INDUSTRIAL WASTE TREATMENT

Volume I

Third Edition

A Field Study Training Program

prepared by

Office of Water Programs
College of Engineering and Computer Science
California State University, Sacramento

in cooperation with

California Water Environment Association

❧ ❧ ❧

Kenneth D. Kerri, Project Director
Bill B. Dendy, Co-Director
John Brady, Consultant and Co-Director
William Crooks, Consultant

❧ ❧ ❧

2005

Funding for the production of this operator training manual was provided by the California State University Sacramento Foundation. Mention of trade names or commercial products does not constitute endorsement or recommendation for use by the project director; the Office of Water Programs; or California State University Sacramento Foundation.

Copyright © 2005 by
California State University Sacramento Foundation

First edition published 1987. Second edition 1994. Third edition 2005.

All Rights Reserved.
Printed in the United States of America

16 15 14 13 12   2 3 4 5 6

ISBN
Volume I: 978-1-59371-028-6
                       1-59371-028-3
Volumes I and II (set): 978-1-59371-029-3
                       1-59371-029-1

http://www.owp.csus.edu

OFFICE OF WATER PROGRAMS

The Office of Water Programs is a nonprofit organization operating under the California State University Sacramento Foundation, to provide distance learning courses for persons interested in the operation and maintenance of drinking water and wastewater facilities. These training programs were developed by people who explain, through the use of our manuals, how they operate and maintain their facilities. The university, fully accredited by the Western Association of Schools and Colleges, administers and monitors these training programs, under the direction of Dr. Ramzi J. Mahmood.

Our training group develops and implements programs and publishes manuals for operators of water treatment plants, water distribution systems, wastewater collection systems, and municipal and industrial wastewater treatment and reclamation facilities. We also offer programs and materials for pretreatment facility inspectors, environmental compliance inspectors, and utility managers. All training is offered as distance learning, using correspondence, video, or computer-based formats with opportunities for continuing education and contact hours for operators, supervisors, managers, and administrators.

Materials and opportunities available from our office include manuals in print, CD, or video formats, and enrollments for courses providing CEU (Continuing Education Unit) contact hours. Here is a sample:

- Industrial Waste Treatment, 2 volumes (print, course enrollment)
- Operation of Wastewater Treatment Plants, 2 volumes (print, CD, course enrollment)
- Advanced Waste Treatment (print, course enrollment)
- Treatment of Metal Wastestreams (print, course enrollment)
- Pretreatment Facility Inspection (print, video, course enrollment)
- Small Wastewater System Operation and Maintenance, 2 volumes (print, course enrollment)
- Operation & Maintenance of Wastewater Collection Systems, 2 volumes (print, course enrollment)
- Collection System Operation & Maintenance Training Videos (video, course enrollment)
- Utility Management (print, course enrollment)
- Manage for Success (print, course enrollment)
- and more

These and other materials may be ordered from:

Office of Water Programs
California State University, Sacramento
6000 J Street
Sacramento, CA 95819-6025
(916) 278-6142 – phone
(916) 278-5959 – FAX

or

visit us on the web at http://www.owp.csus.edu

ADDITIONAL VOLUMES OF INTEREST

Industrial Waste Treatment, Volume II

The Industrial Plant Operator
Fixed Growth Processes
Activated Sludge Process Control
Sequencing Batch Reactors
Enhanced Biological Treatment
Anaerobic Treatment
Residual Solids Management
Maintenance

Operation of Wastewater Treatment Plants, Volume I

The Treatment Plant Operator
Why Treat Wastes?
Wastewater Treatment Facilities
Racks, Screens, Comminutors, and Grit Removal
Sedimentation and Flotation
Trickling Filters
Rotating Biological Contactors
Activated Sludge
Wastewater Stabilization Ponds
Disinfection and Chlorination

Operation of Wastewater Treatment Plants, Volume II

Activated Sludge
Sludge Digestion and Solids Handling
Effluent Disposal
Plant Safety
Maintenance
Laboratory Procedures and Chemistry
Applications of Computers for Plant O & M
Analysis and Presentation of Data
Records and Report Writing
Treatment Plant Administration

Treatment of Metal Wastestreams

Need for Treatment
Sources of Wastewater
Material Safety Data Sheets (MSDSs)
Employee Right-To-Know Laws
Methods of Treatment
Advanced Technologies
Sludge Treatment and Disposal
Operation, Maintenance, and Troubleshooting
Polymers
Oxidation-Reduction Potential (ORP)

Pretreatment Facility Inspection

The Pretreatment Facility Inspector
Pretreatment Program Administration
Development and Application of Regulations
Inspection of a Typical Industry
Safety in Pretreatment Inspection and Sampling Work
Sampling Procedures for Wastewater
Wastewater Flow Monitoring
Industrial Wastewaters
Pretreatment Technology (Source Control)
Industrial Inspection Procedures
Emergency Response

PREFACE TO THE THIRD EDITION

The Office of Water Programs at California State University, Sacramento, continuously strives to make our operator training manuals and programs useful for industrial wastewater treatment plant operators in the field. Many operators enrolled in our courses and studying our manuals will provide us with helpful suggestions regarding how to improve our manuals.

Ms. Judy Bruenjes, PE, Environmental Training Associates, Portland, Maine, provided many valuable suggestions for this Third Edition. Emphasis is placed on the needs of industrial operators at physical–chemical wastewater treatment plants. Information has been added on the description of industrial wastewaters, their sources, their impacts on the environment, and the sampling and monitoring of treatment processes and receiving waters. The chapter on safety is now after the treatment processes, and a new chapter on maintenance has been added to follow the other chapters. The glossary of word definitions in the chapters and at the end has also been revised and updated. Occasionally, words defined in a chapter's glossary or footnotes may contain cross references to other definitions not contained in the chapter. To look up these definitions, please refer to the appendix section, "Industrial Waste Words," at the end of this manual.

The Office of Water Programs requests your assistance in helping us improve our operator training manuals. While you are reading the material in this manual, please make notes of questions and areas where you would improve the material. You may also notice that we have changed the look of this manual, both inside and out. The font was chosen to improve readability, we have highlighted the question sections throughout the chapters, and we have redesigned the cover. We would like to hear what you think of these changes as well. If you have ideas or suggestions on how we can do a better job or produce a better training manual, please contact us. Operators in the future will appreciate your assistance and will benefit from your contributions.

        Kenneth D. Kerri
        Office of Water Programs
        California State University, Sacramento
        6000 J Street
        Sacramento, CA 95819-6025
        (916) 278-6142 – phone
        wateroffice@csus.edu – e-mail

2005

REVIEWERS
INDUSTRIAL WASTE TREATMENT, THIRD EDITION

Russel M. Sanchez Adams
Consulting Sanitary Engineer
Woodland, CA 95776

Bill Garrett
CSDLAC
Industrial Waste Section
Whittier, CA 90607

Julio Guerra
City of Merced
Merced, CA 95340

Peg Hannah
Office of Water Programs
Sacramento, CA 95819-6025

Jon Jewett
Agency of Natural Resources
Waterbury, VT 05671-0406

Patrick Kwok
San Jose/Santa Clara WPC Plant
San Jose, CA 95134

Robert Montgomery
City of Oxnard
Oxnard, CA 93041

Andy Weist
Operations Assistance Section,
 NY DEC
Albany, NY 12233-3506

Vicki Willis
Wastewater Treatment Division
Benicia, CA 94510-3272

TECHNICAL CONSULTANTS

Russ Armstrong
William Garber
George Gardner
Larry Hannah
Mike Mulbarger
Joe Nagano

Carl Nagel
Al Petrasek
Frank Phillips
Warren Prentice
Ralph Stowell
Larry Trumbull

USES OF THIS MANUAL

Originally this manual was developed to serve as a home-study course for operators in remote areas or persons unable to attend formal classes either due to shift work, personal reasons, or the unavailability of suitable classes. This home-study training program used the concepts of self-paced instruction where you are your own instructor and work at your own speed. In order to certify that a person had successfully completed this program, objective tests and special answer sheets for each chapter are provided when a person enrolls in this course.

Once operators started using this manual for home study, they realized that it could serve effectively as a textbook in the classroom. Many colleges and universities have used the manual as a text in formal classes often taught by operators. In areas where colleges were not available or were unable to offer classes in the operation of industrial wastewater treatment plants, operators and utility agencies joined together to offer their own courses using the manual.

Occasionally a utility agency has enrolled from three to over 300 of its operators in this training program. A manual is purchased for each operator. A senior operator or a group of operators are designated as instructors. These operators help answer questions when the persons in the training program have questions or need assistance. The instructors grade the objective tests, record scores, and notify California State University, Sacramento, of the scores when a person successfully completes this program. This approach eliminates any waiting while papers are being graded and returned by CSUS.

This manual was prepared to help operators run their treatment plants. Please feel free to use it in the manner which best fits your training needs and the needs of other operators. We will be happy to work with you to assist you in developing your training program. Please feel free to contact:

       Project Director
       Office of Water Programs
       California State University, Sacramento
       6000 J Street
       Sacramento, CA 95819-6025
       (916) 278-6142 – phone
       (916) 278-5959 – FAX
       wateroffice@csus.edu – e-mail

INSTRUCTIONS TO PARTICIPANTS IN HOME-STUDY COURSE

Procedures for reading the lessons and answering the questions are contained in this section.

To progress steadily through this program, you should establish a regular study schedule. For example, many operators in the past have set aside two hours during two evenings a week for study.

The study material in Volume I is contained in 14 chapters. Some chapters are longer and more difficult than others. For this reason, many of the chapters are divided into two or more lessons. The time required to complete a lesson will depend on your background and experience. Some people might require an hour to complete a lesson and some might require three hours; but that is perfectly all right. The important thing is that you understand the material in the lesson.

Each lesson is arranged for you to read a short section, write the answers to the questions at the end of the section, check your answers against suggested answers; and then *YOU* decide if you understand the material sufficiently to continue or whether you should read the section again. You will find that this procedure is slower than reading a typical textbook, but you will remember much more when you have finished the lesson.

Some discussion and review questions are provided following each lesson in the chapters. These questions review the important points you have covered in the lesson. Write the answers to the discussion and review questions in your notebook.

In the appendix at the end of this manual, you will find some comprehensive review questions and suggested answers. These questions and answers are provided as a way for you to review how well you remember the material. You may wish to review the entire manual before you attempt to answer the questions. Some of the questions are essay-type questions, which are used by some states for higher-level certification examinations. After you have answered all the questions, check your answers with those provided and determine the areas in which you might need additional review before your next certification or civil service examination. Please do not send your answers to California State University, Sacramento.

You are your own teacher in this program. You could merely look up the suggested answers at the end of the chapters or comprehensive review questions or copy them from someone else, but you would not understand the material. Consequently, you would not be able to apply the material to the operation of your plant nor recall it during an examination for certification or a civil service position.

You will get out of this program what you put into it.

SUMMARY OF PROCEDURE

OPERATOR (YOU)

1. Read what you are expected to learn in each chapter; the major topics are listed at the beginning of the chapter.
2. Read sections in the lesson.
3. Write your answers to questions at the end of each section in your notebook. You should write the answers to the questions just as you would if these were questions on a test.
4. Check your answers with the suggested answers.
5. Decide whether to reread the section or to continue with the next section.
6. Write your answers to the discussion and review questions at the end of each lesson in your notebook.

ORDER OF WORKING LESSONS

To complete this program you will have to work all of the lessons. You may proceed in numerical order, or you may wish to work some lessons sooner.

Safety is a very important topic. Everyone working in an industrial wastewater treatment plant must always be safety conscious. Operators daily encounter situations and equipment that can cause a serious disabling injury or illness if the operator is not aware of the potential danger and does not exercise adequate precautions. For these reasons you may decide to work on the chapter on "Safety" early in your studies. In each chapter, safe procedures are always stressed. See Chapter 13, "Safety," for details.

INDUSTRIAL WASTE TREATMENT COURSE OUTLINE

VOLUME I, THIRD EDITION

Chapter	Topic	Page
1	THE INDUSTRIAL PLANT OPERATOR	1
	Ken Kerri, Office of Water Programs, California State University, Sacramento, CA	
	Dan Campbell, New York DEC, Albany, NY	
2	INDUSTRIAL WASTEWATERS	13
	William Crooks, Central Valley Regional Water Quality Control Board, Sacramento, CA	
	Dan Campbell, New York DEC, Albany, NY	
	Richard von Langen, RVL & Associates, Pasadena, CA	
	Ken Kerri, Office of Water Programs, California State University, Sacramento, CA	
3	REGULATORY REQUIREMENTS	69
	Robert Montgomery, City of Oxnard, CA	
4	PREVENTING AND MINIMIZING WASTES AT THE SOURCE	109
	Philip Lo, Sanitation Districts of LA County, Whittier, CA	
	Theresa Dodge, Sanitation Districts of LA County, Whittier, CA	
	Alison Gemmell, Del Monte Foods, Walnut Creek, CA	
	Rob Lamppa, Land O'Lakes, Inc., Arden Hills, MN	
	Mischelle Mische, Sanitation Districts of LA County, Whittier, CA	
	John Polanski, University of Minnesota, Minneapolis, MN	
5	INDUSTRIAL WASTE MONITORING	151
	Larry Bristow, Sacramento County, CA	
6	FLOW MEASUREMENT	207
	John Esler, New York DEC, Albany, NY	
7	PRELIMINARY TREATMENT (EQUALIZATION, SCREENING, AND pH ADJUSTMENT)	249
	Richard von Langen, RVL & Associates, Pasadena, CA	
	D. F. Reddy, HYCOR Corp., Lake Bluff, IL	
	Mark Acerra, Retired	
8	PHYSICAL–CHEMICAL TREATMENT PROCESSES (COAGULATION, FLOCCULATION, AND SEDIMENTATION)	309
	Paul Amodeo, McLaren/Hart, Albany, NY	
	Ross Gudgel, Operational Performance Solutions, Pearl City, HI	
	James L. Johnson, City of Santa Rosa, CA	
	Paul J. Kemp, McLaren/Hart, Albany, NY	
9	FILTRATION	383
	James L. Johnson, City of Santa Rosa, CA	
	Ross Gudgel, Operational Performance Solutions, Pearl City, HI	
	Robert G. Blanck, Koch Membrane Systems, Wilmington, MA	
	Francis J. Brady, Koch Membrane Systems, Wilmington, MA	

| 10 | PHYSICAL TREATMENT PROCESSES (AIR STRIPPING AND CARBON ADSORPTION) | 481 |

Fred Edgecomb, Metcalf & Eddy Services Systems, Riverside, CA

| 11 | TREATMENT OF METAL WASTESTREAMS | 517 |

Bill Strangio, Strangio and Associates, San Jose, CA

| 12 | INSTRUMENTATION | 607 |

Leonard Ainsworth, June Lake PUD, June Lake, CA

| 13 | SAFETY | 665 |

Robert Reed, NEC Electronics, Inc., Roseville, CA
Revised by Russ Armstrong, Sacramento County, CA

| 14 | MAINTENANCE | 739 |

John Brady, Norman Farnum, and Stan Walton, Sacramento County, CA
Revised by Malcolm Carpenter, Sacramento County, CA

Appendix	COMPREHENSIVE REVIEW QUESTIONS AND SUGGESTED ANSWERS	877
	INDUSTRIAL WASTE WORDS	903
	SUBJECT INDEX	949

CHAPTER 1

THE INDUSTRIAL PLANT OPERATOR

by

Ken Kerri

and

Dan Campbell

TABLE OF CONTENTS

Chapter 1. THE INDUSTRIAL PLANT OPERATOR

			Page
OBJECTIVES			4
1.0	WHAT IS AN INDUSTRIAL PLANT OPERATOR?		5
	1.00	What Does an Industrial Plant Operator Do?	5
	1.01	Who Does the Industrial Plant Operator Work For?	5
	1.02	Where Does the Industrial Plant Operator Work?	5
	1.03	What Pay Can an Industrial Plant Operator Expect?	5
	1.04	What Does It Take To Be an Industrial Plant Operator?	6
1.1	YOUR PERSONAL TRAINING COURSE		6
1.2	WHAT DO YOU ALREADY KNOW?		6
1.3	THE WATER QUALITY PROTECTOR: *YOU*		6
1.4	YOUR QUALIFICATIONS		7
	1.40	Your Job	8
1.5	STAFFING NEEDS AND FUTURE JOB OPPORTUNITIES		9
1.6	TRAINING YOURSELF TO MEET THE NEEDS		10
DISCUSSION AND REVIEW QUESTIONS			11
SUGGESTED ANSWERS			11

OBJECTIVES

Chapter 1. THE INDUSTRIAL PLANT OPERATOR

At the beginning of each chapter in this manual you will find a list of objectives. The purpose of this list is to stress those topics in the chapter that are most important. Contained in the list will be items you need to know and skills you must develop to operate, maintain, repair, and manage an industrial wastewater treatment plant as efficiently and as safely as possible.

Following completion of Chapter 1, you should be able to:

1. Explain the type of work done by an industrial wastewater treatment plant operator.

2. Describe where to look for jobs in this profession.

3. Find sources of training and further information on how to do the jobs performed by industrial treatment plant operators.

CHAPTER 1. THE INDUSTRIAL PLANT OPERATOR

Chapter 1 was prepared especially for the new or the potential industrial WASTEWATER[1] treatment plant operator. If you are an experienced operator, you may find some new viewpoints.

1.0 WHAT IS AN INDUSTRIAL PLANT OPERATOR?

Before modern society entered the scene, water was purified in a natural cycle as shown below:

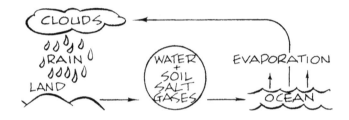

Simplified natural purification cycle

But modern society and the intensive use of the water resource and the resulting water pollution could not wait for sun, wind, and time to accomplish the purification of soiled water; consequently, municipal and industrial treatment plants were built. Thus, nature was given an assist by a team consisting of designers, builders, and treatment plant operators. Designers and builders occupy the scene only for an interval, but operators go on forever. They are the final and essential link in maintaining and protecting the aquatic environment upon which all life depends.

1.00 What Does an Industrial Plant Operator Do?

Simply described, the operator keeps an industrial wastewater treatment plant working. Physically, the operator turns valves, pushes switches, collects samples, lubricates equipment, reads gauges, and records data.

An operator may also maintain equipment and plant areas by painting, weeding, gardening, repairing, and replacing. Mentally, an operator inspects records, observes conditions, makes calculations to determine that the plant is working effectively, and predicts necessary maintenance and facility needs to ensure continued effective operation of the plant. The operator also has an obligation to explain to supervisors, management, civic bodies, and even the general public what the plant does, and most importantly, why the industry and its facilities are vital to the welfare of the community.

1.01 Who Does the Industrial Plant Operator Work For?

An operator's paycheck comes from the industry, city, or municipal sanitation district that employs the operator. As an operator you are responsible to your employer for maintaining an economically and efficiently operating facility. An even greater obligation rests with the operator because the great numbers of people who rely upon downstream water supplies are totally dependent upon the operator's competence and trustworthiness for their welfare. In the final analysis, the operator is really working for these vitally affected downstream water users.

1.02 Where Does the Industrial Plant Operator Work?

Obviously, the operator works in an industrial treatment plant. But the different types and locations of treatment plants offer a wide range of working conditions. From the mountains to the sea, wherever people congregate into communities, industrial treatment plants will be found. From a unit process operator at a complex industrial facility to a one-person manager of a small pretreatment facility, you can select your own special place in industrial plant operation.

1.03 What Pay Can an Industrial Plant Operator Expect?

In dollars? Prestige? Job satisfaction? Community service? In opportunities for advancement? By whatever scale you use, returns are what you make them. If you choose a large industry, the pay is good and advancement prospects are tops. Choose a small facility and pay may not be as good; but job satisfaction, freedom from time-clock hours, service, and prestige may well add up to outstanding personal achievement. Total reward depends on you.

The operator's duties

---

[1] *Wastewater.* A community's used water and water-carried solids (including used water from industrial processes) that flow to a treatment plant. Stormwater, surface water, and groundwater infiltration also may be included in the wastewater that enters a wastewater treatment plant. The term sewage usually refers to household wastes, but this word is being replaced by the term wastewater.

1.04 What Does It Take To Be an Industrial Plant Operator?

Desire. First, you must choose to enter this profession. You can do it with a grammar school, a high school, or a college education. While some jobs will always exist for manual labor, the real and expanding need is for trained operators. New techniques, advanced equipment, and increasing instrumentation require a new breed of operator, one who is willing to learn today, and gain tomorrow, for surely your industrial plant will move toward newer and more effective operating procedures and treatment processes. Indeed, the truly service-minded operator assists in adding to and improving the plant performance on a continuing basis.

Tomorrow's forgotten operator stopped learning yesterday

You can be an industrial operator tomorrow by beginning your learning today; or you can be a better operator, ready for advancement, by accelerating your learning today.

This training course, then, is your start toward a better tomorrow, both for you and for the industry and public who will receive better water from your efforts.

QUESTIONS

Place an "X" by the correct answer or answers. After you have answered all the questions, check your answers with those given at the end of the chapter on page 11. Reread any sections you did not understand and then proceed to the next section. You are your own teacher in this training program, and *YOU* should decide when you understand the material and are ready to continue with new material.

EXAMPLE:

This is a training course on

_____ A. Accounting

_____ B. Engineering

  X   C. Industrial treatment plant operation

_____ D. Salesmanship

1.0A Wastewater is the same thing as

_____ A. Rain

_____ B. Soil

   X   C. Sewage

_____ D. Condensation

1.0B What does an industrial treatment plant operator do?

   X   A. Collect samples

   X   B. Lubricate equipment

   X   C. Record data

1.0C Who employs industrial plant operators?

   X   A. Cities

   X   B. Sanitation districts

   X   C. Industries

1.1 YOUR PERSONAL TRAINING COURSE

Beginning on this page, you are embarking on a training course that has been carefully prepared to allow you to improve your knowledge of and ability to operate an industrial wastewater treatment plant. You will be able to proceed at your own pace; you will have an opportunity to learn a little or a lot about each topic. The course has been prepared this way to fit the various needs of operators, depending on what kind of plant you have or how much you need to learn about it. To study for certification examinations, you will have to cover all the material. You will never know everything about your plant or about the wastewater that flows through it, but you can begin to answer some very important questions about how and when certain things happen in the plant. You can also learn to manipulate your plant so that it operates at maximum efficiency.

1.2 WHAT DO YOU ALREADY KNOW?

If you already have some experience operating an industrial wastewater treatment plant, you may use the first chapter for a review. If you are relatively new to the industrial wastewater treatment field, this chapter will provide you with important background information. The remainder of this introductory chapter describes your role as a protector of water quality, your qualifications to do your job, a little about staffing needs in the industrial wastewater treatment field, and some information on other training opportunities.

Water quality protector

1.3 THE WATER QUALITY PROTECTOR: *YOU*

Historically, Americans have shown a great lack of interest in the protection of their water resources. We have been content to think that "the solution to pollution is dilution." For years we were able to dump our wastes with little or no treatment back into the nearest *RECEIVING WATERS*.[2] As long as there was enough dilution water to absorb the waste material, nature took care of our disposal problems for us. As more and more towns and industry sprang up, waste loads increased until the natural purification processes could no longer do the job. Many waterways were converted into open sewers. Unfortunately, for many areas this did not signal the beginning of a clean-up campaign. It merely increased the frequency of the cry: "We don't have the money for a treatment plant," or the ever-popular, "If we make industries treat their wastes they will move to another state." Thus, the pollution of our waters increased.

Within the last several years, we have seen many changes in this depressing picture. We now realize that we must give nature a hand by treating wastes before they are discharged. Adequate treatment of wastes will not only protect our health and that of our downstream neighbors; it can also increase property values, allow game fishing and various recreational uses to be enjoyed, and attract water-using industries to the area. Today we are seeing massive efforts being undertaken to control water pollution and improve water quality throughout the nation. This includes the efforts not only of your own industry, community, county, and state, but also the federal government. As we continue to learn more about the health effects of contaminants, increasingly stringent wastewater discharge regulations are being developed and enforced to protect public health and the environment.

Great sums of public and private funds are now being invested in large, complex municipal and industrial wastewater treatment facilities to overcome this pollution; and you, the treatment plant operator, will play a key role in the battle. Without efficient operation of your plant, much of the research, planning, and building that has been done and will be done to accomplish the goals of water quality control in your area will be wasted. You are the difference between a facility and a performing unit. You are, in fact, a water quality protector on the front line of the water pollution battle.

The receiving water quality standards and waste discharge requirements that your plant has been built to meet have been formulated to protect the water users downstream or the *PUBLICLY OWNED TREATMENT WORKS (POTW)*[3] to which your facility discharges. Downstream water uses may include domestic water supply, industrial water supply, agricultural water supply, stock and wildlife watering, propagation of fish and other aquatic and marine life, shellfish culture, swimming and other water-contact sports, boating, aesthetic enjoyment, hydroelectric power, navigation, and others. Therefore, you have an obligation to the users of the water downstream, as well as to the people of your district or municipality and industry. You are the key water quality protector and must realize that you are in a responsible position.

QUESTIONS

Write your answers in a notebook and then compare your answers with those on page 11.

1.3A How did many receiving waters become polluted?

1.3B Why must municipal and industrial wastewaters receive adequate treatment?

1.4 YOUR QUALIFICATIONS

The skill and ability required for your job depend to a large degree on the size and type of treatment plant where you are employed. You may work at a large, modern treatment plant

---

[2] *Receiving Water.* A stream, river, lake, ocean, or other surface or groundwaters into which treated or untreated wastewater is discharged.

[3] *POTW.* Publicly Owned Treatment Works. A treatment works that is owned by a state, municipality, city, town, special sewer district, or other publicly owned and financed entity as opposed to a privately (industrial) owned treatment facility. This definition includes any devices and systems used in the storage, treatment, recycling, and reclamation of municipal sewage (wastewater) or industrial wastes of a liquid nature. It also includes sewers, pipes, and other conveyances only if they carry wastewater to a POTW treatment plant. The term also means the municipality (public entity) that has jurisdiction over the indirect discharges to and the discharges from such a treatment works.

treating several complex industrial wastestreams and employing 10 or more operators. In this case, you are probably a specialist in one or more phases of the treatment process.

On the other hand, you may operate a small plant or pretreatment facility. You may be the only operator at the plant or, at best, have only one or two additional employees. If this is the case, you must be a "jack-of-all-trades" because of the diversity of your tasks. You will need to know how each process in your plant works as well as how each process influences the rest of the unit processes.

1.40 Your Job

To describe the operator's duties, let us start at the beginning. Let us say that the need for a new or improved wastewater treatment plant has long been recognized by your industry. The industry has allocated the funds to finance the project, and the consulting engineers have submitted plans and specifications. It is to the best interests of the industry and the consulting engineer that you be in on the ground-floor planning. If it is a new plant, you should be present or at least available during the construction period in order to become completely familiar with the entire plant, including the equipment and machinery and their operation. This will provide you with the opportunity to relate your plant drawings to actual facilities.

You and the engineer should discuss how the treatment plant should best be run and the means of operation the designer had in mind when the plant was designed. If it is an old plant being remodeled, you are in a position to offer excellent advice to the consulting engineer. Your experience provides valuable technical knowledge concerning the characteristics of wastewater, its sources, and the limitations of the present facilities. Together with the consultant, you are a member of an expert team able to advise the industry.

Once the plant is operating, you become an administrator. In a small plant, your duties may not include supervision of personnel, but you are still in charge of records. You are responsible for operating the plant as efficiently as possible, keeping in mind that the primary objective is to protect the receiving water quality by continuous and efficient plant performance. Without adequate, reliable records of every phase of operation, the effectiveness of your operation and your compliance with regulatory requirements have not been documented (recorded).

You may also be the budget administrator. Most certainly you are in the best position to give advice on budget requirements, management problems, and future planning. You should be aware of the necessity for additional expenditures, including funds for plant enlargement, equipment replacement, and laboratory requirements. You should recognize and define such needs in sufficient time to inform the proper officials to enable them to accomplish early planning and budgeting.

You are in the field of public relations and must be able to explain the purpose and operation of your plant to visitors, civic organizations, school classes, representatives of news media, and even to the city council or managers and directors of your industry. Public interest in water quality is increasing, and you should be prepared to conduct plant tours (Figure 1.1) that will con-

Fig. 1.1 Visitors touring a treatment plant

tribute to public acceptance and support. A well-guided tour for officials of regulatory agencies or other operators may provide these people with sufficient understanding of your plant to allow them to suggest helpful solutions to operational problems.

Special care and safety must be practiced when visitors are taken through your treatment plant. An accident could spoil all of your public relations efforts.

The appearance of your plant indicates to the visitor the type of operation you maintain. If the plant is dirty and rundown with tools and litter scattered about, you will be unable to convince your visitors that the plant is doing a good job. Your records showing a high-quality *EFFLUENT*[4] will mean nothing to these visiting citizens and officials unless your plant appears clean and well maintained and the effluent looks good.

Another aspect of your public relations duties is your dealings with the downstream water user. Unfortunately, the operator is often considered by the downstream user as a polluter rather than a water quality protector. Through a good public information program, backed by facts supported by reliable data, you can correct the impression held by the downstream user and establish "good neighbor" relations. This is indeed a challenge. Again, you must understand that you hold a very responsible position and be aware that the sole purpose of the operation of your plant is to protect the downstream user, be that user a private property owner, another city or district, an industry, or a fisherman.

You are required to understand certain laboratory procedures in order to conduct various tests on samples of wastewater and receiving waters. On the basis of the data obtained from these tests, you may have to adjust the operation of the treatment plant to meet stream standards or discharge requirements.

As an operator, you must have a knowledge of the complicated mechanical principles involved in many treatment mechanisms. In order to measure and control the wastewater flowing through the plant, you must have some understanding of hydraulics. Practical knowledge of electrical motors, circuitry, and controls is also essential.

Safety is a very important operator responsibility. Unfortunately, too many operators take safety for granted. This is one reason why the wastewater treatment industry has one of the worst safety records of any industry. *You* have the responsibility to be sure that your treatment plant is a safe place to work and visit. Everyone must follow safe procedures and understand why safe procedures must be followed at all times. All operators must be aware of the safety hazards in and around treatment plants. You should plan or be a part of an active safety program. Chief operators frequently have the responsibility of training new operators and must encourage all operators to work safely.

Clearly then, today's industrial treatment plant operator must possess a broad range of qualifications.

QUESTIONS

Write your answers in a notebook and then compare your answers with those on page 11.

1.4A Why is it important that the operator be present during the construction of a new plant?

1.4B How does the operator become involved in public relations?

1.5 STAFFING NEEDS AND FUTURE JOB OPPORTUNITIES

The industrial treatment field, like so many others, is changing rapidly. New plants are being constructed, and old plants are being modified and enlarged to handle the wastewater from our growing industries and to treat the new chemicals being produced by our space-age technology. Operators, maintenance personnel, lead people, managers, instrumentation experts, and laboratory technicians are sorely needed.

A look at past records and future predictions indicates that industrial wastewater treatment is a rapidly growing field. According to the US Bureau of Labor Statistics (BLS),[5] water and wastewater treatment plant operators held about 98,000 jobs in 1996. Approximately half of the jobs were in the wastewater treatment industry, and the majority of operators worked for local governments. The BLS estimates that the number of jobs in the water and wastewater treatment industries is expected to increase by an average of 46 percent in the period from 2002 to 2012. Factors contributing to the increase include population growth, retirement of many current operators, regulatory requirements, more sophisticated treatment, and operator certification. The factors creating a demand for municipal wastewater treatment plant operators are the same factors that are creating a demand for industrial wastewater treatment plant operators. The need for *trained* operators is increasing rapidly and is expected to continue to grow in the future.

---

[4] *Effluent* (EF-loo-ent). Water or other liquid—raw (untreated), partially treated, or completely treated—flowing *FROM* a reservoir, basin, treatment process, or treatment plant.

[5] Refer to the Bureau of Labor Statistics Website at www.bls.gov for additional information about the types of jobs available in the water and wastewater industries, working conditions, earnings potential, and the job outlook.

1.6 TRAINING YOURSELF TO MEET THE NEEDS

This training course is not the only one available to help you improve your abilities. The states have offered various types of both long- and short-term operator training. Their health departments and water pollution control or environment associations have provided training classes conducted by members of the associations, largely on a volunteer basis. State and local colleges and universities have provided valuable training under their own sponsorship or in partnership with others. Many state, local, and private agencies have conducted both long- and short-term training as well as interesting and informative seminars.

Listed below are several very good references in the field of municipal wastewater treatment plant operation. The name in quotes represents the term usually used by operators when they mention the reference.

1. *"MOP 11." OPERATION OF MUNICIPAL WASTEWATER TREATMENT PLANTS* (MOP 11). Obtain from Water Environment Federation (WEF), Publications Order Department, 601 Wythe Street, Alexandria, VA 22314-1994. Order No. M05110. Price to members, $120.00; nonmembers, $148.00; plus shipping and handling.

2. *"NEW YORK MANUAL." MANUAL OF INSTRUCTION FOR WASTEWATER TREATMENT PLANT OPERATORS* (two-volume set), distributed in New York by the New York State Department of Health, Office of Public Health Education, Water Pollution Control Board. Distributed outside of New York State by Health Education Services, PO Box 7126, Albany, NY 12224. Price, $25.00, includes cost of shipping and handling.

3. *"TEXAS MANUAL." MANUAL OF WASTEWATER TREATMENT.* Obtain from Texas Water Utilities Association, 1106 Clayton Lane, Suite 101 East, Austin, TX 78723-1093. Price to members, $25.00; nonmembers, $35.00; plus $3.50 shipping and handling.

These publications cover the entire field of treatment plant operation. They emphasize treatment of municipal wastewaters. This is the only manual available that stresses the safe operation and maintenance of industrial waste treatment plants. At the end of many of the other chapters in this manual, lists of additional references will be provided.

QUESTION

Write your answer in a notebook and then compare your answer with the one on page 11.

1.6A In addition to this course, what are some other sources of training to become an industrial wastewater treatment plant operator or to improve your knowledge and skills?

Please answer the discussion and review questions next.

DISCUSSION AND REVIEW QUESTIONS
Chapter 1. THE INDUSTRIAL PLANT OPERATOR

The purpose of these questions is to indicate to you how well you understand the material in the chapter. Write the answers to these questions in your notebook.

1. Describe in general terms the types of activities that make up an industrial treatment plant operator's job.

2. List several examples of downstream water uses affected by how well an industrial wastewater treatment plant is operated.

3. Why is recordkeeping an important part of an operator's job?

4. What are the responsibilities of an operator with regard to safety?

SUGGESTED ANSWERS
Chapter 1. THE INDUSTRIAL PLANT OPERATOR

You are not expected to have the exact answer suggested for questions requiring written answers, but you should have the correct idea. The numbering of the questions refers to the section in the manual where you can find the information to answer the questions. For example, answers to questions numbered 1.0 can be found in Section 1.0, What is an Industrial Plant Operator?

Answers to questions on page 6.

1.0A   C

1.0B   A, B, C

1.0C   A, B, C

Answers to questions on page 7.

1.3A   Receiving waters became polluted by a lack of public concern for the impact of waste discharges and by discharging wastewater into a receiving water beyond its natural purification capacity.

1.3B   Municipal and industrial wastewaters must receive adequate treatment to protect receiving water users.

Answers to questions on page 9.

1.4A   The operator should be present during the construction of a new plant in order to become familiar with the plant before the operator begins operating it.

1.4B   The operator becomes involved in public relations by explaining the purpose and operation of the plant to visitors, civic organizations, news reporters, and even city council members and industry managers or directors.

Answer to question on page 10.

1.6A   In addition to this course, other sources of operator training include state health departments and water pollution control or environment associations, state and local colleges, and many state, local, and private training agencies.

CHAPTER 2

INDUSTRIAL WASTEWATERS

by

William Crooks

Dan Campbell

Richard von Langen

Ken Kerri

TABLE OF CONTENTS
Chapter 2. INDUSTRIAL WASTEWATERS

		Page
OBJECTIVES		18
PROJECT PRONUNCIATION KEY		19
WORDS		20

LESSON 1

2.0	PREVENTION OF POLLUTION	27
2.1	WHAT IS PURE WATER?	27
2.2	TYPES OF WASTE DISCHARGES	28
2.3	EFFECTS OF WASTE DISCHARGES	28
	2.30 Sludge and Scum	28
	2.31 Oxygen Depletion	28
	2.32 Human Health	29
	2.33 Other Effects	29
2.4	SOLIDS IN WASTEWATER	30
	2.40 Types of Solids	30
	2.41 Total Solids	30
	2.42 Dissolved Solids	30
	2.43 Suspended Solids	30
	2.44 Organic and Inorganic Solids	31
	2.45 Floatable Solids	31
2.5	NATURAL CYCLES	31
2.6	NEED TO TREAT INDUSTRIAL WASTES by Dan Campbell	32
	2.60 Uses of Water by Industry	32
	2.61 Reasons for Treatment of Industrial Wastes	33
	2.610 Dissolved and Suspended Wastes	33
	2.611 Drinking Water	33
	2.612 Cooking Water	35
	2.613 Water-Contact Recreation	35
	2.614 Non-Body-Contact Water Recreation	36
	2.615 Water for Fish, Wildlife, and Aquatic Vegetation	36

		2.616	Agricultural Use	36
		2.617	Industrial Use	36

DISCUSSION AND REVIEW QUESTIONS 37

LESSON 2

2.7   IMPORTANCE OF UNDERSTANDING INDUSTRIAL WASTEWATERS 38
by Richard von Langen

     2.70   Manufacturing Processes and Wastewater Generation 39

     2.71   Utility Processes and Wastewater Generation 39

     2.72   Maintenance Activities and Wastewater Generation 40

     2.73   IWTS Wastewater Generation 40

2.8   INDUSTRIAL WASTESTREAM VARIABLES 42

     2.80   Compatible and Noncompatible Pollutants 42

     2.81   Dilute Solutions 42

     2.82   Concentrated Solutions 42

     2.83   Concentration Versus Mass of the Pollutant 43

     2.84   Frequency of Generation and Discharge 44

		2.840	Hours of Operation Versus Discharge	44
		2.841	Discharge Variations	45
		2.842	Continuous and Intermittent Discharges	45

2.9   EFFECTS OF INDUSTRIAL WASTEWATERS 46

     2.90   Operator's Responsibility 46

     2.91   Effects on the Collection System 47

		2.910	Hydraulic Capacity Problems	47
		2.911	Plugging	47
		2.912	Odors	47
		2.913	pH Problems	48
		2.914	Flammables	49
		2.915	Temperature	50

     2.92   Effects on the Treatment System 50

		2.920	Hydraulic Overload	50
		2.921	Interference	51
		2.922	Influent Variability	51
		2.923	Slug Loadings	51

     2.93   Effects on Effluent and Sludge Disposal and Reuse 53

     2.94   Effects on the POTW 53

2.10   MANUFACTURING PROCESSES AND WASTEWATER GENERATION 54

     2.100   Metal Finishing Industries 54

		2.1000	Unit Process Description	54
		2.1001	Waste/Wastewater Characteristics	58
	2.101		Printed Circuit Board Manufacturing	60
		2.1010	Unit Process Description	60
		2.1011	Waste/Wastewater Characteristics	62

2.11 REFERENCES 62

2.12 ACKNOWLEDGMENTS 62

DISCUSSION AND REVIEW QUESTIONS 64

SUGGESTED ANSWERS 64

OBJECTIVES
Chapter 2. INDUSTRIAL WASTEWATERS

Following completion of Chapter 2, you should be able to:

1. Give reasons for preventing pollution.
2. Identify various types of waste discharges.
3. Recognize the effects of waste discharges on receiving waters.
4. Describe the different types of solids in wastewater.
5. Explain what happens in a natural cycle.
6. Explain why industrial wastewaters need to be treated.
7. List the general types of manufacturing activities and processes that are sources of industrial wastes or wastewater.
8. Describe the types of industrial wastewaters discharged to sewers.
9. Explain the difference between concentration and mass of pollutants.
10. Describe the effects of industrial wastewaters on wastewater collection, treatment, and disposal systems.

PROJECT PRONUNCIATION KEY

by Warren L. Prentice

The Project Pronunciation Key is designed to aid you in the pronunciation of new words. While this key is based primarily on familiar sounds, it does not attempt to follow any particular pronunciation guide. This key is designed solely to aid operators in this program.

You may find it helpful to refer to other available sources for pronunciation help. Each current standard dictionary contains a guide to its own pronunciation key. Each key will be different from each other and from this key. Examples of the difference between the key used in this program and the *WEBSTER'S NEW WORLD COLLEGE DICTIONARY*[1] "Key" are shown below.

In using this key, you should accent (say louder) the syllable that appears in capital letters. The following chart is presented to give examples of how to pronounce words using the Project Key.

WORD	SYLLABLE				
	1st	2nd	3rd	4th	5th
acid	AS	id			
coliform	KOAL	i	form		
biological	BUY	o	LODGE	ik	cull

The first word, *ACID*, has its first syllable accented. The second word, *COLIFORM*, has its first syllable accented. The third word, *BIOLOGICAL*, has its first and third syllables accented.

We hope you will find the key useful in unlocking the pronunciation of any new word.

---

[1] The *WEBSTER'S NEW WORLD COLLEGE DICTIONARY*, Fourth Edition, 1999, was chosen rather than an unabridged dictionary because of its availability to the operator. Other editions may be slightly different.

WORDS
Chapter 2. INDUSTRIAL WASTEWATERS

At the beginning of each chapter in this manual you will find a glossary containing definitions of words used by industrial wastewater treatment plant operators. The material in each chapter on how to safely operate and maintain industrial treatment plant facilities will be easier to understand if you know and understand the meanings of these words.

ACUTE HEALTH EFFECT

An adverse effect on a human or animal body, with symptoms developing rapidly.

AEROBIC (air-O-bick)

A condition in which atmospheric or dissolved oxygen is present in the aquatic (water) environment.

AEROBIC BACTERIA (air-O-bick back-TEER-e-uh)

Bacteria that will live and reproduce only in an environment containing oxygen that is available for their respiration (breathing), namely atmospheric oxygen or oxygen dissolved in water. Oxygen combined chemically, such as in water molecules (H_2O), cannot be used for respiration by aerobic bacteria.

ALGAE (AL-jee)

Microscopic plants containing chlorophyll that live floating or suspended in water. They also may be attached to structures, rocks, or other submerged surfaces. Excess algal growths can impart tastes and odors to potable water. Algae produce oxygen during sunlight hours and use oxygen during the night hours. Their biological activities appreciably affect the pH, alkalinity, and dissolved oxygen of the water.

ANAEROBIC (AN-air-O-bick)

A condition in which atmospheric or dissolved oxygen (DO) is *NOT* present in the aquatic (water) environment.

ANAEROBIC BACTERIA (AN-air-O-bick back-TEER-e-uh)

Bacteria that live and reproduce in an environment containing no free or dissolved oxygen. Anaerobic bacteria obtain their oxygen supply by breaking down chemical compounds that contain oxygen, such as sulfate (SO_4^{2-}).

ANODIZING

An electrochemical process that deposits a coating of an insoluble oxide on a metal surface. Aluminum is the most frequently anodized material.

APPURTENANCE (uh-PURR-ten-nans)

Machinery, appliances, structures, and other parts of the main structure necessary to allow it to operate as intended, but not considered part of the main structure.

BOD (pronounce as separate letters)

Biochemical Oxygen Demand. The rate at which organisms use the oxygen in water or wastewater while stabilizing decomposable organic matter under aerobic conditions. In decomposition, organic matter serves as food for the bacteria and energy results from its oxidation. BOD measurements are used as a surrogate measure of the organic strength of wastes in water.

BIOCHEMICAL OXYGEN DEMAND (BOD)

See BOD.

BIOCHEMICAL OXYGEN DEMAND (BOD) TEST

A procedure that measures the rate of oxygen use under controlled conditions of time and temperature. Standard test conditions include dark incubation at 20°C for a specified time (usually five days).

BIODEGRADABLE (BUY-o-dee-GRADE-able)

Organic matter that can be broken down by bacteria to more stable forms that will not create a nuisance or give off foul odors is considered biodegradable.

BLOWDOWN

The removal of accumulated solids in boilers to prevent plugging of boiler tubes and steam lines. In cooling towers, blowdown is used to reduce the amount of dissolved salts in the recirculated cooling water.

CENTRIFUGE

A mechanical device that uses centrifugal or rotational forces to separate solids from liquids.

CHELATING (KEY-LAY-ting) AGENT

A chemical used to prevent the precipitation of metals (such as copper).

CHELATION (key-LAY-shun)

A chemical complexing (forming or joining together) of metallic cations (such as copper) with certain organic compounds, such as EDTA (ethylene diamine tetracetic acid). Chelation is used to prevent the precipitation of metals (copper). Also see SEQUESTRATION.

CHRONIC HEALTH EFFECT

An adverse effect on a human or animal body with symptoms that develop slowly over a long period of time or that recur frequently.

COLIFORM (KOAL-i-form)

A group of bacteria found in the intestines of warm-blooded animals (including humans) and also in plants, soil, air, and water. The presence of coliform bacteria is an indication that the water is polluted and may contain pathogenic (disease-causing) organisms. Fecal coliforms are those coliforms found in the feces of various warm-blooded animals, whereas the term "coliform" also includes other environmental sources.

COMPATIBLE POLLUTANTS

Those pollutants that are normally removed by the POTW treatment system. Biochemical oxygen demand (BOD), suspended solids (SS), and ammonia are considered compatible pollutants.

CONDUCTIVITY

A measure of the ability of a solution (water) to carry an electric current.

CONVENTIONAL POLLUTANTS

Those pollutants that are usually found in domestic, commercial, or industrial wastes, including suspended solids, biochemical oxygen demand, pathogenic (disease-causing) organisms, and oil and grease.

DELETERIOUS (DELL-eh-TEER-ee-us)

Refers to something that can be or is hurtful, harmful, or injurious to health or the environment.

DISINFECTION (dis-in-FECT-shun)

The process designed to kill or inactivate most microorganisms in water or wastewater, including essentially all pathogenic (disease-causing) bacteria. There are several ways to disinfect, with chlorination being the most frequently used in water and wastewater treatment plants. Compare with STERILIZATION.

DRAG OUT

The liquid film (plating solution) that adheres to the workpieces and their fixtures as they are removed from any given process solution or their rinses. Drag-out volume from a tank depends on the viscosity of the solution, the surface tension, the withdrawal time, the draining time, and the shape and texture of the workpieces. The drag-out liquid may drip onto the floor and cause wastestream treatment problems. Regulated substances contained in this liquid must be removed from wastestreams or neutralized prior to discharge to POTW sewers.

EFFLUENT (EF-loo-ent)

Water or other liquid—raw (untreated), partially treated, or completely treated—flowing *FROM* a reservoir, basin, treatment process, or treatment plant.

EVAPOTRANSPIRATION (ee-VAP-o-TRANS-purr-A-shun)

(1) The process by which water vapor is released to the atmosphere from living plants. Also called TRANSPIRATION.

(2) The total water removed from an area by transpiration (plants) and by evaporation from soil, snow, and water surfaces.

EXFILTRATION (EX-fill-TRAY-shun)

Liquid wastes and liquid-carried wastes that unintentionally leak out of a sewer pipe system and into the environment.

FACULTATIVE (FACK-ul-tay-tive) POND

The most common type of pond in current use. The upper portion (supernatant) is aerobic, while the bottom layer is anaerobic. Algae supply most of the oxygen to the supernatant.

IMHOFF CONE

A clear, cone-shaped container marked with graduations. The cone is used to measure the volume of settleable solids in a specific volume (usually one liter) of water or wastewater.

IMMISCIBLE (im-MISS-uh-bull)

Not capable of being mixed.

INFILTRATION (in-fill-TRAY-shun)

The seepage of groundwater into a sewer system, including service connections. Seepage frequently occurs through defective or cracked pipes, pipe joints and connections, interceptor access risers and covers, or manhole walls.

INHIBITORY SUBSTANCES

Materials that kill or restrict the ability of organisms to treat wastes.

INORGANIC WASTE

Waste material such as sand, salt, iron, calcium, and other mineral materials that are only slightly affected by the action of organisms. Inorganic wastes are chemical substances of mineral origin; whereas organic wastes are chemical substances usually of animal or plant origin. Also see NONVOLATILE MATTER, ORGANIC WASTE, and VOLATILE SOLIDS.

INTERFERENCE

Interference refers to the harmful effects industrial compounds can have on POTW operations, such as killing or inhibiting beneficial microorganisms or causing treatment process upsets or sludge contamination.

LIPOPHILIC (lie-puh-FILL-ick)

Having a strong affinity for fats. Compounds that dissolve in fats, oils, and greases.

MASS EMISSION RATE

The rate of discharge of a pollutant expressed as a weight per unit time, usually as pounds or kilograms per day.

MERCAPTANS (mer-CAP-tans)

Compounds containing sulfur that have an extremely offensive skunk-like odor; also sometimes described as smelling like garlic or onions.

MILLIGRAMS PER LITER, mg/*L*

A measure of the concentration by weight of a substance per unit volume in water or wastewater. In reporting the results of water and wastewater analysis, mg/*L* is preferred to the unit parts per million (ppm), to which it is approximately equivalent.

MISCIBLE (MISS-uh-bull)

Capable of being mixed. A liquid, solid, or gas that can be completely dissolved in water.

NONBIODEGRADABLE (NON-buy-o-dee-GRADE-uh-bull)

Substances that cannot readily be broken down by bacteria to simpler forms.

NONCOMPATIBLE POLLUTANTS

Those pollutants that are normally *NOT* removed by the POTW treatment system. These pollutants may be a toxic waste and may pass through the POTW untreated or interfere with the treatment system. Examples of noncompatible pollutants include heavy metals, such as copper, nickel, lead, and zinc; organics, such as methylene chloride, 1,1,1-trichloroethylene, methyl ethyl ketone, acetone, and gasoline; or sludges containing toxic organics or metals.

NUTRIENT

Any substance that is assimilated (taken in) by organisms and promotes growth. Nitrogen and phosphorus are nutrients that promote the growth of algae. There are other essential and trace elements that are also considered nutrients. Also see NUTRIENT CYCLE.

NUTRIENT CYCLE

The transformation or change of a nutrient from one form to another until the nutrient has returned to the original form, thus completing the cycle. The cycle may take place under either aerobic or anaerobic conditions.

ORP (pronounce as separate letters)

Oxidation-Reduction Potential. The electrical potential required to transfer electrons from one compound or element (the oxidant) to another compound or element (the reductant); used as a qualitative measure of the state of oxidation in water and wastewater treatment systems. ORP is measured in millivolts, with negative values indicating a tendency to reduce compounds or elements and positive values indicating a tendency to oxidize compounds or elements.

ORGANIC WASTE

Waste material that may come from animal or plant sources. Natural organic wastes generally can be consumed by bacteria and other small organisms. Manufactured or synthetic organic wastes from metal finishing, chemical manufacturing, and petroleum industries may not normally be consumed by bacteria and other organisms. Also see INORGANIC WASTE and VOLATILE SOLIDS.

POTW

Publicly Owned Treatment Works. A treatment works that is owned by a state, municipality, city, town, special sewer district, or other publicly owned and financed entity as opposed to a privately (industrial) owned treatment facility. This definition includes any devices and systems used in the storage, treatment, recycling, and reclamation of municipal sewage (wastewater) or industrial wastes of a liquid nature. It also includes sewers, pipes, and other conveyances only if they carry wastewater to a POTW treatment plant. The term also means the municipality (public entity) that has jurisdiction over the indirect discharges to and the discharges from such a treatment works.

PASSIVATING

A metal plating process that forms a protective film on metals by immersion in an acid solution, usually nitric acid or nitric acid with sodium dichromate.

PATHOGENIC (path-o-JEN-ick) ORGANISMS

Organisms, including bacteria, viruses, or cysts, capable of causing diseases (such as giardiasis, cryptosporidiosis, typhoid, cholera, dysentery) in a host (such as a person). Also called PATHOGENS.

pH (pronounce as separate letters)

pH is an expression of the intensity of the basic or acidic condition of a liquid. Mathematically, pH is the logarithm (base 10) of the reciprocal of the hydrogen ion activity. If $\{H^+\} = 10^{-6.5}$, then pH = 6.5. The pH may range from 0 to 14, where 0 is most acidic, 14 most basic, and 7 neutral.

PICKLE

An acid or chemical solution in which metal objects or workpieces are dipped to remove oxide scale or other adhering substances.

POLLUTION

The impairment (reduction) of water quality by agricultural, domestic, or industrial wastes (including thermal and radioactive wastes) to a degree that the natural water quality is changed to hinder any beneficial use of the water or render it offensive to the senses of sight, taste, or smell or when sufficient amounts of wastes create or pose a potential threat to human health or the environment.

PRIMARY TREATMENT

A wastewater treatment process that takes place in a rectangular or circular tank and allows those substances in wastewater that readily settle or float to be separated from the wastewater being treated. A septic tank is also considered primary treatment.

REAGENT (re-A-gent)

A pure, chemical substance that is used to make new products or is used in chemical tests to measure, detect, or examine other substances.

RECEIVING WATER

A stream, river, lake, ocean, or other surface or groundwaters into which treated or untreated wastewater is discharged.

SECONDARY TREATMENT

A wastewater treatment process used to convert dissolved or suspended materials into a form more readily separated from the water being treated. Usually, the process follows primary treatment by sedimentation. The process commonly is a type of biological treatment followed by secondary clarifiers that allow the solids to settle out from the water being treated.

SEPTIC (SEP-tick) or SEPTICITY

A condition produced by bacteria when all oxygen supplies are depleted. If severe, the bottom deposits produce hydrogen sulfide, the deposits and water turn black, give off foul odors, and the water has a greatly increased oxygen and chlorine demand.

STABILIZATION

Conversion to a form that resists change. Organic material is stabilized by bacteria that convert the material to gases and other relatively inert substances. Stabilized organic material generally will not give off obnoxious odors.

STERILIZATION (STAIR-uh-luh-ZAY-shun)

The removal or destruction of all microorganisms, including pathogens and other bacteria, vegetative forms, and spores. Compare with DISINFECTION.

TOXIC

A substance that is poisonous to a living organism. Toxic substances may be classified in terms of their physiological action, such as irritants, asphyxiants, systemic poisons, and anesthetics and narcotics. Irritants are corrosive substances that attack the mucous membrane surfaces of the body. Asphyxiants interfere with breathing. Systemic poisons are hazardous substances that injure or destroy internal organs of the body. Anesthetics and narcotics are hazardous substances that depress the central nervous system and lead to unconsciousness.

TRANSPIRATION (TRAN-spur-RAY-shun)

The process by which water vapor is released to the atmosphere by living plants. This process is similar to people sweating. Also see EVAPOTRANSPIRATION.

VOLATILE (VOL-uh-tull)

(1) A volatile substance is one that is capable of being evaporated or changed to a vapor at relatively low temperatures. Volatile substances can be partially removed from water or wastewater by the air stripping process.

(2) In terms of solids analysis, volatile refers to materials lost (including most organic matter) upon ignition in a muffle furnace for 60 minutes at 550°C (1,022°F). Natural volatile materials are chemical substances usually of animal or plant origin. Manufactured or synthetic volatile materials, such as plastics, ether, acetone, and carbon tetrachloride, are highly volatile and not of plant or animal origin. Also see NONVOLATILE MATTER.

WASTEWATER

A community's used water and water-carried solids (including used water from industrial processes) that flow to a treatment plant. Stormwater, surface water, and groundwater infiltration also may be included in the wastewater that enters a wastewater treatment plant. The term sewage usually refers to household wastes, but this word is being replaced by the term wastewater.

CHAPTER 2. INDUSTRIAL WASTEWATERS

(Lesson 1 of 2 Lessons)

Chapter 2 was prepared for industrial wastewater treatment plant operators. The chapter explains why industrial and municipal wastewaters must be treated. Words used by all operators are explained. The importance of understanding industrial wastewaters and the effects of industrial wastewaters are presented. This manual emphasizes the treatment of the physical–chemical characteristics of industrial wastewaters, including pH adjustment and control of toxic chemicals. Volume II in this series of manuals stresses the treatment of industrial wastes from industries producing *ORGANIC WASTES*[2] that require biological treatment processes.

2.0 PREVENTION OF POLLUTION

The operator's main job is to protect the many users of receiving waters. As an operator you must do the best you can to remove any substances that will unreasonably affect these users. Industrial wastewater treatment plant operators often operate plants that discharge to municipal wastewater collection systems. If toxic wastes are released to the collection system, then they can flow untreated or inadequately treated through municipal treatment plants, which are not set up to handle them. In addition, toxic industrial wastes can upset municipal plant processes and kill aquatic life in receiving waters.

Many people think *any* discharge of waste to a body of water is pollution. However, with our present system of using water to carry away the waste products of homes and industries, it would be impossible and perhaps unwise to prohibit the discharge of all wastewater to oceans, streams, and groundwater basins. Today's technology is capable of treating wastes in such a manner that existing or potential receiving water uses are not unreasonably affected. Definitions of pollution include any interference with the beneficial reuse of water or failure to meet water quality requirements. Any questions or comments regarding this definition must be settled by the appropriate enforcement agency.

2.1 WHAT IS PURE WATER?

Water is a combination of two parts hydrogen and one part oxygen, or H_2O. This is true, however, only for "pure" water such as might be manufactured in a laboratory. Water as we know it is not "pure" hydrogen and oxygen. Even the distilled water we purchase in the store has measurable quantities of various substances in addition to hydrogen and oxygen. Water picks up dissolved substances as it falls as rain, flows over land, and is used for domestic, industrial, agricultural, and recreational purposes. Rainwater, even before it reaches the earth, contains many substances. These substances, since they are not found in "pure" water, may be considered "impurities." When rain falls through the atmosphere, it gains nitrogen and other gases. As soon as the rain flows over land, it begins to dissolve from the earth and rocks such substances as calcium, carbon, magnesium, sodium, chloride, sulfate, iron, nitrogen, phosphorus, and many other materials. Organic matter (matter derived from plants and animals) is also dissolved by water from contact with decaying leaves, twigs, grass, or small insects and animals. Thus, a fresh flowing mountain stream may pick up many natural "impurities," some possibly in harmful amounts, before it ever reaches civilization or is affected by the waste discharges of society. Many of these substances, however, are needed in small amounts to support life and are useful to humans. Concentrations of impurities must be controlled or regulated to prevent harmful levels in receiving waters.

Water + Impurities

QUESTIONS

Write your answers in a notebook and then compare your answers with those on page 64.

2.1A   How does water pick up dissolved substances?

2.1B   In addition to hydrogen and oxygen, what other dissolved substances may be contained in water?

---

[2] *Organic Waste.*   Waste material that may come from animal or plant sources. Natural organic wastes generally can be consumed by bacteria and other small organisms. Manufactured or synthetic organic wastes from metal finishing, chemical manufacturing, and petroleum industries may not normally be consumed by bacteria and other organisms. Also see INORGANIC WASTE and VOLATILE SOLIDS.

2.2 TYPES OF WASTE DISCHARGES

The waste discharge that first comes to mind in any discussion of stream pollution is the discharge of domestic wastewater. Wastewater contains a large amount of organic waste. Industry also contributes substantial amounts of organic waste. Some of these organic industrial wastes come from vegetable and fruit packing; dairy processing; meat packing; tanning; and processing of poultry, oil, paper, and fiber (wood), and many other industries. All organic materials have one thing in common—they all contain carbon.

Another classification of wastes is *INORGANIC WASTES*.[3] Domestic wastewater contains inorganic material as well as organic, and many industries discharge inorganic wastes that add to the mineral content of receiving waters. For instance, a discharge of salt brine (sodium chloride) from water softening will increase the amount of sodium and chloride in the receiving waters. Some industrial wastes may introduce inorganic substances such as chromium or copper, which are very toxic to aquatic life. Other industries (such as gravel washing plants) discharge appreciable amounts of soil, sand, or grit, which also may be classified as inorganic wastes.

There are two other major types of wastes that do not fit either the organic or inorganic classification. These are heated (thermal) wastes and radioactive wastes. Waters with temperatures exceeding the requirements of the enforcing agency may come from cooling processes used by industry and from thermal power stations generating electricity. Radioactive wastes are usually controlled at their source, but could come from hospitals, research laboratories, and nuclear power plants.

QUESTIONS

Write your answers in a notebook and then compare your answers with those on page 64.

2.2A What are the four types or classifications of waste discharges?

2.2B What are some examples of industries that contribute substantial amounts of organic waste?

2.3 EFFECTS OF WASTE DISCHARGES

Certain substances not removed by wastewater treatment processes can cause problems in receiving waters. This section reviews some of these substances and discusses why they should be treated.

2.30 Sludge and Scum

If certain wastes (including industrial and domestic wastewater) do not receive adequate treatment, large amounts of solids may accumulate on the banks of the receiving waters, or they may settle to the bottom to form sludge deposits or float to the surface and form rafts of scum. Sludge deposits and scum are not only unsightly but, if they contain organic material, they may also cause oxygen depletion and be a source of odors. *PRIMARY TREATMENT*[4] units in the wastewater treatment plant are designed and operated to remove the sludge and scum before they reach the receiving waters.

2.31 Oxygen Depletion

Most living creatures need oxygen to survive, including fish and other aquatic life. Although most streams and other surface waters contain less than 0.001% dissolved oxygen (10 milligrams of oxygen per liter of water, or 10 mg/L[5]), most fish can thrive if there are at least 5 mg/L and other conditions are favorable. When oxidizable (organic) wastes are discharged to a stream, bacteria begin to feed on the waste and decompose or break down the complex substances in the waste into simple chemical compounds. These bacteria also use dissolved oxygen (similar to human respiration or breathing) from the water and are called *AEROBIC BACTERIA*.[6] As more organic waste is added, the bacteria reproduce rapidly; and as their population increases, so does their use of oxygen. Where waste flows are high, the population of bacteria may grow large enough to use the entire supply of oxygen from the stream faster than it can be replenished by natural diffusion from the atmosphere. When this happens, fish and most other living things in the stream that require dissolved oxygen die.

Oxygen depletion

Therefore, one of the principal objectives of wastewater treatment is to prevent as much of this "oxygen-demanding" organic material as possible from entering the receiving water. The treatment plant actually removes the organic material the same way a

---

[3] *Inorganic Waste.* Waste material such as sand, salt, iron, calcium, and other mineral materials that are only slightly affected by the action of organisms. Inorganic wastes are chemical substances of mineral origin; whereas organic wastes are chemical substances usually of animal or plant origin. Also see NONVOLATILE MATTER, ORGANIC WASTE, and VOLATILE SOLIDS.

[4] *Primary Treatment.* A wastewater treatment process that takes place in a rectangular or circular tank and allows those substances in wastewater that readily settle or float to be separated from the wastewater being treated. A septic tank is also considered primary treatment.

[5] *Milligrams per Liter, mg/L.* A measure of the concentration by weight of a substance per unit volume in water or wastewater. In reporting the results of water and wastewater analysis, mg/L is preferred to the unit parts per million (ppm), to which it is approximately equivalent.

[6] *Aerobic Bacteria* (air-O-bick back-TEER-e-uh). Bacteria that will live and reproduce only in an environment containing oxygen that is available for their respiration (breathing), namely atmospheric oxygen or oxygen dissolved in water. Oxygen combined chemically, such as in water molecules (H_2O), cannot be used for respiration by aerobic bacteria.

stream does, but it accomplishes the task much more efficiently by removing the wastes from the wastewater. SECONDARY TREATMENT[7] units are designed and operated to use natural organisms such as bacteria in the plant to STABILIZE[8] and to remove organic material.

Another effect of oxygen depletion, in addition to the killing of fish and other aquatic life, is the problem of odors. When all the dissolved oxygen has been removed, ANAEROBIC BACTERIA[9] begin to use the oxygen that is combined chemically with other elements in the form of chemical compounds, such as sulfate (sulfur and oxygen), which are also dissolved in the water. When anaerobic bacteria remove the oxygen from sulfur compounds, hydrogen sulfide (H_2S), which has a "rotten egg" odor, is produced. This gas is not only very odorous, but it also erodes (corrodes) concrete and can discolor and remove paint from homes and structures. Hydrogen sulfide also may form explosive mixtures with air and is a toxic gas capable of paralyzing your respiratory system. Other products of anaerobic decomposition (putrefaction: PYOO-truh-FACK-shun) also can be objectionable.

QUESTIONS

Write your answers in a notebook and then compare your answers with those on pages 64 and 65.

2.3A   How can oxygen depletion occur when oxidizable (organic) wastes are discharged to a stream?

2.3B   What kind of bacteria cause hydrogen sulfide gas to be produced?

2.32   Human Health

Up to now, we have discussed the physical or chemical effects that a waste discharge may have on the uses of water. More important, however, may be the effect on human health through the spread of disease-causing bacteria and viruses. Initial efforts to control human wastes evolved from the need to prevent the spread of diseases. Although untreated wastewater contains many billions of bacteria per gallon, most of these are not harmful to humans, and some are even helpful in wastewater treatment processes. However, humans who have a disease that is caused by bacteria or viruses may discharge some of these harmful organisms in their body wastes. Many serious outbreaks of communicable diseases have been traced to direct contamination of drinking water or food supplies by the body wastes from a human disease carrier.

Some known examples of diseases that may be spread through wastewater discharges are

Diseases

Fortunately, the bacteria that grow in the intestinal tract of diseased humans are not likely to find the environment in the wastewater treatment plant or receiving waters favorable for their growth and reproduction. Although many PATHOGENIC ORGANISMS[10] are removed by natural die-off during the normal treatment processes, sufficient numbers can remain to cause a threat to any downstream use involving human contact or consumption. If such uses exist downstream, the treatment plant must also include a DISINFECTION[11] process.

The disinfection process most often used is the addition of chlorine. In most cases, proper chlorination of a well-treated waste will result in a complete kill of the pathogenic organisms. Operators must realize, however, that the breakdown or malfunction of equipment could result in the discharge at any time of an effluent that contains pathogenic bacteria. To date, no one working in the wastewater collection or treatment field is known to have become infected by the AIDS virus due to conditions encountered while working on the job. Good personal hygiene is an operator's best defense against infections and diseases.

2.33   Other Effects

Some wastes adversely affect the clarity and color of the receiving waters, making them unsightly and unpopular for recreation.

---

[7] *Secondary Treatment.* A wastewater treatment process used to convert dissolved or suspended materials into a form more readily separated from the water being treated. Usually, the process follows primary treatment by sedimentation. The process commonly is a type of biological treatment followed by secondary clarifiers that allow the solids to settle out from the water being treated.

[8] *Stabilization.* Conversion to a form that resists change. Organic material is stabilized by bacteria that convert the material to gases and other relatively inert substances. Stabilized organic material generally will not give off obnoxious odors.

[9] *Anaerobic Bacteria* (AN-air-O-bick back-TEER-e-uh).   Bacteria that live and reproduce in an environment containing no free or dissolved oxygen. Anaerobic bacteria obtain their oxygen supply by breaking down chemical compounds that contain oxygen, such as sulfate (SO_4^{2-}).

[10] *Pathogenic* (path-o-JEN-ick) *Organisms.*   Organisms, including bacteria, viruses, or cysts, capable of causing diseases (such as giardiasis, cryptosporidiosis, typhoid, cholera, dysentery) in a host (such as a person). Also called PATHOGENS.

[11] *Disinfection* (dis-in-FECT-shun).   The process designed to kill or inactivate most microorganisms in water or wastewater, including essentially all pathogenic (disease-causing) bacteria. There are several ways to disinfect, with chlorination being the most frequently used in water and wastewater treatment plants. Compare with STERILIZATION.

Many industrial wastes are highly acid or alkaline (basic), and either condition can interfere with aquatic life, domestic use, and other uses. The accepted measurement of a waste's acidic or basic condition is its *pH*.[12] Before wastes are discharged, they should have a pH similar to that of the receiving sewer or receiving water.

Waste discharges may contain toxic substances, such as heavy metals (lead, mercury, cadmium, and chromium) or cyanide, which may affect the use of the receiving water for domestic purposes or for aquatic life. Plant effluents chlorinated for disinfection purposes may have to be dechlorinated to protect receiving waters from the toxic effects of residual chlorine.

Taste- and odor-producing substances may reach levels in the receiving water that are readily detectable in drinking water or in the flesh of fish.

Treated wastewaters contain *NUTRIENTS*[13] capable of encouraging excess *ALGAE*[14] and plant growth in receiving waters. These growths hamper domestic, industrial, and recreational uses. Conventional wastewater treatment plants do not remove a major portion of the nitrogen and phosphorus nutrients.

QUESTIONS

Write your answers in a notebook and then compare your answers with those on page 65.

2.3C   Where do the disease-causing bacteria in wastewater come from?

2.3D   What is the term that means "disease-causing"?

2.3E   What disinfection process is most often used at treatment plants?

2.4 SOLIDS IN WASTEWATER (Figure 2.1)

One of the primary functions of a treatment plant is the removal of solids from wastewater.

2.40 Types of Solids

In Section 2.2 you read about the different types of industrial wastes: organic, inorganic, thermal, and radioactive. For a normal municipal wastewater that contains domestic wastewater as well as some industrial and commercial wastes, the concerns of the treatment plant designer and operator usually are to remove the organic and inorganic suspended solids, to remove the dissolved organic solids (the treatment plant does little to remove dissolved inorganic solids, and to kill or inactivate the pathogenic organisms by disinfection. Thermal and radioactive wastes require special treatment processes.

Since the main purpose of the treatment plant is removal of solids from the wastewater, a detailed discussion of the types of solids is in order. Figure 2.1 will help you understand the different terms.

2.41 Total Solids

For discussion purposes, assume that you obtain a one-liter sample of raw wastewater entering the treatment plant. Heat this sample enough to evaporate all the water and weigh all the solid material left (residue); it weighs 1,000 milligrams. Thus, the total solids concentration in the sample is 1,000 milligrams per liter (mg/L). This weight includes both dissolved and suspended solids.

2.42 Dissolved Solids

How much is dissolved and how much is suspended? To determine this, you could take an identical sample and filter it through a very fine mesh filter such as a membrane filter or fiberglass. The suspended solids will be caught on the filter, and the dissolved solids will pass through with the water. You can now evaporate the water and weigh the residue to determine the weight of dissolved solids. In Figure 2.1 the amount is shown as 800 mg/L. The remaining 200 mg/L is suspended solids.

2.43 Suspended Solids

Suspended solids are composed of two parts: settleable and nonsettleable. The difference between settleable and nonsettleable solids depends on the size, shape, and weight per unit volume of the solid particles; large-sized particles tend to settle more rapidly than smaller particles. The amount of settleable solids in the raw wastewater should be estimated in order to design settling basins (primary units), sludge pumps, and sludge handling facilities. Also, measuring the amount of settleable solids entering and leaving the settling basin allows you to calculate the efficiency of the basin for removing the settleable solids. A device called an *IMHOFF CONE*[15] is used to measure settleable solids in milliliters per liter, mL/L. (The example in Figure 2.1 shows a settleable solids concentration of 130 mg/L. The settled solids in the Imhoff cone had to be dried and weighed by proper procedures to determine their weight.)

---

[12] *pH* (pronounce as separate letters).   pH is an expression of the intensity of the basic or acidic condition of a liquid. Mathematically, pH is the logarithm (base 10) of the reciprocal of the hydrogen ion activity. If $\{H^+\} = 10^{-6.5}$, then pH = 6.5. The pH may range from 0 to 14, where 0 is most acidic, 14 most basic, and 7 neutral.

[13] *Nutrient.*   Any substance that is assimilated (taken in) by organisms and promotes growth. Nitrogen and phosphorus are nutrients that promote the growth of algae. There are other essential and trace elements that are also considered nutrients. Also see NUTRIENT CYCLE.

[14] *Algae* (AL-jee).   Microscopic plants containing chlorophyll that live floating or suspended in water. They also may be attached to structures, rocks, or other submerged surfaces. Excess algal growths can impart tastes and odors to potable water. Algae produce oxygen during sunlight hours and use oxygen during the night hours. Their biological activities appreciably affect the pH, alkalinity, and dissolved oxygen of the water.

[15] *Imhoff Cone.*   A clear, cone-shaped container marked with graduations. The cone is used to measure the volume of settleable solids in a specific volume (usually one liter) of water or wastewater.

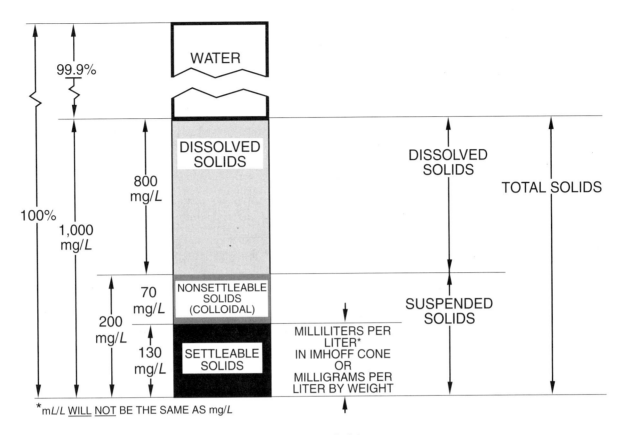

Fig. 2.1 Typical composition of solids in raw wastewater (floatable solids not shown)

You may calculate the weight of nonsettleable solids by subtracting the weight of dissolved and settleable solids from the weight of total solids. In Figure 2.1 the nonsettleable solids concentration is shown as 70 mg/L.

2.44 Organic and Inorganic Solids

For total solids or for any separate type of solids, such as dissolved, settleable, or nonsettleable, the relative amounts of organic and inorganic matter can be determined. This information is important for estimating solids handling capacities and for designing treatment processes for removing the organic portion in waste. The organic portion can be very harmful to receiving waters.

2.45 Floatable Solids

Treatment units are designed to remove the solids in raw wastewater and treated effluent. There is no standard method for the measurement and evaluation of floatable solids. Floatable solids are undesirable in the plant effluent because the sight of floatables in receiving waters indicates the presence of inadequately treated wastewater.

QUESTIONS

Write your answers in a notebook and then compare your answers with those on page 65.

2.4A Total solids consist of _dissolved_ and _suspended_ solids, both of which contain organic and inorganic matter.

2.4B Why is it necessary to determine the amount of settleable solids entering and leaving the settling basin?

2.4C An Imhoff cone is used to measure _settleable_ solids.

2.5 NATURAL CYCLES

When the treated wastewater from a plant is discharged into *RECEIVING WATERS*[16] such as streams, rivers, or lakes, natural cycles in the aquatic (water) environment may become upset. Whether any problems are caused in the receiving waters depends on the following factors:

1. Type or degree of treatment.

2. Size of flow from the treatment plant.

3. Characteristics of wastewater from treatment plant.

---

[16] *Receiving Water.* A stream, river, lake, ocean, or other surface or groundwaters into which treated or untreated wastewater is discharged.

4. Amount of flow in the receiving stream or volume of receiving lake that can be used for dilution.

5. Quality of the receiving waters.

6. Amount of mixing between *EFFLUENT*[17] and receiving waters.

7. Uses of receiving waters.

Natural cycles of interest in wastewater treatment include the natural purification cycles such as the cycle of water from evaporation or *TRANSPIRATION*[18] to condensation to precipitation to runoff and back to evaporation, the life cycles of aquatic organisms, and the cycles of nutrients. These cycles are occurring continuously in wastewater treatment plants and in receiving waters at different rates depending on environmental conditions. Treatment plant operators control and accelerate these cycles to work for their benefit in treatment plants and in receiving waters rather than have these cycles cause plant operational problems and disrupt downstream water uses.

NUTRIENT CYCLES[19] are a special type of natural cycle because of the sensitivity of some receiving waters to nutrients. Important nutrients include carbon, hydrogen, oxygen, sulfur, nitrogen, and phosphorus. All of the nutrients have their own cycles, yet each cycle is influenced by the other cycles. These nutrient cycles are very complex and involve chemical changes in living organisms.

To illustrate the concept of nutrient cycles, a simplified version of the nitrogen cycle will be used as an example (Figure 2.2). A wastewater treatment plant discharges nitrogen in the form of nitrate (NO_3^-) in the plant effluent to the receiving waters. Algae take up the nitrate and produce more algae. The algae are eaten by fish, which convert the nitrogen to amino acids, urea, and organic residues. If the fish die and sink to the bottom, these nitrogen compounds can be converted to ammonium (NH_4^+). In the presence of dissolved oxygen and special bacteria, the ammonium is converted to nitrite (NO_2^-) then to nitrate (NO_3^-), and finally the algae can take up the nitrate and start the cycle all over again.

If too much nitrogen is discharged to receiving waters, too many algae could be produced. Water with excessive algae can be unsightly. Bacteria decomposing dead algae from occasional die-offs can deplete the dissolved oxygen and cause a fish kill. Thus, the nitrogen cycle has been disrupted, as well as the other nutrient cycles. If no dissolved oxygen is present in the water, the nitrogen compounds are converted to ammonium (NH_4^+), the carbon compounds to methane (CH_4), and the sulfur compounds to hydrogen sulfide (H_2S). Ammonia (NH_3) and hydrogen sulfide are odorous gases. Under these conditions, the receiving waters are *SEPTIC*[20]; they stink and look terrible. Operators need to know how to control these nutrient cycles in their treatment plant in order to treat wastes and to control odors, as well as to protect receiving waters.

QUESTIONS

Write your answers in a notebook and then compare your answers with those on page 65.

2.5A Why do treatment plant operators need to have an understanding of natural cycles?

2.5B What can happen when nutrient cycles are disrupted and no dissolved oxygen is present in the water?

2.6 NEED TO TREAT INDUSTRIAL WASTES
by Dan Campbell

2.60 Uses of Water by Industry

Most industrial processes either require water or are made more efficient by the use of water for one purpose or another. As the water is being used, some of the other materials used in the industrial process may become mixed in with the water. That part of the water which is not reused by the industry is discharged as industrial wastewater.

Figure 2.3 shows some of the major potential uses of water by industry. At "A" the water is taken from its source. This could be taken directly from a surface water source, such as a river or lake, directly from a groundwater source, including various types of wells, or from a community water supply. Note that whatever the source, water used by industry may need to be treated by the industry before it is suitable for use.

The water may then be used by industries in a number of ways, as shown at "B." In many cases, water is necessary in the process itself. An obvious example is laundry, but other examples include the pulp and paper industry, dairies, breweries, distilleries, and others. In many cases, water is used to make processing of other materials easier or more efficient. In any of these cases, this is referred to as "process water." Another use of water by industry is for the transport of materials to, through, and from the industrial process, such as in vegetable processing. As shown in Figure 2.3, water also is used for cooling, as for example in refineries and steel mills. Many industries use wash water for both products (fruit and vegetable processing, some metal finishing operations) and facilities (canneries, dairies).

---

[17] *Effluent* (EF-loo-ent). Water or other liquid—raw (untreated), partially treated, or completely treated—flowing *FROM* a reservoir, basin, treatment process, or treatment plant.

[18] *Transpiration* (TRAN-spur-RAY-shun). The process by which water vapor is released to the atmosphere by living plants. This process is similar to people sweating. Also see EVAPOTRANSPIRATION.

[19] *Nutrient Cycle.* The transformation or change of a nutrient from one form to another until the nutrient has returned to the original form, thus completing the cycle. The cycle may take place under either aerobic or anaerobic conditions.

[20] *Septic* (SEP-tick) *or Septicity.* A condition produced by bacteria when all oxygen supplies are depleted. If severe, the bottom deposits produce hydrogen sulfide, the deposits and water turn black, give off foul odors, and the water has a greatly increased oxygen and chlorine demand.

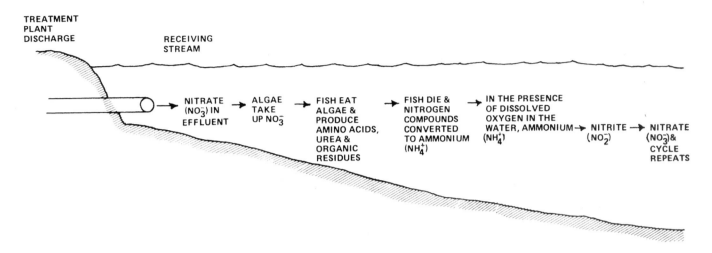

Fig. 2.2 Simplified illustration of nitrogen cycle

After being put to one or more of the above uses, the water may be recycled for repeated use "C" if its characteristics have not been changed so much that it becomes unsuitable. The water may be recycled directly or it may be treated to make reuse possible. When the water is treated, materials are removed. These materials may then be a disposal problem "E." Also, these materials may have value to the industry or other industries either in the same or a different process. An example of this is the use of wastes from steel pickling by municipal wastewater treatment plants "D." When it is uneconomical for the industry to recycle this water, it becomes industrial wastewater "F."

In almost all cases, industrial wastewater is not suitable for direct discharge to a receiving water, but must first be treated to remove pollutants. Figure 2.3 shows three possible alternatives for treatment of industrial wastewater:

(G) Discharge to a Publicly Owned Treatment Works (POTW) to be combined with municipal wastes and possibly wastes from other industries, treated, and returned to the environment.

(H) Pretreatment by industry, followed by discharge to a POTW, as above.

(I) Treatment by industry to the extent required before discharge to a receiving water.

Whatever scheme is used, the ultimate objective is the protection of the receiving water for protection of the public health and for the various uses of the receiving waters.

QUESTIONS

Write your answers in a notebook and then compare your answers with those on page 65.

2.6A What are the major ways water is used by industries?

2.6B What are three possible alternatives for treatment of industrial wastewater?

2.61 Reasons for Treatment of Industrial Wastes

2.610 Dissolved and Suspended Wastes

During any of the four general uses of water by industry, the water may acquire waste materials. These may be in the form of "dissolved" matter, usually defined for wastewater purposes as those wastes that will pass through a fine filter, or those in the form of "suspended" matter, that is, particles that may settle to the bottom or float to the surface if given enough time. More frequently, waste materials are combinations of dissolved and suspended matter. Depending on the specific nature of the matter that is added, the characteristics of the water may be changed, thus affecting its potential for use.

An important basic concept is that water in the environment belongs to all of us, not to any individual or group. Individuals or groups are allowed to use water, but they must use it in such a way as not to interfere with its use by others with rights to use the water. Let us briefly examine some of the uses of water and some ways in which wastes can interfere with the use of water.

2.611 Drinking Water

Use of water for drinking and other domestic purposes is the most basic use of water and generally has priority over any competing use. There are many ways in which water can be rendered unsafe or unfit for drinking. These different types of contamination are discussed in the following paragraphs.

1. *BIOLOGICAL CONTAMINATION.* This problem usually stems from prior use of water to carry human wastes. Sanitary wastewater, domestic wastes, or simply sewage or wastewater are terms used to designate this category of wastes. Bacteria, viruses, and other microorganisms are excreted from the human body and other warm-blooded animals. Some of these organisms may cause disease if they are successful in completing or short-circuiting a cycle through wastewater treatment, discharge to the environment, and

34   Treatment Plants

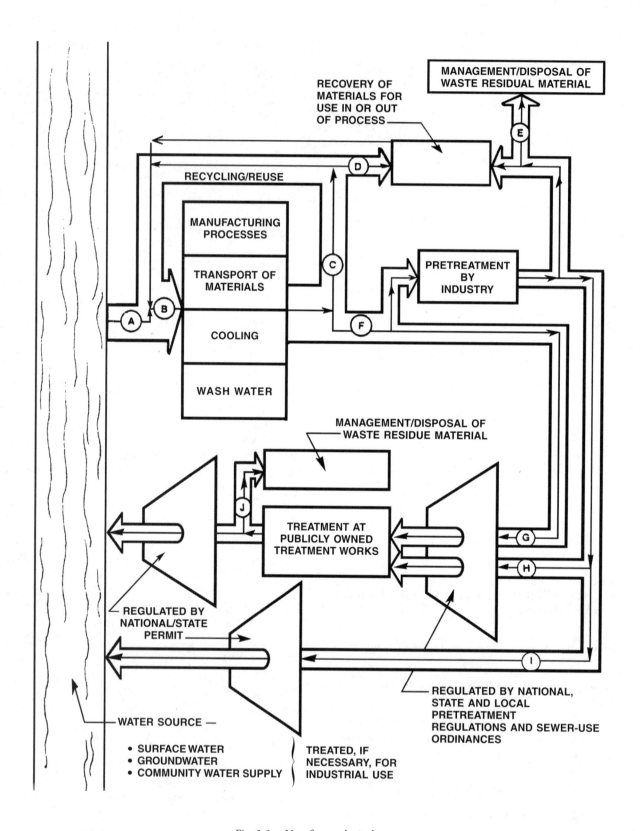

Fig. 2.3   Use of water by industry

treatment for water supply, and are ingested by a human. In addition to being found in sanitary wastewater, such microorganisms may reach environmental waters through stormwater as it washes the land of animal wastes and through certain industries, most notably slaughterhouses and tanneries.

Protection of human populations from this biological contamination has been the historical foundation for public wastewater treatment and water treatment, with disinfection being the particular process most involved. Where wastewater and water treatment are routine practices, diseases that otherwise be transmitted through drinking water, such as typhoid, cholera, and dysentery, are rare. Industries that may potentially discharge pathogenic organisms must be sure that their wastes are adequately disinfected.

2. *CHEMICAL CONTAMINATION.* This is potentially a whole host of problems, rather than a single, neat category. Depending on their nature and concentration, foreign chemicals in drinking water may produce *ACUTE HEALTH EFFECTS*,[21] or may produce long-term *CHRONIC HEALTH EFFECTS*[22] that are not diagnosed for years, but that may have severe effects. This is not to say that all chemicals in water, even drinking water, are bad. Chlorine, for example, has been used for generations as an economical and effective disinfectant, and its use has undoubtedly saved untold numbers of lives. We are now learning, however, that when combined with certain other chemicals, chlorine may cause long-term disease. This serves to emphasize that we are not free to use water to "wash away" compounds we are not sure what else to do with when we want to get rid of them.

Disposing of chemical wastes is made even more complex by the fact that many of the compounds considered toxic are dangerous at even extremely low concentrations, measured as micrograms per liter or parts per billion. Imagine trying to find six black pingpong balls in a room filled with them and 999,999,994 white ones. This is equivalent to six parts per billion and roughly equivalent to 6 micrograms per liter. Some of the categories of chemical contamination are listed below:

a. *TOXIC CHEMICALS.* A great variety of chemicals have toxic effects in drinking water. These range from simple chemical elements, such as mercury, chromium, and other metals, to compounds consisting of numbers of elements, such as pesticides. From an industrial wastewater standpoint, the most likely source of this type of contamination is from the manufacturing processes in which these chemicals are either used or made and from in-plant spills and accidents.

A number of recent episodes with toxic and potentially toxic chemicals have shown us that many of the dangers of toxic compounds are not even suspected at the time we begin using them.

b. *CHEMICALS NOT NORMALLY TOXIC.* Many chemicals are not considered toxic in concentrations normally encountered, but may present other problems if present in drinking water. Oils and greases present in water supplies may pass through water treatment processes in concentrations high enough to cause tastes and odors. Inorganic salts in low concentrations are considered beneficial to water supplies. In higher concentrations, however, they present problems. Sulfate in high concentrations produces a laxative effect. Waters high in inorganic salts cause crusting or deposits on the insides of pipes, thus decreasing the capacity of the pipes.

3. *RADIOACTIVE CONTAMINATION.* Long-term health damage and even death could be caused by the presence of radioactive contaminants in drinking water.

4. *OTHER TYPES OF CONTAMINATION.* Other factors, such as temperature, color, and turbidity, may be problems in water supplies. These factors may not cause health problems by themselves, but may cause people to use a different source of water that may be unsafe because the safe source is made unattractive.

2.612 Cooking Water

Water used for food preparation may be ingested, and of course the food is ingested. Consequently, the potential problems are very much like those listed above for drinking water.

2.613 Water-Contact Recreation

Activities, such as swimming and water skiing, in which the body has direct contact with the water, are referred to as water-contact recreation. Although the health risks from water-contact recreation are not as great as from drinking water, they are real. The most obvious forms of pollution interfering with water-contact recreation are those that make the water unappealing.

1. *BIOLOGICAL CONTAMINATION.* The most likely risk from biological contamination is that of eye, ear, and skin infection, although accidental ingestion of contaminated water is also a risk.

2. *CHEMICAL CONTAMINATION.* Ingestion is considered less of a problem due to the small volume that might be acci-

---

[21] *Acute Health Effect.* An adverse effect on a human or animal body, with symptoms developing rapidly.
[22] *Chronic Health Effect.* An adverse effect on a human or animal body with symptoms that develop slowly over a long period of time or that recur frequently.

dentally swallowed. Acids and other chemicals, if present in water, may cause irritation of the eyes and other sensitive parts of the body. There are many chemicals, such as some pesticides, that have toxic effects that are severe upon contact with the skin.

3. OXYGEN-CONSUMING WASTES. These wastes actually are a form of chemical pollution. However, this problem is so severe that it is considered a problem in its own right. Many organic or oxygen-consuming wastes may be discharged from dairies, breweries, distilleries, pharmaceutical manufacturing, fruit and vegetable processing, textile manufacturing, paper manufacturing, coke and gas industries and municipalities.

These types of wastes are actually treated in a stream or river by the action of the same microorganisms found in a treatment plant. In this treatment process, the organisms "breathe" the oxygen that is present in the water. If all the oxygen is gone, other microorganisms, which can get their oxygen from compounds in the water, continue the treatment process. The by-products of this latter type of treatment (anaerobic decomposition) produce unpleasant odors, colors, and sludges that discourage water-contact recreation.

4. INORGANIC ELEMENTS. This also is a form of chemical pollution, but some types of inorganic elements, such as phosphorus and nitrogen, are nutrients that promote the growth of algae. Algae are plants which, under proper conditions, can multiply to such an extent in water that they interfere with water-contact recreation and other uses of the water. In sunlight, algae add oxygen to water, but in the absence of light and as they decay, dissolved oxygen is removed from the water. Algae reduce the clarity of water and give it a green color. Both of these items are considered undesirable from an aesthetic standpoint. Phosphorus and nitrogen also promote the growth of larger vegetation, such as aquatic weeds.

5. SOLIDS AND VISIBLE POLLUTION. Visible pollution obviously interferes with water-contact and other forms of water recreation. Floating solids, oil, and foam are potential contaminants from industry that would be in this category. Other types of solids may settle to the bottom of the water and form sludge beds. Organic material in the sludge beds decomposes, presenting the problems identified under "Oxygen-Consuming Wastes."

6. OTHER CONTAMINANTS. Increased temperature in water decreases the amount of oxygen available and, at the same time, leads to an increasing rate of microorganism uptake of the oxygen that is present. This compounds the problems of oxygen-demanding wastes described above. Radiation poses a health hazard from contact in which the radioactive material may remain on the skin.

2.614 Non-Body-Contact Water Recreation

Boating and similar non-body-contact recreation activities are interfered with primarily by visible types of pollution. Fishing, of course, would depend on the suitability of water to support fish.

2.615 Water for Fish, Wildlife, and Aquatic Vegetation

Water is necessary for all forms of life. The types of contamination described above, particularly with respect to drinking water, can interfere with the use of the water to support forms of life other than human. Many fish and other organisms are even more susceptible to damage from toxic materials than humans. Many states limit the amount of chlorine that can be added to wastewater effluent because of the sensitivity of aquatic life to chlorine. An additional problem is that some lower forms of life can accumulate some toxic materials. As fish and higher forms of life consume these smaller organisms, toxicity levels rise in the higher forms. This bioaccumulation magnifies the hazard both to the fish and other creatures higher in the food chain, including humans.

Fish also depend on the dissolved oxygen present in water for life. For this reason, many of the standards set for levels of dissolved oxygen in water are based on the needs of fish.

2.616 Agricultural Use

Water for irrigation should be reasonably free from biological contamination. Many chemicals, including some metals and inorganic salts, will retard or prevent crop growth at certain concentrations and would be harmful for livestock.

2.617 Industrial Use

The suitability of water for industrial use depends largely on the industry in question. Water that is safe for drinking sometimes must receive additional treatment for certain industrial uses.

QUESTIONS

Write your answers in a notebook and then compare your answers with those on page 65.

2.6C What are some of the uses of water with which wastes may interfere?

2.6D Name three ways in which water can be biologically contaminated.

2.6E What is the particular treatment process most involved with the protection of human populations from biological contamination?

2.6F From an industrial wastewater standpoint, what is the most likely source of toxic chemical contamination?

2.6G Describe how oxygen-consuming wastes interfere with water use.

END OF LESSON 1 OF 2 LESSONS
on
INDUSTRIAL WASTEWATERS

Please answer the discussion and review questions next.

DISCUSSION AND REVIEW QUESTIONS
Chapter 2. INDUSTRIAL WASTEWATERS
(Lesson 1 of 2 Lessons)

At the end of each lesson in this chapter you will find some discussion and review questions. The purpose of these questions is to indicate to you how well you understand the material in the lesson. Write the answers to these questions in your notebook.

1. What is the main job of an industrial wastewater treatment plant operator?

2. What are the major types of industrial wastes?

3. Before industrial wastes are discharged, what should be the pH level?

4. What are the different types of solids in industrial wastewaters?

5. Identify one advantage and one limitation of the use of chlorine as a disinfectant.

6. What problems could be caused when oils and greases are present in drinking water supplies?

7. If algae add oxygen to water in the presence of sunlight, then why are algae considered undesirable?

CHAPTER 2. INDUSTRIAL WASTEWATERS

(Lesson 2 of 2 Lessons)

2.7 IMPORTANCE OF UNDERSTANDING INDUSTRIAL WASTEWATERS
by Richard von Langen

There are two major reasons for understanding the sources of industrial wastes. First, by understanding the sources and quantities of industrial wastewaters, an operator can identify, define, and solve industrial wastewater treatment system (IWTS) problems caused by industrial discharges. Your IWTS was designed to treat specific wastestreams. Sources of industrial wastewater include the manufacturing process, utility systems maintenance activities, and the industrial wastewater treatment system itself. If the character of the waste changes (type of pollutant, concentration, volume of the discharge, or frequency of the discharge), your system may not operate correctly. The result may be that the discharge goes out of compliance with the discharge limits.

As an operator, you will need to understand how the various industrial wastes interact with each other if you are to deal effectively with the impact of industrial wastewaters on your facilities. You must have a basic understanding of chemistry and chemical reactions, the industrial wastewater generating processes, biological reactions, and your own collection and treatment system.

The second reason for understanding the sources of industrial wastes is to determine their effect on the environment. The very reason why you have an IWTS is to prevent the discharge from your facility from adversely affecting the POTW if you are an indirect discharger. If you are a direct discharger, the purpose of the IWTS is to maintain and preserve the receiving waters. Protection of the POTW and the receiving waters is the basis for the IWTS discharge limits, self-monitoring requirements and reporting requirements in the discharge permit.

Industrial wastewater discharges can have numerous adverse impacts on the wastewater collection system, treatment facilities, and the environment. The discharges can corrode sewers, plug sewers, and release toxic gases and obnoxious odors in the collection system. When these wastewaters reach treatment plants, they can be toxic to the biological processes, they can inhibit microorganisms from performing their intended functions, and they can accumulate to undesirable levels in the microorganisms. After industrial wastes leave the treatment facility in process sludges or the treated effluent, they can be toxic to microorganisms, aquatic life, and people in the environment.

In order to fully understand the effects of industrial wastes on treatment systems and the environment, an operator must be able to accurately characterize the industrial wastestreams and plant discharges. This means the operator must be able to identify and describe not only the types of pollutants in a waste or wastestream, but also the quantities and concentrations of pollutants. The frequency of generation and discharge, whether it be once per year or once per day, is also very important.

Wastes generated by industry may be COMPATIBLE POLLUTANTS,[23] such as biochemical oxygen demand (BOD), suspended solids (SS), ammonia, or oil and grease; or they may be NONCOMPATIBLE POLLUTANTS,[24] such as heavy metals (cadmium, chromium, copper) or organics (acetone, methyl ethyl ketone, toluene, 1,1,1-trichloroethylene). Other chemical constituents, such as sodium, calcium, nitrate, carbonate, sulfide, and pH, in excess can also affect collection and treatment facilities, as well as the potential reuse of the effluent. Physical characteristics of the industrial waste, such as temperature, solubility or insolubility, and viscosity, can affect the chemical, mechanical, and biological activities of the conveyance and treatment systems.

This lesson will present information about the most common types of industrial wastes, explain how they are generated, discuss the characteristics of the wastes, and explain the effects industrial wastes may have on your facilities and those of the POTW collection, treatment, and disposal system. In Section 2.10, "Manufacturing Processes and Wastewater Generation," you will find descriptions of the wastes and wastewaters generated at each step of the manufacturing process for two major in-

---

[23] *Compatible Pollutants.* Those pollutants that are normally removed by the POTW treatment system. Biochemical oxygen demand (BOD), suspended solids (SS), and ammonia are considered compatible pollutants.

[24] *Noncompatible Pollutants.* Those pollutants that are normally *NOT* removed by the POTW treatment system. These pollutants may be a toxic waste and may pass through the POTW untreated or interfere with the treatment system. Examples of noncompatible pollutants include heavy metals, such as copper, nickel, lead, and zinc; organics, such as methylene chloride, 1,1,1-trichloroethylene, methyl ethyl ketone, acetone, and gasoline; or sludges containing toxic organics or metals.

dustries. Even if your industry is not specifically described, try to "read between the lines" as you read this lesson to understand the methodology of investigating industrial waste sources. Look at the examples as a guide to the investigative technique and thought processes. By noting the methodology, you can develop a more complete understanding of industrial wastes and the mix of wastes at your facility.

2.70 Manufacturing Processes and Wastewater Generation

Knowledge of the various manufacturing processes is important in understanding how raw materials are transformed into products, by-products, and waste products, and in understanding the effects of the waste products on collection systems, the industrial wastewater treatment plant, and the environment. Many different industries use the same basic manufacturing processes, including plating, sheet forming, distilling, extraction, screening, grinding, and heating. For example, the metal plating process is used to plate everything from printed circuit boards to pool table legs. This manufacturing process immerses the part into a concentrated chemical solution to clean, activate, plate, or seal it. The part is then removed from the solution and transferred to another solution, usually water, to remove any excess chemical. Subsequently, the part may again be moved and dipped into another concentrated chemical solution. Let us look at some of the many different ways wastewaters can be generated and then discharged to the industrial sewer with this simple but widely used manufacturing process.

1. If the concentrated solution is *VOLATILE*,[25] reactive with the part, or may fume, the dipping of the part into the concentrated solution will produce an air contaminant that may have to be condensed, scrubbed, or removed. The scrubber wastewater is an industrial wastewater.

2. If the concentrated solution tank were to burst due to an accident or tank failure, the concentrated chemical solution could flow into the drain to the plant sewer unless there are provisions for spill containment.

3. The excess concentrated solution (*DRAG OUT*[26]) on the part may fall on the floor during the process of transferring the part from one tank to the next. This liquid may flow into the drain to the plant sewer if not contained.

4. The drag out on the parts must be rinsed off using either still rinses or flowing rinses in order to prevent contamination of subsequent plating baths. Still rinses and the overflow from constant flow rinses are industrial wastewaters.

5. Once the concentrated chemical solution is spent because the active chemical has been used up, it must be disposed of. This spent solution is an industrial wastewater.

6. The plater was adding water to a tank and forgot to turn it off. The tank overflowed to the floor; this flow is an industrial wastewater.

7. The process of stripping a finish of unacceptable quality is a potential problem. Stripping the plated metal with either a cyanide solution or a strong acid or alkaline solution generates a quantity of very concentrated wastewaters, which are considered an industrial wastewater.

Wastes from all of these different sources should flow to the industry's treatment facility before being discharged to a POTW's wastewater collection system or a receiving stream. To understand the full impact of industrial wastes, two other variables must also be considered: concentration of potentially harmful constituents, and the overall volume of waste discharged per unit of time. The concentration of a pollutant from an industrial source may be low, but because of a high overall volume of discharge, the total mass discharged to the sewer may impact your treatment facilities or the POTW. Conversely, even if a discharged concentration is high, there may be no adverse effects if the total flow volume is low.

2.71 Utility Processes and Wastewater Generation

The utility (service) processes are similar in their effects on the IWTS and POTW systems. In the example described below, wastestreams generated by the operation of a boiler illustrate how industrial wastes may be produced by a utility process.

1. Water used in the boiler may be treated to remove suspended solids by filtration. Backwash of the filter will produce an industrial waste.

2. If the water is softened or demineralized, the regeneration of the softener or demineralizer will produce a backwash wastestream with suspended solids, spent regenerant containing high concentrations of the ions that have been removed (such as calcium and magnesium), as well as regenerant (salt, in the case of a softener, and acid and alkali, in the case of demineralizers) and rinse water. Each of the separate steps in the regeneration process produces an industrial waste.

3. Chemical additives to remove oxygen, inhibit corrosion, or reduce scaling are often added to the boiler feedwater. Industrial wastestreams from this operation could include leaks and spills of these chemicals.

---

[25] *Volatile* (VOL-uh-tull). (1) A volatile substance is one that is capable of being evaporated or changed to a vapor at relatively low temperatures. Volatile substances can be partially removed from water or wastewater by the air stripping process. (2) In terms of solids analysis, volatile refers to materials lost (including most organic matter) upon ignition in a muffle furnace for 60 minutes at 550°C (1,022°F). Natural volatile materials are chemical substances usually of animal or plant origin. Manufactured or synthetic volatile materials, such as plastics, ether, acetone, and carbon tetrachloride, are highly volatile and not of plant or animal origin. Also see NONVOLATILE MATTER.

[26] *Drag Out.* The liquid film (plating solution) that adheres to the workpieces and their fixtures as they are removed from any given process solution or their rinses. Drag-out volume from a tank depends on the viscosity of the solution, the surface tension, the withdrawal time, the draining time, and the shape and texture of the workpieces. The drag-out liquid may drip onto the floor and cause wastestream treatment problems. Regulated substances contained in this liquid must be removed from wastestreams or neutralized prior to discharge to POTW sewers.

4. Part of the normal operation of the boiler is to remove the buildup of suspended and dissolved solids from the boiler system. In small systems this is done by periodically discharging a small stream of condensed steam *(BLOWDOWN*[27]*)* to the drain. Larger systems may continuously discharge blowdown. This is an industrial waste.

5. Steam traps are often located within the steam distribution system. Their function is to remove from the steam lines any condensed steam that would impede the flow or decrease the heat value of the steam. The steam trap drains may discharge to a condensate return system or the area drain. If this liquid is discharged to the drain, it is an industrial waste.

6. The condensate return system typically includes a receiver, level control system, and pumps to return the condensate back to the feedwater system for reuse. If the pumps leak or the receiver leaks or overflows to the drain, the released condensate could be an industrial wastestream.

These industrial wastestreams may be continuous or intermittent discharges and follow the general rule that the larger the system, the more likely the discharge will be continuous. Table 2.1 summarizes the sources of utility process wastestreams and indicates the type of discharge (continuous or intermittent).

2.72 Maintenance Activities and Wastewater Generation

Maintenance activities also produce industrial wastestreams that can affect the IWTS, the POTW, or the environment. Both routine and emergency maintenance can contribute concentrated wastes which, if not properly managed, will interfere with the operation of the IWTS.

By definition, routine maintenance activities are planned and therefore can be controlled. Activities that result in the disassembly of equipment (for example, pump packing replacement or in-line pH probe replacement) generate an industrial waste when liquid in the piece of equipment is emptied to work on the equipment. This waste may also be combined with other wastes generated as part of the maintenance activity, such as greases, lubricants, old packing material, soaps, buffer solutions, and cleaning compounds. The result is mixed industrial waste that has characteristics of several different types of wastes (liquid and/or solid) and may require special handling.

Other routine maintenance activities that deal with concentrated processes or utility chemicals and the equipment that handles them may generate wastes that are outside of the standard wastestreams the IWTS is designed to treat. Proper characterization of this waste is necessary to properly pretreat or blend this waste into the IWTS.

Emergency maintenance activities do not occur on a regular basis and, by definition, are not planned. Explosions, fires, and tank ruptures all represent catastrophic emergency situations that result in the generation of industrial wastes that may require treatment. In situations like these, the industrial wastes generated are likely to be mixed wastes. They may range from concentrated to dilute, and the volume may be small or thousands of gallons. For example, a process tank, such as a cadmium cyanide bath, ruptures and overflows to an oily waste trench. This event will generate a concentrated, mixed wastestream that must be treated for oil, cyanide, and heavy metals. Or, suppose a fire occurs in a *MISCIBLE*[28] solvent tank. The water used to suppress and put out the fire may generate a dilute waste. As part of getting the manufacturing process back on line, you will have to handle and possibly treat these wastes.

Both emergency and routine maintenance activities may generate wastes that are not normally a part of the IWTS wastestreams. Manufacturing processes, such as drilling or grinding operations, or utility processes, such as inert gas generation, are normally dry processes. However, maintenance of the equipment for these processes may generate an oily waste that the IWTS does not normally treat or was not designed to treat. As an industrial wastewater treatment plant operator, you need to understand the maintenance functions and activities. A maintenance wastestream that is different from the type of waste (either in pollutant type, concentration, or volume) the system was designed for must be evaluated prior to treatment.

You will also need to plan for the unexpected. You will need to determine how you will evaluate and then handle an industrial waste generated as a result of an emergency maintenance or response activity. Understanding how that waste was generated and properly characterizing the waste will assist you in developing a strategy for treatment and disposal.

2.73 IWTS Wastewater Generation

The last source of industrial wastestreams is the IWTS itself. The IWTS will generate industrial wastestreams that will affect the operation of the IWTS. Recycle streams, such as the backwash from a granular filter or filtrate from the filter press, put a load on the IWTS. While the system should be designed to account for these streams, changes in these operations will impact your IWTS. Examples include:

1. Increases in the backwash flow rate from the filter increase the overall volume that must be treated.

---

[27] *Blowdown.* The removal of accumulated solids in boilers to prevent plugging of boiler tubes and steam lines. In cooling towers, blowdown is used to reduce the amount of dissolved salts in the recirculated cooling water.

[28] *Miscible* (MISS-uh-bull). Capable of being mixed. A liquid, solid, or gas that can be completely dissolved in water.

TABLE 2.1 UTILITY SYSTEM DISCHARGES OF WASTEWATER

Utility System Component	Type of Discharge, C/I[a]	Remarks
Cooling Tower	C	Cooling tower blowdown may contain suspended and dissolved solids, chromium, molybdenum, and biocides. Smaller systems may discharge once per shift.
Boiler	I	Boiler blowdown contains suspended and dissolved solids, heat, and sulfite. Larger systems discharge continuously.
Air Compressor		
Moisture	I	During air compression, moisture is condensed. No pollutants.
Cooling Water	C	Cooling water is required for larger compressors.
Inert Gas	C	Water is usually required to cool the compressor. Some moisture may also condense and contain sulfate or nitrogen oxides.
Pump, Mixer, or Equipment Seals	C	Water may be used on rotating equipment to seal or lubricate. Leaks or normal weeping will contain the product being mixed or pumped.
Demineralization Systems	I	Brine wastes high in total dissolved solids from softeners. Acidic and alkaline wastes from regeneration of demineralizers.
Air Scrubbers	C	Scrubber blowdown contains suspended solids, the solubilized air contaminants, and treatment chemicals (typically alkaline). For smaller scrubbers or systems with high solubility of contaminants, blowdown is intermittent.

[a] C, Continuous discharge
I, Intermittent discharge

2. A decrease in the time between backwashes (or lowering the differential pressure backwash control point) will increase the amount of wastewater that must be treated.

3. Acid cleaning of the filter media will generate an infrequent but concentrated industrial wastestream. Changes in the frequency of cleaning, the type of acid, the concentration of acid, or the amount of material on the filter media will impact this recycle stream.

4. To increase production of a filter press, you decide to install a more permeable filter cloth. This could result in a change in the characteristics of the filtrate. It may contain more suspended solids and, with the higher production rate, a greater volume of filtrate may need to be treated.

Routine maintenance of IWTS equipment, such as cleaning the chemical feed pumps, will produce an industrial waste that will have to be treated. Effects of cleaning the pump include both the downtime of the system and the waste generated by cleaning the pump.

QUESTIONS

Write your answers in a notebook and then compare your answers with those on pages 65 and 66.

2.7A What physical characteristics of industrial wastes might affect the chemical, mechanical, and biological activities of the conveyance and treatment systems?

2.7B Why is a knowledge of the various industrial manufacturing processes important for an industrial wastewater treatment plant operator?

2.7C How could routine maintenance, such as replacing the packing in a pump, be the source of an industrial waste?

2.7D What is a "mixed waste"?

2.8 INDUSTRIAL WASTESTREAM VARIABLES

This section discusses the various types of industrial wastewaters, how they can be generally classified, the importance of knowing the frequency of generation and discharge, and the effects these discharges may have on the industry's treatment system. Examples of each type of industrial waste are given.

2.80 Compatible and Noncompatible Pollutants

Compatible pollutants can be defined as those pollutants that are normally removed by the POTW treatment system. Biochemical oxygen demand (BOD), suspended solids (SS), oil and grease, and ammonia are considered compatible pollutants. The POTW is designed to treat primarily domestic wastewater and the compatible pollutants discharged by industry.

Noncompatible pollutants are defined as those pollutants that are not normally removed by the POTW, may be toxic to a biological IWTS, and may cause pass-through or *INTERFERENCE*[29] with the treatment system. Even some biologically degradable wastes, such as soluble, synthetic cooling oils, may cause interference with the heavy metal removal system by inhibiting floc formation. Other examples of noncompatible pollutants include heavy metals, such as copper, nickel, lead, and zinc; organics, such as methylene chloride, 1,1,1-trichloroethylene, methyl ethyl ketone, acetone, and gasoline; and sludges containing toxic organics or metals.

From the perspective of the POTW, compatible pollutants sometimes exhibit the characteristics of noncompatible pollutants, and vice versa. Soluble BOD from a food industry may have some harmful effects on a POTW's secondary treatment system. The accidental discharge of ammonia by a fertilizer manufacturer may disrupt the nitrification/denitrification or stripping tower processes used by the POTW to treat ammonia. On the other hand, some of the heavy metals (usually classified as noncompatible pollutants) are used as micronutrients to aid in the production of biological mass and the reduction of BOD. Certain organic chemical wastes, such as acetone and isopropanol, are biodegradable and, in dilute solutions, are removed by biological action in secondary treatment.

2.81 Dilute Solutions

The discharges from continuous manufacturing processes are normally dilute solutions of compatible and sometimes noncompatible pollutants. They may be discharged to the industry's pretreatment system or directly to the POTW without any pretreatment. Manufacturing processes such as plating bath rinses, raw food cleaning, and crude oil dewatering are all examples of dilute solutions of pollutants that may be discharged directly to a POTW sanitary sewer. If a problem occurs in the manufacturing process, a probable result is that the quality of wastewater will change; it may be more laden with pollutants. Some wastestreams from utility services, such as cooling tower and boiler blowdown, are continuous and represent the discharge of dilute solutions. (The other characteristics of utility system discharges are discussed in Section 2.71 and listed in Table 2.1.)

Another low-strength wastewater is stormwater runoff from chemical handling and storage areas. Products that may have spilled on the industry's grounds are washed off during a rainstorm or during the spring thaw. The pollutant concentration is usually too dilute to require pretreatment before discharge to the sewer, but exceeds the discharge standards for discharge to surface waters. While the strength of the storm runoff may be low, the volume that must be treated, in addition to normal flow to the pretreatment system or to the POTW, can cause hydraulic capacity problems. Excessive flows can be diverted to storage reservoirs or basins and then gradually discharged to the pretreatment system.

A great deal of attention is presently focused on cleaning up groundwater sources that have been contaminated by leaking underground storage tanks. Cleanup projects of this nature typically involve large quantities of wastes that may contain high concentrations of solvents, fuels, heavy metals, and pesticides. Because of the public attention surrounding groundwater cleanup projects, pretreatment of the contaminated water is almost always required and the result is usually a "high-quality" industrial wastewater.

2.82 Concentrated Solutions

Typically, concentrated solutions are batch-generated and the frequency of generation is usually not daily but weekly, monthly, annually, or even longer. These solutions are process chemicals or products that cannot be reconditioned or reused in the same manufacturing process.

Concentrated solutions, such as spent plating baths, acids, alkalies, static drag out solutions, and reject product, may have concentrations of pollutants hundreds or thousands of times higher than the discharge limits of the POTW or higher than can be adequately treated by the pretreatment system if discharged all at once. Take time to examine and understand each manufacturing process so you can identify these concentrated solutions and take the necessary steps to prevent damage to your facilities.

Some wastes may be considered concentrated by the POTW but not by the industry. For example, the ten percent sulfuric acid solution used for pickling parts is considered "dilute" by comparison to the 98 percent or 50 percent stock solution that the industry uses to make up the pickling solution. When this solution is spent or can no longer be used as a pickling solution, proper treatment and disposal are required. From the industrial manufacturer's point of view, the solution is spent and no longer concentrated. However, from a wastewater treatment point of view, the solution is concentrated since it contains high concentrations of acid (pH less than 1.0) and heavy metals

---

[29] *Interference.* Interference refers to the harmful effects industrial compounds can have on POTW operations, such as killing or inhibiting beneficial microorganisms or causing treatment process upsets or sludge contamination.

(1,000 mg/L) compared to the normal pH of 1.0 to 4.0 and heavy metal concentrations of less than 100 mg/L.

Another source of concentrated solutions is the wastewater from equipment cleanup. While the amount of material in the process chemical bath may be considered dilute by industry standards, it forms a concentrated wastestream when discharged during the cleanup of manufacturing equipment. Cleanup wastestreams contain a high concentration of the product during the first washing of the tank, pipe, or pump. This discharge of concentrated waste is followed by successive rinses that contain fewer and fewer pollutants. If cleanup flow concentrations are not equalized, the cleanup cycle can cause problems in the IWTS.

Spills of process chemicals to the floor, if not contained, can flow directly to the floor drain and the pretreatment or sewer system. The adverse effects on the pretreatment system and POTW are the same as those of any other concentrated solutions. This is why chemical containment areas must not have drains.

2.83 Concentration Versus Mass of the Pollutant

An understanding of the concentration and the mass of a pollutant in an industrial waste is needed to determine the effects on the industry's pretreatment system, the POTW collection, treatment, and disposal systems, and the sampling of the industry's discharge. The concentration of a substance in wastewater is normally expressed as milligrams per liter (mg/L) and is a measurement of the mass per unit of volume. The mass of a substance is normally expressed in pounds or kilograms and is a weight measurement. A mass emission rate is a measurement of weight per unit time and is usually expressed as pounds or kilograms per day.

Many of the electroplating and all of the metal finishing Federal Categorical Standards are written in concentrations, whereas most of the other Categorical Standards are written as mass emission rate standards. Mass emission rate standards are preferred over concentration-based standards because they recognize that with more production and water, the mass of pollutant will also increase. This approach prevents dilution of the pollutant to meet concentration limitations.

The mass emission rate of a substance can be calculated by knowing the concentration of the pollutant in the wastewater and the volume of wastewater. The mass emission rate calculation for an effluent of 10,000 gallons per day containing 4.5 milligrams per liter (mg/L) of copper is shown in Example 1.

EXAMPLE 1

Known	Unknown
Effluent Flow, GPD = 10,000 gal/day	Mass Emission Rate, lbs/day
Waste Conc, mg/L = 4.5 mg/L	

Calculate the mass emission rate in pounds per day.

$$\text{Mass Emission Rate, lbs/day} = \frac{(\text{Flow, gal/day})(\text{Conc, mg}/L)(3.785\ L/\text{gal})}{454,000\ \text{mg/lb}}$$

$$= \frac{(10,000\ \text{gal/day})(4.5\ \text{mg}/L)(3.785\ L/\text{gal})}{454,000\ \text{mg/lb}}$$

$$= 0.38\ \text{lb/day}$$

OR

$$\text{Mass Emission Rate, lbs/day} = (\text{Flow, MGD})(\text{Conc, mg}/L)(8.34\ \text{lbs/gal})$$

$$= (10,000\ \text{gal/day})(1\ \text{MGD}/10^6\ \text{gal/day}) \\ (4.5\ \text{mg}/L)(8.34\ \text{lbs/gal})$$

$$= (0.010\ \text{MGD})(4.5\ \text{mg}/L)(8.34\ \text{lbs/gal})$$

$$= 0.38\ \text{lb/day}$$

NOTE: One liter of water weighs one million milligrams. Therefore, 4.5 mg/L is the same as 4.5 pounds of waste in one million pounds of water.

$$\text{Mass Emission Rate, lbs/day} = (\text{Flow, Mil}\ \tfrac{\text{Gal}}{\text{Day}})(\text{Conc,}\ \tfrac{\text{lbs}}{\text{Mil lbs}})(8.34\ \text{lbs/gal})$$

$$= \text{lbs waste/day}$$

If the industry had a permit limit of 4.12 mg/L, they would be in violation. However, if the permit limit was 0.4 lbs/day, the industry would be in compliance.

The same mass of pollutant contained in different amounts of wastewater can have vastly different effects on the industry's pretreatment system. An unnoticed spill of ten pounds of copper cyanide from a process tank could increase the influent copper concentration to a pretreatment system from 10 mg/L to 500 mg/L and probably cause a pass-through of pollutants to the POTW. If the ten pounds of copper cyanide came from a new process line and was discharged on a continuous basis, the influent copper concentration would only increase to 15 mg/L. This slight increase in influent concentration would not be likely to cause any deterioration in the effluent quality. This example shows how the instantaneous influent mass emission rate can affect a pretreatment plant's performance.

Adverse effects on the pretreatment system can also occur from the same mass loading on the system with only small variations in concentration. For example, assume a pretreatment system was designed to treat 10 mg/L of copper in a flow of 100,000 gallons per day (a mass loading of 8.34 pounds per day). This same system may not be able to treat 5 mg/L of copper in 200,000 GPD (still 8.34 pounds per day) because the solids separation equipment could become hydraulically overloaded (a flow of 200,000 GPD is twice a flow of 100,000 GPD).

The effects of pollutant concentration and mass on the POTW collection, treatment, and disposal systems are generally the same as their effects on the IWTS. However, hydraulic problems in any portion of the POTW system could cause pollutants to pass through the POTW untreated, even though the mass of the pollutant did not change. If the daily mass loading is the same, but the instantaneous mass emission rate is highly variable, the POTW's collection system may not equalize the slug loading of a highly concentrated solution. The result may be interference with the treatment system, causing violations of either or both effluent and sludge disposal limitations.

QUESTIONS

Write your answers in a notebook and then compare your answers with those on page 66.

2.8A What is the difference between a compatible and a noncompatible pollutant?

2.8B List five examples of noncompatible pollutants.

2.8C What are the common units of expression for the following terms: (1) concentration, (2) mass, and (3) mass emission rate?

2.8D How can high flows or concentrations cause interference or pass-through of the pollutant?

2.84 Frequency of Generation and Discharge

Important to both the operation of the industry's pretreatment system and the POTW's collection, treatment, and disposal systems is the frequency of industrial waste generation and discharge. Wastewater sampling to investigate process problems and to determine compliance with the discharge limits are also affected by the hours of discharge. An operator needs to understand when the waste from each process is generated and when it is discharged to the IWTS. Similarly, the operator must know when there is a discharge to the POTW sewer or receiving waters from the IWTS.

2.840 *Hours of Operation Versus Discharge*

Normally, the hours of operation are also the hours of discharge to the IWTS. Thus, the operator can generally expect to receive flow for treatment during the hours of operation. If the production is constant, the discharge volume and chemical constituents will also be constant. For example, a coil coater that runs five million square feet per month from January through December, eight hours per day, five days per week, without any changes in manufacturing processes or chemicals, will have a constant effect on the sewer system and can be sampled in the same manner each time. However, if production increases and a second shift is added, or the work week is extended to six days per week, there is an obvious change in the hours of discharge.

Several common situations where an industrial waste must be treated after the normal production hours are described below:

1. The "wet" processes run for one shift, but the "dry" processes run for two. Compressed air or boilers may be used in the dry processes, each of which could generate a wastewater discharge.

2. In industries with long collection systems, production and wastewater flow to the system may stop, but the IWTS may continue to operate and discharge until the wastewater in the collection system has been processed.

3. Spills, accidental discharges, or stormwater flow that goes to the IWTS may cause the IWTS to operate outside of the normal production hours.

4. A food processing plant operates for one or two shifts, generating some wastewater, but most of the equipment cleaning operations occur on an off shift. The cleaning generates most of the wastewater volume.

5. The IWTS has an equalization tank either at the beginning of the IWTS or at the end of the manufacturing system. Discharge from the equalization tank to the rest of the IWTS may continue after production stops because it is programmed to pump to the next unit process until it reaches its low level.

Equalization of the wastewater is an important factor affecting the actual hours of wastewater discharge to the IWTS and sewer. In order to deliver a relatively constant flow and concentration of pollutants to the IWTS, large wastewater collection sumps, equalization tanks, or storage tanks may be used. As noted above, these equalization devices may also lengthen the time of discharge beyond the actual hours of operation of the manufacturing facility. For instance, an eight-hour work shift generates 100,000 gallons of wastewater, but the pretreatment system is only designed for 150 GPM. Therefore, it will take eleven hours to process the wastewater from the eight-hour production period. Since flow to the IWTS and the sewer span an eleven-hour period, sampling of the discharge will also be necessary over this longer period.

EXAMPLE 2

	Known	Unknown
Wastewater Volume, gal	= 100,000 gal	Process Time, hrs
Treatment Flow, GPM	= 150 GPM	

Calculate the treatment process time in hours.

$$\text{Process Time, hrs} = \frac{\text{Wastewater Vol, gal}}{(\text{Treatment Flow, GPM})(60 \text{ min/hr})}$$

$$= \frac{100{,}000 \text{ gallons}}{(150 \text{ gal/min})(60 \text{ min/hr})}$$

$$= 11.1 \text{ hours}$$

Equalization of industrial wastewater flows can also be beneficial to the POTW. By lengthening the hours of discharge from the industry, there is an effective increase in the available hydraulic capacity of the POTW collection system because of the decreased industrial flow rates. Due to the normal diurnal variation in domestic wastewater flows (peak flows usually occur between 8:00 am and 6:00 pm), the hydraulic capacity of a sewer may be exceeded if a large industrial flow is allowed to be discharged to the sewer during a short period. Therefore, it may be necessary for the industry to discharge only at night. Sampling of this discharge would then be shifted to the nighttime hours. However, the use of flow-proportional sampling will ensure that the composite samples will be collected only during the time the facility is discharging process wastewaters.

2.841 Discharge Variations

Industries that have daily, weekly, or seasonal manufacturing cycles will show variations in wastewater generation. Business cycles for each of the various segments of the industrial community will have an effect on production, and therefore on the generation of wastewater.

The food processing industry provides a good example of daily, weekly, and seasonal variations in discharge quantity and quality. For example, an industry that processes citrus peel to make pectin is dependent on when the peel arrives at the industry's plant. This may mean anywhere from three to six days per week. As the season progresses, the type of peel changes from orange to lemon, and the sugar content changes yielding a slightly different type of wastewater. After the citrus season, the plant is completely shut down.

In certain industries, variations in the quantity of wastewater reflect the nature of the business or the business cycle of the particular business segment. In a small shop producing printed circuit boards, it is typical to have a 30-day turnaround with sales, ordering, and development taking place during the first part of the month. Production is slow while making test boards, but once the board is developed, production proceeds at a rapid pace to produce the boards for shipment in the last week of the month.

The printed circuit board industry is subject to both downturns and upturns in the market. The major pollutant from the industry is copper and, consequently, the quantity of copper discharged to the industrial sewer fluctuates according to market and production cycles.

Variations in the quality of industrial waste can also occur due to market forces or environmental concerns requiring a different type of product. In the metal finishing industry, for example, companies are moving from cadmium-plated metal, an environmentally more hazardous substance with more stringent discharge limitations, to zinc-plated parts. Knowledge of the industry, the manufacturing processes, and market forces are valuable tools needed by the industrial waste treatment plant operator to anticipate variations in industrial discharges.

2.842 Continuous and Intermittent Discharges

Discharges from manufacturing facilities usually reflect the type of manufacturing process used at the facility. Processes that are continuous tend to produce wastewater on a continuous basis, with relatively constant volume and quality. Batch processes, or activities that occur once per shift, per day, or per week, tend to produce an intermittent discharge. Also, as a general rule-of-thumb, the larger the manufacturing process, the more likelihood there is of a continuous discharge.

Examples of manufacturing processes that have continuous discharges include rinsing or cleaning of parts or food, processing of crude oil, either at the well head or refinery, air or fume scrubbing, papermaking, and leather tanning.

Intermittent discharges of wastewater are characterized by discharges of a volume of wastewater separated by a time period between discharges. These typically occur at the beginning or ending of a manufacturing process or during equipment cleanup, a spill, replacement of spent solution, or disposal of a reject product. Intermittent discharges also tend to be more concentrated and of smaller volume than the wastewater normally discharged.

For an industrial pretreatment facility, the intermittent discharges and the variations in waste generation determine the design capacity of the system. A conventional activated sludge wastewater treatment system at an adhesives, resins, and dye chemical plant may be able to handle 8,000 pounds of BOD per day, but the discharge of 100 GPM of a three percent isopropyl alcohol wastestream from a water-cooled contact condenser for one hour every 12 hours will have a significant impact on the pretreatment system. This is because thirty percent of the waste load is discharged to the system during only eight percent of the time.

EXAMPLE 3

Known		Unknown
Waste Flow, GPM	= 100 GPM	1. Waste Load, lbs BOD/day
Discharge Time, min/day	= 1 hr/12 hrs	2. Waste Load, %
or	= 2 hrs/24 hrs	3. Discharge Time, %
	= 120 min/day	
Waste Conc, %	= 3%	
Waste Conc, mg/L	= 30,000 mg/L	
BOD, $\frac{\text{mg BOD}}{\text{mg alcohol}}$	= $\frac{0.8 \text{ mg BOD}}{\text{mg alcohol}}$	
Plant Capacity, lbs BOD/day	= 8,000 lbs BOD/day	

1. Calculate the BOD waste load in pounds of BOD per day.

$$\frac{\text{Waste Load,}}{\text{lbs BOD/day}} = \frac{(\text{Flow, gal/min})(\text{Disch Time, min/day})(\text{Waste Conc, mg}/L)}{(\text{BOD, mg BOD/mg waste})(3.785\ L/\text{gal})(1\ \text{lb}/454,000\ \text{mg})}$$

$$= (100\ \text{gal/min})(120\ \text{min/day})(30,000\ \text{mg}/L)(0.8\ \text{mg BOD/mg}) \frac{(3.785\ L/\text{gal})}{454,000\ \text{mg/lb}}$$

= 2,401 lbs BOD/day

Conversion factors

1%	= 10,000 mg/L
1 gal	= 3.785 liters
1 lb	= 454 grams
	= 454,000 milligrams

2. Determine the percent of waste load discharged during the two hours.

$$\text{Waste Load, \%} = \frac{(\text{Discharge Waste Load, lbs/day})(100\%)}{\text{Plant Capacity, lbs/day}}$$

$$= \frac{(2,401 \text{ lbs BOD/day})(100\%)}{8,000 \text{ lbs BOD/day}}$$

$$= 30\%$$

3. Calculate the percent of time the waste is discharged.

$$\text{Discharge Time, \%} = \frac{(\text{Discharge Time, hours/day})(100\%)}{\text{Total Time, hours/day}}$$

$$= \frac{(2 \text{ hr/day})(100\%)}{24 \text{ hr/day}}$$

$$= 8.3\%$$

QUESTIONS

Write your answers in a notebook and then compare your answers with those on page 66.

2.8E Why is it important for the IWTS operator at an industrial facility to know the facility's hours of operation and discharge?

2.8F Why might an industrial facility temporarily store wastewater in large collection sumps, equalization tanks, or storage tanks?

2.8G When are intermittent wastewater discharges from a manufacturing process most likely to occur?

2.9 EFFECTS OF INDUSTRIAL WASTEWATERS

Some of the effects of industrial wastewater discharges on collection and treatment systems were discussed briefly in Section 2.8, "Industrial Wastestream Variables." This section will describe in greater detail how industrial wastewaters can affect the operation and performance of both an IWTS and a POTW, and how direct discharges to the environment could affect receiving waters.

If an industrial wastestream is discharged to an IWTS that was not designed to handle it, the discharge may cause serious problems. It could interfere with the IWTS processes or pass through untreated to the POTW sewer. Similar effects may occur at the POTW and result in a violation of the discharge permit or prevent the reuse or recycle of water. The untreated industrial discharge could contaminate the industrial wastewater sludge or cause an air emission problem. It potentially could affect maintenance or production personnel working in or around the industrial sewer or treatment system through the generation of a toxic gas.

The seriousness of the effect will depend on the characteristics of the industrial wastestreams, the size and design of the IWTS, and the standards for discharge, recycle, or disposal of wastewater, sludge, or air emissions. Accordingly, the effects of discharging the industrial effluent to the POTW or the environment will depend on the characteristics of the effluent, the type and size of the POTW system, and their standards for sludge and wastewater disposal or reuse. Waste characteristics, such as temperature, pH, odor, toxicity, concentration, and flow, must be evaluated to determine their acceptability to the IWTS. Similarly, understanding these characteristics of the IWTS effluent will also enable you to predict the effect the effluent may have on the POTW system.

The effects of industrial waste discharges are not always negative; some beneficial effects also occur. For example, in a short POTW collection system, such as a small treatment system discharging to a trout stream, a continuous discharge of boiler blowdown from a large power plant can be cause for concern. High temperature discharges to sewers can accelerate (1) biological degradation, (2) slime growths, (3) odor production from anaerobic decomposition, and (4) corrosion of concrete pipe and metal sewer appurtenances. The high temperature wastewater can cause a bacterial population shift in the secondary treatment causing floating sludge and reduced BOD removal efficiency. This in turn would endanger the treatment plant's ability to meet its discharge permit limits. The high temperature wastewater may also cause the plant to exceed its temperature standards to the trout stream. On the other hand, the high temperature wastewater discharge from a power plant in a larger conveyance and treatment system located in a colder climate may, in fact, enhance the POTW secondary treatment processes' removal efficiencies by keeping the wastewater temperature above 65°F (18°C) all year.

When evaluating an industrial wastestream, it is necessary to understand the specific characteristics of the waste and how they may affect each portion of the IWTS and, in turn, how the effluent will affect the POTW's conveyance, treatment, disposal, and reuse facilities.

2.90 Operator's Responsibility

The industrial wastewater treatment plant operator is responsible for operating the IWTS in an efficient manner and ensuring that the discharge complies with the discharge limits stated in the company's permit. The IWTS includes the collection, treatment, and disposal systems and it is necessary for the operator to have a good understanding not only of the IWTS, but of the manufacturing processes and utilities that are the sources of the industrial wastewater. This understanding is necessary so that the operator can optimize the system for efficient operation as well as locate the sources of problems and take appropriate steps to solve them.

2.91 Effects on the Collection System

The IWTS collection system is designed and built to transport the individual and combined industrial wastestreams. If the collection system is not designed, built, or operated correctly or if there is a spill, leak, or accidental discharge of materials, the industrial discharges by themselves, or in combination with other industrial wastewater, can cause plugging, odors, erosion, corrosion, explosions, and numerous other problems. The good news, however, is that some industrial discharges contain substances that have a positive effect on the collection system, which may mitigate (lessen) the effect of another industrial wastewater. The beneficial effects could include in-line neutralization. Large flows may produce scouring velocities in low-flow sewers or dilute a concentrated spill enough to produce a treatable waste within the capabilities of the IWTS.

The next sections discuss the commonly encountered effects of industrial discharges on the collection system and offer suggested solutions to these problems to give you some ideas about how they can be resolved. There are many similarities to the problems and solutions encountered in the POTW collection system.

2.910 Hydraulic Capacity Problems

Hydraulic overload problems can occur if a large slug of wastewater or a continuous flow is discharged to the industrial sewer. The cause of a slug discharge may be a tank rupture or water line break. The cause of a continuous large flow may be a broken valve or one left open by mistake. The result in either case may be a sewer backup or pump station overflow.

The smaller the capacity of the sewer or system, and the larger the contribution by the individual wastestream, the more likely it is this problem will occur. The solution may be to require flow restrictors on water valves or tank level switches to alarm high or low levels. If the condition regularly exists, for example, because of the introduction of a new manufacturing process that discharges a slug, equalization of the discharge may be necessary to store the effluent for off-peak hour discharge.

A hydraulic overload condition may also occur if similar manufacturing processes discharge at the same time. For example, in a food processing industry there may be two sections of the plant that clean tanks, reactors, or cooking pots at virtually the same time. While the discharge from one manufacturing line may not cause a problem, the similar discharge schedule from another line will combine the wastewater flows and cause a hydraulic overload condition. Possible solutions include equalization of flow at the IWTS or at the manufacturing process and scheduling production and cleanup so that both lines are not cleaning at the same time.

2.911 Plugging

If the discharge from a manufacturing process contains large amounts of fibrous or stringy materials, heavy solids, adhesives, or grease, plugging of the sewer system may result. Plugging may occur just downstream of the discharge or in the pumping station. Fibrous or stringy materials get caught on rough surfaces and soon build up by entangling more solids. These types of materials can also wind themselves around pump impellers or shafts causing the pump to fail. If problems are occurring, it may be an indication of a problem with the manufacturing process or that the waste should have been pretreated prior to discharge. Review the manufacturing process to determine if changes in the process or disposal of wastes are required or if the sewer needs to be enlarged to accommodate the materials.

Heavy solids, such as sand, ceramic or porcelain solids, or grindings, can build up in a sewer or pump station wet well and reduce its hydraulic capacity. Solids that are not removed by pretreatment at the process may be discharged during peak wastewater flows during the day and may settle in pump station wet wells or oversized sewers downstream of the actual point of discharge when the flow subsides. The solids then have an opportunity to compact and may not become resuspended when the flow in the sewer returns to its peak flow. This cycle of transporting the solids to a section of the collection system to settle, build up, and compact will eventually cause a restriction.

A complete blockage may also occur if large objects are released to the sewer. Rags, tools, rejected food products, and discarded by-products may accidentally be released to the sewer due to operator carelessness or equipment malfunction. Because of their size, they can easily become wedged or entangled with other waste material and completely block the sewer or lift station pump.

2.912 Odors

Examples of industrial discharges that can be odorous are those from petroleum refining, petrochemical manufacturing, and food processing. Generally, the odors are produced from a compound containing sulfur, such as *MERCAPTANS*[30] or hydrogen sulfide. These compounds in air are detectable in the parts-per-billion range (by volume) and can cause complaints from residents and other industries. While the problem is airborne, the actual cause originates in the industrial discharge. It is even more common to find this problem in the discharge to the POTW system.

The first solution may be to change the manufacturing process. Sour water, which is wastewater containing high concentrations of sulfide from the petroleum refining industry, can be stripped with steam and reduced to elemental sulfur using the Klaus process. This process, and other similar recovery processes, have reduced the odor pollution problem while producing a marketable by-product (sulfur). Another solution may be to oxidize the offending components prior to discharge using air, hydrogen peroxide, or chlorine; or not discharge them at all.

---

[30] *Mercaptans* (mer-CAP-tans). Compounds containing sulfur that have an extremely offensive skunk-like odor; also sometimes described as smelling like garlic or onions.

The wastewater produced during the etherification reaction to make polyester is very odiferous. Because of the quantity of organics in the wastes, it is practical to incinerate the wastes at no net fuel expense and solve the odor problem.

Industrial discharges of sulfide can result in toxic and corrosive conditions. If there is biodegradable material, a source of bacteria and a source of sulfide or sulfate in the industrial wastestreams, hydrogen sulfide gas may be produced under anaerobic conditions in the sewer. Bacteria reduce the inorganic sulfate to sulfide when there is insufficient oxygen in the wastewater (less than 0.1 mg/L), thus producing hydrogen sulfide gas. The sulfide is subsequently oxidized to sulfate by other bacteria under aerobic conditions, producing sulfuric acid, which is extremely corrosive to the crown (upper section) of sewer pipes.

Besides an odor problem, hydrogen sulfide also presents a safety (toxic gas) problem to sewer maintenance personnel and the IWTS operator or, if discharged to the sanitary sewer, the POTW collection system and treatment plant operators. Hydrogen sulfide when dissolved in the wastewater will also produce sulfurous and sulfuric acid, very corrosive materials that attack uncoated metal and concrete surfaces. The anaerobic reduction usually requires a long detention time and an active biological population. Sources of sulfide and sulfate should be identified and recovered or treated prior to discharge. Some suggested solutions to this problem are: require oxygenation or chlorination prior to discharge; aerate the wastewater in the collection system; periodically remove the slime layer of anaerobic growth in the system with a slug loading of alkali or chlorine; or periodically clean the sewer with a high-velocity cleaner or a pig (a sewer-cleaning device).

Industrial discharges to the POTW containing high concentrations of sulfide are normally restricted. Limitations of 5.0 mg/L of total sulfide and 0.5 mg/L of dissolved sulfide are used by the County Sanitation Districts of Orange County, California, and are typical of sulfide limits. The same limits are suggested in an industrial collection system unless it is specifically designed to handle the material, the collection system is relatively short (the detention time is therefore relatively short), or the conditions in the sewer inhibit biological growth.

2.913 pH Problems

The pH of an industrial discharge or the amount of acids and alkalies discharged to an industrial sewer are normally taken into account during design. While older plants in the petroleum, primary metals, and chemical industries have sewers constructed from less corrosion-resistant materials, many of the modern facilities use plastics, fiberglass, or other resin material for the industrial wastewater piping and sewer systems. Difficulties can arise when the manufacturing process changes or new chemicals are used that are not compatible with the existing sewer system. For example, fiberglass piping is an acceptable material of construction for sulfuric acid, but if the plating operation adds a process using hydrofluoric acid, the fiberglass may be severely damaged.

The industrial collection system may be designed to handle strong acids or alkalies, but may not be designed to withstand the heat of solution or reaction. For example, when a concentrated solution of sodium hydroxide (such as a spent alkaline cleaner) is discharged to the sewer, there could be a large temperature rise due to the heat of solution. If there is only a small quantity of stagnant wastewater in the sewer or pump station, the heat of solution may exceed 104°F (40°C), the deformation temperature of PVC. A spill of liquid chlorine can cause a temperature rise sufficient to produce steam resulting in a very toxic gas. Liquid chlorine can also damage plastics directly.

Acids will corrode concrete and cast-iron sewers, concrete wet wells and tanks, the internal steel equipment in the primary and secondary clarifiers, trickling filters, aerators, and pumps. Mineral acids, such as sulfuric, nitric, hydrochloric, and phosphoric acids, are used extensively to clean base metals in the metal finishing industries. The fertilizer, iron and steel, mining, and petroleum industries also use vast quantities of these strong acids. Mineral acids are also used in pretreatment systems for chromium reduction, neutralization of alkalies, and pretreatment of CHELATED [31] metal plating solutions.

Discharge of acid to the sewer from a spill or due to an equipment or control instrumentation failure can cause a pH violation and damage to the collection system. Spill containment provisions are essential in all areas where strong acids or alkalies are being used or stored.

Too high a chlorine concentration is also corrosive to the collection system. Many platers will overchlorinate their cyanide wastewater to ensure they meet the requirements for cyanide concentrations. However, 40 to 50 mg/L excess chlorine can be corrosive to equipment and dangerous to personnel servicing a pump station.

---

[31] *Chelation* (key-LAY-shun). A chemical complexing (forming or joining together) of metallic cations (such as copper) with certain organic compounds, such as EDTA (ethylene diamine tetracetic acid). Chelation is used to prevent the precipitation of metals (copper). Also see SEQUESTRATION.

The organic acids, such as acetic, maleic, benzoic, oxalic and citric acids, are weaker than mineral acids but, nonetheless, can have a pH of 4.0 or less. They too can corrode the sewer or attack the solvent joints of plastic or resin-based sewers. They also represent an organic load to your IWTS. If your pretreatment system does not remove organics, then these acids will represent an organic load to the POTW. Organic acids are typically used in food processing, beverage and consumer product manufacturing, and in the manufacture of chemical intermediates.

Strong alkalies can corrode sewers and pumping stations; aluminum is particularly affected by high pH. High pH may also precipitate metals such as calcium, potentially causing a solids buildup problem in the sewer. The strong alkalies include sodium hydroxide, lime, and ammonia. These are used in the metal finishing industry to clean and chemically mill base metals. The water treatment industry uses significant quantities of lime to soften water, and pretreatment systems use strong alkalies to neutralize industrial wastes. Because of the potential damage these substances may cause, it is important to periodically review the IWTS spill containment measures and check the failure mode of the chemical addition controls.

The acceptable pH range for the discharge of industrial wastewater to the POTW collection system, as regulated in many industrial waste or sewer-use ordinances, is 6.0 to 9.0. In some ordinances, the pH range may be widened. Remembering that a pH of 7.0 is neutral, the trend is to allow more alkaline or basic material in the discharge rather than materials that are more acidic. The construction materials for sewers, pumping stations, treatment equipment, and biological processes all withstand alkaline discharges better than they withstand the discharge of corrosive acids. However, the discharge of strong alkalies to the POTW sewer may actually be beneficial in removing the anaerobic slime layer from the sewer. When this is allowed, it should be done with the POTW's permission and knowledge of each discharge so that the POTW influent and secondary treatment can be monitored to prevent a treatment process upset.

The discharge of out-of-pH-range wastewater will result in damage to the sewer. Over a period of time, such discharges can eventually corrode the pipe completely, causing *EXFILTRATION*[32] and contamination of the groundwater or *INFILTRATION*[33] of the groundwater into the sewer where the groundwater level is above the depth of the sewer. Industrial discharge violations of pH will also increase the maintenance requirements on pumps in the pumping stations. The damage to the pumps could eventually cause their failure, resulting in sewer backups and raw wastewater overflows.

pH problems are most severe nearest the discharge source because the wastes are not diluted by other wastewater. Frequent discharges of acidic wastes will corrode and completely etch through an industry's sewers and may result in heavy metal or solvent contamination of the groundwater. Use of the proper materials of construction for sewers, control of pH, and annual inspection or testing of the pipes and sewers are essential in preventing groundwater contamination from this type of source.

2.914 Flammables

The discharge of flammables is potentially the most damaging industrial discharge to the collection system. Gasoline, aviation fuel, and hexane used in soybean extraction have been responsible for explosions in sewers causing losses of millions of dollars for sewers and businesses, the loss of service to hundreds of people, and loss of life. Industries producing, distributing, and using fuels and solvents are regulated and monitored to prevent discharge of these materials. Generally, fuels and solvents are only slightly soluble in water and have a specific gravity less than water. When accidentally discharged to the sewer, they will float and accumulate in slow-moving sewers and in pump station wet wells. Any source of ignition, such as an arc from tripping a breaker or a motor, or a spark created while removing a manhole cover with a pick, can cause a fire or explosion.

Another concern with flammable wastes is the exposure of industry personnel or the IWTS operator to volatile toxic substances. When solvents are discharged to the sewer and then aerated, they volatilize, thus exposing operators and other personnel to hazardous fumes. This is true of both *IMMISCIBLE*[34] solvents and miscible solvents, such as acetone, methyl ethyl ketone, and isopropyl alcohol. If the concentration of fumes is high enough, an explosive atmosphere may develop.

Solvents can also cause the joints of plastic and resin-type industrial piping systems and sewers to fail. In addition, care must be taken when using these types of piping materials because they are also combustible in the case of a fire or explosion.

---

[32] *Exfiltration* (EX-fill-TRAY-shun). Liquid wastes and liquid-carried wastes that unintentionally leak out of a sewer pipe system and into the environment.

[33] *Infiltration* (in-fill-TRAY-shun). The seepage of groundwater into a sewer system, including service connections. Seepage frequently occurs through defective or cracked pipes, pipe joints and connections, interceptor access risers and covers, or manhole walls.

[34] *Immiscible* (im-MISS-uh-bull). Not capable of being mixed.

The discharge of flammables to the POTW sewer is dangerous for the same reasons noted above. If the concentration of flammables is high enough, an explosive atmosphere can develop, especially if the secondary treatment process is covered or uses pure oxygen. Any hydrocarbon may cause a flammable hazard in a pure oxygen activated sludge system. However, these systems are usually equipped with sensors and purge systems to prevent flammable and explosive conditions from developing.

2.915 Temperature

Heated industrial wastewaters originate from controlling manufacturing process reactions and as a by-product from utilities' production of energy. In manufacturing processes, heat is often used to increase the rate of reaction and thus creates a heated product or waste that must be cooled. Water or steam is often used directly or indirectly (by means of heat exchangers) to heat or cool the product or by-product and to transport it to the next processing step. The metal finishing industry uses steam to heat process solutions.

Accumulated solids must be removed from boilers to prevent plugging of the boiler tubes and steam lines. The discharge is called boiler blowdown. In cooling systems, single-pass cooling water and cooling tower blowdown can also contribute a heat load to the industrial and POTW sewers. Heated industrial discharges can cause many problems in the IWTS collection and treatment systems including evolution of gases and odors, overheating of pump and rotating equipment bearings, shifts in the population of microorganisms used in biological treatment of industrial wastewater, or even sterilization (killing of all organisms) in the wastewater.

Plastic pipe (PVC) has temperature limitations of around 104°F (40°C) and can fail if used for hot water transport. In the POTW sewer laterals, the O-rings may not be designed to withstand a constant high temperature; if they fail, exfiltration or infiltration of the collection system may occur.

The same problems identified in the industrial sewer can also occur in the POTW collection and treatment system. In addition, if the POTW is discharging to a stream or lake with a temperature limit (for example, a trout stream), then a high temperature discharge by an industrial source can cause the POTW to violate its permit limit.

QUESTIONS

Write your answers in a notebook and then compare your answers with those on page 66.

2.9A What types of heavy solids can cause plugging of sewers?

2.9B What types of industrial discharges can cause odor problems?

2.9C What industries use mineral acids and for what purposes?

2.9D What problems could be caused in IWTS collection and treatment systems by the discharge of heated industrial wastewaters?

2.92 Effects on the Treatment System

Industrial waste discharges damage treatment plant equipment in many of the same ways they damage the collection system. High volume discharges can exceed the pumping capacities; plugging of mechanical equipment, such as bar screens or pumps, can occur from a high solids discharge; acids and alkalies will corrode metal parts eventually causing failure; and flammables in the treatment plant are an explosive problem that can cause almost instantaneous damage. The added potential problem with industrial discharges is their effect on the treatment processes, including blinding of filters with oil; plugging microfiltration, nanofiltration, or reverse osmosis membranes; interfering with recovery processes by contaminating the by-product; and overloading or upsetting the aerobic and anaerobic biological treatment processes.

2.920 Hydraulic Overload

A hydraulic overload can cause a decrease in the efficiency of treatment processes, an increase in solids carryover, and possible effluent limit violations at a wastewater treatment plant. Unit processes such as neutralization, sedimentation, filtration, and biological treatment operate best at a constant flow and constant loading conditions. Large changes in the volume of flow or rapid changes in loading will decrease the efficiency of these processes. Hydraulic surges from an industrial process or utility discharge can cause these rapid variations. To compensate, the treatment plant must make a series of changes in their plant operating conditions, such as changing the sludge removal rate, increasing the blower output, or increasing the chemical addition rate. The alternative is to suffer possible effluent limit violations. Equalization of the flow at the source or as a part of the IWTS provides the best means of controlling hydraulic surges and operating the treatment processes at a constant or near-constant flow.

2.921 Interference

EPA defines interference as a discharge which, alone or in conjunction with discharges from other sources, inhibits or disrupts the POTW, its treatment processes or operations, its sludge processes, use or disposal, and is a cause of violation of the NPDES permit or prevents the lawful use or disposal of sludge. This definition of interference applies equally well to discharges by industrial processes to the IWTS. By working closely with the manufacturing and utility operators, the IWTS operator can identify potential interference problems before they cause a discharge violation. Good communication between the operators in the manufacturing facility and the IWTS operator is the most reliable way to identify changes, whether sudden or gradual, in the operation of the plant or quality of the effluent.

For example, a decrease in the cyanide destruction efficiency or BOD removal rate of the anaerobic digester may be caused by a change in the manufacturing process that chelates the cyanide or treatment plant operational changes. Check first with the manufacturing process personnel to see if recent changes have been made in the manufacturing plant processes. If not, an industrial discharge of a noncompatible pollutant to the cyanide process (such as an iron salt) or too much of a compatible pollutant (such as soluble BOD) could have caused the problem. If not corrected at the source (the manufacturing process), the problem could cause violation of the discharge limits.

Discharge of untreated wastes or even large quantities of treated wastes can cause interference with the POTW treatment processes. Table 2.2 illustrates examples of how industrial discharges may cause potential interference with the POTW's treatment processes.

2.922 Influent Variability

Measurements of wastewater flow, pH, temperature, and CONDUCTIVITY[35] are used to detect changes in the influent to the IWTS or POTW. As with hydraulic surges, variability in the chemical composition of the influent wastewater can cause upsets in the treatment processes. The larger the difference between the existing influent composition and the contribution from the industrial discharge, the larger the potential for problems.

Remember that a change of one pH unit represents a 10-fold change in the concentration of acid in the influent. Chemical reactions, precipitation, settleability, and filterability are greatly changed by the pH of the wastewater. For biological treatment systems, both aerobic and anaerobic treatment are inhibited by rapid changes in environmental conditions. Operation outside of the pH range of 7.0 to 8.5 can be toxic to bacteria; however, if the change is gradual the microorganisms can become acclimated to pH levels slightly beyond this range.

Changes in conductivity or ORP[36] normally represent increases or decreases in soluble salts, cyanide, or metals. Inhibition or interference can range from overloading the chemical processes with the mass of metals or cyanide requiring treatment to inhibiting the biological reactions. Changes in soluble salt concentrations alter the rate of oxygen transfer through bacterial cell walls and therefore affect the health and performance of the microorganisms.

2.923 Slug Loadings (also called Shock Loads)

Slug loadings or batch dumps of compatible or noncompatible pollutants from industrial processes, whether accidental or as part of normal production, may cause interference with the treatment processes or pass-through of pollutants. To assess the effect of a slug loading, you will have to consider the mass of the discharge and the resulting concentration at the treatment plant. For example, if a concentrated solution containing 0.1 pound of copper is discharged to a biological treatment system, it may result in a concentration of 5.0 mg/L for a 5-minute period at the treatment plant. While this is a significant variation from a 0.25 mg/L average influent concentration, the effect on the activated sludge treatment system would be minimal, and the sludge reuse potential would not appreciably suffer from a one-time occurrence.

However, if the concentration were to remain at 5.0 mg/L for a one-hour period, the biological treatment system would likely be affected, severely reducing or stopping biological treatment, and the sludge would be contaminated. The concentration of the slug loading as measured at the treatment plant was the same in both examples, but the second example illustrated a batch

---

[35] *Conductivity.* A measure of the ability of a solution (water) to carry an electric current.
[36] *ORP* (pronounce as separate letters). Oxidation-Reduction Potential. The electrical potential required to transfer electrons from one compound or element (the oxidant) to another compound or element (the reductant); used as a qualitative measure of the state of oxidation in water and wastewater treatment systems. ORP is measured in millivolts, with negative values indicating a tendency to reduce compounds or elements and positive values indicating a tendency to oxidize compounds or elements.

TABLE 2.2 INTERFERENCE FROM INDUSTRIAL DISCHARGES

Source	Pollutant	Effect on Treatment System
Metal Finishing and Printed Circuit Board Manufacture	A. Heavy Metals	1. Decrease or stop biological removal rates for secondary and anaerobic treatment. 2. Prevent reuse of sludge or make it a hazardous waste.
	B. Chlorinated Solvents	1. Same effects as A. 2. Exposure of POTW workers to toxic gas.
	C. Acids	1. Destroy microbes, stopping treatment. 2. Upset anaerobic digester reducing gas production. 3. Corrode structures.
Cleaning Operations (Machinery Repair, Food Process, Clean-in-place Operations)	D. Detergents	1. Foam in secondary treatment facilities reduces settling characteristics and dewaterability.
Oil Production, Refining or Dispensing	E. Oil	1. Interferes with settling. 2. Toxic to anaerobic bacteria in large quantities reducing gas production. 3. Explosive when using a pure oxygen activated sludge system.
	F. Flammables	1. Same effects as A. 2. Explosive when it accumulates.
	G. Sulfide (Oil Production)	1. Toxic to treatment plant workers. 2. Odor complaints. 3. Increases oxygen demand and blower requirements.
	H. Salt (Oil Production)	1. Decreases oxygen transfer efficiency. 2. Inhibits biological activity.
Food Processing	I. BOD (Soluble and Insoluble)	1. Increases oxygen demand in secondary treatment. 2. May change microbiology of secondary treatment, causing secondary treatment settling problems. 3. Creates odors.
Organic Chemicals (Ketones, Alcohols)	J. Acetone, Methyl Ethyl Ketone, Isopropanol	1. If biological treatment microorganisms are acclimated, effects same as I-1. 2. If biological treatment microorganisms are not acclimated, effects same as B.
Utilities (Steam, Electricity, Cooling Towers)	K. Temperature[a] (Hot)	1. Depending on discharge point of POTW, exceed temperature limits. 2. Change microbiology or biological treatment efficiency. 3. Accelerate hydrogen sulfide production, which causes odors and corrosion.

[a] Warm wastewaters may improve rather than interfere with treatment. Warm temperatures increase settleability, biological activity, and overall removal rates.

dump that was 12 times more mass than the first. It would have caused discharge violations, sludge contamination, and the biological treatment removal efficiencies would suffer until new bacteria could be cultured to return to the previous efficiency.

If slug loadings, such as in the first example, are allowed to continue on a daily basis, organisms in the activated sludge or trickling filter process may become acclimated and the daily discharges probably will not affect the effluent quality. The sludge, however, will now be more contaminated, possibly affecting its use or disposal.

2.93 Effects on Effluent and Sludge Disposal and Reuse

Industrial discharges which, alone or in conjunction with discharges from other sources, pass through the POTW's facilities to navigable waters and cause a violation of the discharge permit are considered pass-through discharges. Pass-through of compatible and noncompatible pollutants can occur when the POTW treatment system is under stress from hydraulic or compatible waste overloads or shock loadings of toxic pollutants. When the pollutant removal efficiency decreases, the constituents from industrial discharges are found in the effluent.

Excluding slug loadings, the constituents most likely to pass through a biological IWTS are small quantities of the toxic organics that are very miscible (ketones and alcohols, if not stripped, are metabolized by secondary treatment), or immiscible and *LIPOPHILIC*[37] (pesticides or polychlorinated biphenyls), and soluble heavy metals that are not used as micronutrients. The constituents that are likely to pass through a physical–chemical IWTS are small quantities of toxic organics that are miscible solvents or chelated metals.

If the toxic constituents in industrial processes are controlled on site, the level of toxics discharged to the sewer is minimal. This optimizes the recycle and reuse options of both the effluent and sludge. Effluent can be further treated for reclaimed water uses; sludge can be applied to land as fertilizer or mixed with a bulking agent and made into compost if it is biological. If the sludge contains a high percentage of metal, it may be reclaimable as an ore by refining or smelting.

Industrial processes whose discharges upset or pass through the treatment system eventually have an effect on the effluent and sludge quality. In essence, the industrial process has contaminated the wastewater. Instead of being a potential resource, the effluent and sludge become a liability.

2.94 Effects on the POTW

Effects of an industrial discharge on the POTW collection, treatment, and disposal system parallel those of a manufacturing process waste on the IWTS. There are problems with each component of the system. The Federal Pretreatment Regulations were established to remove toxic pollutants at the source and to protect the POTW's collection, treatment, and disposal systems and the environment. The local Industrial Waste Ordinance specifies the exact operating conditions each POTW must observe to prevent pass-through and interference.

The effects of an industrial discharge on the POTW will always depend on the characteristics and flexibility of the system, the level of skill possessed by the POTW inspectors, laboratory analysts, and POTW operators, and the amount and type of industrial flow. Factors such as the size and length of the sewer system also influence how an industrial discharge will affect the POTW collection system. In general, the larger the system, the less effect a single industrial discharge will have on the POTW regardless of whether the industrial discharge is a slug loading or a constant discharge. Dilution and equalization of the industrial discharge occur naturally in the larger collection systems, thereby reducing the effect on the POTW facilities.

As the complexity of the POTW treatment system increases from only primary treatment to tertiary treatment, the effect of an industrial discharge also increases. The higher degrees of treatment are more sensitive to upset from industrial discharges. Secondary and tertiary biological processes, such as activated sludge, nitrification, denitrification, and anaerobic digestion, can be upset by a toxic "overdose" of heavy metals. Tertiary physical–chemical processes, such as sand filtration, can be rendered useless by a pass-through of oil or a carryover of gelatinous (jelly-like) bacteria from an upset biological process.

One duty of the IWTS operator is to *prevent* slug discharges. If a slug discharge is released, the operator must report the incident to the POTW. The POTW must be able to respond. When a POTW is designed with system flexibility, the industry and the POTW operators may be able to mitigate (lessen the impact of) a slug loading that results from an accidental or illegal slug discharge. Some POTWs are equipped with equalization basins, flow control structures, chemical treatment points along the collection system, aeration basins, and adequate aeration equipment, along with return sludge and effluent recycling capabilities. Facilities and equipment such as these enable the POTW to modify the secondary biological treatment systems to cope with the anticipated slug load and treat the wastewater adequately to meet NPDES discharge requirements.

---

[37] *Lipophilic* (lie-puh-FILL-ick).   Having a strong affinity for fats. Compounds that dissolve in fats, oils, and greases.

If the configuration of the treatment system can be easily changed, the effect of an industrial discharge may be lessened. Changing the recycle ratio on a trickling filter or altering the biomass concentration in an activated sludge system could prevent pass-through of noncompatible pollutants or air strip volatile organic compounds. Changing a two-unit process from parallel operation to series operation may help to remove high loadings of compatible pollutants.

The disposal of the POTW effluent and sludge are also affected by industrial discharges. The effluent discharge requirements are more stringent for water reuse than for discharge to receiving waters. POTW sludge being used as a component in compost for resale must meet stricter quality requirements than sludge being landfilled. Toxic components of industrial discharges may limit the recycle and reuse options if the POTW is not properly protected from slug loadings or if contaminated concentrations reach a level that may pass through and be discharged in the effluent or sludge. When certain metals reach high enough concentrations in the sludge, then the sludge must be handled as a hazardous waste under RCRA regulations.

QUESTIONS

Write your answers in a notebook and then compare your answers with those on page 66.

2.9E A hydraulic overload can cause what kinds of problems at a wastewater treatment plant?

2.9F Define interference.

2.9G Changes in conductivity indicate changes in what types of wastestream constituents?

2.10 MANUFACTURING PROCESSES AND WASTEWATER GENERATION

This section presents information on two major industrial waste generators. The information about each industry includes a general description and schematic of the manufacturing process; a description of each of the manufacturing unit processes and the wastes generated; and a schematic diagram of the raw materials used in each unit process and the wastes generated.

Even if the specific industry in which you work is not described here, these two examples may give you an idea of how manufacturing processes generate various types of wastes.

2.100 Metal Finishing Industries (Figures 2.4 and 2.5)

The largest group of industries regulated by Federal Categorical Standards is the metal finishing industry. This industry's distinguishing characteristic is the formation of a surface coating on a base material. Surface coatings are applied to provide corrosion protection, wear or erosion resistance, antifrictional characteristics, or for decorative purposes. The coating of common metals includes the processes in which a ferrous or nonferrous base material is plated with copper, nickel, chromium, brass, bronze, zinc, tin, lead, cadmium, iron, aluminum, or combinations of these metals. Precious metals plating includes the processes in which a base material is plated with gold, silver, palladium, platinum, osmium, iridium, rhodium, indium, ruthenium, or combinations of these metals.

The general processes of the metal finishing industry include machining, cleaning and surface preparation, plating and coating, anodizing, and etching and chemical milling. A schematic of the general and unit processes for the metal finishers is shown in Figure 2.6.

2.1000 Unit Process Description

The following descriptions provide general information for each of the above unit processes. The descriptions apply specifically to the metal finishing industry, but will help you understand similar operations in other industries.

Machining

Machining is the general process of removing stock from a workpiece with a rotating or moving cutting tool. Machining operations such as stamping, turning, milling, grinding, drilling, boring, tapping, sawing, and cut-off are included in this category.

Cleaning and Surface Preparation

In order to prepare the surface of the workpiece to form a good bond, the base metal must be properly prepared and cleaned. The cleaning process involves the removal of oil, grease, and dirt from the surface of the base material using water with or without a detergent or other dispersing agent. Electrolytic and nonelectrolytic alkaline cleaning and acid cleaning are included in this category.

1. Alkaline Cleaning—is used for removal of oily dirt or solids from workpieces. The detergent nature of the cleaning solution provides most of the cleaning action; agitation of the solution and movement of the workpiece increase cleaning effectiveness. Alkaline cleaners are classified into three types: soak, spray, and electrolyte. Soak cleaners are used on easily removed soil, spray cleaners combine the detergent properties of the solution with the impact force of the spray, and electrolytic cleaning produces the cleanest surface by strong agitation of the solution during electrolysis. Also, certain dirt particles become electrically charged and are repelled from the surface. Some alkaline cleaning processes also remove oxide films.

2. Acid Cleaning—is a process in which a solution of an acid, organic acid, or an acid salt, in combination with a wetting agent or detergent, is applied to remove oil, dirt, or oxide from metal surfaces. Acid cleaning can be referred to as pickling, acid dipping, descaling, or desmutting. Heated acid solutions may be used in this process. Acid dip processes may follow alkaline cleaning prior to plating.

3. Paint Stripping—is the process of removing an organic coating from a workpiece, usually by using solvent, caustic, acid, or molten salt solutions.

Fig. 2.4 Typical barrel plating production line. Floor is diked and rinses are plumbed into standpipes. (NOTE: POTW inspectors may be suspicious of white bucket in lower left that could be used to transfer static drag out or floor spills to standpipes.)

4. Solvent Degreasing—is a process for removing oils and grease from the surfaces of a workpiece by the use of organic solvents, such as aliphatic petroleum, aromatics, halogenated hydrocarbons, oxygenated hydrocarbons, and combinations of these solvents. Solvent cleansing can be accomplished by applying solvents in either liquid or vapor form, but solvent vapor degreasing is normally the quicker process. Ultrasonic vibration is sometimes used in conjunction with liquid solvent degreasing processes. Emulsion cleaning is a type of solvent degreasing that uses common organic solvents in combination with an emulsifying agent.

Plating and Coating

1. Electroplating—is the production of a thin surface coating of one metal upon another by electrodeposition. Ferrous or nonferrous base materials may be coated with a variety of common metals (copper, nickel, lead, chromium, brass, bronze, zinc, tin, cadmium, iron, aluminum, or combinations thereof) or precious metals (gold, silver, platinum, osmium, iridium, palladium, rhodium, indium, ruthenium, or combinations of these metals). In electroplating, metal ions supplied by the dissolution of metal from anodes or other pieces are reduced on the workpieces (cathodes) while immersed in either acid, alkaline, or neutral solutions.

The electroplating baths contain metal salts, alkalies, and other bath control compounds in addition to plating metals, such as copper, nickel, silver, or lead. Many plating solutions contain metallic, metallo-organic, and organic additives to induce grain refining, leveling of the plating surface, and deposit brightening. Many plating operations are now using other common metal salts rather than cyanide in their plating baths. However, most precious metal baths contain cyanide salts. Cyanide is commonly used in gold plating, copper plating where a thick coat is required, and cadmium plating.

2. Electroless Plating—uses a chemical reduction/oxidation reaction to provide a uniform plating thickness on all areas of the part regardless of the shape. The most common electroless plating metals are copper and nickel. In electroless plating, the source of the metal is a salt, and a reducer, such as sodium hydrophosphite or formaldehyde, is used to reduce the metal ions to their base state. A complexing agent

Fig. 2.5 Close-up of Figure 2.4 demonstrating high volume of drag out dripping from barrel. Barrels should spend as much time as possible dripping over plating bath, NOT over floor or running rinse tank. "Spill guards" should direct drag out to plating bath.

(C*HELATING AGENT*[38]), such as EDTA or Rochelle salts, holds the metal ions in solution.

3. Mechanical Plating—is the process of depositing a metal coating on a workpiece using a tumbling barrel, metal powder, and an impact medium, usually glass beads. The operation involves cleaning and rinsing processes before and after the plating operation.

4. Chemical Coatings—include chromating, phosphating, metal coloring, and passivating. In chromating, a portion of the base metal is converted to a component of the protective film formed by the coating solutions containing hexavalent chromium and active organic or inorganic compounds. Phosphate coatings are formed by the immersion of steel, iron, or zinc-plated steel in a dilute solution of phosphoric acid plus other *REAGENTS*[39] to condition the surfaces for cold-forming operations, prolong the life of organic coatings, provide good paint bonding, and improve corrosion resistance. Metal coloring chemically converts the metal surface into an oxide or similar metallic compound to produce a decorative finish. Passivating is the process of forming a protective film on metals by immersion in an acid solution, usually nitric acid or nitric acid with sodium dichromate.

5. Hot Dip Coating—is the process of coating a metallic workpiece with another metal by immersion in a molten bath to provide a protective film. The most common process of this type is galvanizing, where workpieces are coated with zinc.

Anodizing

Anodizing is an electrochemical process that converts the metal surface to a coating of an insoluble oxide. Aluminum is the most frequently anodized material. The formation of the oxide occurs when the parts are made anodic in dilute sulfuric or chromic acid solutions. The oxide layer is formed at the extreme outer surface, and as the reaction proceeds, the oxide grows into the metal. Chromic acid anodic coatings are more

---

[38] *Chelating* (KEY-LAY-ting) *Agent.* A chemical used to prevent the precipitation of metals (such as copper).
[39] *Reagent* (re-A-gent). A pure, chemical substance that is used to make new products or is used in chemical tests to measure, detect, or examine other substances.

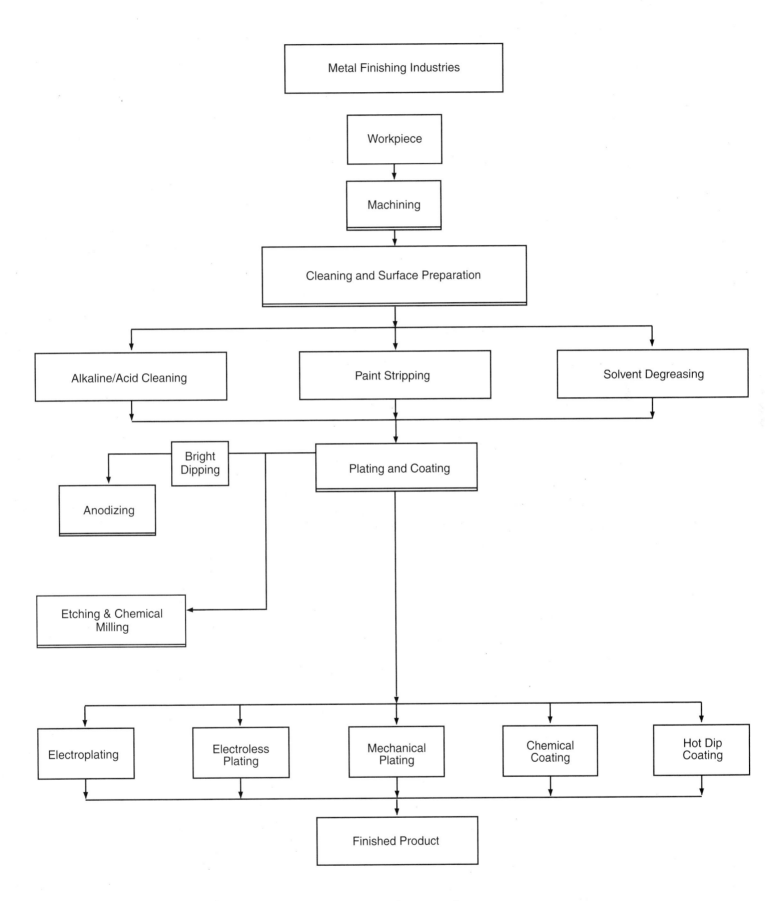

Fig. 2.6 Metal finishing industry manufacturing processes

protective than sulfuric acid coatings and are used if a complete rinsing of the part cannot be achieved.

Etching and Chemical Milling

Etching and chemical milling are processes used to produce specific design configurations or surface appearances on parts by controlled dissolutions with chemical reagents or etchants. Chemical etching is the same process as chemical milling except the rates and depths of metal removal are usually much greater in chemical milling.

Bright Dipping

A specialized form of etching is the bright dipping process. It is used to remove oxide and tarnish from ferrous and nonferrous materials. Bright dip solutions are mixtures of two or more acids: sulfuric, chromic, phosphoric, nitric, or hydrochloric. The process is frequently performed just prior to anodizing, and produces a bright to brilliant finished surface.

2.1001 Waste/Wastewater Characteristics

The raw materials and the waste constituents most commonly found in the waste/wastewater streams generated by the metal finishing industry are presented in Figure 2.7.

Waste Types

- Heavy metals   - cadmium, chromium, copper, lead, nickel, zinc, tin, aluminum, iron

- Precious metals   - gold, silver, platinum, palladium, rhodium, iridium, osmium, ruthenium, indium

- Complexed metals   - complexed wastes containing common and precious metals bonded with complexing solutions, such as formaldehyde, hydrophosphite, ammonia solutions, EDTA, and citrate

- Acid wastewaters   - typically sulfuric acid, others include nitric, hydrochloric, phosphoric, and tri-acid (sulfuric, nitric, hydrofluoric)

- Alkaline wastes   - typically alkaline cleaners containing sodium hydroxide, sodium carbonate, soaps, and surfactants

- Hexavalent chromium - chromium plating solution, such as chromic acid in combination with sulfuric acid or sulfate

- Cyanide wastes   - cyanide plating of copper, cadmium, zinc, brass, gold, silver; electroless plating of gold, silver; immersion of brass, silver, and tin; and cyanide stripping

- Oily wastes   - free or emulsified oil and grease

- Solvent wastes   - common solvents such as aliphatics, aromatics, halogenated hydrocarbons, and oxygenated hydrocarbons

Water Usage/Wastewater Generation

In the metal finishing industry, the water is mostly used for rinsing workpieces, washing away spills, process fluid makeup, cooling, and washing of equipment and parts. The following paragraphs describe the high water usage processes of metal finishing manufacturers.

Rinsing—Rinse water is used to remove the film remaining on the surfaces of the workpieces after removal from the process baths. As a result, the rinse water becomes contaminated with the constituents in the preceding process solution.

Washing Away Spills—The wastewater generated from washing away spills is heavily contaminated with constituents of the process materials and dirt.

Process Fluid Makeup—Due to evaporation, drag out, or spills, process solutions (cleaning and plating solutions) are eventually used up and new process fluids have to be made up. The major component is water. Spent or contaminated process solutions are either collected for off-site disposal or conveyed to treatment facilities.

Cooling—Coolants containing petroleum or synthetic-based oils are required to lubricate and cool equipment and workpieces during many metal machining operations. The film and residues from these fluids are removed during cleaning, washing, or

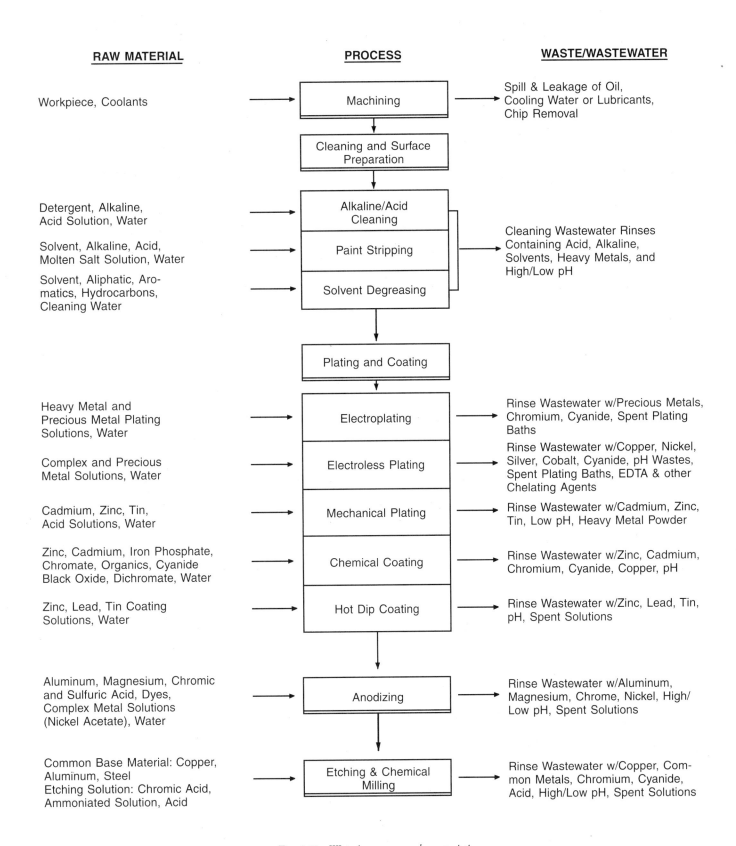

Fig. 2.7 Waste/wastewater characteristics for the metal finishing industry

rinsing operations, and the resulting wastewater becomes contaminated with oil.

Washing—Washing equipment, such as filters, pumps, and tanks, picks up residues of concentrated process solutions, salts, or oils and may contribute a large amount of wastewater.

Deionization Bed Regeneration—Most printed circuit board manufacturers have large deionization plants on site that are regenerated with an acid solution (usually hydrochloric or sulfuric acid) and a caustic solution (usually sodium hydroxide). These regeneration streams should be thoroughly mixed and the pH adjusted to meet effluent limits prior to discharge to the sanitary sewer.

QUESTIONS

Write your answers in a notebook and then compare your answers with those on pages 66 and 67.

2.10A What metals are used by the metal finishing industry to produce a surface coating on a base material?

2.10B List the general processes of the metal finishing industry.

2.10C What types of chemicals are found in electroplating baths?

2.10D List the nine common waste types found in the waste/wastewater streams generated from the metal finishing processes.

2.101 Printed Circuit Board Manufacturing

This industry is classified as a metal finishing industry and represents a significant percentage of the total industry. Printed circuit board manufacturing involves the formation of a circuit pattern of conductive metal (usually copper) on nonconductive board materials such as plastic or glass. The general processes of printed circuit board manufacturing include laminate machining, cleaning and surface preparation, electroless plating, pattern printing and masking, electroplating, and etching. The schematic of the manufacturing processes is shown in Figure 2.8.

2.1010 Unit Process Description

The following is a brief description of the general processes of printed circuit board manufacturing.

Laminate Machining

Laminate machining consists of mechanical processes such as cutting to size, drilling holes, and shaping, by which the circuit boards (made of laminated materials) are prepared for the vital chemical processes. All the machining processes are dry and no liquid wastes are generated.

Cleaning and Surface Preparation

Cleaning and surface preparation are necessary to remove oil, grease, and dirt from the surface of the base material. Water, with or without detergent or other dispersing agent, is commonly used. Cleaning processes could include alkaline cleaners, acid cleaners, vapor degreasing, and scrubbing, with alkaline and acid cleaning being the most common. The alkaline cleaning process removes oil or dirt, and the acid cleaning process removes oil, dirt, and oxide from metal surfaces.

Electroless Plating

Electroless plating provides a uniform plating thickness on all areas of the part by chemical reduction/oxidation. An electroless plated surface is dense and virtually nonporous. Copper and nickel electroless plating for printed circuit boards are the most common operations. The nickel or copper salt in combination with a reducing agent is used to reduce the metal ions to their base state. A chelating agent is used to hold the nickel or copper ions in solution.

Pattern Printing and Masking

This process prints a pattern or the desired circuit configuration onto the board by using a suitable negative-resist pattern for a photoresist. Ink-resist, or positive-resist pattern is used if plating is to be used as an etchant-resist. In the case of negative-resist pattern, gold or solder plating is applied to the non-photoresist areas.

Liquid photosensitive resists are thin coatings produced from organic solutions which, when exposed to light of the proper wavelength, are chemically changed in their solubility to certain solvents (developers). Two types are available: negative-acting and positive-acting.

Electroplating

This process is used after the pattern printing and masking process to deposit an adherent metallic (copper, gold, or nickel) coating upon a negatively charged board by the passage of an electric current in a conducting medium. The electroplating baths contain metal salts, alkalies, and other bath control compounds.

Etching

The last major step in chemical processing is metal removal or etching to achieve the desired circuit pattern. Typical solutions for etching are ferric chloride, nitric acid, ammonium persulfate, chromic acid, cupric chloride, hydrochloric acid, or a combination of these solutions.

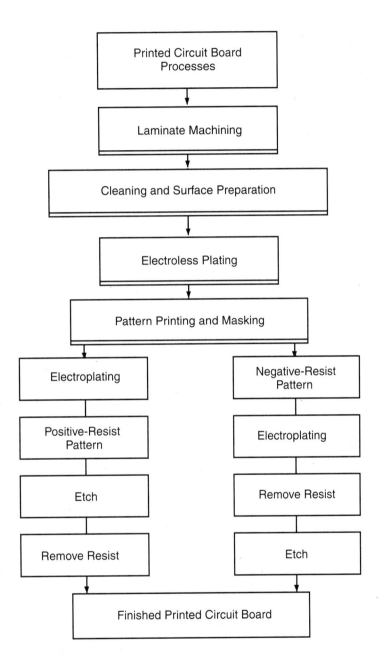

Fig. 2.8 Printed circuit board manufacturing processes

2.1011 Waste/Wastewater Characteristics

Water is used for cleaning, rinsing, spray rinsing, spill clean-up, solution replenishment, and cooling of equipment used in the manufacture of printed circuits. Rinses contain small amounts of the previous chemical bath the board was dipped in, such as cleaner solutions, copper plating bath, and resist. If the chemical bath preceding rinsing was acidic (used to activate the surface for plating), the rinse water may also contain copper or lead that has been solubilized.

Water used in closed-loop systems to cool or heat chemical baths or rinses may have corrosion inhibitors or glycols to protect and aid the heat transfer capabilities. Leaks from these systems, either directly to the sewer or into the rinse water, constitute industrial wastewaters.

Spills to the floor may occur during the transfer of circuit boards from the chemical bath to the rinse tank. In some shops, this floor wastewater is washed to a central sump and metered into the pretreatment system. This waste may contain any of the chemicals used in the facility.

Spent solutions usually make up the second largest volume of wastewaters, rinse wastewaters being first. While these solutions are no longer strong enough to be used in the manufacturing process, they still contain large amounts of alkalies, acids, and surface-active agents.

The sources of waste and wastewater production in the printed circuit board industry are similar to the metal finishing industry described in Section 2.100. Figure 2.9 shows a schematic flow diagram of waste/wastewater generation for the printed circuit board industry.

QUESTIONS

Write your answers in a notebook and then compare your answers with those on page 67.

2.10E Describe a printed circuit board.

2.10F List the general processes of printed circuit board manufacturing.

2.11 REFERENCES

1. *DEVELOPMENT DOCUMENT FOR PROPOSED EFFLUENT LIMITATIONS GUIDELINES, NEW SOURCE PERFORMANCE STANDARDS FOR THE METAL FINISHING POINT SOURCE CATEGORY* (Richard Kinch, August 1982, Effluent Guidelines Division, Office of Water Regulations and Standards, US EPA, Washington, DC 20460. Proposed. EPA No. 440-1-82-091B.) Available from National Technical Information Service (NTIS), 5285 Port Royal Road, Springfield, VA 22161, (800) 553-6847. Order No. PB83-102004. Price, $141.00, plus $5.00 shipping and handling per order.

   Describes unit operations in the industry, water usage by operation and waste type, waste characterization of metal finishing unit operations (includes electroplating, electroless plating, anodizing, conversion coating, etching, cleaning, machining, grinding, polishing, barrel finishing, burnishing, impact formation, heat treating, thermal cutting, welding, electrical discharge, electrochemical machining, laminating, hot dip coating, salt bath descaling, solvent degreasing, paint stripping, painting, testing, mechanical plating), and waste treatment methods.

2. *DEVELOPMENT DOCUMENT FOR EXISTING SOURCE PRETREATMENT STANDARDS FOR THE ELECTROPLATING POINT SOURCE CATEGORY* (J. Bill Hansen, August 1979, Effluent Guidelines Division, Office of Water and Hazardous Materials, US EPA, Washington, DC 20460. EPA No. 440-1-79-003.) Available from National Technical Information Service (NTIS), 5285 Port Royal Road, Springfield, VA 22161, (800) 553-6847. Order No. PB80-196488. Price, $114.00, plus $5.00 shipping and handling per order. Information includes industry characterization, waste characterization, and pretreatment technologies.

2.12 ACKNOWLEDGMENTS

Most of the material in this section originally appeared in the manual on *PRETREATMENT FACILITY INSPECTION*, Chapter 8, "Industrial Wastewaters," by Richard von Langen and Mahin Talebi.

END OF LESSON 2 OF 2 LESSONS
on
INDUSTRIAL WASTEWATERS

Please answer the discussion and review questions next.

Industrial Wastewaters 63

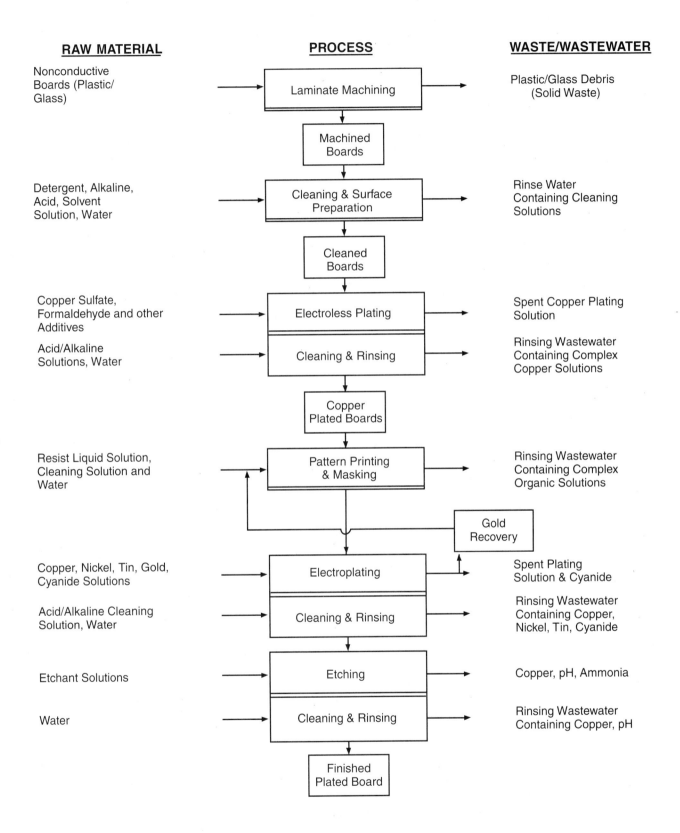

Fig. 2.9 Waste/wastewater characteristics for printed circuit board manufacturing

DISCUSSION AND REVIEW QUESTIONS
Chapter 2. INDUSTRIAL WASTEWATERS
(Lesson 2 of 2 Lessons)

Write the answers to these questions in your notebook. The question numbering continues from Lesson 1.

8. Why is it important for industrial wastewater treatment system operators to understand the sources of industrial wastes at their facilities?
9. List four industrial utility system components that could generate industrial wastewaters.
10. Explain how compatible pollutants can sometimes exhibit the characteristics of noncompatible pollutants, and vice versa.
11. How can spills of process chemicals reach a pretreatment system or POTW?
12. Why are mass emission rate standards preferred over concentration-based standards?
13. Identify several types or sources of intermittent discharges of wastewater.
14. What problems could be caused by discharging an industrial wastestream to the industrial wastewater treatment system if the system was not designed to handle it?
15. What waste characteristics must an IWTS operator evaluate in order to determine the waste's acceptability to the IWTS?
16. How can problems caused by hydrogen sulfide from industrial discharges into a collection system be corrected?
17. What problems can be caused in sewers when an industry discharges out-of-pH-range wastewater?
18. Heated industrial discharges can cause what kinds of problems in IWTS and POTW collection and treatment systems?
19. What is the best means of controlling hydraulic surges and operating the treatment processes at a constant or near-constant flow?
20. Slug loadings or batch dumps cause what kinds of problems?
21. What are pass-through discharges?
22. What can cause a pass-through discharge?
23. What processes are used to clean and prepare the surface of a workpiece for a surface coating?
24. Water is used for what purposes in the metal finishing industry?

SUGGESTED ANSWERS
Chapter 2. INDUSTRIAL WASTEWATERS

ANSWERS TO QUESTIONS IN LESSON 1

Answers to questions on page 27.

2.1A Water picks up dissolved substances as it falls as rain, flows over land, and is used for domestic, industrial, agricultural, and recreational purposes.

2.1B In addition to hydrogen and oxygen, water may contain such dissolved substances as calcium, carbon, magnesium, sodium, chloride, sulfate, iron, nitrogen, phosphorus, and organic matter.

Answers to questions on page 28.

2.2A Four types or classifications of waste discharges are organic, inorganic, heated (thermal), and radioactive discharges.

2.2B Examples of industries that contribute substantial amounts of organic waste include vegetable and fruit packing; dairy processing; meat packing; tanning; and processing of poultry, oil, paper, and fiber (wood).

Answers to questions on page 29.

2.3A When oxidizable (organic) wastes are discharged to a stream, aerobic bacteria begin to feed on the waste. These bacteria also use dissolved oxygen from the water (similar to human respiration or breathing). As more organic waste is added, the bacteria reproduce rapidly; as their population increases, so does their use of oxygen. The population of bacteria may grow large enough to use the entire supply of oxygen from the stream faster than it can be replenished by natural diffusion from the atmosphere.

2.3B   Hydrogen sulfide gas, which has a "rotten egg" odor, is produced when anaerobic bacteria remove the oxygen from sulfur compounds.

Answers to questions on page 30.

2.3C   Disease-causing bacteria in wastewater come from the body wastes of humans who have a disease.

2.3D   Pathogenic means disease-causing.

2.3E   The disinfection process most often used at treatment plants is the addition of chlorine or chlorination.

Answers to questions on page 31.

2.4A   Total solids consist of dissolved and suspended solids, both of which contain organic and inorganic matter.

2.4B   The amount of settleable solids in the raw wastewater should be estimated in order to design settling basins (primary units), sludge pumps, and sludge handling facilities. Also, measuring the amount of settleable solids entering and leaving the settling basin allows you to calculate the efficiency of the basin for removing the settleable solids.

2.4C   An Imhoff cone is used to measure settleable solids.

Answers to questions on page 32.

2.5A   Treatment plant operators need to have an understanding of natural cycles so they can control and accelerate them to work for their benefit in treatment plants and in receiving waters rather than have these cycles cause plant operational problems and disrupt downstream water uses.

2.5B   When nutrient cycles are disrupted and no dissolved oxygen is present in the water, odorous ammonia and hydrogen sulfide gases are produced. Under these conditions, the receiving waters become septic; they stink and look terrible.

Answers to questions on page 33.

2.6A   The major ways water is used by industries are:

1. To make processing of other materials easier or more efficient (this is referred to as "process water").
2. To transport materials to, through, and from the industrial process.
3. For cooling.
4. As wash water for both products and facilities.

2.6B   Three possible alternatives for treatment of industrial wastewater include:

1. Discharge to a Publicly Owned Treatment Works (POTW) to be combined with municipal wastes and possibly wastes from other industries, treated, and returned to the environment.
2. Pretreatment by industry, followed by discharge to a POTW, as above.
3. Treatment by industry to the extent required before discharge to a receiving water.

Answers to questions on page 36.

2.6C   Wastes may interfere with the following uses of water:

1. Drinking
2. Cooking
3. Water-contact recreation
4. Non-body-contact water recreation
5. Fish, wildlife, and aquatic vegetation
6. Agricultural use
7. Industrial use

2.6D   Water can be biologically contaminated by bacteria, viruses, and other microorganisms excreted from the human body and other warm-blooded animals. These excretions may reach the environment through (1) sanitary wastewater, (2) stormwaters washing the land of animal wastes, and (3) through certain industries, most notably slaughterhouses and tanneries.

2.6E   Disinfection is the treatment process most involved with the protection of human populations from biological contamination.

2.6F   From an industrial wastewater standpoint, the most likely source of toxic chemical contamination is from the manufacturing processes in which toxic chemicals are either used or made and from in-plant spills and accidents.

2.6G   Oxygen-consuming wastes are actually treated in a stream or river by the action of the same microorganisms found in a treatment plant. In this treatment process, the organisms "breathe" the oxygen that is present in the water. If all the oxygen is gone, other microorganisms, which can get their oxygen from compounds in the water, continue the treatment process. The by-products of this latter type of treatment (anaerobic decomposition) produce unpleasant odors, colors, and sludges that discourage water-contact recreation.

ANSWERS TO QUESTIONS IN LESSON 2

Answers to questions on page 41.

2.7A   Physical characteristics of the industrial waste, such as temperature, solubility or insolubility, and viscosity, can affect the chemical, mechanical, and biological activities of the conveyance and treatment systems.

2.7B   Knowledge of the various manufacturing processes is important if the operator is to understand how raw materials are transformed into products, by-products, and waste products, and in understanding the effects of the waste products on collection systems, the industrial wastewater treatment plant, and the environment.

2.7C   Routine replacement of pump packing requires the pump to be emptied and disassembled. The liquid that was in the pump is an industrial waste.

2.7D   A mixed waste has the characteristics of several different types of wastes, for example, combinations of liquid and solid wastes, concentrated and dilute wastes, or various types of contaminants.

Answers to questions on page 44.

2.8A   The difference between compatible and noncompatible pollutants is that compatible pollutants are normally removed by the POTW system and noncompatible pollutants are normally *NOT* removed.

2.8B   Examples of noncompatible pollutants include heavy metals, such as copper, nickel, lead, and zinc; organics, such as methylene chloride, 1,1,1-trichloroethylene, methyl ethyl ketone, acetone, and gasoline; and sludges containing toxic organics or metals.

2.8C   The common units of expression for the terms are as follows:

1. Concentration is expressed as milligrams per liter (mg/L) and is a mass per unit of volume measurement.
2. Mass is expressed in pounds or kilograms.
3. Mass emission rate is expressed in pounds or kilograms per day and is a weight per unit time measurement.

2.8D   If the flow of wastewater is large, the hydraulic system may become overloaded and cause pass-through of the pollutant. If a slug load of a highly concentrated solution reaches a treatment plant, the results may be interference with the treatment system and a violation of either or both effluent and sludge disposal limitations.

Answers to questions on page 46.

2.8E   It is important for the IWTS operator at an industrial facility to know the facility's hours of operation because normally, the hours of operation are also the hours of discharge to the IWTS. Thus the operator can generally expect to receive flow for treatment during the hours of operation. Wastewater sampling to investigate process problems and to determine compliance with the discharge limits are also affected by the hours of discharge.

2.8F   An industrial facility might temporarily store wastewater in large collection sumps, equalization tanks, or storage tanks in order to deliver a relatively constant flow and concentration of pollutants to the IWTS.

2.8G   Intermittent wastewater discharges from a manufacturing process typically occur at the beginning or ending of the process or during equipment cleanup, a spill, replacement of spent solution, or disposal of a reject product.

Answers to questions on page 50.

2.9A   Types of heavy solids that can cause plugging of sewers include sand, ceramic or porcelain solids, and grindings, which can build up in a sewer and reduce its hydraulic capacity. Large objects that can plug a sewer include rags, tools, rejected food products, and discarded by-products.

2.9B   Examples of industrial discharges that can be odorous are those from petroleum refining, petrochemical manufacturing, and food processing.

2.9C   Mineral acids such as sulfuric, nitric, hydrochloric, and phosphoric acids are used extensively to clean base metals in the metal finishing industries. The fertilizer, iron and steel, mining, and petroleum industries use vast quantities of strong acids. Mineral acids are also used in pretreatment systems for chromium reduction, neutralization of alkalies, and pretreatment of chelated metal plating solutions.

2.9D   Heated industrial discharges can cause such problems as evolution of gases and odors, overheating of pump and rotating equipment bearings, shifts in the population of microorganisms used in biological treatment, or even sterilization of the wastewater. High-temperature wastewaters may also cause plastic pipes and O-rings to fail thereby permitting exfiltration or infiltration of the collection system.

Answers to questions on page 54.

2.9E   A hydraulic overload can cause a decrease in the efficiency of treatment processes, an increase in solids carryover, and possible effluent limit violations at a wastewater treatment plant.

2.9F   EPA defines interference as a discharge which, alone or in conjunction with discharges from other sources, inhibits or disrupts the POTW, its treatment processes or operations, its sludge processes, use or disposal, and is a cause of violation of the NPDES permit or prevents the lawful use or disposal of sludge.

2.9G   Changes in conductivity normally represent increases or decreases in soluble salts, cyanide, or metals.

Answers to questions on page 60.

2.10A   The common metals used by the metal finishing industry to provide a surface coating on a base material include copper, nickel, chromium, brass, bronze, zinc, tin, lead, cadmium, iron, aluminum, or combinations thereof. In precious metals plating, a base material is plated with gold, silver, palladium, platinum, osmium, iridium, rhodium, indium, ruthenium, or combinations of these metals.

2.10B The general processes of the metal finishing industry include machining, cleaning and surface preparation, plating and coating, anodizing, and etching and chemical milling.

2.10C Types of chemicals found in electroplating baths include metal salts, alkalies, and other bath control compounds in addition to the plating metals. Many plating solutions contain metallic, metallo-organic, and organic additives to induce grain refining, leveling of the plating surface, and deposit brightening.

2.10D The nine common waste types found in the waste/wastewater streams generated from metal finishing processes include: (1) heavy metals, (2) precious metals, (3) complexed metals, (4) acid wastewaters, (5) alkaline wastes, (6) hexavalent chromium, (7) cyanide wastes, (8) oily wastes, and (9) solvent wastes.

Answers to questions on page 62.

2.10E A printed circuit board is a board or plate of nonconductive material such as glass or plastic on which a conductive metal circuit pattern has been printed or etched.

2.10F The general processes of printed circuit board manufacturing include laminate machining, cleaning and surface preparation, electroless plating, pattern printing and masking, electroplating, and etching.

CHAPTER 3

REGULATORY REQUIREMENTS

by

Robert Montgomery

TABLE OF CONTENTS
Chapter 3. REGULATORY REQUIREMENTS

			Page
OBJECTIVES			74
WORDS			75
ABBREVIATIONS			79
3.0	OVERVIEW OF POLLUTION CONTROL AUTHORITY		81
	3.00	Legislative History of Federal Pollution Control Regulations	81
	3.01	Environmental Protection Agency	81
		3.010 Scope of Authority	81
		3.011 Organizational Structure	82
	3.02	Delegation of Federal Authority	82
	3.03	Types of Regulated Dischargers	82
3.1	NPDES PERMIT PROGRAM		83
3.2	NATIONAL PRETREATMENT PROGRAM		84
	3.20	General Pretreatment Regulations	84
	3.21	Prohibited Discharge Standards	87
	3.22	Categorical Pretreatment Standards	88
		3.220 EPA Regulation Development Process	91
		3.221 EPA-Regulated Categories	92
		3.222 Categories Exempt From EPA Categorical Regulations	92
		3.223 Category Determination Requests	92
		3.224 Local Limits for Noncategorical Industries	92
3.3	APPLICATION OF CATEGORICAL PRETREATMENT REGULATIONS		93
	3.30	Types of Categorical Standards	93
		3.300 Industrial Categories With Concentration-Based Standards	93
		3.301 Industrial Categories With Mass-Based Standards	93
		3.302 Industrial Categories With Both Concentration- and Mass-Based Standards	93
	3.31	Types of Wastestreams	94
	3.32	Total Toxic Organics (TTO)	94
		3.320 Regulated Toxic Organics	94
		3.321 TTO Monitoring	94

3.33	EPA Reporting Requirements		94
	3.330	Baseline Monitoring Reports	95
	3.331	Compliance Schedule	95
	3.332	Final Compliance Report	95
	3.333	Periodic Compliance Reports	95
	3.334	Slug Loading Reporting	95
	3.335	Resampling to Confirm Violations	96
	3.336	Notification of Changed Discharge	96
	3.337	Hazardous Waste Disposal Reporting	96
3.34	Modification of Categorical Standards		96
	3.340	Variance From Categorical Standards for Fundamentally Different Factors (FDF)	97
	3.341	Net Gross Calculations	97
	3.342	Removal Credits	97

3.4 LOCAL ORDINANCES AFFECTING INDUSTRIAL DISCHARGERS 98

3.40	Wastewater Ordinances		98
	3.400	Administration	98
	3.401	EPA Minimum Requirements	98
	3.402	General Provisions	99
	3.403	Specific Provisions	99
	3.404	Industrial Wastewater Limitations	99
	3.405	Administrative Fine Penalties	100
	3.406	Misdemeanor and Felony Criminal Actions	100
3.41	Sanitary Sewer Codes		100
3.42	Building Codes		101
3.43	Underground Tank Laws		101
3.44	Land Use Ordinances		101

3.5 OTHER FEDERAL STATUTES AND REGULATIONS 101

3.50	Resource Conservation and Recovery Act (RCRA)	101
3.51	Superfund Amendment Reauthorization Act (SARA)—Title III	102
3.52	Clean Air Act	103
3.53	Occupational Safety and Health Act (OSHA)—Hazard Communication (29 CFR 1910.1200)	103
3.54	Occupational Safety and Health Act (OSHA)—Process Safety Management of Highly Hazardous Chemicals (29 CFR 1910.119)	103
3.55	Clean Water Act—Stormwater Regulations	103
3.56	Department of Transportation—Hazardous Material Regulations (HM 126 and HM 181)	104

3.6	KEEPING CURRENT	104
3.7	ACKNOWLEDGMENTS	104
DISCUSSION AND REVIEW QUESTIONS		105
SUGGESTED ANSWERS		105

OBJECTIVES
Chapter 3. REGULATORY REQUIREMENTS

Following completion of Chapter 3, you should be able to:

1. Describe the role of the Environmental Protection Agency in pollution control efforts.

2. Define the categories or types of industrial dischargers regulated under the Clean Water Act.

3. Explain which types of industrial dischargers are regulated by the National Pollutant Discharge Elimination System (NPDES) permit program.

4. Explain which types of industrial dischargers are regulated by the National Pretreatment Program.

5. List the general types of pollutants governed by the General Pretreatment Regulations.

6. Outline the development of categorical limits.

7. Identify types of industries and facilities subject to the categorical program.

8. Explain the role of local authorities in the control of pollutants.

9. Keep current with changing regulations.

WORDS
Chapter 3. REGULATORY REQUIREMENTS

40 CFR 403

EPA's General Pretreatment Regulations appear in the Code of Federal Regulations under 40 CFR 403. 40 refers to the numerical heading for the environmental regulations portion of the Code of Federal Regulations. 403 refers to the section that contains the General Pretreatment Regulations. Significant amendments to the General Pretreatment Regulations include the PIRT Amendments (*FEDERAL REGISTER*, October 18, 1988) and the DSS Amendments (*FEDERAL REGISTER*, July 24, 1990).

BOD (pronounce as separate letters)

Biochemical Oxygen Demand. The rate at which organisms use the oxygen in water or wastewater while stabilizing decomposable organic matter under aerobic conditions. In decomposition, organic matter serves as food for the bacteria and energy results from its oxidation. BOD measurements are used as a surrogate measure of the organic strength of wastes in water.

BASELINE MONITORING REPORT (BMR)

All industrial users subject to categorical pretreatment standards must submit a baseline monitoring report (BMR) to the control authority (POTW, state, or EPA). The purpose of the BMR is to provide information to the control authority to document the industrial user's current compliance status with a categorical pretreatment standard.

BEST AVAILABLE TECHNOLOGY (BAT)

A level of technology represented by a higher level of wastewater treatment technology than required by Best Practicable Technology (BPT). BAT is based on the very best (state-of-the-art) control and treatment measures that have been developed, or are capable of being developed, and that are economically achievable within the appropriate industrial category.

BEST PRACTICABLE TECHNOLOGY (BPT)

A level of technology represented by the average of the best existing wastewater treatment performance levels within the industrial category.

BIOCHEMICAL OXYGEN DEMAND (BOD)

See BOD.

CATEGORICAL STANDARDS (LIMITS)

Industrial waste discharge standards (limits) developed by EPA that are applied to the effluent from any industry in any category anywhere in the United States that discharges to a Publicly Owned Treatment Works (POTW). These are standards based on the technology available to treat the wastestreams from the processes of the specific industrial category and normally are measured at the point of discharge from the regulated process. The standards are listed in the Code of Federal Regulations.

CODE OF FEDERAL REGULATIONS (CFR)

A publication of the US government that contains all of the proposed and finalized federal regulations, including safety and environmental regulations.

COLIFORM (KOAL-i-form)

A group of bacteria found in the intestines of warm-blooded animals (including humans) and also in plants, soil, air, and water. The presence of coliform bacteria is an indication that the water is polluted and may contain pathogenic (disease-causing) organisms. Fecal coliforms are those coliforms found in the feces of various warm-blooded animals, whereas the term "coliform" also includes other environmental sources.

COMPLIANCE

The act of meeting specified conditions or requirements.

CONVENTIONAL POLLUTANTS

Those pollutants that are usually found in domestic, commercial, or industrial wastes, including suspended solids, biochemical oxygen demand, pathogenic (disease-causing) organisms, and oil and grease.

CRADLE TO GRAVE

A term used to describe a hazardous waste manifest system used by regulatory agencies to track a hazardous waste from the point of generation to the hauler and then to the ultimate disposal site.

DIRECT DISCHARGER

A point source that discharges a pollutant(s) to waters of the United States, such as streams, lakes, or oceans. These sources are subject to the National Pollutant Discharge Elimination System (NPDES) program regulations.

EFFLUENT LIMITS

Pollutant limitations developed by a POTW for industrial plants discharging to the POTW system. At a minimum, all industrial facilities are required to comply with federal prohibited discharge standards. The industries covered by federal categorical standards must also comply with the appropriate discharge limitations. The POTW may also establish local limits more stringent than or in addition to the federal standards for some or all of its industrial users.

EXISTING SOURCE

An industrial discharger that was already in operation when the proposed pretreatment standard for the industrial category was promulgated.

HAZARDOUS WASTE

A waste that possesses any one of the following four characteristics:

(1) Ignitability, which identifies wastes that pose a fire hazard during routine management. Fires not only present immediate dangers of heat and smoke, but also can spread harmful particles over wide areas. A liquid that has a flash point of less than 140°F (60°C).

(2) Corrosivity, which identifies wastes requiring special containers because of their ability to corrode standard materials, or requiring segregation from other wastes because of their ability to dissolve toxic contaminants. An aqueous solution with a pH less than or equal to 2 or a pH greater than or equal to 12.5.

(3) Reactivity (or explosiveness), which identifies wastes that, during routine management, tend to react spontaneously, to react vigorously with air or water, to be unstable to shock or heat, to generate toxic gases, or to explode.

(4) Toxicity, which identifies wastes that, when improperly managed, may release toxicants in sufficient quantities to pose a substantial present or potential hazard to human health or the environment.

INDIRECT DISCHARGER

A nondomestic discharger introducing pollutants to a POTW. These facilities are subject to the EPA pretreatment regulations.

NAICS

North American Industry Classification System. A code number system used to identify various types of industries. This code system replaces the SIC (Standard Industrial Classification) code system used prior to 1997. Use of these code numbers is often mandatory. Some companies have several processes, which will cause them to fit into two or more classifications. The code numbers are published by the US Government Printing Office, Superintendent of Documents, PO Box 371954, Pittsburgh, PA 15250-7954. Stock No. 041-001-00509-9; price, $33.00. There is no charge for shipping and handling.

NPDES PERMIT

National Pollutant Discharge Elimination System permit is the regulatory agency document issued by either a federal or state agency that is designed to control all discharges of potential pollutants from point sources and stormwater runoff into US waterways. NPDES permits regulate discharges into US waterways from all point sources of pollution, including industries, municipal wastewater treatment plants, sanitary landfills, large animal feedlots, and return irrigation flows.

NEW SOURCE

Any building, structure, facility, or installation from which there is or may be a discharge of pollutants. Construction of the facility must have begun after promulgation of the applicable Pretreatment Standards. The building, structure, facility, or installation must also be constructed at a site at which no other source is located; or, must totally replace the existing process or production equipment producing the discharge at the site; or, must be substantially independent of an existing source of discharge at the same site.

NONPOINT SOURCE

A runoff or discharge from a field or similar source, in contrast to a point source, which refers to a discharge that comes out the end of a pipe or other clearly identifiable conveyance. Also see POINT SOURCE.

POTW

Publicly Owned Treatment Works. A treatment works that is owned by a state, municipality, city, town, special sewer district, or other publicly owned and financed entity as opposed to a privately (industrial) owned treatment facility. This definition includes any devices and systems used in the storage, treatment, recycling, and reclamation of municipal sewage (wastewater) or industrial wastes of a liquid nature. It also includes sewers, pipes, and other conveyances only if they carry wastewater to a POTW treatment plant. The term also means the municipality (public entity) that has jurisdiction over the indirect discharges to and the discharges from such a treatment works.

POINT SOURCE

A discharge that comes out the end of a pipe or other clearly identifiable conveyance. Examples of point source conveyances from which pollutants may be discharged include: ditches, channels, tunnels, conduits, wells, containers, rolling stock, concentrated animal feeding operations, landfill leachate collection systems, vessels, or other floating craft. A NONPOINT SOURCE refers to runoff or a discharge from a field or similar source.

RCRA (RICK-ruh)

The Federal Resource Conservation and Recovery Act (10/21/76), Public Law (PL) 94-580, provides technical and financial assistance for the development of plans and facilities for recovery of energy and resources from discarded materials and for the safe disposal of discarded materials and hazardous wastes. This act introduces the philosophy of the "cradle-to-grave" control of hazardous wastes. RCRA regulations can be found in Title 40 of the Code of Federal Regulations (40 CFR) Parts 260-268, 270, and 271.

SIC CODE

Standard Industrial Classification code. A code number system used to identify various types of industries. In 1997, the United States and Canada replaced the SIC code system with the North American Industry Classification System (NAICS); Mexico adopted the NAICS in 1998. Also see NAICS.

SIGNIFICANT INDUSTRIAL USER (SIU)

A significant industrial user includes all categorical industrial users, and any noncategorical industrial user that:

(1) Discharges 25,000 gallons per day or more of process wastewater ("process wastewater" excludes sanitary, noncontact cooling and boiler blowdown wastewaters), or

(2) Contributes a process wastestream that makes up five percent or more of the average dry weather hydraulic or organic (BOD, TSS) capacity of a treatment plant, or

(3) Has a reasonable potential, in the opinion of the control or approval authority, to adversely affect the POTW treatment plant (inhibition, pass-through of pollutants, sludge contamination, or endangerment of POTW workers).

SOLVENT MANAGEMENT PLAN

A strategy for keeping track of all solvents delivered to a site, their storage, use, and disposal. This includes keeping spent solvents segregated from other process wastewaters to maximize the value of the recoverable solvents, to avoid contamination of other segregated wastes, and to prevent the discharge of toxic organics to any wastewater collection system or the environment. The plan should describe measures to control spills and leaks and to ensure that there is no deliberate dumping of solvents. Also known as a TOXIC ORGANIC MANAGEMENT PLAN.

TOXIC ORGANIC MANAGEMENT PLAN (TOMP)

A strategy for keeping track of all solvents delivered to a site, their storage, use, and disposal. This includes keeping spent solvents segregated from other process wastewaters to maximize the value of the recoverable solvents, to avoid contamination of other segregated wastes, and to prevent the discharge of toxic organics to any wastewater collection system or the environment. The plan should describe measures to control spills and leaks and to ensure that there is no deliberate dumping of solvents. Also known as a SOLVENT MANAGEMENT PLAN.

TOXIC POLLUTANT

Those pollutants or combinations of pollutants, including disease-causing agents, that cause death, disease, behavioral abnormalities, cancer, genetic mutations, physiological malfunctions (including malfunctions in reproduction), or physical deformations.

WASTEWATER FACILITIES

The pipes, conduits, structures, equipment, and processes required to collect, convey, and treat domestic and industrial wastes, and dispose of the effluent and sludge.

WASTEWATER ORDINANCE

The basic document granting authority to administer a pretreatment inspection program. This ordinance must contain certain basic elements to provide a legal framework for effective enforcement.

ABBREVIATIONS

Chapter 3. REGULATORY REQUIREMENTS

BAT
Best Available Technology Economically Achievable.

BMR
Baseline Monitoring Report.

BPT
Best Practicable Technology Economically Available.

CBI
Compliance Biomonitoring Inspection.

CERCLA
Comprehensive Environmental Response, Compensation, and Liability Act of 1980.

CFR
Code of Federal Regulations.

CSI
Compliance Sampling Inspection.

EPA
Environmental Protection Agency.

EPCRA
Emergency Planning and Community Right-To-Know Act of 1986.

ERP
Enforcement Response Plan.

FDF
Fundamentally Different Factors.

IPCBM
Independent Printed Circuit Board Manufacturers.

IU
Industrial User.

NAICS
North American Industry Classification System.

NPDES
National Pollutant Discharge Elimination System.

NRDC
Natural Resources Defense Council.

NSPS
New Source Performance Standards.

O & M
Operation and Maintenance.

OWWM
Office of Water and Wastewater Management, EPA.

PAI
Performance Audit Inspection.

PCI
Pretreatment Compliance Inspection.

PDL
Prohibited Discharge Limit.

PEL
Permissible Exposure Level.

POTW
Publicly Owned Treatment Works.

PSES
Pretreatment Standards for Existing Sources.

PSNS
Pretreatment Standards for New Sources.

RCRA
Resource Conservation and Recovery Act.

SARA
Superfund Amendment Reauthorization Act.

SIC
Standard Industrial Classification.

SIU
Significant Industrial User.

TLV
Threshold Limit Value.

TOMP
Toxic Organic Management Plan.

TSCA
Toxic Substances Control Act.

TTO
Total Toxic Organics.

TWA
Time Weighted Average.

XSI
Toxics Sampling Inspection.

USC
United States Code.

CHAPTER 3. REGULATORY REQUIREMENTS

In this chapter we will be discussing the water pollution control statutes and regulations that concern industrial facilities that discharge wastewater directly to the environment or to wastewater collection systems and treatment plants (POTWs).

Black's Law Dictionary, 5th Edition, defines *STATUTE* as an act of the legislature declaring, commanding, or prohibiting something. *REGULATIONS* are issued by various governmental departments to carry out the intent of the statute. Agencies issue regulations to guide the activity of those regulated by the agency and to ensure uniform application of the statute. The Clean Water Act (CWA) is a statute. The Environmental Protection Agency (EPA) has issued regulations relating to the CWA and these are contained in Title 40 of the Code of Federal Regulations (CFR).

The *FEDERAL REGISTER* is a printed record of all activity that takes place in the US Congress. The regulations contained in the Code of Federal Regulations are an annually codified (given the force of law) version of the promulgated regulations recorded in the *FEDERAL REGISTER*. Although the CFR is a comprehensive source of regulations, it does not provide all of the background and implementation information that is provided in the preamble to the regulations contained in the *FEDERAL REGISTER*. Therefore, it is very helpful for the industrial waste treatment operator to become familiar with both of these documents.

3.0 OVERVIEW OF POLLUTION CONTROL AUTHORITY

3.00 Legislative History of Federal Pollution Control Regulations

Even prior to the awakening of the environmental movement in the 1960s and 1970s, many local wastewater treatment agencies had established basic industrial waste control programs. These programs were aimed at gross industrial pollutants that were found to cause operational problems in collection systems and at treatment plants of the agency. While such programs were generally beneficial at local levels, there existed no means to ensure that all communities would be protected or that all industries would be uniformly prohibited from discharging environmentally damaging wastes.

Some efforts had been made to protect the environment prior to 1972. The following is a list of federal pollution control efforts that led to modern water pollution regulations:

1899   The Rivers and Harbors Act

1912   The Public Health Service Act

1965   The Oil Pollution Control Act

1966   The Clean Water Restoration Act

1970   The Water Quality Improvements Act

By the early 1970s, public clamor for effective, comprehensive legislation to control water pollution led Congress to enact landmark legislation in the form of the 1972 Federal Water Pollution Control Act. For the first time in the history of the United States, Congress declared with this Act that the pollution of the nation's waters by either industries or municipalities was unlawful. The Act created a system of uniform controls on discharge of all pollutants and required a federal agency, the Environmental Protection Agency (EPA), to establish effluent limitations and standards. The major goals of this legislation were: (1) to have the nation's waters clean enough for swimming by 1983, and (2) to eliminate the discharge of pollutants by 1985.

The second goal, not yet reached, is an ambitious one, which may not be attainable economically in the near future. However, it remains the goal toward which the nation's efforts are directed. The Federal Water Pollution Control Act has been amended several times since its passage and is now known as the Clean Water Act. Additional legislation further reinforces the mandate to protect the nation's waters. Two significant Acts of Congress are the 1977 Clean Water Act and the 1987 Water Quality Act.

3.01 Environmental Protection Agency

3.010 Scope of Authority

Established in 1970, the Environmental Protection Agency (EPA) is the federal agency responsible for developing and enforcing regulations designed to protect the nation's water, air, and land from various pollutants. The agency is charged with developing pollution control guidelines, implementing programs to achieve the goals established by Congress, and coordinating the various pollution control activities throughout the nation. The Clean Water Act (CWA) provides the EPA with broad authority to:

1. Establish the National Pollutant Discharge Elimination System (NPDES) program and the National Pretreatment Program.

2. Define acceptable pollution control technologies and establish effluent limitations based on these technologies.

3. Obtain information through reports and compliance inspections.

4. Take enforcement actions, both civil and criminal, when violations of the CWA occur.

The National Pollutant Discharge Elimination System (NPDES) permit program regulates the discharge of potential pollutants from POINT SOURCES[1] and some stormwater runoff into US waterways. NPDES permits regulate discharges into US waterways from all point sources of pollution, including those coming from industries, municipal wastewater treatment plants, sanitary landfills, large animal feedlots, return irrigation flows, aquatic production facilities, and mining operations. Regulated discharges of stormwater runoff include municipal separate storm sewer systems, industrial and commercial sites, and construction sites. Each point source discharger is required to obtain an NPDES permit to discharge. The permit contains effluent limits, monitoring and reporting requirements, and any other terms and conditions necessary to protect water quality. Industries may operate under their own NPDES permit if they discharge directly to the environment or, if they discharge to a POTW, they may be affected by the terms of the POTW's NPDES permit. The NPDES permit program is described in Section 3.1, "NPDES Permit Program."

The EPA also developed and administers the National Pretreatment Program. Like the NPDES program, the National Pretreatment Program is designed to protect the nation's water, air, and land from contamination. Industries or other facilities that discharge wastes to a POTW are subject to the regulations of the National Pretreatment Program.

3.011 Organizational Structure

EPA is divided into nine staff offices, six program offices, and ten regional offices. The Office of Water and Wastewater Management (OWWM) is a program office that is responsible for development of effluent limitations and standards. The Office of Water Regulations and Standards is a suboffice of the OWWM. The Industrial Technology Division of the suboffice develops the federal categorical pretreatment standards. The Industrial Technology Division was formerly known as Effluent Guidelines Division.

3.02 Delegation of Federal Authority

The Clean Water Act gives EPA the authority to develop and implement programs to control the flow of toxic pollutants into POTWs or the environment. EPA can delegate its implementation authority to state or local officials and, for reasons of effective management, has chosen to do so. Local industrial source control programs already exist in many areas. The local POTWs are familiar with their industrial dischargers, may already have developed an extensive database about dischargers, and may have ongoing wastewater permit and administration mechanisms. By delegating its regulatory authority, EPA utilizes existing programs rather than replacing them with a costly new bureaucracy. State and local authorities are also better able to respond promptly and effectively to wastewater collection system or treatment plant emergencies; they are better able to resolve problems quickly with industrial dischargers; and greater resources are usually available to local agencies to conduct a pretreatment program.

3.03 Types of Regulated Dischargers

In regulating waste dischargers, EPA has developed specific regulatory requirements for two different types of industrial dischargers: direct and indirect.

DIRECT DISCHARGERS are facilities (such as an industry or POTW) that discharge wastewaters directly into US streams, lakes or other waters, or into the oceans. These facilities are subject to the National Pollutant Discharge Elimination System (NPDES) program regulations.

INDIRECT DISCHARGERS are facilities that discharge wastewaters to a POTW and are subject to the regulations known as Pretreatment Standards for New or Existing Sources (PSNS or PSES). An example of an indirect discharger is a metal finishing company that discharges its industrial wastewaters to the local POTW as opposed to discharging directly to the environment.

---

[1] Point Source. A discharge that comes out the end of a pipe or other clearly identifiable conveyance. Examples of point source conveyances from which pollutants may be discharged include: ditches, channels, tunnels, conduits, wells, containers, rolling stock, concentrated animal feeding operations, landfill leachate collection systems, vessels, or other floating craft. A NONPOINT SOURCE refers to runoff or a discharge from a field or similar source.

Each of these two types of dischargers is further classified on the basis of when the plant came into existence. Dischargers may be referred to as existing or new dischargers and may each be subject to a separate set of effluent standards. Existing and New Sources are defined as follows:

EXISTING SOURCES are industrial dischargers that were in existence at the time of promulgation of the proposed applicable categorical standards.

NEW SOURCES are any building, structure, facility, or installation from which there is or may be a discharge of pollutants. Construction of the facility must have begun after publication of the applicable Pretreatment Standards. The building, structure, facility, or installation must also be constructed at a site at which no other source is located; or, must totally replace the existing process or production equipment producing the discharge at the site; or, must be substantially independent of an existing source of discharge at the same site.

QUESTIONS

Write your answers in a notebook and then compare your answers with those on page 105.

3.0A   What is the difference between a statute and a regulation?

3.0B   What were the two major goals of the 1972 Federal Water Pollution Control Act Amendments?

3.0C   What is the basic mission of the Environmental Protection Agency?

3.0D   EPA has developed specific regulatory requirements for what two different types of industrial dischargers?

3.0E   What is the difference between "new sources" and "existing sources"?

3.1 NPDES PERMIT PROGRAM

The Federal Water Pollution Control Act of 1972, as amended by the Clean Water Act (CWA) of 1977 and by the Water Quality Act of 1987, specifies the objectives of restoring and maintaining the chemical, physical, and biological integrity of the nation's waters. Two programs designed to meet this objective are the National Pollutant Discharge Elimination System (NPDES) permit program and the National Pretreatment Program.

The NPDES permit program regulates the discharge of potential pollutants from point sources and some stormwater runoff into US waterways. NPDES permits regulate discharges into US waterways from all point sources of pollution, including industries, municipal wastewater treatment plants, sanitary landfills, large animal feedlots, and return irrigation flows. Regulated discharges of stormwater runoff include municipal separate storm sewer systems, industrial and commercial sites, and construction sites. If a facility discharges pollutants from any point source into surface waters of the United States, then the facility must apply to its regulatory authority for an NPDES permit for each discharge point (point source). The major elements of an NPDES permit include effluent limits, monitoring and reporting requirements, and any other terms and conditions necessary to protect water quality.

An industrial facility may discharge directly to the environment and be regulated by its own NPDES permit, or the facility may discharge directly or following pretreatment to a POTW that is regulated by an NPDES permit. In either case, the industrial facility must operate in compliance with the terms and conditions of the permit. Permits are issued under five broad categories of programs:

1. Basic municipal and industrial permit program
2. Pretreatment program
3. Federal facilities program
4. General permit program
5. Sludge permit program

The EPA may grant states the authority to administer any or all of these five permit programs.

The NPDES permit program is authorized by Section 402 of the CWA and is implemented through 40 CFR Parts 122 through 124. Other parts of the CFR affecting the NPDES program are listed in Table 3.1.

TABLE 3.1 REGULATIONS AFFECTING NPDES PERMIT PROGRAMS

40 CFR Part 125	Technology-Based Standards
40 CFR Part 129	Toxic Pollutant Standards
40 CFR Part 130	Water Quality Management Plans
40 CFR Part 131	Water Quality-Based Standards
40 CFR Part 133	Sewage Secondary Treatment Regulations
40 CFR Part 135	Citizen Suits
40 CFR Part 136	Analytical Procedures
40 CFR Part 257	Sludge Disposal Regulations
40 CFR Part 401	General Effluent Guidelines Provisions
40 CFR Part 403	General Pretreatment Regulations
40 CFR Parts 405-471	Effluent Limitation Guidelines
40 CFR Part 501	Sludge Permitting Requirements
40 CFR Part 503	Sewage Sludge Disposal Standards[a]

[a] *NOTE:* The Sewage Sludge Disposal Standards include a system of "sludge only" NPDES permits, which apply even when there is no point source discharge of liquid effluent, direct *OR* indirect.

To determine whether NPDES permit conditions are being met, Section 308 of the Act authorizes inspections and monitoring of permittee facilities. Two types of monitoring are required: self-monitoring to be conducted by the permittee, and compliance monitoring to be performed by the permit-issuing agency. According to the Act, an inspection may be conducted wherever there is an existing NPDES permit or where a discharge exists or is likely to exist and no permit has been issued.

Compliance with NPDES permit conditions is often monitored by states. Sections 308 and 402 of the Act allow the transfer of federal program authority to conduct NPDES permit compliance monitoring to state agencies. Currently, most states and territories are approved by EPA to implement state NPDES programs.

As mentioned above, each NPDES permit contains specific, legally enforceable effluent limitations and monitoring requirements. The purpose of the NPDES compliance monitoring program (and the various inspections conducted under the program) is to evaluate the compliance of dischargers with permit limitations and conditions. This evaluation involves two aspects: (1) collection of samples of permittee's effluent by an NPDES inspector, as occurs during a Compliance Sampling Inspection (CSI), a Toxics Sampling Inspection (XSI), or a Compliance Biomonitoring Inspection (CBI); and (2) evaluation of a permittee's self-monitoring procedures, as takes place in a Performance Audit Inspection (PAI). Under certain circumstances, the inspection may also evaluate the industrial monitoring and enforcement efforts conducted as part of a municipality's pretreatment program. This type of inspection is known as a Pretreatment Compliance Inspection (PCI).

An industry discharging into municipal collection and treatment systems need not obtain a permit but must meet certain specified pretreatment standards. These permits may outline a schedule of compliance for a wastewater treatment facility, such as dates for the completion of plant design, engineering, construction, or treatment process changes. Instructions for completing NPDES reporting forms and the necessary forms are available from the regulatory agency issuing the permit (Figure 3.1).

Your main concern as an operator is the effluent (discharge) limitations specified in the NPDES permit for your plant. A municipal wastewater treatment plant permit may specify monthly average and maximum levels of suspended solids, *BIOCHEMICAL OXYGEN DEMAND (BOD)*,[2] and the most probable number (MPN) of *COLIFORM*[3] group bacteria. Larger plants must report effluent temperatures because of the impact of temperature changes on natural cycles. Also, average and maximum flows may be identified as well as an acceptable range of pH values. Almost all effluents are expected to contain virtually no substances that would be toxic to organisms in the receiving waters. NPDES permits have effluent limit restrictions on toxic substances. The NPDES permit will specify the frequency of collecting samples and the methods of reporting the results. Details on how to comply with NPDES permits will be provided throughout this manual.

QUESTIONS

Write your answers in a notebook and then compare your answers with those on page 106.

3.1A What is the National Pollutant Discharge Elimination System (NPDES) permit program designed to regulate?

3.1B What are the major elements of an NPDES permit?

3.1C Under what conditions might an industrial discharger be required to comply with the terms and conditions of the POTW's NPDES permit?

3.2 NATIONAL PRETREATMENT PROGRAM

3.20 General Pretreatment Regulations

At the heart of the National Pretreatment Program is a set of rules and standards known as the General Pretreatment Regulations. The purpose of the General Pretreatment Regulations is to implement the Federal Water Pollution Control Act Amendments of 1972 and the Clean Water Act of 1977 as they apply to industries that discharge nondomestic wastewaters to POTWs. The regulations establish the responsibilities of the federal, state, and local governments, industry, and the public to implement the National Pretreatment Program. These regulations set the ground rules on the types and amounts of pollutants that may be discharged to POTWs; they apply universally to all industrial dischargers who discharge to POTWs. The complete text of the General Pretreatment Regulations is contained in the Code of Federal Regulations under *40 CFR 403*.[4]

The General Pretreatment Regulations regulate pollutants that may:

1. Pass through the POTW's treatment system, untreated or partially treated.

2. Interfere with the POTW's treatment processes.

3. Contaminate the POTW's sludge.

---

[2] *BOD* (pronounce as separate letters). Biochemical Oxygen Demand. The rate at which organisms use the oxygen in water or wastewater while stabilizing decomposable organic matter under aerobic conditions. In decomposition, organic matter serves as food for the bacteria and energy results from its oxidation. BOD measurements are used as a surrogate measure of the organic strength of wastes in water.

[3] *Coliform* (KOAL-i-form). A group of bacteria found in the intestines of warm-blooded animals (including humans) and also in plants, soil, air, and water. The presence of coliform bacteria is an indication that the water is polluted and may contain pathogenic (disease-causing) organisms. Fecal coliforms are those coliforms found in the feces of various warm-blooded animals, whereas the term "coliform" also includes other environmental sources.

[4] *40 CFR 403*. EPA's General Pretreatment Regulations appear in the Code of Federal Regulations under 40 CFR 403. 40 refers to the numerical heading for the environmental regulations portion of the Code of Federal Regulations. 403 refers to the section that contains the General Pretreatment Regulations. Significant amendments to the General Pretreatment Regulations include the PIRT Amendments (*FEDERAL REGISTER*, October 18, 1988) and the DSS Amendments (*FEDERAL REGISTER*, July 24, 1990).

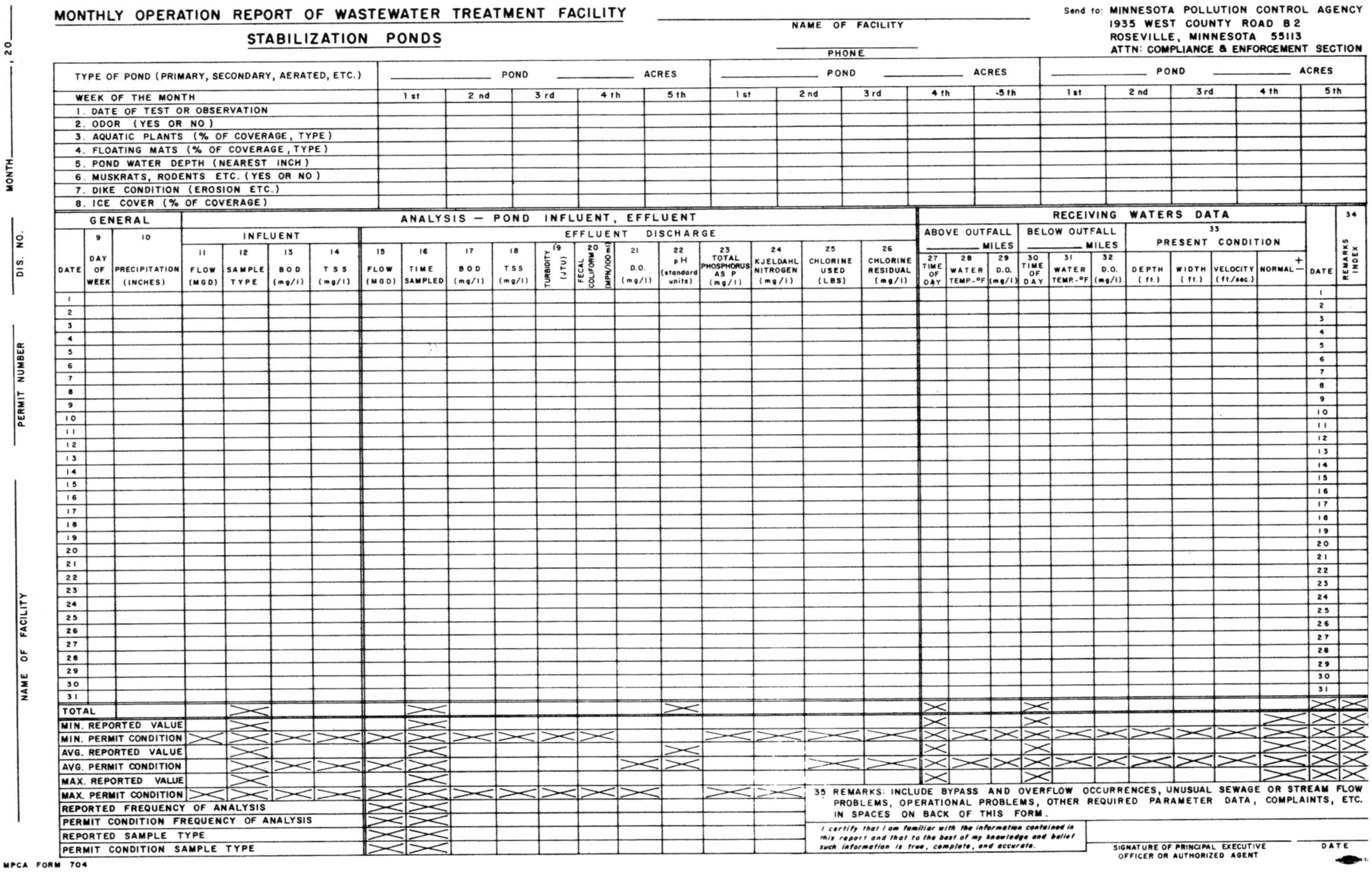

MONTHLY OPERATION REPORT OF WASTEWATER TREATMENT FACILITY

STABILIZATION PONDS

NAME OF FACILITY

PHONE

Send to: MINNESOTA POLLUTION CONTROL AGENCY
1935 WEST COUNTY ROAD B2
ROSEVILLE, MINNESOTA 55113
ATTN: COMPLIANCE & ENFORCEMENT SECTION

MONTH ____, 20__ | DIS. NO. | PERMIT NUMBER | NAME OF FACILITY

TYPE OF POND (PRIMARY, SECONDARY, AERATED, ETC.)	POND ___ ACRES					POND ___ ACRES					POND ___ ACRES				
WEEK OF THE MONTH	1st	2nd	3rd	4th	5th	1st	2nd	3rd	4th	5th	1st	2nd	3rd	4th	5th
1. DATE OF TEST OR OBSERVATION															
2. ODOR (YES OR NO)															
3. AQUATIC PLANTS (% OF COVERAGE, TYPE)															
4. FLOATING MATS (% OF COVERAGE, TYPE)															
5. POND WATER DEPTH (NEAREST INCH)															
6. MUSKRATS, RODENTS ETC. (YES OR NO)															
7. DIKE CONDITION (EROSION ETC.)															
8. ICE COVER (% OF COVERAGE)															

GENERAL: DATE; 9 DAY OF WEEK; 10 PRECIPITATION (INCHES)

ANALYSIS — POND INFLUENT, EFFLUENT

INFLUENT: 11 FLOW (MGD); 12 SAMPLE TYPE; 13 BOD (mg/l); 14 TSS (mg/l)

EFFLUENT DISCHARGE: 15 FLOW (MGD); 16 TIME SAMPLED; 17 BOD (mg/l); 18 TSS (mg/l); 19 TURBIDITY (JTU); 20 FECAL COLIFORM (MPN/100 ml); 21 D.O. (mg/l); 22 pH (standard units); 23 TOTAL PHOSPHORUS AS P (mg/l); 24 KJELDAHL NITROGEN (mg/l); 25 CHLORINE USED (LBS); 26 CHLORINE RESIDUAL (mg/l)

RECEIVING WATERS DATA

ABOVE OUTFALL ___ MILES: 27 TIME OF DAY; 28 WATER TEMP.-°F; 29 D.O. (mg/l)

BELOW OUTFALL ___ MILES: 30 TIME OF DAY; 31 WATER TEMP.-°F; 32 D.O. (mg/l)

33 PRESENT CONDITION: DEPTH (ft.); WIDTH (ft.); VELOCITY (ft./sec.); NORMAL +/−

DATE; 34 REMARKS INDEX

1 2 3 4 5 6 7 8 9 10 11 12 13 14 15 16 17 18 19 20 21 22 23 24 25 26 27 28 29 30 31

TOTAL
MIN. REPORTED VALUE
MIN. PERMIT CONDITION
AVG. REPORTED VALUE
AVG. PERMIT CONDITION
MAX. REPORTED VALUE
MAX. PERMIT CONDITION
REPORTED FREQUENCY OF ANALYSIS
PERMIT CONDITION FREQUENCY OF ANALYSIS
REPORTED SAMPLE TYPE
PERMIT CONDITION SAMPLE TYPE

35 REMARKS: INCLUDE BYPASS AND OVERFLOW OCCURRENCES, UNUSUAL SEWAGE OR STREAM FLOW PROBLEMS, OPERATIONAL PROBLEMS, OTHER REQUIRED PARAMETER DATA, COMPLAINTS, ETC. IN SPACES ON BACK OF THIS FORM.

I certify that I am familiar with the information contained in this report and that to the best of my knowledge and belief such information is true, complete, and accurate.

SIGNATURE OF PRINCIPAL EXECUTIVE OFFICER OR AUTHORIZED AGENT

DATE

MPCA FORM 704

Fig. 3.1 Typical NPDES permit reporting form

The most common of these pollutants are conventional pollutants that are usually found in domestic, commercial, or industrial wastes. These are types of pollutants that a POTW's treatment system is designed to remove. The conventional pollutants are described below.

1. BOD, Biochemical Oxygen Demand, is the rate of oxygen uptake required by the microorganisms to use the organic content of the wastewater. In other words, this is a measure of the organic strength of the wastes in water. BOD content of wastewater is usually reported as mg/L (ppm) of oxygen required by microorganisms commonly found in domestic wastewater to use the organic matter in the wastewater during five days at 20°C (68°F). This measurement may significantly understate the organic strength of many industrial wastewaters.

2. SS, Suspended Solids, also known as TFR or Total Filterable Residue, is a measure of the quantity of suspended materials in the wastewater. POTWs are designed to remove these solids from the wastewater. Suspended solids contents of wastewaters are usually reported as mg/L (ppm).

3. Coliforms are bacteria normally found in the digestive tract of humans and animals. The fecal coliform group of bacteria are used as an indicator organism to reveal the potential presence of pathogenic (disease-causing) bacteria and other harmful organisms. Fecal coliform concentrations are usually reported as Most Probable Number (MPN) of coliforms/100 milliliters.

4. pH is a measure of the basic or acidic condition of the wastewater. A pH of 7.0 is called a neutral pH. pH below 5 is considered an acidic pH while pH above 8 is considered a basic pH.

5. Oil and grease are a measure of the oil and grease content of wastewater and are generally reported as mg/L (ppm).

Any pollutant not classified as a conventional or toxic pollutant is a nonconventional pollutant; for example, ammonia.

The General Pretreatment Regulations apply to pollutants from nondomestic sources that are directly discharged into or transported by truck or rail or otherwise introduced to POTWs. Pollutants that are transported by truck or rail into POTWs, if discharged into the headworks of the treatment plant without prior mixing with domestic wastewater (sewage), would be covered by the *RCRA*[5] regulation (if the pollutant is a listed RCRA hazardous waste). Hazardous wastes delivered to a POTW by truck, rail, or dedicated pipeline must comply with RCRA regulations. POTWs receiving such hazardous wastes are subject to regulation under a RCRA permit.

The Federal Pretreatment Regulations do not intend to affect any pretreatment requirements, including any standards or prohibitions, established by state or local laws as long as the state or local requirements are more stringent than the applicable National Pretreatment Standards or any other requirements or prohibitions under the General Pretreatment Regulations.

The General Pretreatment Regulations require states with NPDES programs to develop a pretreatment program. These regulations also require that a POTW (or combination of POTWs operated by the same authority) with a design flow greater than 5 million gallons per day (MGD) or a significant level of industrial wastewater dischargers must establish a pretreatment program as a condition of its NPDES permit. POTWs with a design flow of less than 5 MGD may also be required to establish a pretreatment program if they have industrial users subject to National Pretreatment Standards or if nondomestic wastes or wastewaters cause upsets, violations of NPDES permit conditions, or sludge contamination. A local pretreatment program must have the legal authority, procedures, funding, and personnel to do at least the following:

1. Deny permission or establish conditions for the discharge of pollutants from Industrial Users (IUs) to the POTW.

2. Require compliance by IUs with applicable EPA and local pretreatment standards.

3. Control through permit, order or similar means the contribution to the POTW by each IU to ensure compliance with applicable pretreatment standards and requirements. In the case of *SIGNIFICANT INDUSTRIAL USERS (SIUs)*,[6] this control can be achieved through permits or equivalent individual control mechanisms issued to each user. Such control mechanisms must be enforceable.

4. Require a compliance schedule from all IUs not in full compliance with the EPA pretreatment standards.

---

[5] *RCRA* (RICK-ruh). The Federal Resource Conservation and Recovery Act (10/21/76), Public Law (PL) 94-580, provides technical and financial assistance for the development of plans and facilities for recovery of energy and resources from discarded materials and for the safe disposal of discarded materials and hazardous wastes. This act introduces the philosophy of the "cradle-to-grave" control of hazardous wastes. RCRA regulations can be found in Title 40 of the Code of Federal Regulations (40 CFR) Parts 260-268, 270, and 271.

[6] *Significant Industrial User (SIU).* A significant industrial user includes all categorical industrial users, and any noncategorical industrial user that:
  (1) Discharges 25,000 gallons per day or more of process wastewater ("process wastewater" excludes sanitary, noncontact cooling and boiler blowdown wastewaters), or
  (2) Contributes a process wastestream that makes up five percent or more of the average dry weather hydraulic or organic (BOD, TSS) capacity of a treatment plant, or
  (3) Has a reasonable potential, in the opinion of the control or approval authority, to adversely affect the POTW treatment plant (inhibition, pass-through of pollutants, sludge contamination, or endangerment of POTW workers).

5. Require an initial *BASELINE MONITORING REPORT (BMR)*[7] plus 90-day compliance reports plus periodic self-monitoring reports from IUs to ensure they are in compliance with all applicable pretreatment standards.

6. Perform inspection and monitoring activities, independent of the above information, that are sufficient to ensure that the IUs are complying with the pretreatment standards (the POTW must be freely able to enter and inspect industrial property in order to perform this function).

7. Obtain remedies for noncompliance by any IU with any pretreatment standard and requirement. All POTWs shall be able to seek injunctive relief for noncompliance by IUs with pretreatment standards and requirements. All POTWs shall also have authority to seek or assess civil or criminal penalties in at least the amount of $1,000 a day for each violation by IUs of pretreatment standards and requirements.

8. Identify and locate IUs affected by the EPA regulations and notify them of the applicability of these regulations.

9. Investigate and remedy instances of noncompliance by IUs in the POTW service area.

10. Obtain effective control of industrial waste discharges that endanger public health, the environment, or the operation of the POTW.

Federal regulations also require that the POTW have an adequate control mechanism (for example, permit system, administrative orders) to implement the above requirements.

One problem that must be resolved by many dischargers is establishing who actually has authority to regulate their waste discharges. This becomes an issue when dischargers are located outside the political boundaries of the POTW to which they discharge. EPA requires the agency (the POTW) holding the NPDES discharge permit to enforce pretreatment requirements throughout the tributary service area. If your facility is located outside your POTW's political boundaries, you may be required to enter into a contractual agreement with the local agency providing your collection system services. Such an agreement would give the POTW legal authority to regulate your waste discharges through its permit system or by an administrative order.

QUESTIONS

Write your answers in a notebook and then compare your answers with those on page 106.

3.2A In general, what types of pollutants are regulated by the General Pretreatment Regulations?

3.2B What are the most common conventional pollutants?

3.2C Under what conditions might a POTW with a design flow of less than 5 MGD be required to establish a pretreatment program?

3.21 Prohibited Discharge Standards

In the 40 CFR 403 regulations (General Pretreatment Regulations), two types of regulatory programs are established: the Prohibited Discharge Standards and the Industrial Categorical Pretreatment Discharge Standards.

The Prohibited Discharge Standards are relatively simple and include two General Prohibitions and eight Specific Prohibitions.

A. *GENERAL PROHIBITIONS* disallow industrial wastewaters introduced into POTWs that:

1. Pass through the POTWs untreated.

2. Interfere with the operation or performance of the POTWs.

B. *SPECIFIC PROHIBITIONS* disallow the introduction of eight specific categories of pollutants into POTWs as follows:

1. Pollutants that create a fire or explosion hazard in the POTW's sewer system or at the treatment plant, including but not limited to, wastestreams with a closed cup flashpoint of less than 140°F or 60°C using the test methods specified in the RCRA regulation (not 40 CFR Part 136!).

2. Pollutants that are corrosive, including any discharge with a pH lower than 5.0, unless the POTW is specifically designed to handle such discharges.

3. Solid or viscous pollutants in amounts that will obstruct the flow in the collection system and treatment plant, resulting in interference with operations.

4. Any pollutant discharged in quantities sufficient to interfere with POTW operations (including BOD).

5. Discharges with temperatures above 104°F (40°C) when they reach the treatment plant, or hot enough to interfere with biological processes at the wastewater treatment plant.

6. Petroleum oil, nonbiodegradable cutting oil, or products of mineral oil origin in amounts that will cause interference or pass-through.

---

[7] *Baseline Monitoring Report (BMR).* All industrial users subject to categorical pretreatment standards must submit a baseline monitoring report (BMR) to the control authority (POTW, state, or EPA). The purpose of the BMR is to provide information to the control authority to document the industrial user's current compliance status with a categorical pretreatment standard.

7. Pollutants that result in the presence of toxic gases, vapors, or fumes within the POTW in a quantity that may cause acute worker health and safety problems.

8. Any trucked or hauled pollutants, except at discharge points designated by the POTW.

The state or local pretreatment program is required to establish specific discharge limits for industrial facilities as needed to enforce the Prohibited Discharge Standards. Local limits are a major component of any local pretreatment program. At a minimum, all industrial facilities must comply with Federal Prohibited Discharge Standards, and categorical companies (described in Section 3.22) must comply with the Categorical Pretreatment Standards. In addition to these standards, however, POTWs may also establish effluent limits that are more stringent than, or in addition to, federal standards.

3.22 Categorical Pretreatment Standards

In addition to the Prohibited Discharge Standards, the EPA has established the Industrial Categorical Pretreatment Standards. These standards were established by EPA in conjunction with its program to regulate the direct discharge to the environment of wastewaters from certain categories of industrial facilities. Categorical Pretreatment Standards apply to both direct discharges to the environment and to indirect discharges passing through a POTW. For many of the categories, several subcategories have been created leading to a number of effluent limitations. The local POTW is required to enforce these regulations at categorical companies within its service area. In some special situations, the state may enforce the regulations in place of the POTW. Table 3.2 lists the industrial categories regulated by the EPA.

Early regulations to implement the Clean Water Act of 1972 originally applied effluent limits only to conventional pollutants such as pH, oil and grease, BOD, and suspended solids. As a result of lawsuits brought against EPA by environmental groups, EPA agreed to undertake a major review of industrial effluent limitations. (This effort became known as the BAT (Best Available Technology Economically Achievable) review.) It was intended to develop more stringent and far-reaching controls on industrial pollution. The net effect, however, was a major shift in EPA's pollution control strategy from the control of conventional pollutants to control of toxic pollutants.

Toxic pollutants are compounds or classes of compounds identified by EPA to be harmful to one or more forms of plant or animal life. Originally, EPA identified 129 such pollutants but subsequently deleted three, leaving 126 substances classified as toxic pollutants. These priority toxic pollutants are subdivided into two categories: inorganic pollutants and organic pollutants. Table 3.3 contains a listing of both the inorganic and organic priority toxic pollutants targeted by EPA for regulation.

POTWs are not designed to remove toxic pollutants, although some toxic pollutants are incidentally removed through the POTW's treatment system. In fact, heavy metals, such as zinc and copper, inhibit sulfide corrosion of the sanitary sewers by combining with the dissolved sulfide in the wastewater. However, excessive concentrations of heavy metals are toxic to the bacteria that purify the wastewater in POTWs. Heavy metals may also make the sludge unusable for land application under 40 CFR 503 regulations and thereby increase disposal costs. According to the EPA, implementation of the categorical pretreatment program will reduce introduction of heavy metals to POTWs from 56 million to 9 million pounds per year (84 percent reduction).

In establishing the pretreatment regulations for industrial categories, EPA performed detailed reviews of the wastes created at various types of industrial sites and evaluated the types of wastewater treatment in use or that could be used. Based on these studies and the projected economic impacts on the affected industries, EPA defined a reasonable expectation for the quality of the treated wastewater.

The resulting categorical standards are technology-based standards, meaning they are based upon the available treatment technologies that could be used to remove pollutants, in contrast to receiving water standards, which are based on the capacity of the stream that receives the wastewater to tolerate the pollutants.

QUESTIONS

Write your answers in a notebook and then compare your answers with those on page 106.

3.2D EPA's General Pretreatment Regulations (40 CFR 403) established what two types of regulatory programs?

3.2E The General Pretreatment Regulations specifically prohibit the introduction of what eight types of pollutants into POTWs?

3.2F What are toxic pollutants?

3.2G What is the difference between technology-based standards and receiving water standards?

TABLE 3.2 PUBLISHED LIST OF INDUSTRIAL CATEGORIES REGULATED BY EPA

Industrial Category	40 CFR Part No.	Year[a]	Industrial Category	40 CFR Part No.	Year[a]
Aluminum Forming	467	1983	Metal Products and Machinery	438	2003
Asbestos Manufacturing	427	1974	Mineral Mining & Processing	436	1977
Battery Manufacturing	461	1984	Nonferrous Metals Forming & Metal Powders	471	1985
Carbon Black Manufacturing	458	1978	Nonferrous Metals Manufacturing	421	1984
Cement Processing	411	1974			
Coal Mining	434	1985	Oil & Gas Extraction	435	1993
Coil Coating	465	1982	Ore Mining & Dressing	440	1982/88
Concentrated Animal Feeding Operations (CAFO)	412	2003	Organic Chemicals & Plastics & Synthetic Fibers	414	1987
Copper Forming	468	1983	Paint Formulating	446	1975
Dairy Products Processing	405	1974	Paving & Roofing Materials (Tars & Asphalts)	443	1975
Electrical/Electronic Components	469	1983	Pesticide Chemicals	455	1993
Electroplating	413	1981	Petroleum Refining	419	1982
Explosives Manufacturing	457	1976	Pharmaceutical Manufacturing	439	1983
Feedlots	412	1974	Phosphate Manufacturing	422	1976
Ferroalloy Manufacturing	424	1974	Photographic Processing	459	1976
Fertilizer Manufacturing	418	1974/79	Plastics Molding & Forming	463	1984
Fruits & Vegetables Processing	407	1974	Porcelain Enameling	466	1982
Glass Manufacturing	426	1974/86	Pulp, Paper & Paperboard	430	1982
Grain Mills Manufacturing	406	1974	Rubber Manufacturing	428	1974
Gum and Wood Chemicals Manufacturing	454	1976	Seafood Processing	408	1974
Hospitals	460	1976	Soap & Detergent Manufacturing	417	1975
Ink Formulating	447	1975			
Inorganic Chemicals	415	1982	Steam Electric Power Generating	423	1982
Iron & Steel Manufacturing	420	1982	Sugar Processing	409	1974
Landfills	445	2000	Textile Mills	410	1982
Leather Tanning & Finishing	425	1982	Timber Products Processing	429	1981
Meat Products	432	1974/76	Transportation Equipment Cleaning	442	2000
Metal Finishing	433	1983	Waste Combustors	444	2000
Metal Molding & Casting (Foundries)	464	1985	Waste Treatment (Centralized)	437	2000

[a] Per EPA Information Pamphlet (EPA-821-F-93-002), 2003.

TABLE 3.3   PRIORITY TOXIC POLLUTANTS

ORGANIC PRIORITY POLLUTANTS

1. Acenaphthene
2. Acrolein
3. Acrylonitrile
4. Benzene
5. Benzidine
6. Carbon tetrachloride (tetrachloromethane)
7. Chlorobenzene
8. 1,2,4-trichlorobenzene
9. Hexachlorobenzene
10. 1,1-dichloroethane
11. 1,2-dichloroethane
12. 1,1,1-trichloroethane
13. Hexachloroethane
14. 1,1,2-trichloroethane
15. 1,1,2,2-tetrachloroethane
16. Chloroethane
17. Bis(2-chloroethyl) ether
18. 2-chloroethyl vinyl ether (mixed)
19. 2-chloronaphthalene
20. 2,4,6-trichlorophenol
21. Parachlorometa cresol
22. Chloroform (trichloromethane)
23. 2-chlorophenol
24. 1,2-dichlorobenzene
25. 1,3-dichlorobenzene
26. 1,4-dichlorobenzene
27. 3,3-dichlorobenzidine
28. 1,1-dichloroethylene
29. 1,2-trans dichloroethylene
30. 2,4-dichlorophenol
31. 1,2-dichloropropane
32. 1,2-dichloropropylene (1,3-dichloropropene)
33. 2,4-dimethylphenol
34. 2,4-dinitrotoluene
35. 2,6-dinitrotoluene
36. 1,2-diphenylhydrazine
37. Ethylbenzene
38. Fluoranthene
39. 4-chlorophenyl phenyl ether
40. 4-bromophenyl phenyl ether
41. Bis(2-chloroisopropyl) ether
42. Bis(2-chloroethoxy) methane
43. Methylene chloride (dichloromethane)
44. Methyl chloride (chloromethane)
45. Methyl bromide (dibromomethane)
46. Bromoform (tribromomethane)
47. Dichlorobromomethane
48. Chlorodibromomethane
49. Hexachlorobutadiene
50. Hexachlorocyclopentadiene
51. Isophorone
52. Naphthalene
53. Nitrobenzene
54. 2-nitrophenol
55. 4-nitrophenol
56. 2,4-dinitrophenol
57. 4,6-dinitro-o-cresol
58. N-nitrosodimethylamine
59. N-nitrosodiphenylamine
60. N-nitrosodi-n-propylamine
61. Pentachlorophenol
62. Phenol
63. Bis(2-ethylhexyl) phthalate
64. Butyl benzyl phthalate
65. Di-n-butyl phthalate
66. Di-n-octyl phthalate
67. Diethyl phthalate
68. Dimethyl phthalate
69. 1,2-benzanthracene (benzo(a)anthracene)
70. Benzo(a)pyrene (3,4-benzopyrene)
71. 3,4-benzofluoranthene (benzo(b)fluoranthene)
72. 11,12-benzofluoranthene (benzo(k)fluoranthene)
73. Chrysene
74. Acenaphthylene
75. Anthracene
76. 1,12-benzoperylene (benzo(ghi)perylene)
77. Fluorene
78. Phenanthrene
79. 1,2,5,6-dibenzanthracene (dibenzo(a,h)anthracene)
80. Indeno(1,2,3-cd)pyrene (2,3-o-phenylene pyrene)
81. Pyrene
82. Tetrachloroethylene
83. Toluene
84. Trichloroethylene
85. Vinyl chloride (chloroethylene)
86. Aldrin
87. Dieldrin
88. Chlordane (technical mixture & metabolites)
89. 4,4-DDT
90. 4,4-DDE (p,p-DDX)
91. 4,4-DDD (p,p-TDE)
92. Alpha-endosulfan
93. Beta-endosulfan
94. Endosulfan sulfate
95. Endrin
96. Endrin aldehyde
97. Heptachlor
98. Heptachlor epoxide (BHC-hexachlorocyclohexane)
99. Alpha-BHC
100. Beta-BHC
101. Gamma-BHC (lindane)
102. Delta-BHC

PCBs - polychlorinated biphenyls:
103. PCB-1242 (Arochlor 1242)
104. PCB-1254 (Arochlor 1254)
105. PCB-1221 (Arochlor 1221)
106. PCB-1232 (Arochlor 1232)
107. PCB-1248 (Arochlor 1248)
108. PCB-1260 (Arochlor 1260)
109. PCB-1016 (Arochlor 1016)

110. Toxaphene
111. 2,3,7,8-tetrachlorodibenzo-p-dioxin (TCDD)

INORGANIC PRIORITY POLLUTANTS

112. Antimony (Total)
113. Arsenic
114. Asbestos
115. Beryllium
116. Cadmium
117. Chromium
118. Copper
119. Cyanide
120. Lead
121. Mercury
122. Nickel
123. Selenium
124. Silver
125. Thallium
126. Zinc

3.220 EPA Regulation Development Process

The Industrial Technology Division of the EPA Office of Water Regulations and Standards, formerly the Effluent Guidelines Division, develops the categorical regulations and limits using the following process.

As previously described, EPA divides industrial dischargers into categories or groups of similar industries for regulatory purposes. Each category is further subdivided by the facility type and point of discharge. Based on point of discharge, industries are referred to as direct or indirect dischargers. In addition, each discharger type has been subdivided depending on when an industrial plant came into existence, and plants are referred to as existing or new dischargers. Each subdivision may be subject to a separate set of effluent standards.

To develop categorical pretreatment standards, EPA first studies a particular category or type of discharger, conducts effluent sampling in various industrial sites, and establishes a database. The data are evaluated by EPA to determine the type and quantity of wastewater generated by an industrial category.

Next, EPA conducts field studies at industrial sites within a category to identify the range and capabilities of various wastewater control systems. Technical and economic analyses are performed to test the effectiveness and economic feasibility of each alternative pollution control system. Based on these studies, EPA identifies two levels of technology:

BEST PRACTICABLE TECHNOLOGY (BPT). BPT level of technology represents the average of the best existing wastewater treatment performance levels within the industrial category.

BEST AVAILABLE TECHNOLOGY (BAT). BAT levels of technology represent a higher level of wastewater treatment technology than those required by Best Practicable Technology (BPT). BAT is based on the very best (state-of-the-art) control and treatment measures that have been developed, or are capable of being developed, and that are economically achievable within the appropriate industrial category.

BAT technology is the basis for categorical effluent standards for direct dischargers and pretreatment standards for indirect dischargers. An additional consideration in the development of pretreatment standards, however, is the capacity of POTWs to remove various pollutants. EPA has developed data on the performance of 50 typical POTWs. It uses the information in this database to determine whether or not a typical POTW is capable of removing any of the pollutants discharged by the categorical dischargers to the same extent as BAT. If it is determined that a typical POTW is capable of removing a pollutant to the same extent as BAT, pretreatment standards for those pollutants are not promulgated for that category.

The detailed technical information compiled by EPA during its initial studies is published as "Development Documents" for the proposed regulations concerning each industrial category. These documents provide detailed technical information about the alternative technologies studied.

Once the EPA has completed its preliminary studies, proposed effluent limits are published in the *FEDERAL REGISTER* as "Proposed Regulations." A brief comment period is provided to consider opinions of the parties affected by the regulations. After review of the public comments and appropriate modifications of the proposed regulations, the regulations are promulgated in final form in the *FEDERAL REGISTER* and codified in the Code of Federal Regulations (CFR).

EPA decides on a model treatment technology for an industrial category and then issues discharge or effluent limitations; these are the categorical standards for the industrial category. Usually, different standards are developed for the various types of dischargers: new and existing direct sources and new and existing indirect sources. Each of these sets of standards is described below.

EXISTING SOURCES—DIRECT DISCHARGERS. The Federal Water Pollution Act of 1972 required that all existing industries discharging industrial wastewater into waters of the United States were to achieve the Best Practicable Technology (BPT) economically available by July 1977, and to achieve the Best Available Technology (BAT) economically achievable by July 1, 1983. The standards applicable to this group of dischargers are referred to as Categorical Standards.

NEW SOURCES—DIRECT DISCHARGERS. The Federal Water Pollution Act of 1972 required that all new industries constructed after the publication of a proposed regulation in the *FEDERAL REGISTER* that discharge industrial wastewaters into waters of the United States were to use the Best Available Technology (BAT). The regulations for this group of dischargers are called the New Source Performance Standards (NSPS).

EXISTING SOURCES—INDIRECT DISCHARGERS. The Pretreatment Standards for Existing Sources (PSES) are the category-specific regulations that apply to existing sources discharging industrial wastewater to POTWs. Table 3.4 is an example of the PSES regulations for the metal finishing category. The PSES for the metal finishing category apply to facilities that began construction before August 31, 1982 (the date for the proposed categorical regulations for the metal finishing category (40 CFR 433)).

NEW SOURCES—INDIRECT DISCHARGERS. The Pretreatment Standards for New Sources (PSNS) are the category-specific regulations that apply to new industrial sources discharging to POTWs. Table 3.5 is an example of the PSNS regulations for the metal finishing category. The PSNS for the metal finishing category apply to facilities that began construction after August 31, 1982.

TABLE 3.4 PRETREATMENT STANDARDS FOR EXISTING SOURCES FOR THE METAL FINISHING CATEGORY

Contaminant	Maximum For Any One Day, mg/L	Monthly Average Limit, mg/L
Cadmium (T)	0.69	0.26
Chromium (T)	2.77	1.71
Copper (T)	3.38	2.07
Lead (T)	0.69	0.43
Nickel (T)	3.98	2.38
Silver (T)	0.43	0.24
Zinc (T)	2.61	1.48
Cyanide (T) a,b	1.20	0.65
TTO	2.13	—

T: Total
TTO: Total Toxic Organics

[a] For industrial facilities with cyanide treatment and upon agreement of the pollution control authority, a one-day maximum limit of 0.86 mg/L and a monthly average limit of 0.32 mg/L of amenable cyanide may apply in place of the total cyanide limit specified above. (Amenable means capable of being treated by alkaline chlorination.)

[b] Sampling for cyanide analyses to meet these standards (as for all other standards for cyanide) must be taken at the completion of the cyanide treatment processes, and prior to intermixing with any other noncyanide wastewaters in the total pretreatment system.

TABLE 3.5 PRETREATMENT STANDARDS FOR NEW SOURCES FOR THE METAL FINISHING CATEGORY

Contaminant	Maximum For Any One Day, mg/L	Monthly Average Limit, mg/L
Cadmium (T)	0.11	0.07
Chromium (T)	2.77	1.71
Copper (T)	3.38	2.07
Lead (T)	0.69	0.43
Nickel (T)	3.98	2.38
Silver (T)	0.43	0.24
Zinc (T)	2.61	1.48
Cyanide (T) a,b	1.20	0.65
TTO	2.13	—

T: Total
TTO: Total Toxic Organics

[a] For industrial facilities with cyanide treatment and upon agreement of the pollution control authority, a one-day maximum limit of 0.86 mg/L and a monthly average limit of 0.32 mg/L of amenable cyanide may apply in place of the total cyanide limit specified above. (Amenable means capable of being treated by alkaline chlorination.)

[b] Sampling for cyanide analyses to meet these standards (as for all other standards for cyanide) must be taken at the completion of the cyanide treatment processes, and prior to intermixing with any other noncyanide wastewaters in the total pretreatment system.

3.221 EPA-Regulated Categories

EPA entered into a Consent Decree in the case of *Natural Resources Defense Council, Inc. (NRDC) v. Train*. This Decree required promulgation of technology-based standards to control toxic pollutants from 21 industrial categories. A 1982 court order modified the NRDC Settlement Agreement creating a total of 38 categories. Twelve categories were exempted under paragraph eight of the Agreement. Paragraph eight of the NRDC Agreement permits EPA to exempt any industrial category that does not warrant regulation after comprehensive evaluations. After these evaluations were conducted, regulations for the remaining categories were promulgated by EPA.

Current categorical standards are prescribed in 40 CFR Parts 400–471 or on the EPA website at www.epa.gov/waterscience/guide. The EPA website contains information regarding final regulations, proposed effluent guidelines, and what is under development.

3.222 Categories Exempt From EPA Categorical Regulations

Currently, paragraph eight of the NRDC Agreement exempts six industrial categories from regulation. Following comprehensive evaluations, EPA has determined that these categories do not warrant regulations. The exempted categories are:

1. Adhesives and Sealants
2. Auto and Other Laundries
3. Explosives Manufacturing
4. Gum and Wood
5. Photographic Equipment and Supplies
6. Printing and Publishing

These industrial categories are still subject to local limits adopted by the POTWs.

3.223 Category Determination Requests

In the course of implementation of categorical pretreatment regulations, if there is a question or disagreement about applicability of a designated industrial category to an industrial discharge, the POTW or the industrial user can request a ruling by the EPA concerning the appropriateness of the industrial category. The decision of the Water Division Director of the EPA Regional Office is final.

3.224 Local Limits for Noncategorical Industries

Local authorities may establish limits for noncategorical industries when necessary. These limits also apply to categorical industries if they are more stringent than the Categorical Standards.

QUESTIONS

Write your answers in a notebook and then compare your answers with those on page 106.

3.2H EPA's categorical regulations identify two levels of treatment technology. What are the two levels?

3.2I What are "Development Documents"?

3.2J What are the regulatory requirements for the control of water pollutants from indirect industrial wastewater dischargers called?

3.2K Who resolves questions or disagreements about the applicability of a designated industrial category between a POTW and an industrial user?

3.3 APPLICATION OF CATEGORICAL PRETREATMENT REGULATIONS

3.30 Types of Categorical Standards

Two different types of categorical standards are used in the EPA pretreatment program: concentration-based standards and production-based standards. The production-based standards are occasionally referred to as mass-based standards.

The concentration-based standards are based on the relative strength of a pollutant in a wastestream, which is usually expressed as mg/L; for example, the pretreatment standard for an existing source in the metal finishing category for cadmium is 0.69 mg/L.

Production-based (mass-based) standards are based on the actual mass of pollutants in a categorical wastewater stream per unit of production. Production-based standards are the most equitable (fairest) type of standard. However, in practice it is difficult to develop, implement, and enforce this type of standard. The production-based or mass-based standards are generally reported as pounds/square foot or milligrams/square meter of production area.

When the limits in a Categorical Pretreatment Standard are expressed only in terms of mass of pollutant per unit of production, the POTW may convert the limits to equivalent limitations expressed either as mass of pollutant discharged per day or effluent concentration for purposes of calculating effluent limitations applicable to individual IUs.

Equivalent, *MASS-PER-DAY LIMITATIONS* are calculated by multiplying the limits in the Categorical Pretreatment Standard by the IU's average rate of production. However, the average rate of production is not the designed production capacity but rather a reasonable measure of the IU's actual long-term daily production, such as the average daily production during a representative year. For new sources, actual production is estimated using projected production.

EQUIVALENT CONCENTRATION LIMITATIONS are calculated by dividing the mass limitations derived above by the average daily flow rate of the IU's regulated process wastewater. This average daily flow rate should be based on a reasonable measure of the IU's actual long-term, average flow rate, such as the average daily flow rate during the representative year.

Equivalent limitations calculated using the procedures described above are considered Pretreatment Standards for the purposes of section 307(d) of the Clean Water Act and the General Pretreatment Standards. If equivalent limitations are specified in a facility's permit, they are enforceable by the POTW in the same manner as the Categorical Pretreatment Standards and local limits.

Many Categorical Pretreatment Standards specify one limit for calculating maximum daily discharge limitations and a second limit for calculating maximum monthly average or 4-day average limitations. Where such standards are being applied, the same production or flow figure must be used in calculating both types of equivalent limitations.

Any IU with a POTW-issued permit using equivalent mass or concentration limits calculated from a production-based standard must notify the POTW within two business days after the IU has a reasonable basis to know that the production level will significantly change within the next calendar month. Any IU not notifying the POTW of such anticipated change will be required to meet the mass or concentration limits in its permit that were based on the original estimate of the long-term average production rate.

For any IU having equivalent mass or concentration limits in its permit, the initial report on compliance and the periodic compliance reports (see Section 3.33) must contain a reasonable measure of the IU's Pretreatment Standards expressed in terms of allowable pollutant discharge per unit of production (or other measure of operation); those reports must include the IU's actual production during the appropriate sampling or reporting period.

3.300 Industrial Categories With Concentration-Based Standards

EPA has promulgated concentration-based pretreatment standards for the following six categories:

1. Electrical and Electronic Component Manufacturing
2. Leather Tanning and Finishing
3. Metal Finishing
4. Organic Chemicals
5. Pharmaceutical Manufacturing
6. Steam Electric Power Generation

3.301 Industrial Categories With Mass-Based Standards

EPA has promulgated production-based (mass-based) pretreatment standards for the following nine categories:

1. Aluminum Forming
2. Battery Manufacturing
3. Coil Coating
4. Copper Forming
5. Iron and Steel Manufacturing
6. Metal Molding and Casting
7. Nonferrous Metals Forming
8. Nonferrous Metals Manufacturing
9. Pesticides

3.302 Industrial Categories With Both Concentration- and Mass-Based Standards

EPA has promulgated both concentration- and mass-based pretreatment standards for the following six categories:

1. Electroplating
2. Inorganic Chemicals Manufacturing
3. Petroleum Refining

4. Porcelain Enameling

5. Pulp, Paper, and Paperboard Manufacturing

6. Timber Products Processing

3.31 Types of Wastestreams

This section defines three different types of wastestreams that are described in regulations.

Regulated streams are industrial process wastewater streams subject to a national categorical pretreatment standard; for example, wastewater produced as a result of rinsing electroplated parts in an electroplating facility.

Unregulated streams are industrial process wastewater streams not subject to a national categorical pretreatment standard; for example, wastestreams resulting as rinses from a galvanizing operation.

Dilution streams are wastewaters from nonregulated process wastewater streams that contain nonregulated pollutants; for example, boiler blowdown, sanitary wastewater, or noncontact cooling water.

Except where expressly authorized to do so by an applicable pretreatment standard or requirement, no IU is ever permitted to increase the use of process water, or in any other way attempt to dilute a discharge, as a partial or complete substitute for adequate treatment to achieve compliance with a pretreatment standard or requirement. The POTW may impose mass limitations on IUs that are using dilution to meet applicable pretreatment standards or requirements, or in other cases where the imposition of mass limitations is appropriate.

3.32 Total Toxic Organics (TTO)

3.320 Regulated Toxic Organics

For each regulated industrial category, EPA specifies which organic compounds it will regulate. The regulated industry then must measure and report the quantities of listed pollutants in its wastestreams. In Table 3.3 (page 90), items 1–111 are the regulated toxic organics for the electroplating and metal finishing categories. (For some industrial categories, the number of toxic organics is fewer than those listed in Table 3.3.) In the electroplating and metal finishing regulations, TTO is defined as the summation (total) of all quantifiable (measureable) values of components 1 through 111 in Table 3.3 in excess of 10 micrograms per liter.

3.321 TTO Monitoring

Requiring analytical determination of all 111 regulated organic compounds (listed in Table 3.3) by all small metal finishers and electroplaters may be unreasonable and uneconomical. Therefore, in place of monitoring for TTO, the control authority may permit the industrial user to submit a *SOLVENT MANAGEMENT PLAN*[8] (also known as a Toxic Organic Management Plan (TOMP)) for the POTW's review and approval.

To request that no monitoring be required, an industrial user must submit a solvent management plan to the POTW. The plan must specify, to the control authority's satisfaction, procedures for ensuring that toxic organics used do not routinely spill or leak into the wastewater and that there is no deliberate dumping of any of the solvents.

If monitoring is necessary to measure compliance with the TTO standard, the industrial discharger needs to analyze for only those pollutants that would reasonably be expected to be present.

Total Toxic Organics (TTO) monitoring is *REQUIRED* for Baseline Monitoring Reports (BMR) and Final Compliance Reports, except where the POTW allows for oil/grease measurements to be substituted.

EPA has developed a guidance manual specifically for application of the TTO standard entitled: Guidance Manual for Implementing Total Toxic Organic (TTO) Pretreatment Standards.

QUESTIONS

Write your answers in a notebook and then compare your answers with those on pages 106 and 107.

3.3A What are the two different types of categorical standards used in the EPA pretreatment program?

3.3B What is a regulated stream?

3.3C What does the phrase "Total Toxic Organics" mean?

3.3D What pollutants must a discharger analyze for in order to comply with the TTO standard?

3.33 EPA Reporting Requirements

The General Pretreatment Regulations require all categorical dischargers to submit a number of reports to the POTW (Control Authority) authorized to implement the EPA categorical program. The reports must be signed by a responsible corporate officer of the industrial user (IU). The reporting requirements for the categorical dischargers are as follows:

1. Baseline Monitoring Report/Compliance Schedule

2. Final Compliance Report/90-Day Report

3. Periodic Compliance Reports

These reports are subject to the provisions of 18 United States Code (USC) section 1001 relating to fraud and false statements; the provisions of sections 309(c)(4) of the Clean Water Act, as

---

[8] *Solvent Management Plan.* A strategy for keeping track of all solvents delivered to a site, their storage, use, and disposal. This includes keeping spent solvents segregated from other process wastewaters to maximize the value of the recoverable solvents, to avoid contamination of other segregated wastes, and to prevent the discharge of toxic organics to any wastewater collection system or the environment. The plan should describe measures to control spills and leaks and to ensure that there is no deliberate dumping of solvents. Also known as a TOXIC ORGANIC MANAGEMENT PLAN.

amended, governing false statements, representation, or certification; and the provisions of section 309(c)(6) of the Clean Water Act regarding responsible corporate officers. Significant criminal and civil penalties await those who submit fraudulent or false statements.

For compliance reports, a minimum of four grab samples must be used for pH, cyanide, total phenols, oil and grease, sulfide, and volatile organics. A minimum of one representative sample (24-hour composite sample obtained through flow proportional composite sampling techniques, where feasible) must be analyzed for all other regulated pollutants. Collection and analysis of all samples must follow the procedures described in 40 CFR Part 136, Guidelines Establishing Test Procedures for the Analysis of Pollutants Under the Clean Water Act as Amended.

The sampling and analysis for compliance reports may be performed by the Control Authority instead of by the IU. Where the POTW performs the required sampling and analysis, the IU is not required to submit a compliance certification stating whether applicable Pretreatment Standards or Requirements are being met on a consistent basis. The purpose behind not requiring this certification language is that the POTW will already know if the IU is in consistent compliance by reviewing the data the POTW collects. In addition, where the POTW itself collects all the information required for the report, including flow data and production data from the IU, the IU is not required to submit a BMR, 90-day Compliance Report, or Periodic Report on Compliance.

3.330 *Baseline Monitoring Reports*

A Baseline Monitoring Report (BMR) is the first report an indirect discharger must file following promulgation of a categorical standard applicable to the category. The BMR is due 180 days after the effective date of the regulations. For new dischargers where the regulation has been in effect for more than 180 days, the BMR must be submitted at least 90 days before the facility begins to discharge wastes. The information required in the BMR includes:

1. Identification of the indirect discharger.
2. A list of environmental control permits.
3. A description of its operations.
4. A report on flows of regulated streams.
5. The results of sampling and analyses of the industrial wastewater discharges to determine levels of regulated pollutants in those streams.
6. A certification statement by the discharger indicating compliance or noncompliance with the applicable pretreatment standards.
7. A description of any additional steps, including a schedule, required to achieve compliance for noncompliant dischargers.

3.331 *Compliance Schedule*

A noncompliant industrial user who is not likely to achieve compliance with the applicable categorical pretreatment standards with minor operation and maintenance (O & M) modifications is required to submit a compliance schedule to the control authority as a part of BMR submission. This schedule must represent the shortest schedule within which the IU can achieve compliance with the applicable pretreatment standards. The final completion date on the compliance schedule must not be later than the applicable categorical compliance date specified in the regulations. If the date specified in the regulations has already passed or is impossible to meet, the final completion date must be as soon as possible. The schedule must contain specific incremental progress dates not exceeding nine months for completion of any increment. In actual practice, the POTW and the industrial user commonly work together to establish an acceptable compliance schedule.

3.332 *Final Compliance Report*

All industrial facilities subject to the pretreatment regulations are required to submit a final compliance report. The final compliance report must be filed within 90 days after the compliance date of the regulations for all existing industrial dischargers. The new categorical dischargers must file a final compliance report as soon as operations begin. A final report must contain the following information:

1. The sampling results for regulated pollutants in the industrial wastewater discharges.
2. Average and maximum daily industrial wastewater flows.
3. A statement of compliance.
4. A statement as to whether additional operation and maintenance modifications and/or pretreatment equipment is necessary to achieve compliance, for noncompliant dischargers.

3.333 *Periodic Compliance Reports*

The General Pretreatment Regulations require all categorical dischargers to submit a minimum of two periodic self-monitoring reports each year to the POTW. Some POTWs require more frequent submission of self-monitoring reports, for example, quarterly. A periodic compliance report must contain the following information:

1. Type of facility.
2. Type of discharge.
3. Type and concentration of pollutants in the discharge.
4. A certification statement concerning the accuracy of the submitted information.

3.334 *Slug Loading Reporting*

All categorical and noncategorical IUs must notify the POTW immediately of all slug discharges that could cause problems to the POTW, including any discharges that would create a fire or explosion hazard; pollutants that would cause corrosive, structural damage to the sewer lines or the treatment plant; solid or viscous pollutants that would cause obstruction of the flow; any pollutant, including oxygen-demanding pollutants released in a discharge at a flow rate and/or pollutant concentration that will cause interference; heat that will inhibit biological activity; petroleum oil, nonbiodegradable cutting oil, or products of mineral oil origin in amounts that will cause interference or pass-through; or pollutants that result in the presence of toxic gases, vapors, or fumes within the POTW in a quantity that may cause acute worker health and safety problems. (Note that these are identical to the first seven of the eight Specific Prohibitions in the Prohibited Discharge Standards of the General Pretreatment Regulations.)

3.335 Resampling to Confirm Violations

If sampling performed by an IU indicates a violation, the user must notify the POTW within 24 hours of becoming aware of the violation. The IU must also repeat the sampling and analysis and submit the results of the repeat analyses to the POTW within 30 days after becoming aware of the violation, except the IU is not required to resample if:

a. The POTW performs sampling at the IU at a frequency of at least once per month.

b. The POTW performs sampling at the IU between the time when the IU performs its initial sampling and the time when the IU receives the results of this sampling.

3.336 Notification of Changed Discharge

All IUs must notify the POTW in advance of any substantial change in the volume or character of pollutants in their discharge, including the listed or characteristic hazardous wastes for which the IU has submitted initial notification and described in Section 3.337 of this manual.

3.337 Hazardous Waste Disposal Reporting

All Significant Industrial Users (SIUs) are required to notify the POTW, the EPA Regional Waste Management Division Director, and the state hazardous waste authorities in writing of any discharge into the POTW of a substance which, if otherwise disposed of, would be a hazardous waste under the Resource Conservation and Recovery Act (RCRA) and specifically contained in EPA's regulations in 40 CFR Part 261 (RCRA regulations). Information in this report must include:

- SIU identifying information.
- The name of the hazardous waste as set forth in 40 CFR Part 261.
- The EPA hazardous waste number.
- The type of discharge (continuous, batch, or other).

If the SIU discharges more than 100 kilograms of such waste per calendar month or any volume of acutely hazardous waste to the POTW, the notification should also contain the following information to the extent such information is known and readily available to the SIU:

- An identification of the hazardous constituents contained in the wastes.
- An estimation of the mass and concentration of such constituents in the wastestream discharged during that calendar month.
- An estimation of the mass of constituents in the wastestream expected to be discharged during the following twelve months.

Any notification needs to be submitted only once for each hazardous waste discharged. However, notifications of *changed* discharge must be submitted in advance whenever there will be a significant change in the volume or characteristics of the wastes being discharged. The notification requirement does not apply to pollutants already reported by the BMR, Final Compliance Report, or the Periodic Compliance Reports and detailed in the POTW permit issued to the SIU.

The SIU is exempt for the requirements of this section during a calendar month in which the facility discharges no more than 15 kilograms of hazardous wastes, unless the wastes are acute hazardous wastes as specified in the RCRA regulation. Discharge of more than 15 kilograms of non-acute hazardous wastes in a calendar month, or any quantity of acute hazardous wastes, requires a one-time notification. Subsequent months during which the IU discharges more than these quantities of any hazardous waste do not require additional notification.

In the case of any new regulations under section 3001 of RCRA identifying additional characteristics of a hazardous waste or listing any additional substance as a hazardous waste, the IU must notify the POTW, the EPA Regional Waste Management Division Director, and state hazardous waste authorities of the discharge of such substance within 90 days of the effective date of such regulations.

When submitting this information, the SIU must certify that it has a program in place to reduce the volume and toxicity of hazardous wastes generated to the degree it has determined to be economically practical.

This report can be signed by a technical representative of the SIU familiar with the source(s) of information contained within. It is not necessary to have the report signed by a responsible corporate officer, general partner or proprietor, or duly authorized representative of the corporate officer, as required of other reports in this section.

3.34 Modification of Categorical Standards

General Pretreatment Regulations state the ground rules for the implementation of EPA's Categorical Pretreatment Program. Although the Categorical Pretreatment Standards apply uniformly to all affected categorical companies that discharge to POTWs, the Pretreatment Regulations specify three ways in which Categorical Pretreatment Standards could be modified:

1. Variance for fundamentally different factors (FDF)
2. Net gross adjustments
3. Removal credits

3.340 Variance From Categorical Standards for Fundamentally Different Factors (FDF)

The General Pretreatment Regulations provide for modification of the Categorical Pretreatment Standards applicable to an industrial discharger if an industrial firm or an interested party can show that factors relating to the industrial user are fundamentally different from factors considered by EPA in establishing the standards. The General Pretreatment Regulations identify six factors that may be considered fundamentally different. These factors are:

1. The nature or quality of pollutants contained in the raw waste.
2. The volume of the industrial user's wastewater and quantity of effluent discharged.
3. Non-water-quality environmental impacts of control and treatment of the waste.
4. Energy requirements for the application of control technology.
5. Age, size, land availability, and configuration as relates to equipment or facilities, processes used, process modifications, and engineering aspects of application of the control technology.
6. Cost of compliance.

3.341 Net Gross Calculations

A net gross credit allows the subtraction of the concentration of regulated pollutants in the intake water from the concentration level in the industrial discharger's effluent. Therefore, a net gross adjustment allows a facility to discharge a pollutant at a concentration level in excess of the applicable Federal Pretreatment Standards if it can be shown that a regulated pollutant is present in the industrial user's incoming water. To obtain a net gross adjustment, the discharger must submit a formal request to EPA certifying that:

1. Intake water is drawn from the same body of water that the POTW discharges to.
2. The pollutants present in the incoming water will not be entirely removed by the industrial discharger's treatment system.
3. The pollutants in the intake water do not vary chemically or biologically from the pollutants limited by the applicable categorical standards.
4. The industrial discharger does not significantly increase the concentration of pollutants in the intake water, even if the total quantity of pollutants remains the same.

3.342 Removal Credits

The Clean Water Act provides that POTWs may grant credit to indirect dischargers, based on the degree of removal actually achieved at the POTW. If such removal credits were applied to the Categorical Pretreatment Standards, the affected industries could discharge regulated pollutants in excess of the amounts permitted by the applicable EPA pretreatment standards.

A POTW with an approved pretreatment program, and approved authority to grant removal credits, has the discretion to grant removal credits to indirect industrial dischargers located within its service area. To date, only a handful of POTWs have been authorized by EPA to grant removal credits.

To justify a removal credit, a POTW must demonstrate the level of pollutant removal it is able to consistently achieve. To determine the consistent removal for a pollutant, the removal credit regulations require a POTW to use the average of the lowest 50 percent of the removals measured based on POTW treatment plant influent and effluent sampling. For example, such sampling at a POTW treatment plant may indicate that its consistent removal for total chromium is 30 percent. The categorical standard for total chromium for a job shop electroplater is normally 7.0 mg/L. However, an electroplater located within the service area of a POTW with an approved 30 percent removal credit for total chromium may discharge up to 10 mg/L without violating the categorical pretreatment standard.

Removal credits have been controversial. The status of removal credits is continually being refined. The current status of removal credits is contained in 40 CFR Part 403.7.

The final sewage sludge regulations (40 CFR Part 257 et al., Technical Standards for the Use or Disposal of Sewage Sludge) were published on February 19, 1993, and contain provisions that would allow the POTW to grant removal credits. However, granting removal credits would be conditional on meeting the requirements of the sewage sludge regulations. Contact your POTW, local regulatory agency, or regional EPA office for further information on removal credits.

QUESTIONS

Write your answers in a notebook and then compare your answers with those on page 107.

3.3E The General Pretreatment Regulations require categorical dischargers to file what types of reports?

3.3F What information is required in a Baseline Monitoring Report?

3.3G List the three ways in which Categorical Pretreatment Standards could be modified?

3.3H What is a net gross credit?

3.4 LOCAL ORDINANCES AFFECTING INDUSTRIAL DISCHARGERS

Thus far in the chapter we have examined the history of pollution control regulations, the general nature of those regulations, and the manner in which legal authority to implement and enforce them can be delegated to local agencies. We will look now at local ordinances, laws, and codes, particularly wastewater ordinances, that affect the disposal of industrial wastes. The wastewater ordinance is the primary instrument enforced by the POTW, but other local laws and regulations may also affect the operation of your facility. In your work as an industrial waste treatment operator, you will need to understand virtually every detail of your local wastewater ordinance.

3.40 Wastewater Ordinances

The basic document granting authority to a local POTW to administer a pretreatment program is the wastewater ordinance or sewer-use ordinance. Wastewater ordinances specify the conditions under which discharges to the POTW's collection system may be made. These ordinances may vary greatly in format depending on the laws of the state within which the POTW operates. However, all such ordinances must contain certain basic elements to provide a legal framework for effective enforcement. In addition, there are specific items that are unique to each POTW.

Most ordinances regulate both domestic and industrial wastewater disposal. For the purpose of this discussion, we will be concerned primarily with the industrial waste control provisions. Essential elements include:

- Administration
- EPA Minimum Requirements
- General Provisions
- Specific Provisions
- Industrial Wastewater Limitations

3.400 Administration

The first portion of a wastewater ordinance usually describes the general purpose of the organization operating the POTW and outlines the intent of the ordinance. The state laws that give the POTW legal authority to regulate wastewater disposal are usually listed along with any regulations under which the POTW is required to operate.

A further statement of scope may be included defining specific limitations and defining terms to be used in interpretation of the ordinance. The scope will outline the major provisions to follow within the document, including the intent to regulate all discharges to the *WASTEWATER FACILITIES*[9] (POTW), the quantity and quality of discharges, degree of required pretreatment, fees and distribution of costs, issuance of permits, and penalties for violations of the ordinances.

The ordinance may set priorities for use of the wastewater collection system. For example, it will commonly say that the primary use of the wastewater facilities is for the collection, conveyance, treatment, and disposal of domestic wastewater, and that industrial discharges will be subject to additional regulation to accomplish the primary objective.

Additional statements may be made limiting wastes accepted by the POTW operator to those that will not (a) damage the system, (b) create nuisances, (c) menace public health, (d) impose unreasonable costs, (e) interfere with wastewater treatment processes, including sludge disposal, (f) violate any quality requirements set by governmental regulatory agencies, or (g) be detrimental to the environment.

Policy statements may also encourage conservation and reuse of reclaimed wastewater and sometimes give priority consideration to those dischargers who practice water conservation. The POTW operator may also reserve the right to restrict the flow of industrial dischargers during periods of upset or repair to the wastewater facilities and may specify when such a discharge may occur.

3.401 *EPA Minimum Requirements*

Every wastewater ordinance specifies pretreatment standards for industrial discharges into the system and specifically incorporates the EPA's Categorical Pretreatment Standards. A POTW operator may also establish local pretreatment standards that are more stringent than, and in addition to, the EPA standards.

The wastewater ordinance describes the control mechanism used to regulate dischargers to the POTW, and specifies the conditions and limitations of discharge. The control mechanism may be a permit, administrative order, or some other type of non-permit arrangement with dischargers. The ordinance also specifies the records IUs must maintain and outlines the notification procedures to be used in the event of process changes or unauthorized discharges in excess of permit limitations.

---

[9] *Wastewater Facilities.* The pipes, conduits, structures, equipment, and processes required to collect, convey, and treat domestic and industrial wastes, and dispose of the effluent and sludge.

The ordinance grants specific inspection authority and gives the POTW the right of access to processes generating industrial wastewaters and also pretreatment facilities. Such authority includes the ability to take samples and to examine records on the operation and maintenance of the pretreatment facility, and other waste handling or chemical records, such as purchases and waste manifests.

The ordinance describes the POTW's enforcement procedures and the notification process that will be used when violations are found. The ordinance also provides penalties for violations. Such penalties may be civil and/or criminal and may include fines, compliance orders, suspension or revocation of a permit to discharge, and disconnection from the POTW system.

The wastewater ordinance must explain how the POTW calculates its wastewater treatment costs and how these costs will be recovered. For example, cost recovery may take place through permit and inspection fees, recovery of capital costs through connection charges and fees, or through charges or surcharges based on the quantity and quality of wastewater discharged into the POTW system.

3.402 General Provisions

The general provisions of the wastewater ordinance specify who is authorized to enforce the ordinance and how this authority may be delegated to subordinate employees. The general provisions also specify penalties for violation of the ordinance.

The EPA general pretreatment regulations require POTWs to verify compliance or noncompliance with the categorical pretreatment standards by the affected dischargers independent of information submitted by the industrial dischargers. This implies that POTWs are required to monitor industrial wastewater dischargers. Since most pretreatment standards are based on daily maximum numbers, POTWs are required to obtain daily composite samples in order to verify compliance or noncompliance with applicable pretreatment standards. In some areas POTWs collect several samples per month or collect monthly samples instead of daily samples. If noncompliance is suspected or discovered, the frequency of sampling is increased.

The general provisions of the ordinance give complete details of the procedures that will be followed when violations are discovered or threatened. For example, the ordinance will provide for the issuance of notification of the discharger; it will specify the means by which notice must be served (consistent with state law); and the ordinance will even specify what information the notice must contain, such as a statement of the nature of the violation and the time limits within which a correction must take place.

The general provisions also spell out the rights of the permittee or discharger when served by a notice. These rights may include ability to rebut or deny such charges by appeal to the administrative authority. The administrative authority, however, may be able to administratively suspend a permit to discharge until the matter is resolved.

The EPA Requirements section of the ordinance describes what fees the POTW will impose but the manner of fee collection and the enforcement authority available to the POTW operator are normally described in the general provisions of the wastewater ordinance.

3.403 Specific Provisions

The sewer-use ordinance may describe in this section specific prohibitions on wastewater discharges. These provisions may prohibit the industrial user from discharging or depositing or causing or allowing to be discharged or deposited into the wastewater collection or treatment system any wastewater containing excessive amounts or concentrations of:

1. Oil and grease
2. Explosive mixtures
3. Noxious materials
4. Improperly shredded garbage
5. Radioactive wastes
6. Solid or viscous wastes
7. Excessive discharge rate
8. Toxic substances
9. Unpolluted waters
10. Discolored material
11. Corrosive wastes

A section in the ordinance would be devoted to each of the above items and describe excessive amounts or concentrations and the problems they can create.

3.404 Industrial Wastewater Limitations

Industrial wastewater limitations commonly refer to the Specific Provisions described in Section 3.403 and describe the limitations in either general limitations and/or specific limitations. The general limitations typically prohibit the discharge or conveyance to public sewers (POTW) of any wastewater containing pollutants of such character or quantity that will:

1. Not be susceptible to treatment or will interfere with the processes or efficiency of the POTW.
2. Constitute a hazard to human or animal life, or to the stream or watercourse receiving the treatment plant effluent.
3. Violate pretreatment standards.
4. Cause the treatment plant to violate its NPDES permit or applicable receiving water standards.

Specific limitations will list the maximum concentrations or mass limitations of pollutants allowable in industrial wastewater discharged to the POTW. This section will also contain a clause stating that dilution of any wastewater discharge for the purpose of satisfying these requirements will be considered a violation of the ordinance.

3.405 Administrative Fine Penalties

A variation of the administrative compliance program in some wastewater agencies involves a schedule of administrative fines. Fines are allowed by many state laws when used to penalize industries for discharging pollutants beyond their permitted limits. A fine (penalty) schedule establishes specific dollar fines for the discharge of excessive quantities of pollutants.

EPA has identified four goals that should be considered in assessing penalties. These goals are:

1. Penalties should recover the economic benefit of noncompliance plus some amount for the seriousness of the violation.

2. Penalties should be large enough to deter future noncompliance.

3. Penalties should be uniform or reasonably consistent for similar instances of noncompliance.

4. A logical basis for the calculation of penalties should exist.

If a company is discharging large quantities of pollutants, fines may be a substantial motivating factor to compel compliance without resorting to severe legal actions. On the other hand, the fines could be thought of as buying pollutant discharges in some contexts. Companies that can afford it could find it less expensive to pay fines than to implement pollution control measures.

3.406 Misdemeanor and Felony Criminal Actions

The option of filing criminal actions against hard-core, noncompliant companies may not be available to all POTWs. State laws must enable the POTW to implement such criminal prosecutions. At most agencies, the law does permit the POTW to take criminal misdemeanor actions against companies that violate the POTW's industrial waste ordinance requirements.

The POTWs use criminal misdemeanor actions against companies that essentially refuse to comply with a POTW's ordinance requirements, in spite of the series of in-house administrative warnings and violation notices. In years past, criminal misdemeanor actions were not taken very seriously either by the lawyers prosecuting the cases or by the criminal justice system. The cases were heard in municipal court where the judges also dealt with felony actions involving crimes such as rape, murder, felonious assault, and major theft crimes. In this context, the criminal justice system tended to believe that the pollution violations were minor, white-collar crimes. In spite of major actions by the agencies involving, in some cases, hundreds of hours of staff time plus thousands of dollars of wastewater analyses, the penalty administered by the court was frequently a few hundred dollars' fine plus six months' to two years' probation. Under these circumstances, going to the final step of a criminal misdemeanor prosecution was not too meaningful and was not undertaken except for the most serious violators.

Within the past few years, however, attitudes about the violation of environmental pollution control laws have changed; violations are now considered a more serious problem by local law enforcement personnel. A District Attorney with the City of Los Angeles instituted some major legal actions against companies found to be purposely violating environmental regulations. The courts ruled in favor of the POTW and imposed stiff penalties. With these penalties has come a new interest in applying the full penalties of the law for companies found in violation of the agency's wastewater ordinance requirements.

The Los Angeles County district attorney's office has established an Environmental Crimes Division where lawyers who specialize in interpreting environmental regulations vigorously prosecute violations of these laws. With the development of this new interest in environmental crimes by the criminal justice system, the use of misdemeanor and felony criminal actions will become increasingly common at POTWs. The EPA regulations require the POTW to pursue every legal alternative necessary to ensure compliance with permit limits. This mandate often includes criminal prosecution if a company persists in violating its discharge requirements.

3.41 Sanitary Sewer Codes

The POTW operating agency may not actually be responsible for maintenance of local sewers nor control connections to the local system. Local ordinances sometimes establish flow restrictions or allocate sewer capacity based on land use (zoning) or actual occupancy. Local sewers may have been financed by bond issues, which may require payment of fees for the right to connect or discharge. Other methods of financing include industrial cost recovery and rate ordinances. The local ordinance may specify which property has the right to discharge to a specific sewer.

Local ordinances may also contain restrictions on entering manholes and require permits and inspection of any work or connection to the system. Such ordinances may require reimbursement to the local wastewater collection system agency for any damage to the system caused by an industrial discharger.

The local wastewater collection system agency may also regulate noncritical industrial wastes that may present local sewer maintenance problems but not necessarily present a problem to the wastewater treatment plant. Such wastes include those where no EPA categorical pretreatment standard has been established. Other items of concern to the local agency include excessive grease from restaurants and food processing industries; high temperatures; low pH; high pH that may cause scaling; flammable, corrosive, or toxic gases; incompatible wastes; excessive solids discharges, which can cause stoppages in sewers; and dissolved and unsettleable solids.

3.42 Building Codes

These laws normally regulate all construction within a private land parcel and specify minimum standards for electrical, mechanical, and plumbing systems. The wastewater ordinance may require the building official to determine that industrial waste discharge approval has been obtained before building permits are issued for any proposed structures. In addition, the actual construction of required facilities may still require standard building and plumbing permits.

Codes administered by the building official usually specify sizes and materials for any plumbing installed within a building and the method of connection to the sewer. Standards may also exist for fixtures such as floor drains, clarifiers, grease traps, pumps, and sampling manholes. Electrical permits are generally required for any electrically operated equipment including pretreatment or monitoring systems.

The industrial waste facility operator should be especially aware of local building code requirements when requesting modifications to existing pretreatment systems. If you have any doubts about what permits are required or if you need help interpreting building codes, contact the appropriate plumbing or building inspectors.

3.43 Underground Tank Laws

Many states have enacted laws regulating the underground storage of hazardous materials and new federal regulations have been developed by EPA. EPA issued regulations on hazardous waste tanks (40 CFR 260-265, 270, and 271) on July 14, 1986 and issued regulations on new and existing underground tanks and associated piping used to store petroleum and hazardous substances (40 CFR 280 and 281) on September 23, 1988.

The 40 CFR 280 technical standards are organized in seven subparts that progressively add requirements. The subparts address:

1. Program scope (applicability, definitions, the interim prohibition).
2. The standards for new and upgraded system design, construction, installation, and notification.
3. General operating requirements.
4. Release detection.
5. Reporting and investigation.
6. Response and corrective action.
7. Closure.

Although these regulations provide very specific criteria at the federal level, they are designed for implementation by state and local agencies.

The applicability of local underground storage tank regulations should be reviewed closely to determine whether or not any portion of your facility's pretreatment system meets the definition of an underground storage tank.

3.44 Land Use Ordinances

These laws, also known as zoning ordinances, regulate a wide variety of activities on property including the type of business allowed. Many existing businesses operate as nonconforming users. That is to say, the zoning of the property they occupy has been changed at some point and, under present zoning restrictions, those businesses would not be permitted to locate there. While nonconforming users may be allowed to continue operations, zoning restrictions may affect the design and placement of any new construction, such as treatment facilities. You should be aware of the requirements of local zoning laws, especially those regulating the storage of chemicals, setbacks that may affect placement of treatment facilities, restrictions on the use of outside areas, such as work space (treatment facilities may have to be enclosed), and the designation of specific areas for employee parking, which may preclude use for storage of items such as sludge containers.

QUESTIONS

Write your answers in a notebook and then compare your answers with those on page 107.

3.4A What is the basic document granting authority to administer a pretreatment program?

3.4B List the essential elements of a wastewater ordinance.

3.4C What are wastewater facilities?

3.4D In addition to wastewater ordinances, list three other types of local ordinances, laws, and codes that may affect industrial wastewater dischargers.

3.4E List five items that might be included in a sanitary sewer code.

3.4F What types of noncritical industrial wastes might be regulated by sanitary sewer codes?

3.5 OTHER FEDERAL STATUTES AND REGULATIONS

3.50 Resource Conservation and Recovery Act (RCRA)

Congress enacted RCRA in 1986 (and subsequently amended it several times) to define a federal role in solid waste and resource management and recovery. The Act's primary goals are:

1. To protect human health and the environment from hazardous and other solid wastes.
2. To protect and preserve natural resources through programs of resource conservation and recovery.

The Act's principal regulatory focus is to control hazardous waste. To this end, RCRA mandates a comprehensive system to identify hazardous wastes and to trace and control their movement from generation through transport, treatment, storage, and ultimate disposal.

Extensive hazardous waste regulations have been promulgated under RCRA's authority. These regulations are codified under 40 CFR 260, 261, 262, 263, 264, 265, 266, and 270. Specifically, RCRA provisions are focused in the following way:

Part 260: General

Part 261: Hazardous Waste Identification and Listing

Part 262: Hazardous Waste Generators

Part 263: Hazardous Waste Transporters

Part 264-265: Owners and Operators of Hazardous Waste Facilities

Part 266: Special Requirements

Part 270: Hazardous Waste Permits

RCRA does contain exceptions and exemptions. Wastes that are normally subject to hazardous waste regulations are exempt in three specific circumstances:

1. Domestic Sewage Exemption
2. On-Site Treatment Exemption
3. Small Quantity Generator Exclusion

DOMESTIC SEWAGE EXEMPTION

Hazardous wastes that are discharged to a POTW and are mixed with domestic sewage are excluded from RCRA control because they are not defined as "solid waste."

ON-SITE TREATMENT EXEMPTION

RCRA regulations governing treatment, storage, or disposal facilities and RCRA permitting regulations contain provisions that exempt owners and operators of the following types of facilities, based on definitions in 40 CFR 260.10:

1. Wastewater Treatment Units
2. Totally Enclosed Treatment Facilities
3. Elementary Neutralization Units

SMALL QUANTITY GENERATOR

EPA does not regulate generators of small quantities of hazardous waste as stringently as it regulates generators of larger quantities. This exclusion reduces the burden of paperwork on small quantity generators, state hazardous waste agencies, and EPA. Generators who generate less than one kilogram per month of acutely hazardous waste and less than 100 kilograms per month of nonacutely hazardous waste are only required to perform a hazardous waste determination and store, treat, or dispose of the hazardous waste on-site in accordance with regulations or ensure its delivery to an authorized hazardous or nonhazardous treatment, storage, or disposal facility.

3.51 Superfund Amendment Reauthorization Act (SARA)—Title III

The Emergency Planning and Community Right-To-Know Act of 1986 (EPCRA) was enacted as a freestanding provision of SARA.[10] This law requires states and local governments to develop plans for responding to unanticipated environmental releases of a number of hazardous substances. In addition, EPCRA requires certain businesses to notify state and local planning entities of the presence of hazardous substances and toxic chemicals at their facilities and to report on the quantities and environmental releases (both planned and unplanned) of these materials. EPCRA also provides a variety of enforcement mechanisms including citizens' suits and criminal, civil, and administrative sanctions.

EPCRA contains three subtitles. Subtitle A creates the framework for local emergency planning and release notifications. Subtitle B provides the mechanisms for community awareness with respect to hazardous chemicals present in their locality. In addition, it requires the submission of material safety data sheets and emergency and hazardous chemical inventory forms to state and local governments, and the submission of toxic chemical release forms to the state and EPA. Subtitle C contains general provisions concerning trade secret protection, enforcement, citizen suits, and public availability of information. The EPCRA regulations are codified in 40 CFR Parts 300, 350, 355, 370, and 372.

---

[10] Enacted by Public Law No. 99-499, October 17, 1986. EPCRA is codified at 42 USC Section 11001, et. seq.

3.52 Clean Air Act

The Clean Air Act Amendments of 1990 were signed into law on November 15, 1990. The original Clean Air Act was less than 50 pages. The 1990 amendment was nearly 800 pages and contained seven major titles.

TITLE I	Attainment of Ambient Air Quality
TITLE II	Mobile Sources
TITLE III	Air Toxics
TITLE IV	Acid Deposition
TITLE V	Permits
TITLE VI	Ozone-Depleting Chemicals
TITLE VII	Enforcement

Title III addresses toxic emissions, which the original Clean Air Act failed to regulate. The amendments take a two-phased approach to regulating 189 air toxics. EPA is required to issue maximum achievable control technology (MACT) standards based on the best demonstrated control technology or practices of the regulated industry.

Title V required EPA to issue a final rule on air pollution permits, creating a program similar to the National Pollutant Discharge Elimination System. Each major source is required to have a five-year permit outlining its compliance requirements.

Title VII provides for serious civil and criminal liabilities for corporations and corporate officials. Criminal penalties for knowing violations, previously misdemeanors, are now felonies. The law also sets new criminal authorities for knowing or negligent endangerment. A conviction of knowingly endangering the public could lead to 15 years in prison. EPA was provided with new authorities to issue administrative penalty orders up to $200,000 and field inspectors can issue on-the-spot citations of up to $5,000. EPA has been given the authority to issue administrative subpoenas to gather compliance data and issue compliance orders and compliance schedules.

Citizen suit provisions in the original Clean Air Act were revised to allow private citizens and groups to seek penalties against violators. The penalties will go to a special fund EPA can use for compliance and enforcement efforts.

3.53 Occupational Safety and Health Act (OSHA)—Hazard Communication (29 CFR 1910.1200)

The purpose of the Hazard Communication Standard (HCS) or Worker Right-To-Know is to ensure that the hazards of all chemicals produced or imported are evaluated (hazard determination), and that information concerning their hazards is transmitted to employers and employees. This transmission of information is accomplished by means of a comprehensive hazard communication program, which is to include a written hazard communication program, container labeling and other forms of warning, material safety data sheets, and employee information training. (Also see Chapter 13, "Safety," for more detailed information about hazard communication programs.)

3.54 Occupational Safety and Health Act (OSHA)—Process Safety Management of Highly Hazardous Chemicals (29 CFR 1910.119)

On February 24, 1992, OSHA issued the final rule on process safety management of highly hazardous chemicals. This rule applies to any process that involves listed quantities of one or more of 132 listed chemicals or 5 or more tons of flammable liquids or gases at one location. The cornerstone of this new rule is the requirement to perform a process hazard analysis. Similar requirements are mandated by the Clean Air Act Amendment of 1990.

A process hazard analysis can be performed in various ways. These vary in complexity from a "what-if" analysis to a Hazard and Operability (HAZOP) study, a Failure Mode Effects and Criticality Analysis (FMECA) or a Fault-Tree Analysis. The HAZOP procedure is the one most commonly used. The goal of a HAZOP is to systematically identify hazards or operational problems at a facility to evaluate the likelihood and consequences of a hazardous material release. The regulations require the employer to establish a system to promptly address and resolve the findings and recommendations of the analysis and ensure that these resolutions are properly documented. In addition, the regulations address written operating procedures, training, contractor requirements, pre-start-up safety review, mechanical integrity, hot work permit, management of change, incident investigation, emergency planning and response, compliance audits, and trade secrets.

3.55 Clean Water Act—Stormwater Regulations

Under the National Pollutant Discharge Elimination System (NPDES) program, efforts to improve water quality focused mainly on industrial discharges and discharges of municipal wastewater since these were thought to be the primary sources of water pollutants. Those efforts have significantly reduced the contamination of surface waters. At the same time, however, it has become apparent that industrial and municipal wastewater discharges are not the only major sources of water pollution. Researchers have found that stormwater picks up and carries unexpectedly large quantities of pollutants as it falls through the atmosphere and flows across the surface of the earth as storm runoff. These pollutants often include heavy metals, such as chromium, cadmium, copper, lead, nickel, and zinc. Runoff from cultivated areas contains pesticides and herbicides, and runoff from paved surfaces may contain fuels, waste oils, solvents, lubricants, and grease.

In urban areas, runoff usually flows either into the municipal wastewater collection system or into a separate stormwater collection system. The water may be treated to remove pollutants before discharge or may be discharged directly to the environment. When runoff is discharged through separate storm sewers or some other type of pipeline or channel, it is considered a point source and may be regulated under the Clean Water Act. Section 402(p) of the Clean Water Act required EPA to set up an NPDES permit program for stormwater discharges in the following situations:

1. Discharges associated with industrial activity.
2. Discharges from large municipal separate storm sewer systems (serving a population of 250,000 or more).
3. Discharges from medium-sized municipal separate storm sewer systems (serving a population of 100,000 or more, but less than 250,000).

The permit application requirements for this program were published on November 16, 1990 (55 *FEDERAL REGISTER* 47990). Runoff from many areas of an industrial facility are regulated by the new permit requirements, including shipping, receiving, manufacturing, storage, and loading areas if they are exposed to stormwater. Operators of industrial facilities should review the requirements of the Storm Water Regulations to determine if a permit is required for their facility. The classification, "storm water discharges associated with industrial activity," is defined in 40 CFR 122.26(b)(14). Additional information is available in EPA's publication, *GUIDANCE MANUAL FOR STORM WATER DISCHARGES ASSOCIATED WITH INDUSTRIAL ACTIVITY.*

3.56 Department of Transportation—Hazardous Material Regulations (HM 126 and HM 181)

On October 1, 1993, new Department of Transportation (DOT) regulations took effect regarding hazardous materials. Two important aspects of these regulations are:

1. Hazard identification (symbols, numbers) and hazard classifications now conform with the international United Nations transportation system's identifications and classifications.
2. The regulatory approach has changed from specifying exact package shape and material and step-by-step training to performance-oriented packaging and training.

These regulations require employers to certify, in writing, that their employees whose jobs directly involve hazardous materials transportation safety have been trained and tested, and retrained at least every two years. How does this affect an industrial waste treatment operator? The DOT training regulations apply to employees who:

- Load, unload, pack, or handle hazardous materials.
- Prepare hazardous materials for transportation.
- Select, classify, test, or recondition hazardous material shipping containers.
- Provide hazard and other information for shipping papers.
- Prepare shipping papers, manifests, or other documentation.
- Select the type of carrier for a hazardous material.
- Operate a vehicle that transports hazardous materials.

In addition, the regulations specify four levels of training:

1. General awareness training to help employees understand DOT's hazardous material shipping requirements and recognize and identify hazardous materials on the job.
2. Safety training to address hazardous material exposure prevention and incident response.
3. Function-specific training to provide knowledge, skills, and abilities for particular tasks.
4. Driver training, which must include the three types of training described above as well as safe hazardous material transport truck information.

3.6 KEEPING CURRENT

These regulations were current when this manual was prepared. Industrial waste treatment operators must be aware that regulations change when better information becomes available, waste treatment technology improves, and results of litigation are implemented. *YOU* have the responsibility to be alert for changes and adjust accordingly.

3.7 ACKNOWLEDGMENTS

Most of the material in this chapter is adapted from Chapter 3, "Development and Application of Regulations," by Eddie Esfandi from the *PRETREATMENT FACILITY INSPECTION* training manual in this series. Portions of the material in the original chapter were prepared by Carl Sjoberg, Jay Kremer, and Robert Steidel. Their contributions are greatly appreciated.

QUESTIONS

Write your answers in a notebook and then compare your answers with those on page 107.

3.5A What are the goals of the Resource Conservation and Recovery Act (RCRA)?

3.5B What types of pollutants are commonly found in stormwater runoff?

3.5C What types of stormwater discharges may be regulated by an NPDES permit?

3.5D What types of training must employers provide for employees whose work directly involves hazardous materials transportation safety?

Please answer the discussion and review questions next.

DISCUSSION AND REVIEW QUESTIONS
Chapter 3. REGULATORY REQUIREMENTS

The purpose of these questions is to indicate to you how well you understand the material in the chapter. Write the answers to these questions in your notebook.

1. What types of discharges are regulated under the National Pollutant Discharge Elimination System (NPDES)?

2. What are the advantages of EPA's delegation of implementation authority to local agencies?

3. What is the difference between a direct and indirect discharger?

4. The General Pretreatment Regulations were established to regulate what types of pollutants?

5. What are the eight specific categories of pollutants regulated by the Prohibited Discharge Standards?

6. What are categorical pretreatment standards?

7. How are categorical pretreatment standards developed?

8. Explain the difference between concentration-based standards and production- or mass-based standards.

9. How can an industrial discharger avoid monitoring for TTO?

10. How can categorical pretreatment standards be modified?

11. What is a wastewater ordinance?

12. What is the principal regulatory focus of the Resource Conservation and Recovery Act (RCRA)?

SUGGESTED ANSWERS
Chapter 3. REGULATORY REQUIREMENTS

Answers to questions on page 83.

3.0A A statute is an act of the legislature declaring, commanding, or prohibiting something. Regulations are issued by various governmental departments to carry out the intent of the statute. Agencies issue regulations to guide the activity of those regulated by the agency and to ensure uniform application of the statute.

3.0B The major goals of the 1972 Federal Water Pollution Control Act Amendments were: (1) to have the nation's waters clean enough for swimming by 1983, and (2) to eliminate the discharge of pollutants by 1985.

3.0C The Environmental Protection Agency is responsible for developing and enforcing regulations designed to protect the nation's water, air, and land from various pollutants. The agency is charged with developing pollution control guidelines, implementing programs to achieve the goals established by Congress, and coordinating the various pollution control activities throughout the nation.

3.0D EPA has developed specific regulatory requirements for industrial facilities that discharge wastewaters directly into US streams, lakes, or other waters, or into the oceans (direct dischargers) and different requirements for industrial facilities that discharge wastewaters to POTWs (indirect dischargers).

3.0E A new source is any building, structure, facility, or installation from which there is or may be a discharge of pollutants. Construction of the facility must have begun after publication of the applicable Pretreatment Standards. The building, structure, facility, or installation must also be constructed at a site at which no other source is located; or, must totally replace the existing process or production equipment producing the discharge at the site; or, must be substantially independent of an existing source of discharge at the same site. Existing sources are industrial dischargers that were in existence at the time of promulgation of the proposed applicable categorical standards.

Answers to questions on page 84.

3.1A The NPDES permit program regulates the discharge of potential pollutants from point sources and some stormwater runoff into US waterways. NPDES permits regulate discharges into US waterways from all point sources of pollution, including industries, municipal wastewater treatment plants, sanitary landfills, large animal feedlots, and return irrigation flows. Regulated discharges of stormwater runoff include municipal separate storm sewer systems, industrial and commercial sites, and construction sites.

3.1B The major elements of a NPDES permit include effluent limits, monitoring and reporting requirements, and any other terms and conditions necessary to protect water quality.

3.1C If an industrial facility discharges wastewater to a POTW, either directly or after pretreatment, the discharger must comply with the terms and conditions of the POTW's NPDES permit.

Answers to questions on page 87.

3.2A The General Pretreatment Regulations regulate pollutants that may:

1. Pass through the POTW's treatment system, untreated or partially treated.
2. Interfere with the POTW's treatment processes.
3. Contaminate the POTW's sludge.

3.2B The most common conventional pollutants are biochemical oxygen demand, suspended solids, fecal coliforms, pH, and oil and grease.

3.2C POTWs with a design flow of less than 5 MGD may be required to establish a pretreatment program if they have industrial users subject to National Pretreatment Standards or if nondomestic wastes or wastewaters cause upsets, violations of NPDES permit conditions, or sludge contamination.

Answers to questions on page 88.

3.2D EPA's General Pretreatment Regulations (40 CFR 403) established two types of regulatory programs, the Prohibited Discharge Standards and the Industrial Categorical Discharge Standards.

3.2E The General Pretreatment Regulations specifically disallow the introduction to POTWs of:

1. Pollutants that create a fire or explosion hazard in the POTW's sewer system or at the treatment plant.
2. Pollutants that are corrosive, including any discharge with a pH lower than 5.0, unless the POTW is specifically designed to handle such discharges.
3. Solid or viscous pollutants in amounts that will obstruct the flow in the collection system and treatment plant, resulting in interference with operations.
4. Any pollutant discharged in quantities sufficient to interfere with POTW operations (including BOD).
5. Discharges with temperatures above 104°F (40°C) when they reach the treatment plant, or hot enough to interfere with biological processes at the wastewater treatment plant.
6. Petroleum oil, nonbiodegradable cutting oil, or products of mineral oil origin in amounts that will cause interference or pass-through.
7. Pollutants that result in the presence of toxic gases, vapors, or fumes within the POTW in a quantity that may cause acute worker health and safety problems.
8. Any trucked or hauled pollutants, except at discharge points designated by the POTW.

3.2F Toxic pollutants are compounds or classes of compounds identified by EPA to be harmful to one or more forms of plant or animal life.

3.2G Technology-based standards are based upon the available treatment technologies that could be used to remove pollutants, whereas receiving water standards are based on the capacity of the stream that receives the wastewater to tolerate the pollutants.

Answers to questions on page 92.

3.2H The levels of treatment technology identified by EPA's categorical regulations are Best Practicable Technology (BPT) and Best Available Technology (BAT).

3.2I Development Documents contain detailed technical information about the treatment technologies studied by EPA during its development of a proposed regulation.

3.2J The regulatory requirements for the control of water pollutants from indirect industrial wastewater dischargers are referred to as Pretreatment Standards for Existing Sources (PSES) and Pretreatment Standards for New Sources (PSNS). The General Pretreatment Regulations also establish several prohibited discharge criteria.

3.2K If there is a question or disagreement about applicability of a designated industrial category to an industrial discharge, the POTW or the industrial user can request a ruling by the EPA concerning the appropriateness of the industrial category.

Answers to questions on page 94.

3.3A The two different types of categorical standards used in the EPA pretreatment program are concentration-based standards and production-based standards (also referred to as mass-based standards).

3.3B A regulated stream is an industrial process wastestream subject to national categorical pretreatment standards.

3.3C The phrase "Total Toxic Organics" refers to the summation (total) of all toxic organic compounds on the EPA Priority Toxic Pollutants list in excess of 10 micrograms per liter.

3.3D If monitoring is necessary to measure compliance with the TTO standard, the industrial discharger needs to analyze for only those pollutants that would reasonably be expected to be present.

Answers to questions on page 97.

3.3E The General Pretreatment Regulations require categorical dischargers to file a Baseline Monitoring Report/Compliance Schedule, a Final Compliance Report/90-Day Report, and Periodic Compliance Reports.

3.3F Information required in a Baseline Monitoring Report (BMR) includes:

1. Identification of the indirect discharger.
2. A list of environmental control permits.
3. A description of its operations.
4. A report on flows of regulated streams.
5. The results of sampling and analyses of the industrial wastewater discharges to determine levels of regulated pollutants in those streams.
6. A certification statement by the discharger indicating compliance or noncompliance with the applicable pretreatment standards.
7. A description of any additional steps required to achieve compliance for noncompliant dischargers.

3.3G Categorical Pretreatment Standards could be modified on the basis of fundamentally different factors (FDF), net gross adjustments, and removal credits.

3.3H A net gross credit allows the subtraction of the concentration of regulated pollutants in the intake water from the concentration level in the industrial discharger's effluent.

Answers to questions on page 101.

3.4A The basic document granting authority to administer a pretreatment program is the wastewater ordinance or sewer-use ordinance.

3.4B The essential elements of a wastewater ordinance include Administration, EPA Minimum Requirements, General Provisions, Specific Provisions, and Industrial Wastewater Limitations.

3.4C Wastewater facilities are the pipes, conduits, structures, equipment, and processes required to collect, convey, and treat domestic and industrial wastes, and dispose of the effluent and sludge.

3.4D In addition to wastewater ordinances, other types of local ordinances, laws, and codes that may affect industrial wastewater dischargers include sanitary sewer codes, building codes, underground tank laws, and land use ordinances.

3.4E Items that might be included in a sanitary sewer code include:

1. Flow restrictions.
2. Connection fees.
3. Which property has the right to discharge to a specific sewer.
4. Restrictions on entering manholes.
5. Requirements for permits.
6. Inspection of any work or connection to the system.
7. Reimbursement for any damage to the system caused by an industrial discharger.
8. Restrictions on wastes not regulated by EPA standards.

3.4F Noncritical industrial wastes that might be regulated by sanitary sewer codes include: wastes for which no EPA categorical pretreatment standard has been established; excessive grease; high temperature wastes; high or low pH wastes; flammable, corrosive, or toxic gases; incompatible wastes; excessive solids; and dissolved and unsettleable solids.

Answers to questions on page 104.

3.5A The goals of the Resource Conservation and Recovery Act are:

1. To protect human health and the environment from hazardous and other solid wastes.
2. To protect and preserve natural resources through programs of resource conservation and recovery.

3.5B Pollutants found in stormwater runoff often include heavy metals, such as chromium, cadmium, copper, lead, nickel, and zinc. Runoff from cultivated areas contains pesticides and herbicides, and runoff from paved surfaces may contain fuels, waste oils, solvents, lubricants, and grease.

3.5C Stormwater discharges may be regulated by an NPDES permit in the following situations:

1. Discharges associated with industrial activity.
2. Discharges from large municipal separate storm sewer systems (serving a population of 250,000 or more).
3. Discharges from medium-sized municipal separate storm sewer systems (serving a population of 100,000 or more, but less than 250,000).

3.5D If an employee's job directly involves hazardous materials transportation safety, the employer must provide:

1. General awareness training to help employees understand DOT's hazardous material shipping requirements and recognize and identify hazardous materials on the job.
2. Safety training to address hazardous material exposure prevention and incident response.
3. Function-specific training to provide knowledge, skills, and abilities for particular tasks.
4. Driver training, which must include the three types of training described above as well as safe hazardous material transport truck information.

CHAPTER 4

PREVENTING AND MINIMIZING WASTES AT THE SOURCE

by

M. Philip Lo

With the Assistance of

Theresa Dodge
Alison Gemmell
Rob Lamppa
Mischelle Mische
John Polanski

TABLE OF CONTENTS

Chapter 4. PREVENTING AND MINIMIZING WASTES AT THE SOURCE

			Page
OBJECTIVES			113
WORDS			114
4.0	POLLUTION PREVENTION—RIGHT FROM THE START		115
	4.00	The Pollution Prevention Strategy	115
	4.01	Switching From Treatment to Prevention	115
	4.02	Regulatory Requirements for Pollution Prevention	116
	4.03	The Role of Industrial Treatment Personnel in Pollution Prevention	116
	4.04	Economic Benefits of Pollution Prevention	117
4.1	GENERAL POLLUTION PREVENTION OPPORTUNITIES		117
	4.10	Good Operating Practices	117
		4.100 Raw Material Purchasing	118
		4.101 Raw Material and Product Storage and Loading	118
		4.102 Material Distribution Systems	118
		4.103 Process Operations	118
		4.104 Waste Segregation	118
		4.105 Pollution Prevention Audits and Top Management Support	118
	4.11	Material Substitution	118
	4.12	Process Modification	119
	4.13	Product Reformulation	119
4.2	INDUSTRY-SPECIFIC POLLUTION PREVENTION OPPORTUNITIES FOR SELECTED INDUSTRIES		120
	4.20	Pollution Prevention Opportunity Checklists	120
	4.21	A Sample "Walk-Through" of the Pollution Prevention Opportunity Checklist for the Metal Finishing Industry	120
	4.22	A Sample "Walk-Through" of the Pollution Prevention Opportunity Checklist for the Chemicals Formulating Industry	121
DISCUSSION AND REVIEW QUESTIONS			121
SUGGESTED ANSWERS			122

APPENDIX 123
    POLLUTION PREVENTION OPPORTUNITY CHECKLISTS
    FOR SELECTED INDUSTRIES 123
      1. CHEMICAL MANUFACTURING 124
      2. CHEMICALS FORMULATING: PESTICIDES, CHEMICALS, AND PAINTS 129
      3. DAIRY PROCESSING 130
      4. DRY CLEANING 132
      5. FLUID MILK PROCESSING 133
      6. MEAT PACKING: BEEF 135
      7. METAL FABRICATION 136
      8. METAL FINISHING 137
      9. OIL AND GAS EXTRACTION 139
    10. PETROLEUM REFINING 140
    11. PHOTO PROCESSING 146
    12. PRINTED CIRCUIT BOARD MANUFACTURING 146
    13. PRINTING 148
    14. PULP AND PAPER MANUFACTURING: KRAFT SEGMENT .. 149
    15. RADIATOR REPAIR 150

OBJECTIVES

Chapter 4. PREVENTING AND MINIMIZING WASTES AT THE SOURCE

Following completion of Chapter 4, you should be able to:

1. Explain the differences between industrial waste treatment and pollution prevention.

2. Describe the industrial treatment facility operator's role in pollution prevention.

3. Identify general types of pollution prevention opportunities in the areas of: (a) good operating practices, (b) material substitution, (c) process modification, and (d) product reformulation.

4. Describe the economic benefits of preventing pollution.

5. Prepare a pollution prevention checklist for your industrial facility.

WORDS
Chapter 4. PREVENTING AND MINIMIZING WASTES AT THE SOURCE

BIOSOLIDS

A primarily organic solid product, produced by wastewater treatment processes, that can be beneficially recycled. The word biosolids is replacing the word sludge.

CONVENTIONAL POLLUTANTS

Those pollutants that are usually found in domestic, commercial, or industrial wastes, including suspended solids, biochemical oxygen demand, pathogenic (disease-causing) organisms, and oil and grease.

DRAG OUT

The liquid film (plating solution) that adheres to the workpieces and their fixtures as they are removed from any given process solution or their rinses. Drag-out volume from a tank depends on the viscosity of the solution, the surface tension, the withdrawal time, the draining time, and the shape and texture of the workpieces. The drag-out liquid may drip onto the floor and cause wastestream treatment problems. Regulated substances contained in this liquid must be removed from wastestreams or neutralized prior to discharge to POTW sewers.

RECYCLE

The use of water or wastewater within (internally) a facility before it is discharged to a treatment system. Also see REUSE.

REUSE

The use of water or wastewater after it has been discharged and then withdrawn by another user. Also see RECYCLE.

SURFACTANT (sir-FAC-tent)

Abbreviation for surface-active agent. The active agent in detergents that possesses a high cleaning ability.

CHAPTER 4. PREVENTING AND MINIMIZING WASTES AT THE SOURCE

4.0 POLLUTION PREVENTION—RIGHT FROM THE START

4.00 The Pollution Prevention Strategy

Pollution prevention is a preventive approach to the protection of human health and the environment. The focus of pollution prevention is on minimizing the amount of waste generated as opposed to treating and recycling wastes after generation. The Pollution Prevention Act of 1990 establishes the Pollution Prevention Hierarchy (Strategy) as a national policy, declaring that:

- Pollution should be prevented or reduced at the source whenever feasible.

- Pollution that cannot be prevented should be *RECYCLED*[1] in an environmentally safe manner whenever feasible.

- Pollution that cannot be prevented or recycled should be treated in an environmentally safe manner whenever feasible.

- Disposal or other release into the environment should be used only as a last resort and should be conducted in an environmentally safe manner.

Pollution prevention generally encompasses the following areas of endeavor:

- Process modification, material substitution, and product reformulation.

- Improved process operation and maintenance.

- Good operating practices (good housekeeping).

- Material recycle, *REUSE*,[2] and recovery for in-process use.

Waste minimization is a related term primarily used in connection with hazardous wastes. Waste minimization is the reduction of hazardous waste that is generated or subsequently treated, stored, or disposed of. It includes any activity that results in the reduction of the total volume or quantity of hazardous waste, or the reduction of toxicity of the hazardous waste. The goal is to minimize present and future threats to human health and the environment. Under waste minimization, activities are distinguished between source reduction and recycling. Source reduction is the reduction or elimination of the generation of waste at the source, usually within a production process. Recycling is the reuse of materials in the original process or in another process, or the reclamation of materials from a waste for resource recovery or by-product production.

4.01 Switching From Treatment to Prevention

Industrial waste treatment and pretreatment practices focus on the reduction of pollutant concentrations in wastewater discharged from an industrial facility to minimize its impact on the wastewater treatment plant or the environment. Treatment is basically a remedial process that satisfies health and safety concerns. However, since many waste treatment processes create residual sludge that needs to be landfilled, treatment of industrial wastewater may simply be shifting pollutants from one medium of the environment to another with no net reduction. Moreover, additional reduction of toxic pollutants discharged to the wastewater collection system may be required in the future to meet more stringent regulations for sludge reuse and disposal, reclaimed water recharge to groundwater, and health risk reductions from volatile organic emissions at municipal wastewater treatment plants.

Pollution prevention is a complementary tool for reducing the discharge of pollutants to a wastewater collection system or the environment. It is not new, and was applied intensively in the 1970s to meet local discharge limits without the installation of pretreatment equipment. The 1980s saw a focus on pretreatment because of the EPA Categorical Pretreatment regulations. However, in the 1990s, pollution prevention was once again the preferred choice for source control and waste management. It has the potential of minimizing or even eliminating a particular pollutant from the wastestream. Good operating practices could minimize the volume of waste entering the treatment system. Material substitution could eliminate a pollutant altogether, and a process change could eliminate a whole wastestream.

---

[1] *Recycle.* The use of water or wastewater within (internally) a facility before it is discharged to a treatment system. Also see REUSE.
[2] *Reuse.* The use of water or wastewater after it has been discharged and then withdrawn by another user. Also see RECYCLE.

Pollution prevention programs are now being implemented in many types of industrial facilities, regardless of whether they generate only CONVENTIONAL POLLUTANTS[3] or toxic/hazardous wastes as well. Food processors, paper manufacturers, and textile dyers have reduced wastewater flow and minimized the amounts of conventional pollutants like suspended solids and chemical oxygen demand (COD) discharged to the sewer. The incentive is for water conservation and cost savings in sewer service charges. Facilities undergoing expansion have an additional incentive to institute pollution prevention. Expanded industrial capacity often creates a need for greater wastewater treatment plant capacity. Construction of the added treatment capacity will usually be financed (at least in part) by higher waste load charges. If the expanding industrial facility is able to increase its size or production level while reducing the waste load it generates by implementing a pollution prevention program, then its waste treatment costs could decline and additional treatment plant capacity may not be needed.

4.02 Regulatory Requirements for Pollution Prevention

The Congress of the United States made waste minimization a national policy in the 1984 Hazardous and Solid Waste Amendments (HSWA) to the Resource Conservation and Recovery Act (RCRA) of 1976. Later, the Pollution Prevention Act of 1990 made pollution prevention a national policy. The policy states that pollution should be prevented or reduced at the source whenever feasible.

Generators of hazardous wastes must certify on their hazardous waste manifests that, "a program is in place to reduce the volume or quantity and toxicity of hazardous waste to the degree it has determined to be economically practicable." In addition, generators are also required to report in their biennial reports to EPA for hazardous wastes all efforts made to reduce the volume of hazardous waste and the reduction achieved. Many, though not all, industrial dischargers to sewer systems are also hazardous waste generators and are covered by these waste minimization requirements. The Pollution Prevention Act of 1990 also directs facilities required to report releases to the US EPA for the Toxic Release Inventory (TRI) to provide documentation of their procedures for preventing the release of those materials.

Many states have also passed legislation requiring pollution prevention or waste minimization. At this time, the requirements primarily apply to hazardous wastes. However, the concept has been extended to include wastewater discharges and air emissions. State legislation generally requires a hazardous waste generator to conduct an audit and prepare a plan for waste minimization. For a few states, a numerical waste reduction goal is also required. Some states also link pollution prevention planning to a related goal of toxic material use reduction. In many states, pollution prevention legislation establishes program offices, advisory boards, or commissions to provide technical assistance and to promote education, training, and research.

Pollution prevention requirements have also been promulgated for wastewater dischargers to municipal wastewater treatment plants (POTWs). A provision of the federal General Pretreatment Regulations requires an industrial discharger to notify the municipal wastewater treatment agency, the state, and the EPA of "any discharge into the POTW (Publicly Owned Treatment Works) of a hazardous substance which, if otherwise disposed of, would be a hazardous waste under 40 CFR (Code of Federal Regulations) Part 261." Further, the industrial discharger must certify that it has developed a program to minimize toxic wastes to the extent it can do so economically. Presumably, this new requirement gives POTWs the regulatory mechanism to inspect and review waste reduction programs of applicable dischargers.

Individual municipal wastewater treatment agencies have also instituted local ordinances or administrative requirements for industrial pollution prevention. The agencies are hoping to use pollution prevention, in addition to pretreatment, to help meet regulations for BIOSOLIDS[4] reuse, effluent discharges, and air emissions. The types of pollution prevention requirements include performing an audit, preparing a plan, and, in enforcement cases, agreeing to install cost-effective pollution prevention equipment or systems. The agencies usually have pollution prevention outreach programs to assist industries.

The present regulatory trend is to integrate and institutionalize pollution prevention into the whole regulatory structure of permitting, inspection, and enforcement. Pollution prevention is beginning to be required for pretreatment system permit approval. Pollution prevention opportunities are being looked for by inspectors on visits. Pollution prevention "Supplemental Environmental Projects" have also been written into enforcement compliance agreements. The ultimate goal is to establish pollution prevention as the preferred choice for pollution control and management.

4.03 The Role of Industrial Treatment Personnel in Pollution Prevention

Industrial treatment personnel should familiarize themselves not only with end-of-pipe treatment technologies, but also with opportunities for pollution prevention. There are opportunities in good housekeeping, process modification, material substitution, and improved operation and maintenance. Treatment personnel need to learn more about the production processes and the chemicals used and generated at their facilities, and should

---

[3] *Conventional Pollutants.* Those pollutants that are usually found in domestic, commercial, or industrial wastes, including suspended solids, biochemical oxygen demand, pathogenic (disease-causing) organisms, and oil and grease.

[4] *Biosolids.* A primarily organic solid product, produced by wastewater treatment processes, that can be beneficially recycled. The word biosolids is replacing the word sludge.

meet and work with production and design personnel to implement pollution prevention strategies. Pollution prevention is another tool for meeting discharge requirements. Information on general pollution prevention opportunities and industry-specific opportunities presented in the following sections may be used as a guide for developing a pollution prevention plan for your facility.

Industrial treatment personnel can also be the "champion" for pollution prevention for their facilities. Treatment personnel generally have the best understanding of the environmental implications of their facilities' discharges to the environment, the costs of wastewater treatment and residual disposal, and the burden of environmental and health and safety requirements that have to be met. Therefore, they are probably in the best position to appreciate the benefits of pollution prevention. Treatment facility personnel can spearhead the move toward pollution prevention in the areas of good housekeeping, improved operation and maintenance, process modification, and material substitution by heading up task force teams to identify pollution prevention opportunities in their facilities. They can provide training for facility personnel on pollution prevention practices. They can also convey information on pollution prevention to facility design and operations staff for process and material input changes. Indeed, treatment facility personnel are key links to their facilities for pollution prevention.

4.04 Economic Benefits of Pollution Prevention

Management often stresses to operators the need to avoid spending money or to save money. In promoting a pollution prevention program, operators could point to the following economic benefits and cost savings that could be achieved by such a program:

- Reduced storage, handling, and treatment costs.
- Reduced raw material costs.
- Reduced production costs through better management and more efficient use of materials.
- Reduced compliance costs for permits and monitoring.
- Reduced transportation and disposal costs for sludges.
- Reduced sewer-use fees from POTWs.
- Lower manifesting and reporting costs.
- Lower risks of emergencies, spills, and accidents.
- Lower health and safety costs.
- Lower long-term environmental liability and insurance costs.
- Reduced liability for landfill disposal of hazardous waste.
- Reduced liability for cleanup of contaminated landfill sites.
- Income from the sale, reuse, or recycle of wastes.

Successful pollution prevention programs provide industry with the opportunity to share the good news with its employees, stockholders, and with the community where its facilities operate.

QUESTIONS

Write your answers in a notebook and then compare your answers with those on page 122.

4.0A Pollution prevention generally encompasses what areas or types of endeavor?

4.0B What is the focus of industrial waste treatment practices?

4.0C Municipal wastewater treatment agencies are hoping to use pollution prevention, in addition to pretreatment, to help meet what regulations?

4.0D What are the opportunities for pollution prevention in an industrial facility?

4.1 GENERAL POLLUTION PREVENTION OPPORTUNITIES

In any industry, general pollution prevention opportunities are available to reduce the toxicity or volume of industrial wastewater. Four general categories of pollution prevention opportunities are:

- Good operating practices (best management practices)
- Material substitution
- Process modification
- Product reformulation

Opportunities in some categories are low-cost, others may be implemented on a short time schedule, and others may require only minimal employee education. Each category will be discussed in terms of relative cost, general implementation time frame, and degree of employee training needed. General examples will be given to illustrate each category.

4.10 Good Operating Practices

Good operating practices (GOPs) are procedures and policies that reduce the quantity or toxicity of wastewater. GOPs are also frequently referred to as best management practices or good housekeeping. In general, GOPs are relatively low capital cost changes that can be implemented in short time frames with regular, repeated employee training. GOPs typically result in immediate waste reduction and often result in the most significant quantity of reduction because they are the easiest and least expensive to implement.

Examples of GOPs for wastewater pollution prevention are explained in the following paragraphs. The GOPs are organized according to steps in a general industry process and include regular inspections and attention to spill prevention, frequent employee training, and properly scheduled maintenance and cleaning.

4.100 Raw Material Purchasing

If possible, raw materials should be purchased in packaging that will not have to be rinsed prior to disposal or reuse. A mechanism should be established for having the signature of a waste disposal person on a purchase order before a new raw material can be purchased. When an industry wants to purchase a new raw material, a waste disposal person should review the material constituents and comment if those constituents or the use of the new material in a process will contribute to greater toxicity or quantity of wastewater. Raw materials should be ordered as needed, rather than in bulk, discounted quantities that may never be used.

4.101 Raw Material and Product Storage and Loading

Storage areas should be protected from rainwater intrusion. Storage containers should be routinely inspected for leaks, corrosion, dents, or other conditions that could lead to spills. Liquid chemical storage areas should be designed with secondary containment. Regular inspections and repairs to ensure the integrity of the containment areas should be performed. Employee traffic patterns should be minimized in storage areas. Employees should be trained to use raw materials on a first-in, first-out basis and routinely check for expiration dates. Fill lines, hoses, and connections should be located where spills will be contained. A quality assurance program should be implemented to ensure that raw materials are not off-specification when they are received from the vendor.

4.102 Material Distribution Systems

As much as possible, material distribution should be done through pumps and pipes rather than through manual pouring from buckets or barrels. This practice reduces the chance of spills. If materials must be poured by hand, employees should be trained to clean up spills using minimal quantities of washdown water, or no washdown water if possible. Employees should be trained to close valves and place drip buckets or absorbent material in locations where drips from valves or faucets are likely to occur.

4.103 Process Operations

Employees should be trained regarding proper process operations and procedures so that off-specification process batches will not be generated or will be reworked rather than disposed of through the wastewater treatment system. Process temperatures and raw material quantities should be closely monitored if an incorrect guideline would generate waste. For example, chemical reactions should be operated so that unwanted by-products are reduced as much as possible and so that decant water contains only trace amounts of soluble contaminants. Equipment cleaning using water-based cleaners should be scheduled as needed, rather than on a set schedule that may be more often than necessary. Dry cleaning techniques, such as rag wipes or squeegees, should be encouraged and implemented where feasible. Maintenance schedules should also be optimized to reduce the number of times wastes are generated.

Employees should be encouraged to investigate recycling or reuse opportunities for process wastes that may be discharged to the wastewater treatment system or the environment and a reward system should be established for employee suggestions that are implemented and successful. Overflow alarms and other liquid level control devices should be routinely inspected and repaired if necessary.

4.104 Waste Segregation

For the wastewater streams that are generated, segregation of concentrated streams from dilute streams or segregation of single, recoverable constituent streams from general wastestreams would improve the recyclability or treatability of the wastestreams.

4.105 Pollution Prevention Audits and Top Management Support

Pollution prevention audits with follow-up employee training should be implemented or incorporated into annual environmental compliance audits. Regular analysis of pollution prevention opportunities, with financial support and employee rewards from top management, will encourage employees to comply with new or improved operating practices.

4.11 Material Substitution

Material substitution is the replacement of one raw material with another raw material in order to reduce the toxicity of wastewater or reduce the quantity of wastewater generated. This pollution prevention technique generally involves low capital costs; however, operation and maintenance expenses may increase. Substituted materials may be more expensive, require more employee attention to a process, and increase the need for quality control testing. Some capital costs may be incurred if a powdered material is substituted for a liquid material. The equipment used for handling one type of material versus another may have to be purchased and installed before the material substitution can be implemented. Material substitutions can be implemented in short time periods if the research, demonstration, and successful performance of the substituted material is well documented. If research and development must be conducted before an appropriate substitute material can be selected, the implementation time frame will increase. Employee education and training will be required, particularly if the new material has different health and safety concerns or requires more operator attention to maintain product quality.

Numerous examples of material substitution are available and are generally industry-specific. Some of these will be mentioned in this general pollution prevention discussion. For more detail, see industry-specific pollution prevention checklists in the Appendix of this chapter. A few technical clearinghouses and pollution prevention research and development organizations are available now and others are being developed so that information about material substitution will be more readily available to industries that look into this opportunity.

Material substitutions can include replacement of cyanide plating baths with noncyanide plating baths. Rinse water tanks located downstream of the plating tanks will then contain less-toxic materials from *DRAG OUT*.[5] In some applications, hexavalent chromium plating baths can be replaced with trivalent chromium plating baths. Rinse waters downstream from the plating tanks will contain the less-toxic form of chromium. Water-based film developing systems in the electronics industry can be replaced with dry systems. Inks may be purchased with lower metal or no metal content so that rinses of ink-containing equipment would contain less toxic material.

A current trend in waste toxicity reduction is to replace solvent-based cleaning processes with water-based cleaning processes. Although this material substitution is encouraged by hazardous waste and industrial health and safety personnel, this practice increases the amount of wastewater generated at a facility. Water-based cleaners may also contain *SURFACTANT*[6] chemicals that would interfere with biological treatment of wastewater. Therefore, the benefits of using a water-based cleaner versus a solvent-based cleaner should be weighed against the potential negative impact of increased wastewater and interference with waste treatment processes.

4.12 Process Modification

Process modification is a change in process operation or equipment. Process modifications can be costly, require long implementation schedules, and require extensive employee training. On the other hand, some rinse water reduction methods and equipment can be designed and installed at relatively low cost, with short implementation schedules and low to no employee training. Simple process operation modifications, such as increased plating bath temperature for water evaporation, are also possibilities. Wastewater segregation, collection, treatment, and reuse in the process can be implemented to reduce the quantity of wastewater discharged to the sewer, if cost effective. Examples of technologies used for recycling and reuse of process wastewater include evaporation, ion exchange, and reverse osmosis.

Employee acceptance of new processes or operating procedures must be achieved through training or reward systems. Implementation schedules may have to include extensive research and development time.

Numerous examples of process modifications are available for the painting and paint and coatings removal processes. Examples include use of powder coat paints, electrostatic painting, laser paint stripping, flashlamp stripping, cryogenic coating removal, and dry ice blasting instead of aqueous alkaline stripping, or use of high-pressure water-jet blasting.

Low-cost, proven process modifications include optimizing process operating guidelines, such as temperature and residence time. This can improve the process reaction so that fewer contaminant by-products are produced.

Different raw materials may be used to produce the same chemical, with fewer by-products, even if the process takes longer or requires more operator attention.

Rinse water reduction methods, such as countercurrent rinsing, spray rinsing, static rinsing, and use of flow restrictors, can be relatively low-cost, low-employee training, pollution prevention opportunities.

The distinction between material substitution and process modification is not always clear; however, process modification generally requires more capital investment and longer implementation schedules.

4.13 Product Reformulation

Product reformulation involves changing a product in such a way that the process for the new product generates less wastewater or less-toxic wastewater. Costs, implementation schedules, and employee training for product reformulations can vary widely depending on the availability of alternative product information and the complexity of the alternative product processes.

Examples of product reformulation include removal of lead from paint formulations and removal of cadmium from ink formulations. Removal of these metals from final products results in less-toxic wastewaters from these industries.

QUESTIONS

Write your answers in a notebook and then compare your answers with those on page 122.

4.1A What are good operating practices?

4.1B What factors should be considered when purchasing raw materials?

4.1C What factors should be considered in raw material storage areas?

4.1D How can material substitution minimize wastes?

4.1E How can product reformulation minimize wastes?

---

[5] *Drag Out.* The liquid film (plating solution) that adheres to the workpieces and their fixtures as they are removed from any given process solution or their rinses. Drag-out volume from a tank depends on the viscosity of the solution, the surface tension, the withdrawal time, the draining time, and the shape and texture of the workpieces. The drag-out liquid may drip onto the floor and cause wastestream treatment problems. Regulated substances contained in this liquid must be removed from wastestreams or neutralized prior to discharge to POTW sewers.

[6] *Surfactant* (sir-FAC-tent). Abbreviation for surface-active agent. The active agent in detergents that possesses a high cleaning ability.

4.2 INDUSTRY-SPECIFIC POLLUTION PREVENTION OPPORTUNITIES FOR SELECTED INDUSTRIES

Pollution prevention opportunity checklists for selected industries were developed to stimulate thinking on the part of industrial wastewater treatment facility personnel to reduce and minimize the volume and toxicity of waste to be treated.

4.20 Pollution Prevention Opportunity Checklists

Examples of pollution prevention opportunity checklists are contained in the Appendix of this chapter to give you some ideas for developing your own checklist. Some of the items on the checklists may be obvious and merely good common sense, while others may need further investigation by facility personnel for appropriateness for your facility. The checklists are by no means all-inclusive; you should feel free to add items that apply to your industry or facility and remove items that do not.

The pollution prevention opportunities on the checklists cover four broad topic areas:

- Process Modification, Materials Substitution, and Product Reformulation
- Process Operation and Maintenance
- Materials Recycle, Reuse, and Recovery
- Good Housekeeping

Individual pollution prevention opportunities are shown in bold type and each is followed by a short explanation of the idea. The author of each checklist is also identified for those who are seeking additional information. Where appropriate, the key to pollution prevention for the particular industry is identified to help industrial facility personnel focus their efforts and resources.

The Appendix contains pollution prevention opportunity checklists for the following industries:

- Chemical Manufacturing
- Chemicals Formulating: Pesticides, Chemicals, and Paints
- Dairy Processing
- Dry Cleaning
- Fluid Milk Processing
- Meat Packing: Beef
- Metal Fabrication
- Metal Finishing
- Oil and Gas Extraction
- Petroleum Refining
- Photo Processing
- Printed Circuit Board Manufacturing
- Printing
- Pulp and Paper Manufacturing
- Radiator Repair

4.21 A Sample "Walk-Through" of the Pollution Prevention Opportunity Checklist for the Metal Finishing Industry

The keys to pollution prevention in metal finishing are to minimize chemical drag out, minimize water use for rinsing, and recover, reuse, and recycle plating chemicals. Substitution of less-toxic chemicals for cyanide and chromate should also be pursued.

Process modification offers the most significant reduction of waste. The checklist highlights efficient rinsing techniques like fog and spray rinses, and countercurrent dip rinses. The checklist also suggests ways to minimize drag out of plating chemicals from the plating bath to the rinse water. Other suggestions include mechanizing drag out control, providing a drip bar, installing static rinses and checking metal parts for pockets that may trap plating solutions.

Better process operation and maintenance helps to minimize the generation of wastes and reduce operating costs. The checklist identifies ways to keep the plating bath in good operating condition and to increase rinsing effectiveness while using less rinse water. Using deionized water, purer chemicals and bath filtering will help to reduce impurities in the plating bath. Flow control devices and conductivity sensors can help to reduce rinse water while ensuring adequate rinsing.

Materials recycling, reuse, and recovery help to reduce the wastes that need to be treated and, in some instances, produce by-products that may have economic value rather than waste sludge. Segregation of wastes is the key to effective recycling and reuse. The checklist identifies various separation and concentration techniques for material reuse and recovery. Plating out of metal ions onto metal sheets is also suggested as a means for reclamation.

Good housekeeping offers simple ideas for waste reduction that can be implemented relatively quickly and inexpensively. The checklist identifies spill containment and prevention, carefully planned chemical inventory, and other preventive steps as good housekeeping ideas that can be implemented easily.

The checklist offers a starting point to stimulate thinking about pollution prevention for your facility. Operators are encouraged to pursue other ideas appropriate for their facilities.

4.22 A Sample "Walk-Through" of the Pollution Prevention Opportunity Checklist for the Chemicals Formulating Industry

The keys to pollution prevention in chemicals formulation are better scheduling to minimize changeover, using better and alternative cleaning methods to minimize cleaning wastes, and avoiding the use of chlorinated solvents.

Process modification in cleaning will help minimize the volume of waste from cleaning. Jet spray, water knife, and steam cleaning methods will reduce the amount of rinse water requiring treatment and disposal. Dedicating transfer piping and mixing tanks also helps eliminate or reduce the need for cleaning. Process operation and maintenance procedures can be adjusted to optimize operation and minimize changeover. Opportunities for materials recycle, reuse, and recovery are also readily available. Rinse water can be mixed into the formulation with proper control, and leaks and spills from filling operations can be recovered for reformulation. Traditional good housekeeping techniques like spill prevention, inventory shelf life control, and the use of bulk containers may also help prevent wastes and pollution.

QUESTIONS

Write your answers in a notebook and then compare your answers with those on page 122.

4.2A What is the purpose of the pollution prevention opportunity checklists?

4.2B What are the keys to pollution prevention in metal finishing?

4.2C What are the keys to pollution prevention in the chemicals formulating industry?

Please answer the discussion and review questions next.

DISCUSSION AND REVIEW QUESTIONS
Chapter 4. PREVENTING AND MINIMIZING WASTES AT THE SOURCE

The purpose of these questions is to indicate to you how well you understand the material in the chapter. Write the answers to these questions in your notebook.

1. What is the meaning of "waste minimization"?
2. How can industrial waste treatment personnel be the "champion" for pollution prevention for their facilities?
3. List five ways pollution prevention could be economically beneficial to an industry.
4. What are four general types of activities in which good operating practices can help prevent pollution?
5. How should employees be trained with regard to off-specification process batches?
6. What are some possible disadvantages of replacing solvent-based cleaning processes with water-based cleaning processes?

SUGGESTED ANSWERS

Chapter 4. PREVENTING AND MINIMIZING WASTES AT THE SOURCE

Answers to questions on page 117.

4.0A Pollution prevention generally encompasses the following areas of endeavor:

1. Process modification, material substitution, and product reformulation.
2. Improved process operation and maintenance.
3. Good operating practices (good housekeeping).
4. Material recycle, reuse, and recovery for in-process use.

4.0B Industrial waste treatment practices focus on the reduction of pollutant concentrations in wastewater discharged from an industrial facility to minimize its impact on the wastewater treatment plant or the environment.

4.0C Municipal wastewater treatment agencies are hoping to use pollution prevention, in addition to pretreatment, to help meet regulations for biosolids reuse, effluent discharges, and air emissions.

4.0D In an industrial facility there are opportunities for pollution prevention in the areas of good housekeeping, process modification, material substitution, and improved operation and maintenance.

Answers to questions on page 119.

4.1A Good operating practices are procedures and policies that reduce the quantity or toxicity of wastewater. Good operating practices are also frequently referred to as best management practices or good housekeeping.

4.1B Factors that should be considered when purchasing raw materials include (1) packaging that will not have to be rinsed prior to disposal or reuse, (2) material that will not contribute to greater toxicity or quantity of wastewater, and (3) ordering materials as needed, rather than in bulk quantities that may never be used.

4.1C Factors that should be considered in raw material storage areas include (1) protection from rainwater intrusion, (2) inspection of storage containers for leaks, corrosion, dents, or other conditions that could lead to spills, and (3) secondary containment for liquids.

4.1D Material substitution can help minimize wastes by the replacement of one raw material with another raw material in order to reduce the toxicity of wastewater or reduce the quantity of wastewater generated.

4.1E Product reformulation minimizes wastes by changing a product in such a way that the process for the new product generates less wastewater or less-toxic wastewater.

Answers to questions on page 121.

4.2A The purpose of the pollution prevention opportunity checklists is to stimulate thinking on the part of industrial wastewater treatment facility personnel to reduce and minimize the volume and toxicity of waste to be treated.

4.2B The keys to pollution prevention in metal finishing are to minimize chemical drag out, minimize water use for rinsing, and recover, reuse, and recycle plating chemicals. Substitution of less-toxic chemicals for cyanide and chromate is also appropriate.

4.2C The keys to pollution prevention in the chemicals formulation industry are better scheduling to minimize changeover, using better and alternative cleaning methods to minimize cleaning wastes, and avoiding the use of chlorinated solvents.

APPENDIX

POLLUTION PREVENTION OPPORTUNITY CHECKLISTS FOR SELECTED INDUSTRIES

	Page
1. Chemical Manufacturing	124
2. Chemicals Formulating: Pesticides, Chemicals, and Paints	129
3. Dairy Processing	130
4. Dry Cleaning	132
5. Fluid Milk Processing	133
6. Meat Packing: Beef	135
7. Metal Fabrication	136
8. Metal Finishing	137
9. Oil and Gas Extraction	139
10. Petroleum Refining	140
11. Photo Processing	146
12. Printed Circuit Board Manufacturing	146
13. Printing	148
14. Pulp and Paper Manufacturing: Kraft Segment	149
15. Radiator Repair	150

124 Treatment Plants

Checklist 1. POLLUTION PREVENTION OPPORTUNITIES FOR THE CHEMICAL MANUFACTURING INDUSTRY*

The keys to pollution prevention are good operating practices and production process modifications. Wastes are usually generated from the mishandling of materials and the inadvertent production of off-spec materials.

Y/N	OPPORTUNITIES	COMMENTS
	I. Good Operating Practices in Material Input, Storage, and Handling	
___	Inventory control	First in, first out to prevent expiration
___	Designate material storage area	Provide protection, spill containment; keep area clean and organized; give one person the responsibility to maintain the area
___	Return obsolete materials to suppliers	Suppliers are the best persons to handle them
___	Segregate wastestreams, especially nonhazardous from hazardous	Prerequisite for recovery and reuse
___	Store packages properly and shelter from weather	To prevent damage, contamination, and product degradation
___	Prevent and contain spills and leaks by proper equipment maintenance and increased employee training and supervision	To prevent the generation of wastes
___	Minimize traffic through material storage area	To reduce contamination and dispersal of materials
___	Improve quality of feed by working with suppliers or installing purification equipment	Impurities in feedstream can be major contributors to waste
___	Reexamine need for each raw material	Need for a raw material that ends up as waste may be reduced or eliminated by modifying the process and control
___	Replace raw material containing hazardous ingredients with nonhazardous ones	To avoid the use of hazardous materials and the generation of hazardous wastes
___	Use off-spec material	Occasionally, a process can use off-spec material because the particular quality that makes the material off-spec is not important to the process
___	Improve product quality	Product impurities may be creating wastes at customers' plants; effect should be discussed with customers
___	Use inhibitors and continuously upgrade	Inhibitors prevent unwanted side reactions or polymer formation
___	Reformulate products from powder to pellet	To reduce dust emissions and waste generation
___	Reuse inert ingredients when flushing solids handling equipment	To minimize need for disposal
___	Change shipping containers, both for raw materials and products	To avoid disposal, change to reusable containers, tote bins, or bulk shipments
___	Recover product from tankcars and tank trucks	To minimize product drained from tanks going to waste
	II. Production Process Modifications	
	Reactors: The reactor is the heart of the process and can be a primary source for waste products. The quality of mixing is the key.	
___	Improve physical mixing in a reactor	Install baffles, a high RPM motor for the agitator, a different mixing blade design, multiple impellers, pump recirculation, or an in-line static mixer

Checklist 1. POLLUTION PREVENTION OPPORTUNITIES FOR THE CHEMICAL MANUFACTURING INDUSTRY*
(continued)

Y/N	OPPORTUNITIES	COMMENTS
___	Distribute feeds better for better yield and conversion, both for inlet and outlet	Add feed distributor to equalize residence time through fixed bed reactor to minimize under- and overreactions that form by-products
___	Improve ways reactants are introduced into the reactor	Get closer to the ideal reactant concentrations before the feeds enter the reactor to avoid secondary reactions that form unwanted by-products in the premixing of reactants
___	Improve catalyst and continuously upgrade	Catalyst has a significant effect on reactor conversion and product mix; changes in the chemical makeup of a catalyst, the method by which it is prepared, or its physical characteristics can lead to substantial improvements in catalyst life and effectiveness
___	Provide separate reactor for recycle streams	The ideal reactor conditions for converting reactor streams to usable products are different from those in the primary reactor; this separation affords optimization for both streams
___	Better heating and cooling techniques for reactors	To avoid hot spots that would give unwanted by-products
___	Consider different reactor design	The classic stirred-tank batch mix reactor is not necessarily the best choice. A plug flow reactor offers the advantage that it can be staged, and each stage can be run at different conditions for optimum product mix and minimum waste generation
___	Improve control to maintain optimal conditions in reactor	To increase yield and decrease by-product; at a minimum, stabilize conditions in reactor operation frequently if advanced computer control is not available
	Heat Exchangers: Heat exchangers can be a source of waste, especially with products that are temperature-sensitive. Reducing tube-wall temperature is the key.	
___	Use lower pressure steam	To reduce tube-wall temperature
___	Desuperheat steam	To reduce tube-wall temperatures and increase the effective surface area of the exchanger because the heat transfer coefficient of condensing steam is ten times greater than that of superheated steam
___	Install a thermocompressor	To reduce tube-wall temperature by combining high and low pressure steam
___	Use staged heating	To minimize degradation, staged heating can be accomplished first using waste heat, then low pressure steam, and finally, desuperheated high-pressure steam
___	Use on-line cleaning techniques for exchangers	Recirculating sponge balls and reversing brushes can be used to reduce exchanger maintenance, and also to keep the tube surface clean so that lower temperature heat sources can be used
___	Use scraped-wall exchanger	To recover saleable products from viscous streams, such as monomers from polymer tar
___	Monitor exchanger fouling	Sometimes an exchanger fouls rapidly when plant operating conditions are changed too fast or when a process upset occurs; monitoring can help to reduce such fouling
___	Use noncorroding tube	Corroded tube surfaces foul more quickly than noncorroded ones

Checklist 1. POLLUTION PREVENTION OPPORTUNITIES FOR THE CHEMICAL MANUFACTURING INDUSTRY*
(continued)

Y/N OPPORTUNITIES	COMMENTS
Pumps: Preventing leaks is the key.	
___ Recover seal flushes and purges	Recycle to the process where possible
___ Use sealless pumps	Use can-type or magnetically driven sealless pumps
Furnaces: Avoiding the hot tube-wall temperature is the key.	
___ Replace coil	Alternative designs should be investigated whenever replacement becomes necessary
___ Replace furnace with intermediate exchanger	Use a high temperature intermediate heat transfer fluid to eliminate direct heat
___ Use existing steam superheat	Sufficient superheat may be available to heat a process stream, avoiding exposure of the fluid to the hot tube-wall temperature of a furnace)
Distillation Column: A distillation column typically produces waste in three ways:	
• Allowing impurities to remain in a product from inadequate separation	
• Forming waste within the column itself through polymerization from the high reboiler temperature in the column	
• Losing products through venting or flaring from inadequate condensing	
___ Increase reflux ratio (if column capacity is adequate) for better separation	Increase the ratio by raising the pressure drop across the column and increasing the reboiler temperature using additional energy
___ Add section to column for better separation	The new section can have a different diameter and can use trays or high efficiency packing
___ Retray or repack column for better separation	Repack to lower pressure drop across a column and decrease the reboiler temperature; large-diameter columns have been successfully packed
___ Change feed tray for better separation	Match the feed conditions with the right feed tray in the column through valving changes
___ Insulate	Good insulation prevents heat losses and fluctuation of column conditions with weather
___ Improve feed distribution	Especially for a packed column
___ Preheat column feed	Preheating improves column efficiency and also requires lower temperatures than supplying the same heat to the reboiler; often the feed can be preheated by cross exchange with another stream
___ Remove overhead products from tray near top of column	To obtain a higher purity product if it contains a light impurity
___ Increase size of vapor line	To reduce pressure drop and decrease the reboiler temperature
___ Modify reboiler design	A falling film reboiler, a pumped recirculation reboiler, or high-flux tubes may be preferred to the conventional thermosiphon reboiler for heat-sensitive fluids

Checklist 1. POLLUTION PREVENTION OPPORTUNITIES FOR THE CHEMICAL MANUFACTURING INDUSTRY*
(continued)

Y/N	OPPORTUNITIES	COMMENTS
___	Reduce reboiler temperature	General temperature reduction techniques include using lower pressure steam or desuperheated steam, installing a thermocompressor, and using an intermediate transfer fluid
___	Lower column pressure	To decrease reboiler temperature; the overhead temperature, however, will also be reduced, which may create a condensing problem
___	Improve overhead condensers	To capture any overhead losses through retubing, condenser replacement or supplemental vent condenser addition
___	Improve column control	Similar to improving reactor control
___	Forward vapor overhead to the next column	Use a partial condenser and introduce the vapor stream to the downstream column
	Piping: A simple piping change can result in a major reduction of waste.	
___	Recover individual wastestream	Segregation is crucial for reuse
___	Avoid overheated lines	Review the amount and temperature of heat-sensitive materials in lines and in vessel tracing and jacketing
___	Avoid sending hot materials to storage	To prevent excessive venting and degradation of products
___	Eliminate leaks	To prevent waste generation
___	Change metallurgy or use lining	Metal may cause a color problem or act as a catalyst for the formation of by-products
___	Monitor major vents and flare system and recover vented products	Storage tanks, tankcars, and tank trucks are common sources of vented products; install a condenser or vent compressor for recovery
___	Consider "pipeless" batch processing	Reactants are transported in process vessels, eliminating the need for pipe cleaning and providing better prevention of contamination and greater flexibility in scheduling
	Process control: Modern technology allows computer control system to respond more quickly and accurately than human beings.	
___	Improve on-line control	Good process control reduces waste by optimizing process conditions and reducing plant trips and wastes
___	Optimize daily operation	A computer can be programmed to analyze the process continually and optimize the conditions to prevent waste
___	Automate start-ups, shutdowns, and product changeover	To bring the plant to stable conditions quickly to minimize the generation of off-spec wastes
___	Program plant to handle unexpected upsets and trips	To minimize downtime, spills, equipment loss, and waste generation
	Miscellaneous:	
___	Avoid unexpected trips and shutdowns	A good preventive maintenance program, adequate spare equipment, and adequate warning system for critical equipment

Checklist 1. POLLUTION PREVENTION OPPORTUNITIES FOR THE CHEMICAL MANUFACTURING INDUSTRY*
(continued)

Y/N	OPPORTUNITIES	COMMENTS
___	Use wastestreams from other plants	Internal waste exchanges are feasible, but wastestreams should be adequately characterized
___	Reduce number and quantity of samples	Review sampling frequency and procedure and recycle the samples
___	Find a market for waste product	Wastes can be converted to saleable by-products with additional processing and creative salesmanship
___	Install reusable insulation	Particularly effective on equipment where the insulation is removed regularly to perform maintenance
	III. Good Operation and Maintenance Practices for Equipment Cleaning and Changeover	
___	Avoid unnecessary equipment cleaning	Explore the feasibility of eliminating cleaning step between batches
___	Maximize equipment dedication	Dedicating tanks to one product will reduce clean-out and save time and labor cost for changeover
___	Recover more products	Scraping down tanks, pigging, or blowing lines can recover more product and reduce wastes
___	Use less cleaner	High-pressure sprays, pressurized air, steam, and heated cleaning bath can reduce the amount of cleaner used and disposed of as waste
___	Reuse cleaner	Reclaim and reuse cleaner if feasible
___	Consider alternative cleaning methods and less hazardous cleaners	Mechanical cleaning, such as plastic media blasting and ultrasonic cleaning, together with more biodegradable cleaner, can reduce waste volume and toxicity
___	Standardize cleaning products used in plant	To maximize recovery potential

* Prepared by Philip Lo, Industrial Waste Section, County Sanitation Districts of Los Angeles County.

ACKNOWLEDGMENT: Materials for production process modification were adapted from Ken Nelson, Dow Chemicals USA, "Use These Ideas to Cut Wastes," *HYDROCARBON PROCESSING*, March 1990.

Checklist 2. POLLUTION PREVENTION OPPORTUNITIES FOR THE CHEMICALS FORMULATING INDUSTRY*

Including Pesticides, Chemicals, and Paints

PROCESS MODIFICATION	PROCESS OPERATION AND MAINTENANCE	MATERIAL RECYCLE, REUSE, AND RECOVERY	GOOD HOUSEKEEPING
Use continuous processes Batch processes involve more frequent mix tank cleaning. **Use pumps and pipes** For raw material and product transfer, use closed systems as much as possible. Reduces potential for spills. **Segregate wastestreams** Increases recovery potential and treatment efficiency. Segregate water-based streams to reduce amount of solvent to wastewater treatment system. **Dedicate equipment** Reduces need for tank rinsing between batches. **Substitute nonhazardous raw materials** Avoids the production of hazardous wastes.	**Raw material purity** Use high-quality raw material in batch to minimize contamination and reduce waste. **Use wiper blades on mix tanks** Physically wiping down sides of mix tanks will reduce amount of rinse water required. **Perform preventive maintenance** Routinely check for leaks in valves and fittings. Repair immediately. **Optimize inventory and production schedule using computer** Minimizes need for changeover and consolidates batch production. **Rinsing** 1. Use jet sprays with pressure booster pump Reduces amount of required rinse water. 2. Use water knife spray Reduces amount of required rinse water. 3. Use steam cleaner Reduces waste use.	**Install drip pan for filling line** Recover and recycle product to filling reservoir. **Reuse mix tank rinse water in next process batch** **Reuse floor wash water** Treat and reuse as wash water or equipment rinse water (including empty drum rinsing). Treatment technologies include chemical precipitation, biological treatment, activated carbon adsorption, air or steam stripping, hydrolysis, chemical oxidation, and resin adsorption. **Reuse container rinse water** Treat and reuse both solvent and water rinses for containers. **Recover materials for reuse from wastestreams** Separate and concentrate materials using membrane separation and evaporative technologies. **Collect and reuse stormwater** Reuse as floor wash water.	**Control inventory** Do not allow material to exceed shelf life. Use materials on a first-in, first-out basis. Do not get rid of expired products by discharging to wastewater treatment system. **Use mop floor washing** Mops and squeegees reduce amount of wash water required. **Reduce use of containers** Less water will be required for rinsing. Have suppliers deliver materials in tank trucks directly to on-site storage tanks, or use returnable tote bins. **Buy appropriate amounts** Buy materials in small quantities if only small amounts are required. Savings on large quantity purchases can be lost if unused material must be disposed of or is discharged to laboratory sink drains. **Manage laboratory samples** Do not allow concentrated lab samples to be discharged to wastewater treatment system. Recycle into process. **Cover outdoor storage** Divert clean stormwater away from material storage and handling areas. **Install spill containment** Spills can be contained and managed appropriately rather than draining to wastewater treatment system and causing system upsets.

* ACKNOWLEDGMENT: Materials for production process modification were adapted from Ken Nelson, Dow Chemicals USA, "Use These Ideas to Cut Wastes," *HYDROCARBON PROCESSING*, March 1990. Reviewed by Charlie Henderson, Rohm and Haas Southern California Incorporated.

Checklist 3. POLLUTION PREVENTION OPPORTUNITIES FOR THE DAIRY PROCESSING INDUSTRY*

Including Fluid Milk, Spread Processing, Cheese/Whey Processing, Process Cheese Manufacturing

PROCESS MODIFICATION	PROCESS OPERATION AND MAINTENANCE	MATERIAL RECYCLE, REUSE, AND RECOVERY	GOOD HOUSEKEEPING
Install water desludging system on milk and whey separators instead of desludging with product Most applicable for non-fluid milk separators (check with USDA). Reduces amount of product to drain. **Review product formulation** Look at all ingredients that make up the product. Evaluate each for impacts from delivery, handling, spillage, cleanup, and ultimate disposal. Consider alternative materials that may have less environmental impact without compromising product quality. **Dedicate equipment as much as possible** Reduces the need for frequent cleanings between batches. For example, a bottling plant could have separate process and filling lines for milk system and juice system.	**Automate cheese salting process** Reduces amount of human error, spillage, and overuse of salt resulting in additional salt drippings for disposal. **Review cleaning chemicals** Check to see environmental impacts of cleaning chemicals. Can alternative chemicals be used that have less environmental impact without compromising cleaning ability? For example, substitute nitric acid for phosphoric acid to reduce plant phosphorus discharge. **Integrate environmental considerations into all aspects of processing** Remember there are always two product streams, one to market, the other to "drain" or disposal; both are equally important and must be considered in making production decisions. Know your processes well; then you can find ways to reduce pollution. **Use high-quality raw materials** So batches will not become contaminated and have to be managed as a waste. **Perform preventive maintenance** Routinely check for leaks in valves and fittings. Repair immediately. **Install CIP (Cleaning In Place) monitoring systems on automated CIP processes** Monitor flow, time, temperature, and conductivity. Can provide information leading to reduction of water and cleaning chemicals, optimization of system.	**Install product reclaim system in milk intake (receiving station)** System involves automated CIP (Cleaning In Place) system, which uses potable water for initial truck, milk line, and silo rinses. **Product reclaim** "Burst" rinse truck tanks with potable water and chase milk/water slurry to collection tank. **Product reclaim** Air blow milk lines to silo and "burst" flush. Chase milk/water slurry to collection tank. **Product reclaim** In non-fluid milk applications, pump water/milk slurry (from collection tank) to milk silos (check with USDA). **Product reclaim** As milk silo is emptied, "burst" rinse silo with potable water and pump milk/water slurry to collection tank. In non-fluid milk applications, pump milk/water slurry from collection tank to manufacturing milk silo (check with USDA). **Install membrane systems for cleaning salt brines** Reduces the number of times salt brine is discharged. The retentate is segregated for alternative disposal. **Install "product reclaim" systems on condensed whey storage and loadout system** Very similar to milk receiving product reclaim system for storage tanks, whey concentrate lines, and truck tanks. **Collect separate "desludge" for alternative disposal** If handled properly, this high-strength waste could be used as a supplement to animal feed, for example, hog feed.	**Covered outdoor storage** Divert clean stormwater away from material storage and handling areas. **Install spill containment** Spills can be contained and managed appropriately rather than draining to wastewater treatment system and causing system upsets. **Spill cleanup procedures** Establish procedures for what to do with a spill. Reduces chance of spill being discharged to wastewater treatment plant. **Provide slope for milk trucks during unloading** Trucks must be emptied as completely as possible before washing. This reduces the potential for product going to the drain. Can be done with ramps, or actually sloping the floor (1/4" per foot is desirable). **Provide high-pressure nozzles on all water hoses** Allows for adequate cleaning with less water. Also, automatically shuts off water stream when not in use. **Perform dry sweepings before floor washing** Picks up loose bits of cheese curd and powder so that they would not get washed down the drain. **Provide drip pans** Place in areas of leaky valves, seals, barrel, or block draining areas to collect product wastes rather than having them run directly onto the floor and the floor drains.

Checklist 3. POLLUTION PREVENTION OPPORTUNITIES FOR THE DAIRY PROCESSING INDUSTRY* (continued)

PROCESS MODIFICATION	PROCESS OPERATION AND MAINTENANCE	MATERIAL RECYCLE, REUSE, AND RECOVERY	GOOD HOUSEKEEPING
	Install computer-controlled processing systems Reduces the potential for "bad" runs that will subsequently need to be disposed of. **Adopt a definite waste prevention program and build an educational program** Helps to ensure that all plant personnel are aware of waste prevention concerns. **Instruct plant personnel completely in proper operation and handling of all dairy plant processing equipment** Major losses are due to poorly maintained equipment and to negligence of inadequately trained and insufficiently supervised personnel. **Repair or replace all worn out and obsolete equipment or parts of equipment, including sanitary valves, fittings, and pumps** To the extent possible, drips and leaks occurring during the processing run should be collected in containers and not allowed to go down floor drains. **Install suitable liquid level controls** Install controls with automatic pump stops, alarms, and other devices at all points where overflows could occur, such as storage tanks, processing tanks, filler bowls, and Cleaning In Place (CIP) tanks. **Use care in materials handling** Avoid spillage of cased, canned, or barreled dairy products and product ingredients.	**Reuse "cow" water (or condensate of whey) instead of potable water for the following applications (USDA dependent):** • Boiler Feed Waste • Floor Washes • External Truck Washes • CIP (Cleaning In Place) • Makeup Water for Intermediate Rinses • Pump Seal Water **Explore possible reverse osmosis polishing of cow water with subsequent disinfection** Could lead to further uses of water in place of well or city water. **Process salt drippings with nanofiltration membrane system** USDA approval needed. Can "desalt" salt drippings, which can then be blended back into sweet whey. **Install reclaim systems for acid/caustic** Can use simple gravity separation techniques, such as cone bottom tanks, to segregate solids and decant chemicals for next day's first wash. Eliminates the need to dump all chemicals each day. **Look at feasibility of dryer, scrubber water reuse** Segregate and reuse this material rather than dumping to drain. Need to review any reuse process with USDA. **Perform dry cleanup on dryers before wet wash** Remove and collect as much dry product as possible before wet washing. Can be used as animal feed. **Study the plant and develop a material balance** Determine where losses occur and take steps to modify and replace unsatisfactory equipment. Where improper maintenance is the cause of losses, a specific maintenance program could be instigated and maintained.	**Screen removable traps in floor drains** By providing screened traps, large particulates are able to be collected for alternative disposal or reuse as "fish bait" or animal feed. Screened traps also prevent undesirable materials from entering the process waste system, such as pump seals, bags, string, and gloves. **Check raw product quality, for example, antibiotics, in small quantity increments** Sample milk on delivery and evaluate for regulated constituents before mixing with large bulk quantities to be processed. Minimizes large quantities of raw product needing disposal. **Develop procedures for handling returned product** By-product outlets minimize waste hauling.

* Prepared by Rob Lamppa, Engineering Department, Land O'Lakes, Inc., Arden Hills, MN 55126.

Checklist 4. POLLUTION PREVENTION OPPORTUNITIES FOR THE DRY CLEANING INDUSTRY*

The keys are to ensure proper PERC (perchloroethylene) separation in the water separator and to warn against illegal disposal of still bottom residuals to the sewer. The preferable program is zero discharge to sewer, with the water and still bottom transported for off-site reclamation by a contract service.

PROCESS MODIFICATION	PROCESS OPERATION AND MAINTENANCE	MATERIAL RECYCLE, REUSE, AND RECOVERY	GOOD HOUSEKEEPING
Convert to "dry-to-dry" machine Reduces solvent vapor losses.	(Mainly for minimizing air emissions)	**Contract collection of cartridge filter, separator water, and still bottom for off-site reclamation** Provides for PERC reclamation and condensate treatment under proper expert supervision.	
Install solvent leak detectors Monitor for vapor losses.	**Keep lids on containers** Reduces evaporation and spills.		
Use refrigeration/condensation system Reduces vapor losses.	**Label all raw material containers** Prevents unnecessary disposal.	**Redistill still bottom with more water following boil-down** Recovers more solvent and reduces solvent content in wastestream.	
Use a reclaiming dryer Reduces solvent losses.	**Store containers shake-proof** Prevents spills in an earthquake.	**Use cartridge stripper to remove solvent from cartridge** Recovers more solvent.	
Redesign separator with baffles and decant taps Reduces PERC entrainment and affords better decanting of water.	**Replace seals regularly on dryer, deodorizer, and aeration valves** Reduces emission leaks.		
	Replace door gasket on button trap		
Allow only batch discharge of decant water from separator after visual inspection Check for inadequate separation before sewer discharge.	**Replace gaskets around cleaning machine door or tighten enclosure**		
	Repair holes in air and exhaust ducts		
	Secure hose connects and couplings		
Steam strip cleaning filter cartridge Permits recovery of solvent.	**Clean lint screens** Avoids clogging fans and condensers.		
Substitute low temperature laundering for dry cleaning for applicable fabric Avoids unnecessary PERC use.	**Open button traps and lint gaskets only long enough to clean**		
	Check baffle assembly in cleaning machine biweekly		
	Check air relief valves for proper enclosure		
	Adjust IN and OUT condensing coil temperatures on heater to within 10°F of each other		

* Prepared by Philip Lo, Industrial Waste Section, County Sanitation Districts of Los Angeles County, Whittier, CA.

ACKNOWLEDGMENT: Some materials were adapted from the Alaskan Health Project, "Waste Reduction Tips For Dry Cleaners," 1987.

Checklist 5. POLLUTION PREVENTION OPPORTUNITIES FOR THE FLUID MILK PROCESSING INDUSTRY*

PROCESS MODIFICATION	PROCESS OPERATION AND MAINTENANCE	MATERIAL RECYCLE, REUSE, AND RECOVERY	GOOD HOUSEKEEPING
Dedicate equipment as much as possible Reduces the need for frequent cleanings between batches. For example, a bottling plant could have separate process and filling lines for milk system and juice system.	**Review cleaning chemicals** Check to see environmental impacts of cleaning chemicals. Can alternative chemicals be used that have less environmental impact without compromising cleaning ability? For example, substitute nitric acid for phosphoric acid to reduce plant phosphorus discharge. **Integrate environmental considerations into all aspects of processing** Remember there are always two product streams, one to market, the other to "drain" or disposal; both are equally important and must be considered in making production decisions. Know your process well; then you can find ways to reduce pollution. **Perform preventive maintenance** Routinely check for leaks in valves and fittings. Repair immediately. **Install CIP (Cleaning In Place) monitoring systems on automated CIP processes** Monitor flow, time, temperature, and conductivity. Can provide information leading to reduction of water and cleaning chemicals, optimization of system. **Install computer-controlled processing system** Reduces the potential for "bad" runs that will subsequently need to be disposed of. **Adopt a definite waste prevention program and build an educational program** Helps to ensure that all plant personnel are aware of waste prevention concerns.	**Install an air blow system in milk intake (receiving station)** System involves automated CIP (Cleaning In Place) system that uses pressurized air to chase milk to storage silos. This removes as much product as possible before cleaning. **Collect separator "desludge" for alternative disposal** If handled properly, this high-strength waste could be used as a supplement to animal feed, for example, hog feed. **Install reclaim systems for acid/caustic** Can use simple gravity separation techniques, such as cone bottom tanks, to segregate solids and decant chemicals for next day's first wash. Eliminates the need to dump all chemicals each day. **Study the plant and develop a material balance** Determine where losses occur and take steps to modify and replace unsatisfactory equipment. Where improper maintenance is the cause of losses, a specific maintenance program could be instigated and maintained.	**Install spill containment** Spills can be contained and managed appropriately rather than draining to wastewater treatment system and causing system upsets. **Spill cleanup procedures** Establish procedures for what to do with a spill. Reduces chance of spill being discharged to wastewater treatment plant. **Provide slope for milk trucks during unloading** Trucks must be emptied as completely as possible before washing. This reduces the potential for product going to the drain. Can be done with ramps or by actually sloping the floor (1/4" per foot is desirable). **Provide high-pressure nozzles on all water hoses** Allows for adequate cleaning with less water. Also, automatically shuts off water stream when not in use. **Provide drip pans** Place in areas of leaky valves and seals to collect product wastes rather than have these run directly on the floor and to the floor drains. **Check raw product quality, for example, antibiotics, in small quantity increments** Sample milk on delivery and evaluate for regulated constituents before mixing with large quantities to be processed. Minimizes large quantities of raw product needing disposal. **Develop procedures for handling return product** By-product outlets minimize waste hauling.

Prevention 133

Checklist 5. POLLUTION PREVENTION OPPORTUNITIES FOR THE FLUID MILK PROCESSING INDUSTRY* (continued)

PROCESS MODIFICATION	PROCESS OPERATION AND MAINTENANCE	MATERIAL RECYCLE, REUSE, AND RECOVERY	GOOD HOUSEKEEPING
	Instruct plant personnel completely in proper operation and handling of all dairy plant processing equipment Major losses are due to poorly maintained equipment and to negligence of inadequately trained and insufficiently supervised personnel. **Install suitable liquid level controls** Use controls with automatic pump stops, alarms, and other devices at all points where overflows could occur, such as storage tanks, processing tanks, filler bowls, and Cleaning In Place (CIP) tanks. **Use care in materials handling** To avoid spillage of cased or packaged dairy products and product ingredients. **Repair or replace all worn out and obsolete equipment or parts of equipment including sanitary valves, fittings, and pumps** To the extent possible, drips and leaks occurring during the processing run should be collected in containers and not allowed to go down floor drains.		

* Prepared by Rob Lamppa, Engineering Department, Land O'Lakes, Inc., Arden Hills, MN 55126.

Checklist 6. POLLUTION PREVENTION OPPORTUNITIES FOR THE MEAT PACKING (BEEF) INDUSTRY*

The keys to pollution prevention are to prevent product and contaminants from entering the wastestream and to reduce water use to a minimum.

MANAGEMENT'S ROLE	PROCESS MODIFICATION	GOOD HOUSEKEEPING	PRETREATMENT OPTIONS
Maintain motivation and support The success or failure of a program will depend on management's attitudes and actions.	**Allow sufficient time for the carcass to bleed after slaughter** Collect as much blood as possible at the slaughter site rather than allowing the blood to enter the packing line.	**Dry clean as much as possible before washdown** Do not use the hose as a broom. When solid material is kept out of the water stream, treatment costs are less because the solids do not have to be removed from the wastewater stream.	**Flow equalization** **Screening** Static, vibrating, and rotary screens can be used to capture solid particles. **Centrifuges** **Grease and suspended solids separation** Settling basins and dissolved air flotation, along with electrocoagulation and lignosulfate, can be used to separate grease and suspended solids from the wastewater. **Biological processes** Aerobic methods include basins, trickling filters, and contact stabilization. Anaerobic methods include basins and digesters.
Educate the employees Provide instruction on proper water use, waste management, and cleaning procedures. Update training on a regular basis.	**Collect the blood** Use troughs and curbs where necessary to direct the flow of blood.	**Maintain a water record system** Monitor water use, and waste load and wastewater discharge quantities. Use only as much water as necessary.	
Appoint a water/waste supervisor Allow the supervisor(s) to control water use.	**Dry clean the paunch** Remove and collect paunch contents.	**Do not let blood coagulate** Coagulated blood requires large amounts of water for removal.	
Develop a job description for all personnel Make waste management and waste reduction part of the job description.	**Transfer paunch contents using a dry system** Use a screw conveyor or an air-energized system.	**Trim all loose particles from the carcass** By removing blood clots from the neck area and loose particles from the carcass, you can prevent these by-products from entering the wastestream when the carcass is rinsed with a high-pressure washer.	
Develop a preventive maintenance program Maintain your plant at maximum efficiency to reduce waste.	**Collect all trimmings** Do not let solids enter the wastestream. Keep all by-products off the floor.	**Segregate and collect all by-products** This includes not only the slaughterhouse but also the holding pens and the retail cutting room.	
Monitor cleanup Strive for an efficient, environmentally conscious sanitation crew.	**Use a high-pressure, low-volume water system for all wet cleaning** Adjust the pressure as needed so minimum water pressure and water volume are used for any cleaning operation.	**Continuously look for areas in the plant where waste reduction can be implemented.** Observing separate processing areas helps you gain insight into how each area contributes to overall waste production. These observations provide opportunities for integrating water use reduction with solid waste reduction.	
Plan a system for by-product recovery Your wastes can become someone else's resources.	**Install mechanical and automatic controls where necessary** In addition to employee monitoring, mechanical and automatic controls can assist in reducing waste.		
Conduct planning sessions Get management and employees together to discuss suggestions for waste reduction.		**Use the correct detergent for cleaning** Using the correct detergent in the correct amount allows for cleaning with minimal rinsing.	

* Prepared by John Polanski, Research Assistant, Minnesota Technical Assistance Program (MnTAP), University of Minnesota, Suite 207, 1313 Fifth Street S.E., Minneapolis, MN 55414 and Dr. Roy E. Carawan, Ph.D., Professor, Department of Food Science, North Carolina State University, Cooperative Extension Service, Box 7624, 129 Schaub Hall, Raleigh, NC 27695-7624. Printed with permission of the Minnesota Technical Assistance Program and North Carolina State University. © Copyright 1993.

Checklist 7. POLLUTION PREVENTION OPPORTUNITIES FOR THE METAL FABRICATION INDUSTRY*

PROCESS MODIFICATION	PROCESS OPERATION AND MAINTENANCE	MATERIAL RECYCLE, REUSE, AND RECOVERY	GOOD HOUSEKEEPING
Change to UV-cured coatings Eliminates use of carrier solvents, maximizes paint transfer efficiency, and minimizes overspray wastes.	**Optimize bath concentrations** Only replace plating chemical when necessary. Lengthens bath life.	**Recycle metalworking fluids** Extend usable life of fluid through filtration, skimming, dissolved air flotation, coalescing, hydrocycloning, centrifugation, and pasteurization.	**Seal and wiper replacement** Reduces chance of oil contamination of metalworking fluid if seal or wiper should fail.
Change to powder coatings Eliminates use of carrier solvents, maximizes paint transfer efficiency, and overspray powder can be collected, filtered, and reused.	**Agitate rinse bath** Agitation promotes better rinsing. Agitate water or part.	**Reuse high-performance fluids** Reuse hydraulic fluids that no longer meet spec for less stringent spec cutting oils.	**Keep fluids from floor drains** Do not allow discharge of spills or spent fluids to sewer. Eliminate floor drains if necessary.
Change to synthetic fluids Synthetic fluids are less susceptible to contamination, therefore have a longer useful life.	**Lengthen drag out time** Allows more chemical to drip back to process tank, so reduces the amount of chemical introduced in rinse water.	**Reuse secondary rinse** Reuse second rinse as primary rinse or makeup for cleaning solutions.	**Install drain boards or drip guards** Boards and guards minimize spillage between tanks and are sloped away from rinse tanks so drag out fluids drain back to plating tanks.
Change to gas coolant Use a gas coolant for certain applications instead of a liquid.	**Establish drag out timing** Post drag out times at tanks to remind employees.	**Reuse deionized rinse water** Depending on product, this rinse water can be reused in a plating bath as evaporated water makeup.	
Mechanize drag out Eliminates possibility of employee using too short a drag out time, maintains product QA/QC standards if timing is set properly.	**Use foot pump or photosensor to activate rinse** These items allow use of water only when processing parts. A photosensor may be used on automatic plating lines.	**Ion exchange on rinse water** Ion exchange can be used to concentrate metals in rinse waters and metal can be recovered from the ion exchange acid regenerant stream.	
Install drip bars Drip bars allow personnel to drain part hands-free without waiting so personnel will not use too short a drag out time.	**Use demineralized water** For mixing purposes, use of high-quality water mitigates contamination problems of machining fluid.	**Reverse osmosis** Concentrate drag out for reuse in plating bath; the water stream can also be reused.	
Reduce pockets on parts Place parts on drag out rack to minimize chances of chemical pooling in corners or in other pockets.	**Install bath filter** Filter can remove particulates and trace contaminant organics in the process bath, lengthens bath life. Use a filter that can be unrolled, cleaned, and reused.	**Electrodialysis** Recover chromium from hard chromium plating baths and rinse waters.	
Use countercurrent rinses These rinses dramatically reduce the amount of water required for rinsing and, therefore, reduce the amount of wastewater to be treated or sent for metal recovery.	**Extend life of fluid** Clean fluid through filtration and clarification and use of specialized biocides.	**Electrowinning** Recover metals from spent plating baths or ion exchange acid regenerant streams.	
	Raw material purity Use high-quality raw materials in bath so bath will not become contaminated as quickly.	**Evaporation** Concentrate drag out for reuse; the water condensate can also be reused.	

Prevention 137

Checklist 7. POLLUTION PREVENTION OPPORTUNITIES FOR THE METAL FABRICATION INDUSTRY* (continued)

PROCESS MODIFICATION	PROCESS OPERATION AND MAINTENANCE	MATERIAL RECYCLE, REUSE, AND RECOVERY	GOOD HOUSEKEEPING
Use spray or fog rinsing Reduces rinse water quantity required and can also be used over plating baths.		**Reuse mild acid rinse water** Use mild acid rinse water as influent to rinse following alkaline cleaning bath. Improves efficiency of rinse so less rinse water is required. **Reuse alkaline rinse water** Reuse rinse water from an alkaline cleaner operation to rinse parts from an acid cleaning operation. **Reuse spent acid/alkaline** Spent acid can be used to neutralize an alkaline wastestream. Spent alkali can be used to neutralize an acid wastestream.	

* Prepared by Alison Gemmell, Del Monte Foods, Walnut Creek, CA.

Checklist 8. POLLUTION PREVENTION OPPORTUNITIES FOR THE METAL FINISHING INDUSTRY*

PROCESS MODIFICATION	PROCESS OPERATION AND MAINTENANCE	MATERIAL RECYCLE, REUSE, AND RECOVERY	GOOD HOUSEKEEPING
Use spray or fog rinsing Reduces rinse water amount required and can also be used over plating baths. **Use countercurrent rinses** These rinses dramatically reduce the amount of water required for rinsing and therefore reduce the amount of wastewater to be treated or sent for metal recovery. **Mechanize drag out** Eliminates possibility of employee using too short a drag out time, maintains product QA/QC standards if timing is set properly. **Convert to dry floor** Reduces chances of spills reaching floor drains or causing upset in wastewater pre-treatment plant.	**Increase bath temperature** Evaporates bath water so relatively clean waste rinse water can be reused as bath makeup water. Reduces solution viscosity so more chemical drains back to process tank during drag out. Do not use on cyanide or hexavalent chromium baths. **Use deionized (DI) water** Use DI water in plating baths, static rinses, and, if practical, in running rinses. DI water reduces impurities in the plating bath to extend its life and minimizes the precipitation of minerals in water as sludge. **Raw material purity** Use high-quality raw materials in bath so bath will not become contaminated as quickly.	**Segregate wastestreams** Increases recovery and treatment technology efficiencies. Acidic/alkaline. Chrome/nonchrome. Concentrated/dilute. Chelated/nonchelated. Cyanide/noncyanide. **Reuse deionized rinse water** Depending on product, this rinse water can be reused in a plating bath as evaporated water makeup. **Ion exchange on rinse water** Ion exchange can be used to concentrate metals in rinse waters and metal can be recovered from the ion exchange acid regenerant stream. **Reverse osmosis** Concentrate drag out for reuse in plating bath; the water stream can also be reused.	**Control inventory** Do not allow material to exceed shelf life. Use materials on a first-in, first-out basis. **Install spill containment** Spills can be contained and managed. Reduces wastewater treatment upsets. **Spill cleanup procedures** Establish procedures for what to do with a spill. Reduces chance of spill being discharged to wastewater treatment plant or environment. **Buy appropriate amounts** Buy materials in small quantities if only small amounts are required. **Perform preventive maintenance** Routinely check for leaks in valves and fittings. Repair immediately.

138 Treatment Plants

Checklist 8. POLLUTION PREVENTION OPPORTUNITIES FOR THE METAL FINISHING INDUSTRY* (continued)

PROCESS MODIFICATION	PROCESS OPERATION AND MAINTENANCE	MATERIAL RECYCLE, REUSE, AND RECOVERY	GOOD HOUSEKEEPING
Use different process Replace toxic cadmium plating with relatively nontoxic aluminum ion vapor deposition to achieve metal hardening properties.	**Optimize bath concentrations** Only replace plating chemical when necessary. Lengthens bath life.	**Reuse spent acid/alkaline** Spent acid can be used to neutralize an alkaline wastestream. Spent alkali can be used to neutralize an acid wastestream.	**Establish drag out timing** Post drag out times at tanks to remind employees.
Eliminate cyanide baths Change to a noncyanide plating bath. Alternative chemistries are available with the exception of copper strike.	**Reduce bath dumps** Optimize bath operation so bath dumps are infrequent.	**Electrowinning** Recover metals from spent plating baths or ion exchange acid regenerant streams.	**Cover outdoor storage** Divert clean stormwater away from storage areas.
Install drip bars Drip bars allow personnel to drain part hands-free without waiting so personnel will not use too short a drag out time.	**Eliminate intermittent jobs** Stop performing small plating operations that generate intermittent wastestreams that personnel are not familiar with treating.	**Evaporation** Concentrate drag out for reuse; the water condensate can also be reused.	**Install drain boards or drip guards** Boards and guards minimize spillage between tanks and are sloped away from rinse tanks so drag out fluids drain back to plating tanks.
Use static rinses Static rinses usually follow the plating bath and capture the most concentrated drag out for return to the plating bath or for metal recovery.	**Agitate rinse bath** Agitation promotes better rinsing. Agitate water or part.	**Electrodialysis** Recover chromium from hard chromium plating baths and rinse waters.	
Reduce pockets on parts Place parts on drag out rack to minimize chance of chemical pooling in corners or in other pockets.	**Lengthen drag out time** Allows more chemical to drip back to process tank, so reduces the amount of chemical introduced in rinse water.	**Reuse mild acid rinse water** Use mild acid rinse water as influent to rinse following alkaline cleaning bath. Improves efficiency of rinse so less rinse water is required.	
	Use foot pump or photosensor to activate rinse These items allow use of water only when processing parts. A photosensor may be used on automatic plating lines.		
	Install bath filter Filter can remove particulates and trace contaminant organics in the process bath; lengthens bath life. Use a filter that can be unrolled, cleaned, and reused.		
	Use conductivity sensor This sensor gives an indication of the cleanliness of the rinse water. Sensor can be designed to trigger clean rinse water flow when the tank water gets too dirty. Also allows better QA/QC.		
	Install flow restrictors		
	Install flow control meter		

* Prepared by Alison Gemmell, Del Monte Foods, Walnut Creek, CA.

Checklist 9. POLLUTION PREVENTION OPPORTUNITIES FOR THE OIL AND GAS EXTRACTION INDUSTRY*

The keys are to maximize oil separation and minimize spillage. Fail-safe devices are necessary since many of these discharge locations are unattended.

Y/N	OPPORTUNITIES	COMMENTS
___	Install free water knock-out tank with sufficient detention time and pressure relief valve	To maximize oil/water separation, especially at high flow
___	Install inlet and outlet baffles in gravity separation tank	To prevent short-circuiting
___	Install high-level alarm, remote dialer, and pump shutoff in separation tank	To prevent tank overflow and to alert operator at remote location
___	Install oil/water interface sensor at the bottom of the oil retention baffle of the separation tank, or install automatic oil skimming equipment	To prevent over-accumulation of oil and carryover to the water discharge
___	Install heater treater if appropriate	To enhance oil/water separation, especially for low-viscosity crude oil
___	Install elevated "gooseneck" surge box at discharge end of the separation tank	To prevent accidental spill of floating oil in tank from excessive drawdown of water
___	Provide steady flow from tank to secondary oil removal units like dissolved air flotation unit or WEMCO depurator	Flotation units work best with uniform flow rate and oil and grease content of a few hundred mg/L
___	Provide buffer storage capacity between separation device and the discharge point to the sewer	To allow for additional storage safeguards for system malfunction, especially for unattended locations, until the next operator visit
___	Add polymer and adjust dosage frequently	To enhance oil and water separation
___	Cover separator and flotation units and capture volatile organics emissions through vacuum suction	Volatile organics may be removed through chilling, activated carbon, or catalytic oxidation
___	Install explosimeter, if available, or oil sensor, together with auto shutoff valve for sewer discharge and flow diversion; provide temporary holding capability	To minimize oil discharge to sewer in the event of accidental spill or overloading of the flotation unit
___	Store treated water for reuse	Use treated water for reinjection
___	Discharge brine after further treatment to salt water channel, if available	High salt content in oil field brine water may affect reuse of treated municipal wastewater

* Prepared by Philip Lo, Industrial Waste Section, County Sanitation Districts of Los Angeles County, Whittier, CA.

Checklist 10. POLLUTION PREVENTION OPPORTUNITIES FOR THE PETROLEUM REFINING INDUSTRY*

The keys to pollution prevention for the petroleum refining industry are, for the short term, waste segregation, good operating practices, and source control for oily wastes. For the medium term, the driving force is probably product reformulation, which has resulted in production changes in meeting limitations for airborne toxic compounds and vapor pressure in fuel products. For the longer term, the keys may be more targeted hydrocarbon rebuilding and reforming to produce the desirable fuel components, while avoiding the undesirable toxic ones. More specifically, removal of undesirable precursors in reforming, isomerization, catalytic conversions, and expanded use of hydrogenation may hold the most promise.

(The following checklist presents a compilation of pollution prevention opportunities. However, since every refinery is unique, some of the opportunities may be more applicable to one refinery than to another. Please use the checklist with caution.)

Y/N	OPPORTUNITIES	COMMENTS
	I. Good Operating Practices	
___	Specify lower bottom sludge and water content for crude oil supply	To reduce wastes in storage tank and desalter through improved separation of water and bottom sludge at extraction
___	Use recycled water as makeup water for crude desalter	Recycled water quality is sufficient for desalting
___	Reroute desalter water with emulsifiers to intermediate tankage	To minimize emulsifier carryover to API separator
___	Segregate and dispose of ballast water to salt water channel, if available	To minimize brine contamination of treated water for reuse
___	Eliminate moisture contact with oxygenates (MTBE and methanol) in storage and process	To minimize water contamination of oxygenates
___	Replace desalting with an aggressive chemical treatment system for applicable situation, through crude oil dehydration in tankage with emulsion breaker, chloride reduction with caustic injection, ammonia replacement with neutralizing amine, film inhibitor feed rate optimization, and anti-foulant injection to debutanizer heat exchanger (*Oil & Gas Journal*, 3/20/1989, pg. 60)	To eliminate desalter water blowdown, which could be high in benzene and emulsifiers, while maintaining corrosion protection
___	Segregate and discharge blowdown and water treatment regenerate to salt water channel or truck to ocean outlet, if available	To minimize brine contamination of treated water for reuse
___	Use corrosion-resistant lines in storage and slop oil tanks	To minimize sludge formation and need for tank cleaning
___	Install agitator in crude oil storage tanks	To minimize sludge accumulation
___	Avoid high shear pumping of oily wastes; use Archimedean screw pumps as appropriate	To minimize emulsion formation
___	Install tank cover and seal	To minimize emission loss and moisture entry
___	Upgrade to non-leaking pump seals	To eliminate leak losses
___	Replace packed pump with mechanical seal pump	To eliminate leak losses
___	Install sealless pump	To eliminate leaks and fugitive emissions, though such pumps have limited service
___	Maintain pump seals regularly	To prevent leaks
___	Recycle seal flushes and purges	To minimize wastes for treatment
___	Recycle seal and bearing cooling water stream	To minimize wastes for treatment

Checklist 10. POLLUTION PREVENTION OPPORTUNITIES FOR THE PETROLEUM REFINING INDUSTRY*
(continued)

Y/N	OPPORTUNITIES	COMMENTS
___	Pave process area	To minimize dirt entry to sewer
___	Install cover for sewer drain	To minimize dirt entry to sewer
___	Load and unload catalytic fines in closed system	To prevent fines from becoming wastes
___	Recover coke fines for sale with coke	To prevent solids entry to sewer
___	Reuse recycled water for washdown if quality is desirable	To minimize need for discharge
___	Integrate process units to pass processing streams from one unit to the next, if appropriate	To avoid intermediate tankage but may lose operational flexibility
___	Blend fuels in-line	To avoid blending tankage
___	Install closed-loop sampling system	To flush materials back to the tank or pipeline and minimize volatile compound emissions and hydrocarbon discharge to wastewater
___	Use computer software to track all hazardous materials and wastes	To better manage virgin materials and wastestreams
___	Return oily wastewater and sludge from distribution and sales terminals to refinery as permitted by federal and state recycling regulations	To afford proper handling of oily wastes
___	Segregate scrap metals for sale	To reclaim metals for reuse
___	Recondition valves and vessels for reuse	To further reduce scrap metal wastes
___	Recover and reuse sandblasting grit as blasting media or as a light aggregate in concrete production	To minimize need for grit disposal, but beware of lead and heavy metal contamination
	Stormwater Management	
___	Selectively cover loading rack and process areas to divert rainwater	To preclude rainwater contamination
___	Segregate stormwater collection system from process drainage	To prevent cross contamination of stormwater
___	Impound rainwater in collection basin or tank as appropriate	To hold water pending determination of treatment needs
___	Sweep streets and provide weirs in catch basin to exclude dirt	To prevent dirt entry to storm drain
___	Keep tank farm and process area clean, including secondary containment areas	To avoid contaminating rainwater
___	Reuse rainwater after gravity recovery of oil and solids	To minimize need for discharge
___	Discharge rainwater to public storm drain system under NPDES permit	To avoid using sewer capacity
___	Dike process area that drains to stormwater collection system as appropriate	To prevent contamination of stormwater
___	Regularly clean out drainage system to remove accumulated dirt	To minimize contamination of stormwater

Checklist 10. POLLUTION PREVENTION OPPORTUNITIES FOR THE PETROLEUM REFINING INDUSTRY*
(continued)

Y/N	OPPORTUNITIES	COMMENTS
	Firefighting Water and Spillage Management	
___	Install tank overfill prevention system	To prevent spills
___	Pave areas under pipe rack	To facilitate leak detection
___	Contain spillage with diking and absorbent materials	To minimize spreading of spillage
___	Recover and reuse spillage	To minimize need for disposal
___	Impound firefighting water in rainwater basins or storage tanks as appropriate	To hold and test before discharge or reuse
___	Prevent automatic overflow of storm drain to wastewater collection system	To prevent spills and firefighting water that entered the storm drain system from overwhelming the wastewater treatment system
	Groundwater and Contaminated Soil Cleanup	
___	Recover floatable oil for reuse	To recover oil at source and avoid entrainment in transport
___	Pretreat and reinject treated groundwater if appropriate	To eliminate need for discharge to sewer
___	Reuse hydrocarbon contaminated soil as filler in asphalt paving manufacture	To avoid need for disposal
___	Reuse soil with mineral contents similar to shale as raw material substitute for cement kiln; reuse in pre-heater and calciner kiln is preferred, to maximize volatile hydrocarbon destruction	To avoid need for disposal
II.	**Production Process Modifications**	
	Separation Process	
___	Improve separation in distillation column through various means including the following:	To increase yield and the separation of volatiles, for example, benzene
	• Increase the reflux ratio • Add a new section to the column • Match feed condition with the right feed tray • Preheat column feed • Install reusable insulation to prevent heat loss and fluctuation of column condition with weather	
___	Lower the reboiler temperature in distillation column through various means including the following:	To minimize degradation and waste generation from high reboiler temperature
	• Retray column to lower pressure drop • Increase size of vapor line to reduce pressure drop • Use lower pressure steam or desuperheated steam • Install a thermocompressor • Lower column pressure	
___	Improve overhead condensers to capture overhead losses	To minimize flaring and emissions

Checklist 10. POLLUTION PREVENTION OPPORTUNITIES FOR THE PETROLEUM REFINING INDUSTRY*
(continued)

Y/N	OPPORTUNITIES	COMMENTS
	Conversion and Upgrading Processes	
___	Improve conversion in reactors through various means including the following:	To improve yield and conversion and minimize the formation of undesirable compounds from side reactions
	• Distribute feeds better at inlets and outlets • Upgrade catalysts continuously • Provide separate reactor for recycled streams for more ideal reactor conditions • Adjust heating and cooling to avoid hot spots • Improve control to maintain optimum conditions in reactor • Use inhibitors to minimize unwanted side reactions	
___	Filter catalyst fines from decanter oil from the fluid catalytic cracking unit (FCCU)	To recover and reuse catalyst
___	Reclaim hydroprocessing catalysts for metals and alumina	To recover the metals on the catalysts like cobalt and molybdenum, as well as those removed from oil like nickel and vanadium; the alumina carrier is also recovered
___	Recycle catalyst for bauxite in cement manufacturing	To minimize need for disposal
___	Recover fluoride from spent caustics from an HF alkylation process by calcium precipitation	To produce calcium fluoride solids for use in cement industry or as fluxing agent in glass and steel industries
___	Reuse spent fluidized catalytic cracking unit (FCCU) catalysts in residue FCCU	To reuse catalysts in another FCCU where higher metal content on the catalysts can be tolerated
___	Reactivate catalysts for reuse	To reuse catalysts after the nickel and vanadium deposits are removed
___	Regenerate spent sulfuric acid by commercial reclaimer using incineration	To regenerate the acid and avoid neutralization
___	Reclaim extraction solvents like sulfolane and sulfinol	To recover solvents for reuse, with the residuals going for feed to a sulfuric acid plant because of their high BTU and sulfur contents
	Product Treatment	
___	Minimize the amount of caustic and rinse water used for product treatment through better contacting and recycling	To minimize need for treatment of wastewater
___	Consider hydrotreating for pollutant removal	To eliminate the use of caustic and water in product treatment
___	Send spent caustics to reclaimer	To reclaim cresylic and naphthenic compounds for sale
___	Reuse spent sulfuric caustics for paper manufacturing	To reuse the caustics if the strength is high enough
___	Regenerate clay from jet fuel filtration by washing with naphtha and drying by steam heating and feeding to furnace	To recycle filter clay

Checklist 10. POLLUTION PREVENTION OPPORTUNITIES FOR THE PETROLEUM REFINING INDUSTRY*
(continued)

Y/N	OPPORTUNITIES	COMMENTS
	Equipment Cleaning—Heat Exchangers	
___	Use lower pressure steam	To reduce tube-wall temperature and sludge formation
___	Desuperheat steam	To reduce tube-wall temperature and increase the effective surface area of the exchanger because the heat transfer coefficient of condensing steam is ten times greater than that of superheated steam
___	Install a thermocompressor	To reduce tube-wall temperature by combining high- and low-pressure stream
___	Use staged heating	To minimize degradation, staged heating can be accomplished first using waste heat, then low-pressure steam and finally, desuperheated high-pressure steam
___	Use on-line cleaning techniques for exchangers	Recirculating sponge balls and reversing brushes can be used to reduce exchanger maintenance and also to keep the tube surface clean so that lower temperature heat sources can be used
___	Use noncorroding tube	Corroded tube surfaces foul more quickly than uncorroded ones
	Waste Gas Treatment	
___	Regenerate di-ethanol-amine (DEA) using slip stream filtration in addition to carbon filtration	To remove degradation products and prolong DEA life
___	Substitute Sulften Sulfur Recovery Process for Beavon Process	To avoid generation of spent Stretford Solution, which contains vanadium
___	Regenerate activated carbon from gas scrubbing	To avoid need for disposal
	Wastewater and Sludge Treatment	
___	Add forebay skimming for API separator	To recover more hydrocarbons for recycle
___	Use floating roof on treatment tanks and drains	To minimize air emissions
___	Use pressurized air in flotation	To minimize air emissions
___	Pretreat desalter water blowdown before co-mingling with other oily wastes, using absorption with light oil, or stripping with steam, nitrogen, methane, or vacuum	To pretreat the high concentration of benzene and, possibly, emulsifiers in the desalter water blowdown
___	Thicken sludge in sludge tank and decant supernatant	To aid in sludge dewatering
___	Treat sludge with heat and chemicals to release more oil and water	To further reduce hydrocarbon content in sludge
___	Dewater sludge to cake form	To minimize water content and remove some oil
___	Reclaim hydrocarbons in sludge by feeding it to a delayed coker that produces fuel grade coke	To dispose of solids and to reclaim hydrocarbon value
___	Use solvent extraction to remove hydrocarbons from sludge	To treat sludge for disposal and recover hydrocarbons

Checklist 10. POLLUTION PREVENTION OPPORTUNITIES FOR THE PETROLEUM REFINING INDUSTRY*
(continued)

Y/N	OPPORTUNITIES	COMMENTS
___	Use high-temperature sludge drying to desorb hydrocarbons	To treat sludge for disposal and recover hydrocarbons
___	Feed sludge cake to cement kiln for energy recovery	To recycle sludge for its energy value
___	Evaluate gasification of oily wastes	To convert waste to usable methane
	Utility Production—Steam, Hydrogen	
___	Use closed-loop cooling water system	To minimize water loss
___	Demineralize cooling tower feedwater	To reduce cleaning and waste generation
___	Use polymers for boiler feedwater treatment	To reduce boiler cleaning
___	Collect condensate for reuse	To avoid sewer discharge
___	Use non-chromate corrosion inhibitor	To minimize chromate emissions and also chromate treatment in blowdown
___	Reclaim hydrogen plant catalysts	To recover materials in catalysts
III.	**Product Reformulation and Material Substitution**	
___	Reformulate leaded gasoline to unleaded alternative with MTBE	To eliminate lead from gasoline and product storage tanks
___	Reduce benzene and other volatile hydrocarbons in gasoline through reblending with oxygenates like MTBE	To decrease emissions of air toxics and smog-forming volatile organics

* Prepared by Philip Lo, Industrial Waste Section, County Sanitation Districts of Los Angeles County, Whittier, CA.

Checklist 11. POLLUTION PREVENTION OPPORTUNITIES FOR THE PHOTO PROCESSING INDUSTRY*

PROCESS MODIFICATION	PROCESS OPERATION AND MAINTENANCE	MATERIAL RECYCLE, REUSE, AND RECOVERY	GOOD HOUSEKEEPING
Change to silver-less film Silver does not end up in rinse waters or spent fix baths when this type of film is used. Examples include vesicular, diazo, and electrostatic films.	**Adjust replenishment** Adjust chemical replenishment rates and wash water flow rates on photoprocessor to optimize bath life and reduce wastewater quantity.	**Metal replacement canister** Install canister on rinse waters and on effluent of wastewater electrowin to recover silver from solution and reduce toxicity of wastewater.	**Control inventory** Do not allow material to exceed shelf life and then have to be discarded as waste. Use materials on a first-in, first-out basis. Do not discharge expired products to wastewater treatment system.
Add ammonium thiosulfate Addition of this chemical to silver-contaminated bath extends the useful life of the bath.		**Install electrowin** Install electrowin unit on first rinse and developer wastestreams to recover silver from solution and reduce toxicity of wastewater.	**Keep lids on solutions** In storage, keep lids on bulk solutions to prevent oxidation and contamination.
Add acetic acid Acetic acid added to the fix bath keeps the pH low to maximize soluble complexes, therefore, extend bath life.		**Install rinse water recycling** Reduces wastewater.	**Spill containment** Store chemicals and metal recovery system in double-wall tanks or in diked area to prevent accidental discharges.
Purchase new machines Purchase new developing machines that use less rinse water (for example, countercurrent rinsing) or have squeegees or air blades to reduce drag out from chemical baths to rinse waters.		**Recover fix bath silver** Install electrowin unit on fix bath of photoprocessor. Extends bath life.	
		Segregate wastestreams Spent fix baths should be segregated from rinse waters and developer solutions because silver recovery is more efficient on the more concentrated spent wastestream.	

* Prepared by Alison Gemmell, Del Monte Foods, Walnut Creek, CA.

Checklist 12. POLLUTION PREVENTION OPPORTUNITIES FOR THE PRINTED CIRCUIT BOARD MANUFACTURING INDUSTRY*

PROCESS MODIFICATION	PROCESS OPERATION AND MAINTENANCE	MATERIAL RECYCLE, REUSE, AND RECOVERY	GOOD HOUSEKEEPING
Mechanize drag out Eliminates possibility of employee using too short a drag out time, maintains product QA/QC standards if timing is set properly.	**Lengthen drag out time** Allows more chemical to drip back to process tank, so reduces the amount of chemical introduced in rinse water.	**Segregate wastestreams** Increases recovery and treatment technology efficiencies. Acidic vs. alkaline. Concentrated metal (spent baths) vs. dilute metal (rinse water streams). Chelated vs. non-chelated streams.	**Install drain boards or drip guards** Boards and guards minimize spillage between tanks and are sloped away from rinse tanks so drag out fluids drain back to plating tanks.
Reduce pockets on parts Place parts on drag out rack to minimize chances of chemical pooling in corners or in other pockets.	**Establish drag out timing** Post drag out times at tanks to remind employees.	**Reuse deionized rinse water** Depending on product, this rinse water can be reused in a plating bath as evaporated water makeup.	**Spill cleanup procedures** Establish procedures for what to do with a spill. Reduces chance of spill being discharged to wastewater treatment plant.
Use static rinses Static rinses usually follow the plating bath and capture the most concentrated drag out for return to the plating bath or for metal recovery.	**Use conductivity sensor** This sensor gives an indication of the cleanliness of the rinse water. Sensor can be designed to trigger clean rinse water flow when the tank water gets too dirty.	**Ion exchange on rinse water** Ion exchange can be used to concentrate metals in rinse waters and metal can be recovered from the ion exchange acid regenerant stream.	**Perform preventive maintenance** Routinely check for leaks in valves and fittings. Repair immediately.
	Use foot pump or photosensor to activate rinse These items allow use of water only when processing parts. A photosensor may be used on automatic plating machines.		**Control inventory** Do not allow material to exceed shelf life. Use materials on a first-in, first-out basis.

Checklist 12. POLLUTION PREVENTION OPPORTUNITIES FOR THE PRINTED CIRCUIT BOARD MANUFACTURING INDUSTRY* (continued)

PROCESS MODIFICATION	PROCESS OPERATION AND MAINTENANCE	MATERIAL RECYCLE, REUSE, AND RECOVERY	GOOD HOUSEKEEPING
Use countercurrent rinses These rinses dramatically reduce the amount of water required for rinsing and therefore reduce the amount of wastewater to be treated or sent for metal recovery. **Use spray or fog rinsing** Reduces rinse water quantity required and can also be used over plating baths. **Eliminate chelated baths** Change to a nonchelated plating bath to improve metal wastewater treatment. Chelated streams make it difficult to precipitate metal in wastewater treatment system. **Buy efficient etch machine** An efficient etch machine results in less copper in rinse water. **Convert to dry floor** Reduces chance of spills reaching floor drains or causing upset in wastewater pretreatment plant. **Electroless tanks** Use continuous passivation on stainless-steel components in electroless plating tanks to prevent copper plate-out. Copper plate-out needs to be stripped with nitric acid. Reduces amount of spent nitric acid that needs to be treated.	**Agitate rinse bath** Agitation of water or part promotes better rinsing. **Reduce bath dumps** Optimize bath operation so bath dumps are infrequent. **Install flow restrictors** Restrictors automatically reduce the amount of rinse water, so operators do not need to adjust inlet valves. **Install drip bars** Drip bars allow personnel to drain parts hands-free without waiting, so personnel will not use too short a drag out time. **Install bath filters** Filters can remove particulates and trace contaminant organics in the process bath. Lengthens bath life. Use filter that can be unrolled, cleaned, and reused. **Raw material purity** Use high-quality raw materials so bath will not become contaminated as quickly. **Increase bath temperature** Evaporates bath water so relatively clean waste rinse water can be reused as bath makeup water. Reduces solution viscosity, so more chemical drains back to process tank during drag out. **Use deionized (DI) water** Use DI water in plating baths, static rinses, and, if practical, in running rinses. DI water reduces impurities in the plating bath to extend its life and minimizes the precipitation of minerals in water as sludge. **Optimize bath concentrations** Only replace plating chemical when necessary. Lengthens bath life. Replace chemicals on electroless copper baths to extend bath life. Automate chemical replacement through on-line analyzers and chemical flowmeters.	**Reuse spent acid/alkaline** Spent acid can be used to neutralize an alkaline wastestream. Spent alkali can be used to neutralize an acid wastestream. **Reverse osmosis** Concentrate drag out for reuse in plating bath; the water stream can also be reused. **Evaporation** Concentrate drag out for reuse; the water condensate can also be reused. **Electrowinning** Recover metals from spent plating baths or ion exchange acid regenerant streams. **Recover particulate copper** On debur operation, recover particulate copper, using centrifuge or paper filter. Reuse water. **Reuse mild acid rinse water** Use mild acid rinse water as influent to rinse following alkaline cleaning bath. Improves efficiency of rinse so less rinse water is required. **Recover copper sulfate** On microetch line, recover copper sulfate crystals directly from etch tank and reuse crystals in copper electroplating baths. **Reclaim etchant** Send etchant to an off-site reclaimer instead of treating etchant in wastewater treatment system. **Recycle photoresist stripper** Decant spent photoresist stripper from polymer residue and recycle stripper. Do not discharge spent stripper to wastewater treatment system.	**Buy appropriate amounts** Buy materials in small quantities if only small amounts are required. **Cover outdoor storage** Divert clean stormwater away from storage areas. **Install spill containment** Spills can be contained and managed. Reduces wastewater treatment upsets.

* Prepared by Alison Gemmell, Del Monte Foods, Walnut Creek, CA.

Checklist 13. POLLUTION PREVENTION OPPORTUNITIES FOR THE PRINTING INDUSTRY*

PROCESS MODIFICATION	PROCESS OPERATION AND MAINTENANCE	MATERIAL RECYCLE, REUSE, AND RECOVERY	GOOD HOUSEKEEPING
Change to silver-less film Silver does not end up in rinse waters or spent fix baths when this type of film is used. Examples include vesicular, diazo, and electrostatic films. **Change to ultraviolet inks** Will not dry in ink fountain overnight. Reduces need for fountain cleaning. **Install laser platemaking** Eliminates need for photoprocessing. Expensive. **Install electronic imaging** Eliminates need for photoprocessing. Expensive. **Use flexographic process** Replaces metal etch processes for plate processing. **Purchase new machines** 1. Purchase new developing machines that use less rinse water (for example, countercurrent rinsing) or have squeegees or air blades to reduce drag out from chemical baths to rinse waters. 2. Purchase waterless paper and film developing machines to reduce the volume of fix waste. **Use non-dry aerosol** Special non-drying aerosol materials can be sprayed on ink fountains to keep them from drying out overnight. Fewer ink fountain cleanings required. **Add acetic acid** Acetic acid added to the fix bath keeps the pH low to maximize soluble complexes, therefore extend bath life. **Add ammonium thiosulfate** Addition of this chemical to silver-contaminated bath extends the useful life of the bath.	**Adjust replenishment** Adjust chemical replenishment rates and wash water flow rates on photoprocessor to optimize bath life and reduce wastewater quantity. **Run similar jobs at once** Minimizes need for cleaning between jobs.	**Recycle inks** Recycle inks to make black ink instead of discharging to sewer. **Segregate wastestreams** Spent fix baths should be segregated from rinse waters and developer solutions because silver recovery is more efficient on the more concentrated spent fix wastestreams. **Install rinse water recycling** Reduces wastewater. **Metal replacement canister** Install canister on rinse waters and on effluent of wastewater electrowin to recover silver from solution and reduce toxicity of wastewater. **Recover fix bath silver** Install electrowin unit on fix bath of photoprocessor. Extends bath life. **Install electrowin** Install electrowin unit on first rinse and developer wastestreams to recover silver from solution and reduce toxicity of wastewater.	**Control inventory** Do not allow material to exceed shelf life and then have to be discarded as waste. Use materials on a first-in, first-out basis. Do not discharge expired products into wastewater treatment system. **Return used ink to vendor** Purchase ink from distributors who will take back unused or spent ink so ink will not be discharged to the sewer. **Keep lids on solutions** In storage, keep lids on bulk solutions to prevent oxidation and contamination. **Minimize spills** Use dry method cleanups. **Spill containment** Store chemicals in diked area to prevent accidental discharges.

* Prepared by Alison Gemmell, Del Monte Foods, Walnut Creek, CA.

Checklist 14. POLLUTION PREVENTION OPPORTUNITIES FOR THE KRAFT SEGMENT OF THE PULP AND PAPER INDUSTRY*

The keys to pollution prevention for the kraft segment of the pulp and paper industry are to purchase preservative-free wood fiber, maintain and operate chippers to produce chips of uniform dimensions, use chlorine dioxide for pulp bleaching, and reuse treated wastewaters as makeup water for the log flumes.

PROCESS MODIFICATION	PROCESS OPERATION AND MAINTENANCE	MATERIAL RECYCLE, REUSE, AND RECOVERY	GOOD HOUSEKEEPING
Process pulp with oxygen in an alkaline environment After conventional or extended pulping processes but prior to bleaching with chlorine or chlorine derivatives.	**Operate recovery boilers within selected operating guidelines** Minimizes total reduced sulfur emissions. Reduces atmospheric emissions of total reduced sulfur. Improves boiler efficiency.	**Reuse treated wastewater as makeup water for the log flumes** Conserves water and reduces discharge loadings of total suspended solids and BOD_5.	**Install retaining walls to contain chips and fugitive dusts from the chip piles** Minimizes chip and dust losses.
Purchase wood fiber (rough wood chips, sawdust) that has not been treated with wood preservatives, particularly chlorophenols Improves product and effluent quality.	**Use relatively low charges of anthraquinone in digester** Accelerates pulping process and increases pulp yield. Reduces effluent color and BOD_5.	**Recycle scrubber water to the causticizing circuit**	**Modify barge unloading operations to use front loader instead of clam shell** Minimizes chip and dust losses.
Modify existing (continuous) digesters or install new (batch or continuous) digesters to produce brownstock pulp of lower Kappa number Accomplishes greater degree of delignification with recoverable pulping chemicals as opposed to delignification with disposable bleach plant chemicals that need to be disposed of.	**Use high-shear mixers for bleach plant chemical additions** Ensures efficient chemical application and minimizes localized overchlorination, which generates unwanted chlorinated compounds.	**Collect and recover steam condensates from pulp dryers and paper machines**	**Designate material storage area** Provide protection, spill containment; keep area clean and organized; give one person the responsibility to maintain the area.
Steam strip digester, evaporator, and turpentine condensates for the removal of total reduced sulfur and BOD_5 (methanol, acetone) Steam-to-feed ratios of 15 to 20 percent are necessary to achieve efficient BOD_5 removal. Stripper overheads are combusted in power boilers. Stripper bottoms are reused for brownstock washing.	**Maintain and operate chippers to produce chips of uniform dimensions with emphasis on chip thickness control** Improves yield and reduces rejects. More uniform pulping occurs, which results in lower bleach plant chemical consumption.	**Reuse acid stage filtrates as dilution and wash water on first bleaching stage and on log sequence bleach lines, and use filtrate from the second extraction stage in the first extraction stage**	**Store packages properly and shelter from weather** Prevents damage, contamination, and product degradation.
Convert recovery boilers to a low odor design by the elimination of direct contact evaporator, modification of secondary and tertiary combustion air systems, and installation of a new economizer section	**Use instrumentation to monitor and control the application of chlorine or chlorine dioxide in the first bleaching stage** Maintain chlorine multiple or Kappa factor within ranges where chlorinated compounds are minimized.		**Minimize traffic through material storage area** Reduces contamination and dispersal of materials.
Use dry drum debarkers instead of hydraulic debarkers	**Install a total reduced sulfur control system for collection and combustion of vent streams from brownstock washers, foam tanks, black liquor filters, oxidation tanks, and storage tanks**		
Install chip screening device designed to produce chips of uniform thickness	**Install new or modify existing brownstock pulp screening and deknotting systems so that black liquid is recovered and recycled to recovery system**		
Use chlorine dioxide instead of elemental chlorine in the first bleaching stage			

* Compiled by Mischelle Mische and Philip Lo, County Sanitation Districts of Los Angeles County, Whittier, CA. Information taken from US EPA Region X's "Model Pollution Plan for the Kraft Segment of the Pulp and Paper Industry," September 1992, EPA#910/9-92-030.

Checklist 15. POLLUTION PREVENTION OPPORTUNITIES FOR THE RADIATOR REPAIR INDUSTRY*

The keys are to minimize drag in and drag out of contaminants to and from the caustic boil-out tank and test tank, thereby reducing the amount of contaminants discharged with the flush booth water. Zero discharge to the sewer is possible; however, additional treatment and specialized equipment may be required. Waste minimization techniques for zero discharge may be somewhat different.

PROCESS MODIFICATION	PROCESS OPERATION AND MAINTENANCE	MATERIAL RECYCLE, REUSE, AND RECOVERY	GOOD HOUSEKEEPING
Use a low-zinc flux Reduces zinc level in discharge; however, it may not produce a satisfactory soldering job. Experiment first. **Use smaller process tanks with ultrasonic cleaning** Reduces volume of waste generated. **Use smaller test tanks** For efficient operation to minimize volume of wastewater.	**Remove as much oil as possible from oil cooler using compressed air** Minimizes drag in to boil-out tank. **Remove and recycle antifreeze, if any** Minimizes drag in. **Use higher pressure and lower water flow in flushing booth** Minimizes water use and need for disposal. **Provide hang bars over caustic boil-out tank and test tank** Allows convenient draining. **Provide drain board between tanks** Minimizes spillage to floor. **Blow out residual test tank or caustic solution to tank using compressed air** Minimizes drag out to flushing booth. **No soldering over the test tank** Provide drip pans to collect excess solder or flux to minimize zinc and lead contamination of test water. **Pre-rinse radiator over boil-out tank using fog spray** Minimizes drag out. **Maintain and monitor boil-out tank** Minimizes need for tank changing. **Upon changing of tank, remove the solids, reuse the liquid, and reconstitute the bath** Minimizes volume of waste upon bath changing.	**Filter solids and reuse test tank water when bath is cloudy** Minimizes tank changing. **Reuse flushing booth rinses** Use for tank makeup to minimize water use. **Minimize use of cleaner or flux containing metal chelating compounds** Reduces interference with wastewater treatment.	**Secondary containment** For test tank, boil-out tank and sludge storage area to prevent spills and leaks from contaminating discharge.

CHAPTER 5

INDUSTRIAL WASTE MONITORING

by

Larry Bristow

TABLE OF CONTENTS

Chapter 5. INDUSTRIAL WASTE MONITORING

	Page
OBJECTIVES	155
WORDS	156

LESSON 1

5.0	NEED FOR INDUSTRIAL WASTE MONITORING		159
	5.00	Objectives of Industrial Waste Monitoring	159
	5.01	Importance of a Monitoring Program	159
	5.02	Objectionable Characteristics of Industrial Wastes	159
5.1	ADMINISTRATION OF A MONITORING PROGRAM		160
	5.10	Organization	160
	5.11	The Database	160
	5.12	Legal Authority	160
	5.13	Dealing With Industry	161
5.2	HOW TO MONITOR DISCHARGES		161
	5.20	Monitoring Programs	161
		5.200 Self-Monitoring	161
		5.201 Continuous Monitoring	161
	5.21	Monitoring Provisions	162
	5.22	Need for Representative Samples	162
	5.23	Sampling Points	162
	5.24	Types of Samples	162
		5.240 Grab Samples	162
		5.241 Composite Samples	163
	5.25	Portable Equipment	163
	5.26	Handling Portable Samplers	163
	5.27	Equipment Problems	166
	5.28	Maintenance	170
DISCUSSION AND REVIEW QUESTIONS			170

LESSON 2

5.3	LOCATING SOURCES OF DISCHARGES		171
	5.30	"Prospecting" in the Sewers	171
	5.31	Identifying Waste Materials	171

5.4	SAMPLE PRESERVATION AND SECURITY		171
	5.40	Storage Time	171
	5.41	Storage Temperature	171
	5.42	Chemical Preservation	171
	5.43	Security	171
5.5	FLOW METERING		175
	5.50	Need for Accurate Measurements	175
	5.51	Water Meters	175
5.6	SAFETY		175
	5.60	Traffic Safety	175
	5.61	Confined Spaces	179
	5.62	Equipment Storage Areas	179
	5.63	Battery Charging	179
5.7	MONITORING STRATEGIES		179
	5.70	Development of Strategies	179
	5.71	Regulation of High Flows	180
	5.72	Decisions on Industrial Discharges	180
	5.73	Reactions to Industrial Discharges	181
	5.74	Warning Systems	181
5.8	ADDITIONAL READING		181
DISCUSSION AND REVIEW QUESTIONS			184
SUGGESTED ANSWERS			184
APPENDIX			187
	A	Industrial Sewer-Use Permit Application	187
	B	Sample Sewer-Use Ordinance	197
	C	Standard Form A—Municipal	204
	D	Safety Orders for Battery Charging	206

OBJECTIVES
Chapter 5. INDUSTRIAL WASTE MONITORING

Following completion of Chapter 5, you should be able to:

1. Develop an industrial waste monitoring program.
2. Justify an industrial waste monitoring program.
3. Administer and manage the program.
4. Locate sources of industrial waste discharges.
5. Collect and preserve representative samples.
6. Document the "chain of custody" of your sampling program.
7. Develop a monitoring strategy.
8. Conduct your duties in a safe manner.

WORDS
Chapter 5. INDUSTRIAL WASTE MONITORING

CHAIN OF CUSTODY

A record of each person involved in the handling and possession of a sample from the person who collected the sample to the person who analyzed the sample in the laboratory and to the person who witnessed disposal of the sample.

COMPOSITE (PROPORTIONAL) SAMPLE

A composite sample is a collection of individual samples obtained at regular intervals, usually every one or two hours during a 24-hour time span. Each individual sample is combined with the others in proportion to the rate of flow when the sample was collected. Equal volume individual samples also may be collected at intervals after a specific volume of flow passes the sampling point or after equal time intervals and still be referred to as a composite sample. The resulting mixture (composite sample) forms a representative sample and is analyzed to determine the average conditions during the sampling period.

CONFINED SPACE

Confined space means a space that:

(1) Is large enough and so configured that an employee can bodily enter and perform assigned work; and

(2) Has limited or restricted means for entry or exit (for example, manholes, tanks, vessels, silos, storage bins, hoppers, vaults, and pits are spaces that may have limited means of entry); and

(3) Is not designed for continuous employee occupancy.

Also see DANGEROUS AIR CONTAMINATION and OXYGEN DEFICIENCY.

DANGEROUS AIR CONTAMINATION

An atmosphere presenting a threat of causing death, injury, acute illness, or disablement due to the presence of flammable and/or explosive, toxic, or otherwise injurious or incapacitating substances.

(1) Dangerous air contamination due to the flammability of a gas, vapor, or mist is defined as an atmosphere containing the gas, vapor, or mist at a concentration greater than 10 percent of its lower explosive (lower flammable) limit (LEL).

(2) Dangerous air contamination due to a combustible particulate is defined as a concentration that meets or exceeds the particulate's lower explosive limit (LEL).

(3) Dangerous air contamination due to the toxicity of a substance is defined as the atmospheric concentration that could result in employee exposure in excess of the substance's permissible exposure limit (PEL).

NOTE: A dangerous situation also occurs when the oxygen level is less than 19.5 percent by volume (OXYGEN DEFICIENCY) or more than 23.5 percent by volume (OXYGEN ENRICHMENT).

GRAB SAMPLE

A single sample of water collected at a particular time and place that represents the composition of the water only at that time and place.

OSHA (O-shuh)

The Williams-Steiger Occupational Safety and Health Act of 1970 (OSHA) is a federal law designed to protect the health and safety of workers, including the operators of water supply and treatment systems and wastewater collection and treatment systems. The Act regulates the design, construction, operation, and maintenance of water and wastewater systems. OSHA regulations require employers to obtain and make available to workers the Material Safety Data Sheets (MSDSs) for chemicals used at industrial facilities and treatment plants. OSHA also refers to the federal and state agencies that administer the OSHA regulations.

OXYGEN DEFICIENCY

An atmosphere containing oxygen at a concentration of less than 19.5 percent by volume.

OXYGEN ENRICHMENT

An atmosphere containing oxygen at a concentration of more than 23.5 percent by volume.

PERISTALTIC (PAIR-uh-STALL-tick) PUMP

A type of positive displacement pump.

REFRACTORY (re-FRACK-toe-ree) MATERIALS

Materials difficult to remove entirely from wastewater, such as nutrients, color, taste- and odor-producing substances, and some toxic materials.

REPRESENTATIVE SAMPLE

A sample portion of material, water, or wastestream that is as nearly identical in content and consistency as possible to that in the larger body being sampled.

SLUG

Intermittent release or discharge of wastewater or industrial wastes.

TRUNK SEWER

A sewer line that receives wastewater from many tributary branches and sewer lines and serves as an outlet for a large territory or is used to feed an intercepting sewer. Also called MAIN SEWER.

CHAPTER 5. INDUSTRIAL WASTE MONITORING

(Lesson 1 of 2 Lessons)

5.0 NEED FOR INDUSTRIAL WASTE MONITORING

The rapid advance of industrial technology has added new concerns for the wastewater treatment field. Chemical compounds of increased complexity and quantities are being discharged into the sewers. The impact of these waste discharges on treatment processes is harmful in some cases and unknown in others. This, coupled with increased knowledge of the effects of these chemicals on health and the environment, creates new or additional problems for many treatment plant operators. This chapter will discuss the problems associated with, and methods of regulating, industrial waste discharges. For information on actually treating industrial wastes, see the remaining chapters of this manual.

5.00 Objectives of Industrial Waste Monitoring

Industrial waste monitoring may be performed by industrial wastewater treatment plant operators to alert them of waste loads flowing into the treatment plant or pretreatment facility. Also, industrial waste monitoring may be performed by POTW operators to alert them of industrial discharges into their collection system and treatment facilities. Industrial wastes often cause shock loads that may disrupt or interfere with wastewater treatment processes. A program for monitoring and regulating industrial discharges is essential to:

1. Prevent serious shock loads and thus ensure that no disruptions of plant processes occur.
2. Provide physical protection of personnel and facilities.
3. Prevent discharge of pollutants from treatment plants.
4. Provide a database for fair and reasonable sewer-use charges and for engineering data.

5.01 Importance of a Monitoring Program

A discharge of improperly pretreated industrial wastes may affect a treatment plant to the extent that it fails to meet the discharge requirements set by the NPDES permit, the regulating agency, or other authority. One cause for the failure may be a process upset occurring due to toxic materials that destroy or affect the metabolism of microorganisms that are part of the treatment process. Another cause may be toxic materials that pass through the treatment process and are present in the plant effluent. Hydraulic overloads can upset treatment processes by "washing out" the microorganisms in the treatment process. Solids carryover to the effluent will result in a process upset, as will odors resulting from upset biological treatment processes.

Industrial discharges may cause problems in both the collection system and treatment plant including excessive solids that may cause stoppages, corrosive substances that may damage the pipes, and flammable materials that may cause fires or explosions. Dangerous gases may result from some discharges. In some cases, the gases will cause an oxygen deficiency. In others, toxic gases, such as hydrogen sulfide or cyanide, may be present. Additionally, some gases may create flammable or explosive conditions.

5.02 Objectionable Characteristics of Industrial Wastes

Some undesirable properties and effects of various industrial waste discharges are presented in this section.

Possible Collection System Problems:

1. Acidic and Alkaline Wastes—cause corrosion of pipes.
2. Thermal Wastes—hot discharges raise the temperature of the wastewater, speeding up decomposition and thus causing septic conditions. Septic conditions can produce odor and corrosion problems as well as toxic, flammable, explosive, and oxygen-deficient atmospheres.
3. Solids—excessive solids may settle and cause stoppages and septic conditions.
4. Oil and Grease—build up on pipe walls, reducing capacity and causing stoppages and septic conditions.
5. Odors—the discharge itself may be odorous, or the odors may result from the conditions caused by the discharge.
6. Toxic Substances—may include infectious wastes, poisons (such as pesticides), or gases (such as cyanide or hydrogen sulfide).
7. Flammable and Explosive Materials—include fuels, paint thinners, and hydrogen sulfide.

NOTE: Hydrogen sulfide results from septic conditions. Hydrogen sulfide will cause odors, corrosion, and can be flammable under certain conditions and toxic to your respiratory system.

Possible Treatment Plant Problems:

1. Organic Overloads—more suspended and dissolved solids than the plant can handle.
2. Toxic Substances—(see previous section on Possible Collection System Problems).

3. Oil and Grease—may flow through plant and form surface scums on receiving waters (see previous section on "Possible Collection System Problems").

4. Acidic and Alkaline Wastes—cause corrosion of pipes and can be toxic to organisms in treatment processes.

5. Hydraulic Overloads—reduce detention or treatment times and cause incomplete treatment of wastes.

6. Refractory Materials—materials that are not removed by plant processes, such as nutrients, color, taste- and odor-producing substances, and some toxic materials.

7. Hazardous Wastes—solvents, fuels, corrosives, poisonous substances, infectious wastes, radioactive wastes, gases (explosive, toxic, or oxygen-displacing).

Thus, it is possible for a seemingly harmless food processing plant, for example, to discharge excessive solids and organic materials and cause intolerable conditions in the collection system and the wastewater treatment plant.

QUESTIONS

Write your answers in a notebook and then compare your answers with those on page 184.

5.0A What are the objectives of an industrial waste monitoring program?

5.0B What types of industrial waste discharges have undesirable properties that could cause problems for:

(1) collection systems?
(2) treatment plants?

5.1 ADMINISTRATION OF A MONITORING PROGRAM

5.10 Organization

The US Environmental Protection Agency (EPA) requires monitoring programs that control problems at their sources. State agencies also have monitoring requirements. In some cases, they may be more stringent than the basic EPA regulations. In order to abide by these regulations and achieve the objectives listed in Section 5.00, certain actions will be necessary. Essential actions include the building and maintenance of a database and the establishment and enforcement of effective local ordinances regulating industrial discharges.

5.11 The Database

A database file should be established and maintained on each discharging industry. The technical data contained in the file are essential for making decisions in areas such as operation and maintenance of treatment plant processes, program monitoring, and engineering design of facilities. The following information should be included in each discharger's file.

1. Industry Identification. Name of firm, products or services provided, raw materials or chemicals used, and a brief description of manufacturing and treatment processes used (for example, batch, continuous, seasonal).[1]

2. Flow Data. Characteristics of the wastewater discharge including flow patterns (peaks and lows), total volume, and BOD, suspended solids, pH, and other critical characteristics.

3. Laboratory Results. Dates, times, and results of laboratory tests on discharges.

4. Billing Information. Calculations used to establish billing rates.

5. Treatment Information. Pretreatment measures used by the industry and any special measures that must be taken by the treatment plant.

6. Reports of Unusual Circumstances or Incidents. Data on spills, product or process changes, or changes in pretreatment efforts by the industry.

7. Industrial Facilities. Layout of the industrial site showing location of sewer lines and connectors, chemical storage areas, spill protection measures provided, monitoring facilities, and name (or title) and location of the person to contact upon arrival for inspection visit.

8. Emergency Information. Names and phone numbers of at least two persons with the industry who have authority to take appropriate action in case of accidents or spills.

5.12 Legal Authority

A good sewer-use ordinance is the proper tool to provide the necessary authority to control industrial wastes. The objectives of the ordinance are to provide for the control of industrial discharges to the degree that (1) no harm is done to personnel or wastewater collection and treatment facilities, and (2) these discharges do not cause the wastewater treatment plant to violate its discharge requirements. Also, it should provide for equitable sewer service charges.

---

[1] EPA's Standard Form A (see Appendix C) may be used for this purpose. This form includes a space for the standard industrial classification code number of the industry involved. The code numbers are published by the US Government Printing Office, Superintendent of Documents, PO Box 371954, Pittsburgh, PA 15250-7954. The stock number of the book is 041-001-00314-2, price is $40.00 and the title is "Standard Industrial Classification (SIC) Manual 1987." Use of these code numbers is often mandatory. (NOTE: Many industries will fit into two or more classifications. The EPA Form A (Appendix C) provides only for the "major" product or service. This is not adequate for monitoring purposes. In these cases, more than one Form A per industry, or special forms will be needed.)

A sewer-use ordinance or a pretreatment facility inspection ordinance should authorize the following activities by the wastewater treatment agency:

1. Inspecting and monitoring.
2. Requiring pretreatment or flow adjustment.
3. Curtailing or halting discharges.
4. Requiring spill protection measures.
5. Issuing permits to dischargers.

   Regulatory agencies usually will require the local agency to use a permit system. This is a good mechanism for inspections and gathering base information. A sample copy of an Industrial Sewer-Use Permit Application is located in Appendix A.

6. Billing for sewer service charges.

   a. Normal sewer-use charges.

   b. Special charges or fines for damages caused by spills.

   Billing rates should be fair and equitable for everyone. Charges to industry for treating flows, including BOD and suspended solids, should be similar to the charges to businesses and homeowners. Sufficient funds should be collected from all dischargers to cover all of the operating and maintenance costs of the wastewater collection and treatment facilities. Fines and extra charges for accidental spills, dumps, or additional loadings should be greater than the extra costs associated with resulting problems and a poor plant effluent in order to discourage future problems resulting from negligence.

A sample ordinance, prepared by the Environmental Protection Agency (EPA), is included in Appendix B.

5.13 Dealing With Industry

Your personal contact with industry representatives can be crucial. You have the authority to seriously affect their operations. However, you must obtain their cooperation. Good communications are a must. If you are going to require an action or response to a problem on the part of an industry, first be sure that you can show the need for the action. Most people will cooperate fully if they are convinced that a real need exists.

QUESTIONS

Write your answers in a notebook and then compare your answers with those on page 184.

5.1A Why should a database file be established and maintained on each discharging industry?

5.1B What information should be included in each discharger's file in developing a database?

5.1C A sewer-use ordinance provides legal authority for the wastewater treatment agency to conduct what activities?

5.2 HOW TO MONITOR DISCHARGES

Monitoring is the key to industrial waste control. There must be a program of surveillance to ensure compliance with prescribed limits. In the larger cities with heavy industrial areas, self-monitoring by the industries, with spot-checks by the local agency, is usually practiced. Smaller communities may do their own monitoring, or at least a part of it.

5.20 Monitoring Programs

5.200 Self-Monitoring

Self-monitoring by industry is common in large industrial areas. The burden of monitoring a large number of industries otherwise would be overwhelming to a local agency. Self-monitoring may be used where the industry has laboratory facilities or access to a commercial laboratory. The laboratories used should be certified by the appropriate health department or other authorized regulatory agency.

Routine inspections should be made, as well as examination of regular reports received from the industry. In addition, random sampling (with the samples analyzed at a different laboratory) should be done to confirm the results of the self-monitoring program. When this is done, the sample should be split so both laboratories might analyze it. This will bring to light any variations in procedure between the laboratories. When making inspections, be sure to abide by all of the security and safety procedures of the firms or companies being inspected.

5.201 Continuous Monitoring

Special sensors are available for continuously monitoring a number of water quality indicators. The most commonly used ones are for temperature, pH, and specific conductance. In some cases, frequent cleaning and calibration of the special sensors will be necessary to maintain their accuracy.

There is considerable research being done in the field of continuous monitoring. A broader range of water quality indicators and increased reliability may be expected in the future.

5.21 Monitoring Provisions

Suitable provisions must be provided for monitoring industrial wastes. In most cases, monitoring facilities are provided by the industry. For monitoring purposes, a flowmeter frequently must be installed. While water meter data can be used to determine total flow in some cases, this does not provide peak flow data. A good sampling location is required, with possible locations ranging from a special manhole to complete monitoring structures.

5.22 Need for Representative Samples[2]

As is true in sampling for treatment plant process control, good sampling of industrial wastes is both difficult and of the utmost importance. The sample should be representative; that is, it should be an accurate representation of the entire wastewater discharge. This is essential if you are to know the actual characteristics of the discharge. No matter how highly trained the laboratory technicians or how sophisticated their equipment, their work can be no better than the samples that have been collected for analysis. Very often, a great deal of emphasis is placed on the accuracy of the laboratory analyses, with only casual attention given to obtaining the samples. The collection of samples must be carefully thought out and performed by responsible, trustworthy individuals. Representative samples of slug discharges of chemicals or wastes from industrial plants are difficult to obtain even when you know the time of discharge.

5.23 Sampling Points

A good sampling point is one that is easily accessible and may be located anywhere that a representative sample of the industrial wastewater discharge may be obtained. Remember that you are sampling only the flow from one industrial waste discharger and not the total flow in a municipal collection system. This usually involves a special manhole or monitoring structure.

The special manhole may simply be an ordinary manhole that is easily accessible. The manhole may have a measuring weir or flume. Where continuous monitoring of certain water quality indicators (such as pH) is needed, a special monitoring station may be necessary. The station would house the monitoring equipment, which would typically be a flowmeter, sampler, refrigerator, and pH or other special monitoring equipment. Sometimes the monitoring station is equipped with an alarm system and provisions to stop or divert the discharge to a holding tank or pond. For example, a pH probe is installed to monitor the discharge from an industry. An accidental spill in the plant produces a discharge with a pH of 4.7. If the minimum pH was 6.0, an alarm would sound. Downstream from the probe, a flow diversion device would divert the discharge to a holding basin or to a pretreatment facility to neutralize the discharge. Other facilities in a sampling station may include a sink and a small workbench.

The sampling point needs to be located where the wastewater flow from the industrial plant is well mixed and before the flow is discharged into the public sewer. Flat, sluggish areas are poor sampling locations because solids will tend to settle to the bottom. Under these circumstances, the sample may contain too many or too few solids, depending on the manner in which the sample was collected. Solids also may settle out upstream from weirs. Bubbling air through the wastewater stream has been used for mixing in channels and above weirs. The solids-settling problem is reduced when flumes are used as primary flow-measuring devices.

Where oil or other floating materials are floating, representative sampling is very difficult. If it is possible to divert the entire wastestream to a large container for a short time period, the approximate quantity of floatables or oil may be determined by measuring the material that rises to the top of the container.

If the industry has two or more outlets that discharge process waste into the sewer, the monitoring problem is compounded. If possible, the outlets should be combined into one discharge. Otherwise, separate facilities will be required on each discharge.

QUESTIONS

Write your answers in a notebook and then compare your answers with those on page 185.

5.2A Why are self-monitoring programs commonly used in large industrial areas?

5.2B What type of maintenance may be needed for the sensors used in continuous monitoring programs?

5.2C Define the term "representative sample."

5.2D To collect a representative sample, where in the industrial plant's wastewater flow should the sampling point be located?

5.24 Types of Samples

5.240 Grab Samples[3]

The simplest and least expensive sampling method is the grab sample. However, grab samples are not useful for discharges that vary in makeup (constituents) or quantity. This is particularly true when the samples are being taken for billing purposes. Grab samples can be useful for qualitative analyses to determine if particular materials (for example, heavy metals) that must be preserved are present. The grab sample should be obtained from the center of the wastewater flow.

To determine the peak concentration and time of occurrence of a toxic waste discharge, collect grab samples hourly or use a composite sampler that collects samples at regular time intervals and does not mix the samples. Analyze the individual samples—do not mix them.

---

[2] *Representative Sample.* A sample portion of material, water, or wastestream that is as nearly identical in content and consistency as possible to that in the larger body being sampled.

[3] *Grab Sample.* A single sample of water collected at a particular time and place that represents the composition of the water only at that time and place.

5.241 Composite Samples[4]

Composite samples will provide the most accurate representation of a wastewater stream that can be obtained at a reasonable cost. The simplest method of preparing a composite sample is by taking grab samples at predetermined times, either by hand or through a clock-controlled sampler. The principal objection to these timed grab samples is that the sample quantities are the same for high- and low-flow periods. This can be overcome if the samples are stored in individual bottles and then proportionally combined based on ratios from a flowmeter chart. A limitation of composite samples is that an isolated or peak discharge of a concentrated toxic waste could be diluted by the other samples taken during the sampling period.

Automatic flow-proportional sampling is the standard method for obtaining representative samples. In this method, the sampler is activated by a flowmeter. A sample is collected for each preselected quantity of flow (for example, one sample for each 10,000 gallons). This results in more frequent samples during periods of high flow. Other flow-proportional sampling systems use uniform timing and vary the quantity of sample according to the flow.

Various methods are used to obtain the samples. Scoops, conveyors with small "buckets" attached, vacuum chambers, and small pumps are the most common. Some samplers obtain the sample directly from the wastewater flow. In others a pump supplies a constant stream of the wastewater to the samplers, with an overflow line carrying the excess back to the sewer. Figure 5.1 shows a sampling device, a peristaltic (a type of positive displacement pump) pump sampler.

Flow-proportional samplers usually are installed in permanent monitoring facilities. Most of these samplers combine the various samples in a refrigerated compartment. This prevents deterioration and change of several characteristics (BOD, acidity-alkalinity, specific conductance) of the samples.

5.25 Portable Equipment

Samplers, flowmeters, and some instruments, such as pH meters and conductivity meters, are available for field work. Portable flowmeters use a weir or Palmer-Bowlus flume for easy installation in a sewer line. The meters themselves use floats, air or nitrogen bubbling, built-in electronic sensing, motor-operated sensing probes, or sonar for sensing the water level behind the weir or in the measuring flume. Recording charts and totalizers usually are part of the equipment so that a record of the flow is obtained. (Portable flowmeters and recorders are shown in Figure 5.2.)

Most portable samplers use peristaltic pumps or vacuum chambers to pick up the samples (Figures 5.1 and 5.3). Power is supplied by rechargeable batteries or, when possible, from a 110-volt AC line using a converter that fits the same space usually occupied by the battery.

The pumping rate must be considered when a sample is collected. A velocity of 2 to 5 feet per second (0.6 to 1.5 m/sec) in the sampler suction tubing produces repeatable results. Lower velocities tend to leave solids behind in the tubing, while higher velocities may pull in large chunks of material, yielding erratic results in the data.

Most portable samplers use individual bottles for each sampling period. The bottles are on a rotating table or plate on a timer that positions a different bottle under the sampler discharge during each sampling time. This permits a sample from a particular sampling time to be individually inspected. The separate samples are combined manually or automatically to make a composite sample.

The usefulness of a portable sampler will depend on the flexibility of its controls. Desirable features include automatic purging before and after the sample is taken, and timing flexibility to allow a range of sample times from every few minutes to a couple of hours between samples. At times, you may want to take more than one sample per bottle or to fill several bottles during each sampling cycle. All of these features, plus other refinements, are provided by several manufacturers in their portable samplers.

Chilling the sample for preservation may be important (see Section 5.4, "Sample Preservation and Security"). An ice compartment is generally provided for this purpose. If the sampler is to be used in one location for some time, it may be practical to use chilled water. A small refrigerated-water chilling unit, a small water circulation pump, and some plastic tubing will do the job. Notch the cover of the sampler to insert the tubing or install short lengths of $1/8$-inch (3-mm) pipe through the bottom of the sampler using locknuts, washers, and a good sealant (Figure 5.4).

5.26 Handling Portable Samplers

Most portable samplers are designed to fit into a manhole and are equipped with harnesses to permit hanging them within the manhole. Figure 5.5 shows a sampler and the gear for supporting it. Usually missing is something to hook the harness to. Also missing is a way to keep the suction hose approximately in the middle of the wastewater flow. This is especially noticeable if you are sampling a *TRUNK SEWER*.[5] The following method

---

[4] *Composite (Proportional) Sample.* A composite sample is a collection of individual samples obtained at regular intervals, usually every one or two hours during a 24-hour time span. Each individual sample is combined with the others in proportion to the rate of flow when the sample was collected. Equal volume individual samples also may be collected at intervals after a specific volume of flow passes the sampling point or after equal time intervals and still be referred to as a composite sample. The resulting mixture (composite sample) forms a representative sample and is analyzed to determine the average conditions during the sampling period.

[5] *Trunk Sewer.* A sewer line that receives wastewater from many tributary branches and sewer lines and serves as an outlet for a large territory or is used to feed an intercepting sewer. Also called MAIN SEWER.

164 Treatment Plants

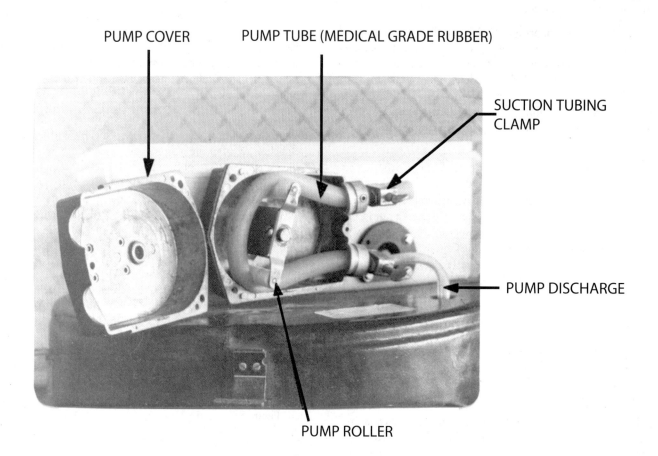

Fig. 5.1 Peristaltic pump sampler

Industrial Monitoring 165

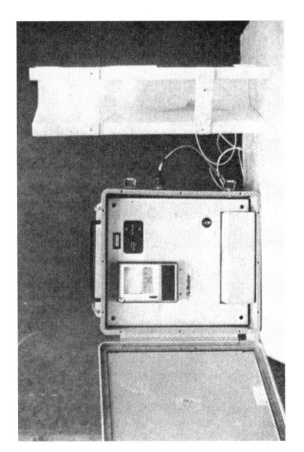
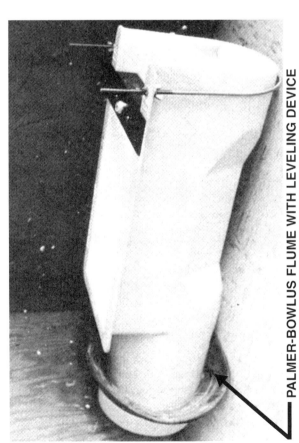

Fig. 5.2 Portable flowmeters and recorders

166 Treatment Plants

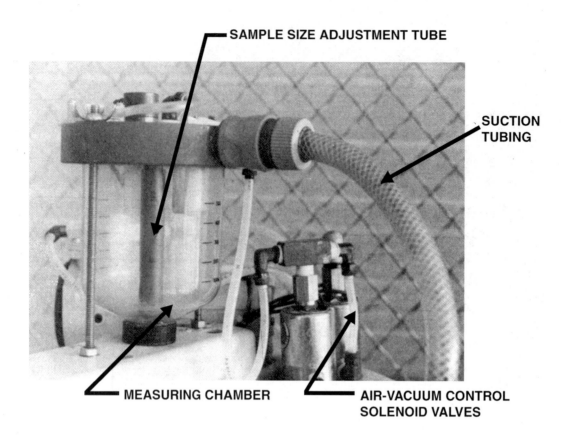

Fig. 5.3 Vacuum chamber sampler

has served well to overcome these problems. Many other ideas also will work with the individual situation and your ingenuity usually determining your approach.

1. A special "manhole ring" (Figure 5.6) is suspended inside the manhole from the inside of the manhole frame. The manhole cover sits on top of the manhole ring. The manhole ring can be made up by a sheet metal shop from 16-gauge galvanized sheet metal. The ring cylinder should be slightly smaller than the inside diameter of the top of the manhole.

2. A harness for the sampler (Figure 5.5) is made of nylon or other synthetic rope and attached to the manhole ring with harness snaps. Several harnesses of various lengths should be made for manholes of various depths. The harness snaps should be inspected frequently for possible replacement because they may corrode rapidly under conditions found in manholes.

3. A knotted rope is attached from the manhole ring to the sampler to raise and lower the sampler.

4. A piece of pipe with a bail on it (Figure 5.5) is suspended from the sampler. This serves as a "hanging wet well." The suction line is taped to the bail to hold it in place, and a piece of nylon cord suspends it to the desired depth.

In a sewer, debris, such as rags, may hang up on this device, but it will swing downstream and the debris will slide off. In practice, this method has yielded good results with very few missed samples.

5.27 Equipment Problems

Cleanliness is a very important part of getting accurate tests. Contamination of sample containers, tubing, and internal parts of samplers will affect the sample, and thereby lead to inaccurate laboratory results.

Another potential problem that can lead to inaccurate test results is interference due to the type of material from which the tubing and containers are made. Depending on the materials, some elements in the wastewater may adsorb on the walls of the

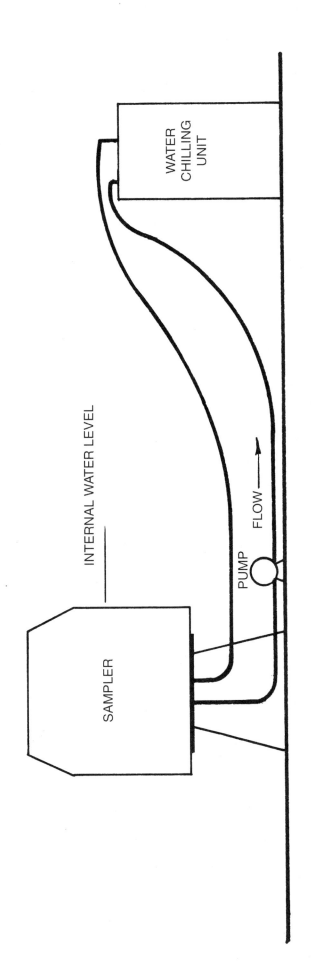
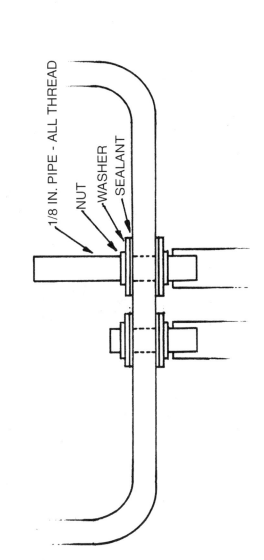

Fig. 5.4 *Sketch of arrangement for chilling sampled waters*

168 Treatment Plants

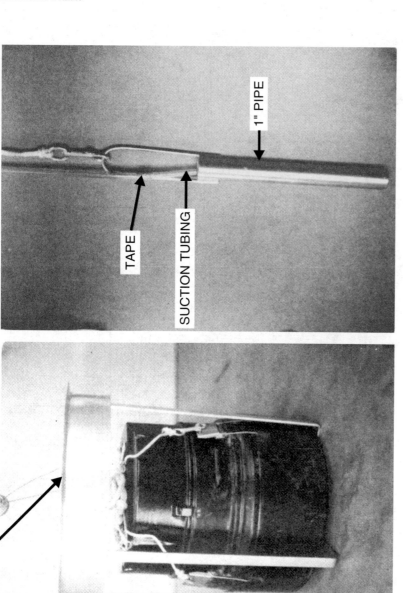

Fig. 5.5 Hanging sampler

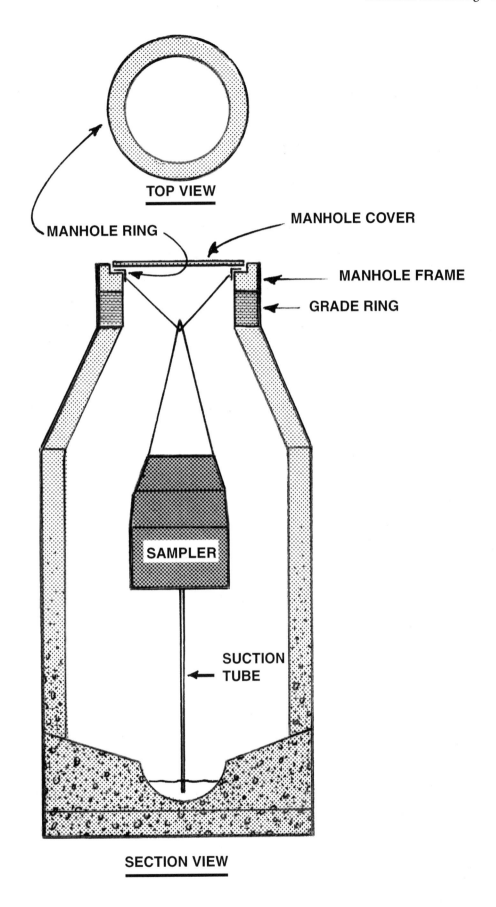

Fig. 5.6 Special "manhole ring" details

tubing or container. Also, some wastewater constituents may react with tube or container materials causing impurities that may dissolve into the sample. This is a greater problem with the containers because the contact time of the sample with the tubing is very limited. While this type of interference normally is not a problem, it is important when testing for very low levels of heavy metals and other toxins (that is, in parts per billion or micrograms per liter). If you are faced with this situation, it is a good idea to obtain sample containers made from Teflon or linear polyethylene. These materials are excellent in that they do not adsorb materials from the sample or contain impurities that can dissolve into the sample.

5.28 Maintenance

The reliability of samplers varies considerably. A unit that is considered poor by one operator may perform well for another. The most reliable equipment will be that used where regular maintenance is performed in accordance with the manufacturer's instructions. Replacement items must also be as specified. For example, if a specific tubing is designated for a peristaltic pump, substitutes may cause malfunctions. Where samplers are equipped with desiccant (drying) cartridges, it is important to keep them changed as instructed. To service sampling equipment, both mechanical and electronic skills are required.

QUESTIONS

Write your answers in a notebook and then compare your answers with those on page 185.

5.2E What are the differences between grab samples and composite samples from the standpoint of industrial waste monitoring? What are the limitations with each type?

5.2F How is power provided for portable sampling equipment in the field?

5.2G Why is the flow velocity in the sampler suction tube important?

5.2H What kind of care and maintenance should be provided monitoring equipment?

END OF LESSON 1 OF 2 LESSONS
on
INDUSTRIAL WASTE MONITORING

Please answer the discussion and review questions next.

DISCUSSION AND REVIEW QUESTIONS
Chapter 5. INDUSTRIAL WASTE MONITORING
(Lesson 1 of 2 Lessons)

At the end of each lesson in this chapter you will find some discussion and review questions. The purpose of these questions is to indicate to you how well you understand the material in the lesson. Write the answers to these questions in your notebook.

1. What problems could a wastewater treatment plant experience if improperly pretreated industrial wastes are discharged?

2. How would you effectively deal with an industrial waste discharger?

3. How could you check or verify the accuracy of an industrial laboratory analyzing samples from its own self-monitoring program?

4. Why must representative samples be collected?

5. How would you collect a representative sample when the wastewater has oil or other floating materials on the surface?

6. How would you attempt to determine the peak concentration and time of occurrence of a toxic waste discharge?

CHAPTER 5. INDUSTRIAL WASTE MONITORING

(Lesson 2 of 2 Lessons)

5.3 LOCATING SOURCES OF DISCHARGES

5.30 "Prospecting" in the Sewers

Often, the problem facing the wastewater treatment plant operator is that of an industrial waste from an unknown source. The magnitude of the problem is, of course, directly proportional to the problems caused by the unknown discharge. When trying to locate the source, established procedures must be followed to obtain accurate results. Some of the equipment used and techniques of locating the sources of discharges are discussed in Section 5.2, "How to Monitor Discharges."

To trace or locate an industrial waste discharge that is causing problems, look for signs or evidence of the waste. Begin at the treatment plant or at a point in the collection system where the discharge has been discovered, then follow signs of the waste back to the source. Look for indicative color or solids and simply inspect the sewer at each sewer line intersection and follow it up to its source. If the discharge is not visible, use portable samplers to follow the waste to its source. If several samplers are available, the process can be speeded up greatly.

If the discharge is intermittent, the task can be very difficult and time-consuming. $SLUGS^6$ can pass by the sampler between sampling periods. If short sampling time intervals are used, the samplers must be inspected every few hours. Solvents float on the surface and will not be picked up by ordinary sampling equipment. If you are looking for petroleum solvents or similar floatable materials, you can fasten the suction tubing to a block of Styrofoam or similar material and thus hold the suction end of the sampling tube very close to the surface.

Techniques other than sampling will be appropriate at times. For instance, a solvent discharge problem might be solved by inspecting all the shops, stores, service stations, and laundries connected to that part of the collection system. A survey of the area may prove to be helpful. Use the permit application forms provided in the Appendix (A and C) as they appear there, or use them as a guide for preparation of your own survey forms.

5.31 Identifying Waste Materials

Once a sample has been collected and analyzed by the laboratory, the next problem is to interpret the results. A list of the common types of discharges from some businesses and industries is contained in Table 5.1. Once the type of discharge is determined, one helpful source of information is the telephone directory. The Yellow Pages can usually produce a list of firms or industries to inspect. Personal visits or inspections can then help to get the problem solved.

5.4 SAMPLE PRESERVATION AND SECURITY

5.40 Storage Time

Laboratory tests are most accurate if a sample is analyzed immediately after it is collected. Prompt testing prevents physical, chemical, or biological changes in the sample. While immediate testing usually is not possible, in most cases refrigeration adequately preserves the quality of the samples (see the sample preservation list in Table 5.2). A related problem arises when samples must be tested within a specified time after collection. For example, practically all BOD samples are from 24-hour composite samples plus additional storage time. As indicated on the list, the maximum storage time for BOD is 48 hours. Even though you will not be able to get perfectly accurate test results under these circumstances, you can minimize the amount of error if you maintain a consistent storage time for each sample tested. This will maintain a constant degree of error on all samples, and thereby reduce errors in interpretation of results.

5.41 Storage Temperature

The recommended temperature for sample storage is 4°C. At this temperature, the rate of chemical reactions and biological activity is significantly reduced. Set the control on the refrigeration unit in permanent installations or chilling water unit for a modified portable sampler. When using portable samplers in the field, put ice on the compartment and hope for the best. If you are working where the wastewater stream is warm, you can use dry ice. Experiment to find out how much to use (too much may freeze the samples). Samples must be kept chilled while they are being transported to the laboratory.

5.42 Chemical Preservation

In some cases, it is appropriate to add chemicals to preserve the samples (see Table 5.2 for detailed recommendations). For portable samplers, the chemical can be measured into each sample bottle in advance.

5.43 Security

Once a sample has been obtained and is ready for laboratory analysis, it is extremely important to ensure that each sample is

---

[6] *Slug.* Intermittent release or discharge of wastewater or industrial wastes.

TABLE 5.1 TYPES OF POLLUTANTS FROM SOME COMMON INDUSTRIES

TYPES OF POLLUTANTS

Type of Industry	BOD	SUSPENDED SOLIDS	DISSOLVED SOLIDS	OIL AND GREASE	pH	TURBIDITY	COLOR	HEAT	HEAVY METALS	CYANIDE	AMMONIA	SULFATE	ALKALINITY	PHENOL	SULFITE	SETTLEABLE SOLIDS	SURFACTANTS	NUTRIENTS
Canneries	X	X			X											X		
Chemical Mfg.	X	X		X	X				X	X			X					
Dairy Products	X	X	X		X											X		X
Detergent Mfg.	X	X		X	X		X					X					X	
Electroplating		X	X		X				X	X								
Glass, Asbestos	X	X	X	X	X	X	X	X				X		X				
Grain Mills	X	X	X		X		X											X
Leather Tanning	X	X	X	X	X		X								X			
Meat Products	X	X	X	X												X		X
Rubber Products	X		X				X							X			X	
Sugar Processing	X	X	X		X			X										X
Paper Mills	X	X	X			X	X									X		X
Plastics	X	X			X				X						X			
Wood Products	X	X	X				X											

carefully labeled. Test results are useless if there is any doubt that they are for the correct sample. A good starting point is to firmly affix a label or tag to the sample container (Figure 5.7). Another reason for careful labeling is so the *CHAIN OF CUSTODY*[7] is traceable. If legal action results, you must be able to prove that the sample was in the possession of responsible people at all times and also that no one tampered with the sample.

A logging or "sign-in" system should be used at the laboratory. This system should include the name of the person delivering the sample as well as the date and time received. The laboratory should then assign a number that identifies the sample and appears on the laboratory report.

QUESTIONS

Write your answers in a notebook and then compare your answers with those on page 185.

5.3A How would you attempt to trace or locate an industrial waste discharge that is causing problems?

5.3B How can the suction end of a sampling tube be adjusted to sample for petroleum solvents or similar floatable materials?

5.4A Why should a sample be analyzed as soon as possible?

5.4B Why are samples stored at 4°C?

---

[7] *Chain of Custody.* A record of each person involved in the handling and possession of a sample from the person who collected the sample to the person who analyzed the sample in the laboratory and to the person who witnessed disposal of the sample.

TABLE 5.2 REQUIRED CONTAINERS, SAMPLE PRESERVATION TECHNIQUES, AND HOLDING TIMES[a]

RESIDUE (SOLIDS) TESTS

Measurement	Container[b]	Preservation[c,d]	Maximum Holding Time[e]
Total Solids	P,G	Cool, 4°C	7 days
Suspended Solids	P,G	Cool, 4°C	7 days
Volatile Suspended Solids	P,G	Cool, 4°C	7 days

CHEMICAL TESTS

Measurement	Container[b]	Preservation[c,d]	Maximum Holding Time[e]
Acidity-Alkalinity	P,G	Cool, 4°C	14 days
Biochemical Oxygen Demand	P,G	Cool, 4°C	48 hours
Calcium	P,G	None required	7 days
Chemical Oxygen Demand	P,G	Add H_2SO_4 to pH <2 at 4°C	28 days
Chloride	P,G	None required	28 days
Color	P,G	Cool, 4°C	48 hours
Cyanide	P,G	Add NaOH to pH >12 at 4°C	14 days
Dissolved Oxygen	G	Determine on site	No holding
Fluoride	P	None required	28 days
Hardness	P,G	HNO_3 to pH <2	6 months
Metals, Total	P,G	HNO_3 to pH <2	6 months
Metals, Dissolved	P,G	Filtrate: 3 mL 1:1 HNO_3 per liter	6 months
Nitrogen, Ammonia	P,G	H_2SO_4 to pH <2 at 4°C	28 days
Nitrogen, Kjeldahl	P,G	H_2SO_4 to pH <2 at 4°C	28 days
Nitrogen, Nitrate-Nitrite	P,G	H_2SO_4 to pH <2 at 4°C	28 days
Oil and Grease	G	H_2SO_4 to pH <2 at 4°C	28 days
Organic Carbon	P,G	H_2SO_4 to pH <2 at 4°C	28 days
pH	P,G	Determine on site	No holding
Phenolics	G	H_2SO_4 to pH <2 at 4°C	28 days
Phosphorous, Total	P,G	H_2SO_4 to pH <2 at 4°C	28 days
Specific Conductance	P,G	Cool, 4°C	28 days
Sulfate	P,G	Cool, 4°C	28 days
Sulfide	P,G	Zn acetate + NaOH to pH >9 at 4°C	7 days
Threshold Odor	P,G	Cool, 4°C	7 days
Turbidity	P,G	Cool, 4°C	48 hours

[a] "Required Containers, Preservation Techniques, and Holding Times," Part 136.3. *CODE OF FEDERAL REGULATIONS*, Protection of the Environment, 40, Parts 136-149, 2004. This publication is available from the US Government Printing Office, Superintendent of Documents, PO Box 371954, Pittsburgh, PA 15250-7954. Order No. 869-052-00157-1. Price, $61.00.

[b] Polyethylene (P) or Glass (G).

[c] Sample preservation should be performed immediately upon sample collection. For composite chemical samples, each aliquot should be preserved at the time of collection. When use of an automated sampler makes it impossible to preserve each aliquot, then chemical samples may be preserved by maintaining at 4°C until compositing and sample splitting is completed.

[d] When any sample is to be shipped by common carrier or sent through the United States Mails, it must comply with the Department of Transportation Hazardous Materials Regulations (49 CFR Part 172).

[e] Samples should be analyzed as soon as possible after collection.

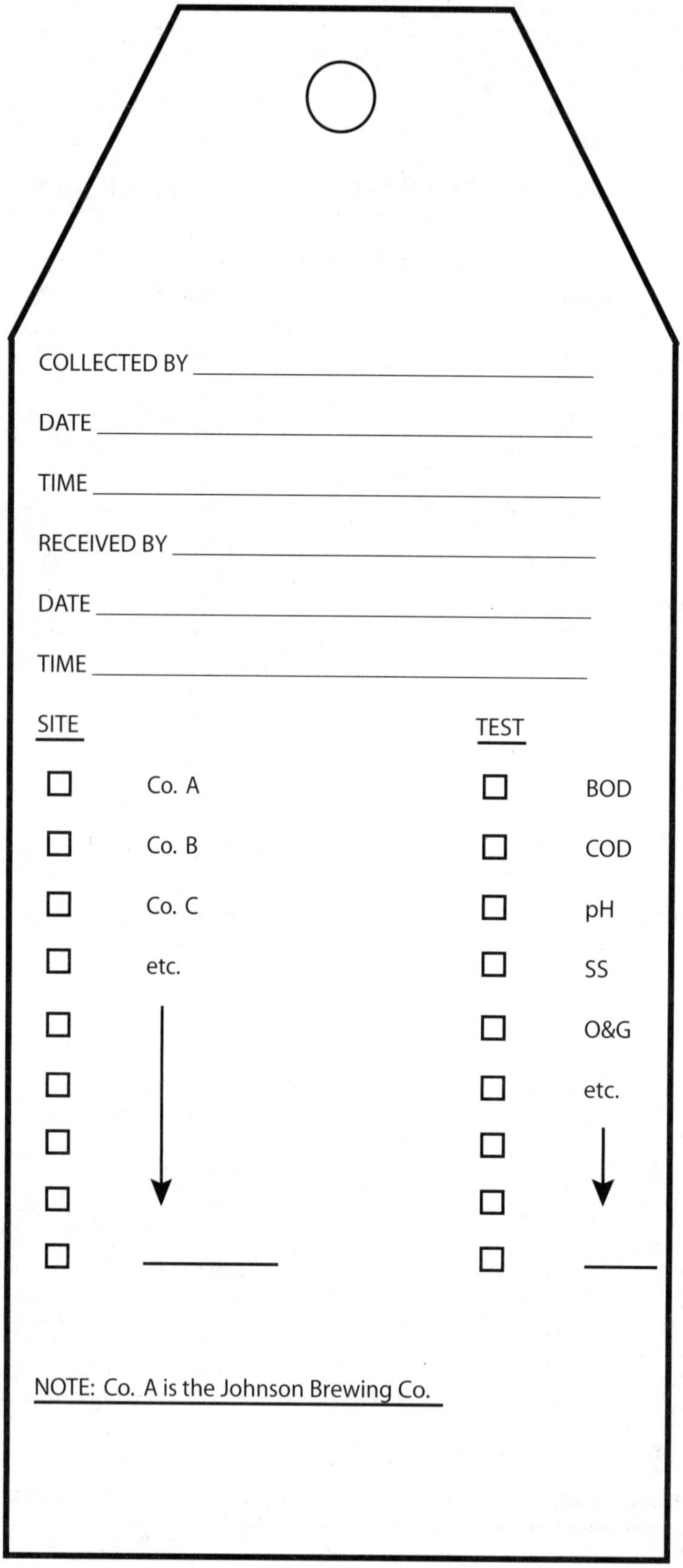

174 Treatment Plants

Fig. 5.7 Label or identification tag for sample container

5.5 FLOW METERING

5.50 Need for Accurate Measurements

Measuring the flow from an industry is equally as important as obtaining good samples. The quantity of pollutants cannot be calculated without accurate flow data. Knowledge of the quantities of pollutants from the various industries is essential to reaching decisions regarding pretreatment requirements to be set, and also to prepare accurate and equitable billing calculations. Flow and pollutant data on industries also may be an important part of the design work for future expansions or new facilities. Flows indicate the hydraulic loading on both the collection system and the wastewater treatment facilities.

5.51 Water Meters

In many cases, the flow to the sewer can be calculated based on data from water meters placed on public water supplies or private wells. When water meter data are used, allowances may have to be made for evaporation, irrigation, or water used in the product (for example, soft drinks). Water meters on boiler feed lines, product lines, or irrigation water will serve to verify this type of usage.

For the larger industries, open-channel flowmeters, such as Parshall flumes, generally are used. For these "fixed" installations, the same type of equipment is used as for treatment plants. Refer to Chapter 6, "Flow Measurement," for detailed information.

For the smaller enterprises, portable metering equipment often is installed in existing collection systems. Many types of portable meters are available. The standard approach is to use a compact flume, such as a Palmer-Bowlus flume, as the primary flow-measuring device. Some compact flumes have provisions for several types of primary devices (flumes, V-notch, and rectangular weirs) for measuring the flows. The level in the primary device is sensed by a float, bubbler, electric probe, sonar (ultrasonic), or built-in capacitance sensor in the flume. The meters generally are equipped with a totalizer and small strip-chart recorder, although at least one manufacturer uses a circular chart. Also, there is usually a connection for controlling a sampler so that flow-proportional sampling can be done in the field.

If you are purchasing a flowmeter and the recorder is an optional feature, you are strongly advised to include the recorder. Knowing the flow pattern can be very helpful in evaluating the discharges by an industry, and perhaps in helping them to correct problems.

5.6 SAFETY

5.60 Traffic Safety

Obtaining a sample from a manhole in the street without getting hit by a car requires some advance preparation and effort. Table 5.3 shows a good example of a traffic safety instruction sheet. Follow these instructions very carefully as they apply to your situation every time you must collect a sample from a manhole in a street. Figure 5.8 shows how cones may be used in a typical layout for blocking one lane of the street when you must work in the street. Channelizing devices and barricades for directing traffic are shown in Figures 5.9 and 5.10.

TABLE 5.3 TRAFFIC SAFETY INSTRUCTIONS

Safe working conditions must be maintained at all times. This is especially important when taking samples from manholes located within normal traffic lanes.

The important considerations are: (1) to be visible to oncoming traffic, and (2) to guide this traffic safely around you. The following basic procedure is for your general guidance. (For unusual situations, look at the situation carefully and use common sense. If you consider a particular location too hazardous, skip that location and contact your supervisor for advice.)

1. Wear fluorescent orange-red or fluorescent yellow-green safety vest, hard hat, and safety glasses.
2. Slow down in advance of work area. Use hand signals and turn on beacon (the rotating yellow light on the vehicle roof). Do not stop suddenly.
3. Park vehicle between you and the prevailing traffic, about 10 feet (3 m) away from work area.
4. Turn wheels away from you and oncoming traffic, set hand brake firmly, and place in parking gear.
5. Leave vehicle carefully—set out traffic cones as quickly and as safely as possible. Figure 5.8 shows how cones may be used for blocking one lane of a street. Channelizing devices and barricades for directing traffic are shown in Figures 5.9 and 5.10.
6. Be alert to the traffic at all times.
7. Finish all work and be ready to go before retrieving traffic cones.

NOTE: Where a hazard exists to employees because of traffic at work sites that encroach upon public streets or highways, a system of traffic control must be used to minimize the hazard. This list is suggested; check with your State Department of Transportation (DOT) regulations for specific requirements.

176 Treatment Plants

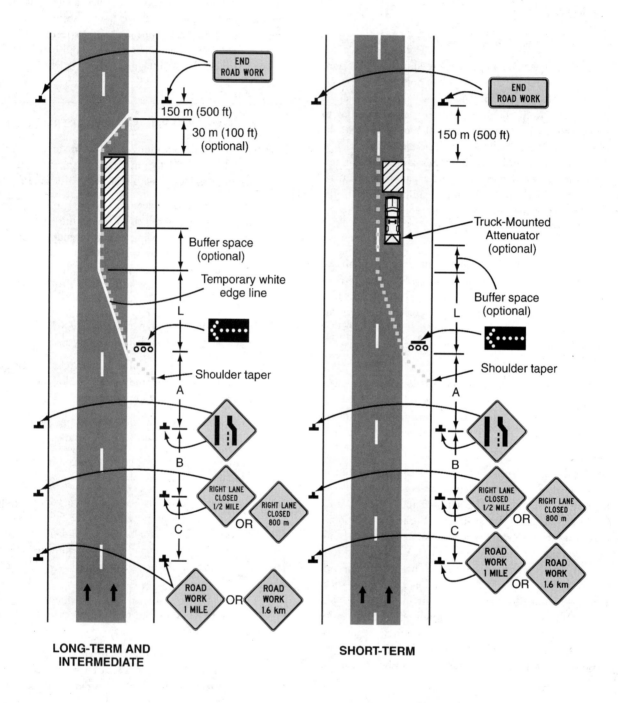

Fig. 5.8 Typical layout for blocking one lane of traffic
(Source: Manual on Uniform Traffic Control Devices, 2003)

Industrial Monitoring 177

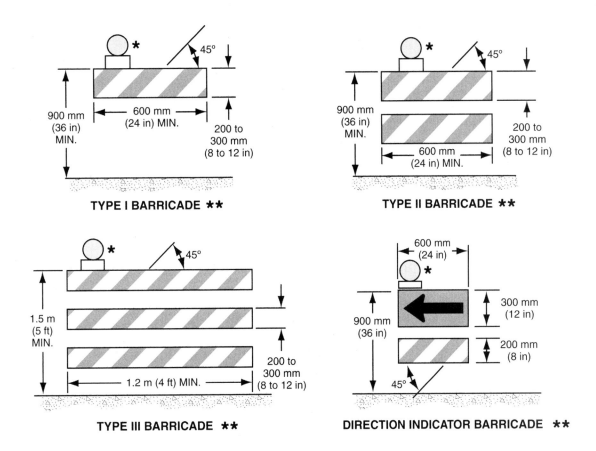

* Warning lights (optional)
** Rail stripe widths shall be 150 mm (6 in), except that 100 mm (4 in) wide stripes may be used if rail lengths are less than 900 mm (36 in). The sides of barricades facing traffic shall have retroreflective rail faces.

Note: If barricades are used to channelize pedestrians, there shall be continuous detectable bottom and top rails with no gaps between individual barricades to be detectable to users of long canes. The bottom of the bottom rail shall be no higher than 150 mm (6 in) above the ground surface. The top of the top rail shall be no lower than 900 mm (36 in) above the ground surface.

Fig. 5.9 Channelizing devices and barricades
(Source: Manual on Uniform Traffic Control Devices, 2003)

178 Treatment Plants

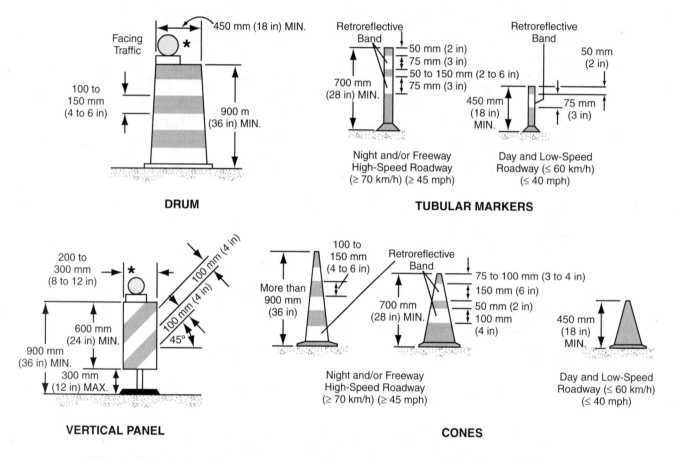

* Warning lights (optional)

Note: If drums, cones, or tubular markers are used to channelize pedestrians, they shall be located such that there are no gaps between the bases of the devices, in order to create a continuous bottom, and the height of each individual drum, cone, or tubular marker shall be no less than 900 mm (36 in) to be detectable to users of long canes.

Fig. 5.10 Delineating and channelizing devices
(Source: Manual on Uniform Traffic Control Devices, 2003)

5.61 Confined Spaces[8]

Manholes are usually considered confined spaces. Use a manhole hook or manhole lift to open manhole covers. Sample the atmosphere in the manhole by inserting a probe through the pick hole prior to removal of the cover to determine if combustible gases are present. When removing the cover you may cause a spark, which could be disastrous if an explosive atmosphere exists in the manhole. When a manhole cover is off, guard the area around the open manhole with traffic cones or a portable guardrail. Remove items from shirt pockets before working over the open manhole. When replacing the cover, remove any dirt or other debris from the base, so that the cover will seat properly.

OSHA[9] regulations (and common sense) dictate several precautionary measures that must be taken before entering manholes or other areas that may be confined spaces. Before entering an area that may be a confined space, the atmosphere in the area must be tested for oxygen deficiency or enrichment, toxic gases (for example, hydrogen sulfide), and explosive conditions, including lower explosive limit (LEL). If you discover explosive conditions, you must not enter. Notify the proper authorities immediately. If there is an oxygen deficiency, you may use a self-contained breathing apparatus only for emergency entry. Correct any situation in which an oxygen deficiency/enrichment exists or toxic gases are encountered by providing adequate ventilation. Do not work in a confined space until all detection devices indicate that the area is safe for entry. The important fact to remember is that you seldom get a second chance. Be sure it is safe before entering confined spaces. Recommended procedures for work in confined spaces are provided in Chapter 14, "Plant Safety," *OPERATION OF WASTEWATER TREATMENT PLANTS*, Volume II.

If there is no oxygen deficiency/enrichment, no toxic gases, no explosive conditions, and no dangerous air contamination or oxygen deficiency/enrichment can occur while the space is occupied, the area or space is considered *not* confined. Before entering the area, you must be wearing a safety harness. Two people must be standing by topside to help the person out of the hole or area in an emergency. The atmosphere in the area must be monitored with sufficient frequency to ensure that the development of dangerous air contamination (toxic gases, explosive conditions) or oxygen deficiency/enrichment does not occur whenever anyone is working in the space. If the monitor alarm goes off, the person in the hole must be removed immediately.

NEVER enter a confined space even with a self-contained breathing apparatus and sufficient help standing by unless you absolutely must enter the space. If at all possible, eliminate hazardous conditions existing in the confined space before entering and working in the area.

For details on safe traffic regulation and manhole entry, see Chapter 4, "Safe Procedures," in *OPERATION AND MAINTENANCE OF WASTEWATER COLLECTION SYSTEMS*, Volume I.

5.62 Equipment Storage Areas

Orderliness is the key to avoiding accidents and equipment damage in the storage area. Common hazards include items protruding from shelves, items falling from shelves, and items placed in normal traffic patterns on the floor. Containers should be cleaned before storage and protected so they will stay clean. Chemicals should be stored separately from other items, with a separate room being preferable. All bottles must be properly labeled and marked with the date opened or the date the reagents were prepared.

5.63 Battery Charging

There are some hazards to consider when using samplers that operate from lead-acid batteries. When a lead-acid battery is being charged, hydrogen gas is given off. Therefore, good ventilation is essential. New batteries must be shipped dry; therefore, you must add the electrolyte (diluted sulfuric acid) yourself. Protective clothing (goggles, gloves, and aprons) must be worn. Eye wash facilities must be provided. Some excerpts from safety orders for battery charging are given in Appendix D.

5.7 MONITORING STRATEGIES

5.70 Development of Strategies

The following discussion is based on the assumption that an industrial discharge is the cause of the particular problem under consideration. This will, of course, be obvious in many cases. However, the "state of the art," particularly regarding activated sludge, is not sufficiently advanced that the operator can always be certain about the effects of a particular industrial discharge. A process that is not completely healthy may be seriously upset by a discharge that would otherwise have little effect. Extremely small quantities of certain exotic chemical compounds (such as pesticides) may upset the process and not be detected at all by

---

[8] *Confined Space.* Confined space means a space that:
   (1) Is large enough and so configured that an employee can bodily enter and perform assigned work; and
   (2) Has limited or restricted means for entry or exit (for example, manholes, tanks, vessels, silos, storage bins, hoppers, vaults, and pits are spaces that may have limited means of entry); and
   (3) Is not designed for continuous employee occupancy.
   Also see DANGEROUS AIR CONTAMINATION and OXYGEN DEFICIENCY.

[9] *OSHA* (O-shuh). The Williams-Steiger Occupational Safety and Health Act of 1970 (OSHA) is a federal law designed to protect the health and safety of workers, including the operators of water supply and treatment systems and wastewater collection and treatment systems. The Act regulates the design, construction, operation, and maintenance of water and wastewater systems. OSHA regulations require employers to obtain and make available to workers the Material Safety Data Sheets (MSDSs) for chemicals used at industrial facilities and treatment plants. OSHA also refers to the federal and state agencies that administer the OSHA regulations.

most monitoring programs. The items discussed are very important when preparing a plan of action on how to operate wastewater treatment plants that must handle industrial wastes.

5.71 Regulation of High Flows

Hydraulic loading is an important factor in plant performance. The flow from an industry may greatly affect a treatment plant, particularly where a sizeable industry has located in a smaller community. Regulation of the industrial flow to coincide with low domestic flow periods may be required. Seasonal industries, such as canneries, or vacation shutdown periods, may have a significant impact on treatment plant operation. To regulate or handle a large, fluctuating industrial flow that hydraulically overloads a treatment plant during the day, try requiring the industry to store a portion of its discharges during peak-flow periods for release during low-flow periods, such as during the graveyard shift.

5.72 Decisions on Industrial Discharges

When a new industry is locating in your area, decisions will have to be made regarding pretreatment, controlled timing or rate of discharge, and installation of monitoring equipment. When reviewing the application or permit of an industrial waste discharger, several basic items need to be considered. Will there be any toxic or hazardous materials in the discharge? Will there be discharges that can cause problems in the sewers (for example, oxygen-consuming wastes, explosive hazards, or solids settling)? Calculate what the BOD loading will be and compare it to the design capacity (less present loading) of the treatment plant. Consider whether there might be clarifier hydraulic overloading or excessive floatable or settleable solids. High solids loading could lead to overloading the solids treatment capacity of the digesters.

If a color is to be discharged, will it be from dyes or food colors? Food coloring usually is no problem, because the colors disappear within the plant processes. Other sources of color will require some testing. Some wastes can be treated by acclimatizing organisms to the specific toxic materials present in the wastes. This is usually best handled by a pretreatment process at the industry.

In general, you would not want to allow one new firm or industry to use up all of the remaining hydraulic, BOD, or solids capacity in your treatment plant. A new industry often will attract more businesses, provide more jobs, and cause an increase in home construction. Some plant capacity should be reserved for these other or secondary activities resulting from a new industry.

Tables 5.1 (page 172), 5.4, and 5.5 summarize helpful information indicating what to expect when evaluating the potential impact of a new industry. These tables indicate the types of wastes that can be expected, whether they might interfere with the treatment processes, and whether the pollutants might pass through a treatment plant untreated.

TABLE 5.4 POLLUTANTS THAT MIGHT INTERFERE WITH WASTEWATER TREATMENT PLANTS

Inorganic Substances	Organic Substances	Other Substances
Acidity, alkalinity, and pH	Alcohols	Corrosive materials
Ammonia	Agricultural chemicals	Explosives and flammable materials
Alkali and alkaline earth metals	Carbon tetrachloride	High temperature wastes
Arsenic	Chlorinated hydrocarbon	Materials that cause blockages
Borate (and other boron species)	Chloroform	
Bromine	Methylene chloride	
Cadmium	Miscellaneous organic chemicals	
Chloride	Oils and grease	
Chlorine	Organic nitrogen compounds	
Chromium	Phenols	
Copper	Surfactants	
Cyanide		
Iron		
Lead		
Manganese		
Mercury		
Nickel		
Silver		
Sulfate		
Sulfide		
Zinc		

Source: *JOINT TREATMENT OF INDUSTRIAL AND MUNICIPAL WASTEWATERS*, Water Environment Federation, Washington, DC.

TABLE 5.5 POLLUTANTS THAT MIGHT PASS THROUGH WASTEWATER TREATMENT PLANTS

Cadmium	Nickel
Chloride	Nitrogen
Chromium	Organic carbon
Copper	Phenols
Cyanide	Phosphorus
Iron	Radioactive wastes
Lead	Suspended solids
Manganese	Zinc
Mercury	

Source: *JOINT TREATMENT OF INDUSTRIAL AND MUNICIPAL WASTEWATERS*, Water Environment Federation, Washington, DC.

5.73 Reactions to Industrial Discharges

Table 5.6 lists some common industrial discharge situations, possible problems that may result, and suggested actions to take. The table is useful primarily as an "idea file." A listing of all the types of discharges that might occur is impossible as is providing a specific corrective action for each possible case or problem. The operator should review the situations listed, plus any others that might be peculiar to the community, and prepare for these possibilities. Preplanning will allow you to implement preventive maintenance measures, such as the purchase of neutralizing chemicals or construction of a holding basin.

5.74 Warning Systems

Continuous monitoring and alarm systems for treatment plant influents are becoming common for potentially harmful conditions. Conductivity, pH, explosive atmosphere, including lower explosive limit (LEL), and oxygen deficiency/enrichment detectors are suitable for continuous monitoring. Special sensors are available for many wastewater constituents; however, few are suitable at present for continuous use. High water level or flow alarms are available to warn of surges or hydraulic overloads.

The operator should keep abreast of progress in this field. Reading the technical journals and attending seminars and conferences will help to keep you current.

5.8 ADDITIONAL READING

1. *NEW YORK MANUAL*, Chapter 8, "Industrial Wastes."

2. *TEXAS MANUAL*, Chapter 26, "Industrial Wastes."

3. *PRETREATMENT FACILITY INSPECTION*. Obtain from the Office of Water Programs, California State University, Sacramento, 6000 J Street, Sacramento, CA 95819-6025. Price, $45.00.
   NOTE: This manual covers all topics presented in this chapter in great detail.

4. *OPERATION AND MAINTENANCE OF WASTEWATER COLLECTION SYSTEMS*, Volumes I and II. Obtain from the Office of Water Programs, California State University, Sacramento, 6000 J Street, Sacramento, CA 95819-6025. Price, $45.00 per volume.

5. *HANDBOOK FOR MONITORING INDUSTRIAL WASTEWATER*, US Environmental Protection Agency, Technology Transfer, Cincinnati, OH 45268. Available from National Technical Information Service (NTIS), 5285 Port Royal Road, Springfield, VA 22151. Ordering Number PB-259146. Price, $59.50, plus $5.00 shipping and handling per order.

6. *STANDARD INDUSTRIAL CLASSIFICATION MANUAL (SIC), 1987*. Obtain from the US Government Printing Office, Superintendent of Documents, PO Box 371954, Pittsburgh, PA 15250-7954. Order No. 041-001-00314-2. Price, $40.00.

7. *DEVELOPING SOURCE CONTROL PROGRAMS FOR COMMERCIAL AND INDUSTRIAL WASTEWATER*. Obtain from Water Environment Federation (WEF), Publications Order Department, 601 Wythe Street, Alexandria, VA 22314-1994. Order No. MM2004. Price to members, $27.74; nonmembers, $37.74; price includes cost of shipping and handling.

QUESTIONS

Write your answers in a notebook and then compare your answers with those on page 185.

5.5A Why should the flow from an industry be measured?

5.6A List the major areas of potential safety hazards while monitoring industrial wastes.

5.7A How would you regulate or handle a large, fluctuating industrial flow that hydraulically overloads a treatment plant during the day?

5.7B For which potentially harmful conditions are continuous monitoring and alarm systems for treatment plant influents suitable?

END OF LESSON 2 OF 2 LESSONS
on
INDUSTRIAL WASTE MONITORING

Please answer the discussion and review questions next.

TABLE 5.6 REACTIONS TO INDUSTRIAL DISCHARGES

Situation	Problem	Action
Gasoline, solvent; other hydrocarbons	Explosive or flammable conditions in wet wells and headworks.	➤ Hold gasoline or solvent on top of primary clarifiers and allow wind to disperse fumes. ➤ Inspect wet well headworks ventilation and provide extra ventilation with blowers only (no suction of vapors into fans). ➤ Divert to holding basin. ➤ Keep personnel clear of area. ➤ Pure oxygen plants: vent and purge reactors with air; shut down oxygen feed until lower explosive limit (LEL) out of alarm range and normal readings are obtained.
	Possible damage to secondary processes.	➤ Watch DO levels, waste and return sludge rates.
Excessive solids	Cause stoppages in sewers. Overload treatment processes.	➤ Flush sewers. ➤ Increase pumping from clarifiers. ➤ Balance feed to digesters. ➤ Watch for septic conditions in primary clarifiers. ➤ Watch for low DO in aeration systems.
Acidic or caustic wastes	Toxic to organisms in biomass.	➤ Neutralize. ➤ If too late and the biomass is damaged, adjust air rates, wasting rates, and foam control devices. Increase chlorine dosage for adequate disinfection.
Color and turbidity	Color and turbidity in effluent.	➤ Trace discharge back to source. ➤ Wastes from different industries may combine to form permanent colors. Control colors at source.
High-temperature waste	Septic conditions in sewers. Upset biological treatment processes. Trickling filter ponding. Increased oxygen demand.	➤ Try prechlorination to correct. ➤ Stop high-temperature discharges at source. ➤ Increase recirculation rates. ➤ More air is required in activated sludge process.

High organic load (suspended and dissolved solids)	Septic conditions in sewers.
	Upset biological treatment processes.
		➤ Try prechlorination to correct.
		➤ Activated sludge:
		  ➤ More air required.
		  ➤ Fast solids buildup. Try increasing wasting.
		  ➤ Foam. Try using sprays and/or defoamers.
		  ➤ Trickling filters. Increase recirculation rates.

TABLE 5.6 REACTIONS TO INDUSTRIAL DISCHARGES (continued)

Situation	Problem	Action
Oil and grease	Grease buildup in sewers.	➤ Clean more frequently.
	Scum blankets in digesters.	➤ Increase mixing.
	Excessive grease in clarifier scum box. Difficulty in pumping.	➤ Try removing grease more frequently.
	Secondary processes. ➤ May carry oil and grease over weirs. ➤ Trickling filters. Plugged orifices and ponding. ➤ Activated sludge. Oil coats biomass and foam problems develop.	➤ Increase maintenance: 　➤ Increase skimming at clarifier surface. 　➤ Increase frequency of cleaning and hosing. 　➤ Skim floating sludge and scum from surface of secondary clarifiers.
Toxic substances (heavy metals and pesticides)	Usually not detected in plant effluent. Plant processes deteriorate. Trickling filter sloughing can occur.	➤ Prevent from entering collection system. ➤ Try to build up biomass by stopping all sludge wasting and return sludge pumping. If necessary, obtain seed sludge from another plant. ➤ If wasting is continued, do not waste toxic substances to another biological process (aerobic or anaerobic digester). Try to minimize wasting of helpful organisms. ➤ Usually less air is required because of reduced biological activity. Excess air rates may strip out some toxic substances. ➤ Increase foam control if necessary. ➤ Extensive lab work to identify material. ➤ Find out what is being discharged and trace to source.
Nutrients (nitrogen and phosphorus)	Poor removals of nutrients by conventional plant processes and may cause aquatic growths in receiving waters.	➤ Source control of nutrients. ➤ Tertiary treatment to remove nutrients.
	Conversion of ammonia nitrogen to nitrite nitrogen will cause high chlorine demands and possibly reduce coliform kills.	➤ Change process to avoid converting ammonia to nitrite by lowering DO levels and MCRT.
Taste and odor problems	Undesirable tastes and odors in drinking waters.	➤ Phenolic compounds good example. They pass through the plant and cause water supply problems. Control at source. ➤ Carefully control chlorine so tastes and odors will not get worse.
	Hydrogen sulfide from: ➤ Septic conditions in sewers. ➤ Tannery and other wastes.	➤ Control hydrogen sulfide: 　➤ Try prechlorination. 　➤ Control at source.
Pathogenic wastes Radioactive wastes	Harmful to humans and usually not detectable at treatment plant.	➤ Control at source, such as hospitals and clinics.
Hydraulic shock load	Overload treatment processes.	➤ Reduce if possible by: 　➤ Diverting to storage basin. 　➤ Restricting influent flow. ➤ Increase chlorination rate.
	Activated sludge process may lose solids.	➤ Stop wasting.

DISCUSSION AND REVIEW QUESTIONS
Chapter 5. INDUSTRIAL WASTE MONITORING
(Lesson 2 of 2 Lessons)

Write the answers to these questions in your notebook. The question numbering continues from Lesson 1.

7. How would you attempt to trace or locate an industrial waste discharge that is causing problems?
8. Why is the "chain of custody" of a sample important?
9. How would you measure the flow from an industry?
10. What are the potential hazards that could be encountered when entering a manhole?
11. What basic items would you consider when reviewing the application or permit of an industrial waste discharger?
12. What are potential problems that could be caused by an industrial discharge of gasoline or solvent and what are some possible actions to take?

SUGGESTED ANSWERS
Chapter 5. INDUSTRIAL WASTE MONITORING

ANSWERS TO QUESTIONS IN LESSON 1

Answers to questions on page 160.

5.0A The objectives of an industrial waste monitoring program are to:
  1. Prevent serious shock loads and thus ensure that no disruptions of plant processes occur.
  2. Provide physical protection of personnel and facilities.
  3. Prevent discharge of pollutants from treatment plants.
  4. Provide a database for fair and reasonable sewer-use charges and for engineering data.

5.0B Certain types of industrial waste discharges have undesirable properties that could cause problems for collection systems and treatment plants. These include:
  1. Collection systems—acidic and alkaline wastes, thermal wastes, solids, oil and grease, odors, toxic substances, and flammable and explosive materials.
  2. Treatment plants—organic overloads, toxic substances, oil and grease, acids and alkalies, hydraulic overloads, refractory materials, and hazardous wastes.

Answers to questions on page 161.

5.1A A database file should be established and maintained on each discharging industry because the technical data contained in the file are essential for making decisions in areas such as operation and maintenance of treatment plant processes, program monitoring, and engineering design of facilities.

5.1B In developing a database, include the following information in each discharger's file: industry identification, flow data, laboratory results, billing information, treatment information, reports of unusual circumstances or incidents, industrial facilities, and emergency information.

5.1C A sewer-use ordinance should provide legal authority for the following activities:
  1. Inspecting and monitoring.
  2. Requiring pretreatment and/or flow adjustment.
  3. Curtailing or halting discharges.
  4. Requiring spill protection measures.
  5. Issuing permits to dischargers.
  6. Billing for sewer service charges.

Answers to questions on page 162.

5.2A Self-monitoring programs are commonly used in large industrial areas because the burden of monitoring a large number of industries would be overwhelming to a local agency.

5.2B Sensors used in continuous monitoring programs may require frequent cleaning and calibration to maintain their accuracy.

5.2C Representative sample: A sample portion of material or wastestream that is as nearly identical in content and consistency as possible to that in the larger body of material or wastestream being sampled.

5.2D To collect a representative sample, the sampling point needs to be located where the wastewater flow from the industrial plant is well mixed.

Answers to questions on page 170.

5.2E Grab samples are the simplest and least expensive sampling method. However, grab samples are not useful for discharges that vary in makeup (constituents) or quantity. Composite samples will provide the most accurate representation of a wastewater stream that can be obtained at a reasonable cost. A limitation of composite samples is that an isolated or peak discharge of a concentrated toxic waste could be diluted by the other samples taken during the sampling period.

5.2F Portable sampling equipment obtains power in the field from rechargeable batteries or, when possible, from a 110-volt AC line using a converter.

5.2G The flow velocity in the sampler suction tube is important in obtaining repeatable results. Lower velocities tend to leave solids behind in the tubing, while higher velocities may pull in large chunks of material, yielding erratic results in the data.

5.2H Cleanliness of monitoring equipment is very important to ensure accurate tests. Contamination of sample containers, tubing, and internal parts of samplers will affect the sample, and thereby lead to inaccurate laboratory results. The most reliable monitoring equipment will be that used where regular maintenance is performed in accordance with the manufacturer's instructions.

ANSWERS TO QUESTIONS IN LESSON 2

Answers to questions on page 172.

5.3A To trace or locate an industrial waste discharge that is causing problems, look for signs or evidence of the waste. Begin at the treatment plant or at a point in the collection system where the discharge has been discovered, then follow signs of the waste back to the source. Look for indicative color or solids and simply inspect the sewer at each sewer line intersection and follow it up to its source. If the discharge is not visible, use portable samplers to follow the waste to its source. If several samplers are available, the process can be speeded up greatly.

5.3B To sample for petroleum solvents or similar floatable materials, you can fasten the suction tubing to a block of Styrofoam or similar material and thus hold the suction end of the sampling tube very close to the surface.

5.4A A sample should be analyzed as soon as possible to prevent physical, chemical, or biological changes in the sample.

5.4B Samples are stored at 4°C to reduce the rate of chemical reactions and biological activity.

Answers to questions on page 181.

5.5A The flow from an industry should be measured to determine the quantity of pollutants discharged and also to indicate the hydraulic loading on the collection system and the wastewater treatment facilities.

5.6A Major areas of potential safety hazards while monitoring industrial wastes include:

1. Traffic safety
2. Confined spaces
3. Equipment storage areas
4. Battery charging

5.7A To regulate or handle a large, fluctuating industrial flow that hydraulically overloads a treatment plant during the day, try requiring the industry to store a portion of its discharges during peak flow periods for release during low flow periods such as during the graveyard shift.

5.7B Continuous monitoring and alarm systems for treatment plant influents are suitable for potentially harmful conditions including conductivity, pH, explosive atmosphere, including lower explosive limit (LEL), and oxygen deficiency/enrichment conditions.

APPENDIX A

INDUSTRIAL SEWER-USE PERMIT APPLICATION

Source: Water Quality Division,
Department of Public Works
Sacramento County,
Sacramento, California

188   Treatment Plants

INDUSTRIAL SEWER USE PERMIT APPLICATION

_____ SANITATION DISTRICT

PAGE 1
GENERAL
(Instructions on reverse)

DISTRICT USE
SUP # _____
AP # _____
CP # _____
SIC # _____

1. NAME OF BUSINESS _____
2. ADDRESS OF PREMISES _____
3. OWNER OF BUSINESS _____
4. OWNER'S ADDRESS (Street) _____
   (City) _____ (Zip) _____
5. OWNER OF PROPERTY (IF DIFFERENT) _____
6. PROPERTY OWNER'S ADDRESS (Street) _____
   (City) _____ (Zip) _____
7. SEND SEWER BILLS TO: ☐ BUSINESS (1&2); ☐ BUSINESS OWNER (3&4); ☐ PROPERTY OWNER (5&6)
8. PERSON TO CONTACT ABOUT THIS APPLICATION _____ Phone _____
9. DESCRIPTION OF BUSINESS _____
10. RATE BASIS: _____ _____
    UNIT   QUANTITY
11. WASTEWATER GENERATING OPERATIONS: _____
    _____
    _____
    _____
    _____
    _____

12. SEASONAL VARIATIONS _____
    _____
    _____
    _____
    _____
    _____

13. ARE ANY OF THE FOLLOWING MATERIALS USED OR STORED ON THE PREMISES?
    1. Flammable or explosive materials. 2. Acid, alkaline, or corrosive material. 3. Pesticides or toxic material such as Aldrin, Dieldrin, Benzidine, Cadmium, Cyanide, DDD, DDE, DDT, Endrin, Mercury, PCBs, Toxaphene, Etc. 4. Oil, grease, or solvents. 5. Metal solutions. 6. Phenols. 7. Large amounts of soaps or detergents. 8. Radioactive material. 9. Dyes.

    ☐ NO   ☐ YES (if yes, please give description, and the approximate quantities used and/or stored.)
    _____
    _____
    _____
    _____

14. PERSON TO CONTACT IN AN EMERGENCY _____
    TITLE _____ Phone _____ Emergency No. _____

INSTRUCTIONS FOR COMPLETING PAGE 1

1. ENTER NAME OR TITLE OF BUSINESS.
2. ENTER FULL STREET ADDRESS OF BUILDING OR PREMISES PRODUCING THE WASTEWATER.
3. ENTER NAME OF INDIVIDUAL OR FIRM THAT IS THE OWNER OF THE BUSINESS.
4. ENTER MAILING ADDRESS OF OWNER OF BUSINESS.
5. ENTER NAME OF LEGAL OWNER OF PROPERTY UPON WHICH THE BUSINESS IS LOCATED, IF IT IS DIFFERENT THAN THE OWNER OF THE BUSINESS.
6. ENTER MAILING ADDRESS OF PROPERTY OWNER.
7. INDICATE WHERE SEWER BILLS SHOULD BE SENT. (UNLESS OTHERWISE INDICATED, THE BILLS WILL BE SENT TO THE PROPERTY OWNER.)
8. IDENTIFY PERSON WHO IS THOROUGHLY FAMILIAR WITH THE FACTS REPORTED ON THESE FORMS AND MAY BE CONTACTED BY THE DISTRICT.
9. DESCRIBE THE PRINCIPAL ACTIVITY ON THE PREMISES, SUCH AS FOOD CANNING, BANK, ETC.
10. DETERMINE FROM TABLE BELOW WHICH UNIT WOULD BE APPLICABLE FOR THIS TYPE OF BUSINESS AND THE NUMBER OF UNITS. IF, FOR EXAMPLE, THE BUSINESS IS A MARKET, ENTER "AREA" FOR UNIT. THEN ENTER THE NUMBER OF SQUARE FEET IN THE STORE FOR QUANTITY.

BUSINESS	UNIT	BUSINESS	UNIT
Auto dealerships	Area (s.f.)	Market	Area (s.f.)
Bakeries	Area (s.f.)	Medical, Dental	Area (s.f.)
Banks and financial	Area (s.f.)	Mortuaries	Slumber rooms
Barber/beauty shops	Chairs	Offices	Area (s.f.)
Bars	Area (s.f.)	Places of Worship	Area (s.f.)
Bowling alleys	Lanes	Public agencies	Area (s.f.)
Car wash (full service)	Flow*	Rest/convalescent homes	Beds
Car wash (self service)	Stalls	Restaurants	Area (s.f.)
Dry cleaners	Area (s.f.)	Retail stores	Area (s.f.)
Garages	Bays	Schools	Attendance
Halls, Lodges	Area (s.f.)	Service stations	Pumps
Health studio, Gym	Area (s.f.)	Theaters	Seats
Hotel, Motel	Sleeping Rooms	Used car lots	Fixture units
Laundry (self service)	Machines	Warehouses	Area (s.f.)
Laundry (commercial)	Flow*	Others	Flow*

* estimate number of gallons of wastewater discharged each month

11. DESCRIBE WASTEWATER GENERATING PROCESS OCCURING ON THE PREMISES, INCLUDING PLANT OPERATIONS, RAW MATERIALS USED, CHEMICALS USED, AND ANY VARIATIONS IN DISCHARGE VOLUMES.
12. INDICATE WHETHER THE BUSINESS ACTIVITY IS CONTINUOUS THROUGHOUT THE YEAR OR IF IT IS SEASONAL. DESCRIBE THE SEASONAL VARIATION, LISTING MONTHS OF SEASONAL ACTIVITY.
13. LIST SIGNIFICANT RAW MATERIALS USED OR STORED ON PREMISES AND INDICATE DURATION OF STORAGE. INCLUDE ALL HAZARDOUS, POISONOUS, OR TOXIC MATERIALS EVEN IF THEY ARE KEPT ON PREMISES ONLY OCCASIONALLY. NEGLECT MATERIALS USED IN LABORATORY OR QUALITY CONTROL OPERATIONS. IF QUANTITIES IN INVENTORY VARY, SELECT AN AVERAGE AMOUNT OR GIVE RANGES. USE ADDITIONAL SHEETS IF NECESSARY.
14. GIVE NAME, TITLE, AND TELEPHONE NUMBER(S) OF A RESPONSIBLE PERSON WHO CAN BE CONTACTED IN CASE OF AN EMERGENCY (e.g. SPILLING OF A TOXIC MATERIAL).

INDUSTRIAL SEWER USE PERMIT APPLICATION

_____ SANITATION DISTRICT

PAGE 2
LAYOUT
(Instructions on reverse)

SITE PLAN OF PREMISES

INDUSTRIAL SEWER USE PERMIT APPLICATION

INSTRUCTIONS FOR COMPLETING PAGE 2

DRAW A SITE PLAN SHOWING LOCATION OF ALL PERTINENT BUILDINGS, SHEDS, WAREHOUSES, LOADING AND UNLOADING FACILITIES, FENCES, PROPERTY LINES, STREETS, AND ROADS. INDICATE ALL SEWERS, STORM DRAINS, DRAINAGE DITCHES, MANHOLES, SAMPLING AND MONITORING LOCATIONS, WATER LINES AND METERS, AND SHOW THE SIZES OF THESE ITEMS. SHOW ALL POINTS OF CONNECTION TO THE PUBLIC SEWER AND DRAIN LINES. INDICATE SCALE AND NORTH ARROW. USE ADDITIONAL SHEETS IF NECESSARY. (BUILDING OR SITE PLANS ACCEPTABLE TO THE DISTRICT MAY BE SUBSTITUTED FOR THIS FORM.)

192 Treatment Plants

INDUSTRIAL SEWER USE PERMIT APPLICATION

_____ SANITATION DISTRICT

PAGE 3
WASTEWATER DATA
(Instructions on reverse)

1. WATER SOURCE
   - ☐ PRIVATE WELL
   - ☐ PUBLIC (METERED)
   - ☐ PUBLIC (UNMETERED)
   - ☐ OTHER _____

2. WASTEWATER FLOW RATE TO THE SEWER:

PEAK HOURLY	MAX. DAILY	AVERAGE DAILY
_____ (gal/hour)	_____ (gal/day)	_____ (gal/day)

3. DO YOU WISH PERMISSION TO DISCHARGE ANY OF THE FOLLOWING

	YES	NO
1. More than 50,000 gallons per day.	☐	☐
2. A "slug" (more than five times the normal flow or strength for longer than 15 minutes).	☐	☐
3. A concentration of BOD or suspended solids in excess of 300 mg/L, or COD in excess of 500 mg/L	☐	☐
4. More than 3,000 lbs of BOD or suspended solids, or 5,000 lbs of COD per day	☐	☐
5. Animal or vegetable derived oil or grease in excess of 300 mg/L, or petroleum derived oil or grease in excess of 100 mg/L.	☐	☐
6. pH between 5.0 and 6.0, or higher than 9.0 (less than 5.0 is prohibited).	☐	☐
7. Metals, metal pickling wastes or plating solutions, phenols or other taste or odor producing substances, or soap or detergent.	☐	☐
8. Temperature over 160°F maximum, or 120°F average.	☐	☐
9. Stormwater, cooling water, etc., which is polluted or otherwise unacceptable for discharge into storm drains or natural outlets.	☐	☐
10. Garbage (except from homes or restaurants).	☐	☐
11. Radioactive waste.	☐	☐
12. Pool water or waste.	☐	☐
13. Materials which cause unusual amounts of inert suspended solids (e.g. soil solids), dissolved solids (e.g. sodium chloride), or discoloration (e.g. dyes).	☐	☐
14. Discharges regulated by the EPA.	☐	☐

If any are marked "YES," give details of each in "REMARKS" below, or attach sheet.
NOTE: THESE ITEMS MAY BE DISCHARGED ONLY IF SPECIFICALLY APPROVED.

4. ATTACH SHEET DESCRIBING ANY PRETREATMENT USED OR PLANNED.

5. REMARKS:

INSTRUCTIONS FOR COMPLETING PAGE 3

1. INDICATE WHERE WATER USED ON THE PREMISES IS OBTAINED.
2. LIST TOTAL DISCHARGES FROM PREMISES TO EACH SEPARATE SEWER.

    PEAK HOURLY—Indicate expected maximum flow in a one-hour period at any time in the year. Do not include batch discharges.

    MAXIMUM DAILY—Indicate expected maximum flow during a 24-hour period.

    AVERAGE DAILY—Estimate the flow during an average 24-hour period.

3. THESE ITEMS ARE REGULATED AND MAY BE DISCHARGED ONLY WITH THE SPECIFIC PERMISSION OF THE DISTRICT. INDICATE WHETHER YOU ANTICIPATE DISCHARGING THESE. IF SO, GIVE DETAILS IN "REMARKS" (#5).
4. DESCRIBE, ON A SEPARATE SHEET, ANY PRETREATMENT GIVEN THE WASTEWATER BEFORE IT IS DISCHARGED INTO THE SEWER. THE TREATMENT FACILITY SHOULD BE DESCRIBED IN SUFFICIENT DETAIL TO ENABLE AN ESTIMATION OF ITS EFFECTIVENESS.
5. GIVE ANY DETAILS REQUIRED BY #3. ALSO PROVIDE ANY INFORMATION ABOUT YOUR OPERATION THAT YOU FEEL MIGHT HAVE AN EFFECT ON THE SEWERAGE SYSTEM.

INDUSTRIAL SEWER USE PERMIT APPLICATION

_____ SANITATION DISTRICT

PAGE 4
CERTIFICATION
(Instructions on reverse)

> WARNING—DISCHARGE OF SUBSTANCES INTO THE PUBLIC SEWER IS REGULATED BY LAW AND IS SUBJECT TO CIVIL AND CRIMINAL PENALTIES. IF YOU ANTICIPATE DISCHARGING ANYTHING OTHER THAN NORMAL DOMESTIC SEWAGE, YOU ARE ADVISED TO READ THE "SEWER USE ORDINANCE" ADOPTED BY THE BOARD OF DIRECTORS OF THE DISTRICT.

The Sewer Use Ordinance prohibits any discharge which would cause a hazard or interfere with the operation of the District's facilities, or would result in contamination, nuisance, or pollution of public waterways (refer to the ordinance for wording). In addition, the ordinance specifically prohibits the discharge to public sewers of the following:

PROHIBITED DISCHARGES—(BRIEF DESCRIPTION, see Sect. 6.4, Regional Sewer Use Ordinance for full description):

1. Unpolluted storm or other waters.
2. Wastewater having pH lower than 5.0, or other corrosive properties.
3. Certain pesticides and other toxic pollutants.
4. Solid or viscous substances capable of causing an obstruction to the flow in sewers, or other interference with the proper operation or maintenance of the sewerage system.
5. Wastes which cannot be treated by the District's processes.
6. Materials prohibited by the EPA.

NOTE—A SEWER USE PERMIT PERTAINS ONLY TO THE DISCHARGE OF WASTEWATER INTO THE PUBLIC SEWERAGE SYSTEM. CONNECTION TO THE PUBLIC SEWER, AND THE INSTALLATION OR MODIFICATION OF ON-SITE PLUMBING, REQUIRES SEPARATE PERMITS.

- -

CERTIFICATION: I certify that the information contained herein is true and correct to the best of my knowledge.

Signature _____   Date _____

Name (type or print) _____   Title _____

SEWER USE PERMIT

The above named applicant is hereby authorized to use the public sewerage system subject to the following conditions:

1. Compliance with applicable sewer use ordinances
2. Payment of all applicable fees and charges
3.

The applicant shall report to the Water Quality Division any changes (permanent or temporary) to the premises or operations that could significantly change the quality or volume of the discharge, or deviate from the conditions under which this permit is granted. This permit is not transferable.

LOCAL APPROVAL   Signed _____   Date _____

INSTRUCTIONS FOR COMPLETING PAGE 4

CERTIFICATION:

    The application must be signed and dated by an officer, employee, or other agent of the business who has legal authority to bind the applicant business. Also type or print the name and title of the person signing the application.

RETURN THE APPLICATION TO:

    Industrial Waste Section
    Water Quality Division

* *

Do not complete the portion below the line. The Water Quality Division will determine what conditions will be applied to the permit, then will sign and return the permit to the applicant.

APPENDIX B

SAMPLE SEWER-USE ORDINANCE

Source: US Environmental Protection Agency

ORDINANCE NO. _____

AN ORDINANCE establishing rules and regulations for the discharge of wastewaters into the wastewater treatment system of the City of _____; and

WHEREAS, the Federal Water Pollution Control Act Amendments of 1972, PL 92-500 (hereinafter referred to as the "Act") have resulted in an unprecedented program of cleaning up our Nation's waters; and

WHEREAS, this City has already made and will continue to make a substantial financial investment in its wastewater treatment system to achieve the goals of the Act; and

WHEREAS, this City seeks to provide for the use of its wastewater treatment system by industries served by it without damage to the physical facilities, without impairment of their normal function of collecting, treating, and discharging domestic wastewater, and without the discharge by this City's wastewater treatment system of pollutants which would violate the discharge allowed under its National Pollutant Discharge Elimination System (NPDES) permit and the applicable rules of all governmental authorities with jurisdiction over such discharges;

NOW, THEREFORE, BE IT ORDAINED AND ENACTED by the City Council of the City of _____, County of _____, State of _____, as follows:

SECTION 1: DEFINITIONS

Unless the context specifically indicates otherwise, the following terms as used in this Ordinance, shall have the meanings hereinafter designated:

(a) *"BIOCHEMICAL OXYGEN DEMAND"* (BOD) means the quantity of oxygen utilized in the biochemical oxidation of organic matter under standard laboratory procedure in five (5) days at 20°C, expressed in terms of weight and concentration (milligrams per liter).

(b) *"COOLING WATER"* means the water discharged from any use such as air conditioning, cooling, or refrigeration, during which the only pollutant added to the water is heat.

(c) *"COMPATIBLE POLLUTANT"* means BOD, suspended solids, pH, and fecal coliform bacteria, and such additional pollutants as are now or may be in the future specified and controlled in this City's NPDES permit for its wastewater treatment works where said works have been designated and used to reduce or remove such pollutants.

(d) *"DIRECTOR/(SUPERINTENDENT)"* means the (director/superintendent of wastewater treatment system/or water pollution control/or public works) of this City or the duly appointed deputy, agent, or representative.

(e) *"DOMESTIC WASTES"* means liquid wastes (i) from the noncommercial preparation, cooking, and handling of food, or (ii) containing human excrement and similar matter from the sanitary conveniences of dwellings, commercial buildings, industrial facilities, and institutions.

(f) *"GARBAGE"* means solid wastes from the domestic and commercial preparation, cooking, and dispensing of food, and from the handling, storage, and sale of food.

(g) *"INCOMPATIBLE POLLUTANT"* means any pollutant which is not a "compatible pollutant" as defined in this section.

(h) *"INDUSTRIAL WASTEWATER"* means the liquid wastes resulting from the processes employed in industrial, manufacturing, trade or business establishments, as distinct from domestic wastes.

(i) *"NATIONAL POLLUTANT DISCHARGE ELIMINATION SYSTEM"* (NPDES) means the program for issuing, conditioning, and denying permits for the discharge of pollutants from point sources into the navigable waters, the contiguous zone, and the oceans pursuant to Section 402 of the Act.

(j) *"PERSON"* means any individual, firm, company, partnership, corporation, association, group, or society, and includes the State of _____, and agencies, districts, commissions, and political subdivisions created by or pursuant to State law.

(k) *"pH"* means the logarithm of the reciprocal of the concentration of hydrogen ions in grams per liter of solution.

(l) *"PRETREATMENT"* means application of physical, chemical, and biological processes to reduce the amount of pollutants in or alter the nature of the pollutant properties in a wastewater prior to discharging such wastewater into the publicly owned wastewater treatment system.

(m) *"PRETREATMENT STANDARDS"* means all applicable Federal rules and regulations implementing Section 307 of the Act, as well as any nonconflicting State or local standards. In cases of conflicting standards or regulations, the more stringent thereof shall be applied.

(n) *"SIGNIFICANT INDUSTRIAL USER"* means any industrial user of the City's wastewater treatment system whose flow exceeds (i) (50,000) gallons per day, or (ii) (five (5)) percent of the daily capacity of the treatment system.

(o) *"STORM WATER"* means any flow occurring during or immediately following any form of natural precipitation and resulting therefrom.

(p) *"SUSPENDED SOLIDS"* means the total suspended matter that floats on the surface of, or is suspended in, water, wastewater, or other liquids, and which is removable by laboratory filtering.

(q) *"UNPOLLUTED WATER"* is water not containing any pollutants limited or prohibited by the effluent standards in effect, or water whose discharge will not cause any violation of receiving water quality standards.

(r) *"USER"* means any person who discharges, causes, or permits the discharge of wastewater into the City's wastewater treatment system.

(s) *"USER CLASSIFICATION"* means a classification of user based on the 1987 (or subsequent) edition of the Standard

Industrial Classification (SIC) Manual prepared by the Office of Management and Budget.

(t) *"WASTEWATER"* means the liquid and water-carried industrial or domestic wastes from dwellings, commercial buildings, industrial facilities, and institutions, together with any groundwater, surface water, and storm water that may be present, whether treated or untreated, which is discharged into or permitted to enter the City's treatment works.

(u) *"WASTEWATER TREATMENT SYSTEM"* (system) means any devices, facilities, structures, equipment, or works owned or used by the City for the purpose of the transmission, storage, treatment, recycling, and reclamation of industrial and domestic wastes, or necessary to recycle or reuse water at the most economical cost over the estimated life of the system, including intercepting sewers, outfall sewers, sewage collection systems, pumping, power, and other equipment, and their appurtenances; extensions, improvements, remodeling, additions, and alterations thereof; elements essential to provide a reliable recycled supply such as standby treatment units and clear well facilities; and any works, including site acquisition of the land that will be an integral part of the treatment process or is used for ultimate disposal of residues resulting from such treatment.

(v) Terms not otherwise defined herein shall be as adopted in the latest edition of *STANDARD METHODS FOR THE EXAMINATION OF WATER AND WASTEWATER*, published by the American Public Health Association, the American Water Works Association, and the Water Pollution Control Federation.

SECTION 2: PROHIBITIONS AND LIMITATIONS OF WASTEWATER DISCHARGES

(a) *PROHIBITIONS ON WASTEWATER DISCHARGES.* No person shall discharge or deposit or cause or allow to be discharged or deposited into the wastewater treatment system any wastewater which contains the following:

(1) *OILS AND GREASE.* (A) Oil and grease concentrations or amounts from industrial facilities violating Federal pretreatment standards. (B) Wastewater from industrial facilities containing floatable fats, wax, grease, or oils (Optional: (C) Wax, grease, or oil concentrations of mineral origin of more than ( ) mg/L whether emulsified or not, or containing substances which may solidify or become viscous at temperatures between 32° and 150°F (0° and 65°C) at the point of discharge into the system.) (Optional: (D) Total fat, wax, grease, or oil concentrations of more than ( ) mg/L, whether emulsified or not, or containing substances which may solidify or become viscous at temperatures between 32° and 150°F (0° and 65°C) at the point of discharge into the system.)

(2) *EXPLOSIVE MIXTURES.* Liquids, solids, or gases which by reason of their nature or quantity are, or may be, sufficient either alone or by interaction with other substances to cause fire or explosion or be injurious in any other way to the sewerage facilities or to the operation of the system. At no time shall two successive readings on an explosion hazard meter, at the point of discharge into the sewer system, be more than five percent (5%) nor any single reading over ten percent (10%) of the Lower Explosive Limit (LEL) of the meter. Prohibited materials include, but are not limited to, gasoline, kerosene, naphtha, benzene, toluene, zylene, ethers, alcohols, ketones, aldehydes, peroxides, chlorates, perchlorates, bromates, carbides, hydrides, and sulfides.

(3) *NOXIOUS MATERIAL.* Noxious or malodorous solids, liquids, or gases, which, either singly or by interaction with other wastes, are capable of creating a public nuisance or hazard to life, or are or may be sufficient to prevent entry into a sewer for its maintenance and repair.

(4) *IMPROPERLY SHREDDED GARBAGE.* Garbage that has not been ground or comminuted to such a degree that all particles will be carried freely in suspension under flow conditions normally prevailing in the public sewers, with no particle greater than one-half (1/2) inch in any dimension.

(5) *RADIOACTIVE WASTES.* Radioactive wastes or isotopes of such half-life or concentration that they do not comply with regulations or orders issued by the appropriate authority having control over their use and which will or may cause damage or hazards to the sewerage facilities or personnel operating the system.

(6) *SOLID OR VISCOUS WASTES.* Solid or viscous wastes which will or may cause obstruction to the flow in a sewer, or otherwise interfere with the proper operation of the wastewater treatment system. Prohibited materials include, but are not limited to, grease, uncomminuted garbage, animal guts or tissues, paunch manure, bones, hair, hides or fleshings, entrails, whole blood, feathers, ashes, cinders, sand, spent lime, stone or marble dust, metal, glass, straw, shavings, grass clippings, rags, spent grains, spent hops, waste paper, wood, plastic, tar, asphalt residues, residues from refining or processing of fuel or lubricating oil, and similar substances.

(7) *EXCESSIVE DISCHARGE RATE.* Wastewaters at a flow rate or containing such concentrations or quantities of pollutants that exceed for any time period longer than fifteen (15) minutes more than five (5) times the average twenty-four (24) hour concentration, quantities or flow during normal operation and that would cause a treatment process upset and subsequent loss of treatment efficiency.

(8) *TOXIC SUBSTANCES.* Any toxic substances in amounts exceeding standards promulgated by the Administrator of the United States Environmental Protection Agency pursuant to Section 307(a) of the Act, and chemical elements or compounds, phenols or other taste- or odor-producing substances, or any other substances which are not susceptible to treatment or which may interfere with the biological processes or efficiency of the treatment system, or that will pass through the system.

(9) *UNPOLLUTED WATERS.* Any unpolluted water including, but not limited to, water from cooling systems or of storm-water origin, which will increase the hydraulic load on the treatment system.

(10) *DISCOLORED MATERIAL.* Wastes with objectionable color not removable by the treatment process.

(11) *CORROSIVE WASTES*. Any waste which will cause corrosion or deterioration of the treatment system. All wastes discharged to the public sewer system must have a pH value in the range of six (6) to nine (9) standard units. Prohibited materials include, but are not limited to, acids, sulfide, concentrated chloride and fluoride compounds, and substances which will react with water to form acidic products.

(b) *LIMITATIONS ON WASTEWATER DISCHARGES.* (Use either Option A or Option B.)

(Option A—General Limitations)

No person shall discharge or convey, or permit or allow to be discharged or conveyed, to a public sewer any wastewater containing pollutants of such character or quantity that will:

(1) Not be susceptible to treatment or interfere with the process or efficiency of the treatment system.

(2) Constitute a hazard to human or animal life, or to the stream or water course receiving the treatment plant effluent.

(3) Violate pretreatment standards.

(4) Cause the treatment plant to violate its NPDES permit or applicable receiving water standards.

(Option B—Specific Limitations)

The following are the maximum concentrations of pollutants allowable in wastewater discharges to the wastewater treatment system. Dilution of any wastewater discharge for the purpose of satisfying these requirements shall be considered a violation of this Ordinance.

Pollutant	**Concentration (mg/L) or Mass Limitation (kg/kg)**
Arsenic	(Options: See, e.g., Federal Guidelines; State and Local Pretreatment Programs, Volume I, Section C)
Barium	
Boron	
Cadmium	
Chromium (Total)	
Chromium (Trivalent)	
Chromium (Hexavalent)	
Chlorinated Hydrocarbons	
Copper	
Cyanide	
Iron	
Lead	
Manganese	
Mercury	
Nickel	
Phenolic Compounds	
Phosphorus	
Selenium	
Silver	
Surfactants	
Zinc	
pH	
Temperature	Not over 150°F (except where higher temperatures are permitted by law).

(c) *SPECIAL AGREEMENTS*. Nothing in this Section shall be construed as preventing any special agreement or arrangement between the City and any user of the wastewater treatment system whereby wastewater of unusual strength or character is accepted into the system and specially treated subject to any payments or user charges as may be applicable.

SECTION 3: CONTROL OF PROHIBITED WASTES

(a) *REGULATORY ACTIONS*. If wastewaters containing any substance described in *SECTION 2* of this Ordinance are discharged or proposed to be discharged into the sewer system of the City or to any sewer system tributary thereto, the Director and (Corporation Counsel/City Attorney) may take any action necessary to:

(1) Prohibit the discharge of such wastewater.

(2) Require a discharger to demonstrate that in-plant modifications will reduce or eliminate the discharge of such substances in conformity with this Ordinance.

(3) Require pretreatment, including storage facilities, or flow equalization necessary to reduce or eliminate the objectionable characteristics or substances so that the discharge will not violate these rules and regulations.

(4) Require the person making, causing, or allowing the discharge to pay any additional cost or expense incurred by the City for handling and treating excess loads imposed on the treatment system.

(5) Take such other remedial action as may be deemed to be desirable or necessary to achieve the purpose of this Ordinance.

(b) *SUBMISSION OF PLANS*. Where pretreatment or equalization of wastewater flows prior to discharge into any part of the wastewater treatment system is required, plans, specifications, and other pertinent data or information relating to such pretreatment or flow-control facilities shall first be submitted to the Director for review and approval. Such approval shall not exempt the discharge or such facilities from compliance with any applicable code, ordinance, rule, regulation, or order of any governmental authority. Any subsequent alterations or additions to such pretreatment or flow-control facilities shall not be made without due notice to and prior approval of the Director.

(c) *PRETREATMENT FACILITIES OPERATIONS*. If pretreatment or control of waste flows is required, such facilities shall be maintained in good working order and operated as efficiently as possible by the owner or operator at his own cost and expense, subject to the requirements of these rules and regulations and all other applicable codes, ordinances, and laws.

(d) *ADMISSION TO PROPERTY*. Whenever it shall be necessary for the purposes of these rules and regulations, the Director, upon the presentation of credentials, may enter upon any property or premises at reasonable times for the purpose of (1) copying any records required to be kept under the provisions of this Ordinance, (2) inspecting any monitoring equipment or method, and (3) sampling any discharge of wastewater to the treatment works. The Director may enter upon the property at any hour under emergency circumstances.

(e) *PROTECTION FROM ACCIDENTAL DISCHARGE.* Each industrial user shall provide protection from accidental discharge of prohibited materials or other wastes regulated by this Ordinance. Facilities to prevent accidental discharge of prohibited materials shall be provided and maintained at the owner or operator's own cost and expense. Detailed plans showing facilities and operating procedures to provide this protection shall be submitted to the Director for review, and shall be approved by him before construction of the facility. Review and approval of such plans and operating procedures shall not relieve the industrial user from the responsibility to modify the facility as necessary to meet the requirements of this Ordinance.

(f) *REPORTING OF ACCIDENTAL DISCHARGE.* If, for any reason, a facility does not comply with or will be unable to comply with any prohibition or limitations in this Ordinance, the facility responsible for such discharge shall immediately notify the Director so that corrective action may be taken to protect the treatment system. In addition, a written report addressed to the Director detailing the date, time, and cause of the accidental discharge, the quantity and characteristics of the discharge and corrective action taken to prevent future discharges, shall be filed by the responsible industrial facility within five (5) days of the occurrence of the noncomplying discharge.

SECTION 4: INDUSTRIAL WASTEWATER MONITORING AND REPORTING

(a) *DISCHARGE REPORTS.*

(1) Every significant industrial user shall file a periodic Discharge Report at such intervals as are designated by the Director. The Director may require any other industrial users discharging or proposing to discharge into the treatment system to file such periodic reports.

(2) The discharge report shall include, but, in the discretion of the Director, shall not be limited to, nature of process, volume, rates of flow, mass emission rate, production quantities, hours of operation, concentrations of controlled pollutants, or other information which relates to the generation of waste. Such reports may also include the chemical constituents and quantity of liquid materials stored on site even though they are not normally discharged. In addition to discharge reports, the Director may require information in the form of (Industrial Discharge Permit Applications and (optional)) self-monitoring reports.

(b) *RECORDS AND MONITORING.*

(1) All industrial users who discharge or propose to discharge wastewaters to the wastewater treatment system shall maintain such records of production and related factors, effluent flows, and pollutant amounts or concentrations as are necessary to demonstrate compliance with the requirements of this Ordinance and any applicable State or Federal pretreatment standards or requirements.

(2) Such records shall be made available upon request by the Director. All such records relating to compliance with pretreatment standards shall be made available to officials of the US Environmental Protection Agency upon demand. A summary of such data indicating the industrial user's compliance with this Ordinance shall be prepared (quarterly) (optional) and submitted to the Director.

(3) The owner or operator of any premises or facility discharging industrial wastes into the system shall install at his own cost and expense suitable monitoring equipment to facilitate the accurate observation, sampling, and measurement of wastes. Such equipment shall be maintained in proper working order and kept safe and accessible at all times.

(4) The monitoring equipment shall be located and maintained on the industrial user's premises outside of the building. When such a location would be impractical or cause undue hardship on the user, the Director may allow such facility to be constructed in the public street or sidewalk area, with the approval of the public agency having jurisdiction over such street or sidewalk, and located so that it will not be obstructed by public utilities, landscaping, or parked vehicles.

(5) When more than one user can discharge into a common sewer, the Director may require installation of separate monitoring equipment for each user. When there is a significant difference in wastewater constituents and characteristics produced by different operations of a single user, the Director may require that separate monitoring facilities be installed for each separate discharge.

(6) Whether constructed on public or private property, the monitoring facilities shall be constructed in accordance with the Director's requirements and all applicable construction standards and specifications.

(c) *INSPECTION, SAMPLING, AND ANALYSIS.*

(1) *COMPLIANCE DETERMINATION.* Compliance determinations with respect to *SECTION 2* prohibitions and limitations may be made on the basis of either instantaneous grab samples or composite samples of wastewater. Composite samples may be taken over a 24-hour period, or over a longer or shorter time span, as determined necessary by the Director to meet the needs of specific circumstances.

(2) *ANALYSIS OF INDUSTRIAL WASTEWATERS.* Laboratory analysis of industrial wastewater samples shall be performed in accordance with the current edition of "*STANDARD METHODS,*" "*METHODS FOR CHEMICAL ANALYSIS OF WATER AND WASTE,*" published by the US Environmental Protection Agency, or the "*ANNUAL BOOK OF STANDARDS, PART 23, WATER, ATMOSPHERIC ANALYSIS,*" published by the American Society for Testing and Materials. Analysis of those pollutants not covered by these publications shall be performed in accordance with procedures established by the (State Department of Environmental Health).

(3) *SAMPLING FREQUENCY* (Optional). Sampling of industrial wastewater for the purpose of compliance determination with respect to *SECTION 2* prohibitions and limitations will be done at such intervals as the Director may designate. However, it is the intention of the Director to conduct compliance sampling or to cause such sampling to be conducted for all major contributing industries at least once in every (1 year) (optional) period.

SECTION 5: INDUSTRIAL DISCHARGE PERMIT SYSTEM (OPTIONAL)

(a) *WASTEWATER DISCHARGE PERMITS REQUIRED.* All significant industrial users proposing to connect to or discharge into any part of the wastewater treatment system must first obtain a discharge permit. All existing significant industrial users connected to or discharging to any part of the City system must obtain a wastewater discharge permit within ninety (90) (optional) days from and after the effective date of this Ordinance.

(b) *PERMIT APPLICATION.* Users seeking a wastewater discharge permit shall complete and file with the Director an application on the form prescribed by the Director, and accompanied by the application fee. In support of this application, the user shall submit the following information:

(1) Name, address, and SIC number of applicant.

(2) Volume of wastewater to be discharged.

(3) Wastewater constituents and characteristics including, but not limited to, those set forth in *SECTION 2* of this Ordinance as determined by a reliable analytical laboratory.

(4) Time and duration of discharge.

(5) Average and (thirty (30)) (optional) minute peak wastewater flow rates, including daily, monthly, and seasonal variations, if any.

(6) Site plans, floor plans, mechanical and plumbing plans, and details to show all sewers and appurtenances by size, location, and elevation.

(7) Description of activities, facilities, and plant processes on the premises including all materials and types of materials which are, or could be, discharged.

(8) Each product produced by type, amount, and rate of production.

(9) Number and type of employees, and hours of work.

(10) Any other information as may be deemed by the Director to be necessary to evaluate the permit application.

The Director will evaluate the data furnished by the user and may require additional information. After evaluation and acceptance of the data furnished, the Director may issue a wastewater discharge permit subject to terms and conditions provided herein.

(c) *PERMIT CONDITIONS.* Wastewater discharge permits shall be expressly subject to all provisions of this Ordinance and all other regulations, user charges, and fees established by the City. The conditions of wastewater discharge permits shall be uniformly enforced in accordance with this Ordinance, and applicable State and Federal regulations. Permit conditions will include the following:

(1) The unit charge or schedule of user charges and fees for the wastewater to be discharged to the system.

(2) The average and maximum wastewater constituents and characteristics.

(3) Limits on rate and time of discharge or requirements for flow regulations and equalization.

(4) Requirements for installation of inspection and sampling facilities, and specifications for monitoring programs.

(5) Requirements for maintaining and submitting technical reports and plant records relating to wastewater discharges.

(6) Daily average and daily maximum discharge rates, or other appropriate conditions when pollutants subject to limitations and prohibitions are proposed or present in the user's wastewater discharge.

(7) Compliance schedules.

(8) Other conditions to ensure compliance with this Ordinance.

(d) *DURATION OF PERMITS.* Permits shall be issued for a specific time period, not to exceed (five (5)) (optional) years. A permit may be issued for a period of less than (one (1)) (optional) year, or may be stated to expire on a specific date. If the user is not notified by the Director (thirty (30)) (optional) days prior to the expiration of the permit, the permit shall automatically be extended for ( ) months. The terms and conditions of the permit may be subject to modification and change by the (responsible official) during the life of the permit, as limitations or requirements as identified in *SECTION 2* are modified and changed. The user shall be informed of any proposed changes in the permit at least (thirty (30)) (optional) days prior to the effective date of change. Any changes or new conditions in the permit shall include a reasonable time schedule for compliance.

(e) *TRANSFER OF A PERMIT.* Wastewater discharge permits are issued to a specific user for a specific operation. A wastewater discharge permit shall not be reassigned or transferred or sold to a new owner, new user, different premises, or a new or changed operation.

(f) *REVOCATION OF PERMIT.* Any user who violated the following conditions of the permit or of this Ordinance, or of applicable State and Federal regulations, is subject to having the permit revoked. Violations subjecting a user to possible revocation of the permit include, but are not limited to, the following:

(1) Failure of a user to accurately report the wastewater constituents and characteristics of the discharge;

(2) Failure of the user to report significant changes in operations, or wastewater constituents and characteristics;

(3) Refusal of reasonable access to the user's premises for the purpose of inspection or monitoring; or

(4) Violation of conditions of the permit.

SECTION 6: ENFORCEMENT PROCEDURES

(a) *NOTIFICATION OF VIOLATION.* Whenever the Director finds that any person has violated or is violating the Ordinance, or any prohibition, limitation, or requirement contained herein, the Director may serve upon such person a written notice stating the nature of the violation and providing a reasonable time, not to exceed thirty (30) days, for the satisfactory correction thereof.

(b) *SHOW CAUSE HEARING.*

(1) If the violation is not corrected by timely compliance, the Director may order any person who causes or allows an unauthorized discharge to show cause before the (hearing authority) why service should not be terminated. A notice shall be served on the offending party, specifying the time and place of a hearing to be held by the (hearing authority) regarding the violation, and directing the offending party to show cause before (said authority) why an order should not be made directing the termination of service. The notice of the hearing shall be served personally or by registered or certified mail (return receipt requested) at least 10 (ten) days before the hearing. Service may be made on any agent or officer of a corporation.

(2) The (hearing authority) may itself conduct the hearing and take the evidence, or may designate any of its members or any officer or employee of the (assigned department) to:

(A) Issue in the name of the (hearing authority) notices of hearings requesting the attendance and testimony of witnesses and the production of evidence relevant to any matter involved in any such hearings.

(B) Take the evidence.

(C) Transmit a report of the evidence and hearing, including transcripts and other evidence, together with recommendations to the (hearing authority) for action thereon.

(3) At any public hearing, testimony taken before the hearing authority or any person designated by it, must be under oath and recorded stenographically. The transcript, so recorded, will be made available to any member of the public or any part to the hearing upon payment of the usual charges therefor.

(4) After the (hearing authority) has reviewed the evidence, it may issue an order to the party responsible for the discharge directing that, following a specified time period, the sewer service be discontinued unless adequate treatment facilities, devices, or other related appurtenances shall have been installed or existing treatment facilities, devices, or other related appurtenances are properly operated, and such further orders and directives as are necessary and appropriate.

(c) *LEGAL ACTION.* Any discharge in violation of the substantive provisions of this Ordinance or an Order of the (hearing authority) shall be considered a public nuisance. If any person discharges sewage, industrial wastes, or other wastes into the City treatment system contrary to the substantive provisions of this Ordinance or any Order of the (hearing authority), the (Corporation Counsel/City Attorney) shall commence an action for appropriate legal and/or equitable relief in the (Circuit) Court of this County.

SECTION 7: PENALTIES AND COSTS

Any person who is found to have violated an Order of the (hearing authority) or who willfully or negligently failed to comply with any provisions of this Ordinance, and the orders, rules, and regulations issued hereunder, shall be fined not less than (one hundred dollars ($100.00)) (optional) nor more than (one thousand dollars ($1,000.00)) (optional) for each offense. Each day on which a violation shall occur or continue shall be deemed a separate and distinct offense. In addition to the penalties provided herein, the City may recover reasonable attorneys' fees, court costs, court reporters' fees, and other expenses of litigation by appropriate suit at law against the person found to have violated this Ordinance or the orders, rules, and regulations issued hereunder.

SECTION 8: SAVINGS CLAUSE

If any provision, paragraph, word, section, or article of this Ordinance is invalidated by any court of competent jurisdiction, the remaining provisions, paragraphs, words, sections, and articles shall not be affected and shall continue in full force and effect.

SECTION 9: CONFLICT

All Ordinances and parts of ordinances inconsistent or conflicting with any part of this Ordinance are hereby repealed to the extent of such inconsistency or conflict.

SECTION 10: EFFECTIVE DATE

This Ordinance shall be in full force and effect (Option A) from and after its passage, approval, and publication, as provided by law. (Option B)

on the _____ day of _____, 20 ___.

INTRODUCED the _____ day of _____, 20 ___.

FIRST READING: _____, 20 ___.

SECOND READING: _____, 20 ___.

PASSED THIS _____ day of _____, 20 ___.

AYES:

NAYS:

ABSENT:

NOT VOTING:

APPROVED by me this _____ day of _____, 20 ___.

_____

MAYOR

ATTEST:_____ (Seal) City Clerk

Published the _____ day of _____, 20 ___.

APPENDIX C

STANDARD FORM A—MUNICIPAL

Section IV. Industrial Waste Contribution to Municipal System

Industrial Monitoring 205

Form Approved OMB No. 158-R0100

STANDARD FORM A - MUNICIPAL

SECTION IV. INDUSTRIAL WASTE CONTRIBUTION TO MUNICIPAL SYSTEM

Submit a description of each major industrial facility discharging to the municipal system, using a separate Section IV for each facility description. Indicate the 4 digit Standard Industrial Classification (SIC) Code for the industry, the major product or raw material, the flow (in thousand gallons per day), and the characteristics of the wastewater discharged from the industrial facility into the municipal system. Consult Table III for standard measures of products or raw materials. (See instructions)

1. Major Contributing Facility
   (See instructions)
   - Name — 401a _____
   - Number & Street — 401b _____
   - City — 401c _____
   - County — 401d _____
   - State — 401e _____
   - Zip Code — 401f _____

2. Primary Standard Industrial Classification Code (See instructions) — 402 _____

3. Principal Product or Raw Material (See instructions)

		Quanity	Units (see Table III)
Product	403a _____	403c _____	403e _____
Raw Material	403b _____	403d _____	403f _____

4. Flow   Indicate the volume of water discharged into the municipal system in thousand gallons per day and whether this discharge is intermittent or continuous
   - 404a _____ Thousand gallons per day
   - 404b ____ Intermittent (int)   ____ Continuous (con)

5. Pretreatment Provided   Indicate if pretreatment is provided prior to entering the municipal system.
   - 405 ____ Yes   ____ No

6. Characteristics of Wastewater (See instructions)

	Parameter Name							
406a	Parameter Number							
406b	Value							

EPA Form 7550-22 (7-73)   IV-2   This section contains 1 page

APPENDIX D

SAFETY ORDERS FOR BATTERY CHARGING

Source: California Code of Regulations, Title 8, Section 5185

SAFE PROCEDURES FOR CHANGING AND CHARGING STORAGE BATTERIES

Battery charging installations shall be located in areas designated for that purpose. Employees assigned to work with storage batteries shall be instructed in emergency procedures such as dealing with accidental acid spills.

The area shall be adequately ventilated to prevent concentrations of flammable gases exceeding 20 percent of the lower explosive limit, and to prevent harmful concentration of mist from the electrolyte.

While batteries are being charged, smoking and open flame shall be prohibited in battery charging area and signs stating that prohibition shall be posted in the area.

Electrolyte (acid or base, and distilled water) for battery cells shall be mixed in a well-ventilated room. Acid or base shall be poured gradually into the water while stirring. Water shall never be poured into concentrated (greater than 75 percent) acid solutions.

When charging batteries, the vent caps shall be kept firmly in place to avoid electrolyte spray. Care shall be taken to assure that vent caps are functioning.

Fire extinguishing equipment adequate to cope with the hazards which may be encountered shall be provided and maintained close at hand.

Eye protection devices which provide side as well as frontal eye protection for employees shall be provided when measuring storage battery specific gravity or handling electrolyte, and the employer shall ensure that such devices are used by the employees. The employer shall also ensure that acid-resistant gloves and aprons shall be worn for protection against spattering.

CHAPTER 6

FLOW MEASUREMENT

by

John Esler

TABLE OF CONTENTS
Chapter 6. FLOW MEASUREMENT

			Page
OBJECTIVES			212
WORDS			213
6.0	NEED FOR FLOW MEASUREMENT		217
6.1	BASICS OF FLOW MEASUREMENT		217
	6.10	Flow is…	217
	6.11	Types of Flow Systems	218
		6.110 Open Channel Flow	218
		6.111 Closed Channel (Pressure Pipe) Flow	218
	6.12	Head	218
	6.13	Flow Measurement Devices	218
6.2	OPEN CHANNEL FLOW MEASUREMENT		221
	6.20	Primary Elements	221
		6.200 Weirs	221
		6.201 Flumes	224
		6.202 Flow Nozzles	226
		6.203 Operation and Maintenance of Weirs and Flumes	226
	6.21	Secondary Elements/Devices	230
		6.210 Staff Gauge	230
		6.211 Stilling Well	230
		6.212 Floats	230
		6.213 Ultrasonic Devices	232
		6.214 Bubblers	232
		6.215 Submerged Pressure Transducers	233
		6.216 Electrical Capacitance Strips	233
6.3	CLOSED PIPE FLOW MEASUREMENT		233
	6.30	Differential Pressure Devices	233
		6.300 Venturi Nozzle	233
		6.301 Dall Flow Tube	235
		6.302 Orifice Plate	235

	6.31		Mechanical Devices	235
		6.310	Electromagnetic/Magnetic Flowmeter (Average Velocity)	235
		6.311	Rotameter	235
		6.312	Displaced Volume Devices	235
		6.313	Propeller Meter	238
	6.32		Acoustic Devices	238
		6.320	Doppler Effect	238
		6.321	Transit Time (Time-Travel)	238
6.4			SPECIAL CASES	238
	6.40		Very Low Flows	238
	6.41		Stormwater	238
6.5			METHODS OF CHECKING THE ACCURACIES OF FLOWMETERS	240
	6.50		Hydraulic Calibration	240
	6.51		Instrument Calibration	240
	6.52		Comparative Depth Measurements	240
	6.53		Dye Dilution Method	242
	6.54		Calibration Cylinder	243
	6.55		Comparison of Typical Flowmeter Accuracies	243
6.6			FLOWS TO MEASURE	243
	6.60		Influent/Effluent	243
	6.61		Equalization Basin Flow	244
	6.62		Primary Sludge Flow	244
	6.63		Activated Sludge Flows To or From	244
		6.630	Aeration Basin	244
		6.631	Pass or Compartment	244
		6.632	Final Clarifier	245
	6.64		Return Activated Sludge (RAS)	245
		6.640	To Each Aeration Tank	245
		6.641	From Each Final Clarifier	245
	6.65		Waste Activated Sludge (WAS)	245
		6.650	Wasting to an Aerobic Digester	245
		6.651	Wasting to a Thickener	246
	6.66		Filter Backwash Rate	246
	6.67		Chemical Feed Rates	246

6.7	FLOW PROPORTIONING COMPOSITE SAMPLES	246
6.8	ACKNOWLEDGMENT	246
DISCUSSION AND REVIEW QUESTIONS		247
SUGGESTED ANSWERS		247

OBJECTIVES
Chapter 6. FLOW MEASUREMENT

Following completion of Chapter 6, you should be able to:

1. List the reasons for measuring flows.
2. Explain the functions of primary and secondary elements.
3. Describe the various types of open channel flow measurement devices.
4. Determine the accuracy of open channel flowmeters.
5. Identify the various types of closed pipe flow metering systems.
6. Identify appropriate flow measurement devices for a variety of treatment plant flows.
7. Prepare a flow-proportioned composite sample.

WORDS
Chapter 6. FLOW MEASUREMENT

ANALOG READOUT

The readout of an instrument by a pointer (or other indicating means) against a dial or scale. Also, the continuously variable signal type sent to an analog instrument (for example, 4–20 mA). Also see DIGITAL READOUT.

COMPOSITE (PROPORTIONAL) SAMPLE

A composite sample is a collection of individual samples obtained at regular intervals, usually every one or two hours during a 24-hour time span. Each individual sample is combined with the others in proportion to the rate of flow when the sample was collected. Equal volume individual samples also may be collected at intervals after a specific volume of flow passes the sampling point or after equal time intervals and still be referred to as a composite sample. The resulting mixture (composite sample) forms a representative sample and is analyzed to determine the average conditions during the sampling period.

CONDUCTIVITY

A measure of the ability of a solution (water) to carry an electric current.

DENSITY

A measure of how heavy a substance (solid, liquid, or gas) is for its size. Density is expressed in terms of weight per unit volume, that is, grams per cubic centimeter or pounds per cubic foot. The density of water (at 4°C or 39°F) is 1.0 gram per cubic centimeter or about 62.4 pounds per cubic foot.

DIGITAL READOUT

The readout of an instrument by a direct, numerical reading of the measured value or variable. The signal sent to such instruments is usually an analog signal.

FRICTION LOSS

The head, pressure, or energy (they are the same) lost by water flowing in a pipe or channel as a result of turbulence caused by the velocity of the flowing water and the roughness of the pipe, channel walls, or restrictions caused by fittings. Water flowing in a pipe loses head, pressure, or energy as a result of friction losses. Also called HEAD LOSS.

HEAD

The vertical distance, height, or energy of water above a reference point. A head of water may be measured in either height (feet or meters) or pressure (pounds per square inch or kilograms per square centimeter). Also see DISCHARGE HEAD, DYNAMIC HEAD, STATIC HEAD, SUCTION HEAD, SUCTION LIFT, and VELOCITY HEAD.

HEAD LOSS

The head, pressure, or energy (they are the same) lost by water flowing in a pipe or channel as a result of turbulence caused by the velocity of the flowing water and the roughness of the pipe, channel walls, or restrictions caused by fittings. Water flowing in a pipe loses head, pressure, or energy as a result of friction losses. The head loss through a filter is due to friction losses caused by material building up on the surface or in the top part of a filter. Also called FRICTION LOSS.

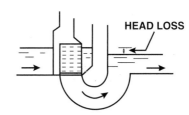

HYDRAULIC JUMP

The sudden and usually turbulent abrupt rise in water surface in an open channel when water flowing at high velocity is suddenly retarded to a slow velocity.

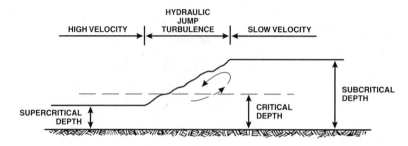

INTEGRATOR

A device or meter that continuously measures and calculates (adds) a process rate variable in cumulative fashion; for example, total flows displayed in gallons, million gallons, cubic feet, or some other unit of volume measurement. Also called a TOTALIZER.

MANOMETER (man-NAH-mut-ter)

An instrument for measuring pressure. Usually, a manometer is a glass tube filled with a liquid that is used to measure the difference in pressure across a flow measuring device, such as an orifice or a Venturi meter. The instrument used to measure blood pressure is a type of manometer.

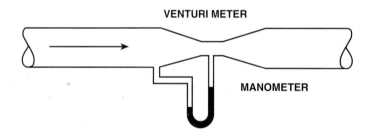

NAPPE (NAP)

The sheet or curtain of water flowing over a weir or dam. When the water freely flows over any structure, it has a well-defined upper and lower water surface.

ORIFICE (OR-uh-fiss)

An opening (hole) in a plate, wall, or partition. An orifice flange or plate placed in a pipe consists of a slot or a calibrated circular hole smaller than the pipe diameter. The difference in pressure in the pipe above and at the orifice may be used to determine the flow in the pipe. In a trickling filter distributor, the wastewater passes through an orifice to the surface of the filter media.

PRIMARY ELEMENT

(1) A device that measures (senses) a physical condition or variable of interest. Floats and thermocouples are examples of primary elements. Also called a SENSOR.

(2) The hydraulic structure used to measure flows. In open channels, weirs and flumes are primary elements or devices. Venturi meters and orifice plates are the primary elements in pipes or pressure conduits.

RECEIVER

A device that indicates the result of a measurement, usually using either a fixed scale and movable indicator (pointer), such as a pressure gauge, or a moving chart with a movable pen like those used on a circular flow-recording chart. Also called an INDICATOR.

RECORDER

A device that creates a permanent record, on a paper chart or magnetic tape, of the changes in a measured variable.

REPRESENTATIVE SAMPLE

A sample portion of material, water, or wastestream that is as nearly identical in content and consistency as possible to that in the larger body being sampled.

SECONDARY ELEMENT

The secondary measuring device or flowmeter used with a primary measuring device (element) to measure the rate of liquid flow. In open channels, bubblers and floats are secondary elements. Differential pressure measuring devices are the secondary elements in pipes or pressure conduits. The purpose of the secondary measuring device is to (1) measure the liquid level in open channels or the differential pressure in pipes, and (2) convert this measurement into an appropriate flow rate according to the known liquid level or differential pressure and flow rate relationship of the primary measuring device. This flow rate may be integrated (added up) to obtain a totalized volume, transmitted to a recording device, or used to pace an automatic sampler.

SENSOR

A device that measures (senses) a physical condition or variable of interest. Floats and thermocouples are examples of sensors. Also called a PRIMARY ELEMENT.

STILLING WELL

A well or chamber that is connected to the main flow channel by a small inlet. Waves and surges in the main flow stream will not appear in the well due to the small-diameter inlet. The liquid surface in the well will be quiet, but will follow all of the steady fluctuations of the open channel. The liquid level in the well is measured to determine the flow in the main channel.

SURCHARGE

Sewers are surcharged when the supply of water to be carried is greater than the capacity of the pipes to carry the flow. The surface of the wastewater in manholes rises above the top of the sewer pipe, and the sewer is under pressure or a head, rather than at atmospheric pressure.

TIMER

A device for automatically starting or stopping a machine or other device at a given time.

TOTALIZER

A device or meter that continuously measures and calculates (adds) a process rate variable in cumulative fashion; for example, total flows displayed in gallons, million gallons, cubic feet, liters, cubic meters, or some other unit of volume measurement. Also called an INTEGRATOR.

TRANSDUCER (trans-DUE-sir)

A device that senses some varying condition measured by a primary sensor and converts it to an electrical or other signal for transmission to some other device (a receiver) for processing or decision making.

VENTURI (ven-TOOR-ee) METER

A flow measuring device placed in a pipe. The device consists of a tube whose diameter gradually decreases to a throat and then gradually expands to the diameter of the pipe. The flow is determined on the basis of the difference in pressure (caused by different velocity heads) between the entrance and throat of the Venturi meter.

NOTE: Most Venturi meters have pressure sensing taps rather than a manometer to measure the pressure difference. The upstream tap is the high pressure tap or side of the manometer.

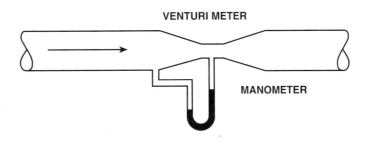

CHAPTER 6. FLOW MEASUREMENT

6.0 NEED FOR FLOW MEASUREMENT

Of all the things an industrial wastewater treatment plant operator must know or be able to do, the accurate use of the appropriate flow measurement devices ranks high on the list of essential skills. Nearly every facet of your job involves flow measurement devices in one way or another. For example, to determine how much flow is being contributed from the different process lines, you must be able to measure those flows. To calculate the amount of chemicals needed for neutralization or for nutrient addition, you must know the wastestream rate of flow over a period of time. To set up accurate flow-proportioning of a COMPOSITE SAMPLE,[1] you must know the rate of flow during each sample period. All of these activities rely heavily on the use of flow measurement devices.

Many industries that discharge to municipal wastewater treatment systems are required to install effluent flow measurement devices on their discharge lines to determine their wastewater contribution to the POTW (Publicly Owned Treatment Works). This procedure allows for the assessment of user charges based on flow. In some cases, this flow measurement is made by the POTW agency. In any event, this measurement is often a major factor in determining the potential impact of the industrial discharge on the municipal collection and treatment systems.

Waste discharge requirements for most effluent water quality indicators are issued either on a mass or concentration basis. Concentration-based limits simply require that the amount of pollutant in a discharge does not exceed a certain mass per unit volume, such as milligrams per liter (mg/L); no attempt is made to determine the particular quantity of the pollutant discharged. However, with a mass-based limit, the measured concentration of a pollutant in wastewater (generally from a composite sample taken over a period of time) is applied against a corresponding measured flow rate so that the quantity of the pollutant in question can be limited. This procedure produces a calculated value. Without reliable flow rate data, mass-based limits cannot be applied.

For industries with fluctuating or unusual waste discharge flow rates, composite samples taken over time are most representative if they are acquired using flow proportioning techniques. In such instances, discrete (separate) time-interval samples, collected either manually or by automatic equipment, are proportioned according to the prevailing flow rate to make up a single sample that is representative of the discharge conditions. For example, if the flow rate during time interval "A" averaged twice that measured during time interval "B," the amount of sample "A" in the flow-proportioned composite would be twice that of sample "B." However, the preferred way of collecting flow-proportioned composite samples involves connecting an automatic wastewater sampler to a flow measurement device so that the individual samples in the sampler are precisely and uniformly taken at predetermined, fixed flow intervals, such as one sample every 10,000 gallons. In either instance, a properly operating flowmeter is required in order for flow-proportioned samples to be taken.

Although the focus of flow measurement is often on effluent flow rate, in this chapter we are equally concerned with the measurement of flow streams throughout the industrial wastewater treatment plant. Indeed, to plant operations staff, control of a process begins and ends with the measurement of flow. The ability to use and evaluate existing flowmeters is an important skill for the industrial wastewater treatment plant operator.

6.1 BASICS OF FLOW MEASUREMENT

6.10 Flow is …

Flow is the amount of water (or another fluid) going past a particular reference point over a certain period of time. It is measured in units of volume per unit of time. Typical units are cubic feet per second (CFS), gallons per minute (GPM), or million gallons per day (MGD). Flow is usually calculated by the equation:

$Q = AV$ where Q = the quantity of flow

A = the cross-sectional area of the flow

V = the velocity of the flow

For instance, Q (cubic feet per second) equals V (feet per second) times A (square feet) for the liquid area in a pipe or channel. From this basic equation, you can see that the flow can be calculated by measuring both the average velocity of the water and the area that the water is flowing through. Some types of flow measurement systems determine the flow by directly measuring these quantities; other types of systems measure a depth or pressure that is related to the flow.

---

[1] *Composite (Proportional) Sample.* A composite sample is a collection of individual samples obtained at regular intervals, usually every one or two hours during a 24-hour time span. Each individual sample is combined with the others in proportion to the rate of flow when the sample was collected. Equal volume individual samples also may be collected at intervals after a specific volume of flow passes the sampling point or after equal time intervals and still be referred to as a composite sample. The resulting mixture (composite sample) forms a representative sample and is analyzed to determine the average conditions during the sampling period.

6.11 Types of Flow Systems

There are two basic types of flow systems in wastewater treatment plants: "open channel" flow and "closed channel" (or "closed pipe") flow. Each of these flow systems has certain characteristics that differentiate it from the other; each requires a different system for measuring flow (Figure 6.1).

6.110 Open Channel Flow

Open channel flow conditions occur whenever the flow stream has a free or unconfined surface that is open to the atmosphere (that is, the flow is not pressurized). Examples of this are streams and canals, distribution channels, gravity sewer lines, and open units, such as chlorine contact chambers. In open channel flow, the total energy contained in the flow is in the form of velocity (called the "velocity head") and pressure energy, which consists of the depth of the flow.

6.111 Closed Channel (Pressure Pipe) Flow

Closed pipe flow occurs whenever the conduit or pipe is completely filled and under pressure. Examples of this are force mains, pump discharge lines, sludge lines, and closed process lines. In closed pipe flow, the energy is present in the forms of velocity (the "velocity head"), pressure (called the "pressure head"), and elevation or distance above some reference point.

6.12 Head

Each of these forms of energy (head) can be expressed as a single linear quantity that is an equivalent height of a liquid. For example, the velocity component of a flow stream can be converted to the "velocity head" using the formula:

Head, feet = $V^2/2g$

where V = the velocity in fps

g = the acceleration due to gravity (32 feet/second^2)

Similarly, the pressure component (in closed pipe flow) can be converted to the "pressure head" using the formula:

Head, feet = p/w

where p = the fluid pressure in pounds per square foot (psf)

w = the weight of the fluid in pounds per cubic foot

QUESTIONS

Write your answers in a notebook and then compare your answers with those on page 247.

6.0A Why does an operator need to be able to measure flows?

6.1A What are the typical units of flow measurement?

6.1B What do the letters Q = AV mean?

6.1C What are the two basic types of flow systems?

6.13 Flow Measurement Devices

There are several ways to measure the rate of flow in a system. The two most accurate methods are the "timed volumetric" (or "bucket and stopwatch"), where the volume of water is measured over a set time, and "dye dilution," where a known amount of dye is diluted by the unknown amount of water. However, neither of these methods lends itself to the type of continuous measurement of flow that is essential to the control of an industrial facility. Other more commonly used methods of measuring plant flows include the "velocity-area" method, where the average velocity of the fluid is multiplied by the cross-sectional area of the flow (and applying the relationship of Q = AV), and the use of a "hydraulic structure" in the flow stream. The purpose of the hydraulic structure is to restrict the flow stream and produce a change in the level or pressure of the flowing liquid that can be related to the quantity of flow.

In discussing flow measurement systems, the terms primary element and secondary element are commonly used. The hydraulic structures that are placed in the flow streams are known as primary elements. In open channel flow, the primary element generally consists of either a flume or a weir (Figure 6.2). In closed pipe flow, the primary element consists of some type of device that creates a pressure condition that can be related to the velocity of flow. These types of primary elements often consist of either a flow tube (such as a VENTURI[2]) or an ORIFICE[3] plate.

The secondary element for a flow measurement device (the primary element) in an open channel consists of some type of device for measuring the elevation of the water surface above a reference point on the flume or weir. The secondary element usually has an additional function, which is to convert the eleva-

---

[2] *Venturi* (ven-TOOR-ee) *Meter.* A flow measuring device placed in a pipe. The device consists of a tube whose diameter gradually decreases to a throat and then gradually expands to the diameter of the pipe. The flow is determined on the basis of the difference in pressure (caused by different velocity heads) between the entrance and throat of the Venturi meter.

NOTE: Most Venturi meters have pressure sensing taps rather than a manometer to measure the pressure difference. The upstream tap is the high pressure tap or side of the manometer.

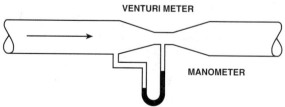

[3] Orifice (OR-uh-fiss). An opening (hole) in a plate, wall, or partition. An orifice flange or plate placed in a pipe consists of a slot or a calibrated circular hole smaller than the pipe diameter. The difference in pressure in the pipe above and at the orifice may be used to determine the flow in the pipe. In a trickling filter distributor, the wastewater passes through an orifice to the surface of the filter media.

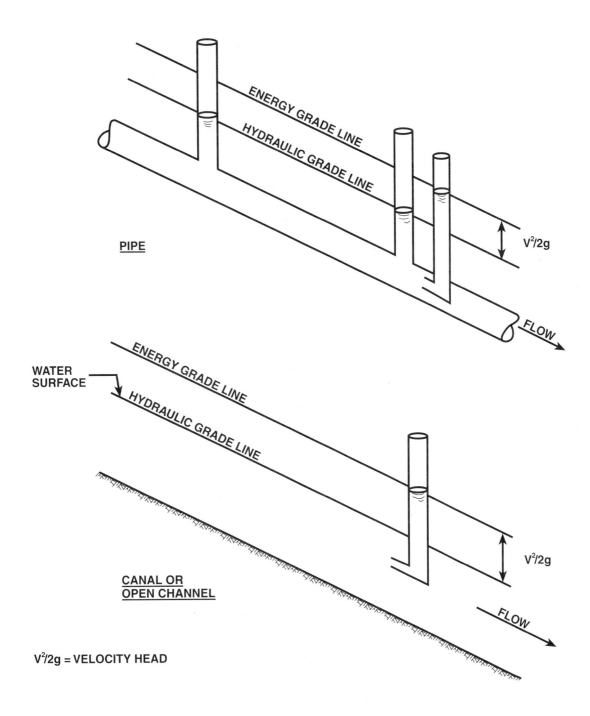

Fig. 6.1 *Flow and energy conditions in a pipe and a canal*

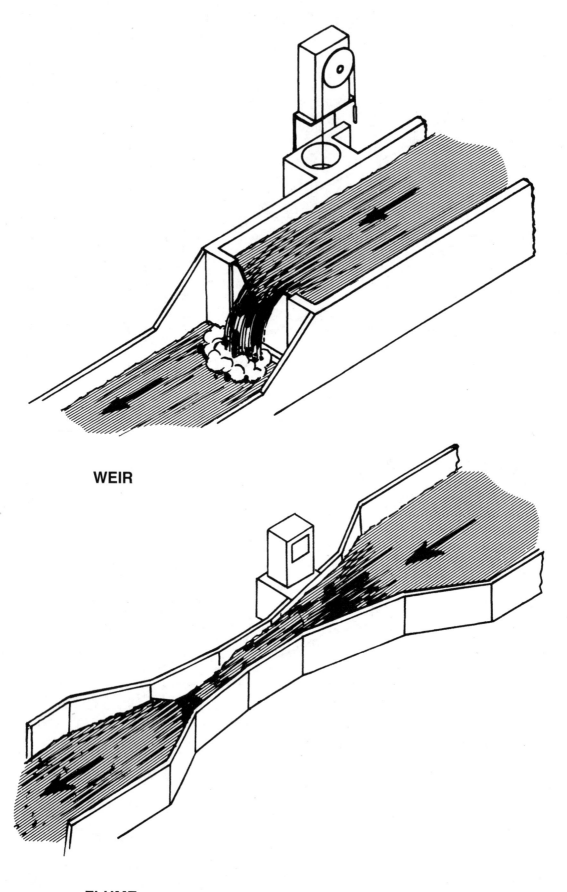

WEIR

FLUME

Fig. 6.2   Weirs and flumes are primary flow measuring devices (hydraulic structures)

tion (head) reading to the appropriate rate of flow for that device. In a closed pipe system, the secondary element measures the pressure change brought about by the liquid flowing through the flow tube or orifice plate and converts the pressure difference into a rate of flow.

The instrumentation of an automatic flowmeter usually contains three integrated devices to obtain data for describing the flow. A totalizer is used to continuously sum the flow (flow integrator) over an extended period; it is essentially like an odometer on a vehicle. Totalizers typically have a digital display indicating flow in tens, hundreds, or thousands of gallons. A second device, the flow indicator, shows the instantaneous flow rate. These indicators may have either an analog or digital display. Some instrumentation may also incorporate a third device that produces a printed and graphed display to record the instantaneous flow measurements. Graphical displays (flowcharts) are desirable when measuring industrial effluent flows because they can reveal times and magnitudes of batch discharges or dumps. The instrumentation may also be equipped to pace an automatic sampler, operate an external recorder, or trigger alarms.

This chapter presents the information an operator needs to properly operate and maintain existing flow metering systems. It is not intended as a guide for selecting or designing flow metering systems since these tasks are usually performed by a qualified design engineer. However, an understanding of the principles on which flow metering systems are based, combined with some practical knowledge of them, will enable you to properly operate and maintain the flow metering equipment at your industrial wastewater treatment facility.

The flow metering devices described in this chapter are often installed to function as part of a total system for flow measurement and process control. In addition to the actual flow measurement devices, such systems usually also incorporate various types of instrumentation, such as totalizers and recorders. Please refer to Chapter 11, "Instrumentation," for detailed information about how the flow measurement devices described here fit into the overall system of instrumentation for process control.

6.2 OPEN CHANNEL FLOW MEASUREMENT

6.20 Primary Elements

The purpose of an open channel primary element is to create flow conditions that produce a known relationship between flow and depth. When these conditions are created, the channel width is known and the velocity of the water does not need to be measured. Instead, the secondary element senses the depth at the established measurement point and converts it to flow based on the known relationship between depth and flow.

The primary element for an open channel flow condition may be either a weir or a flume (Figure 6.2). Each device has certain characteristics that make it the most suitable device for measuring flows in specific situations.

6.200 Weirs

A weir may be defined as an overflow structure built across an open channel. The discharge rate of a weir is determined by measuring the vertical distance from the crest of the overflow portion of the weir to the surface of the upstream pool of water and referring to tables or computations that apply to that particular size and shape of weir.

Weirs are either broad-crested (wide in the direction of flow) or sharp-crested (Figure 6.3). The most common types of sharp-crested weirs are V-notch, rectangular, Cipolletti, and rectangular with end contractions (suppressed) (Figure 6.4). A sharp-crested weir has certain standard characteristics and dimensions that permit it to be used for flow measurement with the aid of simple weir discharge tables (Table 6.1). These conditions include:

- A sharp crest that the water flows over.
- A minimum crest height of 2 to 3 times the expected "head" measurement.
- A point to measure the head that is 3 to 4 times the expected head upstream of the weir.

Each type of weir has characteristics that make it suitable for particular operating conditions. In general, for best accuracy, a weir should be selected to provide the following conditions:

- The minimum head should be at least 0.2 foot to prevent the *NAPPE*[4] from clinging to the crest.
- The length of rectangular weirs should be at least three times the head.
- The crest should be placed high enough so that the downstream water elevation is at least 0.2 foot below the crest.

TABLE 6.1 TYPICAL WEIR DISCHARGE TABLE FOR 90° V-NOTCH WEIR

Formula $Q = 2.500\ H^{2.5}$
where H is in feet and Q is in CFS.

H, ft	Q, CFS	Q, MGD	Q, GPM
0.10	0.008	0.005	3.54
0.20	0.045	0.029	20.0
0.30	0.125	0.080	55.2
0.40	0.253	0.163	113.3
0.50	0.442	0.285	198
0.60	0.697	0.450	312
0.70	1.025	0.661	459
0.80	1.431	0.925	641
0.90	1.921	1.239	860
1.0	2.500	1.613	1,119
1.5	6.889	4.444	3,085
2.0	14.142	9.124	6,332
2.5	24.705	15.939	11,062
3.0	38.971	25.143	17,449

---

[4] *Nappe* (NAP). The sheet or curtain of water flowing over a weir or dam. When the water freely flows over any structure, it has a well-defined upper and lower water surface.

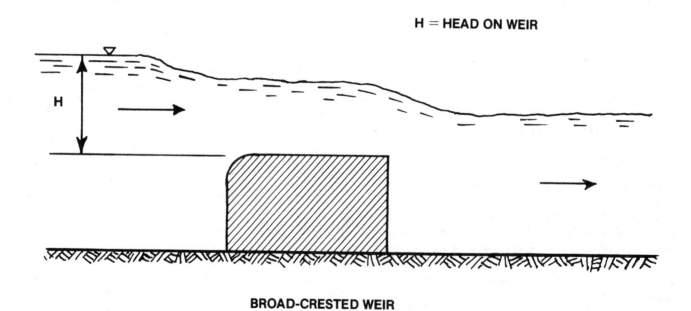

Fig. 6.3 Sharp-crested and broad-crested weirs

Flow Measurement 223

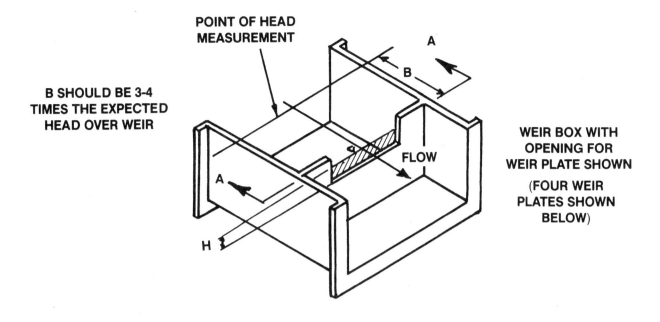

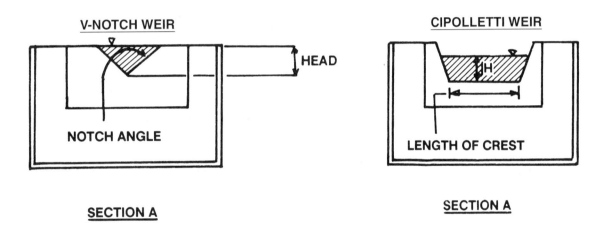

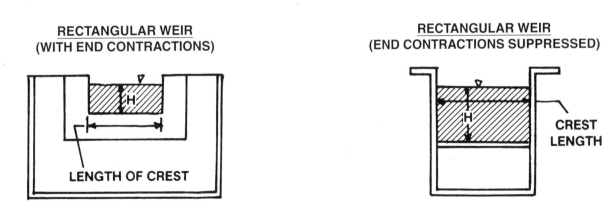

H = HEAD OR DEPTH BEHIND WEIR CREST
MEASURE H AT A DISTANCE B UPSTREAM IN A STILLING WELL

Fig. 6.4 Weirs

V-NOTCH WEIRS

A V-notch weir (or triangular weir) is a form of sharp-crested weir where the flow passes through a triangular notch in the weir. The angle of this notch may be 22$\frac{1}{2}$, 30, 45, 60, or 90 degrees. Because a V-notch weir has no crest length, the head required for a small flow through it is greater than that required for other types of weirs. This is an advantage for small discharges in that the nappe will spring free of the crest when it would cling to the crest of another type of weir and make the measurement meaningless (Figure 6.5).

RECTANGULAR WEIRS WITHOUT END CONTRACTIONS (SUPPRESSED RECTANGULAR WEIRS)

A standard rectangular weir consists of a thin, horizontal plate. The weir crest extends to the full width of the approach channel so that there is no lateral contraction of the water passing through the weir. Special care must be taken with this type of weir to provide aeration underneath the sheet of water flowing over the crest. This is usually accomplished with vents under the nappe on both sides of the weir (Figure 6.5).

RECTANGULAR WEIRS WITH END CONTRACTIONS

A standard contracted weir has its crest and sides both acting as weirs with the sheet of water forming a jet narrower than the weir opening. These end contractions also provide the ventilation needed for the weir crest.

6.201 Flumes

A flume is a specially shaped open channel section that can be installed in an open channel or an unpressurized pipeline. As previously illustrated in Figure 6.2, each flume has an entrance (converging) section, a throat section, and a discharge (diverging) section. These sections are designed to constrict the flow and expand it again in a predictable manner. Constriction of the flow causes changes in head (measured as elevation of the water surface), which can be converted to flow rates.

Flumes are generally used instead of weirs where head loss is a concern, for larger flows, or for flows that may contain solids or debris that would tend to clog a V-notch weir. The distinct advantage of a flume is that it can function as a flowmeter over a wide range of flows with only a minimum loss of head. The main disadvantage of flumes is the initial cost as compared to weirs.

PARSHALL FLUMES

The Parshall flume is the most popular type of flume for measuring flows at treatment plants and is available in throat sizes from 1 inch to greater than 20 feet. They are designated by throat width and are generally furnished as prefabricated fiberglass structures. Its developer, Ralph Parshall of the Department of the Interior, intended them to be used primarily for placement in the beds of small rivers and streams, so they are ideally suited for flow measurement in entrance or exit channels at treatment plants.

Head must be measured in a very specific location on any flume because flumes are designed to produce a specific relationship between depth of flow and rate of flow at the specified measurement point. For Parshall flumes, the measurement for head must be made at the "$\frac{2}{3}$A" point (Figure 6.6)(in flumes up to 8 feet in width). Measuring the head upstream or downstream from this point can result in flow indications that are greater or less than the actual flow.

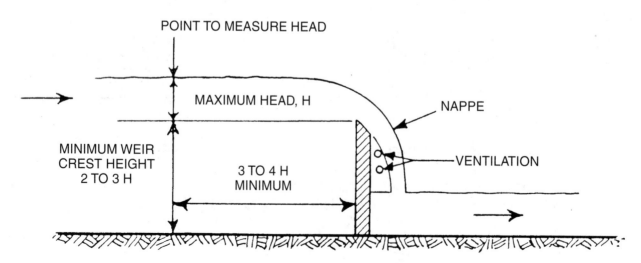

Fig. 6.5 Weir dimensions and nappe

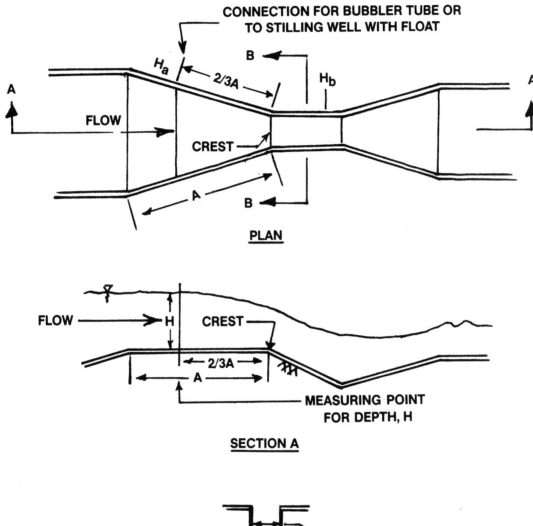

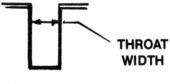

Fig. 6.6 Parshall flume

Palmer-Bowlus Flumes/Leopold-Lagco Flumes

The Palmer-Bowlus flume configuration was developed to provide a simple means for measuring flows in existing sewers. This type of flume actually can be provided in several different configurations. However, most Palmer-Bowlus flumes now conform to a more standardized size and shape (Figure 6.7). The size of the flume is generally determined by the size of the sewer; a 6-inch flume would be inserted into a 6-inch sewer line, a 12-inch flume would be inserted into a 12-inch sewer line.

Because of its simplicity and its ease of insertion into an existing sewer line, the Palmer-Bowlus flume is most often used to provide a measuring point for flow where none existed previously. This includes a flow line from an individual process, as well as a line tributary to a municipal system.

As with the Parshall flume, the head determination for a Palmer-Bowlus flume must be made at a specific location. In this case, the measurement is made at a point that is D/2 upstream from the transition section (Figure 6.7).

The dimensions to which a Palmer-Bowlus flume is constructed have been standardized, but in a generic sense the term Palmer-Bowlus-type flume can apply to any flume of this general shape. Be aware, however, that head-to-flow tables are not identical for different manufacturers due to slight differences in styles. For instance, another similar type of flume, the Leopold-Lagco flume, also is occasionally installed in an existing line. It has a rectangular cross section rather than a trapezoid cross section and, consequently, produces different head-to-flow readings than a Palmer-Bowlus flume of the same nominal size.

The primary concern with a Palmer-Bowlus flume is that it produces an obstruction in the invert (lowest point) of the sewer that becomes a place for grit and other settleables to collect. Any deposits that raise the surface of the water at the measuring point would cause the head measurement to indicate higher than actual. A good practice is to set up a sewer flushing schedule for each flume location that is based on the experience at that site.

6.202 Flow Nozzles

A Kennison nozzle and a parabolic nozzle (Figure 6.8) are devices that connect to the free-flowing end of a gravity sewer to measure flow under low head conditions with the line partly filled. By providing a constricted throat section, the surface of the water is raised within the nozzle, providing a head (elevation change) that can be related to the flow. One major advantage of the Kennison nozzle is that its change in head is directly proportional to the flow, providing a generous change in head for equal flow rate increments. It also has an unobstructed, self-cleaning flow path, which makes it more suitable for measuring solids-bearing fluids, such as raw influent or sludge, with reasonable accuracy.

6.203 Operation and Maintenance of Weirs and Flumes

All open channel primary elements create readily observed flow profile characteristics by manipulating the channel slope and size. The flow is constricted and made to drop through a steep and precisely dimensioned section (the primary element) before flow through the regular channel is resumed. A known, repeatable relationship between depth and flow results in open channel flumes.

A flume or weir that is properly sized, constructed, installed, and maintained will generate the correct relationship between depth and flow. Unfortunately, they are not always installed properly, are not always maintained properly, do not last forever, and occasionally the flows for which they are sized change. Changing conditions in sewer lines downstream from the flow metering system can also affect the accuracy of the flume or weir by backing up and flooding the flume or weir and producing erroneously high flow readings during flooded conditions. It is usually easy to identify a problem that would cause an incorrect relationship between depth and flow to occur. Some problems are more difficult to detect but it is often obvious when a more thorough evaluation is needed.

The characteristics of a properly operating flume are distinct and easy to observe. Starting some distance upstream of the flume, the water will be relatively deep and slow moving. In comparison, as it passes through the throat of the flume it will become much shallower and faster (Figure 6.9). Downstream from the throat, the water will return to its deeper and slower moving condition. A visible hydraulic drop occurs as the flow approaches the throat of the flume and speeds up. When the flow leaves the throat section, a *HYDRAULIC JUMP*[5] occurs (see Figure 6.9). The water surface will often show a "V" section formed by standing waves as the water enters the flume. The hydraulic jump also often has a "V" shape to it. A hydraulic jump that occurs upstream of the flume may be an indication that the upstream piping was laid at too steep a slope or that accumulated debris needs to be removed.

---

[5] *Hydraulic Jump.* The sudden and usually turbulent abrupt rise in water surface in an open channel when water flowing at high velocity is suddenly retarded to a slow velocity.

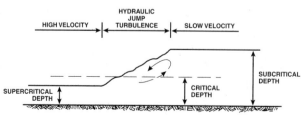

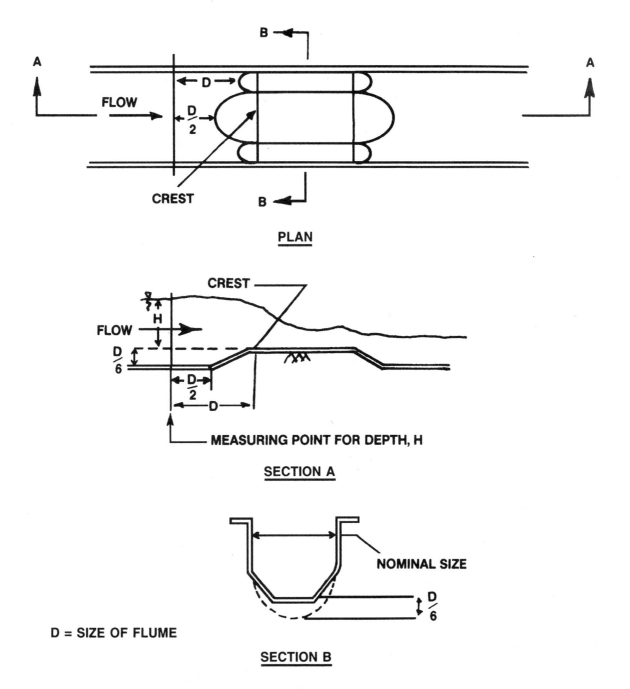

Fig. 6.7 Palmer-Bowlus flume

228 Treatment Plants

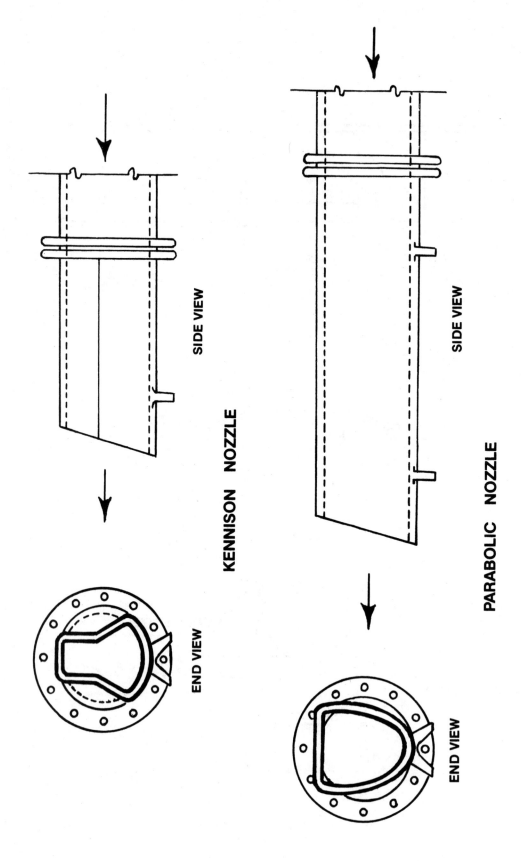

Fig. 6.8 Kennison nozzle and parabolic nozzle

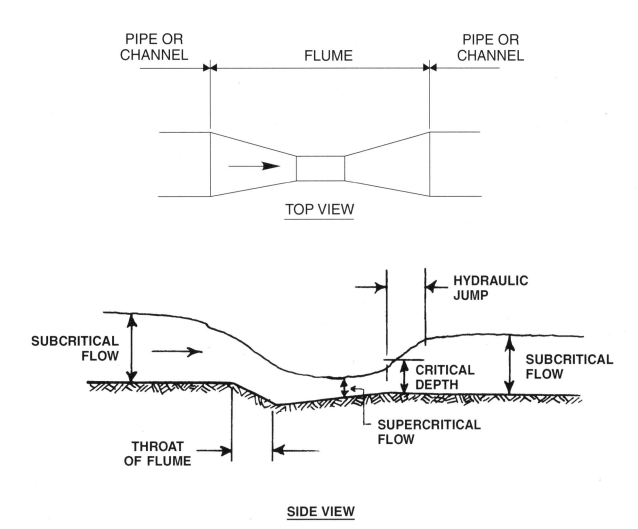

Fig. 6.9 Flow through a flume

Weirs operate on the same principles as flumes; however, they can look quite different. The approach section is sized so that the approach velocity is minimal. The water speeds up as it pours through the weir notch and then returns to a slower, deeper flow in the afterbay of the weir.

Under normal conditions, you will see that the nappe (the flow through the notch) springs away from the weir plate. This means that the weir is operating with a free discharge and that the nappe is well ventilated, or aerated; that is, air can move freely beneath the nappe. Only at low flows should the water cling to the face of the weir plate.

A weir cannot be operated in a submerged condition (Figure 6.10). The nappe of the water must fall freely into the weir afterbay. If the level in the afterbay rises too high, aeration of the nappe may cease and the measured discharge will be greater than the actual discharge. A weir should be constructed with several inches' clearance between the crest of the weir (the bottom of the

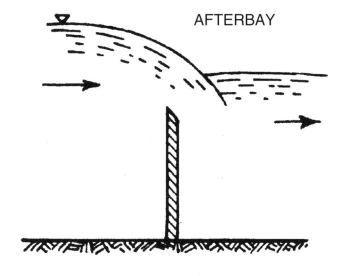

Fig. 6.10 Submerged weir

notch) and the afterbay level. In general, a weir should be constructed with the top of the downstream pipe at least six inches below the crest of the weir. If the discharge pipe is not visible and the afterbay level is approaching the crest of the weir, it is likely that the proper depth-to-flow relationship does not exist.

In contrast to weirs, however, the discharge of a flume can occur under two conditions of flow. The first, free flow, occurs when there is insufficient backwater depth downstream to reduce the discharge rate. This is generally in the range of 65 percent of submergence for Parshall flumes (Figure 6.6). For free flow, only the upstream head must be measured to determine the flow. The second condition, submerged flow, (where the submergence is greater than 50 to 80 percent, depending on the flume size), requires the measurement of both the upstream and downstream heads to determine the flow. This condition may occur, for example, if a bar screen is placed just downstream of the flume, or if there is a restriction that causes flooding of the flume. This condition should be avoided if at all possible in wastewater treatment plants.

You will need to clean the flumes and weirs regularly to maintain their proper dimensions. Flumes are described as self-cleaning, but that does not mean that they do not need occasional maintenance. Use a brush, high-pressure water hose, or other similar device, to remove accumulated solids. Some types of wastewater, especially oily wastewater, can rapidly coat a flume and cause significant errors to occur. Oils and greases can coat the pipe from the throat to the stilling well, thus hindering flow and creating inaccurate flow readings. Cans and rags sometimes lodge in narrow flumes. Excessive debris in the approach channel may create waves and may even cause an improper water surface profile to occur. These are often clues that the flume needs to be cleaned. Some debris can accumulate ahead of flumes and weirs without affecting their accuracy but excessive debris in a weir increases the approach velocity of the flow, which affects the weir's accuracy. The accuracy of a weir is also sensitive to rounding of the edges of the notch. Keep the edges clean and sharp and keep the upstream face of the weir plate clean. For dependable, long-term service, develop a regular schedule of cleaning and maintaining flumes and weirs.

QUESTIONS

Write your answers in a notebook and then compare your answers with those on page 247.

6.2A List the two most accurate methods of measuring the rate of flow in a system.

6.2B What is the secondary element for a flow measurement device in an open channel?

6.2C What are the three different sections of a flume?

6.2D Why must head measurements be made at a specific location on Parshall or Palmer-Bowlus flumes?

6.21 Secondary Elements/Devices

The purpose of the secondary measuring device in an open channel is to measure or indicate the liquid level in the primary device. The secondary device is also used in conjunction with an instrumentation system to convert this "head" to a flow measurement. The selection of the type of secondary device to be used is based on a consideration of many factors, such as the location, type of information required, and cost.

6.210 Staff Gauge

A staff gauge is a ruler mounted on or near the primary device to indicate the elevation (head) of the water at that point (Figure 6.11). Most staff gauges are graduated to read in feet and tenths and hundredths of feet (to correspond with weir or flume flow tables) and are usually made of a corrosion-resistant plastic or have an enameled surface. If the primary device is a prefabricated fiberglass flume, the staff gauge can be included in the manufacturing process to provide a permanent mounting at the correct location for measuring the head. This simple staff gauge is sometimes all that is needed to provide a convenient indication of flow rate.

6.211 Stilling Well

The flow of wastewater through a primary device is often quite unsteady and can be laden with debris. A stilling well is a small reservoir that is directly connected to the primary device at the proper point for head measurement; its surface elevation is the same as that in the primary device (Figure 6.11). The purpose of the stilling well is to provide a steadier indication of the flow depth and to avoid the problems associated with the debris and solids in the wastewater flow. In some installations, another connection is made to a point downstream of the flume or weir in order to be able to flush out the stilling well. This connection must be valved closed except during cleaning of the stilling well when the flow reading from the instrument will not be used. The advantage of a stilling well is that it is not affected by wave action in a flume or weir. Also, a stilling well eliminates the effects of foam in the wastewater discharge for the users of ultrasonic depth sensors and also the effects of rags for the users of sensors fouled by rags. However, frequent maintenance is often necessary to keep both the stilling well and the connection to the flume or weir clean.

6.212 Floats

SEPARATE FLOAT IN STILLING WELL

A simple float, either connected directly to an indicating rod or connected to a cable and pulley, is the simplest method for measuring a head elevation. As long as the stilling well is kept free of debris and silt, the float will provide an accurate indication of the head (Figure 6.12).

Flow Measurement 231

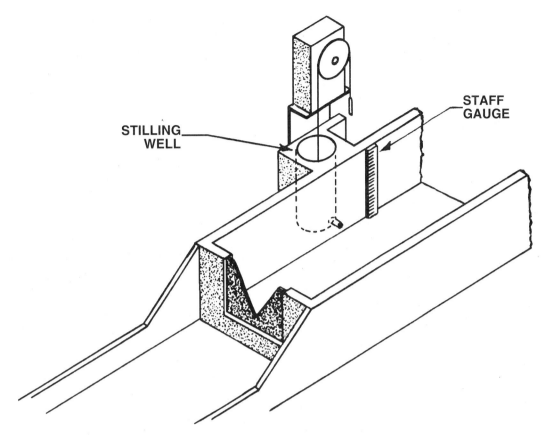

Fig. 6.11 Staff gauge and stilling well. Both used to measure head on weir (shown) and flumes

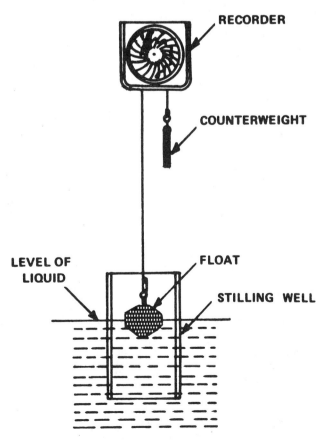

Fig. 6.12 A float is used to measure liquid level and convert the level reading to a flow measurement

BALL OR SCOW FLOATS ON PIVOTED ARMS

A float may also be attached to a pivoting arm to measure changes in the surface level of flowing water, such as in a flume or weir. The pivoting arm permits the float to remain at the proper measuring point while the liquid level rises and falls with changing flow rates. Use of a round ball or streamlined "scow" float minimizes disturbance of the water surface and accumulation of debris on the float.

Measurement errors in a float-operated system arise from float lag; water leaking into float; errors in manufacturing tolerances; the buildup of grease on the float, float feed cables, and stilling well walls; and wear and tear. The effect of float lag can be minimized when taking a comparative depth measurement by taking the measurement when the flow is steady. Leaking floats can be avoided by using leakproof materials and construction. Much greater errors result from debris and grease in stilling well lines. Excessive error in manufacturing tolerances and errors due to gradual wear and tear should show up and be compensated for during calibrations of the meter. A kink in the float cable or a broken mechanical device should show as an obvious error. An accumulation of debris on the float or a foam layer on the water surface may affect the accuracy of the system. As with all flow metering systems, some error will result if the meter drifts out of adjustment.

Float measuring systems should be calibrated monthly or quarterly to ensure they are being properly maintained. Maintenance includes flushing the stilling well, checking for free movement of float and cable, and proper lubrication.

6.213 Ultrasonic Devices

An electronic device, referred to as an "ultrasonic" flowmeter, that emits a beam of sound waves, similar to "sonar," can be used to measure water depth by determining the time it takes an acoustic pulse generated in a transducer or transmitter to reach and be reflected from the surface of the wastewater stream (Figure 6.13). The receiver picks up the signal from the transmitter and converts this signal to a linear (analog) output that can be transmitted to a remote instrument such as a recorder or controller.

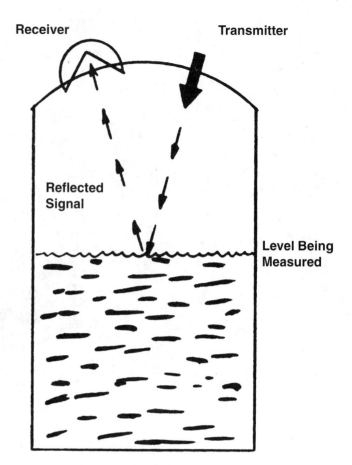

Fig. 6.13 Ultrasonic sound is used to measure liquid level and convert the level reading to a flow measurement

Ultrasonic devices offer the accuracy of other secondary devices without disturbing the water surface. Another advantage is that they can be installed in a variety of locations. This permits calibration of the instrument without entering a confined space, such as a manhole or metering pit.

6.214 Bubblers

A bubbler is a system that provides a constant flow of air through a small tube that is submerged at the selected measuring point in the primary device. The change in water elevation is determined by measuring the change in air pressure required to push the air bubbles out of the tube (Figure 6.14).

Measurement errors arise from lag effects, although these errors should be minor. Temperature changes may affect the pressure sensing element if high-temperature wastestreams are discharged from the manufacturing processes. The *DENSITY*[6] of the liquid can also change and this changes the pressure sensed by the bubbler. However, this source of error should usu-

---

[6] *Density.* A measure of how heavy a substance (solid, liquid, or gas) is for its size. Density is expressed in terms of weight per unit volume, that is, grams per cubic centimeter or pounds per cubic foot. The density of water (at 4°C or 39°F) is 1.0 gram per cubic centimeter or about 62.4 pounds per cubic foot.

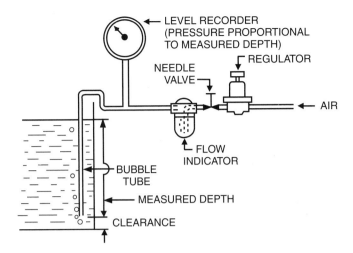

Fig. 6.14 A bubbler tube is used to measure liquid level and convert the depth reading to a flow measurement

ally be minor. Significant errors can result if the end of the bubbler tubing becomes clogged from solids in the wastestream. Errors can also result if condensate forms and freezes in any dips or low points in the piping or tubing; this error invariably causes the meter to read high. This circumstance can be prevented with frequent maintenance or by installing an automatic purging system to blow out accumulated material in the tube. A split in the bubbler tubing will cause a low reading and will be indicated by an unsteady bubble rate. A kink in the bubbler tubing may also cause an inaccurate reading. Too low a bubble rate may cause the meter to fail to properly respond to changing flows; the meter should be set to the manufacturer's recommended bubble rate. Do not turn up the bubble rate to keep the bubbler tubing clean because a higher than necessary bubble rate could produce high pressure or depth readings.

6.215 Submerged Pressure Transducers

This device is similar in function to a bubbler, except that the change in water surface elevation is sensed and measured by a sealed pressure transducer that is submerged at the point of measurement.

6.216 Electrical Capacitance Strips

The liquid elevation can also be determined by inserting an electrical conductor in the water. As the surface level moves up or down on the strip, the CONDUCTIVITY[7] of the strip increases or decreases in direct proportion to the amount of wetted surface.

QUESTIONS

Write your answers in a notebook and then compare your answers with those on page 248.

6.2E What is the purpose of a secondary measuring device in an open channel?

6.2F What factors are considered when selecting the type of secondary device?

6.2G What is the purpose of a stilling well?

6.2H How does an ultrasonic flowmeter measure water depth?

6.2I How does a bubbler measure flow depth?

6.3 CLOSED PIPE FLOW MEASUREMENT

The measurement of flows in closed pipes requires devices that measure the velocity directly or convert the velocity head to a pressure head by restricting the flow in the pipe.

Flowmeters in pipes will produce accurate flowmeter readings when the meter is located at least five pipe diameters distance downstream from any pipe bends, elbows, or valves and also at least two pipe diameters distance upstream from any pipe bends, elbows, or valves. Flowmeters also should be calibrated in place to ensure accurate flow measurements.

6.30 Differential Pressure Devices

Differential pressure devices measure flow in a pipe by measuring the change in pressure that occurs when the diameter of the pipe is reduced. As with flumes and weirs, the velocity increases and the pressure drops in predictable ways as wastewater enters the constricted section. Differential pressure devices measure the decrease in pressure, compare it to the original line pressure, and relate the pressure drop to the known change in pipe area. The velocity can then be determined and converted to a flow measurement.

A general disadvantage of measuring industrial wastewater flows in closed pipes is the difficulty of determining if the meter is clean. The disadvantages include the materials present in some wastewaters, which can coat, clog, or corrode a meter in an undesirably short period of time. The pressure taps on differential pressure devices must be kept clean to obtain accurate measurements, and they must be calibrated regularly (every six months) after installation.

6.300 Venturi Nozzle

The standard Venturi nozzle (Figure 6.15) consists of an entrance section, a throat section, and a discharge section. As the liquid passes through the reduced throat section, the velocity in-

---

[7] *Conductivity.* A measure of the ability of a solution (water) to carry an electric current.

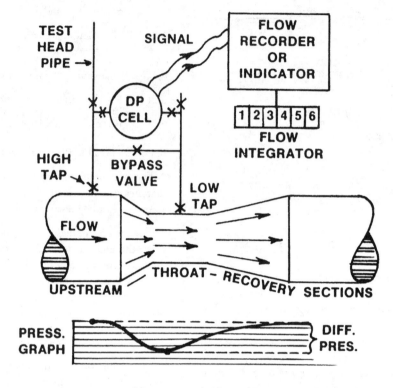

Venturi system (flow rate)

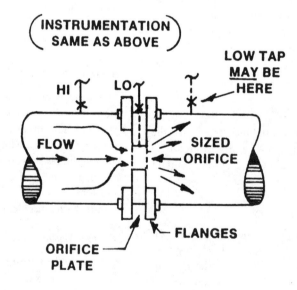

Orifice plate installation (flow rate)

Fig. 6.15 Schematic diagrams differential pressure flow metering for Venturi systems and orifice plates

creases and results in a pressure differential between the entrance and the throat sections. This pressure differential can be measured directly, using a *MANOMETER*,[8] or translated into a flow rate.

6.301 Dall Flow Tube

The Dall flow tube is a modification of the Venturi nozzle that has a greatly reduced inlet section and outlet cone, resulting in a considerably shorter unit. Because of its short construction, it has a lower permanent head loss than a Venturi while providing a considerably higher differential pressure for the same entrance and throat sizes.

6.302 Orifice Plate

An orifice plate flowmeter consists of a thin, flat plate with a precision-bored hole through it. Measuring taps are located on either side of the orifice plate to measure the pressure drop through the plate (Figure 6.15). Orifice plate meters are especially sensitive to deposits on the edge of the hole in the plate.

6.31 Mechanical Devices

6.310 Electromagnetic/Magnetic Flowmeter (Average Velocity) (Figure 6.16)

When a conductor (in this case, wastewater) moves through a magnetic field, a voltage is generated in this conductor. Two coils mounted in the meter section generate the magnetic field, and the induced voltage is received by two insulated electrodes.

Since the voltage induced is proportional to the strength of the magnetic field, the distance between the electrodes, and the velocity of the flow, the rate of flow can be determined using the continuity equation, Q = AV.

The advantages of this type of meter include an obstructionless design, high accuracy, and an output signal that is directly proportional to the flow rate, that is, a linear output.

The accuracy and linearity of an electromagnetic meter are affected by nonuniform flow patterns. Sudden changes in flow direction near the meter produce nonuniform flow patterns. For this reason, the meter should be installed where there will be three to five pipe diameters of straight pipe upstream of the meter and two diameters of straight pipe downstream.

6.311 Rotameter

Small flows in pipes, such as chlorination systems and chemical feed systems, can often be determined by measuring the drag effect on an almost buoyant float or ball suspended in a vertically mounted tube (Figures 6.17 and 6.18). As the rate of flow around the float increases, the drag forces around the float increase, causing the float to rise in the tube. The height of the float in the tube is taken as an indication of the flow rate.

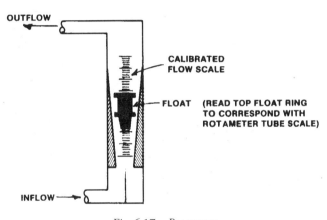

Fig. 6.17 Rotameter

6.312 Displaced Volume Devices

Large positive displacement pumps, such as piston/plunger pumps, are generally used as flowmeters by calculating the volume of liquid displaced by each stroke of the pump (Volume = "bore" × "stroke") and multiplying that by the rate of pumping (strokes per minute or strokes per day) (Figure 6.19). The volume is calculated from catalog information, but may also be measured on the pump; the number of strokes pumped is usually obtained from a "stroke counter" that is attached to the pump frame.

Chemical feeder pumps are often of the positive displacement type to ensure accurate measurement of chemicals (Figure 6.20).

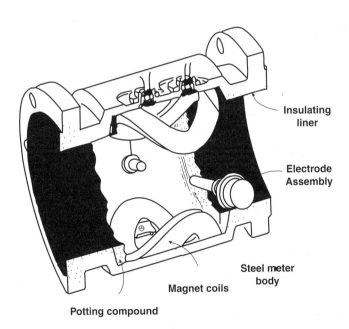

Fig. 6.16 Magnetic flowmeter
(Permission of Fischer and Porter Co.)

---

[8] *Manometer* (man-NAH-mut-ter). An instrument for measuring pressure. Usually, a manometer is a glass tube filled with a liquid that is used to measure the difference in pressure across a flow measuring device, such as an orifice or a Venturi meter. The instrument used to measure blood pressure is a type of manometer.

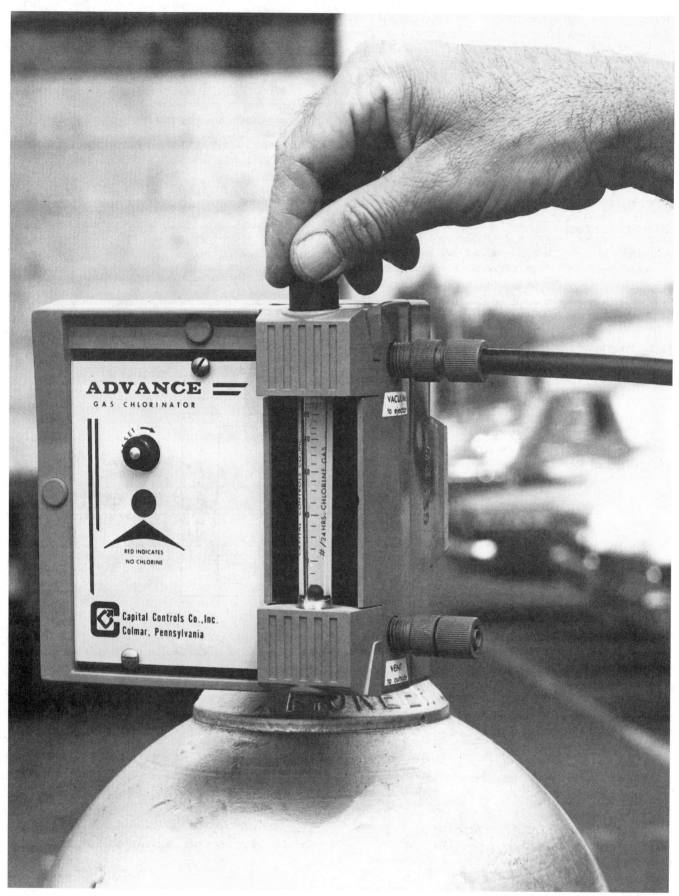

*Fig. 6.18 Chlorinator with rotameter showing
feed rate #/24 hr (lbs/day) chlorine gas*
(Permission of Capital Controls Company, Colmar, PA)

Flow Measurement 237

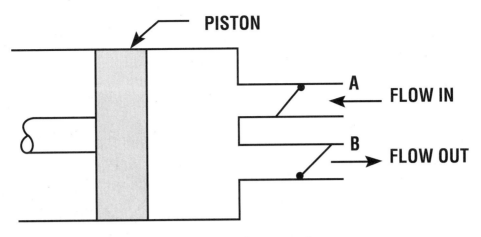

Fig. 6.19 Piston/plunger type of pump

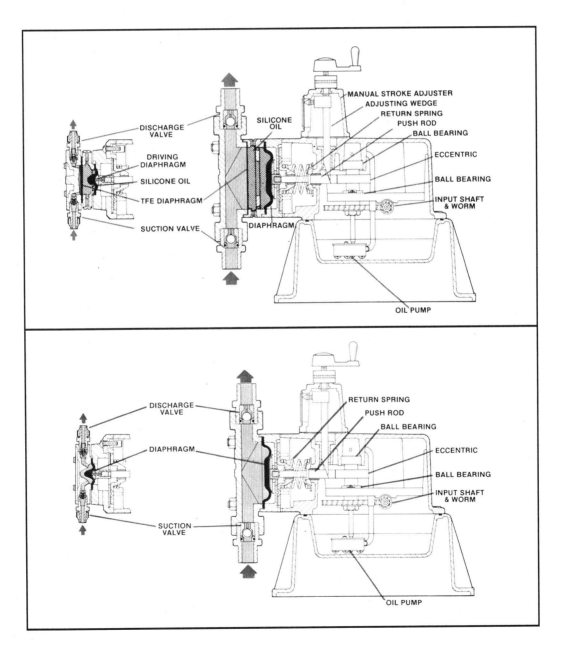

Fig. 6.20 Positive displacement diaphragm pumps
(Permission of Wallace & Tiernan Division, Pennwalt Corporation)

6.313 Propeller Meter

Although not favored by most designers because of the possibility of stringy material collecting on the rotating propeller, this type of mechanical meter has been used successfully, even in solids-bearing wastestreams (Figure 6.21).

6.32 Acoustic Devices

Acoustic, or ultrasonic, flowmeters used in measuring flow in closed pipes are based on either of two principles: transit time (time-travel) measurement or the Doppler effect.

6.320 Doppler Effect

If a sound wave is reflected off a moving object (in this case, a particle in the flow stream), a change in frequency occurs. The velocity of the particle can be determined from this change in frequency and applied to the continuity equation (Q = AV) to calculate the flow rate. The Doppler meter uses a sensor (or sensors) mounted on the outside of the pipe that transmits an ultrasonic beam through the pipe wall into the fluid at an angle to the flow stream. The same sensors receive the sound signal and compare its frequency to that of the transmitted signal. It is essential in this type of system that the fluid that is being measured contain particles from which the sound waves can be reflected; this is generally the case in all but the cleanest of industrial wastewater flow streams.

6.321 Transit Time (Time-Travel)

The velocity of a sound wave in a moving fluid is also dependent on the velocity of the fluid. If two sensors are mounted directly across from each other and at an angle to the direction of flow, the velocity of flow can be determined by the difference of the two sound waves; the velocity of sound waves transmitted in the direction of flow will be increased while the velocity of sound waves transmitted against the direction of flow will be decreased. Since this principle does not depend on the reflection of sound waves, it is effective in virtually all moving fluids. As with the Doppler meter, the velocity is related to the area of the pipe using the continuity equation (Q = AV) to calculate the flow rate.

Transit time meters are affected by nonuniform flow patterns produced by pipe elbows or constrictions. The usual recommendation is to leave at least ten diameters of straight pipe before the meter and three to five diameters of straight pipe after the meter.

6.4 SPECIAL CASES

6.40 Very Low Flows

Very low flows (flows that would be below the effective range of V-notch weirs) are often the most difficult to measure using conventional primary devices. One special flowmeter, called a weighted gate, uses a suspended flapper valve to restrict the flow in a pipeline. This restriction of flow causes a change in water surface upstream of the gate that is proportional to the weight of the gate and the rate of flow.

6.41 Stormwater

New requirements for monitoring stormwater flows require the measurement of highly variable flow volumes. If the flows are conveyed through sewer pipes, the flow conditions can vary from normal open channel flow to SURCHARGED[9] conditions. Primary devices that are suitable for open channel flow will not function under surcharged conditions, while devices that measure velocity only are not suitable for measuring open channel flows. Special types of flowmeters have been developed that measure both velocity and head, thereby enabling the flow to be calculated under the full range of conditions using the continuity equation.

QUESTIONS

Write your answers in a notebook and then compare your answers with those on page 248.

6.3A How are flows measured in closed pipes?

6.3B What are the advantages of an electromagnetic/magnetic flowmeter?

6.3C How do positive displacement pumps meter flows?

---

[9] *Surcharge.* Sewers are surcharged when the supply of water to be carried is greater than the capacity of the pipes to carry the flow. The surface of the wastewater in manholes rises above the top of the sewer pipe, and the sewer is under pressure or a head, rather than at atmospheric pressure.

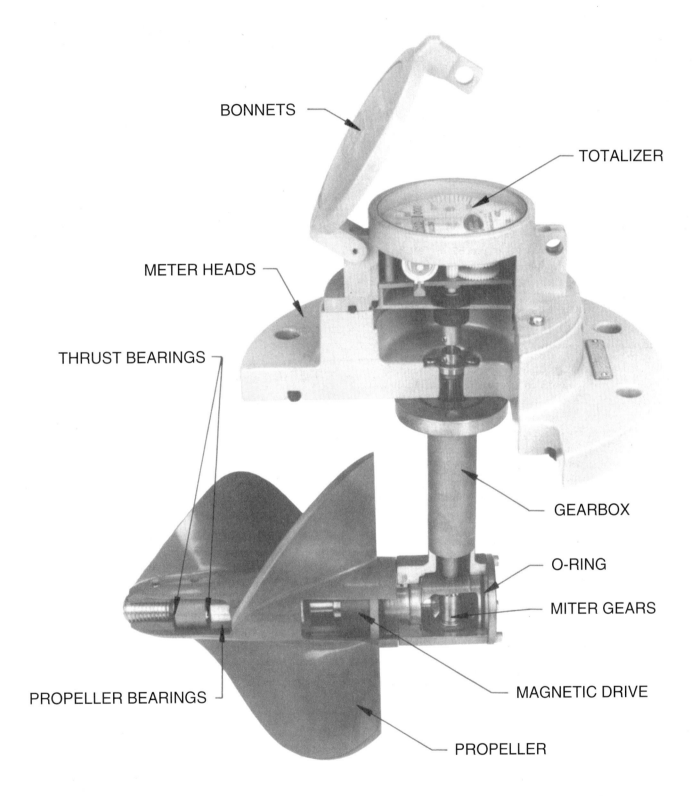

Fig. 6.21 Propeller meter
(Courtesy of Water Specialties Corporation)

6.5 METHODS OF CHECKING THE ACCURACIES OF FLOWMETERS

The basic methods to check the accuracy of flow metering systems are hydraulic calibration, instrument calibration, and comparative depth measurements. Hydraulic calibration checks both the flume or weir and the instrumentation as a system. Instrument calibration checks whether the instrumentation is operating correctly over the range of the system. A comparative depth measurement is an approximate check of a meter's accuracy at a particular flow. Two other ways to calibrate flow metering systems are the dye dilution method and the use of a calibration cylinder. Both of these methods will be discussed briefly at the end of this section.

Hydraulic calibrations and instrument calibrations are usually performed under the supervision of a qualified engineer. An operator should have a general understanding of the procedures involved. However, since both types of procedures require the use of extensive auxiliary equipment or a partial dismantling of the flow metering system, neither calibration procedure is likely to be performed by the industrial wastewater treatment plant operator.

6.50 Hydraulic Calibration

When a flow metering system is first installed, a hydraulic calibration is performed in order to verify that the system follows the standard rating curve for the system or to develop a nonstandard rating curve. Known flows are introduced through the system and the accuracy of the system is compared to the calibration standard. Typically, it is recommended that flow measurements be obtained for five known flows over the normal operating range of the system. The corresponding depths are also obtained. Once these five flow rate relationships are obtained, they can be graphed, thereby providing the necessary rating curve. This procedure is fairly standard for all open channel flow systems. A hydraulic calibration of the system should be performed every six months. Newer units are capable of simulating flows for hydraulic calibration. This is necessary to verify that the depth-to-flow relationship of the flume or weir has remained constant. An occasional system will develop a nonstandard curve over time due to wear and tear or ground settlement. The nonstandard curve can be generated by performing a hydraulic calibration. Many types of instruments will accept nonstandard curves when converting depths of water to flows.

Most flowmeters are in installations that can be calibrated by measuring the rate of flow into or out of a container of a known volume. This is referred to as the "timed volumetric" method. Depending on the size of container, a high degree of accuracy can be obtained by either repeating the test or by taking repetitive readings over a period of time. For example, if a 3- or 5-gallon pail or a 22-gallon container is used, the process is generally repeated a minimum of three times, with the flow rate determined from an average of the three readings. If a larger container is used, say a final clarifier, accuracy is obtained by measuring the rate of filling or emptying over a period of time, with elevations determined at one-minute intervals. The accuracy of this method is improved by averaging a series of successive readings.

The flow rate of a positive displacement piston pump is usually determined by multiplying the volume displaced per pump stroke by the number of strokes. This method, however, assumes that the pump is operating with no internal leakage. In actual practice, the amount pumped per stroke is reduced by leakage between the piston and the cylinder wall, as well as by leakage in the check valves. Accordingly, their accuracy is best checked by pumping from or to a tank of known volume for a set time period.

6.51 Instrument Calibration

An instrument calibration is done in order to check the performance of the instrumentation. This procedure does not require any flow through the system but instead allows for the secondary device to be removed and tested separately. Some types of instrumentation can be calibrated in place electronically.

An instrument calibration for a bubbler system, capacitance probe, or pressure sensor is usually performed with a column of water into which the bubbler tube, probe, or sensor is placed (Figure 6.22). The depth of the column of water (standpipe) is varied so as to simulate the flow through the primary element. The instrument should be checked at several depth levels, ranging from zero to the most common flow to the maximum head required for the system. The totalizing system of the bubbler should also be checked during this procedure to ensure its accuracy.

For an ultrasonic transducer, an instrument calibration can be performed by removing the transducer from its mounting bracket and aiming it against an adjustable target plate. The meter should be checked at several distances, especially at maximum distance (zero flow) and minimum distance (100 percent flow). Also check the totalizing system of the meter.

6.52 Comparative Depth Measurements

Of more importance to the operator are the techniques needed to make comparative depth measurements to verify the accuracy of an operating flowmeter. Provided a confined space entry is not required, one person can check a flowmeter if the weir or flume and instrumentation are close together or if the flow is steady. However, simultaneous readings of depth measurements and indicated instrumentation levels are preferable and two persons are usually involved. The following are some of

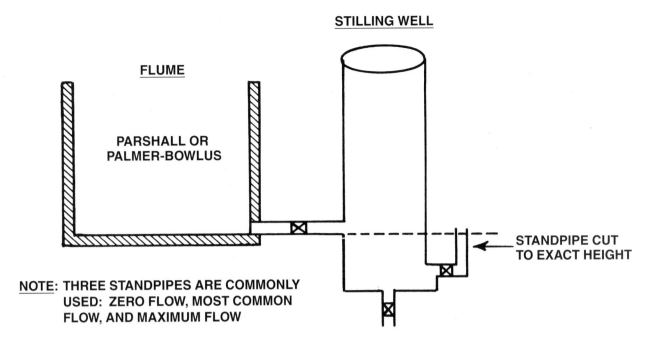

Fig. 6.22 Instrument calibration for depth measurement

the required items that you should be familiar with in order to make these measurements.

It is important to check the flow rate during an average discharge period. The accuracy of a meter can be determined at any flow rate; however, observing a flow rate during an average discharge period will also allow you to observe the normal flow conditions for that particular system and to verify that the characteristics of the flow through the primary element are correct.

Some manufacturers of prefabricated open channel elements install staff gauges on the primary element at the measuring point. A staff gauge is simply a ruler, and may be graduated in feet, inches, or units of flow. If an installation has such a staff gauge, then a quick check of the meter's accuracy can be performed by reading the level at the staff gauge and comparing it with the level indicated by the instantaneous flow rate shown on the instrumentation. It is difficult, however, to read staff gauges located in deep manholes or vaults. Also, solids tend to cover the gauges or obliterate the markings. It is often not possible to see exactly where the water touches the gauge and some allowance for this should be made when checking the depth of the water.

If a staff gauge has not been installed on the primary device or if it cannot be easily read, then a manual measurement needs to be taken. This type of measurement can be made with a thin steel rule or a measuring tape. First, mark a reference point above the measuring point of the flume or weir and determine the distance from the zero flow level of the flume or weir to the reference point. Then, preferably during a steady flow period, measure the distance from the reference point to the surface of the water. The difference between these two distances is the depth in the depth-to-flow relationship (see Figures 6.6 and 6.7).

In the above method of measuring the depth at the reference point, the flow is not disturbed while the measurement is being taken. When the measurement to the water's surface is made, wave action may make it difficult to take a precise measurement. Flowmeter instrumentation is usually able to dampen out such wave action.

In order for accurate depths to be measured, care must be taken in determining the distance from the reference point to the water surface. Take the measurement to the water's surface by lowering a thin steel rule[10] into the water and noting the interference waves that form if it is not possible to clearly see the rule touch the water. When the surface of the water is touched, a drop of water will often remain on the end of the measuring stick. This drop should be knocked off before another measurement is taken or a false indication of the water surface may result.

In some situations it may not be possible to set a reference point or get close enough to the water to measure the depth with a thin steel rule. An alternative method for checking the accuracy is to use a long, thin pole that has been tapered on the lower end (with knife edges on both sides) to minimize turbulence and waves when inserting it into the flow. The wetted section of the pole can then be measured and checked against the instanta-

---

[10] Rain gauge sticks make excellent rules to measure water depths. These sticks are specially calibrated and will "wet" to the point of the water surface.

neous reading of the flowmeter. This method is not as precise as the methods previously mentioned, but provides a fairly good check of a meter's accuracy.

A number of manufacturers have developed portable equipment for checking the accuracy of an existing meter. Portable, automatic flowmeters or velocity meters can be installed to provide continuous flow readings, and other types of equipment can be used to determine the instantaneous level or flow.

There are some important differences in the procedures used to take depth measurements from different types of primary elements. With Palmer-Bowlus flumes and V-notch and rectangular weirs, the measurement point is located a short distance upstream of the flume throat or weir plate. Also, as previously discussed, the desired depth is not the full depth of the water at the measurement point. To set a reference point in these cases, it is necessary to place a carpenter's level between the measurement point and the weir or flume. The reference measurement is set above the floor of the flume, or crest (notch) of the weir to the level, then the depth is determined from the distance down to the water surface at the measuring point. The depth or head is the distance from the reference point to the water surface. For weirs, the full measured depth is the head on the weir. With Parshall flumes, the depth of the flow is the full depth of the water at the measurement point.

If it is only possible to take a depth measurement at a Palmer-Bowlus flume by measuring the full depth of water, the assumed reference distance D/6 (see Figure 6.6) must be subtracted. However, as mentioned earlier, the actual distance from the floor of the channel to the floor of the flume may not be precisely D/6.

A reference point at a V-notch weir is usually set by placing a ruler into the notch to set the level. Due to the thickness of the ruler, it may not quite touch the very bottom of the notch and will cause a wave, thus producing an inaccurate measurement.

In many installations, it is not always possible to take the depth measurement precisely at the measuring point. With Palmer-Bowlus flumes and Parshall flumes, the measurement can be taken in the channel a short distance upstream or downstream of the measurement point. Since weirs form a large stilling pool, there is considerable latitude in where the measurement can be taken. In general, the depth measurement should be taken as close to the proper measuring point as conditions will allow.

A small amount of error will be associated with any method of measuring the depth of water behind a weir or flume. The magnitude of the error in the comparison check should be considered in the evaluation of a permanent flowmeter's accuracy.

The methods described above are not equally accurate. Errors related to the reading of a staff gauge are assumed to be minor and therefore this means of determining a flow rate should be considered very accurate, provided the staff gauge is properly installed and can be accurately read. Errors related to the determination of head by means of a reference point should be considered minor as long as the flow rate remains fairly constant during the check. Errors related to the use of a long tapered pole should be considered to be the greatest since the insertion of any obstruction into the flow can affect the flow conditions.

QUESTIONS

Write your answers in a notebook and then compare your answers with those on page 248.

6.5A What are the basic methods to check the accuracy of a flow metering system?

6.5B What factors cause problems or inaccuracies when determining the flow rate of a positive displacement piston pump?

6.5C How can an operator check the accuracy of a flowmeter?

6.53 Dye Dilution Method

In this calibration method, the flow to be measured is injected with a known concentration of dye (such as a fluorescent dye) or chemical tracer at a known, constant rate. The tracer is then diluted by the unknown flow rate to a dilution that can be measured by some instrument (such as a fluorometer). Dye injection is continued for a period of time to enable the downstream solution concentration to reach a steady state. This is a simple mass balance technique that determines the flow rate by amount of dilution present when the tracer concentration remains at a steady level or value. The method is described by the formula:

$$q_i \times C_i = Q_t \times c_f$$

This formula can be rearranged to isolate the value to be determined:

$$Q_t = (q_i \times C_i)/c_f$$

where Q_t = the total flow being measured

q_i = the rate of injection of the tracer

C_i = the initial concentration of the tracer solution

c_f = the final concentration of the tracer as diluted by the unknown flow stream

The accuracy of this method is dependent on the accuracy with which the injection rate can be determined and the accuracy by which the initial and final dye concentrations can be measured. It is especially useful for measuring large flow rates that are not readily determined by other methods; it is not useful for continuous flow monitoring.

6.54 Calibration Cylinder

The calibration of chemical metering pumps and meters measuring small pumped flow rates can be accomplished periodically with the use of a calibration cylinder (Figure 6.23). This method makes use of a cylinder that is calibrated in known volume increments and attached to the suction line of the pump. The cylinder is filled by gravity from the supply tank and then emptied by the metering pump over a known time interval. This is a built-in version of the "bucket and stopwatch" method that was discussed earlier.

6.55 Comparison of Typical Flowmeter Accuracies

Table 6.2 presents useful information about suitable applications for various types of flowmeters, their typical accuracies, and commonly available sizes.

QUESTIONS

Write your answers in a notebook and then compare your answers with those on page 248.

6.5D When is the dye dilution method most useful in measuring flows?

6.5E How are chemical metering pumps and meters measuring small pumped flow rates calibrated?

6.6 FLOWS TO MEASURE

At any industrial wastewater treatment facility various flow streams need to be measured for permit reporting purposes, and there are other flows that are either crucial or good to know for process control purposes. Whether these flow rates are continuously indicated, recorded, or totalized—or any combination of these functions—is determined by the use for the information.

6.60 Influent/Effluent

Whether required by permit or for process control, the total flow into or out of a treatment plant or plant site is usually needed. If it is important to know what the instantaneous rates

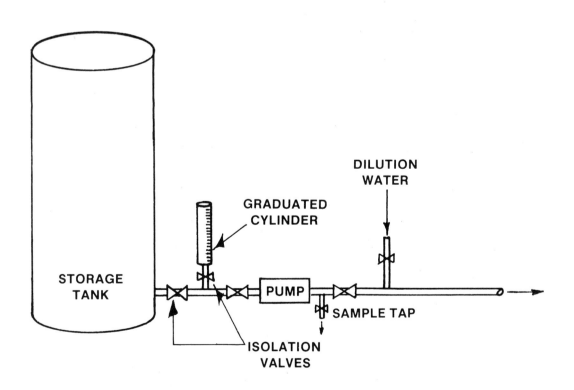

Fig. 6.23 Calibration cylinder system

TABLE 6.2 COMPARISON OF TYPICAL FLOWMETERS*

	Good for Clean Liquids	Good for Dirty Liquids	Typical Accuracy Percent	Usable Accurate Range	Typical Sizes, Inches
Flume	yes	yes	2–5 fs[a]	>10 to 1	3–120
Weir	yes	maybe[b]	2–5 fs	>10 to 1[c]	3–120
Venturi flow tube	yes	maybe[d]	1–2 fs	4 to 1[e]	3–72
Insert flow tube	yes	no	1–2 fs	4 to 1[e]	3–48
Magnetic	yes	yes	1 (rate)	10 to 1	1–96
Transit-time ultrasonic	yes	maybe[f]	1 (rate)	10 to 1	1–96
Doppler ultrasonic	no	yes	2–5 fs	10 to 1	1–48
Propeller	yes	no	2 (rate)	10 to 1	2–48
Turbine	yes	no	1.5 (rate)	>25 to 1	2–16

* Table 6.2 is adapted from, "Flow meters for water- and wastewater-treatment plants," by Patrick Moorman, WATERWORLD REVIEW, Jan/Feb 1994, page 27, and is reprinted with permission of WATERWORLD REVIEW.
[a] fs = full scale; rate = percent of actual rate
[b] Weirs can be used for dirty fluids if the solids particles stay in suspension as flow crosses the weir.
[c] Range can exceed 100 to 1 for V-notch weirs.
[d] Venturi tubes can be used for dirty fluids if appropriate seals or back-flushing devices are used to prevent dirt particles from plugging the pressure ports.
[e] Range can exceed 10 to 1 if a dual transmitter scheme is used.
[f] Transit-time meters can tolerate some dirty liquids as long as the suspended particles do not obstruct sound-wave transmission.

of flow, or flows, during a particular time period are, the influent flow to the plant should be measured. Most often this is required in order to signal an automatic sampler for the influent flow stream.

If it is only important to determine the rate or amount of flow leaving a facility, an effluent flow measurement will be adequate. Unless there is an equalization basin at the plant, the effluent flow rate will follow the influent flow rate. It will have approximately the same peaks and valleys in rates, but will be offset by a time delay determined by the nature of the process tankage. The delay could be as short as five minutes or as long as three hours.

The lack of solids and debris, or interferences by flows backed up or released by influent bar screens, are some of the considerations favoring effluent flow measurement.

6.61 Equalization Basin Flow

If a facility has an equalization basin, it is important to be able to monitor flows before and after the basin in order to control its operation and optimize its effectiveness. The meter for the influent flow can be the plant's influent flowmeter, while the rate of flow from the equalization basin can be determined by a typical effluent flowmeter. If a separate flowmeter is required for process control, it is usually only necessary to indicate a rate of flow for the effluent.

6.62 Primary Sludge Flow

Primary sludge flow rates are usually monitored for the purpose of keeping track of the overall process performance. A change in influent quality is sometimes first indicated by a noticeable increase or decrease in primary (raw) sludge production. Primary sludge pumping rates are usually monitored in terms of the amount of time required to empty the sludge hopper with a check on production rate from a periodic recording of the stroke count indicator.

Occasionally, the stroke counter is supplemented or replaced by some type of full-pipe flowmeter, especially when a continuous indication of the rate of sludge flow is required. However, the nature of primary sludge is such that meters that are directly connected to the flow stream, like magnetic flowmeters or Venturi-type meters, are prone to require more frequent maintenance than other types of measuring devices might require. Hence, the reliance on stroke counting. Remember, though, if you need an accurate measure of the primary sludge flow, compare the volume of sludge pumped as determined from the stroke counter against an actual volumetric test.

6.63 Activated Sludge Flows To or From

6.630 Aeration Basin

The rate of flow to or from each aeration basin is generally not measured. Flow rates are usually assumed to be approximately equal if the basins are arranged to receive equal flows by a symmetrical arrangement in the basin layout. Imbalances in basin flows are indicated, however, when there is a continual imbalance in the mixed liquor concentrations.

6.631 Pass or Compartment

Similarly, flow rates to the individual compartments or "passes" in a step-feed aeration system are rarely measured. Flow distribution is largely determined by the hydraulic design of the system and by the operation of any feed control valves.

6.632 Final Clarifier

The ability to measure flow to or from each clarifier in an activated sludge plant is essential for sludge blanket control and to achieve good effluent quality. It is especially critical when settling conditions are poor or if the plant is approaching a hydraulic overload. Although the flow to the secondary clarifiers may be controlled by a distribution system designed to deliver equal flows, the actual distribution is often unequal.

Flow distribution to clarifiers can either be measured in the feed channel to each clarifier, using a low-head primary device, such as a Parshall flume, or as it exits the clarifier through the effluent launder system. One way to monitor effluent flow from a circular clarifier is to block off the effluent flow from one side of the clarifier. The entire flow can then be routed to the other side and measured using a simple (but usually nonstandard) weir or a Parshall or Palmer-Bowlus flume. The primary function of these flowmeters is to provide a measure of the relative amount of flow to each clarifier. Using this information, the operator can balance the flow to various clarifiers.

In monitoring flows from rectangular clarifiers, the same principles apply. Determine the relative amount of flow either by blocking flow in some of the launders and shunting it to a nonstandard weir inserted in one representative effluent launder, or by inserting the same weir in each launder and measuring the head over one representative weir in each clarifier.

Keep in mind that the purpose of monitoring these flows is not to be able to measure the exact flow but to be able to have control over its distribution.

6.64 Return Activated Sludge (RAS)

The amount of return activated sludge is usually only of general interest in terms of the amount of flow, such as "30 percent of plant flow." In optimizing plant performance, however, it may be necessary to monitor and control the rate with more accuracy. The importance of this measurement is sometimes affected by the mode of operation; for example, the amount may be more important when operating a contact stabilization process than an extended aeration process.

6.640 To Each Aeration Tank

Mixed liquor concentrations depend on the amount of influent flow to an aeration tank as well as on the rate of the RAS flow. Return sludge flows are usually measured by meters on the sludge force mains or by weir boxes located at each aeration tank. A weir box equipped with an internal baffle to reduce the agitation from the air lift provides a simple visual measure of the flow rate; this arrangement is especially useful with air lift pumps.

6.641 From Each Final Clarifier

Even though most plants have meters for the total RAS flow, it is equally important to measure the rate of flow from each clarifier. Unless this capacity was provided in the design, it can be difficult to find a location for installing nonobtrusive devices, such as an ultrasonic meter, with the recommended lengths of straight pipe sections before and after the meter.

Since the control of RAS flow rates can be critical to the process, these meters should be calibrated periodically. This can be done most effectively by isolating (stopping) the influent to a clarifier and performing a draw-down test using the RAS pump to be calibrated. Knowing the volume of the clarifier and the rate of drop in the clarifier's water surface during a brief time interval (for example, 10 minutes), the rate of pumping can be determined with a high degree of accuracy.

EXAMPLE 1

A 50-foot diameter clarifier drops 3.0 inches in 10 minutes during a calibration test on the RAS pump. Calculate the RAS pumping rate in gallons per minute (GPM).

Known	Unknown
Diameter, ft = 50 ft	Pumping Rate, GPM
Drop, in = 3.0 in	
Time, min = 10 min	

Calculate the RAS pumping rate in GPM.

$$\text{Pumping Rate, GPM} = \frac{\text{Volume Pumped, gal}}{\text{Time Pumped, min}}$$

$$= \frac{(0.785)(\text{Diameter, ft})^2(\text{Drop, in})(7.48 \text{ gal/cu ft})}{(\text{Time, min})(12 \text{ in/ft})}$$

$$= \frac{(0.785)(50 \text{ ft})^2(3.0 \text{ in})(7.48 \text{ gal/cu ft})}{(10 \text{ min})(12 \text{ in/ft})}$$

$$= 367 \text{ GPM}$$

6.65 Waste Activated Sludge (WAS)

Controlling the amount of activated sludge that is wasted from the process is one of the most critical tasks in operating an activated sludge process. Whether the process control strategy is based on mixed liquor concentration, a food to microorganism ratio (F:M), or a selected cell residence time (MCRT or SRT), the key control activity is determining the amount of sludge wasted from the process. The ability to waste the proper amount of sludge is controlled by an accurate measurement of the waste sludge concentration along with the wasting flow rate. Although the capability to measure the waste sludge flow rate should be provided for in the design of the plant, it is often necessary to find other methods for measuring this flow.

6.650 Wasting to an Aerobic Digester

Wasting sludge to an aerobic digester provides an opportunity to use the level change in the digester to calculate the actual volume that is wasted. When this method is used, it is usually based on the volume of the digester per inch of depth.

6.651 Wasting to a Thickener

If sludge is wasted to a thickener, the wasting rate can be checked periodically by using a weir placed in the effluent channel of a gravity thickener or in the supernatant trough of a flotation thickener. Although these will normally not be "standard" weirs, they can be made accurate enough for this determination by calibrating the weirs based on the rate at which the thickener fills.

6.66 Filter Backwash Rate

Sand and multi-media filters require a periodic backwash in order to flush out the filtered solids and maintain their throughput capability. In gravity flow filters, the backwash flow rate can be monitored by periodically measuring the rate of rise of the backwash flow within the filter. In continuous backwash filters, this rate is determined by the capacity of the backwash pump and the action of a flow control valve on the pump's discharge. The backwash flow rate can be determined directly, however, by inserting either a weir or a flume at the end of the discharge trough. Since the backwash often contains sand particles that would tend to settle out in a channel controlled by a weir, a flume such as a Montana flume would allow for a good visual indication of the flow rate without any obstruction in the invert of the trough. (A Montana flume consists of the entrance section of a Parshall flume, without any throat or discharge section.)

6.67 Chemical Feed Rates

Diluted or concentrated chemical solutions have many applications in the industrial waste treatment process. Chemicals in liquid form are used to provide acidity or alkalinity, for nutrient addition, and for coagulation and precipitation processes.

Chemicals are usually fed from a storage tank or mixing tank in small quantities at either a constant rate or at a rate that varies with process requirements. These varying rates are usually flow-paced or paced by a specific chemical requirement, such as would be indicated by a pH probe.

Depending on the sensitivity of the process demands, chemical feed pumping rates are monitored by observing the rate of change of the liquid elevation in the storage tank or by use of a calibration cylinder. The storage tank level can be monitored over a period of time with a sight glass attached to the tank, or if the tank is translucent (semitransparent), by observing the liquid level through the tank wall. A calibration cylinder, on the other hand, enables the operator to check the feed rate at any time simply by filling up the calibration cylinder and timing the rate of emptying over a matter of minutes.

6.7 FLOW PROPORTIONING COMPOSITE SAMPLES

Samples for reporting or for process control purposes may have to be composited over time in order to provide a sample that is more representative. If the wastewater characteristics are relatively constant for a given flow period, samples can be time-weighted, such as one per hour for six consecutive hours, or one every 15 minutes for four hours. However, if the waste characteristics are variable, a more representative sample would be obtained from a flow-proportioned sample (a sample composited based on flow).

A flow-proportioned composite sample consists of a number of grab samples, each representing a single instant. For example, grab samples of 500 mL are collected every hour during a 24-hour period and then returned to the lab. The lab, using the flows recorded on a chart during the sampling period, transfers a portion of each grab sample to one large sample container. The size of each grab sample transferred is proportional to the flow at the time the sample was collected (as shown below) by adding 10 mL of grab sample for every 100 gallons per day of flow rate.

TIME	FLOW, GPD	SAMPLE, mL
1000	600	60
1200	800	80
1400	1,000	100
1600	1,600	160
1800	2,000	200
2000	1,700	170
2200	1,200	120
2400	1,000	100
0200	800	80
0400	700	70
0600	500	50
0800	600	60

Composite Sample Size, mL = **1,250**

Mixing the sample produces a single sample equal to the average of the pollutants or constituents in all of the grab samples. If enough grab samples are taken, the composite sample will contain constituents approximately equal to the average of the wastewater that was discharged.

QUESTIONS

Write your answers in a notebook and then compare your answers with those on page 248.

6.6A Why should flows be measured before and after an equalization basin?

6.6B How can the accuracy of a primary sludge pump be determined?

6.6C How are return activated sludge flows measured?

6.6D How can the actual volume of waste activated sludge pumped to an aerobic digester be measured?

6.7A How can a representative sample be obtained if the waste characteristics are variable?

6.8 ACKNOWLEDGMENT

Drawings in this chapter were obtained from Chapter 7, "Wastewater Flow Monitoring" by Lory Rising, Rob Wienke, and Paul Martyn in *PRETREATMENT FACILITY INSPECTION* in this series of operator training manuals. For additional information, please review Chapter 7, "Wastewater Flow Monitoring."

Please answer the discussion and review questions next.

DISCUSSION AND REVIEW QUESTIONS
Chapter 6. FLOW MEASUREMENT

The purpose of these questions is to indicate to you how well you understand the material in the chapter. Write the answers to these questions in your notebook.

1. How can a representative sample be obtained from an industry with a fluctuating or unusual waste discharge flow rate?
2. When do open channel flow conditions occur?
3. What are the commonly used methods of measuring plant flows?
4. Where are flumes used instead of weirs to measure flows?
5. What is the primary concern when using a Palmer-Bowlus flume?
6. How does a differential pressure meter measure flow in a pipe?
7. How are the accuracy and flow rate from a piston pump best determined?
8. How can imbalances in aeration basin flows be detected?
9. How can an RAS pump be calibrated?

SUGGESTED ANSWERS
Chapter 6. FLOW MEASUREMENT

Answers to questions on page 218.

6.0A An operator needs to know how to measure flows in order to:
1. Determine how much flow is being contributed from the different process lines.
2. Calculate the correct amount of chemicals needed for treatment processes such as neutralization or nutrient addition.
3. Prepare composite samples that accurately reflect the wastestream.
4. Determine compliance with mass-based limits.

6.1A Typical units of flow measurement are expressions of volume per unit of time, such as CFS, GPM, or MGD.

6.1B The letters Q = AV mean the flow, Q, is equal to the area, A, times the volume, V.

6.1C The two basic types of flow systems are "open channel" flow and "closed channel" (or "closed pipe") flow.

Answers to questions on page 230.

6.2A The two most accurate ways to measure the rate of flow in a system are the "timed volumetric" (or "bucket and stopwatch") and "dye dilution" methods.

6.2B The secondary element for a flow measurement device in an open channel consists of some type of device for measuring the elevation of the water surface above the measurement reference point of the flume or weir.

6.2C The three different sections of a flume are the (1) entrance (converging) section, (2) throat section, and (3) discharge (diverging) section.

6.2D Head measurements must be made at a specific location on Parshall or Palmer-Bowlus flumes because the flumes are designed to produce a specific relationship between depth of flow and rate of flow at the specified measurement point. Measuring the head upstream or downstream from the proper point can result in inaccurate flow indications.

248 Treatment Plants

Answers to questions on page 233.

6.2E The purpose of the secondary measuring device in an open channel is to measure or indicate the liquid level in the primary device. The secondary device is also used to convert this "head" to a flow measurement.

6.2F The selection of the type of secondary device is based on consideration of many factors, such as location, type of information required, and cost.

6.2G The purpose of a stilling well is to provide a steadier indication of the flow depth and to avoid the problems associated with the debris and solids in the wastewater flow.

6.2H An electronic device, referred to as an "ultrasonic" flowmeter, that emits a beam of sound waves, similar to "sonar," can be used to measure water depth by determining the time it takes an acoustic pulse generated in a transducer or transmitter to reach and be reflected from the surface of the wastewater stream. The receiver picks up the signal from the transmitter and converts this signal to a linear (analog) output that can be transmitted to a remote instrument such as a recorder or controller.

6.2I A bubbler is a system that measures flow depth by providing a constant flow of air through a small tube that is submerged at the selected measuring point in the primary device. The change in water elevation is determined by measuring the change in air pressure required to push the air bubbles out of the tube.

Answers to questions on page 238.

6.3A Flows are measured in closed pipes by devices that measure the velocity directly or convert the velocity head to a pressure head by restricting the flow in the pipe.

6.3B The advantages of an electromagnetic/magnetic flowmeter include an obstructionless design, high accuracy, and an output signal that is directly proportional to the flow rate.

6.3C Positive displacement pumps meter flows by calculating the volume of liquid displaced by each stroke of the pump.

Answers to questions on page 242.

6.5A The basic methods to check the accuracy of a flow metering system are hydraulic calibration, instrument calibration, comparative depth measurements, the dye dilution method, and use of a calibration cylinder.

6.5B The flow rate of a positive displacement piston pump is usually determined by multiplying the volume displaced per pump stroke by the number of strokes. Problems or inaccuracies with this method can occur if there is leakage between the piston and the cylinder wall or leakage in the check valves.

6.5C An operator can check the accuracy of a flowmeter by determining the depth of flow on a staff gauge or by manually measuring the depth of flow.

Answers to questions on page 243.

6.5D The dye dilution method is most useful for measuring large flow rates that are not readily determined by other methods.

6.5E Chemical metering pumps and meters measuring small pumped flow rates are calibrated with the use of a calibration cylinder.

Answers to questions on page 246.

6.6A Flows should be measured before and after an equalization basin in order to control the basin's operation and optimize its effectiveness.

6.6B The accuracy of a primary sludge pump can be determined by comparing the volume of sludge pumped as determined from the stroke counter against an actual volumetric test.

6.6C Return activated sludge flows are usually measured by meters on the sludge force mains or by weir boxes located at each aeration tank.

6.6D The actual volume of waste activated sludge pumped to an aerobic digester can be measured by the level change in the digester, assuming nothing leaves the digester during the pumping time period.

6.7A A representative sample can be obtained if waste characteristics are variable by collecting a flow-proportioned sample (a sample composited based on flow).

CHAPTER 7

PRELIMINARY TREATMENT
(EQUALIZATION, SCREENING, AND pH ADJUSTMENT)

by

Richard von Langen

D. F. Reddy

Mark Acerra

TABLE OF CONTENTS

Chapter 7. PRELIMINARY TREATMENT
(EQUALIZATION, SCREENING, AND pH ADJUSTMENT)

	Page
OBJECTIVES	254
WORDS	255

LESSON 1

		Page
7.0	WHAT IS PRELIMINARY TREATMENT?	257
7.1	EQUALIZATION OF INDUSTRIAL WASTEWATER FLOWS	257
	by Richard von Langen	
	7.10 What Is Equalization?	257
	7.11 When Is Equalization Needed?	257
	7.12 Beneficial Effects of Equalization	259
	7.13 Where to Locate Equalization in the Industrial Wastewater Treatment System	260
	7.130 Equalization Basins	260
	7.131 Equalization in the Collection System	260
	7.132 Equalization at the Manufacturing Process	261
7.2	FLOW EQUALIZATION TANKS	261
	7.20 Flow Equalization Volume	261
	7.21 Other Factors for Sizing an Equalization Tank	264
	7.22 Mixing	265
	7.23 Tank Construction	265
	7.24 Pumps and Flow Controls	266
	7.25 Strategy for Operation	266
7.3	SUMMARY	267
7.4	ADDITIONAL READING	267
DISCUSSION AND REVIEW QUESTIONS		267

LESSON 2

		Page
7.5	SCREENING	268
	by D. F. Reddy	
	7.50 Why Screen?	268
	7.51 Types of Screens	268
	7.510 Coarse Screens (Bar Screens)	268
	7.5100 Description	268

		7.5101 Start-Up/Shutdown Procedures	269
		7.5102 Troubleshooting Automatically Cleaned Bar Screens	271
	7.511	Fine Screens	273
		7.5110 Static Screens	273
		7.5111 Rotating Screens	275
		7.5112 In-Channel or In-Tank Screens	282
7.52	Maintenance		282
	7.520	Coarse Screens	283
	7.521	Fine Screens	283
7.53	Safety		284

DISCUSSION AND REVIEW QUESTIONS 284

LESSON 3

7.6 NEED FOR pH ADJUSTMENT OF INDUSTRIAL WASTEWATERS 285
 by Richard von Langen and Mark Acerra

7.7 DEFINITION OF pH AND TITRATION CURVES 286

    7.70 Definition of pH 286

    7.71 Chemicals Used for pH Adjustment 287

    7.72 Titration Curve 287

7.8 pH SENSORS 290

    7.80 Use of pH Sensors 290

    7.81 Measurement Electrode 291

    7.82 Reference Electrode 291

    7.83 Factors Affecting pH Sensors 293

        7.830 Temperature 293

        7.831 Absorption Effects 293

        7.832 Face Velocity, Fouling, and Response Time 293

        7.833 Alkalinity Error 294

        7.834 Acidity Error 294

7.9 STRATEGY FOR A SUCCESSFUL pH CONTROL SYSTEM 294

    7.90 Equalization 294

    7.91 Mixing, Dead Time, and Residence Time 296

7.10 CONTROL SYSTEM 298

    7.100 Primary Element 298

    7.101 pH Transmitter 298

    7.102 pH Controller 298

    7.103 Final Element 299

Preliminary Treatment 253

7.11		COMMON pH ADJUSTMENT SYSTEMS	299
	7.110	pH Adjustment for Discharge	300
	7.111	pH Adjustment for Biological Treatment	300
	7.112	pH Adjustment for Heavy Metal Removal	301
	7.113	pH Adjustment for Oxidation and Reduction	301
7.12		MAINTAINING pH ADJUSTMENT EQUIPMENT	302
7.13		TROUBLESHOOTING pH ADJUSTMENT PROBLEMS	302
	7.130	Mechanical Operation	302
	7.131	pH Electrode	302
	7.132	Manufacturing Process Changes	304
	7.133	Back to the Basics	304
7.14		SUMMARY	304
7.15		ADDITIONAL READING	304
DISCUSSION AND REVIEW QUESTIONS			306
SUGGESTED ANSWERS			306

OBJECTIVES
Chapter 7. PRELIMINARY TREATMENT
(EQUALIZATION, SCREENING, AND pH ADJUSTMENT)

Following completion of Chapter 7, you should be able to:

1. Explain the benefits of flow and waste load equalization.
2. Determine whether equalization would benefit your plant.
3. Calculate the minimum volume of an equalization tank needed to accommodate your plant's wastewater flows.
4. Develop a strategy for operating an equalization tank.
5. Explain the importance of screening industrial wastes before treatment.
6. Describe the various types of coarse and fine screens.
7. Start up, shut down, maintain, and troubleshoot wastewater screens.
8. Explain why pH adjustment is necessary.
9. Describe the major components of a pH probe.
10. List and discuss the factors that influence pH sensor accuracy.
11. Prepare a titration curve.
12. Troubleshoot pH adjustment problems.

WORDS
Chapter 7. PRELIMINARY TREATMENT
(EQUALIZATION, SCREENING, AND pH ADJUSTMENT)

ANION (AN-EYE-en)

A negatively charged ion in an electrolyte solution, attracted to the anode under the influence of a difference in electrical potential. Chloride ion (Cl^-) is an anion.

BOD_5

BOD_5 refers to the five-day biochemical oxygen demand. The total amount of oxygen used by microorganisms decomposing organic matter increases each day until the ultimate BOD is reached, usually in 50 to 70 days. BOD usually refers to the five-day BOD or BOD_5.

BIOSOLIDS

A primarily organic solid product produced by wastewater treatment processes that can be beneficially recycled. The word biosolids is replacing the word sludge when referring to treated waste.

BLINDING

The clogging of the filtering medium of a microscreen or a vacuum filter when the holes or spaces in the media become clogged or sealed off due to a buildup of grease or the material being filtered.

BUFFER CAPACITY

A measure of the capacity of a solution or liquid to neutralize acids or bases. This is a measure of the capacity of water or wastewater for offering a resistance to changes in pH.

CATION (KAT-EYE-en)

A positively charged ion in an electrolyte solution, attracted to the cathode under the influence of a difference in electrical potential. Sodium ion (Na^+) is a cation.

CHELATING (KEY-LAY-ting) AGENT

A chemical used to prevent the precipitation of metals (such as copper).

COAGULATION (ko-agg-yoo-LAY-shun)

The clumping together of very fine particles into larger particles (floc) caused by the use of chemicals (coagulants). The chemicals neutralize the electrical charges of the fine particles, allowing them to come closer and form larger clumps.

DECANT (de-KANT) WATER

Water that has separated from sludge and is removed from the layer of water above the sludge.

DOCTOR BLADE

A blade used to remove any excess solids that may cling to the outside of a rotating screen.

ELECTROLYTE (ee-LECK-tro-lite)

A substance that dissociates (separates) into two or more ions when it is dissolved in water.

EMULSION (e-MULL-shun)

A liquid mixture of two or more liquid substances not normally dissolved in one another; one liquid is held in suspension in the other.

EQUIVALENCE POINT

The point on a titration curve where the acid ion concentration is equal to the base ion concentration.

FLOCCULATION (flock-you-LAY-shun)

The gathering together of fine particles after coagulation to form larger particles by a process of gentle mixing. This clumping together makes it easier to separate the solids from the water by settling, skimming, draining, or filtering.

HEAD LOSS

The head, pressure, or energy (they are the same) lost by water flowing in a pipe or channel as a result of turbulence caused by the velocity of the flowing water and the roughness of the pipe, channel walls, or restrictions caused by fittings. Water flowing in a pipe loses head, pressure, or energy as a result of friction losses. The head loss through a filter is due to friction losses caused by material building up on the surface or in the top part of a filter. Also called FRICTION LOSS.

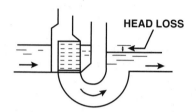

ION

An electrically charged atom, radical (such as SO_4^{2-}), or molecule formed by the loss or gain of one or more electrons.

MICRON (MY-kron)

μm, Micrometer or Micron. A unit of length. One millionth of a meter or one thousandth of a millimeter. One micron equals 0.00004 of an inch.

N or NORMAL

A normal solution contains one gram equivalent weight of reactant (compound) per liter of solution. The equivalent weight of an acid is that weight which contains one gram atom of ionizable hydrogen or its chemical equivalent. For example, the equivalent weight of sulfuric acid (H_2SO_4) is 49 (98 divided by 2 because there are two replaceable hydrogen ions). A one N solution of sulfuric acid would consist of 49 grams of H_2SO_4 dissolved in enough water to make one liter of solution.

NEUTRALIZATION (noo-trull-uh-ZAY-shun)

Addition of an acid or alkali (base) to a liquid to cause the pH of the liquid to move toward a neutral pH of 7.0.

OXIDATION-REDUCTION POTENTIAL (ORP)

The electrical potential required to transfer electrons from one compound or element (the oxidant) to another compound or element (the reductant); used as a qualitative measure of the state of oxidation in water and wastewater treatment systems. ORP is measured in millivolts, with negative values indicating a tendency to reduce compounds or elements and positive values indicating a tendency to oxidize compounds or elements.

PRECIPITATE (pre-SIP-uh-TATE)

(1) An insoluble, finely divided substance that is a product of a chemical reaction within a liquid.

(2) The separation from solution of an insoluble substance.

SEPTIC (SEP-tick) or SEPTICITY

A condition produced by bacteria when all oxygen supplies are depleted. If severe, the bottom deposits produce hydrogen sulfide, the deposits and water turn black, give off foul odors, and the water has a greatly increased oxygen and chlorine demand.

SHORT-CIRCUITING

A condition that occurs in tanks or basins when some of the flowing water entering a tank or basin flows along a nearly direct pathway from the inlet to the outlet. This is usually undesirable since it may result in shorter contact, reaction, or settling times in comparison with the theoretical (calculated) or presumed detention times.

CHAPTER 7. PRELIMINARY TREATMENT
(EQUALIZATION, SCREENING, AND pH ADJUSTMENT)

(Lesson 1 of 3 Lessons)

7.0 WHAT IS PRELIMINARY TREATMENT?

Preliminary treatment consists of a series of steps that prepare industrial wastewaters for the actual treatment processes that follow. Typically, preliminary treatment includes equalization, screening, and, if necessary, pH adjustment. In certain industries, oil and grease removal may also be necessary. This chapter focuses on the processes most commonly encountered by industrial wastewater treatment facility operators and does not include oil and grease removal.

7.1 EQUALIZATION OF INDUSTRIAL WASTEWATER FLOWS
by Richard von Langen

7.10 What Is Equalization?

The influent to an industrial wastewater treatment system (IWTS) varies in flow and concentration of pollutants because the discharges from the manufacturing and utility processes are not constant. This variation affects the operation of the IWTS and could adversely affect the effluent quality from the plant.

Equalization reduces the effects of fluctuations in flow volumes and pollutant concentrations by evening out the variations. There are two common methods of flow equalization: in-line equalization or side-line equalization (Figure 7.1). Side-line equalization diverts flows that are greater than the average flow volume or concentration to a holding tank for subsequent release when flows are below average. In-line equalization blends all incoming flows in the holding tank and discharges at the average flow rate. Flow equalization also evens out the loading (flow times the concentration, or mass) of pollutants to the downstream treatment processes. Ideally, the equalization tank is large enough to equalize the load to the rest of the IWTS and thus permit all of the treatment systems to operate at their optimum loading rates.

In-line and side-line equalization systems are both effective ways to equalize flow volumes, but in-line equalization is more effective for leveling out the variations in influent concentration because the entire flow is blended with the entire contents of the holding tank. With side-line equalization, only wastewater at high flow conditions is stored so there is little blending with other wastewaters. Side-line equalization is therefore not as effective in reducing the concentration variations. In either case, equalization of IWTS influent wastewater often significantly improves the performance of an existing treatment system or reduces the size of a new IWTS.

This lesson will describe the beneficial effects of equalization, explain how to determine whether equalization would improve an existing treatment system, describe the characteristics of a properly designed equalization tank, and suggest some strategies for operating an equalization tank.

7.11 When Is Equalization Needed?

Because of the nature of industrial processes, equalization will nearly always benefit the downstream processes of the IWTS. The more variable the flow and concentration coming into the IWTS, the more equalization will help. The influent concentration may vary because of batch discharges, such as spent plating baths, or the batch release of process wastewater, such as from the receiving flumes in a tomato processing plant. The periodic blowdown of a cooling tower may not add a significant load of biochemical oxygen demand (BOD), but will certainly add a large quantity of flow to the influent. Spills and dumps can also cause variations. The variability of the influent wastewater flow and pollutant concentration will generally increase as the size of the plant decreases because each manufacturing operation is a larger percentage of the total. The same situation occurs when there are many different manufacturing or utility processes.

To determine whether equalization would be beneficial for a specific treatment system, measure the variability of the influent flow or loading. Take hourly samples (or a minimum of 20 samples at regular time intervals over the course of the production

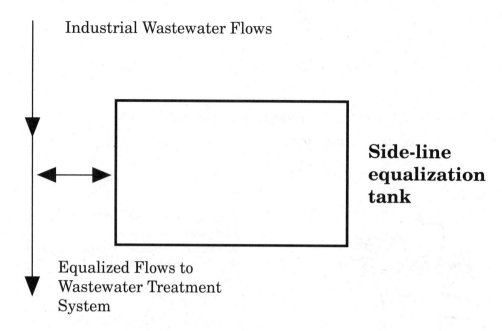

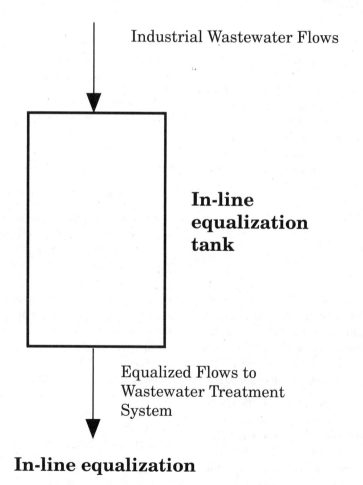

Fig. 7.1 Methods of flow equalization

day) and compute the coefficient of variation. This is the standard deviation divided by the mean and multiplied by 100 to get a percentage.[1] If the coefficient of variation of flow rate or loading rate is greater than 25 to 30 percent, then equalization probably would benefit the IWTS. Likewise, if samples were collected from the effluent of an existing equalization tank and the same calculations were made, the coefficient of variation should be less than 25 to 30 percent.

EXAMPLE 1

The mean flow to an industrial wastewater treatment system is 20 GPM and the standard deviation of the flows is 8 GPM. Calculate the coefficient of variation to determine if flow equalization would help the IWTS.

Known	Unknown
Mean Flow, GPM = 20 GPM	Coefficient of Variation, %
Standard Deviation of Flow, GPM = 8 GPM	

Calculate the coefficient of variation.

$$\text{Coefficient of Variation, \%} = \frac{(\text{Standard Deviation, GPM})(100\%)}{\text{Mean, GPM}}$$

$$= \frac{(8\ \text{GPM})(100\%)}{20\ \text{GPM}}$$

$$= 40\%$$

Flow equalization would benefit the IWTS because the calculated coefficient of variation of 40 percent is greater than the 30 percent that equalization should provide.

7.12 Beneficial Effects of Equalization

The result of equalization is the ability to operate the downstream processes in the IWTS at a constant flow with less variation in loading. The primary benefit is an improvement in the effluent quality. Shock loads from the manufacturing process that may overload or even toxify the IWTS are reduced in magnitude (volume or concentration) and can be treated more easily. Recycle streams, such as filter backwash, filter press filtrate, or sludge holding tank overflow, can be routed to the equalization tank to prevent overloading the IWTS with a batch load of concentrated pollutants.

Some of the chemical additions in the IWTS are flow- and load-paced. For heavy metal precipitation systems, polymer is usually added in proportion to the flow going to the flocculator and solids removal equipment. Even if the polymer feed pump is turned on whenever wastewater is pumped to the clarifier, the loading of suspended solids and therefore the amount of polymer required is not constant if the loading has not been equalized. For a pH adjustment system feeding a biological reactor, the addition of the neutralizing agent is based on a pH of the solution. By equalizing the flow to the pH adjustment tank, the variation in the pH is reduced and the system can adjust the pH to within a small range. Limiting pH variations to a small range improves the performance of the biological system because the organisms are not constantly adapting to new pH conditions.

Pumps and other mechanical equipment downstream of the equalization tank that are controlled by the flow and load to the system can operate at nearly constant conditions and save energy costs. When large level-controlled sump pumps feeding the IWTS turn on, there is a large start-up current load or energy demand. This results in a large electrical "demand" cost. If the flows are equalized and the pumps remain on all the time at a lower electrical load, the demand charge will be lower.

Equalization of flows to a biological IWTS is particularly important. Variable loadings (BOD) to the activated sludge process can seriously affect the effluent quality. Studies of transient loadings (a sudden increase or decrease) on an activated sludge process found the following:

1. An increase in the BOD level in the influent was seen almost immediately and peaked two to five hours after the step increase.

2. The ability of the sludge to settle decreased with an increase or decrease in the BOD loading.

---

[1] *Mean and Standard Deviation.* The mean, $\bar{X}$, is an indication of the average or mean value of the flows or a group of measurements (flows). The standard deviation and variance are measures of the variability, dispersion or spread of the flows or a group of measurements (flows). Some computer programs that will compute standard deviation and mean are commercially available or you may wish to review the explanation of these terms in OPERATION OF WASTEWATER TREATMENT PLANTS, Volume II, Chapter 18, "Analysis and Presentation of Data," in this series of manuals.

$$\text{Mean, } \bar{X} = \frac{\text{Sum of Measurements, x}}{\text{Number of Measurements, n}}$$

$$\text{Variance, } S^2 = \frac{\text{Sum of Square of Each Measurement, } x^2 - (\text{Sum of Measurements, x})^2/n}{\text{Number of Measurements, n} - 1}$$

$$\text{Standard Deviation, } S = (S^2)^{1/2}$$

3. For a constant influent BOD level, the effluent BOD level increased more with an increase in flow (loading) than with a decrease in flow.

4. If the loading of BOD was constant, the process could tolerate a 100 percent increase or decrease in flow without a serious degradation (reduction) of effluent quality.

5. Gradual flow increases and decreases had less effect on the effluent quality than sudden changes.

Equalization of flows going to a biological IWTS will usually produce the following benefits:

1. Improvement in primary settling because the *BIOSOLIDS*[2] will tend to flocculate during the mixing that occurs in the equalization tank. (This advantage is reduced if the feed from the equalization tank is by a centrifugal pump that will tend to break up the flocs.)

2. Improved settling in the final clarifier, which reduces the washout of solids. With a higher solids inventory, the food-to-microorganism ratio (F/M) will decrease and the sludge retention time (SRT) will increase. This may result in an increased level of nitrification and a decrease in sludge production.

3. Improved response to shock loads. In a simulation of a municipal activated sludge system subjected to a shock load, the coefficient of variation for the BOD load going to the aeration tank was reduced from 73 percent to 28 percent with an equalization basin just large enough to equalize the flow. The effluent BOD coefficient of variation was reduced from 72 percent to 15 percent and the flow-weighted average BOD load was reduced by 15 percent because of the equalization tank.

4. Some BOD reduction is likely to occur in an aerated equalization tank. A 10 to 20 percent reduction is predicted, but is dependent on the detention time, the amount of aeration provided, temperature, and other factors.[3]

If the IWTS does not have an equalization tank, the addition of an equalization tank to the system will improve the performance of an overloaded IWTS. If a new IWTS is being designed, the addition of an equalization tank will reduce the size of all of the other processes. The pH adjustment tanks, clarifiers, and activated sludge aeration tank can be designed for the average flow rather than peak flows. With peak to average flows typically ranging from 2:1 to 4:1 and peak to average loads of a pollutant typically around 10:1, it is not difficult to understand that an equalization basin can greatly assist in reducing the size and operating range of the IWTS.

In addition, difficult-to-treat wastewaters or difficult-to-meet effluent limits may be less of a problem when equalization allows the IWTS to operate at its optimum flow and load.

QUESTIONS

Write your answers in a notebook and then compare your answers with those on page 306.

7.1A What is the difference between side-line and in-line flow equalization?

7.1B What could cause the influent to an IWTS to vary in flow and concentration of pollutants?

7.1C What is the primary benefit of flow equalization for an industrial wastewater treatment system?

7.1D Why should IWTS recycle streams be routed to the equalization tank?

7.13 Where to Locate Equalization in the Industrial Wastewater Treatment System

7.130 Equalization Basins

Equalization is often the first operation in the IWTS. This reduces the sizing of all of the remaining unit processes and allows them to operate at a constant flow or a constant load. However, for some wastewater streams, it may be advantageous to have equalization follow preliminary or primary treatment. For example, if the wastewater contains a high concentration of solids or solvents, it would be better to remove these prior to equalization. If the solids are not removed, the mixing requirements may become excessive or interfere with the next unit operation. Inert solids (such as sand or grit) may build up in the aeration tank of an activated sludge process or the solids may cause excessive wear on the equipment. If the solvents are not removed prior to equalization, they may become volatile in the equalization tank causing a potential violation of air pollution regulations. Mixing in the equalization tank could *EMULSIFY*[4] the solvents, making them harder to remove in the next treatment process.

7.131 Equalization in the Collection System

Equalization could also be accomplished within the wastewater collection system. Typically, it would be done with larger industrial wastewater treatment systems or systems having sumps and pumping stations where flows could be discharged at controlled rates. Care must be taken while storing the flow in the collection system; the velocity in the industrial sewer should not drop so low as to cause deposition of solids. Solids accumulation will reduce the capacity of the collection system, or worse yet, plug an industrial sewer causing wastewater to back up into buildings. An overflow protection device, such as an overflow weir or an audible alarm, should be installed. Finally, if the collection system is used as the means of equalizing the flow, reex-

---

[2] *Biosolids.* A primarily organic solid product produced by wastewater treatment processes that can be beneficially recycled. The word biosolids is replacing the word sludge when referring to treated waste.

[3] EPA Technology Transfer publication, *FLOW EQUALIZATION*, Environmental Protection Agency, May 1974.

[4] *Emulsion* (e-MULL-shun). A liquid mixture of two or more liquid substances not normally dissolved in one another; one liquid is held in suspension in the other.

amine all discharges into the sewer to make sure that they are not concentrated or incompatible wastes that could cause a violent reaction when combined in the sewer.

7.132 Equalization at the Manufacturing Process

A means to partially equalize the wastewater flows to the IWTS is to control the discharge at the manufacturing process. Examples include changing the periodic blowdown of cooling tower water to continuous flow release, restricting the drain valve opening when cleaning a process tank to discharge at a slower rate, installing a separate bleed tank for highly concentrated solutions to meter the wastewater into the sewer, or using an existing sump to equalize the flow from one area of the plant. These methods will help reduce shock loadings to the IWTS. However, the best method for equalizing the flow is the in-line equalization tank.

QUESTIONS

Write your answers in a notebook and then compare your answers with those on page 306.

7.1E Why should solvents be removed from the wastestream before it enters an equalization tank?

7.1F What precautions should be taken when a collection system is used for equalization of flows?

7.2 FLOW EQUALIZATION TANKS

This section discusses the characteristics of a properly designed flow equalization tank system. As an industrial wastewater treatment plant operator, you can compare your existing equalization tank to the one described below to note its advantages, disadvantages, and what simple things you can do to make it operate better. There are five areas for comparison: volume needed for flow equalization, other factors that affect size, mixing requirements, tank construction factors, and pumping operation and controls. If your facility does not presently use an equalization system, the information provided in the next several paragraphs can help you determine whether installation of an equalization system would be beneficial and cost-effective.

7.20 Flow Equalization Volume

The volume required for an in-line and a side-line equalization tank is the same, and the equipment required for either system is nearly the same. Because the in-line type of system gives better load equalization, it is preferred. If side-line equalization is currently being used, you may want to convert it to an in-line system to take advantage of the benefits.

To determine the size (capacity) of a flow equalization tank for your facility, prepare a table similar to Table 7.1. Begin by measuring or estimating the average IWTS influent flow at small time intervals, such as half-hour (30-minute) intervals, during the industrial wastewater flow cycle. In selecting the time interval you will measure, try to choose a length of time that will give you at least 20 time intervals during the manufacturing cycle.

The manufacturing cycle may be any length, but is typically 8 to 24 hours or more depending on the processes involved. The total length of the cycle may also depend on how long the wastewater treatment system is designed to operate, which may be the same as, shorter than, or longer than the manufacturing cycle. In selecting the cycle to measure, choose a cycle that represents the typical maximum industrial wastewater flow for the plant.

EXAMPLE 2

In this example we are going to assume that the wastewater flow from manufacturing starts at 0745 and that each half-hour flow measurement represents the average flow from 15 minutes before the measurement to 15 minutes after the measurement. Therefore, a flow measurement of 40 GPM at 0800 assumes that 40 GPM is the average flow from 0745 to 0815.

The flow volume for each interval is computed by multiplying the average flow rate during the time interval by the time interval itself (30 minutes). Therefore, the flow volume during the first time period in our example is:

Volume, gal = (Flow, GPM)(Time, minutes)

= (40 GPM)(30 minutes)

= 1,200 gallons

Each 30-minute volume should be added to the total of the other previous volumes to obtain the cumulative (total) flow volume. As you can see in Table 7.1, the cumulative total at 0745 is zero; at 0815 (which is the end of the 0745 to 0815 time interval) the cumulative total is 1,200 gallons. When the volume during the next interval (1,800 gal, from 0815 to 0845) is added to the previous cumulative total (1,200 gal), the new cumulative volume is 3,000 gallons. Continue making these calculations for one complete manufacturing cycle.

Once you have recorded the flow data in a table similar to Table 7.1, you have enough information to calculate the average rate of wastewater flow to the IWTS during the manufacturing cycle. This average flow rate will become the rate of discharge from the equalization tank. To calculate the average flow rate, divide the total flow volume by the total release time for the flow.

TABLE 7.1 EXAMPLE CALCULATION OF CUMULATIVE FLOW VOLUMES FOR DETERMINING THE SIZE OF AN EQUALIZATION TANK

Time	Avg Flow, GPM	Volume, gal	Cumulative Volume, gal
0745			0
0800[a]	40	1,200	
0815			1,200
0830	60	1,800	
0845			3,000
0900	60	1,800	
0915			4,800
0930	30	900	
0945			5,700
1000	30	900	
1015			6,600
1030	30	900	
1045			7,500
1100	60	1,800	
1115			9,300
1130	40	1,200	
1145			10,500
1200	90	2,700	
1215			13,200
1230	90	2,700	
1245			15,900
1300	110	3,300	
1315			19,200
1330	110	3,300	
1345			22,500
1400	90	2,700	
1415			25,200
1430	70	2,100	
1445			27,300
1500	40	1,200	
1515			28,500
1530	30	900	
1545			29,400
1600	40	1,200	
1615			30,600
1630	20	600	
1645			31,200
1700	30	900	
1715			32,100
1730	30	900	
1745			33,000
1800	50	1,500	
1815			34,500
1830	60	1,800	
1845			36,300
1900	50	1,500	
1915			37,800
1930	50	1,500	
1945			39,300
2000	20	600	
2015			39,900

Total Flow Volume = 39,900 gallons

[a] 0800 is the middle of the time interval from 0745 to 0815. The cumulative volume for this time interval should be plotted on Figure 7.2 at the end (0815) of the time interval.

Preliminary Treatment 263

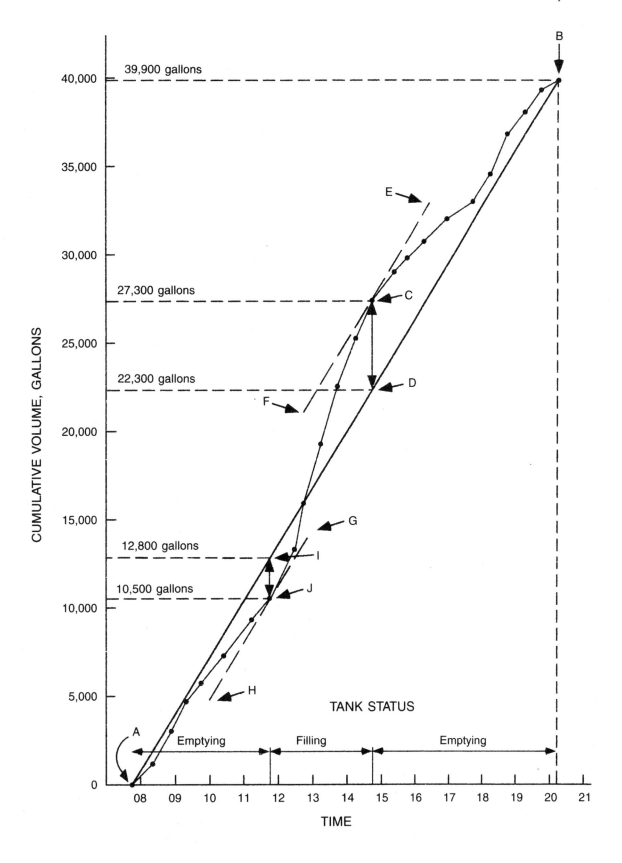

Fig. 7.2 Cumulative flow hydrograph used to determine average flow and equalization tank volume

$$\text{Avg Flow, gal/hr} = \frac{\text{Cumulative Volume, gal}}{\text{Time, hr}}$$

$$= \frac{39{,}900 \text{ gal}}{2015 - 0745}$$

$$= \frac{39{,}900 \text{ gal}}{12.5 \text{ hr}}$$

$$= 3{,}192 \text{ gal/hr}$$

$$\text{Avg Flow, GPM} = \frac{\text{Avg Flow, gal/hr}}{60 \text{ min/hr}}$$

$$= \frac{3{,}192 \text{ gal/hr}}{60 \text{ min/hr}}$$

$$= 53.2 \text{ GPM}$$

Based on these calculations, wastewater arrives at the IWTS at an average rate of 53.2 GPM. It will therefore be discharged from the equalization basin at a steady rate of 53.2 GPM or 1,596 gallons per 30-minute time interval (53.2 GPM × 30 min = 1,596 gallons).

Next, prepare a hydrograph similar to the one shown in Figure 7.2 by plotting the data you have collected for Table 7.1. Begin by indicating cumulative flow volumes along the left (vertical) edge of the graph and indicating time intervals along the bottom (horizontal) edge. Now plot the cumulative volumes at the end of each 30-minute time interval as shown in Figure 7.2 and connect the plotting points. If the flows start at 0745, then the first plotting point for zero (0) cumulative volume is 0745. The next cumulative volume should be plotted at 0815, the end of that measurement interval. If the flows start at 0800, the first plotting point for zero cumulative volume would be 0800 and the next would be at 0830 (the end of the time interval with the average flow being measured at 0815).

The first plotting point (point A in Figure 7.2) represents the beginning of the cycle where there is no flow at 0745. The last plotting point (point B) represents the cumulative flow of 39,900 gallons at the end of the last time interval (2015). Draw a line from A to B; this line represents the average cumulative flow into and, therefore, the cumulative flow out of the equalization basin. The line connecting all of the plotting points from A to B shows the variations in flow coming into the equalization tank in relation to the cumulative volume discharged.

Holding a ruler parallel to line A-B and above the average line, find the point farthest from line A-B and draw a line that is tangent to A-B. Do the same thing for the point farthest from A-B below the average line. In Figure 7.2, the two farthest points are at C and J and the tangents are the two broken lines E-F and G-H. Points C and J represent the greatest variations from the average cumulative flow.

Next, we need to figure out what size tank will be needed to accommodate the maximum flow variations plotted as points C and J. Draw a vertical line from point C to the average flow line; this is line C-D on Figure 7.2. Repeat the same step below the average line by drawing a line from point J to the average line (line I-J on Figure 7.2).

Read the volumes on the left of the chart for points C (27,300 gal) and D (22,300 gal) and subtract volume D from volume C (27,300 gal – 22,300 gal = 5,000 gal). Do this again for points I and J (12,800 gal – 10,500 gal = 2,300 gal). Add the results of these two calculations to find the minimum storage tank volume needed to handle the typical flow variations plotted in the hydrograph.

$$\text{Equalization Vol, gal} = (27{,}300 \text{ gal} - 22{,}300 \text{ gal}) +$$
$$(12{,}800 \text{ gal} - 10{,}500 \text{ gal})$$

$$= 5{,}000 \text{ gal} + 2{,}300 \text{ gal}$$

$$= 7{,}300 \text{ gal}$$

In addition to showing the flow variations and minimum equalization tank volume needed, the hydrograph also visually illustrates the filling and emptying of the tank. To interpret the hydrograph, assume the equalization tank is half full at 0745 (point A). Whenever the curving influent line intersects or crosses line A-B (the cumulative discharge volume) the tank will be half full. All of the plotting points from 0745 through 1245 fall below the cumulative discharge line. As the curving line moves away from line A-B, the tank is emptying; as it moves toward line A-B, the tank is filling. At 1245 the two lines cross, meaning the cumulative volume received is equal to the cumulative volume discharged; the tank is again half full. After 1245, the curved line moves above the cumulative discharge line A-B. The tank is filling as the lines move apart and emptying as they move closer together. At point B, the curving influent line intersects the cumulative discharge line A-B; the tank is once again half full. The volume received equals the volume discharged. (When interpreting a hydrograph, remember that when the cumulative influent line is *below* the cumulative discharge line A-B, the tank is emptying as the two lines move apart and filling as they move closer together. The reverse is true *above* line A-B: the tank is filling as the two lines move apart and emptying as they move closer together.)

7.21 Other Factors for Sizing an Equalization Tank

While the above discussion provides a method to evaluate whether the existing equalization tank is large enough to equalize the incoming flows, there are other factors that influence the volume needed to fully equalize the flow and dampen the variations in load to the rest of the system. Plant recycle streams, the operating levels of the mixing or aeration equipment, and a safety factor are common elements that will add to the minimum volume calculated in Section 7.20 above.

Flows from the filter backwash will introduce a significant flow with a high concentration of suspended solids. *DECANT WATER* [5] from the digester or sludge holding tank will intro-

---

[5] *Decant* (de-KANT) *Water*. Water that has separated from sludge and is removed from the layer of water above the sludge.

duce a concentrated solution of pollutant into the system. These are good candidates for discharge to the equalization tank. Therefore, IWTS recycle flows need to be included in the calculation of the volume of the equalization tank.

As will be discussed later, mixing is an important factor in an effective equalization tank. The actual working volume of the tank (that volume available for equalization) is determined by the operating levels of the mixing or aeration equipment. If the six-inch tall mixer blade is located two feet above the bottom of the tank and requires at least one foot of liquid on top of it before it is started, the working or fluctuating volume of the tank starts three and one-half feet above the bottom of the tank. If floating aerators are used, the working volume of the tank is affected by the low-level shutoff point on the aerators. For smaller systems, recirculation and mixing of the equalization tank using pumps may offer more working volume for the same depth of tank. However, this arrangement does sacrifice the effectiveness and efficiency of a mechanical mixer.

Finally, the size of the equalization tank should include a safety factor for the unknown peak flow occurrences. These may occur on start-up or shutdown of a manufacturing process or during the periodic cleanout of a large tank. The size of the safety factor depends on how much is known about the process and how accurately it has been characterized. A typical safety factor may be 10 to 20 percent of the calculated volume of the tank.

QUESTIONS

Write your answers in a notebook and then compare your answers with those on page 307.

7.2A If the cumulative influent flow volume to an equalization tank during a 15-hour manufacturing cycle is 47,250 gallons, what is the average tank discharge rate in gallons per minute?

7.2B In addition to the volume determined using the hydrograph in Figure 7.2, what other factors must be considered when sizing an equalization tank?

7.22 Mixing

Successful operation of an equalization tank requires proper mixing and, if the waste is biodegradable, adequate aeration. The two purposes of mixing are to blend the entire contents of the tank and to prevent solids deposition in the tank. Without complete mixing, the equalization tank will not act effectively to dampen shock loads; without sufficient agitation to prevent solids deposition, the solids will build up in the tank and reduce its effective volume. Typical energy requirements for mixing a large equalization basin are 0.02 to 0.04 horsepower per 1,000 gallons of stored equalization wastewater.[6] Smaller tanks may require more energy simply because of available motor sizes and the tank's size and shape. Smaller tanks may also lend themselves to using compressed air or a recirculating pump to mix the tank.

If the waste is biodegradable and may become SEPTIC[7] while in the equalization tank, aeration is required. Aeration may be used to mix the contents of the tank as well as provide oxygen. Aerobic conditions require approximately 1.25 to 2 cubic feet of air per minute per 1,000 gallons of storage.[8] The use of mechanical aerators is one method to provide both mixing and aeration. Surface aerators can also be used, but have the disadvantage of needing a minimum of five feet of water depth. If surface aerators are used in an earthen basin, a low-level shutoff is needed to prevent scouring the bottom of the basin.

When mixing to prevent solids deposition is the controlling factor for an equalization tank rather than blending or oxygen transfer, a two-prong approach may be used. Mixing to prevent solids deposition can be done with mechanical mixers and oxygen transfer can be done with a diffused air system or a surface aerator blade on the mixer.

Other factors also influence the size, type, quantity, and placement of aeration and mixing equipment. These include the maximum operating depth of the tank and tank configuration. Questions about whether the existing equipment is adequate can be resolved by consulting with the equipment vendor.

7.23 Tank Construction

Tanks for new facilities can be made of almost any material compatible with the wastewater including, but not limited to, steel, concrete, earthen basins, plastics, or reinforced fiberglass. Equalization for existing facilities may be implemented with relative ease by converting abandoned tanks, such as clarifiers, aeration tanks, digesters, or large sumps.

---

[6] EPA Technology Transfer publication, *FLOW EQUALIZATION*, Environmental Protection Agency, May 1974.

[7] *Septic* (SEP-tick) *or Septicity.* A condition produced by bacteria when all oxygen supplies are depleted. If severe, the bottom deposits produce hydrogen sulfide, the deposits and water turn black, give off foul odors, and the water has a greatly increased oxygen and chlorine demand.

[8] See reference No. 8 in the EPA Technology Transfer publication, *FLOW EQUALIZATION*, Environmental Protection Agency, May 1974.

Tanks should be equipped with baffles to promote complete mixing and prevent SHORT-CIRCUITING.[9] Square tanks provide more volume per area, but without baffles will develop "dead" areas in the corners. Circular tanks need baffles to prevent the wastewater from just moving in the same direction without mixing. Long rectangular or long oval tanks also should be avoided because the wastewater tends to flow through in a plug flow rather than dispersing and mixing.

Sloping sides prevent solids deposition but reduce the effective volume of the tank. For larger tanks, compartmentalization or a means to flush or clean solids from the bottom of the tank may be required for maintenance purposes. If an earthen basin is used, concrete scour pads should be installed beneath the aerators or mixers.

Piping to and from the equalization tank is also important. The inlet should be near the mixer(s) to disperse the flow as quickly as possible, and the outlet should be configured so as to minimize short-circuiting. Also, check to see that the mixer rotation does not promote short-circuiting by moving the wastewater toward the outlet. The tank should be equipped with an alarm for high level and with either a bypass around the tank or a means to shut down the incoming wastewater. If the wastewater has a tendency to foam, a high water takeoff should be provided for withdrawing floating material and foam.

7.24 Pumps and Flow Controls

Flow equalization causes a certain amount of HEAD LOSS[10] as the wastewater passes through the system. Some head loss occurs as wastewater flows into and out of the tank; this is the dynamic energy of flowing liquids. Also, the normal variations in the surface level of the tank cause additional loss of pressure head. At a minimum, therefore, the head loss is equal to the sum of the dynamic losses and the losses from pressure variations. If the next process in the treatment sequence is upgradient from the equalization tank, it may be necessary to make up these head losses using pumps.

The two most typical flow arrangements are: (1) a gravity feed into the equalization tank with a pumped outlet, or (2) a pumped inlet with a gravity-controlled outlet. With gravity flow into the equalization basin, the tank is equipped with a low level shutoff and a high level alarm. Typically, wastewater is pumped to the next treatment process through a meter and flow control valve. If the flow is pumped into the equalization tank, the tank is equipped with a low level shutoff, a high level alarm, and a bypass or overflow. Wastewater flows by gravity through a meter and flow control valve to the next treatment process.

7.25 Strategy for Operation

If the flow for the day matches the design flow for the equalization tank, the pumping rate is equal to the average flow for the design manufacturing cycle. While this is only sometimes the case, the system generally can be run at this average flow rate without encountering problems. Experience has shown that the larger the flow into the system or the more consistent the manufacturing process, the more frequently the design average flow rate will be sufficient to operate the equalization tank successfully.

If the flows are less than the design, the tank will be drawn down to its low level and stop until enough flow enters the tank to turn the pump back on. In this sense the equalization tank acts as a large sump. If this ON/OFF sequence occurs more than two or three times for each manufacturing cycle, then the operator should reduce the flow rate out of the equalization tank unless it is below the optimum range of the rest of the downstream equipment. The operator needs to decide which is better for the downstream treatment processes, no flow or below-optimum flow.

If the high level alarm comes on at regular intervals, the discharge rate can be increased to accommodate the larger-than-design condition for the equalization tank. Another option would be to collect concentration and flow data and then construct a loading curve of the main water quality indicator of concern versus time. Locate the low loading time periods and increase the flow rates to the rest of the IWTS at those times in an attempt to equalize the load rather than the flow. Another solution for higher-than-design average flows is to operate the IWTS longer than the manufacturing cycle. This becomes attractive if the manufacturing cycle is, for example, eight hours and the IWTS can be operated in an automatic mode either before you arrive or after you leave.

If the pumping rate from the equalization tank needs to be changed, it should be done in small increments, typically five to

---

[9] *Short-Circuiting.* A condition that occurs in tanks or basins when some of the flowing water entering a tank or basin flows along a nearly direct pathway from the inlet to the outlet. This is usually undesirable since it may result in shorter contact, reaction, or settling times in comparison with the theoretical (calculated) or presumed detention times.

[10] *Head Loss.* The head, pressure, or energy (they are the same) lost by water flowing in a pipe or channel as a result of turbulence caused by the velocity of the flowing water and the roughness of the pipe, channel walls, or restrictions caused by fittings. Water flowing in a pipe loses head, pressure, or energy as a result of friction losses. The head loss through a filter is due to friction losses caused by material building up on the surface or in the top part of a filter. Also called FRICTION LOSS.

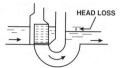

ten percent of the present rate. This is to minimize hydraulic surges through the system and provide an opportunity for the chemical feed systems to smoothly respond to a change in load. Remember that if you change the flow (and loading rate), the processes that are flow-paced, such as polymer addition, also will need to be changed.

7.3 SUMMARY

Equalization of incoming wastewater flows provides an economical means of improving the performance of your IWTS or downsizing the rest of the equipment in an IWTS that is being designed. Flow equalization also provides load equalization, which is important when trying to optimize the operation of the IWTS. Equalization protects the IWTS from shock loadings that might otherwise cause pass-through of pollutants or interfere with the treatment system processes.

The minimum recommended size of an equalization tank is the volume needed to equalize the flow. Additional increments of volume should be added for the recycle streams from the IWTS, the minimum operating volumes of the mixing and aeration equipment, and a safety factor. Mixing is very important to receive the maximum benefits from equalization. Baffles, proper placement of the inlet and outlet, compartmentalization, and the ability to flush or clean out any solids are also considerations.

7.4 ADDITIONAL READING

Environmental Protection Agency, EPA Technology Transfer publication, *FLOW EQUALIZATION*, May 1974.

QUESTIONS

Write your answers in a notebook and then compare your answers with those on page 307.

7.2C How can short-circuiting be controlled in equalization tanks?

7.2D How can higher-than-design average flows be treated?

7.2E What precautions should be taken to protect downstream processes and equipment when changing the equalization tank discharge pumping rate?

END OF LESSON 1 OF 3 LESSONS
on
PRELIMINARY TREATMENT

Please answer the discussion and review questions next.

DISCUSSION AND REVIEW QUESTIONS
Chapter 7. PRELIMINARY TREATMENT
(EQUALIZATION, SCREENING, AND pH ADJUSTMENT)
(Lesson 1 of 3 Lessons)

At the end of each lesson in this chapter, you will find some discussion and review questions. The purpose of these questions is to indicate to you how well you understand the material in the lesson. Write the answers to these questions in your notebook.

1. Why is in-line equalization more effective than side-line equalization in leveling influent concentration variations?

2. Why is the equalization of industrial wastewater flows important?

3. What factors should be considered when determining the size of an equalization tank?

4. Why is adequate mixing important in an equalization tank?

CHAPTER 7. PRELIMINARY TREATMENT
(EQUALIZATION, SCREENING, AND pH ADJUSTMENT)

(Lesson 2 of 3 Lessons)

7.5 SCREENING
by D. F. Reddy

7.50 Why Screen?

At its very simplest, screening of industrial wastestreams captures solids that may have value and allows downstream treatment processes to function more effectively. Screening is performed in an open system (without a vacuum or at atmospheric pressure).

When solids or by-products of industrial operations are captured early in the wastewater treatment process, they are relatively easy to remove because the particles are larger and fresher. (Solids particles tend to disintegrate and fracture the farther they are pumped.) Solids recovered by screening can often be sold or recycled into the plant's process; that is, meat packing waste solids are recovered for rendering, fibers recovered from paper mill waste can be recycled, vegetable processing wastes can be used as animal feed or a soil enhancer, and PVC beads or fines can be recycled back into the plant process.

Screening reduces the overall solids loading on downstream processes. Floatable, suspended, or settleable solids can be effectively screened from wastewater. This protects downstream processes and equipment. Pumps, valves, pipes, aeration headers, heat exchangers, and the like often become clogged with nuisance solids if they are not protected by screens. Reducing the solids load on downstream equipment also enhances the performance of that equipment. Fewer chemicals, less power, and less maintenance downtime are the secondary process results of prescreening.

If the industrial plant does not have secondary wastewater treatment, but discharges its wastewater to a POTW (Publicly Owned Treatment Works), screening wastewater prior to discharge normally lowers sewer-use fees. Reduction in the strength of wastewater discharged to the city is often used to cost-justify the installation of screens.

Many plants with primary and secondary wastewater treatment use fine screens in place of primary sedimentation, saving about 50 percent of the capital cost of clarifiers. Screens are also very conservative of space when used in place of primary sedimentation. Because screening is a very quick operation, high operating temperatures can be maintained; this is particularly important if subsequent treatment involves heat exchangers.

Significant BOD_5[11] reduction can be achieved by screening, but this is a consequence of solids removal rather than the goal of screening, since most BOD is dissolved and cannot be removed mechanically. Biological treatment of wastewater reliably removes BOD_5 and other pollutants when this phase of treatment is not burdened by large quantities of solids.

QUESTIONS
Write your answers in a notebook and then compare your answers with those on page 307.

7.5A Why is screening used to pretreat industrial wastewater?

7.5B What types of solids can be effectively screened from industrial wastewater?

7.5C What is the impact of screening on sewer-use fees?

7.51 Types of Screens

7.510 Coarse Screens (Bar Screens)

7.5100 DESCRIPTION

Coarse screens, also known as trash racks or bar screens, generally have clear openings of $3/4$ inch to 3 inches. They are used to capture coarse solids that would damage downstream equipment. Hand-raked coarse screens are seldom installed today; almost all are automatically self-cleaning.

---

[11] BOD_5. BOD_5 refers to the five-day biochemical oxygen demand. The total amount of oxygen used by microorganisms decomposing organic matter increases each day until the ultimate BOD is reached, usually in 50 to 70 days. BOD usually refers to the five-day BOD or BOD_5.

Automatically cleaned bar screens normally consist of fixed bars of steel or stainless steel mounted vertically in an open wastewater influent channel. A framework of painted carbon or stainless steel extending above grade supports the main drive and cleaning mechanisms. Bar screen designs without underwater drive components are preferred at most industrial plants due to reliability and ease of maintenance. Debris trapped by the submerged bars in the channel restricts the flow of wastewater through the bars and raises the liquid level upstream of the screen. When the water level rises above a preset point, a level-sensing device (float system, differential level control) triggers the descent of a cleaning rake into the channel to remove the debris from the bars. The rake lifts the debris up out of the channel and drops it into a container or receiving conveyor. The cleaning operation begins whenever the preset upstream water level is exceeded and continues until the water drops to an acceptable level. A cleaning cycle can also be initiated by a timer or manual controls. (See Figure 7.3.)

Many automatic bar screens have the ability to "walk over" objects trapped by the bars that are too large to be removed by the cleaning mechanism. After such an object is bypassed, the cleaning mechanism will once again mesh with the bars and continue its upward cleaning motion. Most automatic bar screens are equipped with torque overload switches that will stop the unit and sound an alarm if an object is too large to be bypassed.

Bar screens differ in whether they are front cleaned (upstream side) or back cleaned (downstream side), and in their drive mechanisms, that is, cog rail, pin rack, cable, or chain, to name a few. The Grabber automatically cleaned bar screen shown in Figure 7.4 is of the front-cleaned, cog-rail type. Figure 7.5 shows the rack and cog mechanism.

7.5101 START-UP/SHUTDOWN PROCEDURES

Always refer first to manufacturer's instruction manual.

1. Start-Up

   a. Clean the channel of all obstructions and ensure no concrete or gravel remains between the bars.
   b. Ensure gear motor is properly filled with oil and check all grease points to make sure all are properly lubricated.
   c. Check for proper motor rotation and cleaning rake movement.
   d. Make a test run with the empty channel to ensure that all parts mesh properly, nothing jams, and no unusual noises or movements are apparent. Diagnose and correct any problems noticed.
   e. Start the influent to the channel and start the unit.

2. Shutdown

   a. Close or insert inlet gate to stop flow through the channel.
   b. Turn the mechanical unit off.
   c. Drain the channel.
   d. Thoroughly hose down and clean the bars and cleaning mechanism.

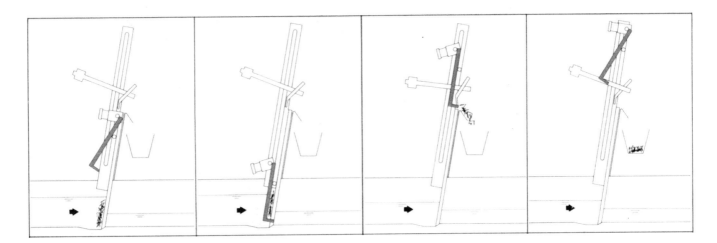

OPERATION OF A MECHANICAL BAR SCREEN

1. The cleaning rake is activated by a bubbler control system, which senses differential level, or is triggered by an automatic timer. The rake assembly moves downward toward the open channel and the rake arm swings outward.

2. When the rake reaches its lowest point, it swings inward to engage the bar screen. As the rake travels upward out of the channel, it automatically cleans the bar screen and drops the debris into a hopper.

3. The rake assembly continues upward to its highest point at which it turns and the machine shuts down automatically. The rake arm remains in this home position until the next cleaning cycle.

Fig. 7.3 Automatically raked bar screen
(Diagram courtesy of Hycor Corporation)

270   Treatment Plants

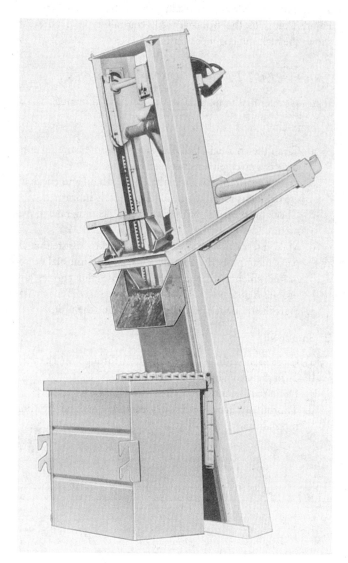

Fig. 7.4   Typical bar screen
(Grabber® Screen from Hycor Corporation)

Special cog wheel system moves the cleaning rake along.

Fig. 7.5   Cog wheel drive system
(Courtesy of Hycor Corporation)

7.5102 TROUBLESHOOTING AUTOMATICALLY CLEANED BAR SCREENS

COMPONENTS: Carriage and Cleaning Rake Assembly

PROBLEM	PROBABLE CAUSE	REMEDY
EXCESSIVE NOISE OR VIBRATION	Dry pin or cog rail track	Lubricate rails and cog wheels.
	Worn pin or cog rails or cog wheel	Check tooth profile and replace pins, rail, and wheels as necessary.
	Drive shaft end bearings rubbing on side frame	Check drive shaft position and mounting bearings. Adjust as required. Check locking collars on the flanged bearings. Tighten as necessary.
	Carriage frame slide bushing rubbing	Check carriage position and flanged drive shaft mounting bearings. Adjust as required.
	Cleaning rake teeth touching deadplate	Check wear shoes and adjust or replace as necessary.
	Faulty cleaning rake teeth/bar rack engagement	Check position of rake teeth and bar rack assemblies. Adjust as necessary. Check position of flanged drive shaft mounting bearings and adjust as necessary.
	Too much compression on rake arm springs or dry guide rails	Check rake pressure against guide rails and adjust compression springs or clean and grease guide rails.
FAULTY BAR RACK ENGAGEMENT BY CLEANING RAKE	Carriage position changed	Check carriage alignment and readjust.
	Bars or bar rack damaged	Repair or replace damaged bar rack.
	Loose bar rack section	Check bar rack and adjust and tighten fasteners as necessary.
	Loose or damaged cleaning rake teeth	Check rake teeth position. Tighten fasteners as necessary, repair or replace.
	Bent cleaning rake arm	Check cleaning arm alignment and repair or replace as necessary.
SCREENINGS DROPPING OFF CLEANING RAKE SHELF	Wear shoes incorrectly adjusted	Check wear shoes and adjust as necessary.
	Rake arm compression too weak	Adjust compression mechanism as necessary.
	Bent rake arm	Check rake arm alignment. Repair or replace as necessary.

COMPONENTS: Motor and Reducer Bar Screen Drive

PROBLEM	PROBABLE CAUSE	REMEDY
NO ROTATION	Loss of power	Check power distribution. Check electrical connections.
	Power overload	Check starter overloads.

7.5102 TROUBLESHOOTING AUTOMATICALLY CLEANED BAR SCREENS (continued)

COMPONENTS: Motor and Reducer Bar Screen Drive (continued)

PROBLEM	PROBABLE CAUSE	REMEDY
NO ROTATION (continued)	Torque overload	Check machine for obstacle or hang-up and reset.
	Loss of control power and auto signal	Check control power fuses and transformer. Check control relays and PC interface (if applicable).
	Motor windings bad	Check motor windings, repair or replace.
EXCESSIVE NOISE	Motor bearing failure	Check or replace motor bearings.
	Brake pads	Check pads, adjust or replace.
	Low oil in reducer	Check oil level and fill as necessary.
TOO HIGH TEMPERATURE	Supply voltage change	Check input power.
	Bad motor winding	Check motor. Repair or replace.
	Low oil in reducer	Check oil levels and fill.
	Reducer vent plugged	Check reducer vent fill plug.
	Unit overload	Check material being discharged.

COMPONENT: Wiper Assembly

PROBLEM	PROBABLE CAUSE	REMEDY
EXCESSIVE BOUNCING AND NOISE	Main counterweight positioned incorrectly	Reposition main counterweight to balance wiper assembly. Retighten fasteners equally. Check alignment of wiper assembly, adjust as necessary.
	Main counterweight positioned incorrectly	Reposition main counterweight to balance wiper assembly. Retighten fasteners equally. Check alignment of wiper assembly, adjust as necessary.
	Rear counterweight positioned incorrectly	Reposition rear counterweight. Manually check balance of rocker arm before running unit, adjust as necessary.
	Shock absorbers not functioning	Adjust or replace as necessary.
NOT WIPING CLEANING RAKE PROPERLY	Wiper assembly not properly balanced	Adjust position of main counterweight.
	Rocker arm not returning to cleaning position	Check counterweights, guides, and return bumpers. Adjust or replace as needed.
	Wiper blade not cleaning entire rake shelf	Check alignment and condition of wiper blade. Realign or replace as needed.
	Rocker arm position shifting	Adjust position of rocker arm cleaning stroke in relation to the discharge chute.

QUESTIONS

Write your answers in a notebook and then compare your answers with those on page 307.

7.5D Why are bar screen designs without underwater drive components preferred at most industrial plants?

7.5E What procedures should be followed when shutting down a mechanical bar screen?

7.5F List the major components of an automatically cleaned coarse screen (bar screen) that might be sources of poor operation or breakdowns.

7.511 Fine Screens

Fine screens generally incorporate equipment having clear openings of 75 *MICRONS*[12] to 0.025 inch. These screens capture solids that pass through coarse screens. The small particles captured by fine screens are solids that may have value for resale or recycling, or solids that would clog or overload subsequent treatment processes.

7.5110 STATIC SCREENS

Static screens, also known as "cascade type" or "run-down" screens, are popular in industrial wastewater treatment because they have no moving parts and a relatively low capital cost (Figure 7.6). They are especially appropriate for the pretreatment of fibrous and non-grease-bearing wastes.

Fig. 7.6 The Hydroscreen static screen has no motors or moving parts and is easy to operate.
(Courtesy of Hycor Corporation)

All static screens feature a stainless-steel screen panel, in a flat or curved form, that fits into a cabinet containing a simple influent chamber and usually an effluent chamber as well. The cabinet is normally made of stainless steel or FRP (fiber reinforced plastic). The screen panel (Figure 7.7) is typically fabricated from horizontal wedgewire bars with clean openings of 0.010 to 0.100 inch. In larger models the panel is sometimes hinged to make cleaning the underside easier.

 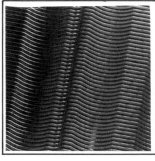

Fig. 7.7 The wedgewire screen is a highly effective dewatering screen. The wider "wedge" of the front surface prevents solids from catching and allows liquids to drain.
(Diagrams courtesy of Hycor Corporation)

Untreated wastewater enters the static screen cabinet from the top or rear. As the headbox (influent area) of the cabinet fills, the wastewater rises over a weir and cascades over the upper lip of the static screen panel. (See Figure 7.8.)

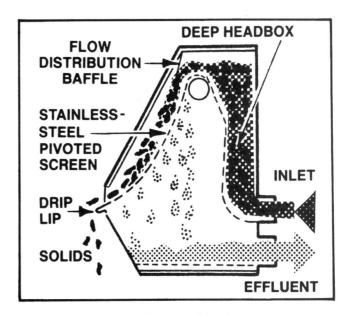

Fig. 7.8 Operation of a static screen
(Diagram courtesy of Hycor Corporation)

Solids larger than the screen openings are trapped on the outside face of the screen and roll downward to discharge at the lower lip of the screen panel. Liquid drains through the clear openings of the panel and is freely discharged by gravity into the effluent collection area of the cabinet, or into a catchment tank or channel beneath the screen. (See Figures 7.9 and 7.10.)

---

[12] *Micron* (MY-kron).   μm, Micrometer or Micron. A unit of length. One millionth of a meter or one thousandth of a millimeter. One micron equals 0.00004 of an inch.

Fig. 7.9 High solids captured on a static screen
(Photograph courtesy of Hycor Corporation.)

Fig. 7.10 A row of static screens in wastewater process line
(Photograph courtesy of Hycor Corporation)

Static screens are not self-cleaning and require periodic manual cleaning with a brush or hose. Cleaning frequency depends upon the nature of the wastewater being treated. Care should be taken to ensure that both the front face of the screen panel and the underside or back are clear of debris that could interfere with the free flow of liquid through the screen. Screen blinding (solids covering the clear screen openings) reduces hydraulic throughput capability of the screen and results in wetter captured solids.

START-UP/SHUTDOWN PROCEDURES

Always refer first to manufacturer's instruction manual.

1. Start-Up. Slowly start the influent. Inspect the unit and piping for leaks. If everything is satisfactory, increase flow at a moderate rate to the desired level and check the system operation.

2. Shutdown. If the static screen must be shut down for a period of time, solids can dry on the screen and create problems when the unit is later restarted. The following procedure will eliminate these problems:

a. Shut off the influent.

b. Hose down the screen to remove solids trapped between screen wires. A brush or broom can also be used to remove solids from the screen panel. Do not clean the screen panel or cabinet with metallic brushes other than stainless steel. Other types of metallic brushes may leave free iron deposits on the unit which will result in rusting and pit formation.

c. Drain the headbox and clean as required.

QUESTIONS

Write your answers in a notebook and then compare your answers with those on page 307.

7.5G What is the purpose of fine screens?

7.5H Static screens are especially appropriate for the pretreatment of what types of wastes?

7.5I How are static screens cleaned?

TROUBLESHOOTING STATIC SCREENS

PROBLEM	PROBABLE CAUSE	REMEDY
SCREEN BLINDING	Solids built up on screen face	Shut off influent. Hose down screen or use a portable pressure spray cleaner. Scrub screen with hard bristle brush (never use a carbon steel wire brush).
	Solids on underside of screen	Remove screen panel and clean underside.
HEADBOX DEPOSITS	Heavy solids in influent flow	Shut off influent. Manually remove debris.
SWING BAFFLE BINDING, NOT MOVING FREELY	Solids caught on baffle or tie rods too tight	Manually remove solids caught on baffle. Loosen tie rods if required.

7.5111 ROTATING SCREENS

Rotating screens represent an advance in fine screening in that they are self-cleaning or largely self-cleaning. They are made of stainless-steel wedgewire or perforated metal, or even stainless-steel or synthetic fabrics. Only the most common types will be discussed here.

1. *Externally fed rotating screens* feature a slowly rotating cylinder mounted horizontally and supported by bearings outside the liquid flow. The supporting frame contains a headbox (Figure 7.11) and sometimes incorporates a liquid collection base for screened water beneath the cylinder. (See Figures 7.12 and 7.13.)

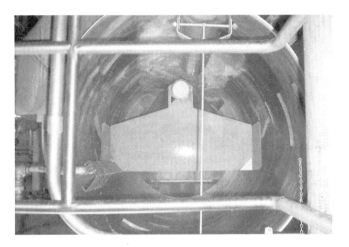

Close-up of headbox of externally fed rotating screen

Fig. 7.11 Rotating screen in operation
(Courtesy of Hycor Corporation)

Various flows of 50 to 12,000 GPM (190 to 45,400 *L*PM) can be accommodated with cylinders of diameters from 25 up to 48 inches (60 up to 120 cm) and lengths from 12 up to 144 inches (30 up to 365 cm). Wedgewire cylinders with clear openings of 0.010 to 0.100 inch (0.02 to 0.20 cm) are most commonly used. Most externally fed rotating screens use low-horsepower, direct drive, fixed-speed gear motors.

Wastewater is delivered from the headbox over a weir made of synthetic, replaceable material, to the horizontal cylinder. Influent solids larger than the clear openings in the cylinder are retained on the outside surface. As the cylinder slowly rotates, these solids are carried out of the liquid flow and around to the other side of the cylinder, where they are scraped off by a *DOCTOR BLADE*[13] (Figures 7.14 and 7.15). Liquid falls by gravity through the cylinder twice, once from the outside to the inside and then again as it falls from the inside to the outside. It is this "double pass" that creates the self-cleaning action for which rotating screens are so popular. Spray bars can also be installed within the cylinder for additional automatic cleaning when required by the application.

Externally fed rotating screens efficiently handle a wide variety of industrial wastewaters including even fatty, greasy effluents. They are generally unsuitable, however, for wastewaters containing significant amounts of fiber, such as textile mill or industrial laundry wastes. Periodic inspection of the replaceable seals and doctor blades is recommended to ensure optimum performance.

2. *Internally fed rotating screens* (Figures 7.16 and 7.17) commonly feature wedgewire or perforated metal cylinders of various diameters from 24 to 80 inches (60 to 200 cm) and lengths of 48 to 144 inches (120 to 365 cm). The stainless-steel screen cylinder is supported by trunnions, or rollers, on a frame of painted carbon or stainless steel. Screen rotation is normally achieved by a chain and sprocket arrangement fixed to the influent end of the cylinder although some equipment is belt-driven.

Influent wastewater flow is delivered to an internally fed rotating screen by a headbox inside the cylinder. Wastewater cascades by gravity over one or two weirs on the headbox and falls onto the slowly rotating cylinder. Solids are captured inside the cylinder and are gently shifted toward the discharge end by means of flights welded within the cylinder. Liquid is discharged by gravity through the clear cylinder openings (Figure 7.18).

The scouring action of the solids trapped within the cylinder provides a cleaning effect, and internal and external spray systems assist in keeping the clear openings of the screen clean of obstructions. Automatic spraying cycles are recommended, and care should be taken to match the spray frequency and duration to the demands of the application.

Internally fed rotating screens are quite versatile in the pretreatment of industrial wastewater and are very popular with food processors. They are particularly appropriate for wastewater flows with very large solids or with a heavy concentration of solids as well as for wastewater containing coarse, fibrous materials.

3. *Disk filters* (Figure 7.19) feature mesh-covered pairs of disks, or wheels, which rotate on a common shaft in a contoured chamber. Stainless or synthetic screen mesh commonly used by industry has 75- to 240-micron (0.003- to 0.010-inch) clear openings. The disks supporting the screen mesh can be up to 60 inches (150 cm) in diameter, with up to 6 pairs of disks in a single machine.

Disk filters are especially suited to fibrous solids separation, such as might be required in pulp and paper, textile, laundry, or similar plant wastewaters (Figure 7.20). They are generally inappropriate for greasy applications.

In disk filter operation (Figure 7.21), wastewater is introduced between a pair or pairs of disks and rises to a maximum level of about one-third of the disk diameter. The disks build up a precoat of solids and a plug of solids settles

---

[13] *Doctor Blade.* A blade used to remove any excess solids that may cling to the outside of a rotating screen.

Fig. 7.12   Externally fed rotating screen (Rotostrainer ®)
(Courtesy of Hycor Corporation)

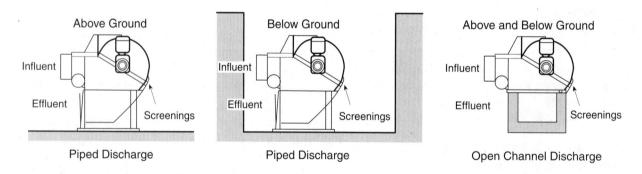

Rotary screens can be positioned above or below ground with gravity or pumped flow.

Fig. 7.13   Rotary screen installations
(Courtesy of Hycor Corporation)

Preliminary Treatment 277

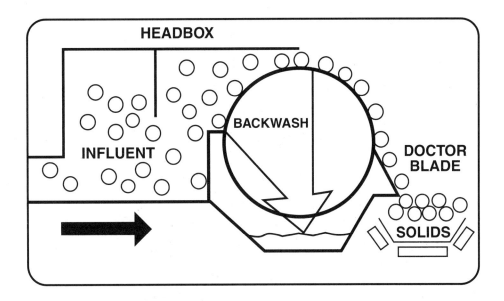

Wastewater influent flows through the cylindrical screen twice while solids are captured on the screen.

Fig. 7.14 Operation of cylindrical screen
(Courtesy of Hycor Corporation)

Doctor blades keep externally fed rotating screens clean.

Fig. 7.15 Rotating screen in operation
(Courtesy of Hycor Corporation)

278 Treatment Plants

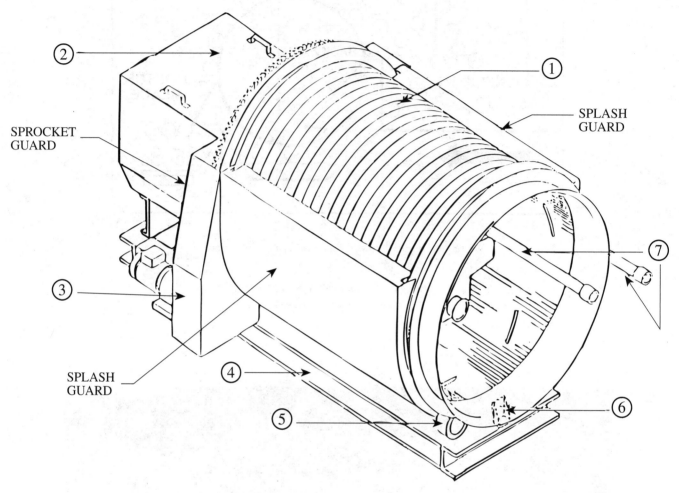

1. SCREEN ELEMENT
2. HEADBOX AND DISTRIBUTION SYSTEM
3. DRIVE SYSTEM
4. FRAME
5. TRUNNIONS
6. STABILIZERS
7. SPRAY WASH SYSTEM

Fig. 7.16 Major components of an internally fed rotating screen (Rotoshear ®)
(Courtesy of Hycor Corporation)

Preliminary Treatment 279

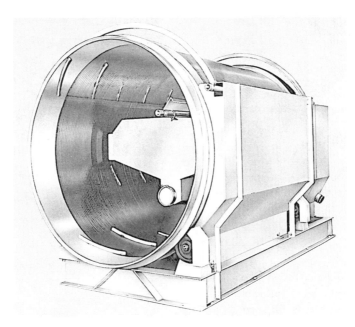

Fig. 7.17 Internally fed rotating screen
(Courtesy of Hycor Corporation)

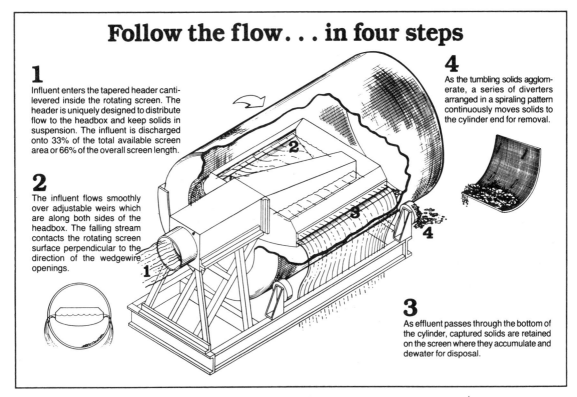

Fig. 7.18 Internally fed screen operation
(Courtesy of Hycor Corporation)

280 Treatment Plants

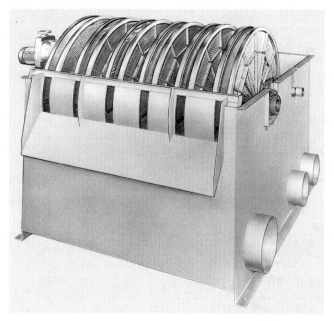

Hycor Discostrainer's® mesh screen removes fibrous solids from wastestreams by combining the principles of screening, sedimentation, and precoat filtration.

Fig. 7.19   Discostrainer ® screen
(Courtesy of Hycor Corporation)

Fine mesh disk screens yield extremely dry solids.

Fig. 7.20   Disk filters separating lint from plant wastewater
(Courtesy of Hycor Corporation)

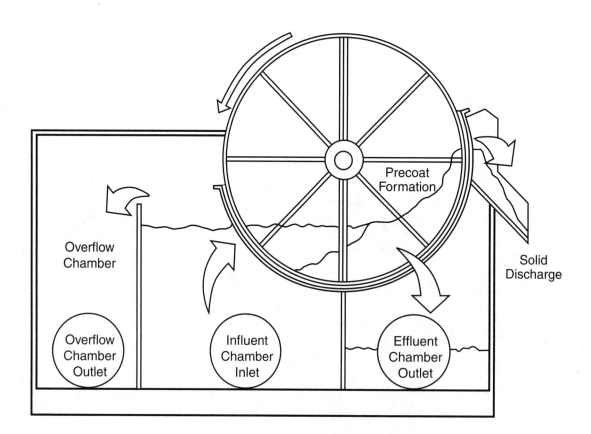

Cross section of a Discostrainer® screen. Influent is moved between the rotating disks. Solids are separated from the liquid and discharged.

Fig. 7.21   Diagram of disk filter operation
(Courtesy of Hycor Corporation)

and forms between them while liquid constantly drains through the clear openings in the mesh covering the disks. The turning action of the disks moves this plug of solids upward toward a discharge slot in the contoured chamber and solids are automatically ejected. Liquid that has passed through the mesh-covered disks to the inside of the disk pair is collected in a separate compartment of the cabinet of the unit and is piped out for discharge or further treatment.

Spray bars mounted above the water level wash entrained (trapped) solids off the disks. The sprayers are used on an intermittent or continuous basis to clean the screen, as needed. Disk filters are equipped with variable-speed gear motors so that the disks can be turned at a speed appropriate to the flow rate and solids loading of the application.

The disks rotate against replaceable seals in the contoured chamber. Periodic inspection of these seals and of the screen mesh itself (which can be rather delicate) is important to ensure that no wastewater bypasses the screening process.

START-UP/SHUTDOWN PROCEDURES

Always refer first to manufacturer's instruction manual.

1. Start-Up

   a. Inspect all piping connections and ensure gear motor is properly filled with oil.

   b. Check cylinder or disk support bearings to ensure all bolts are secure.

   c. Visually inspect the unit for proper cylinder or disk alignment.

   d. Jog (quickly start and stop) the motor to check for proper machine rotation. If rotating in reverse, turn off all power and switch any two incoming power leads to the motor.

   e. Check for proper spray bar orientation and spray pattern.

   f. Ensure all guards are in place.

   g. Run the machine to observe overall mechanical operation. Stop the machine and correct any misalignments.

   h. Turn the machine on (with variable-speed gear motor, begin on low speed) and slowly start the influent. Recheck all piping connections and observe solids separation, which will be immediate with rotating cylinder-type screens. If everything is satisfactory, increase flow at a moderate rate to desired level and recheck system operation.

2. Shutdown

   a. Shut off influent.

   b. Allow the machine to continue to turn with spray bars operating until entire machine is rinsed down. This process ensures no solids will be left on the screen to harden.

   c. Rinse away any accumulated solids from headbox, guards, or spray bars.

   d. Turn off spray bars and stop machine rotation. Drain the headbox and rinse away any residual solids.

   e. For long-term shutdown, always consult the manufacturer's instructions for special procedures.

TROUBLESHOOTING ROTATING SCREENS

PROBLEM	PROBABLE CAUSE	REMEDY
SCREEN DRUM OR DISK DOES NOT TURN	No power to motor	Check circuit breakers. Check connections.
	Broken chain drive	Replace chain.
	Obstructed drum or disk	Remove obstruction.
	Zero motion switch turns off power	Find obstruction or overload condition. Clear machine, restart.
SCREEN DRUM OR DISK TURNS ERRATICALLY	Chain stretched	Adjust chain tension. Replace chain.
	Sprocket stripped	Replace sprocket.
	Frozen or broken trunnion	Replace trunnion or parts of trunnion assembly affected.
	Screen overload	Clear screen.
	Broken gear	Check reducer output shaft. Repair gear reducer.
SPRAY NOZZLES OBSTRUCTED	Solids dried on outside surface of nozzles	Rinse spray bars thoroughly and clean with soft wire brush.

TROUBLESHOOTING ROTATING SCREENS (continued)

PROBLEM	PROBABLE CAUSE	REMEDY
AFTER CLEANING, NOZZLES STILL OBSTRUCTED	Nozzle orifice plugged	Remove spray bar, remove nozzles, clean nozzles, flush spray bar and supply lines. Reassemble and test.
UNEVEN INFLUENT FLOW TO SCREEN	Weirs out of adjustment or covered with solids	Clean headbox assembly. Adjust weirs if needed.
	Erratic feed condition	Check supply pumps or plumbing.
INFLUENT ESCAPING SCREENING UNIT	Guards loose or out of adjustment	Tighten or readjust guards.
	Screen plugged or overloaded with solids	Clear solids from drum. Rinse screen completely. Steam clean screen.
	Seals worn	Replace seals.
TRUNNION NOT TURNING	Lack of lubrication	Lubricate bearings.
	Frozen bearing	Replace bearing.
OVERFLOWS DURING NORMAL OPERATION	Spray cleaning not frequent enough	Change spray cleaning cycle.
	Solids composition or flow changed	Review system process and adjust operation of unit.
SOLIDS BYPASS	Seals or doctor blade worn	Replace seals or doctor blade.

QUESTIONS

Write your answers in a notebook and then compare your answers with those on page 307.

7.5J What is the major advantage of rotating fine screens over static screens?

7.5K Internally fed rotating screens treat what types of wastestreams?

7.5L Disk filters treat what types of wastestreams?

7.5112 IN-CHANNEL OR IN-TANK SCREENS

These types of screens are common in domestic wastewater treatment plants and are not common in physical–chemical industrial wastewater treatment systems (IWTSs). They will not be discussed in this section.

7.52 Maintenance

Coarse and fine screens are known as rugged workhorses. Proper maintenance ensures peak screening performance, minimizes emergencies, and increases the useful life of any equipment. Due to the many types, unique features, and options available for mechanical screens, all maintenance steps for every type of unit cannot be covered here.

Always refer to the manufacturer's operation and maintenance manual for detailed and specific instructions. If the O & M manual for any equipment is missing, contact the

manufacturer for a replacement. Most manufacturers will be pleased to provide copies and are eager to share the information needed to keep their equipment in good working order.

Screens should be visually checked at least once per shift for proper liquid/solid separation performance. More frequent observation may be needed depending on the nature of the wastewater and solids and the type of screen installed. Static screens, for example, require more cleaning maintenance than rotating screens to operate at peak performance. Any abnormal noises or movements associated with motorized equipment should be diagnosed and corrected immediately.

Maintenance records should be kept for all screening equipment recording daily, weekly, monthly, and yearly maintenance operations performed. This will safeguard the equipment warranty, make spare parts requirements predictable, and highlight any recurring problems so they can be logically addressed. Accurate, up-to-date maintenance records are an important method of communication between changing shifts of operating personnel to ensure that important maintenance procedures are not skipped or needlessly duplicated. (See Volume II, Chapter 8, "Maintenance," for a description of an equipment maintenance recordkeeping system that is simple and easy to use.)

Always turn off and lock out the main power switch or remove the power fuse when working on any motorized screening unit. This is crucial, particularly when working on equipment that is automatically activated by level sensing or timer devices.

7.520 Coarse Screens

1. Daily

Observe the drive, cleaning rake, and solids discharge functions of the screen, noting any unusual sound, movement, or function. Diagnose and correct as needed.

2. Weekly

Inspect and lubricate pin rack or cog rail and gears, or chains and sprockets.

Always shut the unit off first, *NEVER* reach into the operating range of machinery while it is running. Slow-moving equipment is especially hazardous. Because it moves slowly, it does not appear dangerous. However, most geared-down machinery is so powerful that it can crush almost any obstruction. A human hand, for instance, offers little resistance to this type of equipment.

3. Monthly

  a. Inspect screening unit for any signs of abnormal operation. Diagnose and correct as needed.

  b. Inspect control system to ensure proper operation.

  c. Inspect power cables, motor, and motor brake assembly for possible damage or need for adjustment.

  d. Inspect cleaning rake, guide rollers, chains, sprockets, side rails, and frames for damage or excessive wear, or need for adjustment.

  e. Inspect solids discharge assembly for proper screening removal or wear.

  f. Inspect and tighten all fasteners as required.

  g. Inspect and lubricate bearings.

  h. Inspect gear box oil level and refill if needed.

  i. Check for loose or damaged wiring, connections, or conduit.

4. Yearly

  a. Clean pin rack, cog rails, gears, chains, and sprockets.

  b. Check all moving parts for wear or damage. Repair or replace as needed.

  c. Inspect and tighten all fasteners as required.

  d. Inspect and clean the motor/brake assembly and gear drive. This may need to be done more frequently according to manufacturer's instructions. Verify all electrical connections are secure and not corroded.

  e. Change lubricating oil and clean gear box.

  f. Test run in accordance with start-up procedures.

7.521 Fine Screens

Static Screens: only cleaning maintenance is normally required.

Rotating, in-channel or in-tank screens:

1. Daily

Observe operation, noting any unusual sound, movement, or performance. Diagnose and correct as needed.

2. Weekly

  a. Inspect and lubricate bearings, chains, or sprockets. Replace worn bearings and tighten loose chain tension as needed.

  b. Inspect all seals or doctor blades to ensure untreated wastewater is not bypassing the screening unit. Adjust or replace seals or doctor blades as needed.

  c. Inspect condition of any cleaning brushes fixed to the unit. Replace as needed.

  d. Visually check spray nozzle performance. If nozzles are clogged, remove and clean thoroughly, reinstall, and check operation.

  e. Check liquid/solid separation performance. Ensure there are no flow obstructions and that solids are being properly removed from the machine. Correct any areas of concern.

  f. Clean the screening unit as a part of normal good housekeeping.

3. Monthly

  a. Check gear box oil level and fill if needed.

  b. Check all fasteners and tighten as needed.

  c. Inspect the screening surface for damage. Repair or replace as needed.

d. Check chain-driven units for proper tension. Tighten as needed.

e. Check any trunnions, rollers, or wear pads or rails for excessive wear. Replace as needed.

4. Yearly

a. Inspect and clean the gear motor and reducer. (This may need to be done more frequently, according to manufacturer's instructions.) Change the oil.

b. Verify all electrical connections are secure and not corroded.

7.53 Safety

Use standard safety procedures and good common sense in working around moving equipment and maintaining the area around it. Slow-moving equipment is particularly hazardous because it moves slowly and does not appear dangerous. However, most geared-down machinery is so powerful that it can crush almost any obstruction, including a human hand. Whenever you work around open channels or tanks, be careful not to trip or slip and fall into the wastewater or moving machinery. Stay behind guardrails whenever possible, and do not operate screening equipment without them. Walkways and guardrails must be kept clean, in good repair, and free of obstructions.

The entire screening area should be well lit and ventilated. Post "No Smoking" signs throughout the area because industrial discharges may contain explosive materials and gases.

Any time you must work on mechanical equipment, be sure to first shut off, lock out, and tag the power and operating controls to prevent accidental start-up of the equipment.

QUESTIONS

Write your answers in a notebook and then compare your answers with those on page 307.

7.5M Why are maintenance records important?

7.5N What types of hazards might an operator be exposed to when working around open channels or tanks?

END OF LESSON 2 OF 3 LESSONS
on
PRELIMINARY TREATMENT

Please answer the discussion and review questions next.

DISCUSSION AND REVIEW QUESTIONS
Chapter 7. PRELIMINARY TREATMENT
(EQUALIZATION, SCREENING, AND pH ADJUSTMENT)
(Lesson 2 of 3 Lessons)

Write the answers to these questions in your notebook. The question numbering continues from Lesson 1.

5. What types of downstream processes and equipment are protected by screening?

6. Externally fed rotating screens can handle what types of wastes and should not treat what other types of wastes?

7. Internally fed rotating screens are suitable for treating what types of wastestreams?

8. Why should coarse and fine screens have proper maintenance?

9. What makes slow-moving equipment particularly hazardous?

CHAPTER 7. PRELIMINARY TREATMENT

(EQUALIZATION, SCREENING, AND pH ADJUSTMENT)

(Lesson 3 of 3 Lessons)

7.6 NEED FOR pH ADJUSTMENT OF INDUSTRIAL WASTEWATERS
by Richard von Langen and Mark Acerra

pH adjustment is one of the most commonly used treatment processes in an industrial wastewater treatment system (IWTS). Its purpose is primarily to optimize chemical reactions, such as those associated with precipitating heavy metals, oxidizing cyanide, reducing hexavalent chromium, optimizing biological activity, preventing corrosion, reducing compounds (sulfate to hydrogen sulfide), or NEUTRALIZING[14] an acidic or basic solution to bring it within discharge limits.

Industrial wastes usually contain acidic or alkaline (caustic) materials that require neutralization before biological treatment or discharge to receiving waters or a POTW (Publicly Owned Treatment Works) collection system. The neutralization of water is measured in terms of pH. pH is measured on a scale from 0 to 14 with 7 being neutral. Levels below 7 are acidic and above 7 are caustic. In practice, industrial wastewaters are rarely truly neutralized to a pH of 7. They are adjusted to within an acceptable pH range. The range is determined by either the water quality criteria of the receiving stream, the waste treatment process being used, or the physical integrity of the wastewater collection system.

Water quality criteria are determined by the needs of the receiving body of water and are usually determined by the appropriate state and federal regulatory agencies. Receiving water quality criteria are written into NPDES discharge permits under which the process discharges and are independent of the process being used to treat the wastewater. pH is an important factor in the chemical and biological systems of natural waters. For example, cyanide toxicity to fish increases as the pH is lowered. Ammonia has been shown to be 10 times as toxic at pH 8.0 as it is at pH 7.0. The solubility of metal compounds contained in bottom sediments or as suspended material is affected by pH. The following pH ranges from "Quality Criteria for Water" have been adopted (July 1986) by the US Environmental Protection Agency.

Range	Beneficial Water Uses
5–9	Domestic water supplies
6.5–9.0	Freshwater aquatic life
6.5–8.5	Marine aquatic life

NPDES permits for discharges into receiving waters with these beneficial water uses require effluent pH values within these ranges. In addition to meeting the water quality criteria of the receiving stream, industrial wastewaters usually require treatment for specific pollutants such as oil, grease, metals, suspended solids, organic materials, and other polluting compounds. The various waste treatment processes commonly used are pH-dependent. These processes work or work best within a specific pH range. Some wastewater streams require several pH adjustments to accommodate process steps as well as final discharge to the receiving stream. When pH makes the difference in whether or not a process will work, pH control is critically important and must be continuously monitored. When pH determines the relative efficiency of a process, pH control is not critical if the use of excess treatment chemicals or longer mixing times are available. However, this will result in higher operating costs and decreased ability to handle process upsets.

Neutralization can also be important in corrosion control. The lower the pH, the greater the rate of corrosion. If the pH of a process solution is allowed to drift outside the designed range, corrosion of equipment and piping can start. Some process wastestreams require pH adjustment so that they can be discharged through an existing wastewater collection system without damaging it.

---

[14] *Neutralization* (noo-trull-uh-ZAY-shun). Addition of an acid or alkali (base) to a liquid to cause the pH of the liquid to move toward a neutral pH of 7.0.

pH adjustment is one of the most difficult chemical reactions to control. pH is measured on a logarithmic scale, meaning that for every one unit of change on the pH scale, there is a ten-fold difference in the hydrogen ion (acid) concentration. Changes in the concentrations of the other constituents, such as calcium carbonate or calcium sulfate, affect the *BUFFERING CAPACITY*[15] and the amount of acid or base needed to change the pH. Usually, with any industrial wastewater there are variations in both of these, making pH adjustment a set target under variable conditions. In addition, the instrument used to measure and guide us in adjusting the pH (a pH probe in the wastestream) has problems of its own and can rarely come within 0.25 pH unit of an accurate reading.

This section provides a definition of pH and the problems with pH measurement, a description of a good pH adjustment system, some common pH adjustment processes, an approach to solving pH control problems, and ways to optimize pH control. The information on the definition of pH and the problems with pH measurement will give you an understanding of the problems encountered with measuring and controlling pH. By studying what is involved in a good pH control system and some of the common pH control systems, you will be able to compare it to your pH control system and understand the strengths and weaknesses of your own system. Finally, the section on solving pH control problems and optimizing the system you have will help you solve those everyday problems that you are now facing.

7.7 DEFINITION OF pH AND TITRATION CURVES

7.70 Definition of pH

The textbook definition of pH is that pH is an expression of the intensity of the basic or acidic condition of a liquid. Mathematically, pH is the logarithm (base 10) of the reciprocal of the hydrogen *ION*[16] activity.

$$pH = \text{Log}\frac{1}{[H^+]}$$

pH may range from 0 to 14 with 0 the most acidic, 14 the most basic, and 7 as neutral.

Wastewaters with high pH levels are adjusted with the addition of acids such as sulfuric acid (H_2SO_4) or hydrochloric acid (HCl) (also sometimes known as muriatic acid). Also, carbon dioxide (CO_2) and sulfur dioxide (SO_2) can be used in a gaseous form to lower the pH of wastewater.

Low pH wastewaters are adjusted with calcium oxide or lime (CaO), hydrated lime ($Ca(OH)_2$), sodium hydroxide (NaOH), ammonia (NH_3), magnesium oxide (MgO), magnesium hydroxide ($Mg(OH)_2$), or sodium carbonate (soda ash) (Na_2CO_3). Table 7.2 shows the relative basicity of these compounds.

Concentrated solutions of strong acids and bases are used to bring pH within a general range, for example 6 to 9 pH. Weaker acids and bases, or diluted solutions of strong acids, are used to

TABLE 7.2 LIST OF NEUTRALIZING REAGENTS SUGGESTED FOR WASTEWATER TREATMENT

Neutralizing Reagent	Chemical Formula	Basicity Factor[a]
Potassium Hydroxide	KOH	0.89
Trisodium Phosphate	Na_3PO_4	0.92
Sodium Carbonate (Soda Ash)	Na_2CO_3	0.94
Calcium Carbonate (Limestone)	$CaCO_3$	1.00
Calcium Magnesium Carbonate (Dolomite)	$(Ca\text{-}Mg)CO_3$	1.09±
Sodium Hydroxide (Caustic Soda)	NaOH	1.25
Calcium Hydroxide (Hydrated Lime)	$Ca(OH)_2$	1.35
Ammonium Hydroxide	NH_4OH	1.43
Magnesium Hydroxide	$Mg(OH)_2$	1.72
Calcium Oxide (Quicklime)	CaO	1.78
Magnesium Oxide	MgO	2.48
Sodium Sulfide	Na_2S	[b]
Potassium Permanganate	$KMnO_4$	[b]

[a] Grams of Calcium Carbonate ($CaCO_3$) equivalent per gram of reagent.
[b] Basicity factor will vary depending on cations present in wastewater.

---

[15] *Buffer Capacity.* A measure of the capacity of a solution or liquid to neutralize acids or bases. This is a measure of the capacity of water or wastewater for offering a resistance to changes in pH.
[16] *Ion.* An electrically charged atom, radical (such as SO_4^{2-}), or molecule formed by the loss or gain of one or more electrons.

"fine-tune" the pH to an optimum point for heavy metal precipitation.

Acids and bases mixed with water produce solutions composed partially or wholly of ions. Solutes, those compounds dissolved in water and that exist solely as ions in solution, are termed strong ELECTROLYTES.[17] Solutes that do not react completely form neutral molecules and the ions formed from them are termed weak electrolytes.

Water is an example of a neutral solution. It has an equal amount of hydrogen ions (H^+) and hydroxide ions (OH^-). If a solution has a pH of 5 then it has 100 times more acid in it than a solution with a pH of 7. (pH 7 − pH 5 = 2. The log of 2 is 10^2 or 10 × 10 = 100.) With deionized (DI) water being used in more and more manufacturing processes, pH adjustment of wastewater containing large quantities of DI water approaches this theoretical relationship between hydrogen ions (H^+) and hydroxide ions (OH^-). However, most of the pH adjustment situations that an industrial wastewater treatment operator encounters are solutions that are buffered with other compounds.

pH adjustment is also dependent on chemical reaction rate, temperature, mixing conditions, and the starting pH. Because these factors vary with the wastestream and with time, different amounts of neutralizing agents may be required to change one pH unit in different wastestreams or from point to point in the same wastestream.

7.71 Chemicals Used for pH Adjustment

Process water pH adjustment is achieved by adding available acid or alkaline substances. Alkaline substances commonly used are:

1. CaO (calcium oxide or lime), MgO (magnesium oxide), $Ca(OH)_2$ (calcium hydroxide, a hydrated form of lime) or $Mg(OH)_2$ (magnesium hydroxide), are the most commonly used chemicals because of availability, low cost, and high capacity. Sludge bulk (volume) is a major problem, but recovery is possible. Lime is either high calcium or dolomitic and comes either as quicklime or hydrated. It comes in dry form and is usually mixed with water before use.

2. Sodium hydroxide (caustic soda). Sodium hydroxide (NaOH) is a convenient, controllable, and commonly available chemical, but expensive. It is generally used for small or occasional applications or where limitation of sludge deposits is sought. Caustic soda is available in liquid form in two concentrations, 50 percent NaOH, which begins to crystallize at 54°F (12°C), and 73 percent NaOH, which begins to crystallize at 145°F (63°C). Therefore, they must be properly stored or diluted prior to use. Caustic soda is also available in an anhydrous or dry state (solid, flake, ground, or powdered) at 100 percent concentration. In the dilution process, considerable heat is generated. Therefore, the rate of dilution and method of cooling must be carefully controlled so that there is no boiling or splattering.

The Alkali Conversion Tables (Table 7.3) and Alkali Neutralization Graph (Figure 7.22 on page 289) show the equivalent amounts of sodium hydroxide and the various forms of lime that must be used to accomplish the same degree of neutralization. The Alkali Neutralization Graph also shows the weights of alkalis required to neutralize a given weight of any of the acids indicated.

Acid substances commonly used are:

1. Sulfuric acid (H_2SO_4) is the cheapest and most readily available. It is strongly corrosive, dense, oily, and clear or dark brown (depending on purity). Sulfuric acid should be of the USP (United States Pharmaceutical) grade, free of heavy metals; it is available in a number of grades containing 60 to 94 percent H_2SO_4. In the 93 percent grade, it is noncorrosive to steel drums; however, upon dilution it is highly corrosive.

2. Hydrochloric acid (HCl) or muriatic acid is a clear or slightly yellow, fuming, pungent liquid. It is poisonous and may contain iron or arsenic. Hydrochloric acid should be obtained in the purified form (USP) and is shipped in glass bottles, carboys, and rubber-lined steel drums, tank cars, or trucks. It contains approximately 35 percent available hydrogen chloride. Fuming can be reduced by dilution to 20 percent HCl.

3. Where available, carbon dioxide (CO_2) or sulfur dioxide (SO_2) may be applied in gaseous form. Flue gases are accessible and economical for neutralization of alkaline waters in certain industries.

7.72 Titration Curve

The laboratory test used to measure changes in pH involves a procedure called titration. A titration is the drop-by-drop addition of one solution (called a reagent or titrant) to another solution (the wastewater sample) until there is a measurable change in the sample. This change is frequently a color change or the formation of a PRECIPITATE.[18] When the color change occurs

---

[17] *Electrolyte* (ee-LECK-tro-lite). A substance that dissociates (separates) into two or more ions when it is dissolved in water.
[18] *Precipitate* (pre-SIP-uh-TATE). (1) An insoluble, finely divided substance that is a product of a chemical reaction within a liquid. (2) The separation from solution of an insoluble substance.

TABLE 7.3   ALKALI CONVERSION TABLES[a]

NaOH	CaO	Ca(OH)$_2$	Ca - MgO	Na$_2$CO$_3$	Na$_2$CO$_3$	CaO	Ca(OH)$_2$	CaO - MgO	NaOH
1	.70	.93	.60	1.32	1	.53	.70	.45	.75
2	1.40	1.85	1.20	2.65	2	1.06	1.40	.91	1.51
3	2.10	2.78	1.81	3.97	3	1.59	2.10	1.36	2.26
4	2.80	3.70	2.41	5.30	4	2.12	2.80	1.82	3.02
5	3.50	4.63	3.01	6.62	5	2.65	3.49	2.27	3.77
6	4.21	5.56	3.61	7.95	6	3.17	4.19	2.73	4.53
7	4.91	6.48	4.22	9.27	7	3.70	4.89	3.18	5.28
8	5.61	7.41	4.82	10.60	8	4.23	5.59	3.64	6.04
9	6.31	8.33	5.42	11.92	9	4.76	6.29	4.09	6.79
10	7.01	9.26	6.02	13.25	10	5.29	6.99	4.55	7.55
15	10.51	13.89	9.04	19.87	15	7.94	10.48	6.82	11.32
20	14.02	18.52	12.05	26.50	20	10.58	13.98	9.09	15.10
25	17.52	23.15	15.06	33.12	25	13.23	17.47	11.37	18.87
30	21.03	27.78	18.07	39.75	30	15.87	20.97	13.64	22.64
35	24.53	32.41	21.08	46.37	35	18.52	24.46	15.91	26.42
40	28.04	37.04	24.10	53.00	40	21.16	27.96	18.19	30.19
45	31.54	41.67	27.11	59.62	45	23.81	31.45	20.46	33.97
50	35.05	46.30	30.12	66.24	50	26.45	34.95	22.73	37.74
55	38.55	50.93	33.13	72.87	55	29.10	38.44	25.01	41.51
60	42.05	55.57	36.15	79.49	60	31.74	41.94	27.28	45.29
65	44.56	60.20	39.16	86.12	65	34.39	45.43	29.56	49.06
70	49.06	64.83	42.17	92.74	70	37.03	48.93	31.83	52.83
75	52.57	69.46	45.18	99.37	75	39.68	52.42	34.10	56.61
80	56.07	74.09	48.19	105.99	80	42.32	55.92	36.38	60.38
85	59.58	78.72	51.21	112.62	85	44.97	59.41	38.65	64.16
90	63.08	83.35	54.22	119.24	90	47.61	62.91	40.92	67.93
95	66.59	87.98	57.23	125.86	95	50.26	66.40	43.20	71.70
100	70.09	92.61	60.24	132.49	100	52.90	69.90	45.47	75.48

[a] From *CHEMICAL LIME FACTS*, permission of National Lime Association, Washington, DC, 1951.

or the precipitate forms, the end point of the titration has been reached.

The amount of titrant added to reach the end point can be plotted as a graph (see Figure 7.23) called a titration curve. The titration curve graphically represents the amount of neutralizing agent needed to adjust the pH from its starting point to the one desired to optimize the downstream reaction. The curve presents the amount of the titrant, the acid or base solution, added to the raw wastewater to obtain a given pH. The key to preparing a titration curve is to use the chemical and concentration that you are currently using, or proposing to use, at the same temperature as the raw wastewater and with the same contact time.

Figure 7.23 illustrates the titration of a strong base (NaOH) with a strong acid (H$_2$SO$_4$) in an unbuffered solution, such as DI water. Initial additions of the titrant may have a minor effect upon pH because of the buffer capacity of the water being treated. The curve is almost flat for each addition of titrant (until the buffer capacity is exceeded) prior to the inflection (sharp curve). The *EQUIVALENCE POINT*[19] is graphically located halfway along the straight line on the graph between the upper and

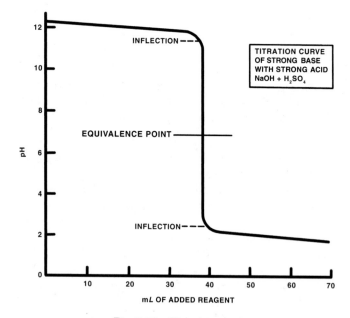

Fig. 7.23   Titration curve
(From *PHYSICAL CHEMICAL TREATMENT TECHNOLOGY*, US Environmental Protection Agency, Washington, DC, 1972)

---

[19] *Equivalence Point.*   The point on a titration curve where the acid ion concentration is equal to the base ion concentration.

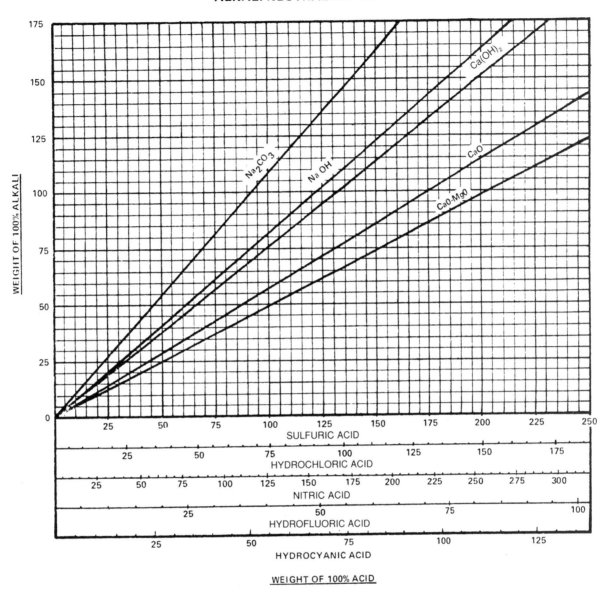

Fig. 7.22 Alkali neutralization graph
(From *CHEMICAL LIME FACTS*, permission of National Lime Association, Washington, DC, 1951)

lower inflections. Strong acid and base equivalence points commonly occur near pH 7.0. The product of the volume times the concentration of added titrant at the equivalence point is an estimate of sample basicity.

Figure 7.24 shows the effect of adding 4 percent (1 N NaOH) or 4.9 percent (1 N H_2SO_4) to a sample of industrial wastewater. (See footnote 20 for an explanation of N in 1 N H_2SO_4 and 1 N NaOH.) From the curve, the operator can readily determine the amount of caustic or acid needed to adjust the pH to whatever value best suits the desired result. The graphs in Figures 7.23 and 7.24 differ significantly in curvature, but provide the same basic information.

If the industrial wastewater in Figure 7.24 were rinse water from a hexavalent chromium plating bath and needed the pH adjusted to 2.2, 13,000 gallons of 4.9 percent sulfuric acid should be added for every 1,000,000 gallons of wastewater to be treated. If the wastewater were from a zinc plating solution that was going to be treated for cyanide oxidation and needed the pH adjusted to 11.0, 3,000 gallons of 4 percent sodium hydroxide would need to be added.

---

[20] *N* or *Normal*. A normal solution contains one gram equivalent weight of reactant (compound) per liter of solution. The equivalent weight of an acid is that weight which contains one gram atom of ionizable hydrogen or its chemical equivalent. For example, the equivalent weight of sulfuric acid (H_2SO_4) is 49 (98 divided by 2 because there are two replaceable hydrogen ions). A one *N* solution of sulfuric acid would consist of 49 grams of H_2SO_4 dissolved in enough water to make one liter of solution.

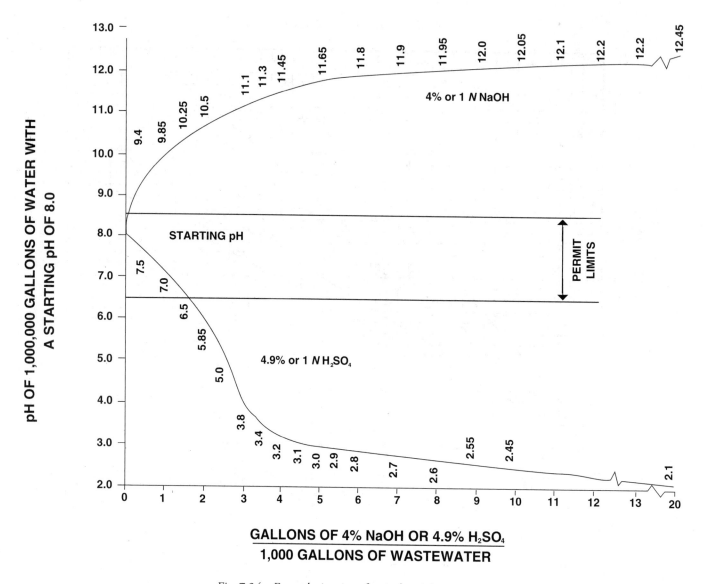

Fig. 7.24 Example titration of an industrial wastewater

QUESTIONS

Write your answers in a notebook and then compare your answers with those on page 308.

7.6A Industrial wastewaters require pH adjustment for what purposes?

7.7A Define pH.

7.7B pH adjustment depends on what factors?

7.8 pH SENSORS

7.80 Use of pH Sensors

In order to control pH, you must be able to measure the pH of the solution requiring adjustment as well as the changes in pH when you add the neutralizing or pH adjustment chemicals. The optimum pH for a process is usually defined as a specific pH to the nearest 0.1 pH unit. The critical part of the control system is the pH electrode.

For most industrial applications, the actual accuracy of pH electrodes falls far short of the operator's expectations. With one electrode, some operators may believe they have achieved an accuracy of 0.1 pH unit or better. Some systems have three electrodes and use the middle electrode value for control. Even with three pH probes in the reactor, the best accuracy recorded to date is 0.25 pH unit. It is said that if all three electrodes agree to within 0.1 pH unit for more than a few minutes, the electrodes are probably coated, broken, or still have their protective caps in place.

To understand why pH measurement and control are so difficult, we will start with the sources of pH electrode error. The operator can then assess how to compensate for the expected performance due to these errors. The pH sensor (Figure 7.25) is composed of a measuring electrode and a reference electrode. The next paragraphs will briefly explain how the pH electrodes work and what happens to the electrode's response while in an industrial wastewater.

7.81 Measurement Electrode

pH measurement is the sensing of an electrical potential in comparison to a reference potential. The measurement potential is the result of a transfer of protons [H⁺] from the process solution to an external gel layer, the glass layer, the hydrated internal gel layer, and the internal buffer layer to create a millivolt electrical potential. This electrical potential is proportional to the difference between the known pH in the fill of the electrode and the unknown pH in the process solution. Figure 7.26 illustrates how a pH electrode senses potential.

7.82 Reference Electrode

The reference electrode is designed to provide a constant reference potential; there are three general types. The first type of electrode is a flowing electrode. The internal liquid junction of a flowing electrode changes quickly with a change in the concentration of ions in the process stream. This type of electrode was used in the United States decades ago (and is still used in Europe). The difficulty with these electrodes is that they can become poisoned with a high concentration of strong acids or bases in the process stream.

A second type of reference electrode is a sealed electrode. The sealed electrodes are slow to reach equilibrium due to the fact that they are designed to resist process contamination. Combination electrodes, where the reference electrode surrounds the measuring electrode, are also slow to reach equilibrium. These electrodes use potassium chloride-filled wooden dowels to reduce the contamination rate. Whether this is a problem is dependent on the application. There is some difficulty in calibrating them and using them in batch processes because they are slow to equilibrate. Batch operations where the pH changes throughout the treatment (two-stage cyanide treatment) may not allow the reference to reach equilibrium with the changing pH so an inaccurate reading may result.

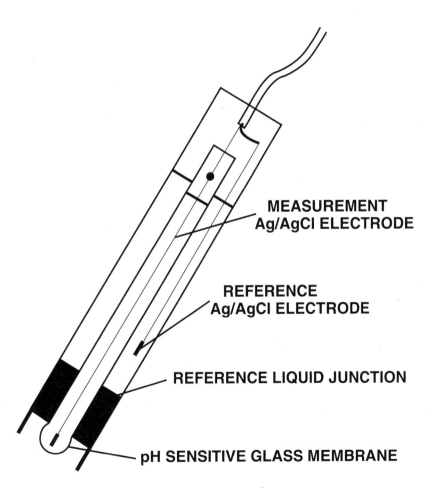

Fig. 7.25 pH probe
(Permission of Signet Scientific)

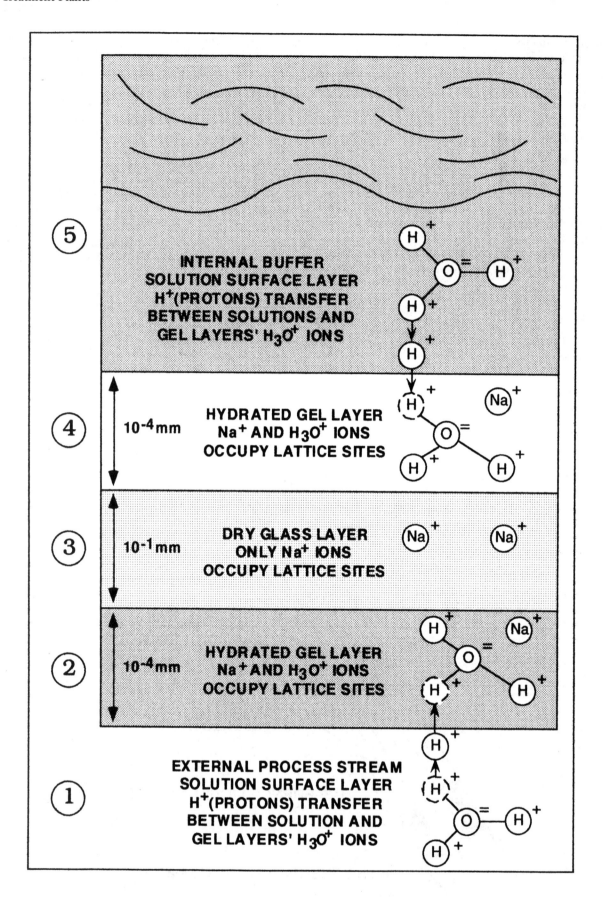

Fig. 7.26 *How a pH electrode senses potential*
(Reproduced with permission of the American Institute of Chemical Engineers.
©1991 AIChE. All rights reserved.)

The third type of reference electrode or technique that can be used is a differential electrode. This uses another glass measurement electrode behind a second junction surrounded by buffer solution as shown in Figure 7.27. This type of reference electrode is not susceptible to sulfide attack and is widely used in wastewater applications where the control range is a pH of 6 to 9. The differential pH electrode uses only a small reference area exposed to the process solution. This reduces the poisoning problems, but differential electrodes are more easily clogged than other types of electrodes.

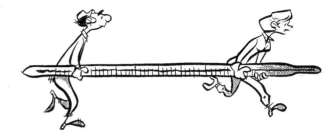

7.830 Temperature

The change in pH to temperature is zero as the pH approaches its isopotential point (the point where levels of potential are equal). This point is generally accepted as being a pH of 7 but, in fact, this is rarely the case. However, using this premise, if a wastewater containing strong electrolytes shows a pH of 8 at 60°C (140°F), it would change to a pH of 9 if the wastewater were at 25°C (77°F). pH changes in the opposite direction of the temperature change.

Temperature compensation is slow because the temperature sensor is imbedded in the interior of the electrode. The farther the sensor is from the tip of the electrode, the longer it takes to sense the change in temperature. Generally, the temperature changes in a wastewater stream are relatively slow and of little consequence. However, there are two situations where temperature compensation needs to be seriously considered. The first is when there is a significant and rapid change in the wastewater temperature and the second is in the laboratory.

For example, the manufacture of barium ferrite (used in making rubberized and ceramic magnets) uses hot concentrated hydrochloric acid to dissolve the impurities and divide paired crystals of barium ferrite. The discharge of the 40°C (109°F) wastestream into a small pH control receiving pit dramatically affects the temperature of the wastewater. The second situation is one that occurs more frequently. It is in the laboratory where a pH electrode is moved between solutions of different temperatures. Remember to allow enough time for the temperature compensator to reach the solution temperature.

7.831 Absorption Effects

Absorption affects the alkaline buffer solutions and, therefore, the calibration of the pH electrode. An exposed buffer solution of 10 pH can drop 0.1 pH unit per day through the absorption of carbon dioxide in the air. While the rate is much lower, carbon dioxide can also be absorbed through plastic bottle walls. For this reason, it is a good idea to make up a fresh buffer solution each time you calibrate your pH electrodes. pH 9 buffer solution is preferred in some cases because it is much less affected by CO_2 absorption than a pH 10 buffer solution.

7.832 Face Velocity, Fouling, and Response Time

A slow response time to a change in pH in the process solution means poor control of the process. As the velocity past the face of the electrode increases, the boundary layer thickness decreases and the response time to a pH change decreases, eventually becoming almost instantaneous. Good mixing to create a high flow velocity past the electrode or the use of flow-through

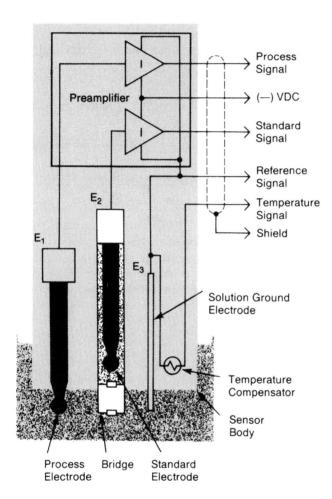

Fig. 7.27 How a differential electrode technique works
(Courtesy of Great Lakes Instruments)

7.83 Factors Affecting pH Sensors

The performance of a pH electrode can be adversely affected by a number of physical and chemical factors. Harsh wastewater characteristics can damage the external gel layer of the electrode; this will shorten the electrode's measurement span and slow its response. Damage to the gel layer could occur if the electrode remains dry or is abraded by slurries, etched by hydrofluoric acid (below pH 3) or sodium hydroxide (above 12 pH), or dehydrated by low pH (below pH 1) or low water content in the wastestream (40 percent). Temperature, absorption, face velocity and fouling of the probe, buffering of the solution, alkalinity, and acidity all affect the pH electrode accuracy and response time.

pH electrodes will reduce the response time of a pH electrode. Velocities up to 7 feet/second (fps) past the probe will improve the response time and help to keep the electrode clean. Velocities greater than 10 fps can cause excessive signal noise and wear. Please note that when calculating the face velocity, we must take into account the reduction in area due to the electrode being in the stream.

A slime layer or coating of one millimeter on an electrode can severely slow the response time from 10 seconds to 7 minutes. A coating on an electrode may trap the proton [H$^+$] concentration and the pH reading freezes. The thickness of the glass electrode also affects the response time. Some of the more rugged electrodes have response times as high as 100 seconds. Air bubbles clinging to or passing by a pH electrode can cause signal noise or shift. Particulates can also cause a similar shift or noise.

Response of a pH electrode is highly nonlinear and depends on the degree of buffering, the size of the pH change and the direction of change. In general the response is slower as the buffering decreases and the size of the pH change increases. The response time is also slower in the negative direction. For example, the response time may increase up to 100 seconds with a decrease from 1.55 to 1.38 pH and up to 200 seconds from 1.55 to 1.36 pH.

7.833 Alkalinity Error

The glass electrode reads low if the concentration of alkali ions is high enough to penetrate the gel layer and create a millivolt potential like a proton. The smaller the size of the alkali ion, the more error it can cause. Thus, at the same concentration, lithium will show a greater error than sodium and sodium will give a greater error than potassium. Alkalinity error also increases with a higher pH and a rise in temperature. Care must also be taken with solutions that may be neutral or even slightly acidic if there is a high salt concentration because of the high sodium or potassium concentration.

Alkalinity error can be confusing. If the electrode is removed from the highly alkaline solution, the electrode will typically return to normal condition when immersed in buffer solutions or brought down to a lower temperature.

7.834 Acidity Error

pH electrodes will read high if the pH is below 1. You will need to reduce the pH reading to get the true value. This error increases with time. Short immersions in hydrochloric acid will strip away coatings or the old hydrated gel layer, but repeated cleaning will eventually "kill" the electrode.

QUESTIONS

Write your answers in a notebook and then compare your answers with those on page 308.

7.8A What could be happening if all three pH measurement electrodes in a system agree to within 0.1 pH unit for more than a few minutes?

7.8B What factors influence pH electrode accuracy and response time?

7.9 STRATEGY FOR A SUCCESSFUL pH CONTROL SYSTEM

Three important elements contribute to a successful pH control strategy:

1. Equalize the waste before trying to adjust it. If you cannot equalize or there is still a large pH adjustment to be made, more than one tank in series may be necessary.

2. Provide enough residence time in the tank for the reaction and ensure good mixing.

3. Optimize your controller by minimizing dead time and by understanding how the controller responds.

The operator should evaluate the existing system using the information below to identify deficiencies and recognize potential limitations. Improvements in the pH adjustment system can be identified and sometimes accomplished with minimal changes to the system when you understand the things that can affect the performance of the pH adjustment system.

7.90 Equalization

The purpose of equalization, as explained in Lesson 1, is to reduce variations in concentration and flow (that is, the waste loading) to the subsequent processes. Generally, the larger the equalization tank, the better. For pH, this means reducing the variation in the loading of hydrogen ions to the pH adjustment tank. In some cases, wastestreams may self-neutralize if there is enough detention time in the equalization tank. An example would be the equalization of wastes from the regeneration of the CATION[21] and ANION[22] exchange columns.

If equalization is not available, if the variation in pH is still too great, or if the titration curve(s) is very steep (more like Figure 7.23 than Figure 7.24), two or more pH adjustment tanks may be needed. The first tank requires agitation and, using one of several control schemes (as shown in Figure 7.28), adjusts the pH to within a small range, say two pH units. The second tank is also likely to be agitated and used to fine-tune the

---

[21] *Cation* (KAT-EYE-en). A positively charged ion in an electrolyte solution, attracted to the cathode under the influence of a difference in electrical potential. Sodium ion (Na$^+$) is a cation.

[22] *Anion* (AN-EYE-en). A negatively charged ion in an electrolyte solution, attracted to the anode under the influence of a difference in electrical potential. Chloride ion (Cl$^-$) is an anion.

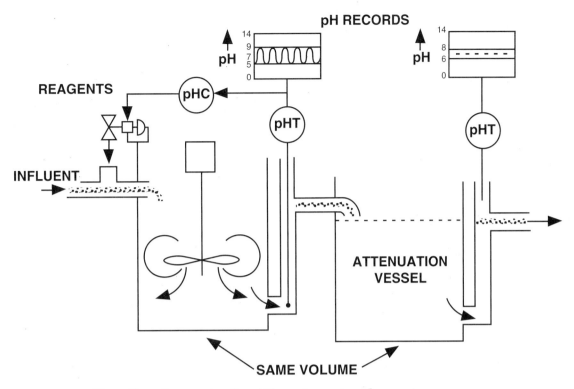

Feedback control with a two-tank system

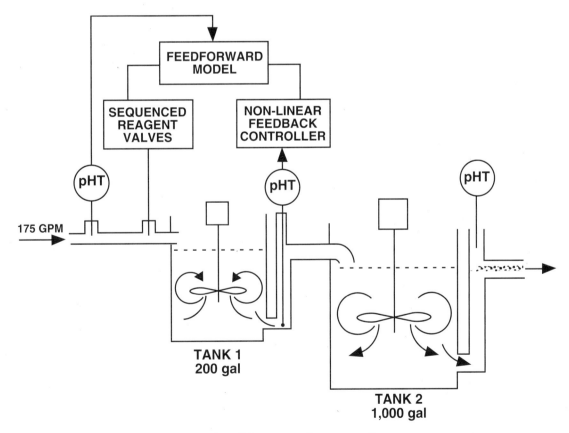

A wide-range feedforward control system

Fig. 7.28 pH adjustment systems with two tanks

(Reprinted by special permission from *CHEMICAL ENGINEERING,* November 8, 1976.
Copyright © 1976, by McGraw-Hill, Inc., New York, NY 10020)

pH. A third tank may be needed to level out any variations and as a point of measurement for the feedback loop in the control system. The first tank is by far the most important tank and really determines the controllability of the process.

7.91 Mixing, Dead Time, and Residence Time

Mixing is probably the most important, but often forgotten, part of pH adjustment. Without it, the reaction does not occur in a timely manner, there may be large pH variations in the reaction tank so the pH electrode reading will be erratic, the process control is poor, and the faulty pH reading could lead to a violation of the discharge limits. Good mixing is required if good pH adjustment and control are to be accomplished.

Dead time is the time interval between the addition of the pH adjustment chemical (reagent) and the first observable change in the effluent measurement. As the operator, you want to minimize the dead time and have a quick response to a change in the conditions in the reactor. Residence time is the volume of the reactor divided by the flow rate through the reactor. If there is no mixing, the dead time will approach the retention time.

Dead time is affected by agitation, tank geometry (size and shape), and retention time. For agitated vessels, the mathematical relationship between dead time and residence time is:

$$\frac{T_d}{V/F} = \frac{F}{F_a}$$

where: T_d = dead time

V = volume of the reactor

F = influent flow rate

F_a = pumping capacity of the agitator

The pumping capacity of an agitator is available from the equipment manufacturer. For pH applications, the dead time to residence time ratio should be about 0.05.

There are two types of mixing: intermixing, to get the reagent dispersed and reacted with the incoming wastestream, and backmixing, to reduce dead time and react any remaining reagent. For good intermixing, the reagent should be added to the wastewater before it enters the reactor. The reagent should be dispersed in the influent stream and turbulence provided to react it with the wastewater. Turbulence can be created by putting the reagent injector into the wastestream, using a static mixer, installing one or two 90-degree bends in the piping, or causing the influent to free fall one or two feet (0.3 to 0.6 m) to the surface of a stirred tank. The reagent feed piping should be equipped with a loop seal or some other means of preventing the line from draining freely into the wastestream. Poor intermixing results in an erratic or noisy signal observed in the effluent pH measurement.

Backmixing is more important in pH adjustment applications and is accomplished through the use of baffles and proper agitator selection and rotation, tank configuration, and vessel inlet and outlet placement. The preferred configuration of a pH adjustment reactor is cubic, meaning that the width (or diameter) and the working volume depth are about equal.

The residence time is dependent on the reagent, the rate of reaction, and the amount of solids that may interfere with the reaction. Liquid (for example, NaOH) or gaseous (for example, NH_3) reagents will need less time to react than solid reagents (for example, CaO) or slurries (for example, $Ca(OH)_2$). The more solids present or created by the pH adjustment, the longer the residence time should be. The precipitates trap unreacted reagent or influent material and cause an extended reaction time since the trapped material must separate from the precipitate before the reaction can occur. If the purpose of the reactor is just pH adjustment, then the reaction time is shorter than if the reaction is to oxidize cyanide or reduce chromium, for example. Typical residence times for industrial wastewater undergoing pH adjustment and treatment are shown in Table 7.4.

TABLE 7.4 TYPICAL RESIDENCE TIMES FOR pH ADJUSTMENT OF INDUSTRIAL WASTEWATER[a]

Process	Suggested Residence Time (minutes)
1. Batch pH Adjustment Only	3–10
2. Continuous pH Adjustment Only	5–7
3. Chromium Treatment	
a. Reduction	15–30
b. Hydroxide Precipitation	10–20
4. Cyanide Treatment	
a. First Stage Oxidation	20–45
b. Second Stage Oxidation	45–90

[a] From *TREATMENT OF METAL WASTESTREAMS*

Baffling should be provided and the agitator positioned to prevent a whirlpool effect. Baffles should extend from near the top to near the bottom of the working volume of the tank. The width of each baffle should equal approximately ten percent of the tank diameter or width. For circular tanks, four baffles should be placed 90 degrees apart, perpendicular to the tank sides. For square tanks, the baffles should be in the center of each side.

The agitator for small tanks can be side-mounted, but center-mounted is preferred and should be used for larger tanks. The normal flow pattern desired is to force the liquid down the center and up the sides of the tank. Thus, a propeller or axial-flow[23] impeller should be used and flat-bladed or radial-flow[23] impellers should be avoided. Impeller diameter to tank diameter should range from 0.25 to 0.40. The agitator should rotate in a direction opposite to the inlet-to-outlet wastewater flow direction to produce good mixing.

The recommended minimum agitator net shaft horsepower per 1,000 gallons is presented in Figure 7.29 and shows a de-

---

[23] Axial flow means the liquid moves in a direction parallel to the shaft of the impeller. In radial flow, the liquid moves perpendicular to the shaft.

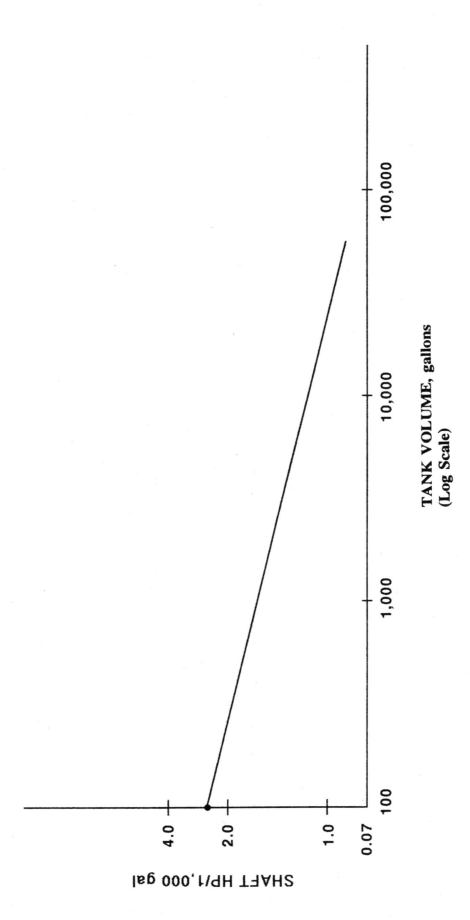

Fig. 7.29 Recommended minimum agitator shaft horsepower per 1,000 gallons (Reprinted by special permission from CHEMICAL ENGINEERING, November 8, 1976. Copyright © 1976, by McGraw-Hill, Inc., New York, NY 10020)

crease with an increase in the size of the tank. Therefore, as the tank gets larger, more horsepower is needed, but the extra horsepower for an increase in tank size gets smaller. The shaft horsepower as well as the pumping capacity need to be evaluated to ensure that the fluid is moving fast enough to achieve the desired reduction in dead time from inlet to outlet.

QUESTIONS

Write your answers in a notebook and then compare your answers with those on page 308.

7.9A List three elements of a successful pH control strategy.

7.9B What is dead time?

7.10 CONTROL SYSTEM

We are not going to try to make you an instrumentation and control expert, but this section will provide you with some fundamentals and things to look for when that expert is not immediately available. If you know things that affect the performance of the control loop, then you can minimize the bad effects and maximize the good things about your pH adjustment system. As discussed elsewhere in this book (Chapter 12, "Instrumentation"), an instrumentation and control system consists of:

1. The primary element (pH electrode).

2. The transmitter, usually directly associated and coupled to the pH electrode.

3. The pH controller, which may be a proportional, integral (reset), and derivative (PID) controller for ON/OFF control of a chemical metering pump, or an adaptive feedforward/feedback controller for a ganged set of three different sizes of control valves.

4. The final element or control variable, which is the chemical metering pump with ON/OFF or variable speed control or a control valve(s) on a reagent loop.

In this section, we will discuss the characteristics of these four elements as they may relate to the operator. We will concentrate on things that you can do without having to "tune the loop" or take apart the controller. Unless it is your job to also maintain the instruments and you know what you are doing, it is best to request assistance.

7.100 Primary Element

Most of the potential problems with a pH electrode already have been discussed in the previous section. However, there are a few of the mechanical aspects about pH electrodes that should be noted. The placement of the pH electrode in the control system is critical in getting a representative measurement of the pH in the reactor. pH electrodes in the reactor should be placed at a point between 30 to 60 percent of the depth of the working volume of the tank on the side where the effluent pipe is located.

An alternative is to put the pH electrode in a pipeline that is being pumped and recirculated to the reactor. This has been used successfully where the recirculation pump is also the method of mixing the contents of the reactor.

If you will be using more than one pH electrode, it is suggested that you use three electrodes. If only two are installed, there will be perpetual disagreement and only a fortune teller would know which one is correct. With three electrodes, you would choose the middle (not mean) value as the basis for control. These electrodes would be evenly spaced around the tank between the 30 and 60 percent depth levels.

Using a three-electrode system allows the following advantages in identification of electrode problems:

1. Failure of one electrode does not disable the system.

2. On-line washing or calibration is possible without disabling the system.

3. Measurement noise is reduced.

4. Measurement accuracy is improved.

5. Confidence that the system works properly is enhanced.

7.101 pH Transmitter

In many cases, the pH transmitter is coupled directly to the pH element. As an operator, there are a few items you can check that may help in making sure the system is working properly.

For submersible pH electrodes and transmitters, you may want to consider putting a small amount of compressed air into the conduit line to provide slight positive pressure. This will help prevent leaks from ruining the electronics of the electrode and transmitter. If your system already is equipped like this, check to make sure that the compressed air pressure is at the recommended level (1 to 3 psi (70 to 200 gr/sq cm)). You may also wish to check for leaks by turning off the agitator after the end of the day to see if there are bubbles coming from the probe assembly.

7.102 pH Controller

The pH controller can be as simple as turning on the chemical feed pump when it is below a certain range (assuming that the wastewater is acidic and needs a base added to it to bring it into range), and turning it off when it reaches a set point. These systems are usually applied when the wastewater has been equalized or varies over a small range and the allowable discharge range is relatively large (6 to 9 pH).

A pH control system also can be very complicated when there is a lot of variation in flow and concentration in the incoming wastewater, and the system has a minimum amount of residence time to accomplish a tight pH range (6 to 6.5 pH). This system may include a feedforward pH electrode, which senses the incoming load and is the main signal to the controller, and then a feedback pH electrode located in the reactor itself, which acts to fine-tune the amount of caustic added. The controller has to add these two signals together to allow the correct amount of caustic for pH adjustment.

There are two things that the operator can do to help ensure the good performance of the controller. The most helpful is to make sure that the dead time in the system is as low as possible. Reduction of dead time was discussed in the previous sections.

Periodically (once per year), you should review the pH adjustment system to ensure that conditions that would increase the dead time have not changed. Second, check to see if there are changes to the characteristics of the wastewater and, therefore, the titration curve. This is important because pH is a nonlinear function, meaning that for a unit change in pH you may need a 15-fold change in reagent at one starting pH and a 100-fold change at another starting pH. Some controllers today can be programmed to take the inverse of the pH titration curve and divide it into many small line segments so that there is a linear relationship over a certain range in pH.

7.103 Final Element

The final element in the control system is usually a metering pump or a control valve. The metering pump may turn on or off, increase or decrease speed, or increase flow rate based on the signal from the controller. The control valve is either at the end of the delivery system or is at one end of a recirculation system. A recirculating reagent system is recommended because the feed pump is always on to provide a relatively constant condition for the operation of the control valve.

The two main things the operator can do to help the performance of the final element are to perform the normal maintenance functions on the pumps and control valves and to make sure there is enough reagent. Periodically, you should also check to verify that the reagent pump turns on or the valve opens when the pH is low and base needs to be added. This can be done while the system is working or by temporarily adjusting the pH to a higher set point.

QUESTIONS

Write your answers in a notebook and then compare your answers with those on page 308.

7.10A List the components of an instrumentation and control system for pH adjustment.

7.10B What can an operator do to help ensure reliable performance of a pH controller?

7.11 COMMON pH ADJUSTMENT SYSTEMS

Almost every wastewater treatment process used to treat industrial wastewaters involves pH adjustment and the final effluent usually requires neutralization. The need for an optimum pH may be determined by a chemical reaction being used (such as cyanide destruction), the manufacturer's specifications with respect to a piece of equipment (such as to protect a reverse osmosis membrane), or by a physical process (such as coagulation of wastewater and sludge). An industrial wastewater treatment facility may receive streams separated as acid and alkaline or separated by pollutant. Process dynamics and economics often dictate preliminary treatment of each stream, including pH adjustment. The streams are then combined for additional treatment including pH adjustment and neutralization.

pH is considered to be the single most important variable in the *COAGULATION*[24] process. Jar testing as well as full-scale operation have clearly shown that there is at least one pH range for any given water or wastewater within which good coagulation and *FLOCCULATION*[25] occur in the shortest time with a given coagulant dose. The pH range is affected by the chemical composition of the wastewater and types of coagulant and coagulant aids used as well as by the concentrations. Alum or aluminum sulfate ($Al_2(SO_4)_3 \cdot 18\ H_2O$), ferric chloride ($FeCl_3$), and ferric sulfate ($Fe_2(SO_4)_3$) are the most commonly used coagulants. Although alum is cheaper, more readily available, and easier to handle, iron salts have the advantage of offering good coagulation over a broader pH range. This is most important where wastewater streams vary significantly in pH and pH control does not exist.

Commonly used methods of wastewater treatment, such as biological processes, reverse osmosis, ozonation, carbon adsorption, and ultrafiltration, are either pH-dependent or have a net effect on wastewater pH. Figure 7.30 shows phosphorus removal relative to alum dosage with almost zero phosphorus remaining at an alum dosage of 400 mg/L. However, the addition of alum also lowers the wastewater pH from above 7 to below 6. This shows that a chemical process sometimes will affect the wastewater pH even though pH adjustment was not needed. The amount of pH change is dependent on the buffering capacity of the wastestream.

Four common pH adjustment systems are discussed below in order to familiarize the operator with the different types of systems encountered. The first example is a pH adjustment system to neutralize an acidic waste with sodium hydroxide up to the discharge range of 6 to 9 pH. The second is adjusting the pH to a tighter pH range for optimizing the following downstream biological treatment system. Heavy metal precipitation at a specific pH is the third example. The final example is a discussion of how pH adjustment is used in the treatment of cyanide and chromium.

---

[24] *Coagulation* (ko-agg-yoo-LAY-shun). The clumping together of very fine particles into larger particles (floc) caused by the use of chemicals (coagulants). The chemicals neutralize the electrical charges of the fine particles, allowing them to come closer and form larger clumps.

[25] *Flocculation* (flock-you-LAY-shun). The gathering together of fine particles after coagulation to form larger particles by a process of gentle mixing. This clumping together makes it easier to separate the solids from the water by settling, skimming, draining, or filtering.

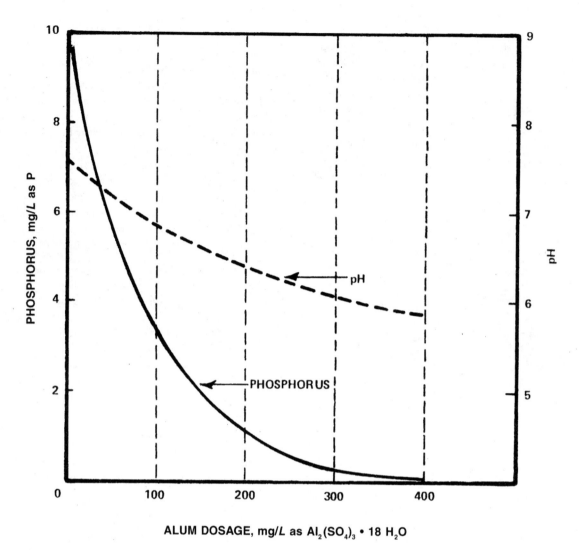

Fig. 7.30 Alum coagulation of wastewater
(From HANDBOOK OF ADVANCED WASTEWATER TREATMENT, 2nd Edition, by Russell L. Culp, George M. Wesner, and Gordon C. Culp, © 1978 by Litton Educational Publishing, Inc. Reprinted by permission of Van Nostrand Reinhold)

7.110 pH Adjustment for Discharge

This type of system is common where pH is the only water quality indicator that must be treated. This may occur when there are only acids or bases in the wastewater. It also is frequently used when the wastewater has been treated, say, for heavy metal removal and the pH must be reduced to bring it within the discharge limits of 6 to 9 pH. The system may consist of a single reaction tank with an agitator, a combination pH electrode, and a PID controller for ON/OFF operation of a chemical metering pump. For this example, we will assume that the pH is acidic and that the control range is 6 to 9 pH.

The objective is to be in compliance all of the time and to add the minimum amount of sodium hydroxide to minimize the operating costs. For this example, you may wish to put the pH set point between 6.5 and 6.8 pH. Much of this will depend on the titration curve, amount of equalization ahead of the reactor, and the variation in the influent and, therefore, the effluent pH. The reason for putting it 0.5 pH unit above the minimum of 6.0 is because the best a pH electrode can measure is a pH of 0.25 unit. From the regulatory agency's point of view, it is better to be high than low at this pH level. This is because the bacteria used in a municipal biological wastewater treatment plant can tolerate a slightly basic condition better than an acidic one.

7.111 pH Adjustment for Biological Treatment

Discharge from a petrochemical plant goes through an API separator for removal of free oil, solvents, and solids. It then flows to a pH adjustment system prior to going to an aerobic lagoon using a rotary brush system for oxidation. The pH adjustment system is designed to treat wastewater that can be either more acidic or basic than the desired range of 7.0 to 8.5 pH. Most of the time, the wastestream is acidic. The system consists of an agitated equalization tank followed by a pH adjustment reactor equipped with an agitator, a pH measuring and reference electrode, a PID controller, and two variable speed chemical metering pumps (one for ammonia and one for sulfuric acid). Ammonia is used because it is a strong base and because it helps in providing a source for one of the deficient nutrients, nitrogen, for the biological system.

The most important factor for biological treatment is to keep the pH as consistent as possible and, for reasons similar to the one described above, it is better to be slightly alkaline than acidic. If the discharge from the pH adjustment system were going to an anaerobic digester (such as the discharge from a brewery or citrus food processor), the preferred pH adjustment reagent may be lime. This reagent will buffer the influent to the digester. Buffering the influent is particularly important because the first step in anaerobic digestion is the production of acid-forming bacteria, which then provide food for the methane formers. Without the buffering and strict control of the influent pH, the acid formers may overpopulate, which causes the digester to "turn sour" and not function properly.

Another factor that may influence the choice of reagent for reducing the pH is the level of phosphorus in the biological system. If it is deficient, you may want to consider using either phosphoric acid or a combination of phosphoric acid and sulfuric acid to reduce the pH. Calculation should be made to determine that the amount of phosphoric acid that is to be used will supply enough phosphorus to meet the biological needs and not be so much as to cause a violation of a phosphorus limit. If either is in question, it may be better to use a neutral phosphate salt for the nutrient requirements.

Another option for neutralizing reagents is to use waste acids or bases that are generated on site. Be aware, however, that some of the acids and bases may contain toxic organics or heavy metals, which will pass through to the effluent or interfere with the treatment process if final pH adjustment is needed prior to discharge to the sewer or receiving water.

7.112 pH Adjustment for Heavy Metal Removal

Most metals are soluble at lower pH and will precipitate when the pH is increased. Precipitation is followed by coagulation, flocculation, and gravity settling or filtration. Each metal has its own optimum pH level for precipitation and many will dissolve at a pH on either side of this optimum. Co-precipitates, *CHELATES*[26] and other interferences affect the optimum pH for removing heavy metals.

In a typical situation, the wastewater contains multiple heavy metals, each with a different optimum pH. The strategy here is to optimize the pH for the heavy metal with the lowest discharge limit and determine if the other metals will be removed to below their limits. If this does not work, you may have to use a two-stage system to precipitate and remove one metal and then adjust the pH to the second optimum for the other metals.

When selecting the optimum pH control point, remember that the field conditions vary and you are not able to measure pH accurately. Test the effects of pH around the optimum to determine how sharp the inflection point is when the metals go back into solution. The field control point may be higher or lower than that found in the laboratory. Also, remember to examine the optimum pH set point and the type of wastewater you are dealing with to determine the effects that may occur to the pH measuring and reference electrodes.

7.113 pH Adjustment for Oxidation and Reduction

Cyanide (CN) oxidation and hexavalent chromium reduction are two common pretreatment methods. Both use an optimum pH and *OXIDATION-REDUCTION POTENTIAL (ORP)*[27] to complete the reaction in a reasonable length of time. If the conditions are not correct, the reaction time can go up by a factor of two or three, or the reaction may go out of control.

CN OXIDATION. pH is a critical factor in the oxidation of cyanide. The rate of reaction as well as the end products depend on maintaining proper pH values. For the first stage reaction, cyanide (CN) is converted to cyanogen chloride and then to cyanate. Cyanogen chloride is a gas with limited solubility at pH less than 10. Under those conditions, the cyanogen chloride, tear gas, is liberated and the reaction of cyanogen chloride to cyanate is greatly reduced.

pH also affects the amount of chlorine or sodium hypochlorite that must be added to maintain the required ORP. At a pH of 10.5 to 11.0, the ORP must be +550 millivolts (mV) with a residence time of 20 to 40 minutes, depending on the type of cyanide and the ability to control the pH. If the pH drops below 9.5, the oxidizing potential must be +650 mV. Below that pH, the reaction liberation of cyanogen chloride essentially stops the reaction, so a higher ORP has little effect. If the pH drops below 7.0, there is a potential to create hydrogen cyanide gas, which is highly toxic. For this reason, a low pH level alarm will sound to halt the process if the pH reaches 8.5.

At a pH above 11.5, there is excessive alkali and a suppression of ORP values. The ORP value is selected so that excess hypochlorite will be present to dominate the reaction. The difficulty is that when the controller requests additional hypochlorite, the hypochlorite also contains free sodium hydroxide and the pH can rise because of this simultaneous addition of alkali. When the pH rises, the ORP value decreases and the system is out of control as it requests additional hypochlorite, perpetuating the problem when there is no cyanide present. This situation is corrected when acid is added to reduce the pH to the optimum range.

The second stage reaction converts cyanate to carbon dioxide (CO_2) and nitrogen gas. The recommended optimum pH is 8.5 to 9.5 pH with an ORP above +800 mV and a residence time of 45 minutes. However, if the pH drops below 7.0, a different reaction takes place and the nitrogen hydrolyzes (combines chemically) with water to produce ammonia compounds.

---

[26] *Chelating* (KEY-LAY-ting) *Agent.* A chemical used to prevent the precipitation of metals (such as copper).

[27] *Oxidation-Reduction Potential (ORP).* The electrical potential required to transfer electrons from one compound or element (the oxidant) to another compound or element (the reductant); used as a qualitative measure of the state of oxidation in water and wastewater treatment systems. ORP is measured in millivolts, with negative values indicating a tendency to reduce compounds or elements and positive values indicating a tendency to oxidize compounds or elements.

CHROMIUM REDUCTION. The optimum pH for hexavalent chromium reduction varies with different wastes, reducing agents, temperatures, and chromium concentrations. Reducing agents include sulfur dioxide, sodium bisulfite, sodium metabisulfite, and ferrous sulfate. The hexavalent chromium is reduced to trivalent chromium and then precipitated as chromic hydroxide. The first reaction takes place best around a pH of 2.0 with an ORP of +250 mV and a residence time of 15 to 30 minutes. If the pH is above 2.0, the reaction rate decreases with increasing pH and is extremely slow for pH levels above 5.0.

Sodium hydrosulfite can also be used as a reducing agent for the reduction of chromium. This reducing agent has the advantage of being used at a higher pH (around 4.0), but it is more expensive than sodium bisulfite and oxidizes when it is in solution and allowed to stand. Its use is restricted mainly to manual chemical addition.

The second stage reaction to precipitate the trivalent chromium to the insoluble hydroxide takes place at a pH of 8.0 to 8.5 with a residence time of 10 to 20 minutes. The optimum pH should be determined experimentally, using the actual waste, because the solubility of chromium changes dramatically with pH on either side of the minimum point.

QUESTIONS

Write your answers in a notebook and then compare your answers with those on page 308.

7.11A List the common pH adjustment systems.

7.11B What is the most important factor an operator should keep in mind when adjusting the pH of a wastewater prior to biological treatment?

7.11C How is pH adjusted when the wastewater contains more than one type of heavy metal?

7.12 MAINTAINING pH ADJUSTMENT EQUIPMENT

pH adjustment systems consist of parts found in other locations of the treatment plant as well as the industrial site being served. Pumps are pumps and recorders are recorders. For the process to work, they must all function together as a system. The breakdown of one part will either shut the process down or necessitate manual operation. If certain parts are allowed to drift or to go out of calibration due to poor maintenance, process control will suffer. Because a chemical reaction is taking place and is usually being controlled by sensitive instrumentation, neutralization systems are trouble-prone and maintenance "hogs" (time consumers). An ounce of preventive maintenance is often worth a pound of breakdown maintenance. Points of particular concern are:

1. Primary sensors, such as level probes and pH probes
2. Chemical feed pump bearings and seals
3. Chemical feed pump stroke controllers or positioners
4. Automatic valves and limit switches
5. Screw feeders or vibrators for dry chemicals
6. Mixers

Maintenance procedures should be set up and be as formal and detailed as practical. Most industrial sites are staffed to set up such programs. The operators should work with these people to see that a usable program is implemented. The treatment plant should maintain a complete set of drawings (updated) and equipment manuals. Operating logs showing all aspects of process, system, and equipment operation should be maintained.

7.13 TROUBLESHOOTING pH ADJUSTMENT PROBLEMS

This section describes a general methodology for troubleshooting pH adjustment systems. The first thing to do in resolving a problem with your system is to clearly identify the problem. This involves four basic steps: check the mechanical operation of the equipment, check the operation of the pH electrodes, determine if the process wastewater has changed, and, if all else fails, go back to the basics of a good pH control system.

7.130 Mechanical Operation

The mechanical operation of the pH adjustment system equipment is the easiest and normally the first thing that you should check. Other sections in this manual will explain in greater detail how to identify problems with each piece of equipment, but Table 7.5 provides a summary of items to check to identify a problem.

7.131 pH Electrode

The second place to look for the source of the problem is the pH electrode or reference electrode. Unless physically abused by handling, by solids in the wastewater, or by being subjected to chemical attack by the contents of the wastewater and reagent, the pH probe should last nine months to two years in most systems. To keep the handling to a minimum, you might use the following method for checking the calibration of the pH electrode:

1. Take a sample of the wastewater noting the pH reading by the existing pH electrode.

2. Immediately measure the pH in the laboratory with an electrode that has been calibrated with two buffer solutions.

3. Standardize the field pH using the differential between the previously noted measurement and the laboratory reading.

If a new electrode is needed, the same procedure can be used except that before calibration, the new electrode should be immersed in the process liquid for no less than twenty minutes and up to two or three hours for solid-state reference electrodes in a highly ionic solution.

If you have a three-electrode system, it is easier to tell if there is a problem with an electrode. With three electrodes, you would remove one only if it differed by more than 1.0 pH unit for more than 10 minutes after repeated process calibration. If one electrode is coated, it will lag behind the other two. If two

TABLE 7.5 TROUBLESHOOTING PROBLEMS OF A pH ADJUSTMENT SYSTEM

Equipment	Items to Check
Mixer	a. Proper Rotation—Rotation should be in the direction away from the point of discharge and that the wastewater is being pushed downward. b. Speed—Fast enough to provide good mixing, but slow enough to prevent splashing. c. Pumping Capacity—Determine if it is sufficient to provide a dead time of 0.05. d. Blades—Should be propeller type. Check to make sure that all of them are on the shaft. e. Power—If the mixer is not operating, make sure that there is power to the mixer.
Chemical Feed Pumps	a. Proper Rotation and Speed—Check the rotation of the pump to make sure that it is pulling the influent into the impeller and discharging through the discharge pipe. Check the speed of the pump versus the required dosage amount. b. Power—If the pump is not pumping, make sure that there is power to the pump. c. Receiving Signal—Check to determine if the pump/controller valve is receiving a signal from the controller. Use a buffer solution or change the pH set point to determine if the controller is sending a signal to the pump/controller valve. d. Poor Pump Performance—If the discharge pressure is low, this may be an indication of a clogged inlet or outlet condition.
Control Valves	a. Respond to Signal—Test to determine if the control valve is receiving a signal from the controller. Adjust either the pH to above the existing measured pH if an alkali is being used (or below if an acid is being used) and observe whether the control valve attempts to open. b. Leaking—Change the set point to completely close the valve and observe whether the valve is leaking. An obstruction in the valve or the corrosion of the valve may be the reason for the leak. The leak may be the source of error in adding too much reagent and reducing the ability to accurately control the pH. c. Clogged—Another reason for the control valve not responding to a call for an increase in the amount of reagent is that the valve is clogged. With the pump off and all of the pressure upstream and downstream of the valve released, remove the control valve and inspect for debris. Corrosion may also be the cause of the valve not opening at all or to the extent required by the controller. d. Draining—The control valve should be as close to the delivery point as possible. Check the reagent delivery piping arrangement to determine if the end section is free draining.
Reagents	a. Inventory—If the pH is not being adjusted, there may not be any solution in the reagent tank. Check to determine if there is enough solution in the tank. b. Concentration—Check the concentration of the reagent to determine if it has been changed, particularly if it is made up from a concentrate that is diluted with water. If there is a question, it may be easier to just remake the solution. If the solution was made more dilute than normal, the chemical feed pump or control valve will operate at its highest capacity and not be able to keep up with controlling the pH. If the solution is too concentrated, the pump will operate at the low end or nearly closed and not be able to control the pH properly, because a small amount of reagent dramatically changes the pH. c. Proper Reagent—Check to make sure that the reagent that you are pumping is the one that is needed for the process. It is possible that the wrong reagent was put into the tank.

or all three are coated, there will be a loss of movement of the readings with respect to each other. Table 7.6 summarizes the different conditions and symptoms for many of the more common sources of pH probe errors. Just the sheer quantity of sources shows you that pH measurement accuracy can easily deteriorate and that pH measurement troubleshooting can be difficult. By identifying the symptoms, the problems that do develop can be remedied.

Chunks of solids striking the electrode, chemical attack of glass by hydrofluoric acid, or just old age can lead to electrode breakage. The measurement electrode gel layer wears away and penetrates deeper into the glass with time. Even under the best of conditions, the useful life of a measurement electrode varies from nine months to two years. If the measurement electrode bulb breaks, process fluid will be in contact with both the inside and outside pH-sensitive glass layers. The hydrogen activity, and hence the potential, will be about equal so that the difference in potentials is nearly zero, which corresponds to a constant 7 pH reading.

7.132 Manufacturing Process Changes

The third area to check is to see if the manufacturing process may have changed. Changes in chemicals, the amounts of acid or base, changes in the amount of production, or changes to the manufacturing procedures can create a change in the characteristics of the wastewater and, therefore, a possible change in the titration curve(s) and the controller. Check with your contacts in the production department to see if any changes were made.

One of your tools to determine if a change has been made or if the wastewater has changed in its characteristics is the titration curve. If different processes discharge different acids and bases or different ionic strength, then more than one curve will have to be created and compared. If the titration curve is different from the one that was originally used to program or set constants on the controller, you may need to tune the loop using the new titration curves. (Be sure to request assistance unless you are qualified and authorized to adjust the controller.)

7.133 Back to the Basics

If all else fails, go back to the basics of what makes a good pH control system. Examine whether the equalization tank is correctly sized and that mixing is sufficient and in the proper direction. If you do not have an equalization tank, then you should have at least two reactors in series with mixing in the first and probably mixing in the second. Check the dead time for the reactor. Check the electrode, transmitter, and controller dead time. This is to determine if there is a built-in lag between the time of reagent addition brought about by a change in the influent pH and the time it is noted by the pH meter.

Check to see if the concentration of the reagent has increased or if it is a different chemical altogether. Either of these could require a longer residence time to react. Check to see if the electrode is being affected by its environment. Is the pH so high that it causes an alkalinity error or so low that it causes an acidity error? Has the temperature of the wastewater changed? Is the velocity past the pH electrode fast enough to provide a quick response time?

A review of all of these factors should then reveal the problem. With the problem identified, you proceed in a step-wise manner to correct the problem or modify the equipment to meet the new conditions.

7.14 SUMMARY

pH adjustment is a common method of either directly meeting discharge limits, optimizing the process reaction that follows, or is part of optimizing the ongoing reaction. While common, it is one of the most difficult processes to control. Extremes in temperature, concentration of acids, alkali, or salt, velocity of the wastewater past the electrode, variation of the influent pH and buffering capacity, or steepness of the titration all interact and cause problems with controlling pH. Adequate mixing, equalization or two or more tanks in series, reactor residence time, and the proper controller influence the adequacy and responsiveness of the pH adjustment system.

There are four basic types of pH adjustment systems. They include simple neutralization of acidic and basic wastes, adjustment of the pH to optimize a following downstream treatment process, pH adjustment for the precipitation of heavy metals, and optimizing the reaction time for a pH dependent conversion of cyanide to carbon dioxide and nitrogen and hexavalent chromium to trivalent chromium hydroxide. As the operator, you are now equipped with an understanding of the basic requirements of the pH adjustment system and the effects of the various operating guidelines. You now have a logical approach to identify and solve a pH problem.

7.15 ADDITIONAL READING

1. *TREATMENT OF METAL WASTESTREAMS* in this series of training manuals. Available from Office of Water Programs, California State University, Sacramento, 6000 J Street, Sacramento, CA 95819-6025. Price $25.00.

2. McMillan, Gregory K., *ADVANCED pH MEASUREMENT AND CONTROL*, Third Edition. Available from Instrument Society of America, 67 Alexander Drive, PO Box 12277, Research Triangle Park, NC 27709. Order No. 1556174837. Price: ISA members, $72.00; nonmembers, $79.00.

QUESTIONS

Write your answers in a notebook and then compare your answers with those on page 308.

7.13A List the four basic steps for troubleshooting pH adjustment systems.

7.13B What might cause one electrode to lag behind the other two in a three-electrode system?

7.13C How might changes in the manufacturing processes affect pH adjustment?

END OF LESSON 3 OF 3 LESSONS
on
PRELIMINARY TREATMENT

Please answer the discussion and review questions next.

TABLE 7.6 SOURCES OF pH ELECTRODE ERRORS AND RESPONSE SYMPTOMS[a,b]

Source of Error	Response Symptom
Measurement Electrode:	
Bulb broken	No response (4 to 6 pH)
Fill contamination	No response (4 to 6 pH)
Bulb abrasion	Slow, erratic, shortened span and upscale pH
Bulb dehydration	Slow, erratic, shortened span and upscale pH
Bulb etching	Slow, erratic, shortened span and upscale pH
Partial bulb coating	Very slow
Complete bulb coating	No response
Low temperature	pH increases as temperature decreases
Reference Electrode:	
Bulb broken	Drift upscale or downscale
Fill contamination	Drift upscale or downscale
Partial bulb coating	Drift upscale (typically)
Complete bulb coating	Offscale up or down, depending on meter type
Thermistor:	
Open circuit	Shortened span
Shorted circuit	Lengthened span
Solution:	
Acidic solvent (no water)	pH offscale downward
Basic solvent (no water)	pH offscale upward
Alcohol solvent (no water)	pH upscale
Hydrocarbon solvent (no water)	pH upscale and decreased lower and increased upper limits
Pure water	pH upscale and erratic
Composition changes	Drift upscale or downscale
Gas bubbles	Upscale noise
Low pH (acid error)	Shortened span at low pH end
High pH (alkalinity or sodium ion error)	Shortened span at high pH end
Low temperature	Shortened span
High temperature	Lengthened span
Terminals:[c]	
Short from M to R	Fixed at 7 pH
Broken electrode wire	Fixed at 7 pH
Short from M to ground	Fixed at 7 pH
Moisture on M	Stays near 7 pH
Moisture on R	Upscale pH

[a] Adapted from McMillan, Gregory K., *pH CONTROL*, Instrument Society of America, Research Triangle Park, NC, 1984.
[b] Due to numerous possible errors, proper diagnosis requires knowledge of the electrical symptoms and response symptoms.
[c] M = Meter; R = Electrode Resistance

DISCUSSION AND REVIEW QUESTIONS
Chapter 7. PRELIMINARY TREATMENT
(EQUALIZATION, SCREENING, AND pH ADJUSTMENT)
(Lesson 3 of 3 Lessons)

Write the answers to these questions in your notebook. The question numbering continues from Lesson 2.

10. How are strong and weak acids and bases used to adjust the pH of industrial wastewaters?

11. Why do some pH adjustment systems use three pH electrodes?

12. What are the main elements of a successful pH control strategy?

13. Why might two or more pH adjustment tanks be needed?

SUGGESTED ANSWERS
Chapter 7. PRELIMINARY TREATMENT
(EQUALIZATION, SCREENING, AND pH ADJUSTMENT)

ANSWERS TO QUESTIONS IN LESSON 1

Answers to questions on page 260.

7.1A Side-line equalization diverts flows that are greater than the average flow volume or concentration to a holding tank for subsequent release when flows are below average. In-line equalization blends all incoming flows in the holding tank and discharges at the average flow rate.

7.1B Influent to an IWTS may vary in flow and concentration of pollutants because of batch discharges, such as spent plating baths, or the batch release of process wastewater. The periodic blowdown of a cooling tower may not add a significant load of biochemical oxygen demand (BOD), but will certainly add a large quantity of flow to the influent. Spills and dumps can also cause variations.

7.1C The primary benefit of flow equalization for an industrial wastewater treatment system is an improvement in the effluent quality.

7.1D IWTS recycle streams should be routed to the equalization tank so they will not overload the IWTS with a batch load of concentrated pollutants.

Answers to questions on page 261.

7.1E If solvents are not removed prior to equalization, they may become volatile in the equalization tank causing a potential violation of air pollution regulations. Mixing in the tank could emulsify the solvents, making them harder to remove in the next treatment process.

7.1F Precautions that should be taken when using a collection system for equalization of flows include:

1. Maintaining adequate flow velocity to prevent deposition of solids.
2. Installing an overflow protection device and alarm.
3. Making sure that discharges into the sewer do not contain concentrated or incompatible wastes that could cause a violent reaction when combined in the sewer.

Answers to questions on page 265.

7.2A If the cumulative influent flow volume to an equalization tank is 47,250 gallons during a 15-hour manufacturing cycle, the average discharge rate is:

$$\text{Avg Flow, gal/hr} = \frac{\text{Cumulative Volume, gal}}{\text{Time, hr}}$$

$$= \frac{47,250 \text{ gal}}{15 \text{ hr}}$$

$$= 3,150 \text{ gal/hr}$$

$$\text{Avg Flow, GPM} = \frac{\text{Avg Flow, gal/hr}}{60 \text{ min/hr}}$$

$$= \frac{3,150 \text{ gal/hr}}{60 \text{ min/hr}}$$

$$= 52.5 \text{ GPM}$$

7.2B Additional factors that must be considered when determining the volume of an equalization tank include plant recycle streams, operating levels of the mixing or aeration equipment, and a safety factor.

Answers to questions on page 267.

7.2C Equalization tanks should be equipped with baffles to promote complete mixing and prevent short-circuiting. Piping to and from the equalization tank is also important. The inlet should be near the mixer(s) to disperse the flow as quickly as possible, and the outlet should be configured so as to minimize short-circuiting. Also, check to see that the mixer rotation does not promote short-circuiting by moving the wastewater toward the outlet.

7.2D A solution for higher-than-design average flows is to operate the treatment system longer than the manufacturing cycle.

7.2E Precautions that should be taken when changing the pumping rate from the equalization tank include making the change in small increments of five to ten percent of the current rate. Gradual changes minimize hydraulic surges downstream and allow the chemical feed systems to smoothly respond to a change in load. The operator will also need to adjust the settings on any flow-paced processes, such as polymer addition.

ANSWERS TO QUESTIONS IN LESSON 2

Answers to questions on page 268.

7.5A Screening is used to pretreat industrial wastewater because screening captures solids that may have value and allows downstream treatment processes to function more effectively.

7.5B Floatable, suspended, or settleable solids can be effectively screened from wastewater.

7.5C Screening wastewater prior to discharge to a POTW normally lowers sewer-use fees.

Answers to questions on page 273.

7.5D Bar screen designs without underwater drive components are preferred at most industrial plants due to reliability and ease of maintenance.

7.5E Shutdown procedures for a mechanical bar screen include stopping flow through the channel, shutting off the unit, draining the channel, and thoroughly hosing down and cleaning the bars and cleaning mechanism.

7.5F The major components of an automatically cleaned coarse screen (bar screen) that might be sources of poor operation or breakdowns include (1) carriage and cleaning rake assembly, (2) motor and reducer bar screen drive, and (3) wiper assembly.

Answers to questions on page 274.

7.5G Fine screens capture small particles that have value for resale or recycling, or solids that would clog or overload subsequent treatment processes.

7.5H Static screens are especially appropriate for the pretreatment of fibrous and non-grease-bearing wastes.

7.5I Static screens require periodic manual cleaning with a brush or hose.

Answers to questions on page 282.

7.5J The major advantage of rotating fine screens over static fine screens is that rotating fine screens are self-cleaning or largely self-cleaning.

7.5K Internally fed rotating screens are widely used by food processors. They are particularly appropriate for wastestreams with very large solids or with a heavy concentration of solids, and they are also suitable for coarse, fibrous wastes.

7.5L Disk filters are especially suited to fibrous solids separation, such as might be required in pulp and paper, textile, laundry, or similar wastewaters. They are generally inappropriate for greasy applications.

Answers to questions on page 284.

7.5M Maintenance records are important to safeguard the equipment warranty, make spare parts requirements predictable, and highlight any recurring problems so they can be logically addressed.

7.5N Hazards that an operator may be exposed to when working around open channels or tanks include slipping or tripping and falling into the wastewater or moving machinery, and exposure to explosive materials or gases.

ANSWERS TO QUESTIONS IN LESSON 3

Answers to questions on page 290.

7.6A Industrial wastewaters require pH adjustment to optimize the chemical reaction; for example, precipitate heavy metals, oxidize cyanide, reduce hexavalent chromium, optimize biological activity, prevent corrosion, reduce compounds (sulfate to hydrogen sulfide), or neutralize an acidic or basic solution to bring it within discharge limits.

7.7A pH is an expression of the intensity of base or acid condition of a liquid. Mathematically, pH is the logarithm (base 10) of the reciprocal of the hydrogen ion activity,

$$pH = \log \frac{1}{[H^+]}$$

7.7B pH adjustment depends on amount of buffering with other compounds, chemical reaction rate, temperature, mixing conditions, and the starting pH.

Answers to questions on page 294.

7.8A If all three pH measurement electrodes in a system agree to within 0.1 pH unit for more than a few minutes, there is the possibility that the electrodes are coated, broken, or still have their protective caps in place.

7.8B Factors that influence pH electrode accuracy and response time include temperature, absorption, face velocity and fouling of the probe, buffering of the solution, alkalinity, and acidity.

Answers to questions on page 298.

7.9A Three elements of a successful pH control strategy include:

1. Equalize the waste before trying to adjust it.
2. Provide enough residence time in the tank for the reaction and ensure good mixing.
3. Optimize your controller by minimizing dead time and by understanding how the controller responds.

7.9B Dead time is the time interval between the addition of the pH adjustment chemical (reagent) and the first observable change in effluent measurement.

Answers to questions on page 299.

7.10A A pH adjustment instrumentation and control system consists of:

1. The primary element (pH electrode)
2. The transmitter
3. The pH controller
4. The final element or control variable

7.10B To help ensure reliable performance of a pH controller, the operator should reduce dead time as much as possible and periodically check the characteristics of the wastewater to make sure the titration curve reflects the actual wastewater composition.

Answers to questions on page 302.

7.11A Common pH adjustment systems include:

1. pH adjustment for discharge
2. pH adjustment for biological treatment
3. pH adjustment for heavy metal removal
4. pH adjustment for oxidation and reduction

7.11B The most important factor for biological treatment is to keep the pH as consistent as possible, preferably slightly alkaline.

7.11C When the wastewater contains multiple heavy metals, each with a different optimum pH, the operator should optimize the pH for the heavy metal with the lowest discharge limit and determine if the other metals will be removed below their limits. If this approach does not remove all of the heavy metals, a two-stage system may be necessary to precipitate and remove one metal and then adjust the pH to the second optimum pH for the other metals.

Answers to questions on page 304.

7.13A The four basic steps for troubleshooting a pH adjustment system are:

1. Check the mechanical operation of the equipment.
2. Check the operation of the pH electrodes.
3. Determine whether the process wastewater has changed.
4. Review the basics of a good pH control system.

7.13B If one electrode is coated, it will lag behind the other two in a three-electrode system.

7.13C Changes in the manufacturing processes could change the characteristics of the process wastewaters. If this occurs, new pH titration curves will need to be prepared and the pH controller may need to be adjusted using the new titration curves.

CHAPTER 8

PHYSICAL–CHEMICAL TREATMENT PROCESSES
(COAGULATION, FLOCCULATION, AND SEDIMENTATION)

by

Paul Amodeo

Ross Gudgel

James L. Johnson

Paul J. Kemp

TABLE OF CONTENTS

Chapter 8. PHYSICAL–CHEMICAL TREATMENT PROCESSES (COAGULATION, FLOCCULATION, AND SEDIMENTATION)

		Page
OBJECTIVES		314
WORDS		315

LESSON 1

8.0	NEED TO REMOVE SOLIDS FROM INDUSTRIAL WASTESTREAMS	321
8.1	SOLIDS REMOVAL FROM WASTESTREAMS USING CHEMICALS	321
8.2	HOW COAGULATION/FLOCCULATION WORKS	322

DISCUSSION AND REVIEW QUESTIONS 326

LESSON 2

8.3	CHEMICALS USED TO IMPROVE SETTLING		327
	8.30	Aluminum Sulfate (Dry) ($Al_2(SO_4)_3 \cdot 14\ H_2O$)	327
	8.31	Aluminum Sulfate (Liquid)	327
	8.32	Ferric Chloride	328
	8.33	Lime	328
	8.34	Polymeric Flocculants	328
8.4	SELECTING CHEMICALS AND DETERMINING DOSAGES		331
	8.40	The Jar Test	331
	8.41	Jar Testing Procedures	333
		8.410 Preparation	333
		8.411 General Setup Considerations	333
	8.42	Example Calculations	334
	8.43	Sampling	335
	8.44	Addition of Test Chemicals	336
	8.45	Specific Tests for Chemical Selection, Dosage, and Process Control	336
		8.450 Preliminary Screening Test	336
		8.451 Dosage Testing	338
		8.452 Full-Scale Trial	339
		8.453 System Performance Optimization	339

312 Treatment Plants

8.46	Procedure for Plants Without Laboratory Facilities	339
8.47	Phosphate Monitoring	340
8.48	Safe Working Habits	340
8.49	Summary	341

DISCUSSION AND REVIEW QUESTIONS ... 341

LESSON 3

8.5 PHYSICAL–CHEMICAL TREATMENT PROCESS EQUIPMENT ... 342

- 8.50 Chemical Storage and Mixing Equipment ... 342
- 8.51 Chemical Feed Equipment ... 342
  - 8.510 Types of Chemical Metering Equipment ... 342
  - 8.511 Selecting a Chemical Feeder ... 351
  - 8.512 Reviewing Chemical Feed System Designs ... 351
  - 8.513 Chemical Feeder Start-Up ... 355
  - 8.514 Chemical Feeder Operation ... 355
  - 8.515 Shutting Down Chemical Systems ... 355
- 8.52 Coagulant Mixing Units ... 358
- 8.53 Flocculators ... 358
- 8.54 Clarifiers ... 358
  - 8.540 Types of Clarifiers ... 358
  - 8.541 Operating Guidelines ... 358
  - 8.542 Clarifier Efficiency ... 366
- 8.55 Dissolved Air Flotation (DAF) Thickeners ... 367
  - 8.550 Factors Affecting Dissolved Air Flotation ... 369
  - 8.551 Operating Guidelines ... 369
    - 8.5510 Solids and Hydraulic Loadings ... 369
    - 8.5511 Air to Solids (A/S) Ratio ... 370
    - 8.5512 Recycle Rate and Sludge Blanket ... 371
  - 8.552 Normal Operating Procedures ... 371
  - 8.553 Typical Performance ... 372
  - 8.554 Troubleshooting ... 373

8.6 OPERATION, START-UP, AND MAINTENANCE ... 374

- 8.60 Operating Strategy ... 374
- 8.61 Start-Up and Maintenance Inspection ... 374
- 8.62 Actual Start-Up ... 375

	8.63	Normal Operation	376
	8.64	Abnormal Operation	376
	8.65	Troubleshooting	377
	8.66	Safety	377

DISCUSSION AND REVIEW QUESTIONS 379
SUGGESTED ANSWERS 379

OBJECTIVES

Chapter 8. PHYSICAL–CHEMICAL TREATMENT PROCESSES (COAGULATION, FLOCCULATION, AND SEDIMENTATION)

Following completion of Chapter 8, you should be able to:

1. Explain the role of coagulation and flocculation processes in removing suspended solids from industrial wastestreams.
2. Describe various types of sedimentation basins and how they work.
3. Develop an operational strategy for physical–chemical treatment processes.
4. Safely start up, operate, maintain, and shut down the processes.
5. Perform a jar test.
6. Recognize abnormal operating conditions and troubleshoot to determine the cause.
7. Select the most cost-effective chemicals and determine proper dosage.
8. Start up, operate, maintain, and shut down a chemical feed system.
9. Adjust chemical feed rates.
10. Troubleshoot a chemical feed system.
11. Safely store and handle chemicals.

WORDS

Chapter 8. PHYSICAL–CHEMICAL TREATMENT PROCESSES
(COAGULATION, FLOCCULATION, AND SEDIMENTATION)

AGGLOMERATION (uh-glom-er-A-shun)

The growing or coming together of small scattered particles into larger flocs or particles, which settle rapidly. Also see FLOC.

ALKALI (AL-kuh-lie)

Any of certain soluble salts, principally of sodium, potassium, magnesium, and calcium, that have the property of combining with acids to form neutral salts and may be used in chemical processes such as water or wastewater treatment.

ALKALINITY (AL-kuh-LIN-it-tee)

The capacity of water or wastewater to neutralize acids. This capacity is caused by the water's content of carbonate, bicarbonate, hydroxide, and occasionally borate, silicate, and phosphate. Alkalinity is expressed in milligrams per liter of equivalent calcium carbonate. Alkalinity is not the same as pH because water does not have to be strongly basic (high pH) to have a high alkalinity. Alkalinity is a measure of how much acid must be added to a liquid to lower the pH to 4.5.

ANHYDROUS (an-HI-drous)

Very dry. No water or dampness is present.

ANION (AN-EYE-en)

A negatively charged ion in an electrolyte solution, attracted to the anode under the influence of a difference in electrical potential. Chloride ion (Cl^-) is an anion.

ASPIRATE (AS-per-rate)

Use of a hydraulic device (aspirator or eductor) to create a negative pressure (suction) by forcing a liquid through a restriction, such as a Venturi tube. An aspirator may be used in the laboratory in place of a vacuum pump; sometimes used instead of a sump pump.

BOD (pronounce as separate letters)

Biochemical Oxygen Demand. The rate at which organisms use the oxygen in water or wastewater while stabilizing decomposable organic matter under aerobic conditions. In decomposition, organic matter serves as food for the bacteria and energy results from its oxidation. BOD measurements are used as a surrogate measure of the organic strength of wastes in water.

BASE

(1) A substance that takes up or accepts protons.

(2) A substance that dissociates (separates) in aqueous solution to yield hydroxyl ions (OH^-).

(3) A substance containing hydroxyl ions that reacts with an acid to form a salt or that may react with metals to form precipitates.

BULKING

Clouds of billowing sludge that occur throughout secondary clarifiers and sludge thickeners when the sludge does not settle properly. In the activated sludge process, bulking is usually caused by filamentous bacteria or bound water.

COD (pronounce as separate letters)

Chemical Oxygen Demand. A measure of the oxygen-consuming capacity of organic matter present in wastewater. COD is expressed as the amount of oxygen consumed from a chemical oxidant in mg/L during a specific test. Results are not necessarily related to the biochemical oxygen demand (BOD) because the chemical oxidant may react with substances that bacteria do not stabilize.

CAVITATION (kav-uh-TAY-shun)

The formation and collapse of a gas pocket or bubble on the blade of an impeller or the gate of a valve. The collapse of this gas pocket or bubble drives water into the impeller or gate with a terrific force that can cause pitting on the impeller or gate surface. Cavitation is accompanied by loud noises that sound like someone is pounding on the impeller or gate with a hammer.

CLARIFICATION (klair-uh-fuh-KAY-shun)

Any process or combination of processes the main purpose of which is to reduce the concentration of suspended matter in a liquid.

CLARIFIER (KLAIR-uh-fire)

A tank or basin in which water or wastewater is held for a period of time during which the heavier solids settle to the bottom and the lighter materials float to the surface. Also called settling tank or SEDIMENTATION BASIN.

COAGULANT (ko-AGG-yoo-lent)

A chemical that causes very fine particles to clump (floc) together into larger particles. This makes it easier to separate the solids from the liquids by settling, skimming, draining, or filtering.

COAGULANT (ko-AGG-yoo-lent) AID

Any chemical or substance used to assist or modify coagulation.

COAGULATION (ko-agg-yoo-LAY-shun)

The clumping together of very fine particles into larger particles (floc) caused by the use of chemicals (coagulants). The chemicals neutralize the electrical charges of the fine particles, allowing them to come closer and form larger clumps.

COLIFORM (KOAL-i-form)

A group of bacteria found in the intestines of warm-blooded animals (including humans) and also in plants, soil, air, and water. The presence of coliform bacteria is an indication that the water is polluted and may contain pathogenic (disease-causing) organisms. Fecal coliforms are those coliforms found in the feces of various warm-blooded animals, whereas the term "coliform" also includes other environmental sources.

COLLOIDS (KALL-loids)

Very small, finely divided solids (particles that do not dissolve) that remain dispersed in a liquid for a long time due to their small size and electrical charge. When most of the particles in water have a negative electrical charge, they tend to repel each other. This repulsion prevents the particles from clumping together, becoming heavier, and settling out.

COMPOSITE (PROPORTIONAL) SAMPLE

A composite sample is a collection of individual samples obtained at regular intervals, usually every one or two hours during a 24-hour time span. Each individual sample is combined with the others in proportion to the rate of flow when the sample was collected. Equal volume individual samples also may be collected at intervals after a specific volume of flow passes the sampling point or after equal time intervals and still be referred to as a composite sample. The resulting mixture (composite sample) forms a representative sample and is analyzed to determine the average conditions during the sampling period.

DETENTION TIME

(1) The time required to fill a tank at a given flow.
(2) The theoretical (calculated) time required for water to pass through a tank at a given rate of flow.
(3) The actual time in hours, minutes, or seconds that a small amount of water is in a settling basin, flocculating basin, or rapid-mix chamber. In septic tanks, detention time will decrease as the volumes of sludge and scum increase. In storage reservoirs, detention time is the length of time entering water will be held before being drafted for use (several weeks to years, several months being typical).

$$\text{Detention Time, hr} = \frac{(\text{Basin Volume, gal})(24 \text{ hr/day})}{\text{Flow, gal/day}}$$

or

$$\text{Detention Time, hr} = \frac{(\text{Basin Volume, m}^3)(24 \text{ hr/day})}{\text{Flow, m}^3/\text{day}}$$

DIAPHRAGM PUMP

A pump in which a flexible diaphragm, generally of rubber or equally flexible material, is the operating part. It is fastened at the edges in a vertical cylinder. When the diaphragm is raised, suction is exerted, and when it is depressed, the liquid is forced through a discharge valve.

ELECTROLYTE (ee-LECK-tro-lite)

A substance that dissociates (separates) into two or more ions when it is dissolved in water.

FLOC

Clumps of bacteria and particles, or coagulants and impurities, that have come together and formed a cluster. Found in flocculation tanks, sedimentation basins, aeration tanks, secondary clarifiers, and chemical precipitation processes.

FLOCCULATION (flock-you-LAY-shun)

The gathering together of fine particles after coagulation to form larger particles by a process of gentle mixing. This clumping together makes it easier to separate the solids from the water by settling, skimming, draining, or filtering.

HYDROLYSIS (hi-DROLL-uh-sis)

(1) A chemical reaction in which a compound is converted into another compound by taking up water.

(2) Usually a chemical degradation of organic matter.

LINEAL (LIN-e-ul)

The length in one direction of a line. For example, a board 12 feet (meters) long has 12 lineal feet (meters) in its length.

MATERIAL SAFETY DATA SHEET (MSDS)

A document that provides pertinent information and a profile of a particular hazardous substance or mixture. An MSDS is normally developed by the manufacturer or formulator of the hazardous substance or mixture. The MSDS is required to be made available to employees and operators or inspectors whenever there is the likelihood of the hazardous substance or mixture being introduced into the workplace. Some manufacturers are preparing MSDSs for products that are not considered to be hazardous to show that the product or substance is not hazardous.

MICRON (MY-kron)

μm, Micrometer or Micron. A unit of length. One millionth of a meter or one thousandth of a millimeter. One micron equals 0.00004 of an inch.

NAMEPLATE

A durable, metal plate found on equipment that lists critical operating conditions for the equipment.

NEUTRALIZATION (noo-trull-uh-ZAY-shun)

Addition of an acid or alkali (base) to a liquid to cause the pH of the liquid to move toward a neutral pH of 7.0.

OVERFLOW RATE

One factor of the design flow of settling tanks and clarifiers in treatment plants used by operators to determine if tanks and clarifiers are hydraulically (flow) over- or underloaded. Also called SURFACE LOADING.

$$\text{Overflow Rate, GPD/sq ft} = \frac{\text{Flow, gallons/day}}{\text{Surface Area, sq ft}}$$

or

$$\text{Overflow Rate, } \frac{m^3/day}{m^2} = \frac{\text{Flow, } m^3/day}{\text{Surface Area, } m^2}$$

POLYELECTROLYTE (POLY-ee-LECK-tro-lite)

A high-molecular-weight (relatively heavy) substance, having points of positive or negative electrical charges, that is formed by either natural or synthetic (manmade) processes. Natural polyelectrolytes may be of biological origin or obtained from starch products or cellulose derivatives. Synthetic polyelectrolytes consist of simple substances that have been made into complex, high-molecular-weight substances. Used with other chemical coagulants to aid in binding small suspended particles to larger chemical flocs for their removal from water. Often called a POLYMER.

POLYMER (POLY-mer)

A long-chain molecule formed by the union of many monomers (molecules of lower molecular weight). Polymers are used with other chemical coagulants to aid in binding small suspended particles to larger chemical flocs for their removal from water. Also see POLYELECTROLYTE.

PRECIPITATE (pre-SIP-uh-TATE)

(1) An insoluble, finely divided substance that is a product of a chemical reaction within a liquid.

(2) The separation from solution of an insoluble substance.

ROTARY PUMP

A type of displacement pump consisting essentially of elements rotating in a close-fitting pump case. The rotation of these elements alternately draws in and discharges the water being pumped. Such pumps act with neither suction nor discharge valves, operate at almost any speed, and do not depend on centrifugal forces to lift the water.

SCFM

Standard Cubic Feet per Minute. Cubic feet of air per minute at standard conditions of temperature, pressure, and humidity (0°C, 14.7 psia, and 50 percent relative humidity).

SEPTIC (SEP-tick) or SEPTICITY

A condition produced by bacteria when all oxygen supplies are depleted. If severe, the bottom deposits produce hydrogen sulfide, the deposits and water turn black, give off foul odors, and the water has a greatly increased oxygen and chlorine demand.

SHORT-CIRCUITING

A condition that occurs in tanks or basins when some of the flowing water entering a tank or basin flows along a nearly direct pathway from the inlet to the outlet. This is usually undesirable since it may result in shorter contact, reaction, or settling times in comparison with the theoretical (calculated) or presumed detention times.

SLAKE

To mix with water so that a true chemical combination (hydration) takes place, such as in the slaking of lime.

SURFACE LOADING

One factor of the design flow of settling tanks and clarifiers in treatment plants used by operators to determine if tanks and clarifiers are hydraulically (flow) over- or underloaded. Also called OVERFLOW RATE.

$$\text{Surface Loading, GPD/sq ft} = \frac{\text{Flow, gallons/day}}{\text{Surface Area, sq ft}}$$

or

$$\text{Surface Loading, } \frac{m^3/\text{day}}{m^2} = \frac{\text{Flow, } m^3/\text{day}}{\text{Surface Area, } m^2}$$

SUSPENDED SOLIDS

(1) Solids that either float on the surface or are suspended in water, wastewater, or other liquids, and that are largely removable by laboratory filtering.

(2) The quantity of material removed from water or wastewater in a laboratory test, as prescribed in *STANDARD METHODS FOR THE EXAMINATION OF WATER AND WASTEWATER*, and referred to as Total Suspended Solids Dried at 103–105°C.

TRUE COLOR

Color of the water from which turbidity has been removed. The turbidity may be removed by double filtering the sample through a Whatman No. 40 filter when using the visual comparison method.

TURBID

Having a cloudy or muddy appearance.

VISCOSITY (vis-KOSS-uh-tee)

A property of water, or any other fluid, that resists efforts to change its shape or flow. Syrup is more viscous (has a higher viscosity) than water. The viscosity of water increases significantly as temperatures decrease. Motor oil is rated by how thick (viscous) it is; 20 weight oil is considered relatively thin while 50 weight oil is relatively thick or viscous.

WEIR (WEER)

(1) A wall or plate placed in an open channel and used to measure the flow of water. The depth of the flow over the weir can be used to calculate the flow rate, or a chart or conversion table may be used to convert depth to flow. Also see PROPORTIONAL WEIR.

(2) A wall or obstruction used to control flow (from settling tanks and clarifiers) to ensure a uniform flow rate and avoid short-circuiting.

CHAPTER 8. PHYSICAL–CHEMICAL TREATMENT PROCESSES
(COAGULATION, FLOCCULATION, AND SEDIMENTATION)

(Lesson 1 of 3 Lessons)

8.0 NEED TO REMOVE SOLIDS FROM INDUSTRIAL WASTESTREAMS

As increasing demands are placed upon our nation's receiving waters, it has become necessary to set industrial wastewater treatment plant discharge standards at a level that cannot be consistently met by conventional secondary wastewater treatment plants or by many industrial metal removal facilities.

Some locations have stringent discharge requirements because more and more wastes are being discharged into the receiving waters and increasing demands are being placed on the waters by water users. At some locations, very stringent National Pollutant Discharge Elimination System (NPDES) permit discharge requirements are being imposed. To comply with these requirements, the effluent from a standard secondary treatment plant must receive additional or tertiary treatment. Improving solids removal from the effluent of secondary wastewater treatment plants may be accomplished by chemical addition or by several filtration processes. This chapter will review physical–chemical methods presently in use for improving solids removal from the effluent of secondary treatment plants and the use of these methods to remove metals from industrial wastestreams.

QUESTIONS

Write your answers in a notebook and then compare your answers with those on page 379.

8.0A Why do some locations have stringent discharge requirements?

8.0B What do the initials NPDES stand for?

8.1 SOLIDS REMOVAL FROM WASTESTREAMS USING CHEMICALS

Physical–chemical treatment is a three-step process consisting of (1) *COAGULATION*,[1] (2) *FLOCCULATION*,[2] and (3) liquid/solids separation. The three steps must occur in the proper sequence. During the coagulation phase, chemicals are added to the wastewater and rapidly mixed with the process flow. At this time, certain chemical reactions occur quickly, resulting in the formation of very small particles, usually called "pinpoint floc."

Flocculation follows coagulation and consists of gentle mixing of the wastewater. The purpose of the gentle or slow mixing is to produce larger, denser floc particles that will settle rapidly. The liquid is agitated slowly to ensure contact of coagulating chemicals with particles in suspension. Floc growth is accelerated by controlled particle collisions. Suspended particles gather together and form larger particles with higher settling velocities.

The liquid/solids separation step follows flocculation and is almost always conventional sedimentation by gravity settling, although other processes, such as dissolved air flotation, are used occasionally.

Sedimentation can be defined in a broad sense as those operations performed in which a suspension of particles is separated into a clarified liquid and a more concentrated suspension. Sedimentation can be physically located downstream of any process in which such a suspension is generated (such as a biological treatment process) or it may be autonomous (by itself), representing the bulk of the treatment process of a plant. The latter is most common in the industrial waste treatment facilities in which metals and large quantities of suspended matter exist prior to any treatment. In most cases, allowing the solids to settle at their own pace takes too long to be cost-effective. Consequently, chemical coagulation is generally used to enhance the settling quality of the suspension, thereby decreasing the detention time required to achieve the desired liquid clarification.

Care should be taken not to use the words flocculation and coagulation interchangeably. Coagulation is the act of adding and mixing a coagulating chemical to destabilize the suspended particles allowing the particles to collide and "stick" together, forming larger particles. Flocculation is the actual gathering together of smaller suspended particles into flocs, thus forming a more readily settleable mass.

---

[1] *Coagulation* (ko-agg-yoo-LAY-shun). The clumping together of very fine particles into larger particles (floc) caused by the use of chemicals (coagulants). The chemicals neutralize the electrical charges of the fine particles, allowing them to come closer and form larger clumps.

[2] *Flocculation* (flock-you-LAY-shun). The gathering together of fine particles after coagulation to form larger particles by a process of gentle mixing. This clumping together makes it easier to separate the solids from the water by settling, skimming, draining, or filtering.

A separate chemical treatment process can be added on to an existing primary or secondary treatment plant as a tertiary treatment process. Chemical treatment performed in this manner requires the construction of additional basins or tanks, which may significantly increase the capital cost of the treatment plant. However, chemical treatment can also be practiced by adding chemicals at specific locations in existing primary or secondary treatment plants (Figure 8.1). This approach is often called chemical addition, and it eliminates the need for constructing additional clarifiers.

Regardless of the form of the chemical treatment process (tertiary or chemical addition), the most important process control guidelines are:

1. Providing enough energy to completely mix the chemicals with the wastewater.

2. Controlling the intensity of mixing during flocculation.

3. Controlling the chemical(s) dose.

Filters (Chapter 9) often are installed after chemical treatment to produce a highly polished effluent. Also, chemicals may be added to reduce emergency problems such as those created by sludge *BULKING*[3] in the secondary clarifier, upstream equipment failure, accidental spills entering the plant, and seasonal overloads. Chemicals can be used effectively as a "band-aid" during problem situations with relatively minor capital expense.

Keep in mind that the addition of chemicals is usually meant to capture some additional solids; therefore, more sludge must be handled. Care must be taken in controlling dosage into the secondary system because large chemical additions may be toxic to the organisms in biological treatment processes. This will reduce the activity or even kill the organisms treating the wastes in the system.

Whenever applying chemicals, it is important to always know, understand, and carefully control the dosage. You must understand each chemical's characteristics so the chemical will be properly stored and safely handled. Many of the chemicals used are harmful, especially to the eyes. Safety is of the utmost importance when chemicals are stored or applied.

QUESTIONS

Write your answers in a notebook and then compare your answers with those on page 379.

8.1A What is coagulation?

8.1B What is flocculation?

8.1C Why might chemicals be used in a wastewater treatment plant in addition to removing solids from secondary effluents?

8.1D What precaution must be exercised when adding chemicals upstream from a biological treatment process?

8.2 HOW COAGULATION/FLOCCULATION WORKS

The coagulation needed to efficiently remove solids from a suspension such as wastewater involves both chemical (destabilization) and physical (mixing and *AGGLOMERATION*[4]) processes. Coagulation amounts to collecting together a number of small suspended solids particles in such a way that they form a larger unit that is big enough to fall out of the wastewater flow stream (settle) or be trapped within filtering media (filtration) without slipping through. It is critical that coagulation be accomplished in such a way that the solids particles form units strong enough to withstand subsequent handling without breaking up (shearing apart) again.

For suspended solids particles in the ultra-fine or *COLLOIDAL*[5] size range (1 to 2 *MICRONS*[6]), random physical contact between particles within the suspension would normally cause a "snowball" type of effect commonly known as coagulation. However, fine suspended solids particles tend to accumulate an electrical charge. Whether the charge is positive (+) or negative (–) depends on the composition of the particles involved. This accumulated charge has two effects: (1) electrostatic repulsion, and (2) coordination sphere formation. Electrostatic repulsion is the tendency of particles with the same electrical charge to push away from or repel each other. Coordination sphere formation results from the characteristic of water molecules to line up in an orderly manner in the vicinity of an electrical charge. With a colloidal-sized charged particle, the water molecules line up in semi-attached, onion-like layers, surrounding the particle with "shells" of water that tend to hold their physical shape.

In order to get coagulation to proceed, the electrostatic charges of the particles must be modified to reduce their tendency to repel each other. This is the process of destabilization by ionic charge neutralization (sometimes also referred to as electrostatic charge reduction). It is accomplished by adding chemicals with a charge opposite to the charge of the suspended particles. Wastewater solids particles tend to have a negative ionic charge and the wastewater tends to have a pH on the low side of pH 6.5. Sometimes, the chemicals added for pH adjustment will produce ions (truly dissolved charged atoms or molecules) having a charge opposite to the one that tends to disperse solids particles. The chemicals used for pH adjustment are com-

---

[3] *Bulking.* Clouds of billowing sludge that occur throughout secondary clarifiers and sludge thickeners when the sludge does not settle properly. In the activated sludge process, bulking is usually caused by filamentous bacteria or bound water.

[4] *Agglomeration* (uh-glom-er-A-shun). The growing or coming together of small scattered particles into larger flocs or particles, which settle rapidly. Also see FLOC.

[5] *Colloids* (KALL-loids). Very small, finely divided solids (particles that do not dissolve) that remain dispersed in a liquid for a long time due to their small size and electrical charge. When most of the particles in water have a negative electrical charge, they tend to repel each other. This repulsion prevents the particles from clumping together, becoming heavier, and settling out.

[6] *Micron* (MY-kron).   µm, Micrometer or Micron. A unit of length. One millionth of a meter or one thousandth of a millimeter. One micron equals 0.00004 of an inch.

Physical–Chemical Treatment 323

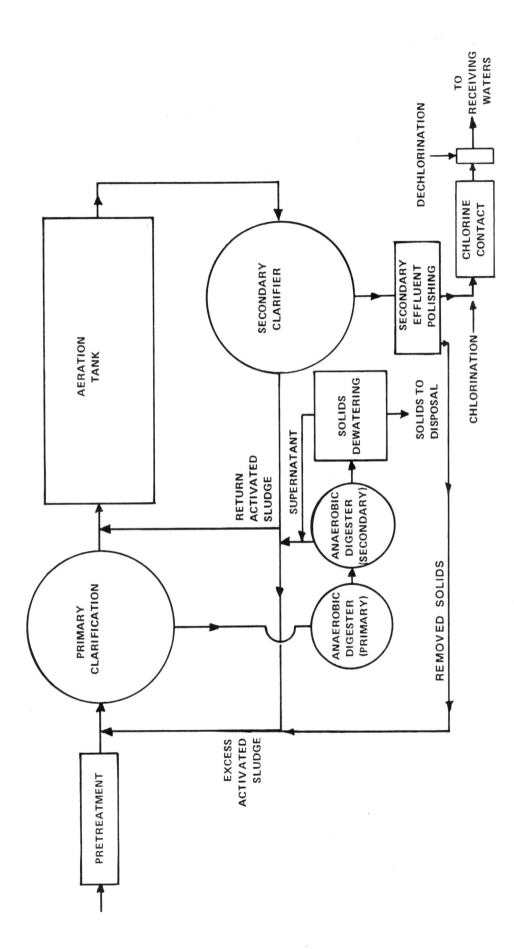

Fig. 8.1 Plan layout of a typical activated sludge plant with secondary effluent polishing process

monly a caustic (alkali or base) such as sodium hydroxide, soda ash or lime, or an acid such as sulfuric or muriatic acid. When the wastewater has a high phosphate content, the addition of lime ($Ca(OH)_2$) may be sufficient to both destabilize and coagulate the solids particles. Lime reacts with phosphate to precipitate calcium phosphate that collects extra calcium ions (Ca^{2+}). The positively charged phosphate precipitate adsorbs onto (sticks on the surface of) the negatively charged solids particles in the wastewater, neutralizing the electrostatic charge(s) and eliminating the repulsion effect. When lime addition is sufficient for coagulation requirements, the low cost of lime makes its use attractive.

Unfortunately (from a chemical conditioning standpoint), high phosphate wastewaters are not common. More often, aluminum (3+) (alum) or iron (3+) (ferric) metal salts are used to destabilize and coagulate wastewater solids. The cationic (positively charged) metal salts adsorb onto negatively charged wastewater solids and neutralize their negative charge(s). In addition, chemical reactions involving the metal salts produce insoluble (they do not dissolve easily) PRECIPITATES,[7] which also carry positive charges. The precipitates further assist in the destabilization reactions.

Once a colloidal suspension has been destabilized, coagulation can begin. Since this process mainly involves interparticle collisions, energetic mixing or stirring generally enhances the coagulation process. The physical turbulence created by rapid mixing helps collect the colloidal solids in a way that resembles churning butter or rolling up a snowball.

The initial process of rapid mix is important in two regards. First it ensures that there is a homogeneous mixture (complete mix) of suspended particles and coagulating chemicals. Second, it causes the needed contact between particles. The rapid mix operation is typically a relatively quick process. It should continue only long enough to create a homogeneous mixture because too long a rapid mix may break up and separate the forming floc. Therefore, the speed of the paddles becomes very important. Too rapid a speed may mechanically break up floc. On the other hand, too slow a speed may not provide the needed mixing and may promote dead spots within the tank where mixing does not occur. Floc remaining too long in a dead spot may begin to settle out in the mixing tank before it can be effectively handled by removal equipment in the downstream processes. The amount of energy applied to optimize coagulation will vary from system to system and may also vary with time in an individual system. Routine testing is the only means available to keep the performance optimized as operations proceed.

Proper coagulation produces agglomerated solids particle units that tend to hold together and do not easily redisperse. Depending on the concentration of suspended solids particles, their composition and the destabilization methods used, coagulation may or may not be sufficient by itself to allow satisfactory solids removal from wastestreams. For example, coagulation alone may be insufficient when suspended solids loadings are high (as with waste activated sludge and anaerobic digester sludge) or when a high-quality effluent must be produced (as in the case of wastewater reuse systems). In these circumstances, it is often desirable or necessary to go one step further using a process called flocculation to precondition the solids particles and enhance the sludge dewatering processes or the tertiary filtration processes.

Flocculation is a process of further collecting together the coagulated solids particles into still larger aggregated units (floc). This process most resembles stringing beads together on a long thread. It is most easily accomplished using synthetic POLYMERS,[8] which are, by comparison, much lighter and more fragile than the solids particles (floc) they gather together. Flocculation, therefore, is a more placid process of gentle mixing that requires special care. The results, however, can be spectacular. Flocculation can produce floc that is visible to the naked eye from a coagulated mixture that only looked cloudy or turbid before.

Stirring during flocculation is for the purpose of promoting maximum contact between suspended particles so that they will gather together or coagulate to form a larger floc mass. Paddle configuration (layout) and speed are such that the water and floc are encouraged to move slowly through the tank. The aggregates or flocs have a greater overall density after flocculation and can be more readily separated from the liquid portion. As in the case of the rapid mix, two dangers must be avoided during flocculation. Paddle speed must be sufficient to keep the floc from settling while at the same time it must not be so great as to shear and break up the floc formed.

---

[7] *Precipitate* (pre-SIP-uh-TATE). (1) An insoluble, finely divided substance that is a product of a chemical reaction within a liquid. (2) The separation from solution of an insoluble substance.

[8] *Polymer* (POLY-mer). A long-chain molecule formed by the union of many monomers (molecules of lower molecular weight). Polymers are used with other chemical coagulants to aid in binding small suspended particles to larger chemical flocs for their removal from water. Also see POLYELECTROLYTE.

By far the most precise application of flocculation is in the preparation of secondary effluents for tertiary filtration (often used to "polish" the effluent). Commonly, simple coagulation is insufficient to enlarge the suspended solids particles enough to meet pre-filtration conditioning requirements. This is especially true when the concentration of suspended solids in the wastewater leaving the secondary treatment process is very low (15 mg/L or less). Polymers are often used under these circumstances. Selection of the best type of polymer for a specific application depends on factors such as pH, conductivity (dissolved solids content), type and concentration of suspended solids, particle size ranges, type and amount of coagulant(s) applied, and what the next stage of treatment will be.

Two other important physical–chemical reactions occur during coagulation and flocculation that also contribute to the formation of floc that can be easily removed by sedimentation or other processes. The two reactions are: interparticle bridging and physical enmeshment.

Some polymers have the capacity to adsorb (stick) to sites on suspended particles and act as a bridge between them, thus enhancing floc formation in two ways: (1) the positively charged polymer neutralizes the negatively charged wastewater particle, and (2) the polymer acts as a bridge that connects solids particles. The effects of interparticle bridging are: (1) particles are attached into larger floc, and (2) the larger floc traps (physical enmeshment) additional solids within its mass as it collides with them during flocculation or sedimentation. Figure 8.2 illustrates how a positively charged polymer adsorbs onto negatively charged solids particles creating interparticle bridging.

Physical enmeshment also takes place when cationic (positively charged) metal salts of iron, aluminum, or (at high pH) magnesium are used for destabilization. These cations will combine with hydroxyl ions (OH$^-$) found in water from its natural dissociation as shown below.

$$H_2O \rightleftharpoons H^+ + OH^-$$

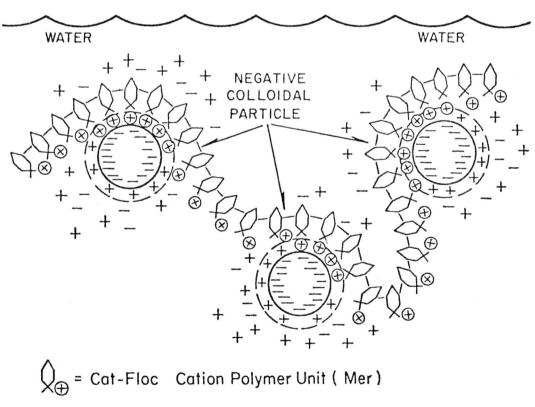

Fig. 8.2 Drawing illustrating the coagulation-flocculation reaction produced by Cat-Floc

This natural dissociation of water comes about either as a result of the alkalinity of the water or as a result of increasing the pH by the use of lime or soda ash. The metal ions combine with the hydroxide ions to form precipitates that tend to have an extended open flake form, commonly referred to as gelatinous. This means that they form in soft strings and flakes containing a lot of incorporated water and "void" space. As this gelatinous precipitate forms, it also collects any other solids in the vicinity as a co-precipitate, thus enlarging the forming floc. The gelatinous flake structure can be destroyed by severe turbulence, but vigorous mixing during coagulation produces a more granular gelatinous formation, which is more physically stable.

There is no substitute for testing the wastewater to select the right combination of chemicals and the proper dosage rates for effective coagulation and flocculation.

QUESTIONS

Write your answers in a notebook and then compare your answers with those on page 379.

8.2A   What is destabilization and why is it necessary?

8.2B   What is the typical particle charge when a wastewater tends to have a pH on the low (acidic) side of pH 6.5?

8.2C   What is the importance of mixing in the coagulation process?

END OF LESSON 1 OF 3 LESSONS
on
PHYSICAL–CHEMICAL TREATMENT PROCESSES

Please answer the discussion and review questions next.

DISCUSSION AND REVIEW QUESTIONS
Chapter 8.   PHYSICAL–CHEMICAL TREATMENT PROCESSES
(COAGULATION, FLOCCULATION, AND SEDIMENTATION)
(Lesson 1 of 3 Lessons)

At the end of each lesson in this chapter you will find some discussion and review questions. The purpose of these questions is to indicate to you how well you understand the material in the lesson. Write the answers to these questions in your notebook.

1. What is the difference between coagulation and flocculation?

2. What are the three most important process control guidelines in any chemical treatment process?

3. What is the meaning of "electrostatic repulsion"?

4. How do the precipitates of metal salts assist in the destabilization of suspended particles?

5. Why is paddle speed important during flocculation?

CHAPTER 8. PHYSICAL–CHEMICAL TREATMENT PROCESSES
(COAGULATION, FLOCCULATION, AND SEDIMENTATION)
(Lesson 2 of 3 Lessons)

8.3 CHEMICALS USED TO IMPROVE SETTLING

Secondary effluent quality may be improved by adding coagulant aids ahead of the secondary settling tanks or filters. Chemicals usually added are alum, ferric chloride, lime, or polyelectrolytes. Other useful chemicals may include sodium aluminate, ferric sulfate, ferrous chloride, and ferrous sulfate. These chemicals may be used alone or in combinations as determined by laboratory testing (Section 8.4, "Selecting Chemicals and Determining Dosages") and the results obtained in actual plant operation.

8.30 Aluminum Sulfate (Dry) ($Al_2(SO_4)_3 \cdot 14\ H_2O$)

Alum may be purchased in varying grades identified as lump, ground, rice, and powdered. Lump alum consists of lumps varying in size from 0.8 inch to 8 inches (2 to 20 cm) in diameter and is rarely used due to its irregularity in size and the difficulties of applying and achieving a satisfactory dose. Ground (granulated) alum is a mixture of rice-size material and some fines (very small particles). This form of alum is used by the majority of the water and wastewater plants. Ground alum feeds easily and does not bulk (stick together) in the hoppers if kept free of moisture or water. Also, ground alum does not require special protection of the hopper interiors from corrosion and wear.

Commercial filter alum (ground alum) has a bulk density of 60 to 70 lbs/cu ft (1.0 to 1.1 gm/cu cm) and is shipped in 100-pound (45-kg) bags or in large quantities (20 tons or 18,000 kg) by bulk trucks and railroad hopper cars. Special care should be taken to prevent alum from getting damp or it will cake into a solid lump. All mechanical equipment, such as conveyors, should be run until well cleaned of all alum before shutting down because the alum can harden and jam the equipment. Keep alum dry by storing it inside a well-ventilated location. Storage bins should have a 60-degree slope to the bottom to ensure complete emptying. Be sure the alum will not get wet when hosing down equipment or washing floors.

Both dry dust and liquid forms of alum are irritating to the skin and mucous membranes and can cause serious eye injury. Wear protective clothing to protect yourself from dust, splashes, or sprays. Proper clothing consists of a face shield, rubber or plastic gloves, rubber shoes, and rubber clothing when working around alum dust. Prevent inhalation by the use of local exhaust or approved respiratory protection.

8.31 Aluminum Sulfate (Liquid)

Alum is also available as a liquid. One gallon weighs about 11 pounds and contains the equivalent of 5.4 pounds (2.45 kg) of dry aluminum sulfate (49% as $Al_2(SO_4)_3 \cdot 14\ H_2O$). Obtain a chemical analysis from the supplier for each delivery to determine the exact content. Liquid alum is preferred by operators because of its ease of handling; however, you must pay shipping costs for transporting the water portion.

Liquid alum is shipped in 2,000- to 4,000-gallon tank trucks or 55- to 110-ton railroad tank cars.

Alum becomes very corrosive when mixed with water; therefore, dissolving tanks, pumps, and piping must be protected. Liquid storage tanks must be constructed of corrosion-resistant material such as rubber-lined steel or fiberglass. Bulk liquid alum storage tanks must be protected from extreme cold because normal commercial concentrations will crystallize at temperatures below 32°F (0°C) and freeze at about 18°F (–8°C).

Alum will support a bacterial growth and cause sludge deposits in feed lines if wastewater is used to transport the alum to the point of application. These growths and deposits can completely plug the chemical feed line. This problem can be reduced by maintaining a high velocity to scour the line continuously. Also, a concentrated alum solution will not support the bacterial growth so reducing the amount of carrier water helps.

Alum reduces the alkalinity in the water being treated during the coagulation process. Hydrated lime, soda ash, or caustic soda may be required if there is not enough natural alkalinity present to satisfy the alum dosage.

When added to water, alum is acidic; a 1 percent solution will have a pH of 3.5. Overdosing of alum may depress the pH to a point that it will reduce the biological activity in the secondary system. Also, this lowered pH may allow the chlorine added as a disinfectant to further depress the pH and affect the aquatic life in the receiving waters. This, along with chemical costs, emphasizes the need to maintain proper chemical dosages and closely monitor effluent quality.

Regularly analyze the bulk chemicals to determine if the concentration has changed. If the concentration has changed, you will need to adjust the chemical feed rate. Also, test the effluent quality to determine if sufficient solids are being removed or if an adjustment in the chemical feed rate might be helpful.

Liquid alum can be very hazardous. A face shield and gloves should be worn around leaking equipment. The eyes or skin should be flushed and washed upon contact with liquid alum. Liquid alum becomes very slick upon evaporation and therefore spills and leaks should be avoided.

QUESTIONS

Write your answers in a notebook and then compare your answers with those on page 380.

8.3A What are the four most common chemicals added to improve settling of solids?

8.3B Why should alum be kept dry?

8.3C Why should all mechanical equipment, such as conveyors, be run until well cleaned of all alum before shutting down?

8.32 Ferric Chloride

Ferric chloride is available in three forms—*ANHYDROUS*,[9] crystal hydrated, and liquid. The dry forms will absorb enough moisture from the air to quickly form highly corrosive solutions.

Anhydrous ferric chloride is shipped in 150- and 350-pound drums. Once these drums are opened, they should be completely emptied to prevent the formation of corrosive solutions. Care must be taken when making up solutions because the temperature of the solution will rise as the chemical dissolves.

Crystal ferric chloride is shipped in 100-, 400-, or 450-pound drums. Store the crystals in a cool, dry place and always completely empty any opened containers. The heat rise in dissolving crystal ferric chloride is much lower than that of anhydrous ferric chloride and is not a problem.

Liquid ferric chloride (35 to 45% $FeCl_3$) is shipped in rubber-lined tank cars or trucks (3,000 to 10,000 gallons). This chemical must be stored in corrosion-resistant tanks. If storage occurs at temperatures below 30°F (–1°C), it may be necessary to provide tank heaters or insulation to prevent crystallization.

Positive displacement metering pumps should be used for accurate measurements. Both the feeder and the lines must be corrosion-resistant.

All forms of ferric chloride will cause bad stains. This staining will occur on almost every material including walls, floors, equipment, and even operators.

Safety precautions required for handling ferric chloride in concentrated forms should be the same as those for acids. Wear protective clothing, chemical goggles, rubber or plastic gloves, and rubber shoes. Flush all splashes off clothing and skin immediately.

8.33 Lime

Hydrated lime (calcium hydroxide or $Ca(OH)_2$) is used to coagulate solids or adjust the pH to improve the coagulation process of other chemicals. Lime may be purchased in 50- or 100-pound bags or in bulk truck or railroad car loads. Lime should be stored in a dry place to avoid absorbing moisture. Bulk bin outlets should be provided with non-flooding rotary feeders. Hopper slopes may vary from 60 to 66 degrees.

Lime also may be purchased as anhydrous or quicklime, but must be *SLAKED*[10] before it can be used. Quicklime is more difficult to store because it will easily absorb moisture and cake into a solid clump. Quicklime is less expensive to purchase than lime; however, the added equipment for slaking and the requirement for increased operational safety must be considered.

Heat is generated when water is added to quicklime. If the controlled water supply fails and the water is shut off while the lime feed continues, boiling temperatures can be reached quickly. If a boiling reaction results, hot lime may cause the slaker to erupt and spew out hot lime. If high-temperature controls are properly installed, they should activate an alarm or shut down the unit. Mixers and pumps should be inspected frequently (daily) for wear because the lime slurry will rapidly erode or wear moving parts.

When transporting concentrated lime slurries in pipelines, a scale will build up on the inside of the pipe and eventually plug the line. A 2- to 3-inch (50- to 60-mm) diameter pipe may need replacing every year or two due to this scale. Rubber or flexible piping with easy access and short runs will permit cleaning by squeezing the walls and washing out the broken scale. Standby lines should be provided for use during the cleaning operation.

Lime is irritating to the skin, the eyes, the mucous membranes, and the lungs. Protect your eyes and lungs with approved full-face respiratory protection devices and wear protective clothing when working around lime.

8.34 Polymeric Flocculants

Polymeric flocculants are high-molecular-weight organic compounds with the characteristics of both polymers and *ELECTROLYTES*.[11] They are commonly called *POLYELECTROLYTES*.[12] These flocculants may be of natural or synthetic origin.

---

[9] *Anhydrous* (an-HI-drous). Very dry. No water or dampness is present.

[10] *Slake*. To mix with water so that a true chemical combination (hydration) takes place, such as in the slaking of lime.

[11] *Electrolyte* (ee-LECK-tro-lite). A substance that dissociates (separates) into two or more ions when it is dissolved in water.

[12] *Polyelectrolyte* (POLY-ee-LECK-tro-lite). A high-molecular-weight (relatively heavy) substance, having points of positive or negative electrical charges, that is formed by either natural or synthetic (manmade) processes. Natural polyelectrolytes may be of biological origin or obtained from starch products or cellulose derivatives. Synthetic polyelectrolytes consist of simple substances that have been made into complex, high-molecular-weight substances. Used with other chemical coagulants to aid in binding small suspended particles to larger chemical flocs for their removal from water. Often called a POLYMER.

Technically speaking, a polymer is any material that is composed of one single base unit that is chemically linked together with many more base units of the same type. In many ways, a polymer's base unit (properly named its "monomer") resembles a link in a chain. When a large number of these "links" are assembled into one single unit, the entire assembly is an entity with entirely different characteristics from the individual monomer "links" in their detached or unassembled form. It is this similarity to a chain with which we commonly have personal experience that has generated much of the polymer terminology, such as references to "polymer chains" when discussing individual polymer molecules or talking about "chain lengths" when describing the size of a polymer molecule.

All synthetic polyelectrolytes are classified on the basis of the type of charge on the polymer chain, the molecular weight (or length) of the polymer chain, and the charge density (spacing of the charges) along the chain. Negative-charge polymers are called "anionic" and positive-charge polymers are called "cationic." Polymers carrying no free electrical charge are "nonionic polyelectrolytes."

The sizes of polyelectrolytes are designated by their molecular weights. This is a chemical term used to describe the length of a polymer chain and has the units of gm/mole or MW/mole (Molecular Weight/mole). A medium-molecular-weight polymer will weigh <100,000 gm/mole. A high-molecular-weight polymer will range from 100,000 to 1,000,000 gm/mole, a very high-molecular-weight polymer will be from 1,000,000 to 10,000,000 gm/mole and ultrahigh-molecular-weight polymer will be >10,000,000 gm/mole.

Charge densities refer to the relative number of locations along the polymer chain that actually carry a charge. The more precise specification reports the percentage of charged sites out of the maximum number of possible locations. Polymers are also often grouped as low charge (<25% charge), medium charge (26–50% charge), or high charge (>50% charge).

A great assortment of polyelectrolytes are available to the wastewater treatment facility operator. They may be applied alone or in combination with other chemicals to aid coagulation. With this large selection of polymers available, it is possible to find a beneficial combination for almost all conditions.

Because of the variety of polymers available, a method of cataloging polymers has been developed called a polymer map. This can be in the form of a table or an actual graphical "map" (Figure 8.3) that organizes polymers according to their design or production specifications. The polymer map is actually a logarithmic graph of polymer specifications. This kind of map is particularly useful when an initial set of polymer selection studies is undertaken because it helps in determining which polymers may work well in a particular application.

Across the upper portion of the map (horizontal "x" axis) is shown the range of polymer ionic charges. These charges range from −100% (minus 100 percent) to +100% (plus 100 percent). The definition of a −100% value is that every available site along a polymer chain in this group has an anionic (negative) charge when the polymer is properly dissolved in water. Conversely, a +100% value indicates that every available site along a polymer chain in this group has a cationic (positive) charge when the polymer is properly dissolved in water.

The center of the polymer map (vertical "y" axis) displays the uncharged (0%) designation assigned to nonionic polymers. Nonionic polymers do not develop a "free" ionic charge in solution. However, they have the ability to form bonds with both positively and negatively charged suspended solids particles.

Along the vertical "y" axis are marked the molecular weights (MW) of polymers. The values are presented in orders of magnitude (powers of 10) of the molecular weight (for example 10^4 = 10,000; 10^5 = 100,000), thus the scale is logarithmic.

For you to prepare a polymer map, you will need to know both the "charge density" (± % charge) and the molecular weight for each polymer you will be investigating. With such information, each polymer to be tested can be plotted as a point on the map. A completed polymer map will resemble a map of the stars in the sky—a scattering over all the possible range of polymers of interest to you. In Figure 8.3, some of the typical molecular weight "zones" of polymer types suitable for various process applications are identified (the irregularly shaped boxes).

Because of the wide selection of available products and the different and changing chemical characteristics of water being treated, extensive laboratory testing should be conducted before treating the entire plant effluent. The selection of a polymer for any individual application is highly dependent on the characteristics of the wastewater to be treated. Specific influences (factors) that require attention are: pH, conductivity (dissolved solids content), type and concentration of suspended solids, particle size ranges, type and amount of coagulant(s) applied, and what the next stage of treatment will be. Most polyelectrolyte suppliers have field representatives who will assist with the testing of their products at the treatment plant.

Polyelectrolytes are commonly used in very small doses, usually less than 1 mg/L. The effective dosage range is limited. An overdose can be worse than no polymer addition at all.

Polyelectrolytes are available as dry powder, as a liquid suspension, or as a true solution. Care must be taken when storing powders because they may quickly absorb moisture and become ineffective.

330 Treatment Plants

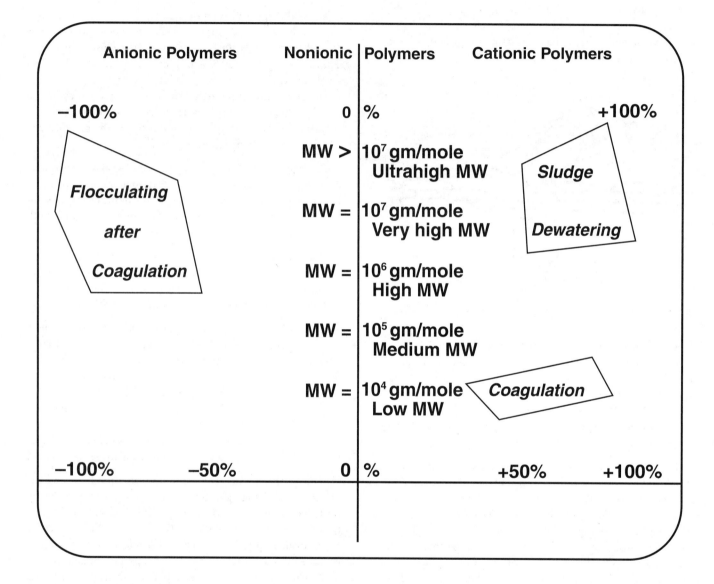

Definitions:
1. MW = Molecular Weight.
2. gm/mole = molecular weight of a substance expressed in grams.

Fig. 8.3   Polymer map

Solutions for treating wastewater must be made up in water prior to their use. This step is especially important with high- and ultrahigh-molecular-weight polymers supplied as powders or suspensions. With these materials, a minimum two-hour hydration time is necessary to get the polymer fully dissolved in water. True solution polymers do not require this kind of preparation and they are applied with dilution water in order to disperse or mix them with the wastewater to be treated. When mixing a batch, care is required to add a powder or a suspension liquid slowly while continuously mixing. If care is not taken, useless lumps will form that can clog feed pumps and lines.

Polyelectrolytes are considered nonhazardous to handle; however, good housekeeping must be practiced because polyelectrolytes will create an extremely slippery surface when wet. Clean up spills immediately. Salts, drying agents, and highly concentrated solutions will break down polyelectrolytes and help with a cleanup effort. There are commercially available polymer cleaning solutions based on isopropyl (rubbing) alcohol but a variety of alternatives are equally effective. Polymer manufacturers' Material Safety Data Sheets (MSDSs) recommend using inert absorbent materials to clean up polymer spills, such as sand or earth. Some polyelectrolytes have a low pH and can be corrosive to the makeup day or age tank (tank used to store solution).

QUESTIONS

Write your answers in a notebook and then compare your answers with those on page 380.

8.3D What safety precautions are required for handling ferric chloride in concentrated solutions?

8.3E How can the scale of lime that builds up on the inside of pipe be cleared?

8.3F What problems can be created when a polyelectrolyte is spilled?

8.3G How would you clean up a polyelectrolyte spill?

8.4 SELECTING CHEMICALS AND DETERMINING DOSAGES

In order to select the chemical products to be used in any treatment process application, a systematic accumulation of data is required. Factors that should be considered when selecting a chemical include minimum dosage, upper dosage limit, volume of sludge, cost, safety, availability of chemical supply, and reliability of chemical supply. When preparing to set up a chemical treatment program, a three-stage series of tests is used. These stages are: (1) preliminary screening, (2) dosage testing, and (3) full-scale trial. The common element through all stages is the jar test procedure, although specific details and objectives are different for each stage. Once the correct combination of chemicals has been selected for actual process use, jar test procedures for system performance optimization (similar to full-scale trial tests) are regularly used by the operator to maintain the best possible chemical use efficiency during routine process operation.

One invariable requirement in all jar test procedures is that the tests only have meaning if the tested wastewater exactly resembles the flow stream that will be ultimately treated by the chemicals being evaluated. The bottom line is that for the tests to be truly valid, they need to be performed at the chemical application location using freshly drawn samples from the actual flow stream. A delay of as little time as 30 minutes between sample collection time and jar testing can significantly affect the test results. It is equally important that all other test conditions match the actual system (plant) conditions.

1. *Preliminary Screening*

   Preliminary screening tests cover a wide variety of possible chemicals for treatment use and are more qualitative than quantitative. These tests are meant to show obvious differences between potential treatment chemicals and narrow the possibilities down to a few choices.

2. *Dosage Testing*

   Dosage testing is used to identify (bracket) the minimum-to-maximum amounts of potential treatment chemicals that will be needed. Performance results for a range of application rates are measured. These tests are used to predict the costs associated with the use of each potential treatment chemical.

3. *Full-Scale Trial*

   Following completion of dosage testing, precise estimates of chemical application guidelines for treating the process flow stream are available and can be used to begin treatment. While the chemicals are being applied, their performance is measured through the actual treatment process, along with a parallel series of jar tests. This allows two essential sets of data to be collected. First, the actual performance of the treatment process can be measured against minor, fine-tuning adjustments of the chemical(s) being applied. Second, the actual treatment results will show the operator how closely the jar tests predicted what would happen. Knowing this helps the operator interpret and use future test results.

4. *System Performance Optimization*

   In order to maintain optimum, cost-effective treatment process performance, regular monitoring is essential. In addition to regularly testing the effluent quality from the treatment process, regular jar testing of the influent just upstream of the chemical application point will enable the operator to confirm chemical treatment dosage rates and provide required dosage rate adjustment information when flow stream conditions change.

8.40 The Jar Test

Probably the single most valuable tool in operating and controlling a chemical treatment process is the variable speed, multiple station (or "gang") jar test unit (Figure 8.4). Various types of chemicals or different doses of a single chemical are added to sample portions of wastewater and all portions of the sample are rapidly mixed. After rapid mixing, the samples are slowly mixed

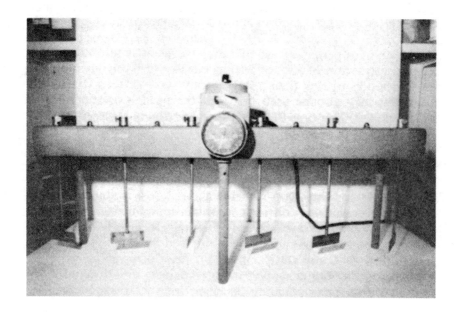

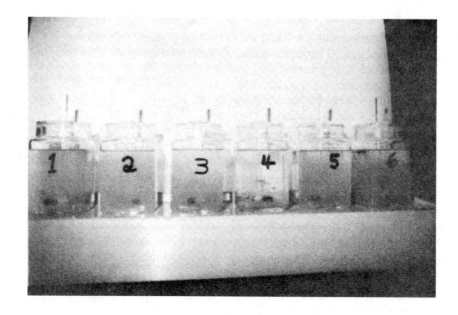

Fig. 8.4 Jar test units with mechanical (top) and magnetic (bottom) stirrers
(Source: EPA *PROCESS DESIGN MANUAL FOR SUSPENDED SOLIDS REMOVAL*)

to approximate the conditions in the plant. Mixing is then stopped and the floc formed is allowed to settle. The appearance of the floc, the time required to form a floc, and the settling conditions are recorded. The supernatant (liquid above the settled sludge) is analyzed for turbidity, suspended solids, and pH. With this information, the operator selects the best chemical or best dosage to feed on the basis of clarity of effluent and minimum cost of chemicals.

Jar test units vary somewhat in configuration depending on the manufacturer. These differences, such as the number of test stations, the size and shape of the test jars (round or square), stirrer controls, portability, internal illumination, and method of mixing, do not affect the performance of the device. However, two things must be known before a particular jar test unit is used: (1) the volume of the test jars, and (2) the speed rates of the stirrers. Some jar test units are supplied with a gauge that displays the turning rate of the stirrers in revolutions per minute (RPM). Other units have a control dial calibrated in RPM, and some are marked in percent of maximum speed (0–100%). In the latter case, it will be necessary for the operator to perform a calibration check so that the percentage numbers can be associated with the RPM of the stirrers. When testing flow streams containing low suspended solid concentrations, any type of stirring option is acceptable: rotating paddles, magnetically driven stirring bars, or vertical plungers. For flow streams containing high suspended solid concentrations (for example, waste activated sludge, digested sludge), magnetically driven stirring bars will be unsuitable because the resistance of the solids within the sample can decouple the stirring bar from the drive magnet.

QUESTIONS

Write your answers in a notebook and then compare your answers with those on pages 380.

8.4A What are the three stages of testing used to select chemical products for a treatment process?

8.4B Briefly describe the jar test procedure.

8.4C What two features must be known about a jar test unit before it is used?

8.41 Jar Testing Procedures

8.410 Preparation

In many cases, the preparation of the chemical(s) to be tested will be virtually the same. In some cases, especially with treatment system performance optimization testing, normal, "convenient" concentrations of the chemical(s) to be used may be selected. The descriptions that follow are intended as examples and are meant to be typical rather than specific. Always perform jar tests using the same chemicals that will be used to treat the water. Do not use laboratory grade chemicals because they may contain fewer impurities and therefore could produce slightly different results.

BASIC GUIDELINES

1. Jar test volume: 1,000 mL (= 1.0 L = 1,000 gm).

2. Test chemical stock solution strength: 0.5% (= 5,000 mg/L).

3. Working chemical solution strength: 0.1% (= 1,000 mg/L).

4. Rapid mix time at 140 to 160 RPM for 10 to 15 seconds for polymers; 3 to 5 minutes for aluminum or iron metal salts.

5. Slow mix time at 15 to 20 RPM = 1 to 3 minutes for polymers; slow mix is not used for aluminum or iron metal salts alone.[13]

6. One gallon of water weighs 8.34 lbs.

NOTE: The ranges given for items 4 and 5 above represent the ranges of normal operating conditions. The actual values selected for the tests will be fixed at single values based on experience with the type of jar test equipment used and the composition of the flow stream being examined.

8.411 General Setup Considerations

THE JARS

If the jars used in the jar test unit have a calibration mark at the 1,000 mL level, this is usually satisfactory for most testing applications. If there is no such mark on the jars, obtain and use a measuring container that is calibrated for 1,000 mL to fill the test jars. For precision testing, it is preferable to measure sample volumes using a graduated cylinder.

---

[13] Slow mix will be used with aluminum and/or iron metal salts during the second stage of a combination test where a flocculant (polymer) is added as a second chemical after the initial metal salt coagulation period.

TEST CHEMICALS

To prepare stock solutions of the chemicals to be tested:

1. Prepare clean 250-mL (half-pint) storage bottles with appropriate labels and the date of stock solution preparation.

2. Into each storage bottle, accurately measure 199 mL of deionized or distilled water.

3. Weigh exactly 1.0 gm of the selected chemical(s) into the respective storage bottle(s).[14]

4. Thoroughly mix the polymer and deionized or distilled water with constant stirring for at least 10 minutes. Polymer solutions will require an additional aging period of at least 2 hours with occasional stirring at 10 to 15 minute intervals.[15]

5. Store the chemical stock solutions in a cool place, convenient to the jar testing work area.[16]

To prepare a working solution of a chemical to be tested:

1. Prepare a clean, temporary container and mark it for identification.

2. Into the container, add exactly 20 mL of the chemical stock solution.

3. Into the same container, add 80 mL deionized or distilled water.

4. Thoroughly mix the contents of the container for at least 5 minutes, and let the solution stand for at least 5 minutes before use.

5. Discard any unused working solution at the end of the sequence of jar tests.

To calculate trial dosage levels for a jar test a generally applicable formula is:

$$\begin{pmatrix}\text{Volume}\\\text{(of stock}\\\text{solution)}\end{pmatrix}\begin{pmatrix}\text{Concentration}\\\text{(of working}\\\text{solution)}\end{pmatrix} = \begin{pmatrix}\text{Final Volume}\\\text{(of working}\\\text{solution)}\end{pmatrix}\begin{pmatrix}\text{Final Concentration}\\\text{(of working}\\\text{solution)}\end{pmatrix}$$

Which can be solved to find the final concentration as:

$$\begin{matrix}\text{Final Concentration}\\\text{(of working solution)}\end{matrix} = \frac{\begin{pmatrix}\text{Volume}\\\text{(of stock solution)}\end{pmatrix}\begin{pmatrix}\text{Concentration}\\\text{(of stock solution)}\end{pmatrix}}{\begin{matrix}\text{Final Volume}\\\text{(of working solution)}\end{matrix}}$$

Or, which can be solved to find the needed amount of stock solution as:

$$\begin{matrix}\text{Volume}\\\text{(of stock solution)}\end{matrix} = \frac{\begin{pmatrix}\text{Final Volume}\\\text{(of working solution)}\end{pmatrix}\begin{pmatrix}\text{Final Concentration}\\\text{(of working solution)}\end{pmatrix}}{\begin{matrix}\text{Concentration}\\\text{(of stock solution)}\end{matrix}}$$

EXAMPLES:

1. Using 1.0 mL of stock solution (5,000 mg/L) in a 1,000-mL jar, what is the dosage?

$$\begin{matrix}\text{Final Concentration}\\\text{(of jar test solution)}\end{matrix} = \frac{(1.0\text{ m}L)(5{,}000\text{ mg}/L)}{1{,}000\text{ m}L}$$

$$= 5.0\text{ mg}/L\text{ (in the jar)}$$

2. How much working solution is needed to provide 5.0 mg/L in the test jar?

$$\begin{matrix}\text{Volume}\\\text{(of working solution)}\end{matrix} = \frac{(1{,}000\text{ m}L)(5.0\text{ mg}/L)}{5{,}000\text{ mg}/L}$$

$$= 1.0\text{ m}L\text{ (into the jar)}$$

8.42 Example Calculations

When conducting jar tests and setting chemical feed rates, the following standard units of measurement are used:

- Jar test results are expressed in milligrams of chemical per liter of sample (mg/L) or parts chemical per million parts sample (ppm).

- Chemical feed rates are expressed in pounds of chemical delivered per hour (lbs/hr) or pounds of chemical delivered per day (lbs/day).

- Chemical feed pump settings are expressed in gallons of chemical fed per hour (gal/hr).

- Chemical pumping rates are expressed in gallons of chemical fed per hour (gal/hr).

- Chemical application rates are expressed in pounds of chemical applied per hour (lbs/hr) or pounds of chemical applied per day (lbs/day).

- Chemical dosage rates are expressed in milligrams of chemical per liter of wastewater (mg/L), parts of chemical per mil-

---

[14] When preparing solutions from liquid polymer supplies, a positive displacement measuring device will be needed, such as a hypodermic syringe, or a positive displacement micropipet. For preliminary screening tests, it is adequate to dispense 1.0 mL of the polymer supply into the storage bottle. For more precise testing, it is necessary to correct the volume for the density of the liquid polymer supply. (For example, if a polymer liquid has a density of 8.8 lbs/gallon (consult the polymer MSDS or other data sheet), the accurate volume of 1.0 gm of the polymer can be found using the ratio of the density of water (8.34 lbs/gallon) divided by the density of the polymer (8.8 lbs/gallon) = 0.94 mL.)

[15] When mixing a dry polymer into water, the water must be stirred rapidly while gradually sprinkling the polymer granules or powder over the water surface. Care must be taken to be sure that the granules of powder do not pile up in lumps on the water surface or cling to the sides of the mixing container, and that each granule of polymer is wetted individually.

[16] Generally, stock solutions of polymer are stable for 5 to 7 days and stock solutions should only be used for that length of time. After the designated period, leftover polymer stock solutions should be discarded. Metal salt solutions and soda ash are stable indefinitely and may be kept until the solutions are exhausted.

lion parts wastewater (ppm), or pounds of chemical per million pounds of wastewater (lbs/M lbs equivalent process flow).

The basic formula for calculating a chemical feed is:

Chem Feed Rate, lbs/day = (Flow, MGD)(Conc, mg/L)(8.34 lbs/gal).

EXAMPLE 1

A process flow rate is 347 GPM. What is the equivalent process flow rate in M lbs/day?

Known	Unknown
Flow Rate, GPM = 347 GPM	Flow Rate, M lbs/day

Convert the flow rate from GPM to M lbs/day.

1. Flow Rate, GPD = (Flow Rate, GPM)(1,440 min/day)
   = (347 GPM)(1,440 min/day)
   = 499,680 GPD (or 0.5 MGD)

2. Flow Rate, lbs/day = (Flow Rate, GPD)(8.34 lbs/gal)
   = (499,680 GPD)(8.34 lbs/gal)
   = 4,167,331 lbs/day

3. Flow Rate, M lbs/day = $\dfrac{\text{Flow Rate, lbs/day}}{1{,}000{,}000/M}$
   = $\dfrac{4{,}167{,}331 \text{ lbs/day}}{1{,}000{,}000/M}$
   = 4.167 M lbs/day

Therefore, 347 GPM = 4.167 M lbs/day equivalent process flow rate.

EXAMPLE 2

Results of jar testing indicate that a concentration of 5.0 mg/L of chemical should be used to treat a process flow of 0.5 MGD. What is the chemical application rate in lbs/day and lbs/hr? (*NOTE:* For each 1,000,000 lbs of water, it takes 1 lb of chemical to produce a 1 mg/L (1 ppm) chemical concentration in that water.)

Known	Unknown
Flow Rate, MGD = 0.5 MGD	Chemical Application Rate, lbs/day and lbs/hr
Chem Conc, mg/L = 5.0 mg/L	

1. Calculate the chemical application rate in lbs/day.

   Application Rate, lbs/day = (Flow, MGD)(Chem Conc, mg/L)(8.34 lbs/gal)
   = (0.5 MGD)(5.0 mg/L)(8.34 lbs/gal)
   = 20.85 lbs/day

2. Convert the application rate from lbs/day to lbs/hr.

   Application Rate, lbs/day = $\dfrac{\text{Application Rate, lbs/day}}{24 \text{ hr/day}}$
   = $\dfrac{20.85 \text{ lbs/day}}{24 \text{ hr/day}}$
   = 0.869 lbs chemical/hr

Chemicals are supplied or prepared in solutions of various strengths. Polymers are usually limited to 1.0 percent maximum solution strength so that the solution can be pumped (metered) easily.

EXAMPLE 3

Based on jar test results, 5.0 mg/L of chemical is applied to 0.5 MGD equivalent process flow rate (from Example 1 flow of 347 GPM). If the chemical feed solution strength is 0.4 percent, what is the chemical pumping rate in gallons per hour?

Known	Unknown
Chem Dosage, mg/L = 5.0 mg/L	Chem Pumping Rate, gal/hr
Equiv Process Flow Rate, MGD = 0.5 MGD	
Solution Strength, % = 0.4%	

Calculate the chemical pumping rate for a 0.4% solution.

Chem Pumping Rate, gal/hr = $\dfrac{(\text{Flow, MGD})(\text{Dosage, mg/}L)(8.34 \text{ lbs/gal})(100\%)}{(24 \text{ hr/day})(8.34 \text{ lbs/gal})(\text{Sol Strength, \%})}$

= $\dfrac{(0.5 \text{ MGD})(5.0 \text{ mg/}L)(8.34 \text{ lbs/gal})(100\%)}{(24 \text{ hr/day})(8.34 \text{ lbs/gal})(0.4\%)}$

= 26.05 gal/hr

8.43 Sampling

Before starting the jar test procedure, outline the objectives of your test and list the information you need to collect to reach your objectives. Choose a sampling point that is as near as possible to the location where treatment chemicals will be or are being fed into the flow stream. When testing a treatment process that has an existing chemical feeding system, take care to obtain the water test samples far enough upstream of the chemical injection point(s) so that there will not be any added chemicals in the test sample.

Next, use a container that will hold enough sample to fill all the jars in the jar test unit. If for some reason you can only collect small portions of the sample, all of the scoops of sample should be placed in a single container and stirred thoroughly before the sample is put into the individual jars. When running jar tests on high solids solutions (for example, waste activated sludge or digested sludge), stirring of the sample must be continuous in the holding container. It may be necessary to work with a partner to accomplish the stirring while the samples are being measured into the jars. Retain a portion of the untreated sample for laboratory analysis uses.

8.44 Addition of Test Chemicals

Performance of the treatment chemicals depends on chemical concentration, method of application, and time of reaction. It is therefore important to add the treatment chemicals to all of the test jars as nearly simultaneously as possible. To do this, prefill the chemical measuring device with the required amount of chemical working solution before beginning chemical addition. For *VISCOUS*[17] polymer solutions, a positive displacement device such as a hypodermic syringe or positive displacement micropipet is needed. For nonviscous solutions such as aluminum or iron metal salts, an ordinary pipet or graduated cylinder is satisfactory.

It is usually preferable to obtain assistance when injecting the chemical working solution into the jars. One person can handle up to two samples simultaneously, so it takes two people for four jars and three people for six jars. Simultaneous injection is most important with flocculant addition where the rapid mixing time is only a few seconds. When all the chemical working solutions are ready for injection, start the jar test unit stirrer at rapid mix speed and proceed with the chemical injection. As soon as the chemicals are in the test jars, begin the appropriate timing sequence and data logging. Some jar test devices have a chemical holder attached to each stirrer so all chemical doses can be applied simultaneously when the stirrers are placed in the jars.

8.45 Specific Tests for Chemical Selection, Dosage, and Process Control

8.450 Preliminary Screening Test

Objective: To select a limited number of potentially suitable chemical treatment products from a large number of choices.

In order to properly run a preliminary screening test series on polymer samples, it is essential to make use of the polymer map (Figure 8.3). The polymers to be screened are first marked on the map, and then a number of polymers are chosen with substantially different properties. An initial screening dosage level suitable to the flow stream to be examined is then selected. For secondary effluent flow streams, a dosage of 3.0 mg/L is usually a surplus and makes a suitable initial dosage. This will require application of 3.0 mL of polymer working solution (1,000 mg/L). For sludge samples, dosage ranges of 75 mg/L or higher are not uncommon, for which 15 mL of chemical stock solution (5,000 mg/L) will be needed. To calculate trial dosage levels for a jar test, refer to TEST CHEMICALS in Section 8.411.

1. Prepare a "jar test bench sheet" (similar to the one shown in Figure 8.5) on which to record specific data including, but not limited to:

   a. Source of water sample.

   b. Name of each treatment product to be tested.

   c. Dosage of treatment chemicals.

   d. Rapid mixing speed and duration.

   e. Slow mixing speed and duration.

   f. The sequence in which visible flocculation appears (which jar is first, which is second).

   g. The relative apparent size of the floc that forms.

   h. How quickly the floc settles to the bottom of the jar after stirring stops.

   i. How deep the floc is at the bottom of the jar after settling.

   j. The relative clarity of the supernatant water above the settled floc after settling.

2. Collect a suitable amount of sample and fill all the jars with sample.

3. Start the jar test unit stirrer and set it at the rapid mixing speed.

4. Fill the chemical measuring device with the appropriate amount of chemical working solution.

5. Add the chemical working solution to the test jars and start the timer. (Coagulation tests skip to step 7.)

6. Reduce the stirring speed after the appropriate interval. (Flocculation tests only.)

7. Record data items 1-f and 1-g during the stirring period(s).

8. At the end of the timing sequence, stop the stirrer and record data items 1-h, 1-i, and 1-j on the jar test bench sheet.

Before going on to additional preliminary screening tests, compare the results obtained so far with the polymer map layout and make a note of what kind of chemical gave the best kind of result. It is often useful to mark the relative performance of each polymer screened by rank (for example: 1 = best, 2 = next best) next to the associated polymer location on the polymer map. This will help you see the comparative effectiveness of the various polymer types used in your investigation. Using this as your guide to predicting results, select another set of test chemicals for the next stage of screening. It is usually a good idea to carry along the chemical having the best performance into the next round of screening to provide a cross reference between tests. Using the newly selected test chemicals, repeat the screening procedure beginning at step 2 above.

Use each set of preliminary screening tests to help you decide which, if any, new chemicals look promising. After a few screenings, most of the ineffective chemicals will be eliminated and four to six chemicals will have begun to show consistently favorable results. If there is a clear performance difference in the last few "best performers," especially if there is a single obvious best

---

[17] *Viscosity* (vis-KOSS-uh-tee). A property of water, or any other fluid, that resists efforts to change its shape or flow. Syrup is more viscous (has a higher viscosity) than water. The viscosity of water increases significantly as temperatures decrease. Motor oil is rated by how thick (viscous) it is; 20 weight oil is considered relatively thin while 50 weight oil is relatively thick or viscous.

JAR TEST BENCH SHEET

PERFORMED BY: _____   DATE: _____

SAMPLE DATA

SOURCE: _____
TIME TAKEN: _____ AM/PM
FLOW: _____ pH: _____
TEMPERATURE: _____
TURBIDITY: _____
SUSPENDED SOLIDS: _____
C.O.D.: _____
CONDUCTIVITY: _____

PROCESS DATA

RAPID MIX
SPEED (RPM): _____
TIME (MINUTES): _____

SLOW MIX
SPEED (RPM): _____
TIME (MINUTES): _____

CHEMICALS USED: _____

JAR NO.	CHEMICALS (PPM)				FLOC CHARACTERISTIC AND TIME OF APPEARANCE	ANALYSIS OF SUPERNATANT					SLUDGE	
	ALUM	LIME	FERRIC	POLYMER	CAUSTIC		pH	TURB	C.O.D.	SUSP. SOLIDS	COAG RESIDUAL	VOLUME	SETTLING TIME
UNTREATED													
1													
2													
3													
4													
5													
6													

REMARKS: _____

CHEMICAL(S) AND DOSAGE TO BE USED: _____

Fig. 8.5 Jar test bench sheet

choice, the next stage of testing, dosage testing, may be started. If there are two or more apparent best performers, repeat the preliminary screening test with these products at reduced chemical levels. For secondary effluent wastewater, drop the test dosage from 3.0 mg/L to 1.0 mg/L. For sludges, drop the dosage to 50 mg/L. If this reduced dosage does not reveal noticeable differences between the choices, try again at still lower application rates, or go on to dosage testing with all of the "winners" of the screening test.

8.451 Dosage Testing

Objective: To determine the chemical dosages required to achieve the desired treatment performance.

The information collected in this and subsequent test methods must be more detailed and more precise than for the preliminary screening test procedure above. What is required are reliable numbers that will allow you to set up a dependable chemical addition program at the lowest practical cost. The term practical is used here rather than "lowest possible cost" because as regulatory requirements become more and more stringent, and enforcement penalties become more costly, the value of reliability is substantial.

Along with dosage testing, a detailed analysis of the untreated flow stream is needed. In addition to a number of particular factors that will be measured after chemical treatment (for example, suspended solids, turbidity, and BOD/COD[18]), a number of other factors directly influence the performance of the treatment chemicals. These include pH, conductivity (dissolved solids), calcium and magnesium hardness, phosphate, sulfate, sulfide, sodium, potassium, nitrogen, and chloride. For polymers, the most critical of these are conductivity (dissolved solids) and pH. For inorganic coagulants, those two, along with phosphate and sulfate, are also critical. For iron metal salts, nitrogen (particularly ammonia nitrogen) and sulfide are very important as well.

The dosage tests follow the same general procedures as the preliminary screening tests. The main difference is that only one chemical is tested at a time, and a different amount of it is used in each jar. Also some additional tests need to be made on the "clarified" water.

1. Prepare a jar test bench sheet similar to the one shown in Figure 8.5 on which to record specific data including, but not limited to:

    a. Source of water sample.

    b. Name of each treatment product to be tested.

    c. Dosage of treatment chemicals.

    d. Rapid mixing speed and duration.

    e. Slow mixing speed and duration.

    f. The sequence in which visible flocculation appears (which jar is first, which is second).

    g. The relative apparent size of the floc that forms.

    h. How quickly the floc settles to the bottom of the jar after stirring stops.

    i. How deep the floc is at the bottom of the jar after settling.

    j. The turbidity of the clarified liquid.

    k. The suspended solids in the clarified liquid.

    l. The *TRUE COLOR*[19] of the clarified liquid.

    m. The residual BOD or COD (whichever is part of normal discharge guideline testing) of the clarified liquid.

    n. The volatile and nonvolatile suspended solids.

2. Collect a suitable amount of flow stream sample. Using a graduated cylinder, accurately fill all the jars with their respective sample volumes.

3. Start the jar test unit stirrer and set it at the rapid mixing speed.

4. Fill the chemical measuring devices with the different dosages of the same chemical working solution.

5. Add the various dosages of chemical working solution to the respective test jars and start the timer. (Coagulation tests skip to step 7.)

6. Reduce the stirring speed after the appropriate interval. (Flocculation tests only.)

7. Record data items 1-f and 1-g during the stirring period(s).

8. At the end of the timing sequence, stop the stirrer and record data items 1-h through 1-n on the jar test bench sheet.[20]

Because of the time required to analyze the results of the chemical treatment on the flow stream samples in each test jar, these jar tests are run "semi-blind." That is, variations in performance at near-optimum conditions may not be visible to the eye of the observer (this is the reason for the laboratory analysis). However, the judgment of the testing personnel should be applied wherever there is an obviously inferior test. Discard any such samples and note this on the jar test bench sheet.

Two very important pieces of information should become evident following the dosage testing: (1) there will likely be a mini-

---

[18] *COD* (pronounce as separate letters). Chemical Oxygen Demand. A measure of the oxygen-consuming capacity of organic matter present in wastewater. COD is expressed as the amount of oxygen consumed from a chemical oxidant in mg/L during a specific test. Results are not necessarily related to the biochemical oxygen demand (BOD) because the chemical oxidant may react with substances that bacteria do not stabilize.

[19] *True Color.* Color of the water from which turbidity has been removed. The turbidity may be removed by double filtering the sample through a Whatman No. 40 filter when using the visual comparison method.

[20] To complete the data collection for items numbered 1-j through 1-n, it will be necessary to carefully remove the clarified liquid from each test jar using a pipet or other suitable extraction device taking care not to disturb either the sediment at the bottom of the jar or anything that might be floating on the surface or sticking to the sides.

mum dosage that will give suitable performance in meeting treatment specifications and above which better than necessary treatment process performance results can be obtained, and (2) nearly always there will be an upper dosage limit beyond which performance will diminish. (This is one classic example of "if a little bit is good, a lot is not necessarily better.") These two dosage limits are the upper and lower ends of the safe operating dosage range for each chemical product. If more than one treatment chemical is carried through this dosage testing procedure, the test exhibiting the widest safe operating dosage range will most likely handle the greatest variation in treatment process operating conditions. This factor needs to be included in the considerations used to select a treatment chemical. Two other considerations should be evaluated along with volume of sludge, cost, and safe operating range; they are: (1) availability of the chemical supply, and (2) reliability of the chemical supply (which are not necessarily the same issues.)

8.452 Full-Scale Trial

Objective: To confirm the performance and fine-tune the dosage rates for the selected chemical(s).

By the time dosage testing has been completed, the performance of one or two chemicals will stand out as clearly superior. Because the decision about which chemical to use has far-reaching effects for both plant operation and future planning, the final stage is to try the selected chemicals in the process flow stream itself. The jar tests that will be run during this process are identical to those for dosage testing.

Use a set of chemical dosages that start below the level being used in the process flow stream and extending well above it. During the full-scale trial period, detailed analyses (as described in dosage testing) for the untreated flow, as well as the chemically treated flow, should be run at various times of the day in order to record any significant changes in flow composition as treatment process flow rates vary. The range of dosages for these jar tests should be run at about the same times that the samples for untreated and chemically treated flows are taken. By doing this, the results from the jar tests and the results in the treatment process flow can be measured and confirmed and any correlations or variations can be noted. When this procedure is followed, the jar tests can then be applied with full confidence to the fourth phase of testing described in the following section.

8.453 System Performance Optimization

Objective: To keep treatment process performance within the desired limits while minimizing the chemical costs involved.

The required frequency of system performance optimization testing will be determined by a number of factors. These include the stability (or variability) of treatment process flow rates, the reliability of the treatment process itself, and the nature and frequency of process monitoring. A secondary objective of system performance optimization testing is to detect and compensate for potential treatment process upsets before they become an urgent problem. A treatment plant that operates with virtually no variation in effluent quality will need optimization testing very infrequently, perhaps monthly or quarterly and then mainly to ensure that chemical treatment dosages are not excessive. The more common situations, however, involve systems that operate near or at (or even significantly above) their design capacities on a daily basis. In these systems, jar testing should coincide with peak flows in order to obtain the information necessary to keep the treatment process operating satisfactorily.

The system performance optimization test procedure is identical to the ones described above for dosage testing and for full-scale trial testing except that the detailed analysis of the untreated water may be considered a matter of historical information rather than an essential guideline. The second difference is that, where at all possible, an attempt should be made to bracket the "safe operating dosage range" of the chemical(s) in such a way that the limits are apparent to the operator's naked eye. This is so that the time delay between the jar testing and completion of the laboratory analyses does not completely limit the operator's ability to adjust operation of the treatment process to keep it within performance specifications.

A third variation is that the jar test results (all of them), along with chemical system feed rates, should be recorded in a log format that can be used for tracking and projection purposes.

8.46 Procedure for Plants Without Laboratory Facilities

Source: *PROCEDURE FOR DETERMINING THE ALUM DOSAGE.* Permission of Industrial Chemicals Division, Allied Chemical Corporation, Solvay, NY

In case a laboratory is unavailable, it is possible to maintain reasonable process control by conducting coagulation tests using a simple hand-stirring method.

Clear glass fruit jars, one- to two-quart capacity, are a good substitute for beakers and are easily obtainable. If necessary, the local druggist or high-school chemistry teacher will usually assist in preparing alum solutions and lime suspensions. If a pipet is unavailable, approximately 20 drops from a common medicine dropper is roughly equivalent to one milliliter. A calibrated dropper (1.0 mL) may be obtained from the drugstore.

Procedure:

1. Dissolve 9.46 grams of alum and dilute to 1 liter (8.95 grams of alum to one quart). One mL of this solution will provide a treatment of 10 mg/L of alum when added to one quart (946 mL) of the water sample.

2. With the pipet or medicine dropper, add 1 mL of the alum solution to one quart of the sample and stir rapidly for approximately two minutes. Actual rapid mixing time should

be similar to the actual detention time of the water being treated in the flash mixer. This actual detention time may vary from 50 seconds to six minutes depending on design and flows. Then stir gently for at least 15 minutes to permit floc particles to form. Again, this actual gentle mixing time should be similar to the actual process flocculation time. During gentle mixing, the speed of the paddles in the jars should be similar to the speed of the actual flocculator paddles. When running the jar test, try adjusting the paddle speed to produce the best floc particle growth and then operate the flocculator paddles at this speed. Under some conditions the best floc can be produced by stopping the flocculators. Avoid violent agitation during this floc conditioning stage to prevent the breakup of the floc.

3. Observe the quality of the floc, the rate of settling, and the clarity of the settled water.

4. Repeat the above steps using a higher alum dose until the desired floc and clarity are achieved.

Helpful conversion factors:

1 liter	= 1,000 mL
1 quart	= 946 mL
1 grain/US gal	= 17.1 mg/L
1 grain/US gal	= 143 lbs/million US gallons
1 mg/L	= 8.34 lbs/million US gallons

8.47 Phosphate Monitoring

When the coagulant is being used to precipitate phosphate as well as to remove solids, coagulant dosage control may be obtained by automatically analyzing the incoming wastewater for soluble orthophosphate. The coagulant feeder is set to maintain a selected ratio of coagulant-to-phosphate either automatically or manually. Equipment is available that will automatically do this type of coagulant feeding.

Coagulant dosages that produce good phosphate removal will generally produce good solids removal if polymers are used to aid in flocculating the fine, precipitated matter. Usually, the polymer feed should be flow-paced; however, the polymer feeder may function with manual adjustments.

8.48 Safe Working Habits

Chemical feed equipment and chemical handling areas have safety hazards that each operator should become aware of for each plant. In addition to the usual electrical and mechanical hazards associated with automatic equipment, chemical treatment hazards include:

1. Strong acids
2. Strong caustics
3. High pressures
4. High temperatures
5. Dust in the air
6. Slippery walk areas

Avoid getting inorganic coagulants on the skin or in the eyes. These chemicals are either corrosive or caustic and cause irritation and possibly permanent damage. Wear protective gloves, chemical goggles, and a face shield, and ensure that an emergency eye wash and deluge shower are immediately available. Use local exhaust or approved respiratory protection to prevent inhalation of mists and dusts. Areas contacted should be flushed immediately with water. If a dry coagulant chemical comes in contact with your skin, attempt to mechanically wipe or brush off as much as possible before flushing. Check the appropriate *MATERIAL SAFETY DATA SHEET (MSDS)*[21] for specific health and safety information.

Polymers mixed with water are extremely slippery and can cause a fall hazard if left unattended on walkways. Spills should be cleaned immediately, using inert absorbent materials, such as sand or earth.

Develop safe working habits by always wearing proper safety equipment and protective clothing. If you do not have the required safety equipment or are not trained in its use, do not use the chemical until you have both the equipment and the training.

Good housekeeping is a part of the total plant operation. Good housekeeping around the chemical feed systems is very

---

[21] *Material Safety Data Sheet (MSDS).* A document that provides pertinent information and a profile of a particular hazardous substance or mixture. An MSDS is normally developed by the manufacturer or formulator of the hazardous substance or mixture. The MSDS is required to be made available to employees and operators or inspectors whenever there is the likelihood of the hazardous substance or mixture being introduced into the workplace. Some manufacturers are preparing MSDSs for products that are not considered to be hazardous to show that the product or substance is not hazardous.

important to good operations and safety. A dry chemical feeder that weighs its output will change its feed rate if chemicals are allowed to build up on the scales. Good housekeeping will reduce the hazard of slipping around chemical handling areas and will keep the dust down in work areas. Good housekeeping is a daily duty and must not be neglected.

8.49 Summary

Treatment chemical testing is most effective when it proceeds from a general review (preliminary screening) through more precise analyses as optimum chemical and treatment process operating performance is approached. Evaluation of test results will initially lead to another set of tests, which, in turn, will indicate further testing. Simply running a single grab batch of tests will rarely provide information of sufficient reliability to operate a treatment process. As enough information accumulates, the number of key guidelines that are significant will become evident, and those of less importance can be reduced in test frequency while any that prove to be virtually unimportant can be discontinued. Testing is not only valuable in controlling treatment process performance, but regular review of recorded test, application, and performance data will also serve to evaluate the needed testing as well.

QUESTIONS

Write your answers in a notebook and then compare your answers with those on page 380.

8.4D How are stock solutions of polymers prepared?

8.4E What is the maximum solution strength for a polymer?

8.4F Where should the sampling point be located to collect samples for a jar test?

END OF LESSON 2 OF 3 LESSONS
on
PHYSICAL–CHEMICAL TREATMENT PROCESSES

Please answer the discussion and review questions next.

DISCUSSION AND REVIEW QUESTIONS
Chapter 8. PHYSICAL–CHEMICAL TREATMENT PROCESSES
(COAGULATION, FLOCCULATION, AND SEDIMENTATION)

(Lesson 2 of 3 Lessons)

Write the answers to these questions in your notebook. The question numbering continues from Lesson 1.

6. What precautions should be taken when using alum?

7. What is a polymer map?

8. Why should extensive laboratory testing be conducted before treating the entire plant effluent with a particular polyelectrolyte?

9. What factors influence the selection of a polymer?

10. What factors should be considered when selecting a chemical?

11. Explain how to run the jar test.

12. What would you do if a dry coagulant chemical came in contact with your skin?

CHAPTER 8. PHYSICAL–CHEMICAL TREATMENT PROCESSES
(COAGULATION, FLOCCULATION, AND SEDIMENTATION)

(Lesson 3 of 3 Lessons)

8.5 PHYSICAL–CHEMICAL TREATMENT PROCESS EQUIPMENT

8.50 Chemical Storage and Mixing Equipment

Coagulants can be divided physically into two categories, liquid and solid. The physical category will necessarily dictate the type of equipment needed to accommodate it.

Solids are often dissolved before actual use and therefore use some equipment similar to that for liquid coagulants. Storage of all inorganic solids should be in dry tanks or bins because almost all inorganic solids exhibit caustic or corrosive properties if they become moist. Bins for powdered solids generally will be provided with a dust collector to keep the material ventilated as well as contained as a precaution against flash combustion. A shaker or vibrator is often installed at the feed system exit to prevent caking of the material. In the case of some limes, a slaker is often found at this point.

Chemical mixing equipment (Figure 8.6) is needed to prepare a solution of known concentration that can be metered (measured) into the water being treated. Polyelectrolytes can be difficult to dissolve. Also, polyelectrolytes may need a period of aging prior to application. A dissolving tank with a mechanical mixer is used to prepare the solution for feeding. The resulting solution is stored in a day tank (holding tank) from which it is metered out at the proper dosage into the water being treated.

8.51 Chemical Feed Equipment

Chemical feeders (metering equipment) are required to accurately control the desired dosage. The chemical to be used and the form in which it will be purchased must be determined first because chemicals used for solids removal in wastewater treatment usually can be purchased in either solid or liquid form.

After metering, solid chemicals are generally converted into a solution or a slurry (watery mixture) before being fed to the wastewater stream. Flushing water is often used with both slurries and liquids to rapidly carry the chemical to the point of application. This is especially true with lime (Figure 8.7), caustic (Figure 8.8), alum (Figure 8.9) and organic polymers (Figure 8.10). In such cases, measured amounts of the dry chemical are passed to a makeup or slurry tank in which a desired fluid dilution is made.

Dry chemicals may also be added directly under the assistance of water jets spraying at the points of entry. The latter method is an acceptable means of delivering dry polymer.

8.510 Types of Chemical Metering Equipment

To maintain accurate feed rates, there cannot be any slippage in the metering equipment; therefore, most liquid feeders are of the positive displacement type. The quality of the water used for both mixing and flushing the polymer system is important. Use of poor-quality plant effluent for either purpose may cause clumps ("fish eyes") to form, which will plug feeders, small orifices (openings), and even piping.

POSITIVE DISPLACEMENT PUMPS

A plunger pump (Figure 8.11) is used for metering chemicals due to the accuracy of the positive displacement stroke and the ease of adjusting the piston stroke to regulate the chemical feed rate. With each stroke, a fixed amount or volume of chemical or solution is discharged. By knowing the amount discharged per stroke and the number of strokes per minute, it is easy to calculate the chemical output.

Other positive displacement pumps besides the plunger pump include the gear pump (Figure 8.11) and the diaphragm pump (Figures 8.12 and 8.13). Each of these pumps will produce a constant chemical flow rate for a specific setting. Another type of positive displacement pump is the progressive cavity pump.

The feed rate for dry chemicals must also be accurately controlled. Typical dry chemical feeders include the screw feeder, vibrating trough, rotating feeder, and belt-type gravimetric feeder.

SCREW FEEDER

A screw feeder unit (Figure 8.14) maintains a desired output by varying the speed or the amount of time the screw rotates as it moves chemicals out the discharge port. Care must be taken that the chemical does not cake up in the hopper and stop feeding the screw. Also, the screw must be kept clean or the amount discharged per revolution will change.

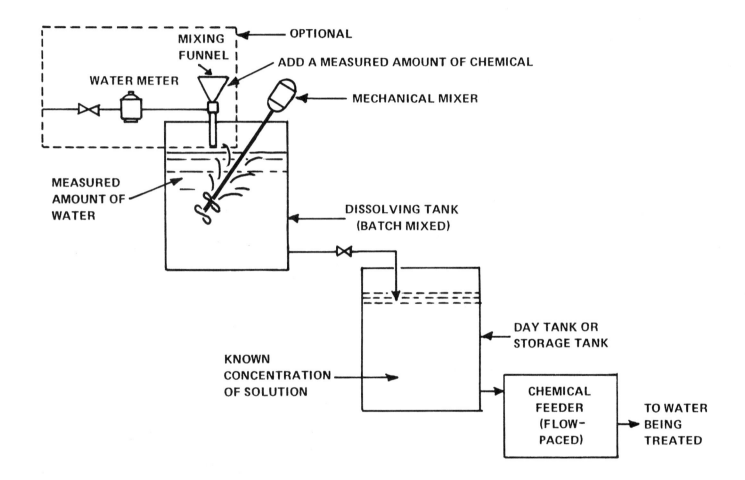

Fig. 8.6 Dry chemical dissolver, day tank, and feeder

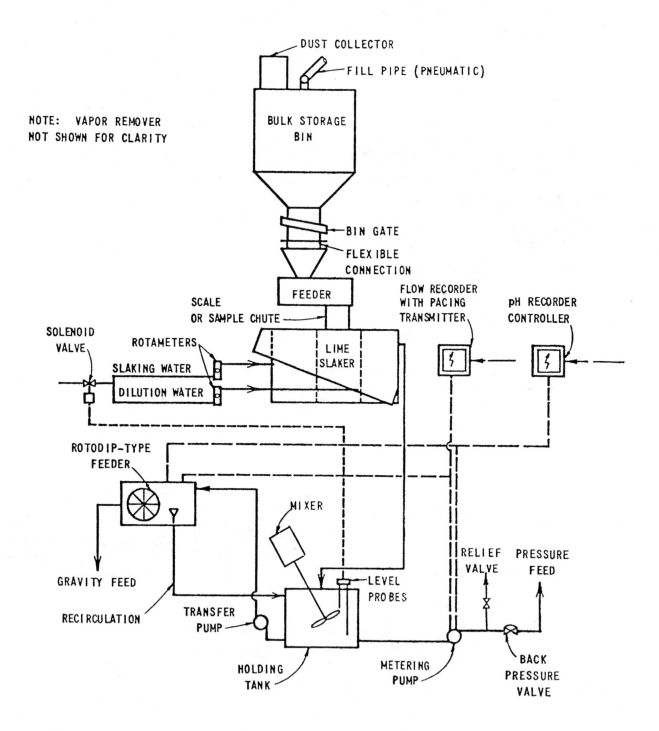

Fig. 8.7 Typical lime feed system

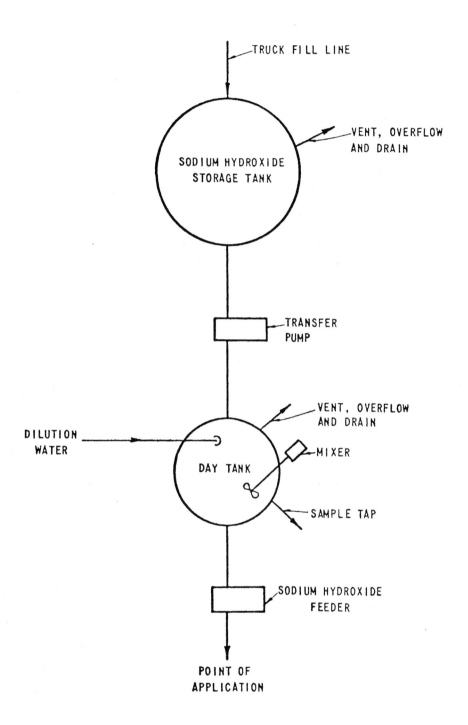

Fig. 8.8 Typical schematic of a caustic soda feed system

346 Treatment Plants

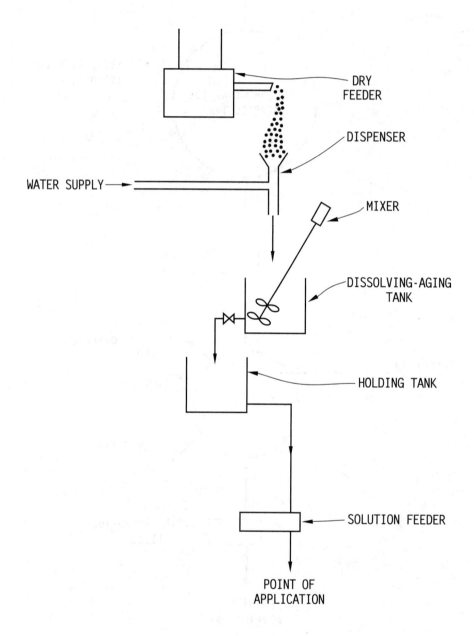

Fig. 8.9 Typical schematic of a dry alum feed system

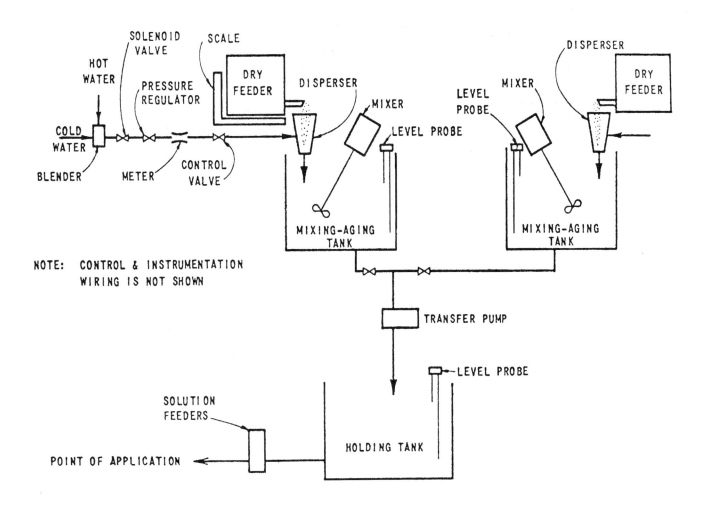

Fig. 8.10 Typical schematic of an automatic dry polymer feed system

348 Treatment Plants

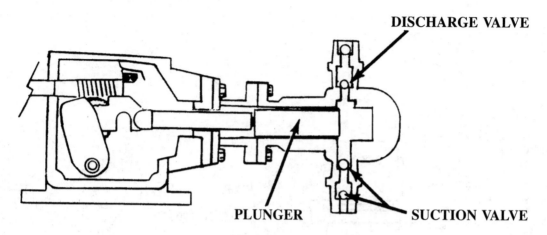

PLUNGER PUMP
(Courtesy of Wallace & Tiernan)

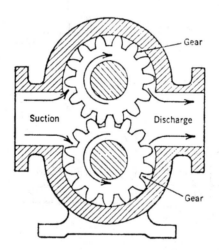

GEAR PUMP
(Courtesy of *CHEMICAL ENGINEERING*, 76 8, 45 (April 1969))

Fig. 8.11 Plunger and gear pumps

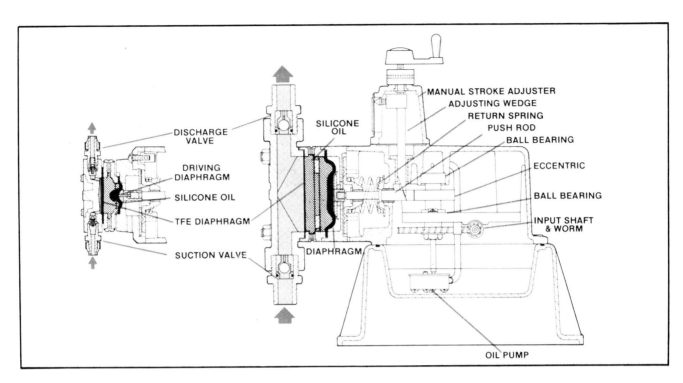
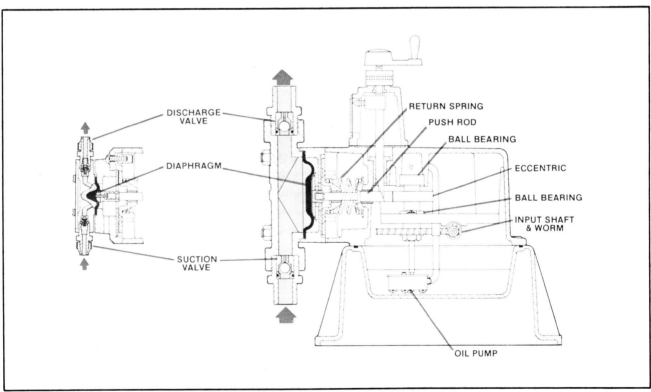

Fig. 8.12 Positive displacement diaphragm pumps
(Permission of Wallace & Tiernan Division, Pennwalt Corporation)

350 Treatment Plants

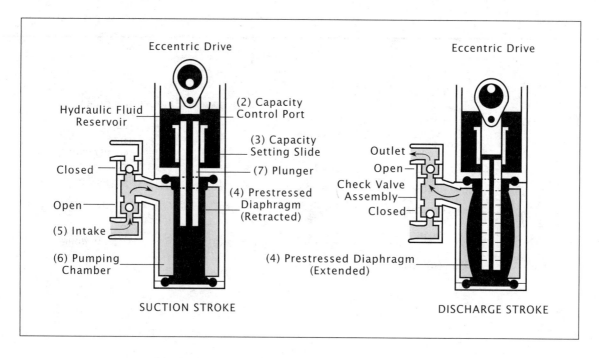

Fig. 8.13 Positive displacement diaphragm metering
(Permission of BIF, a Unit of General Signal)

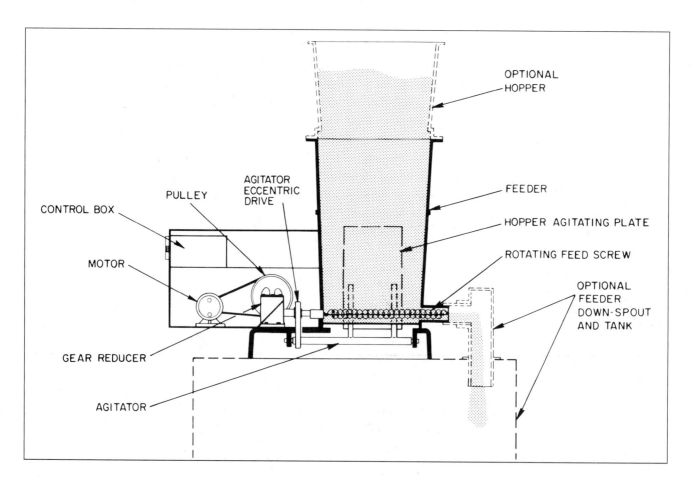

Fig. 8.14 Volumetric screw feeder
(Permission of Wallace & Tiernan Division, Pennwalt Corporation)

VIBRATING TROUGH FEEDER

The vibrating trough maintains a constant depth of chemical discharged and controls its chemical output by the magnitude and the duration (length of time) of the vibrations.

Care must be taken that the chemical does not cake in the hopper and stop feeding into the trough. Also, caking on the trough will prevent an even flow of chemical, which could change the output volume.

ROTARY FEEDER

Rotary feeders are similar to the positive displacement gear pump because a fixed amount of chemical is discharged from between each tooth (Figure 8.15). The output can be controlled by the speed or running time of the rotor. Care must be taken to maintain the rotor lobes clean and free of buildup that will change the chemical output volume.

BELT-TYPE GRAVIMETRIC FEEDER

A gravimetric belt feeder (Figure 8.16) maintains a constant chemical weight on a revolving belt. This is accomplished using a vibrating trough and a balance system. The chemical output is controlled by the amount of chemical on the belt and the speed and time the belt travels. The amount of chemical is varied by the opening or closing of a feed gate or as a weighing deck moves up or down.

Care must be taken with this unit to ensure that chemicals do not build up on the balance because this will change the chemical output. By catching and weighing the chemical discharged at a constant speed over a measured amount of time, the feeder output can be verified.

Table 8.1 lists various types of chemical feeders, their uses, and limitations.

QUESTIONS

Write your answers in a notebook and then compare your answers with those on page 380.

8.5A How are chemical solutions prepared for feeding?

8.5B List the most common types of chemical feeders or metering equipment.

8.511 Selecting a Chemical Feeder

When you must decide which chemical feeder to purchase for your situation, include the following considerations:

1. *TOTAL OPERATING RANGE*

    Will the unit run at today's lowest expected chemical output as well as the future required output?

2. *ACCURACY*

    Will the unit maintain the same feed rate after it has been installed, calibrated, and operated?

3. *REPEATABILITY*

    Can you return to previous settings and obtain the same feed rates as before?

4. *RESISTANCE TO CORROSION*

    Will the equipment, including electrical components, withstand the corrosive environment to which they may be exposed?

5. *DUST CONTROL*

    Is a means provided to control dust, if needed?

6. *AVAILABILITY OF PARTS*

    Are replacement parts readily available?

7. *SAFETY*

    Is the system designed with safety of both operation and maintenance in mind?

8. *ECONOMICS*

    What are the costs of purchase, installation, operation, maintenance, and replacement and energy requirements?

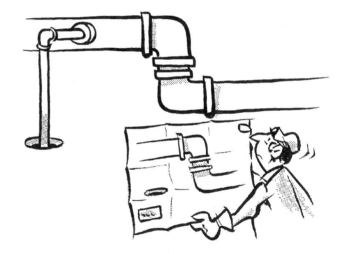

8.512 Reviewing Chemical Feed System Designs

When reviewing chemical feed system designs and specifications, the operator should check the following items:

1. Review the results of pre-design tests to determine the chemical feed rate for both the present and future. The chemical feeders should be sized to handle the full range of chemical doses or provisions should be made for future expansion.

2. Determine if sampling points are provided to measure chemical feeder output.

3. Be sure provisions are made for standby equipment in order to maintain uninterrupted dosages during equipment maintenance.

4. Look for adequate valving to allow bypassing or removing equipment for maintenance without interrupting the chemical dosage.

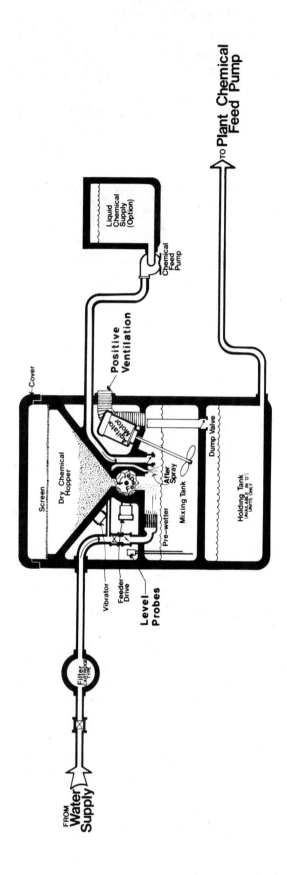

Fig. 8.15 Rotary feeder
(Permission of Neptune Microfloc)

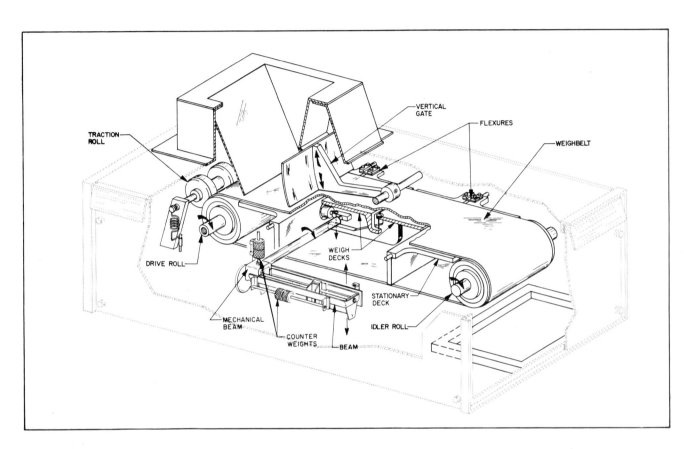

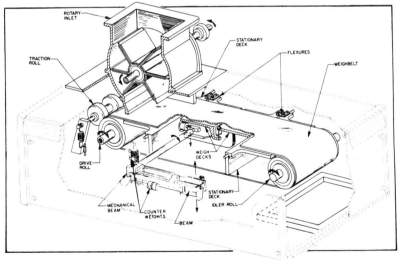

Fig. 8.16 Gravimetric belt feeders
(Permission of Wallace & Tiernan Division, Pennwalt Corporation)

TABLE 8.1 TYPES OF CHEMICAL FEEDERS

(Source: EPA *PROCESS DESIGN MANUAL FOR SUSPENDED SOLIDS REMOVAL*)

Type of Feeder	Use	Limitations General	Capacity, cu ft/hr	Range
Dry feeder:				
Volumetric:				
Oscillating plate	Any material, granules or powder.		0.01 to 35	40 to 1
Oscillating throat (universal)	Any material, any particle size.		0.02 to 100	40 to 1
Rotating disk	Most materials including NaF, granules, or powder.	Use disk unloader for arching.	0.01 to 1.0	20 to 1
Rotating cylinder (star)	Any material, granules or powder.		8 to 2,000 or 7.2 to 300	10 to 1 or 100 to 1
Screw	Dry, free flowing material, powder or granular.		0.05 to 18	20 to 1
Ribbon	Dry, free flowing material, powder, granular, or lumps.		0.002 to 0.16	10 to 1
Belt	Dry, free flowing material up to 1 1/2-inch size, powder or granular.		0.1 to 3,000	10 to 1 or 100 to 1
Gravimetric:				
Continuous-belt and scale	Dry, free-flowing, granular material, or floodable material.	Use hopper agitator to maintain constant density.	0.02 to 2	100 to 1
Loss in weight	Most materials, powder, granular, or lumps.		0.02 to 80	100 to 1
Solution feeder:				
Nonpositive displacement:				
Decanter (lowering pipe)	Most solutions or light slurries.		0.01 to 10	100 to 1
Orifice	Most solutions.	No slurries	0.16 to 5	10 to 1
Rotameter (calibrated valve)	Clear solutions.	No slurries	0.005 to 0.16 or 0.01 to 20	10 to 1
Loss in weight (tank with control valve)	Most solutions.	No slurries	0.002 to 0.20	30 to 1
Positive displacement:				
Rotating dipper	Most solutions or slurries.		0.1 to 30	100 to 1
Proportional pump:				
Diaphragm	Most solutions. Special unit for 5% slurries.[a]		0.004 to 0.15	100 to 1
Piston	Most solutions, light slurries.		0.01 to 170	20 to 1
Gas feeder:				
Solution feed	Chlorine		8,000 lb/day max	20 to 1
	Ammonia		2,000 lb/day max	20 to 1
	Sulfur dioxide		7,600 lb/day max	20 to 1
	Carbon dioxide		6,000 lb/day max	20 to 1
Direct feed	Chlorine		300 lb/day max	10 to 1
	Ammonia		120 lb/day max	7 to 1
	Carbon dioxide		10,000 lb/day max	20 to 1

[a] Use special heads and valves for slurries.

5. Examine plans for valving to allow flushing the system with water before removing from service.

6. Be sure corrosion-resistant drains are provided to prevent chemical leaks from reaching the floor; for example, drips from pump packing.

7. Check for corrosion-resistant pumps, piping, valves, and fittings as needed.

8. Determine the amount of maintenance required. The system should require a minimum of maintenance. Equipment should be standard, with replacement parts readily available.

9. Consider the effect of changing head conditions, both suction and discharge, on the chemical feeder output. Changing head conditions should not affect the output if the proper chemical feeder has been specified.

10. Determine whether locations for monitoring readouts and dosage controls are convenient to the operation center and are easy to read and record.

8.513 Chemical Feeder Start-Up

After the chemical feed system has been purchased and installed, the operator must carefully check it out before starting it up. Even if the contractor who installed the system is responsible for ensuring that the equipment operates as designed, the operation by plant personnel, the functioning of the equipment, and the results from the process are the responsibility of the chief operator. Therefore, before start-up, check the following items:

1. Inspect the electrical system for proper voltage, for properly sized overload protection, for proper operation of control lights on the control panel, for proper safety lockout switches and operation, and for proper equipment rotation.

2. Confirm that the manufacturer's lubrication and start-up procedures are being followed. Equipment may be damaged in minutes if it is run without lubrication.

3. Examine all fittings, inspection plates, and drains to ensure that they will not leak when placed in service.

4. Determine the proper positions for all valves. A positive displacement pump will damage itself or rupture lines in seconds if allowed to run against a closed valve or system.

5. Be sure that the chemical to be fed is available. A progressive cavity pump will be damaged in minutes if it is allowed to run dry.

6. Inspect all equipment for binding or rubbing.

7. Confirm that safety guards are in place.

8. Examine the operation of all auxiliary equipment including the dust collectors, fans, cooling water, mixing water, and safety equipment.

9. Check the operation of alarms and safety shutoffs. If it is possible, operate these devices by manually tripping each one. Examples of these devices are alarms and shutoffs for high water, low water, high temperature, high pressure, and low chemical levels.

10. Be sure that safety equipment, such as eye wash facilities, deluge showers, approved respiratory protection, face shields, gloves, and vent fans, is in place and functional.

11. Record all important *NAMEPLATE*[22] data and place it in the plant files for future reference.

8.514 Chemical Feeder Operation

Once the chemical feed equipment is in operation and the major bugs are worked out, the feeder will need to be fine-tuned. To aid in fine-tuning and build confidence in the entire chemical feed system, the operator must maintain accurate records (Figures 8.17 and 8.18). These records will include the flows and characteristics of the wastewater before treatment, the dosage and conditions of the chemical treatment, and the results obtained after treatment. A comment section should be used to note abnormal conditions, such as a feeder plugged for a short time, a sudden change in the characteristics of the influent waste, and related equipment that malfunctions. Daily logs should be summarized into a form that operators can use as a future reference.

8.515 Shutting Down Chemical Systems

If the equipment is going to be shut down for an extended length of time, it should be cleaned out to prevent corrosion or the solidifying of the chemical. Lines and equipment could be damaged when restarted if chemicals left in them solidify. Operators could be seriously injured if they open a chemical line that has not been properly flushed out.

The following items should be included in your checklist for shutting down the chemical system:

1. Shut off the chemical supply.

2. Run chemicals completely out of the equipment and clean the equipment.

3. Flush out all the solution lines.

4. Shut off the electric power.

5. Shut off the water supply and protect from freezing.

6. Drain and clean the mix and feed tanks.

QUESTIONS

Write your answers in a notebook and then compare your answers with those on page 380.

8.5C List the items that should be considered when selecting a chemical feeder.

8.5D What information should be recorded for a chemical feeder operation?

---

[22] *Nameplate.* A durable, metal plate found on equipment that lists critical operating conditions for the equipment.

SODIUM HYDROXIDE LOG

	TANK #1 GAL.	TANK #2 GAL.	TANK #3 GAL.	GAL. #3 BEFORE TRANS.	GAL. #3 AFTER TRANS.	H2O TO NAOH	DILUTE GAL. USED	TOTAL GAL REC.	REMARKS
1	880	-0-	1280				300		
2	880	-0-	990				290		
3	880	-0-	700				290		
4	880	-0-	580				120		
5	880	-0-	300				280		
6	-0-	-0-	1840	200	2000	1-1	260		
7	-0-	-0-	1600				240		
8	-0-	-0-	1400				200		
9	-0-	-0-	1240				160	4,000	
10	1990	1940	1050				190		
11	1990	1940	850				200		
12	1990	1940	650				200		
13									
14									
15									
16									
17									
18									
19									
20									
21									
22									
23									
24									
25									
26									

Fig. 8.17 Typical record of chemical feeder operation

CHEMICAL FEED RECORD

CHEMICAL ALUM   10 GALLONS PER INCH   LOCATION SECONDARY

DATE	TIME	CHEMICAL TANK LEVEL			TREATED FLOW X 1,000 METER READINGS			PUMP SET A/M	CHEM USED mg/l	OPER.	REMARKS
		PREVIOUS	PRESENT	AMT. USED	PRESENT	PREVIOUS	TOTAL				
5/31/04		—	123"		105,376	—					
6-1	0800	123"	100"	23"	115,376	105,376	10,000	AUTO	22.0	T.J.	O.K.
6-2	0800	100"	75"	25"	125,026	115,376	9,650	A	25.9	T.S.	O.K.
6-3	0800	75"	51"	24"	134,574	125,026	9,548	A	25.1	AL	O.K.
6-3	1030	51"	48"	3"	135,876	134,574	1,302	A	23.0	B.C.	REFILLED TANK
6-3	1030		200"								
6-4	0800	200"	184"	16"	144,806	135,876	8,930	M/30%	17.9	B.C.	TREATMENT BETTER
6-5	0800	184"	162"	22"	154,466	144,806	10,660	A	20.6	B.C.	"
6-5	1500	162"	154"	8"	158,140	154,466	3,674	A	21.8	T.J.	CHECKING PUMP

TOTAL / MAX. / MIN. / AVG.   CHEMICAL COST $/MG

Fig. 8.18   *Typical form for chemical feeder operation*

8.52 Coagulant Mixing Units

Rapid mixing of coagulants may be accomplished in one of three modes: (1) high-speed mixers (impeller or turbine), (2) in-line blenders and pumps, and (3) baffled compartments or pipes (static mixers). The use of high-speed mechanical mixers is most common. They are often seen in parallel to increase residence time. Static mixers make use of fluid passing through baffled chambers at high velocities to bring about turbulence and mixing. In-line blenders and pumps accomplish the same by virtue of a high velocity through pipelines and pumps.

8.53 Flocculators

Mechanical flocculating units (Figure 8.19) may be rotary, horizontal shaft-reel type, rotary-shaft turbine, or rotary reciprocating. All three systems possess vertical shafts. Standard rotary and rotary reciprocating units use paddle impellers; reciprocating types have two or more shafts rotating in opposite directions.

In all flow-through flocculators, tapered flocculation is found to be most beneficial. By this method, a small, dense floc is formed initially followed by aggregation to form a denser, larger floc. This is accomplished on single shafts by variation of the paddle sizes. On multiple-shaft units, variation of the speed of the individual units or the number of paddles per shaft is effective.

8.54 Clarifiers (Also see Chapter 5, "Sedimentation and Flotation," Volume I, *OPERATION OF WASTEWATER TREATMENT PLANTS*)

8.540 Types of Clarifiers

Clarifiers can take on two basic configurations based upon the flow character, that is, vertical or horizontal flow. Horizontal flow is the most common in both rectangular (Figure 8.20) and circular (Figure 8.21) clarifiers.

Rectangular clarifiers with horizontal flow have the influent entering at one end. Flow generally hits a baffle and moves by gravity to the opposite end where the effluent overflows the outlet weirs. A surface skimmer made up of flights pushes oil and floating debris to a spiral collector located at one end. Settled sludge is moved by flights along the bottom to a sludge hopper where it is collected and pumped to a dewatering facility.

Circular clarifiers (Figure 8.21) with horizontal flow take on one of three configurations:

1. Center influent with radial effluent
2. Radial influent with center effluent
3. Radial influent and effluent

In each case, sludge is collected at the center of the conical (cone-shaped) base. Oil and scum are skimmed by a radial arm at the surface of the water, which deposits the material into a sump.

Vertical-flow units (Figure 8.22) have the general distinction of the influent flowing along the bottom and rising toward the top to be discharged over the effluent weir. One advantage of vertical-flow clarifiers is that flow can be forced up through the sludge blanket, thus aiding in solids retention and improving flow control. Both rectangular and circular configurations exist.

Circular configurations have solids contact units in which all three activities leading up to and including coagulation and precipitation take place (Figure 8.23). Influent becomes rapidly mixed with the coagulant at the influent discharge and flows down through a center baffle. Flocculation occurs in this zone. Flow proceeds radially upward through the sludge blanket and clarification occurs. Effluent discharge is radial.

Rectangular configurations may have tube and lamella separators in which settling becomes compartmentalized. Tube settlers (Figures 8.24, 8.25, and 8.26) consist of a collection of closely packed small-diameter tubes placed at an angle. Flow proceeds upward as sludge settles downward. A lamella separator uses parallel plates rather than tubes. The wastewater flows upward and sludge moves downward similar to tube settlers (Figure 8.24). Both configurations depend on the assumption that the paths of all particles with the same settling velocities will be straight parallel lines. In the above two configurations, the surface area of the clarifier is effectively increased without increasing the actual size.

8.541 Operating Guidelines

Four major factors are used to predict the performance of clarifiers: detention time, weir overflow rate, surface loading rate, and solids loading. If any one or more of these factors exceeds design or expected values, we can expect facility performance to deteriorate. Each of these factors is readily calculated and will give the operator an indication of expected clarifier efficiency. These four factors are also used to design clarifiers.

Detention time is the length of time it would take a plug of water to enter a clarifier and to exit in the effluent. This is related to clarifier efficiency in that particles should be allowed ample time to settle out. If the detention time is less than the settling rate, then there will be a carryover of particles into the effluent. Detention time is calculated knowing the flow and tank dimensions as follows:

Tank Volume, cu ft = Length, ft × Width, ft × Depth, ft

or

Tank Volume, m³ = Length, m × Width, m × Depth, m

and

$$\text{Detention Time, hr} = \frac{\text{Tank Vol, cu ft} \times \frac{7.48 \text{ gal}}{\text{cu ft}} \times \frac{24 \text{ hr}}{\text{day}}}{\text{Flow, gal/day}}$$

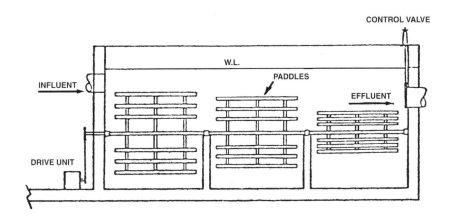

Mechanical Flocculation Basin
Horizontal Shaft—Reel Type

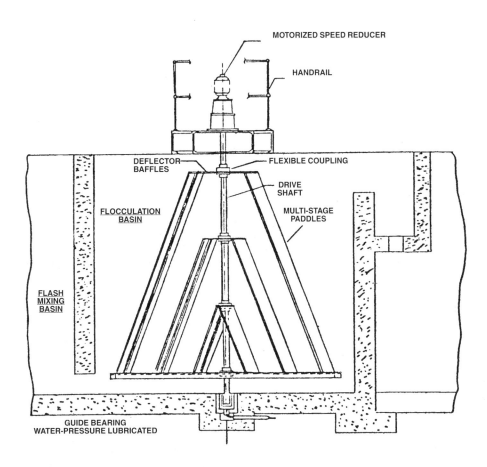

Mechanical Flocculator
Vertical Shaft—Paddle Type
(Courtesy of Ecodyne Corp.)

Fig. 8.19 Mechanical flocculators

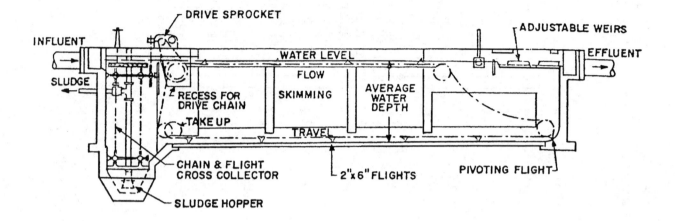

A. WITH CHAIN AND FLIGHT COLLECTOR

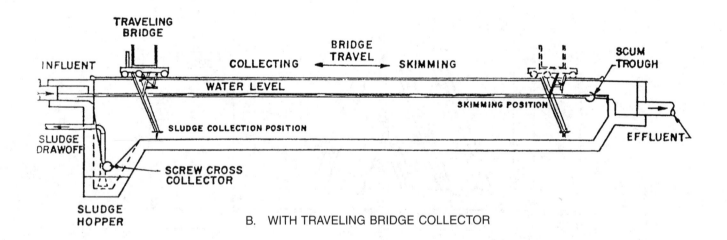

B. WITH TRAVELING BRIDGE COLLECTOR

Fig. 8.20 Rectangular sedimentation tanks
(Courtesy of FMC Corp.)

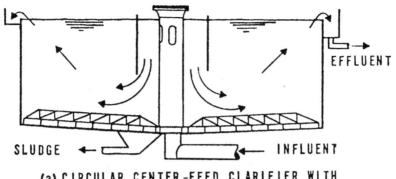

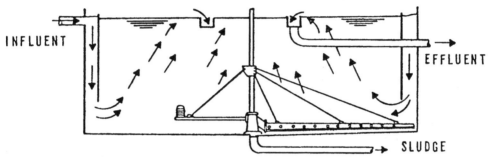

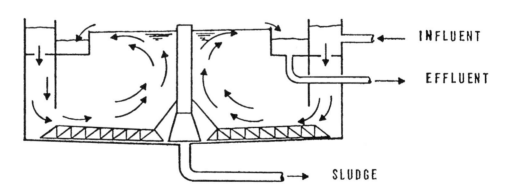

Fig. 8.21 Typical circular clarifier

362 Treatment Plants

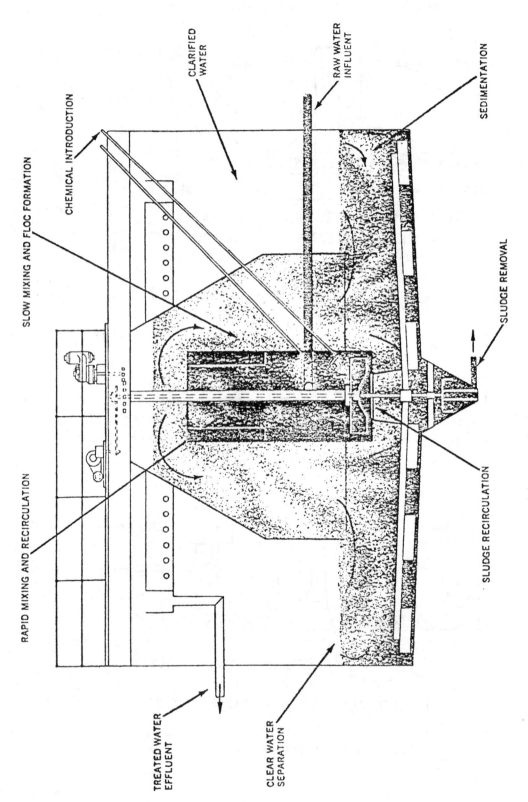

Fig. 8.22 Solids contact clarifier without sludge blanket filtration
(Courtesy of Econodyne Corp.)

Physical–Chemical Treatment 363

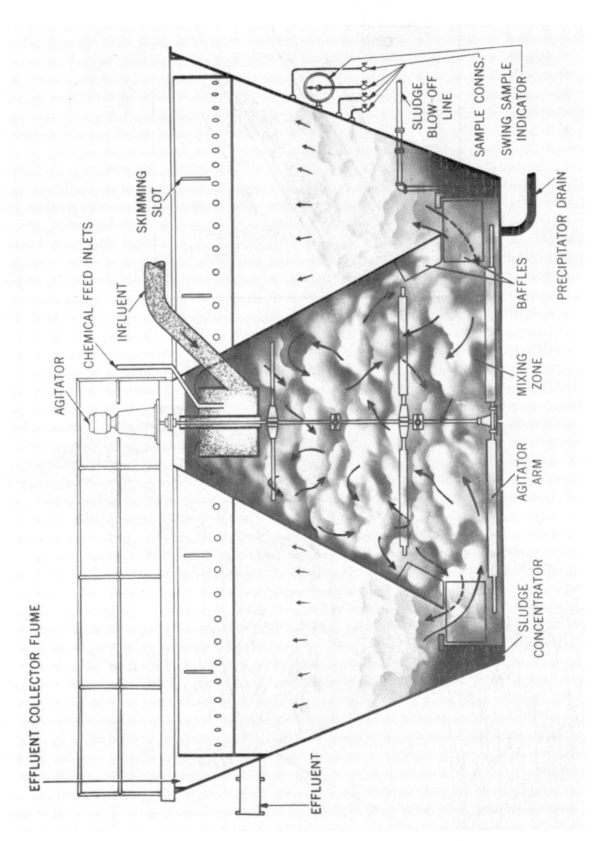

Fig. 8.23 Solids contact clarifier with sludge blanket filtration
(Courtesy of the Permutit Co.)

364 Treatment Plants

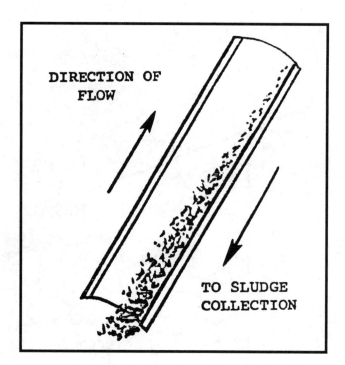

Fig. 8.24 Tube settlers—flow pattern

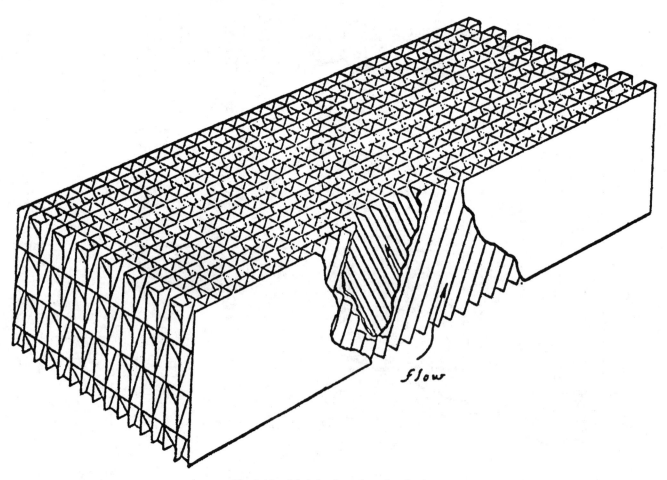

Fig. 8.25 Module of steeply inclined tubes
(Courtesy Neptune Microfloc, Inc.)

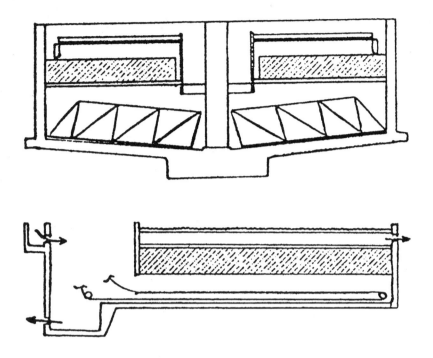

TUBE SETTLERS IN EXISTING CLARIFIER

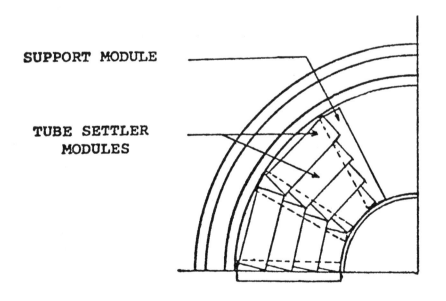

Fig. 8.26 Plan view of modified clarifier

or

$$\text{Detention Time, hr} = \frac{\text{Tank Vol, m}^3 \times 24 \text{ hr/day}}{\text{Flow, m}^3/\text{day}}$$

If the detention time proves to be less than the settling rate (as shown by results of laboratory tests), then it may be necessary to increase the clarifier capacity by placing other basins into operation.

The weir overflow rate expresses the quantity of water that passes out of the clarifier in relation to the *LINEAL*[23] feet of weir available. Weir overflow rate is calculated as follows:

$$\text{Weir Overflow Rate, GPD/ft} = \frac{\text{Flow, gal/day}}{\text{Length of Weir, lineal feet}}$$

or

$$\text{Weir Overflow Rate, m}^3/\text{day/m} = \frac{\text{Flow, m}^3/\text{day}}{\text{Length of Weir, m}}$$

Surface loading rate expresses the quantity of water being treated in relation to the available clarifier surface. As previously stated, sedimentation efficiency increases with increased surface area. Surface loading rate may be calculated as follows:

$$\text{Surface Loading Rate, GPD/sq ft} = \frac{\text{Flow, gal/day}}{\text{Clarifier Surface Area, sq ft}}$$

or

$$\text{Surface Loading Rate, m}^3/\text{day/sq m} = \frac{\text{Flow, m}^3/\text{day}}{\text{Clarifier Surface Area, sq m}}$$

Another very important loading guideline for clarifiers is the solids loading. Solids loadings are especially important for industrial waste treatment because the solids carried by industrial wastewater may be significantly different from solids in municipal wastewaters.

$$\text{Solids Loading, lbs/hr/sq ft} = \frac{\text{Solids, lbs/hr}}{\text{Clarifier Surface Area, sq ft}}$$

or

$$\text{Solids Loading, kg/hr/sq m} = \frac{\text{Solids, kg/hr}}{\text{Clarifier Surface Area, sq m}}$$

The above characteristics are useful when compared with the original design considerations and specifications in attempting to diagnose the possible causes for clarifier inefficiency. These, coupled with the results of laboratory tests, such as dye tracer studies, will aid the operator in troubleshooting clarifier performance problems.

Other physical characteristics will also affect clarifier efficiency. One important feature is the settling characteristics of the particles. Settling rate may be affected by many factors including particle size, shape, temperature of the surrounding water, and particle density in relationship to that of the surrounding water. Particles of greater weight and density will settle faster than those of lesser density. The horizontal velocity and the tank depth will also come into play in determining how long it takes a particle to settle.

A particle settling quiescently (in still water) will tend to settle in a path perpendicular to the settling surface. With flow through a clarifier, force is applied to a particle causing it to settle in a plane diagonal to the settling surface. Since it now has a longer path to travel, the particle will take slightly longer to settle.

The ultimate settling velocity of the particle will be affected by the flow rate, the viscosity of the water, and the settling characteristics of the particle. Temperature may play a vital role in the settling characteristics of the particle. With decreasing temperature, water becomes more dense and more viscous. Consequently, the space between the water molecules becomes more restricting on the particle's ability to settle. Increased temperature has the opposite effect. Settling rate is therefore greater at a higher temperature than it is at a lower temperature.

SHORT-CIRCUITING[24] of flow can adversely affect clarifier efficiency. This occurs when the flow is not homogeneous (completely uniform) throughout the tank. That is to say, there exist zones or layers where the flow is either faster or slower than the surrounding areas. If velocity is too high in an area, suspended material may pass out of the clarifier without settling. Too slow a velocity will cause dead spaces and may produce a *SEPTIC*[25] condition if organic or other biologically degradable material is present.

Short-circuiting may be caused by differences in water density due to different temperatures existing at the surface and the bottom of the clarifier. This is especially true in temperate climates during the winter and summer seasons. Density differences may also be caused by a high suspended solids content in the influent. Short-circuiting may be made worse by high inlet velocities, high outlet weir rates, and strong winds blowing along the tank surface. In all cases, the most effective solution is the use of weir plates, baffles, and port openings to produce a flow velocity throughout the clarifier that is as even as possible.

8.542 Clarifier Efficiency

Clarifier efficiency may be defined as the percent of a pollutant removed by the clarifier. Four major factors are used to

---

[23] *Lineal* (LIN-e-ul). The length in one direction of a line. For example, a board 12 feet (meters) long has 12 lineal feet (meters) in its length.
[24] *Short-Circuiting*. A condition that occurs in tanks or basins when some of the flowing water entering a tank or basin flows along a nearly direct pathway from the inlet to the outlet. This is usually undesirable since it may result in shorter contact, reaction, or settling times in comparison with the theoretical (calculated) or presumed detention times.
[25] *Septic* (SEP-tick) or *Septicity*. A condition produced by bacteria when all oxygen supplies are depleted. If severe, the bottom deposits produce hydrogen sulfide, the deposits and water turn black, give off foul odors, and the water has a greatly increased oxygen and chlorine demand.

measure performance or efficiency: flows, suspended solids, settleable solids, and floatable solids. Efficiency should be based upon the analyses of both inlet and outlet COMPOSITE SAMPLES[26] over a 24-hour period. Calculation is as follows:

$$\text{Efficiency, \%} = \frac{\text{In} - \text{Out}}{\text{In}} \times 100\%$$

QUESTIONS

Write your answers in a notebook and then compare your answers with those on pages 380 and 381.

8.5E List the three common modes of rapid mixing of coagulant chemicals.

8.5F What is tapered flocculation?

8.5G What is an advantage of vertical-flow clarifier units?

8.5H List the four major factors that are used to predict the performance of clarifiers.

8.5I What happens if the detention time in a clarifier is too short?

8.5J List the possible causes of short-circuiting in a clarifier.

8.5K List the four major factors that are used to measure clarifier performance or efficiency.

8.55 Dissolved Air Flotation (DAF) Thickeners (Figure 8.27)

The objective of flotation thickening is to separate solids from the liquid phase in an upward direction by attaching air bubbles to particles of suspended solids. Four general methods of flotation are commonly employed. These include:

1. Dispersed air flotation where bubbles are generated by mixers or diffused aerators.

2. Biological flotation where gases formed by biological activity are used to float solids.

3. Dissolved air (vacuum) flotation where water is aerated at atmospheric pressure and released under a vacuum.

4. Dissolved air (pressure) flotation where air is put into solution under pressure and released at atmospheric pressure.

Flotation by dissolved air (pressure) is the most commonly used procedure for wastewater sludges and will be the topic of discussion in this section. Flotation units may be either rectangular or circular in design. The dissolved air system uses either a compressed air supply or an ASPIRATOR-TYPE[27] air injection assembly to obtain a pressurized air-water solution. The key components of dissolved air flotation thickener units are (1) air injection equipment (located on and within pressurized retention tank), (2) agitated or unagitated pressurized retention tank, (3) recycle pump, (4) inlet or distribution assembly, (5) sludge scrapers, and (6) an effluent baffle.

The sludge to be thickened is either introduced to the unit at the bottom through a distribution box and blended with a pre-pressurized effluent stream or the influent stream is saturated with air, pressurized, and then released to the inlet distribution assembly. Total wastestream pressurization may shear (break up) flocculated sludges and seriously reduce process efficiency. Direct saturation and pressurization of the sludge stream is not the preferred mode of operation where primary sludges are to be thickened. Primary sludges often contain stringy material that can clog or "rag-up" the aeration equipment in a pressurized retention tank. Flotation thickening of excess biological solids may use air saturation and pressurization of the wastestream with less possibility of clogging the air addition and dissolution equipment.

The preferred mode of operation from a maintenance standpoint is the use of a recycle stream to serve as the air carrying medium. Referring again to Figure 8.27, the operation of dissolved air flotation (DAF) units that incorporate recycle techniques is as follows. A recycled DAF effluent, primary or secondary effluent stream, is introduced into a retention tank to dissolve air into the liquid. The retention tank is maintained at a pressure of 45 to 70 psig (3.2 to 4.9 kg/sq cm). Compressed air is either introduced into the retention tank directly or at some point upstream of the retention tank or an aspirator assembly is used to draw air into the stream.

The pressurized air-saturated liquid then flows to the distribution or inlet assembly and is released at atmospheric pressure through a back pressure-relief valve. The decrease in pressure causes the air to come out of solution in the form of thousands of minute air bubbles. The bubbles make contact with the influent sludge solids in the distribution box and attach to the solids

---

[26] *Composite (Proportional) Sample.* A composite sample is a collection of individual samples obtained at regular intervals, usually every one or two hours during a 24-hour time span. Each individual sample is combined with the others in proportion to the rate of flow when the sample was collected. Equal volume individual samples also may be collected at intervals after a specific volume of flow passes the sampling point or after equal time intervals and still be referred to as a composite sample. The resulting mixture (composite sample) forms a representative sample and is analyzed to determine the average conditions during the sampling period.

[27] *Aspirate* (AS-per-rate). Use of a hydraulic device (aspirator or eductor) to create a negative pressure (suction) by forcing a liquid through a restriction, such as a Venturi tube. An aspirator may be used in the laboratory in place of a vacuum pump; sometimes used instead of a sump pump.

368 Treatment Plants

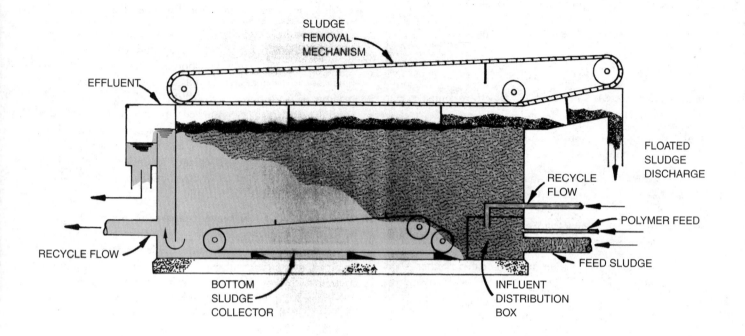

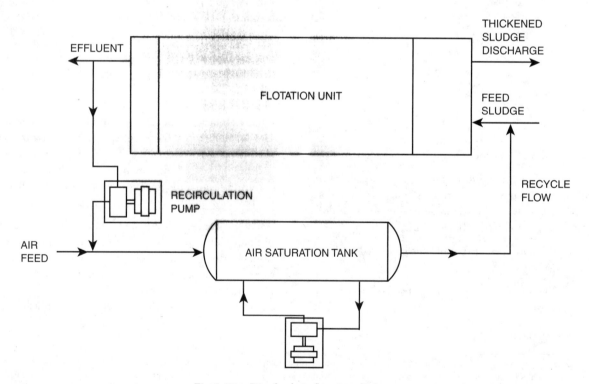

Fig. 8.27 Dissolved air flotation thickener

causing them to rise to the surface. These concentrated solids are then removed from the surface. An effluent baffle is provided to keep the floated solids from floating out with the effluent.

The effluent baffle extends approximately 2 to 3 inches (5.0 to 7.5 cm) above the water surface and down to within 12 to 18 inches (0.3 to 0.45 m) of the bottom of the tank. The effluent baffle is provided to keep the floated solids from contaminating the effluent. Clarified effluent flows under the baffle and leaves the unit through an effluent weir. If air is introduced or aspirated upstream of the retention tank, it is usually done on the suction side of the recycle pump to use the pump as a driving force for dissolving air into the liquid. The main disadvantage associated with introducing air to the suction side of pumps is the possibility of pump CAVITATION[28] and the subsequent loss of pump capacity. Systems that add compressed air directly to the retention tank commonly use a float control mechanism to maintain a desired air-liquid balance. A sight glass should be provided to periodically check the level of the air-liquid interface (point of contact) because if the float mechanisms fail, the retention tank may fill completely with either liquid or air. In both cases, the net effect will be a drastic reduction in flotation efficiency.

8.550 *Factors Affecting Dissolved Air Flotation*

The performance of dissolved air flotation units depends on (1) type of sludge, (2) age of the feed sludge, (3) solids and hydraulic loadings, (4) air to solids (A/S) ratio, (5) recycle rate, and (6) sludge blanket depth.

As is the case with gravity thickeners, the type and age of sludge applied to flotation thickeners will affect the overall performance. Primary sludges are generally heavier than excess biological sludges and are not as easy to treat by flotation concentration. If enough air is introduced to float the sludge mass, the majority of the primary sludge solids will float to the surface and be removed by the skimming mechanisms. Gritty or heavy primary sludge particles will settle and be deposited on the floor of the flotation unit and provisions should be made to remove these settled solids. If a flotation unit is used for primary sludge thickening, the flotation cell is usually equipped with sludge scrapers to push the settled solids to a collection hopper for periodic removal. Problems will arise when concentrating primary sludges or combinations of primary sludge and waste activated sludge if the flotation chamber is not equipped with bottom sludge scrapers and sludge removal equipment. Solids buildup will result in a decrease in flotation volume and a reduction in thickener efficiency.

Excess biological sludges are easier to treat by flotation thickening than primary sludges because they are generally lighter and thus easier to float. Bottom sludge scrapers should still be incorporated in the design of units used solely for biological sludge because a small fraction of solids will settle. These settled solids will eventually become anaerobic and rise due to gasification. If these solids are deposited at the effluent end of the unit, solids may be carried under the effluent baffle and exit the unit with the effluent.

Sludge age usually does not affect flotation performance as drastically as it affects gravity concentrators. A relatively old sludge has a natural tendency to float due to gasification and this natural buoyancy will have little or no negative effect on the operation of flotation thickeners. However, rising sludge does create problems in primary and final sedimentation processes and should be avoided by controlling the sludge withdrawal rate from these unit processes.

Solids and hydraulic loadings, A/S (air to solids) ratios, recycle rate, and sludge blanket depth are normal operational guidelines and are discussed in the following paragraphs.

8.551 *Operating Guidelines*

The size of dissolved air flotation units is determined by the engineers who design them. The operator has control over A/S ratio, recycle rate, and the blanket thickness and can optimize performance by properly adjusting these variables. Before discussing the control variables, the operator should be familiar with determining applied loading rates.

8.5510 SOLIDS AND HYDRAULIC LOADINGS

Solids and hydraulic loadings for flotation units are based on the same calculations used to determine loading rates for gravity thickeners. If either the solids or hydraulic loading becomes excessive, effluent quality declines and thickened sludge concentrations are reduced. The following example shows how to calculate loading rates.

EXAMPLE 4

Given: A dissolved air flotation unit receives 100 GPM of waste activated sludge with a suspended solids concentration of 8,000 mg/L. The rectangular flotation unit is 40 feet long and 15 feet wide.

Find: The hydraulic loading (GPM/sq ft) and solids loading (lbs/hr/sq ft).

Solution:

Known		**Unknown**
Flow, GPM	= 100 GPM	1. Hydraulic Loading, GPM/sq ft
Sus Sol, mg/L	= 8,000 mg/L	2. Solids Loading, lbs/hr/sq ft
, %	= 0.8%	
Flotation Unit		
Length	= 40 ft	
Width	= 15 ft	

---

[28] *Cavitation* (kav-uh-TAY-shun). The formation and collapse of a gas pocket or bubble on the blade of an impeller or the gate of a valve. The collapse of this gas pocket or bubble drives water into the impeller or gate with a terrific force that can cause pitting on the impeller or gate surface. Cavitation is accompanied by loud noises that sound like someone is pounding on the impeller or gate with a hammer.

1. Determine the hydraulic loading, GPM/sq ft.

$$\text{Hydraulic Loading, GPM/sq ft} = \frac{\text{Flow, GPM}}{\text{Liquid Surface Area, sq ft}}$$

$$= \frac{100 \text{ gal/min}}{40 \text{ ft} \times 15 \text{ ft}}$$

$$= \frac{100 \text{ gal/min}}{600 \text{ sq ft}}$$

$$= 0.2 \text{ GPM/sq ft}$$

2. Determine the solids loading, lbs/hr/sq ft.

$$\text{Solids Loading, lbs/hr/sq ft} = \frac{\text{Flow, GPM} \times 60 \frac{\text{min}}{\text{hr}} \times 8.34 \frac{\text{lbs}}{\text{gal}} \times \text{SS,} \frac{\%}{100\%}}{\text{Liquid Surface Area, sq ft}}$$

$$= \frac{100 \frac{\text{gal}}{\text{min}} \times 60 \frac{\text{min}}{\text{hr}} \times 8.34 \frac{\text{lbs}}{\text{gal}} \times \frac{0.8\%}{100\%}}{40 \text{ ft} \times 15 \text{ ft}}$$

$$= \frac{400 \text{ lbs/hr}}{600 \text{ sq ft}}$$

$$= 0.67 \text{ lb/hr/sq ft} = 16 \text{ lbs/day/sq ft}$$

8.5511 AIR TO SOLIDS (A/S) RATIO

The quantity of air introduced and dissolved into the recycle or wastestream is critical to the operation of flotation thickeners. Enough air has to be added and dissolved to float the sludge solids. The most effective method of accomplishing this is to introduce air into a pressurized retention tank along with the wastestream to be thickened or along with a portion of the thickener effluent stream. Air also can be dissolved in primary or secondary effluent, thus avoiding solids recycling in the DAF unit. Mixing of the retention tank contents should also be used to increase the amount of air that can be put into solution. In unmixed pressure retention tanks, only about 50 percent of the injected air will dissolve while 90 percent saturation can be obtained by vigorous agitation of the tank contents. As previously discussed, following a short detention time in the pressurized retention tank, the saturated air/liquid stream is pumped to the inlet side of the flotation unit where it enters a distribution assembly by operation of a back pressure-relief valve. The release of the saturated air stream to atmospheric pressure causes the air to come out of solution in the form of very small bubbles. Thousands of these minute bubbles attach to particles of suspended solids. The solids float to the surface, concentrate, and are removed by the sludge skimming mechanism. The more air you have dissolved in the retention tank, the greater the number of minute air bubbles that will be released in the distribution assembly. And, the more bubbles you produce in the distribution assembly, the more efficient your operation will be.

The amount of air supplied to the unit is usually controlled by an air rotameter and compressor assembly, which are activated by a liquid level indicator in the retention tank. The most important operational concern is to ensure that the air rotameter, compressor, and the float mechanism to actuate air injection are in proper working order.

Example 5 shows how to calculate the quantity of air applied to the system.

EXAMPLE 5

Given: An air rotameter and compressor provide for 10 cubic feet per min (SCFM)[29] of air to be injected into a pressurized retention tank.

Find: The pounds of air applied to the unit per hour (lbs/hr).

Solution:

Known	Unknown
Air Flow, SCFM = 10 SCFM	Air Applied, lbs/hr

Calculate the air applied in pounds of air per hour.

$$\text{Air Applied, lbs/hr} = \text{Air Flow,} \frac{\text{cu ft}}{\text{min}} \times 60 \frac{\text{min}}{\text{hr}} \times 0.075 \frac{\text{lb air}}{\text{cu ft air}}$$

$$= 10 \frac{\text{cu ft}}{\text{min}} \times 60 \frac{\text{min}}{\text{hr}} \times 0.075 \frac{\text{lb}}{\text{cu ft}}$$

$$= 45 \text{ lbs/hr}$$

NOTE: The conversion factor of 0.075 pound of air per cubic foot of air will change with temperature and elevation or barometric pressure.

The ratio between air supplied and the quantity of solids applied to the flotation unit is then the air to solids (A/S) ratio. The following example illustrates the determination of air/solids (A/S) ratio.

EXAMPLE 6

Given: A dissolved air flotation unit receives 100 GPM of waste activated sludge at a concentration of 9,000 mg/L (0.9% solids). Air is supplied at a rate of 5.0 cu ft/min.

Find: The air to solids (A/S) ratio.

Solution:

Known		Unknown
Solids Flow, GPM	= 100 GPM	Air to Solids (A/S) Ratio
Sl Conc, mg/L	= 9,000 mg/L	
, %	= 0.9% Solids	
Air, cu ft/min	= 5.0 cu ft/min	

---

[29] *SCFM*. Standard Cubic Feet per Minute. Cubic feet of air per minute at standard conditions of temperature, pressure, and humidity (0°C, 14.7 psia, and 50 percent relative humidity).

Calculate the air to solids (A/S) ratio.

$$\frac{\text{Air, lbs}}{\text{Solids, lbs}} = \frac{\text{Air, cu ft/min} \times 0.075 \text{ lb/cu ft}}{\text{Solids, GPM} \times 8.34 \frac{\text{lbs}}{\text{gal}} \times \text{Sl Conc,} \frac{\%}{100\%}}$$

$$= \frac{5.0 \text{ cu ft/min} \times 0.075 \text{ lb/cu ft}}{100 \frac{\text{gal}}{\text{min}} \times 8.34 \frac{\text{lbs}}{\text{gal}} \times \frac{0.9\%}{100\%}}$$

$$= \frac{0.375 \text{ lb air}}{7.5 \text{ lbs solids}}$$

$$= 0.05 \text{ lb air/lb solids}$$

8.5512 RECYCLE RATE AND SLUDGE BLANKET

Both the rate of effluent recycle and the thickness of the sludge blanket are operational controls available to optimize DAF performance. Typically, recycle rates of 100 to 200 percent are used. A recycle rate of 100 percent means that for every gallon of influent sludge there is one (1) gallon of DAF effluent recycled to the DAF inlet works.

The following example illustrates the determination of recycle rate.

EXAMPLE 7

Given: A dissolved air flotation unit receives waste activated sludge flow of 50 GPM. The recycle rate is set at 75 GPM.

Find: Percentage of recycle.

Solution:

Known	**Unknown**
Waste Flow, GPM = 50 GPM	Percentage of Recycle, %
Recycle Flow, GPM = 75 GPM	

Calculate the percentage of recycle.

$$\text{Recycle, \%} = \frac{\text{Recycle Flow, GPM} \times 100\%}{\text{Waste Flow, GPM}}$$

$$= \frac{75 \text{ GPM}}{50 \text{ GPM}} \times 100\%$$

$$= 150\%$$

The optimum recycle rate for a particular unit will vary from one treatment plant to the next and it is impossible to define that rate for every application. The important point is that the recycle stream carries the air to the inlet of the unit. Obviously, as the rate of recycle increases, the potential to carry more air to the inlet also increases. The term "potential" is used here because the recycle rate and the quantity of air dissolved and released in the inlet assembly are dependent on one another because of what happens in the retention tank. DAF recycle pumps are usually centrifugal pumps, which means that as the pressure upstream (retention tank) increases, the output (flow) decreases. Therefore, the rate of recycle is directly dependent on the pressure maintained within the retention tank. As stated previously, retention tank pressures of 45 to 70 psi (3.2 to 4.9 kg/sq cm) are commonly used. As the pressure within the retention tank is increased or decreased by closing or opening the back pressure-relief valve, the recycle rate will decrease or increase. The optimum recycle rate and retention tank pressure are usually determined by experimentation.

The thickness of the floating sludge blanket can be varied by increasing or decreasing the speed of the surface sludge scrapers. Increasing the sludge scrapers speed usually tends to thin out the floated sludge while decreasing the scrapers speed will generally result in a more concentrated sludge.

8.552 Normal Operating Procedures

Typically, the flow through the thickener is continuous and should be set as constant as possible. Monitoring of the influent, effluent, and thickened sludge streams should be done at least once per shift and composite samples should be taken for later laboratory analysis.

Under normal operating conditions, the effluent stream should be relatively free of solids (less than 100 mg/L suspended solids) and will resemble secondary effluent. The float (thickened sludge) solids will have a consistency resembling cottage cheese. The depth of the float solids should extend approximately 6 to 8 inches (15 to 20 cm) below the surface. The surface farthest from the float solids collection and the discharge point should be scraped clean of floating solids with each pass of the sludge collection scrapers. If the sludge blanket is allowed to build up (too thick) and drop too far below the surface, thickened (floated) solids will be carried under the effluent baffle and contaminate the effluent.

Normal start-up and shutdown procedures are outlined below:

START-UP

- Fill the unit with fresh water, primary effluent, or secondary effluent.

- Open the inlet and discharge valves on the recycle pump and turn on the recycle pump only when thickener is full.

- Adjust the retention tank pressure to the desired pressure (45 to 70 psig or 3.2 to 4.9 kg/sq cm) by opening or closing the pressure regulating valve.

- Open the inlet and discharge valves on the reaeration pump (if provided) and turn on the reaeration pump. If a mechanical mixer is used instead of a reaeration pump, turn on the mixer in the retention tank. Mixing in the retention tank may be accomplished by methods other than the use of mechanical mixers.

- Open the appropriate air injection valves and turn on the air compressor.

- Open the appropriate chemical addition valves and turn on the chemical pump if chemicals are used.

- Wait until the surface of the DAF thickener is covered by a uniform pattern of small bubbles.

- Open the inlet and discharge valves on the sludge feed pump and start the feed pump.
- Allow floated sludge mat to build up, then turn on sludge collection scrapers.
- Turn on the thickened sludge pump and adjust the withdrawal rate as required.

If the thickener is not operated in a continuous mode and daily or frequent shutdowns are required, the following procedures should be followed:

SHUTDOWN

- Turn off the sludge inlet pump and close the appropriate valves.
- Turn off the chemical pump and close appropriate valves.
- Turn on the fresh water supply to the unit and allow it to run on fresh water until the surface is free of floating sludge.
- Turn off the air compressor and close appropriate air valves.
- Turn off the reaeration and recycle pumps and close appropriate valves.
- Turn off the sludge collectors.
- Turn off the thickened sludge pump.
- Hose down and clean up as required.

8.553 Typical Performance

Typical operating guidelines as well as thickened sludge concentration and suspended solids removals for waste activated sludge are presented in Table 8.2

TABLE 8.2 OPERATIONAL AND PERFORMANCE GUIDELINES FOR FLOTATION THICKENERS

	Without Polymer Addition	With Polymer Addition
Solids Loading,		
lbs/hr/sq ft[a]	0.4–1	1–2
lbs/day/sq ft	9.6–24	24–48
Hydraulic Loading, GPM/sq ft[b]	0.5–1.5	0.5–2.0
Recycle, %	100–200	100–200
Air/Solids, lb/lb	0.01–0.10	0.01–0.10
Minimum Influent Solids Concentration, mg/L	5,000	5,000
Float Solids Concentration, %	2–4	3–5
Solids Recovery, %	50–90	90–98

[a] lbs/hr/sq ft × 4.883 = kg/hr/sq m
[b] GPM/sq ft × 0.679 = L/sec/sq m

The determination of solids recovery in the operation of the DAF unit is based on laboratory analysis and the following calculations.

EXAMPLE 8

Given: A 100-foot diameter dissolved air flotation unit receives 750 GPM of waste activated sludge at a concentration of 0.75% (7,500 mg/L) sludge solids. The effluent contains 50 mg/L of suspended solids. The float or thickened sludge is at a concentration of 3.3 percent.

Find: The solids removal efficiency (%) and the concentration factor (cf).

Solution:

Known	Unknown
Dissolved Air Flotation Unit	1. Solids Removal Efficiency, %
Infl Solids, mg/L = 7,500 mg/L	2. Concentration Factor (cf)
, % = 0.75%	
Effl Solids, mg/L = 50 mg/L	
Effl Sludge, % = 3.3% (Thickened Sludge)	

1. Determine the solids removal efficiency, %.

$$\text{Solids Removal Efficiency, \%} = \frac{(\text{Infl Solids, mg}/L - \text{Effl Solids, mg}/L)\,100\%}{\text{Infl Solids, mg}/L}$$

$$= \frac{(7{,}500 \text{ mg}/L - 50 \text{ mg}/L)\,100\%}{7{,}500 \text{ mg}/L}$$

$$= 99.3\%$$

2. Calculate the concentration factor (cf) for the thickened sludge.

$$\text{Concentration Factor (cf)} = \frac{\text{Thickened Sludge Concentration, \%}}{\text{Influent Sludge Concentration, \%}}$$

$$= \frac{3.3\%}{0.75\%}$$

$$= 4.4$$

QUESTIONS

Write your answers in a notebook and then compare your answers with those on pages 381 and 382.

8.5L List the main components of dissolved air flotation (DAF) units.

8.5M Discuss the function of the distribution box, the retention tank, and the effluent baffle.

8.5N Why should a sight glass be provided on the retention tank?

8.5O List the factors that affect the performance of DAF thickeners.

8.5P What effect does sludge age have on the performance of DAF thickeners?

8.5Q Determine the hydraulic loading (GPD/sq ft) for a 20-foot diameter DAF unit. The influent flow is 100 GPM. The formula for surface area of a circular tank is:

$$\text{Area} = \frac{\pi}{4} \text{Diameter}^2 \text{ or Area} = 0.785 \text{ Diameter}^2$$

8.5R For the above problem, determine the solids loading, A/S ratio, and recycle flow rate (GPM), if the influent sludge has a suspended solids concentration of 0.75% (7,500 mg/L), and is supplied at a rate of 2.5 cu ft/min. Air is supplied at a rate of 0.75 cu ft/min and a recycle ratio of 100 percent is required.

8.5S Determine the suspended solids removal efficiency (%) and the concentration factor (cf) if a DAF unit receives an influent sludge at 1.0 percent (10,000 mg/L) suspended solids. The effluent is at 50 mg/L suspended solids and the float or thickened sludge is at a concentration of 3.8 percent.

8.554 Troubleshooting

Visual inspection of the dissolved air flotation unit in conjunction with a working knowledge of the operating techniques is the operator's biggest asset in ensuring efficient operation. The specific areas that the operator should be concerned with are: (1) effluent quality, and (2) thickened sludge (float) characteristics. The effluent from DAF units should be relatively clear (less than 100 mg/L suspended solids). Well-operated units should produce effluents equivalent in appearance to secondary clarifier effluent. If the effluent from the unit contains an unusually high amount of suspended solids, the problem may be related to: (1) sludge blanket thickness, (2) chemical conditioning, (3) A/S ratio, (4) recycle rate, (5) solids or hydraulic loading, or any combination of these factors.

If the float solids appear to be well flocculated and concentrated (resembling cottage cheese), the speed of the sludge scrapers should be increased. Poor effluent quality in conjunction with a concentrated float sludge usually results from allowing the sludge blanket to develop too far below the surface. When this happens, the undermost portions of the blanket will break off and be carried under the effluent baffle. Increasing the sludge collector speed will result in a decrease in blanket thickness and prevent solids from flowing under the baffle.

If the scrapers are already operating at full speed and the blanket level is below the effluent baffle, the unit is probably overloaded with regard to solids. In this case, the influent flow rate and concentration should be checked and the flow rate should be decreased, if possible.

High solids carryover with the effluent, in conjunction with lower than normal float solids concentrations, usually indicates that problems exist with the air system, chemical conditioning system, or the loading rates. The operator should systematically check the retention tank pressure and sight glass, the recycle pump, the air compressor assembly, the reaeration pump, the chemical conditioning equipment, and the influent flow.

Equipment malfunctions are quickly revealed by checking the retention tank pressure and the sight glass. Higher than desired pressures will result in decreased recycle rates and the back pressure-relief valve should be opened somewhat to decrease the pressure and increase the recycle rate. Lower than normal pressures will result in higher recycle rates. In this case, the pressure-relief valve should be closed somewhat to decrease the recycle rate, and allow more time for air to dissolve in the retention tank.

Malfunctions in the retention tank liquid level indicator and air compressor activation assembly will also cause drastic decreases in flotation efficiency. The liquid level in the sight glass is the best indicator of this problem.

If the liquid level in the retention tank is lower or higher than normal, either the float mechanism to activate the air inlet valve or control is malfunctioning or the air compressor and solenoid valves are not operating correctly. If the liquid level in the retention tank is not at the desired level, shut off the DAF unit, open the hatch on the retention tank, and clean the liquid level indicator probes.

If everything (air, recycle, and retention pressure) seems to be in proper order, but the DAF effluent is still high in solids and the float solids are at a low concentration, check the retention tank mixer (reaeration pump), the chemical conditioning system, and the loading rates.

If chemical conditioners are used, they must be prepared properly and applied at the desired dosage. Chemical conditioning is discussed in Section 3.3 of *ADVANCED WASTE TREATMENT* in this series of operator training manuals. Proper operation of the chemical conditioning system will greatly help the performance of the DAF unit. The chemical mixing and delivery systems should be carefully watched and calibrated because of the high cost of chemicals.

If all the equipment is operating properly and the problem still exists, check the hydraulic and solids loading according to Example 4 and adjust flow rates as required.

Table 8.3 summarizes problems that may arise when operating a dissolved air flotation system and the corrective measures that might be taken.

TABLE 8.3 TROUBLESHOOTING DISSOLVED AIR FLOTATION

Operational Problem	Possible Cause	Check or Monitor	Possible Solution
1. Solids carryover with effluent but good float concentration	1. Float blanket too thick	1. a. Flight speed b. Solids loading	1. a. Increase flight speed b. Lower flow rate to unit if possible
2. Good effluent quality but float sludge thin (dilute)	2. Float blanket too thin	2. a. Flight speed b. Solids loading	2. a. Decrease flight speed b. Increase flow rate if possible
3. Poor effluent quality and thin (dilute) float sludge	3. a. A/S low b. Pressure too low or too high c. Recycle pump inoperative d. Reaeration pump inoperative e. Chemical addition inadequate f. Loading excessive	3. a. (1) Air rate (2) Compressor b. Pressure gauge c. Pressure gauge and pump d. Pump pressure e. Chemical system f. Loading rates	3. a. (1) Increase air input (2) Repair or turn on compressor b. Open or close valve c. Turn on recycle pump d. Turn on reaeration pump e. Increase chemical dosage f. Lower flow rate

QUESTION

Write your answer in a notebook and then compare your answer with the one on page 382.

8.5T On a routine check of a dissolved air flotation unit, the operator notices high suspended solids in the effluent and a thinner than normal sludge. Discuss the possible causes and solutions to the problem.

8.6 OPERATION, START-UP, AND MAINTENANCE

8.60 Operating Strategy

The development of an operational strategy for a physical–chemical treatment process will prepare you to deal with sudden changes in the water being treated, to train new operators, and to plan for the future. Procedures that should be part of your strategy include:

1. Set up your laboratory so jar tests can be run quickly and easily. The jar test is the most important control test for chemical treatment. Accurate jar tests can result in significant chemical and cost savings. For example, if the dosage of a polyelectrolyte costing $2.00 per pound could be reduced by 0.5 mg/L in a 10 MGD plant, the cost savings would be $83.40 per day.

2. Monitor chemical feeders closely to ensure proper output. On new equipment, measure the actual feed rates and compare them with feed settings at least weekly.

3. Adjust chemical dosages whenever the flow rate changes. Long detention times during low flows may not require as high a chemical dosage as shorter detention times during high flows.

4. Monitor water conditions and quality at least daily for alkalinity, pH, temperature, turbidity, and suspended solids because these water quality indicators may signal a need for a chemical dosage change. If you are removing phosphorus, measure soluble phosphorus also.

5. Consider in-plant conditions when collecting samples for jar tests and adjusting chemical dosages. For example, if one-half of the primary clarifiers in a plant are out of service or if a digester is upset, these situations can affect required chemical doses.

8.61 Start-Up and Maintenance Inspection

The items listed below should be checked during your general start-up inspection of a physical–chemical treatment process. This checklist can also be used as the basis for an overall maintenance inspection. Because of the wide variety of industrial wastewater treatment plants and the broad differences from one manufacturer's equipment to another, it is suggested that the manufacturer's instructions be consulted for more in-depth maintenance procedures. For simplicity and to avoid duplication, some items in this checklist have been grouped under equipment type rather than process stage.

1. General

   a. Determine that all tanks, basins, and piping are clean and free of debris.

   b. Ensure that all drawings, equipment manufacturer's specifications, and operating manuals are complete, up to date, and available.

   c. Verify that the correct spare parts are on hand.

d. Ensure the proper operation of all Start-Emergency-Stop controls both on-site and at the control panel. Check all electrical connections and power supplies.

e. Ensure that all piping and valves are properly installed and adequately braced.

2. Motors and Drives

   a. Ensure that all motors and drives are securely fastened. Check bearing supports, shaft alignments to drive motors, and belts (for both condition and tightness).

   b. Ensure proper lubrication of motors, drives, shafts, chains, and bearings according to the manufacturer's specifications.

   c. Verify that motors run at the speed and rotational directions prescribed and that all voltage requirements are satisfied.

   d. Ensure that chains move freely without binding.

   e. Ensure that all chain and other equipment guards are in place.

3. Pumps

   a. Piston Pumps. Check ball seatings, packing, shear pin, drive belts, and hydraulic fluid.

   b. Centrifugal Pumps. Check impeller for wear or plugging. Check for prime. Check packing.

   c. Progressive Cavity Pumps. Check rotor and stator for wear or plugging. Check prime and packing.

   d. Diaphragm pumps. Be sure that diaphragm is intact and working properly.

4. Chemical Feed Systems

   a. Ensure that all level alarms in tanks are functioning properly.

   b. Ensure proper calibration of all flow and metering systems.

   c. Ensure that water blenders operate properly and check dilution mixers for proper placement and installation.

   d. Ensure proper temperature and pressure for dilution water.

   e. Ensure in-line mixers are in place, properly braced with accommodation for proper bypass.

5. Gates for Control of Flow

   a. Ensure all gates are properly aligned in angles and for travel clearance.

   b. Ensure proper lubrication of wheels and rising stems.

   c. Ensure proper operation in both automatic and manual modes.

6. Rapid Mix

   a. Inspect impeller conditions. Ensure that impellers are free from obstructions.

   b. Check motors and drives (item 2 above).

7. Flocculators

   a. Check motors and drives (item 2 above).

   b. Ensure baffles are correctly set and securely anchored.

   c. Ensure that drive stuffing box is properly placed and grouted.

   d. Ensure that drive bearing and sprocket are complete.

   e. Ensure that mixers rotate freely before coupling to the gear drive.

   f. Ensure that sump pumps are in place and operational.

8. Clarifiers

   a. Check motors and drives (item 2 above).

   b. Ensure that all sprockets and shafts are in alignment and free for the rotation of the sludge collectors.

   c. Lube all rails. Run collectors in empty tank (dry) for two hours before allowing the water into the flocculation tank.

   d. Ensure all flights are connected to the chain and that all shoes are attached.

   e. Check that all drive sprockets are operational. Ensure that shear pins are installed in all sprockets.

   f. Check the operation of limit switches to ensure that they stop the motor.

   g. Check cross-collector travel.

9. Scum Collectors

   a. Ensure that the collectors and trough are properly secured and aligned.

   b. Ensure that the wiper blades on the spiral collectors have proper uniform contact with the breaching plate.

8.62 Actual Start-Up

The following list is a general pattern of procedures to be followed when placing the entire system on-line. Sections may be applied to the start-up of an individual part.

1. Chemical Feed System

   a. Ensure proper temperature for dilution waters, if applicable.

   b. Check that proper chemical strengths are set on automatic feeds.

   c. Open all manual valves, as appropriate.

   d. Set proper flow rate.

376 Treatment Plants

2. Mixing Tank

   a. Open effluent gates.

   b. Allow tanks to fill.

   c. Start mixers and adjust speed.

3. Flocculation Tanks

   a. Open influent gates.

   b. Turn on and adjust paddle/turbine drives.

4. Clarifiers

   a. Turn on cross-collectors and longitudinal collectors.

   b. Start scum collectors.

   c. Allow basins to fill.

8.63 Normal Operation

A general procedure for the normal operation of a chemical coagulation-precipitation system is outlined below.

1. Chemical Feed System

   a. Perform a jar test to determine the proper chemical dosage.

   b. Ensure proper chemical dilution.

   c. Set controls for the proper feed rate.

   d. Report in the operating log the amount of chemical used per unit time.

2. Rapid Mixing Tanks

   a. Ensure proper mixing speed by observing the floc formed.

   b. Check for scum formation. If scum accumulates in the influent, open the scum gates. If scum is floating in the tanks, adjust the mixer speed. If the condition does not improve, open the slide gates and allow the scum to pass through with the effluent.

   c. Only operate mixers when the tank is filled to capacity.

   d. Do not allow mixers to be off for an extended period while material is still in the tank.

3. Flocculation Tanks

   a. Adjust paddle/turbine speed so that particles receive just enough agitation to remain suspended.

   b. Check to see that chemical is being added if no floc forms.

   c. Adjust all paddles to the same speed unless:

      (1) Sludge formation occurs at one point. Increase the paddle speed for this area.

      (2) Coagulant is added at the floc tanks. Increase the tip speed of the mixers immediately preceding the point of discharge.

      (3) Coagulant is added in the influent channel. Increase the tip speed of the first mixer.

4. Clarifiers

   a. If possible, all tanks should be kept in operation since the best sedimentation occurs with the greatest amount of surface area.

   b. Sludge control. Level should be kept at a minimum.

   c. Never store sludge in the clarifiers. Move it to the thickeners for storage.

   d. Scum Removal. Check periodically.

      Skimmer. Clean daily.

      Scum pit. Clean after each pumping.

   e. Clean weirs, scum baffles, and launders daily.

8.64 Abnormal Operation

This section contains a list of abnormal conditions that could occur at any time during the operation of a wastewater treatment plant. Included are recommendations that should help you adjust the chemical treatment system in order to maintain a high-quality effluent. Whenever you detect indications of abnormal conditions, increase the frequency of process monitoring.

1. High solids concentrations in effluent leaving the secondary clarifiers due to bulking sludge, rising sludge, or solids washout.

   a. Inspect chemical feeders for proper output.

   b. Run jar tests to determine if dosage requirements have changed.

   c. Examine overall plant operations to locate the cause of high solids.

   d. Increase sludge removal rates from clarifiers.

2. Unusually low suspended solids in effluent leaving the secondary clarifiers.

   a. Inspect chemical feeders for proper output.

   b. Run jar tests to determine if dosage requirements have changed.

   c. Record in log book conditions and dosage that produced low solids in the effluent. You need to know how you produced a good quality effluent.

3. High flows passing through the treatment plant.

   a. Prepare to feed a greater quantity of chemicals.

b. Run jar tests to determine the dosage that will produce rapid settling rates when detention times are reduced.

c. Be sure jar test flash mixing and flocculation times are similar to actual detention times through these units during the high flows.

4. Low flows passing through the treatment plant.

    a. Run jar tests to determine optimum dosage because longer detention times may allow a reduction in chemical dosage, which will reduce chemical costs.

    b. Watch for chemical overdoses that could produce toxic conditions in biological treatment processes or in the receiving waters.

5. A change in the pH of the water being treated by one or more units.

    a. Inspect chemical feeders to determine if the chemicals being added are causing the pH change.

    b. Run jar tests to determine if chemical feed rates need adjusting.

    c. Extreme pH changes may affect biological activity and effectiveness of disinfection. Try to control chemical feeders to minimize chemical changes.

    d. If existing chemical dosages will not cause coagulation, new chemicals may be required. For example, you may have to switch from one type of polyelectrolyte to another type.

6. A change in water temperature resulting from seasonal weather conditions, groundwater infiltration, or wet weather inflows.

    a. Run a jar test to determine if new chemical feed rates are required. Coagulation and settling rates change when the temperature changes.

8.65 Troubleshooting

Two common problems in physical–chemical treatment systems are foaming in the rapid-mix tanks and flocculators and no coagulation of suspended particles. Foaming can be controlled by the use of water spray from hoses or by surface spray nozzles installed on the tanks.

Any number of conditions could lead to a failure of the particles to coagulate properly. First, run a jar test to determine the proper chemical dosage for the wastewater being treated. Be sure the jar test chemicals are the proper strength. If the chemical dosage is correct, inspect the following items:

1. Chemical feed pump operation.
2. Chemical supply and valve positions.
3. Solution carrier water flow and valve positions.
4. Applied water for a significant change.
5. Actual feeder output by catching a timed sample.
6. Feed chemical strength.

Table 8.4 lists several other problems operators should watch for when operating a physical–chemical treatment system and suggests possible causes and solutions.

8.66 Safety

Most of the chemicals used in coagulation and flocculation processes are either caustic or corrosive. Operators who work with these chemicals must make it a habit to use the safe working procedures discussed in Section 8.48, "Safe Working Habits."

In addition to being exposed to dangerous chemicals, however, operators of coagulation, flocculation, and sedimentation systems also must be aware of the dangers associated with working around large tanks or basins. Drowning in a basin filled with wastewater is not a common occurrence, but it could happen at any time. Spilled polymers tend to be extremely slippery. If not promptly cleaned up, an operator could easily slip and fall into an open tank or basin. Always approach basins through areas that have appropriate walkways and be sure life preservers are readily available in work areas around open tanks or channels. Keep walkways clear of clutter and free of grease, oil, and chemicals. Be sure that approved guardrails are provided around all tanks and channels and make use of handrails in all work locations.

Before working on a piece of mechanical or electrical machinery, be sure all power is turned off, locked out, and properly tagged to prevent accidental start-up. If possible, physically block rotating arms or any part of a machine that could suddenly move or swing free. Avoid placing your arms or legs into any moving part of machinery. If you must work on a piece of equipment while it is operating, consider using an extension tool if the work can be done safely using one. Do not remove guards or safety shields around moving equipment unless absolutely necessary and be sure to replace them immediately when the maintenance work or repair is complete.

QUESTIONS

Write your answers in a notebook and then compare your answers with those on page 382.

8.6A   What water quality indicators should be monitored when operating a physical–chemical treatment process?

8.6B   List the items you would check during the start-up inspection of a chemical feed system.

8.6C   List the procedures you would follow for the normal operation of a chemical feed system in a coagulation-precipitation system.

8.6D   What abnormal conditions could be encountered in the water being treated when operating a physical–chemical treatment process?

8.6E   What are two common problems that could occur when operating a physical–chemical treatment process?

END OF LESSON 3 OF 3 LESSONS
on
PHYSICAL–CHEMICAL TREATMENT PROCESSES

Please answer the discussion and review questions next.

TABLE 8.4 TROUBLESHOOTING A PHYSICAL–CHEMICAL TREATMENT PROCESS

Problem	Cause	Solution
Excess scum buildup	Scum collection device	Inspect scum trough, spiral screw, and scum pumps.
Floc too small	Improper chemical dosage	Check dosage with jar test.
	Low chemical metering	Adjust metering.
	Chemical feed pump adjusted too low	Adjust feed pumps.
	Paddle speed in flocculators or rapid mix too fast	Decrease paddle speed.
	Short-circuiting	Baffling changes, adjust weir plates or port openings.
	Change in pH	Neutralize pH.
Floc too large, settles too soon	Improper chemical dosage	Check dosage with jar test.
	Metering setting	Check and adjust metering.
	Chemical make-up too strong	Check make-up and adjust feed.
	Too little dilution water	Check and adjust metering of dilution water.
	Paddle speed in rapid mix or flocculator too slow	Increase paddle speed.
	Coagulant aid added at wrong point	Optimize point of coagulant aid addition (in rapid mix, before flocculator, in flocculator).
Floating sludge	Sludge collectors not functioning properly	Check motors, chains, drives, and belts for smooth operation.
	Sludge pumping system malfunction	Check sludge pumps for operation, pipes for debris. Switch to standby pump.
	Change in influent character or flow rate	Check dosage with jar test and adjust.
	Excess coagulant aid	Check dosage with jar test and adjust.
Loss of solids over effluent weir	See "floc too small" problem above	
	Improper or misaligned baffling	Adjust baffling at inlet and outlet.
Thin sludge with deep sludge blanket	Cross-collectors not functioning	Check motors, drives, and chains for cross-collectors.
Sludge collector, jerky operation or inoperable	Broken sprocket, chain link, flight, or shear pin	Inspect and repair.
	Sludge blanket too deep	Pump out sludge. May have to drain basin and remove manually.

DISCUSSION AND REVIEW QUESTIONS

Chapter 8. PHYSICAL–CHEMICAL TREATMENT PROCESSES

(COAGULATION, FLOCCULATION, AND SEDIMENTATION)

(Lesson 3 of 3 Lessons)

Write the answers to these questions in your notebook. The question numbering continues from Lesson 2.

13. Why are plunger pumps used for metering chemicals?

14. What economic factors should be considered when selecting a chemical feeder?

15. Why should chemical feed equipment be cleaned before being shut down for an extended length of time?

16. Why are solids loadings especially important in the design of clarifiers treating industrial wastes?

17. What is the most effective solution for short-circuiting?

18. Why should you develop an operational strategy for a physical–chemical treatment process?

SUGGESTED ANSWERS

Chapter 8. PHYSICAL–CHEMICAL TREATMENT PROCESSES

(COAGULATION, FLOCCULATION, AND SEDIMENTATION)

ANSWERS TO QUESTIONS IN LESSON 1

Answers to questions on page 321.

8.0A Some locations have stringent discharge requirements because more and more wastes are being discharged into the receiving waters and increasing demands are being placed on the waters by water users.

8.0B NPDES stands for National Pollutant Discharge Elimination System.

Answers to questions on page 322.

8.1A Coagulation is the act of adding and mixing a coagulating chemical to destabilize the suspended particles allowing the particles to collide and "stick" together, forming larger particles.

8.1B Flocculation is the actual gathering together of smaller suspended particles into flocs, thus forming a more readily settleable mass.

8.1C Chemicals may be added to improve filter performance as well as to reduce emergency problems such as those created by sludge bulking in the secondary clarifier, upstream equipment failure, accidental spills entering the plant, and seasonal overloads.

8.1D When adding chemicals upstream from a biological treatment process, be sure that the chemical or its concentration is not toxic to the organisms treating the wastewater in the biological process.

Answers to questions on page 326.

8.2A When most of the solids particles in a wastestream have the same electrical charge (usually negative), the particles repel each other. The electrostatic repulsion prevents the particles from clumping into floc large enough and dense enough to settle out. Destabilization is the addition of chemicals to change or neutralize the charge of the particles so they will coagulate and can be separated from the wastestream.

8.2B Particles tend to have a negative ionic charge when wastewater has a pH on the low (acidic) side of pH 6.5.

8.2C Vigorous mixing during coagulation is important to ensure that the coagulants are thoroughly mixed into the wastestream and that the solids particles make physical contact with each other.

ANSWERS TO QUESTIONS IN LESSON 2

Answers to questions on page 328.

8.3A The four most common chemicals added to improve settling are alum, ferric chloride, lime, and polyelectrolytes (polymers).

8.3B Alum should be kept dry to prevent it from caking into a solid lump.

8.3C All mechanical equipment, such as conveyors, should be run until well cleaned of all alum before shutting down because the alum can harden and jam the equipment.

Answers to questions on page 331.

8.3D Safety precautions required for handling ferric chloride in concentrated forms should be the same as those for acids. Wear protective clothing, chemical goggles, rubber or plastic gloves, and rubber shoes. Flush all splashes off clothing and skin immediately.

8.3E Rubber or flexible piping with easy access and short runs will permit cleaning by squeezing the walls and washing out the broken scale. Solid piping that is plugged by scale usually requires replacement.

8.3F Polyelectrolytes will create an extremely slippery surface when wet. Clean up spills immediately.

8.3G Manufacturers' MSDSs recommend the use of inert absorbent material to clean polymer spills, such as sand or earth.

Answers to questions on page 333.

8.4A The three stages of testing used to select chemical products for a treatment process are: (1) preliminary screening, (2) dosage testing, and (3) full-scale trial.

8.4B Various types of chemicals or different doses of a single chemical are added to sample portions of wastewater in a jar test unit and all portions of the sample are rapidly mixed. After rapid mixing, the samples are slowly mixed to approximate the conditions in the plant. Mixing is stopped and the floc formed is allowed to settle. The appearance of the floc, the time required to form a floc, and the settling conditions are recorded. The supernatant is analyzed for turbidity, suspended solids, and pH. With this information, the operator selects the best chemical or best dosage to feed on the basis of clarity of effluent and minimum cost of chemicals.

8.4C Before a jar test unit is used, it is essential to know the volume of the test jars and the speed rates of the stirrers.

Answers to questions on page 341.

8.4D When preparing stock solutions of polymers, carefully mix the dry polymer into water. The water must be stirred rapidly while gradually sprinkling the polymer granules or powder over the water surface. Care must be taken to be sure that the granules of powder do not pile up in lumps on the water surface or cling to the sides of the mixing container, and that each granule of polymer is wetted individually.

8.4E Polymers are usually limited to 1.0 percent maximum solution strength so that the solution can be pumped (metered) easily.

8.4F The sampling point for a jar test should be located as near as possible to the location where treatment chemicals will be or are being fed into the flow stream.

ANSWERS TO QUESTIONS IN LESSON 3

Answers to questions on page 351.

8.5A Chemical solutions are prepared for feeding by mixing known amounts of chemicals and water together using a mechanical mixer. The resulting solution is stored in a day tank (holding tank) from which it is metered out at the proper dosage into the water being treated.

8.5B Common types of chemical feeders or metering equipment include:

1. Positive displacement pumps such as the plunger pump, gear pump, diaphragm pump, and progressive cavity pump.
2. Screw feeder.
3. Vibrating trough feeder.
4. Rotary feeder.
5. Belt-type gravimetric feeder.

Answers to questions on page 355.

8.5C Items that should be considered when selecting a chemical feeder include:

1. Total operating range
2. Accuracy
3. Repeatability
4. Resistance to corrosion
5. Dust control
6. Availability of parts
7. Safety
8. Economics

8.5D The following information regarding a chemical feeder operation should be recorded:

1. Flows
2. Characteristics of wastewater before treatment
3. Dosage and conditions of chemical treatment
4. Results after treatment
5. Abnormal conditions observed during treatment

Answers to questions on page 367.

8.5E Rapid mixing of coagulant chemicals may be accomplished in any of three modes: (1) high-speed mixers (impeller or turbine), (2) in-line blenders and pumps, and (3) baffled compartments or pipes (static mixers).

8.5F In tapered flocculation, a small dense floc is formed initially followed by aggregation to form a denser, larger floc.

8.5G An advantage of vertical-flow clarifier units is that flow is forced up through a sludge blanket, thus aiding in solids retention and improving flow control.

8.5H The four major factors used to predict the performance of clarifiers include: detention time, weir overflow rate, surface loading rate, and solids loading.

8.5I If the detention time in a clarifier is too short, there will be a carryover of particles into the effluent.

8.5J Short-circuiting in a clarifier may be caused by:

1. Differences in water density due to different temperatures existing at the surface and bottom of the clarifier.
2. Density differences due to suspended solids.
3. High inlet and outlet velocities.
4. Strong winds blowing along the surface of the tank.

8.5K The four major factors used to measure clarifier performance or efficiency include: flows, suspended solids, settleable solids, and floatable solids.

Answers to questions on page 373.

8.5L The main components of dissolved air flotation (DAF) units are: (1) air injection equipment, (2) agitated or unagitated pressurized retention tank, (3) recycle pump, (4) inlet or distribution assembly, (5) sludge scrapers, and (6) an effluent baffle.

8.5M 1. The function of the distribution box is to allow the air to come out of solution in the form of minute air bubbles that attach to the solids and cause them to rise to the surface.
2. The retention tank provides a location to dissolve air into the liquid.
3. The effluent baffle is provided to keep the floated solids from contaminating the effluent.

8.5N A sight glass should be provided on the retention tank to periodically check the level of the air-liquid interface because on occasion the float mechanisms may fail and the retention tank will either fill completely with liquid or fill completely with air.

8.5O The performance of DAF thickeners depends on (1) type of sludge, (2) age of the feed sludge, (3) solids and hydraulic loadings, (4) air to solids (A/S) ratio, (5) recycle rate, and (6) sludge blanket depth.

8.5P The age of the sludge usually does not affect flotation performance as drastically as it affects gravity concentrators. A relatively old sludge has a natural tendency to float due to gasification and this natural buoyancy will have little or no adverse effect on the operation of flotation thickeners.

8.5Q Determine the hydraulic loading (GPD/sq ft) for a 20-foot diameter DAF unit. The influent flow is 100 GPM. The formula for surface area of a circular tank is:

$$\text{Area} = \frac{\pi}{4} \text{Diameter}^2 \quad \text{or} \quad \text{Area} = 0.785 \text{ Diameter}^2$$

Known
DAF Diameter, ft = 20 ft
Flow, GPM = 100 GPM

Unknown
Hydraulic Surface Loading, GPD/sq ft

$$\text{Area} = \frac{\pi}{4} \text{Diameter}^2 \quad \text{or} \quad \text{Area} = 0.785 \text{ Diameter}^2$$

1. Determine the liquid surface area, sq ft.

$$\text{Surface Area, sq ft} = \frac{\pi}{4} \times (\text{Diameter, ft})^2$$

$$= 0.785 \times (20 \text{ ft})^2$$

$$= 314 \text{ sq ft}$$

2. Calculate the hydraulic surface loading, GPD/sq ft.

$$\text{Surface Loading, GPD/sq ft} = \frac{\text{Flow, GPD}}{\text{Surface Area, sq ft}}$$

$$= \frac{100 \text{ gal/min} \times 1{,}440 \text{ min/day}}{314 \text{ sq ft}}$$

$$= 459 \text{ GPD/sq ft}$$

8.5R For Problem 8.5Q, determine the solids loading, A/S ratio, and recycle flow rate (GPM), if the influent sludge has a suspended solids concentration of 0.75% (7,500 mg/L), and is supplied at a rate of 2.5 cu ft/min. Air is supplied at a rate of 0.75 cu ft/min and a recycle ratio of 100 percent is required.

Known
Conditions in Problem 8.5Q
Influent Sludge, % = 0.75%
, mg/L = 7,500 mg/L
Influent Sludge Flow, CFM = 2.5 cu ft/min
Air, CFM = 0.75 cu ft/min
Recycle Ratio, % = 100%

Unknown
1. Solids Loading, lbs/day/sq ft
2. A/S Ratio, lbs air/lbs solids
3. Recycle Flow Rate, GPM

1. Determine solids applied, lbs/day.

$$\text{Solids Applied, lbs/day} = \text{Flow} \frac{\text{cu ft}}{\text{min}} \times 1{,}440 \frac{\text{min}}{\text{day}} \times 62.4 \frac{\text{lbs}}{\text{cu ft}} \times \text{SS}, \frac{\%}{100\%}$$

$$= 2.5 \frac{\text{cu ft}}{\text{min}} \times 1{,}440 \frac{\text{min}}{\text{day}} \times 62.4 \frac{\text{lbs}}{\text{cu ft}} \times 0.75 \frac{\%}{100\%}$$

$$= 1{,}685 \text{ lbs/day}$$

2. Calculate solids loading, lbs/day/sq ft.

$$\text{Solids Loading, lbs/day/sq ft} = \frac{\text{Solids Applied, lbs/day}}{\text{Surface Area, sq ft}}$$

$$= \frac{1{,}685 \text{ lbs/day}}{314 \text{ sq ft}}$$

$$= 5.4 \text{ lbs/day/sq ft}$$

3. Determine the air supply in pounds per hour.

$$\text{Air Supply, lbs/hr} = \text{Air Flow, } \frac{\text{cu ft}}{\text{min}} \times 60 \frac{\text{min}}{\text{hr}} \times 0.075 \frac{\text{lb air}}{\text{cu ft air}}$$

$$= 0.75 \frac{\text{cu ft}}{\text{min}} \times 60 \frac{\text{min}}{\text{hr}} \times 0.075 \frac{\text{lb}}{\text{cu ft}}$$

$$= 3.375 \text{ lbs/hr}$$

4. Determine the solids applied in pounds per hour.

$$\text{Solids Applied, lbs/hr} = \frac{\text{Solids Applied, lbs/day}}{24 \text{ hr/day}}$$

$$= \frac{1{,}685 \text{ lbs/day}}{24 \text{ hr/day}}$$

$$= 70.2 \text{ lbs/hr}$$

5. Calculate the pounds of air to pounds of solids (A/S) ratio.

$$\frac{\text{Air, lbs}}{\text{Solids, lbs}} = \frac{\text{Air Supply, lbs/hr}}{\text{Solids Applied, lbs/hr}}$$

$$= \frac{3.375 \text{ lbs air/hr}}{70.2 \text{ lbs solids/hr}}$$

$$= 0.05 \text{ lb air/lb solids}$$

6. Determine the recycle flow rate, GPM.

$$\text{Recycle Flow Rate, GPM} = \text{Inflow, GPM} \times \text{Recycle Ratio, } \frac{\%}{100\%}$$

$$= 2.5 \frac{\text{cu ft}}{\text{min}} \times 7.5 \frac{\text{gal}}{\text{cu ft}} \times \frac{100\%}{100\%}$$

$$= 18.8 \text{ GPM}$$

$$\text{or } = 19 \text{ GPM for pumping rate}$$

8.5S Determine the suspended solids removal efficiency (%) and the concentration factor (cf) if a DAF unit receives an influent sludge at 1.0 percent (10,000 mg/L) suspended solids. The effluent is at 50 mg/L suspended solids and the float or thickened sludge is at a concentration of 3.8 percent.

Known		Unknown
DAF Unit		1. Solids Removal Eff, %
Infl Solids, %	= 1.0%	2. Concentration Factor (cf)
, mg/L	= 10,000 mg/L	
Effl Solids, % (Thickened Sludge)	= 3.8%	
Effl Liquid SS, mg/L	= 50 mg/L	

1. Determine the suspended solids removal efficiency.

$$\text{SS Efficiency, \%} = \frac{(\text{SS Infl, mg/}L - \text{SS Effl, mg/}L)}{\text{SS Infl, mg/}L} \times 100\%$$

$$= \frac{(10{,}000 \text{ mg/}L - 50 \text{ mg/}L)}{10{,}000 \text{ mg/}L} \times 100\%$$

$$= 99.5\%$$

2. Determine the concentration factor for the thickened solids.

$$\text{Concentration Factor (cf)} = \frac{\text{Thickened Solids Concentration, \%}}{\text{Influent Solids Concentration, \%}}$$

$$= \frac{3.8\%}{1.0\%}$$

$$= 3.8$$

Answer to question on page 374.

8.5T *PROBLEM*. Poor effluent quality (high suspended solids) and thinner than normal sludge.

Possible Causes	Possible Solutions
1. A/S low.	1. Increase air input. Repair or turn on compressor.
2. Pressure too low or too high.	2. Open or close valve.
3. Recycle pump inoperative.	3. Turn on recycle pump.
4. Reaeration pump inoperative.	4. Turn on reaeration pump.
5. Chemical addition inadequate.	5. Increase dosage.
6. Loading excessive.	6. Lower flow rate.

Answers to questions on page 377.

8.6A Water quality indicators that should be monitored when operating a physical–chemical treatment process include alkalinity, pH, temperature, turbidity, and suspended solids.

8.6B Items to be checked during the start-up inspection of a chemical feed system include:

1. Level alarms in tanks.
2. Calibration of flow and metering systems.
3. Water blenders and dilution mixers.
4. Temperature and pressure of dilution water.
5. In-line mixers.

8.6C Procedures for the normal operation of a chemical feed system in a coagulation-precipitation system include:

1. Perform a jar test to determine the proper chemical dosage.
2. Ensure proper chemical dilution.
3. Set controls for the proper feed rate.
4. Report in the operating log the amount of chemicals used per unit time.

8.6D Abnormal conditions that could be encountered in the water being treated when operating a physical–chemical treatment process include high solids, high or low flows, and changes in pH and temperature.

8.6E Two common problems that could occur when operating a physical–chemical treatment process are foaming in the rapid-mix tanks and flocculators and no coagulation of suspended particles.

CHAPTER 9

FILTRATION

by

Ross Gudgel

James L. Johnson

Robert G. Blanck

Francis J. Brady

TABLE OF CONTENTS
Chapter 9. FILTRATION

	Page
OBJECTIVES	389
WORDS	390

LESSON 1

- 9.0 TYPES OF FILTRATION SYSTEMS ... 395
- 9.1 SOLIDS REMOVAL FROM INDUSTRIAL WASTESTREAMS USING GRAVITY FILTERS ... 395
  by Ross Gudgel and James L. Johnson
  - 9.10 Gravity Filters ... 395
    - 9.100 Use of Filters ... 395
    - 9.101 Description of Filters ... 395
    - 9.102 Filtering Process ... 395
    - 9.103 Backwashing Process ... 396
  - 9.11 Methods of Filtration ... 396
    - 9.110 Filter Types ... 396
    - 9.111 Surface Straining ... 396
    - 9.112 Depth Filtration ... 396
  - 9.12 Location of Filters in a Treatment System ... 400
  - 9.13 Major Parts of a Filtering System ... 400
    - 9.130 Inlet ... 400
    - 9.131 Filter Media ... 400
    - 9.132 Filter Underdrains ... 400
    - 9.133 Filter Media Scouring Systems ... 400
    - 9.134 Wash Water Troughs ... 400
    - 9.135 Backwash Water Drain ... 400
    - 9.136 Backwash Water Supply ... 402
    - 9.137 Backwash Water Rate Control ... 402
    - 9.138 Used Backwash Water Holding Tank ... 402
    - 9.139 Effluent Rate-Control Valve ... 405
  - 9.14 Filter System Instrumentation ... 405
    - 9.140 Head Loss ... 405
    - 9.141 Filter Flow Rate Indicator and Totalizer ... 405

		9.142	Applied Turbidity	405
		9.143	Effluent Turbidity	406
		9.144	Indicator Lights	406
		9.145	Alarms	406
	9.15	Operation of Gravity Filters		406
		9.150	Pre-Start Checklist	406
		9.151	Normal Operation	406
			9.1510 Filtering	406
			9.1511 Backwashing	406
		9.152	Abnormal Operations	408
		9.153	Operational Strategy	410
		9.154	Shutdown of a Gravity Filter	412
	9.16	Troubleshooting		412
	9.17	Safety		414
	9.18	Review of Plans and Specifications		414
DISCUSSION AND REVIEW QUESTIONS				415

LESSON 2

9.2	SOLIDS REMOVAL FROM WASTESTREAMS USING INERT-MEDIA PRESSURE FILTERS			416
	by Ross Gudgel			
	9.20	Use of Inert-Media Pressure Filters		416
	9.21	Pressure Filter Facilities		416
		9.210	Holding Tank (Wet Well)	416
		9.211	Filter Feed Pumps	418
		9.212	Chemical Feed Systems	418
		9.213	Filters	422
		9.214	Backwash System	423
		9.215	Decant Tank (Backwash Recovery)	425
	9.22	Operation		426
		9.220	Operational Strategy	426
		9.221	Abnormal Operation	426
	9.23	Maintenance		426
	9.24	Safety		427
	9.25	Review of Plans and Specifications		429
	9.26	Acknowledgments		429
	9.27	Additional Reading		429
DISCUSSION AND REVIEW QUESTIONS				429

LESSON 3

9.3 SOLIDS REMOVAL FROM WASTESTREAMS USING CONTINUOUS BACKWASH, UPFLOW, DEEP-BED SILICA SAND MEDIA FILTERS 430
by Ross Gudgel

- 9.30 Use of Continuous Backwash, Upflow, Deep-Bed Silica Sand Media Filters 430
- 9.31 Auxiliary Equipment 433
  - 9.310 Channel or Piping 433
  - 9.311 Filter Influent and Effluent Turbidity Metering 433
  - 9.312 Filter Influent and Effluent Flow Metering 436
  - 9.313 Coagulation and Flocculation Equipment 436
- 9.32 Operation of Continuous Backwash, Upflow, Deep-Bed Silica Sand Media Filters 436
  - 9.320 Normal Operation 436
  - 9.321 Operating Strategy 439
  - 9.322 Abnormal Operation 440
- 9.33 Maintenance 440
- 9.34 Safety 445
- 9.35 Review of Plans and Specifications 445

DISCUSSION AND REVIEW QUESTIONS 446

LESSON 4

9.4 CROSS FLOW MEMBRANE FILTRATION 447
by Robert G. Blanck and Francis J. Brady

- 9.40 Types of Membrane Filtration Processes 447
  - 9.400 Microfiltration 447
  - 9.401 Ultrafiltration 447
  - 9.402 Nanofiltration 449
  - 9.403 Reverse Osmosis 449
- 9.41 Membrane Materials 449
- 9.42 Process Time and Membrane Life 449
- 9.43 Membrane Configurations 449
  - 9.430 Tubular Membranes 449
  - 9.431 Hollow Fiber Membranes 449
  - 9.432 Spiral Membranes 451
  - 9.433 Plate and Frame Membranes 452
- **9.5 BASIC ELEMENTS OF A MEMBRANE FILTRATION PROCESS** 453
  - 9.50 Concentrating Components in the Wastewater 453
  - 9.51 Permeate Discharge to Drain 453
  - 9.52 Membrane Flux 453
  - 9.53 Retention 453

		9.54	Calculating the Concentration of Waste Components	454
		9.55	Transmembrane Pressure	455
		9.56	Recirculation Flow	455
		9.57	Temperature	456
		9.58	Concentration-Dependent Flux	456
		9.59	Membrane Fouling	456
	9.6		OPERATION OF A CROSS FLOW MEMBRANE SYSTEM	456
		9.60	Filter Staging	456
		9.61	Operating Modes	456
		9.62	Feed Pretreatment	458
		9.63	System Operation	458
		9.64	Cleaning Procedures	460
		9.65	Water Flux Measurements	461
		9.66	Sampling	462
		9.67	Recordkeeping	463
	9.7		SAFETY PRECAUTIONS WITH MEMBRANE SYSTEMS	463
	9.8		REVERSE OSMOSIS	463
		9.80	What Is Reverse Osmosis?	463
		9.81	Reverse Osmosis Membrane Structure and Composition	465
		9.82	Membrane Performance and Properties	465
		9.83	Definition of Flux	466
		9.84	Mineral Rejection	467
		9.85	Effects of Feedwater Temperature and pH on Membrane Performance	469
		9.86	Recovery	469
	9.9		REFERENCE	474
DISCUSSION AND REVIEW QUESTIONS				476
SUGGESTED ANSWERS				476

OBJECTIVES
Chapter 9. FILTRATION

Following completion of Chapter 9, you should be able to:

1. Identify and describe the components of inert-media gravity and pressure filters.
2. Explain how membrane filters operate.
3. Start up and shut down filters.
4. Safely operate and maintain filters.
5. Troubleshoot a filtration system.
6. Develop operational strategies for inert-media and membrane filtration systems.
7. Review plans and specifications for filter systems.

WORDS
Chapter 9. FILTRATION

atm

The abbreviation for atmosphere. One atmosphere is equal to 14.7 psi or 100 kPa.

AIR BINDING

The clogging of a filter, pipe, or pump due to the presence of air released from water. Air entering the filter media is harmful to both the filtration and backwash processes. Air can prevent the passage of water during the filtration process and can cause the loss of filter media during the backwash process.

ALGAE (AL-jee)

Microscopic plants containing chlorophyll that live floating or suspended in water. They also may be attached to structures, rocks, or other submerged surfaces. Excess algal growths can impart tastes and odors to potable water. Algae produce oxygen during sunlight hours and use oxygen during the night hours. Their biological activities appreciably affect the pH, alkalinity, and dissolved oxygen of the water.

ANNULAR (AN-yoo-ler) SPACE

A ring-shaped space located between two circular objects. For example, the space between the outside of a pipe liner and the inside of a pipe.

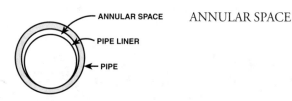

BOD (pronounce as separate letters)

Biochemical Oxygen Demand. The rate at which organisms use the oxygen in water or wastewater while stabilizing decomposable organic matter under aerobic conditions. In decomposition, organic matter serves as food for the bacteria and energy results from its oxidation. BOD measurements are used as a surrogate measure of the organic strength of wastes in water.

BULKING

Clouds of billowing sludge that occur throughout secondary clarifiers and sludge thickeners when the sludge does not settle properly. In the activated sludge process, bulking is usually caused by filamentous bacteria or bound water.

COD (pronounce as separate letters)

Chemical Oxygen Demand. A measure of the oxygen-consuming capacity of organic matter present in wastewater. COD is expressed as the amount of oxygen consumed from a chemical oxidant in mg/L during a specific test. Results are not necessarily related to the biochemical oxygen demand (BOD) because the chemical oxidant may react with substances that bacteria do not stabilize.

COLIFORM (KOAL-i-form)

A group of bacteria found in the intestines of warm-blooded animals (including humans) and also in plants, soil, air, and water. The presence of coliform bacteria is an indication that the water is polluted and may contain pathogenic (disease-causing) organisms. Fecal coliforms are those coliforms found in the feces of various warm-blooded animals, whereas the term "coliform" also includes other environmental sources.

CONCENTRATION POLARIZATION

(1) A buildup of retained particles on the membrane surface due to dewatering of the feed closest to the membrane. The thickness of the concentration polarization layer is controlled by the flow velocity across the membrane.

(2) Used in corrosion studies to indicate a depletion of ions near an electrode.

(3) The basis for chemical analysis by a polarograph.

CONFINED SPACE

Confined space means a space that:

(1) Is large enough and so configured that an employee can bodily enter and perform assigned work; and

(2) Has limited or restricted means for entry or exit (for example, manholes, tanks, vessels, silos, storage bins, hoppers, vaults, and pits are spaces that may have limited means of entry); and

(3) Is not designed for continuous employee occupancy.

Also see DANGEROUS AIR CONTAMINATION and OXYGEN DEFICIENCY.

DALTON

A unit of mass designated as one-sixteenth the mass of oxygen-16, the lightest and most abundant isotope of oxygen. The dalton is equivalent to one mass unit.

DECANT (de-KANT)

To draw off the upper layer of liquid (water) after the heavier material (a solid or another liquid) has settled.

DELAMINATION (DEE-lam-uh-NAY-shun)

Separation of a membrane or other material from the backing material on which it is cast.

FILTER AID

A chemical (usually a polymer) added to water to help remove fine colloidal suspended solids.

FLUIDIZED (FLOO-id-i-zd)

A mass of solid particles that is made to flow like a liquid by injection of water or gas is said to have been fluidized. In water and wastewater treatment, a bed of filter media is fluidized by backwashing water through the filter.

FLUX

A flowing or flow.

HEAD LOSS

The head, pressure, or energy (they are the same) lost by water flowing in a pipe or channel as a result of turbulence caused by the velocity of the flowing water and the roughness of the pipe, channel walls, or restrictions caused by fittings. Water flowing in a pipe loses head, pressure, or energy as a result of friction losses. The head loss through a filter is due to friction losses caused by material building up on the surface or in the top part of a filter. Also called FRICTION LOSS.

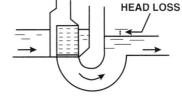

HYDROLYSIS (hi-DROLL-uh-sis)

(1) A chemical reaction in which a compound is converted into another compound by taking up water.

(2) Usually a chemical degradation of organic matter.

INDICATOR

(1) (Chemical indicator) A substance that gives a visible change, usually of color, at a desired point in a chemical reaction, generally at a specified end point.

(2) (Instrument indicator) A device that indicates the result of a measurement, usually using either a fixed scale and movable indicator (pointer), such as a pressure gauge, or a moving chart with a movable pen like those used on a circular flow-recording chart. Also called a RECEIVER.

INTEGRATOR

A device or meter that continuously measures and calculates (adds) a process rate variable in cumulative fashion; for example, total flows displayed in gallons, million gallons, cubic feet, or some other unit of volume measurement. Also called a TOTALIZER.

JAR TEST

A laboratory procedure that simulates coagulation/flocculation with differing chemical doses. The purpose of the procedure is to estimate the minimum coagulant dose required to achieve certain water quality goals. Samples of water to be treated are placed in six jars. Various amounts of chemicals are added to each jar, stirred, and the settling of solids is observed. The lowest dose of chemicals that provides satisfactory settling is the dose used to treat the water.

LAUNDERS

Sedimentation basin and filter discharge channels consisting of overflow weir plates (in sedimentation basins) and conveying troughs.

MATERIAL SAFETY DATA SHEET (MSDS)

A document that provides pertinent information and a profile of a particular hazardous substance or mixture. An MSDS is normally developed by the manufacturer or formulator of the hazardous substance or mixture. The MSDS is required to be made available to employees and operators or inspectors whenever there is the likelihood of the hazardous substance or mixture being introduced into the workplace. Some manufacturers are preparing MSDSs for products that are not considered to be hazardous to show that the product or substance is not hazardous.

MIL

A unit of length equal to 0.001 of an inch. The diameter of wires and tubing is measured in mils, as is the thickness of plastic sheeting.

MUDBALLS

Material, approximately round in shape, that forms in filters and gradually increases in size when not removed by the backwashing process. Mudballs vary from pea-sized up to golf-ball-sized or larger.

NPDES PERMIT

National Pollutant Discharge Elimination System permit is the regulatory agency document issued by either a federal or state agency that is designed to control all discharges of potential pollutants from point sources and stormwater runoff into US waterways. NPDES permits regulate discharges into US waterways from all point sources of pollution, including industries, municipal wastewater treatment plants, sanitary landfills, large animal feedlots, and return irrigation flows.

NEPHELOMETRIC (neff-el-o-MET-rick)

A means of measuring turbidity in a sample by using an instrument called a nephelometer. A nephelometer passes light through a sample and the amount of light deflected (usually at a 90-degree angle) is then measured.

ORIFICE (OR-uh-fiss)

An opening (hole) in a plate, wall, or partition. An orifice flange or plate placed in a pipe consists of a slot or a calibrated circular hole smaller than the pipe diameter. The difference in pressure in the pipe above and at the orifice may be used to determine the flow in the pipe. In a trickling filter distributor, the wastewater passes through an orifice to the surface of the filter media.

PATHOGENIC (path-o-JEN-ick) ORGANISMS

Organisms, including bacteria, viruses, or cysts, capable of causing diseases (such as giardiasis, cryptosporidiosis, typhoid, cholera, dysentery) in a host (such as a person). Also called PATHOGENS.

PERMEATE (PURR-me-ate)

(1) To penetrate and pass through, as water penetrates and passes through soil and other porous materials.

(2) The liquid (demineralized water) produced from the reverse osmosis process that contains a low concentration of dissolved solids.

POLYELECTROLYTE (POLY-ee-LECK-tro-lite)

A high-molecular-weight (relatively heavy) substance, having points of positive or negative electrical charges, that is formed by either natural or synthetic (manmade) processes. Natural polyelectrolytes may be of biological origin or obtained from starch products or cellulose derivatives. Synthetic polyelectrolytes consist of simple substances that have been made into complex, high-molecular-weight substances. Used with other chemical coagulants to aid in binding small suspended particles to larger chemical flocs for their removal from water. Often called a POLYMER.

POLYMER (POLY-mer)

A long-chain molecule formed by the union of many monomers (molecules of lower molecular weight). Polymers are used with other chemical coagulants to aid in binding small suspended particles to larger chemical flocs for their removal from water. Also see POLYELECTROLYTE.

POTTING COMPOUNDS

Sealing and holding compounds (such as epoxy) used in electrode probes.

PROGRAMMABLE LOGIC CONTROLLER (PLC)

A small computer that controls process equipment (variables) and can control the sequence of valve operations.

RECEIVER

A device that indicates the result of a measurement, usually using either a fixed scale and movable indicator (pointer), such as a pressure gauge, or a moving chart with a movable pen like those used on a circular flow-recording chart. Also called an INDICATOR.

RISING SLUDGE

Rising sludge occurs in the secondary clarifiers of activated sludge plants when the sludge settles to the bottom of the clarifier, is compacted, and then starts to rise to the surface, usually as a result of denitrification, or anaerobic biological activity that produces carbon dioxide and/or methane.

ROTAMETER (ROTE-uh-ME-ter)

A device used to measure the flow rate of gases and liquids. The gas or liquid being measured flows vertically up a tapered, calibrated tube. Inside the tube is a small ball or bullet-shaped float (it may rotate) that rises or falls depending on the flow rate. The flow rate may be read on a scale behind or on the tube by looking at the middle of the ball or at the widest part or top of the float.

SPC CHART

Statistical Process Control chart. A plot of daily performance such as a trend chart.

SHORT-CIRCUITING

A condition that occurs in tanks or basins when some of the flowing water entering a tank or basin flows along a nearly direct pathway from the inlet to the outlet. This is usually undesirable since it may result in shorter contact, reaction, or settling times in comparison with the theoretical (calculated) or presumed detention times.

SPECIFIC GRAVITY

(1) Weight of a particle, substance, or chemical solution in relation to the weight of an equal volume of water. Water has a specific gravity of 1.000 at 4°C (39°F). Particulates with specific gravity less than 1.0 float to the surface and particulates with specific gravity greater than 1.0 sink.

(2) Weight of a particular gas in relation to the weight of an equal volume of air at the same temperature and pressure (air has a specific gravity of 1.0). Chlorine gas has a specific gravity of 2.5.

SURFACTANT (sir-FAC-tent)

Abbreviation for surface-active agent. The active agent in detergents that possesses a high cleaning ability.

TOTALIZER

A device or meter that continuously measures and calculates (adds) a process rate variable in cumulative fashion; for example, total flows displayed in gallons, million gallons, cubic feet, liters, cubic meters, or some other unit of volume measurement. Also called an INTEGRATOR.

TRAMP OIL

Oil that comes to the surface of a tank due to natural flotation. Also called free oil.

TURBIDITY (ter-BID-it-tee)

The cloudy appearance of water caused by the presence of suspended and colloidal matter. In the waterworks field, a turbidity measurement is used to indicate the clarity of water. Technically, turbidity is an optical property of the water based on the amount of light reflected by suspended particles. Turbidity cannot be directly equated to suspended solids because white particles reflect more light than dark-colored particles and many small particles will reflect more light than an equivalent large particle.

TURBIDITY (ter-BID-it-tee) UNITS (TU)

Turbidity units are a measure of the cloudiness of water. If measured by a nephelometric (deflected light) instrumental procedure, turbidity units are expressed in nephelometric turbidity units (NTU) or simply TU. Those turbidity units obtained by visual methods are expressed in Jackson turbidity units (JTU), which are a measure of the cloudiness of water; they are used to indicate the clarity of water. There is no real connection between NTUs and JTUs. The Jackson turbidimeter is a visual method and the nephelometer is an instrumental method based on deflected light.

WATER HAMMER

The sound like someone hammering on a pipe that occurs when a valve is opened or closed very rapidly. When a valve position is changed quickly, the water pressure in a pipe will increase and decrease back and forth very quickly. This rise and fall in pressures can cause serious damage to the system.

CHAPTER 9. FILTRATION

(Lesson 1 of 4 Lessons)

9.0 TYPES OF FILTRATION SYSTEMS

The purpose of filtration in industrial wastewater treatment facilities is to remove particulate impurities so that the treated water can be disposed of or recycled. Solids particles are commonly removed from wastestreams by passing the wastewater through a bed of granular media (such as sand), which traps the solids and permits the cleaned water to pass through. When the contaminants to be removed are too small to be captured by granular media, membrane filters are frequently used. In this type of process, wastewater flows under pressure across the surface of a porous material (the membrane). The pores (openings) in the membrane separate the contaminants from the wastestream by permitting the liquid portion to pass through the membrane but blocking the passage of the solids.

9.1 SOLIDS REMOVAL FROM INDUSTRIAL WASTESTREAMS USING GRAVITY FILTERS
by Ross Gudgel and James L. Johnson

9.10 Gravity Filters

9.100 Use of Filters

The use of gravity filtration is second only to gravity sedimentation for the separation of wastewater solids. This same process, using deep-bed filtration and granular media, has long been used in municipal and industrial water supplies. However, gravity filter systems are more frequently used for domestic water supplies that have much lower suspended solids concentrations than are found in industrial wastewater treatment facilities.

The following specific applications have been observed in industrial settings:

1. Removal of residual biological floc in settled effluents from secondary treatment by trickling filters or activated sludge processes.
2. Removal of residual chemical-biological floc after alum, iron, or lime precipitation of phosphate in secondary settling tanks of biological treatment processes.
3. Removal of solids remaining after the chemical coagulation of wastewaters from tertiary or independent physical–chemical wastewater treatment processes.

9.101 Description of Filters

Applied water generally flows through wastewater filters from top to bottom. The applied water is distributed evenly over the surface of the filter media through an inlet distribution system. This may be the same system that is used later to uniformly collect the dirty backwash water.

The water travels through the filter media where the solids are trapped. The filter bed may be made up of one or several materials or several grades of materials. This is determined by the designer based primarily upon the quality of the applied water.

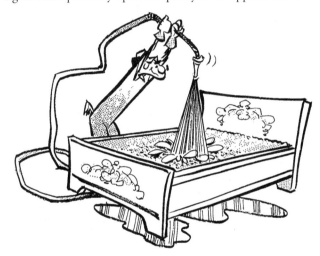

The underdrain system is designed to collect the filtered water uniformly throughout the bed. It also is used to uniformly apply the backwash water during backwashing. The system design must prevent the filter media from passing into the underdrain system, thereby being lost from the bed.

A surface wash system is beneficial during backwashing to scrub the surface mat of accumulated solids, thus breaking it up with minimum amounts of water. In some installations, air is used in place of surface washing as a means of breaking up the accumulated solids.

Valves control the volume, direction, and duration of flows through the unit, and instruments are used to monitor and record the volumes and quality of the water being processed through the filtering system.

9.102 Filtering Process

Water to be treated enters at the top of the filter bed through an inlet valve and is distributed over the entire filter surface. The water passes evenly down through the media (sand) and leaves the solids behind. Filtered water then travels out the bottom of the filter and into the underdrain collection system, which is designed to uniformly collect the flow. Once inside the underdrain collection system, the water passes through a flowmeter and rate-control valve. The rate-control valve maintains the desired flow through the filter and prevents backwash water from mixing with the filtered water during backwashing.

Most gravity filters operate on a batch basis whereby the filter operates until its capacity to remove solids is nearly reached but before solids break through into the effluent. At this time it is completely removed from service and cleaned. Other designs are available that filter continuously with a portion of the media always undergoing cleaning. The cleaning of the media may take place either externally or in place.

9.103 Backwashing Process

As suspended solids are removed from water, the filter media becomes clogged. This is indicated by a *HEAD LOSS*[1] reading. Through operating experience, the maximum head loss before backwashing will be determined. The filter should be backwashed after the solids capacity of the media has been reached, but before solids pass through the filter and begin to appear in the effluent (a condition known as breakthrough).

Backwashing consists of closing valves to stop influent flow and to protect the filtered water. Backwash water either flows by gravity or is pumped to the filter. This water flows through the underdrain system and back through the media. As the water flows through the media, the sand particles are lifted and are cleaned by rubbing against each other. The solids retained by the media are washed away with the backwash water and the media has been cleaned.

When the media is cleaned externally, it is removed from the filter bed, cleaned in a separate system, and recycled back into the bed. In-process cleaning involves washing a small section of the filter bed with a traveling backwash water or air-pulsing system while the remainder of the bed remains in service.

9.11 Methods of Filtration

9.110 Filter Types

Most gravity filters used in wastewater treatment are "rapid sand" filters (Figure 9.1(a)). They also may be called "downflow" (water flows down through the bed) or "static bed" (bed does not move or expand when filtering) filters. These filters operate continuously until they must be shut down for backwashing. Other designs, such as the upflow and the biflow, are on the market. Both of these designs are attempts to use more of the filter media, thereby removing and holding more solids per filter run.

In the upflow filter (Figures 9.1(b) and 9.2), water enters at the bottom of the filter and is removed from the top. The biflow system has water applied at both the top and bottom and water is withdrawn from the interior of the bed. Filters are always backwashed in an upflow direction regardless of the operating flow direction.

9.111 Surface Straining

Downflow filters are designed to remove suspended solids by either the surface-straining method or the depth-filtration method. In surface straining, the filter is designed to remove the solids at the very top of the media. The fine grade-sized media is uniform throughout the bed. Because of this conformity, surface-straining systems will have a rapid head loss buildup, short filter runs, and they must be backwashed frequently. There are, however, no problems with breakthrough of solids. The solids compress into a mat at the surface, which aids in removing solids; however, the mat is difficult to remove during backwashing. Backwashing a surface-straining system, although needed more frequently, requires less water per wash than does a depth-filtration system.

9.112 Depth Filtration

Depth filtration is designed to permit the solids to penetrate deep into the media, thereby capturing the solids within as well as on the surface of the media. Depth filtration will have a slower buildup of head loss, but solids will break through more readily than with surface straining.

To reduce breakthrough, yet retain depth filtering, the multi-media design is used. This combines a fine, denser media (sand) on the bottom with a coarse, lighter media (anthracite coal) on the top (Figures 9.1(d) and 9.3). The coarse media remove large solids that would quickly clog finer media. The fine media will surface-strain solids that penetrate the full depth of the coarse media bed thereby prevent a breakthrough of solids. The filter is designed to prevent the fine media from escaping unless the head loss becomes too great.

QUESTIONS

Write your answers in a notebook and then compare your answers with those on page 476.

9.1A Do most gravity filters operate on a batch or on a continuous basis?

9.1B When should a gravity filter be cleaned?

9.1C What is meant by the following terms that are used to describe "rapid sand" filters:
1. Downflow?
2. Static bed?

9.1D From what part of the filter are solids removed by
1. Surface straining?
2. Depth filtration?

---

[1] *Head Loss.* The head, pressure, or energy (they are the same) lost by water flowing in a pipe or channel as a result of turbulence caused by the velocity of the flowing water and the roughness of the pipe, channel walls, or restrictions caused by fittings. Water flowing in a pipe loses head, pressure, or energy as a result of friction losses. The head loss through a filter is due to friction losses caused by material building up on the surface or in the top part of a filter. Also called FRICTION LOSS.

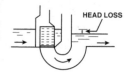

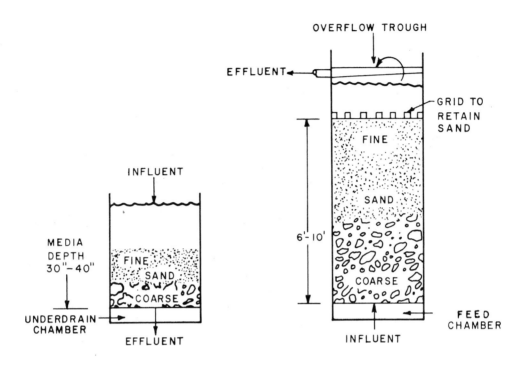

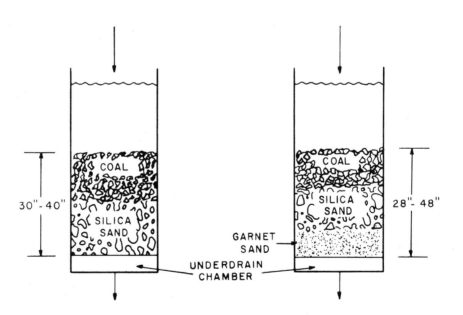

Fig. 9.1 Filter configurations
(Source: EPA *PROCESS DESIGN MANUAL FOR SUSPENDED SOLIDS REMOVAL*)

398 Treatment Plants

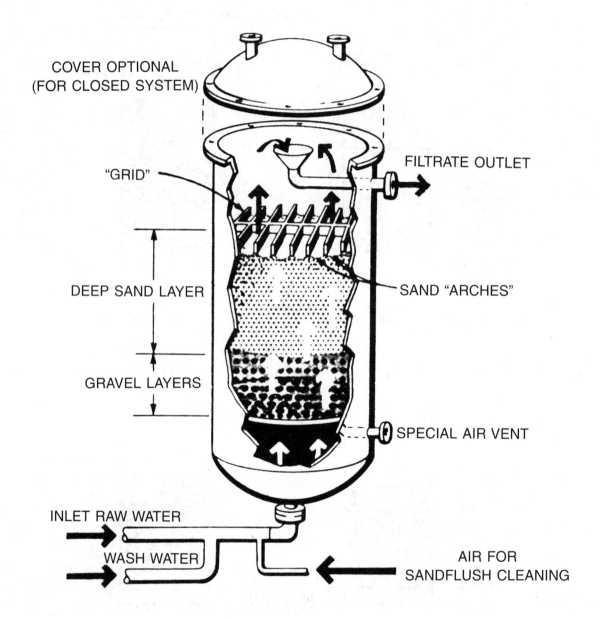

Fig. 9.2 Cross section of upflow filter
(Source: EPA *PROCESS DESIGN MANUAL FOR SUSPENDED SOLIDS REMOVAL*)

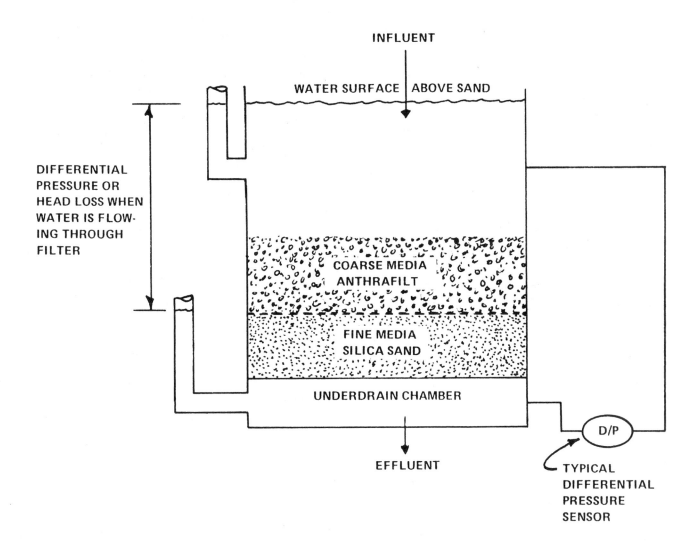

Fig. 9.3 Differential pressure through a sand filter

9.12 Location of Filters in a Treatment System

In wastewater treatment, the filters may be used in the following modes (Figure 9.4):

1. To polish secondary effluent without the addition of chemicals as filter aids just ahead of the filters.
2. To polish secondary effluent with the addition of chemicals as filter aids just ahead of the filters.
3. To polish secondary effluent that has been chemically pretreated and settled.
4. To polish raw wastewater that has undergone coagulation, flocculation, and sedimentation in a physical-chemical treatment system.

9.13 Major Parts of a Filtering System (Figure 9.5)

This section describes the major parts of a filtering system and also how each part works or functions during the filtration process.

9.130 Inlet

The filter inlet gate allows the applied water to enter the top of the filter media. When closed, it will permit emptying the filter for backwashing or maintenance.

9.131 Filter Media

The filter media selection is one of the most important design considerations. Filter beds are made up of silica sand, anthracite coal, garnet, or ilmenite. Garnet and ilmenite are commonly used in multi-media beds.

Because of rapid plugging, the conventional single media filter bed commonly used in potable (drinking) water systems is generally unsatisfactory for removing solids from industrial wastewater. To lengthen filter runs and use the full bed depth, the dual- and multi-media filters are used. A layer of coarse media (anthracite) is placed over finer, dense material (sand or garnet). The coarse layer allows deep penetration of the solids into the bed causing a minimum of head loss. The fine material prevents breakthrough of solids into the effluent.

9.132 Filter Underdrains (Figure 9.6)

The filter underdrain system is designed to contain the filter media within the bed and to maintain uniform water flows through the entire bed during both filtering and backwashing.

9.133 Filter Media Scouring Systems

If the filter media is not cleaned thoroughly at each backwashing, a buildup of solids will occur. The end result of incomplete cleaning is the formation of *MUDBALLS*[2] within the bed. These mudballs settle to the filter bottom and eventually make it necessary to rebuild the entire bed. "Surface wash" and "air scour" are two systems used to improve cleaning of the media.

The surface wash system consists of either fixed or rotating nozzles installed just above the media. During a backwash, the sand expansion places the nozzles within the media where high-pressure water jetting out of the nozzles will agitate and clean the surface. Because these wash systems are designed primarily to break up the surface mat, deep filtering beds need nozzles placed deeper within the media.

The air scour system injects air into the bottom of the bed. This agitates the entire bed, yet requires no additional wash water. Care must be taken to prevent air and water flowing at the same time or the media will be washed out and lost.

9.134 Wash Water Troughs

During backwashing, the accumulated solids strained out by the filter media are carried out of the filter bed by means of the backwash water troughs. The troughs must be level to uniformly collect and withdraw the backwash water. This will help prevent dead spots (areas where no water circulates) during the backwash operation.

A smooth trough surface, such as fiberglass, will reduce routine cleaning; however, fiberglass troughs may be damaged more easily by the weight of the backwash water than steel or concrete troughs. Filter troughs, particularly fiberglass, must be well anchored to ensure that they will not warp or attempt to float during backwashing. They also must be designed to withstand the weight of water if filled when there is no water over the bed.

9.135 Backwash Water Drain

The filter drain allows the backwash water to leave the filter and return to the plant headworks for reprocessing. This drain must be opened before the backwash water flow begins, but not before the filtered water has dropped below the level of the backwash trough. If the drain opens while the filter is still full of applied water, the water above the troughs will needlessly be recycled back through the plant.

The drain should be closed completely before the inlet valve is opened or applied water again will be wasted.

---

[2] *Mudballs.* Material, approximately round in shape, that forms in filters and gradually increases in size when not removed by the backwashing process. Mudballs vary from pea-sized up to golf-ball-sized or larger.

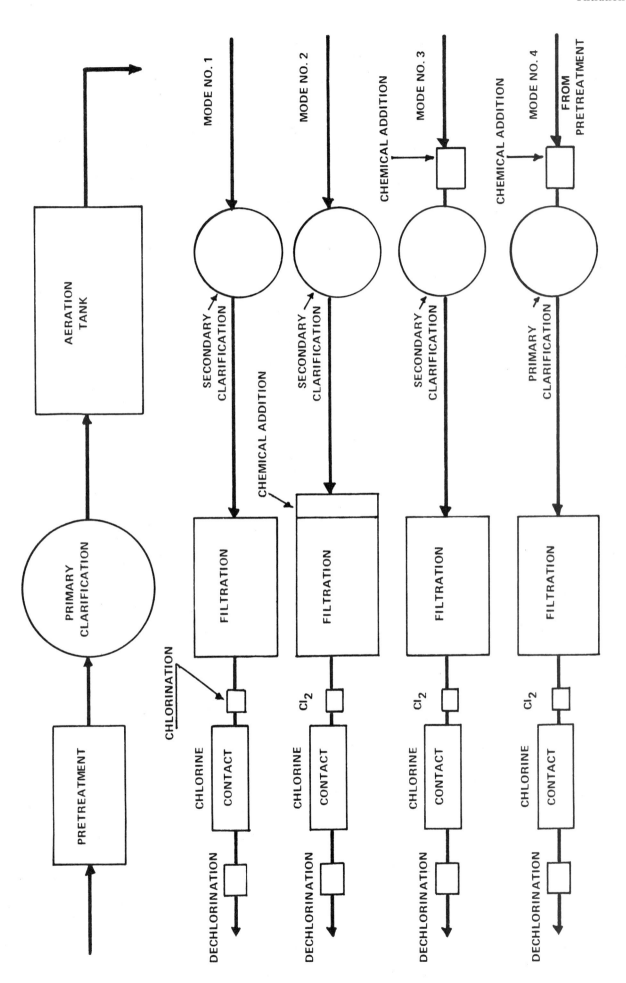

Fig. 9.4 Four possible modes of using filters to remove solids

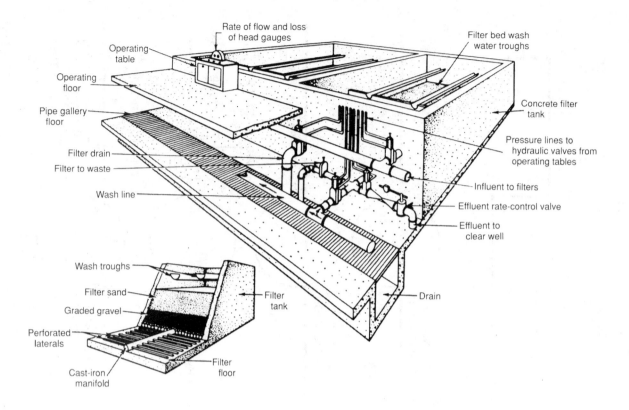

Fig. 9.5 Typical rapid sand filter
(From WATER SUPPLY AND TREATMENT by C.P. Hoover, permission of National Lime Association)

9.136 Backwash Water Supply

The backwash water is usually water that has gone through the complete treatment process and is of the best quality available. If unfiltered water is supplied to the backwash system, clogging of the underdrain system may occur.

Filter backwashes require large volumes of water over a short period of time; therefore, small- to medium-size plants need a wash water storage reservoir. Water from the chlorine contact tank commonly is used.

Large filters are often split in half to reduce the size of pumps and piping required for backwashing. This also can reduce the water storage requirements because a pause between washing the two halves will provide time to refill the storage reservoir.

Sectional filters (Figure 9.7), designed to backwash one small section at a time, do not require a large backwash water storage supply. These filters use pumped water as it is being filtered through other sections.

9.137 Backwash Water Rate Control

The backwash water may be supplied through pumps or by gravity from a storage tank. Both methods require careful control of the flow rate.

Backwash water supplied through pumps will maintain a more constant flow over the entire wash cycle than wash water from gravity storage. Water supplied from storage tanks may require adjustment of the rate-control valve to maintain constant flows as the storage tank level drops, due to a decrease in the available pressure head on the backwash water.

9.138 Used Backwash Water Holding Tank

The filter backwash water contains solids concentrated from many gallons of applied water. Because of the high solids concentration, this water must be retreated in the treatment process. Since the backwash flow rates are very high, they must be dampened through a holding tank to avoid hydraulic overloads on the treatment plant. A holding tank is generally provided to equalize flows and prevent plant overloads. The tank is filled during the backwash operation and slowly emptied into the plant headworks between washings. (Some improvement in primary settling efficiency may be noted because of this recycled water.) If these high flows were returned directly to the headworks, a hydraulic overload would occur that would upset the treatment process and flow pacing systems in all but the largest plants. A filter system designed to backwash like the sectional filter may avoid the need for a used backwash water holding tank.

Filtration 403

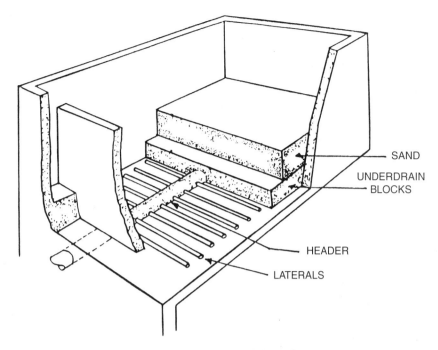

A. HEADER LATERALS
(Courtesy of the AWWA)

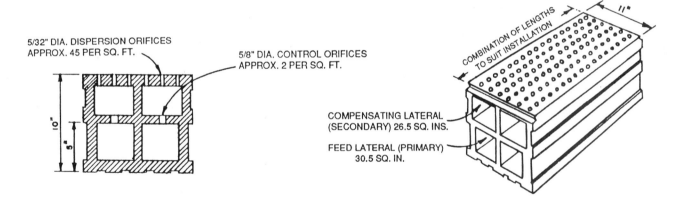

B. LEOPOLD BLOCK SYSTEM
(Courtesy of F. B. Leopold Co.)

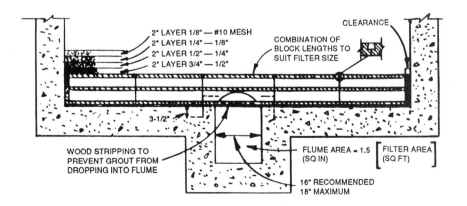

Fig. 9.6 Underdrains
(Source: EPA *PROCESS DESIGN MANUAL FOR SUSPENDED SOLIDS REMOVAL*)

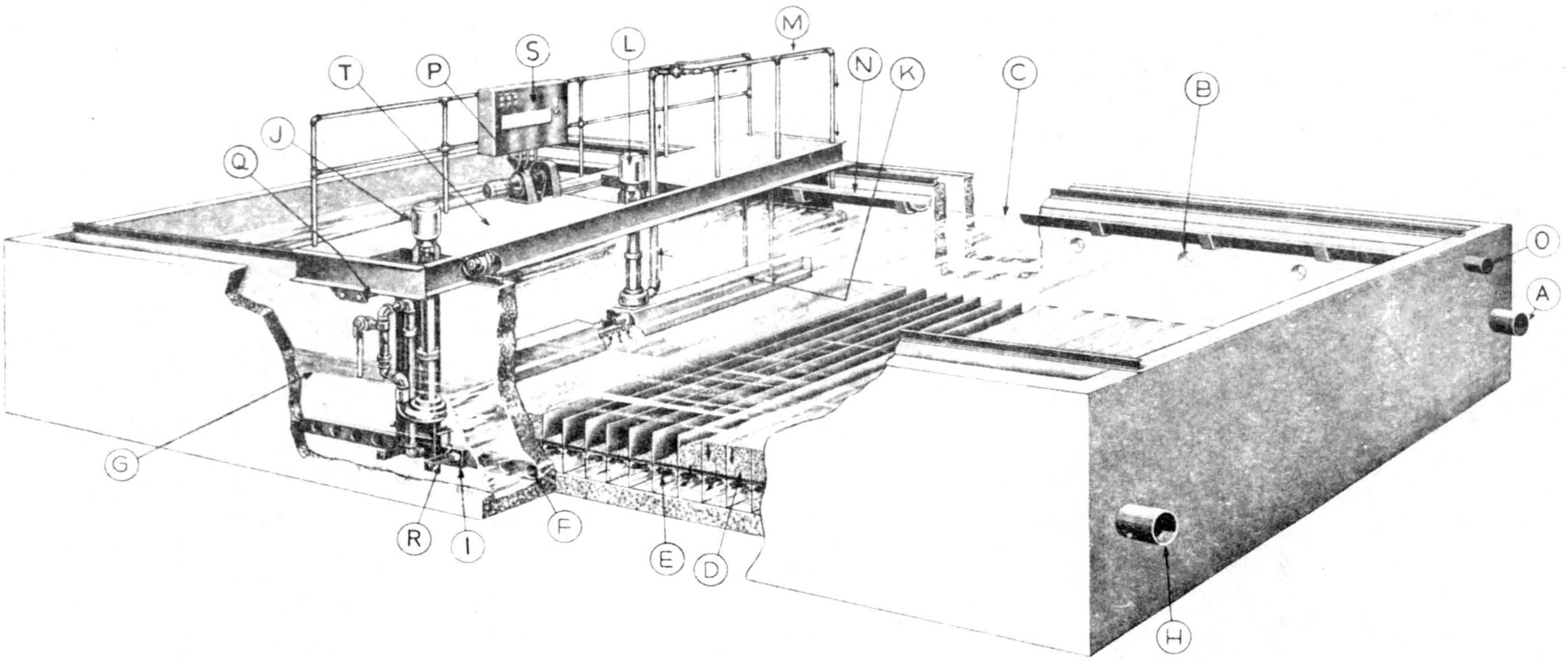

404 Treatment Plants

A. Influent line
B. Influent ports
C. Influent channel
D. Compartmented filter bed
E. Sectionalized underdrain

F. Effluent and backwash ports
G. Effluent channel
H. Effluent discharge line
I. Backwash valve
J. Backwash pump assembly

K. Wash water hood
L. Wash water pump assembly
M. Wash water discharge pipe
N. Wash water trough
O. Wash water discharge

P. Mechanism drive motor
Q. Backwash support retaining springs
R. Pressure control springs
S. Control instrumentation
T. Traveling backwash mechanism

Fig. 9.7 Sectional filter
(Permission of Environmental Elements Corporation)

9.139 Effluent Rate-Control Valve

A valve automatically controls the filtered water flow leaving the bed. The effluent rate-control valve is designed to maintain a constant water level in the filter. When operating a clean filter, this valve will be closed down to restrict the flow. As the head loss through the media increases, this valve must open more to maintain a constant flow. The effluent rate-control valve must be closed during filter backwash to prevent backwash water from mixing with previously filtered water.

QUESTIONS

Write your answers in a notebook and then compare your answers with those on page 477.

9.1E   What types of materials are used for filter media?

9.1F   What can happen if the filter media is not thoroughly cleaned during each backwashing?

9.1G   Why should the backwash water be of the best quality available?

9.1H   What is the purpose of a used backwash water holding tank?

9.14 Filter System Instrumentation

Instrumentation is essential for all but the small package plant installations. Instrumentation associated with filtering is used to monitor the plant performance, to operate the plant in the absence of the operator, and to trigger an alarm if abnormal conditions develop. The system may be simple or very complex depending on the facilities; each has its place.

As with any equipment, the usefulness of instrumentation is limited by the quality of the maintenance it receives. Stated another way, if there are intermittent errors in a flowmeter signal and they are not corrected, then the operator cannot trust any of the readings and must disregard all of them. The usefulness of the instrument is then very limited.

Comments regarding instrumentation in the following sections apply to plants of all sizes. (Also see Chapter 12, "Instrumentation," for more detailed information.)

9.140 Head Loss

Head loss is one of the most important control guidelines in the operation of the rapid sand filter. Each filter or filter half requires a head loss (differential pressure) indicator, preferably one with a readout chart. This will indicate the present condition of the bed, its ability to remove solids, and the effectiveness of the backwash operation.

Head loss is determined by measuring the water pressure above and below the filter media (see Figure 9.3, page 399). With the filter out of service, the pressures will be the same (zero difference).

When water flows through the bed of a typical downflow gravity filter, the pressure below the media will be less than the pressure above the media (when the pressure levels are measured or read at the same elevation). Measured in feet (or meters) of water, the difference becomes the head loss.

As the media bed becomes filled with solids, the head loss becomes greater. There is a point at which little or no water can pass through the filter. The operator wants the head loss to always be less than at that point; therefore, the filter backwash control point must be less than the maximum design head loss.

9.141 Filter Flow Rate Indicator[3] and Totalizer[4]

Each filter or filter half requires a flow indicator and totalizer on the filtered water line. This is needed to determine proper filtering rates (gal/min/sq ft or liters/sec/sq meter). Also, knowing the total volume filtered and the volume of backwash water used, the percent of production (filtered) water used for backwashing can be calculated. This is important because excessive wash water usage is costly and must be controlled. The backwash water should average 5 to 10 percent of total water production.

9.142 Applied Turbidity[5]

A continuous-reading turbidimeter with readout chart on the applied water is useful in monitoring the performance of the secondary settling tanks. This readout will alert the operator to developing problems if the turbidity suddenly increases. With experience, chemical dosages can be adjusted as turbidity changes.

---

[3] *Indicator.* (1) (Chemical indicator) A substance that gives a visible change, usually of color, at a desired point in a chemical reaction, generally at a specified end point. (2) (Instrument indicator) A device that indicates the result of a measurement, usually using either a fixed scale and movable indicator (pointer), such as a pressure gauge, or a moving chart with a movable pen like those used on a circular flow-recording chart. Also called a RECEIVER.

[4] *Totalizer.* A device or meter that continuously measures and calculates (adds) a process rate variable in cumulative fashion; for example, total flows displayed in gallons, million gallons, cubic feet, liters, cubic meters, or some other unit of volume measurement. Also called an INTEGRATOR.

[5] *Turbidity* (ter-BID-it-tee). The cloudy appearance of water caused by the presence of suspended and colloidal matter. In the waterworks field, a turbidity measurement is used to indicate the clarity of water. Technically, turbidity is an optical property of the water based on the amount of light reflected by suspended particles. Turbidity cannot be directly equated to suspended solids because white particles reflect more light than dark-colored particles and many small particles will reflect more light than an equivalent large particle.

9.143 Effluent Turbidity

A continuous reading of turbidity with a chart on the filter effluent will monitor the filter performance. A sudden increase may indicate a filter breakthrough (cracked bed) and may be used or instrumented to set off alarms if specified limits are exceeded. One turbidimeter unit, with proper valving, may be used to monitor more than one filter.

9.144 Indicator Lights

Indicator lights are beneficial to operators in keeping track of the filter system. Lights can easily indicate which filter is in service, out of service, or backwashing. They can indicate if filter pumps, wash water pumps, or air blowers are running, out of service, or on standby and ready to run. Indicator lights are also used with the alarm system to show abnormal conditions.

9.145 Alarms

Alarms needed to alert the operator should include high applied water level, high turbidity, and pump malfunctions. Backwash water supply and holding tanks both need high water level alarms. All alarms should be tested for proper functioning at least every 60 days.

QUESTIONS

Write your answers in a notebook and then compare your answers with those on page 477.

9.1I How is the head loss through the filter media determined?

9.1J How often should filter system alarms be tested for proper operation?

9.15 Operation of Gravity Filters

9.150 Pre-Start Checklist

Before starting up any major system, such as gravity filters, a thorough check of each component must be made to prevent damage to the equipment or injury to personnel. The following items should be included in your checklist for starting filtering systems.

1. Be sure all construction debris has been removed. Wood scraps, concrete chips, nails, and other trash can damage equipment such as pumps and valve seats. Trash dropped into the filter media will work its way to the bottom, thus reducing the effective area of the filter.

2. Inspect the electrical installation for completeness. Check safety lockouts, fuse sizes, safety covers, and equipment overload protections.

3. Check motors and drives for proper alignment, for proper safety guards, and for free rotation.

4. Examine motors, drive units, and bearings for proper lubrication.

5. Check motors for proper rotation. (A three-phase motor may run in either direction.)

6. Inspect pumps and motors for excessive vibration.

7. Fill tanks and piping and look for leaks.

8. Open and close valves manually and run each valve through a complete cycle to check limit setting.

9. Put the automatic controls through a "dry run."

10. Inspect the total system for safety hazards.

11. Backwash the media several times. Skim the fines (tiny particles that tend to float) from the surface between each washing prior to placing filter into service. After the final pre-start backwashing sequence, fill the filter with wash water up to the level of the wash water troughs.

9.151 Normal Operation

Since most wastewater gravity filters are deep-bed, downflow, rapid sand-type filters, this section will present information based on them. Nevertheless, most of the information can be applied to other filter designs with some possible modifications.

9.1510 FILTERING

The applied wastewater enters at the top of the filter bed through an inlet valve and is distributed over the entire filter surface. The water passes evenly down through the media and leaves the solids behind. Filtered water then travels out the bottom of the filter and into the underdrain collection system, which is designed to uniformly collect the flow. Once inside the underdrain collection system, the water passes through a flowmeter and rate-control valve. The rate-control valve maintains the desired flow through the filter and prevents backwash water from mixing with the filtered water during backwashing. Successful filter operation depends on effective backwashing of the filter media.

9.1511 BACKWASHING

As suspended solids are removed from wastewater, the filter media becomes clogged. This is indicated by the head loss reading (Figure 9.3, page 399). The filter should be backwashed after the capacity of the media to hold solids is nearly used up, but before solids break through into the effluent. Operating experience will be the best guide to determining the maximum head loss before backwashing is required.

A typical set point to start backwashing is at 7.0 feet (2.0 m) of head loss. If a filter is operating with a 6.0-foot (1.8-m) head loss, the operator knows the filter will need washing soon. If the head loss is 4.0 feet (1.2 m) after washing, this indicates a very poor washing or it may indicate a malfunctioning instrument. After a proper washing, the head loss should be less than 0.5 foot (0.15 m) at start-up. The head loss will then slowly increase to the point where backwashing is required again.

Backwashing a filter manually, although sometimes necessary, is very time-consuming; moreover, manual backwashings are inconsistent. Automatic backwashing, on the other hand, can be a simple procedure that requires a minimum of operator time.

To maintain smooth operations, the automatic backwash cycle should be initiated by the operator. This mode of operation permits the operator to backwash at a convenient time thereby allowing time for keeping records current and completing the necessary maintenance duties. Automatically starting backwashes, although workable in a large system, can be very inconvenient to the operation of a small system.

At the start of the backwash cycle, the rate-control valve must be opened slowly to a low rate of backwash. This prevents damaging the underdrain system or disturbing the rock and gravel layers of the bed. This damage can occur when an empty bed has high backwash water flows suddenly injected into it or if trapped air in the piping and underdrain system is violently forced into the bed. After the air has been slowly purged and the water level is up to the wash water troughs, the bed can no longer be damaged by high backwash rates.

Some plants use an air scouring system to clean the filter media. The air scour system injects air into the bottom of the media bed. This agitates the entire bed, yet requires no additional wash water. Care must be taken to prevent air and water flowing at the same time, or the media will be washed out and lost.

To prevent the loss of filter media into the backwash troughs:

1. Draw the water level in the filter down to within a few inches over the top of the filter media.

2. If air scouring is used, pause a moment after air washing before starting the water wash.

3. Wash with a low water flow rate until the trapped air has escaped the filter media.

4. Never backwash a filter with water containing large quantities of air.

The various types of media become intermixed during the high agitation of air scrubbing or high-rate backwashing. With proper control, however, the media will automatically regrade due to the difference in *SPECIFIC GRAVITIES*[6] of the particles.

By design, the filter media is prevented from escaping into the underdrain system; nevertheless, operational care must be taken to prevent damaging the underdrains while backwashing or the filter media will be lost into the underdrain collection system.

Uniform water flow through the filter bed is important to prevent the breakthrough of solids in the effluent due to localized high velocities. Also, high velocities will cause the media to be disturbed and relocated if the backwash flow is not uniform throughout the bed.

The following situations indicate a disturbed or damaged filter underdrain:

1. Consistently poor-quality effluent (high suspended solids levels) while there is little buildup of the filter head loss.

2. Boiling areas and very quiet ("dead") areas of the filter media during backwashing. This is most noticeable during high wash rates in a nearly clean filter.

3. Filter media in the effluent.

Improper control of the system during backwashing is generally the cause of damaged filter bottoms, providing they were properly installed. Damage to the filter bottom could result if:

1. The maximum backwash rate is allowed to enter an empty filter.

2. A large volume of air preceded the maximum backwash rate causing *WATER HAMMER*.[7]

---

[6] *Specific Gravity.* (1) Weight of a particle, substance, or chemical solution in relation to the weight of an equal volume of water. Water has a specific gravity of 1.000 at 4°C (39°F). Particulates with specific gravity less than 1.0 float to the surface and particulates with specific gravity greater than 1.0 sink. (2) Weight of a particular gas in relation to the weight of an equal volume of air at the same temperature and pressure (air has a specific gravity of 1.0). Chlorine gas has a specific gravity of 2.5.

[7] *Water Hammer.* The sound like someone hammering on a pipe that occurs when a valve is opened or closed very rapidly. When a valve position is changed quickly, the water pressure in a pipe will increase and decrease back and forth very quickly. This rise and fall in pressures can cause serious damage to the system.

The only way to correct a damaged filter bottom is to remove the media and rebuild the bed. A bed with the media displaced to a minor extent may be corrected by extended and properly controlled backwashing. This will regrade the media.

After the filter media is clean, slowly reduce the backwash water flow. This permits the media to regrade itself through gravity settling. The heavier particles (gravel, garnet, sand) will settle to the bottom first. Then, as the uplift velocities decline, the lighter particles (anthracite coal) will settle, thereby regrading the filter bed back to its original placement. This regrading must occur at the end of each backwash cycle.

After backwashing, the filter normally contains water up to the sides of the troughs. To fill the remaining portion of the filter, open the inlet valve. Be sure to waste some of the filtered water at the start until completely filtered and clear water is leaving the filter.

If the filter has been drained for maintenance, fill the filter as if you were starting to backwash, up to the top of the sides of the troughs, and then fill the filter using the inlet valve. An empty filter should not be filled through the inlet valve because the water falling onto the media will disturb the bed and result in uneven filtering. Also, filling the backwash troughs with water in an empty filter will place an unnecessary load (weight of water) on the troughs.

When the used backwash water holding tank is empty, the tank should be inspected. An observation of the solids settled on the bottom of the empty holding tank will alert the operator to any loss of filter media due to improper backwash procedures, such as an excessively high flow rate or SHORT-CIRCUITING.[8]

By analyzing the records and observing the complete wash cycle, the operator can determine if the backwashing sequence is adequate. If highly turbid water is still in the bed at the end of the cycle, experiment with one or all of the following:

1. Adjust the media scouring time.
2. Adjust the low wash rate.
3. Adjust the high wash rate.
4. Adjust the time of regrading the media.
5. Backwash more frequently by beginning to wash at a lower head loss.

QUESTIONS

Write your answers in a notebook and then compare your answers with those on page 477.

9.1K Why should a pre-start check be conducted before starting filtering systems?

9.1L What is the purpose of the rate-control valve?

9.1M When should a filter be backwashed?

9.152 Abnormal Operations

Following is a list of conditions that are not normally found in the day-to-day operation of filtration systems; however, these conditions could occur at almost any time. Recommendations are added to aid you in adjusting for the situation.

1. High solids in the applied water due to *BULKING*[9] sludge, *RISING SLUDGE*,[10] or solids washout in the secondary clarifier.

   a. Run *JAR TESTS*[11] and adjust chemical dosage as needed. (Jar test procedures are described in Chapter 8, "Physical-Chemical Treatment Processes.")

   b. Place more filters in service to prevent breakthrough.

   c. Prepare to backwash more frequently.

2. Low suspended solids in applied water; however, solids pass through filter.

   a. Run jar tests and adjust chemical dosage as needed. Test a combination of chemicals and *POLYELECTROLYTES*.[12]

   b. Place more filters in service to reduce velocity through the media.

   c. Backwash filter and precoat clean filter with *FILTER AID*.[13]

---

[8] *Short-Circuiting.* A condition that occurs in tanks or basins when some of the flowing water entering a tank or basin flows along a nearly direct pathway from the inlet to the outlet. This is usually undesirable since it may result in shorter contact, reaction, or settling times in comparison with the theoretical (calculated) or presumed detention times.

[9] *Bulking.* Clouds of billowing sludge that occur throughout secondary clarifiers and sludge thickeners when the sludge does not settle properly. In the activated sludge process, bulking is usually caused by filamentous bacteria or bound water.

[10] *Rising Sludge.* Rising sludge occurs in the secondary clarifiers of activated sludge plants when the sludge settles to the bottom of the clarifier, is compacted, and then starts to rise to the surface, usually as a result of denitrification, or anaerobic biological activity that produces carbon dioxide and/or methane.

[11] *Jar Test.* A laboratory procedure that simulates coagulation/flocculation with differing chemical doses. The purpose of the procedure is to estimate the minimum coagulant dose required to achieve certain water quality goals. Samples of water to be treated are placed in six jars. Various amounts of chemicals are added to each jar, stirred, and the settling of solids is observed. The lowest dose of chemicals that provides satisfactory settling is the dose used to treat the water.

[12] *Polyelectrolyte* (POLY-ee-LECK-tro-lite). A high-molecular-weight (relatively heavy) substance, having points of positive or negative electrical charges, that is formed by either natural or synthetic (manmade) processes. Natural polyelectrolytes may be of biological origin or obtained from starch products or cellulose derivatives. Synthetic polyelectrolytes consist of simple substances that have been made into complex, high-molecular-weight substances. Used with other chemical coagulants to aid in binding small suspended particles to larger chemical flocs for their removal from water. Often called a POLYMER.

[13] *Filter Aid.* A chemical (usually a polymer) added to water to help remove fine colloidal suspended solids.

3. Loss of filter aid chemical feed.
   a. Place more filters in service to reduce velocity through media.
   b. Backwash more frequently.
   c. Precoat clean filters by hand feeding chemicals into them when first placed into service.
4. High wet weather peak flows.
   a. Place more filters in service.
   b. Run jar tests and adjust chemical dosage as needed.
   c. Prepare for peak daily flows by backwashing early.
5. Low applied water flows.
   a. Reduce number of filters in service. Run one-half of a filter at a time.
   b. Prepare to take one filter out of service and backwash when flow or head loss increases, thereby preventing breakthrough.
6. High color loading.
   a. Run jar tests and adjust chemical dosage as needed.
   b. Add chlorine to applied water.
   c. Usually color cannot be removed with filtration; consequently the problem must be corrected at the source.
7. High water temperature.
   a. Run jar tests and adjust chemical dosage as needed.
   b. Prepare for *AIR BINDING*[14] of filters because water will release gases more readily at higher temperatures.
   c. Place more filters in service to reduce head loss through the media.
   d. Increase backwash water flow rates to obtain the same bed expansion as used when backwashing with colder water.
8. Low water temperature.
   a. Run jar tests and adjust chemical dosage as needed.
   b. Prepare for air binding of filters as cold water will carry more gases to the filters. Backwash more frequently if air binding occurs.
   c. Place more filters in service to reduce head loss through filter media.
   d. Reduce backwash water flow rates to obtain the same bed expansion as used when backwashing with warmer water.
9. Air binding.
   a. Backwash at a lower head loss.
   b. Place more filters on line to reduce head loss through media.
   c. Take filter out of service and allow air to escape to the atmosphere. This will reduce head loss; however, if placed back into service without backwashing, solids will likely be drawn through the media and into the effluent. These solids may or may not cause a problem.
10. Negative pressure in the filter.
    a. Reduce flow through the filter by adding additional units.
    b. Backwash at a lower head loss.
    c. Skim surface of media (about one-half inch or 1.3 cm) to remove fines.
    d. Prevent filter from running at a low filtration rate. This builds a mat on the media surface and then sharply increases the head loss if a higher rate of water flows through the filter.
    e. A negative pressure within the filter will cause a false reading from the differential pressure sensor.
11. Chlorine in the applied water.
    a. Discontinue adding polyelectrolytes as chlorine will interfere with them.
    b. Run jar tests and adjust chemical dosage as needed.
12. pH change in the applied water.
    a. Run jar tests and adjust chemical dosage as needed.
    b. Change type of filter aid if necessary.
13. High quantities of grease and oil in the applied water.
    a. If in solution, they will pass through media.
    b. If not in solution, they will be trapped in the bed, thus requiring extra hosedown during each backwash.

QUESTIONS

Write your answers in a notebook and then compare your answers with those on page 477.

9.1N List at least five of the various types of abnormal operating conditions that could occur while operating a filtration system.

9.1O How would you adjust for a situation in which you were treating a high solids content in the water applied to a filter?

---

[14] *Air Binding.* The clogging of a filter, pipe, or pump due to the presence of air released from water. Air entering the filter media is harmful to both the filtration and backwash processes. Air can prevent the passage of water during the filtration process and can cause the loss of filter media during the backwash process.

9.153 Operational Strategy

The development of an operational strategy for the filtration of wastewater will aid in dealing with situations such as sudden changes in applied water, in training new operators, or in planning for the future. Following are points to consider when developing or reviewing your plans.

1. Maintain the filtering rates within the design limits. Add units or remove them from service as needed. Very low filtering rates will produce matting on the surface. This matting will cause breakthroughs if the flows are increased sharply. Excessively high rates will pull the solids through the filter and into the effluent.

2. Each backwash must be a complete cleaning of the media or solids will build up and form mudballs, or cause the media to crack.

3. To remove mudballs, first backwash thoroughly. Then, superchlorinate manually and draw the chlorinated water into the filter media. Allow this chlorinated water to stand for 24 or 48 hours to soak the mudballs and then backwash thoroughly again.

4. Run jar tests to maintain optimum chemical dosages. As the applied water quality changes (solids, alkalinity, temperature), the filter aid requirements will change. The operator must be aware of the changes and the effectiveness of the chemicals applied.

5. With complete backwashing, a high-quality effluent can be maintained without filtering to waste before placing the filter back into service.

6. If the effluent turbidity reaches 3 to 4 *TURBIDITY UNITS*,[15] a change should be made to correct the problem. Either adjust chemical dosage, adjust flow rate, or backwash the filter.

7. Filter walls that are constructed with a smooth surface (sacked) or painted with a good sealant are easy to keep clean. A rough surface provides an excellent area for *ALGAE*[16] and slimes to grow.

8. Controls and instrumentation must be protected from the elements. Cabinets that are opened to adjust instruments must be out of the rain, dust, and extreme heat.

9. Air used to operate instruments or transmit signals must be cleaned and dried to prevent damaging the equipment.

10. Every three or four months, measure and record the freeboard to the filter media surface (Figure 9.8). A small amount of media loss is normal, but an excessive amount (2 to 3 inches or 5 to 7 centimeters) indicates operational problems.

11. After the filters have been in service for some time, obtain a profile of the media to determine if it is being displaced. A plug sample (Figure 9.8) will show if the media are being regraded after each backwash.

12. When landscaping around uncovered filters, keep trees and shrubs that will drop leaves into the bed away from the filter because leaves are very difficult to backwash out of the media.

13. Never throw trash such as cigarette butts into the filter media. Trash may not backwash out and instead may work its way deep into the media.

14. Occasionally, chlorinate ahead of the filters to control algal and slime growths on the walls and within the media. There will be a short period of discolored effluent after the initial application, but the water will run clear in a short while.

15. Calculate the unit cost to treat wastewater. Apply this cost to the volume of water used per backwash. Inform all operators of this because it may easily cost in excess of $100 per filter wash.

16. Always fill an empty filter bed through the backwash system to prevent disturbing the media surface. If an empty filter is filled through the influent valve, water will flow into the wash water troughs, over the top edges and onto the top of the media. The force of this falling water will disturb the media.

---

[15] *Turbidity* (ter-BID-it-tee) *Units (TU)*. Turbidity units are a measure of the cloudiness of water. If measured by a nephelometric (deflected light) instrumental procedure, turbidity units are expressed in nephelometric turbidity units (NTU) or simply TU. Those turbidity units obtained by visual methods are expressed in Jackson turbidity units (JTU), which are a measure of the cloudiness of water; they are used to indicate the clarity of water. There is no real connection between NTUs and JTUs. The Jackson turbidimeter is a visual method and the nephelometer is an instrumental method based on deflected light.

[16] *Algae* (AL-jee). Microscopic plants containing chlorophyll that live floating or suspended in water. They also may be attached to structures, rocks, or other submerged surfaces. Excess algal growths can impart tastes and odors to potable water. Algae produce oxygen during sunlight hours and use oxygen during the night hours. Their biological activities appreciably affect the pH, alkalinity, and dissolved oxygen of the water.

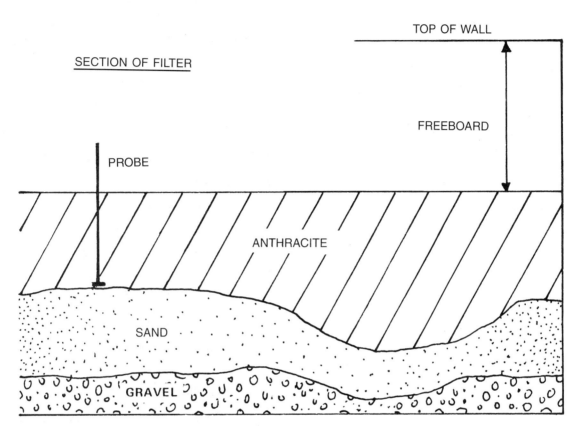
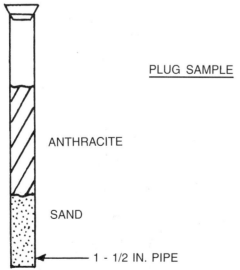

Fig. 9.8 Section of filter and plug sample

17. Allow dry filter media to soak several hours before backwashing. Dry media will tend to float out with the backwash water.

18. Maintain a log (Figure 9.9) of the filtering operation that includes the following:

    a. Time filter was placed into service and total hours run between washings.
    b. Volume of water processed between washings.
    c. Applied water rate at start and end of filter run.
    d. Head loss at start and end of filter run.
    e. Applied suspended solids.
    f. Effluent suspended solids.
    g. Percent removal of suspended solids.
    h. Chemicals added as filter aids, mg/L.
    i. Chlorine added to applied water, mg/L.
    j. Remarks of special observations and maintenance.
    k. Backwash water flow rates and duration.
    l. Surface wash flow rate and duration.
    m. Influent and effluent turbidity.

9.154 Shutdown of a Gravity Filter

If the filter is to be out of service more than a week, it should be backwashed, dewatered, and air dried. This will help control slime and algal growth on the walls, troughs, and within the media. Dried algae can be hosed from the walls prior to backwashing and returning to service.

To remove a filter from service, first backwash the filter. Then switch controls to the manual mode of operation and:

1. Close the influent valve.
2. Filter all the water possible through the rate-control valve.
3. Open the drain valve.

Hose down and backwash the filter again before returning it to service.

QUESTIONS

Write your answers in a notebook and then compare your answers with those on page 477.

9.1P   How would you determine if media are being lost from a filter?

9.1Q   Why should trees and shrubs be kept away from uncovered filters?

9.16 Troubleshooting

PROBLEM: HIGH TURBIDITY AND SUSPENDED SOLIDS IN THE EFFLUENT.

1. Check for excessive head loss. Breakthrough will occur at a high head loss.
2. Look for fluctuating flows. Widely varying flows will cause breakthrough.
3. Determine filter aid dosages.
4. Examine backwash cycle for complete wash.
5. Inspect for damaged bed due to backwashing.

PROBLEM: RAPID BUILDUP OF HEAD LOSS.

1. Check applied water suspended solids.
2. Check filter aid dosage.
3. Determine applied water flow rate.
4. Check backwash cycle for complete wash.
5. Inspect head loss differential pressure sensor for air in one side. This will give a false reading.

PROBLEM: INSIGNIFICANT BUILDUP OF HEAD LOSS.

1. Check applied water suspended solids.
2. Check applied water flow rate.
3. Determine filter aid dosages.
4. Check head loss differential pressure sensor for air in one side. This will give a false reading.
5. Examine filter effluent for suspended solids going out (filter breakthrough).
6. Inspect for damaged bed due to backwashing.
7. Backwash and check for complete cycle.

FILTER LOG

MONTH JAN 20 04   FILTER NUMBER 1

START FILTER					STOP FILTER							FILTER WASH			
DATE	TIME	FILTER RATE MGD		HEAD LOSS FEET		DATE	TIME	FILTER RATE MGD		HEAD LOSS FEET		HRS.	NTU	DATE	TIME	REMARKS	OPER.
		A	B	A	B			A	B	A	B						
1-4-04	1500	1.5	1.5	—	—	1-3-01	0100	—	—	—	—	—	—	—	—	BACKWASHED SEMI-AUTOMATIC	A.S.
1-6	2100	1.0	1.0	—	—	1-6	2000	1.5	1.5	7	7	—	—	1-6	2000	MANUAL BACKWASH	R.N.N.
1-9-04	1500					1-9	1300	1.1	1.1	6.5	6.5	64		1/9	1300	AUTO	B.T.
1-12-04	0900	1.4	1.4	0.4	0.4	1-11	1100	2.2	2.4	6.0	9.0	44		1-11	1500	AUTO	L.S.
1-14-04	1300	1.4	1.4	0.4	0.4	1-16	1700	1.7	1.7	8.8	8.6	52	.8	1-16	1700	MANUAL	RICK
1-16	1800	1.8	1.8	.2	.2	1-17	1700	2.0	2.0	7.5	6.5	23	.7	1-17	1700	MANUAL	RICK
1-18	1100	1.3	1.3	—	—	1-20	0300	1.9	1.9	7.0	6.5	60		1-20	0500	AUTO	L.S.
1-25	2000	1.2	1.2	—	—	1-27	2100	1.8	1.0	6.9	7.4	49	.5				

Fig. 9.9 Log of filter operation

PROBLEM: RAPID LOSS OF FILTER MEDIA.

1. Look for washout during backwash cycle.
2. Examine for media in effluent indicating a damaged filter underdrain.
3. Check for excessive scouring during backwash cycle time. Excessive scouring will grind the media into fines.

PROBLEM: HIGH HEAD LOSS THROUGH CLEANED FILTER.

1. Inspect differential pressure sensor for air in one side.
2. Check for incomplete backwash cycle.
3. Look for mudballs in filter media. Take a plug sample (Figure 9.8, page 411).

PROBLEM: FLOW INDICATED WHEN EFFLUENT VALVE IS CLOSED.

1. Inspect differential pressure sensor for air in one side.
2. Check instrumentation loop for calibration.
3. Examine valve for proper position.

PROBLEM: BACKWASH STOPS BEFORE COMPLETING CYCLE.

1. Look for sticking valve.
2. Inspect for electrical relay hang-up.
3. Check for timer out of sequence.
4. Examine backwash water supply.
5. Inspect electrical control (pump lockout).

9.17 Safety

Always think safety when working around moving equipment and motors with automatic controls. Filtration systems have electrical, chemical, and mechanical safety hazards. Operators are usually well protected from electrical hazards; however, there are times when opening a panel to look for trouble or to adjust a timer may expose you to electrical hazards. Always use safety equipment (rubber electrical gloves, multimeters, fuse pullers, and lockout switches) and approved safety procedures when working with electricity. REMEMBER: *Only trained and qualified individuals should be allowed to work on electrical systems.*

Chemical hazards include chemical burns and skin irritation from direct contact with chemicals. Also, there is a hazard of slipping and falling caused by chemical spills. Review appropriate *MATERIAL SAFETY DATA SHEETS (MSDSs)*[17] if any doubt exists as to the safety precautions required when dealing with specific chemicals. Good housekeeping will reduce the safety hazards caused by chemicals.

Mechanical hazards associated with filters are similar to those found throughout the treatment plant. Safety guards must be in place, equipment operated automatically must be identified by warning signs, and work areas should be well lighted.

9.18 Review of Plans and Specifications

While reviewing the plans and specifications of a gravity filtration system, you should consider the items listed in this section.

1. Filters require regular servicing; therefore, provisions must be made to handle the normal flows during periods of servicing. Regular maintenance includes servicing of valves, instruments, and filter media.
2. The quality of the water applied to the filters must be considered when a filtering media is specified. A high suspended solids content in the applied water will quickly plug a fine-media filter, thereby requiring frequent backwashing.
3. Install sufficient instrumentation to adequately monitor the process and to determine operating efficiencies. Include instrumentation to measure and record applied flows, backwash flows, head loss, and water quality before and after filtration.
4. Be sure that each step in the automatic system is complete before the following or next step can begin.
5. Provide a means to reset the automatic system if the backwash cycle is interrupted.
6. Keep the automatic backwashing system uncomplicated, especially in small plants. The operator should be on hand at the start of a filter backwash cycle. Housekeeping chores

---

[17] *Material Safety Data Sheet (MSDS).* A document that provides pertinent information and a profile of a particular hazardous substance or mixture. An MSDS is normally developed by the manufacturer or formulator of the hazardous substance or mixture. The MSDS is required to be made available to employees and operators or inspectors whenever there is the likelihood of the hazardous substance or mixture being introduced into the workplace. Some manufacturers are preparing MSDSs for products that are not considered to be hazardous to show that the product or substance is not hazardous.

can be performed while keeping an eye on the filter washing process.

7. Install instruments out of the weather and well protected from the weather. Even weather-proof cabinets must be opened during the maintenance and servicing of instruments and equipment.

8. Install the instruments' readout meters, charts, and gauges in a convenient and centralized location.

9. Separate and shield instrumentation signals from all voltages 110 volts and higher and from other equipment noise that may be picked up by the instruments as a false signal.

10. Provide adequate storage for chemicals. A minimum supply of chemicals must be on hand even while waiting for a full shipment.

11. Provide adequate storage for both the backwash water supply and the used backwash water.

12. When reviewing designs for the future, keep today's flows in mind. Equipment operating below 10 percent capacity may be useless for years.

13. Visit similarly designed plants that are currently in operation and talk to the operators regarding possible design improvements.

QUESTIONS

Write your answers in a notebook and then compare your answers with those on page 477.

9.1R What are the three main types of safety hazards around filtration systems?

9.1S When reviewing plans and specifications for a filtration system, instrumentation should be available to measure and record what items?

9.1T Where should the instruments' readout meters, charts, and gauges be installed?

END OF LESSON 1 OF 4 LESSONS

on

FILTRATION

Please answer the discussion and review questions next.

DISCUSSION AND REVIEW QUESTIONS

Chapter 9. FILTRATION

(Lesson 1 of 4 Lessons)

At the end of each lesson in this chapter you will find some discussion and review questions. The purpose of these questions is to indicate to you how well you understand the material in the lesson. Write the answers to these questions in your notebook.

1. Why are multi-media filters used?

2. What is the purpose of instrumentation used with a filter system?

3. How does a rapid sand filter work?

4. How can a filter bottom be damaged?

5. Why should you attempt to maintain filtering rates within the design limits?

6. How would you remove a gravity filter from service?

CHAPTER 9. FILTRATION

(Lesson 2 of 4 Lessons)

9.2 SOLIDS REMOVAL FROM WASTESTREAMS USING INERT-MEDIA PRESSURE FILTERS
by Ross Gudgel

The pressure filters described in this section are similar in many ways to the gravity filters discussed earlier in this chapter. In both types of filtration systems, the filter media consists of one or more inert materials, that is, materials that do not react chemically with the wastewater being filtered. Silica sand, anthracite coal, and garnet sand are three types of inert media widely used in both gravity and pressure filters.

Inert-media gravity and pressure filters both remove suspended particles in basically the same way. Solids are trapped in the spaces between media particles and on the surface of the media bed. When the filter approaches the limit of its capacity to remove suspended solids, both types of filters must be backwashed.

As the name implies, pressure is the main difference between an inert-media pressure filter and an inert-media gravity filter. The pressure filter uses a closed tank or vessel that allows the operator to apply pressure to force the wastewater through the media at a faster rate than would be achieved by gravity flow.

9.20 Use of Inert-Media Pressure Filters

Inert-media pressure filters remove suspended solids, associated BOD, and turbidity from the secondary effluent after the addition of chemical coagulants such as a *POLYMER*[18] or alum. The filtration process is used to meet waste discharge requirements for final effluent suspended solids and turbidity limits established by an *NPDES PERMIT*[19] when these limits cannot be met by secondary treatment processes. Filtration also will have a direct bearing on the disinfection of the final effluent by the removal of more solids from the water to be disinfected. Fewer solids will reduce the amount of disinfectant necessary to meet the NPDES permit bacteriological requirements.

The filter system usually consists of:

1. A holding tank or wet well for secondary effluent storage.
2. Filter feed pumps, which pump the secondary effluent from the holding tank to the filters.
3. A chemical coagulant feed pump system, which injects the necessary coagulants into the influent line to the filters.
4. Single, dual, or multi-media filters that trap the suspended solids and remove the turbidity.
5. A filter backwash wet well for clean backwash water storage.
6. Filter backwash pumps, which pump clean water back through the filter to remove the trapped suspended solids.
7. A *DECANT*[20] tank that provides for holding the spent backwash water to allow the suspended solids to settle or rise while the clarified water is either directly recycled to the filters or is returned to the headworks.

Figure 9.10 shows a schematic view of the items outlined above and they are further discussed in the following sections.

QUESTIONS

Write your answers in a notebook and then compare your answers with those on page 477.

9.2A What is the purpose of the inert-media pressure filter?

9.2B What chemicals are commonly used with the filtration process and why?

9.2C List the major components of a pressure filter system.

9.21 Pressure Filter Facilities

The following sections describe facilities that are typical for a filter plant with a capacity of 5 MGD. Facilities at larger or smaller plants would be quite similar but might differ significantly in the numbers and sizes of the various components.

9.210 Holding Tank (Wet Well)

Secondary effluent from the treatment plant's secondary sedimentation tanks is conducted through a channel or pipe to a holding tank. The purpose of this tank is to store water and to allow additional settling of the suspended solids before the water is applied to the filters. Most tanks of this type are very similar to secondary clarifiers. They have flights or scrapers to move the

---

[18] *Polymer* (POLY-mer). A long-chain molecule formed by the union of many monomers (molecules of lower molecular weight). Polymers are used with other chemical coagulants to aid in binding small suspended particles to larger chemical flocs for their removal from water. Also see POLYELECTROLYTE.

[19] *NPDES Permit*. National Pollutant Discharge Elimination System permit is the regulatory agency document issued by either a federal or state agency that is designed to control all discharges of potential pollutants from point sources and stormwater runoff into US waterways. NPDES permits regulate discharges into US waterways from all point sources of pollution, including industries, municipal wastewater treatment plants, sanitary landfills, large animal feedlots, and return irrigation flows.

[20] *Decant* (de-KANT). To draw off the upper layer of liquid (water) after the heavier material (a solid or another liquid) has settled.

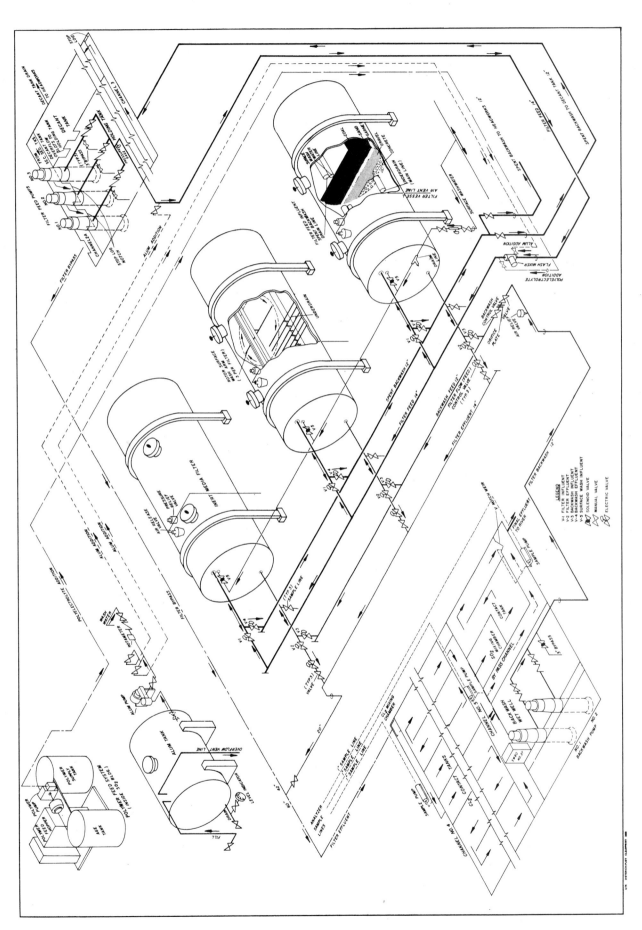

Fig. 9.10 Schematic view of pressure filter system

settled solids toward a sludge hopper for return to the solids handling facility.

A bypass structure should be provided to permit secondary effluent to bypass the pressure filters during emergency conditions, such as equipment failures or clogged filters. Bypassed flows should go into emergency holding basins or into the chlorine contact tank for final treatment before discharge. An alternative emergency storage procedure would be to divert secondary effluent into the decant tank.

Spent backwash water may also be returned to the decant tank. Both flows receive some settling and the clarified effluent then overflows into the holding tank through weir slots between the two tanks for recycle to the filters. In either method of operation, the floatable materials in the holding tank are collected, using a pan-type skimmer, and discharged to the solids handling section of the plant for disposal. For additional information on clarifier operation and maintenance, see Chapter 8, "Physical–Chemical Treatment Processes," in this manual and Chapter 5, "Sedimentation and Flotation," Volume I, *OPERATION OF WASTEWATER TREATMENT PLANTS*.

9.211 Filter Feed Pumps (Figure 9.11)

Filter feed pumps lift the secondary effluent from the holding tank and pump it through the filters. Generally, they are of the vertical-turbine wet-pit type pump with either a closed or semi-open impeller. The pumps are driven by either fixed-speed, multi-speed (two speed), variable-speed motors, or a combination of these. Each pump should be equipped with a manually adjusted bypass valve to avoid the possibility of the system operating at the shutoff pressure of the pumps. If this happens, the pumps could be damaged because no water would flow through the pumps. Each valve should be adjusted to allow a given bypass flow as recommended by the manufacturer.

The water level in the holding tank may be sensed by a level transmitter. The transmitter produces a signal to start and stop the pumps and a set point signal for the controller, which controls filter flow.

Starting and stopping of the pumps is controlled by a HAND-OFF-AUTO (HOA) switch for each pump. Another switch is used to select the sequence of automatic starting (lead or lag). Normal automatic start/stop control of the pumps may be by means of current trips using the signal representing the water level in the holding tank. A low-water probe in the holding tank will stop all pumps if the water level drops below a preset elevation. For additional information on the operation and maintenance of pumps, see Chapter 8, "Maintenance," in Volume II of this manual.

9.212 Chemical Feed Systems

Various types of chemicals may be added to the filter influent flow to ensure coagulation and flocculation of the suspended material. This coagulation and flocculation aids the filtering process by joining many of the finely divided and colloidal suspended solids into a floc mass that is easily trapped on or in the filter media, thus allowing clear water to pass through the filter. Alum and polymers are the chemicals most often used as filter aids. The following paragraphs briefly describe the reasons for using chemicals to aid filtration and the methods used to apply the chemicals to water. For a more detailed description of the coagulation and flocculation processes, see Chapter 8, "Physical–Chemical Treatment Processes."

ALUM (ALUMINUM SULFATE) (Figure 9.12)

Alum is a coagulant that produces a hydrated (containing water) oxide floc. This floc causes suspended material to stick together by electrostatic or interionic force when contact of the

Fig. 9.11 Filter feed pumps

Fig. 9.12 Alum storage tank and feed pump

chemical and a solid particle is made in the filter influent flow. The effective performance of alum is critically affected by the pH of the water in which it is used. The ideal condition is to maintain the value of the pH between 6.5 and 7.5. Economics may dictate that the working range is as much as 1.5 pH units above or below this "perfect" range, depending on the "natural" (or unregulated) pH of the system and the cost of pH control.

The alum may be pumped by a mechanical diaphragm, positive displacement pump. The dosage is manually adjusted by adjustment of the pump stroke length. Motor speed may be controlled by a silicon controlled rectifier (SCR) drive unit that uses a filter flow signal to pace the pump in the automatic mode. The SCR drive is also equipped with a manual potentiometer for manual speed control and a meter indicating percentage of total or full motor speed.

The pump discharge check valve has a built-in back pressure device to prevent uneven delivery due to low discharge pressure and to prevent siphoning. All wetted parts of the pump are selected for their chemical resistance.

POLYMERS (POLYELECTROLYTES)

Polymers are flocculation aids that are classified on the basis of the type of electrical charge on the polymer chain, the molecular weight (or length) of the polymer chain, and the charge density (spacing of the charges) along the chain.

Polymers possessing negative charges are called "anionic," positive-charge polymers are called "cationic," and polymers that carry no free electric charge are called "nonionic." Polymers cause the suspended material to stick together by chemical bridging or chemical enmeshment when contact is made in the filter influent flow.

Anionic polymers are most commonly used with the application of alum but systems' performances vary. The optimum alum–polymer combination can only be selected by testing the system water under actual system conditions.

Polymer usually is injected into the influent line of the filters far enough downstream from the point of the alum injection for the alum floc to form properly and far enough upstream of the filters for final coagulation to become complete.

The polymer may be prepared for use (dilution and aging) by a polyelectrolyte mixer unit (Figure 9.13). This unit consists of a dry polymer feeder with storage hopper, a solution water flowmeter with regulating valve, pressure regulating valve, pressure gauge, solenoid valves, dilution water flowmeter with regulating valve, polymer wetting cones, a mixing/aging tank, slow-speed mixer, transfer pump, metering/storage tank, and a metering pump with SCR drive.

A variable-area flowmeter (ROTAMETER [21]) is provided to indicate flow of dilution water to the metered polymer. The polymer is mixed automatically by the polyelectrolyte mixer. The polymer feeder is calibrated to dispense a metered quantity of dry polymer to obtain a desired solution concentration. The polymer drops from the feeder hopper to the wetting cones where it is spread on a high-velocity water surface and the individual grains or droplets of polymer are wetted to form a polymer solution. This solution then flows to the aging tank where it is mixed and aged. On completion of the aging cycle, the polymer solution is pumped to the metering/storage tank by the transfer pump. When the aging tank empties, the polymer preparation and mixing cycle begins again. The metering pump, calibrated to deliver a desired dosage, draws the polymer solution from the metering/storage tank and delivers it to the influent line of the filters.

The dry feeder is a screw-type feeder capable of metering dry polymer of densities ranging from 14 to 42 lbs/cu ft (225 to 675 kg/m^3) at an adjustable rate to the wetting cones in order to prepare various solution concentrations.

Application of liquid polymers is essentially the same. Instead of the dry screw feeder, a high-viscosity chemical metering pump is used. The pump may be connected directly to the containers in which the polymer is supplied or to an intermediate holding tank (in place of the hopper).

MIXING AND ADDING CHEMICALS

The mixing/aging tank (day tank) and the metering/storage tank are sized based on the projected use of polymer. The tanks may be of molded polyethylene, fiberglass reinforced polyester, stainless steel, mild steel with a plastic liner or, in some circumstances, unlined mild steel. The mixer, a low-shear type (to avoid breaking up floc), is fitted with a stainless-steel shaft and impellers.

The slow-speed transfer pump (either a progressive cavity or a gear type is recommended) conveys the mixed polymer solution from the mixing/aging tank to the metering/storage tank with minimal polymer shear. The metering pump is capable of delivering various amounts of polymers at various percent solutions. Flow pacing, adjustment of the dosage rate, and operation of the

---

[21] *Rotameter* (ROTE-uh-ME-ter). A device used to measure the flow rate of gases and liquids. The gas or liquid being measured flows vertically up a tapered, calibrated tube. Inside the tube is a small ball or bullet-shaped float (it may rotate) that rises or falls depending on the flow rate. The flow rate may be read on a scale behind or on the tube by looking at the middle of the ball or at the widest part or top of the float.

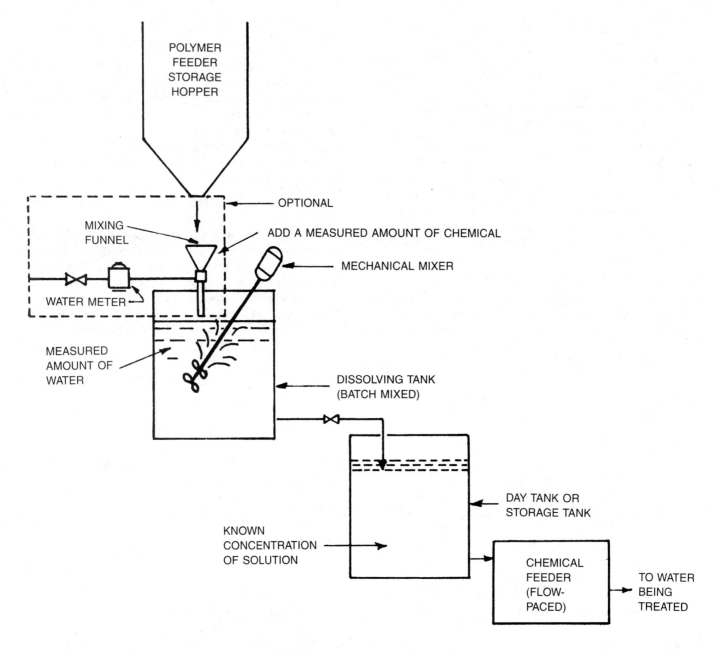

Fig. 9.13 *Polymer dissolver, day tank, and feeder*

polymer metering pump are the same as for the alum feed pump.

Steps to calculate polymer and alum dosage are outlined in the following examples. Information concerning the concentration of chemical (lbs/gal) delivered to your plant may be obtained from the chemical manufacturer or supplier.

EXAMPLE 1

Known		**Unknown**
Polymer Delivered, lbs/day	= 72 lbs polymer/day	Polymer Dose, mg/L
Flow Through Filter, GPM	= 6,000 GPM	

Determine the pounds of polymer used per million pounds of water, which is the same as mg per million mg or mg/L.

$$\text{Polymer Dose, mg}/L = \frac{\text{Polymer Delivered, lbs polymer/day}}{\text{Flow Through Filter, M lbs water/day}}$$

$$= \frac{72 \text{ lbs polymer/day}}{6,000 \text{ gal/min} \times 8.34 \text{ lbs/gal} \times 60 \text{ min/hr} \times 24 \text{ hr/day}}$$

$$= \frac{72 \text{ lbs polymer/day}}{72,057,600 \text{ lbs water/day}}$$

$$= \frac{72 \text{ lbs polymer/day}}{72 \text{ M lbs water/day}}$$

$$= 1 \text{ mg polymer/liter water}$$

$$= 1 \text{ mg}/L$$

EXAMPLE 2

Polymer is supplied to your plant at a concentration of 0.5 pound polymer per gallon (60 gm/L or 60 kg/m^3). The polymer feed pump delivers a flow of 0.10 GPM (0.0063 L/sec) and the flow to the filters is 3,000 GPM (190 L/sec). Calculate the concentration or dose of polymer in mg/L in the water applied to the filter.

Known	Unknown
Polymer Conc, lbs/gal = 0.5 lb/gal	Polymer Dose, mg/L
Polymer Pump, GPM = 0.1 GPM	
Flow to Filter, GPM = 3,000 GPM	

ENGLISH

Calculate the polymer dose, mg/L.

$$\text{Dose, mg}/L = \frac{\text{Pump, gal/min} \times \text{Conc, lbs polymer/gal}}{\text{Flow, gal/min} \times 8.34 \text{ lbs water/gal}}$$

$$= \frac{0.1 \text{ gal/min} \times 0.5 \text{ lb polymer/gal}}{3,000 \text{ gal/min} \times 8.34 \text{ lbs water/gal}}$$

$$= \frac{0.05 \text{ lb polymer}}{25,020 \text{ lbs water}} \times \frac{1,000,000^*}{1 \text{ M}}$$

$$= \frac{2.0 \text{ lbs polymer}}{1 \text{ M lbs water}}$$

$$= \frac{2.0 \text{ mg polymer}}{1 \text{ M mg water}^{**}}$$

$$= 2.0 \text{ mg}/L$$

---

* We multiplied the top and the bottom by the same number, 1,000,000 or 1 M. This is similar to multiplying the top and bottom by 1, you do not change the equation, only the units.
** 1 M mg water = 1 liter.

METRIC

$$\text{Dose, mg}/L = \frac{\text{Flow, } L/\text{sec} \times \text{Conc, gm polymer}/L \times 1,000 \text{ mg/gm}}{\text{Flow, } L/\text{sec}}$$

$$= \frac{0.0063 \text{ } L/\text{sec} \times 60 \text{ gm polymer}/L \times 1,000 \text{ mg/gm}}{190 \text{ } L/\text{sec}}$$

$$= 2 \text{ mg}/L$$

EXAMPLE 3

Liquid alum usually is supplied at a concentration of 5.4 pounds alum per gallon (650 gm/L or 650 kg/m^3). In this example, the alum feed pump delivers 88 mL per minute and the flow to the filter is 3,000 GPM (190 L/sec). Calculate the concentration or dose of alum in mg/L in the water applied to the filter.

Known	Unknown
Alum Conc, lbs/gal = 5.4 lbs/gal	Alum Dose, mg/L
Alum Pump, mL/min = 88 mL/min	
Flow to Filter, GPM = 3,000 GPM	

ENGLISH

Calculate the alum dose, mg/L.

$$\text{Dose, mg}/L = \frac{\text{Pump, m}L/\text{min} \times \text{Conc, lbs alum/gal} \times 0.00026 \text{ gal/m}L^*}{\text{Flow, gal/min} \times 8.34 \text{ lbs water/gal}}$$

$$= \frac{88 \text{ m}L/\text{min} \times 5.4 \text{ lbs alum/gal} \times 0.00026 \text{ gal/m}L}{3,000 \text{ gal/min} \times 8.34 \text{ lbs water/gal}}$$

$$= \frac{0.124 \text{ lb alum}}{25,020 \text{ lbs water}} \times \frac{1,000,000}{1 \text{ M}}$$

$$= \frac{5 \text{ lbs alum}}{1 \text{ M lbs water}}$$

$$= 5 \text{ mg}/L$$

---

* Conversion factor. 1 mL = 0.00026 gallon

METRIC

$$\text{Dose, mg}/L = \frac{\text{Pump, m}L/\text{min} \times \text{Conc, gm alum}/L \times 1,000 \text{ mg/gm}}{\text{Flow, } L/\text{sec} \times 60 \text{ sec/min} \times 1,000 \text{ m}L/L}$$

$$= \frac{88 \text{ m}L/\text{min} \times 650 \text{ gm alum}/L \times 1,000 \text{ mg/gm}}{190 \text{ } L/\text{sec} \times 60 \text{ sec/min} \times 1,000 \text{ m}L/L}$$

$$= \frac{950 \text{ mg/sec}}{190 \text{ } L/\text{sec}}$$

$$= 5 \text{ mg}/L$$

QUESTIONS

Write your answers in a notebook and then compare your answers with those on page 478.

9.2D What is the purpose of the holding tank located just ahead of a filter?

9.2E Cross out the incorrect words within the following parentheses in order to make the statement correct.

Alum is used for *(COAGULATION OR FLOCCULATION)* while polymers are used for *(COAGULATION OR FLOCCULATION)*.

9.2F Polymer is supplied at a concentration of 0.6 pound polymer per gallon (72 gm/L or 72 kg/m^3). The polymer feed pump delivers a flow of 0.15 GPM (0.0095 L/sec) and the flow to the filters is 5,000 GPM (315 L/sec). Calculate the concentration or dose of polymer in mg/L in the water applied to the filter.

9.213 Filters (See Figure 9.10, page 417)

This section discusses the parts of a pressure filter and the purpose of each part.

VESSELS (Figures 9.14 and 9.15)

Each pressure vessel containing filter media consists of a cylindrical shell closed at both ends. Accessways are provided to allow entry to the vessel for media installation and maintenance work. Pressure gauges are attached to the accessway covers to facilitate monitoring of the vessel pressure.

A direct, spring-loaded pressure relief valve is installed on top of the filter and is set to release at a preset pressure. The relief valve is provided to prevent vessel rupture in case effluent flow is restricted or stopped while influent flow continues.

A combination-type air release valve with a large orifice (opening) is installed on top of the filter to permit air to exhaust when the filter vessel is charged with water and to allow air to reenter when the filter vessel is drained. A small orifice is also provided to exhaust small pockets of air that may collect during operation of the filter.

INTERIOR PIPING (Figure 9.10, page 417)

Interior vessel surfaces, influent and effluent headers, and supports are painted with a protective coating to inhibit corrosion. The influent header is suspended and supported from the upper side of the vessel by lugs. Each filter is equipped with rotary surface wash arms that are installed and supported just beneath the influent header. These are self-propelling, revolving "straight line" wash arms.

The surface wash piping consists of an influent water line, solenoid valve, a central bearing of all-bronze construction, a bronze tee with a water nozzle to spray water directly below and from the center of the tee, and arms extending laterally from the tee. The lateral arms are fitted with numerous brass nozzles located at double-angle positions to most effectively cover the area of the filter bed to be cleaned. Each nozzle is fitted with a synthetic rubber cap slit or grooved to act as a check valve to keep filter media away from the nozzle.

Water to the wash arms is supplied from an external source, usually from the treatment plant wash water system. Water from the surface wash arms quickly breaks up the mat of suspended material that has accumulated in and on the top layer of filter media. This occurs during the first portion of the backwash cycle.

The effluent header is encased in concrete fill in the lower section of the filter. PVC underdrain laterals are attached to the effluent header. Each lateral has numerous small-diameter holes facing toward the bottom of the filter. The ends of the laterals are capped. The filtered water is collected by the underdrain laterals, which pass the water to the effluent header for discharge from the filter.

UNDERDRAIN GRAVEL (Support Media)

The inert filtering media is supported by underdrain gravel consisting of specifically sized, hard, durable, rounded stones with an average specific gravity of not less than 2.5. The gravel is placed in the filter in many specific layers starting with the larger stones (2-inch or 5-cm diameter) on the bottom and progressing to the smallest stones ($1/4$-inch or 0.7-cm diameter) on top. The depth of each layer, specific stone sizes, and overall gravel depth will depend on the application, type, and quantity of inert media that will be used in a filter.

INERT MEDIA

Granular filter media commonly used in wastewater filtration include anthracite coal, silica sand, and garnet sand. These

Fig. 9.14 Filter vessels

Fig. 9.15 Filter vessels

filter media range in size from 0.20 mm to 1.20 mm and specific gravities range from 1.35 to 4.5. The largest media, anthracite coal, has the lowest specific gravity. Conversely, the smallest media, garnet sand, has the highest specific gravity.

Inert-media filter configurations vary according to the specific characteristics of the water to be filtered. The common applications use either silica sand or garnet sand in a single-media filter; anthracite coal and silica sand or garnet sand in dual-media filters; and anthracite coal, silica sand, and garnet sand in multimedia or mixed media filters.

In most filter applications, the various types of media are placed in the filter in the following order and proportions:

1. The smaller size, higher specific gravity media is placed on the bottom of the filter first and makes up about 10 percent of the total media depth.

2. Next, the medium size, medium density media material is added on top of the small, dense layer. This middle layer equals approximately 30 percent of the total media depth.

3. The top and final layer is made up of the larger size, lower specific gravity media material. This layer equals about 60 percent of the total media depth. (*NOTE:* Total filter media depth varies with the application.)

Due to the size and density ratio of the media and its placement in the filter, the larger size, lower specific gravity media stay at the top and the smaller size, higher specific gravity media remain at the bottom. Most dual-media filters are designed to keep the media separated after backwashing.

FLOW CONTROL METHOD (Figure 9.16)

In filter operation, the rate of flow through a filter is expressed in gallons per minute per square foot:

$$\frac{\text{Rate of Flow,}}{\text{GPM/sq ft}} \sim \frac{\text{Driving Force}}{\text{Filter Resistance}} \sim \frac{\text{Total Available Head}}{\text{Total Head Loss}}$$

Therefore, as the total head loss increases, the rate of flow decreases. The driving force refers to the pressure drop across the filter, which is available to force the water through the filter. At the start of the filter run, the filter is clean and the driving force only needs to overcome the resistance of the clean filter media. As filtration continues, the suspended solids removed by the filter collect on the media surface or in the filter media, or both, and the driving force must overcome the combined resistance of the filter media and the solids removed by the filter.

The filter resistance (head loss) refers to the resistance of the filter media to the passage of water. The head loss increases during a filter run because of the accumulation of solids removed by the filter. The head loss increases rapidly as the pressure drop across the suspended solids mat increases because the suspended solids already removed compress and become more resistant to flow. As the head loss increases, the driving force across the filter must increase proportionally to maintain a constant rate of flow.

The constant rate method of filtration is commonly used for pressure filters. In this method, a constant pressure is supplied to the filter and the filtration rate is then held constant by the action of a manually or automatically operated filter rate-of-flow controller. At the beginning of the filter run, the filter is clean and has little resistance. If the maximum available water pressure was applied to the filter, and the effluent flow was not restricted, the flow rate would be very high. To maintain a constant flow rate, the available pressure is dampened or reduced by the rate-of-flow controller (RFC). At the start of the filter run, the RFC is nearly closed to provide the additional head loss needed to maintain the desired flow rate. As filtration continues, the filter gradually becomes clogged with suspended solids and the RFC opens proportionally. When the valve is fully opened, any further increase in the head loss will not be balanced by a corresponding decrease in the head loss of the RFC. Thus, the ratio of pressure to filter resistance will decrease, and the flow rate will decrease. This action is also known as filter differential pressure. When the flow rate decreases, filter differential pressure increases and this is an indication that the filter run must be terminated and a filter backwash should be initiated.

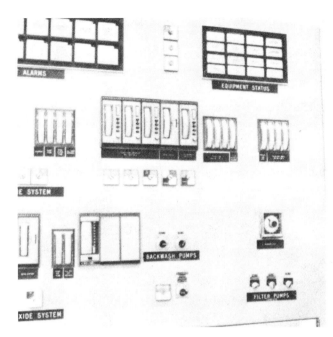

Fig. 9.16 Filter controls

In general, a pressure filter should run for a minimum of 6 to 8 hours during peak load conditions before backwashing is required. Under average flow conditions, this will mean a filter run length of about 24 hours.

9.214 Backwash System

As the suspended material accumulates on the filter media surface, or in the filter media bed, or both, the differential pressure across the filter increases, flow through the filter decreases, and filter effluent quality deteriorates. The filter backwash cycle removes the suspended solids accumulation from the filter, thus restoring the filter efficiency.

WET WELL

The backwash wet well is used to store a large volume of filtered and chlorinated wastewater to backwash the filters. The water, usually from the chlorine contact tank, flows to the wet well until it is filled up and then it flows to further final treatment processes. This flow method ensures a continuous water supply to the wet well.

PUMPS (Figure 9.17)

The filter backwash pumps lift the filtered and chlorinated wastewater from the backwash wet well and pump it through the filters to remove the trapped suspended material. The pumps are generally of the vertical-turbine wet-pit type with either a closed or semi-open impeller. The pumps may be driven by fixed-speed motors.

Fig. 9.17 Backwash pumps

The pump system is equipped with a solenoid-operated bypass valve installed on the main discharge line to prevent the possibility of the system operating at the shutoff pressure of the pumps. A pressure switch with an adjustable operating range (psi) will cause the valve to open in response to rising pressure. The bypass flow is returned to the backwash wet well. The main discharge line is also equipped with an air relief valve that purges air from the system to prevent air slugs from disturbing the filter media bed.

Normal starting and stopping of the pumps may be controlled by means of current switches in the backwash program unit. Lead-lag position selector switches provide the means of selecting the sequence of starting for the pumps. A low-water probe in the backwash wet well will stop the pumps in the event that the water level drops below a preset elevation.

The backwash pump main discharge line is provided with an ORIFICE [22] plate and flow-control valve. Flow control is accomplished by means of a cascade control system using a cam programmer to provide a set point signal. A cam is cut so as to gradually introduce the backwash flow to the filters, thereby avoiding sudden disturbance or uneven expansion of the filter media bed.

BACKWASH CYCLE

Whenever possible, the filters should be backwashed during the plant's low-flow hours when the full capacity of the filters is not needed. The backwash cycle may be activated either manually, automatically by a preset filter differential-pressure level, or automatically by a programmed timer. In the manual mode, one or more filters may be backwashed as needed. In the automatic mode, all filters in the system that are on line will be washed when the differential pressure reaches the preset level. Upon completion of backwash of one filter, the next filter on line will begin to backwash.

The total backwash duration per filter usually is adjustable. The total backwash flow and duration should be adequate to FLUIDIZE [23] and expand the media bed. The largest media size and the warmest expected water temperature will dictate the maximum filter backwash rates required.

When the backwash cycle is manually or automatically activated, the following sequence occurs:

1. Filter influent valve (V-1) and effluent valve (V-2) close to terminate filter feed flow. (See Figure 9.10, page 417, for locations of valves.)

2. Backwash influent valve (V-3) and effluent valve (V-4) open to allow backwash flow into and out of the filter.

3. The surface wash arms' influent water line solenoid valve (V-5) opens, allowing the wash arms to function in initially breaking up the mat of suspended material that has accumulated on the top layer of filter media.

4. The backwash pumps start pumping against a closed backwash control valve. The backwash flow rate is brought up to full rate in one to two minutes as determined by the cam programmer transmitting a gradual open signal to the backwash flow-control valve operator.

. As the surface wash continues to operate, the backwash flow gradually enters the filter from the bottom. As the flow increases, the bed fluidizes and expands upward (about 20 percent of the total media depth) allowing a uniform rolling action of the filter media bed, which results in cleaning of the media due to the hydrodynamic shear (water causes grains to clean each other) that occurs. The media bed expands upward

---

[22] *Orifice* (OR-uh-fiss). An opening (hole) in a plate, wall, or partition. An orifice flange or plate placed in a pipe consists of a slot or a calibrated circular hole smaller than the pipe diameter. The difference in pressure in the pipe above and at the orifice may be used to determine the flow in the pipe. In a trickling filter distributor, the wastewater passes through an orifice to the surface of the filter media.

[23] *Fluidized* (FLOO-id-i-zd). A mass of solid particles that is made to flow like a liquid by injection of water or gas is said to have been fluidized. In water and wastewater treatment, a bed of filter media is fluidized by backwashing water through the filter.

and into the rotating surface wash arms. The arms now aid in breaking up the suspended material and mudballs that have accumulated in the top section of the media.

5. After two to five minutes of surface wash, the surface wash influent water line solenoid valve closes. Surface wash is discontinued two to ten minutes before the backwash ends so that the surface of the filter media will be smooth and level at the beginning of the cleaned filter run cycle.

6. After seven to twenty minutes of backwash, the backwash flow-control valve gradually begins to close. Shortly after the backwash flow-control valve is fully closed, the backwash pumps stop.

7. Backwash influent valve (V-3) and effluent valve (V-4) close.

8. Filter influent valve (V-1) and effluent valve (V-2) open.

NOTE: When the backwash cycle is activated, the filter flow-control valve fully closes. Upon completion of the cycle, the valve opens slightly.

The two valve sequences indicated in items 1 and 2, and items 7 and 8 occur simultaneously to ensure that the filter does not become "air bound" (clogged by air released from water). Air binding will reduce or block filter influent flow or create media bed disturbance when filter backwash begins.

9.215 *Decant Tank (Backwash Recovery) (Figures 9.18 and 9.19)*

Most decant tanks are very similar to secondary clarifiers because they have flights or scrapers to collect settled material toward a sludge hopper.

Filter backwash effluent leaves the filter and may be discharged to the decant tank. The suspended material in the backwash water is dosed with a cationic polymer and allowed to rise to the surface. The clarified effluent overflows to the holding tank through weir slots between the two tanks for recycle to the filters. The floating solids are collected toward a skimmings trough and the settled material is collected toward a hopper in the tank for periodic discharge to the solids handling facility.

If poor solids capture occurs in the decant tank, all spent backwash flow may be returned to the plant headworks through the tank drain line. The drain line may be equipped with a propeller meter and a motor-operated butterfly valve. The opening limit of the valve should be set to discharge tank flow at a rate that will not hydraulically overload the plant.

The tank may be equipped with high water level probes that will open the motor-operated valve fully to allow a predetermined volume of water to leave the tank rapidly. This may be necessary if the tank becomes surcharged (overloaded) due to frequent filter backwashes.

QUESTIONS

Write your answers in a notebook and then compare your answers with those on page 478.

9.2G List the major components of pressure filters.

9.2H How is the mat of suspended material on the media surface initially broken up during a backwash?

9.2I What is the source of water used to backwash the filter?

9.2J What is the purpose of the decant tank?

Fig. 9.18 Decant tank drain line automatic valves

Fig. 9.19 Backwash recovery tank and feed pumps

9.22 Operation

9.220 Operational Strategy

This lesson has covered some of the basic concepts of inert-media pressure filters used to remove suspended solids, associated BOD and turbidity from secondary effluents before chlorination.

If the filters become overloaded due to high suspended solids concentrations, excessive plant flows, high chemical concentrations, or exposure to very cold temperatures, be prepared for the problems discussed in this section.

1. High suspended solids concentrations will cause a filter to plug up fast, thus requiring very frequent filter backwashes. This will result in high recycle flow rates through the plant and eventually the filters. This problem may be eliminated or reduced by having adequate spent backwash storage capacity or by having a closed filter system that will allow for clarification and reuse of spent backwash water for subsequent backwashes.

2. If no backwash storage or a closed system is provided, hydraulic surcharge (overload) on the upstream side of the filters will result. Provisions must be made for filter bypass or storage. If bypass is the only alternative, you should anticipate increased disinfection demands at the disinfection injector point as a result of the increase in unfiltered suspended material. Adjust the disinfection rate to compensate for the greater demands.

3. By allowing suspended material to bypass the filters and enter chlorine contact tanks, more frequent cleaning of these tanks will be required.

4. If higher than normal plant flows can be anticipated (rain), it would be a good idea to operate the filter holding tank/wet well at a lower water level to provide for additional water storage. This action will reduce the surcharge possibility on the upstream side of the filters. This preventive action should also be used if a filter must be taken out of service for repairs or inspection.

5. If liquid alum is used and it is exposed to cold temperatures, the liquid alum will start to crystallize and the delivery of alum to the filter influent flow will be seriously impaired. If climatic conditions of this type are common in your area, consider storing the alum in an enclosed, warm space or providing insulated storage tanks and heat-traced piping to prevent the alum from crystallizing.

6. If chemical feed pump check valves or antisiphon devices fail, large quantities of chemical will be drawn into the filters. This will result in short filter run times due to increased differential pressure across the filter when a polymer is the chemical involved. When excessive alum concentrations are involved, the alum will pass through the filter media and filter effluent turbidity and suspended solids values will increase due to the alum breakthrough.

7. Filter flow and differential pressure valves may be sensed by differential pressure cells. These cells are water activated and are fed through small-diameter piping. During periods of extremely cold weather, these cells could freeze and prevent proper functioning of the filter and control instruments. Heavy insulation or heat tape will prevent the water in the cell piping from freezing.

9.221 Abnormal Operation

Efficient filter operation is essential if your final effluent quality is to comply with the waste discharge requirements established for your plant. Table 9.1 lists a few of the more common pressure filter operational problems and suggestions on how to correct them.

QUESTIONS

Write your answers in a notebook and then compare your answers with those on page 478.

9.2K What happens when large quantities of alum or polymer accidentally reach the filter?

9.2L What precautions should be taken in regions where freezing temperatures occur?

9.2M What could cause high operating filter differential pressures?

9.23 Maintenance

A comprehensive preventive maintenance program is an essential part of plant operations. Good maintenance will ensure longer and better equipment performance. The following may be used as a guideline in performing the required maintenance on the pressure filter system.

A filtration system performance test should be conducted monthly. This test will enable you to evaluate performance and determine if the pumps, valves, filters, and control instruments are functioning properly. If they are not, the proper corrective action must be taken. A sample performance test chart for three filters is shown in Figure 9.20.

Filter media and interior vessel surfaces should be inspected quarterly. The filter should be backwashed just prior to the inspection. Some of the items to look for are:

1. Is the media surface fairly flat and level? If not, the surface wash time should be reduced.

2. Are there mudball formations on or in the media? If so, an increased surface wash time in conjunction with a lower backwash rate should bring the media back to a clean condition.

3. Are very small quantities of mid-filter media particles visible on the top layer media surface? This condition is normal.

4. Do the surface wash arms rotate freely and in the proper direction? If not, the trouble could be a defective central bearing and tee. Are any nozzles plugged? If so, they must be cleaned.

CAUTION: Wear goggles when observing the operation of the surface wash arms. The velocity of water produced from the wash arms is great and will kick up surface media.

TABLE 9.1 ABNORMAL PRESSURE FILTER OPERATION

Abnormal Condition	Possible Cause	Operator Response
PUMPS Do not meet pumping requirements.	Inappropriate valve positioning.	Adjust valves to proper position.
	Insufficient motor speed.	Install higher RPM motors.
	Pump impeller improperly set in bowl of pump.	Set impeller as per manufacturer's instructions.
	Excessive filter system head losses.	Air in filter system. Analyze problem and take corrective action such as install higher RPM motors, redesign pump station, redesign force main, redesign orifice plates.
	Broken pump shaft.	Replace.
FILTERS (GENERAL) High operating filter differential pressure.	Filled with suspended material.	Backwash filter at least once every 24 hours.
	Excessive chemical feed "blinding" media.	Evaluate and reduce dosage. Backwash filter.
Water discharges from pressure-relief valve.	Effluent valve(s) blocked or closed.	Investigate and correct valve problem.
	Foreign object lodged between valve and seat.	Secure filter and clean valve seat.
Water discharges from air-relief valve.	Air pocket in underdrain system (most common after a backwash).	Secure filter for 2 to 3 minutes to allow vessel water level to stabilize, return filter to service. Adjust filter feed and backwash valves to open and close simultaneously to keep vessel full of water.
MEDIA Support media upset.	Air slug forced out by the backwash flow.	Install air-relief valve on backwash influent line.
	Backwash flow pumped too suddenly.	Install flow-control valve for regulated flow rates.
	Backwash flow rate too high.	Install valve stops or limiting orifices.
Mudball formation. Media surface cracks.	Inadequate surface wash time.	Increase surface wash time. Check to ensure arms are operating.
Backwash water dirty at end of wash cycle.	Insufficient backwash time or flow rate.	Increase time or flow rate until clean water appears.
Media surface uneven after backwash.	Surface wash too long.	Decrease surface wash time.
Algal growth in media bed.	Nutrient-rich water being filtered.	Prechlorinate continuously at low rates.

5. Is there a small amount of foreign matter on the media surface (plastic, cigarette butts)? This condition is fairly common and most often occurs at the extreme effluent end of the backwash effluent header. The foreign matter is carried away by the backwash water during the next backwash cycle. If a large amount of foreign matter accumulates, it will have to be removed by manual means.

6. Inspect all interior metal surfaces to ensure that the corrosion-inhibitive protective coating is in good condition. If not, prepare the affected surface and reapply the proper coating. An epoxy tar is frequently used for this purpose.

At least once monthly, the backwash rate should be observed to ensure that the flow rate is correct as specified by the manufacturer's backwash rate/flow curve and that the backwash flow is allowed to enter the filter at a regulated rate. Observe the backwash effluent flow. The water should be clear at the end of the wash cycle. If it is not, an increase in the wash time is indicated.

The flow rates for the chemical feed pumps should be checked at least every two weeks. Corrective adjustments should be made to maintain the proper flow rates.

9.24 Safety

Safety precautions for working in confined spaces and working around mechanical equipment should be observed when operating and maintaining this equipment in the pressure filter system. Chapter 13, "Safety," discusses these precautions in detail.

VALENCIA WRP						FILTER AND FILTER FEED PERFORMANCE TESTS										DATE: 7-21-04 BY: REDNER & KETTLE		
Test	Pumps Running			Filter on			Discharge Through Filters, GPM				Holding Tank Level	Pressure at Pump Discharge	Filter Differential Pressure			Filter Effluent Flow Control Valve, % Open		
#	1	2	3	1	2	3	1	2	3	TOTAL	FEET	PSI	1	2	3	1	2	3
1	on	off	off	on	off	off	1400			1400	7.8	1.5	6.5			100	0	0
2	off	on	off	on	off	off	1100			1100	8.3	3.0	8.3			100	0	0
3	off	off	on	on	off	off	970			970	8.5	3.5	9.5			100	0	0
4	on	on	off	on	on	off	850	1400		2250	8.5	—	9.0	5.0		100	100	0
5	on	on	off	on	on	on	725	1224	1000	2950	8.5	5.0	8.5	5.0	5.0	100	100	100
6	on	off	on	on	on	on	700	1200	1000	2900	8.5	5.2	8.5	5.0	5.0	100	100	100
7	off	off	on	on	on	on	625	1075	925	2625	8.5	3.0	8.5	5.5	5.5	100	100	100
8	on	on	on	on	on	on	900	1275	1150	3325	7.7	8.0	9.5	6.0	6.0	100	100	100
9	on	off	off	off	off	on			1150	1150	7.7	3.0			7.0	0	0	100
10	on	off	off	off	on	off		1300		1300	7.9	3.0		6.5		0	100	0
11	on	off	off	on	off	off	975			975	8.0	3.2	10.0			100	0	0

Note: The filter effluent flow control valve should be 100 percent open during the monthly test.
REMARKS:

Fig. 9.20 *Filtration system performance test chart*

In addition, the following safety precautions should be observed:

1. Wear safety goggles and gloves when working with alum or polymers. Wear approved respiratory protection if inhalation of dry alum or polymer is possible. Flush away any alum or polymer that comes in contact with your skin with cool water for a few minutes.

2. Be very careful when walking in an area where polymer mixing takes place. When a polymer is wet, it is very slippery. Clean up polymer spills with an inert absorbent material such as sand or earth.

3. When inspecting the interior of a filter vessel, *ALWAYS* follow confined space entry procedures, and:

   a. Ensure that all flow-control instruments are in the OFF position and that all valves are in the MANUAL or OFF position. Position all valves to prevent flow from entering the filter. Lock out and tag any controls or valves as well as the power supply to the equipment to prevent accidental start-up.

   b. Always ventilate vessel. Open two access covers. Install and start an exhaust blower in one of the two openings to provide fresh air circulation before entering the filter.

   c. Check vessel atmosphere for toxic gases (hydrogen sulfide), explosive conditions, and sufficient oxygen.

   d. Entering a filter vessel is a three-person operation. Two must be outside the vessel whenever one person is inside.

   e. Wear a hard hat when working inside a filter or around the filter vessel piping to protect your head from injury.

9.25 Review of Plans and Specifications

As an operator, you can be very helpful to design engineers in pointing out some design features that would make your job easier and safer. This section attempts to point out some of the items that you should look for when reviewing plans and specifications for expansion of existing facilities or construction of new pressure filter systems.

1. The variable hydraulic and suspended solids load in secondary effluents must be considered in the design to avoid short filter runs and excessive backwash water requirements.

2. A filter that allows penetration of suspended solids (a coarse-to-fine filtration system) is essential to obtain reasonable filter run lengths. The filter media on the influent side should be at least 1 to 1.2 mm in diameter.

3. Auxiliary agitation of the media is essential for proper backwashing. Surface washers should be installed.

4. The effect of recycling used backwash water through the plant on the filtration rate and filter operation must be considered in predicting peak loads on the filters and resulting run lengths.

5. The filtration rate and head loss should be selected to achieve a minimum filter run length of 6 to 8 hours during peak load conditions. This requirement will mean an average filter run length of 24 hours. Estimates of head loss development and filtrate quality should be based on pilot-scale observations of the proposed facility conducted at the treatment plant before the full-scale facility is designed.

6. Accessways should be sized large enough to allow operators and equipment ease of entering and leaving the filter.

7. A media core sample port(s) should be provided to allow evaluation of the entire media depth.

8. Ladders and walkways should be provided to allow easy access to vessels, pipes, and valves.

9. Filter flow charts should be provided to aid in monitoring filter performance.

9.26 Acknowledgments

1. County Sanitation Districts of Los Angeles, Valencia Water Reclamation Plant.

2. Jerry Schmitz, Draftsman, County Sanitation Districts of Los Angeles.

9.27 Additional Reading

The two EPA publications listed below are also available from the National Technical Information Service (NTIS), 5285 Port Royal Road, Springfield, VA 22161.

1. *WASTEWATER FILTRATION DESIGN CONSIDERATIONS*, Technology Transfer, US Environmental Protection Agency, July 1974. Order No. PB-259448. Price, $33.50, plus $5.00 shipping and handling per order.

2. *PROCESS DESIGN MANUAL FOR SUSPENDED SOLIDS REMOVAL*, Technology Transfer, US Environmental Protection Agency. Order No. PB-259147. Price, $78.50, plus $5.00 shipping and handling per order.

QUESTIONS

Write your answers in a notebook and then compare your answers with those on page 478.

9.2N How frequently should a filter system performance test be conducted?

9.2O What caution should be exercised when observing the operation of the surface wash arms?

9.2P What safety precautions should be taken when working with alum or polymers?

END OF LESSON 2 OF 4 LESSONS

on

FILTRATION

Please answer the discussion and review questions next.

DISCUSSION AND REVIEW QUESTIONS
Chapter 9. FILTRATION
(Lesson 2 of 4 Lessons)

Write the answers to these questions in your notebook. The question numbering continues from Lesson 1.

7. How are floatable and settleable solids removed from a holding tank?

8. How would you attempt to control corrosion of the interior surfaces of a pressure filter vessel?

9. During what time of the day should the filters be backwashed?

10. What would be the impact of allowing suspended material to bypass the filters and enter chlorine contact tanks?

CHAPTER 9. FILTRATION

(Lesson 3 of 4 Lessons)

9.3 SOLIDS REMOVAL FROM WASTESTREAMS USING CONTINUOUS BACKWASH, UPFLOW, DEEP-BED SILICA SAND MEDIA FILTERS
by Ross Gudgel

9.30 Use of Continuous Backwash, Upflow, Deep-Bed Silica Sand Media Filters

As water demands for agricultural, industrial, and municipal use increase, wastewater reuse has become more important, especially as our nation implements essential water conservation programs. In many parts of the country, a high level of water quality is required and specific treatment criteria must be met before treated wastewaters can be reused. Continuous backwash, upflow, deep-bed silica sand media filters consistently produce high-quality effluents that do not exceed average operating values of 2 turbidity units (NTUs). This high level of water quality significantly reduces demands on the disinfection process because *PATHOGENS*[24] and viruses can be removed or inactivated much more effectively and at lower cost when turbidity levels are low.

The actual filtering process, as with most conventional granular media type filters, involves the capture of previously coagulated and flocculated suspended solids within the voids (spaces) between the granules of media that make up the filter bed. Continuous backwash, upflow, deep-bed silica sand media filters continuously remove solids from the influent while at the same time cleaning and recycling the filter media internally through an airlift pipe and sand washer. The cleaned sand is redistributed on top of the filter media bed, which allows for uninterrupted flow of filtered effluent and backwash reject water. Therefore, this type of filter has two distinct advantages over other types of granular media filters.

First, these filters do not need to be taken out of service for backwashing. This is of particular operational importance with regard to the addition of coagulants during periods of low filter influent suspended solids. In most conventional downflow filters, coagulants added to the filter influent flow when the suspended solids are low tend to accumulate and create a mat of coagulant on top of and within the uppermost portions of the granular media. When the filter influent suspended solids subsequently increase, virtually all of the incoming suspended solids are captured on top of and within the coagulant mat. The filter then blinds, filter head loss increases dramatically, and the filter must be taken out of service for backwashing.

Second, because these filters do not need to be taken out of service for backwashing, full design flow processing capacity can be provided with fewer filters or smaller-sized filters.

Continuous backwash, upflow, deep-bed silica sand media filters are supplied in two basic styles, bottom feed cylindrical (Figure 9.21) and bottom feed concrete basin (Figure 9.22). The internal design and the operation of the two styles are virtually identical. The main difference between the two types is the outer housing of the filter. Therefore, the following section will use the bottom feed cylindrical filter to explain this type of filtration.

The filter system usually consists of:

1. A channel or pipe from the upstream treatment process to conduct flow to the filtration system. Alternatively, a holding tank or wet well may be provided for storage of treatment process effluent. Filter feed pumps may also be provided, which pump the effluent from the holding tank or wet well to the filtration system.

2. Filter influent and effluent turbidity metering to continuously monitor and record filter influent and effluent water quality.

3. Filter influent and effluent flow metering to quantify and record flow values and provide dosage control signals to metering pumps and disinfection chemical addition control systems.

4. A chemical coagulant (ferric chloride or alum) dosage control and metering pump system, which injects coagulant solution into the filter influent flow.

5. A chemical flocculant (polyelectrolyte) mixing, dosage control, and metering pump system, which injects flocculant solution into the filter influent flow.

6. A flocculation tank to enhance the development of large floc particles (the sticking together of chemically treated suspended solid and dissolved materials).

---

[24] *Pathogenic* (path-o-JEN-ick) *Organisms.* Organisms, including bacteria, viruses, or cysts, capable of causing diseases (such as giardiasis, cryptosporidiosis, typhoid, cholera, dysentery) in a host (such as a person). Also called PATHOGENS.

Filtration 431

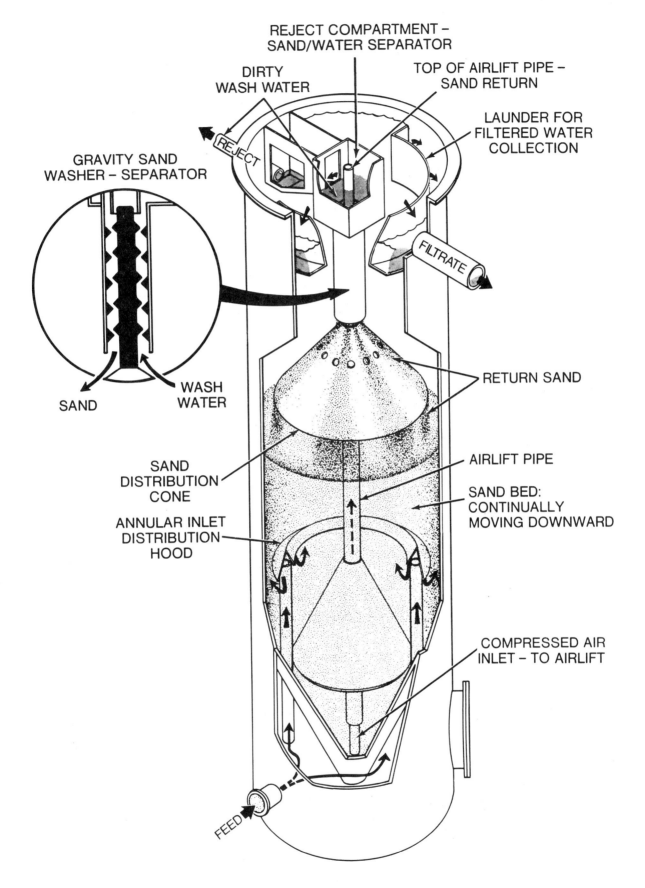

Fig. 9.21 Bottom feed cylindrical filter

(Reproduced by permission. DynaSand® Filter is a registered
trademark of Parkson Corporation.)

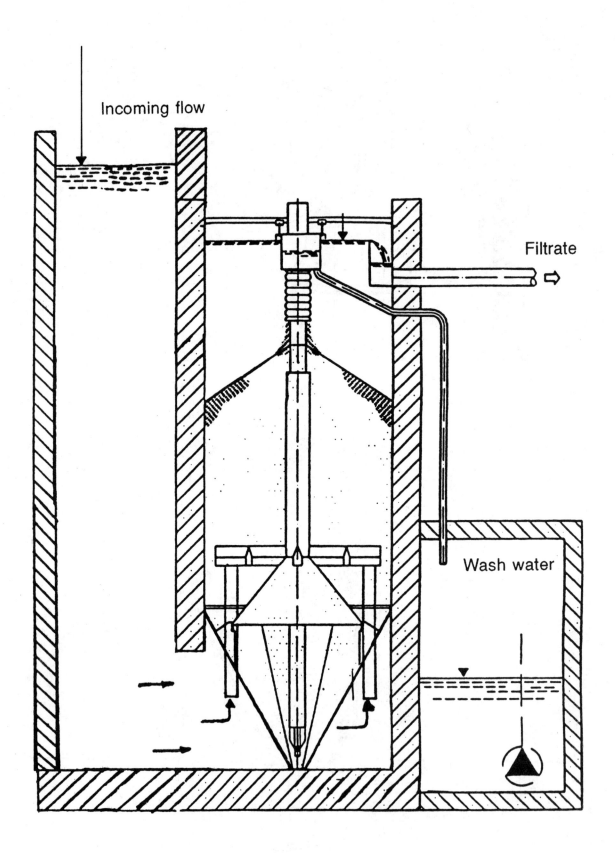

Fig. 9.22 Bottom feed concrete basin filter
(Reproduced by permission. DynaSand® Filter is a registered trademark of Parkson Corporation.)

7. Continuous backwash, upflow, deep-bed silica sand media filters that trap suspended solids and turbidity and remove them from the treatment flow stream.
8. A filter reject (backwash) water sump or wet well equipped with pumps to collect and return suspended solids and backwash water to the plant for further treatment.

Figure 9.23 shows a schematic view of the filter components described above.

QUESTIONS

Write your answers in a notebook and then compare your answers with those on page 478.

9.3A How does a continuous backwash, upflow, deep-bed silica sand media filter work?

9.3B What is the purpose of the filter influent and effluent turbidity meters?

9.3C Why are the influent and effluent flows metered?

9.31 Auxiliary Equipment

The following sections describe facilities that are typical for a filtration system plant using bottom feed cylindrical filters with a rated flow capacity of 0.5 MGD. Facilities at larger plants would be quite similar but would differ significantly in the numbers and sizes of the various components.

9.310 Channel or Piping

Effluent, normally from the treatment plant's secondary sedimentation tanks, is typically conducted to the filtration system facilities through a covered channel or through piping. Where channels are used, covered channels are preferred in order to limit the risk of having foreign objects get into the flow stream. Such objects can damage flocculation equipment or plug the airlift pump in a filter.

Features to bypass the filtration system facilities should be provided to permit diverting treatment process effluent to holding or equalization basins during emergency conditions, such as extremely high flows, equipment failures, or clogged filters. Following resolution of the emergency, flow can then be regulated back into the filter influent flow stream from the holding or equalization basin for normal and complete treatment through the filtration system.

9.311 Filter Influent and Effluent Turbidity Metering

Low range turbidity meters are normally used to continuously analyze and record the turbidity (clarity) of the filter influent and effluent flows.

Portions of the respective flow streams are continuously withdrawn from the flow channel or pipe by a sample pump. The sample portions first pass through a bubble trap to remove the air bubbles (air bubbles interfere with turbidity measurement). The bubble trap also serves as a head regulator to dampen fluctuations in flow due to pulsations from the sample pump.

The wastewater being sampled enters the bottom of the bubble trap assembly and rises toward the top. As the flow rises, most of the air bubbles are released and vented to the atmosphere. The wastewater flows out of the trap by gravity through an overflow discharge pipe, which is fitted with a shutoff/flow control valve. The valve allows the operator to regulate flow to the turbidity meter; a typical flow range is 3.5 to 11.5 gallons per hour. Additional flow control/head regulation of the flow fed to the turbidity meter can be obtained by raising or lowering the bubble trap assembly within two pipe hanger straps that support the assembly.

Wastewater entering the turbidity meter (Figure 9.24) passes through another bubble trap baffle network that forces a downward flow of the water. This downward water flow (relatively slow) allows any remaining air bubbles to rise and either cling to surfaces of the baffle or rise to the surface and vent to the atmosphere. At the bottom of the bubble trap baffle network, flow enters a center column and rises up into the turbidity measuring chamber head where turbidity is measured. The flow then spills over a weir contained within the turbidity measuring chamber head and is returned to the process flow stream through an overflow discharge pipe.

Turbidity is measured by directing a strong beam of light from the optical section of the turbidity measuring chamber head down through the surface of the passing water flow. Light is scattered at a 90-degree angle when it strikes suspended solids particles (turbidity) in the water. The scattered light is detected by a photocell located just below the surface of the passing water flow.

The amount of light scattered is proportional to the turbidity of the water. If the turbidity of the water is low, little light will be scattered to the photocell and the turbidity reading will be low. High turbidity, on the other hand, will cause a high degree of light scattering and will result in a high turbidity reading.

A control unit for the turbidity meter has a keyboard, microprocessor board, and power supply components. Operating controls and indicators located on the control unit keyboard are used to program the turbidity meter for recorder output minimums and maximums and for turbidity level alarm set points as well as to perform a number of diagnostic self-tests and programming operations.

434 Treatment Plants

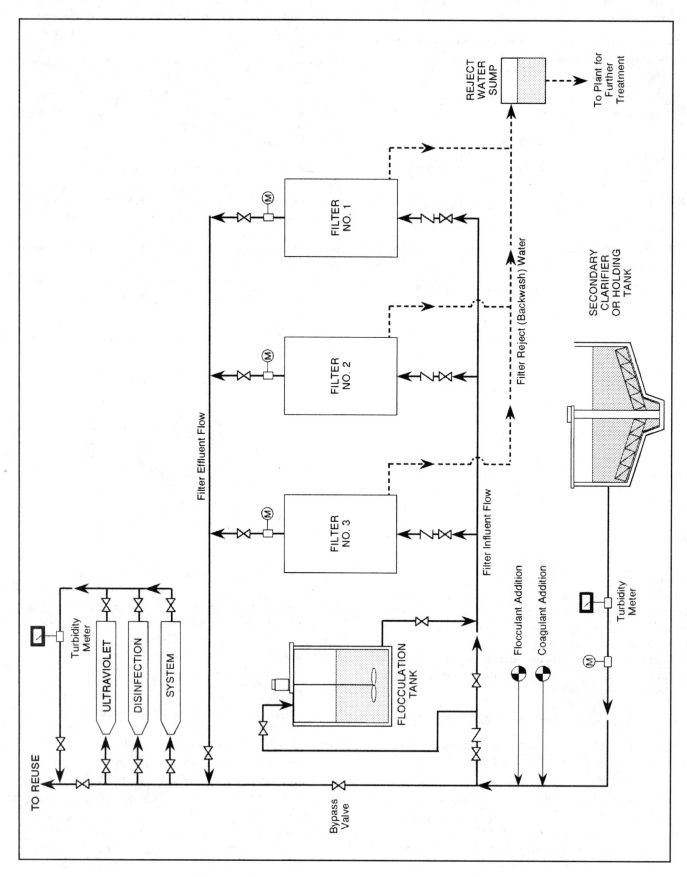

Fig. 9.23   Schematic view of continuous backwash, upflow, deep-bed silica sand media filter system

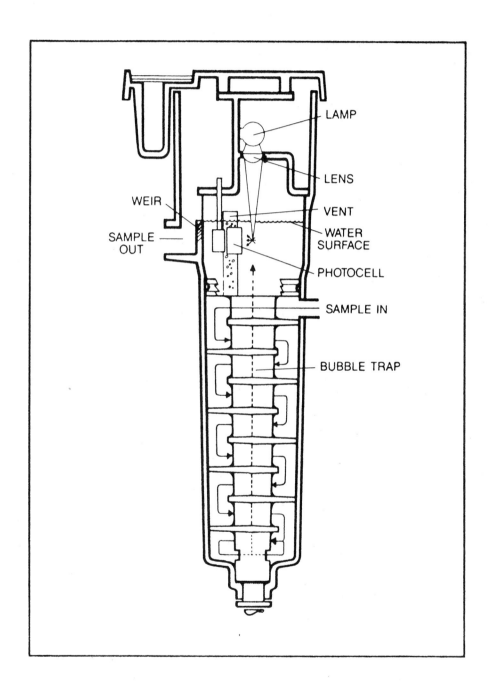

Fig. 9.24 Turbidity meter
(Reproduced by permission of HACH Company)

Turbidity levels are displayed continuously by a four-digit, light emitting diode (LED) display during normal operations. Alarm conditions and certain critical system malfunctions or impending malfunctions are also indicated on the keyboard.

Programmable circuits provide relay closures, both normally open and normally closed, for two turbidity level set points (high turbidity and low turbidity). Alarm set points can be programmed by the operator anywhere within the overall turbidity measuring range of 0.001 to 100.0 NEPHELOMETRIC [25] Turbidity Units (NTUs). High turbidity alarms are transmitted to an alarm monitoring center in the treatment plant, which in turn notifies the operator that a high turbidity condition exists.

NTU values are continuously transmitted from the control unit to a circular chart recorder. The recorder may be a single pen, 10-inch round chart recorder. Chart rotation can be selected for 24-hour or 7-day rotation. Recorder span settings (minimum and maximum values) in NTUs are programmed by the operator at the turbidity meter control unit keyboard.

Most of the information used to develop this section was obtained from the "Hach Model 1720C Low Range Process Turbidimeter Instruction Manual."

9.312 Filter Influent and Effluent Flow Metering

Various types of flow measuring devices suitable for use in these applications are described in Chapter 6, "Flow Measurement." Typical flowmeters include the Venturi and orifice types.

9.313 Coagulation and Flocculation Equipment

The suspended solids removal rates of most types of filters, including continuous backwash filters, can be greatly improved by the addition of chemicals before filtration. Coagulants such as ferric chloride and alum cause the suspended and colloidal (finely divided solids that will not settle by gravity) particles to clump together into bulkier hydroxy precipitates.

To ensure maximum solids removal during filtration, the wastewater is next treated with another type of chemical, usually a POLYMER,[26] in a process called flocculation. This process gathers still more of the suspended and colloidal solids into larger (visible to the naked eye), stronger clumps (floc), which can be efficiently removed by filtration.

Figure 9.25 shows one example of the types of equipment used for flocculation by polymer addition. However, Chapter 8, "Physical–Chemical Treatment Processes," describes the range of equipment and chemicals available for coagulation and flocculation processes. In addition, the chapter explains how to perform a jar test to select the proper chemicals and determine dosages. That information will not be repeated here, so please refer to Chapter 8 to learn about chemical addition and the coagulation and flocculation processes.

9.32 Operation of Continuous Backwash, Upflow, Deep-Bed Silica Sand Media Filters

9.320 Normal Operation

As shown in Figures 9.21 and 9.26, Part A, filter influent flow (feed) is introduced into the bottom of the filter through a feed inlet where it enters a section of the filter called the plenum area. It flows upward through a series of riser tubes and is evenly distributed into the sand bed through the open bottom of the ANNULAR [27] inlet distribution hood. The rate of flow through the filter is expressed as the surface loading rate, in gallons per minute per square foot of filter surface area:

$$\text{Loading Rate, GPM/sq ft} = \frac{\text{Flow, GPM}}{\text{Filter Surface Area, square feet}}$$

The wastewater flows upward through the downward-moving sand bed. The coagulated solids in the wastewater are trapped in the spaces between the granules of sand. A head loss gauge indicates the water pressure (head in inches of water) on the influent side of the filter relative to the atmospheric pressure at the LAUNDER [28] for filtered water collection.

The now cleaned (suspended solids free) influent continues to pass through the sand bed, overflows a launder for filtered water collection, and is discharged from the filter as filtrate (tertiary treated effluent).

Filtrate leaving each filter flows through an effluent pipe and a filtrate flowmeter. The meter indicates and totalizes flow locally and also transmits a flow signal to a filter effluent flow metering system. The flow then passes through an effluent gate valve into a larger discharge pipe that directs flow from each filter to the disinfection system.

As the filter is removing the coagulated solids from the influent flow, the silica sand media is continuously cleaned by recycling the sand internally through an airlift pipe and gravity sand washer-separator. The sand bed, along with the solids trapped among the sand particles, is drawn downward into the suction to the airlift pipe (Figure 9.26, Part B), which is positioned at the bottom and center of the filter. A moderate volume of com-

---

[25] *Nephelometric* (neff-el-o-MET-rick). A means of measuring turbidity in a sample by using an instrument called a nephelometer. A nephelometer passes light through a sample and the amount of light deflected (usually at a 90-degree angle) is then measured.

[26] *Polymer* (POLY-mer). A long-chain molecule formed by the union of many monomers (molecules of lower molecular weight). Polymers are used with other chemical coagulants to aid in binding small suspended particles to larger chemical flocs for their removal from water. Also see POLYELECTROLYTE.

[27] *Annular* (AN-yoo-ler) *Space*. A ring-shaped space located between two circular objects. For example, the space between the outside of a pipe liner and the inside of a pipe.

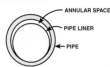

[28] *Launders.* Sedimentation basin and filter discharge channels consisting of overflow weir plates (in sedimentation basins) and conveying troughs.

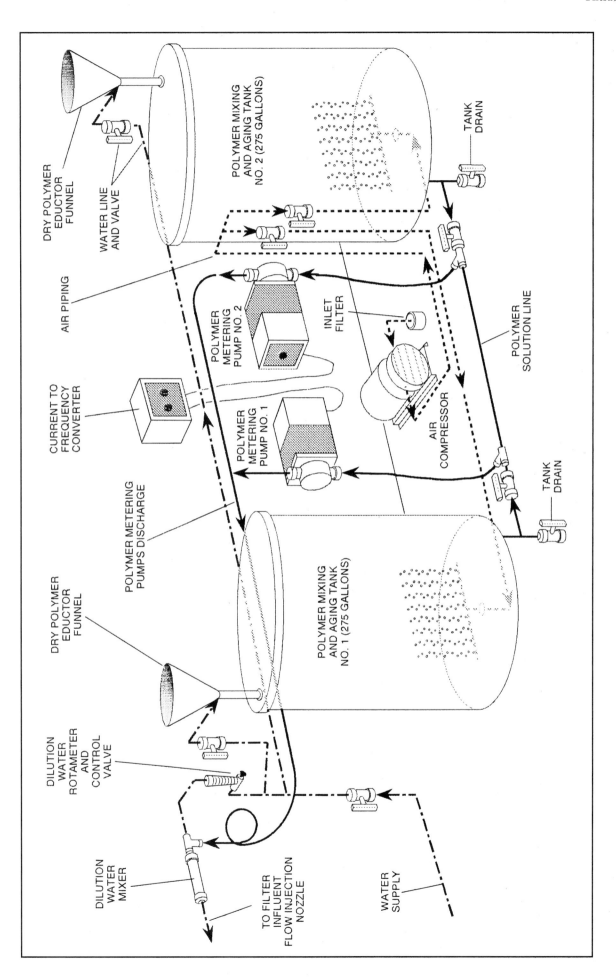

Fig. 9.25 Polymer mixing and aging system

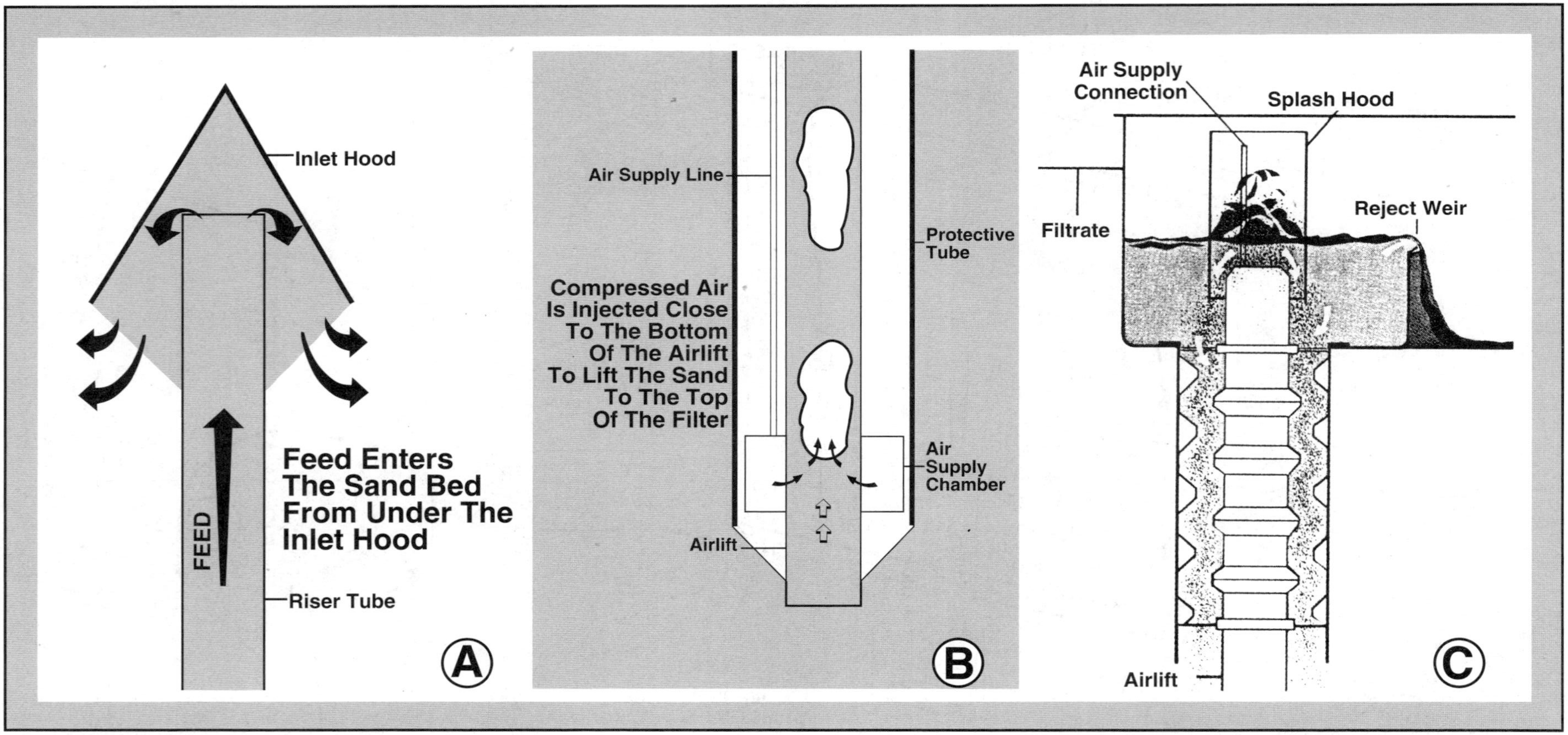

Fig. 9.26 Internal detail of continuous backwash, upflow, deep-bed silica sand media filter (Reproduced by permission. DynaSand® Filter is a registered trademark of Parkson Corporation.)

pressed air (supplied by a compressed air system) is introduced into the bottom of the airlift pipe by an air control panel located at the top of each filter. The air lifts the sand and trapped solids up the airlift pipe into the gravity sand washer-separator. The violently turbulent flow that is created as the sand and trapped solids move up the airlift pipe and into the gravity sand washer-separator scours the trapped solids loose from the sand particles. A typical air flow rate is 100 to 150 standard cubic feet of air per minute per square foot (SCFM/sq ft).

As the sand reaches the top of the airlift pipe (Figure 9.26, Part C), the loosened solids and dirty wash water (reject water) spill over into the reject compartment-sand/water separator. The sand passes downward through the gravity sand washer-separator (Figure 9.21) while a small volume of the filtrate is passing up through the gravity sand washer-separator. This counterflow of rising filtrate and falling sand washes any remaining solids away from the sand.

The cleaned sand is returned to the sand bed through the gravity sand washer-separator and is redistributed over the top of the sand bed by means of gravity, in conjunction with the sand distribution cone.

The small volume of filtrate water, which has now become dirty wash water (reject water) in the gravity sand washer-separator, carries the remaining solids out of the top of the gravity sand washer-separator (Figure 9.26, Part C) into the reject compartment-sand/water separator. From there, the reject water with solids joins with the other reject water and solids exiting at the top of the airlift pipe.

The combined reject waters and solids pass over an adjustable reject weir at a rate of 10- to 15-gallons per minute (GPM). The reject water flow rate leaving the filter is one of the most important determinants for controlling filter operation.

Reject waters and solids leaving the filters are directed to a reject water sump or wet well. A float switch-activated pump returns the reject water and solids to the treatment plant for further treatment.

QUESTIONS

Write your answers in a notebook and then compare your answers with those on page 479.

9.3D   Why should features to bypass filtration system facilities be provided?

9.3E   What is the purpose of adding chemical coagulants to the filter influent?

9.3F   What is the purpose of adding flocculant solution to the filter influent?

9.321 *Operating Strategy*

Normal operation should require very little attention from the operator. This section describes basic requirements for developing an operational strategy. If these requirements are not met, the filters may not consistently perform optimally.

1. When influent water or supplemental water is being processed through the filter, the airlift must be pumping sand. If not, the sand bed will not be cleaned and will eventually become plugged with suspended solids and chemicals.

2. When influent water or supplemental wash water is not being processed through the filter, the airlift must not be operating. If the airlift is allowed to operate without water flow to the filter, no water will be available to the gravity sand washer-separator. This condition will allow dirty sand to be returned to the top of the sand bed. When normal flow and operating conditions are resumed, it will take many hours before clean filtrate is produced.

3. Whenever the filter is operating, there must be sufficient reject flow going out over the reject weir (10 to 15 GPM or a crest of water over the top weir plate of $1/4$ to $1/2$ inch in depth). Without this reject water flow rate, the sand will not be washed away fully and properly. Instead, it will be returned to the top of the sand bed only partially clean and the result will be a low-quality filtrate.

4. There is a minimum (100 SCFM) and maximum (150 SCFM) value for the rate at which sand can be pumped through the airlift. Between these values the rate can be adjusted to match the typical range at which influent suspended solids will be optimally removed. If the sand pumping rate is too low, then the sand will only be partially cleaned. This will result in decreased flow of the influent water through the filter (as the result of increased head loss) and associated poor filtrate quality.

5. The sand pumping rate should be adjusted so that the head loss gauge reading remains stable. The air rate should be high enough to cope with all normal operating flow and suspended solids ranges so that constant air adjustment is not required. Specific head loss is determined by the influent flow fed to the filter and the amount of suspended solids in that flow. Since flow and solids guidelines vary with each filter installation, the equipment manufacturer will provide the specific operating head loss values for your particular filtration system.

   An operating log should be maintained so that head loss readings, influent flow rates, suspended solids values, air flow rates, and back pressure rates for different operating conditions can be recorded. This will allow the operator to evaluate the various filter operating conditions and determine the optimum air rate setting for each filter.

6. If there are problems with filter performance, investigate these questions first:

   a. Is sand being pumped to the top of the airlift and at the proper rate?

   b. Is there enough reject water flow?

c. Does the influent flow rate exceed the design flow and loading rate?

d. Are the chemical feed systems functioning properly and are the chemical dosage rates being applied at the proper concentrations based on jar tests?

e. Do influent suspended solids exceed the design solids loading rate?

7. It is very important not to drop anything into the filter. If this should happen, immediately stop the air flow to the airlift and remove the object from the top of the sand bed. If not removed, the object will move downward with continually moving sand. It will eventually block the bottom of the airlift and prevent the filter from operating properly.

8. Calcium scaling can occur within the filter and on the surface of the sand media granules when calcium ions and sulfate or carbonate ions are present in the filter influent flow. Usually, this happens when lime or sulfuric acid are used for pH control. Prevention or control of scaling can be accomplished in several ways:

   a. Use caustic in place of lime, or hydrochloric acid in place of sulfuric acid in neutralization.

   b. Use scale retardant/dispersant available from water treatment chemical suppliers.

   c. Keep the sand bed moving whenever the influent flow is off by introducing clean water, or wash the sand with clean water for four hours before securing the filter.

9.322 Abnormal Operation

Efficient filter operation is essential if your final effluent quality is to comply consistently with effluent water quality criteria established for your plant. Table 9.2 lists some of the more common filter operational problems and suggestions on how to correct them.

9.33 Maintenance

A comprehensive preventive maintenance program is an essential part of plant operations. Good maintenance will ensure longer and better equipment performance. The following procedures may be used as a guideline in performing the required maintenance on the continuous backwash, upflow, deep-bed silica sand media filters.

1. A filter sand movement evaluation test should be done monthly. The test will enable you to evaluate and determine if the sand granules have "cemented" together, if the entire sand bed is moving downward equally throughout the inner diameter of the filter, and if the rate of downward sand movement is in accordance with manufacturer's specifications.

   a. Make a sand measuring device using a length of $1/2$-inch schedule 80 PVC pipe. Glue caps on both ends and mark one-inch intervals using tape or a permanent-ink marking pen.

   b. Place the sand measuring device on top of the filter sand bed at location "A" shown on Figure 9.27.

   c. Ensure that the end of the PVC pipe sand measuring device only contacts the filter sand bed surface and not any supporting structure within the filter.

   d. Enter the "Start Time" and "Start Inches" on the "Filter Sand Movement Evaluation" form shown as Figure 9.28.

   e. Leave the measuring device in place for 15 minutes with the filter in full operation.

   f. At 5-minute intervals during the 15-minute evaluation period, record (on a separate piece of paper) air flow, filter effluent flow, influent turbidity, effluent turbidity, and head loss values associated with the filter being evaluated.

   g. At the end of the 15-minute evaluation period, enter the "End Time," "End Inches," and "Total Inches" values on the "Filter Sand Movement Evaluation" form.

   h. Calculate the "Inches/Hour" value by multiplying the "Total Inches" value by 4. Sand movement should be 12 to 16 inches per hour. This represents 7 to 10 sand bed turnovers per day on a 24-hour per day basis.

   i. Calculate the "Average Air Flow," "Average Flow," "Average Turbidities," and "Average Head Loss" readings using the data you recorded at 5-minute intervals during the 15-minute evaluation (item "f" above). Enter these average values in the appropriate columns on the "Filter Sand Movement Evaluation" form.

   j. Using the "Average Flow" value, calculate the "Average Loading Rate" value as follows:

   $$\text{Average Loading Rate, GPM/sq ft} = \frac{\text{Average Flow, GPM}}{\text{Filter Surface Area, sq ft}}$$

   k. Repeat steps "b" through "j" for locations "B" and "C" as shown on Figure 9.27.

2. Clean the head loss measuring tube and the wetted areas around the top of the filter as necessary to remove algal growth.

TABLE 9.2 TROUBLESHOOTING GUIDE FOR CONTINUOUS BACKWASH, UPFLOW, DEEP-BED SILICA SAND MEDIA FILTER

Abnormal Condition	Possible Cause	Operator Response
Poor Filtrate Quality	Reject water flow rate too low	Ensure that reject water flow rate is 10 to 15 GPM (reject water crest over the top weir plate is ¼- to ½-inch).
	Influent flow rate or influent solids exceed design range	Calculate the loading rate (GPM/sq ft) and adjust flow to within design range. Evaluate solids concentration in the influent flow and (a) correct upstream process performance, or (b) adjust chemical addition process following performance of jar test. Inspect chemical addition system for proper equipment operation.
	Filter has just experienced solids loading in excess of maximum design loading	After a major solids overload incident, it may take a few hours before filtrate quality returns to normal. Use clean supplemental backwash water, if available, to speed up the sand media washing process.
High or Increasing Head Loss Reading	Sand wash (airlift) air rate too low	Adjust the air rate to clean the sand bed at a higher rate for an hour or two while reducing the filter influent flow rate until filtrate quality returns to normal.
	Air bubbles or foreign matter in or obstructing head loss gauge	Check air relief vent pipe for proper operation. Check head loss gauge inlet for obstruction.
	Airlift is not pumping sand	Check and determine if air is being supplied to the airlift and that all pneumatic systems and control instruments are operating properly. If pneumatic systems are OK, secure filter and (a) check airlift intake for foreign object, or (b) secure filter and remove airlift and clean screen in airlift.
	Excessive chemical application dosages	Adjust chemical addition process following performance of jar test. Inspect chemical addition system for proper equipment operation.
Airlift Functioning Incorrectly	Top of the airlift and the bottom of the splash hood are piled up with sand:	
	1. Sand bed has become plugged with solids due to overloading or improper filter operation, or	1. Reduce or stop filter influent flow until the airlift is operating properly (use clean, supplemental backwash water, if provided).
	2. Airlift lifts only water or just a small amount of sand, or	2. Check that the required air flow is getting through to the airlift. Wearing eye protection, disconnect the air hose from the top of the airlift to determine if the problem is in the airlift itself or the air supply to it. Next, turn off the air and "bump" the airlift by suddenly turning the air on at maximum flow. Try this repeatedly. Wait for one minute between "bumps." Continue with this procedure for some time before moving on to the next troubleshooting step (if your filter is supplied with an "air burst" feature, use it to get maximum air flow for the "bumps").
	3. Airlift is plugged with solids, or	3. Make up an air lance from two sections of ½-inch CPVC pipe. Glue a female threaded fitting to one end of one pipe section, and a male and female threaded fitting on the respective ends of the other pipe section. This will provide you with a two-part air lance. Next, gather together the appropriate pipe fittings that will enable you to connect an air line to the female threaded fitting on the pipe section that also has the male threaded fitting attached. Put on safety goggles or other eye protection. Using both pipe sections screwed together, insert the air lance down into the airlift. With a strong volume of air blowing through the air lance, push the lance down into the airlift as far as it will go. By working the lance up and down, it should be possible to reach all the way down to the bottom of the airlift. Secure the air flow to the lance, remove the lance, and try to get the airlift operating again. If sand pumps for a while and then stops, repeat the air lancing procedure.
	4. Airlift screen is plugged, or	4. Secure the filter and remove the airlift to inspect, clean, repair, or replace the screen as required.
	5. Airlift is plugged	5. If none of the above procedures corrects the problem, it is likely that an object has gotten into the filter and is blocking the bottom of the airlift. Secure and drain the filter. When the filter is drained of water, remove the accessway cover to the plenum area and the inner cone inspection port plate. *WARNING!* The plenum area of the filter is a confined space. Confined space procedures must be followed prior to entry while working with and when exiting the plenum area.

TABLE 9.2 TROUBLESHOOTING GUIDE FOR CONTINUOUS BACKWASH, UPFLOW, DEEP-BED SILICA SAND MEDIA FILTER (continued)

Abnormal Condition	Possible Cause	Operator Response
Airlift Functioning Incorrectly (continued)	5. Airlift is plugged (continued)	*DANGER! Before entering the plenum area, check the inner cone for signs of damage.* In no event should the plenum area be entered if damage to the inner cone is suspected without first removing all sand from the filter. If sand backflow has occurred as the result of improper filter operation, sand accumulation piles will usually be noted just below each vertical riser tube in the plenum area. Sand accumulation in other areas of the plenum may indicate a damaged inner cone. Once the inner cone inspection port plate has been removed, it is possible to dig through the sand with one hand to reach the area around the bottom of the airlift pipe (wear leather gloves). Because the sand will be falling through the inspection port as fast as you dig, it is best to make a sleeve out of a piece of thin-wall PVC pipe that will fit through the inspection port opening and through which you can slide your hand and arm. Cut one end of the pipe at a 45-degree angle and slide this end through the inspection port. Using this device will help keep the weight of the sand off your arm and will make it possible to reach all around the bottom of the airlift to search for the foreign object. This job will take a great deal of determination to do properly and safely.
Plugged Sand Washer	Washer plugged with trash	Because it may be difficult to see down into the sand washer to view possible trash accumulations, use a water hose with significant water pressure to clear the obstruction. Wear eye protection. Insert water hose down into the washer, turn water supply on, and flush the area around the full circumference of the washer to dislodge the obstruction.
	Washer assembly has been displaced	Check to see that the wash rings and spider rings (consult manufacturer's drawings) for the sand washer are in place.
	Sand bed level too high	Check to see if the sand bed level has reached the underside of the washer. If it has, it is most likely that the sand bed is expanding due to calcium scaling. Correction of the scaling problem can be accomplished in one of two ways: (1) while following proper safety precautions, acid wash the sand by maintaining the pH of the water in the filter between 3.0 and 4.0 with inhibited hydrochloric or muriatic acid while the sand is circulated with the airlift. This process will take 8 to 12 hours. Progress can be monitored by probing the sand bed with CPVC pipe and air lancing as necessary. Upon completion of acid cleaning, the filter should be flushed with clean water until the reject water is clear, or (2) remove the contaminated sand from the filter following the manufacturer's instructions and discard the sand in a legal and environmentally approved manner.
Filter Flow Capacity Greatly Reduced	Sand has backflowed into the plenum area and is blocking the riser tubes or the influent pipe	Check the height of the sand bed and compare with height values recorded at initial start-up to determine if sand bed has dropped. It may also be necessary to secure and drain the filter, open the accessway, and inspect for and remove sand accumulation in the plenum area. Follow the safety precautions and procedures described on the previous page (Airlift Functioning Incorrectly, response item #5) before proceeding with the tasks described here. Wear eye protection, use a water hose with significant water pressure, insert water hose up into each riser tube, turn water on, and flush each riser tube to clear any sand from tube. Return removed sand to the filter sand bed.
	Sand bed is plugged due to severe solids overloading or due to operating the filter without operating the airlift	Using the air lance described on the previous page (Airlift Functioning Incorrectly, response item #3), wear eye protection and air scour the sand bed. Secure influent flow to the filter and apply clean supplemental backwash water. As the supplemental backwash water is applied, air lance the sand bed by working the lance up and down through the sand bed from the top of the filter for 15 to 30 minutes. Work the air lance through the lower part of the sand bed as well as the upper part. The reject water will be very dirty after this operation so it is advisable to continue applying the supplemental backwash water until the reject water runs clean. *CAUTION!* Upper surfaces of the filter unit for which catwalks or walkways have not been constructed are not safe to walk on or work from. Obtain and install proper temporary safety planking to walk on and work from, wear a safety harness, and have an assistant tend the safety line, which is attached to the harness you are wearing.
High or Increasing Back Pressure at the Air Control Panel	Screen in airlift is becoming restricted with solids, trash, or scale	Secure filter, remove airlift, and clean screen.

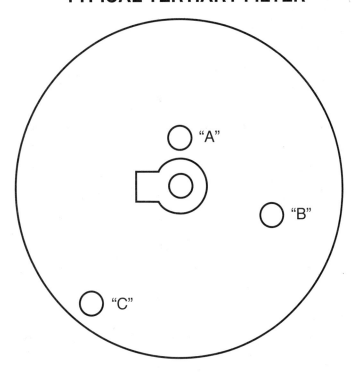

Fig. 9.27 Sand movement evaluation test locations (typical)

TERTIARY FILTERS SAND MOVEMENT EVALUATION

Evaluator: _____ Evaluation Date: ___/___/___

D S F Model: 64, Serial No: D S F - -1970, Sand Size: 0.9 to 1.0 mm, Sand Depth: 40-inches

Filter No: _____

Test Point "A"

Start Time	Start Inches	End Time	End Inches	Total Inches	Factor Inches/Hour	Average Air Flow, SCFH	Average Flow, GPM	Average Loading Rate, GPM/SQ FT	Average Influent Turbidity, NTU	Average Effluent Turbidity, NTU	Average Headloss Reading, Inches
					X-4						
					X-4						
					X-4						
					X-4						

Test Point "B"

Start Time	Start Inches	End Time	End Inches	Total Inches	Factor Inches/Hour	Average Air Flow, SCFH	Average Flow, GPM	Average Loading Rate, GPM/SQ FT	Average Influent Turbidity, NTU	Average Effluent Turbidity, NTU	Average Headloss Reading, Inches
					X-4						
					X-4						
					X-4						
					X-4						

Fig. 9.28 Filter sand movement evaluation form

3. In some applications, sludge can accumulate in the plenum area of the filter. If necessary, remove the filter from service, drain the filter, and remove the accessway cover. Hose out the plenum area to remove accumulations. Because the filter plenum drain outlet is located slightly above the absolute bottom of the filter, an industrial vacuum cleaner designed for both wet and dry work may be used to remove all accumulations and water.

4. The airlift should be removed from the filter once a year for inspection and cleaning. The air chamber at the bottom of the airlift should be disassembled and the screen removed, inspected, cleaned, or replaced. This action is of particular importance if ferric chloride (which is very corrosive to most metals), lime, or sulfuric acid (which causes calcium scaling) are used in your process.

9.34 Safety

Safety precautions described in Chapter 13, "Safety," for sedimentation tanks and pumps should be observed when operating and maintaining this equipment in the continuous backwash, upflow, deep-bed silica sand media filter system.

In addition, the following safety precautions should be observed:

1. Great care must be taken to avoid the contact of ferric chloride with any part of the body, *especially* with eyes. Ferric chloride must be handled with the same care as acid solutions since ferric chloride causes burns similar to those caused by acids. If ferric chloride comes in contact with any part of the body, flush the body area with large amounts of cool water for at least 15 minutes. Seek medical attention immediately.

    a. Before handling ferric chloride solutions or any system components, operators must wear all of the following personal protective equipment. *REMEMBER! No condition or situation involving ferric chloride is more important than your personal safety:*

    - Safety goggles.
    - Face shield with hard hat.
    - Rubber apron properly secured to the body.
    - Rubber gloves of the full gauntlet (flared cuff) type.
    - Rubber boots. Alternatively, regular work shoes with rubber soles and heels and waterproofed leather uppers may be worn (ferric chloride deteriorates leather rapidly).

2. Wear safety goggles and gloves when working with alum or polymers. Wear approved respiratory protection if inhalation of dry alum or polymer is possible. Flush away any alum or polymer that comes in contact with your skin using cool water for a few minutes.

3. Be very careful when walking in an area where polymer is mixed or used. When polymer is wet, it is very slippery. Clean up polymer spills using a highly concentrated salt water solution made with rock salt or use inert absorbent material such as earth or sand.

4. The plenum area of a filter is a *CONFINED SPACE*.[29] Confined space procedures must be followed prior to entry, while working within, and when exiting the plenum area. A confined space permit may be required and anyone working in or around confined spaces must be trained to recognize safety hazards and use safe working procedures. Chapter 13, "Safety," Section 13.12, "Confined Spaces," describes the dangerous conditions associated with confined spaces as well as recommended procedures for confined space work.

5. Before entering the plenum area of a filter, check the inner cone for signs of damage. In no event should the plenum area be entered if damage to the inner cone is suspected without first removing all sand from the filter. The weight of sand in a filter is 20 tons or greater. If sand backflow has occurred as the result of improper filter operation, sand accumulation piles will usually be noted just below each vertical riser tube in the plenum area. Sand in other areas of the plenum may indicate a damaged inner cone.

9.35 Review of Plans and Specifications

As an operator, you can be very helpful to design engineers by suggesting some design features that would make your job easier. This section attempts to point out some of the items that you should look for when reviewing plans and specifications for expansion of existing facilities or construction of new continuous backwash, upflow, deep-bed silica sand media filters.

1. The effect of recycling filter reject (backwash) water through the treatment plant on the filtration rate and filter operation must be considered in predicting peak loads on the filters.

2. Estimates of solids loading, chemical conditioning, head loss development, and filtrate quality should be based on bench-scale tests before the full-scale facility is designed. Bench-scale testing may be performed by the filtration equipment manufacturer.

---

[29] *Confined Space.* Confined space means a space that:
  (1) Is large enough and so configured that an employee can bodily enter and perform assigned work; and
  (2) Has limited or restricted means for entry or exit (for example, manholes, tanks, vessels, silos, storage bins, hoppers, vaults, and pits are spaces that may have limited means of entry); and
  (3) Is not designed for continuous employee occupancy.
Also see DANGEROUS AIR CONTAMINATION and OXYGEN DEFICIENCY.

3. Ladders and walkways should be provided to allow easy access to filters, pipes, and valves.

4. Adequate quantities of spare parts for the filtration system as a whole should be provided for in the construction contract.

5. A source of clean, supplemental water should be piped into the influent line (immediately upstream of the filter) for bottom feed cylindrical filters. For bottom feed concrete basin filters, supplemental water piping should be routed into the top of the filter. In each case, solenoid valve controls on the supplemental water piping may be interlocked with filter influent feed controls (if provided) so that supplemental water will flow to the filter to provide reject water whenever filter influent flow is stopped.

6. Sand backflow from the filter will occur if water within the filter flows backward. To prevent this from happening, check valves must be installed on the influent feed or bypass piping to each filter, or the feed and bypass piping should be installed at an elevation above the level of the launder for filtered water collection.

7. Influent flow should be delivered to the filter at a pressure equal to 3 to 4 feet of head.

8. Piping, pumps, valves, and other ferric chloride handling equipment should be made with, lined with, or coated with Kynar® vinylidene plastic, polyvinyl chloride, rubber, glass, Bakelite, Haveg, ceramic materials, or various other plastics that have given good service with all concentrations of ferric chloride at normal temperatures. The only metals suitable for use with ferric chloride are titanium and tantalum. Additional materials will become available over time. Consult your chemical supplier for up-to-date recommendations.

QUESTIONS

Write your answers in a notebook and then compare your answers with those on page 479.

9.3G What happens if the airlift is allowed to operate without water flow to the filter?

9.3H What should be done if a tool is dropped into the filter?

9.3I Why must ferric chloride be handled with care?

9.3J What safety precautions should be taken before entering the plenum area of a filter?

END OF LESSON 3 OF 4 LESSONS

on

FILTRATION

Please answer the discussion and review questions next.

DISCUSSION AND REVIEW QUESTIONS
Chapter 9. FILTRATION
(Lesson 3 of 4 Lessons)

Write the answers to these questions in your notebook. The question numbering continues from Lesson 2.

11. What are the advantages of continuous backwash, upflow, deep-bed silica sand media filters over other types of granular media filters?

12. How is the silica sand media cleaned?

13. Why should a filter sand movement evaluation test be done monthly?

14. What conditions might be the cause of high or increasing head loss readings?

CHAPTER 9. FILTRATION

(Lesson 4 of 4 Lessons)

9.4 CROSS FLOW MEMBRANE FILTRATION
by Robert G. Blanck and Francis J. Brady

"Cross flow" or "tangential flow" membrane filtration is an effective unit operation for treating industrial, municipal, and food processing wastes. Like all filtration processes, cross flow filtration separates the components of a wastestream. Unlike the upflow and downflow filters described earlier in this chapter, wastewater flows across the surface of a membrane rather than through a granular media. The membrane permits water (permeate) to pass through the membrane to be discharged or recycled, but prevents the passage of solids particles (reject or retentate). The concentration of solids in the retentate, therefore, increases as the wastewater progresses through the membrane filtration system. A major advantage of membrane filters over granular media filters is their relatively long cycle time of 5 to 7 days compared to 8 to 24 hours for granular media filters. Membranes are also easy to clean in a very short period of time (2 to 4 hours) and they are reusable.

Membrane filtration processes are classified on the basis of the size of particle they separate from the wastestream. These processes are: microfiltration (MF); ultrafiltration (UF); nanofiltration (NF); and reverse osmosis (RO). All are rapidly gaining acceptance as alternatives to dissolved air flotation (DAF), biological treatment, chemical treatment, settling ponds, and other conventional techniques. Common commercial membrane applications areas include:

- Food processing wastes
- Metal working wastes
- Primary metal wastes
- Parts washers
- Pulp and paper wastes
- Landfill leachates
- Plating wastes
- Truck wash water waste
- Die casting wastes
- Corrugated box plant wastes
- Flexographic printing wastes
- Laundry wastes
- Textile plant wastes
- Chemical process plant wastes

The separation technique involves a thin, semipermeable membrane that acts as a selective barrier that separates particles on the basis of molecular size. The membrane blocks passage of large particles and molecules, while letting water and smaller constituents pass through the membrane. Membranes made of a variety of materials are available with various pore sizes. Membranes are selected for particular applications on the basis of their differential separation properties. Figure 9.29 shows the particle size and molecular weight retention capacities of one manufacturer's line of membranes.

The driving force of the filtration process is a pressure differential between the wastewater side of the membrane and the effluent side of the membrane. The process differs from conventional filtration in that the waste is pumped at high flow velocities along the surface of the membrane preventing a cake buildup and minimizing a loss of filter capacity.

9.40 Types of Membrane Filtration Processes

9.400 Microfiltration

Microfiltration (MF) membranes have pores ranging from 0.1 to 2.0 microns. MF processes are less common with waste treatment processes because permeate from a microfilter is generally unacceptable for discharge. In some cases, this membrane process may be used in conjunction with settling agents, polymers, activated carbon, and other chemicals that assist in the retention of waste constituents. When using this membrane type, care must be taken to prevent membrane pore blockage by wastestream components by selecting the proper membrane type for the specific plant wastes.

9.401 Ultrafiltration

The process of ultrafiltration (UF) is the most common membrane-based wastewater treatment process; it uses a membrane with pore sizes ranging from 0.005 to 0.1 micron. Particles larger than the pores in the membrane, such as emulsified oils, metal hydroxides, proteins, starches, and suspended solids, are retained on the feed side of the membrane. Molecules smaller than the pores in the membrane, such as water, alcohols, salts, and sugars pass through the membrane. This filtrate (treated water passing through filter) is often referred to as permeate.

Ultrafiltration membranes are sometimes rated on the basis of molecular weight cut-off (MWCO) and range from 1,000 to 500,000 MWCO. For example, a membrane rated at 100,000 MWCO will retain most molecules 100,000 *DALTONS*[30] or

---

[30] *Dalton.* A unit of mass designated as one-sixteenth the mass of oxygen-16, the lightest and most abundant isotope of oxygen. The dalton is equivalent to one mass unit.

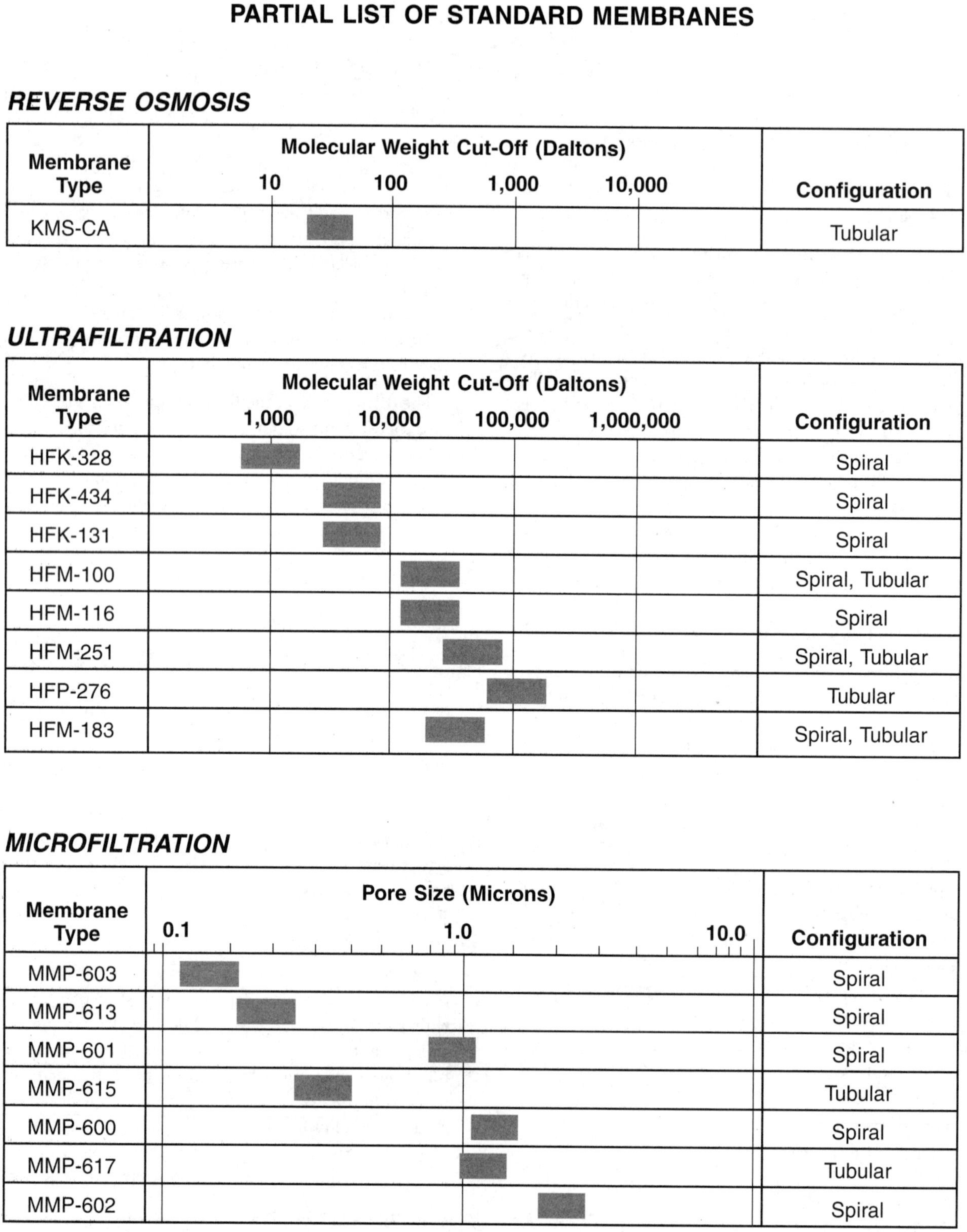

PARTIAL LIST OF STANDARD MEMBRANES

REVERSE OSMOSIS

Membrane Type	Molecular Weight Cut-Off (Daltons) 10 — 100 — 1,000 — 10,000	Configuration
KMS-CA	▮ (≈10–100)	Tubular

ULTRAFILTRATION

Membrane Type	Molecular Weight Cut-Off (Daltons) 1,000 — 10,000 — 100,000 — 1,000,000	Configuration
HFK-328	▮ (≈1,000–3,000)	Spiral
HFK-434	▮ (≈5,000–10,000)	Spiral
HFK-131	▮ (≈5,000–10,000)	Spiral
HFM-100	▮ (≈10,000–30,000)	Spiral, Tubular
HFM-116	▮ (≈10,000–30,000)	Spiral
HFM-251	▮ (≈30,000–80,000)	Spiral, Tubular
HFP-276	▮ (≈80,000–200,000)	Tubular
HFM-183	▮ (≈30,000–80,000)	Spiral, Tubular

MICROFILTRATION

Membrane Type	Pore Size (Microns) 0.1 — 1.0 — 10.0	Configuration
MMP-603	▮ (≈0.1–0.2)	Spiral
MMP-613	▮ (≈0.2–0.3)	Spiral
MMP-601	▮ (≈0.8–1.2)	Spiral
MMP-615	▮ (≈0.3–0.5)	Tubular
MMP-600	▮ (≈0.8–1.0)	Spiral
MMP-617	▮ (≈0.8–1.0)	Tubular
MMP-602	▮ (≈1.5–2.0)	Spiral

Fig. 9.29 Cross flow membrane filtration
(Source: Koch Membrane Systems, Inc.)

larger. The use of MWCO is only an approximate indication of membrane retention capabilities and should be used with guidance of the membrane manufacturer.

9.402 Nanofiltration

Nanofiltration (NF) uses a membrane pore size between UF and RO. These membranes are effective in removing salts from a wastestream by allowing them to pass into the permeate while concentrating other components such as sugars, nitrogen components, and other waste constituents causing high BOD/COD in wastestreams.

9.403 Reverse Osmosis

Reverse osmosis (RO) is the tightest membrane process in that it allows only water to pass through the membrane, retaining salts and higher molecular weight components. RO membranes are used for tertiary treatment producing water with low BOD/COD and of near-potable water quality. RO permeate may be recycled throughout the plant and reused for various plant processes. RO is normally used as a post-treatment process following coarser filtration processes such as ultrafiltration.

9.41 Membrane Materials

Membranes are made from durable polymers. They are reusable after regular cleaning cycles with cleaning chemicals designed to remove foulants from the membrane surface without damaging the membrane. Typical polymers are polysulfone, polyvinylidene fluoride, polyacrylonitrile, and polyamides. The membrane polymer is chosen for its compatibility with the wastestream, ability to retain particles, and ability to show reproducible performance over a long life. Other membrane types are manufactured from ceramics, sintered stainless steel, and carbon. They are used for specialty applications such as high-temperature streams.

9.42 Process Time and Membrane Life

Membranes are designed to process waste and then be cleaned on a periodic basis. Typical membrane processes operate over 5- to 7-day cycles, depending on the membrane type and process design of the system, followed by short cleaning cycles. Membrane filters have a usable life of 1 to 5 years when in daily operation. Most membrane systems are designed to operate 24 hours per day, 7 days per week. Shorter duty cycles may be designed, depending on plant requirements.

9.43 Membrane Configurations

Membranes are housed in various types of modular units. The basic types of membrane configurations are:

Tubular
Hollow Fiber
Spiral
Plate and Frame
Ceramic Tube or Monolith

9.430 Tubular Membranes

Tubular membranes are available in either single- or multi-tube configurations of various diameters. Common tube diameters are 1 inch and $^1/_2$ inch. Figure 9.30 illustrates a 1-inch tube where the membrane is cast on a rigid, porous tube and mounted in a PVC, CPVC, or stainless-steel support shell. Figure 9.31 shows a $^1/_2$-inch multitube configuration mounted in a similar PVC or CPVC housing. Tubular membranes are a nominal 10 feet long and are connected together by U-bends to form a number of tubes in series. The number of tubes in series ranges from 4 to 16, depending on the system design. A number of "series passes" may be present on a commercial system.

Tubular membranes offer advantages in their ability to concentrate to high solids levels without plugging and their ability to operate at pressures as high as 90 psi. Some tubular membranes, such as one-inch diameter tubes, have the ability to be mechanically cleaned with spongeballs. The spongeballs are recirculated through the membrane system along with cleaning solutions to mechanically scour the membrane surface.

9.431 Hollow Fiber Membranes

Hollow fiber membranes, shown in Figure 9.32, are thin tubules of membrane polymer, with the membrane surface normally on the inside of the hollow fiber. The hollow fibers are *POTTED*[31] in a plastic support structure called a cartridge. Hollow fibers offer a high density of membrane packed into a cartridge. This type of membrane produces a high rate of permeate flow (high process fluxes), but is limited to about 30 psi pressure.

Hollow fiber cartridges are available in various fiber diameters ranging from 20 to 106 *MILS*.[32] For waste treatment, common fiber diameters are 43, 75, and 106 mils. The larger diameter fibers are used for concentrating waste to higher solids levels and are more resistant to plugging. The smaller fibers are used for lower concentrations and reduced suspended solids levels.

The hollow fiber cartridge housing is usually constructed of polysulfone, PVC, or other plastic material and may be found in diameters of 3 inches or 5 inches. The length of a hollow fiber typically is approximately 36 inches.

---

[31] *Potting Compounds.* Sealing and holding compounds (such as epoxy) used in electrode probes.
[32] *Mil.* A unit of length equal to 0.001 of an inch. The diameter of wires and tubing is measured in mils, as is the thickness of plastic sheeting.

The Koch one-inch-diameter "FEG" tube is one of the most rugged membrane configurations. This, coupled with the ability to mechanically clean the tubes with spongeballs, makes the FEG tube the membrane configuration that is the most forgiving to system upsets. Wastestreams that would cause severe fouling or plugging of other membrane configurations can be processed with the FEG tube.

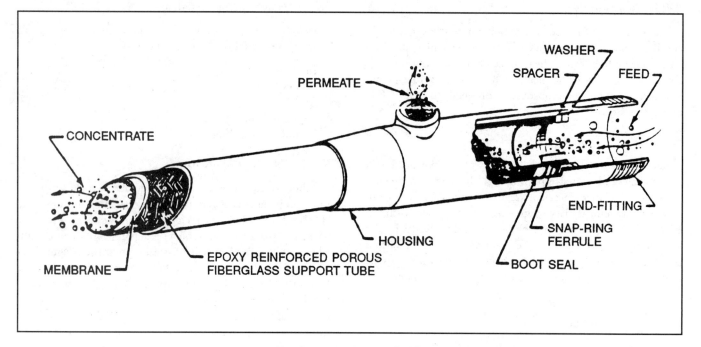

Fig. 9.30   FEG one-inch tube
(Source: Koch Membrane Systems, Inc.)

The Koch ULTRA-COR® VII module packs more membrane area per module than the FEG tube. However, the ULTRA-COR® VII modules are more limited in achieving maximum yields, are less durable and forgiving to upsets, and cannot be mechanically cleaned with spongeballs.

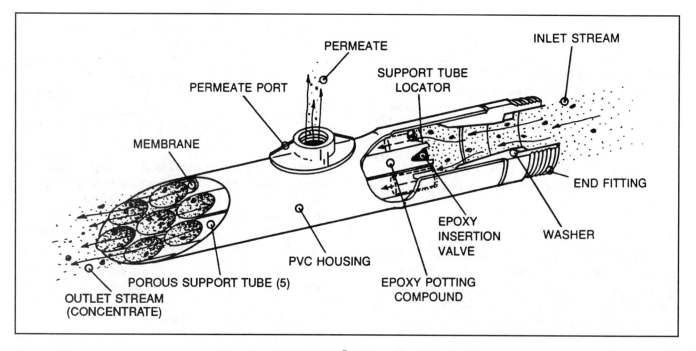

Fig. 9.31   ULTRA-COR® VII ½-inch tubules module
(Source: Koch Membrane Systems, Inc.)

The self-supported hollow fiber membranes are cast in a spinning operation. Since the membrane has no support backing, the cartridges can be backflushed to provide for longer process runs in some applications. With proper operation and waste pretreatment, this configuration can be the most economical choice for many applications.

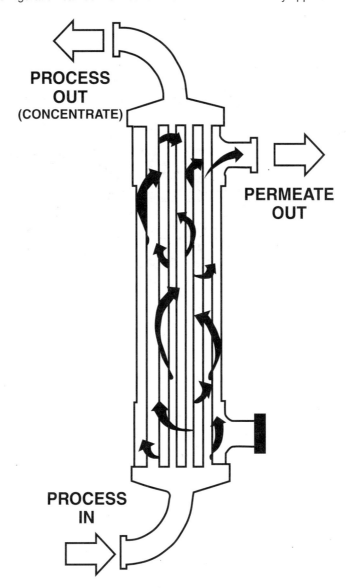

Fig. 9.32   Hollow fiber cartridge
(Source: Koch Membrane Systems, Inc.)

A major advantage of a hollow fiber cartridge is its ability to be backwashed or backflushed without damage to the membrane. In this operation, the permeate is pressurized and made to flow in the opposite direction through the membrane. This technique pushes fouling material off the membrane surface, resulting in increased flow of permeate.

9.432   Spiral Membranes

Membranes can be cast on materials such as nonwoven polyester and manufactured into spiral modules. Spirals also offer a high packing of membrane area into a compact cartridge. They may operate at pressures up to 150 psi for UF and as high as 1,200 psi for RO. A spiral module, shown in Figure 9.33, is composed of the flat sheet membrane, a permeate carrier, a feed spacer, and glue to isolate the feed from the permeate side.

Spiral modules are available with various feed spacer thicknesses that serve to maintain a distance between the membrane sheets. Selection of the correct spacer for a particular feed stream is critical for proper filter operation. In general, high solids wastewaters are more easily processed with thicker feed spacers. Spacer thicknesses range from 28 to 80 mils.

Spiral modules are available in nominal diameters of 2, 4, 6, and 8 inches. Large-diameter spiral modules are used for high-volume wastestreams because they provide large membrane surface areas for a given system size.

Spiral membrane modules are capable of packing a large surface area of membrane into a compact design. For proper operation, these modules require the highest degree of pretreatment of all membrane configurations.

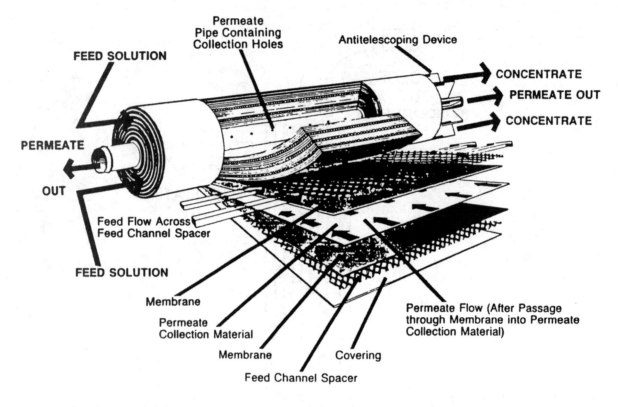

Fig. 9.33 Spiral wound module
(Source: Koch Membrane Systems, Inc.)

The spiral is mounted into either a plastic or stainless-steel shell. This may be replaceable or permanently mounted in the shell. Figure 9.34 illustrates a permanently mounted SPIRAPAK™ module used for wastewater treatment. Most wastewater RO applications are spiral modules.

While offering a large membrane surface area, spiral modules are limited in their ability to process wastes with large amounts of suspended solids and particulates.

9.433 Plate and Frame Membranes

Plate and frame configurations, less common in wastewater treatment, are a series of flat sheet membranes mounted on a frame. These membrane systems resemble a pressure leaf filter, with the membranes fitted in a series of plates. Permeate is diverted out of the plates by a series of small tubes.

QUESTIONS

Write your answers in a notebook and then compare your answers with those on page 479.

9.4A How do cross flow filtration processes differ from conventional filtration?

9.4B How are membrane processes classified?

9.4C List the basic membrane configurations.

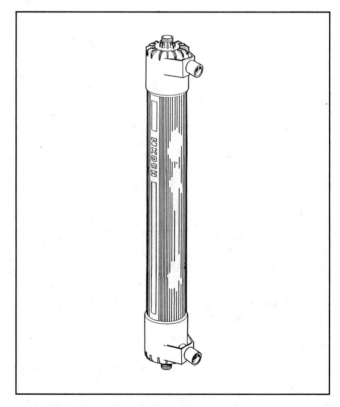

Fig. 9.34 SPIRAPAK™ module
(Source: Koch Membrane Systems, Inc.)

9.5 BASIC ELEMENTS OF A MEMBRANE FILTRATION PROCESS

9.50 Concentrating Components in the Wastewater

Membrane systems concentrate components in the wastestream such as oil, grease, fats, BOD, COD, metal hydroxides, suspended solids, particulates, starches, SURFACTANTS,[33] bacteria, and other microorganisms. The concentrated feed (known as the retentate or concentrate) may be reduced in volume 20- to 100-fold compared to the original feed volume. The level of concentration depends on the original concentration of components in the waste feed. Retentate from an ultrafilter may be as high as 60 percent total solids, depending on the nature of the incoming feed. The level of concentration is also determined by the choice of membrane configuration. Tubular membranes may concentrate to 50 to 60 percent total solids, while spiral and hollow fiber membrane configurations may be limited to 10 to 30 percent, again depending on the type of wastewater.

9.51 Permeate Discharge to Drain

Permeate is made up of the solvent and low-molecular-weight constituents that pass through the membrane. In some plants, the permeate from an ultrafiltration system meets local and state discharge requirements and can be sent directly to the drain. More stringent discharge requirements may require post-treatment of the UF permeate by processes such as pH adjustment, carbon filtration, ozone treatment, and reverse osmosis. The permeate from an ultrafilter is typically clear (non-turbid) and without suspended solids. In some cases, dye or color bodies will tint the permeate. The amount of dissolved substances in the permeate will depend on the retention properties of the membrane.

9.52 Membrane Flux

The rate of mass flow of permeate passing through the membrane is called flux. This measurement is the basis for sizing membrane equipment and is an essential value to evaluate membrane performance. Flux (J) is expressed on the basis of membrane area:

$$\text{Flux, J} = \frac{\text{Permeate Flow Rate}}{\text{Membrane Area}}$$

Units for flux are "gallons/sq ft-day" (GFD) or "liters/sq m-hour" (LMH). Flux values for typical ultrafiltration waste treatment processes are approximately 25 to 50 GFD (43 to 85 LMH). Reverse osmosis and nanofiltration fluxes typically range from 10 to 30 GFD (17 to 51 LMH). Conversion of GFD to LMH is $1.7 \times \text{GFD} = \text{LMH}$.

The amount of flux across a membrane is dependent on a number of factors:

- Transmembrane pressure (driving force).
- Flow rate across the membrane surface (turbulence on the membrane).
- Concentration of waste material.
- Temperature of concentrate.
- Viscosity of concentrate.
- Cleanliness of the membrane surface.

To calculate membrane flux of a commercial system, one needs to know the permeate flow rate measured with a flowmeter, rotameter, or bucket and stopwatch, and the membrane area of the system.

EXAMPLE 4

A membrane filtration system is operating with 1,200 square feet of membrane area and the permeate flow is 25 gallons per minute. What is the flux (J) in GFD (gallons/sq ft-day)?

Known	Unknown
Membrane Area, sq ft = 1,200 sq ft	J (Flux), GFD
Flow, gal/min = 25 gal/min	

Calculate the flux in GFD.

$$J, \text{GFD} = \frac{(\text{Flow, gal/min})(60 \text{ min/hr})(24 \text{ hr/day})}{\text{Membrane Area, sq ft}}$$

$$= \frac{25 \text{ gal/min} \times 60 \text{ min/hr} \times 24 \text{ hr/day}}{1{,}200 \text{ sq ft}}$$

$$= 30 \text{ GFD}$$

9.53 Retention

A membrane filter achieves a separation of various components in the wastewater stream. The ability of a membrane to retain particles is known as retention or rejection. A membrane's retention coefficient (R) is a numerical expression of the concentration of a component in the retentate (concentrate) relative to its concentration in the permeate. The retention coefficient is:

$$\text{Retention, R} = 1 - \left(\frac{\text{Permeate}}{\text{Concentrate}}\right)$$

---

[33] *Surfactant* (sir-FAC-tent). Abbreviation for surface-active agent. The active agent in detergents that possesses a high cleaning ability.

Retention is often expressed as a percent by multiplying by 100%.

Retention, % = Retention Coefficient × 100%

Retention ranges from 0 to 100 percent, depending on the molecular or particle size of the wastewater component being measured. Retentions may be obtained on commercial systems by sampling the concentrate and permeate, and submitting samples to a local lab for component analysis. For most wastestreams, analysis is suggested for freon-extractable oil and grease *(STANDARD METHODS FOR THE EXAMINATION OF WATER AND WASTEWATER, METHOD 5520B, 20th Edition, 1998)*, true oil and grease *(STANDARD METHODS 5520F)*, total soluble solids, total suspended solids, heavy metals, BOD, and COD. Other components analyses for fats, proteins, and sugars may be required for food processing wastes.

EXAMPLE 5

Samples of retentate and permeate from an ultrafiltration system were sent to an analytical lab for oil and grease analysis. The retentate has a concentration of 5,000 mg/L oil and grease and the permeate has a concentration of 25 mg/L oil and grease. What is the percent retention of oil and grease?

Known	Unknown
Permeate Conc, mg/L = 25 mg/L	Retention, %
Retentate Conc, mg/L = 5,000 mg/L	

Calculate the percent retention.

$$R, \% = \left(1.00 - \frac{(25 \text{ mg}/L)}{(5,000 \text{ mg}/L)}\right)(100\%)$$

$$= (1.00 - 0.005)(100\%)$$

$$= 99.5\%$$

9.54 Calculating the Concentration of Waste Components

The wastestream is split into the retentate stream and the permeate stream. A mass balance around the membrane system gives us the volumes and concentrations of permeate and concentrate.

Feed Volume = Permeate Volume + Concentrate Volume

or

$V_f = V_p + V_c$

Membrane filtration data are often presented in terms of the volumetric concentration factor (CF). This value is the amount the waste material is concentrated and can also be expressed as "X." For example, a system concentrating waste 50 times has a CF = 50×. The concentration factor of a system is calculated by the following equation:

$$CF = \frac{\text{Feed Volume}}{\text{Concentrate Volume}}$$

or

$CF = V_f/V_c$

EXAMPLE 6

A waste treatment plant has an original waste volume of 20,000 gallons. After ultrafiltration, a concentrate volume of 400 gallons remains. What is the volumetric concentration factor?

$$CF = \frac{20,000 \text{ gallons}}{400 \text{ gallons}} = 50\times$$

Another way to express the volume balance of a membrane system is "percent volume reduction" or "percent water removed."

Percent volume reduction is determined by:

$$\text{Volume Reduction, \%} = \left(\frac{\text{Feed Volume} - \text{Concentrate Volume}}{\text{Feed Volume}}\right)(100\%)$$

or

$$= \left(\frac{V_f - V_c}{V_f}\right)(100\%)$$

or

$$= \frac{V_p}{V_f}(100\%)$$

Where V_p = Permeate Volume

EXAMPLE 7

From above, 20,000 gallons of waste are reduced to 400 gallons of concentrate. What is the volumetric reduction of the wastewater feed?

Known	Unknown
Feed Vol, gal = 20,000 gal	Volume Reduction, %
Concentrate Vol, gal = 400 gal	

Calculate the percent volume reduction.

$$\text{Volume Reduction, \%} = \left(\frac{V_f - V_c}{V_f}\right)(100\%)$$

$$= \left(\frac{20,000 \text{ gal} - 400 \text{ gal}}{20,000 \text{ gal}}\right)(100\%)$$

$$= 98\%$$

The concentration of components in the concentrated wastewater can be determined by:

Concentration in Retentate = Original Feed Concentration × $(CF)^R$

where R = Retention Coefficient

EXAMPLE 8

A concentrate and permeate sample were collected from an ultrafilter and analyzed for BOD. The retention coefficient was 0.70 (R, % = 70.0%). The original feed BOD was 5,000 mg/L and the concentration factor of the UF system was 50×. What is the concentration of BOD at the end of the batch?

Known	Unknown
BOD Feed, mg/L = 5,000 mg/L	Ending BOD Conc, mg/L
CF (Concentration Factor) = 50×	
BOD Retention, R = 0.70 or % = 70%	

Calculate the ending BOD concentration in mg/L.

$$\text{BOD Conc, mg}/L = (\text{BOD Feed, mg}/L)(\text{CF})^R$$
$$= (5{,}000 \text{ mg}/L)(50)^{0.7}$$
$$= 77{,}312 \text{ mg}/L$$

For UF and tighter membranes, components like true oil and grease, suspended solids, proteins, and fats have a retention close to 1.0. There is little or no passage of these components into the permeate. This means that the concentration of components becomes:

Retentate Concentration = Original Feed Concentration × CF

9.55 Transmembrane Pressure

The permeate rate of a membrane system is dependent on a pressure differential between the feed side and the permeate side of the membrane. This pressure, known as transmembrane pressure, is the driving force of the separation process. Transmembrane pressure typically ranges from 20 to 80 psi for ultrafiltration, 300 to 500 psi for nanofiltration, and 400 to 1,000 psi for reverse osmosis. Process fluxes are a function of pressure; however, as concentration of waste increases in the retentate, fluxes will become independent of pressure. At this point, increasing transmembrane pressures show no increase in process flux. This is due to formation of a resistant gel layer on the membrane surface. Once the gel layer forms, the mass flow of particles toward the membrane surface is equal to the diffusion of particles away from the membrane. Fluxes are most pressure-dependent on water feed rate or when low concentrations of waste are present in the feed.

The average transmembrane pressure (Ptm) is a measure of the driving force of the membrane process exactly in the middle of the system. One can calculate the average transmembrane pressure by knowing the inlet pressure (Pin) and outlet pressures (Pout) of the waste feed recirculating in the membrane system and the permeate pressure.

The inlet pressure is measured on the discharge side of the recirculation pump, just prior to the entrance to the modules or tubes. The outlet pressure is measured at the point where the waste leaves the bank of modules. The permeate pressure may be measured on the permeate manifold, but in most cases is assumed to be zero.

To calculate the average transmembrane pressure of a system:

$$\text{Ptm} = \frac{\text{Pin} + \text{Pout}}{2} - P_{permeate}$$

EXAMPLE 9

The inlet pressure of a commercial system is 64 psi and the outlet pressure is 20 psi. The permeate pressure is zero. What is the average transmembrane pressure?

Known	Unknown
Inlet Pressure, psi = 64 psi	Average Transmembrane Pressure (Ptm), psi
Outlet Pressure, psi = 20 psi	
Permeate Pressure, psi = 0 psi	

Calculate the transmembrane pressure in psi.

$$\text{Ptm, psi} = \frac{\text{Pin} + \text{Pout}}{2} - P_{permeate}$$
$$= \frac{(64 \text{ psi} + 20 \text{ psi})}{2} - 0$$
$$= 42 \text{ psi}$$

9.56 Recirculation Flow

All cross flow membrane systems recirculate the waste feed across the membrane to prevent buildup of particles on the surface. This tangential flow or cross flow creates high turbulence and a sweeping effect on the membrane and enhances the movement of molecules away from the membrane surface. Increasing the recirculation rate usually increases the flux rate, especially when wastes are highly concentrated. Most systems are designed at flow rates that optimize flux and choice of hardware (pumps and pipework).

Tangential flow across the membrane surface is created by recirculation pumps on the system. The amount of recirculation flow is selected by the membrane manufacturer based on the unit's design.

Recirculation flow is normally constant throughout the process run and must not be confused with permeate flow. Usually, recirculation flow is not directly measured in commercial systems. An indirect measurement of recirculation flow is the drop in pressure between the inlet (Pin) and outlet (Pout) of the membrane system. Delta pressure (dP) is measured as:

Delta pressure, dP = 64 psi – 20 psi = 44 psid or psi delta pressure

Delta pressure is an important guideline to monitor in a system because it indicates if the membrane modules have sufficient cross flow velocity to meet design fluxes or are becoming plugged with waste. Normally, delta pressure is relatively constant throughout the process run. The delta pressure might increase when the viscosity of the feed is increasing or when material is accumulating in the membrane modules. (One can imagine two pressure gauges on either side of a valve. When the valve closes, the dP across the valve increases.) Operators should use delta pressure as an indicator of major system problems. Major changes in dP signal the operator to investigate problems in prefiltration or overconcentration in the membrane system. When the delta pressure increases to the shutoff pressure of the pump, there is little or no flow across the surface of the membrane. This can result in significant fouling and eventually will permanently plug the modules.

9.57 Temperature

Increasing the temperature of the feed increases flux through the membrane. This is mainly due to a reduction in viscosity of the waste feed. Membranes have maximum temperature limits based on their materials of construction. The design temperature of a membrane system should be optimized for flux performance and membrane temperature compatibility. Always refer to the system operating manual to determine the temperature limits of the membrane.

Daily temperature variations in the waste feed will cause the membrane system to vary in productivity. For example, the flux at 50 degrees F will be approximately one-half of the flux at 100 degrees F.

9.58 Concentration-Dependent Flux

As materials such as oils, grease, fats, particulate matter, and proteins are concentrated on the feed side of the membrane, the flux of the membrane will drop. This is caused by the formation of the gel layer or CONCENTRATION POLARIZATION[34] layer at the membrane surface. Initially, when wastewater concentrations are low, the flux is high. But, while concentrating over the course of a process run, the productivity of permeate will progressively decline as the concentration increases. Eventually the increasing concentration will drive the membrane flux to zero, although a properly operated membrane system is stopped before the maximum concentration is reached. The typical maximum oil and grease concentration in the retentate on a tubular commercial system is about 50 percent.

9.59 Membrane Fouling

Under normal operation, the surface of the membrane becomes fouled with oil, grease, and other foulants. This fouling results in a decline in the flux of the membrane, however, it is different from the concentration-dependent flux decline described above. Fouling is a molecular attraction between the components in the feed stream and the membrane material.

Two kinds of membrane fouling occur: reversible and irreversible. Reversible fouling is expected. It occurs in all membrane systems and is cleanable with cleaning chemicals. A regular cleaning with caustic, acids, surfactants, chlorine, or hydrogen peroxide (as recommended by the membrane manufacturer) returns the membrane flux very nearly to original values. Irreversible fouling is rare and is a result of incompatible components in the feed stream that cannot be cleaned with normal cleaning chemicals. An example of a common irreversible foulant is silicone. The operator must be aware of potential irreversible foulants in the wastestream and isolate them from the waste at their point sources.

QUESTIONS

Write your answers in a notebook and then compare your answers with those on page 479.

9.5A The amount of flux across a membrane is dependent upon what factors?

9.5B How do membrane systems prevent the buildup of particles on the surface?

9.5C Under what conditions might the delta pressure increase?

9.5D What causes fouling of the membrane surface?

9.6 OPERATION OF A CROSS FLOW MEMBRANE SYSTEM

9.60 Filter Staging

Membrane modules or tubes are mounted on a frame and fitted with a pump, valves, and instrumentation. This functional unit is termed a membrane stage. A membrane system includes a membrane stage or stages, a feed pump to bring waste to the stage, a feed tank, a cleaning station including pump and tank, and control panel. Prefiltration (pretreatment) may be required on some systems.

The amount of membrane area on a stage is a function of the plant effluent requirements and may range from as little as 4 sq ft to 1,200 sq ft or more. Pump and line sizes on the stage increase correspondingly when plant capacities increase. Membranes are placed on the stage in a series and parallel arrangement to maximize the efficiency of the design. For example, a tubular system with 248 tubes on the stage may have them arranged 8-in-series with 31 parallel passes. Figure 9.35 shows the series arrangement of tubes.

9.61 Operating Modes

Most waste treatment systems are designed to operate in a batch or modified batch process mode. A third, less frequently used process mode is a continuous, stages-in-series design. The choice of process mode depends on the volume of waste to be processed and specific plant requirements.

A batch process is the most efficient mode in terms of membrane area requirements and system design. It requires only a feed tank and the membrane stage. As seen in Figure 9.36, the waste feed from the batch tank is recirculated through the mem-

---

[34] *Concentration Polarization.* (1) A buildup of retained particles on the membrane surface due to dewatering of the feed closest to the membrane. The thickness of the concentration polarization layer is controlled by the flow velocity across the membrane. (2) Used in corrosion studies to indicate a depletion of ions near an electrode. (3) The basis for chemical analysis by a polarograph.

CROSS FLOW MEMBRANE FILTRATION
Industrial Wastewater Ultrafiltration 8-in-Series One-Inch Tubular Arrangement

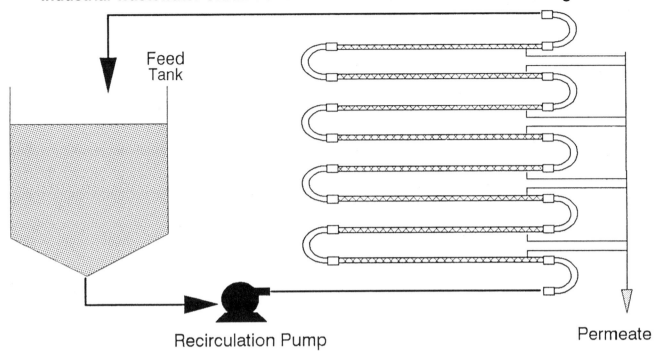

Fig. 9.35 Series tube arrangement
(Source: Koch Membrane Systems, Inc.)

CROSS FLOW MEMBRANE FILTRATION
Batch Process

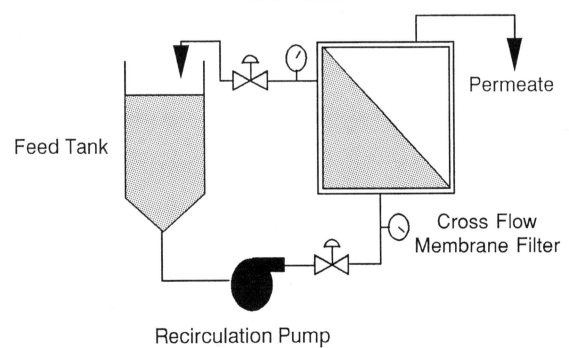

Fig. 9.36 Batch mode
(Source: Koch Membrane Systems, Inc.)

brane stage and back to the feed tank. Permeate is discharged from the membranes and is sent to drain or to further processing. The recirculation rate across the membrane surface is very large compared to the permeate flow off the system. For example, on a tubular design, the recirculation flow is approximately 100 times the permeate rate. The feed material as it exits the stage and returns back to the feed tank is slightly concentrated. The feed is continuously recirculated back through the membranes until the desired concentration of waste is achieved in the feed tank. Since waste treatment batches run for as long as 4 to 7 days, the feed tank size must be large enough to accommodate all the waste over this time period. A disadvantage of a batch design is the large tank requirements for large waste volumes.

In a modified batch mode (Figure 9.37), a smaller feed tank is used and the waste is continuously fed into the feed tank at a rate equal to the permeate production. Material is concentrated in the tank over a 4- to 7-day period, followed by turning off the feed into the tank and batching down the waste volume in the feed tank to its final desired concentration. The final batch-down normally takes 2 to 8 hours, depending on the size of the feed tank. This modified batch mode with end-of-run batch-down is the most common operating mode found in wastewater treatment applications.

Continuous processing uses multiple stages placed in series with each other, as illustrated in Figure 9.38. Feed from a small feed tank enters the first stage of the system where it is slightly concentrated. The waste then enters a second stage where further concentration takes place. After successive concentration in multiple stages (usually 2 to 6), the concentrated waste is bled off the system at a fraction of the feed rate. The system is continuous in that it produces concentrate at maximum concentration during the length of the process run. This differs from a batch system where concentrate reaches maximum concentrations at the end of the batch. Continuous stages-in-series processing is used for RO plants and larger UF plants.

9.62 Feed Pretreatment

Membrane filtration requires proper pretreatment of the feed to optimize the process flux and prevent fouling or plugging of the membrane. For ultrafiltration systems processing primarily industrial waste, the following pretreatment processes are common (see Figure 9.39):

1. Equalization of wastes from throughout the plant combined with surge capacity.

2. Prescreening (prefiltration) to remove large suspended material.

3. Free oil (*TRAMP OIL*[35]) and settleable solids removal.

4. pH treatment (to alkaline conditions) to stabilize the oil/water emulsion.

Equalization of wastes is necessary to level out differences in waste concentrations from different parts of the plant. The equalization tank will also allow for some settling of sludge that can be removed on a periodic basis from the bottom of the tank. A typical equalization tank detention time is one day's waste volume. An alternative to using a large equalization tank is the addition of chemicals to enhance the formation of easily settled floc.

The prefiltration requirements depend on the membrane configuration used and the particle size in the wastestream. Tubular systems require loose prefiltration (5/64th screen), whereas hollow fibers and spirals require prefilter pore sizes as low as 10 microns. Operators must always maintain the prefilter and never bypass it because irreversible membrane plugging or other membrane damage may occur.

Free oil removal is essential for proper membrane operation. Free oil will cause fouling of the membrane surface and reduce the capacity of the membrane system. Oil skimmers are normally placed on the equalization tank to remove free oil from the feed.

In an oily waste system, pH control at the ultrafilter is important for two reasons. Flux rates are improved at the ultrafilter when the wastewater is pH 9 to 10.5 and fouling of the membrane by oil is minimized by stabilizing the oil emulsion. Lower pH values make the oil emulsion unstable and break off oil in the feed tank. Free oil will coat the membrane and reduce process flux. Operators should look for free oil on the surface of the feed tank and, if present, improve the skimming of surface oil on the equalization tank.

9.63 System Operation

Membrane system operation may be manual or controlled by a *PLC*.[36] The steps listed below are a typical operating sequence for processing wastewater in a clean membrane system operating in a batch mode.

---

[35] *Tramp Oil.* Oil that comes to the surface of a tank due to natural flotation. Also called free oil.
[36] *Programmable Logic Controller (PLC).* A small computer that controls process equipment (variables) and can control the sequence of valve operations.

CROSS FLOW MEMBRANE FILTRATION
Modified Batch Process

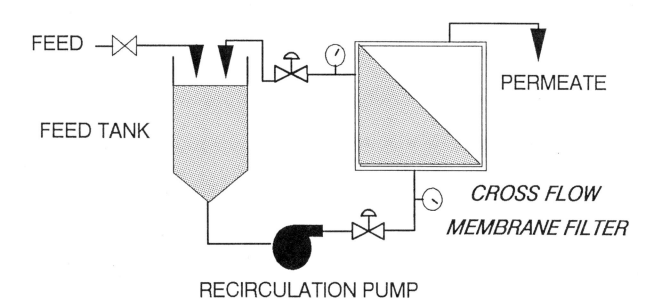

Fig. 9.37 *Modified batch mode*
(Source: Koch Membrane Systems, Inc.)

CROSS FLOW MEMBRANE FILTRATION
Stages-in-Series Process

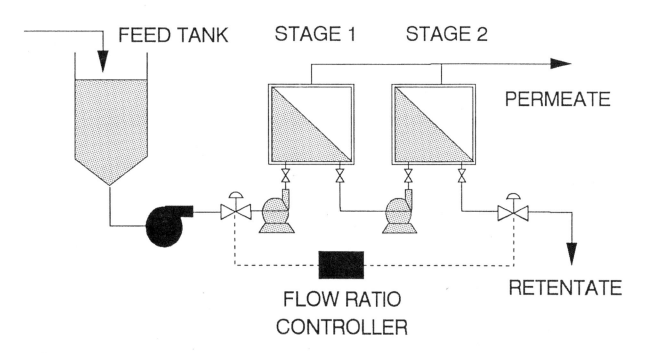

Fig. 9.38 *Continuous stages-in-series*
(Source: Koch Membrane Systems, Inc.)

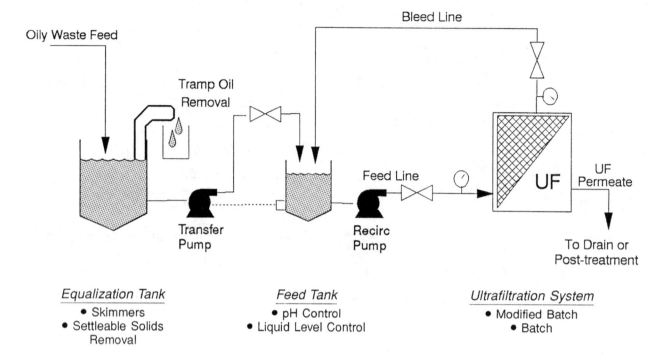

Fig. 9.39 Typical ultrafiltration wastewater treatment flow schematic
(Source: Koch Membrane Systems, Inc.)

START-UP

1. Pretreated waste is introduced into the feed tank.
2. The return valve to the feed tank is opened fully.
3. The recirculation pump is turned on.
4. The discharge valve on the pump is opened slowly.
5. Inlet and outlet pressures are adjusted to the proper delta pressure.
6. System heat exchangers are turned on (if present).
7. For modified batch systems, the rate of feed into the feed tank is adjusted to equal the permeate flow rate.
8. The system is now at steady state.

SHUTDOWN

During batchdown, the following sequence will normally take place:

1. The wastewater to the feed tank is shut off and the UF continues to operate.
2. The feed tank is concentrated down to the final desired concentration factor and the system is stopped.
3. The concentrate in the system is displaced with water.
4. The waste concentrate is transferred to a sludge tank.
5. Cleaning of the system is initiated.

During normal, steady state operation, conditions should be monitored on a regular basis over the course of a process run. Data collected every few hours is usually adequate to ensure proper system operation.

9.64 Cleaning Procedures

Membrane systems are cleaned on a regular basis with chemicals on cycles prescribed by the membrane supplier, typically every 4 to 7 days on wastewater treatment systems. The cleaning cycle usually takes approximately 2 to 4 hours, depending on the required wash cycles. A typical cleaning will include the following cycles:

WASH CYCLES

1. Displacement of waste from the system with water.
2. Washing the membranes with a caustic and surfactant to remove oils and grease. Typical wash cycles are performed at 0.1 percent surfactant, pH 10.5, for 1 hour at 100 to 120 degrees F (40 to 50°C).

3. Flushing the surfactant from the membrane with warm water.

4. Washing the membrane with an acid cleaner to remove salt buildup. This cycle is typically 1 hour at 100 to 120 degrees F (40 to 50°C).

5. Flushing the acid from the membrane with warm water.

6. Recording clean membrane water flux as a check of cleaning effectiveness.

All wash cycles are performed with transmembrane pressures and recirculation flow across the membrane similar to process conditions.

For some tubular membrane systems, the chemical cleaning may be supplemented with spongeballs circulated through the tubes. Spongeballs are sometimes required when membrane foulants are difficult to remove, such as with metal hydroxides or waste latex. Spongeballs may also be used to speed up the cleaning procedure in some plants.

Some membrane configurations, such as hollow fibers, may be backflushed to assist in removal of particles from the membrane surface. Backflushing is the reverse flow of water or permeate from the permeate side of the membrane to the feed side. This technique is most common during cleaning; however, it may also be used during the treatment process on high-fouling wastewater streams. Most tubular and spiral membranes cannot be backflushed due to potential DELAMINATION[37] of the membrane.

Exposure to aggressive cleaning chemicals decreases the life of the membrane and is the principal reason membranes must be replaced periodically. Overuse of chemicals accelerates the failure of membranes and operators must take care not to overclean membranes.

9.65 Water Flux Measurements

After cleaning, the rate of water passage through a membrane in flux, or permeate flow, Q_p, is measured to assess the effectiveness of the cleaning cycle. To measure water flux or permeate flow, all dirty wash water is drained from the system. Clean water is placed in the cleaning tank and then pumped to the membrane system. Permeate flow is measured under standard pressure conditions. Typical transmembrane conditions for UF are 20 or 50 psi (depending on the membrane type) at 77 degrees F (25°C). Permeate flow, Q_p, taken at other temperature and pressure conditions must be corrected to these standard conditions. Correction is performed as follows:

$$Q_{P_{CORRECTED}} = \frac{Q_{P_{ACTUAL}} \times 50 \text{ psi}}{P_{tm_{avg}}} \times F$$

where F = Temperature correction from standard 77 degrees F (see Table 9.3).

TABLE 9.3 WATER FLUX TEMPERATURE CORRECTION FACTOR (F)[a]

Flux (25°C) = Flux (T°C) × F
or Flux (77°F) = Flux (T°F) × F

Temp (°F)	Temp (°C)	F	Temp (°F)	Temp (°C)	F	Temp (°F)	Temp (°C)	F
125.6	52	0.595	96.8	36	0.793	68.0	20	1.125
123.8	51	0.605	95.0	35	0.808	66.2	19	1.152
122.0	50	0.615	93.5	34	0.825	64.4	18	1.181
120.2	49	0.625	91.4	33	0.842	62.6	17	1.212
118.4	48	0.636	89.6	32	0.859	60.8	16	1.243
116.6	47	0.647	87.8	31	0.877	59.0	15	1.276
114.8	46	0.658	86.0	30	0.896	57.2	14	1.320
113.0	45	0.670	84.2	29	0.915	55.4	13	1.346
111.2	44	0.682	82.4	28	0.935	53.6	12	1.383
109.4	43	0.694	80.6	27	0.956	51.8	11	1.422
107.6	42	0.707	78.8	26	0.978	50.0	10	1.463
105.8	41	0.720	77.0	25	1.000	48.2	9	1.506
104.0	40	0.734	75.2	24	1.023	46.4	8	1.551
102.2	39	0.748	73.4	23	1.047	44.6	7	1.598
100.4	38	0.762	71.6	22	1.072	42.8	6	1.648
98.6	37	0.777	69.8	21	1.098	41.0	5	1.699

[a] Based on water fluidity relative to 25°C (77°F) fluidity value.
  $F = (\mu_{T°C/25°C})$, or
  $F = (\mu_{T°F/77°F})$

---

[37] *Delamination* (DEE-lam-uh-NAY-shun). Separation of a membrane or other material from the backing material on which it is cast.

EXAMPLE 10

The clean water productivity of a system is 50 GPM at 100 degrees F, with an average transmembrane pressure of 40 psi. What is the corrected permeate flow, Qp, at 50 psi, 77 degrees F (standard conditions)?

Known	Unknown
Actual Qp, GPM = 50 GPM	Corrected Qp, GPM
Avg Pressure (Ptm_{avg}), psi = 40 psi	
Temp, degrees F = 100°F	

Calculate the corrected permeate flow in gallons per minute.

$$Q_{P\,CORRECTED}, \text{GPM} = \frac{Q_{P\,ACTUAL} \times 50\text{ psi}}{Ptm_{avg}} \times F$$

$$= \frac{(50\text{ GPM})(50\text{ psi})(0.762)}{40\text{ psi}}$$

$$= 47.6\text{ GPM}$$

To calculate clean membrane water productivity in terms of flux, J, simply calculate the permeate flow on a daily basis and divide by the system's membrane area.

EXAMPLE 11

If the membrane system in Example 10 has 545 sq ft of membrane area, what is the system's clean water productivity in GFD (gallons/sq ft-day)?

Known	Unknown
$Q_{P\,CORRECTED}$, GPM = 47.6 GPM	Flux, J, GFD
Membrane Area, sq ft = 545 sq ft	

Calculate the corrected flux, J, in GFD.

$$J_{CORRECTED}, \text{GFD} = \frac{Q_{P\,DAILY\,CORRECTED}}{\text{Membrane Area, sq ft}}$$

$$= \frac{(47.6\text{ GPM})(1{,}440\text{ min/day})}{545\text{ sq ft}}$$

$$= 126\text{ GFD}$$

9.66 Sampling

Samples of feed, concentrate, and permeate should be collected from the membrane system on a regular basis to assess system performance and discharge levels. For most industrial wastestreams, these analyses are recommended:

Total Solids

Total Soluble Solids

Suspended Solids

True Hydrocarbon Oil and Grease
(STANDARD METHOD 5520F)

Freon-Extractable Oil and Grease (Includes fats, oils, grease surfactants, and fatty acids)
(STANDARD METHOD 5520B)

pH

BOD

COD

Metals

Color

For food processing wastewater, these analyses are recommended:

Total Solids

Total Soluble Solids

Total Suspended Solids

pH

Freon-Extractable Oil and Grease (Includes fats, oils, grease surfactants, and fatty acids)
(STANDARD METHOD 5520B)

COD

BOD

TKN[38] Protein

True Protein

Additional analyses may be required when specific discharge limits must be met.

The analysis of permeate on a regular basis allows the operator to monitor membrane life. As membranes degrade, the retention values decrease as more of the retained particles pass into the permeate. Concentrate analysis and the permeate values allow the operator to calculate retention levels with the formulas presented earlier. Feed analysis is recommended on a regular basis to assess variations in the feed, which may alter the performance of the membrane filter. Frequency of analysis must be determined case by case. Good documentation of analyses in the form of SPC[39] charts and log sheets is recommended.

---

[38] Total Kjeldahl Nitrogen
[39] SPC Chart. Statistical Process Control chart. A plot of daily performance such as a trend chart.

9.67 Recordkeeping

Recording of data is important for system performance evaluation. A daily log sheet should be set up for recording process and analytical data. This log should include the following information:

1. Date.
2. Operator's name.
3. Feed description.
4. Upstream conditions within the plant.
5. Time of data collection.
6. Flow to the unit.
7. Permeate flow from the unit.
8. Tank levels.
9. Inlet and outlet pressures.
10. Feed temperature.
11. pH of the feed.
12. Sample ID numbers of feed, permeate, and concentrate.
13. Comment section to record problems.

A separate cleaning data sheet is suggested:

1. Cleaning chemical concentrations (pH, amounts of chemicals added, measured chlorine concentration).
2. Cleaning cycle times.
3. Inlet and outlet pressures.
4. Fluxes during cleaning.
5. Water flux after cleaning, corrected to standard conditions.
6. Pressure and temperature conditions of water flux.
7. Comment section to record problems.

Charts or diagrams of the membrane installation should be readily available. Membrane serial numbers are found on individual modules for identification purposes. All maintenance and repairs should also be recorded.

9.7 SAFETY PRECAUTIONS WITH MEMBRANE SYSTEMS

Each manufacturer will provide a list of safety precautions with the equipment to protect the membrane and ensure safe operation to the plant operators and other employees. Basic safety precautions include:

1. Never expose the membrane to unknown or incompatible materials or chemicals. Examples are silicone, polar hydrocarbons, and organic solvents. Consult the membrane supplier for chemical compatibility.
2. Do not run the system at temperatures, pressures, or flow rates exceeding the manufacturer's recommendations.
3. Do not overconcentrate the waste feed beyond the design of the system.
4. Do not overclean the membrane system; this will shorten the membrane life.
5. Do not modify or bypass prefiltration of the feed.

QUESTIONS

Write your answers in a notebook and then compare your answers with those on page 479.

9.6A List the common pretreatment processes used for ultrafiltration systems processing primarily industrial waste.

9.6B In an oily waste system, why is pH control at the ultrafilter important?

9.6C List the steps for cleaning (washing) a membrane.

9.6D When would spongeballs be used to remove membrane foulants?

9.6E How is the effectiveness of a cleaning cycle assessed?

9.8 REVERSE OSMOSIS

9.80 What Is Reverse Osmosis?

Osmosis can be defined as the passage of a liquid from a weak solution to a more concentrated solution across a semipermeable membrane. The membrane allows the passage of the water (solvent) but not the dissolved solids (solutes).

Osmosis plays a vital role in many biological processes. Nutrient and waste minerals are transported by osmosis through the cells of animal tissues, which show varying degrees of permeability to different dissolved solids. A striking example of a natural osmotic process is the behavior of blood cells placed in pure water. Water passes through the cell walls to dilute the solution inside the cell. The cell swells and eventually bursts, releasing its red pigment. If the blood cells are placed in a concentrated sugar solution, the reverse process occurs; the cells shrink and shrivel up as water moves out into the sugar solution.

The top half of Figure 9.40 illustrates osmosis. The transfer of the water (solvent) from the fresh water side of the membrane continues until the level (shown in shaded area) rises and the head or pressure is large enough to prevent any net transfer of the solvent (water) to the more concentrated solution. At equilibrium, the quantity of water passing in either direction is equal; the difference in water level between the two sides of the membrane is defined as the osmotic pressure of the solution.

If a piston is placed on the more-concentrated solution side of the semipermeable membrane (Figure 9.40) and a pressure, P, is applied that is greater than the osmotic pressure, water flows from the more concentrated solution to the fresh water side of the membrane. This condition illustrates the process of reverse osmosis.

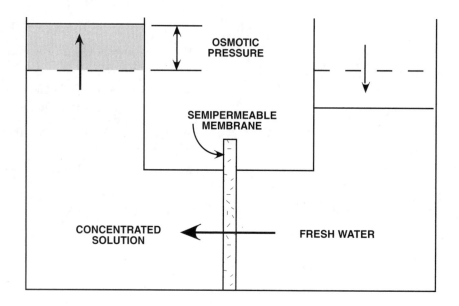

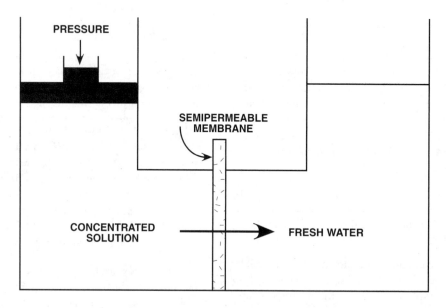

Fig. 9.40 Flows through a semipermeable membrane

9.81 Reverse Osmosis Membrane Structure and Composition

The two types of semipermeable membranes that are used most often for demineralization are cellulose acetate and thin film composites. Cellulose acetate, the first commercially available membrane (since the mid-1960s) is made by casting a cellulose acetate/solvent solution onto a porous support material. After quenching and annealing operations, the cellulose acetate membrane, often referred to as CA, exhibits the structure shown in Figure 9.41. The cellulose acetate membrane is asymmetric, meaning that one side is different from the other. The total cellulose acetate layer is 50 to 100 microns thick; however, a thin, dense layer approximately 0.2 micron thick exists at the surface. This thin, dense layer serves as the rejecting barrier of the membrane. The characteristics of a CA membrane are given in Table 9.4.

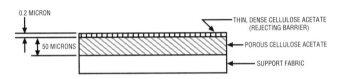

Fig. 9.41 Cellulose acetate membrane cross section

TABLE 9.4 CELLULOSE ACETATE MEMBRANE TYPICAL CHARACTERISTICS

Net Driving Pressure:	400 psi
NaCl Rejection:	92–97%
Flux Rate at 400 psi, 77°F:	25 GFD[a]
Operating pH range:	4.0–6.0
Cleaning pH range:	3.0–6.0
Cost relative to thin film composite membrane:	lower
Allowable feedwater chlorine concentration:	1.0 mg/L
Maximum operating temperature:	104°F (40°C)
Subject to biological attack	
Subject to hydrolysis	
Salt passage rate typically doubles after three years	
Most suitable for treatment of municipal wastes and some heavily pretreated surface supplies (due to lower fouling rate vs. thin film)	

[a] gallons of flux per square foot per day

During the 1970s, researchers realized the need for a membrane with better *FLUX*[40] and rejection characteristics than those of cellulose acetate. The basic approach to developing a better membrane was to improve the efficiencies of the thin rejecting layer and the porous substrate by casting each layer separately during the manufacturing process, hence the term "thin film composite." A typical thin film composite membrane is shown in Figure 9.42. The first step in preparing a thin film composite membrane is the casting of the porous support, usually a polysulfone solution, onto the support fabric. The next step is to contact the composite support with the polymers that will actually form the semipermeable thin film rejecting barrier.

The thin film rejecting barrier is formed *in situ*, or in position, by interfacial polymerization. It is the ability to form the semipermeable membrane separate from the support layers that has enabled membrane manufacturers to select polymers that will produce membranes with optimum dissolved solids rejection and water flux rates. The characteristics of thin film composite membranes are given in Table 9.5.

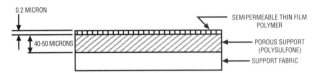

Fig. 9.42 Thin film composite membrane cross section

TABLE 9.5 THIN FILM COMPOSITE MEMBRANE TYPICAL CHARACTERISTICS

Net Driving Pressure:	200 psi
NaCl Rejection:	98–99%
Flux Rate at 200 psi, 77°F:	25–30 GFD[a]
Operating pH range:	3.0–10.0
Cleaning pH range:	2.0–12.0
Cost relative to cellulose acetate membrane:	higher
Allowable feedwater chlorine concentration:	none
Maximum operating temperature:	113°F (45°C)
Salt passage increase after three years:	≤30%
Not subject to biological attack, hydrolysis, or compaction	
Higher rejection and flux rates than cellulose acetate membrane	
Higher fouling rates than cellulose acetate on certain waste and surface water supplies	
Sensitive to oxidants in feedwater	

[a] gallons of flux per square foot per day

9.82 Membrane Performance and Properties

The basic behavior of semipermeable cellulose acetate reverse osmosis membranes can be described by two equations. The product water flow through a semipermeable membrane can be expressed as shown in Equation 1.

EQUATION 1

$$F_w = A(\Delta P - \Delta \pi)$$

Where

F_w = Water flux (gm/sq cm-sec)

A = Water permeability constant (gm/sq cm-sec atm[41])

ΔP = Pressure differential applied across the membrane (atm)

$\Delta \pi$ = Osmotic pressure differential across the membrane (atm)

Note that the water flux is the flow of water in grams per second through a membrane area of one square centimeter. Think of this as similar to the flow through a rapid sand filter in gallons per minute through a filter area of one square foot (GPM/sq ft).

---

[40] *Flux.* A flowing or flow.
[41] *atm.* The abbreviation for atmosphere. One atmosphere is equal to 14.7 psi or 100 kPa.

The mineral (salt) flux (mineral passage) through the membrane can be expressed as shown in Equation 2.

EQUATION 2

$F_s = B(C_1 - C_2)$

Where

F_s = Mineral flux (gm/sq cm-sec)

B = Mineral permeability constant (cm/sec)

$C_1 - C_2$ = Concentration gradient across the membrane (gm/cu cm)

The water permeability (A) and mineral permeability (B) constants are characteristics of the particular membrane that is used and the processing it has received.

An examination of Equations 1 and 2 shows that the water flux (the rate of water flow through the membrane) is dependent on the applied pressure, while the mineral flux is *not* dependent on pressure. As the pressure of the feedwater is increased, the flow of water through the membrane increases while the flow of minerals remains essentially constant. Therefore, both the quantity and the quality of the purified product (the *PERMEATE*[42]) should increase with increased pressure. This occurs because there is more water to dilute the same amount of mineral.

The water flux (F_w) *decreases* as the mineral content of the feed increases because the osmotic pressure contribution ($\Delta\pi$) increases with increasing mineral content. In other words, since $\Delta\pi$ increases, the term ($\Delta P - \Delta\pi$) decreases, which results in a decrease in F_w, the water flux. Further, as more and more feedwater passes through the membrane, the mineral content of the feedwater becomes higher and higher (more concentrated). The osmotic pressure contribution ($\Delta\pi$) of the concentrate increases, resulting in a lower water flux.

Finally, since the membrane rejects a constant percentage of mineral, product water quality decreases with increased feedwater concentration. Also, note that Equation 2 reveals that the greater the concentration gradient ($C_1 - C_2$) across the membrane, the greater the mineral flux (mineral flow). Therefore, the greater the feed concentration, the greater the mineral flux and also mineral concentration in the product water.

Wastewater treatment plant operators must have a basic understanding of these mathematical relationships that describe RO (reverse osmosis) membrane performance. To help develop a better understanding of the interrelationships of flux, rejection, time, temperature, pH, and recovery, further explanation of these variables continues in the next section.

EXAMPLE 12

Convert a water flux of 5×10^{-4} gm/sq cm-sec[43] to gallons per day per square foot.

Known	Unknown
Water Flux, gm/sq cm-sec = 5×10^{-4} gm/sq cm-sec	Flow, GPD/sq ft

Convert the water flux from gm/sq cm-sec to flow in GPD/sq ft.

$$\text{Flow, GPD/sq ft} = \frac{(\text{Water Flux, gm/sq cm-sec})(2.54 \text{ cm/in})^2(12 \text{ in/ft})^2(60 \text{ sec/min})(60 \text{ min/hr})(24 \text{ hr/day})}{(1{,}000 \text{ gm}/L)(3.785 \, L/\text{gal})}$$

$$= \frac{(0.0005 \text{ gm/sq cm-sec})(2.54 \text{ cm/in})^2(12 \text{ in/ft})^2(60 \text{ sec/min})(60 \text{ min/hr})(24 \text{ hr/day})}{(1{,}000 \text{ gm}/L)(3.785 \, L/\text{gal})}$$

$$= 10.6 \text{ GPD/sq ft}$$

QUESTIONS

Write your answers in a notebook and then compare your answers with those on pages 479 and 480.

9.8A What is the osmotic pressure of a solution?

9.8B What are the two types of semipermeable membranes that are used most often for demineralization?

9.8C What is the meaning of water flux and of mineral flux? What units are used to express measurement of these quantities?

9.8D When additional pressure is applied to the side of a membrane with a concentrated solution, what happens?

9.8E When higher mineral concentrations occur in the feedwater, what happens to the permeate (or product water)?

9.83 Definition of Flux

The term flux is the expression used to describe the rate of water flow through the semipermeable membrane. Flux is usually expressed in gallons per day per square foot of membrane surface or in grams per second per square centimeter.

The average membrane flux rate of a reverse osmosis system is an important operating guideline. In practice, most reverse osmosis systems will require periodic cleaning. It has been demonstrated that the cleaning frequency can be dependent on the average membrane flux rate of the system. Too high a flux rate may result in excessive fouling rates requiring frequent cleaning. Some general industry guidelines for acceptable flux rates are:

Feedwater Source	Flux Rate, GFD
Industrial/Municipal Waste	8–12
Surface (river, lake, ocean)	8–14
Well	14–20

EXAMPLE 13

The permeate flow through an arrangement (or array) of RO membrane pressure vessels is 1,330,000 gallons per day (GPD).

---

[42] *Permeate* (PURR-me-ate). (1) To penetrate and pass through, as water penetrates and passes through soil and other porous materials. (2) The liquid (demineralized water) produced from the reverse osmosis process that contains a low concentration of dissolved solids.

[43] 5×10^{-4} is the same as 0.0005.

Feedwater first flows to 33 vessels operating in parallel. The concentrate from the 33 first-pass vessels is combined and sent to a set of 11 second-pass vessels. Each vessel (or tube) contains six membrane elements. Each element contains many fiber membranes, which provide a total of 325 square feet of membrane surface area per element. Calculate the average membrane flux rate for the system in gallons per day per square foot (GFD).

Known		Unknown
Permeate Flow, GPD	= 1,330,000 GPD	Average Flux Rate, GFD
No. of Vessels	= 44 (33 + 11)	
No. of Elements per Vessel	= 6	
Membrane Area per Element, sq ft	= 325 sq ft	

1. Determine the total membrane area in the system.

    Membrane Area, sq ft = (No. Vessels)(No. Elements/vessel)(Surface Area/element)

    = (44 vessels)(6 elements/vessel)(325 sq ft/element)

    = 85,800 sq ft

2. Calculate the average membrane flux rate for the system.

    $$\text{Average Flux Rate, GFD} = \frac{\text{Permeate Flow, GFD}}{\text{Membrane Area, sq ft}}$$

    $$= \frac{1,330,000 \text{ GPD}}{85,800 \text{ sq ft}}$$

    $$= 15.5 \text{ GFD}$$

Even under ideal conditions (pure feedwater and no fouling of the membrane surface), there is a decline in water flux with time. This decrease in flux is due to membrane compaction. This phenomenon is considered comparable to "creep" observed in other plastics or even metals when subjected to compressing stresses (pressure).

The term "flux decline" is used to describe the loss of water flow through the membrane due to compaction plus fouling. In the real world, feedwaters are never pure and contain suspended solids, dissolved organics and inorganics, bacteria, algae, and other potential foulants. These impurities can be deposited or grow on the membrane surface, thus hindering the flow of water through the membrane.

9.84 Mineral Rejection

The purpose of demineralization is to separate minerals from water; the ability of the membrane to reject minerals is called the mineral rejection. Mineral rejection is defined as:

EQUATION 3

$$\text{Rejection, \%} = \left(1 - \frac{\text{Product Concentration}}{\text{Feedwater Concentration}}\right) \times 100\%$$

Mineral rejections can be determined by measuring the TDS and using Equation 3. Rejections also may be calculated for individual constituents in the solution by using their concentrations.

The basic equations that describe the performance of a reverse osmosis membrane indicate that rejection decreases as feedwater mineral concentration increases. Remember, this is because the higher mineral concentration increases the osmotic pressure. Figure 9.43 illustrates the rejection performance for a typical RO (reverse osmosis) membrane operating on three different feedwater solutions. This figure shows that as feed mineral concentration increases (TDS in mg/L), rejection decreases at a given feed pressure. Notice also that rejection improves as feed pressure increases.

Typical rejection for most commonly encountered dissolved inorganics is usually between 92 and 99 percent. Divalent ions like calcium and sulfate are better rejected than monovalent ions such as sodium or chloride. Table 9.6 lists the typical rejection of an RO membrane operating on a brackish feedwater.

TABLE 9.6 TYPICAL REVERSE OSMOSIS REJECTIONS OF COMMON CONSTITUENTS FOUND IN BRACKISH WATER

Contaminant	Units	Feedwater Concentration	Percent Removal
EC[a]	μmhos	1,400	92
TDS[a]	mg/L	900	92
Calcium	mg/L	100	99
Chloride	mg/L	120	92
Sulfate	mg/L	338	99
Sodium	mg/L	158	92
Ammonia	mg/L	22.5	94
Nitrate	mg/L	2.9	55
COD[a]	mg/L	12.5	95
TOC[a]	mg/L	6.0	88
Silver	μg/L	1.2	88
Arsenic	μg/L	<5.0	—
Aluminum	μg/L	71.0	93
Barium	μg/L	24.0	96
Beryllium	μg/L	<1.0	—
Cadmium	μg/L	3.4	98
Cobalt	μg/L	4.6	>90
Chromium	μg/L	3.6	80
Copper	μg/L	12.7	63
Iron	μg/L	24.0	91
Mercury	μg/L	0.8	41
Manganese	μg/L	1.0	85
Nickel	μg/L	2.5	88
Lead	μg/L	<1.0	—
Selenium	μg/L	<5.0	—
Zinc	μg/L	<100.0	—

[a] EC, Electrical Conductivity; TDS, Total Dissolved Solids; COD, Chemical Oxygen Demand; and TOC, Total Organic Carbon.

EXAMPLE 14

Estimate the ability of a reverse osmosis plant to reject minerals by calculating the mineral rejection as a percent. The feedwater contains 1,500 mg/L TDS and the product water TDS is 150 mg/L.

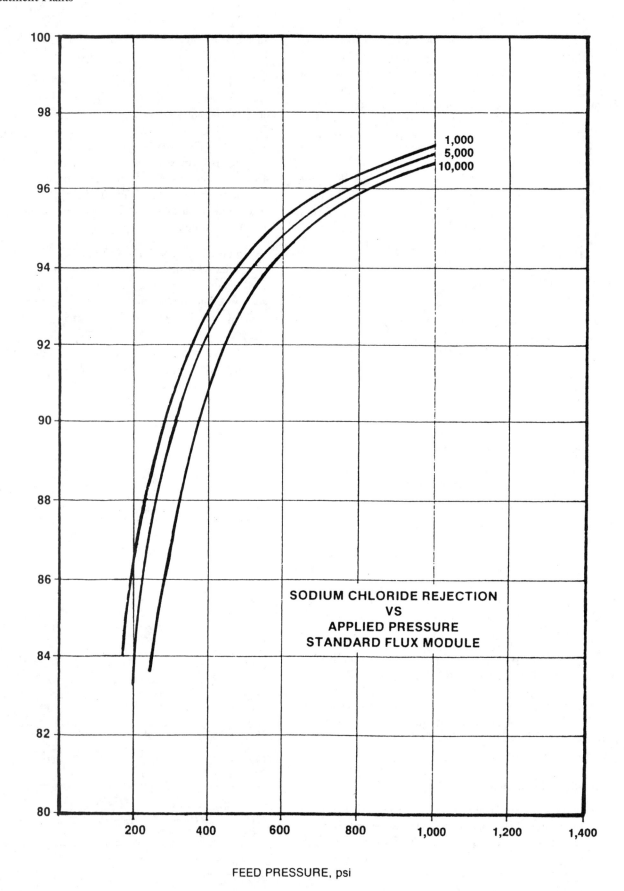

Fig. 9.43 Typical RO rejection for three different feedwater concentrations of TDS in mg/L
(Source: REVERSE OSMOSIS PRINCIPLES AND APPLICATIONS,
by Fluids Systems, Division of UOP, October 1970)

Known		Unknown
Feedwater TDS, mg/L =	1,500 mg/L	Mineral Rejection, %
Product Water TDS, mg/L =	150 mg/L	

Calculate the mineral rejection as a percent.

$$\text{Mineral Rejection, \%} = \left(1 - \frac{\text{Product TDS, mg}/L}{\text{Feed TDS, mg}/L}\right)(100\%)$$

$$= \left(1 - \frac{150 \text{ mg}/L}{1{,}500 \text{ mg}/L}\right)(100\%)$$

$$= (1 - 0.1)(100\%)$$

$$= 90\%$$

While most demineralization applications require the use of a membrane with high rejection rates (greater than 95 percent), some applications can use a membrane with lower rejection rates (80 percent) and lower operating pressures (less than 150 psi). The membranes that fit this classification are commonly referred to as softening or nanofiltration membranes. These membranes produce the same quantity of water as standard RO membranes but at lower operating pressures. A comparison of standard and softening RO membranes is given below:

	Standard	Softening (Nanofiltration)
Flux	25–30 GFD	25–30 GFD
Applied Pressure	225 psi	150 psi
Minimum Salt Rejection	97–98%	75–80%
Hardness Rejection	>99%	>95%

Softening or nanofiltration membranes are seeing widespread use for demineralization of wastewaters that require high rejection rates for hardness and THM formation potential, and moderate TDS rejection.

Reverse osmosis membrane manufacturers provide computer software to project the permeate water quality of a reverse osmosis system. Figures 9.44 and 9.45 are examples of computer printouts of permeate projections showing the expected initial performance of a thin film composite membrane (Figure 9.44) and a cellulose acetate membrane (Figure 9.45). The same feedwater was used for both projections.

QUESTIONS

Write your answers in a notebook and then compare your answers with those on page 480.

9.8F Water flux is usually expressed in what units?

9.8G What is "flux decline"?

9.8H How is mineral rejection measured?

9.8I What are the two major differences between standard RO membranes and softening membranes?

9.85 Effects of Feedwater Temperature and pH on Membrane Performance

In reverse osmosis operation, feedwater temperature has a significant effect on membrane performance and must therefore be taken into account in system design and operation. Essentially, the value of the water permeation constant is only constant for a given temperature. As the temperature of the feedwater increases, flux increases. Usually, flux is reported at some standard temperature reference condition, such as 25°C. Figure 9.46 illustrates the increase in flux for a standard RO module over a range of operating temperatures when 400 psi (2,758 kPa or 28 kg/sq cm) net operating pressure is applied.

Cellulose acetate membrane is subject to long-term *HYDROLYSIS*.[44] Hydrolysis results in a lessening of mineral rejection capability. The rate of hydrolysis is accelerated by increased temperature, and is also a function of feed pH (Figure 9.47). Slightly acidic pH values (5 to 6) ensure a lower hydrolysis rate, as do cooler temperatures. Therefore, to ensure the longest possible lifetime of the cellulose acetate membrane and to slow hydrolysis, acid is added as a pretreatment step before demineralization. Table 9.7 indicates the relative time for mineral passage to increase 200 percent at different feedwater pH levels. Thin film composite membranes are not subject to hydrolysis but pH adjustment of the feedwater may be required for scale control.

TABLE 9.7 TIME REQUIRED TO ACHIEVE A 200 PERCENT INCREASE IN MINERAL PASSAGE AT 23°C AT VARIOUS pH LEVELS

pH 5.0	6 years
6.0	3.8 years
7.0	1 year
8.0	0.14 year = 51 days
9.0	0.01 year = 3.6 days

9.86 Recovery

Recovery is defined as the percentage of feed flow that is recovered as product water. Expressed mathematically, recovery can be determined by Equation 4.

EQUATION 4

$$\text{Recovery, \%} = \left(\frac{\text{Product Flow}}{\text{Feed Flow}}\right)(100\%)$$

The recovery rate is usually determined or limited by two considerations. The first is the desired product water quality.

---

[44] *Hydrolysis* (hi-DROLL-uh-sis). (1) A chemical reaction in which a compound is converted into another compound by taking up water. (2) Usually a chemical degradation of organic matter.

```
HYDRANAUTICS DESIGN PROGRAM -   VERSION 4.01 (1991)           05-26-04
Calculation was made by: CWC

Project name : SANTA ANA WATERSHED            Permeate flow :   1330000 GPD

Feedwater temperature :    21.0 C       Recovery :       77.0%
Raw water pH :              7.20        Element age :     0.0 years
Acid dosage, ppm(100%):    32.7 H2SO4   Flux decline coefficient :       -0.025
Acidified feed CO2,ppm :   66.7         3-yr salt passage increase factor :1.2


Feed Pressure : 157.4 psi               Concentrate Pressure : 118.1 psi

Pass   Feed Flow       Conc. Flow      Beta    Conc.      Element       Element   Array
       Total Vessel    Total Vessel            Press.     Type          No.
       gpm    gpm      gpm    gpm              psi

 1    1199.5  36.3    454.7   13.8     1.22    142.7    8040-LSY-CPA2    198     33x6
 2     454.7  41.3    275.9   25.1     1.08    118.1    8040-LSY-CPA2     66     11x6
```

Ion	Raw water mg/l	Raw water ppm*	Feed water mg/l	Feed water ppm*	Permeate mg/l	Permeate ppm*	Concentrate mg/l	Concentrate ppm*
Ca	140.0	349.1	140.0	349.1	1.8	4.6	602.5	1502.5
Mg	42.0	172.8	42.0	172.8	0.6	2.3	180.8	743.8
Na	168.0	365.2	168.0	365.2	10.5	22.8	695.4	1511.7
K	3.8	4.9	3.8	4.9	0.3	0.4	15.5	19.9
NH4	0.0	0.0	0.0	0.0	0.0	0.0	0.0	0.0
Ba	0.0	0.0	0.0	0.0	0.0	0.0	0.0	0.0
Sr	0.0	0.0	0.0	0.0	0.0	0.0	0.0	0.0
CO3	0.3	0.5	0.1	0.2	0.0	0.0	2.1	3.5
HCO3	367.0	300.8	326.6	267.7	18.8	15.4	1357.2	1112.5
SO4	243.0	253.1	275.1	286.5	2.3	2.3	1188.4	1237.9
Cl	162.0	228.5	162.0	228.5	5.2	7.4	686.8	968.7
F	0.4	1.1	0.4	1.1	0.0	0.1	1.7	4.4
NO3	93.0	75.0	93.0	75.0	5.9	4.8	384.5	310.1
SiO2	40.0		40.0		0.7		171.7	
TDS	1259.5		1251.0		46.1		5286.5	
pH	7.2		6.9		5.7		7.5	

Notes: *ppm as CaCO3.   Calculated concentrations are accurate to   +/- 10%

	Raw water	Feed water	Concentrate
CaSO4/Ksp*100,%	7.9	8.8	55.2
SrSO4/Ksp*100,%	0.0	0.0	0.0
BaSO4/Ksp*100,%	0.0	0.0	0.0
SiO2 sat.,%	34.2	34.2	146.8
Langelier ind.	0.21	-0.14	1.69
Stiff & Davis ind.	0.20	-0.15	1.44
Ionic strength	0.03	0.03	0.11
Osmotic press.,psi	9.2	9.1	38.5

Fig. 9.44 Computer printout of permeate projections for a thin film composite membrane
(Source: Hydranautics)

```
HYDRANAUTICS DESIGN PROGRAM -   VERSION 4.01 (1991)         05-26-04
Calculation was made by: CWC

Project name : SANTA ANA WATERSHED           Permeate flow :   1330000 GPD

Feedwater temperature :      21.0 C     Recovery :       77.0%
Raw water pH :                7.20      Element age :     0.0 years
Acid dosage, ppm(100%):       0.0 H2SO4 Flux decline coefficient :     -0.035
Acidified feed CO2,ppm :      37.9      3-yr salt passage increase factor :2.0


Feed Pressure : 297.9 psi                Concentrate Pressure : 257.5 psi
```

Pass	Feed Flow Total gpm	Feed Flow Vessel gpm	Conc. Total gpm	Flow Vessel gpm	Beta	Conc. Press. psi	Element Type	Element No.	Array
1	1199.5	36.3	480.9	14.6	1.21	282.8	8040-MSY-CAB1	198	33x6
2	480.9	43.7	275.9	25.1	1.11	257.5	8040-MSY-CAB1	66	11x6

Ion	Raw water mg/l	Raw water ppm*	Feed water mg/l	Feed water ppm*	Permeate mg/l	Permeate ppm*	Concentrate mg/l	Concentrate ppm*
Ca	140.0	349.1	140.0	349.1	2.5	6.2	600.4	1497.3
Mg	42.0	172.8	42.0	172.8	0.7	3.1	180.1	741.2
Na	168.0	365.2	168.0	365.2	27.9	60.6	637.2	1385.1
K	3.8	4.9	3.8	4.9	0.8	1.0	13.8	17.7
NH4	0.0	0.0	0.0	0.0	0.0	0.0	0.0	0.0
Ba	0.0	0.0	0.0	0.0	0.0	0.0	0.0	0.0
Sr	0.0	0.0	0.0	0.0	0.0	0.0	0.0	0.0
CO3	0.3	0.5	0.3	0.5	0.0	0.0	4.6	7.7
HCO3	367.0	300.8	367.0	300.8	21.3	17.5	1524.3	1249.4
SO4	243.0	253.1	243.0	253.1	1.8	1.9	1050.5	1094.2
Cl	162.0	228.5	162.0	228.5	22.6	31.8	628.8	886.9
F	0.4	1.1	0.4	1.1	0.0	0.1	1.6	4.3
NO3	93.0	75.0	93.0	75.0	24.3	19.6	323.1	260.6
SiO2	40.0		40.0		10.4		139.0	
TDS	1259.5		1259.5		112.3		5103.5	
pH	7.2		7.2		6.0		7.8	

Notes: *ppm as CaCO3.   Calculated concentrations are accurate to +/- 10%

	Raw water	Feed water	Concentrate
CaSO4/Ksp*100,%	7.9	7.9	50.1
SrSO4/Ksp*100,%	0.0	0.0	0.0
BaSO4/Ksp*100,%	0.0	0.0	0.0
SiO2 sat.,%	34.2	34.2	118.8
Langelier ind.	0.21	0.21	2.04
Stiff & Davis ind.	0.20	0.20	1.79
Ionic strength	0.03	0.03	0.11
Osmotic press.,psi	9.2	9.2	37.0

Fig. 9.45 Computer printout of permeate projections for a cellulose acetate membrane
(Source: Hydranautics)

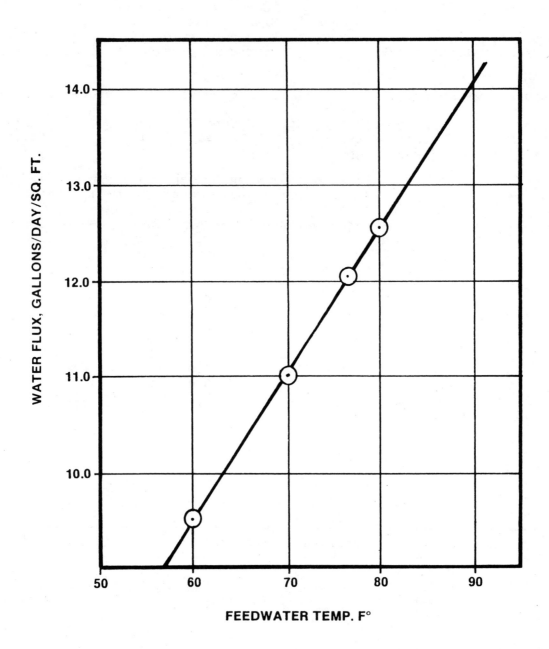

Fig. 9.46 Effect of temperature on water flux rate, cellulose acetate membrane operating pressure at 400 psi (2,758 kPa or 28 kg/sq cm) net
(Source: REVERSE OSMOSIS PRINCIPLES AND APPLICATIONS, by Fluids Systems, Division of UOP, October 1970)

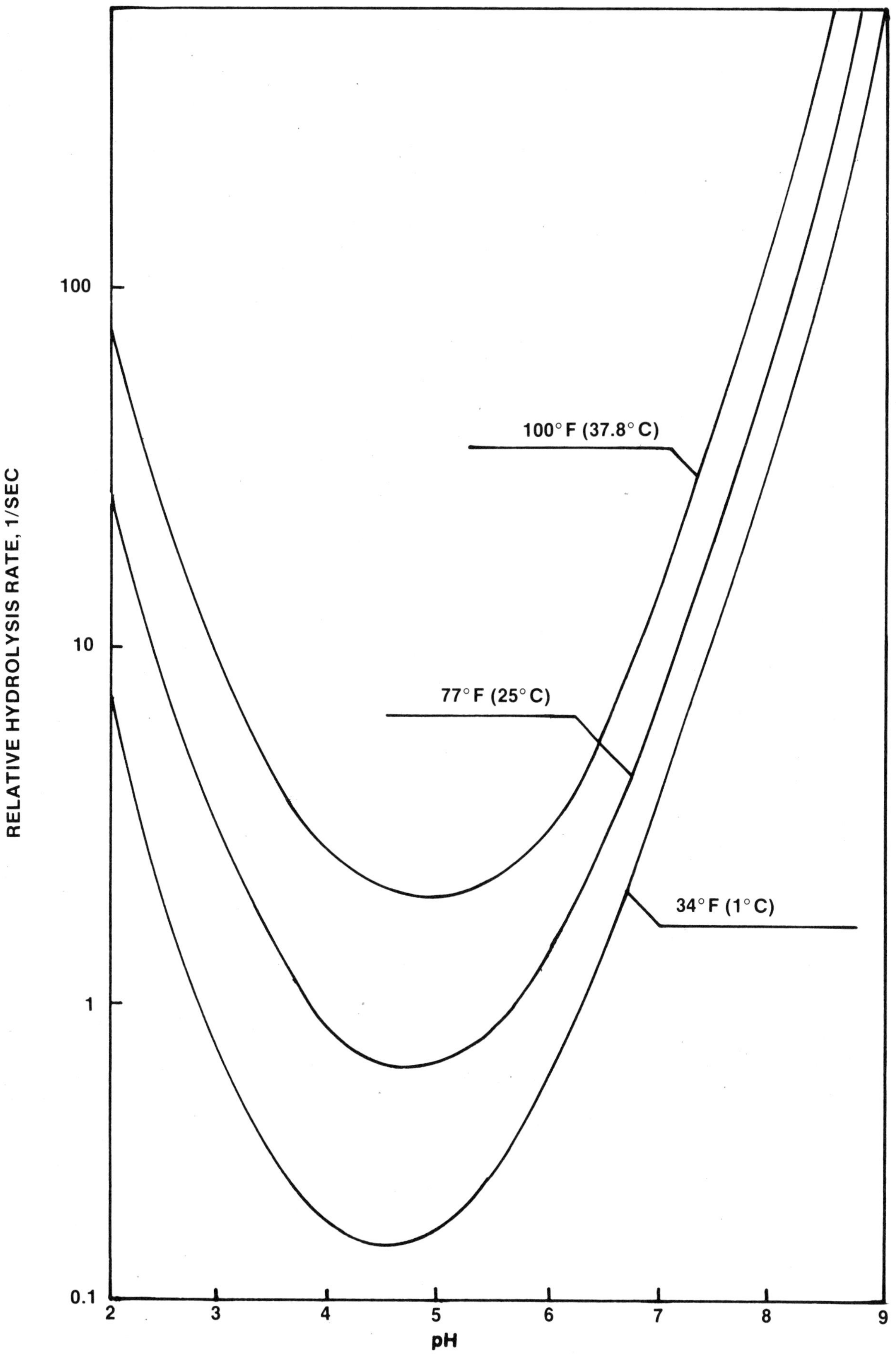

Fig. 9.47 Effect of temperature and pH on hydrolysis rate for cellulose acetate membrane

Since the amount of mineral passing through the membrane is influenced by the concentration differential between the brine and product, there is a possibility of exceeding product quality criteria with excessive recovery. The second consideration concerns the solubility limits of minerals in the brine. One should not concentrate the brine to a degree that would precipitate minerals on the membrane. This effect is commonly referred to as CONCENTRATION POLARIZATION.[45]

The most common and serious problem resulting from concentration polarization is the increasing tendency for precipitation of sparingly soluble salts and the deposition of particulate matter on the membrane surface.

In any flowing hydraulic system, the fluid near a solid surface travels more slowly than the main stream of the fluid. In other words, there is a liquid boundary layer at the solid surface. This is also true at the surface of the membrane in a spiral-wound element or in any other membrane packaging configuration. Since water is transmitted through the membrane at a much more rapid rate than minerals, the concentration of the minerals builds up in the boundary layer (concentration polarization) and it is necessary for the minerals to diffuse back into the flowing stream. Polarization will reduce both the flux and rejection of a reverse osmosis system. Since it is impractical to totally eliminate the polarization effect, it is necessary to minimize it by good design and operation.

The boundary layer effect can be minimized by increased water flow velocity and by promoting turbulence within the RO elements. Brine flow rates can be kept high as product water is removed by staging (reducing) the module pressure vessels. This is popularly referred to as a "Christmas Tree" arrangement. Typical flow arrangements such as 4 units - 2 units - 1 unit (85 percent recovery) or 2 units - 1 unit (75 percent recovery) are used most often (Figure 9.48).

These configurations consist of feeding water to a series of pressure vessels in parallel where about 50 percent of the water is separated by the membrane as product water and 50 percent of the water is rejected. The reject is then fed to half as many vessels in parallel where again about 50 percent is product water and 50 percent rejected. This reject becomes the feed for the next set of vessels. By arranging the pressure vessels in the 4-2-1 arrangement, it is possible to recover over 85 percent of the feedwater as product water and to maintain adequate flow rates across the membrane surface to minimize polarization. For example, a system consisting of a total of 35 vessels would have a 20-10-5 pressure vessel arrangement for an 85 percent recovery.

EXAMPLE 15

Estimate the percent recovery of a reverse osmosis unit with a 4-2-1 arrangement if the feed flow is 5.88 MGD and the product flow is 5.0 MGD.

Known	Unknown
Product Flow, MGD = 5.0 MGD	Recovery, %
Feed Flow, MGD = 5.88 MGD	

Calculate the recovery as a percent.

$$\text{Recovery, \%} = \frac{(\text{Product Flow, MGD})(100\%)}{\text{Feed Flow, MGD}}$$

$$= \frac{(5.0 \text{ MGD})(100\%)}{5.88 \text{ MGD}}$$

$$= 85\%$$

QUESTIONS

Write your answers in a notebook and then compare your answers with those on page 480.

9.8J How will an increase in feedwater temperature influence the water flux?

9.8K How does hydrolysis influence the mineral rejection capability of a membrane?

9.8L How is recovery defined?

9.8M Recovery rate is usually limited by what two considerations?

9.8N Define concentration polarization.

9.9 REFERENCE

STANDARD METHODS FOR THE EXAMINATION OF WATER AND WASTEWATER, 20th Edition, 1998. Obtain from Water Environment Federation (WEF), 601 Wythe Street, Alexandria, VA 22314-1994. Order No. S82010. Price to members, $164.75; nonmembers, $209.75; price includes cost of shipping and handling.

END OF LESSON 4 OF 4 LESSONS

on

FILTRATION

Please answer the discussion and review questions next.

---

[45] *Concentration Polarization.* (1) A buildup of retained particles on the membrane surface due to dewatering of the feed closest to the membrane. The thickness of the concentration polarization layer is controlled by the flow velocity across the membrane. (2) Used in corrosion studies to indicate a depletion of ions near an electrode. (3) The basis for chemical analysis by a polarograph.

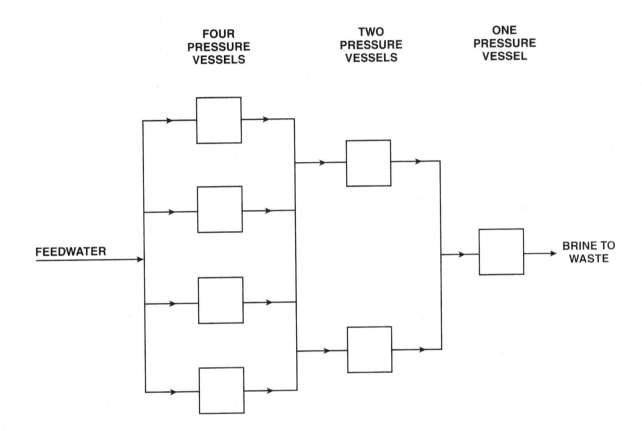

Fig. 9.48 Typical 4-2-1 "Christmas Tree" arrangement

DISCUSSION AND REVIEW QUESTIONS

Chapter 9. FILTRATION

(Lesson 4 of 4 Lessons)

Write the answers to these questions in your notebook. The question numbering continues from Lesson 3.

15. Membrane filtration processes are gaining acceptance as alternatives for what other treatment processes?

16. How does the membrane process differ from conventional filtration?

17. Stringent discharge requirements may require post-treatment of the ultrafiltration (UF) permeate by which treatment processes?

18. What major membrane system problems can be revealed by analyzing delta pressure measurements?

19. How can fouled membranes be cleaned?

20. Why does membrane filtration require proper pretreatment of the feed?

21. Why must membranes be replaced periodically?

22. What is reverse osmosis?

23. Indicate what will happen to both the water flux and mineral flux when:

   a. Pressure differential applied across the membrane (ΔP) increases.

   b. Osmotic pressure differential across the membrane ($\Delta \pi$) increases.

   c. Concentration gradient across the membrane ($C_1 - C_2$) increases.

24. What usually happens to water flux with time? Explain.

25. How does fouling develop on membranes?

26. What factors influence the rate of hydrolysis of a membrane? Explain.

27. What is the most common and serious problem resulting from concentration polarization?

28. Why do demineralization plants use a pressure vessel Christmas Tree configuration?

SUGGESTED ANSWERS

Chapter 9. FILTRATION

ANSWERS TO QUESTIONS IN LESSON 1

Answers to questions on page 396.

9.1A Most filters operate on a batch basis whereby the filter operates until its capacity to remove solids is nearly reached. At this time it is completely removed from service and cleaned.

9.1B A gravity filter should be cleaned when the solids capacity of the media has nearly been reached but before solids break through into the effluent.

9.1C Meanings of the following terms are:
  1. Downflow. Water flows down through the bed.
  2. Static bed. Bed does not move or expand while water is being filtered.

9.1D
  1. In surface straining, the filter is designed to remove the solids at the very top of the media.
  2. Depth filtration is designed to permit the solids to penetrate deep into the media, thereby capturing the solids within as well as on the surface of the media.

Answers to questions on page 405.

9.1E Materials used for filter media include silica sand, anthracite coal, garnet, or ilmenite. Garnet and ilmenite are commonly used in multi-media beds.

9.1F If the filter media is not thoroughly cleaned during each backwashing, a buildup of solids will occur. The end result of incomplete cleaning is the formation of mudballs within the bed.

9.1G If unfiltered water is supplied to the backwash system, clogging of the underdrain system may occur.

9.1H Used backwash water holding tanks are needed to equalize flows and prevent hydraulically overloading the treatment plant when backwash waters are returned to the headworks.

Answers to questions on page 406.

9.1I The head loss through the filter media is determined by measuring the water pressure above and below the filter media. When water flows through the media, the pressure below the media will be less than the pressure above the media (when the pressure levels are measured or read at the same elevation). The difference between the two readings is the head loss.

9.1J Filter system alarms should be tested for proper functioning at least every 60 days.

Answers to questions on page 408.

9.1K A pre-start check should be conducted before starting filtering systems to prevent damage to the equipment or injury to personnel.

9.1L The purpose of the rate-control valve is to maintain the desired flow through the filter and prevent the backwash water from mixing with the filtered water during backwashing.

9.1M A filter should be backwashed after the capacity of the media to hold solids is nearly used up, but before solids break through into the effluent.

Answers to questions on page 409.

9.1N Abnormal operating conditions include:

1. High solids in applied water due to bulking sludge, rising sludge, or solids washout in the secondary clarifier.
2. Low suspended solids in applied water; however, solids pass through filter.
3. Loss of filter aid chemical feed.
4. High wet weather peak flows.
5. Low applied water flows.
6. High color loading.
7. High water temperature.
8. Low water temperature.
9. Air binding.
10. Negative pressure in filter.
11. Chlorine in applied water.
12. pH change in applied water.
13. High grease and oil in applied water.

9.1O To treat a high solids content in the water applied to a filter:

1. Run jar tests and adjust chemical dosage as needed.
2. Place more filters in service to prevent breakthrough.
3. Prepare to backwash more frequently.

Answers to questions on page 412.

9.1P To determine if media are being lost, every three or four months, measure and record the freeboard to the filter media surface. A small amount of media loss is normal, but an excessive amount (2 to 3 inches or 5 to 7 centimeters) indicates operational problems.

9.1Q Trees and shrubs should be kept away from uncovered filters because leaves will drop into the filter and they are very difficult to backwash out of the media.

Answers to questions on page 415.

9.1R The three main types of safety hazards around filtration systems are electrical, chemical, and mechanical.

9.1S Filtration instrumentation should measure and record applied flows, backwash flows, head loss, and water quality before and after filtration.

9.1T Install all readout meters, charts, and gauges of instruments in a convenient and centralized location.

ANSWERS TO QUESTIONS IN LESSON 2

Answers to questions on page 416.

9.2A The purpose of the inert-media pressure filter is to remove suspended solids, associated BOD, and turbidity from secondary effluents to meet waste discharge requirements established by NPDES permits.

9.2B Chemicals commonly used with the filtration process are polymers and alum. The chemicals are used as coagulants for the solids and turbidity to aid in their removal by filtration.

9.2C Major components of a pressure filtration system include:

1. A holding tank or wet well
2. Filter feed pumps
3. Chemical coagulant feed pump system
4. Filters
5. Filter backwash wet well
6. Filter backwash pumps
7. Decant tank

Answers to questions on page 421.

9.2D The purpose of the holding tank is to store water and to allow additional settling of the suspended solids before the water is applied to the filters.

9.2E Alum is used for *COAGULATION* while polymers are used for *FLOCCULATION*.

9.2F Polymer is supplied at a concentration of 0.6 pound polymer per gallon (72 gm/L or 72 kg/m^3). The polymer feed pump delivers a flow of 0.15 GPM (0.0095 L/sec) and the flow to the filters is 5,000 GPM (315 L/sec). Calculate the concentration or dose of polymer in mg/L in the water applied to the filter.

Known	Unknown
Polymer Conc, lbs/gal = 0.6 lb/gal	Polymer Dose, mg/L
Polymer Pump, GPM = 0.15 GPM	
Flow to Filter, GPM = 5,000 GPM	

ENGLISH

Calculate the polymer dose, mg/L.

$$\text{Dose, mg}/L = \frac{\text{Flow, gal/min} \times \text{Conc, lbs polymer/gal}}{\text{Flow, gal/min} \times 8.34 \text{ lbs water/gal}}$$

$$= \frac{0.15 \text{ gal/min} \times 0.6 \text{ lb polymer/gal}}{5,000 \text{ gal/min} \times 8.34 \text{ lbs water/gal}}$$

$$= \frac{0.09 \text{ lb polymer}}{41,700 \text{ lbs water}} \times \frac{1,000,000}{1 \text{ M}}$$

$$= 2.2 \text{ mg}/L$$

METRIC

$$\text{Dose, mg}/L = \frac{\text{Flow, }L/\text{sec} \times \text{Conc, gm polymer}/L \times 1,000 \text{ mg/gm}}{\text{Flow, }L/\text{sec}}$$

$$= \frac{0.0095 \, L/\text{sec} \times 72 \text{ gm polymer}/L \times 1,000 \text{ mg/gm}}{315 \, L/\text{sec}}$$

$$= 2.2 \text{ mg}/L$$

Answers to questions on page 425.

9.2G Major components of pressure filters include:
1. Vessels
2. Interior piping
3. Underdrain gravel (support media)
4. Inert media
5. Flow controls

9.2H Water from the surface wash arms initially breaks up the mat of suspended material on the media surface.

9.2I The water used to backwash the filter comes from the chlorine contact tank (filtered and chlorinated) to the backwash wet well before it is used for backwashing.

9.2J The decant tank receives the backwash water from the filters. The backwash water is allowed to settle and the clarified effluent is recycled to the filters. The settled material is collected and discharged to the solids handling facility.

Answers to questions on page 426.

9.2K When a large quantity of polymer reaches a filter, short filter run times will result due to increased differential pressure across the filter. When excessive alum concentrations are involved, the alum will pass through the filter media and filter effluent turbidity and suspended solids will increase due to alum breakthrough.

9.2L In areas where freezing temperatures occur, heavy insulation or heat tape will prevent the water in the cell piping from freezing. Also, liquid alum should be stored in an enclosed, warm space.

9.2M High operating filter differential pressures could occur if either (1) the media is filled with suspended material; or (2) excessive chemical feed is "blinding" the media.

Answers to questions on page 429.

9.2N Filter system performance tests should be conducted monthly.

9.2O Wear goggles when observing the operation of the surface wash arms.

9.2P Safety precautions that should be taken when working with alum or polymers include:

1. Wear safety goggles and gloves when working with alum or polymers. Flush away any alum or polymer that comes in contact with your skin with cool water for a few minutes. Wear approved respiratory protection if inhalation of dry alum or polymer is possible.
2. Be very careful when walking in an area where polymer mixing takes place. When a polymer is wet, it is very slippery.

ANSWERS TO QUESTIONS IN LESSON 3

Answers to questions on page 433.

9.3A Continuous backwash, upflow, deep-bed silica sand media filters continuously and simultaneously filter solids from the influent flow, while cleaning and recycling the filter media internally through an airlift pipe and sand washer. The cleaned sand is redistributed on top of the filter media bed, which allows for continuous uninterrupted flow of filtered effluent and backwash reject water.

9.3B Filter influent and effluent turbidity meters continuously monitor and record filter influent and effluent water quality.

9.3C Filter influent and effluent flows are metered to quantify and record flows and provide dosage control signals to metering pumps and disinfection chemical addition control systems.

Answers to questions on page 439.

9.3D Features to bypass the filtration system facilities should be provided to permit diverting treatment process effluent to holding or equalization basins during emergency conditions, such as extremely high flows, equipment failures, or clogged filters.

9.3E The purpose of adding chemical coagulants to the filter influent is to cause the suspended and colloidal (finely divided solids that will not settle by gravity) particles to clump together into bulkier hydroxyl precipitates.

9.3F The purpose of adding flocculant solution to the filter influent is to gather still more of the suspended and colloidal solids into larger (visible to the naked eye), stronger clumps (floc), which can be efficiently removed by filtration.

Answers to questions on page 446.

9.3G If the airlift is allowed to operate without water flow to the filter, no water will be available to the gravity sand washer-separator. This condition will allow dirty sand to be returned to the top of the sand bed.

9.3H If a tool is dropped into the filter, *IMMEDIATELY* stop the air flow to the airlift and remove the tool from the top of the sand bed.

9.3I Ferric chloride must be handled with the same care as acid solutions since ferric chloride causes burns similar to those caused by acids.

9.3J Before entering the plenum area of a filter, observe all confined space procedures and check for possible damage to the inner cone. In no event should the plenum area be entered if damage to the inner cone is suspected without first removing all sand from the filter.

ANSWERS TO QUESTIONS IN LESSON 4

Answers to questions on page 452.

9.4A In cross flow filtration, wastewater flows across the surface of a membrane rather than through a bed of granular media. The membrane permits water to pass through but blocks the passage of particles. Other differences include the length of the filter run and the ease of cleaning the membranes.

9.4B Membrane processes are classified on the basis of the size of particle they separate from the wastestream.

9.4C The basic membrane configurations are tubular, hollow fiber, spiral, plate and frame, and ceramic tube or monolith.

Answers to questions on page 456.

9.5A The amount of flux across a membrane is dependent upon transmembrane pressure (driving force), flow rate across the membrane surface (turbulence on the membrane), concentration of waste material, temperature, viscosity, and cleanliness of the membrane surface.

9.5B All membrane systems recirculate the waste feed across the membrane to prevent buildup of particles on the surface.

9.5C The delta pressure might increase when the viscosity of the feed is increasing or when material is accumulating in the membrane modules.

9.5D The surface of the membrane can become fouled with oil, grease, and other foulants.

Answers to questions on page 463.

9.6A For ultrafiltration systems processing primarily industrial waste, the following pretreatment processes are common:

1. Equalization of wastes.
2. Prescreening to remove large suspended material.
3. Free oil and settleable solids removal.
4. pH treatment (to alkaline conditions) to stabilize the oil/water emulsion.

9.6B In an oily waste system, pH control at the ultrafilter is important for two reasons. Flux rates are improved at the ultrafilter when the wastewater pH is 9 to 10.5 and fouling of the membrane by oil is minimized by stabilizing the oil emulsion.

9.6C The steps for cleaning (washing) a membrane are as follows:

1. Displacement of waste from the system with water.
2. Washing the membranes with a caustic and surfactant to remove oils and grease.
3. Flushing the surfactant from the membrane with warm water.
4. Washing the membrane with an acid cleaner to remove salt buildup.
5. Flushing the acid from the membrane with warm water.
6. Recording clean membrane water flux as a check on cleaning effectiveness.

9.6D Spongeballs are sometimes required when membrane foulants are difficult to remove, such as with metal hydroxides or waste latex. Spongeballs may also be used to speed up the cleaning procedure in some plants.

9.6E After cleaning, a water flux is measured to assess the effectiveness of the cleaning cycle.

Answers to questions on page 466.

9.8A The osmotic pressure of a solution is the difference in water level between the two sides of a membrane.

9.8B The two types of semipermeable membranes that are used most often for demineralization are cellulose acetate and thin film composites.

9.8C The water flux is the flow of water in grams per second through a membrane area of one square centimeter (or gallons per day per square foot) while the mineral flux is the flow of minerals in grams per second through a membrane area of one square centimeter.

9.8D When additional pressure is applied to the side of a membrane with a concentrated solution, the water flux (rate of water flow through the membrane) will increase, but the mineral flux (rate of flow of minerals) will remain constant.

9.8E When higher mineral concentrations occur in the feedwater, the mineral concentrations will increase in the permeate (or product water).

Answers to questions on page 469.

9.8F Water flux is usually expressed in gallons per day per square foot (or grams per second per square centimeter) of membrane surface.

9.8G The term "flux decline" is used to describe the loss of water flow through the membrane due to compaction plus fouling.

9.8H Mineral rejection is defined as:

$$\text{Rejection, \%} = \left(1 - \frac{\text{Product Concentration}}{\text{Feedwater Concentration}}\right)(100\%)$$

Mineral rejection can be determined by measuring the TDS and using the above equation. Rejections also may be calculated for individual constituents in the solution by using their concentrations.

9.8I Softening membranes operate at lower pressures and lower rejection rates than standard reverse osmosis membranes.

Answers to questions on page 474.

9.8J An increase in feedwater temperature will increase the water flux.

9.8K Hydrolysis of a membrane results in a lessening of mineral rejection capability.

9.8L Recovery is defined as the percentage of feed flow that is recovered as product water.

$$\text{Recovery, \%} = \frac{(\text{Product Flow})(100\%)}{\text{Feed Flow}}$$

9.8M Recovery rate is usually limited by (1) desired product water quality, and (2) the solubility limits of minerals in the brine.

9.8N Concentration polarization is a buildup of retained particles on the membrane surface due to dewatering of the feed closest to the membrane. The thickness of the concentration polarization layer is controlled by the flow velocity across the membrane.

CHAPTER 10

PHYSICAL TREATMENT PROCESSES
(AIR STRIPPING AND CARBON ADSORPTION)

by

Fred Edgecomb

TABLE OF CONTENTS

Chapter 10. PHYSICAL TREATMENT PROCESSES
(AIR STRIPPING AND CARBON ADSORPTION)

	Page
OBJECTIVES	486
WORDS	487

LESSON 1

10.0	AIR STRIPPING OF VOLATILE ORGANICS		489
	10.00	Purpose of Air Stripping	489
	10.01	Types of Air Stripping Systems	489
	10.02	Principles of Air Stripping	490
		10.020 Chemical Characteristics	490
		10.021 Effects of Air Pressure	490
		10.022 Effects of Temperature	490
	10.03	Air Stripping Equipment	491
	10.04	Methods for Controlling Discharge Contaminants	491
		10.040 Vapor Phase Activated Carbon	491
		10.041 Incineration	492
		10.042 Product Recovery	492
	10.05	Operation of Air Stripping Systems	492
		10.050 Start-Up	492
		10.051 Normal Operation	492
	10.06	Troubleshooting	493
	10.07	Preventive Maintenance	493
	10.08	Safety	494
DISCUSSION AND REVIEW QUESTIONS			494

LESSON 2

10.1	ACTIVATED CARBON ADSORPTION		495
	10.10	How Does Carbon Adsorption Work?	495
	10.11	Purpose of Carbon Adsorption	495
	10.12	The Manufacture of Activated Carbon	495
10.2	THE CARBON ADSORPTION PROCESS		496
	10.20	General Physical Principles	496

	10.21	Equipment Necessary for Carbon Adsorption	496
		10.210 Fixed Beds in Series	497
		10.211 Fixed Beds in Parallel	497
		10.212 Moving Bed Systems	497
	10.22	Pre-Start-Up	498
	10.23	Placing Carbon Adsorption Units Into Operation	500
	10.24	Adsorber Loading Guidelines	505
		10.240 Hydraulic Loading Rates	505
		10.241 Chemical Loading Rates	505
	10.25	Operating Procedures	505
	10.26	Shutting Down a Carbon Adsorption Unit	506
	10.27	Abnormal and Emergency Conditions	506
		10.270 Low COD Removal Efficiencies	506
		10.271 High Head Losses	507
		10.272 Fouling of Activated Carbon Granules	507
		10.273 Plugged Screens	507
		10.274 Collapsed Screens	507
		10.275 Air Pockets	507
		10.276 Deterioration of Tank and Pipe Coatings	507
		10.277 Failure of Upstream Processes	507
10.3	ACTIVATED CARBON REGENERATION	508
10.4	SAMPLING AND ANALYSIS	508
	10.40	Sampling	508
	10.41	Sampling Ports	508
	10.42	Analysis	508
10.5	SAFETY	509
	10.50	Carbon Transfer	509
	10.51	Working Inside a Carbon Adsorption Vessel	509
	10.52	Carbon Dust	509
	10.53	Excessive Pressures Within Carbon Contactor Tank	510
	10.54	Handling Spent Activated Carbon	510
	10.55	Disposal of Activated Carbon	510
10.6	REVIEW OF PLANS AND SPECIFICATIONS	510
	10.60	Unloading Station for Truck or Train Delivery of Fresh Activated Carbon	510
	10.61	Valving Placed at Easy-To-Reach Locations	510
	10.62	Dust Control for Unloading Fresh Carbon	510
	10.63	Proper Ventilation in Carbon Regeneration Furnace Room	511
	10.64	Scaffolding and Catwalks	511

	10.65	Warning Alarms and Signs	511
	10.66	Upstream Processes	511
10.7	ADDITIONAL READING		511
10.8	ACKNOWLEDGMENT		511

DISCUSSION AND REVIEW QUESTIONS ... 512

SUGGESTED ANSWERS ... 512

APPENDIX: ACTIVATED CARBON ADSORPTION LABORATORY PROCEDURES ... 515

OBJECTIVES

Chapter 10. PHYSICAL TREATMENT PROCESSES
(AIR STRIPPING AND CARBON ADSORPTION)

Following completion of Chapter 10, you should be able to:

AIR STRIPPING

1. Explain the purpose of air stripping in industrial wastewater treatment.
2. Describe how a countercurrent packed tower air stripper removes volatile organic compounds from water or wastewater.
3. List and describe three methods for controlling the stripped volatile organics discharged from an air stripper.
4. Safely start up, operate, and maintain an air stripping unit.
5. Perform routine maintenance on an air stripper.
6. Troubleshoot air stripper performance problems.

ACTIVATED CARBON ADSORPTION

1. Describe how activated carbon removes contaminants from a wastestream.
2. Start up, operate, and shut down a carbon adsorption unit.
3. Identify and correct abnormal operating conditions.
4. Operate and maintain an activated carbon adsorption system in a safe manner.
5. Review plans and specifications for a carbon adsorption system.

WORDS

Chapter 10. PHYSICAL TREATMENT PROCESSES

(AIR STRIPPING AND CARBON ADSORPTION)

BIOMASS (BUY-o-MASS)　　　　　　　　　　　　　　　　　　　　　　　　　　　　　　　　BIOMASS

A mass or clump of organic material consisting of living organisms feeding on wastes, dead organisms, and other debris. Also see ZOOGLEAL MASS and ZOOGLEAL MAT (FILM).

COD (pronounce as separate letters)　　　　　　　　　　　　　　　　　　　　　　　　　　　　　　　COD

Chemical Oxygen Demand. A measure of the oxygen-consuming capacity of organic matter present in wastewater. COD is expressed as the amount of oxygen consumed from a chemical oxidant in mg/L during a specific test. Results are not necessarily related to the biochemical oxygen demand (BOD) because the chemical oxidant may react with substances that bacteria do not stabilize.

CHELATING (KEY-LAY-ting) AGENT　　　　　　　　　　　　　　　　　　　　　　　　　　　CHELATING AGENT

A chemical used to prevent the precipitation of metals (such as copper).

EDUCTOR (e-DUCK-ter)　　　　　　　　　　　　　　　　　　　　　　　　　　　　　　　　EDUCTOR

A hydraulic device used to create a negative pressure (suction) by forcing a liquid through a restriction, such as a Venturi. An eductor or aspirator (the hydraulic device) may be used in the laboratory in place of a vacuum pump. As an injector, it is used to produce vacuum for chlorinators. Sometimes used instead of a suction pump.

MESH　　　MESH

One of the openings or spaces in a screen or woven fabric. The value of the mesh is usually given as the number of openings per inch. This value does not consider the diameter of the wire or fabric; therefore, the mesh number does not always have a definite relationship to the size of the hole.

ORGANIC WASTE　　　　　　　　　　　　　　　　　　　　　　　　　　　　　　　　　　ORGANIC WASTE

Waste material that may come from animal or plant sources. Natural organic wastes generally can be consumed by bacteria and other small organisms. Manufactured or synthetic organic wastes from metal finishing, chemical manufacturing, and petroleum industries may not normally be consumed by bacteria and other organisms. Also see INORGANIC WASTE and VOLATILE SOLIDS.

SLURRY　　　　　　　　　　　　　　　　　　　　　　　　　　　　　　　　　　　　　　　SLURRY

A watery mixture or suspension of insoluble (not dissolved) matter; a thin, watery mud or any substance resembling it (such as a grit slurry or a lime slurry).

TURBIDITY (ter-BID-it-tee)　　　　　　　　　　　　　　　　　　　　　　　　　　　　　　　　TURBIDITY

The cloudy appearance of water caused by the presence of suspended and colloidal matter. In the waterworks field, a turbidity measurement is used to indicate the clarity of water. Technically, turbidity is an optical property of the water based on the amount of light reflected by suspended particles. Turbidity cannot be directly equated to suspended solids because white particles reflect more light than dark-colored particles and many small particles will reflect more light than an equivalent large particle.

VOID　　　VOID

A pore or open space in rock, soil, or other granular material, not occupied by solid matter. The pore or open space may be occupied by air, water, or other gaseous or liquid material. Also called an INTERSTICE, PORE, or void space.

VOLATILE (VOL-uh-tull)   VOLATILE

(1) A volatile substance is one that is capable of being evaporated or changed to a vapor at relatively low temperatures. Volatile substances can be partially removed from water or wastewater by the air stripping process.

(2) In terms of solids analysis, volatile refers to materials lost (including most organic matter) upon ignition in a muffle furnace for 60 minutes at 550°C (1,022°F). Natural volatile materials are chemical substances usually of animal or plant origin. Manufactured or synthetic volatile materials, such as plastics, ether, acetone, and carbon tetrachloride, are highly volatile and not of plant or animal origin. Also see NONVOLATILE MATTER.

CHAPTER 10. PHYSICAL TREATMENT PROCESSES

(AIR STRIPPING AND CARBON ADSORPTION)

(Lesson 1 of 2 Lessons)

10.0 AIR STRIPPING OF VOLATILE[1] ORGANICS[2]

10.00 Purpose of Air Stripping

The purpose of air stripping in industrial wastewater treatment facilities is to remove volatile organic compounds (VOCs) (and sometimes ammonia) from the plant effluent. Where not prohibited by air management regulations, the stripped volatile organics are released directly into the atmosphere. However, in most industrial applications of air stripping, the organic vapors stripped from wastewater normally cannot be discharged directly. Instead, they must be directed to an additional treatment system before release. Alternatives for treating air stripper discharges include vapor phase activated carbon, incineration, and product recovery processes.

In many parts of the country there are areas in which the groundwater has been contaminated by volatile organics. Drinking water quality standards require even slight contamination to be cleaned up to make the water potable, that is, acceptable for drinking. For these areas, air stripping of the volatile organic material is a cost-effective practice.

Lesson 2 of this chapter describes another method for removing volatile organics from plant effluent streams: adsorption using activated carbon. One advantage of air stripping versus the use of activated carbon is that no wastestream of spent activated carbon is produced. Spent carbon must be regenerated or disposed of and replaced. For this reason and others, air stripping is seen as an economical alternative to the use of activated carbon in some facilities. Other facilities operate both processes but use air stripping as a way to reduce the use of activated carbon.

10.01 Types of Air Stripping Systems

A wide variety of devices may be used to air strip VOCs from plant process water. These devices include diffused aeration, the coke tray aerator, the countercurrent packed tower, and the cross flow tower.

In diffused aeration, air is bubbled through the contaminated water by a diffuser. This process is similar to the activated sludge aeration process.

A coke tray aerator is a simple, low-maintenance design without air blowers. The water being treated trickles through several layers of trays producing a large surface area for gas transfer. Coke tray aerators are fairly popular for the removal of hydrogen sulfide, iron, and manganese. Recent improvements in the design of these units have led to their use in VOC removal processes.

In countercurrent packed towers (Figure 10.1), packing materials are used to provide high *VOID*[3] space and high surface

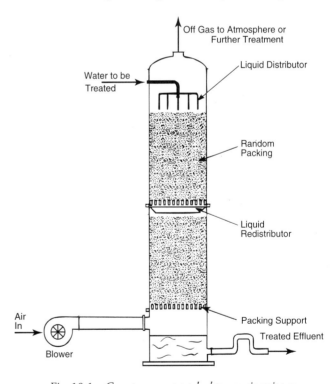

Fig. 10.1 Countercurrent packed tower air stripper

---

[1] *Volatile* (VOL-uh-tull). (1) A volatile substance is one that is capable of being evaporated or changed to a vapor at relatively low temperatures. Volatile substances can be partially removed from water or wastewater by the air stripping process. (2) In terms of solids analysis, volatile refers to materials lost (including most organic matter) upon ignition in a muffle furnace for 60 minutes at 550°C (1,022°F). Natural volatile materials are chemical substances usually of animal or plant origin. Manufactured or synthetic volatile materials, such as plastics, ether, acetone, and carbon tetrachloride, are highly volatile and not of plant or animal origin. Also see NONVOLATILE MATTER.

[2] *Organic Waste.* Waste material that may come from animal or plant sources. Natural organic wastes generally can be consumed by bacteria and other small organisms. Manufactured or synthetic organic wastes from metal finishing, chemical manufacturing, and petroleum industries may not normally be consumed by bacteria and other organisms. Also see INORGANIC WASTE and VOLATILE SOLIDS.

[3] *Void.* A pore or open space in rock, soil, or other granular material, not occupied by solid matter. The pore or open space may be occupied by air, water, or other gaseous or liquid material. Also called an INTERSTICE, PORE, or void space.

area. The untreated water enters the tower through nozzles located at the top and flows downward by gravity. At the same time, air is blown upward from the bottom of the tower.

In a cross flow tower, the water flows down through the packing as in other packed towers, but the air is pulled into the tower from the sides. Cross flow towers have a lower profile than countercurrent towers and are sometimes less costly to build and operate. Cross flow towers are widely used as cooling towers but are not commonly used for air stripping.

Countercurrent air stripping towers are by far the most common type of stripper used for VOC removal. For this reason, the remainder of this section will discuss the countercurrent packed tower type of air stripper.

10.02 Principles of Air Stripping

The mechanism of air stripping of organics can be simply thought of as one of evaporation. All liquids have a tendency to go from the liquid phase into the gaseous phase. In a closed container, a chemical (water, for example) will vaporize, transferring some water molecules into the air in the container. The amount of liquid transferred to the air depends on the properties of the liquid chemical, the air pressure, and the temperature.

The purpose of a countercurrent packed bed contactor is to bring the chemicals in the water into contact with air so that they can evaporate into the air stream. The packing in the column can be thought of as acting somewhat like a waterfall. As the water falls, it forms sheets and droplets that have a lot of surface area. In a waterfall, the water itself evaporates into the air. In a waterfall, the force of air on the falling water and the surface tension of the water itself cause it to form sheets and droplets. In a packed bed contactor, the water continually falls on the surfaces of the packing material and forms and reforms new sheets and droplets of water. Air is blown up through the packing and flows past the droplets and sheets of water evaporating both the chemicals and some water from the water surfaces. The efficiency of a packed bed contactor is attributable to the continuous regeneration of new sheets and droplets forming new surfaces from which the chemicals can evaporate.

10.020 *Chemical Characteristics*

Each chemical has a different tendency to transfer from the liquid to the gaseous phase. This tendency is measured by a property called vapor pressure. The higher a chemical's vapor pressure, the greater is its tendency to transfer from the liquid to the gaseous phase.

The solubility of a chemical (its ability to dissolve in water) has a very strong effect on the vapor pressure of the chemical when it is present in a mixture. For example, acetone is a volatile chemical that evaporates very rapidly when exposed to air; it evaporates much more rapidly than water. When water and acetone are mixed together, however, the vapor pressure of acetone approaches that of water. This is due to the complete solubility of acetone in water. Another compound that is very volatile is trichloroethylene (TCE). However, it is practically insoluble in water. This compound is very easy to remove from water by air stripping, whereas a chemical such as acetone is difficult to remove.

Nonvolatile organic compounds such as phenols and ketones are soluble in water and, therefore, are not readily removed by air stripping. Granular activated carbon is effective in removing nonvolatile organic compounds from the water.

10.021 *Effects of Air Pressure*

In a countercurrent packed bed contactor, the clean air is blown into the bottom of the column and the contaminated water enters at the top. The rate (or speed) with which a chemical will vaporize is related to how close the air is to being saturated with the chemical. For example, on a very humid day, water evaporates very slowly because the air already contains almost all of the water vapor it can hold. On a very hot, dry day, water evaporates quickly because there is very little water in the air. Thus, the "driving force" for evaporation is greater on the hot, dry day.

Air is blown into the bottom of the column to take advantage of this driving-force principle. By the time the contaminated water reaches the bottom of the column, most of the contaminants will have been stripped out. Therefore, the driving force for the contaminants will be greatest at the bottom of the column where clean air is introduced without any of the chemicals in it. As the air flows up through the column, it picks up the chemicals from the water and the driving force for evaporation decreases the farther up the column the air goes. If an excess of air is used, the air will not become saturated with the chemicals and the contaminants will be transferred from the water to the air stream throughout the length of the column.

10.022 *Effects of Temperature*

Most VOCs are relatively insoluble in water and are generally easy to remove. However, because of the very high removal efficiencies required, it is sometimes necessary to raise the water temperature. The increased temperature raises the volatility of the contaminants and the water alike. Since the volatility is increased, the amount of contaminant removed by a certain amount of air is also increased. The amount of water evaporated also increases dramatically at the elevated temperature. The balance between the amount of air used in the stripping process and the temperature of operation of the stripper is selected to minimize the quantity of water vapor evaporated while still removing the required amount of contaminants.

Computer aided design is used to identify the best conditions at which to run the stripper. The results of this computer modeling show the approximate CFM (Cubic Feet of air per Minute) of air along with an inlet water temperature to determine the optimum conditions for a given air to water inlet flow rate ratio.

QUESTIONS

Write your answers in a notebook and then compare your answers with those on page 512.

10.0A Why is air stripping considered an economical means of controlling volatile organic compounds (VOCs)?

10.0B How does the solubility of chemicals in a wastestream influence the effectiveness of the air stripping process?

10.03 Air Stripping Equipment

AIR STRIPPING TOWER

Air stripping towers are constructed from fiberglass, polypropylene, or other corrosion-resistant materials. Spray nozzles at the top of the tower distribute water over the top of the packing. The water flows through the packing and into the wet well. The air flow enters the tower at the bottom of the packing and flows upward through the packing. A demister pad is commonly installed above the spray nozzles. These pads are designed to prevent moisture in the air stream from flowing through the ducting to the atmosphere or downstream treatment processes.

A wet well located at the base of the air stripping tower receives the treated water and provides for surge capacity. The wet well is normally equipped with a level indicator/transmitter, which sends a signal to a level controller. The level controller controls the operation of the pumps, which pump the treated water to the next process or to the effluent receiving location.

PACKING MATERIAL

The packing material used in air strippers is often made of plastic but many other materials, such as stainless steel, are also used. A wide variety of different shapes and sizes are available. Normally, the packing in a stripper consists of a random assortment of pieces ranging from marble size to softball size. The individual pieces may be virtually any shape. Swirls and cylinder-shaped packing are commonly used, but spheres, squares, and rectangles are also widely available.

AIR BLOWERS

Air blowers used in the air stripping process are pressure blowers. They may be of the lobe, vane, or centrifuge type. One blower is normally provided for routine service while another serves as a standby. A blower air pressure switch, located on the common blower discharge header, detects a low air blower discharge pressure in the header.

FLOW INSTRUMENTATION

Flow instrumentation should be provided for both the water flow and the air flow. This allows the operating guidelines to be monitored and adjusted as needed to operate the process at the maximum efficiency and the lowest cost. A differential pressure gauge is also necessary to determine the air pressure drop across the air stripper. This will provide an early indication of when the air stripper media is becoming fouled.

10.04 Methods for Controlling Discharge Contaminants

In most industrial applications, it is necessary to control the discharge of stripped VOCs to the atmosphere. Vapor phase carbon, incineration, and product recovery are three processes commonly used to control these discharges.

10.040 Vapor Phase Activated Carbon

Vapor phase carbon refers to the use of activated carbon to control the release of hazardous materials into the atmosphere. The carbon used for this application is virtually identical to the carbon used in liquid phase operations. Vapor phase carbon is widely used in filter systems for respirators to remove hazardous airborne materials when it is necessary to work in a hazardous atmosphere.

One advantage of using vapor phase carbon is that it can be impregnated with chemicals to enhance the removal of specific contaminants from an air stream. By impregnating the carbon with a caustic material, such as sodium hydroxide or potassium hydroxide, acidic materials, such as hydrogen sulfide, can be removed from the air stream. When the caustic material has been neutralized, the carbon can be chemically reactivated by soaking it in a caustic solution for a given time period.

Vapor phase carbon systems that are used to remove volatile substances from air streams are sometimes designed to be regenerated in place through the use of steam. The steam strips the organics from the carbon. After stripping, the organics are either destroyed by incineration or recovered for reuse.

Pretreatment for vapor phase systems consists of reducing the humidity of the air stream to the lowest degree possible. Humidity adversely affects the operation of vapor phase systems.

10.041 Incineration

Another option for the control of VOCs from air strippers is to incinerate the vapors. This option is selected when the concentration of VOCs in the off gas is high enough that the fuel value of the VOCs becomes significant. Modern incinerators or combusters are designed to destroy a minimum of 99.99 percent of the volatile organics removed by the air stripper and then recover the maximum amount of heat possible.

The combuster system raises the air stripper off gas temperature to a minimum of 1,800°F (980°C) and maintains that temperature for a minimum of two seconds. These conditions will guarantee 99.99 percent destruction of the volatile contamination. The contaminants contain chlorine and are difficult to burn, so high temperatures are required to ensure that the contaminants are burned completely, forming carbon dioxide, water, and hydrochloric acid. If the temperature is decreased or the holding time is decreased, the contaminants might not burn completely. Incomplete combustion of the contaminants could result in the discharge of volatile organics to the atmosphere.

The heat from the combuster can be recovered and used to preheat the water to improve the air stripping process.

10.042 Product Recovery

Product recovery systems are designed to recover VOCs from processes in which they are lost to wash water or enter the liquid stream by accidental means. Activated carbon is sometimes used in the vapor phase to trap the VOCs. Steam is then used to purge the VOCs from the carbon for recovery.

QUESTIONS

Write your answers in a notebook and then compare your answers with those on page 512.

10.0C Why are demister pads installed in air strippers?

10.0D What is the purpose of having vapor phase carbon on the discharge of an air stripper?

10.0E Why would you want to recover the heat from an incinerator system?

10.05 Operation of Air Stripping Systems

10.050 Start-Up

1. Check each piece of equipment to be used.

    Plant log sheets will indicate which equipment was being used when the plant was last operated and the status of that equipment. If it is time to rotate equipment, this is a good time to switch equipment from the lead to standby position. Mechanical equipment, instruments, and electrical switches should be checked for repair or warning tags.

2. Check valve and electrical lineups.

    Ensure equipment and valve positions are correct as indicated in the manufacturer's O & M manual or Standard Operating Procedures (SOP).

3. Check plant effluent or extraction well operation flow set points based on the latest plant operational or hydraulic data.

After completing the pre-start checks, start the air flow through the stripper and then start the water flow to the top of the air stripper. Next, adjust the amount of air flowing to the stripping tower. The design air/water ratio should be maintained at all times. If the flow must vary from the set point, consult the manufacturer or engineering support personnel for advice as soon as possible.

A greater volume of air will result in better removal of VOCs. However, the greater the air flow through the stripper, the less time it will take the air stream to pass through the stripper. Take care not to exceed the design air flow rate specified by the manufacturer.

Changes in the influent temperature will affect the operation of the air stripper. A temperature drop of as little as 10°F (6°C) below the design influent temperature may severely reduce the performance of the air stripper. Low temperature may be compensated for by either increasing the air flow through the stripper to the maximum allowable or, if possible, recycling water flow to the air stripper.

With these steps completed, the system is ready for normal operations.

10.051 Normal Operation

Normal operation of the air stripping system will consist of the following routine checks:

1. Check the control panel for alarms or other unusual conditions.

2. Check the stripper feed pump's operation and flow set point.

3. Check to verify that the wet well level control valve is operating correctly. If not, take manual control of the valve and adjust it until the level is acceptable. Continue to control the valve manually until repairs can be made.

4. Check the blower system for accurate air flow, and look for any frozen (stuck) blower vanes or loose vane adjusters.

5. Take all required samples and review sampling results.

6. Check the differential pressure across the air stripper. A high differential pressure indicates fouling of the packing media.

10.06 Troubleshooting

When an air stripper is not working properly, the most common symptoms are likely to be a high differential pressure reading or low removal efficiencies. Most problems in the operation of air strippers relate to the scaling or biological fouling of the stripper packing material. This fouling is normally caused by the oxidation of dissolved iron and calcium, and the stimulation of biological activity. Field studies indicate that there are three main causes of scaling and fouling in air stripping systems: carbonate scaling, iron oxidation, and microbial action leading to iron biofouling.

The scaling of packing from carbonate (mainly calcium carbonate) is caused by the pressure changes in water that cause carbon dioxide to be stripped from the water. When water enters the packed section of the tower, there is a substantial drop in air pressure, which causes the stripping of the carbon dioxide and the formation of calcium carbonate. Calcium carbonate deposits within the packing material usually form a brittle, cement-like scale.

The presence of iron in the water results in the oxidation of dissolved iron to form insoluble iron compounds. These compounds take the form of a rust-colored, shapeless mass or a black-colored gel.

Microbial activity by iron bacteria is normally found when treating groundwater. Under aerobic conditions found in the air stripper, and with iron to feed on, the population of these organisms increases dramatically. The *BIOMASS*[4] and slimy sheaths of these bacteria have a clogging effect and are difficult to remove.

The best way to prevent the fouling of packing media is to start with a media that resists fouling. Since this is not always known or possible, there are some preventive measures that will help keep scaling to a minimum:

1. Pretreat the water being air stripped by chemical or physical means including preaeration, oxidation, and filtering to remove oxidized solids.

2. Add *CHELATING AGENTS*,[5] such as hexametaphosphate or phosphonates, to inhibit the formation of oxidation products.

3. Periodically clean the unit with acids and disinfectants to remove the scale and biofouling.

10.07 Preventive Maintenance

Preventive maintenance for an air stripping column involves only two major activities: (1) visually inspect the packing material on a monthly basis, and (2) monitor the pressure drop within the packing section on a daily basis. The purpose of a visual inspection is to determine when an acid wash of the packing media is necessary to remove biological growths or solid precipitate deposits. If brown or red discoloration is visible on the packing, then further inspection of the media is required. Remove the top flange and inspect the packing media. Also, inspect the distributor to check for openings blocked by solid precipitates or biological growths. In addition, inspect all pipe flanges and connections for leaks.

The results of laboratory tests on the air stripper effluent may provide an indication of when an air stripping column requires an acid wash. If volatile organics in the treated water increase above the baseline established by past operation, a wash cycle is probably required.

Other inspection and maintenance tasks associated with air stripping towers should be performed at the intervals listed below.

Weekly

1. Check spray nozzle distribution pattern. It may be necessary to turn the blower off while checking.

2. Check blower equipment and pumps by checking operating temperature, listening for noise, and checking valve positions for normal position.

Monthly

1. Inspect packing media.
2. Lubricate fans and pumps.
3. Check blower belts for wear and proper tension.
4. Check the calibration on the differential pressure meter.
5. Calibrate the wet well level indicator.
6. Calibrate the influent water and air flowmeters.

Annually

1. Conduct a thorough inspection of the stripper for air leaks.
2. Monitor the air in the area of the stripper for the release of volatile substances.

---

[4] *Biomass* (BUY-o-MASS). A mass or clump of organic material consisting of living organisms feeding on wastes, dead organisms, and other debris. Also see ZOOGLEAL MASS and ZOOGLEAL MAT (FILM).

[5] *Chelating* (KEY-LAY-ting) *Agent*. A chemical used to prevent the precipitation of metals (such as copper).

3. Clean spray nozzles and distribution lines.
4. Check air ducts, plenums, fan housings, and mist eliminators for corrosion and leakage.
5. Clean sump as needed. Use approved confined space entry procedures if you must enter a sump.

10.08 Safety

The key to working safely around an air stripper is knowing what substances the wastewater contains. Many of the volatile organic compounds an air stripper is designed to remove are cancer-causing agents. Operators should wear respiratory equipment approved for the types of vapors that are likely to be present when investigating leaks or performing maintenance tasks that could expose them to vapors.

Cleaning the sump of an air stripper exposes an operator to the hazards associated with confined space work. Above all, remember that no one should ever work alone in a confined space. Review the discussion of confined space hazards in Chapter 13, "Safety," and follow your facility's confined space procedures.

QUESTIONS

Write your answers in a notebook and then compare your answers with those on page 512.

10.0F What is the probable cause of a high differential pressure across an air stripper?

10.0G List three ways to prevent the fouling of air stripper media.

END OF LESSON 1 OF 2 LESSONS

on

PHYSICAL TREATMENT PROCESSES

Please answer the discussion and review questions next.

DISCUSSION AND REVIEW QUESTIONS

Chapter 10. PHYSICAL TREATMENT PROCESSES

(AIR STRIPPING AND CARBON ADSORPTION)

(Lesson 1 of 2 Lessons)

At the end of each lesson in this chapter you will find some discussion and review questions. The purpose of these questions is to indicate to you how well you understand the material in the lesson. Write the answers to these questions in your notebook.

1. What are the alternative treatment methods for treating discharges of stripped organics from an industrial air stripping process?

2. What are the most common symptoms of operational problems in an air stripper?

3. How can the fouling of packing media be prevented?

4. How can an operator determine if an air stripping column requires an acid wash?

CHAPTER 10. PHYSICAL TREATMENT PROCESSES

(AIR STRIPPING AND CARBON ADSORPTION)

(Lesson 2 of 2 Lessons)

10.1 ACTIVATED CARBON ADSORPTION

10.10 How Does Carbon Adsorption Work?

Activated carbon is a material of remarkably high surface area relative to its volume. This surface area results from the activation process, which produces millions of pores in the base material. Activated carbon can remove trace amounts of organic contaminants from air and liquid wastestreams by the attraction and accumulation of these substances onto the surface of the carbon. This is the definition of adsorption. Particles adhere (stick) to the surface of the activated carbon very much like a magnet attracts particles of iron. As wastewater passes through a field of activated carbon, contaminants are adsorbed onto the carbon's surface and internal pore spaces. This process will continue until the activated carbon can no longer adsorb the material. The carbon is then said to be exhausted or "spent."

The term ADSORPTION should not be confused with the similar sounding term ABSORPTION. Adsorption refers to the gathering or accumulation of one substance on the surface of another substance, whereas absorption refers to the incorporation of one substance into the interior of another, such as the absorption of water into a sponge.

10.11 Purpose of Carbon Adsorption

The use of carbon for the purification of water and chemical substances dates to Ancient Egypt where carbon was used for the purification of oils and for medicinal purposes. By the early 19th century, both wood and bone charcoal were in large scale use for the decolorization and purification of cane sugar.

The development of chemical warfare in World War I led to the development of activated carbon as a filtering media to protect soldiers from these new toxic gases. By the late 1930s, activated carbon was being used extensively for industrial purposes.

The purpose of activated carbon adsorption in industrial wastewater plants is to remove organic pollutants from the industrial wastewater and from stripping tower off gases. The advent of toxicity-based National Pollution Discharge Elimination System (NPDES) permit regulations and local sewer-use ordinances has made activated carbon a necessary part of the treatment process at many industrial facilities.

Activated carbon is used by many industries to achieve this toxicity reduction. In the petroleum industry, activated carbon is used to reduce concentrations of phenols, heavy metals, and volatile and semi-volatile organics. In the chemical industry, activated carbon is used to remove traces of products, such as pesticides, alcohols, and solvents, from the water leaving the plant. Uses in the textile industry include the removal of dyes, colors, and total organic carbon.

The use of activated carbon for foul-air scrubbing is discussed in Chapter 1, "Odor Control," *ADVANCED WASTE TREATMENT*, in this series of manuals. Use of carbon for the removal of organic pollutants from air streams is discussed in Section 10.040 of this chapter, "Vapor Phase Activated Carbon."

10.12 The Manufacture of Activated Carbon

Activated carbon can be produced from wood, coal, nutshells, bone, petroleum residues, or even sawdust. The raw material used to make activated carbon must be a carbonaceous (containing carbon) material. The material must also be available in large quantities as ten tons of raw material will yield only about one ton of good quality, usable activated carbon. The base material used directly influences the quality of the finished product. Activated carbon manufactured from coconut shells is generally regarded as the best carbon available for most purposes, while that manufactured from sawdust is generally of low quality.

Carbon is activated by drying the raw material and slowly heating it to a high temperature. The amount of air is controlled to keep the base material from burning. Extreme heat is required to burn off the remaining residue to produce the activated carbon. It is then exposed to an oxidizing gas at a high temperature. This gas develops a porous structure in the carbon creating a large internal surface area.

QUESTIONS

Write your answers in a notebook and then compare your answers with those on pages 512 and 513.

10.1A   Why is the activated carbon adsorption process used to treat industrial wastewaters?

10.1B   What materials may be used to make activated carbon?

10.1C   How is activated carbon made?

10.2 THE CARBON ADSORPTION PROCESS

10.20 General Physical Principles

During the production of activated carbon, the carbon particles may be put into contact with carbon dioxide or steam while the carbon particles are still very hot. The result is a fracturing of the carbon to produce small cracks or fissures within each of the small particles. These fissures are used in the attraction and adsorption process.

The surface properties of activated carbon are a function of both the initial material used and the exact preparation procedure, so many variations are possible. The two size classifications generally used in wastewater treatment are powdered (diameter of less than 200 *MESH*[6]) and granular (diameter greater than 0.1 mm).

Powdered activated carbon (PAC) is used to enhance the operation of chemical or biological treatment systems. The powdered carbon is mixed with water to form a *SLURRY*,[7] which can be added to the rapid mix of chemical treatment systems or to the aeration basin of an activated sludge system. In activated sludge systems, PAC addition provides for adsorption of organic substances that resist biological treatment or are harmful to the bacteria, thereby allowing the survival of nitrifying bacteria. PAC also increases the settling velocity and thickening characteristics of the suspended solids.

Granular activated carbons (GACs) typically have surface areas of 500 to 1,400 square meters per gram (a carbon granule the size of a quarter has a surface area of approximately $1/2$ square mile). Molecules of organic materials that pass by the carbon particles are attracted to and held on the surface of the carbon. Some smaller molecules may find their way into the pores or the fissures (cracks) that were produced during the quenching or cooling cycle of the carbon manufacturing process. Other particles may be trapped on the surface or wedge themselves into the larger part of the fissure opening at the surface of the carbon particles.

The ability of activated carbon to adsorb a given contaminant depends on the chemical nature of the contaminant, the pH of the water, and the contact time allowed for the contaminant to be adsorbed by the carbon. The chemical composition of a compound refers to its molecular weight (the size of its molecules) and its electrical charge. In general terms, the higher the molecular weight of a substance and the lower its electrical charge, the better it will be adsorbed by the carbon. Organics such as methanol, which have both a high electrical charge and a low molecular weight, adsorb very poorly and may be "bumped off" the surface of the carbon by higher weight organics such as benzene. There is an upper limit of molecular weight above which the adsorption is adversely affected. At some point the molecules will become too large to fit into the available space in the carbon. The polymer molecule is an example of a large molecule that is not easily adsorbed by carbon.

The pH of the water and the contaminant's solubility also affect its adsorbability.

How well any substance or mixture of substances is adsorbed onto activated carbon is determined by conducting a series of laboratory tests. The results of these tests are used to produce a curve referred to as an "isotherm." The isotherm for a particular mixture of contaminants will predict how long the carbon's ability to adsorb the contaminant will last at a constant flow rate, pressure, and temperature.

The isotherm also identifies the minimum contact time required for the adsorption of a given contaminant. Contact times for industrial wastewaters average around 60 minutes. This is twice the time required for municipal wastewater treatment plant tertiary effluent. Contact time can range from as little as 15 minutes to as long as 150 minutes for industrial wastewaters.

Several tests are used to characterize the adsorptive capacity of activated carbon. The phenol number is used as an index of a carbon's ability to remove taste and odor compounds. The iodine number relates to the ability of the activated carbon to adsorb low molecular weight substances (micropores having an effective diameter of less than $4 \mu m$[8]). The molasses number relates to the carbon's ability to adsorb high molecular weight substances (pores ranging from 1 to 50 μm). In general, high iodine number carbons will be most effective for wastewaters with mostly high molecular weight organics.

Newly manufactured carbon is referred to as virgin carbon. Carbon that has been reactivated in a furnace or kiln is referred to as regenerated carbon.

10.21 Equipment Necessary for Carbon Adsorption

The most commonly used equipment for carbon adsorption is a pressure tank or a carbon column (Figure 10.2) filled with activated carbon. The pressure tanks are also known as adsorbers; they contain the carbon for the adsorption process. The tanks are normally 8 to 15 feet (2.4 to 4.5 m) in diameter and can be as tall as 60 feet or, for small treatment systems, as small as 4 feet. Inside the containers are screens that keep the carbon within the pressure vessel and prevent it from flowing in either direction with the water.

Carbon adsorbers are lined with an epoxy or resin to prevent corrosion from the contact of the carbon with the surface of the vessel. Wet carbon can be very corrosive and the lining protects the steel pressure vessel.

Generally, systems used for industrial treatment are a variation of one of three designs: fixed beds in series, fixed beds in parallel, or moving bed systems (Figure 10.3).

---

[6] *Mesh.* One of the openings or spaces in a screen or woven fabric. The value of the mesh is usually given as the number of openings per inch. This value does not consider the diameter of the wire or fabric; therefore, the mesh number does not always have a definite relationship to the size of the hole.

[7] *Slurry.* A watery mixture or suspension of insoluble (not dissolved) matter; a thin, watery mud or any substance resembling it (such as a grit slurry or a lime slurry).

[8] μm = micrometer = one millionth of a meter or = 1 micron; $4 \mu m$ = 4 millionths of a meter.

Fig. 10.2   Carbon columns

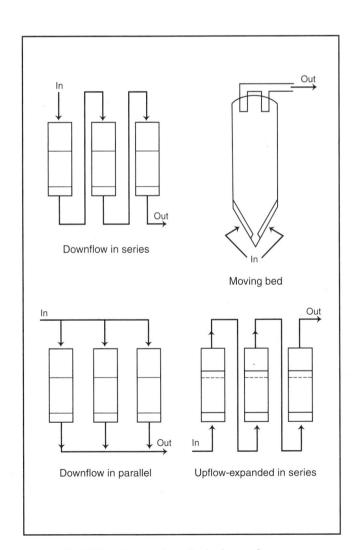

Fig. 10.3   Commonly used adsorber configurations

All of the systems operate on the same general principle of countercurrent flow. That is, wastewater being treated passes through the oldest and dirtiest carbon first. As the water moves through the column, it is treated by successively cleaner activated carbon, which has a higher adsorption capacity. In this way, the water is "polished" as it nears the effluent end of the column.

10.210   Fixed Beds in Series

A continuous flow adsorption system may consist of two or more upflow or downflow fixed bed adsorbers arranged in series. The first or lead adsorber in the series receives the influent wastewater and removes the highest concentration of contaminants. Effluent from the first vessel is then passed on to the successive number of units needed to provide the total carbon depth required to meet the discharge requirements. Once the lead vessel cannot adsorb any more contaminants, it is taken out of service while the spent carbon is replaced. In some systems, a spare vessel is provided to contain the spent carbon until the replacement carbon can be loaded. After refilling, the recharged unit is placed back in service at the end of the series where it is sequentially advanced to the lead position again. This arrangement allows continuous service to consistently meet effluent limitations.

10.211   Fixed Beds in Parallel

Fixed beds can also be operated in parallel. The mode of operation can be upflow or downflow.

10.212   Moving Bed Systems

The moving bed system is one of the earliest systems used and is often more economical and efficient in large plants where large volumes of carbon must be handled on a regular basis. In a moving bed adsorber, water enters the bottom of the vessel and flows up through the carbon bed. Treated water is then removed from the top of the adsorber. In this type of system, the highest concentration of contaminants is adsorbed by the carbon at the bottom of the vessel. To make recharging easier, moving bed contactors are constructed with a cone-shaped bottom that enables rapid discharge of slugs of spent carbon by gravity. When a slug of spent carbon is discharged, the rest of the carbon bed shifts downward and the bed can be recharged by adding fresh carbon to the top of the bed. Also, flow is easily regulated in the upflow mode and it is possible to fill the carbon contactors with new granular carbon even while flow continues to the vessel.

10.22 Pre-Start-Up

1. Specific activities to complete before operating the adsorption equipment should include the following:

   a. Check all piping connections for proper installation and tightness.

   b. Ensure that all gauges and instruments are functional and installed correctly.

   c. Close the valves in the lines around the adsorbers.

   When the preliminary steps are completed, the adsorbers are ready to be filled with granular activated carbon.

2. Transfer of Carbon Into the Adsorbers. The means of transferring activated carbon into the adsorber will vary with the size of the system. In smaller systems, carbon is loaded from bags directly into the adsorber tank. Activated carbon is available in a variety of containers ranging from 50-pound bags to 1,000-pound "super sacks." A small amount of water is added to the adsorber to soften the impact of the carbon on the bottom of the adsorber and to help control dust.

   In medium size systems (10,000- to 20,000-pound capacity), carbon may be delivered by service trailers. These trailers are pulled by large trucks and contain up to 20,000 pounds of carbon. The carbon is unloaded directly into the adsorbers. The carbon is transferred from the service trailers to the adsorber by pressurizing the trailers with air and water systems and moving the carbon as a slurry through flexible hoses into the empty adsorber. To transfer the spent carbon from the adsorber to the empty service trailer, the adsorber is pressurized and the slurry is transferred through a flexible hose to the service trailer.

   Larger systems (20,000 to 50,000 pounds) normally transfer carbon directly from bulk trucks into the adsorbers using an *EDUCTOR*.[9]

   After the desired amount of carbon has been unloaded into the adsorber, check the level of carbon by measuring from the top of the carbon adsorber to the surface of the granular carbon within the pressure vessel.

   When the carbon level is correct, begin a wash cycle to remove carbon fines. Fines are the tiny particles of carbon that have broken off the granular material during shipping and loading. The flow should continue until the fines can no longer be seen in the wash water with the naked eye.

3. Valves and Piping. The standard schematic diagram illustrating piping is shown in Figure 10.4. Valving and piping in and around a carbon vessel must be flexible. Three-way valves are usually installed in order to provide the option for both upflow and downflow conditions as necessary for the activated carbon treatment, and also to allow for backwashing of filter screens. Before placing the carbon vessel on line, the valving should be checked to ensure that the flow direction is correct for the operation desired. The operator should know where each pipe leads and the direction of flow through the pipe that is desired.

   The influent piping around an upflow carbon adsorber is at the bottom of the pressure vessel (see Figure 10.4). Valves are usually located at the bottom of the vessel to allow the normal upflow conditions through the carbon. The valve allows flow to be equally distributed into ports, which are protected by screens within the container itself. In this manner, a uniform flow is usually achieved through the carbon for proper adsorption.

   The effluent piping of an upflow carbon adsorber is located at the top of the container (Figures 10.5 and 10.6). As with the influent piping, screens are located on the ends of several ports that discharge effluent from the top of the container. Uniform flow is achieved throughout the vessel so that short-circuiting is prevented by having a number of ports in the bottom as well as the top of the effluent side of the carbon adsorber. Figure 10.7 is a detail of the combination pressure air and air-vacuum relief valve shown in Figure 10.5

   The piping is versatile in that flow can be up (Figure 10.8) or down (Figure 10.9) through the carbon adsorber. In this manner, the screens that hold the carbon within the vessel can be backwashed to prevent clogging with carbon fines or the small particles of carbon that stick to the surface of the screens.

QUESTIONS

Write your answers in a notebook and then compare your answers with those on page 513.

10.2A   What is the average contact time in an activated carbon pressure vessel treating industrial wastewaters?

10.2B   What is the countercurrent flow principle?

10.2C   Why are three-way valves used in activated carbon piping?

10.2D   How is activated carbon kept in the pressure vessel?

---

[9] *Eductor* (e-DUCK-ter).   A hydraulic device used to create a negative pressure (suction) by forcing a liquid through a restriction, such as a Venturi. An eductor or aspirator (the hydraulic device) may be used in the laboratory in place of a vacuum pump. As an injector, it is used to produce vacuum for chlorinators. Sometimes used instead of a suction pump.

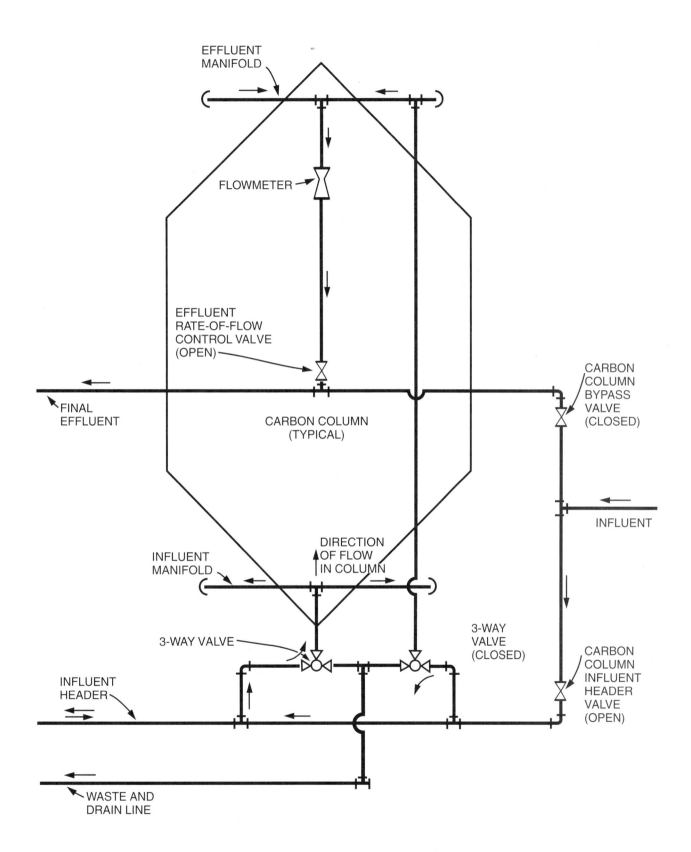

*Fig. 10.4 Upflow carbon column schematic, normal operation
Orange County, CA*

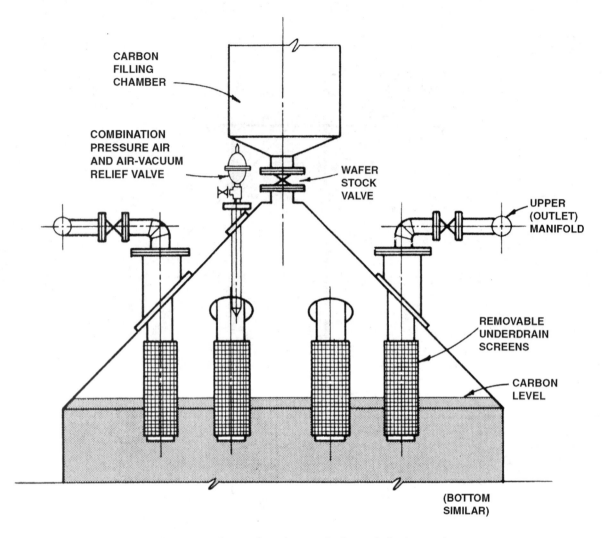

*Fig. 10.5 Section through top underdrain of adsorber vessel
Orange County, CA*

10.23 Placing Carbon Adsorption Units Into Operation

1. *Starting Flow Into the Carbon Adsorber.* The first step in placing a carbon adsorber into operation is to adjust the valving of the adsorber so that effluent from either filtration units or other treatment units that precede the carbon adsorbers will enter the desired influent port of the adsorber. Three-way valves must be aligned so that wastewater will flow through the influent piping of the carbon container and out the effluent piping.

2. *Adjusting Flow Rate Through the Adsorption System.* Flow rates through the carbon system can be controlled in one of two ways: by controlling the pumping rate or by throttling the effluent valve on an individual adsorber.

   In systems where the flow through the adsorber is controlled by pumping, the rate may be set at a constant rate or may fluctuate with the flow of wastewater from the system.

   In systems where the flow is controlled by valving, a control valve regulates the flow through the carbon adsorber. This control valve allows the operator to set the desired gallons-per-minute flow rate through the carbon so that proper adsorption can take place. After the operator has opened the influent valves to allow flow into the carbon adsorber, the flow control valve should be adjusted until the exact flow rate desired is flowing through the carbon and out the effluent piping.

3. *Head Loss Measurements.* There is a head loss or loss of pressure between the influent of a carbon adsorber and the effluent side of the container. This head loss is expected since the wastewater must pass through carbon and thereby create a pressure drop or head loss. When head losses are too high (check the system operations manual to determine the optimum head loss), backflushing of the screen must take place in order to cut down the pressure drop. Pressure sensors are located both at the top and bottom of carbon adsorbers. These instruments may also send a signal to a remote location, such as a control room. The sensors help the operator determine the head loss readings so as to know when a backwash cycle must take place. Carbon adsorber screens are backwashed whenever the head loss exceeds the manufacturer's recommendation.

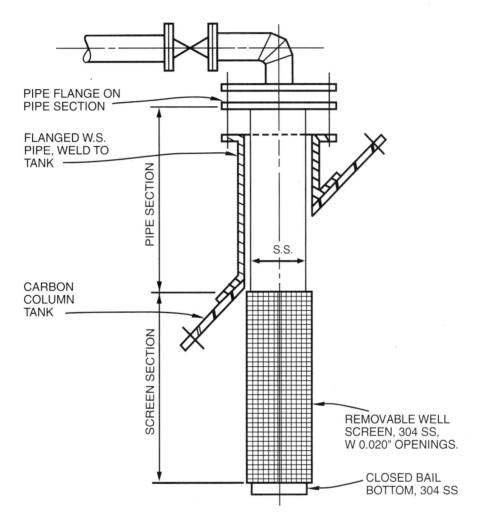

*Fig. 10.6   Top and bottom underdrains on adsorber vessel
Orange County, CA*

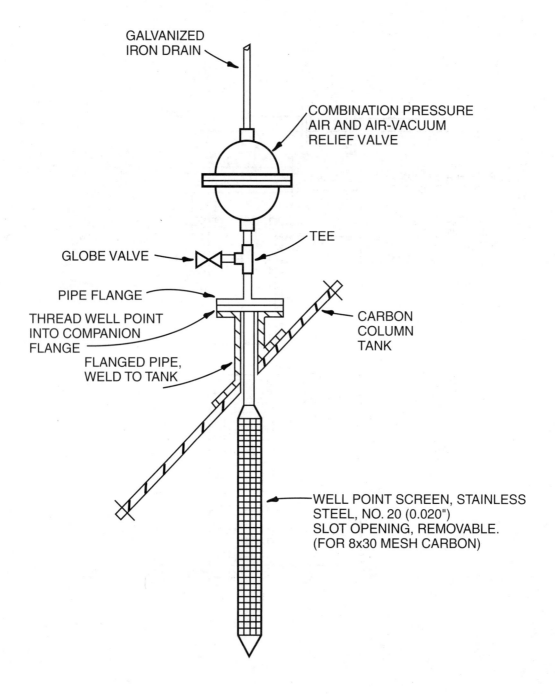

Fig. 10.7 *Combination pressure air and air-vacuum relief valve detail Orange County, CA*

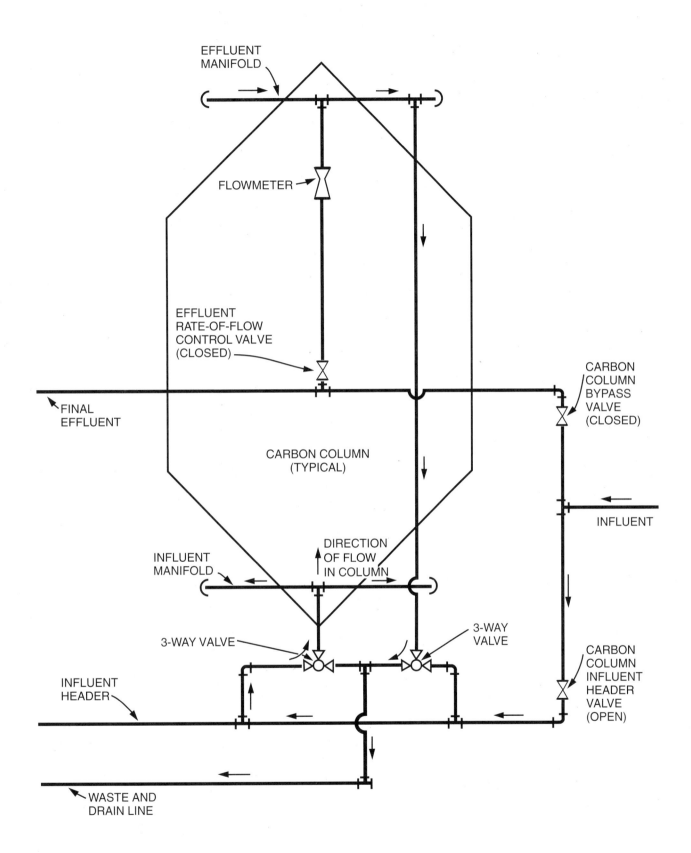

*Fig. 10.8 Upflow carbon column schematic, upflow to waste
(used after adding carbon to columns to flush out lines)
Orange County, CA*

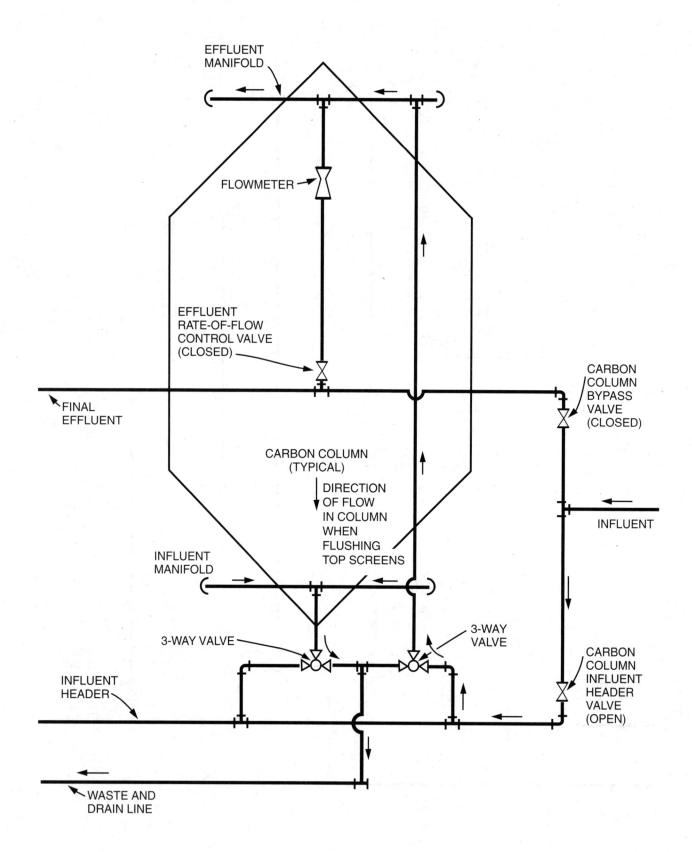

Fig. 10.9 Upflow carbon column schematic—reverse flow
(used to flush top screens)
Orange County, CA

4. Sampling Effluent To Determine Amount of Carbon Fines. To determine whether the screens in the carbon adsorber are doing an efficient job, sample a portion of the effluent from the carbon adsorber. The effluent should be clear with relatively few specks of carbon in the sample. If excessive carbon fines are found, the operator can assume that the carbon is rubbing too hard against the screens causing particles (fines) to break off and flow out in the effluent or that a hole in one of the screens is allowing substantial amounts of carbon to flow out in the effluent. A daily check on the relative amount of carbon fines found in the effluent from the carbon adsorber may be necessary. In many treatment plants, carbon fines are considered suspended solids and a reduction in the amount of carbon fines will help reduce the suspended solids that must be reported to regulatory agencies.

10.24 Adsorber Loading Guidelines

Two types of loading rates are used for activated carbon adsorption processes: (1) hydraulic loading rates, and (2) chemical oxygen demand (COD) or organic loading rates.

10.240 Hydraulic Loading Rates

The typical hydraulic loading rates entering a carbon adsorption column range from 2 to 10 gallons per minute per square foot (1.3 to 6.8 liters per second per square meter). The average hydraulic loading rate (or surface loading rate) is 5 gallons per minute per square foot (3.4 liters per second per square meter).

10.241 Chemical Loading Rates

Chemical loading rates for activated carbon systems will depend on the nature and chemical composition of the wastestream being treated and the discharge requirements. The development of isotherms for the application will be useful in determining the proper chemical loading rates.

10.25 Operating Procedures

1. Routine Backflushing of Fine-Mesh Screens To Reduce Head Losses Through Carbon (Figure 10.9). The screens on each carbon adsorber should be backflushed whenever the pressure drop across the adsorber exceeds the manufacturer's recommendation. This will help reduce the amount of carbon fines found in the effluent as well as reduce head losses through the activated carbon treatment.

If a carbon vessel requires an abnormally high number of backflushing cycles during a given time period, check the instruments and prepare the vessel for an inspection to determine if any damage has taken place inside the vessel. Backflushing and flow reversal will help eliminate air pockets within the carbon adsorber to prevent short-circuiting and a loss of efficiency in removal of organics.

2. Chemical Testing To Determine Removal Efficiencies. Sampling and analysis are important to determine the efficiency of the carbon adsorption process. Testing for the chemicals being removed must be performed on a regularly scheduled basis. For some installations, the carbon must be replaced as soon as a regulated contaminant can no longer be removed and is detected in the effluent. This condition is referred to as breakthrough. The premature breakthrough of organics through the carbon bed normally involves either the failure of upstream pretreatment systems or poor-quality regenerated carbon. Extensive testing of the influent and effluent of upstream processes may help to determine the cause of the failure and proper corrective action. Poor-quality regenerated carbon leaves no practical solution for the system operator.

Some carbon adsorbers are equipped with additional sample points (between the influent and effluent sampling points), which are useful to determine the rate of carbon exhaustion through the bed. If multiple fixed beds in series are used, sample points between each of the beds will help in determining the rate of exhaustion. With properly regenerated activated carbon and appropriate flow rates and loading conditions, the carbon process will efficiently remove a substantial percentage of the chemical pollutants. If the removal efficiencies for any carbon adsorber are lower than normal, regeneration or addition of new carbon may be required in order to bring the removal efficiencies back to the proper levels. The removal efficiency also depends on flow rates and detention times within the carbon adsorber.

3. Turbidity Measurements. If the effluent turbidity is regulated by a discharge standard, measurement of the effluent turbidity should be conducted continuously or as required by the discharge permit. Although effluent from a carbon adsorber will usually be very clear and free from color, a turbidity measurement is important as a check on the operating efficiency. A turbidimeter should be used to determine whether excess carbon fines are causing high turbidity in the effluent. The turbidimeter can also be used to indicate whether color is being removed efficiently by the activated carbon process.

4. Checks To Determine Level of Carbon Remaining in Reactor. For systems in which carbon is frequently transferred for regeneration, the operator should check at least weekly to determine whether an adequate amount of carbon remains in each of the carbon adsorbers. If additional carbon is needed, makeup or virgin carbon should be added to achieve the proper detention time and ratio of pounds of carbon to chemical load removed. Check the operation and maintenance

manual to determine the exact carbon level recommended for each of the adsorbers. The weekly check on the level of carbon will make sure that enough activated carbon is in contact with the wastewater to do an adequate job in removing organics.

5. Process Monitoring. Once each shift, inspect the adsorption system to determine if any abnormal or unusual conditions may exist that could cause trouble in the activated carbon treatment process. The system review includes: (a) checking head losses for abnormally high head losses through the system; (b) checking for excessive carbon fines in the effluent from the carbon treatment process; (c) backflushing the carbon whenever necessary but at least once a day; (d) checking turbidity of effluent from the carbon adsorption process; (e) checking that the flow scheme is correct and the proper valves are opened or closed; (f) checking the flow rate through the carbon adsorber for proper detention time and adequacy of activated carbon adsorption; and (g) checking to be sure daily influent and effluent samples are collected.

6. Recordkeeping. Keep a daily log of adsorption system activities and data. Record the details of any problems that occur, such as plugged screens, excessive head losses, problems with the regeneration furnace, plugged lines, or any other abnormal conditions.

QUESTIONS

Write your answers in a notebook and then compare your answers with those on page 513.

10.2E   How are flow rates through a carbon adsorption system controlled?

10.2F   How could an operator determine the optimum head loss through an adsorber?

10.2G   What conditions might cause an excessive amount of fines in the adsorber effluent?

10.2H   What are the most likely causes of premature breakthrough of contaminants?

10.26 Shutting Down a Carbon Adsorption Unit

If it becomes necessary to shut down the carbon adsorption operation, shut off the flow into the carbon adsorber by closing the influent valve on the pressure vessel. If all processes except the activated carbon adsorption process are to continue operation, be certain to turn the three-way valves properly to divert the flow from various preceding processes to other areas besides the carbon adsorption process.

If one of the adsorbers in a multiple fixed bed system must be shut down, the flow can either be switched to the second adsorber in the train or the flow to the train can be stopped. If a problem develops with the lead (first) adsorber in a train, the flow can be directed to the last polishing vessel without reducing the effluent quality. However, if the polishing adsorber must be shut down, flow can be maintained through the lead vessel only until contamination breaks through the lead (first) bed. At this stage, the flow to the train should be halted to prevent the possibility of exceeding the plant effluent specifications. If a major leak or other similar problem develops, flow to the adsorber should be shut down immediately and steps taken to correct the problem. Proper safety procedures should be observed at all times to prevent damage to the equipment or injury to personnel. See Section 10.5, "Safety."

If carbon adsorption processes must be shut down, the operator must notify supervisors or the regulatory agency. If discharge standards are so strict that activated carbon is required in order to meet the standards, any deviation from the standard flow through the carbon adsorber may result in a violation of the discharge limitations. The regulatory agency must be notified so that emergency procedures can be implemented, if necessary.

Some treatment plants have a storage reservoir available that can be used if final treatment processes must be shut down. Water stored in these reservoirs should be brought back through the processes for final treatment prior to discharge to any lake or stream that requires the strict standards.

For short duration shutdowns lasting less than one or two weeks, little needs to be done. The valves in the water lines to and from the adsorber train should be closed and the valves in the vent lines on each vessel should be opened.

For extended shutdown periods, the adsorbers should be drained and refilled with plant water. All the valves in the water lines to and from the train should be closed and the valves in the adsorber vent lines opened. If the entire plant is shut down, the feed pumps should be shut down and the valves closed in the lines to and from the pumps. Any drain valves in the pump casing should be opened for the duration of the shutdown.

10.27 Abnormal and Emergency Conditions

10.270 Low COD[10] Removal Efficiencies

If low chemical oxygen demand removal efficiencies are recorded for the activated carbon process, the operator must prepare for regeneration. The common reason for low removal efficiencies of organic matter is plugging of the fissures within the granular activated carbon. By sampling and analyzing COD removal efficiencies for each carbon contact reactor, you will know when the regeneration process should begin. As soon as low removal efficiencies of organic matter are reported by the laboratory, begin preparation for the regeneration cycle. Take carbon columns off line, remove carbon from the bottom, and prepare for regeneration in the multiple hearth carbon furnace. Prepare the furnace, bringing it up to the proper temperature

---

[10] *COD* (pronounce as separate letters). Chemical Oxygen Demand. A measure of the oxygen-consuming capacity of organic matter present in wastewater. COD is expressed as the amount of oxygen consumed from a chemical oxidant in mg/L during a specific test. Results are not necessarily related to the biochemical oxygen demand (BOD) because the chemical oxidant may react with substances that bacteria do not stabilize.

and adjusting both the carbon feed rate and the steam quantities for the regeneration process.

During the process of regeneration, 5 to 10 percent of the carbon mass is lost. These losses will lower the amount of available activated carbon in the adsorption process. To counteract the low quantities, be prepared to add fresh makeup carbon to any carbon column whose total quantity of granular activated carbon is low. COD removal efficiencies should improve after fresh carbon is added to the system.

10.271 High Head Losses

During the operation of the activated carbon adsorption processes, the head loss will increase. However, if head losses continue to remain high even after backflushing, be prepared to inspect the carbon columns to determine if a screen is completely plugged or collapsed. High head losses may also indicate that the amount of carbon in the adsorption reactor is too high and therefore causes plugging of the screens.

10.272 Fouling of Activated Carbon Granules

The carbon adsorption process usually follows other wastewater treatment units. Occasionally, the other treatment units may fail and organic matter and suspended matter in large concentrations may enter the carbon reactors. This can cause fouling of the activated carbon granules, which will reduce the unit's adsorption capabilities and increase head loss.

When you suspect a fouling of the carbon granules with suspended solids, immediately backflush the carbon contactor in an attempt to dislodge and flush out as much of the suspended solids as possible. Because most of the suspended solids will be trapped in the lower portions of an upflow carbon reactor, an extra regeneration cycle may be required to remove the excess suspended solids.

10.273 Plugged Screens

Because of the upflow configuration for normal operation of an activated carbon column, the bottom screens on the tank will seldom plug with debris and suspended solids. The screens that do tend to plug with carbon fines and suspended solids are the top screens located at the highest point or effluent end of the carbon reactor.

When screens are plugged, the head losses will increase substantially through the carbon reactor. Backflush the screens in order to dislodge the plugged portions. Occasionally, backflushing will not be adequate to dislodge material trapped in the screens. When this happens, enter the carbon column using approved confined space entry procedures and wire brush the screens to correct the plugging problem.

10.274 Collapsed Screens

Do not attempt to force more flow through the carbon column reactor than the design specifications allow. Extreme pressures can build up within the carbon column and cause severe damage to the carbon tank or to the screens within the vessel.

The screens located within the vessel may collapse if excess pressure from within builds up due to negligence on the part of the operator. The high pressure from within forces against the zero or negative pressure on the other side of the screens or effluent side. This pressure can cause the screen to collapse and it will have to be replaced.

10.275 Air Pockets

Air pockets can develop when the carbon column reactor is filled too quickly. Whenever the carbon column is empty, fill it slowly to allow air to escape through the automatic air escape valves at the top of the vessel. By filling the tank slowly, air is pushed ahead of the water as it enters the bottom of the tank and will escape without forming air pockets within the body of the granular carbon. Air pockets must be eliminated from the carbon column reactors to prevent short-circuiting. If air pockets develop in spite of slow filling rates, backflushing will sometimes eliminate them.

10.276 Deterioration of Tank and Pipe Coatings

Because of the corrosive nature of activated carbon in its granular form, interior pipe and tank surfaces are sealed with protective coatings. At least once every six months, check the inside of the activated carbon tank to be certain that the coating has not scaled or chipped away from the interior metal surface. If you find a chipped spot, contact the supervisor and have the spot repaired using an epoxy to seal the interior lining.

10.277 Failure of Upstream Processes

Because the carbon adsorption process is usually the last process in the wastewater treatment plant, the efficiency is highly dependent on the proper operation of any processes that precede the carbon units. When upstream processes fail, a serious reduction in efficiency of the carbon process may result. If the treatment plant is equipped with an emergency holding pond, it may be in the best interest of the operation of the treatment plant to temporarily store inadequately treated wastewater rather than pumping it through the activated carbon columns. The better the quality of water that enters the carbon column reactors, the more efficient the adsorption process will be. If wastewater is fouled with suspended solids and other matter, the adsorption process will be impaired and the operation will become more expensive because the activated carbon will have to be regenerated.

QUESTIONS

Write your answers in a notebook and then compare your answers with those on page 513.

10.2I   List the possible causes of high head losses through an activated carbon process.

10.2J   Why should the activated carbon column reactor be filled slowly?

10.2K   What would you do if a wastewater treatment process upstream from a carbon adsorption process failed?

10.3 ACTIVATED CARBON REGENERATION

As described previously in this chapter, activated carbon is produced by heating carbonaceous materials to a high temperature and using a gas or steam to help form the porous structure necessary for the adsorption process. Regeneration of activated carbon is a process whereby the pores and the surface of the carbon are cleansed of the molecular organic material that has been adsorbed on the surface.

Activated carbon must be disposed of or regenerated after the carbon has adsorbed all of the molecular organic material that can be adsorbed on the surface and within the pore structure. Some discharge permits and local ordinances strictly regulate the discharge of specific toxic substances. If breakthrough occurs and a restricted substance is detected in the effluent, the bed must be taken out of service for replacement or regeneration.

The purpose of the regeneration process is to drive off the molecular organic material and to allow a clean surface to be readied for additional adsorption of organic material. The most common method for regeneration of activated carbon is through heat treatment in a multiple hearth furnace or a rotary kiln. The carbon is subjected to heat up to 1,750 degrees Fahrenheit (950°C). Oxygen content in the furnace is normally controlled at less than one percent to oxidize the organics without burning the carbon. Steam is added to control the oxidation process. Scrubbers and afterburners are used to cleanse potential air pollutants from the furnace exhaust gases. The adsorption capacity of regenerated carbon is slightly less than that of virgin carbon.

Several types of regeneration furnaces exist. They are the multiple hearth furnace, the fluid bed furnace, the direct fired rotary kiln, and the infrared furnace. Some industrial wastes containing halogens have required special materials of construction in the kiln or furnace to avoid corrosion problems.

Multiple hearth furnaces have been used to reactivate carbon for nearly sixty years. This is particularly true in the large capacity ranges, greater than 10,000 pounds per day of carbon reactivated. At this writing, the largest furnace installed had a capacity of 120,000 pounds per day.

Fluid bed furnaces have been used to reactivate carbon in wastewater and sweetener decolorizing applications.

Direct fired rotary kilns are generally fired countercurrent to the carbon flow; thus, the ability to vary the temperature profile is limited. However, since carbons used in potable (drinking) water are generally easily reactivated, the direct fired rotary kiln is used in this application.

Indirect fired kilns have been used mainly in the metallurgical industry where the reactivation temperatures are less than 1,000°F.

Infrared furnaces have been used successfully to regenerate water treatment plant activated carbon.

Non-thermal reactivation of activated carbon has been attempted using chemicals to oxidize or recover the carbon contaminants. In general, these systems have seen little success. Reactivation of caustic and acid impregnated carbon has been successful in vapor phase operations.

10.4 SAMPLING AND ANALYSIS

10.40 Sampling

Sampling is very important in the operation of the activated carbon process. The operator should sample both the influent and effluent to help determine organic contaminant removal efficiencies and to determine the quantities of carbon fines remaining in the effluent from the process.

10.41 Sampling Ports

Sample valves are usually found at the inlets and outlets of the adsorbers from which the operator can sample the water flowing through the carbon. In this way, the operator can obtain analyses of the influent and effluent of the adsorption process. This will be important in determining the efficiency of the organics removal through adsorption.

10.42 Analysis

Determination of removal efficiencies of organics is accomplished by running chemical oxygen demand tests and TURBIDITY[11] measurements. Other tests are run to help determine the status of the carbon in the carbon column contactors. Influent and effluent samples should be taken at least once a day on each carbon column. The types of lab analyses that must be performed in order to achieve the highest efficiency of operation with the activated carbon regeneration process include apparent density of regenerated carbon, total ash of regenerated carbon, iodine number, molasses number, hardness number, abrasion number, and sieve analyses. A brief description of each of these tests is presented in the Appendix at the end of this chapter.

---

[11] *Turbidity* (ter-BID-it-tee). The cloudy appearance of water caused by the presence of suspended and colloidal matter. In the waterworks field, a turbidity measurement is used to indicate the clarity of water. Technically, turbidity is an optical property of the water based on the amount of light reflected by suspended particles. Turbidity cannot be directly equated to suspended solids because white particles reflect more light than dark-colored particles and many small particles will reflect more light than an equivalent large particle.

QUESTIONS

Write your answers in a notebook and then compare your answers with those on page 513.

10.3A   What is regeneration of activated carbon?

10.4A   What tests are used to evaluate the organics removal efficiency of a carbon adsorption unit?

10.5 SAFETY

10.50 Carbon Transfer

The carbon system is a relatively trouble-free system. When properly operated, the carbon system will present relatively few hazards to the operations and maintenance personnel. Use caution whenever the carbon vessels are pressurized and when carbon is being transferred. The adsorbers and carbon transfer systems will be pressurized to 20 to 25 psi (140 to 170 kPa). Although this is a relatively low pressure, care should be exercised to prevent accidental injury to the operator. Whenever carbon transfer takes place, observe the following precautions:

1. Monitor the pressure of the pressurized vessel carefully. Plugging of the transfer lines may result in overpressurizing the vessel.

2. Ensure that camlock fittings are tightly fastened.

3. When transferring spent carbon, be aware of the fact that this carbon is contaminated by hazardous materials; wear protective clothing and use appropriate safety devices.

10.51 Working Inside a Carbon Adsorption Vessel

Working inside a closed container such as an activated carbon adsorption vessel exposes operators to some very serious types of hazards. Specific safety regulations for confined space work must be observed and a permit may be required. Exact procedures for work in confined spaces may vary with different agencies and geographical locations and must be confirmed with the appropriate safety regulatory agency. For a detailed description of the hazards involved in confined space work and safety precautions that should be taken, see Chapter 13, "Safety," Section 13.12, "Confined Spaces."

SAFETY NOTICE

Before anyone ever enters a tank for any reason, these safety procedures must be followed:

1. Test the atmosphere in the tank for toxic and explosive gases and for adequate oxygen. Contact your local safety equipment supplier for the proper types of atmospheric testing devices. These devices should have alarms that are activated whenever an unsafe atmosphere is encountered.

2. Provide adequate ventilation. A self-contained, positive-pressure breathing apparatus may be necessary.

3. All persons entering a tank must wear a safety harness.

4. One person must be at the tank entrance and observing the actions of all people in the tank. An additional person must be readily available to help the person at the tank entrance with any rescue operation.

In addition to the usual dangers associated with confined spaces, entry into or work in an activated carbon adsorption vessel presents a greater than usual risk of encountering an oxygen-deficient atmosphere. The oxygen content in normal breathing air is 20.9 percent; an oxygen-deficient atmosphere is one that contains less than 19.5 percent oxygen by volume. Appropriate ventilation must be provided inside the tank and use of a self-contained breathing apparatus is required when the oxygen level falls below 19.5 percent. This is of particular importance in activated carbon adsorbers because studies indicate that wet carbon adsorbs oxygen molecules from the atmosphere, thereby lowering the oxygen available for breathing by anyone working in the reactor.

Confined space procedures should be established for any facility using carbon in confined vessels. Also, any operator who must work in or around activated carbon equipment should be trained to use safe procedures and all appropriate safety equipment.

10.52 Carbon Dust

Fresh or dry carbon tends to shed a certain amount of dust particles when moved or handled. Be very careful when handling dry activated carbon to avoid inhaling the dust. Wear an approved respirator to prevent the carbon dust from entering your lungs.

Carbon dust can be particularly hazardous if it is allowed to accumulate in a confined space. The dust can ignite with explosive force if a spark or even a hot surface is present.

10.53 Excessive Pressures Within Carbon Contactor Tank

Make every effort to prevent excessive pressure buildup within the carbon contactor vessels. Excessive pressure may be caused by too high a flow entering the vessel or by collapsed or plugged screens. If pressures are too high within the carbon vessel, the vessel itself could explode or rupture. The air-vacuum relief device (Figure 10.7) installed to prevent overpressurization should be checked periodically as part of the routine preventive maintenance procedures.

10.54 Handling Spent Activated Carbon

Spent carbon may be a hazardous waste due to the quantity and nature of adsorbed materials and must be handled accordingly. If it ever becomes necessary to enter the carbon vessels for maintenance, use an approved respirator and wear suitable protective clothing.

10.55 Disposal of Activated Carbon

For the majority of industrial users, on-site regeneration of activated carbon is not economically feasible. For those users, a decision must be made regarding the disposal of the spent activated carbon. Most carbon supply companies will remove the carbon they have delivered for a fixed price included in the initial purchase price. This allows the user to be sure that proper disposal procedures are followed when the carbon's useful life has been exhausted.

Before agreeing to such an arrangement, however, the supplier will require a spent carbon acceptance test. This will ensure that the regeneration of the carbon by the supplier will not cause an air pollution problem and that the spent carbon can be properly manifested for disposal or regeneration.

If arrangements for the disposal or regeneration of the carbon have not been made, the user must make a determination as to the proper means of disposal of the spent carbon. In some cases, the spent carbon will be nonhazardous and may be disposed of at a municipal landfill. To use this disposal option, the carbon must undergo extensive testing to prove that it is nonhazardous.

Unfortunately, the spent carbon from most industrial uses will be judged a hazardous waste. This carbon must either be manifested to a licensed regeneration facility, incinerated, or transported to a hazardous waste disposal facility.

Facilities using carbon service trailers will transfer carbon directly from the adsorbers to the service company's trailer. The trailer is typically sized to receive 20,000 pounds of spent carbon based on a dry adsorbate-free carbon weight. The actual weight of the spent carbon, including the water remaining on the carbon after the trailer is drained, should not exceed 40,000 pounds. When all the carbon is transferred, the adsorber vessel should be flushed with water to remove the last remaining amount of carbon.

10.6 REVIEW OF PLANS AND SPECIFICATIONS

10.60 Unloading Station for Truck or Train Delivery of Fresh Activated Carbon

Because it is important to add makeup or fresh activated carbon periodically, an unloading station is needed to receive the fresh carbon for the adsorption process. The unloading station should have a proper turnaround if it is a truck station. The station should be close enough to the rail center for easy unloading if the carbon is delivered by rail.

Bagged carbon can sometimes be unloaded rapidly by machine if the proper suction equipment is available to pull carbon from the bags. Otherwise, operators must empty each bag into the fresh carbon handling station on an individual basis.

If bulk carbon is to be unloaded at the treatment plant, it is best to have a suction device to unload the carbon from the bulk container so that manpower can be cut to a minimum. By providing a suction device and adequate local exhaust in the unloading area, dust and other hazards can be reduced or eliminated.

10.61 Valving Placed at Easy-To-Reach Locations

Check the design of the activated carbon adsorption process station to be sure that all valves indicated on the plans are easy to reach. The valves, if overhead, should have operators no higher than six feet (1.8 m). Valves that are at the top of carbon columns should be easy to reach from a catwalk or other handrail-protected operation viewing platform.

10.62 Dust Control for Unloading Fresh Carbon

When unloading fresh carbon, be certain that exhaust fans are operating and that all staff are wearing approved respiratory protection to prevent inhaling the carbon dust. The operator should check the plans to be certain that dust control is adequate to protect all personnel.

10.63 Proper Ventilation in Carbon Regeneration Furnace Room

Because of the heat and dust that can be emitted from a carbon regeneration furnace, check the plans to be certain that adequate ventilation is provided within the furnace room. Proper ventilation and air circulation is mandatory in order to protect the personnel from excessive heat and exposure to fine particles of carbon, smoke, and other hazardous materials.

10.64 Scaffolding and Catwalks

Check plans to be certain that all catwalks are protected with guardrails that not only comply with OSHA standards but that actually ensure safety, particularly when the catwalks are located at the top of the carbon column reactors. Entrance and exit from access ports into carbon column reactors should be protected to eliminate the chances of an operator falling.

10.65 Warning Alarms and Signs

Check the plans to be certain that alarms are adequate to notify personnel of excessive temperatures or lack of oxygen within a building or vessel. Be certain that the plans call for warning signs to be posted at conspicuous locations so that all personnel are aware of the dangers associated with working around activated carbon and carbon handling equipment. REMEMBER: *Confined space procedures must be implemented to enter carbon columns.*

10.66 Upstream Processes

Filtration is usually an upstream process prior to carbon adsorption. A holding tank or reservoir is desirable upstream from the carbon adsorption process in order to provide temporary storage in case of failure of other systems. Check plans to be certain that such a reservoir is provided for temporary or emergency storage. Failure of upstream processes would mean excessive organic or suspended solids loading on the activated carbon process, thereby reducing its effectiveness.

10.7 ADDITIONAL READING

1. *HANDBOOK OF PUBLIC WATER SYSTEMS*, Second Edition, HDR Engineering, Inc. Available from John Wiley & Sons, Customer Care Center - Consumer Accounts, 10475 Crosspoint Boulevard, Indianapolis, IN 46256. ISBN 0-471-29211-7. Price, $199.00, plus shipping and handling.

2. *PROCESS DESIGN MANUAL FOR CARBON ADSORPTION*, US Environmental Protection Agency. Available from National Technical Information Service (NTIS), 5285 Port Royal Road, Springfield, VA 22161. Order No. PB-227157. Price, $68.50, plus $5.00 shipping and handling per order.

10.8 ACKNOWLEDGMENT

Portions of the material in this section were originally prepared by John Gonzales and appeared in the First Edition of *INDUSTRIAL WASTE TREATMENT*.

QUESTIONS

Write your answers in a notebook and then compare your answers with those on page 513.

10.5A   What are the major types of hazards an operator is exposed to when working in a carbon column reactor vessel?

10.5B   How could activated carbon cause an oxygen deficiency in a carbon column reactor vessel?

10.6A   List the items you would consider when reviewing the plans and specifications for an activated carbon adsorption process.

10.6B   What kinds of warning signs and alarms should be provided with an activated carbon adsorption process?

END OF LESSON 2 OF 2 LESSONS

on

PHYSICAL TREATMENT PROCESSES

Please answer the discussion and review questions next.

DISCUSSION AND REVIEW QUESTIONS
Chapter 10. PHYSICAL TREATMENT PROCESSES
(AIR STRIPPING AND CARBON ADSORPTION)
(Lesson 2 of 2 Lessons)

Write the answers to these questions in your notebook. The question numbering continues from Lesson 1.

5. How does activated carbon remove organic material from wastewater?
6. What industries use activated carbon to treat industrial wastestreams?
7. What would you do if the results of lab tests showed low COD removal efficiencies from a carbon adsorption unit?
8. How can extreme pressures build up in carbon column reactors?
9. Why must activated carbon be regenerated?
10. What tests should be run on the effluent of the activated carbon process and why?
11. Where could an operator find out what specific safety regulations must be observed before entering or working in a carbon column reactor vessel?

SUGGESTED ANSWERS
Chapter 10. PHYSICAL TREATMENT PROCESSES
(AIR STRIPPING AND CARBON ADSORPTION)

ANSWERS TO QUESTIONS IN LESSON 1

Answers to questions on page 491.

10.0A Air stripping of VOCs is more economical than carbon adsorption because there is no recurring cost for the replacement and disposal of carbon.

10.0B The solubility of a chemical has a strong influence on air stripping effectiveness. Highly soluble chemicals mix thoroughly with the water and are difficult to remove while less soluble chemicals are effectively removed by air stripping.

Answers to questions on page 492.

10.0C Demister pads are installed in air strippers to reduce the amount of water carried to the air discharge or downstream processes.

10.0D Vapor phase carbon is installed on the discharge of air strippers to prevent the release of VOCs to the atmosphere.

10.0E The heat from the incinerator can be used to raise the water temperature to improve VOC removal.

Answers to questions on page 494.

10.0F A high differential pressure across an air stripper indicates fouling of the packing media.

10.0G Three ways to prevent the fouling of air stripper media are:
1. Pretreat the water being air stripped by chemical or physical means including preaeration, oxidation, and filtering to remove oxidized solids.
2. Add chelating agents to inhibit the formation of oxidation products.
3. Periodically clean the unit with acids and disinfectants to remove the scale and biofouling.

ANSWERS TO QUESTIONS IN LESSON 2

Answers to questions on page 495.

10.1A The activated carbon adsorption process is used by industries to remove organic pollutants, especially toxic organics, from industrial wastewaters.

10.1B Activated carbon can be produced from wood, coal, nutshells, bone, petroleum residues, or even sawdust.

10.1C Activated carbon is made by drying carbonaceous material and slowly heating it to a high temperature in a controlled atmosphere, and then exposing it to an oxidizing gas at a high temperature.

Answers to questions on page 498.

10.2A The average contact time in an activated carbon pressure vessel is 60 minutes.

10.2B Carbon adsorption systems operate on the same general principle of countercurrent flow. That is, wastewater being treated passes through the oldest and dirtiest carbon first. As the water moves through the column, it is treated by successively cleaner activated carbon, which has a higher adsorption capacity. In this way, the water is "polished" as it nears the effluent end of the column.

10.2C Three-way valves are used in activated carbon piping to provide both the upflow conditions necessary for the activated carbon treatment, and downflow conditions needed for backwashing of filter screens.

10.2D Activated carbon is kept in the pressure vessel by screens located at both the top and bottom of the vessel.

Answers to questions on page 506.

10.2E Flow rates through a carbon adsorption system can be controlled in one of two ways: by controlling the pumping rate or by throttling the effluent valve on an individual adsorber.

10.2F To find out the optimum head loss through an adsorber, the operator should check the operations manual supplied by the equipment manufacturer.

10.2G An excessive amount of fines in the adsorber effluent could be caused by the activated carbon rubbing too hard against the screens and breaking off particles, or a hole in one of the screens may be permitting carbon to flow out in the effluent.

10.2H The most likely causes of premature breakthrough of contaminants are failure of an upstream pretreatment system or the use of poor quality regenerated carbon.

Answers to questions on page 507.

10.2I Possible causes of high head losses through an activated carbon process include:
  1. Need for backflushing.
  2. Fouling of activated carbon granules with suspended solids.
  3. Screen plugged or collapsed.
  4. Amount of carbon in the adsorption reactor is too high.

10.2J The activated carbon column reactor should be filled slowly to prevent air pockets from forming in the reactor and causing short-circuiting.

10.2K If a wastewater treatment process upstream from a carbon adsorption process fails, try to prevent suspended solids from reaching the carbon adsorption process. If an emergency holding pond is available, temporarily store the inadequately treated wastewater until the upstream process is working again.

Answers to question on page 509.

10.3A Regeneration of activated carbon is a process whereby the pores and the surface of the carbon are cleansed of the molecular organic material that has been adsorbed on the surface.

10.4A The chemical oxygen demand test and turbidity measurements are used to evaluate the organics removal efficiency of a carbon adsorption unit.

Answers to questions on page 511.

10.5A The major hazards associated with working in a carbon column reactor vessel are confined space hazards and, in particular, oxygen-deficient atmospheres.

10.5B Activated carbon could cause an oxygen deficiency in a carbon column reactor vessel by adsorbing oxygen molecules from the air.

10.6A Items that should be considered when reviewing plans and specifications for an activated carbon adsorption process include:
  1. Unloading station for delivery of fresh activated carbon.
  2. Location of valves within easy reach.
  3. Dust control for unloading fresh carbon.
  4. Proper ventilation in carbon regeneration furnace room.
  5. Scaffolding and catwalks.
  6. Warning alarms and signs.
  7. Upstream processes.

10.6B Alarms should be provided to notify personnel of excessive temperatures or lack of oxygen within a building. Signs should warn personnel of dangers associated with working around activated carbon and carbon handling equipment.

APPENDIX

Chapter 10. PHYSICAL TREATMENT PROCESSES
(AIR STRIPPING AND CARBON ADSORPTION)
ACTIVATED CARBON ADSORPTION
LABORATORY PROCEDURES

A. INFLUENT AND EFFLUENT TESTS

1. Chemical Oxygen Demand (COD)

   A measure of the oxygen-consuming capacity of organic matter present in wastewater, COD is expressed as the amount of oxygen consumed from a chemical oxidant in mg/L during a specific test. Results are not necessarily related to the biochemical oxygen demand (BOD) because the chemical oxidant may react with substances that bacteria do not stabilize. See Chapter 16, "Laboratory Procedures and Chemistry," Volume II, *OPERATION OF WASTEWATER TREATMENT PLANTS*, for test procedures.

2. Turbidity

   Turbidity units, if measured by a nephelometric (reflected light) instrumental procedure, are expressed in nephelometric turbidity units (NTU) or simply TU. Those turbidity units obtained by visual methods are expressed in Jackson Turbidity Units (JTU) and sometimes as Formazan Turbidity Units (FTU). The FTU nomenclature comes from the Formazan polymer used to prepare the turbidity standards for instrument calibration. Turbidity units are a measure of the cloudiness of water.

B. STATUS OF ACTIVATED CARBON

1. Abrasion Number (Ro-Tap)

   The abrasion number of granular carbon is a measure of the resistance of the particles to degrading on being mechanically abraded. This number is measured by contacting a carbon sample with steel balls in a pan on a Ro-Tap machine. The abrasion number is the ratio of the final average (mean) particle diameter to the original average (mean) particle diameter (determined by screen analysis) times 100.

2. Abrasion Number (NBS)

   Similar to 1 above except different equipment is used.

3. Apparent Density

   The weight per unit volume of a dry homogeneous (uniform) activated carbon. To ensure uniform packing of a granular carbon during measurement, a vibrating trough is used to fill the measuring device.

4. Decolorizing Index

   Molasses solution is treated with different weights of a standard carbon of known decolorizing index. The optical densities of the filtrate are measured and plotted with the known decolorizing index values to obtain a standard curve. A molasses solution is then treated with pulverized activated carbon of unknown decolorizing capacity. The optical density of the filtrate is measured and the decolorizing index is determined from the standard curve.

5. Effective Size and Uniformity Coefficient

   Effective size is the size of the particle that is coarser than 10 percent, by weight, of the material. That is, it is the size sieve that will permit 10 percent of the carbon sample to pass but will retain the remaining 90 percent. Effective size is usually determined by the interpolation of a cumulative particle size distribution.

   Uniformity coefficient is obtained by dividing the sieve opening in millimeters that will pass 60 percent of a sample by the sieve opening in millimeters that will pass 10 percent of the sample. These values are usually obtained by interpolation on a cumulative particle size distribution.

6. Hardness Number

   The hardness number is a measure of the resistance of a granular carbon to the degradation action of steel balls in a pan in a Ro-Tap machine. This number is calculated by using the weight of granular carbon retained on a particular sieve after the carbon has been in contact with steel balls. This is the Chemical Warfare Service (CWS) test.

7. Iodine Number

   The iodine number is the milligrams of iodine adsorbed by one gram of carbon at an equilibrium filtrate concentration of $0.02N$ iodine. The number is measured by contacting a single sample of carbon with an iodine solution and extrapolating to $0.02N$ by an assumed isotherm slope.

Iodine number can be correlated with ability to adsorb low molecular weight substances.

8. Methylene Blue Number

    The methylene blue number is the milligrams of methylene blue adsorbed by one gram of carbon in equilibrium with a solution of methylene blue having a concentration of 1.0 mg/L.

9. Moisture

    Moisture is the percent by weight of water adsorbed on activated carbon.

10. Molasses Number

    The molasses number is calculated from the ratio of the optical densities of the filtrate of a molasses solution treated with a standard activated carbon and the activated carbon in question. This is a test method of Pittsburgh Activated Carbon Company.

11. Sieve Analysis (Dry)

    The distribution of particle sizes in a given sample is obtained by mechanically shaking a weighed amount of material through a series of test sieves and determining the weight retained by or passing given sieves.

12. Total Ash of Regenerated Carbon

    The total ash of a carbon is a measure of the amount of inorganic matter present. This test is accomplished by a combustion process in which organic matter is converted to carbon dioxide and water at a controlled temperature to prevent decomposition and volatilization of inorganic substances as much as is consistent with complete oxidation of organic matter.

CHAPTER 11

TREATMENT OF METAL WASTESTREAMS

by

Bill Strangio

TABLE OF CONTENTS

Chapter 11. TREATMENT OF METAL WASTESTREAMS

	Page
OBJECTIVES	522
WORDS	523

LESSON 1

- 11.0 METHODS OF TREATMENT 527
- 11.1 SOURCES OF METAL WASTESTREAMS 528
- 11.2 TYPES OF WASTESTREAMS 531
  - 11.20 Chelating Agents 531
  - 11.21 Spent Baths 531
  - 11.22 Dilute Rinse Waters (Without Chelators) 532
  - 11.23 Dilute Rinse Waters (With Chelators) 532
  - 11.24 Gold, Silver, and Other Metals 532
  - 11.25 Organically Contaminated Wastewaters 532
  - 11.26 Cyanide and Hexavalent Chromium 532
- 11.3 TREATMENT PROCESSES 533
  - 11.30 Batch and Continuous Process Modes 533
  - 11.31 Neutralization (pH Adjustment) 537
  - 11.32 Common Metals Removal 538
  - 11.33 Complexed Metals Removal 546
  - 11.34 Reduction of Hexavalent Chromium 547
  - 11.35 Cyanide Destruction by Oxidation 549
    - 11.350 Safety 549
    - 11.351 Cyanide Sources and Treatment 549
    - 11.352 Chemistry Involved 549
    - 11.353 Batch or Continuous Treatment 550
  - 11.36 Precious Metals Recovery 553
    - 11.360 Recovery Treatment Processes 553
    - 11.361 Troubleshooting Ion Exchange Systems 555
  - 11.37 Oily Waste Removal 557
  - 11.38 Solvent Control 558
  - 11.39 Control of Toxic Organics 558

DISCUSSION AND REVIEW QUESTIONS 559

LESSON 2

11.4 PROCESS INSTRUMENTATION AND CONTROLS 560
- 11.40 pH and ORP Probes 560
  - 11.400 Description 560
  - 11.401 Cleaning 560
  - 11.402 Calibration 564
  - 11.403 Troubleshooting pH and ORP Probes 565
- 11.41 Flow Measurement/Totalization Devices 566
- 11.42 Level Controls 567
- 11.43 Programmable Controllers 567
- 11.44 Voltage, Current, and Resistance Measurements 567

11.5 METALLIC SLUDGE DEWATERING 568
- 11.50 Reduction of Sludge Volumes 568
- 11.51 Methods to Dewater Sludges 568
  - 11.510 Centrifuges 568
  - 11.511 Vacuum Filters 569
  - 11.512 Bag Filters 569
  - 11.513 Filter Press 569
- 11.52 Sludge Drying Methods 574
  - 11.520 Air Drying 574
  - 11.521 Induced Air Drying 574
  - 11.522 Heating 574
  - 11.523 Reduced Pressure, Low Heat Drying (Vacuum Drying) 574

DISCUSSION AND REVIEW QUESTIONS 575

LESSON 3

11.6 OPERATION, MAINTENANCE, AND TROUBLESHOOTING 576
- 11.60 Know the Processes Involved 576
- 11.61 Knowledge and Abilities Required 576
- 11.62 Normal Operation 577
  - 11.620 Visual Observations 577
  - 11.621 Typical Operator Duties 577
  - 11.622 Sampling 579
    - 11.6220 Importance of Sampling 579
    - 11.6221 Representative Sampling 579
    - 11.6222 Time of Sampling 579
    - 11.6223 Types of Samples 579
    - 11.6224 Sampling Devices 580
    - 11.6225 Preservation of Samples 580
    - 11.6226 Quality Control in the Wastewater Laboratory 581

			11.6227	Summary	581
			11.6228	Additional Reading	581
		11.623	Flow Measurement		582
			11.6230	Operators' Responsibilities	582
			11.6231	Types of Flow Measurement Devices	582
			11.6232	Location of Measuring Devices	584
			11.6233	Additional Information	584
		11.624	Ventilation and Exhaust Systems		584
		11.625	Pumps		585
			11.6250	Types of Pumps	585
			11.6251	Starting a New Pump	585
			11.6252	Long-Term Pump Shutdowns	585
			11.6253	Pump-Driving Equipment	587
			11.6254	Electrical Controls	587
			11.6255	Procedures for Starting and Stopping Pumps During Normal Operation	587
	11.63	Maintenance			590
		11.630	Safety		590
		11.631	Preventive Maintenance Program		590
		11.632	Spare Parts Inventory		591
	11.64	Troubleshooting			591
		11.640	pH Meters		591
		11.641	ORP Meters		592
		11.642	Instrumentation		592
		11.643	Centrifugal Pump		593
		11.644	Diaphragm Pump		593
		11.645	Other Pump Problems		594
		11.646	Waste Is Not Flocculating		595
11.7	SAFETY				595
	11.70	Beware of Hazardous Chemicals and Wastes			595
	11.71	Storage and Handling of Chemicals			597
	11.72	First Aid			598
DISCUSSION AND REVIEW QUESTIONS					601
SUGGESTED ANSWERS					601

OBJECTIVES

Chapter 11. TREATMENT OF METAL WASTESTREAMS

Following completion of Chapter 11, you should be able to:

1. Explain the need to treat metal wastestreams.
2. Identify the sources of metal wastestreams.
3. Operate and maintain neutralization, metal precipitation, cyanide destruction, and complexed metal treatment facilities.
4. Collect, treat, and dispose of sludges generated by these treatment processes.
5. Troubleshoot treatment facilities described in this chapter.
6. Safely perform the duties of an operator of these facilities.

WORDS

Chapter 11. TREATMENT OF METAL WASTESTREAMS

ACUTE HEALTH EFFECT

An adverse effect on a human or animal body, with symptoms developing rapidly.

AMBIENT (AM-bee-ent) TEMPERATURE

Temperature of the surroundings.

AMPHOTERIC (AM-fuh-TUR-ick)

Capable of reacting chemically as either an acid or a base.

ANION (AN-EYE-en)

A negatively charged ion in an electrolyte solution, attracted to the anode under the influence of a difference in electrical potential. Chloride ion (Cl⁻) is an anion.

ANNULAR (AN-yoo-ler) SPACE

A ring-shaped space located between two circular objects. For example, the space between the outside of a pipe liner and the inside of a pipe.

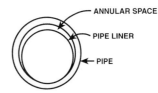

BATCH PROCESS

A treatment process in which a tank or reactor is filled, the water (or wastewater or other solution) is treated or a chemical solution is prepared, and the tank is emptied. The tank may then be filled and the process repeated. Batch processes are also used to cleanse, stabilize, or condition chemical solutions for use in industrial manufacturing and treatment processes.

BUFFER CAPACITY

A measure of the capacity of a solution or liquid to neutralize acids or bases. This is a measure of the capacity of water or wastewater for offering a resistance to changes in pH.

CATION (KAT-EYE-en)

A positively charged ion in an electrolyte solution, attracted to the cathode under the influence of a difference in electrical potential. Sodium ion (Na⁺) is a cation.

CHELATING (KEY-LAY-ting) AGENT

A chemical used to prevent the precipitation of metals (such as copper).

CHRONIC HEALTH EFFECT

An adverse effect on a human or animal body with symptoms that develop slowly over a long period of time or that recur frequently.

COLLOIDS (KALL-loids)

Very small, finely divided solids (particles that do not dissolve) that remain dispersed in a liquid for a long time due to their small size and electrical charge. When most of the particles in water have a negative electrical charge, they tend to repel each other. This repulsion prevents the particles from clumping together, becoming heavier, and settling out.

COLORIMETRIC MEASUREMENT

A means of measuring unknown chemical concentrations in water by measuring a sample's color intensity. The specific color of the sample, developed by addition of chemical reagents, is measured with a photoelectric colorimeter or is compared with color standards using, or corresponding with, known concentrations of the chemical.

COMMON METAL

Aluminum, cadmium, chromium, copper, iron, lead, nickel, tin, zinc, or any combination of these elements are considered common metals.

COMPOSITE (PROPORTIONAL) SAMPLE

A composite sample is a collection of individual samples obtained at regular intervals, usually every one or two hours during a 24-hour time span. Each individual sample is combined with the others in proportion to the rate of flow when the sample was collected. Equal volume individual samples also may be collected at intervals after a specific volume of flow passes the sampling point or after equal time intervals and still be referred to as a composite sample. The resulting mixture (composite sample) forms a representative sample and is analyzed to determine the average conditions during the sampling period.

CONTINUOUS PROCESS

A treatment process in which water is treated continuously in a tank or reactor. The water being treated continuously flows into the tank at one end, is treated as it flows through the tank, and flows out the opposite end as treated water.

DPD METHOD

A method of measuring the chlorine residual in water. The residual may be determined by either titrating or comparing a developed color with color standards. DPD stands for N,N-diethyl-p-phenylenediamine.

DEWATERABLE

This is a property of sludge related to the ability to separate the liquid portion from the solid, with or without chemical conditioning. A material is considered dewaterable if water will readily drain from it.

DIATOMACEOUS (DYE-uh-toe-MAY-shus) EARTH

A fine, siliceous (made of silica) earth composed mainly of the skeletal remains of diatoms.

DIATOMS (DYE-uh-toms)

Unicellular (single cell), microscopic algae with a rigid, box-like internal structure consisting mainly of silica.

DOCTOR BLADE

A blade used to remove any excess solids that may cling to the outside of a rotating screen.

DRAG OUT

The liquid film (plating solution) that adheres to the workpieces and their fixtures as they are removed from any given process solution or their rinses. Drag-out volume from a tank depends on the viscosity of the solution, the surface tension, the withdrawal time, the draining time, and the shape and texture of the workpieces. The drag-out liquid may drip onto the floor and cause wastestream treatment problems. Regulated substances contained in this liquid must be removed from wastestreams or neutralized prior to discharge to POTW sewers.

ELECTROLYTE (ee-LECK-tro-lite)

A substance that dissociates (separates) into two or more ions when it is dissolved in water.

EXOTHERMIC (EX-o-THUR-mick)

A chemical change that produces heat.

HYGROSCOPIC (hi-grow-SKOP-ick)

Absorbing or attracting moisture from the air.

IDLH

Immediately Dangerous to Life or Health. The atmospheric concentration of any toxic, corrosive, or asphyxiant substance that poses an immediate threat to life or would cause irreversible or delayed adverse health effects or would interfere with an individual's ability to escape from a dangerous atmosphere.

ION

An electrically charged atom, radical (such as SO_4^{2-}), or molecule formed by the loss or gain of one or more electrons.

ION EXCHANGE

A water or wastewater treatment process involving the reversible interchange (switching) of ions between the water being treated and the solid resin contained within an ion exchange unit. Undesirable ions are exchanged with acceptable ions on the resin or recoverable ions in the water being treated are exchanged with other acceptable ions on the resin.

ION EXCHANGE RESINS

Insoluble polymers, used in water or wastewater treatment, that are capable of exchanging (switching or giving) acceptable cations or anions to the water being treated for less desirable ions or for ions to be recovered.

MOTHER CIRCUIT BOARD

The base circuit board or the main circuit board.

NEUTRALIZATION (noo-trull-uh-ZAY-shun)

Addition of an acid or alkali (base) to a liquid to cause the pH of the liquid to move toward a neutral pH of 7.0.

ORP (pronounce as separate letters)

Oxidation-Reduction Potential. The electrical potential required to transfer electrons from one compound or element (the oxidant) to another compound or element (the reductant); used as a qualitative measure of the state of oxidation in water and wastewater treatment systems. ORP is measured in millivolts, with negative values indicating a tendency to reduce compounds or elements and positive values indicating a tendency to oxidize compounds or elements.

PICKLE

An acid or chemical solution in which metal objects or workpieces are dipped to remove oxide scale or other adhering substances.

POLYMER (POLY-mer)

A long-chain molecule formed by the union of many monomers (molecules of lower molecular weight). Polymers are used with other chemical coagulants to aid in binding small suspended particles to larger chemical flocs for their removal from water. Also see POLYELECTROLYTE.

POTTING COMPOUNDS

Sealing and holding compounds (such as epoxy) used in electrode probes.

PRECIOUS METAL

Metal that is very valuable, such as gold or silver.

PRECIPITATE (pre-SIP-uh-TATE)

(1) An insoluble, finely divided substance that is a product of a chemical reaction within a liquid.
(2) The separation from solution of an insoluble substance.

SLUDGE (SLUJ)

(1) The settleable solids separated from liquids during processing.
(2) The deposits of foreign materials on the bottoms of streams or other bodies of water or on the bottoms and edges of wastewater collection lines and appurtenances.

SPECIFIC GRAVITY

(1) Weight of a particle, substance, or chemical solution in relation to the weight of an equal volume of water. Water has a specific gravity of 1.000 at 4°C (39°F). Particulates with specific gravity less than 1.0 float to the surface and particulates with specific gravity greater than 1.0 sink.

(2) Weight of a particular gas in relation to the weight of an equal volume of air at the same temperature and pressure (air has a specific gravity of 1.0). Chlorine gas has a specific gravity of 2.5.

WATER HAMMER

The sound like someone hammering on a pipe that occurs when a valve is opened or closed very rapidly. When a valve position is changed quickly, the water pressure in a pipe will increase and decrease back and forth very quickly. This rise and fall in pressures can cause serious damage to the system.

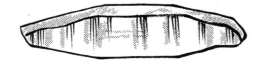

CHAPTER 11. TREATMENT OF METAL WASTESTREAMS

(Lesson 1 of 3 Lessons)

11.0 METHODS OF TREATMENT

Metal wastestreams cannot be treated to the degree that is required for discharge by using only biological processes. Biological processes such as activated sludge can remove some metals by adsorbing them on the flocculated sludge. The metals removed by that process are sent to the anaerobic digesters and processed into a more concentrated form. This accumulation effect, in turn, makes the final disposal of the digested sludge a difficult operational problem. To safeguard the discharge quality of the local POTW, industrial dischargers are legally required to adequately pretreat metal-bearing wastewaters before discharge to POTW sewers. This is done by the use of physical–chemical treatment processes. These are methods specifically designed to remove contaminants based on the physical–chemical nature of the contaminant.

It is not unusual that a combination of processes would be used to treat a given facility's wastestream. Even the least complicated production facility will use many chemicals to make up the solutions used in the production area. While many of the chemicals do not have metallic components, their presence affects the use of the metallic chemicals and their eventual treatment and reuse or disposal. Certain chemicals are "refractory," that is, they are difficult to treat. Therefore, to avoid creating a larger problem by mixing the refractory chemical into the main wastestream, the refractory chemical is treated as a separate wastestream, perhaps as a *BATCH*[1] treatment whose effluent is then safe to route through the main waste treatment system.

The final treatment process in most facilities consists of precipitation followed by filtration or sedimentation to remove the *PRECIPITATED*[2] contaminants. The hydroxide form of most metals is relatively insoluble (it does not dissolve easily); therefore, once the *PRECIPITATE*[2] has been formed, it is fairly easily removed from the wastestream. Processes of this type produce an adequately treated water and a *DEWATERABLE*[3] *SLUDGE*[4] that meets most state guidelines for disposal purposes.

The industrial wastewater treatment facility operator has the job of operating the facility so that it meets discharge requirements and minimizes the amount of sludge produced. The Publicly Owned Treatment Works' (POTW) goal is to enforce the regulations so that their effluent and sludge meet all quality guidelines, while the states and the Environmental Protection Agency (EPA) oversee regional water quality and verify that the sludges produced during treatment are disposed of in a manner that avoids their reintroduction into the environment. Therefore, an industrial wastewater treatment plant operator has to be knowledgeable about regulatory agency reporting and record-keeping requirements as well as about the wastewater treatment processes.

Treatment processes that the operator must be familiar with include the following:

- Oil/water, solvent/water separation
- pH–ORP adjustment and control
- Ion exchange
- Reverse osmosis
- Activated carbon
- Air stripping
- Plate out techniques
- Precipitation
- Filtration

---

[1] *Batch Process.* A treatment process in which a tank or reactor is filled, the water (or wastewater or other solution) is treated or a chemical solution is prepared, and the tank is emptied. The tank may then be filled and the process repeated. Batch processes are also used to cleanse, stabilize, or condition chemical solutions for use in industrial manufacturing and treatment processes.

[2] *Precipitate* (pre-SIP-uh-TATE). (1) An insoluble, finely divided substance that is a product of a chemical reaction within a liquid. (2) The separation from solution of an insoluble substance.

[3] *Dewaterable.* This is a property of sludge related to the ability to separate the liquid portion from the solid, with or without chemical conditioning. A material is considered dewaterable if water will readily drain from it.

[4] *Sludge* (SLUJ). (1) The settleable solids separated from liquids during processing. (2) The deposits of foreign materials on the bottoms of streams or other bodies of water or on the bottoms and edges of wastewater collection lines and appurtenances.

Many of the processes listed above actually are "concentrating" processes. Each of these processes is used to solve a particular treatment problem. They may produce a very high-quality effluent that can be reused or disposed of, but they also frequently produce a concentrated waste that requires further handling or treatment. This would be typical of ion exchange, reverse osmosis, precipitation, and filtration processes. The advantages of these methods are that they may allow the use of smaller sized final treatment units, they may produce reusable water, and the concentrated waste (from ion exchange) may be reusable if the unit is correctly located within the production facility.

11.1 SOURCES OF METAL WASTESTREAMS

Certain types of industries produce metallically contaminated wastestreams. These industries include, but are not limited to:

- Job shop platers
- Hard chrome platers
- Bumper remanufacturers
- Printed circuit board manufacturers
- Hard disk manufacturers
- Metal pickling operations
- Galvanizers
- Metal strut manufacturers

Other industries produce metallically contaminated wastes even though their main product is not obviously metallic in nature; they include:

- Rubber glove manufacturers
- Photo processors
- Fungicide manufacturers
- Algicide manufacturers

Figure 11.1 illustrates a typical wastestream schematic for a printed circuit board manufacturer. In addition to the wastes shown in Figure 11.1, there may be additional wastes resulting from water treatment operations that need to be handled to ensure high-quality deionized water within the production area. Review of the treatment processes shows that there are:

- At least four batch treatments.
- At least two ion exchange processes.
- Several chemicals hauled off site.
- Several chemicals used to treat the wastewater and to produce a dewatered sludge.

Manufacturing processes change frequently to meet new product demands. When this happens, the chemicals used in the processes also change. It is your responsibility to know about any such changes because they will have an effect on the operation of the treatment system. As the operator of an industrial wastewater treatment facility, you will have to be knowledgeable, organized, inquisitive, and capable of performing many separate tasks at the same time.

Before continuing with the discussion of how to treat metal wastestreams, it is important to remember that an effective source control program can significantly reduce not only the amount of wastewater requiring treatment but also the cost and complexity of the treatment systems needed. Plating processes can be changed or altered to reduce the volumes of process wastewaters and to control amounts of wastes that must be treated. In the metal plating industry, volumes of wastewater to be treated can be reduced by countercurrent rinsing processes (Figure 11.2). Rinse flows can be controlled and reduced by using conductivity meters, spray rinses, and proper racking to reduce *DRAG OUT*.[5] Some companies install static rinse tanks, which have no running flow. These tanks are used for rinsing gross contamination. The temperature of static rinse tanks is frequently kept at an elevated level to encourage evaporation. The contents of these tanks are sometimes disposed of by transferring the water to plating tanks as makeup water. Good housekeeping practices, prevention of accidental spills, and control of leaks and drag out are other procedures used to reduce the amount of waste that must be treated. Table 11.1 summarizes procedures for reducing waste treatment costs. Also see Chapter 4, "Preventing and Minimizing Wastes at the Source," for additional information.

TABLE 11.1 PROCEDURES FOR REDUCING WASTE TREATMENT COSTS[a]

1. Reduction of flow volumes
2. Segregation of different wastestreams
3. Recycling of water
4. Changing production processes to decrease generation of wastes
5. Effective rinsing techniques
6. Elimination of batch dumps of process wastes
7. Controlling wastes at source
8. Good housekeeping and control of accidental spills
9. Good solvent management plan

[a] Provided by Patrick Kwok

QUESTIONS

Write your answers in a notebook and then compare your answers with those on page 601.

11.0A Why are physical–chemical processes used to treat metal wastestreams rather than biological processes?

11.1A What types of industries produce metallically contaminated wastestreams?

---

[5] *Drag Out.* The liquid film (plating solution) that adheres to the workpieces and their fixtures as they are removed from any given process solution or their rinses. Drag-out volume from a tank depends on the viscosity of the solution, the surface tension, the withdrawal time, the draining time, and the shape and texture of the workpieces. The drag-out liquid may drip onto the floor and cause wastestream treatment problems. Regulated substances contained in this liquid must be removed from wastestreams or neutralized prior to discharge to POTW sewers.

Metal Wastestreams 529

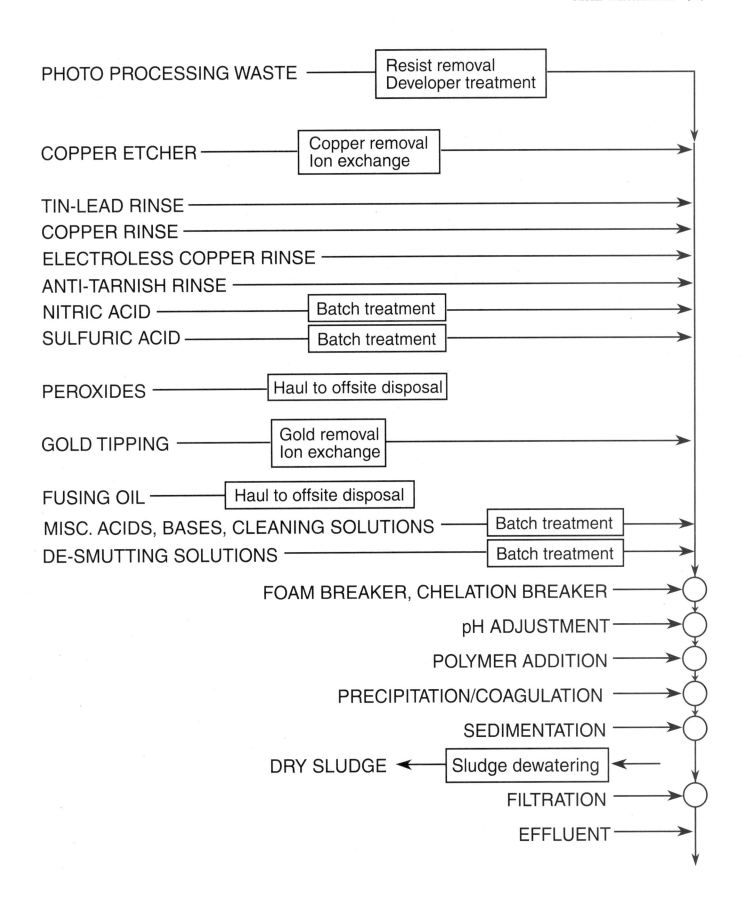

Fig. 11.1 Schematic of wastestream generation and treatment for printed circuit board manufacturing processes

530 Treatment Plants

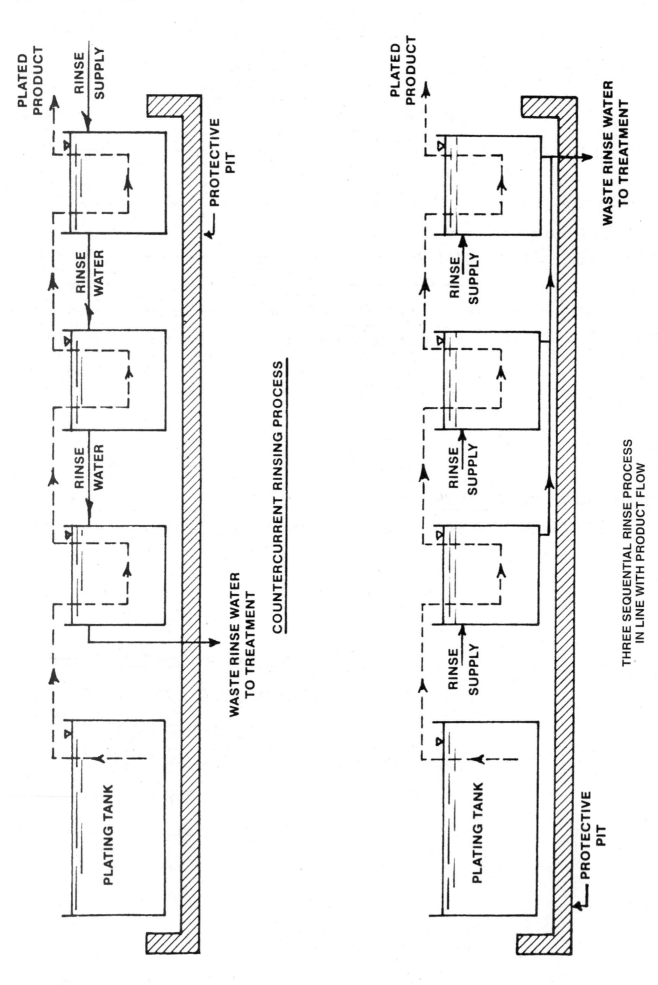

Fig. 11.2 *Countercurrent rinsing and three sequential rinse process in line with product flow*

11.2 TYPES OF WASTESTREAMS

The size and complexity of the production activities at a given facility have a direct bearing on the complexity of the waste treatment system or systems that will be used. It is often necessary to treat some wastestreams individually before combining them with other types of wastes for further treatment. Other types of wastestreams may be blended or grouped for treatment purposes based on the outcome of a thorough review (and possibly laboratory analyses) to ensure chemical compatibility. For example, wastestreams might be grouped as follows for purposes of treatment:

- Spent baths (actually chemically concentrated plating solutions or acids/bases)
- Dilute rinse waters (with no *CHELATING AGENTS*[6])
- Dilute rinse waters (with chelating agents)
- Gold, silver, and some other metals
- Organically contaminated waters (glycol ether, resists, alcohols)
- Cyanide and hexavalent chromium

11.20 Chelating Agents

If acids, bases, and metallic salts were the only chemicals used in all metal solutions used in industry, treatment would be relatively easy. However, that is the exception. Most manufacturers improve their processes in order to produce a better manufactured product. Chelating agents and other dispersants are added to the plating solutions to improve the quality of plated surfaces and to extend the useful life of the plating bath. Tank rinse waters and drag out containing the chelating agents and dispersants eventually find their way to the treatment system. Their presence there causes metals to remain in solution, rather than precipitating, causing the system to violate its contaminant discharge limits.

The "electroless" plating solutions used in industries such as printed circuit board manufacturing and hard disk manufacturing have chelating systems that are extremely effective. Additional chemicals (other than those used to vary pH) must be used to defeat the chelating effect. This adds cost to the treatment process because the extra chemicals have to be purchased, and the additional sludge that is produced has to be dewatered and hauled to a landfill that is certified for toxic/hazardous waste disposal.

Chemicals that have been used by themselves or in combination to overcome the effects of the chelating agents include:

- Ferric chloride
- Ferrous sulfate
- Calcium chloride
- Phosphoric acid
- Calcium hydroxide

These compounds increase the bulk of the sludge and, in some cases, have reduced the metal content below the levels that recyclers deem feasible for reclamation.

A chemical system adopted by several facilities uses a carbamate additive controlled by *ORP*[7] instrumentation. The additive lowers (reduces) the ORP level and allows the metals to precipitate. Little or no polymer is added to the mixture. This reduces the amount of sludge that must be dewatered (50 percent by some estimates). The precipitation step must be followed by a filtration system to remove the suspended fine particles.

11.21 Spent Baths

From a production standpoint, a bath is spent when it no longer is suitable for the intended purpose. This does not mean that the bath is weak from a chemical standpoint. It may be contaminated with other materials or it may be discarded so as to reduce the possibility of producing workpiece defects. Many chemicals are changed at the start of a new shift without regard to the use or lack of use of the chemical.

Spent plating baths can have concentrations in the 75,000 mg/L range. When this is compared to typical rinse waters in the 10 to 20 mg/L range, it is easy to realize how concentrated a spent bath can be.

The operator typically has two options available to treat the spent baths. They are: (1) to meter the spent bath into the rinse water system at a slow rate thereby diluting it, or (2) batch treating it in its concentrated form. The first method, dilution, may not be practical if the bath contains chemicals that will cause treatment problems, or if the rinse flows are not sufficient to dilute the waste to treatable levels. The second method, batch treatment, can be hazardous if heat is generated because of *EXOTHERMIC*[8] reactions or if there is evolution of gases, such

---

[6] *Chelating* (KEY-LAY-ting) *Agent.* A chemical used to prevent the precipitation of metals (such as copper).

[7] *ORP* (pronounce as separate letters). Oxidation-Reduction Potential. The electrical potential required to transfer electrons from one compound or element (the oxidant) to another compound or element (the reductant); used as a qualitative measure of the state of oxidation in water and wastewater treatment systems. ORP is measured in millivolts, with negative values indicating a tendency to reduce compounds or elements and positive values indicating a tendency to oxidize compounds or elements.

[8] *Exothermic* (EX-o-THUR-mick). A chemical change that produces heat.

as hydrogen gas, from the batch treatment process. In general, the second method of treatment is preferred because it is usually easier and cheaper to treat a small amount of concentrated waste than a large volume of less concentrated waste.

11.22 Dilute Rinse Waters (Without Chelators)

Dilute rinse waters can be treated by several methods; the most common method is by coagulation and precipitation. If the waste is or could be contaminated with solvents, oils, and greases, membrane filtration and ion exchange processes are usually not used. Where the wastewater is free of those contaminants, or can be treated for them ahead of time, membrane filtration and ion exchange processes produce excellent effluent and a more concentrated wastestream (blowdown), which will then have to be treated.

The absence of chelating agents allows the operator to use a process consisting of pH adjustment, polymer injection, flocculation, and sedimentation. A final step of pH control is usually provided to make sure that discharge guidelines are met.

11.23 Dilute Rinse Waters (With Chelators)

The presence of chelating agents complicates the treatment of the wastewater. As mentioned above, one or more chemicals will need to be added to overcome the dispersal effect. In most cases, the effluent must also be filtered to remove particles that did not agglomerate (clump together) and settle in the sedimentation tank.

11.24 Gold, Silver, and Other Metals

These valuable metals are used at various locations in the facility; therefore, the wastewaters are typically treated at the location of use by the ion exchange process. The captured metals are then retrieved and reused. In most instances, the ion exchange resin is sent out of the facility to a metal reclaimer, and new resin cylinders are installed. The same technique is sometimes used even when the metal is not as costly (such as copper). The goal is to capture the metal at the source (copper etching machine) and thereby produce less sludge at the final treatment step. Sludge disposal is a significant portion of waste treatment operating costs.

Electroplating and other techniques can be used to remove wastestream metals in a solid form. These processes are usually applied to spent baths and wastestreams where the metal content is relatively high. So-called "dead rinses" are routed to a plate out tank where the metal contaminant (usually copper) is plated onto another object. The metal can thus be removed as a solid and the treated water can be returned to the rinse tank for reuse.

A similar process can be used to greatly reduce the metal content of spent electroless plating solutions. A ball of steel wool is dipped into the reaction vessel containing the metal-bearing wastewater. The metal (typically nickel) plates out on the steel wool and can then be removed. Caution should be exercised when using this technique because hydrogen gas may be produced during the plating reaction. This technique does not affect any chelating agents that might be present in the wastewater; they will still need to be treated if the waste is to be discharged to the main wastewater treatment system.

11.25 Organically Contaminated Wastewaters

The presence of organically contaminated wastewaters can also complicate wastewater treatment. Many organics are regulated and cannot be discharged since they pose a problem to the local POTW or because they represent an *ACUTE*[9] or *CHRONIC HEALTH THREAT*[10] if they are discharged to the environment. When the facility goes through its initial permitting, and also on its re-permitting, the chemicals in use in the facility will be reviewed to ensure that regulated chemicals are identified and controlled. Testing of the effluent for their presence will be required. The operator has to be aware of the requirements and frequency of the tests.

Problems can be caused by various organics, even if they are not on a list of regulated organics. Excessive foaming of the wastewater can occur. Coating of instrumentation probes, with consequent loss of accuracy, as well as clogging of membranes and resins, can occur. Swelling of membranes and resins due to organic solvent exposure can lead to reduced treatment capacities and reduced service life. Tank linings and containment area coatings may also be affected.

Refractory organics are best treated at their point of use before they are diluted by the main wastestream. They can be treated using a method that best suits the type of contamination. These include:

- foam suppression
- air stripping
- gravity separation
- emulsion cracking
- flotation
- activated carbon adsorption

Operators of industrial wastewater treatment facilities need to find out what types of organics are being used in the facility and determine if any problems are being caused at the treatment plant by their use. The operator will have to take an active part in seeing that the problem is recognized by the supervisor and in devising a solution.

11.26 Cyanide and Hexavalent Chromium

Both cyanide and hexavalent chromium require special treatment prior to the precipitation reaction or the metals will not precipitate. Cyanide must be oxidized to at least the cyanate form to lose its toxicity, while hexavalent chromium must be reduced to the trivalent form before it can be precipitated. The two reactions are quite different in their conditions (that is, pH and ORP) and in the speed of the reaction. The hexavalent reaction is very rapid (a few seconds) while cyanide destruction is

---

[9] *Acute Health Effect.* An adverse effect on a human or animal body, with symptoms developing rapidly.
[10] *Chronic Health Effect.* An adverse effect on a human or animal body with symptoms that develop slowly over a long period of time or that recur frequently.

slow (approximately 10 minutes). It is important to note that the reactions take place within a relatively narrow pH range and that the ORP reading is only meaningful as a control value within that pH range. It is the operator's responsibility to keep the instrumentation calibrated correctly and to check the readings frequently to ensure that the system is within the prescribed range.

Each type of waste must be captured at its source and piped to its specific treatment unit. The operator should review production procedures to make sure the chromium solution is not being "dragged into" the cyanide containment area, and vice versa. Hexavalent chromium would not be affected by the cyanide treatment step and would proceed through the precipitative step without being removed, thereby causing discharge violations.

QUESTIONS

Write your answers in a notebook and then compare your answers with those on page 601.

11.2A What chemicals have been used to overcome the effects of chelating agents?

11.2B From a production standpoint, when is a bath considered spent?

11.2C What process is used to treat wastestreams containing gold and silver?

11.3 TREATMENT PROCESSES

There are certain basic treatment principles that the operator must keep in mind regarding the chemical reactions involved in physical–chemical treatment processes. Whenever chemicals are used, adequate mixing is essential to ensure completion of the chemical reactions. Also, remember that not all chemical reactions are instantaneous; some processes require specific detention times or reaction periods. Another important aspect of physical–chemical treatment is precise and accurate control of operating factors such as pH, ORP, and chemical doses. Use of too much chemical is not only wasteful and expensive, but may also contribute to poor performance of downstream processes. It is also important for an operator to know the design limits of the treatment processes and equipment. Excessive concentrations of waste materials can overwhelm the chemical metering capacity of a system resulting in incomplete chemical reactions and inadequately treated wastestreams.

The remainder of this lesson describes several of the most widely used processes for treating metal wastestreams.

11.30 Batch and Continuous[11] Process Modes

Many of the methods used to treat metal wastestreams may be operated as either a batch or a continuous type of treatment process. Usually wastewaters are treated by the batch process when dealing with small wastewater flows because storage requirements for wastewaters prior to treatment are minimal. Batch processes are easier to control and the water being treated is not discharged to sewers or the environment until after a satisfactory level of treatment is achieved.

Many operators consider the batch treatment process the simplest and most dependable. A typical hydroxide precipitation metals removal process can accomplish pH adjustment, mixing, flocculation, and clarification successively in a single tank (Figure 11.3). Usually two or more batch tanks are necessary for efficient operation. Metal wastes can flow into one batch tank while the other tank is treating a full tank of wastes. If only cyanide or hexavalent chromium is present in a separate wastestream, it can be pretreated on a separate batch basis also (Figure 11.4). Batch treatment is cost effective when the operation and waste flows are relatively small. Most industries do not use batch treatment processes, however, because of high wastewater flows and the space required for batch treatment facilities. In shops using continuous processing, batch treatment is an effective way to handle acid dumps, spent plating solutions, alkaline cleaners, and other concentrated solutions that would overwhelm the continuous process systems.

Most total treatment systems separate (segregate) flowing rinses from wastewaters that result from periodic dumping of process solutions that have reached the end of their useful life. The batch dumps are (1) low in volume, and (2) high in contaminant concentration, the two major factors that favor segregation for treatment.

Closer control of treatment processes is possible when the contamination level of the raw waste does not fluctuate (vary) widely. If a spent process solution (a *PICKLE*,[12] for example) is dumped directly into rinse water, the contamination level can increase 10,000 times for a short time. The control system for rinse water treatment cannot respond to this surge without losing the close control required for normal discharge of effluent.

The other factor favoring batch treatment concerns the handling of suspended solids from waste treatment processes. The sludges resulting from batch treatment vary somewhat in solids content, depending upon what is being treated. On the average, however, the sludge solids concentration from a batch treatment process is in the range of 5 to 10 percent (50,000 to 100,000 mg/L) solids. Suspended solids in rinse waters are usually around 100 to 200 mg/L. After the suspended solids have been removed from rinse waters, they produce a sludge in the range of 0.5 to 1.0 percent solids. It is much easier for an operator to deal with a small volume of 5 to 10 percent solids from a batch process than a large volume of 0.5 to 1.0 percent solids if the wastes are bled into the rinse waters for treatment.

Continuous flow treatment facilities are used to treat relatively high flows of metal wastes (Figure 11.5). In the continuous flow treatment mode, several tanks or treatment units are used. If the wastewater flow rate or metal concentrations fluctuate, a flow equalization tank will be needed before the treatment processes.

---

[11] *Continuous Process.* A treatment process in which water is treated continuously in a tank or reactor. The water being treated continuously flows into the tank at one end, is treated as it flows through the tank, and flows out the opposite end as treated water.

[12] *Pickle.* An acid or chemical solution in which metal objects or workpieces are dipped to remove oxide scale or other adhering substances.

534 Treatment Plants

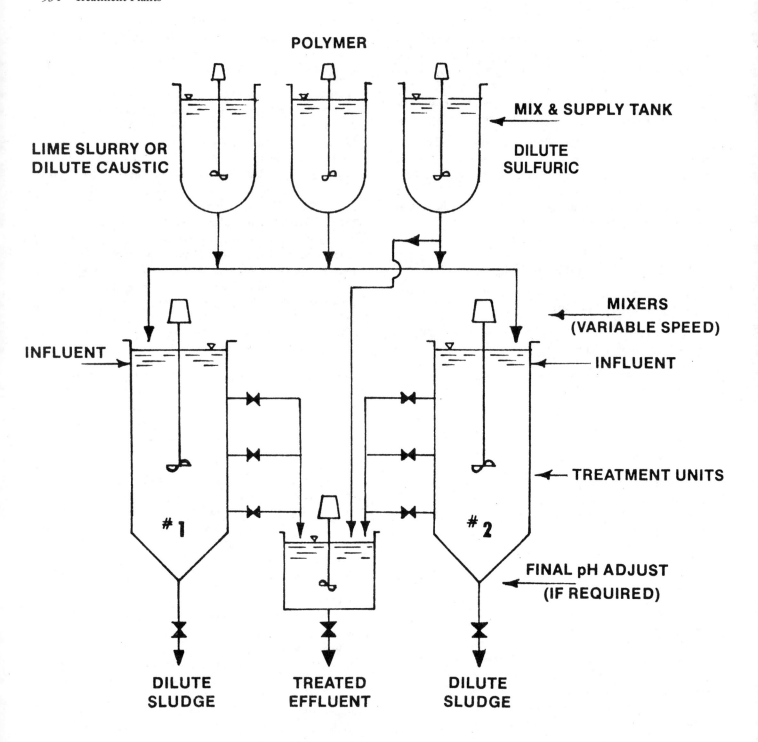

1. Tank #1 fills while Tank #2 is being treated.
2. Lime or caustic, and/or acid, is added to adjust pH for metal hydroxide precipitation.
3. Rapid mix (3-10 min. depending on size) to mix chemicals added for pH adjustment.
4. Polymer is added to improve flocculation.
5. Slow mixing for 20-30 min. to promote floc formation.
6. No mixing for 1-3 hours to allow floc to settle.
7. Decant clarified water to second pH adjustment tank.
8. After decanting, remove settled solids as dilute sludge.
9. If required, add sulfuric to correct pH, mix for 10 min., then discharge treated effluent.

Fig. 11.3 Batch treatment units for common metal removal by hydroxide precipitation and sedimentation (by Joe Shockcor)

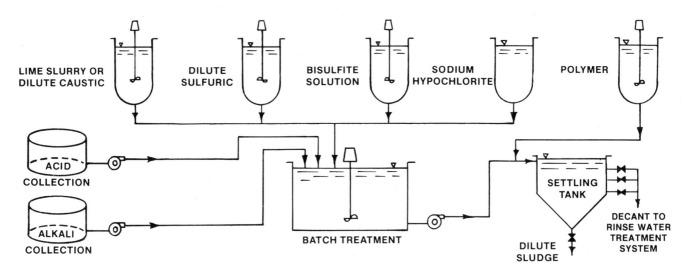

Fig. 11.4 Batch treatment units for common metal removal by hydroxide precipitation and sedimentation for concentrated wastes (by Joe Shockcor)

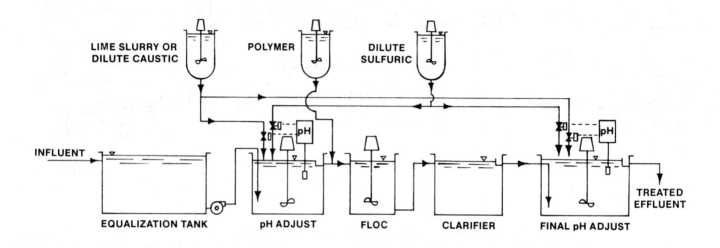

CONTINUOUS TREATMENT PROCESS FOR RINSE WATER

1. Equalization of influent flows and concentrations.
2. Pump equalized raw waste to subsequent treatment at uniform rate.
3. Adjust pH by adding alkali or acid to precipitate heavy metals using rapid mixing for 5 to 7 minutes.
4. During gravity transfer to Floc Tank, add polymer to improve flocculation.
5. Slow mix in Floc Tank for 5 to 7 minutes to promote growth of large settleable floc.
6. Allow suspended solids to settle by gravity in Clarifier during 1 to 3 hr retention period. (Periodically, remove dilute sludge from bottom of tank.)
7. Adjust final pH (if required) by controlled addition of acid or alkali for 5 to 7 minutes.

Fig. 11.5 Continuous treatment units for common metal removal by hydroxide precipitation and sedimentation for rinse water (by Joe Shockcor)

The sequence of processes would then consist of a rapid chemical mixing and precipitation process, flocculation, and final clarification. Metal precipitates are removed from the clarifier as sludge and are then usually dewatered and containerized for disposal.

11.31 Neutralization[13] (pH Adjustment)

Neutralization of strongly acidic (low pH) or strongly basic (high pH) waters is required before they are discharged to a sewer or the environment. Very high or low pH wastestreams may originate at cleaning, manufacturing, or waste treatment processes.

Wastewaters with high pH levels are neutralized by the addition of acids, such as sulfuric acid (H_2SO_4) and hydrochloric acid (HCl). Carbon dioxide (CO_2) and sulfur dioxide (SO_2) can be applied in the gaseous form to lower the pH of liquids. Some industries use flue gases (high in carbon dioxide or sulfuric acid fumes) to lower the pH of waters when the gas is available and this is an economical solution.

Low pH wastewaters are neutralized by adding calcium oxide or lime (CaO), magnesium oxide (MgO), calcium hydroxide ($Ca(OH)_2$) (a hydrated form of lime), magnesium hydroxide ($Mg(OH)_2$), sodium hydroxide (NaOH, caustic soda), or soda ash (Na_2CO_3).

Neutralization may be accomplished by either batch or continuous processes, depending on the amount of wastewater to be treated. In many industrial situations, acidic wastewaters may be generated from one source and basic wastewater is produced by another source. Under these conditions, the wastewaters are frequently stored in separate containers and mixed together at appropriate times to neutralize each other (provided the wastes are compatible). Both acidic and basic chemicals are needed to treat the wastewater if the flow from one source becomes excessive or is drastically reduced. When precise control of pH is essential, weak acids and bases may be used to adjust or "fine-tune" the pH.

When mixed with water, acids and bases produce solutions made up wholly or partially of ions. Solutes (dissolved substances) that exist almost completely as ions in solution are called strong *ELECTROLYTES*.[14] Solutes that do not react completely with the water form both neutral molecules and ions and are called weak electrolytes. The degree of ionization is determined by measuring the electrical conductivity of the solution.

Chemical equilibrium exists when the reactants are forming as rapidly as the products, so that the composition of the mixture remains constant and does not change with time. At any one temperature, the equilibrium constant K has a fixed numerical value characteristic of the particular chemical equation. However, in industrial wastewater treatment, the stream being treated is made up of water plus many contaminants that provide competing equilibria. Furthermore, reaction rate, temperature, and mixing conditions vary as does the waste loading. Therefore, the pure chemistry involved is not directly used by the operator in day-to-day activities.

The laboratory test used to measure changes in pH involves a procedure called titration. A titration is the drop-by-drop addition of one solution (called a reagent or titrant) to another solution (the wastewater sample) until there is a measurable change in the sample. This change may be a color change or the formation of a precipitate. When the change occurs, the end point of the titration has been reached.

A titration curve can be prepared to illustrate how different amounts of a titrant will affect pH. Such curves (graphs) are often designed to indicate other process variables, such as contact time, temperature, nature and amount of solids, and handling characteristics in relation to the system pH.

The following titration curves illustrate terminology, characteristics, and changes with respect to a few commonly encountered adjustments. Figure 11.6 illustrates titration of a strong base (NaOH) with a strong acid (H_2SO_4). Initial additions of the titrant have a minor effect upon pH because NaOH plus product Na_2SO_4 has little *BUFFER CAPACITY*.[15] The curve is almost flat for each addition of titrant prior to the inflection. The inflection indicates an approach to the equivalence point. The equivalence point is graphically located halfway along the straight line on the graph between the upper and lower inflections. Strong acid and base equivalence points commonly occur near pH 7.0. The product of volume and concentration of added titrant at the equivalence point is an estimate of sample basicity.

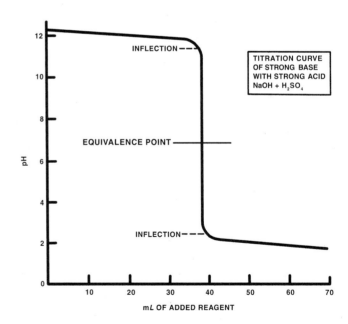

Fig. 11.6 Titration curve
(From *PHYSICAL CHEMICAL TREATMENT TECHNOLOGY*, US Environmental Protection Agency, Washington, DC, 1972)

---

[13] *Neutralization* (noo-trull-uh-ZAY-shun). Addition of an acid or alkali (base) to a liquid to cause the pH of the liquid to move toward a neutral pH of 7.0.

[14] *Electrolyte* (ee-LECK-tro-lite). A substance that dissociates (separates) into two or more ions when it is dissolved in water.

[15] *Buffer Capacity*. A measure of the capacity of a solution or liquid to neutralize acids or bases. This is a measure of the capacity of water or wastewater for offering resistance to changes in pH.

Figure 11.7 shows the effect of adding 4 percent or 1 N NaOH to a sample of industrial wastewater made up primarily of contact cooling water, machining rinse waters, and miscellaneous combined process discharges. From the curve, the operator can readily determine the amount of caustic needed to adjust the pH to whatever value best suits the process used.

Figure 11.7 also shows the effect of adding 4.9 percent or 1 N H_2SO_4 to the same sample of industrial wastewater. The curves in Figures 11.6 and 11.7 differ significantly but provide the same basic information. Five gallons of 4 percent NaOH in 1,000 gallons of wastewater will result in a pH of 11.65, and 5 gallons of 4.9 percent H_2SO_4 in the same 1,000 gallons of wastewater will result in a pH of 3.0.

Similar curves can and should be developed by the operator as a means of understanding any waste that must be neutralized. They are of further use in monitoring processes and calibrating automatic instrumentation. The steepness of the curves near the equivalence points is a good indication of the difficulty of pH control, especially when mixing times are short, tankage is too small, titrant concentrations are too high, and feed equipment, including pumps as well as instrumentation, are marginally sized.

Be very careful whenever strongly acidic and basic solutions are mixed because of the potential for splattering, the generation of heat, and the production of toxic gases. Highly acidic or basic solutions are very harmful to your skin and can cause loss of eyesight; the fumes are extremely irritating to your lungs and the mucous membranes in your body. To avoid splattering always slowly *ADD ACID TO WATER or BASE TO WATER*, never the reverse.

QUESTIONS

Write your answers in a notebook and then compare your answers with those on pages 601 and 602.

11.3A   What types of wastewater flows are treated by the batch process?

11.3B   How are wastewaters with high pH levels neutralized?

11.3C   How are wastewaters with low pH levels neutralized?

11.3D   What safety precautions must an operator take when working with strongly acidic and strongly basic solutions?

11.32   Common Metals[16] Removal

Common metals can be removed from wastestreams by either hydroxide precipitation or sulfide precipitation. Hexavalent chromium and cyanide wastes must be treated before attempting to remove the other common metals. Hexavalent chromium is not removed by hydroxide precipitation and cyanide will interfere with the precipitation processes' ability to remove other dissolved metals.

The first step in hydroxide precipitation of common metals is to determine the optimum pH for precipitating each type of metal you wish to remove. Figure 11.8 contains "idealized"[17] curves of metal concentrations versus pH for the common metals. Assume your wastestream contains trivalent chromium and zinc. Looking at the idealized curves in Figure 11.8, you will see that the optimum pH for trivalent chromium removal would be around 7.5 and the trivalent chromium concentration remaining in the treated water would be about 0.2 mg/L. Optimum pH for zinc removal would be around 10.5 and the remaining zinc concentration would be 0.2 mg/L.

When treating wastestreams containing two or more common metals with significantly different optimum pH levels, a two-stage operation is commonly used. First, the wastewaters are adjusted to the lowest desired pH level (7.5 for trivalent chromium) and the common metals are allowed to form precipitates and settle out by precipitation. Then, the remaining wastewater is transferred to another tank, the pH is readjusted (to 10.5 for zinc removal) and the remaining common metals are allowed to form precipitates and to settle out. Usually, the pH is increased to optimum levels by the addition of a lime slurry or sodium hydroxide (caustic soda).

From a practical standpoint, in many situations it may be necessary and possible to remove the common metals present in a wastestream at a single pH level due to the mix of chemicals in the metal wastes being treated. The optimum pH level can be determined by a series of jar tests[18] or bench tests (small scale laboratory tests) on the wastestream. The pH is changed in increments of 0.5 pH unit in each jar. The jar that requires the minimum pH adjustment and meets pretreatment standards indicates the optimum pH.

When a lime slurry or caustic soda is added to a wastestream containing common metals, these chemicals need to be mixed with the wastestream by some type of mixing device. Since metal hydroxide precipitates are often a *COLLOIDAL*[19] type of suspended solid, a coagulating chemical or polymer is added to encourage coagulation, flocculation, and sedimentation.[20] Final pH adjustment may be required after the precipitated solids have been removed to reduce the high pH created by the basic treatment chemicals.

Hydroxide precipitation produces large quantities of sludge containing toxic metals, which must be collected, treated, dewatered, and disposed of. Filter presses are commonly used to

---

[16] *Common Metal.* Aluminum, cadmium, chromium, copper, iron, lead, nickel, tin, zinc, or any combination of these elements are considered common metals.

[17] *NOTE:* Each of the curves in Figure 11.8 was drawn assuming pure compounds and deionized water. There were no "common ion" effects or chelates considered when the data were developed for plotting the curves.

[18] See Chapter 8, Section 8.4, "Selecting Chemicals and Determining Dosages," for jar test procedures.

[19] *Colloids* (KALL-loids).   Very small, finely divided solids (particles that do not dissolve) that remain dispersed in a liquid for a long time due to their small size and electrical charge. When most of the particles in water have a negative electrical charge, they tend to repel each other. This repulsion prevents the particles from clumping together, becoming heavier, and settling out.

[20] See Chapter 8, "Physical–Chemical Treatment Processes."

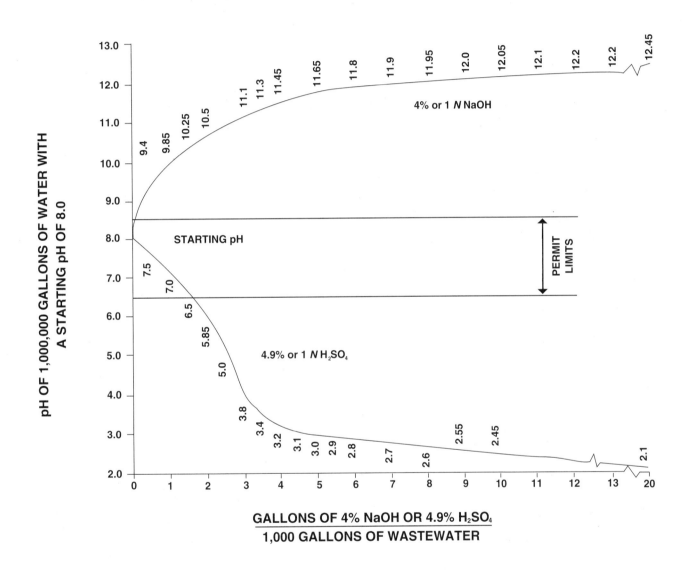

Fig. 11.7 Example titration of an industrial wastewater

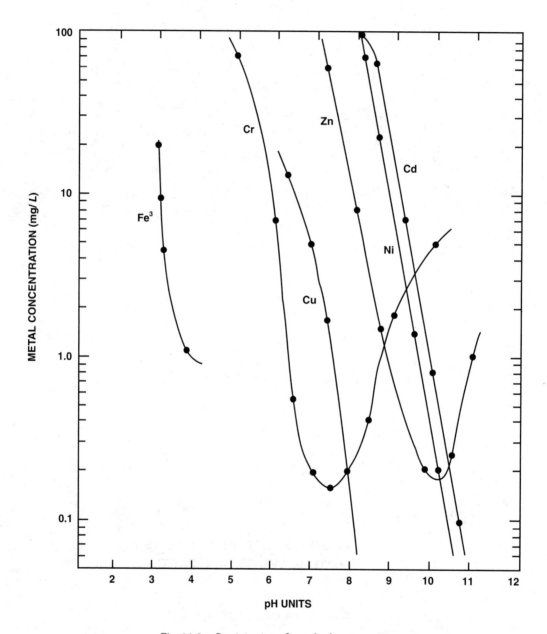

Fig. 11.8   Precipitation of metal salts versus pH
(From *METAL FINISHING WASTES*, EPA Technology/Transfer, US Environmental Protection Agency, Washington, DC)

dewater metal sludges. Metal hydroxide sludges are considered hazardous wastes and must be handled accordingly. To further reduce the suspended metal hydroxide precipitates in the effluent from a clarifier, filtration procedures, such as microscreening, ultrafiltration, and sand filtration, may be used. Figure 11.9 shows a suspended solids removal process using a Lamella settler (Figures 11.10 and 11.11) and a sand filter (Figures 11.12 and 11.13).

Limitations of the hydroxide precipitation process include:

1. The presence of complexing ions (chelating agents) such as EDTA, phosphate, and ammonia commonly found in plating formulations may adversely affect metal removal efficiencies.

2. Hydroxide precipitates tend to resolubilize (go back into solution) if the solution pH is either increased or decreased from the minimum solubility point of the metal, thereby reducing the removal efficiency (pH must be carefully controlled).

3. Different metals have different theoretical minimum solubilities at different pHs.

An effective alternative method for removing many heavy metal compounds is the addition of hydrogen sulfide or soluble sulfide salts, such as ferrous or sodium sulfide, to cause precipitation of the metals as sulfide compounds. This procedure is capable of achieving greater heavy metal removal than hydroxide precipitation because most metal sulfides are less soluble than metal hydroxides at high pH levels. The solubility of metal sulfide compounds depends on the pH, just like the metal hydroxide compounds; however heavy metal sulfides are insoluble over a much wider pH range.

Due to the extremely low solubilities of most metal sulfide compounds, very high metal removal efficiencies can be achieved. The sulfide precipitation process can remove hexavalent chromium without preliminary reduction of hexavalent chromium to trivalent chromium. Also, most complexed metals can be precipitated by the sulfide process.

Unfortunately, the sulfide precipitation process has several severe limitations. If the pH drops below 8, toxic hydrogen sulfide gas can be produced and released to the atmosphere. However, if the source of sulfide is ferrous sulfide, the problem of hydrogen sulfide gas production can be minimized.

SULFIDE IS TOXIC. THE PROCESS EFFLUENT MUST NOT CONTAIN ANY SULFIDE. The use of sulfide for heavy metal precipitation may require add-on effluent oxidation steps involving the addition of chlorine or peroxide to oxidize any residual sulfide. The costs of sulfide precipitating chemicals are high in comparison with hydroxide precipitation chemicals. Another major concern when using the sulfide precipitation process is the disposal of toxic metal sulfide sludges. The cost of placing the sludges in suitable containers, transporting them to an acceptable disposal site, and the long-term risks associated with the ultimate disposal of toxic waste must be weighed against the high-quality effluent achieved by sulfide precipitation. On the positive side, however, heavy metal sulfide sludges are much less soluble over a broad pH range than hydroxide precipitated sludges. When disposed of in a landfill, the sulfide sludges are less likely to leach out and contaminate groundwater supplies or adjacent soil than hydroxide sludges. Table 11.2 is a comparison of the hydroxide and sulfide metal precipitation processes.

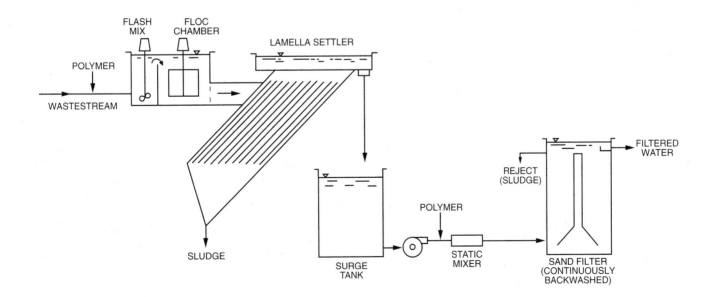

Fig. 11.9 Suspended solids removal using Lamella settler and sand filter

Fig. 11.10   Lamella settler (behind Parkson sign)
(Permission of New England Plating Co., Inc.)

Metal Wastestreams 543

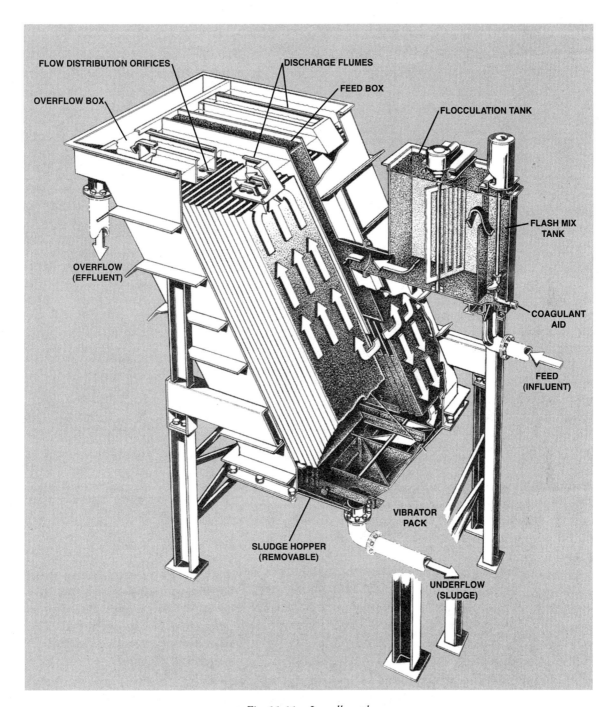

Fig. 11.11   Lamella settler
(Permission of Parkson Corporation. Lamella® is a registered
trademark of Parkson Corporation of Fort Lauderdale, Florida)

544 Treatment Plants

Fig. 11.12   Sand filter
(Permission of New England Plating Co., Inc.)

Metal Wastestreams 545

The DynaSand Filter is a continuous backwash, upflow, deep-bed granular media filter. The filter media is continuously cleaned by recycling the sand internally through an airlift pipe and sand washer. The regenerated sand is redistributed on the top of the sand bed allowing for a continuous, uninterrupted flow of filtrate and reject water.

Feed is introduced into the bottom of the filter, then flows upward through a series of riser tubes and is evenly distributed into the sand bed through the open bottom of an inlet distribution hood (A). The influent flows upward through the downward moving sand bed (B) with the solids being removed. The clean filtrate exits from the sand bed, overflows a weir (C), and is discharged from the filter (D). Simultaneously, the sand bed, along with the accumulated solids, is drawn downward into the suction of an airlift pipe (see below), which is positioned in the center of the filter. A small volume of compressed air is introduced into the bottom of the airlift (E). The sand, dirt, and water are transported upward through the pipe at a rate of about 200 GPM/sq ft. The impurities are scoured loose from the sand during this violently turbulent, upward flow. Upon reaching the top of the airlift (F), the dirty slurry spills over into the central reject compartment (I). The sand is returned to the sand bed through the gravity washer/separator (G), which allows the fast settling sand to penetrate, but not the dirty liquid. The washer/separator is placed concentrically around the part of the airlift and consists of several stages to prevent short-circuiting. By setting the filtrate weir (C) above the reject weir (J), a steady stream of clean filtrate flows upward, countercurrent to the sand, through this washer section and acts as a liquid barrier that carries away the dirt and reject water (K). Since the sand has a higher settling velocity than the dirt particles, it is not carried out of the filter. The sand is redistributed by the means of a sand distribution cone (H). The sand bed is continuously cleaned while both a continuous filtrate and a continuous reject stream are produced.

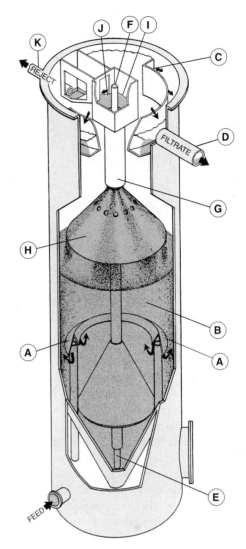

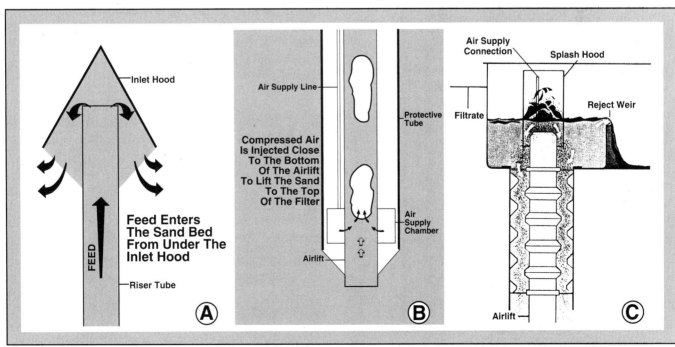

Fig. 11.13 Sand filter
(Permission of Parkson Corporation. DynaSand® Filter is a registered trademark of Parkson Corporation of Fort Lauderdale, Florida)

TABLE 11.2 COMPARISON OF HYDROXIDE AND SULFIDE PRECIPITATION PROCESSES

Item	Hydroxide	Sulfide
1. Effluent quality	Satisfactory metal concentrations	Very low metal concentrations; toxic sulfide may be present
2. Treatment removal efficiency	Satisfactory	Very high
3. pH range for precipitation	Narrow	Wider
4. Hexavalent chromium	Not removed	Effective removal without reduction to trivalent state
5. Complexed metals	Will not precipitate	Will precipitate
6. Hydrogen sulfide gas (toxic)	Not a problem	Generated at pH 8 or lower
7. Cost of precipitating chemicals	Low	High
8. Sludge volume	Large quantity	Smaller quantity
9. Sludge disposal	Dewatered, containerized, and disposed of in approved landfills	Dewatered, containerized, and disposed of in approved landfills
10. Frequency of use	Common	Rare

11.33 Complexed Metals Removal

Complexed metals are metals with a tendency to remain in solution rather than form precipitates and settle out. These metals have reacted with or are tied up with chemical complexing agents, such as ammonia or citrates, tartrates, quadrol, and EDTA. Complexed metal wastes are found in wastestreams from electroless plating, immersion plating, etching, and printed circuit board manufacturing. Cadmium, copper, lead, nickel, and zinc are the most common complexed metals found in wastestreams. These complexed metals will not be effectively precipitated out by the hydroxide precipitation process. Therefore, complexed metal wastestreams must be kept segregated from other metal wastestreams and treated separately. Complexed metal wastestreams may be treated by (1) high pH precipitation, (2) chemical reduction, (3) sulfide precipitation, or several other processes (Table 11.3).

TABLE 11.3 pH RANGES FOR VARIOUS COMPLEXED METAL WASTE TREATMENT PROCESSES

Process	pH Range[a]
Ammonia Stripping	11–12
Ferrous Sulfide	8–9
Ferrous Sulfate	11–12
(with Formaldehyde)	3
Sodium Borohydride	5–7
Ion Exchange	7
EDTA Precipitation	2–3
Starch Xanthate	[b]
Ferric Chloride & Calcium Chloride	11.6–12.5[c]

[a] pH ranges depend on many variables including complexing and batch or continuous treatment processes.
[b] The removal of hexavalent chromium is achieved by lowering the pH to below 3 and then raising it to above 7.
[c] Treatment of solutions of complexed copper.

High-pH precipitation is a process requiring the addition of chemicals that drastically increase the pH of the wastestream to approximately 12.0. This very high pH produces a shift in the chemical equilibrium point, breaks the complexing bond, and results in the production of free metal ions. These metal ions can then be precipitated by available hydroxide ions and removed by sedimentation.

Complexed metals can also be removed from wastestreams by the chemical reduction process. This treatment process lowers the pH of the wastestream to break up the various metal complexes. The pH level depends on the complexed metals. Some require a pH in the 4.0 to 6.0 range while others require a pH of 2.0 to 2.5 or lower. After the pH has been lowered sufficiently, a reducing agent (sulfur dioxide, SO_2; sodium bisulfite, $NaHSO_3$; sodium metabisulfite, $Na_2S_2O_5$; or ferrous sulfate $FeSO_4$) is added to reduce the metals to an oxidation state that permits the precipitation of metals. Basic chemicals are then added to increase the pH to a level (Figure 11.8) where metallic hydroxide precipitation will occur. The precipitates are removed from the wastestream by sedimentation. Polymers or coagulating agents may be added to enhance sedimentation. Bench tests are run on complexed metal wastestreams to determine the optimum reducing agent, doses, pH, and polymer.

QUESTIONS

Write your answers in a notebook and then compare your answers with those on page 602.

11.3E Why are two different pH levels sometimes necessary for effective hydroxide precipitation of metals?

11.3F List the major limitations of the sulfide precipitation process.

11.3G Why are complexed metals difficult to treat?

11.3H How can complexed metal wastewaters be treated?

11.34 Reduction of Hexavalent Chromium

Hexavalent chromium-bearing wastewaters are produced in chromium electroplating, chromium conversion coatings, etching with chromic acid, and in metal finishing operations carried out on chromium as a basis material. Many common plating solutions contain hexavalent chromium at concentrations between 10 and 500 mg/L; however, concentrations over 100,000 mg/L may be encountered. Hexavalent chromium may be present in the form of chromic acid, chromate, or dichromate in an acid solution.

To remove chromium from wastestreams, highly toxic hexavalent chromium (Cr^{6+}) must be reduced to trivalent chromium (Cr^{3+}), which can then be removed from the wastestream by hydroxide precipitation. Strong reducing agents used in this chemical treatment process include sulfur dioxide, sodium bisulfite, sodium metabisulfite, sodium hydrosulfite, and ferrous sulfate. Gaseous sulfur dioxide is the most economical reducing agent used to reduce hexavalent chromium to trivalent chromium for large flow applications. This reaction takes place at a very low pH (around 2.0 for complete reduction). The reduction rate decreases with increasing pH and is extremely slow for pH levels above 5.0.

Chromium wastes are commonly treated by a two-step batch process. The first step is to reduce the highly toxic hexavalent chromium to the less toxic trivalent form. The trivalent form is removed by hydroxide precipitation in the second step.

STEP 1

Hexavalent chromium (Cr^{6+}) is reduced to trivalent chromium (Cr^{3+}). Sulfuric acid is added to lower the pH to 2.0 or lower. In an acid solution, the following reduction reactions will occur, depending on the reducing agent.

Using sulfur dioxide, Cr^{6+} to Cr^{3+}

$$SO_2 + H_2O \rightarrow H_2SO_3$$

Sulfur Dioxide / Water / Sulfurous Acid

$$3H_2SO_3 + 2H_2CrO_4 \rightarrow Cr_2(SO_4)_3 + 5H_2O$$

Sulfurous Acid / Chromic Acid / Chromic Sulfate / Water

Using bisulfite, Cr^{6+} to Cr^{3+}

$$2H_2CrO_4 + 3NaHSO_3 + 3H_2SO_4 \rightarrow Cr_2(SO_4)_3 + 3NaHSO_4 + 5H_2O$$

Chromic Acid / Sodium Bisulfite / Sulfuric Acid / Chromic Sulfate / Sodium Bisulfate / Water

When using sodium bisulfite, the pH is dropped to 2.0 or lower to speed up the reduction of hexavalent chromium by the bisulfite.

Some shops use sodium hydrosulfite for manual additions of chemicals where it is not practical to lower the pH to 2.0. Hydrosulfite works well at a pH of around 4.0, which is a pH range where hexavalent chromium reduction by bisulfite is far too slow. Hydrosulfite is not practical to use on continuous processes because it is more expensive than bisulfite and is oxidized by air once it is in solution and allowed to stand. Sodium hydrosulfite is very convenient to use when chemicals are added manually.

Using ferrous sulfate, Cr^{6+} to Cr^{3+}

$$2H_2CrO_4 + 6FeSO_4 + 6H_2SO_4 \rightarrow 3Fe_2(SO_4)_3 + Cr_2(SO_4)_3 + 8H_2O$$

Chromic Acid / Ferrous Sulfate / Sulfuric Acid / Ferric Sulfate / Chromic Sulfate / Water

STEP 2

The trivalent chromium (Cr^{3+}) compounds formed in Step 1 are removed by alkaline precipitation at a pH between 8.0 and 8.5. Caustic soda (NaOH) is the preferred base, but hydrated lime (Ca(OH)$_2$) is also used.

Using caustic soda to precipitate chromic hydroxide

$$6NaOH + Cr_2(SO_4)_3 \rightarrow 2Cr(OH)_3\downarrow + 3Na_2SO_4$$

Sodium Hydroxide / Chromic Sulfate / Chromic Hydroxide / Sodium Sulfate

Using lime to precipitate chromic hydroxide

$$3Ca(OH)_2 + Cr_2(SO_4)_3 \rightarrow 2Cr(OH)_3\downarrow + 3CaSO_4$$

Lime / Chromic Sulfate / Chromic Hydroxide / Calcium Sulfate

Reduction of hexavalent chromium requires an equalization tank if the flows fluctuate considerably; two reaction tanks can be connected in series (Figure 11.14). Sulfuric acid is added to maintain a pH below 2.0. Gaseous sulfur dioxide or sodium bisulfite is added to each reaction tank to produce an ORP (oxidation-reduction potential) around +250 mV (millivolts) or less. Use a color comparator or hand-held colorimeter kit to de-

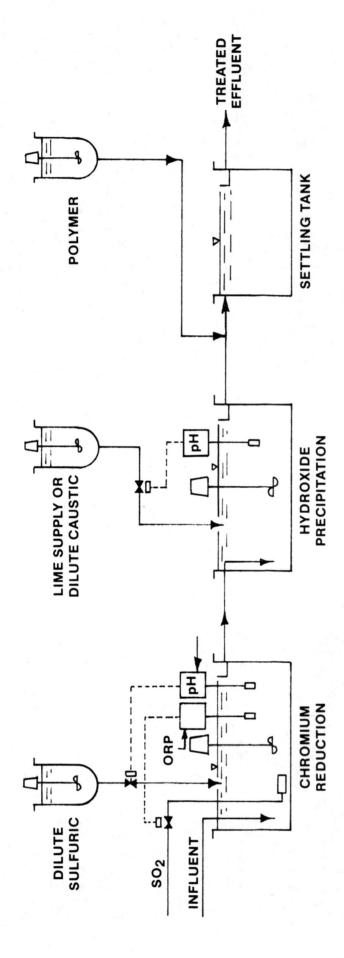

Fig. 11.14 *Reduction of hexavalent chromium and hydroxide precipitation for rinse waters (by Joe Shockcor)*

termine if all of the hexavalent chromium has been converted to the trivalent form. If not, increase the sulfur dioxide dose. Observe and record the ORP value at which all hexavalent chromium has been converted to the trivalent form. Set the sulfonator (SO_2) or chemical feeder of sulfur dioxide to maintain the desired ORP level. Regardless of whether sulfur dioxide or sulfite is used, the tanks should be fitted with an exhaust hood discharging to the atmosphere to prevent people being exposed to sulfur dioxide gas (it can be produced from bisulfite).

Each reaction tank has a propeller mixer and the detention time in each tank is from 15 to 45 minutes. Reaction times vary for different wastes, reducing agents, temperatures, pH, and chromium concentrations.

After the hexavalent chromium has been reduced to trivalent chromium, the wastestream can be combined with other wastestreams containing common metals. These combined wastestreams can be treated by the hydroxide precipitation process to remove chromium and the other common metals.

NOTE: pH and ORP levels given in this manual are approximate. Values cited in other references may vary slightly. Optimum levels for your metal waste treatment facilities may be slightly different due to the mix of wastes being treated.

QUESTIONS

Write your answers in a notebook and then compare your answers with those on page 602.

11.3I   List the possible forms of hexavalent chromium in an acid solution.

11.3J   Why does hexavalent chromium have to be converted to trivalent chromium?

11.3K   How can hexavalent chromium be reduced to trivalent chromium?

11.3L   What is the function of each step in the two-step batch process by which chromium wastes are commonly treated?

11.35 Cyanide Destruction by Oxidation

11.350 Safety

The cyanide ion (CN^-) is extremely toxic and must be removed from metal wastes before discharge to sewers or the environment. Cyanide toxicity (poisoning) in humans is caused by an irreversible reaction with the iron in hemoglobin that results in loss of the blood's ability to transport oxygen. If there is an equipment failure during the treatment of wastes containing cy-

anide, extremely toxic gases could be released. Therefore, all tanks or pits used for cyanide destruction must be properly located and ventilated so that any gases produced will never enter an area occupied by people.

11.351 Cyanide Sources and Treatment

Cyanide compounds get into metal wastestreams as a result of their use as metal salts for plating and conversion coating or they are active components in plating and cleaning baths. Cyanide compounds are used in copper, zinc, cadmium, silver, and gold plating solutions, in the immersion stripping of various electrodeposits, and in some activating solutions. Flowing rinse waters downstream of these production operations can become contaminated with cyanide. Cyanide may be present as the simple alkali cyanides of sodium or potassium, or complexed with heavy metals, such as zinc, cadmium, silver, gold, copper, nickel, or iron. For waste treatment control purposes, the two major groupings of cyanide compounds are segregated into those that can be oxidized by chlorination (also known as cyanides amenable to chlorination) and those that resist oxidation by chlorination (also known as refractory cyanides). Both types are considered to be toxic and must be removed or treated to lower their concentrations below toxic levels prior to discharge of the rinse waters.

Sodium, potassium, cadmium, and zinc cyanides are readily oxidized by chlorination. The copper cyanide complex is also considered amenable to chlorination, although longer reaction times are required; the same is true of silver and gold cyanide complexes. The nickel cyanide complex is more resistant to chlorination than the copper complex, yet can be considered as amenable to chlorination under extreme oxidizing conditions. The iron complexes (most commonly the sodium ferrocyanide salt) are not considered to be oxidizable by chlorination.

Pretreatment permits specify maximum discharge limits for cyanides oxidizable by chlorination and for total cyanide (the sum of those that can be chlorinated and those that cannot). The average limit for amenable cyanides is 0.05 mg/L and the average limit for total cyanide is 0.28 mg/L. The maximum concentrations for amenable and total cyanide are 0.65 mg/L and 1.2 mg/L, respectively.

11.352 Chemistry Involved

The waste treatment systems are designed to oxidize cyanide compounds with sodium hypochlorite (NaOCl). The hypochlorite is purchased at a 12 to 15 percent by weight concentration. This solution has been produced by the reaction between chlorine and sodium hydroxide according to the reaction:

Cl_2 + 2NaOH → NaOCl + NaCl + H_2O

Chlorine Sodium   Sodium   Sodium   Water
          Hydroxide   Hypochlorite   Chloride

Sodium cyanate (NaCNO) is formed by the reaction of hypochlorite and cyanide. This reaction is actually a simplification of the chlorination of cyanide. Intermediate reactions that are very rapid take place. Chlorine is freed from the hypochlorite by dissociation (a reversal of the formation of hypochlorite shown above). The chlorine actually reacts with the cyanide to form cyanogen chloride according to:

$$\text{NaCN} + \text{Cl}_2 \rightarrow \text{CNCl} + \text{NaCl}$$

Sodium Cyanide / Chlorine / Cyanogen Chloride / Sodium Chloride

The cyanogen chloride then reacts with the sodium hydroxide to form the cyanate according to:

$$\text{CNCl} + 2\text{NaOH} \rightarrow \text{NaCNO} + \text{NaCl} + \text{H}_2\text{O}$$

Cyanogen Chloride / Sodium Hydroxide / Sodium Cyanate / Sodium Chloride / Water

The cyanogen chloride is a gas that has limited solubility at neutral or low pH values. At pH values of 10 or higher, the reaction producing the cyanate is quite rapid for cyanides that are not too tightly complexed with heavy metals. At lower pH values, not only is the reaction too slow, but there is a good probability of liberating the cyanogen chloride gas to the atmosphere. This gas is a lacrymator (produces tears) and its liberation is readily apparent to those nearby.

Oxidation of cyanate is possible by chlorination where the reaction is:

$$2\text{NaCNO} + 3\text{NaOCl} + \text{H}_2\text{O} \rightarrow 2\text{CO}_2 + \text{N}_2 + 3\text{NaCl} + 2\text{NaOH}$$

Sodium Cyanate / Sodium Hypochlorite / Water / Carbon Dioxide / Nitrogen Gas / Sodium Chloride / Sodium Hydroxide

The carbon dioxide formed will react with alkali to produce carbonate. Any excess is liberated into the atmosphere. The nitrogen has a limited solubility and also escapes into the air. This second stage oxidation reaction is faster at pH values slightly lower than the initial reaction of cyanide to cyanate. If the pH drops too low (below 7.0), then another reaction takes place with the cyanate and water producing ammonia compounds.

The important point of this discussion is that pH is a critical factor in cyanide destruction. The rate of reaction as well as the end products depend upon maintaining proper pH values. Also, the escape of cyanogen chloride is inhibited by high pH values. Another impact of pH is shown by the oxidizing potential required to chlorinate cyanide to cyanate. At a pH of 10.5 to 11.0, the oxidizing potential must be at +550 mV for the reaction to go to completion. If the pH drops to 9.5, a potential of +650 mV must be maintained to get the same results. If the pH drops much lower, the liberation of cyanogen chloride as an escaping gas essentially stops the reaction so higher potentials (above the +650 mV) do not solve the problem.

Operators must fully understand the extreme hazards associated with cyanide destruction processes. Whenever cyanide treatment processes are accidentally carried on at an acidic pH where significant levels of cyanides exist, extremely hazardous conditions can develop very quickly. Also, when treating cyanides, be sure to determine the pH at the start of the treatment process. The adjustment of pH and the various set points for the treatment system are to some extent dependent upon the concentration of cyanide and cyanate and the initial pH of the wastestreams. Adjusting the pH downward to 9.0 for a treatment process when the initial pH of a cyanide solution is significantly higher could result in the evolution of toxic cyanide gas.

Cyanides complexed with zinc and cadmium react at the same rate as the simple sodium and potassium cyanides. Other metal complexes are more resistant to oxidation by chlorination and long contact times are required. In order to maintain a fast reaction rate, higher oxidizing potentials (additional excess chlorine) are often used. This is most important for the complexes of silver and nickel.

When cadmium cyanide complex is oxidized, the cadmium precipitates from solution as cadmium oxide. When copper, nickel, and zinc cyanide complexes are oxidized, these metals precipitate from solution as the hydroxides. However, the presence of silver cyanide presents an additional concern. The silver cyanide complex, when oxidized, can precipitate as silver chloride (a white precipitate) and silver oxide (a black precipitate). Silver oxide is preferred because it is denser than the silver chloride and is therefore easier to remove from the wastewater. Higher oxidizing conditions (higher pH) favor the oxide formation; lower conditions, the chloride.

In practice, the process control devices are set to control the pH within the ranges of 10.5 to 11.5 for the first stage reaction (cyanide to cyanate) and 8.5 to 9.0 for the second stage reaction (cyanate to carbon dioxide and nitrogen). The control devices are set to maintain minimum oxidizing potentials of +600 mV in the first stage and +850 mV in the second stage.

Another control phenomenon influences the success of this process. For the first stage reaction, a minimum pH must be maintained (as discussed earlier). The upper pH limit (11.5) is established for a different reason. Excessive alkali (as measured by pH) results in suppression of ORP values. The ORP value selected considers the need to have excess hypochlorite present to dominate the reaction. Sodium hypochlorite always contains free sodium hydroxide. When the ORP controller requests the addition of hypochlorite, the pH can rise because of this simultaneous addition of alkali. When the pH rises, the ORP value decreases, establishing an out-of-control condition. Adding more hypochlorite results in the instrumentation requesting still more hypochlorite, even though there is no cyanide present. This phenomenon is eliminated by the addition of acid whenever the pH rises too high.

11.353 Batch or Continuous Treatment

The most practical and economical method of treating metal wastestreams containing cyanide is the alkaline chlorination treatment process. Cyanide destruction by the alkaline chlorination oxidation process may be accomplished in either a batch (Figure 11.3, page 534) or a continuous treatment process (Figure 11.15). If the cyanide wastes are maintained separate from other wastes, the batch process is usually the most cost-effective treatment mode.

In the batch process, the cyanide waste flows into one tank. When the tank is full or at an appropriate time interval, the

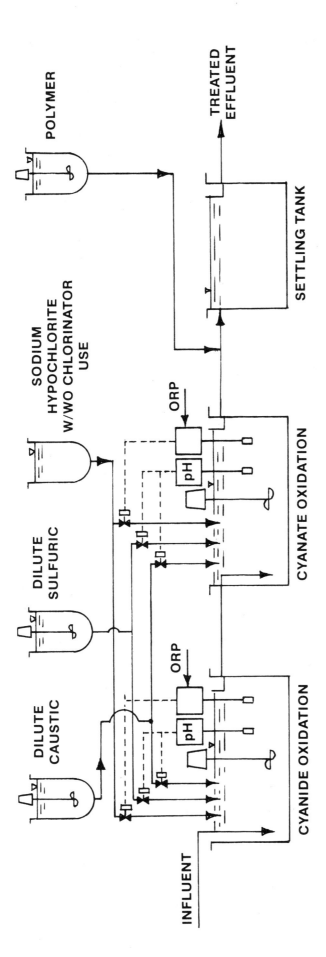

Fig. 11.15 *Continuous flow oxidation of cyanide by alkaline chlorination and hydroxide precipitation for rinse waters (by Joe Shockcor)*

waste is either stopped or diverted to a second tank, while the cyanide waste is treated in the first tank. You will need a pH controller with two set points (for the two steps) and one pH sensor. You will also need an ORP or chlorine controller with two set points and one ORP sensor. The tank must be equipped with a mixer, and a sequential programmer is used to add chlorine and sodium hydroxide at the correct times as determined by the pH and ORP controllers.

The high set point for the pH controller is at a pH of 10.5 and the low set point is at a pH of 8.8. The high set point for the ORP controller is at +790 mV and the low point is at +670 mV. The ORP set point may need to be set at a higher or lower level to maintain the desired chemical concentration, but the difference between the ORP set points should be 120 mV.

The chemical addition and adjustment sequence proceeds in a programmed fashion. Add sodium hydroxide until the pH reaches 10.5. Then add chlorine to produce an ORP level of +670 mV for 10 to 30 minutes. Next, allow the pH to drop to 8.8 (acid may have to be added to lower the pH) and adjust the ORP to +790 mV by the regulation of chlorine. The reactions are usually complete after 90 minutes.

Two tanks are used to treat cyanide in the continuous flow process shown in Figure 11.15. The first tank will have a retention time of 20 to 45 minutes and the second tank requires 45 to 90 minutes' retention time. Some shops use a rule of thumb that the retention time in the second tank should be at least two times the retention time in the first tank. Actual retention times to achieve desired results depend on metal concentrations and the mixture of metals in the wastestream being treated. Each tank should have a mixer capable of completely mixing the tank. Both tanks must have pH and ORP sensors and controllers.

In order to have sufficient chlorine present for the cyanide oxidation reactions in both the batch and continuous processes, at least 10 mg/L of free chlorine must be present as analyzed by the *DPD METHOD*[21] or by using PAO or a similar titration. The oxidation-reduction potential (ORP) is used to control the chlorine concentration. A platinum electrode with a reference electrode is used to sense ORP. These two electrodes are usually combined in one sensor probe. In water containing chlorine, the platinum electrode is sensitive to both chlorine and pH. For every tenfold increase in chlorine concentration, the oxidation-reduction potential increases by approximately 60 mV. For every one unit of pH increase, the oxidation-reduction potential decreases by approximately 60 mV.

At a pH of 8.0 and a free chlorine concentration of 10 mg/L, the ORP will be approximately +790 mV when a silver chloride reference electrode is used. Thus, at a pH of 10 and a free chlorine concentration of 10 mg/L, the oxidation-reduction potential will be +670 mV (a reduction of 120 mV).

The response of the electrode to changes in pH is almost immediate. The response of the electrode to changes in free chlorine concentration occurs within 5 to 15 minutes. To minimize chlorine waste and to ensure complete destruction of cyanide, use the DPD method to analyze the water being treated to be sure a free chlorine residual is being maintained. After the chlorine concentration has been determined, adjust the ORP to the correct level based on the effects of chlorine concentration and pH level. The pH level must be carefully controlled because if the pH is allowed to drift up or down by 1.0 unit, the chlorine concentration will be adjusted up or down by the controller by a factor of 10. Do not rely on the chlorine residual as an indication that all cyanide has been oxidized to cyanate. The only way to know for sure is to collect a sample and analyze for cyanide.

Efficient operation of a cyanide oxidation system requires exact control of the pH. If the pH control is not accurate, there will be excessive use of chlorine or the system will not completely destroy the cyanide. A practical approach is to automatically correct the ORP when the pH values fluctuate. By the use of a glass pH electrode and a microprocessor, the ORP signal can be converted to a chlorine concentration signal that may be read out in milligrams per liter of free chlorine.

A problem develops with the alkaline chlorination treatment of cyanide if soluble iron or certain other transition metal ions are present. The iron forms very stable ferrocyanide complexes that prevent the cyanide from being oxidized. The best solution to this problem is to segregate the wastestreams to keep the cyanide and soluble iron separated.

If the pH is allowed to drop too low in the second stage, all of the cyanate is not converted to carbon dioxide and nitrogen gas. Ammonium compounds are formed (reacting with the nitrogen), which can complex copper and prevent the subsequent precipitation of copper as a hydroxide.

Chlorine is a very toxic gas and must be handled with extreme care. Always follow safe procedures for operating and maintaining chlorination facilities.

Alternative treatment techniques for the destruction of cyanide include oxidation by ozone, oxidation by ozone combined with ultraviolet radiation, oxidation by hydrogen peroxide, or electrolytic oxidation.

QUESTIONS

Write your answers in a notebook and then compare your answers with those on page 602.

11.3M What safety hazard could occur if there is an equipment failure during the treatment of metal wastes containing cyanide?

11.3N How do cyanide compounds get into metal wastestreams?

11.3O What is the most practical and economical method of treating metal wastestreams containing cyanide?

11.3P Why does efficient operation of a cyanide oxidation system require exact control of the pH?

---

[21] *DPD Method.* A method of measuring the chlorine residual in water. The residual may be determined by either titrating or comparing a developed color with color standards. DPD stands for N,N-diethyl-p-phenylenediamine.

11.36 Precious Metals[22] Recovery

11.360 Recovery Treatment Processes

Precious metals recovery is achieved by evaporation, ion exchange, reverse osmosis, and electrolytic metal waste treatment processes. Each of these processes separates or recovers the precious metal from wastestreams.

Evaporation is used to recover precious metals by boiling off the wastewater portion of the wastestream containing the precious metal and removing the metal. The evaporation process allows for the recovery of several process chemicals. The greatest limitation of the evaporation process is the high cost of energy to evaporate the wastewater. New *AMBIENT TEMPERATURE*[23] evaporation processes are being developed, which could be considerably less expensive.

Ion exchange is the process in which ions, held by electrostatic forces to charged functional groups on the surface of an ion exchange resin, are exchanged for ions of similar charge from the solution in which the resin is immersed. In this treatment process, a precious metal in a wastestream can be exchanged for another similar, but harmless and less precious, ion of the resin. Ion exchange units are used not only to recover precious metals, but also to recover other metal plating chemicals and to concentrate and purify plating baths.

Precious metals recovered by the ion exchange process include copper, molybdenum, cobalt, nickel, gold, and silver. Metal plating chemicals recovered by ion exchange include chromium, nickel, phosphate solution, sulfuric acid from anodizing, and chromic acid. Ion exchange units also have been used to treat metal wastes containing aluminum, arsenic, cadmium, chromium (hexavalent and trivalent), cyanide, iron, lead, manganese, selenium, tin, and zinc. The advantage of using the ion exchange process is that a specific metal can be recovered, whereas in the precipitation processes all of the metals are precipitated together and removed in the sludge.

The chemistry involved in the ion exchange process for chromium recovery can be illustrated by the following reactions. The influent plating rinse waters contain metal ions (M^{n+}) and chromate ions (CrO_4^{2-}). The R^- symbol represents the resin of a cation exchanger. This means the resin will exchange one type of cation (such as H^+) for the cation to be removed from the wastestream (such as M^{n+}). The R^+ symbol indicates an anion exchanger.

CATION EXCHANGER

M^{n+} + CrO_4^{2-} + R^-H^+ → R^-M^{n+} + H^+ + CrO_4^{2-}

Metal Ions — Chromate Ions — Cation Exchanger — Cation Exchanger — Hydrogen Ions — Chromate Ions

ANION EXCHANGER

H^+ + CrO_4^{2-} + R^+OH^- → $R^+CrO_4^{2-}$ + H_2O

Hydrogen Ions — Chromate Ions — Anion Exchanger — Anion Exchanger — Product Water

ANION EXCHANGER REGENERANT TO CATION EXCHANGER (CHROMIC ACID RECOVERY)

Na_2CrO_4 + R^-H^+ → R^-Na^+ + H_2CrO_4

Sodium Chromate (Regenerant) — Cation Exchanger — Cation Exchanger — Chromic Acid (Recovered)

Figure 11.16 illustrates the application of an ion exchange system for treating metal wastes for the recovery of hexavalent

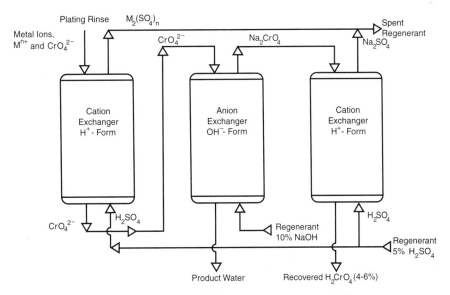

Fig. 11.16 Schematic illustration of plating waste treatment and chromium recovery
(Reproduced by permission of Wiley-Interscience, Copyright 1972, from *PHYSICOCHEMICAL PROCESSES FOR WATER QUALITY CONTROL* by Walter J. Weber, ed.)

---

[22] *Precious Metal.* Metal that is very valuable, such as gold or silver.
[23] *Ambient* (AM-bee-ent) *Temperature.* Temperature of the surroundings.

chromium (Cr^{6+}) as chromic acid (H_2CrO_4) and producing treated water for reuse. The metal wastestream is first passed through a cation exchanger to remove other metals such as iron, copper, zinc, nickel, and trivalent chromium. The hexavalent chromium in the metal wastes passes through the cation exchanger as the chromate ion (CrO_4^{2-}) and is subsequently removed in the anion exchanger. The effluent from the anion exchanger is a demineralized water suitable for reuse. To recover the hexavalent chromium (Cr^{6+}) from the anion exchanger, the water being treated must be stopped, the anion exchanger regenerated with sodium hydroxide, and sodium chromate (Na_2CrO_4) is released. This sodium chromate solution is then passed through another cation exchanger, which exchanges sodium for hydrogen, thus releasing chromic acid (H_2CrO_4) in the effluent for recovery.

When an ion exchange resin loses its exchange capacity, the resin must be regenerated. Regeneration consists of passing a chemical solution through the unit and exchanging or replacing ions removed from the wastestream with ions that can later be exchanged with precious metals from wastestreams after the ion exchange resin has been regenerated. Ion exchange resins must be cleaned when they occasionally become clogged or fouled.

Concentrated regeneration streams require special treatment for recovery of wastes and safe disposal. The regeneration flows from ion exchange units, especially for deionized waters, are frequently very acidic and may require pH adjustment prior to discharge to sewers. These flows may also be high in salt content. Regeneration flows must be properly treated to remove any toxic substances or excess salts. Spent or degraded ion exchange materials (resins) must be properly cleaned to remove any hazardous substances present before disposal. If degraded resins contain toxic metals, the resins must be properly containerized and then disposed of in an approved landfill. Disposal of these wastes must conform with both federal and state laws regulating the disposal of hazardous wastes.

Certain plating operations in the electronics and high-tech industries require water of extremely high purity (very low dissolved solids). Water of high quality can be obtained by using a series of water treatment processes consisting of a reverse osmosis unit, degassifier, strong-acid cation exchanger, strong-base anion exchanger, and, finally, a mixed-bed ion exchange unit (Figure 11.17).

Electrolytic recovery is particularly applicable to precious metals recovery because the precious metals offer a faster payback on equipment and energy costs than other recovery processes. In this process, an electrochemical reduction of metal ions at the cathode changes these ions to the elemental state of the metal. This process is used to recover copper, silver, and tin from plating and etching bath drag out.

Table 11.4 is a summary of recovery techniques.

QUESTIONS

Write your answers in a notebook and then compare your answers with those on page 602.

11.3Q How can precious metals be recovered from metal wastestreams?

11.3R List the major uses of ion exchange units in treating metal wastestreams.

11.3S What precious metals may be recovered by the ion exchange process?

11.3T What metal plating chemicals can be recovered by ion exchange?

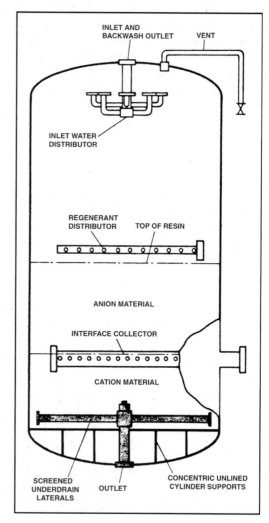

Fig. 11.17 Mixed-bed ion exchange unit showing internal distributors

(Reproduced by permission of Betz Laboratories, Inc. Copyright 1980, from *BETZ HANDBOOK OF INDUSTRIAL WATER CONDITIONING*, Eighth Edition, 1980)

TABLE 11.4 SUMMARY OF RECOVERY TECHNIQUES[a]

Technique	Common Applications	Comments
I. Ion exchange	Gold, silver, chromium, nickel plating, phosphoric acid, anodizing solutions, aluminum etchants.	Suited for metal complexed solution, low concentration removal, and selective process bath stabilization.
II. Electrolyte	Gold, silver, tin, copper, chromium plating solutions, cadmium.	Suited for metal recovery from concentrated wastestream.
III. Electrodialysis	Gold, silver, copper, chromic acid, nickel, zinc, cobalt plating.	Ion exchange membrane allows minimal objectionable inorganic return.
IV. Evaporation	Chromic acids, chrome, lead-tin, nickel, copper, cadmium, brass, bronze, gold, zinc plating.	Efficient for low volume, high concentration recovery of plating or process solutions.
V. Reverse osmosis	Nickel plating.	Recovers brighteners and other organics. Suited for plating or process solution and rinse water recovery.

[a] Reproduced by permission of TMSI Contractors from *METAL FINISHING PLATING BUYERS GUIDE.*

11.361 Troubleshooting Ion Exchange Systems

This section summarizes the three major problems operators may encounter in the day-to-day operation of ion exchange systems and suggests some practical ways operators can prevent or correct most problem situations. The three primary problems operators encounter are: throughput capacity loss, failure to produce specified water quality, and excessive pressure drop across the ion exchange system.

1. Throughput Capacity Loss

To investigate throughput capacity loss, begin by accurately measuring the depth of the resin bed, because loss of resin leads to a loss in throughput. Strong electrolyte ion exchange resins should be measured in the fully exhausted state, while weak electrolyte ion exchange resins should be measured in the regenerated state. If there is no resin loss and no change in water quality, check the following regeneration conditions:

- Check integrator flowmeters to determine whether the throughput measurement is accurate.

- Evaluate the end points (bed exhaustion). Conductivity or photometric methods usually are used to determine end points, so check the accuracy of the analytical instrumentation. In the case of weak-acid cation units, the end point is measured by pH.

- If the integrator flowmeter and end points are within specification limits, evaluate a core sample of the resin. A good practice is to take a core sample (using a grain thief) from each quadrant (quarter) of the resin bed. Combine and mix the core samples. Analyze a sample portion of the mixture (one half to one liter) to determine if the resin's properties are within specifications.

2. Failure to Produce Specified Water Quality

Ion exchange fouling can take various forms: organic fouling, which usually affects anion exchange resins; iron fouling, which affects both cation and anion exchange resins; oil fouling; mud or polyelectrolyte fouling; algal or bacterial fouling; calcium sulfate precipitation; or barium sulfate fouling. Each condition reduces a resin's ability to produce the specified water quality.

Organic fouling. While organics can foul cation exchange resins, the main concern lies in their effect on anion exchange resin performance. The symptoms of organically fouled anion resins are:

- Low pH of treated water (sometimes as low as 5.5)

- Increasing conductivity of treated water

- Increased silica leakage

- Increased rinse volumes

- Loss of throughput capacity

Organic fouling usually can be avoided by choosing the appropriate organically resistant resin, such as gelular acrylic or macroreticular styrene anion exchange resins. In any case, the resin should be cleaned periodically with a brine-and-caustic solution to remove as much organic matter as possible. How often a system requires cleaning depends on influent conditions and the anion exchange resin; some need cleaning once a month, while others must be cleaned only once a year.

Iron fouling. Iron can occur in water supplies as ferrous iron, ferric iron, ferrous hydroxide, ferric hydroxide, and organic iron complexes. It will be present in one of three forms:

- Suspended insoluble iron will be filtered out by the ion exchange bed and can be removed from the resin by vigorous backwashing before each regeneration.

- Soluble ferrous iron that is oxidized to the ferric form before it reaches the ion exchange unit and becomes a gelatinous ferric hydroxide precipitate can coat the resin beads and cause a decrease in capacity. This form of iron sticks to the resin beads and is difficult to remove by backwashing, but can be cleaned with sodium hydrosulfite (a reducing agent). Sodium hydrosulfite reduces the iron to the more soluble ferrous form, which is less likely to stick to the resin.

- Soluble iron that is oxidized to the ferric form after adsorption on a cation exchange resin is the most troublesome type of iron fouling. The recommended treatment is a 10 percent hydrochloric acid solution. Before using this treatment, however, consider whether the acid could corrode the ion exchange unit (check construction material). If corrosion could be a problem, use an inhibited hydrochloric acid solution.

Oil fouling. Oil fouling can occur when oil enters a compressed air source through a leaking seal or when the wastewater contains oil. In these situations, an oil film forms around the resin beads and interferes with the ion exchange reactions. Because lubrication oils contain fluorescent additives, this type of fouling can be detected by ultraviolet irradiation.

Resin fouling caused by oil leaks from malfunctioning equipment usually can be avoided by proper pump and compressor maintenance. When a source of water becomes contaminated, however, pretreatment is required. Pretreatment typically involves the use of coalescing filters with precoat materials or surfactant-treated cation exchange resins. With or without pretreatment, the solution to this problem is a low-foaming, nonionic surfactant.

Mud or polyelectrolyte fouling. Mud or polyelectrolyte fouling can occur in the lead unit of an ion exchange system when wastewater is not pretreated or the pretreatment facilities are not working properly. In these situations, suspended solids accumulate on the ion exchange bed, forming a filter cake (interference with plug flow) and causing excessive pressure drop. Accumulated material cannot be removed under design backwash procedures; extended backwash times (30 minutes) at higher flow rates are required, sometimes with air agitation. If this procedure is inadequate, use a detergent-dispersant cleanup technique.

Algal or bacterial fouling. For reasons that are largely unknown, some ion exchange systems develop algal and bacterial growths that can interfere with flow patterns and internal distribution systems. When this occurs, periodic bed sterilization is required using sodium hypochlorite. This procedure is a highly aggressive oxidative cleanup technique that can damage strong base anion resins if used repeatedly. Dilute solutions of slow-chlorine-release agents are also effective but are more expensive.

Calcium sulfate precipitation. Calcium sulfate fouling typically occurs as a result of faulty regeneration techniques when using sulfuric acid on strong or weak acid cation exchange units. One technique for removing the calcium sulfate involves backwashing the unit with large amounts of soft water, followed by repeated cleaning with 10 percent hydrochloric acid at a rate of 0.25 gallon per minute per cubic foot of resin. Repeated cleanup stages are needed due to the slow reaction rate between hydrochloric acid and calcium sulfate.

Barium sulfate fouling. Barium sulfate fouling occurs when trace amounts of barium are present in the water being treated in strong acid cation exchange resin units regenerated with sulfuric acid. This type of fouling typically occurs when water sources are influenced by runoff from oil field drilling processes in which the drilling muds contain barium.

Barium sulfate is highly insoluble and usually collects around underbed distributors and strainers. It is almost impossible to remove, although prolonged treatment with 10 percent hydrochloric acid is reasonably effective. Certain proprietary (trademarked or patented) methods involving EDTA formulations also seem to be effective.

3. Excessive Pressure Drop

All packed beds of ion exchange resins have a resistance to water flow that causes a head loss or pressure drop as water flows through the resin. This pressure drop can be related to the vessel or the resin. Pressure drop related to the vessel is caused by the bed walls and head loss across the inlet distribution and underbed collection system.

Pressure drop related to the resin depends on water temperature, water density, water viscosity, resin void volume, resin particle size and shape, solids content of the resin, the nature of the resin surface, and wall effects.

It is essential to locate pressure gauges at the top and bottom levels of the resin bed so that the total pressure drop across the vessel can be determined from the inlet to the outlet. This enables operators to determine if the problem is in the equipment or in the resin. Gauges also must be properly calibrated to ensure correct readings.

Equipment problems. Equipment problems contributing to excessive pressure drop can be related to various components, including the inlet distribution system, underbed collection system, flowmeter, or vessel walls. Check each of these items to determine whether one of them is the possible cause of an ion exchange system problem.

Ion exchange resin problems. Generally speaking, good quality ion exchange resins are shaped like beads, and each batch of resin contains particle sizes from 0.297 millimeter to 1.19 millimeters (16 to 50 US mesh). This size range is optimal for bead-shaped resins because it allows close hexagonal packing that yields good hydraulic performance at normal service flow rates.

Typically, ion exchange resin pressure-drop problems relate to external factors such as the following:

- Oxidative conditions. The development of oxidative conditions, such as the inflow of free chlorine residual greater than 0.1 mg/L, or inflow of trace metals (iron, copper, manganese, or chromium), or an increase in temperature, can cause the ion exchange resin to begin to deteriorate. Generally, this occurs in the lead cation exchange resin units. As the resin polymers break down, water-soluble organic by-products are produced. When these organic particles are swept forward to downstream anion exchange units, they may cause fouling that cannot be corrected.

- Resin deposits. Resin deposits develop because of external factors not related to the resin.

    — Iron fouling. Iron fouling can accumulate and lead to pressure-drop problems.

    — Mud fouling. Mud fouling also can lead to pressure-drop problems if not corrected.

    — Calcium sulfate. Calcium sulfate precipitation usually is caused by inappropriate regenerant dilution conditions. This problem can be corrected, but removal of the precipitate deposits can be time-consuming and expensive.

- Resin fines. If a resin is subjected to physical shock (osmotic shock), mechanical abrasion or wear and tear, freeze and thaw conditions, or thermal shock effects, bead breakdown will occur. As the resin beads break down, extremely small pieces of resin (fines) begin to fill the open spaces between the resin particles in the bed and there is an increased pressure drop. If shock conditions occur again, further breakdown will occur, leading to even greater pressure drop. This cycle can produce even more resin fines, which, if not removed by backwashing, will filter out on top of the bed as filter cake and lead to channeling. This, in turn, affects throughput and water quality and also causes a pressure drop.

QUESTIONS

Write your answers in a notebook and then compare your answers with those on pages 602 and 603.

11.3U What are the three primary problems operators encounter with ion exchange systems?

11.3V What are the various forms of ion exchange fouling?

11.3W What are the various forms of iron found in water to be treated?

11.3X What is the solution to problems caused by algal and bacterial growths in ion exchange resins that interfere with flow patterns and internal distribution systems?

11.3Y What can cause ion exchange resin bead breakdown?

11.37 Oily Waste Removal [24]

The principal reasons to minimize the discharge of oily wastes from a wastestream are to prevent the fouling of sensors and to comply with pretreatment discharge limitations. Oily wastes come from process coolants and lubricants, wastes from cleaning operations, wastes from painting processes, and machinery lubricants. Techniques commonly used by electroplaters and metal finishers to remove oils include skimming, coalescing, emulsion breaking, flotation, centrifugation, ultrafiltration, and reverse osmosis. Treatment of oily wastes is most efficient and cost effective if oils are segregated from other wastes and treated separately. The process of separation varies depending on the type of oil involved.

In general, three types of oily wastes are encountered by operators: free oils, emulsified or water soluble oils, and greases. Manufacturing areas can produce segregated oily wastes, sometimes in concentrations as high as 400,000 mg/L. Lower oil and grease concentrations are found in combined oily wastes from the washing or rinsing of oily parts, spills, and leaks.

Treatment of oily wastes depends on the types of wastes. Free or floating oils can be separated from wastestreams by the use of gravity API (American Petroleum Institute) oil separators, centrifuges, dissolved air flotation (DAF) units, and nitrification. Emulsified oils are treated with emulsion breaking procedures such as steam, heat cracking, acids, or the use of emulsion breakers (polymers) with the addition of chemicals to cause the necessary separation of the oils. Oils must be removed from wastestreams before any further processing of the wastestreams occurs. Their presence can disrupt any subsequent operation in the treatment process train having a sensor feedback loop, which can fail due to sensor fouling. Also, regulatory pretreatment standards have an oil limitation.

Oil removal may be achieved by skimming. Once the oils and greases have been removed from the wastestream, the wastewater can be combined with other wastestreams containing common metals. The waste metals may then be removed by the hydroxide precipitation processes followed by sedimentation to remove the precipitates.

---

[24] American Petroleum Institute has some excellent references and literature. Contact American Petroleum Institute, 1220 L Street, NW, Washington, DC 20005-4070. Phone (202) 682-8000. Also, a publication, "A Guide to the Treatment of Oily Wastes," may be obtained from AFL Industries, Inc., 1751 West 10th Street, Riviera Beach, FL 33404. Phone (800) 807-2709.

QUESTIONS

Write your answers in a notebook and then compare your answers with those on page 603.

11.3Z What are the principal reasons to minimize the discharge of oily wastes from a wastestream?

11.3AA List the possible sources of oily wastes in wastestreams.

11.3BB What techniques are commonly used to remove oily wastes from metal wastestreams?

11.3CC What treatment processes are used to treat (a) free or floating oil, and (b) emulsified oil?

11.38 Solvent Control

Spent degreasing solvents must be segregated from other process wastewaters to maximize the value of the recoverable solvents, to avoid contamination of other segregated wastes, and to prevent the discharge of toxic organics to any wastewater collection systems or the environment. To encourage operators to segregate these wastes, provide clearly identified storage containers, establish clear disposal procedures, and train operators in the use of the proper techniques. Check periodically (monthly) to ensure that proper segregation is occurring. Segregated waste solvents are appropriate for on-site solvent recovery or may be contract hauled to another site for disposal or reclamation.

Alkaline cleaning (sodium hydroxide baths) is a feasible substitute for solvent degreasing. The major advantage of alkaline cleaning over solvent degreasing is the elimination or reduction in the quantity of pollutants being discharged. Major disadvantages include high energy consumption and the tendency to dilute oils removed and to discharge these oils as well as the cleaning additive. The aerospace industry is using aqueous cleaners (concentrated soap solutions) instead of degreasers for small flat parts such as circuit boards. When slug discharges of these alkaline cleansers have occurred, foaming problems have developed at small wastewater treatment plants.

QUESTIONS

Write your answers in a notebook and then compare your answers with those on page 603.

11.3DD Why must spent degreasing solvents be segregated from other process wastewaters?

11.3EE How can you encourage operators to segregate spent degreasing solvents from other process wastewaters?

11.3FF List two alternative cleaning methods that are being used instead of solvent degreasing procedures.

11.39 Control of Toxic Organics

Toxic organics may be found with oily wastes and solvents used for cleaning. In manufacturing operations, the toxic organics combine with oils, such as process coolants and lubricants, wastes from cleaning operations, wastes from painting processes, and machinery lubricants. Toxic organics can be controlled by proper storage of the chemicals. When they are used, they should be kept separated from other wastes that will enter wastestreams and must be treated. Spent degreasing solvents may be segregated from other wastes by providing and identifying the necessary storage containers. Operators must be trained in the proper use, collection, and storage of toxic organics. Periodic inspections (monthly) must be conducted to be sure that proper segregation is occurring. The containers holding the different waste solvents can then be taken to recovery facilities or hauled to disposal sites.

The quantity of toxic organics reaching wastestreams can be reduced by proper housekeeping procedures and by using cleaning techniques that require no solvents. These cleaning techniques include wiping, spraying with water, immersing in water, using alkaline or acid mixtures, or using cleansing emulsions. Toxic organics can be kept segregated from other wastestreams by parts racking to avoid pockets of solvents, air drying to cause volatilization (conversion to a vapor) before the next step in the process, and physical separation/segregation/containment to avoid contamination due to spills.

Toxic organics that enter wastestreams can be removed by treatment processes used for the control of other pollutants. Toxic organics tend to be more soluble in oil and grease than in water. Therefore, the removal of oil and grease from wastestreams will reduce the discharge of toxic organics. Aeration and carbon adsorption are the processes commonly used to remove toxic organics from wastestreams. Other treatment processes used to remove toxic organics include settling and volatilization, which can occur during the treatment of metals, cyanide, and oil and grease. In some situations, the activated sludge process may be used if a biological culture can be developed in the aeration tanks that is acclimated (accustomed) to the wastes. Aerobic decomposition, bacterial decomposition, and ozonation have also been used to treat toxic organics.

QUESTIONS

Write your answers in a notebook and then compare your answers with those on page 603.

11.3GG How can the quantity of toxic organics reaching wastestreams be reduced?

11.3HH How can toxic organics that enter wastestreams be removed?

END OF LESSON 1 OF 3 LESSONS

on

TREATMENT OF METAL WASTESTREAMS

Please answer the discussion and review questions next.

DISCUSSION AND REVIEW QUESTIONS
Chapter 11. TREATMENT OF METAL WASTESTREAMS
(Lesson 1 of 3 Lessons)

At the end of each lesson in this chapter you will find some discussion and review questions. The purpose of these questions is to indicate to you how well you understand the material in the lesson. Write the answers to these questions in your notebook.

1. How can the costs and complexity of metal waste treatment facilities be reduced?
2. What are some specific procedures for reducing waste treatment costs?
3. What options are available for treating spent baths?
4. What problems can be caused by organics in metal wastestreams?
5. What basic treatment principles must the operator keep in mind regarding the chemical reactions involved in physical–chemical treatment processes?
6. Why should a spent process solution (a pickle, for example) be treated as a batch process and not dumped directly into the rinse water for treatment?
7. Why must operators be very careful whenever strongly acidic and basic solutions are mixed?
8. How are chromium wastes commonly treated?
9. How can the ion exchange process be used to treat metal wastes?
10. Why must spent degreasing solvents be segregated from other process wastewaters?
11. How can toxic organics be controlled?

CHAPTER 11. TREATMENT OF METAL WASTESTREAMS

(Lesson 2 of 3 Lessons)

11.4 PROCESS INSTRUMENTATION AND CONTROLS

The instrumentation and controls that the operator will need to be familiar with to treat metal wastestreams are the following:

- pH and ORP probes
- Flow measurement/totalization devices
- Level controls
- Programmable controllers
- Voltage, current, and resistance measuring devices

To understand the interrelationship of the various instrumentation and control features, the operator must become familiar with the engineering schematics and wiring diagrams for the systems that must be operated. While the operator does not have to be an electrician, the operator should know enough to determine if it is a major problem or a minor one that must be corrected. The operator has to be able to tell the electrician what appears to be wrong and assist the electrician so that the repairs can be completed as quickly as possible.

Without adequate backup information, such as schematic diagrams, controller hookup information, and other relevant written documentation, even fully qualified instrumentation repair technicians will be greatly handicapped in resolving problems. The operator should have at least two sets of such information. One set would be the master set and should be kept under lock and key. The other set could be kept in a loose-leaf binder and used in the day-to-day operation of the facility.

QUESTION

Write your answer in a notebook and then compare your answer with the one on page 603.

11.4A List the waste treatment process instrumentation and controls that the operator treating metal wastestreams must be familiar with.

11.40 pH and ORP Probes

11.400 Description

pH probes consist of a glass electrode and a reference electrode (Figures 11.18 and 11.19). Both of these electrodes may be filled with a potassium chloride (KCl) electrolyte solution or a silver chloride (AgCl) solution. A temperature thermocouple also may be a part of the electrode assembly. The pH probes require the presence of potassium chloride, but if large reservoirs of KCl solution were provided, the probes would be large and clumsy to use. Therefore, a compromise must be achieved by manufacturers between reducing the size of probes (which causes a shorter probe life) and the use of a gel (which decreases probe sensitivity) instead of KCl. Probes currently being manufactured usually last between six months and two years depending on the type of service. Some manufacturers of pH and ORP electrodes are producing easily refillable reference cells because the electrolyte is continually being depleted.

The four electrodes shown in the upper left photo of Figure 11.18 are described below.

1. Epoxy bodied electrode (sensing cell only).

2. Combination electrode (section shown in drawing on Figure 11.18). This is both a sensing and double reference electrode. The rings (junction) at the top serve as a filter-protection device between the reference electrode and the process solution that is being sensed.

3. Combination electrode. This is both a sensing and double reference electrode. A single junction is shown at the bottom above the bulb.

4. Combination electrode cartridge. The cartridge contains the electrode and facilitates the replacement of electrodes. This cartridge is very easy to insert in and remove from tank walls and wastestreams.

ORP probes (Figures 11.20 and 11.21) commonly consist of a platinum electrode and a reference electrode identical to the pH electrode.

11.401 Cleaning

Operators must clean and calibrate pH and ORP probes on a regularly scheduled basis because the probes must have a good contact with the fluid they are measuring. Probes should be cleaned weekly unless special circumstances require more frequent cleaning. pH probes usually have more fouling problems than ORP probes. However, since ORP probes are always used with pH probes, the recommended procedure is to clean both probes at the same time.

Both pH and ORP probes are cleaned by using the same procedures. Remove the probes from their working location in the wastestream when they are to be cleaned. Do not clean probes in a cyanide destruct system with acid when the probes are on line. If acid spills into a tank containing cyanide, toxic hydrogen cyanide gas could be released into the atmosphere. You must know which cleaning chemicals, process control chemicals, and chem-

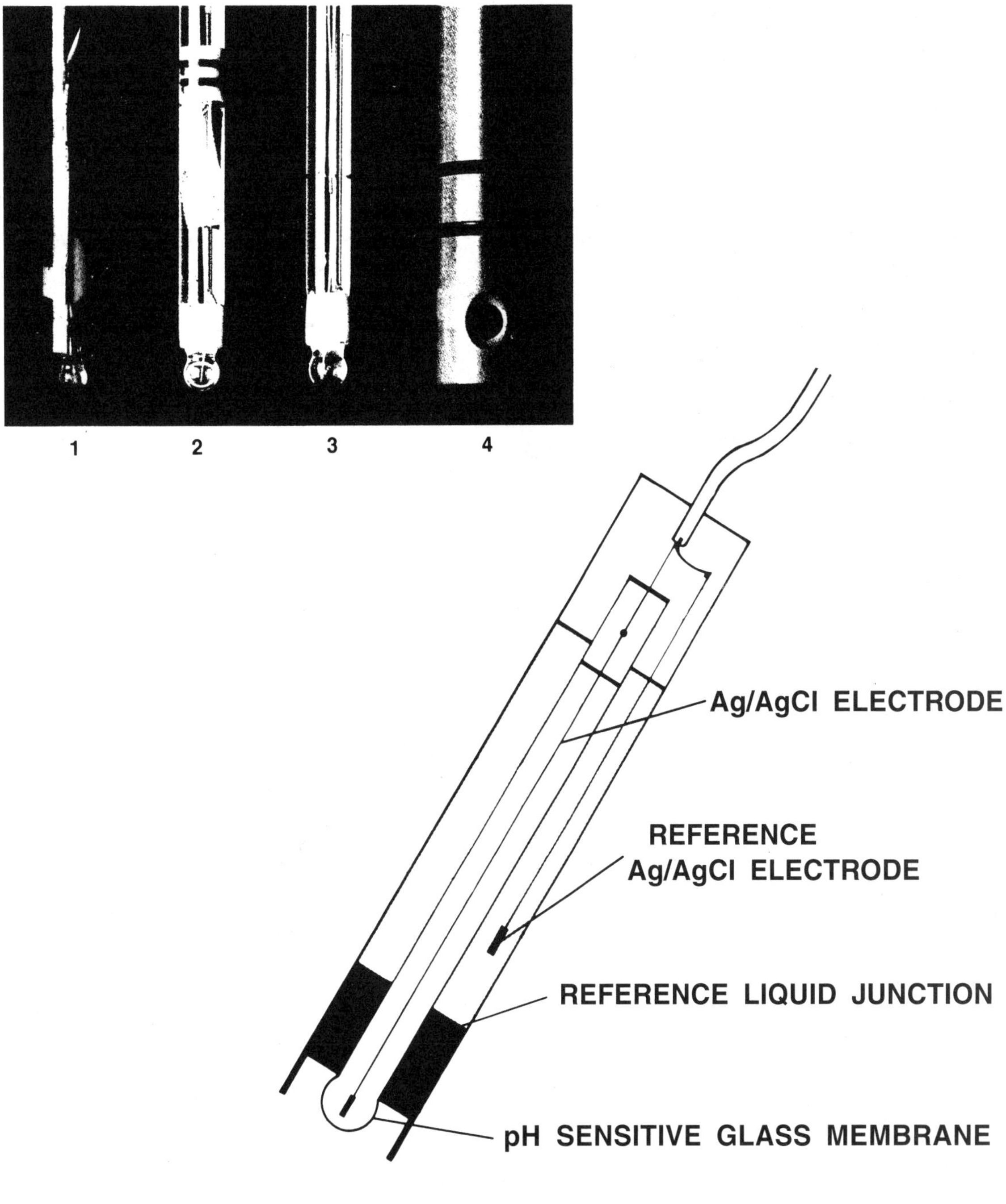

Fig. 11.18 Combination electrode pH probe
(Permission of Signet Scientific)

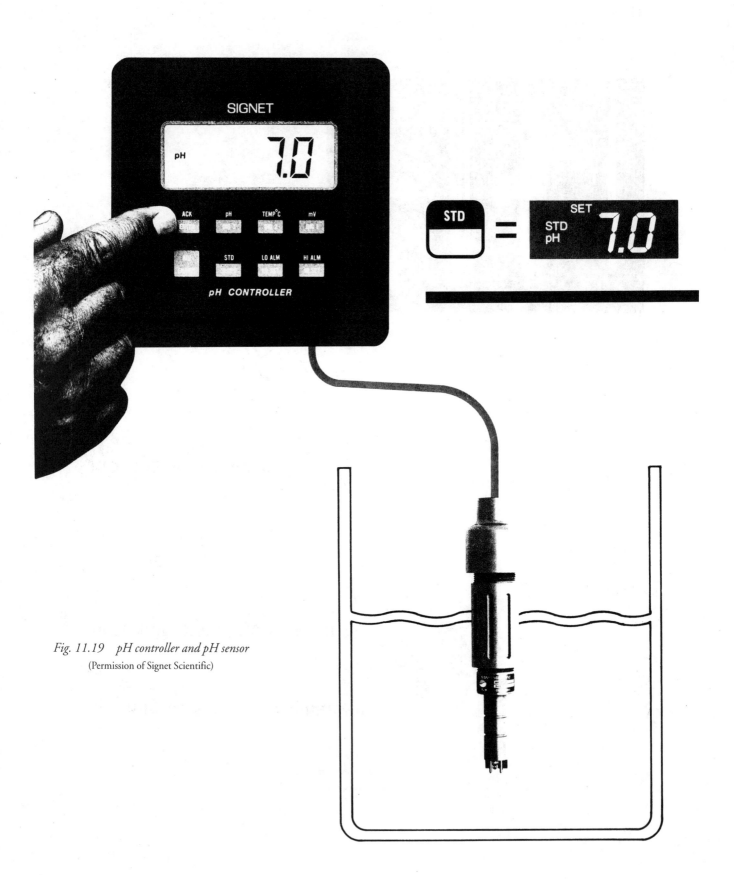

Fig. 11.19 pH controller and pH sensor
(Permission of Signet Scientific)

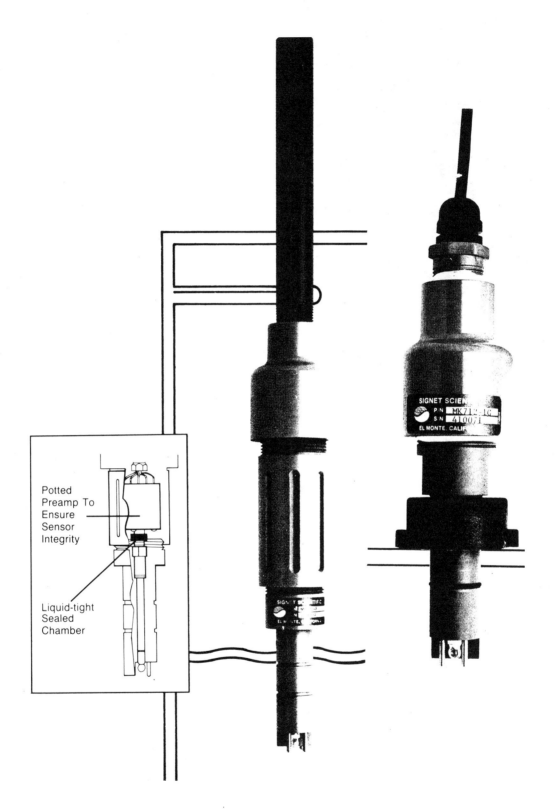

Fig. 11.20 ORP sensors
(Permission of Signet Scientific)

Fig. 11.21 ORP monitor
(Permission of Signet Scientific)

icals in wastestreams are compatible and which chemicals are not compatible.

1. Precipitate Removal

   Precipitates are the most common type of electrode fouling material. All high pH systems tend to precipitate metallic hydroxides as well as calcium and magnesium salts. These precipitates are easily removed by briefly (2 to 3 seconds) dipping a dry probe in a concentrated hydrochloric acid (HCl) solution. The short dipping time should be sufficient to clean the electrode. Longer dipping times may result in damage to the electrode. Some shops clean probes using a 10 percent hydrochloric acid solution and cotton swabs (such as Q-Tips®). Next, rinse the probe in clean water and then allow the probe to equalize (soak in clean water and rehydrate) for five minutes or so before calibrating the probe.

2. Oil and Grease Removal

   Removal of oil and grease from electrodes is not as common a problem as precipitate removal in treatment systems for most industrial wastestreams. Many oils exist as an emulsion (suspended in the wastestream but not dissolved) due to high pH conditions and therefore do not form a coating on probes.

   If the oils and greases that form the coatings are from a known source, it may be possible to find an appropriate solvent that will dissolve the residues on the probe. Before applying any solvent to a probe, contact the probe manufacturer to be sure the solvent will not damage the probe's components since many of the *POTTING COMPOUNDS*[25] and O-rings are made of rubber or plastic.

If you are unable to determine the source or type of oils or greases, try cleaning the probe with 100 percent isopropanol. If isopropanol does not work, use a soft brush and dishwashing liquid. Lightly scrub both the glass electrode and the reference electrode. If the greasy scum is still present, soak the probe in a 33 percent sodium hydroxide (NaOH) solution for five minutes. Next, scrub the probe lightly with a soft brush, rinse the probe, and then repeat the detergent cleaning process. If none of these procedures work, dip the probe in a concentrated hydrochloric acid solution used to remove precipitates.

11.402 Calibration

1. pH Probes

   Two standard buffer solutions are commonly used to calibrate pH probes. The buffer solutions can be purchased in bottles in the already-mixed or ready-to-use form from a chemical supply house. Buffer solutions also can be prepared when they are needed by purchasing the buffer in packet or tablet form and then mixing the dry chemical with distilled water. A pH 7.0 buffer is used as the standard buffer, while a pH of 4 or 10 is used as the "slope" buffer. The choice between the 4 or 10 buffer depends on where the system pH will usually operate. If you are measuring the pH in a metal precipitation process (high pH) or the cyanide destruction process (high pH), use the 10 pH buffer as the slope. Low pH processes, such as chrome reduction, require the use of a low pH (4.0) buffer. pH 10 buffer will absorb carbon dioxide from the atmosphere and must be replaced frequently. Some instrument technicians prefer pH 9 buffer instead because it is much more stable than pH 10 buffer.

   To calibrate pH probes, remove them from the reaction tanks and clean them according to the instructions in Section 11.401, "Cleaning." Place each of the two buffer solutions in a 250- to 500-mL beaker (use three beakers if both high and low pH probes are being calibrated). Also, place a one-liter beaker full of clean water near the probe. Use the following procedure to calibrate the probes.

   a. Place the probe in the pH 7 buffer.

      NOTE: All pH buffers must be temperature correct. For example, if the temperature of the buffer solution is 25°C, the actual pH may be 6.85 instead of the expected 7.0. The buffer bottle or packet will indicate the appropriate pH values of various temperatures. Allow sufficient time for the probes to adjust to the buffers (a few minutes) before adjusting the pot (potentiometer) or slope to help ensure accurate pH values.

   b. The standard pot is used to adjust the pH reading on the pH meter to agree with the pH of the buffer. The pot is an adjustable resistor, which is the adjustment mechanism or device in the pH meter. This pot may be in the form of

---

[25] *Potting Compounds.* Sealing and holding compounds (such as epoxy) used in electrode probes.

a dial or a screw. Some manufacturers require that a non-metallic screwdriver be used to make the pH adjustment.

c. Rinse the probe with deionized water from a squeeze bottle and then place the probe in the slope buffer. Vapors from processes will often form a film on the connectors causing poor response. This film can be removed by isopropanol on a tissue or a cotton swab.

d. Adjust the slope pot so the pH reading will agree with the pH buffer. The slope pot may not be able to adjust the pH on the meter all the way to the desired calibration or buffer pH value on the first try. This problem is not too unusual. If the pH meter cannot be adjusted to read the calibration pH value, go on to step e.

e. Rinse the probe in clean water and then place the probe in the standard buffer.

f. The meter pH value will no longer read 7.0, the standard value. Readjust the pH meter reading to agree with the standard buffer value using the standard pot. The standard and slope pH meter reading values are interconnected by the electronics of the pH meter system. By making these adjustments you are actually "warping" an electronic circuit to fit your probe system.

g. Rinse the probe in clean water and then place the probe in the slope buffer solution.

h. Readjust the slope pH meter reading using the slope pot so the meter reading will agree with the pH of the slope buffer solution. This time the pH meter should be adjustable to the expected value or very close to it.

i. Repeat these steps until the pH meter correctly reads both the standard and slope pH values without any adjustment.

2. ORP Probes

ORP probes are calibrated using a different procedure than the pH probe calibration procedures. Only one standard solution is needed to calibrate an ORP probe. A typical ORP probe calibration procedure uses the following steps.

a. Switch the ORP system from "Read" to "Standby." In most ORP systems this "grounds out" the input to the millivolt meter. The standard pot is then used to precisely adjust the needle or solid-state readout to zero. Operators should realize that it is possible to have negative millivolt readings on an ORP scale. Most ORP scales are capable of reading ORP values of ± 1,400 mV (the device in Figure 11.21 measures ± 1,000 mV).

Some ORP systems do not have "Read" and "Standby" switches. When calibrating such a system, you must ground out the probe leads using a paper clip. The clip is inserted to bridge between the two wires that lead from the ORP unit to where they connect with the probe wires.

b. After adjusting the scale to read exactly zero, the probe is reattached or placed in "Read" and put in the buffer solution.

The buffer solution for ORP probes is not as easy to obtain as pH buffer solutions due to the instability of ORP buffer solutions. Whether the ORP buffer solution is mixed from liquids on site or from a chemical manufacturer's powder packet, the ORP buffer solution must be used within 10 to 30 minutes of its preparation. Discard the solution after 30 minutes.

Some ORP meter manufacturers recommend the use of a pH 7 standard buffer to which a certain amount of hydroquinone is added. The exact proportions of hydroquinone and the ORP value that results varies according to each manufacturer's recipe. One manufacturer, Horiba, Inc.,[26] makes a very convenient powder packet that, when mixed with 250 mL of distilled water, produces an ORP value of 262.5 mV.

c. The slope pot is used to adjust the meter reading to the buffer ORP value.

d. Repeat the procedure to be sure that both the standard and slope values are realized. The standard is the zero reading and the slope reading is the 262.5 mV reading.

11.403 Troubleshooting pH and ORP Probes

If problems develop when using either pH or ORP probes, use the following checklist to locate and correct the causes of the problems.

1. Indicator does not move off one position

a. Check Read-Standby switch. The switch may be on Standby.

b. Clean the probe and calibrate the assembly. The reference electrode may need to be changed.

c. Inspect the probe connections; they may be disconnected.

d. Cross check the pH with a pH test strip or a lab pH meter. The waste characteristics may be unusually consistent.

---

[26] May be obtained by contacting Instrumentation Northwest, Inc., 4620 Northgate Boulevard, Suite 170, Sacramento, CA 95834, phone (916) 922-2900. Pack of 10 buffers, Model No. 160-22, Part No. 350065. Cost, $53.00, plus shipping and handling.

2. Indicator has no reading

   a. Inspect panel circuit breaker.

   b. Examine fuses and overloads.

   c. Check LED (Light Emitting Diode or red light) to be sure it is on or check indicator connector to the unit's circuits.

   d. Inspect probe connections.

3. Indicator has a very low reading

   a. Cross check with pH test strips or lab pH meter.

   b. Check the glass electrode. Clean and calibrate the assembly.

4. Indicator moves very slowly

   a. Clean the probe assembly (clean the glass and reference probes).

   b. If a clean probe is still sluggish or halting, one of the electrodes is failing, probably the glass electrode. The reference electrode usually lasts longer. Replace the electrode.

   c. If a clean probe cannot be adjusted to both the standard and slope values, both electrodes will probably have to be changed. However, always change one electrode at a time (if possible), starting with the glass electrode.

5. Indicator moves erratically

   If a pH reading is behaving erratically, look for:

   a. Electrode contamination.

   b. Ground-loop problem.

   To determine the cause of the problem, measure the pH of a solution in the laboratory. Next, measure the pH of the same solution with the electrode.

   a. If the electrode does not give the same pH, then the problem is probably a contaminated electrode that must be replaced.

   b. If the electrode gives the same pH as measured in the lab, then look for a ground-loop problem. A ground-loop problem is caused by stray electric currents in the wastestream being treated. The problem may be detected during start-up or in a system that has been functioning perfectly for some time, but a new pump or other facilities have been connected to the system. These stray electric currents may not be present all of the time and may fluctuate considerably, thus causing pH readings to fluctuate radically.

   A ground-loop problem may be corrected by any of three approaches.

   a. Find the source of the stray current and prevent the current from flowing through the wastestream.

   b. Place a ground-loop into the tank by inserting a copper rod into the wastestream flowing through the tank. Run a wire from the copper rod to a rod in the ground to complete the ground-loop.

   c. If the copper rod cannot be tied to an earth ground, tie it back to the shield of the instrument.

6. Probe exposed to air

   If a probe used in a batch process is exposed to air when a tank is emptied, the probe must be recalibrated before the next batch is added to the tank. This recalibration problem may be avoided by placing the electrode in a position where it will always be submerged. The electrode could be placed in a shallow cup that will keep the electrode submerged, but will expose the electrode to the contents of the wastestream when it flows through the tank. If tank levels fluctuate considerably, the electrode may be placed on a styrofoam float with the tip submerged below the wastestream's surface. You may encounter problems with this approach if there are density currents through the tank or a scum layer on the surface of the wastestream in the tank.

QUESTIONS

Write your answers in a notebook and then compare your answers with those on page 603.

11.4B   Why must pH and ORP probes be cleaned and calibrated on a weekly basis or more frequently?

11.4C   What types of materials must be removed from pH and ORP probes?

11.4D   What metal waste treatment processes require a high pH slope buffer and which one requires a low pH slope buffer?

11.4E   List the problems that could develop with the indicator when using pH and ORP probes.

11.41 Flow Measurement/Totalization Devices

Flow measurement/totalization is important for the following reasons:

- Dosages of various chemicals, such as polymers, flocculation aids, chelation breakers, and defoaming solutions, are based on flow rates.

- Dilution of concentrated chemicals is based on the rate of flow.

- Sewer fees are based on the total amount of flow that occurs, and in some cases also on the peak flows that can occur.

- Treatment processes and treatment tanks have fixed capacities, therefore flows need to be evaluated and additional capacity provided when it is necessary.

- Samples taken by automated samplers that are initiated by a flow signal are more representative than a strictly time-based sample.

The operator should check the accuracy of the flow measuring devices by physically measuring the quantity of fluid pumped during a known time period. It should be pointed out that centrifugal pumps, and even some positive displacement metering pumps, will pump varying amounts depending on the variations in the level of the fluid being pumped. In addition, variations of the level to which the liquid is being pumped may cause uneven pumping rates. The best results are obtained by keeping these conditions normal and running three or four repetitions of each test.

Small flows, such as metering pumps, can be tested by the use of a graduated cylinder and a stopwatch. Flows up to about 60 GPM can be tested using a 5-gallon bucket and a stopwatch. Above that rate of flow, it will be necessary to lower the level in the receiving tank and pump into it with each foot of height (or less) being measured against time. This method can provide an accurate test of the totalizer as well as the rate of flow. If the totalizer reads in increments of 100 gallons, it will be necessary to pump at least 500 to 600 gallons to obtain a reasonable result. Record the starting time as the totalizer actuates after the pump is turned on. Note the running times for each 100 gallons, as well as the rate of flow values. Repeat the test several times so that an accurate result is obtained. If the results indicate a problem, contact the manufacturer's representative to recalibrate the meter to actual conditions. It is not unusual for meters to be installed and forgotten. Like any other piece of equipment, flow instrumentation needs to be maintained.

11.42 Level Controls

Level controls are available in many different types and configurations. Each different type was developed to overcome some factor that was viewed as a problem. The limitations of each type are described in Chapter 6, "Flow Measurement," and will not be repeated here. From the operator's viewpoint, it will be evident when there is a problem since something will or will not happen (no high-level alarm, pumps that will not shut off at low levels, alarms when there is no cause).

The operator must periodically inspect the sensors to see if they are operating as they should. For instance, in systems using float controls:

- Are the cords tied in a knot (by turbulence)?
- Are the floats held up by a scum blanket?
- Have the cords slipped (changing the position of the float)?
- When the floats are tilted by hand, does the pump respond as it should? Do the alarm level floats activate alarms at the proper time?

11.43 Programmable Controllers

Many metal wastewater treatment systems are using programmable controllers as a means of replacing relays within instruments and control panels. This saves space and money. Some of the more sophisticated units accept sensor signals, directly or through interface (connector) devices, process the signal, and then make control decisions based on a programmed set of instructions. An operator is somewhat at the mercy of the "black box." If you do not understand the device or its programming, and do not have an input keyboard to change control points, you will eventually have to deal with the unit going out of service. If you do not have an identical, programmed unit available, the wastewater treatment plant will be 100 percent out of service until the controller company responds to the emergency. Be sure that all the backup information and hardware are on hand and that a sufficient number of personnel are familiar with the system so that a breakdown can be handled in a timely manner.

11.44 Voltage, Current, and Resistance Measurements

An operator does not have to be an electrician; however, operators work in wet areas where electrical motors, control panels, sensors, and other electrical items are common. For your safety, to ensure the continuous operation of the plant, to analyze electrical problems, and to communicate effectively with electricians, you should have a fundamental knowledge of electrical power and how it is controlled. You should also pay constant attention to electrical safety.

In most facilities, motors use 480-volt, three-phase AC (alternating current) power. Fractional horsepower motors are run with 120-volt, single-phase circuits or lesser voltages. These alternating currents can be dangerous when your body serves as a pathway for the current. The higher the voltage the greater the danger! However, it is the amount of current (the amperage) that goes through the pathway (the body) that actually does the damage.

Voltage can be viewed as the "pressure," while current can be viewed as the "flow." Voltage is measured using a voltmeter, typically called a multimeter, since one meter is used to measure AC, DC, and resistances. Current is typically measured by using an "amp probe," which measures the electrical field caused by the current flow through the conductors and converts that to a current reading in amperes (amps).

The resistance portion of the multimeter is used to see if fuses, relays, and solenoids are burned out or if there is continuity. If the wire is burned out, there should be infinitely high resistance. If there is no resistance between a wire and the case of the device, it means that the wire is "shorted out" to the case.

The operator can use the multimeter to see if there is a voltage reading between the case of a motor and the nearest ground. This would indicate that the insulation is defective. The multimeter can also be used to test the voltage in each leg of a three-phase system to see if they are even (in some systems with a "stinger leg" the readings will not be equal). Low or high voltage readings may indicate an inadequacy of the electrical system.

The amperage readings for the motors indicate whether the motor is overloaded, whether the "electrical heaters" (overload devices) are properly sized, and if the amp readings are reasonably close on all legs of the three-phase circuit supplying the motor.

Since most operators do not work with electrical troubleshooting on a frequent basis, exercise caution when doing so. Become familiar with the above-mentioned equipment and review any items with which you are not familiar with the plant electrician. Practice and exposure to the electrical devices ahead of time will go a long way to building your confidence when there are breakdowns.

QUESTIONS

Write your answers in a notebook and then compare your answers with those on pages 603 and 604.

11.4F How can an operator check the accuracy of the flow measuring devices?

11.4G What should an operator be prepared to do when a programmable controller goes out of service?

11.4H Why should an operator have a fundamental knowledge of electrical power and how it is controlled?

11.5 METALLIC SLUDGE DEWATERING

11.50 Reduction of Sludge Volumes

Because of the expense of disposing of the metallic sludges produced in industrial wastewater treatment systems, many different approaches to the sludge disposal problem have been tried. A very effective method is to not produce (or at least minimize) the production of metallic waste in the first place. Other techniques include:

- Allowing sufficient drip time.
- Using countercurrent rinses with ion removal or plate out techniques.
- Installing air knives to retain drag out.
- Using foggers rather than spray rinses.

These techniques keep the plating solutions in their respective baths and out of the wastewater.

Another way to reduce the amount of sludge is to minimize the use of bulking agents to a practical minimum. Chemicals of this type include lime, calcium chloride, magnesium hydroxide, ferric salts, alum, and phosphoric acid. While sometimes needed to gain a good separation/sedimentation, bulking agents add a considerable volume to the sludge. In many cases, the amount of the sludge from these sources is many times the amount that is caused by the metals that are the object of the removal process. This added bulk, in many cases, causes the metal concentration in the sludge to be below the level that is attractive to metal recyclers.

The presence of organic compounds in the sludge, typically resulting from chelation compounds contained in electroless plating solutions, can have an adverse impact on dewatering operations.

11.51 Methods to Dewater Sludges

Metallic sludges can be dewatered in several ways, and to different degrees of dryness. The methods that are chosen depend on the characteristics of the sludge as well as the amount of sludge to be handled. The drier the sludge, the less it weighs and the less space it takes up. Settled metallic sludges will contain 1 to 4 percent solids by weight.

That same sludge, if it is amenable to being filter pressed, will be in the 25 to 35 percent solids range. This represents a reduction close to a factor of 10. By heating the sludge at relatively low temperatures, a final solids content of 55 to 70 percent (by weight) can be obtained. The approximate reduction in volume is another factor of 4 to 6.

Common methods used to dewater metallic sludges include:

- Centrifuges
- Vacuum filters
- Bag filters
- Filter presses

11.510 Centrifuges

Because of the abrasive nature of most metallic sludges, the use of centrifuges for sludge dewatering has been limited. For sludges that contain oils and grease, or other agents that would blind or clog filter fabrics, centrifuges may be the method of choice. Typical dewatered sludges would be in the 20 percent solids range. The solids capture percentage on most centrifuges is in the 80 percent range, which means that the supernatant is very dirty and must be returned to the process.

Centrifuges are expensive to purchase and operate (high energy costs). If the sludge is abrasive, the maintenance costs will be significant and the down time will also be a critical factor.

From the operator's perspective, the operation of centrifuges is not complicated. The machines usually function automatically and run with little operator attention. The operator is responsible for pumping the waste to the machine, making sure there are sufficient supplies of treatment chemicals/additives (if any are used), and making sure that the chemicals are pumped at the proper rate. At the end of the daily use of the centrifuge, the operator must shut off any chemical feed pumps and clean out the system so that it does not become plugged.

There are several different types of centrifuges. The machines are designed to solve material separation problems caused by the physical characteristics of the mixture. Most metal sludge dewatering centrifuges are of the intermittent discharge (basket) type.

11.511 Vacuum Filters

This type of filter is typically configured as a segmented drum covered by a filter cloth. The drum rotates in a bath of sludge. While rotating in the sludge, that segment of the drum under water/sludge is subjected to a vacuum; as it rotates out of the sludge, the water is pulled from the sludge by the vacuum and the dried sludge is lifted from the cloth by a *DOCTOR BLADE*.[27] Typical results are in the 15 to 18 percent solids range. This type of filter may have a precoat system to enhance the solids retention. Typically, this would be a *DIATOMACEOUS EARTH*[28] application (in depth) on the filter cloth. The complexity, expense (both operating and maintenance), and low capacity of a vacuum filter compared to space required has limited its use to wastes that have special characteristics.

The vacuum is typically supplied by a liquid ring vacuum pump. Since filtrate containing solids passes through the pump, excessive wear can result. Therefore, the operator must check the adequacy of the vacuum and the condition of the pump on a regular basis. The operator must also maintain the polymer system (if used) and any other supplies that are used (such as diatomaceous earth).

11.512 Bag Filters

This low-tech method of dewatering sludge consists of fabric bags, which are attached to a distribution system. Sludge is pumped into the bags, where the sludge is retained and the water that has passed through is caught in a pan and routed back to the treatment process. With a sufficient number of bags, a dewatered sludge in the 13 to 20 percent range can be realized. The significant problems with this method are that there is a great deal of manual labor involved and there is some degree of operator exposure to the heavy metals in the sludge. In general, this method is only suitable for the smallest systems.

11.513 Filter Press

When the term filter press is used, what is being described is a plate and frame filter press. As shown in Figures 11.22 to 11.24, the press consists of plates (covered with filter fabric), which are held in a frame, typically by a hydraulic ram. Sludge is pumped into a center passageway where the solids are filtered out by the fabric and the filtrate is returned to the process. When the press has been filled with sludge, the sludge is pressurized with air to get the last of the free water out. The hydraulic pressure is then released and the sludge is removed (Figure 11.25). Typical solids concentrations are in the 25 to 35 percent range.

Filter presses are available in a large variety of sizes and with varying degrees of automation. The machine is simple in concept, and maintenance and operating costs are relatively low. The purchase cost is a function of how large the press is and what degree of sophistication is desired.

The operator will operate the system on a cycle that lasts for about $2^{1}/_{2}$ hours. On this basis, two full loads per shift are easily possible. The steps in the process are as follows:

- The press is closed by operating the hydraulic ram.

- The hydraulic pressure is increased to the design level (probably in the 3,000 to 6,000 psi range, depending on the specific design).

- The precoat system is activated (if used) and the filter media is coated.

- The transfer pump is activated and the sludge starts to fill the press. The pump, an air-operated diaphragm type, is allowed to run for approximately two hours. Prior to that time the operation of the pump will become intermittent as the pump stalls out due to increased resistance caused by the increasing thickness of the dewatered sludge. After the pump has stalled out completely, or the two hours has expired, the transfer pump is turned off. An air line attached to the input manifold is opened and the sludge is air dried for 15 to 30 minutes. The air application pushes the water out of the center feed channel sludge so that when the press is opened the sludge moisture content should be consistently dry.

- After turning off the air and making sure that all internal pressures have been relieved, the hydraulic system can be switched to the open position, and the press opened for cleaning.

The operator is responsible for providing for precoat materials and such items as polymers, if they are needed at a particular location. Depending on the degree of automation, the operator will be responsible for starting the dewatering cycle and for the cleaning part of the cycle. The cleaning cycle may just consist of opening the press and letting the cake drop out. If the cake has sticky qualities, it may have to be scraped out. Since this is very time consuming, it may be worthwhile to precoat the filter cloth so that the cake does not stick.

Besides the operational tasks described above, cleanup and maintenance of the dewatering area are essential to maintain a well-run and safe working area. Chronic exposure to heavy metals in the form of dust can lead to long-term health problems.

Maintenance will be required on the transfer pump. Wear occurs on both the check valve assemblies and diaphragms. If the pump continues to operate without stalling and the cake is not

---

[27] *Doctor Blade.* A blade used to remove any excess solids that may cling to the outside of a rotating screen.
[28] *Diatomaceous* (DYE-uh-toe-MAY-shus) *Earth.* A fine, siliceous (made of silica) "earth" composed mainly of the skeletal remains of diatoms.[29]
[29] *Diatoms* (DYE-uh-toms). Unicellular (single cell), microscopic algae with a rigid, box-like internal structure consisting mainly of silica.

Fig. 11.22 *Plate and frame filter press* (Permission of New England Plating Co., Inc.)

Fig. 11.23 Plate and frame filter press
(Permission of Lee, Strangio & Associates, Inc.)

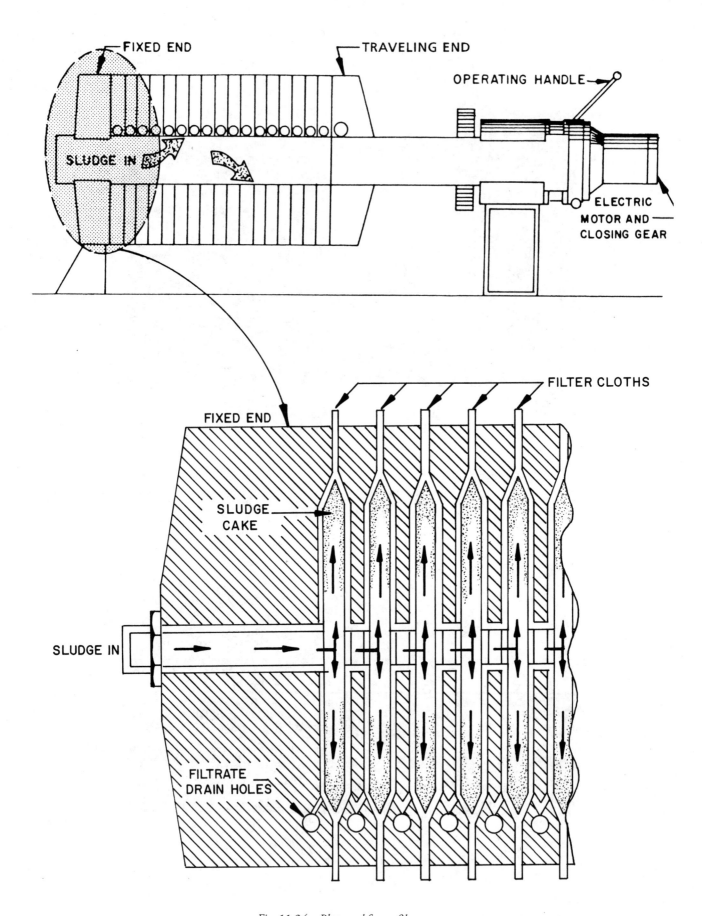

Fig. 11.24 *Plate and frame filter press*

(Permission of Lee, Strangio and Associates, Inc.)

(Permission of Edwards and Jones, Inc.)

Fig. 11.25   Filter press cake

solid, it is likely that the check valves are not operating correctly and should be replaced. If the filter cloths have to be scraped to get the sludge off, they will eventually have to be replaced.

11.52 Sludge Drying Methods

As mentioned previously, further reductions in moisture content result in reduction in the volume of sludge to be disposed of. The methods by which this can be obtained are:

- Air drying
- Induced air drying
- Heating (with or without additional air)
- Reduced pressure, low heat drying (vacuum drying)

11.520 Air Drying

This is simple drying using sunlight and exposure of the sludge to the ambient environment. This method is limited in use because of the potential for the loss of material to the surrounding area. Many small facilities store their sludge in open drums and bins and then refill them as the sludge dries and shrinks. This method is not suitable for rainy or cold climates.

11.521 Induced Air Drying

This is a mechanically induced breeze, which speeds up the drying action. Careful control of dry metallic dusts is essential since such heavy metals are hazardous and strictly regulated.

11.522 Heating

This method uses a heat source to speed up the drying process. Methods such as infrared heaters, hot air, heat exchange tanks (double walled tanks with hot water circulated through the annular space between the walls), and other ways of transferring the heat have been used. The sludge mass is usually mixed so that an optimum transfer of heat is obtained, and the maximum amount of moisture is released. The end product is typically a granular powder; however, some systems include a pelletizer to further consolidate the metallic powder. Control of the dust that may be produced is essential for the safe, long-term use of this method. The economic benefit of using this method of sludge processing should be proved by detailed analysis prior to adopting its use.

As mentioned previously, the presence of certain organic compounds typically associated with electroless plating compounds may make the application of heat an impossibility because of off-gassing, solidification (vulcanization), and potential combustion. Lab tests should be performed before a new sludge is used in an existing drying device.

The operator must be familiar with the operating characteristics and limits of the drying system. Each mechanical system has its own unique layout and responses. Control systems can vary from ON/OFF to fully automated. Temperature levels and safety devices must be checked frequently (on a routine basis) and the equipment must be kept in good operating order.

11.523 Reduced Pressure, Low Heat Drying (Vacuum Drying)

This method has been used primarily on spent plating baths. It relies on the principle of boiling away the bath's water content while the liquid is subjected to a reduced pressure. The partial vacuum lowers the boiling point allowing the vapor to be condensed in a refrigerated zone, while the reject heat from the compressed refrigerant serves as a heat source for the boiling container. The end products are distilled water and metallic crystals. The presence of the organic compounds associated with electroless plating also causes problems with this method. The solid material is difficult to remove from the device since it is not in crystalline form but is more like rubber. Lab tests should be undertaken before this method is adopted for use.

These devices usually operate in a batch fashion. The operator typically has to remove the dewatered end product by hand. The units usually have a storage tank that holds the waste to be treated. The waste is transferred into the treatment unit automatically when the cycle is started by the operator. The operator must be careful to avoid exposure to the material as it is transferred to a disposal drum or bin.

QUESTIONS

Write your answers in a notebook and then compare your answers with those on page 604.

11.5A List the methods used to dewater metallic sludges.

11.5B How long should the transfer pump operate to pump metallic sludge to a plate and frame filter press?

11.5C How can sticky sludge be dewatered using a plate and frame filter press?

11.5D How can dewatered sludge moisture content be further reduced thus reducing the volume of sludge to be disposed of?

END OF LESSON 2 OF 3 LESSONS

on

TREATMENT OF METAL WASTESTREAMS

Please answer the discussion and review questions next.

DISCUSSION AND REVIEW QUESTIONS
Chapter 11. TREATMENT OF METAL WASTESTREAMS
(Lesson 2 of 3 Lessons)

Write the answers to these questions in your notebook. The question numbering continues from Lesson 1.

12. Under what conditions would you use a high pH slope buffer and also a low pH slope buffer?

13. Why is flow measurement/totalization important for the treatment of metal wastestreams?

14. Why should an operator have a fundamental knowledge of electrical power and how it is controlled?

15. What are the advantages and limitations of using centrifuges to dewater metallic sludges?

CHAPTER 11. TREATMENT OF METAL WASTESTREAMS

(Lesson 3 of 3 Lessons)

11.6 OPERATION, MAINTENANCE, AND TROUBLESHOOTING

11.60 Know the Processes Involved

The actual metal wastestream treatment processes used in your facility to obtain the desired effluent results will depend on a combination of factors involved in your specific industrial situation. The factors that influence the types and complexity of the processes include:

1. The particular wastestream characteristics
2. Space available for treatment
3. Costs of the processes
4. Effluent discharge requirements

Frequently, industrial treatment plant operators are required to operate treatment processes that they know little about or have no experience operating. Under these circumstances, the ingenuity of the operator is challenged. Helpful resources that are available to the operator include:

1. This training manual.
2. System O & M manuals (all manufacturers should provide detailed procedures on how to operate and maintain their equipment).

3. System construction drawings (especially the electrical, mechanical, and instrumentation drawings).
4. Manufacturers' technical representatives (troubleshooters).
5. Instrumentation service company representatives.
6. Wastewater treatment specialists and consultants.
7. Operators who treat wastewater for other companies or shops.

After you have become familiar with your systems and processes, you must continue to develop your knowledge of how these facilities work and how to fine-tune the treatment processes. You can expect the characteristics of the wastewaters being treated to change hourly, daily, weekly, and seasonally due to changes in the production processes or maintenance procedures, thus requiring continuous monitoring and adjustment of the treatment processes. You must be aware of the variations in the wastes and be prepared to change the settings of treatment systems to achieve the best possible effluent at the lowest possible cost.

11.61 Knowledge and Abilities Required

Successful operation and maintenance of industrial wastewater treatment facilities demands highly skilled and knowledgeable operators.

Tasks performed by operators require:

1. Ability to recognize safety hazards and to follow safe procedures at all times.
2. Knowledge of the chemical processes involved.
3. Knowledge of how the equipment is supposed to function.
4. Knowledge of waste treatment facility operations and maintenance (O & M) manual.
5. Skill in conducting methodical, periodic observations of the system.
6. Ability to collect samples, analyze results, and verify expected performance.
7. Ability to use laboratory analytical support facilities.
8. Ability to develop and implement a routine preventive maintenance program.

QUESTIONS

Write your answers in a notebook and then compare your answers with those on page 604.

11.6A   List the factors that influence the types and complexity of the processes used to treat metal wastes.

11.6B   Why do metal waste treatment processes require continuous monitoring and adjustment of the treatment processes?

11.62 Normal Operation

For effective operation of a wastewater treatment facility, operators must visually observe the wastewater being treated and the treatment processes; read, record, and analyze instrument readouts and printouts; collect, analyze, and interpret sample test results; and physically adjust process control settings.

11.620 Visual Observations

Operators must learn how the treatment system and effluent should look when everything is operating correctly. For instance, if precipitation is one of the treatment processes, then the water in the precipitation phase should look cloudy or full of solids. A clear wastewater would indicate that the precipitation tank is not receiving treatment chemicals, or that some component of the wastewater (ammonia, chelates, or acids) is defeating the precipitation process. An alert operator will then follow a logical process of deduction to determine what is causing the problem.

Operators should be present at all times during the operation of batch and continuous wastestream treatment facilities. Every time an operator inspects or walks through an industrial wastewater treatment facility, the visual observations in the following list should be performed. These observations should be made at least daily. At many industrial treatment facilities, the operator will walk through the facility and perform the visual observations at the start of a shift and also at the end of a shift. The end of the shift tour prepares the departing operator to inform the incoming operator of the current status of the processes. The incoming shift operator should also walk through the facility at the start of the shift to confirm existing conditions and to take action on any items requiring operator attention. The following is a list of items for operators to check when they walk through their facilities.

1. Survey control panel for alarm conditions (Figure 11.26).

2. Examine effluent to determine if it appears normal (for example, is it clear? color free? is there evidence of floc or sludge carryover?).

3. Look at the flocculation tank to see that the waste is flocculating properly (large floc and clear water surrounding the floc). Observe the settling characteristics (settleability) of the floc.

4. Observe the precipitation processes for the proper appearance and check pH readings.

5. Review the control readings in the preliminary treatment processes to be sure they are in the proper range. For example:

   Cyanide   pH = 8.5 and ORP = +650 mV
   Chromium   pH = 2.5 and ORP = +250 mV

6. Look at the clarifier or settling tank to be sure there is no pin floc or solids washout in the effluent.

7. Inspect sludge levels in the clarifiers and sludge storage tanks. Use a sludge judge or some other device to determine the level of the sludge blanket.

8. Determine polymer tank levels.

9. Check chemical tank levels.

10. Survey the area for flooding, leaks, spills, or equipment problems.

11. Scan motor control and instrumentation panels to determine if:

    a. Switches are in correct position

    b. All equipment is operating properly

    c. Proper lights are on

    d. No circuit breakers are tripped

12. Look for signs of turbulence in oil separation facilities. Turbulent conditions are an indication of overloading.

11.621 Typical Operator Duties

In addition to visual observations, which must be performed at the beginning and end of each shift, as well as whenever an operator walks through treatment facilities, there are many physical tasks that must be performed on a regular basis. How often a task should be performed depends on each particular situation (volumes of waste flows, wastes being treated, and their concentrations) and the importance of the task in the production of an acceptable effluent quality. Typical tasks and recommended frequencies are listed in the remainder of this section.

DAILY

1. Exercise H-O-A (Hand-Off-Automatic) switches to determine that all motors and pumps are working.

2. Clean all screens and filters.

3. Exercise all solenoid valves.

4. Calibrate polymer pump rates. See Section 11.646, "Waste Is Not Flocculating," item 2, "Polymer flow rate," for detailed procedures.

5. Verify instrumentation values using pH test strips, hand-held pH meter, or take sample to laboratory for immediate pH measurement using laboratory pH meter.

6. Determine performance of waste treatment units by: (a) Using colorimetric test kits (HACH kits) or *COLORIMETRIC*[30] test procedures with portable colorimetric/spectrophotometer and prepared testing reagents; (b) Collect-

---

[30] *Colorimetric Measurement.* A means of measuring unknown chemical concentrations in water by measuring a sample's color intensity. The specific color of the sample, developed by addition of chemical reagents, is measured with a photoelectric colorimeter or is compared with color standards using, or corresponding with, known concentrations of the chemical.

Fig. 11.26 Control panels
(Permission of New England Plating Co., Inc.)

ing *COMPOSITE SAMPLES*[31] and sending them to an approved commercial laboratory for analysis; or (c) Verifying performance on site by the use of your laboratory's atomic absorption spectrophotometer to measure metal concentrations.

7. Keep neat and accurate records of:

    a. Test results, including time, date, and comments

    b. Amount of each chemical used:

        (1) Polymer rate and renewal

        (2) Caustics and acids

        (3) Other chemicals

    c. Frequency and volume of sludge hauled

WEEKLY

1. Push all Press-to-Test motor running lights and alarm lights to determine light bulb condition.

2. Fill all oilers.

3. Inspect water traps and regulators on air lines.

4. Measure sump pump rates.

5. Clean pH and ORP probes to remove precipitates that occur in the precipitation steps and the cyanide destruction step. Also, remove any accumulated oily wastes. See Section 11.401, "Cleaning," for detailed procedures.

6. Calibrate pH and ORP probes and meters. See Section 11.402, "Calibration," for detailed procedures.

MONTHLY

Prepare and submit monthly reports.

QUESTIONS

Write your answers in a notebook and then compare your answers with those on page 604.

11.6C What activities are required of operators for effective operation of a wastewater treatment facility?

11.6D What can an operator determine about a precipitation process by visual observations?

11.6E How should the effluent from metal waste treatment processes normally appear?

11.622 Sampling

11.6220 IMPORTANCE OF SAMPLING

The basis for any treatment plant monitoring program is information obtained by sampling. Decisions based upon incorrect data may be made if sampling is performed in a careless and thoughtless manner. Obtaining good results will depend to a great extent upon the following factors:

1. Ensuring that the sample taken is truly representative of the wastestream.

2. Using proper sampling techniques.

3. Protecting and preserving the samples until they are analyzed. Most errors produced in laboratory tests are caused by improper sampling, poor preservation, or lack of enough mixing during compositing and testing.

11.6221 REPRESENTATIVE SAMPLING

Always remember that wastestream flows can vary widely in quantity and composition over a 24-hour period. Also, composition can vary within a given stream at any single time due to partial settling of solids or floating of light materials. Samples should therefore be taken from the wastestream where it is well mixed. Obtaining a representative sample should be a major concern in any sampling and monitoring program.

Laboratory equipment, in itself, is generally quite accurate. Analytical balances weigh to 0.1 milligram. Graduated cylinders, pipets, and burets usually measure to 1 percent accuracy, so that the errors introduced by these items should total less than 5 percent, and under the worst possible conditions only 10 percent.

Even with the most accurate laboratory equipment, test results will be meaningless for process control purposes unless the test sample is representative of the wastestream. To efficiently operate a wastewater treatment plant, the operator must rely on test results that indicate what is happening in the treatment process. Sampling stations, therefore, must be located where the most representative samples can be collected, where mixing is thorough, and where the wastestream quality is uniform.

11.6222 TIME OF SAMPLING

Let us consider next the time and frequency of sampling. In carrying out a testing program, particularly where personnel and time are limited due to the press of operational responsibilities, testing may necessarily be restricted to about one test day per week.

The time and day of sampling are quite important. Samples should be taken to represent typical weekdays or even varied from day to day within the week for a good indication of the characteristics of the wastestreams. The time of sampling should be based on production, start-up, shutdown, and cleaning and maintenance schedules.

11.6223 TYPES OF SAMPLES

The two types of samples collected in treatment plants are known as (1) grab samples, and (2) composite samples, and either may be obtained manually or automatically.

---

[31] *Composite (Proportional) Sample.* A composite sample is a collection of individual samples obtained at regular intervals, usually every one or two hours during a 24-hour time span. Each individual sample is combined with the others in proportion to the rate of flow when the sample was collected. Equal volume individual samples also may be collected at intervals after a specific volume of flow passes the sampling point or after equal time intervals and still be referred to as a composite sample. The resulting mixture (composite sample) forms a representative sample and is analyzed to determine the average conditions during the sampling period.

GRAB SAMPLES

A grab sample is a single sample of wastewater collected at a particular time and place that represents the waste characteristics only at that time and place. A grab sample may be preferred over a composite sample when:

1. The wastewater to be sampled does not flow on a continuous basis.

2. The wastewater characteristics and flow rate are relatively constant.

3. You wish to determine whether or not a composite sample obscures (hides or evens out) extreme conditions of the waste.

4. The wastewater is to be analyzed for dissolved gases (H_2S), residual chlorine, temperature, volatile organic compounds, and pH. (*NOTE:* Grab samples for these water quality indicators may be collected at set times or specific time intervals.)

PROPORTIONAL COMPOSITE SAMPLES

Since the wastewater quality changes from moment to moment and hour to hour, the best results would be obtained by using some sort of continuous sampler-analyzer. However, since operators are usually the sampler-analyzer, continuous analysis would leave little time for anything but sampling and testing. Except for tests that cannot wait due to rapid chemical or biological change of the sample, such as tests for temperature, pH, and sulfide, a fair compromise may be reached by taking samples throughout the day at hourly or two-hour intervals.

When the samples are taken, they should be refrigerated immediately to preserve them from any chemical decomposition. When all of the samples have been collected for a 24-hour period, the samples from a specific location should be combined or composited together according to the rate of flow to form a single, 24-hour composite sample.

To prepare a proportional composite sample, (1) the rate of wastewater flow must be known, and (2) each grab sample must then be taken and measured out in direct proportion to the volume of flow at that time. For example, Table 11.5 illustrates the hourly flow and sample volume to be measured out for a 12-hour proportional composite sample.

A very important point should be emphasized. During compositing and at the exact moment of testing, the samples must be vigorously remixed[32] so that they will be of the same composition and as well mixed as when they were originally sampled. Sometimes, operators become lax about such remixing, with the result that all the solids are not uniformly suspended. Lack of mixing can cause low results in samples of solids that settle out rapidly. Samples must therefore be mixed thoroughly and poured quickly before any settling occurs. If this is not done, errors of 25 to 50 percent may easily occur.

TABLE 11.5 DATA COLLECTED TO PREPARE PROPORTIONAL COMPOSITE SAMPLE[a]

Time	Flow, MGD	×	Factor	=	Sample Volume, mL
6 am	0.2		100		20
7 am	0.4		100		40
8 am	0.6		100		60
9 am	1.0		100		100
10 am	1.2		100		120
11 am	1.4		100		140
12 noon	1.5		100		150
1 pm	1.2		100		120
2 pm	1.0		100		100
3 pm	1.0		100		100
4 pm	1.0		100		100
5 pm	0.9		100		90
				Total =	1,140

[a] A sample composited in this manner would total 1,140 mL.

11.6224 SAMPLING DEVICES

Automatic sampling devices are wonderful timesavers and should be used where possible. However, as with anything automatic, problems do arise and the operator should be aware of potential difficulties. Sample lines to auto-samplers may build up with precipitates, which may periodically slough off and contaminate the sample. Very regular cleanout of the intake line is required.

Manual sampling equipment includes dippers, weighted bottles, hand-operated pumps, and cross-section samplers. Dippers consist of wide-mouth, corrosion-resistant containers (such as cans or jars) on long handles that collect a sample for testing. A weighted bottle is a collection container that is lowered to a desired depth. At this location, a cord or wire removes the bottle stopper so the bottle can be filled. Sampling pumps allow the inlet to the suction hose to be lowered to the sampling depth.

For sample containers, wide-mouth, plastic bottles are recommended. Plastic bottles, although somewhat expensive initially, not only greatly reduce the problem of breakage and metal contamination, but are much safer to use. The wide-mouth bottles ease the washing problem. For regular samples, sets of plastic bottles bearing identification labels should be used. For some organic constituents, glass bottles must be used.

11.6225 PRESERVATION OF SAMPLES

Sample deterioration starts immediately after collection for most wastewaters. The shorter the time that elapses between collection and analysis, the more reliable will be the analytical results. In many instances, however, laboratory analysis cannot be started immediately due to the remoteness of the laboratory or workload. A summary of acceptable EPA (US Environmental

---

[32] *NOTE:* If the sample has a low buffer capacity and the real pH is 6.5 or less, vigorous shaking can cause a significant change in pH level.

Protection Agency) methods of preservation appears in Table 11.6.

11.6226 QUALITY CONTROL IN THE WASTEWATER LABORATORY

Having good equipment and using the correct methods are not enough to ensure correct analytical results. Each operator must be constantly alert to factors in the plant that can cause poor data quality. Such factors include: sloppy laboratory technique, deteriorated reagents, poorly operating instruments, and calculation mistakes.

11.6227 SUMMARY

1. Representative samples must be taken before any tests are made.
2. Select a good sampling location.
3. Collect samples and, if necessary, properly preserve them.
4. Mix samples thoroughly before compositing and at the time of the test.

11.6228 ADDITIONAL READING

1. *HANDBOOK FOR MONITORING INDUSTRIAL WASTEWATER*, US Environmental Protection Agency. Obtain from National Technical Information Service (NTIS), 5285 Port Royal Road, Springfield, VA 22161. Order No. PB-259146. Price, $59.50, plus $5.00 shipping and handling per order.

2. *HANDBOOK FOR ANALYTICAL QUALITY CONTROL IN WATER AND WASTEWATER LABORATORIES*, US Environmental Protection Agency. Obtain from National Technical Information Service (NTIS), 5285 Port Royal Road, Springfield, VA 22161. Order No. PB-297451. Price, $52.00, plus $5.00 shipping and handling per order.

QUESTIONS

Write your answers in a notebook and then compare your answers with those on page 604.

11.6F What are the causes of most errors produced in laboratory tests?

11.6G Why must a test sample be representative of the wastestream?

11.6H How would you prepare a 12-hour proportional composite sample?

TABLE 11.6 US EPA REQUIRED PRESERVATION METHODS FOR WATER AND WASTEWATER SAMPLES[a]

Test[b]	Preservation Method	Max. Recommended Holding Time
Acidity, Alkalinity	Store at 4°C	14 days
Ammonia	Add H_2SO_4 to pH <2 Store at 4°C	28 days
BOD	Store at 4°C	48 hours
COD	Add H_2SO_4 to pH <2 Store at 4°C	28 days
Chloride	None required	28 days
Chlorine, residual	Det. on site	No holding
Cyanide	Add[c] NaOH to pH >12 Store at 4°C	14 days
Dissolved Oxygen[d]	Det. on site	No holding
Fluoride[e]	None required	28 days
Mercury	Add HNO_3 to pH <2	28 days
Metals	Add HNO_3 to pH <2	6 months
Nitrate	Store at 4°C	48 hours
Nitrite	Store at 4°C	48 hours
Oil & Grease[f]	Add H_2SO_4 to pH <2 Store at 4°C	28 days
Organic Carbon	Add H_2SO_4 to pH <2 Store at 4°C	28 days
Orthophosphate	Filter on site Store at 4°C	48 hours
pH	Det. on site	No holding
Phenols[f]	Add 0.008% $Na_2S_2O_3$ to pH <2 Store at 4°C	7/40 days[g]
Phosphorus, total	Add H_2SO_4 to pH <2 Store at 4°C	28 days
Residue, total	Store at 4°C	7 days
Specific Conductance	Store at 4°C	28 days
Sulfate	Store at 4°C	28 days
Sulfide	Add 2 mL 1 M zinc acetate & 1 N NaOH to pH >9 Store at 4°C	7 days
Temperature	Det. on site	No holding
T. Kjeldahl Nitrogen	Add H_2SO_4 to pH <2 Store at 4°C	28 days
Turbidity	Store at 4°C	48 hours

[a] 40 CFR Parts 136–149, Ch. 1 (7-1-05 Edition), paragraph 136.3. For sale by US Government Printing Office, Superintendent of Documents, PO Box 371954, Pittsburgh, PA 15250-7954. Price, $61.00. Order No. 869-056-00159-2.
[b] Use polyethylene or glass containers except as noted below.
[c] Add 0.6 gm ascorbic acid if residual chlorine is present.
[d] Glass bottle and top.
[e] Polyethylene.
[f] Glass.
[g] 7 days before extraction; 40 days after extraction.

11.623 Flow Measurement

Flow measurement is the determination of the quantity of a mass in movement within a known length of time (Figure 11.27). Flow is measured and recorded as a quantity (gallons or cubic feet) moving past a point during a specific time interval (seconds, minutes, hours, or days). Thus, we obtain a flow rate or quantity in cu ft/sec or MGD. The mass may be solid, liquid, or gas and is usually contained within physical boundaries such as tanks, pipelines, and open channels or flumes. The limits of such physical or mechanical boundaries provide a measureable dimensional *AREA* that the mass is passing through. The speed at which the mass passes through these boundaries is related to dimensional distance and units of time; it is referred to as *VELOCITY*. Therefore, we have the basic flow formula:

$$\text{Quantity} = \text{Area} \times \text{Velocity}$$
$$Q = AV$$

or

$$Q, \text{cu ft/sec} = (\text{Area, sq ft})(V, \text{ft/sec})$$

or

$$Q, \text{m}^3/\text{sec} = (\text{Area, sq m})(V, \text{m/sec})$$

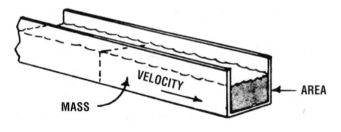

Fig. 11.27 Flow mass

The performance of a treatment facility cannot be evaluated or compared with other plants without flow measurement. Individual treatment units or processes in a treatment plant must be observed in terms of flow to determine their efficiency and loadings. Flow measurement is important to plant operation as well as to records of operation. The devices used for such measurement must be understood, used properly, and most important, maintained so that information obtained is accurate and dependable.

11.6230 OPERATORS' RESPONSIBILITIES

Instrumentation and flow measurement devices are fragile mechanisms. Rough handling will damage the units in as serious a manner as does neglect. Treat the devices with care, keep them clean, and they will perform their designed functions with accuracy and dependability.

11.6231 TYPES OF FLOW MEASUREMENT DEVICES

The selection of a type of flow metering device and its location are decisions made by the facility designer when a new plant is built. However, it is also possible that a metering device will have to be added to an existing facility. In both cases, the various types available, their limitations, and criteria for installation should be known. Often, the criteria for installation must be understood for the proper use and maintenance of a fluid flowmeter. Metering devices commonly used in treatment facilities are listed in Table 11.7 and briefly described in the following paragraphs.

CONSTANT DIFFERENTIAL—A mechanical device called the "float" is placed in a tapered tube in the flow line (Figure 11.28). The difference in pressures above and below the float causes the float to move with flow variations. Instantaneous rate of flow is read out directly on a calibrated scale attached to the tube. Read the scale behind the top of the float to obtain the flow rate.

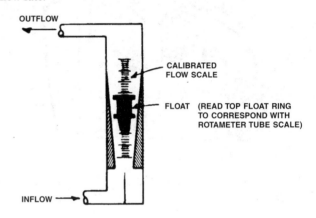

Fig. 11.28 Rotameter

HEAD AREA—A mechanical constriction or barrier is placed in the open flow line causing an upstream rise in liquid level (Figures 11.29, 11.30, and 11.31). The rise of head (H) is mathematically related to the velocity (speed) of the flow. The head measurement can be used in a formula to calculate flow rate.

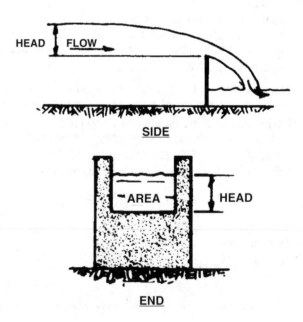

Fig. 11.29 Rectangular weir

TABLE 11.7 SUMMARY OF FLOW METERING DEVICES

Type[a]	Common Name	Application
Constant Differential	Rotameter	Liquids and Gases a. Chlorination b. Purge air for bubblers

Head Area	Weirs
	   Rectangular
	   Cipolletti
	   V-Notch
	   Proportional
		Liquids—partially filled channels, basins, or clarifiers
		a. Influent
		b. Basin control
		c. Effluent
		d. Effluent

	Flumes
	   Parshall
	   Palmer-Bowlus
	   Nozzles
		Liquids—partially filled pipes and channels
		a. Influent
		b. Basin control
		c. Effluent
		d. Distribution

Velocity Meter	Propeller	Liquids—channel flow, clean water piped flow
	Magnetic	Liquids and sludge in closed pipes a. Influent b. Basin control c. Sludge d. Distribution

Differential Head	Venturi Tube
	   Flow Nozzle
	   Orifice
		Gases and liquids in closed pipes
		a. Influent
		b. Basin control
		c. Effluent
		d. Gas
		e. Distribution

Displacement	Piston Diaphragm	Gases and liquids in closed pipes a. Plant water b. Chemical feeders

[a] A description of how each device works is in reality a definition of the meter type.

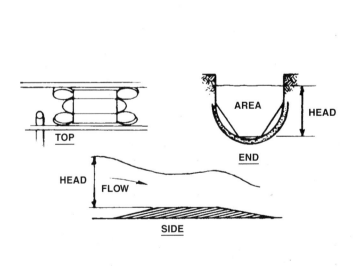

Fig. 11.30 Palmer-Bowlus flume

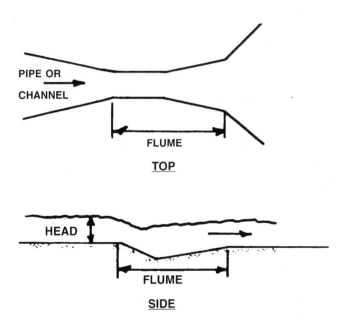

Fig. 11.31 Parshall flume

VELOCITY METERS—The velocity of the liquid flowing past the measurement point through a given area gives a direct relation to flow rate. Several types of meters are commonly used to measure the velocity of a wastestream or chemical solution. The propeller type is turned by fluid flow past propeller vanes, which move gear trains. These gear trains are used to indicate the fluid velocity or flow rate. The velocity of liquid flow past the probes of a magnetic meter is related to flow of electric current between the probes and is read out as the flow rate through secondary instrumentation. Pitot tubes are used to measure the velocity head (H) in flowing water to give the flow velocity ($V = \sqrt{2gH}$). (Figure 11.32)

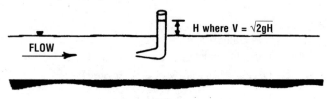

Fig. 11.32 Pitot tube

DIFFERENTIAL PRODUCERS—A mechanical constriction such as an orifice plate is placed in the flow line to reduce the pipe diameter (Figure 11.33). The restriction causes the velocity of flow to increase; when the velocity increases, a pressure drop is created at the restriction. The difference between line pressure at the meter inlet and reduced pressure at the throat section is used to determine the flow rate (which is indicated by a secondary instrument).

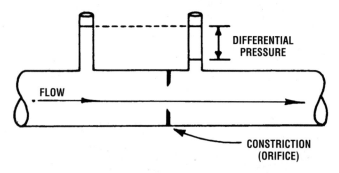

Fig. 11.33 Differential producer

DISPLACEMENT UNITS—Liquid or gas enters and fills a tank or chamber of known dimensions, activates a mechanical counter, and empties the tank in readiness for another filling. As the chamber fills and empties, mechanical gearing activates a counter, which measures the amount of time the cycle takes. Flow rate can then be calculated using the size of the tank and the time factor.

11.6232 LOCATION OF MEASURING DEVICES

The selection of a particular type of meter or measuring device and its location in a particular flow line or treatment facility is usually a decision made by the plant designer. Ideally, the flow should be in a straight section before the meter. The device must be accessible for servicing. In open channels, the flow should not be changing directions, nor should waves be present in the metering section above the measuring device. Valves, elbows, and other fixtures that could disrupt the flow ahead of a meter can upset the accuracy and reliability of a flowmeter. Most flowmeters are calibrated (checked for accuracy) in the factory, but they also should be checked in their actual field installation. When a properly installed and field-calibrated meter starts to give strange results, check for obstructions in the flow channel and the flow metering device.

11.6233 ADDITIONAL INFORMATION

For more information about flow metering devices and procedures, refer to Chapter 6, "Flow Measurement."

> ### QUESTIONS
>
> Write your answers in a notebook and then compare your answers with those on page 604.
>
> 11.6I What is flow measurement?
>
> 11.6J Write the basic flow formula.
>
> 11.6K Why should flow be measured?
>
> 11.6L List several types of flow measuring devices.
>
> 11.6M When a properly installed and field-calibrated meter starts to give strange results, what items should be checked?

11.624 Ventilation and Exhaust Systems

Properly operating ventilation systems are of greatest importance in metal plating and metal finishing waste treatment facilities. The possibility exists that acid mists and extremely toxic gases, such as cyanide, hydrogen sulfide, and chlorine, may be released or escape into the work environment. Toxic gas monitors and their alarm systems must be inspected and calibrated on a daily basis. The actual ventilation systems must be installed, operated, and maintained in strict accordance with the manufacturer's recommendations.

The Uniform Fire Code, Article 80, Hazardous Materials, requires that exhaust ventilation for chlorine and sulfur dioxide areas must be taken from a point within 12 inches (30.5 centimeters) of the floor. Mechanical ventilation should be at a rate of not less than one cubic foot of air per minute per square foot (0.00508 cubic meter of air per second per square meter) of floor area of storage handling area. Normally, ventilation from these areas is discharged to the atmosphere, but when a leak occurs, the ventilated air containing the chlorine or sulfur dioxide should be routed to a treatment system to remove the chemical. A caustic scrubbing system can be used to treat the air from a leak. Treatment systems must be designed to reduce the maximum allowable discharge concentration of chlorine or sulfur dioxide to one-half the *IDLH* [33] (Immediately Dangerous to Life or Health) at the point of discharge to the atmosphere. The

---

[33] *IDLH.* Immediately Dangerous to Life or Health. The atmospheric concentration of any toxic, corrosive, or asphyxiant substance that poses an immediate threat to life or would cause irreversible or delayed adverse health effects or would interfere with an individual's ability to escape from a dangerous atmosphere.

IDLH for chlorine is 10 ppm. For sulfur dioxide the IDLH is 100 ppm. A secondary standby source of power is required for the detection, alarm, ventilation, and treatment systems.

QUESTIONS

Write your answers in a notebook and then compare your answers with those on page 605.

11.6N Why are properly operating ventilation systems particularly important in metal plating and metal finishing facilities?

11.6O How often should toxic gas monitors and their alarms be inspected and calibrated?

11.625 Pumps

11.6250 TYPES OF PUMPS

Pumps serve many purposes in treatment plants. They may be classified by the character of the material handled, such as influent or treated wastewater. Or, they may relate to the conditions of pumping: high lift, low lift, or high capacity. They may be further classified by principle of operation, such as centrifugal, propeller, reciprocating, and turbine (Figure 11.34).

The type of material to be handled and the function or required performance of the pump vary so widely that the designing engineer must use great care in preparing specifications for the pump and its controls. Similarly, the operator must operate and maintain pumps in accordance with the manufacturer's instructions.

11.6251 STARTING A NEW PUMP

The initial start-up work described in this section should be done by a competent and trained person, such as a manufacturer's representative, consulting engineer, or an experienced operator. An inexperienced operator can learn a lot about pumps and motors by accompanying and helping a competent person put new equipment into operation.

Before starting a pump, lubricate it according to the lubrication instructions. Turn the shaft by hand to see that it rotates freely. Then, check to see that the shafts of the pump and motor are aligned and the flexible coupling adjusted. Check the electric voltage with the motor characteristics and inspect the wiring. See that thermal overload units in the starter are set properly. Turn on the motor just long enough to see that it turns the pump in the direction indicated by the rotational arrows marked on the pump. If separate water seal units or vacuum primer systems are used, these should be started. Finally, make sure lines are open. Sometimes, there is an exception (see following paragraph) in the case of the discharge valve.

A pump should not be run without first having been primed. To prime a pump, the pump must be completely filled with water. In some cases, automatic primers are provided. If they are

not, it is necessary to vent the casing. Most pumps are provided with a valve to accomplish this. Allow the trapped air to escape until water flows from the vent; then replace the vent cap. In the case of suction-lift applications, the pump must be filled with water unless a self-primer is provided. In nearly every case, you may start a pump with the discharge valve open. Exceptions to this, however, are where WATER HAMMER[34] or pressure surges might result, or where the motor does not have a sufficient margin of safety or power. Sometimes there are no check valves in the discharge line. In this case (with the exception of positive displacement pumps), it is necessary to start the pump and then open the discharge lines. Where there are common discharge headers, it is essential to start the pump and then open the discharge valve. A positive displacement pump (reciprocating or piston types) should never be operated against a closed discharge line.

After starting the pump, again check to see that the direction of rotation is correct. Packing gland boxes (stuffing boxes) should be observed for slight leakage (approximately 60 drops per minute). Check to see that the bearings do not overheat from over- or underlubrication. The flexible coupling should not be noisy; if it is, the noise may be caused by misalignment or improper clearance or adjustment. Check to be sure pump anchorage is tight. Compare delivered pump flows and pressures with pump performance curves. If pump delivery falls below performance curves, look for obstructions (including trapped air) in the pipelines or inadequate priming (vapor lock) and inspect piping for leaks.

11.6252 LONG-TERM PUMP SHUTDOWNS

When shutting down a pump for a long period, the motor disconnect switch should be opened, locked out, and tagged with the reason for the tag noted. If the electric motor is equipped with winding heaters, check to be sure they are turned on. This helps to prevent condensation from forming, which can weaken the insulation on the windings. All valves on the

---

[34] *Water Hammer.* The sound like someone hammering on a pipe that occurs when a valve is opened or closed very rapidly. When a valve position is changed quickly, the water pressure in a pipe will increase and decrease back and forth very quickly. This rise and fall in pressures can cause serious damage to the system.

586 Treatment Plants

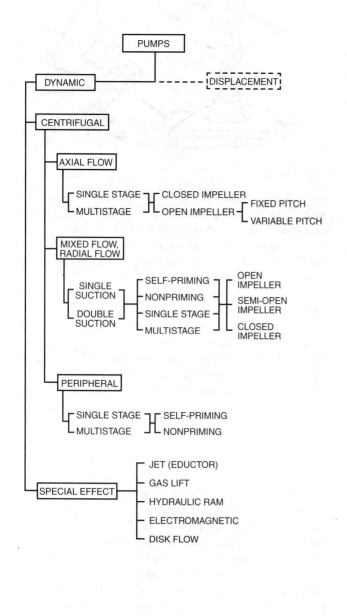

Dynamic types of pumps

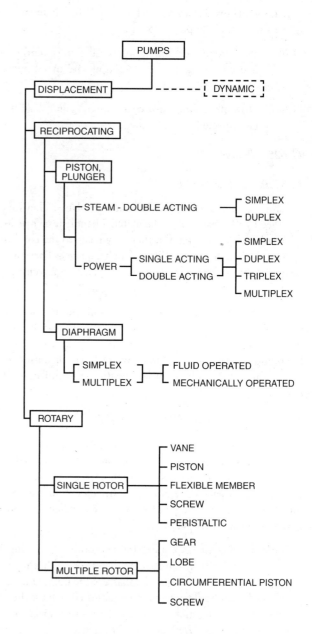

Displacement types of pumps

Fig. 11.34 *Classification of pumps*

suction, discharge, and water seal lines should be shut tightly. Completely drain the pump by removing the vent and drain plugs.

Inspect the pump and bearings thoroughly so that all necessary servicing may be done during the inactive period. Drain the bearing housing and then add fresh lubricant. Follow any additional manufacturer's recommendations.

11.6253 PUMP-DRIVING EQUIPMENT

Driving equipment used to operate pumps includes electric motors and internal combustion engines. In rare instances, pumps are driven with steam turbines, steam engines, air, and hydraulic motors.

In all except the large installations, electric motors are used almost exclusively, with synchronous and induction types being the most commonly used. Synchronous motors operate at constant speeds and are used chiefly in large sizes. Three-phase, squirrel-cage induction motors are most often used in treatment plants. These motors require little attention and, under average operating conditions, the factory lubrication of the bearing will last approximately one year. (Check with the manufacturer for average number of operating hours for bearings.) When lubricating motors, remember that too much grease may cause bearing trouble or damage the winding.

Clean and dry all electrical contacts. Inspect for loose electrical contacts. Make sure that hold-down bolts on motors are secure. Check voltage while the motor is starting and running. Examine bearings and couplings.

11.6254 ELECTRICAL CONTROLS

A variety of electrical equipment is used to control the operation of pumps or to protect electric motors. If starters, disconnect switches, and cutouts are used, they should be installed in accordance with the local regulations (city or county codes) regarding this equipment. In the case of larger motors, the power company often requires starters that do not overload the power lines.

The electrode-type, bubbler-type, and diaphragm-type water level control systems are all similar in effect to the float-switch system. Scum is a problem with many wastestream level controls that operate pumps and it must be removed on a regular basis. Bubbler systems are much less likely to become fouled by scum than other systems.

QUESTIONS

Write your answers in a notebook and then compare your answers with those on page 605.

11.6P What items should an operator check before starting a pump?

11.6Q What could be the cause of a noisy flexible coupling?

11.6R Where would you find out how to lubricate a pump?

11.6S What problems can develop if too much grease is used in lubricating a motor?

11.6255 PROCEDURES FOR STARTING AND STOPPING PUMPS DURING NORMAL OPERATION

The operator must determine what treatment processes will be affected by either starting or stopping a pump. The pump discharge point must be known and valves either opened or closed to direct flows as desired by the operator when a pump is started or stopped.

CENTRIFUGAL PUMPS

Basic rules for the operation of centrifugal pumps include the following items:

1. Do not operate the pump when safety guards are not installed over or around moving parts.

2. Do not start a pump that has been locked or tagged out for maintenance or repairs. An operator working on the pump could be seriously injured or the equipment could be damaged.

3. Never run a centrifugal pump when the impeller is dry. Always be sure the pump is primed.

4. Never attempt to start a centrifugal pump whose impeller or shaft is spinning backward.

5. Do not operate a centrifugal pump that is vibrating excessively after start-up. Shut the unit down and isolate the pump by closing the pump suction and discharge valves. Look for a blockage in the suction line and the pump impeller.

There are several situations in which it may be necessary to start a centrifugal pump against a closed discharge valve. Once the pump is primed, running, and indicating a discharge pressure, slowly open the pump discharge valve until the pump is fully on line. This procedure is used with treatment processes or piping systems with vacuums or pressures that cannot be dropped or allowed to fluctuate greatly while an alternate pump is put on the line.

Most centrifugal pumps used in treatment plants are designed so that they can be easily started even if they have not been primed. This is accomplished with a positive static suction head or a low suction lift. On most of these arrangements, the pump will not require priming as long as the pump and the piping system do not leak. Leaks would allow the water to drain out of the pump volute. When pumps in water systems lose their prime, the cause is often a faulty check valve on the pump discharge line. When the pump stops, the discharge check valve will not seal (close) properly. Water previously pumped then flows back through the check valve and through the pump. The pump is drained and has lost its prime.

About 95 percent of the time, the centrifugal pumps in treatment plants are ready to operate with suction and discharge valves open and seal water turned on. When the automatic start or stop command is received by the pump from the controller, the pump is ready to respond properly.

When the pumping equipment must be serviced, take it off the line by locking and tagging out the pump controls until all service work is completed.

QUESTIONS

Write your answers in a notebook and then compare your answers with those on page 605.

11.6T Why should a pump that has been locked or tagged out for maintenance or repairs not be started?

11.6U Under what conditions might a centrifugal pump be started against a closed discharge valve?

STOPPING PROCEDURES

This section contains a typical sequence of procedures to follow to stop a centrifugal pump. Exact stopping procedures for any pumping system depend upon the condition of the discharge system. The sudden stoppage of a pump could cause severe water hammer problems in the piping system.

1. Inspect process system affected by pump, start alternate pump, if required, and notify supervisor or log action.

2. Before stopping an operating pump, check its operation. This will give an indication of any developing problems, required adjustments, or problem conditions of the unit. This procedure only requires a few minutes. Items to be inspected include:

   a. Pump packing gland.
      (1) Seal water pressure.
      (2) Seal leakage (too much, sufficient, or too little leakage).
      (3) Seal leakage drain flowing clear.
      (4) Mechanical seal leakage (if equipped).

   b. Pump operating pressures.
      (1) Pump suction (pressure vacuum).

         A higher vacuum than normal may indicate a partially plugged or restricted suction line. A lower vacuum may indicate a higher suction water level or a worn pump impeller or wearing rings.

      (2) Pump discharge pressure.

         System pressure is indicated by the pump discharge pressure. Lower than normal discharge pressures can be caused by:

         (a) Worn impeller or wearing rings in the pump.
         (b) A different point of discharge can change discharge pressure conditions.
         (c) A broken discharge pipe can change the discharge head.

   NOTE: To determine the maximum head (shutoff head) a centrifugal pump can develop, slowly close the discharge valve at the pump. Read the pressure gauge between the pump and the discharge valve when the valve is fully closed. This is the maximum pressure the pump is capable of developing. Do not operate the pump longer than a few minutes with the discharge valve closed completely because the energy from the pump is converted to heat and water in the pump can become hot enough to damage the pump.

   c. Motor temperature and pump bearing temperature.

      If motor or bearings are too hot to touch, further checking is necessary to determine if a problem has developed or if the temperature is normal. High temperatures may be measured with a thermometer.

   d. Unusual noises, vibrations, or conditions about the equipment.

      If any of the above items indicate a change from the pump's previous operating condition, additional service or maintenance may be required during shutdown.

3. Actuate stop switch for pump motor and lock out switch. If possible use the lockout/stop switch next to the equipment so that you may observe the equipment stop. Observe the following items:

   a. Check valve closes and seats.

      Valve should not slam shut, or discharge piping will jump or move in their supports. There should not be any leakage around the check valve shaft. If check valve is operated automatically, it should close smoothly and firmly to the fully closed position.

   NOTE: If the pump is not equipped with a check valve, close discharge valve before stopping pump.

   b. Motor and pump should wind down slowly and not make sudden stops or noises during shutdown.

   c. After equipment has completely stopped, pump shaft and motor should not start backspinning. If backspinning is observed in a pump with a check valve or foot valve, close the pump discharge valve slowly. Be extra careful if there is a plug valve on a line with a high head because when the discharge valve is partially closed, the plug valve could slam closed and damage the pump or piping.

4. Go to the power control panel (Motor Control Center or MCC) containing the pump motor starters just shut down and open motor breaker switch, lock out, and tag.

5. Return to pump and close:
   a. Discharge valve.
   b. Suction valve.
   c. Seal water supply valve.
   d. Pump volute bleed line (if so equipped).

6. If required, close and open appropriate valves along piping system through which pump was discharging.

STARTING PROCEDURES

This section contains a typical sequence of procedures to follow to start a centrifugal pump.

1. Check Motor Control Center (MCC) for lock and tags. Examine tags to be sure that no item is preventing start-up of equipment.

2. Inspect equipment.

    a. Be sure lockout/stop switch is locked out at equipment location.

    b. Guards over moving parts must be in place.

    c. Cleanout on pump volute and drain plugs should be installed and secure.

    d. Valves should be in closed position.

    e. Pump shaft must rotate freely.

    f. Pump motor should be clean and air vents clear.

    g. Pump, motor, and auxiliary equipment lubricant levels must be at proper elevations.

    h. Determine if any special considerations or precautions are to be taken during start-up.

3. Follow pump discharge piping route. Be sure all valves are in the proper position and that the pump flow will discharge where intended.

4. Return to Motor Control Center.

    a. Remove tag.

    b. Remove padlock.

    c. Close motor main breaker.

    d. Place selector switch to manual (if you have automatic equipment).

5. Return to pump equipment.

    a. Open seal water supply line to packing gland. Be sure seal water supply pressure is adequate.

    b. Open pump suction valve slowly.

    c. Bleed air out of top of pump volute in order to prime pump. Some pumps are equipped with air relief valves or bleed lines back to the wet well for this purpose.

    d. When pump is primed, slowly open pump discharge valve and recheck prime of pump. Be sure no air is escaping from volute.

    e. Unlock stop switch and actuate start switch. Pump should start.

6. Inspect equipment.

    a. Motor should come up to speed promptly. If ammeter is available, test for excessive draw of power (amps) during start-up and normal operation. Most three-phase induction motors used in treatment plants will draw 5 to 7 times their normal running current during the brief period when they are coming up to speed.

    b. No unusual noise or vibrations should be observed during start-up.

    c. Check valve should be open and no chatter or pulsation should be observed.

    d. Pump suction and discharge pressure readings should be within normal operating range for this pump.

    e. Packing gland leakage should be normal.

    f. If a flowmeter is on the pump discharge, record pump output.

7. If the unit is operating properly, return to the Motor Control Center and place the motor mode of operation selector in the proper operating position (H-O-A).

8. The pump and auxiliary equipment should be inspected routinely after the pump has been placed back into service.

QUESTIONS

Write your answers in a notebook and then compare your answers with those on page 605.

11.6V   What should be done before stopping an operating pump?

11.6W   After equipment has completely stopped, what should be done if a pump shaft or motor starts backspinning in a pump with a check valve or foot valve?

11.6X   Why should the position (open or closed) of all valves be checked before starting a pump?

POSITIVE DISPLACEMENT PUMPS

Steps for starting and stopping positive displacement pumps are outlined in this section. There are two basic differences in the operation of positive displacement pumps as compared with centrifugal pumps. Centrifugal pumps (due to their design) will permit an operator error, but a positive displacement pump will not and someone will have to pay for correcting the damages.

The most important rule regarding the operation of positive displacement pumps is to *NEVER* start or operate the pump against a closed valve, especially a discharge valve. If a positive displacement pump is started or operated against a closed discharge valve, the pipe, valve, or pump could rupture from excessive pressure. The rupture will damage equipment and possibly seriously injure or kill someone standing nearby.

A positive displacement pump should also never be operated when it is dry or empty, especially the progressive-cavity types that use rubber stators. A small amount of liquid is needed for lubrication in the pump cavity between the rotor and the stator.

Also, remember that positive displacement pumps are used to pump sludge and to meter and pump chemicals. Care must be exercised to avoid venting chemicals to the atmosphere.

In addition to *NEVER* closing a discharge valve on an operating positive displacement pump, the only other difference (when compared with a centrifugal pump) may be that the positive displacement pump system may or may not have a check valve in the discharge piping after the pump. Installation of a check valve depends upon the designer and the material being pumped.

Other than the specific differences mentioned in this section, the starting and stopping procedures for positive displacement pumps are similar to the procedures for centrifugal pumps.

must familiarize themselves with the electrical diagrams of their system. *WARNING:* If you do not understand electricity and electronic systems, are not qualified or are not authorized to do electrical maintenance or repairs, then *DO NOT DO IT.* Many facilities have an electrician who repairs all electrical and electronic problems. However, operators should learn as much as possible about electricity *FOR THEIR OWN SAFETY* and to enable them to analyze electrical problems and to communicate effectively with electricians.

11.631 *Preventive Maintenance Program*

All waste treatment facilities need a preventive maintenance program. The purpose of this program is to keep the equipment and treatment processes working as intended, to minimize breakdowns, failures, and repairs, and also to keep overall O & M costs down both today and in the future. An effective preventive maintenance program consists of procedures that notify maintenance people when tasks should be performed, records who performed the maintenance, and records when it was done. Many operators use computers to notify them of scheduled maintenance activities and also to record information on tasks completed.

Consult your equipment manufacturers' maintenance manuals for instructions on equipment maintenance procedures and frequencies. Use this information to develop a specific preventive maintenance program for your facilities. Typical preventive maintenance tasks and their frequencies are listed in the remainder of this section.

DAILY

1. Fix all leaks when they occur. Slow leaks typically leave a deposit or crust near the leak.

2. Change all mechanical seals when they start to leak.

WEEKLY

1. Clean or change any meter faces that are becoming cloudy (opaque) or hard to read.

2. Clean or change any clear hoses or pipes that are becoming opaque. You need to be able to see if anything is moving

QUESTIONS

Write your answers in a notebook and then compare your answers with those on page 605.

11.6Y   What is the most important rule regarding the operation of positive displacement pumps?

11.6Z   What could happen if a positive displacement pump is started against a closed discharge valve?

11.63 Maintenance

11.630 *Safety*

Maintenance of equipment and treatment facilities requires special and diverse talents. Frequently, operators are also expected to perform maintenance duties. Operators must become very familiar with how their treatment facilities work so they can understand what is happening when there is a breakdown or failure. This means that both operators and maintenance people

through these hoses when the pumps are supposed to be operating.

3. Clean pH and ORP probes (if not done by operators). See Section 11.401, "Cleaning."

4. Calibrate pH and ORP probes and meters (also see Section 11.402, "Calibration").

MONTHLY

NOTE: Some manufacturers recommend a more frequent time interval.

1. Grease bearings.

2. Inspect drive belts, starter contacts, and other consumable items and change when necessary.

SEMIANNUALLY

Change lubricant in gear boxes.

ANNUALLY

Clean, neutralize, prime, and paint all painted items. If a paint coating becomes damaged due to splashes, clean, neutralize, prime, and paint as soon as possible to protect the equipment.

11.632 Spare Parts Inventory

An important part of every preventive maintenance program is an inventory of spare parts for repairs of equipment needed for critical service. Many manufacturers recommend a spare parts inventory for their equipment. Computers are being used today to maintain an inventory of all equipment being used as well as spare parts and substitute equipment for critical service. Typical items that should be included in an inventory are listed below:

1. Spare pump for each critical service.
2. Spare seals.
3. Motor heaters.
4. Fuses.
5. Solenoid valves.
6. Instrumentation circuit boards.
7. pH and ORP probes (these have a limited shelf life, so do not overstock).
8. Paint.
9. Motor starter contacts.
10. Critical valves.
11. Belts, gears, and specialized parts, such as check valves and diaphragms.
12. Running lights.
13. Recorder charts and ink.
14. Tools.

QUESTIONS

Write your answers in a notebook and then compare your answers with those on page 605.

11.6AA What is the purpose of a preventive maintenance program?

11.6BB How can an operator obtain instructions on how to maintain equipment and the frequency of each maintenance task?

11.6CC What precaution should you take when stocking extra pH and ORP probes?

11.64 Troubleshooting

Operators performing the visual observations listed in Section 11.620, "Visual Observations," may discover that a treatment process is upset. Process upsets may be caused by equipment, instrumentation, or control system failure, or operator error. Operators should prepare troubleshooting checklists for their facilities to help both new and experienced operators to correct the cause of the upset. This section contains some typical troubleshooting checklists.

11.640 pH Meters

If pH meter readings appear to be incorrect, check the chemical feed systems.

DIRECT CHEMICAL FEED SYSTEM

1. Chemical pump is not pumping.
   a. Pump is turned off.
   b. Supply line is clogged.
   c. Tank shutoff valve is closed.
   d. Pump impeller is jammed.
   e. Pump diaphragm is ruptured.
   f. Pump rate controls have been turned off or set very low.
   g. Pump check valves are stuck open or closed.
   h. Pump is plugged.
   i. Pump has burned its fuse or tripped its circuit breaker.
   j. Chemical feed tank is empty.

INDIRECT CHEMICAL FEED SYSTEM

1. Transfer pump is not working.
   a. See all of the above reasons for "Chemical pump is not pumping."
2. Level controls are not working.
   a. Probes are coated with precipitates.
   b. Probes are shorted out with moisture and chemicals.
   c. Probe rods have vibrated loose.

d. Chemical encrustations (coatings) limit probe movement (float switches or magnetic reed switches).

e. Power is off.

3. Solenoid valve is not working.

Check valve action by exercising the valve with the H-O-A switch. If the valve works in the *HAND* position, the valve is not the problem. If the valve does not work in either the *HAND* or *AUTOMATIC* position, disconnect the coil and test it with a multimeter for continuity, resistance, and shorts in the circuit. If the coil is OK, check the circuit for:

a. Blown fuses.

b. Shorted or loose wires at the panel switch.

c. Tripped circuit breakers.

d. Continuity from the panel to the valve location. Also, check for shorts due to moisture in junction boxes and conduits.

e. Bad switch.

If the valve does not work in the *AUTO* position, check the trip point setting (the setting at which the control relay is supposed to work). Vary the trip point setting above and below the pH or ORP that the instrument is reading. If the relay then works and the solenoid or pump works, then the problem is in the chemical supply and not in the instrumentation. If the relay will not operate when the switch is in the *AUTO* position, the problem could be:

a. A broken or blown relay.

b. Blown fuses.

c. A burned circuit board.

d. A broken or loose connection or wire on the board.

Operators should not attempt repairs beyond this level of difficulty. Problems of this type require the services of a specialist who is familiar with the specific instrumentation.

OTHER REASONS FOR pH METER READINGS OUT OF SCALE

1. Extra high concentrations of chemicals in the waste discharge. Check to see if there is a tank leak in the plant or on the treatment pad.

2. Chemical mixer may not be working properly due to:

   a. Broken belt.

   b. Stripped gears.

   c. Loose impeller.

   d. Tripped circuit breaker.

11.641 ORP Meters

ORP (oxidation-reduction potential) is considered an independent operating guideline. However, for an ORP value to be meaningful for treatment process control purposes, ORP readings must be taken at a relatively fixed or constant pH level. Therefore, if the pH varies, the ORP reading will also vary either up or down depending on the chemical reaction in question (either oxidation or reduction). If an ORP reading is out of its proper range, the first thing you should do is measure the pH. If the pH value is not in its proper range, make the necessary adjustments to get the pH back to its proper level; then observe the ORP values to determine if they are at the expected values. If ORP readings are still out of the proper range, check the items listed below:

1. Supply tank chemical level.

   a. Refill empty tanks.

   b. Inspect shutoff valves or switches.

2. Defective pumps.

3. Clogged chemical supply lines.

4. Chemical supply rate to the reactor tank.

5. Leaks from tanks within the plant (inspect specialized use tanks, such as chrome or cyanide).

In order to get the ORP reaction back to the proper state, you may have to hand transfer chemicals. If hand transfer must be done, *USE EXTREME CAUTION*. Always wear appropriate protective gear and use the proper equipment to transfer any chemical.

11.642 Instrumentation

If none of the instrumentation is functioning, check the following items:

1. Inspect the instrumentation circuit breaker. If the circuit breaker is not tripped, use a multimeter to test for incoming power to the panel circuit breaker.

2. If incoming power is not present at the panel circuit breaker, examine the circuit breakers at the facility distribution panels.

3. Reset any circuit breakers that have been tripped to the OFF position, after checking for shorts (why did it trip?).

4. If power is present and the instrumentation is not working, check for:

   a. Blown fuses on *MOTHER CIRCUIT BOARDS*.[35]

   b. Blown fuses on instrumentation power packs (transformers).

   c. Blown transformers.

   d. Loose wires.

If one specific item of instrumentation is not functioning, inspect the following:

1. Probes for damage.
2. Probe assembly and junction boxes for moisture.
3. Individual unit fuses.
4. Circuit boards for burned condition (examine the traces on the boards to see if they are intact).
5. Wires and components for looseness.
6. Voltages at locations specified by the manufacturer.

Replace defective components on the circuit boards.

If an instrumentation problem cannot be identified and corrected by these procedures, request specialized help.

QUESTIONS

Write your answers in a notebook and then compare your answers with those on pages 605 and 606.

11.6DD What major factors could cause pH meter readings to appear to be incorrect (assume the pH meter works properly) in (1) direct chemical feed systems, and (2) indirect chemical feed systems?

11.6EE What factors could cause level controls not to work properly?

11.6FF What items should be checked if ORP readings are out of the proper range?

11.643 Centrifugal Pump

If a centrifugal pump is not operating properly, check the following items:

1. Pump is on (motor is running).
2. Fluid level of supply.
3. Vapor lock conditions in pump and discharge piping.
4. All valves on supply and discharge sides of the pump are open.
5. Concentration level of chemicals. If the concentration level of a chemical being pumped has increased, this may cause an increase in the *SPECIFIC GRAVITY*[36] of the chemical solution, which could require a larger horsepower motor. Magnetically coupled pumps may not be physically capable of pumping the heavier or more viscous (thick) chemical solution, so the magnetic coupling may stall or slip out. This problem can sometimes be corrected by reducing the diameter of the pump impeller. However, reducing the impeller diameter will also reduce the chemical pumping rate.
6. If the pump's motor switch is on but the motor is not running, turn the motor switch off immediately. The motor is in a tripped condition. Reset starter. If the starter will not reset, inspect the motor heaters to see if they are burned. If they are burned, disconnect the motor and examine the windings for shorts. Replace all defective items.
7. If the pump's impeller is jammed, the cause may be due to:

   a. Stress on the pump body that has warped the pump.

   b. Impurities in the liquid being pumped (clean the pump).

   c. Accumulation of a crust within the pump body (clean the pump).

11.644 Diaphragm Pump

If a diaphragm pump is not working properly, check the following items:

1. Power supply (electricity or air).
2. Air pressure.
3. Fuses or circuit breakers.
4. Shut valves on the delivery side of the pump.
5. Stuck supply valve on air-powered pumps. To fix a stuck supply valve, spray the valve with WD-40® and make sure the air line oil mister has a supply of low-viscosity oil.

If a diaphragm pump is operating but not pumping, check to make sure:

1. Supply line is not clogged.
2. Valves are open on both the supply and discharge sides of the pump.
3. Check valves are operational.
4. Diaphragm is not ruptured.

---

[35] *Mother Circuit Board.* The base circuit board or the main circuit board.

[36] *Specific Gravity.* (1) Weight of a particle, substance, or chemical solution in relation to the weight of an equal volume of water. Water has a specific gravity of 1.000 at 4°C (39°F). Particulates with specific gravity less than 1.0 float to the surface and particulates with specific gravity greater than 1.0 sink. (2) Weight of a particular gas in relation to the weight of an equal volume of air at the same temperature and pressure (air has a specific gravity of 1.0). Chlorine gas has a specific gravity of 2.5.

11.645 Other Pump Problems

The following list of operating troubles includes most of the causes of failure or reduced pump operating efficiency. The remedy or cure is either obvious or may be identified from the description of the cause.

SYMPTOM A—PUMP WILL NOT START

CAUSES:

1. Blown fuses or tripped circuit breakers due to:
   a. Rating of fuses or circuit breakers not correct.
   b. Switch (breakers) contacts corroded or shorted.
   c. Terminal connections loose or broken somewhere in the circuit.
   d. Automatic control mechanism not functioning properly.
   e. Motor shorted or burned out.
   f. Wiring hookup or service not correct.
   g. Switches not set for operation.
   h. Contacts of the control relays dirty and arcing.
   i. Fuses or thermal units too warm.
   j. Wiring short-circuited.
   k. Shaft binding or sticking due to rubbing impeller, tight packing glands, or clogging of pump.
2. Loose connections, fuse, or thermal unit.

SYMPTOM B—REDUCED RATE OF DISCHARGE

CAUSES:

1. Pump not primed.
2. Air in the water.
3. Speed of motor too low.
4. Improper wiring.
5. Defective motor.
6. Discharge head too high.
7. Suction lift greater than anticipated.
8. Impeller clogged.
9. Discharge line clogged.
10. Air bubble trapped in a section of the discharge line.
11. Pump rotating in wrong direction.
12. Air leaks in suction line or packing box.
13. Inlet to suction line too high, permitting air to enter.
14. Valves partially or entirely closed.
15. Check valves stuck or clogged.
16. Incorrect impeller adjustment.
17. Impeller damaged or worn.
18. Packing worn or defective.
19. Impeller turning on shaft because of broken key.
20. Flexible coupling broken.
21. Loss of suction during pumping may be caused by leaky suction line, ineffective water or grease seal.
22. Belts slipping.
23. Worn wearing ring.

SYMPTOM C—HIGH POWER REQUIREMENTS

CAUSES:

1. Speed of rotation too high.
2. Operating heads lower than rating for which pump was designed, resulting in excess pumping rates.
3. Sheaves on belt drive misaligned or maladjusted.
4. Pump shaft bent.
5. Rotating elements binding.
6. Packing too tight.
7. Wearing rings worn or binding.
8. Impeller rubbing.

SYMPTOM D—NOISY PUMP

CAUSES:

1. Pump not completely primed.
2. Inlet clogged.
3. Inlet not submerged.
4. Pump not lubricated properly.
5. Worn impellers.
6. Strain on pumps caused by unsupported piping fastened to the pump.

7. Foundation insecure.
8. Mechanical defects in pump.
9. Misalignment of motor and pump where connected by flexible shaft.
10. Rocks in the impeller.
11. Cavitation.

CAVITATION

Cavitation is the formation and collapse of a gas pocket or bubble on the blade of an impeller or the gate of a valve. The collapse of this gas pocket or bubble drives water into the impeller or gate with a terrific force that can cause pitting on the impeller or gate surface. Cavitation is a condition that can cause a drop in pump efficiency, vibration, noise, and rapid damage to the impeller of a pump. Cavitation occurs due to unusually low pressures within a pump. These low pressures can develop when pump inlet pressures drop below the design inlet pressures or when the pump is operated at flow rates considerably higher than design flows. When the pressure within the flowing water drops very low, the water starts to boil and vapor bubbles form. These bubbles then collapse with great force which knocks metal particles off the pump impeller. Excessive suction lift is a frequent cause of cavitation.

11.646 Waste Is Not Flocculating

If the waste being treated is not flocculating, check the following items.

1. pH. The pH must be in the appropriate range for the waste being treated. In most of the waste treatment systems being discussed in this chapter, the pH range is near 10.5; however, desired pH levels can vary from 7.0 to 12.0.

2. Polymer flow rate. The polymer flow rate can be measured by diverting the flow into a graduated cylinder and measuring the time with a stopwatch.

    EXAMPLE. The polymer feed pump fills a 500 milliliter graduated cylinder in 2 minutes and 12.2 seconds. What is the polymer feed rate in gallons per day (GPD)?

    $$\text{Flow Rate, GPD} = \frac{(\text{Volume Pumped, m}L)(60 \text{ sec/min})(60 \text{ min/hr})(24 \text{ hr/day})}{(\text{Time, sec})(1{,}000 \text{ m}L/L)(3.785 \text{ }L/\text{gal})}$$

    $$= \frac{(500 \text{ m}L)(60 \text{ sec/min})(60 \text{ min/hr})(24 \text{ hr/day})}{[(2 \text{ min})(60 \text{ sec/min}) + 12.2 \text{ sec}](1{,}000 \text{ m}L/L)(3.785 \text{ }L/\text{gal})}$$

    $$= 86 \text{ GPD}$$

3. Adequacy of the polymer flow rate (chemical dosage). Add more polymer by hand and observe whether flocculation improves. Adjust the polymer feed pump if an improvement is observed. Alternatively, run a series of jar tests with different polymer dosages.

4. Change in waste characteristics causing interference with the polymer reactions.

QUESTIONS

Write your answers in a notebook and then compare your answers with those on page 606.

11.6GG What items should be checked if a centrifugal pump is not operating properly?

11.6HH What items would you check if a pump will not start?

11.6II How would you attempt to increase the discharge from a pump if the flow rate is lower than expected?

11.6JJ What is cavitation?

11.6KK If a waste being treated is not flocculating properly, what items should be checked?

11.7 SAFETY

11.70 Beware of Hazardous Chemicals and Waste

Treatment of metal wastestreams involves a number of physical and chemical treatment processes. As an operator of these facilities, you must know how to store and handle the chemicals used in the treatment processes. You must be aware of the hazards related to the chemicals you are using as well as the hazards associated with the wastes you are treating. The manufacturer or formulator of hazardous substances or mixtures must provide you with a Material Safety Data Sheet (MSDS) for the material. This document must contain information on how to handle and store the material. In industrial wastewater treatment, mixtures of several different hazardous substances are common. It is not always possible to determine the hazards associated with mixtures by reading the MSDS for each of its components. Some of the typical chemicals and waste materials you may encounter in the treatment area will be discussed in this section.

Chemicals in treatment processes are used to raise or lower the pH or to serve as oxidizing or reducing agents in chemical reactions. The effectiveness of the chemicals in treatment processes can make them very hazardous or harmful whenever any one of them comes in contact with a human being. Therefore, operators must know how to properly handle each chemical, the hazards associated with the chemical, and the response if an accident occurs. Table 11.8 summarizes a few of the most common treatment process chemicals.

TABLE 11.8 COMMON TREATMENT PROCESS CHEMICALS

Chemical Name (Common Name)	Form	Strength	Use	Safety Warning[a]
1. Sodium Hydroxide (Caustic)	Liquid Pellet and Flake Form	20–50% Solution	Raise pH.	Causes painful burns if it enters breaks in the skin or gets under skin around fingernails. Will cause blindness if it gets into eyes.
2. Calcium Oxide (Quicklime)	Powder and Pellet		Raise pH.	Same as sodium hydroxide.
3. Calcium Hydroxide (Lime)	Powder or Slurry		Raise pH.	Same hazards as sodium hydroxide.
4. Sulfuric Acid	Liquid	90–96% Solution	Lower pH in chrome reduction process. Lower pH of final effluent.	Causes severe burns and blindness. Nylon and rayon fall apart if subject to mist. Most cloth is badly damaged by slight exposure.
5. Hydrochloric Acid	Liquid	37% Solution	Clean probes. Lower pH.	Same hazards as sulfuric acid. Attacks many types of stainless steel. Vapors.
6. Chlorine	Gas, Liquid, or Solid	100% gas; 12.5% liquid industrial bleach.	Oxidizing agent.	Toxic gas that damages respiratory system.
7. Sulfur Dioxide and Sulfide Compounds	Gas or Powders	100% gas	Reducing agent. Hexavalent chromium treatment.	Toxic gas that can cause respiratory edema[b] and pneumonia.
8. Polymers and Polyelectrolytes			Organic compounds used to speed up the removal of precipitated material.	Polymers spilled on surfaces present an extreme slipping hazard.

[a] See Section 11.72, "First Aid," for procedures.
[b] Edema. The flow of fluids into the body tissues or cavities.

Some of the more common industrial waste materials are listed below to give you an indication of the types of hazardous materials that operators should be aware of and could encounter.

1. Spent plating solutions and other process chemicals
   a. Cyanide
   b. Heavy metals
   c. Hydroxide sludge
   d. Spent acids (chromic, hydrochloric, hydrofluoric, nitric, sulfuric)
   e. Cyanide tank sludges

2. Spent caustics
   a. Degreasers
   b. Reverse current alkaline cleaners

3. Organics
   a. Photo resist films
   b. Fusing oil (such as glycol solutions)
   c. Ethanol (photo process)
   d. Detergents
   e. Trichloroethylene (TCE)

QUESTIONS

Write your answers in a notebook and then compare your answers with those on page 606.

11.7A How can an operator be exposed to hazardous chemicals?

11.7B For what purposes are chemicals used in waste treatment processes?

11.71 Storage and Handling of Chemicals

All chemical storage and handling facilities must have proper spill containment measures. These measures must prevent the mixing of acid and cyanide wastes as well as avoid major sewer problems as a result of accidents and spills. Absolutely no floor drains connected to public sewers should be allowed in warehouses and other facilities that store toxic or flammable materials. Spill containment systems may consist of diking or self-contained tanks.

Diking must be capable of containing 105 percent of the volume of the largest tank within a containment area plus an allowance for stormwater. Diking usually consists of concrete blocks and concrete curbing. The concrete or mortar surfaces exposed to a spilled chemical must be coated or otherwise protected against deterioration. Absolutely no openings, manual or electrical discharge gates, or valves of any kind are allowed in spill containment diking.

A self-contained (double wall) tank is a tank with two independent structural shells with the outer shell being capable of containing any leakage from the inner one. A one-inch air gap must be provided between the walls and bottoms of these tanks and the top of the inner tank should be at least one inch lower than the top of the outer tank. Absolutely no discharge valves are allowed in the outer shell. Pits that are constructed or coated with an acid-resistant material and without sewer access are acceptable as self-containment if the containment volume is adequate.

See Material Safety Data Sheets (MSDSs) for specific storage and handling instructions for each chemical.

Most chemicals are stored and used in a concentrated form. Lime and sodium hydroxide (caustic) are often mixed with water at the treatment site. The heat generated during the mixing reaction can be sufficient to weaken plastic tanks and cause them to split open. Dilution of concentrated acids or the mixing of acids and caustics also generates heat and can lead to splatter or explosions, which can be extremely hazardous to operators. Remember, *SAFETY COMES FIRST*.

Acids must be contained in rubber, glass, or plastic-lined equipment. Store acids in clean, cool, well-ventilated areas. The areas should have an acid-resistant floor and adequate containment. Keep acids away from oxidizing agents, alkaline or basic materials, hypochlorite solutions, and cyanides. Protect the containers from damage or breakage. Avoid contact with skin and provide emergency neutralization materials and safety equipment in use areas. Vinegar can be used to neutralize bases and baking soda will neutralize acids.

When you are handling acids, wear protective clothing and equipment to prevent body contact with the acid. Wear rubber gloves, and safety goggles or a face shield for eye protection against splashing. Also, wear a rubber apron, rubber boots, and a long-sleeved, polyester shirt. An eye wash station and safety shower must be located nearby where any acids are being handled. A respirator may be needed in some situations. Although such precautions may seem like too much bother or the safety equipment may be uncomfortable, a serious acid burn is even more uncomfortable. Protect yourself by using proper equipment. *REMEMBER, always pour acid into water, never the reverse.* You pour acid *into* a swimming pool, but you never pour the pool water into an acid bottle.

Sodium hydroxide (caustic soda) is one of the most common and most dangerous bases used in waste treatment processes. Sodium hydroxide is available in pellet, flake, and solution forms; caustic soda solutions usually come as 20 to 50 percent solution of sodium hydroxide. This base is a strong alkali and is very hazardous to operators. Sodium hydroxide is extremely reactive; it reacts violently or explosively with acid and a number of organic compounds. Caustic soda (1) dissolves human skin, (2) when mixed with water causes heat, and (3) reacts with *AMPHOTERIC*[37] metals (such as aluminum) generating hydrogen gas, which is flammable and may explode if ignited. Sodium hydroxide can be dissolved in water and the solution used for the adjustment of pH because it is a liquid and easy to feed. *ALWAYS POUR CAUSTIC INTO WATER,* just like pouring acid into water. If the solution being mixed bumps or splatters, the dilute solution of mostly water will splatter and cause less damage than a concentrated solution.

Operators should be very careful when working with sodium hydroxide because this chemical causes painful burns if it enters breaks in the skin or gets under skin around fingernails. Caustic will cause blindness if it gets into your eyes. Special precautions to be taken when handling or storing caustic soda include: (1) prevent eye and skin contact, (2) do not breathe dusts, particulates, or mists, and (3) avoid storing this chemical next to strong acids or cyanide. Dissolving sodium hydroxide in water or other substances generates excessive heat, causes splattering and mists. Solutions of sodium hydroxide are viscous and slippery and only trained and protected operators should undertake spill cleanup. The operator must act cautiously, dilute the spill with water, and neutralize with a dilute acid, preferably acetic. Neutralizing absorbents can also be used.

---

[37] *Amphoteric* (AM-fuh-TUR-ick).   Capable of reacting chemically as either an acid or a base.

If a strong alkaline solution floods the eyes, it is practically impossible to prevent serious impairment of vision no matter how fast the first aid is applied. Know the location of safety showers and eye wash stations *before* starting to work with dangerous chemicals.

When handling caustic soda, control the mists with good ventilation. Protect your nose and throat with an approved respiratory system. For eye protection, you must wear chemical worker's goggles or a full face shield to protect your eyes. There must be an eye wash and safety shower at or near the work station for this chemical. Protect your body by being fully clothed and by using impervious (leakproof) gloves, boots, apron, and face shield.

If a bucket of dry cleaner salts is accidentally dumped into a hot, alkaline cleaner solution, the resulting eruption (explosion) can flood the operator with a hot, alkali solution. The result is a very serious injury and, in some cases, there has been loss of life. YOU MUST KNOW WHAT YOU ARE DOING WHEN WORKING WITH CHEMICALS.

Certain dry chemicals, such as alum, ferric chloride, and soda ash, are HYGROSCOPIC.[38] These chemicals require special considerations to protect them from moisture during storage. Dry quicklime (calcium oxide) should be kept dry because of the tremendous heat that is generated when it comes in contact with water. This heat is sufficient to cause a fire. Dry forms of sodium hydroxide can cause operators to be exposed to dust.

Some liquid chemicals, such as sodium hydroxide (caustic soda), should not be exposed to air because they may react with the carbon dioxide and form calcium carbonate (a solid). Also, some liquid chemicals may freeze. A 50-percent sodium hydroxide solution becomes crystallized (forms a solid) at temperatures below 55°F (13°C). Therefore, a heater may be required to keep the storage area warm or the solution may need to be diluted down to a 25-percent solution, which remains liquid at temperatures below 32°F (0°C). Some shops order 50-percent sodium hydroxide and add the correct amount of water to the storage tank before delivery to produce a 25-percent dilution after the order has been delivered. The resulting 25-percent dilution will raise the temperature of the mixture to about 160°F (71°C). Therefore, fill the chemical feed tank with 25-percent sodium hydroxide before delivery to give the new mix time to cool down.

Polymer solutions can be degraded (lose their strength) by biological contamination. Clean polymer tanks before a new shipment is delivered to the plant. The tanks may be cleaned by hosing or washing them out with water. Polymer is viscous and slippery when spilled. Spills may be cleaned up using inert absorbent materials, such as sand or earth.

Liquid chemical storage tanks should have a leakproof berm or earth bank around the tanks to contain any chemicals released if the tank fails due to an earthquake, corrosion, or any other reason.

The long-term storage of hypochlorite salts in a tightly closed container could produce a hazardous situation. These salts tend to gasify in time, thus creating the possibility of sufficient pressure in the container to cause an explosion.

Some chemicals, such as chlorine and fluoride compounds, are harmful to the human body when they are released as the result of a leak. Continual surveillance and maintenance of the storage and feeding systems are required. If acid is added to a hypochlorite solution, a large amount of moist chlorine gas will be released. Moist chlorine gas is much more corrosive and much more damaging to the human respiratory system than dry chlorine gas. Mixing ammonia solution and hypochlorite solution results in the release of nitrogen trichloride, a toxic gas.

These guidelines for handling and storage of hazardous chemicals are only general guidelines. Your chemical suppliers should be able to provide you with more detailed information about suitable storage conditions and safe handling procedures. Ask your supplier for this information and then take the time to read and follow the manufacturers' suggestions.

QUESTIONS

Write your answers in a notebook and then compare your answers with those on page 606.

11.7C What problems can be caused by the mixing of caustics?

11.7D How must acids be contained and stored?

11.7E Why should operators be very careful when working with sodium hydroxide (caustic)?

11.7F What special precautions should an operator take when handling or storing caustic soda?

11.72 First Aid

By definition, first aid means emergency treatment for injury or sudden illness before regular medical treatment is available. Everyone in an organization should be able to give some degree of prompt treatment and attention to an injury.

First-aid training in the basic principles and practices of lifesaving steps that can be taken in the early stages of an injury are available through the local Red Cross, Heart Association, local fire departments, and other organizations. Such training should periodically be reinforced so that the operator has a complete understanding of water safety, cardiopulmonary resuscitation (CPR), and other life-saving techniques. All operators need training in first aid, but it is especially important for those who regularly work with electrical equipment or must handle acids, bases, chlorine, and other dangerous chemicals.

---

[38] *Hygroscopic* (hi-grow-SKOP-ick). Absorbing or attracting moisture from the air.

First aid has little to do with preventing accidents, but it has an important bearing upon the survival of the injured patient. A well-equipped first-aid chest or kit is essential for proper treatment. The kit should be inspected regularly by the safety officer to ensure that supplies are available when needed. First-aid kits should be prominently displayed throughout the treatment plant and in company vehicles. Special consideration must be given to the most hazardous areas of the plant, such as shops, laboratories, and chemical handling facilities.

Have immediately available for each supervisor the first-aid instructions for each chemical used in the shop. This information can be copied from the Material Safety Data Sheets (MSDSs) supplied by the chemical manufacturers or suppliers.

ACIDS

The antidote to all acids is neutralization. However, one must be careful in how this is performed. Most often, large amounts of water will serve the purpose, but if the acid is ingested (swallowed), then lime water or milk of magnesia may be needed. If vapors are inhaled, first aid usually consists of providing fresh air, artificially restoring breathing (CPR), or supplying oxygen. In general, acids are neutralized by a base or alkaline substance. Baking soda is often used to neutralize acids on skin because it is not harmful on contact with your skin.

BASES

First aid for the eyes consists of irrigating the eyes immediately and continuously with flowing water for at least 30 minutes. Vinegar (dilute acetic acid) is a good neutralizer on humans. Prompt medical attention is essential. For skin burns, immediate and continuous thorough washing in flowing water for 30 minutes is important to prevent damage to the skin. Consult a physician if required. In case of inhalation, move the victim to fresh air, call a physician, or transport the injured person to a medical facility; *DO NOT INDUCE VOMITING.*

You may also have occasion to use sodium hydroxide as flakes or pellets. All of the precautions stated for liquid caustic also apply for the flake and pellet forms.

EYE BURNS (GENERAL)

1. Apply a steady flow of water to eyes for at least 30 minutes.
2. Call a physician immediately.
3. *DO NOT* remove burned tissue from the eyes or eyelids.
4. *DO NOT* apply medication (except as directed by a physician).
5. *DO NOT* use compresses.

SKIN BURNS (GENERAL)

1. Remove contaminated clothing immediately (preferably in a shower).
2. Flush affected areas with generous amounts of water.
3. Call a physician immediately.
4. *DO NOT* apply medication (except as directed by a physician).

SWALLOWING OR INHALATION (GENERAL)

1. Call a physician immediately.
2. Read antidote on label of any chemical swallowed. For some chemicals, vomiting should be induced, while for other chemicals, vomiting should not be induced. Follow the instructions on the label.

CHLORINE GAS CONTACT

1. If the victim is breathing, place on back with head and back in a slightly elevated position. Keep the victim warm and comfortable. *CALL A PHYSICIAN IMMEDIATELY.*
2. To check for breathing, tilt the head back (tilting the head back opens the airway and may in itself restore breathing), put your ear over the victim's mouth and nose and listen and feel for air. Look at the victim's chest and see if it is rising and falling. Watch for breathing for three to five seconds. If there is no breathing, perform mouth-to-mouth resuscitation.

    a. Tilt victim's head back and lift the chin. Be sure victim's mouth-throat airway is open.

    b. Gently pinch the victim's nose shut with the thumb and index finger, take a deep breath, seal your lips around the outside of the victim's mouth, create an air-tight seal and give victim two full breaths at the rate of 1 to $1^{1}/_{2}$ seconds per breath. Watch for the chest to rise while you breathe into the victim. If you feel resistance when you breathe into the victim, and air will not go in, the most likely cause is that you may not have tilted the head back far enough and the tongue may be blocking the airway. Re-tilt the head and give two full breaths.

    c. Put your ear over the victim's mouth and nose and listen and feel for air. Check for pulse for five to ten seconds.

    d. If victim is not breathing and there is no pulse, give victim 15 compressions (CPR or Cardiopulmonary

Resuscitation) and then two breaths. (Check the pulse at the side of the neck. This pulse is called the carotid pulse.) Feel for the carotid pulse for at least 5 seconds, but no more than 10 seconds.

e. Repeat step (d) four times and then check for breathing and pulse. Do this after giving the two breaths at the end of the fourth cycle of 15 compressions. Tilt the victim's head back and check the carotid pulse for 5 seconds.

If you do not find a pulse, then check for breathing for 3 to 5 seconds. If breathing is present, keep the airway open and monitor breathing and pulse closely. This means that you should look, listen, and feel for breathing while you keep checking the pulse. If there is no breathing, perform rescue breathing and keep checking the pulse.

f. Continue to give CPR until one of the following things happens:

- The heart starts beating again and the victim begins breathing.
- A second rescuer trained in CPR takes over for you.
- Emergency Medical Service (EMS) personnel arrive and take over.
- You are too exhausted to continue.

g. *DO NOT ATTEMPT TO PERFORM CPR UNLESS YOU ARE QUALIFIED.*

3. Eye irritation caused by chlorine gas should be treated by flushing the eyes with generous amounts of water for not less than 15 minutes. Hold eyelids apart to ensure maximum flushing of exposed areas. *DO NOT* attempt to neutralize with chemicals. *DO NOT* apply any medication (except as directed by a physician).

4. Minor throat irritation can be relieved by drinking milk. *DO NOT* give the victim any drugs (except as directed by a physician).

CHLORINE SOLUTION OR HYPOCHLORITE SOLUTION CONTACT

1. Flush the affected area with water. Remove contaminated clothing while flushing (preferably in a shower). Wash affected skin surfaces with soap and water while continuing to flush. *DO NOT* attempt to neutralize with chemicals. Call a physician. *DO NOT* apply medication (except as directed by a physician).

2. If chlorine solution has been swallowed, immediately give the victim large amounts of water or milk, followed with milk of magnesia, vegetable oil, or beaten eggs. *DO NOT* give sodium bicarbonate. *NEVER* give anything by mouth to an unconscious victim. Call a physician immediately.

EMERGENCY PHONE NUMBERS

A list of emergency phone numbers should be located near a telephone that is unlikely to be affected in an emergency. This list should include phone numbers for:

1. Hospital or Physician
2. Ambulance Service
3. Poison Control Center
4. Police Department
5. Fire Department
6. Responsible Plant Officials
7. Local Emergency Disaster Office
8. CHEMTREC (800) 424-9300
9. Emergency Teams (if your plant has one)
10. Regulatory Agencies

QUESTIONS

Write your answers in a notebook and then compare your answers with those on page 606.

11.7G Define first aid.

11.7H What should be done if acid comes in contact with your skin?

11.7I What should be done if your eyes become irritated by chlorine gas?

END OF LESSON 3 OF 3 LESSONS

on

TREATMENT OF METAL WASTESTREAMS

Please answer the discussion and review questions next.

DISCUSSION AND REVIEW QUESTIONS

Chapter 11. TREATMENT OF METAL WASTESTREAMS

(Lesson 3 of 3 Lessons)

Write the answers to these questions in your notebook. The question numbering continues from Lesson 2.

16. What skills and knowledge are required of operators for successful operation and maintenance of industrial wastewater treatment facilities?

17. How would you attempt to determine the cause of the problem if a waste being treated is not flocculating properly?

18. How would you safely store acids?

19. When handling acids, what type of protective clothing would you wear?

20. Why is sodium hydroxide (caustic soda) very hazardous to operators?

21. Why is a knowledge of first aid important?

SUGGESTED ANSWERS

Chapter 11. TREATMENT OF METAL WASTESTREAMS

ANSWERS TO QUESTIONS IN LESSON 1

Answers to questions on page 528.

11.0A   Metal wastestreams cannot be treated to the degree that is required for discharge by using only biological methods. Also, contaminants tend to accumulate in the sludge from biological processes, making the sludge difficult to dispose of. Physical–chemical processes more effectively remove metals from wastestreams because these methods are specifically based on the physical–chemical nature of the contaminant.

11.1A   Industries that produce metallically contaminated wastestreams include job shop platers, hard chrome platers, bumper remanufacturers, printed circuit board manufacturers, hard disk manufacturers, metal pickling operations, galvanizers, metal strut manufacturers, photo processors, and manufacturers of rubber gloves, fungicides, and algicides.

Answers to questions on page 533.

11.2A   Chemicals that have been used to overcome the effects of chelating agents include ferric chloride, ferrous sulfate, calcium chloride, phosphoric acid, and calcium hydroxide.

11.2B   From a production standpoint, a bath is considered spent when it no longer is suitable for the intended purpose.

11.2C   Wastestreams containing gold and silver are treated by the ion exchange process.

Answers to questions on page 538.

11.3A   Usually, small wastewater flows are treated by the batch process because the storage requirements for the wastewater prior to treatment are minimal. Also, concentrated wastes that cannot be handled otherwise are treated by the batch method.

11.3B   Wastewaters with high pH levels are neutralized by the addition of acids such as sulfuric acid and hydrochloric acid. Carbon dioxide and sulfur dioxide can be applied in the gaseous form to lower the pH of liquids. Some industries use flue gases (high in carbon dioxide or sulfuric acid fumes) to lower the pH of waters when the gas is available and this is an economical solution.

11.3C   Low pH wastewaters are neutralized by the addition of calcium oxide or lime, magnesium oxide, calcium hydroxide, magnesium hydroxide, sodium hydroxide, or soda ash.

11.3D   Operators must be very careful whenever strongly acidic and basic solutions are mixed because of the potential for splattering, the generation of heat, and the production of toxic gases. Highly acidic or basic solutions are very harmful to your skin, can cause loss of eyesight, and the fumes are extremely irritating to your lungs and to the mucous membranes in your body. To avoid splattering always slowly ADD ACID TO WATER or BASE TO WATER, never the reverse.

Answers to questions on page 546.

11.3E   Two different pH levels may be necessary for hydroxide precipitation of metals when the metals present have considerably different pH levels for optimum hydroxide precipitation.

11.3F   The major limitations of the sulfide precipitation process include:

1. Generation of toxic hydrogen sulfide gas if the pH drops below 8. Production of hydrogen sulfide gas can be minimized by using ferrous sulfide as the source of sulfide.
2. Disposal of toxic metal sulfide sludges.

11.3G   Complexed metals are difficult to treat because they have a tendency to remain in solution rather than form precipitates and settle out.

11.3H   Complexed metal wastestreams may be treated by (1) high pH precipitation, (2) chemical reduction, or (3) sulfide precipitation.

Answers to questions on page 549.

11.3I   Hexavalent chromium may be present in the form of chromic acid, chromate, or dichromate in an acid solution.

11.3J   Hexavalent chromium must be converted to trivalent chromium to reduce the toxicity level and so chromium can be removed from the wastestream by hydroxide precipitation.

11.3K   Hexavalent chromium can be reduced to trivalent chromium by strong reducing agents, such as sulfur dioxide, sodium bisulfite, sodium metabisulfite, sodium hydrosulfite, and ferrous sulfate.

11.3L   Chromium wastes are commonly treated by a two-step batch process. The first step is to reduce the highly toxic hexavalent chromium to the less toxic trivalent form. The trivalent form is removed by hydroxide precipitation in the second step.

Answers to questions on page 552.

11.3M   If there is an equipment failure during the treatment of metal wastes containing cyanide, extremely toxic cyanide gases could be released. Therefore, all tanks or pits used for cyanide destruction must be properly located and ventilated so that any gases produced will never enter an area occupied by people.

11.3N   Cyanide compounds get into metal wastestreams as a result of their use as metal salts for plating and conversion coating or they are active components in plating and cleaning baths.

11.3O   The most practical and economical method of treating metal wastestreams containing cyanide is the alkaline chlorination treatment process.

11.3P   Efficient operation of a cyanide oxidation system requires exact control of pH to prevent excessive use of chlorine and to ensure complete destruction of cyanide.

Answers to questions on page 554.

11.3Q   Precious metals can be recovered from metal wastestreams by evaporation, ion exchange, reverse osmosis, and electrolytic metal waste treatment processes.

11.3R   Ion exchange units are used to treat metal wastestreams to recover precious metals and other plating chemicals and to concentrate and purify plating baths.

11.3S   Precious metals recovered by the ion exchange process include copper, molybdenum, cobalt, nickel, gold, and silver.

11.3T   Metal plating chemicals that can be recovered by ion exchange include chromium, nickel, phosphate solution, sulfuric acid from anodizing, and chromic acid.

Answers to questions on page 557.

11.3U   The three primary problems operators encounter with ion exchange systems are: (1) throughput capacity loss, (2) failure to produce specified water quality, and (3) excessive pressure drop across the ion exchange system.

11.3V   The various forms of ion exchange fouling are:

1. Organic fouling
2. Iron fouling
3. Oil fouling
4. Mud or polyelectrolyte fouling
5. Algal or bacterial fouling
6. Calcium sulfate precipitation
7. Barium sulfate fouling

11.3W   The three forms of iron found in water to be treated are: (1) suspended insoluble iron, (2) soluble ferrous iron, and (3) soluble iron that is oxidized to the ferric form.

11.3X The solution to problems caused by algal and bacterial growths in ion exchange resins is periodic bed sterilization using sodium hypochlorite.

11.3Y Ion exchange resin bead breakdown can be caused by physical shock (osmotic shock), mechanical abrasion or wear and tear, freeze and thaw conditions, and thermal shock effects.

Answers to questions on page 558.

11.3Z The principal reasons to minimize the discharge of oily wastes from a wastestream are to prevent the fouling of sensors and to comply with pretreatment discharge limitations.

11.3AA Possible sources of oily wastes in wastestreams include process coolants and lubricants, wastes from cleaning and painting processes, and machinery lubricants.

11.3BB Oily wastes can be removed from metal wastestreams by skimming, coalescing, emulsion breaking, flotation, centrifugation, ultrafiltration, and reverse osmosis.

11.3CC a. Free or floating oils can be separated from wastestreams by the use of gravity API oil separators, centrifuges, dissolved air flotation thickeners, and nitrification.

b. Emulsified oils are treated by the use of emulsion breaking procedures such as steam, heat cracking, acids, or the use of emulsion breakers (polymers) with the addition of chemicals to cause the necessary separation and removal of oils. Oil removal may be achieved by skimming.

Answers to questions on page 558.

11.3DD Spent degreasing solvents must be segregated from other process wastewaters to maximize the value of the recoverable solvents, to avoid contamination of other segregated wastes, and to prevent the discharge of toxic organics to any wastewater collection systems or the environment.

11.3EE Encourage operators to segregate spent degreasing solvents from other process wastewaters by providing and identifying the necessary storage containers, establishing clear disposal procedures, training operators in the use of these techniques, and checking periodically (monthly) to ensure that proper segregation is occurring.

11.3FF Instead of using solvent degreasing procedures, possible alternatives include alkaline cleaning (sodium hydroxide baths) and the use of aqueous cleaners (concentrated soap solutions).

Answers to questions on page 558.

11.3GG The quantity of toxic organics reaching wastestreams can be reduced by proper housekeeping procedures and by using cleaning techniques that require no solvents. These cleaning techniques include wiping, immersion, spray techniques using water, alkaline and acid mixtures, and solvent emulsions.

11.3HH Toxic organics that enter wastestreams can be removed by treatment processes used for the control of other pollutants. Aeration and carbon adsorption are the processes commonly used. Other treatment processes include settling and volatilization. In some situations, the activated sludge process is appropriate if a biological culture can be developed that is acclimated to the wastes. Aerobic decomposition, bacterial decomposition, and ozonation have also been used to treat toxic organics.

ANSWERS TO QUESTIONS IN LESSON 2

Answer to question on page 560.

11.4A The waste treatment process instrumentation and controls that an operator must be familiar with include: (1) pH and ORP probes, (2) flow measurement/totalization devices, (3) level controls, (4) programmable controllers, and (4) voltage, current, and resistance measurement devices.

Answers to questions on page 566.

11.4B pH and ORP probes must be cleaned and calibrated on a weekly basis or more frequently because the probes must have a good contact with the fluid they are measuring.

11.4C Both pH and ORP probes must be cleaned to remove precipitates as well as oil and grease.

11.4D Treatment processes requiring the use of a high pH slope buffer include metal precipitation and cyanide destruction processes, while chrome reduction requires the use of a low pH slope buffer.

11.4E Problems that could develop with the indicator when using pH and ORP probes include:

1. Indicator does not move off one position
2. Indicator has no reading
3. Indicator has a very low reading
4. Indicator moves very slowly
5. Indicator moves erratically

Answers to questions on page 568.

11.4F An operator should check the accuracy of the flow measuring devices by physically measuring the quantity of fluid pumped during a known time period.

11.4G When a programmable controller goes out of service, the operator should be prepared to replace the unit with an identical, programmed unit. The operator must ensure that all the backup information and hardware are on hand and that a sufficient number of personnel are familiar with the system to handle the breakdown.

11.4H An operator should have a fundamental knowledge of electrical power and how it is controlled for the operator's safety, to ensure the continuous operation of the plant, to analyze electrical problems, and to communicate effectively with electricians. An operator should also pay constant attention to electrical safety.

Answers to questions on page 574.

11.5A Methods used to dewater metallic sludges include centrifuges, vacuum filters, bag filters, and filter presses.

11.5B The transfer pump should pump metallic sludge to a plate and frame filter press for approximately two hours or until the pump has stalled out completely.

11.5C Sticky sludge may have to be scraped off the filter cloth or it may be worthwhile to precoat the filter cloth so the cake does not stick.

11.5D Dewatered sludge moisture content can be further reduced by air drying, induced air drying, heating, and reduced pressure, low heat drying.

ANSWERS TO QUESTIONS IN LESSON 3

Answers to questions on page 576.

11.6A The factors that influence the types and complexity of the processes used to treat metal wastes include:

1. The particular wastestream characteristics
2. Space available for treatment
3. Costs of the processes
4. Effluent discharge requirements

11.6B Metal waste treatment processes require continuous monitoring and adjustment of the treatment processes because the flows and characteristics of the wastewaters being treated change hourly, daily, weekly, and seasonally due to changes in the production processes and maintenance procedures.

Answers to questions on page 579.

11.6C For effective operation of a wastewater treatment facility, operators must visually observe the wastewater being treated and the treatment processes; read, record, and analyze instrument readouts and printouts; collect, analyze, and interpret sample test results; and physically adjust process control settings.

11.6D An operator can determine if a precipitation process is working properly by visual observations. The water in the precipitation phase of a precipitation process should look cloudy or full of solids if the process is functioning properly. A clear wastewater would indicate that the precipitation tank is not receiving treatment chemicals, or that some component of the wastewater (ammonia, chelates, or acids) is defeating the precipitation process.

11.6E The effluent from metal waste treatment processes should normally appear clear, color free, and with no evidence of floc or sludge carryover.

Answers to questions on page 581.

11.6F Most errors produced in laboratory tests are caused by improper sampling, poor preservation, or lack of enough mixing during compositing and testing.

11.6G Test results will be meaningless for process control purposes unless the test sample is representative of the wastestream. To efficiently operate a wastewater treatment plant, the operator must rely on test results that indicate what is happening in the treatment process. Sampling stations, therefore, must be located where the most representative samples can be collected, where mixing is thorough, and where the wastestream quality is uniform.

11.6H A 12-hour proportional composite sample may be prepared by collecting a sample every hour for 12 hours. The size of each sample is proportional to the flow when the sample is collected. All of these proportional samples are mixed together to produce a proportional composite sample. If an equal volume of sample was collected each hour and mixed, this would simply be a composite sample.

Answers to questions on page 584.

11.6I Flow measurement is the determination of the quantity of a mass in movement within a known length of time. Flow is measured and recorded as a quantity (gallons or cubic feet) moving past a point during a specific time interval (seconds, minutes, hours, or days). Thus, we obtain a flow rate or quantity in cu ft/sec or MGD.

11.6J Quantity = Area × Velocity, or $Q = AV$.

11.6K Flow should be measured in order to determine wastewater treatment plant loadings and efficiencies.

11.6L Different types of flow measuring devices include constant differential, head area, velocity meter, differential head, and displacement.

11.6M When a properly installed and field-calibrated meter starts to give strange results, check for obstructions in the flow channel and the flow metering device.

Answers to questions on page 585.

11.6N Properly operating ventilation systems are of utmost importance in metal plating and metal finishing facilities because the possibility exists that extremely toxic gases, such as cyanide, hydrogen sulfide, and chlorine, may be released or escape into the work environment.

11.6O Toxic gas monitors and their alarms should be inspected and calibrated on a daily basis.

Answers to questions on page 587.

11.6P Before starting a pump, the operator should check:

1. Lubrication.
2. Shaft rotation (should move freely) and direction.
3. Alignment of motor and pump shafts and adjustment of flexible coupling.
4. Electric voltage compared to motor characteristics.
5. Condition of wiring.
6. Thermal overload units in the starter.
7. Water seal units or vacuum primer (if used).
8. Lines are open, as needed.

11.6Q A noisy flexible coupling could be caused by misalignment or improper clearance or adjustment.

11.6R Pumps must be lubricated in accordance with the manufacturer's recommendations.

11.6S In lubricating motors, too much grease may cause bearing trouble or damage the winding.

Answers to questions on page 588.

11.6T If a pump that has been locked or tagged out for maintenance or repairs is started, an operator working on the pump could be seriously injured or the equipment could be damaged.

11.6U There are several situations in which it may be necessary to start a centrifugal pump against a closed discharge valve. Once the pump is primed, running, and indicating a discharge pressure, slowly open the pump discharge valve until the pump is fully on line. This procedure is used with treatment processes or piping systems with vacuums or pressures that cannot be dropped or allowed to fluctuate greatly while an alternate pump is put on the line.

Answers to questions on page 589.

11.6V Before stopping an operating pump:

1. Start an alternate pump (if required).
2. Inspect the operating pump by looking for developing problems, required adjustments, and problem conditions of the unit.

11.6W If backspinning is observed in a pump with a check valve or foot valve, close the pump discharge valve slowly. Be extra careful if there is a plug valve on a line with a high head because when the discharge valve is partially closed, the plug valve could slam closed and damage the pump or piping.

11.6X The position of all valves should be checked before starting a pump to ensure that the water being pumped will go where intended.

Answers to questions on page 590.

11.6Y The most important rule regarding the operation of positive displacement pumps is to *NEVER* operate the pump against a closed discharge valve.

11.6Z If a positive displacement pump is started against a closed discharge valve, the pipe, valve, or pump could rupture from excessive pressure. The rupture will damage equipment and possibly seriously injure or kill someone standing nearby.

Answers to questions on page 591.

11.6AA The purpose of a preventive maintenance program is to keep the equipment and treatment processes working as intended, to minimize breakdowns, failures, and repairs, and also to keep overall O & M costs down both today and in the future.

11.6BB Operators must consult equipment manufacturers' maintenance manuals for instructions regarding how to maintain equipment and the frequency of each task.

11.6CC When stocking pH and ORP probes, remember that they have a limited shelf life and do not overstock these items.

Answers to questions on page 593.

11.6DD Major factors that could cause pH meter readings to appear to be incorrect (assuming the pH meter works properly) include:

(1) *DIRECT CHEMICAL FEED SYSTEMS*
   a. Chemical pump is not pumping
(2) *INDIRECT CHEMICAL FEED SYSTEMS*
   a. Transfer pump is not working
   b. Level controls are not working
   c. Solenoid valve is not working
(3) *OTHER REASONS*
   a. Extra high concentrations of chemicals in the waste discharge
   b. Chemical mixer not working properly

11.6EE Level controls may not work properly due to:

1. Probes coated with precipitates.
2. Probes shorted out with moisture and chemicals.
3. Probe rods have vibrated loose.
4. Chemical encrustations (coatings) limit probe movement (float switches or magnetic reed switches).

11.6FF If ORP readings are out of the proper range, the following items should be checked:

1. pH values.
2. Supply tank chemical level.
3. Defective pumps.
4. Clogged chemical supply lines.
5. Chemical supply rate to the reactor tank.
6. Leaks from tanks within the plant (inspect specialized use tanks, such as chrome or cyanide).

Answers to questions on page 595.

11.6GG If a centrifugal pump is not operating properly, check the following items:

1. Pump is on (motor is running).
2. Fluid level of supply.
3. Vapor lock conditions in pump.
4. All valves on supply and discharge sides of the pump are open.
5. Concentration level (specific gravity) of chemicals.
6. Pump's motor switch, motor heaters and windings.
7. Impeller.

11.6HH If a pump will not start, check for blown fuses or tripped circuit breakers and the cause. Also, check for a loose connection, fuse, or thermal unit.

11.6II To increase the rate of discharge from a pump, you should look for something causing the reduced rate of discharge, such as pumping air, motor malfunction, plugged lines or valves, impeller problems, or other factors.

11.6JJ Cavitation is the formation and collapse of a gas pocket or bubble on the blade of an impeller or the gate of a valve. The collapse of this gas pocket or bubble drives water into the impeller or gate with a terrific force that can cause pitting on the impeller or gate surface.

11.6KK If a waste being treated is not flocculating properly, check:

1. pH.
2. Polymer flow rate.
3. Adequacy of polymer flow rate (chemical dosage).
4. Waste characteristics (for interference with chemical reactions).

Answers to questions on page 596.

11.7A An operator can be exposed to hazardous chemicals by: (1) storing and handling chemicals used in the treatment processes, and (2) treating hazardous chemicals in wastestreams.

11.7B Chemicals are used in waste treatment processes to raise or lower the pH and to serve as oxidizing or reducing agents in chemical reactions.

Answers to questions on page 598.

11.7C The mixing of caustics can generate sufficient heat to weaken plastic tanks and cause them to break open. Mixing of acids and caustics can lead to splatter or explosions, which can be extremely hazardous to operators.

11.7D Acids must be contained in rubber, glass, or plastic-lined equipment. Store acids in clean, cool, well-ventilated areas. The areas should have an acid-resistant floor and adequate containment. Keep acids away from oxidizing agents, alkaline or basic material, hypochlorite solutions, and cyanide.

11.7E Operators should be very careful when working with sodium hydroxide (caustic) because this chemical causes painful burns if it enters breaks in the skin or gets under skin around fingernails. Caustic will cause blindness if it gets into your eyes.

11.7F Special precautions to be taken when handling or storing caustic soda include: (1) prevent eye and skin contact, (2) do not breathe dusts or mists, and (3) avoid storing caustic next to strong acids or cyanide. Also, protect your nose and throat with an approved respiratory system, protect your eyes by wearing chemical workers' goggles or a full face shield, and protect your body by being fully clothed and by using impervious gloves, boots, apron, and face shield.

Answers to questions on page 600.

11.7G First aid means emergency treatment for injury or sudden illness before regular medical treatment is available.

11.7H If acid comes in contact with your skin, wash the area with large amounts of water. Baking soda may be used to neutralize acid on skin because it is not harmful on contact with skin.

11.7I Eye irritation caused by chlorine gas should be treated by flushing the eyes with generous amounts of water for not less than 15 minutes.

CHAPTER 12

INSTRUMENTATION

by

Leonard Ainsworth

TABLE OF CONTENTS

Chapter 12. INSTRUMENTATION

	Page
OBJECTIVES	611
WORDS	612
ABBREVIATIONS AND SYMBOLS	616

LESSON 1

- 12.0 MEASUREMENT AND CONTROL SYSTEMS ... 621
  - 12.00 Importance and Nature of Measurement and Control Systems ... 621
  - 12.01 Importance to the Industrial Waste Treatment Operator ... 621
  - 12.02 Nature of the Measurement Process ... 622
  - 12.03 Explanation of Control Systems ... 623
    - 12.030 Modulating Control Systems ... 623
    - 12.031 Motor Control Stations ... 623
- 12.1 SAFETY HAZARDS OF INSTRUMENTATION AND CONTROL SYSTEMS ... 628
  - 12.10 General Precautions ... 628
  - 12.11 Electrical Hazards ... 628
  - 12.12 Mechanical and Pneumatic Hazards ... 629
  - 12.13 Vaults and Other Confined Spaces ... 630
  - 12.14 Falls and Associated Hazards ... 630
- 12.2 MEASURED VARIABLES AND TYPES OF SENSORS/TRANSDUCERS ... 630
  - 12.20 General Principles of Sensors/Transducers ... 630
  - 12.21 Pressure Measurements ... 630
  - 12.22 Level Measurements ... 632
  - 12.23 Flow (Rate of Flow and Total Flow) ... 636
  - 12.24 Chemical Feed Rate ... 641
  - 12.25 Process Instrumentation ... 642
  - 12.26 Signal Transmitters/Transducers ... 642

DISCUSSION AND REVIEW QUESTIONS ... 644

LESSON 2

- 12.3 CATEGORIES OF INSTRUMENTATION ... 645
  - 12.30 Measuring Elements ... 645
  - 12.31 Panel Instruments ... 645
    - 12.310 Instrumentation ... 645

610 Treatment Plants

		12.311	Indicators	645
		12.312	Recorders	646
		12.313	Totalizers	646
		12.314	Alarms	649
	12.32	Automatic Controllers		649
	12.33	Pump Controllers		650
	12.34	Air Supply Systems		650
	12.35	Laboratory Instruments		652
	12.36	Test and Calibration Equipment		652
	12.37	Computerization		655
		12.370	Basic Principles of Computerization	655
		12.371	Video Display of Data	655
		12.372	Process Control Using Microprocessors	656
12.4	OPERATION AND PREVENTIVE MAINTENANCE			656
	12.40	Proper Care of Instruments		656
	12.41	Indications of Proper Function		657
	12.42	Start-Up/Shutdown Considerations		657
	12.43	Preventive Maintenance		659
	12.44	Operational Checks		660
12.5	ADDITIONAL READING			661
DISCUSSION AND REVIEW QUESTIONS				661
SUGGESTED ANSWERS				662

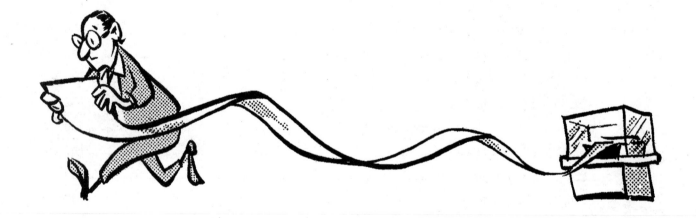

OBJECTIVES
Chapter 12. INSTRUMENTATION

Following completion of Chapter 12, you should be able to:

1. Explain the purpose and nature of measurement and control systems.

2. Identify, avoid, and correct safety hazards associated with instrumentation work.

3. Recognize various types of sensors and transducers.

4. Read instruments and make proper adjustments in the operation of industrial wastewater treatment facilities.

5. Identify symptoms of measurement and control system problems.

WORDS
Chapter 12. INSTRUMENTATION

ACCURACY
How closely an instrument measures the true or actual value of the process variable being measured or sensed.

ALARM CONTACT
A switch that operates when some preset low, high, or abnormal condition exists.

ANALOG READOUT
The readout of an instrument by a pointer (or other indicating means) against a dial or scale. Also, the continuously variable signal type sent to an analog instrument (for example, 4–20 mA). Also see DIGITAL READOUT.

ANALYZER
A device that conducts periodic or continuous measurement of some factor such as chlorine, fluoride, or turbidity. Analyzers operate by any of several methods including photocells, conductivity, or complex instrumentation.

CALIBRATION
A procedure that checks or adjusts an instrument's accuracy by comparison with a standard or reference.

CONTACTOR
An electric switch, usually magnetically operated.

CONTROL LOOP
The path through the control system between the sensor, which measures a process variable, and the controller, which controls or adjusts the process variable.

CONTROL SYSTEM
An instrumentation system that senses and controls its own operation on a close, continuous basis in what is called proportional (or modulating) control.

CONTROLLER
A device that controls the starting, stopping, or operation of a device or piece of equipment.

DESICCANT (DESS-uh-kant)
A drying agent that is capable of removing or absorbing moisture from the atmosphere in a small enclosure.

DESICCATION (dess-uh-KAY-shun)
A process used to thoroughly dry air; to remove virtually all moisture from air.

DETECTION LAG
The time period between the moment a process change is made and the moment when such a change is finally sensed by the associated measuring instrument.

DIGITAL READOUT

The readout of an instrument by a direct, numerical reading of the measured value or variable. The signal sent to such instruments is usually an analog signal.

DRIFT

The difference between the actual value and the desired value (or set point); characteristic of proportional controllers that do not incorporate reset action. Also called OFFSET.

EFFECTIVE RANGE

That portion of the design range (usually from 10 to 90+ percent) in which an instrument has acceptable accuracy. Also see RANGE and SPAN.

FAIL-SAFE

Design and operation of a process control system whereby failure of the power system or any component does not result in process failure or equipment damage.

FEEDBACK

The circulating action between a sensor measuring a process variable and the controller that controls or adjusts the process variable.

HERTZ (Hz)

The number of complete electromagnetic cycles or waves in one second of an electric or electronic circuit. Also called the frequency of the current.

INTEGRATOR

A device or meter that continuously measures and calculates (adds) a process rate variable in cumulative fashion; for example, total flows displayed in gallons, million gallons, cubic feet, or some other unit of volume measurement. Also called a TOTALIZER.

INTERLOCK

An electric switch, usually magnetically operated. Used to interrupt all (local) power to a panel or device when the door is opened or the circuit is exposed to service.

LINEARITY (lin-ee-AIR-it-ee)

How closely an instrument measures actual values of a variable through its effective range; a measure used to determine the accuracy of an instrument.

MEASURED VARIABLE

A factor (flow, temperature) that is sensed and quantified (reduced to a reading of some kind) by a primary element or sensor.

OFFSET

(1) The difference between the actual value and the desired value (or set point); characteristic of proportional controllers that do not incorporate reset action. Also called DRIFT.

(2) A pipe fitting in the approximate form of a reverse curve or other combination of elbows or bends that brings one section of a line of pipe out of line with, but into a line parallel with, another section.

(3) A pipe joint that has lost its bedding support, causing one of the pipe sections to drop or slip, thus creating a condition where the pipes no longer line up properly.

ORIFICE (OR-uh-fiss)

An opening (hole) in a plate, wall, or partition. An orifice flange or plate placed in a pipe consists of a slot or a calibrated circular hole smaller than the pipe diameter. The difference in pressure in the pipe above and at the orifice may be used to determine the flow in the pipe. In a trickling filter distributor, the wastewater passes through an orifice to the surface of the filter media.

PRECISION

The ability of an instrument to measure a process variable and repeatedly obtain the same result. The ability of an instrument to reproduce the same results.

PRIMARY ELEMENT

(1) A device that measures (senses) a physical condition or variable of interest. Floats and thermocouples are examples of primary elements. Also called a SENSOR.

(2) The hydraulic structure used to measure flows. In open channels, weirs and flumes are primary elements or devices. Venturi meters and orifice plates are the primary elements in pipes or pressure conduits.

PROCESS VARIABLE

A physical or chemical quantity that is usually measured and controlled in the operation of a water, wastewater, or industrial treatment plant.

RANGE

The spread from minimum to maximum values that an instrument is designed to measure. Also see EFFECTIVE RANGE and SPAN.

READOUT

The reading of the value of a process variable from an indicator or recorder (or on a computer screen).

RECEIVER

A device that indicates the result of a measurement, usually using either a fixed scale and movable indicator (pointer), such as a pressure gauge, or a moving chart with a movable pen like those used on a circular flow-recording chart. Also called an INDICATOR.

RECORDER

A device that creates a permanent record, on a paper chart or magnetic tape, of the changes in a measured variable.

REFERENCE

A physical or chemical quantity whose value is known exactly, and thus is used to calibrate instruments or standardize measurements. Also called a STANDARD.

ROTAMETER (ROTE-uh-ME-ter)

A device used to measure the flow rate of gases and liquids. The gas or liquid being measured flows vertically up a tapered, calibrated tube. Inside the tube is a small ball or bullet-shaped float (it may rotate) that rises or falls depending on the flow rate. The flow rate may be read on a scale behind or on the tube by looking at the middle of the ball or at the widest part or top of the float.

SCALE

(1) A combination of mineral salts and bacterial accumulation that sticks to the inside of a collection pipe under certain conditions. Scale, in extreme growth circumstances, creates additional friction loss to the flow of water. Scale may also accumulate on surfaces other than pipes.

(2) The marked plate against which an indicator or recorder reads, usually the same as the range of the measuring system. See RANGE.

SENSITIVITY

The smallest change in a process variable that an instrument can sense.

SENSOR

A device that measures (senses) a physical condition or variable of interest. Floats and thermocouples are examples of sensors. Also called a PRIMARY ELEMENT.

SET POINT

The position at which the control or controller is set. This is the same as the desired value of the process variable. For example, a thermostat is set to maintain a desired temperature.

SOFTWARE PROGRAM

Computer program; the list of instructions that tell a computer how to perform a given task or tasks. Some software programs are designed and written to monitor and control treatment processes.

SOLENOID (SO-luh-noid)

A magnetically operated mechanical device (electric coil). Solenoids can operate small valves or electric switches.

SPAN

The scale or range of values an instrument is designed to measure. Also see RANGE.

STANDARD

A physical or chemical quantity whose value is known exactly, and thus is used to calibrate instruments or standardize measurements. Also called a REFERENCE.

STANDARDIZE

To compare with a standard.

(1) In wet chemistry, to find out the exact strength of a solution by comparing it with a standard of known strength. This information is used to adjust the strength by adding more water or more of the substance dissolved.

(2) To set up an instrument or device to read a standard. This allows you to adjust the instrument so that it reads accurately, or enables you to apply a correction factor to the readings.

STARTERS (MOTOR)

Devices used to start up large motors gradually to avoid severe mechanical shock to a driven machine and to prevent disturbance to the electrical lines (causing dimming and flickering of lights).

TELEMETRY (tel-LEM-uh-tree)

The electrical link between a field transmitter and the receiver. Telephone lines are commonly used to serve as the electrical line.

THERMOCOUPLE

A heat-sensing device made of two conductors of different metals joined at their ends. An electric current is produced when there is a difference in temperature between the ends.

TIMER

A device for automatically starting or stopping a machine or other device at a given time.

TOTALIZER

A device or meter that continuously measures and calculates (adds) a process rate variable in cumulative fashion; for example, total flows displayed in gallons, million gallons, cubic feet, liters, cubic meters, or some other unit of volume measurement. Also called an INTEGRATOR.

TRANSDUCER (trans-DUE-sir)

A device that senses some varying condition measured by a primary sensor and converts it to an electrical or other signal for transmission to some other device (a receiver) for processing or decision making.

VARIABLE, MEASURED

A factor (flow, temperature) that is sensed and quantified (reduced to a reading of some kind) by a primary element or sensor.

VARIABLE, PROCESS

A physical or chemical quantity that is usually measured and controlled in the operation of a water, wastewater, or industrial treatment plant.

ABBREVIATIONS AND SYMBOLS
Chapter 12. INSTRUMENTATION

Special symbols are used for simplicity and clarity on circuit drawings for instruments. Usually, instrument manufacturers and design engineers provide lists of symbols they use with an explanation of the meaning of each symbol. This section contains a list of typical instrumentation abbreviations and symbols used in this chapter and also used by the industrial wastewater treatment profession.

ABBREVIATIONS

A — Analyzer, such as a device used to measure a water quality indicator (pH, temperature).

C — Controller, such as a device used to start, operate, or stop a pump.

D — Differential, such as a "differential pressure" (DP) cell used with a flowmeter.

E — Electrical or Voltage.
Element, such as a primary element.

F — Flow rate (*NOT* total flow).

H — Hand (manual operation).
High as in high-level.

I — Indicator, such as the indicator on a flow recording chart.
$I = E/R$ where I is the electric current in amps.

L — Level, such as the level of water in a tank.
Low, as in a low-level switch.
Light, as in indicator light.

M — Motor.
Middle, as in a mid-level switch.

P — Pressure (or vacuum).
Pump.
Program, as in a software program.

Q — Quantity, such as a totalized volume (Σ for summation is also used).

R — Recorder (or printer), such as a chart recorder.
Receiver.
Relay.

S — Switch.
Speed, such as the RPM (revolutions per minute) of a motor.
Starter, such as a motor starter.
Solenoid.

T — Transmitter.
Temperature.
Tone.

V — Valve.
Voltage.

W — Weight.
Watt.

X — Special or unclassified variable.

Y — Computing function, such as a square root ($\sqrt{}$) extraction.

Z — Position, such as a percent valve opening.

TYPICAL PROCESS AND ELECTRICAL SYMBOLS

1. (PT / 1) — Pressure transmitter #1

2. (LIR / 2) — Level indicator/recorder #2

3. (FIC / 3) (FSH / 3) (FSL / 3) — Flow indicator/controller #3 with high-low control switches in the same instrument

Instrumentation 617

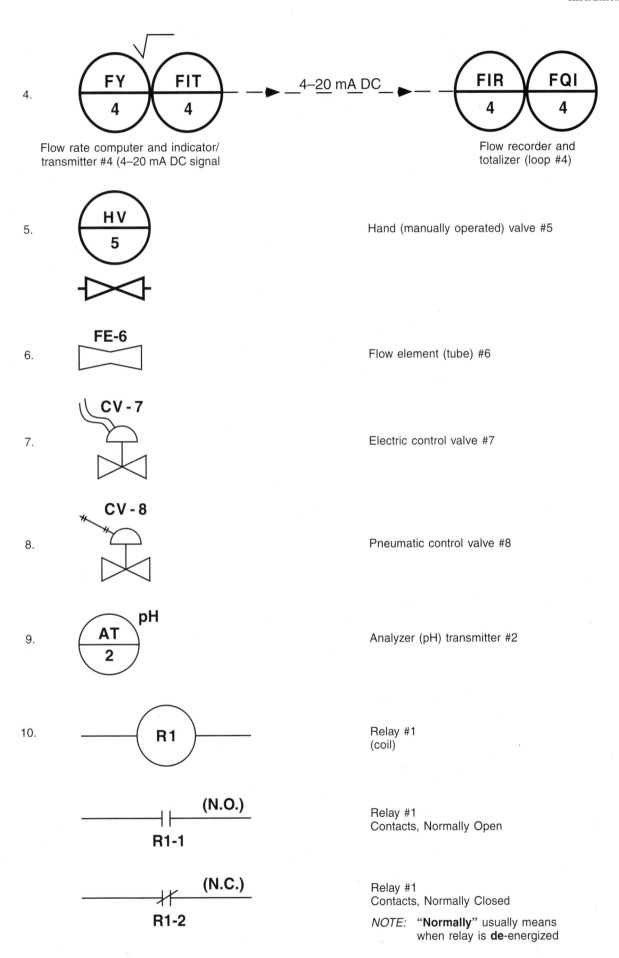

618 Treatment Plants

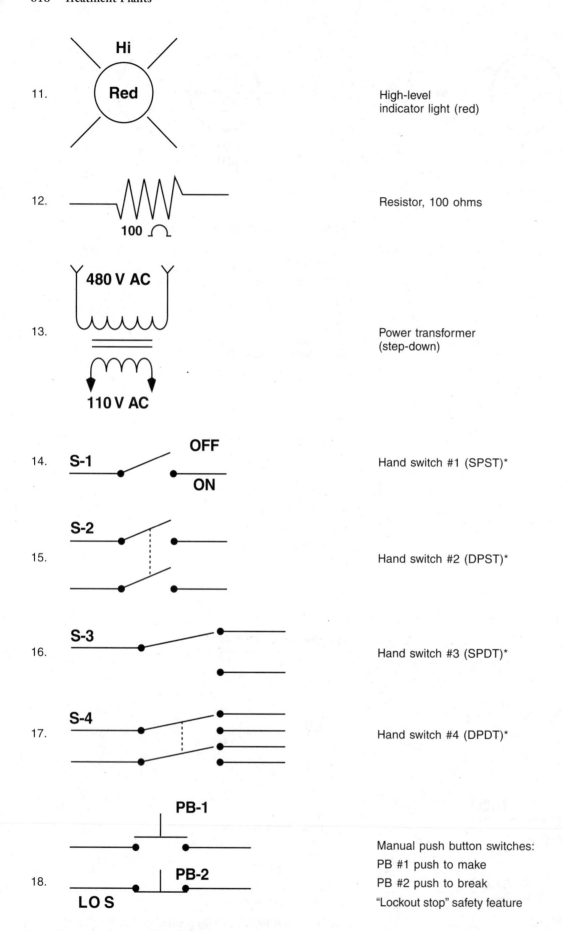

11. High-level indicator light (red)

12. Resistor, 100 ohms

13. Power transformer (step-down)

14. Hand switch #1 (SPST)*

15. Hand switch #2 (DPST)*

16. Hand switch #3 (SPDT)*

17. Hand switch #4 (DPDT)*

18. Manual push button switches:
PB #1 push to make
PB #2 push to break
"Lockout stop" safety feature

* SPST means single-pole, single-throw; DPST means double-pole, single-throw; SPDT means single-pole, double-throw; DPDT means double-pole, double-throw

Instrumentation 619

FUSES/CIRCUIT BREAKERS

19. 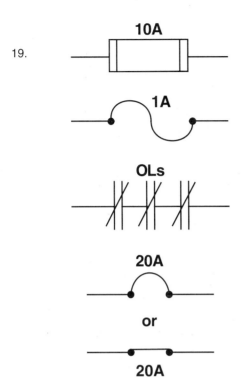 10-amp cartridge fuse

1-amp in-line fuse

Thermal overload contacts (motor)

20-amp circuit breaker

20. 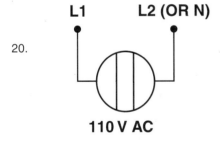 Line 1 and Line 2 (neutral) to standard duplex wall plug outlet

H - O - A Function Switch

21. 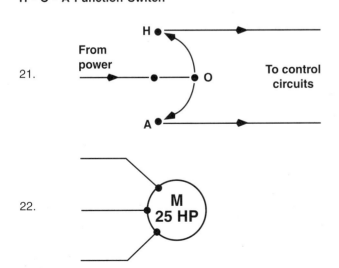 (Hand - Off - Automatic)

22. Electric motor, 3-phase power, 25 horsepower

TYPICAL INDUSTRIAL WASTE TREATMENT PLANT Process and Instrumentation Diagram (P & I.D.)

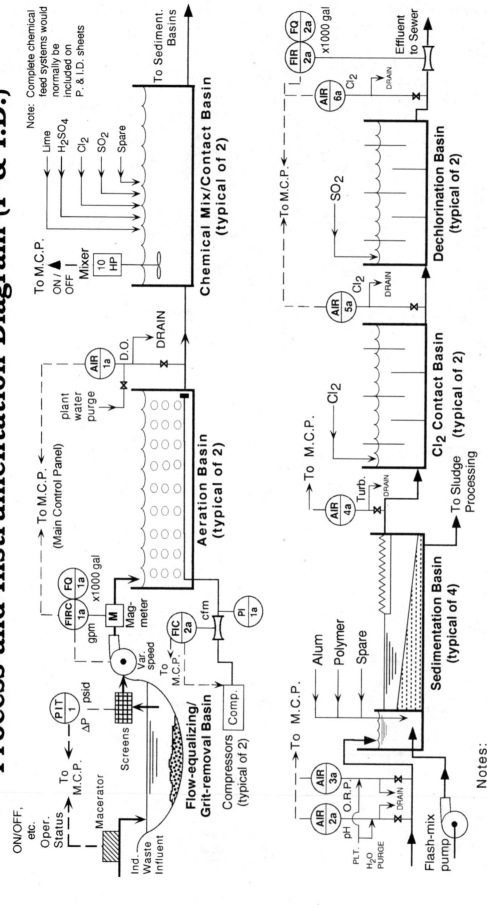

Notes:
1) See Abbreviations and Symbol sheets to interpret this diagram.
2) Lower case ("small") $\underline{a}$ denotes instrumentation for "a" process train, other sets denoted by "b" (and also "c" and "d" for sed. basins).
3) All analytical instruments piped to plant drain for return to aeration basins; instruments noted have plant (fresh) water purge connection.
4) This sheet for instructional purposes only and does not represent an actual process; diagram is simplified in certain respects, for example, no electrical controls or chemical feed systems included.

CHAPTER 12. INSTRUMENTATION

(Lesson 1 of 2 Lessons)

12.0 MEASUREMENT AND CONTROL SYSTEMS

12.00 Importance and Nature of Measurement and Control Systems

In this chapter you will learn about some important concepts and practices regarding industrial wastewater treatment instrumentation measurement systems and their associated control systems. The industrial plant operator must have a good working familiarity with measurement and associated control systems to properly monitor and control the treatment processes. You will learn the what, how, and, to some extent, the why of measurements and automatic controls in industrial wastewater treatment. This chapter is not intended to teach you how to fix malfunctioning instruments; instrument repair requires specialized training and equipment. However, some general information is given on instrument standardization and preventive maintenance measures.

Your informed attitude about measurement essentials can only enhance the effective and efficient operation of your facility. Specifically, if you can recognize that an indicator or recorder is not operating properly (by its erratic readout, for example), your process decision can be based on that knowledge rather than a blind faith in the instruments one might otherwise rely on. This is especially true in our age of black boxes (computers), which tend to be trusted implicitly. Typically, there is no simple way to adjust or bypass their operation, as was possible with most of the older control systems. The days of tapping the gauge, pulling out a relay, or tweaking an errant controller are rapidly coming to an end.

Be this as it may, there are still many measurement and control systems of conventional design the trained operator can and should perform some minor corrective maintenance on, as well as preventive maintenance tasks. The operator who knows enough to free or unclog a balky recorder pen, safely replace a glass fuse after a power failure, or drain an instrument air line can avoid a lot of personal worry and the trouble of the instrument maintenance person service call-out, while protecting the operational integrity of the industrial wastewater process. Thus, the more you know about your plant's instrumentation, the better operator you become.

12.01 Importance to the Industrial Waste Treatment Operator

In a very real sense, measurement instruments can be considered extensions of and improvement to your senses of vision, touch, hearing, and even smell. Not only can instruments provide continuous and simultaneous monitoring of the many process variables throughout the plant, but they do so in a more precise and consistent manner than the human senses. In addition, instruments often provide a permanent record of measurements taken. The associated control systems, in turn, are much like the operator having many far-reaching and powerful hands to constantly manipulate switches, valves, motors, and pumps in specific ways. In effect then, instrumentation provides you with a staff of obedient and hard-working assistants, always on the job with you to provide help so you can operate your process more easily and efficiently. To appreciate these great advantages of automation, consider what a treatment plant would be like without modern instrumentation. In the not-too-distant past, and even in some industrial waste treatment operations today, the situation described in the paragraph below was and is the daily *modus operandi* for some operations:

> Your plant has a complete power failure in the electrical circuit supplying the main control panel power. Resetting the proper circuit breaker does not restore the power. As the operator, you must try to keep the plant on line manually by attempting to control all process variables without any functional measurement and control systems. That is, flows and levels have to be estimated visually, valves turned manually, and pumps started and stopped on "HAND" to control them. Additionally, you are forced to observe and try to regulate the other process variables using only your eyes, ears, sense of touch, and probably, in time, even your sense of smell.

Impossible, you say? No, it can be, has been, and even is being done here and there in old or poorly maintained plants. But even if you could do it (and on some shifts it does seem like this *is* what you have to do), you certainly could not exercise close control or do it for any extended period. Operating a larger or sophisticated plant would probably be impossible without its process instrumentation, even with a sizeable crew.

Accordingly, you, as an operator, would do well to familiarize yourself with your extra "super" eyes, ears, nose, fingers, hands, and arms so as to take full advantage of these miracles of modern technology, for your benefit and that of the public ultimately better served by instrumentation.

QUESTIONS

Write your answers in a notebook and then compare your answers with those on page 662.

12.0A Why must an industrial wastewater treatment plant operator have a good working familiarity with a plant's measurement instrumentation and associated control systems?

12.0B What kind of information is presented in this chapter with regard to instrumentation?

12.0C Measurement instruments can serve as an extension of and improvement to which of the operator's physical senses?

12.02 Nature of the Measurement Process

Our senses provide us only with qualitative (for example, color, sound quality, odor) and relatively short-term quantitative (for example, brighter, louder, stronger) information; measurements provide us with exact quantitative data. That is to say, measurements give us numbers. Without this objective quantification of our environment (that is, the numbers we use to describe it), we could accomplish very little—just as primitive tribes with no real number systems have no technology worthy of the term. Their processing of natural materials is very limited because the information they use consists only of the relative terms "more than" or "less than." Without objective measurements, they have little control over their environment. Modern industrial plants, on the other hand, depend absolutely upon accurate and reproducible measurements of chemical and physical processes, which is what good instrumentation provides.

A measurement is, by definition, the comparison of a quantity with a standard unit of measure. Thus, a tape measure is marked in feet, a standard unit of length; clocks are marked in hours, minutes, and seconds as the standard units of time; and a small container may have fluid ounces marked on its side. These units, usually of a convenient magnitude (size), have been agreed upon internationally to serve as the accepted standard units. The primary standard for mass (weight) is the weight of an actual physical object, which is located in Paris, France; a standard kilogram is defined as the weight of this one specific object. The primary standards for time are defined in wavelengths of light, and the primary standard for length is the wavelength of a certain spectrum line of the element cesium. Other primary standards can be set up for certain quantities—for example, solution strengths—but generally, industrial measurements depend on secondary standards that are based on more or less exact comparisons to a primary standard. A machinist's caliper is calibrated against precision measuring blocks, which are such secondary standards.

Of course, all the measurements used in industrial waste treatment ultimately refer back to a few primary and secondary standards. Measurements of length, and the related calculations for area and volume, are in effect comparisons to the standard foot (meter). Weights of chemicals are traceable to the standard pound (kilogram). Timepieces are designed around the standard second. All the other measurements such as flow rate (volume per unit of time), pressure (weight per unit of area), chemical concentration/dosage (weight of chemical per unit of liquid volume), chemical feed (weight of chemical per unit of time), and others, are derived from the fundamental units of measure.

Some important terms directly relating to the measurement process also need to be defined: a *VARIABLE* (process variable in our case) is a physical or chemical quantity, such as flow rate or pH, that is measured or controlled in an industrial waste treatment process. *ACCURACY* refers to how closely an instrument measures the true or actual value of a process variable; it is usually expressed as within plus or minus a given small percent of the true value, for example, ±2 percent is a typical instrument accuracy. Accuracy depends partly upon the *PRECISION* of an instrument, which amounts to how closely the device can reproduce a given reading (or measurement) time after time (a good example of a *precise* but *inaccurate* instrument is a precision electrical gauge—with a bent needle). *SENSITIVITY* refers to the smallest change in a process variable able to produce an appropriate response from an instrument. It is usually expressed as a percent of the full scale value of the instrument. Typical values are ±0.1 percent of full scale for a high-sensitivity instrument and ±2 percent of full scale for an average-sensitivity instrument. *CALIBRATION* is the complete test of an instrument by measuring several (secondary) standards and its complete adjustment (if required) to read the standard values at several points in its range. *STANDARDIZATION* is an abbreviated calibration where only one standard is measured and the instrument is adjusted to read the proper value, such as standardizing a pH meter with pH 7 buffer. The *RANGE* of an instrument is the spread between the minimum and maximum values it is designed to measure, most often from zero to a maximum value, as with, say, a 0–100 psi pressure gauge. And the *EFFECTIVE RANGE* is that portion of an instrument's complete range within which the instrument has acceptable accuracy, commonly between 10 percent and 90+ percent of its design range. *SPAN* is a closely related term to express that procedure in an instrument's calibration that adjusts it to read both low and high standards accurately; an improperly spanned 0–100 psi pressure gauge might read 10 psi "right on" but then indicate a true 80 psi as 70 psi. *LINEARITY* is the term to describe how well a device tracks a series of standards throughout its effective range—the preceding poorly spanned gauge is off in linearity by −12.5 percent at 80 percent of scale (off 10 psi at 80 psi). It is evident that to be truly accurate, an instrument must have acceptable precision, be in calibration, and be designed with an effective range that is wide enough to measure the range of the treatment plant's process variables.

Finally, two other common terms need be defined. *ANALOG* instruments yield a reading as a pointer (or other cursor) against a marked scale, such as simple pressure or level gauges. *DIGITAL* displays provide direct numerical readouts, as with most modern instrumentation. The best example of both nowadays is a watch that has both readouts; the hands give a quick analog reference and the digital time display gives a more precise reading. The type of signal sent to both analog and digital displays is usually analog; for example, a continuously variable 4–20 mA (milliamp) signal from a transducer.

QUESTIONS

Write your answers in a notebook and then compare your answers with those on page 662.

12.0D What is the main difference between measurements by our senses and by instruments?

12.0E What is the definition of a measurement?

12.0F Explain the difference between precision and accuracy.

12.0G What is the difference between analog and digital readouts?

12.03 Explanation of Control Systems

Control systems are the means by which such process variables as pressure, level, weight, or flow are controlled. The terms "controller" and "control systems" are used to refer to two different types of instrumentation control systems: (1) *MODULATING SYSTEMS*, which sense and control their own operation on a close, continuous basis, and (2) *MOTOR CONTROL STATIONS*, which control only the ON/OFF operation of motors and other devices.

12.030 Modulating Control Systems

The technically proper use of the terms "controller" and "control systems" refers to those systems that provide modulating control of process variables. Examples of modulating control systems include: (1) chlorine residual analyzers/controllers; (2) flow-paced (open-loop) chemical feeders; (3) pressure- or flow-regulating valves; (4) continuous level control of process basins; and (5) variable-speed pumping systems for flow/level control.

In order for a process variable, whether pressure, level, weight, or flow, to be closely controlled, it must be measured precisely and continuously. In a modulating control system, the measuring device or primary element sends an electrical or pneumatic signal, proportional to the value of the variable, to the actual system controller. Within the controller, the signal is compared to the desired or set point value. A difference between the actual and desired values results in the controller sending out a command signal to the controlled element, usually a valve, pump, or chemical feeder. Such an error signal produces an adjustment in the system that causes a corresponding change in the original measured variable, making it more closely match the set point. This continuous "cut-and-try" process can result in very fine ongoing control of variables requiring constant values, such as some flow rates, pressures, levels, or chemical feed rates. The term applied to this circulating action of the variable in such a controller is *FEEDBACK*. The path through the control system is the *CONTROL LOOP*. A diagram of such a control loop, measuring the process variable of wastewater flow, is shown in Figure 12.1. Figure 12.2 is a photo of an actual flow recorder/controller.

The internal settings of the modulating controller can be quite critical since close control depends on sensitive adjustments. Thus, you should not try to adjust any such control system unless you know exactly what you are doing. Many plant and system operations have been drastically upset due to such efforts, however well intentioned, by unqualified personnel.

12.031 Motor Control Stations

A motor control station or panel (Figure 12.3) and the related circuitry (Figure 12.4) essentially provide only ON/OFF operation of an electric motor or other device. The electric motor or device, in turn, might power a pump, control valve, or chemical feeder. With this system, control could be manual or automatic, with a switch responding to a preset value of level, pressure, or some other variable. Motor control stations are typically made up of a standard electric power control panel with manual push buttons, overload relays, and Hand-Off-Automatic (H-O-A) or similar On-Off-Remote switch.

Additionally, they may include, in good electrical *FAIL-SAFE*[1] design practice, provisions for power failure or loss-of-phase, and such protective devices as high or low pressure/temperature/level cutoff switches. For this type of panel to be considered a real controller (within our secondary meaning of the term), its operation

---

[1] *Fail-Safe.* Design and operation of a process control system whereby failure of the power system or any component does not result in process failure or equipment damage.

624 Treatment Plants

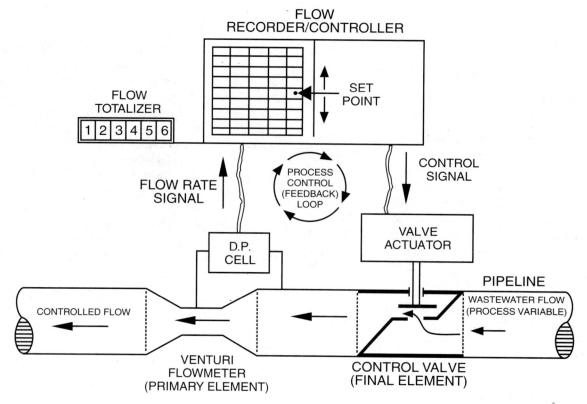

NOTE: Measurement and control system instruments may be electric or pneumatic.

Fig. 12.1 Automatic control system diagram (closed-loop modulating control)

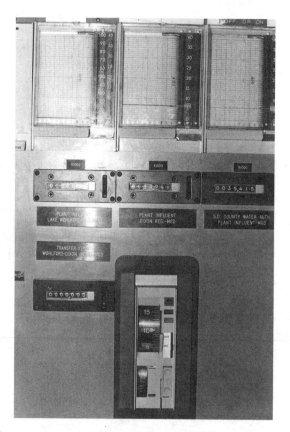

NOTE: Controller in photo is the lower instrument with set point at 10 MGD; recorder for the variable is the middle one above.

Fig. 12.2 Photo of flow recorder/controller in Figure 12.1

Instrumentation 625

Note: "REM" on switch above means "remote", which is automatic operation in this application

*Determines which is "lead" (first to come on) and "lag" (next to come on) pump in automatic operation

Fig. 12.3 *Typical duplex (two-pump) motor control panel*

626 Treatment Plants

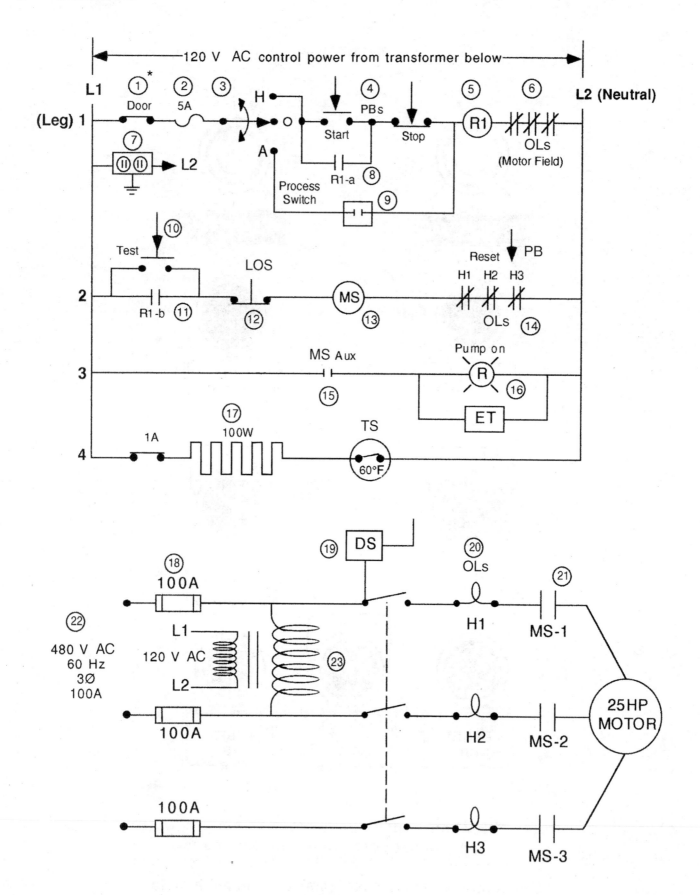

Fig. 12.4 Typical motor control elementary diagram circuitry (do not use for design)
(simplified, self-explanatory ladder diagram)
(see page 627 for parts legend)

FIGURE 12.4   PARTS LEGEND

1. Door or "interlock" switch to kill control power if panel door is opened.
2. Control circuit fuse, 5 amp.
3. Hand-Off-Auto (H-O-A) function switch for manual or automatic operation (shown Off).
4. Momentary contact (spring-loaded) push buttons (Start-Stop), "PB station."
5. Control relay coil (other relays, if used, would be shown as R2, R3, etc.)
6. Motor thermal protection elements embedded in 3 phase motor field ("stator") coils.
7. Duplex 120 V AC outlet within panel (for test equipment usage).
8. "Holding contacts" of control relay; start PB energizes relay coil to close contacts, which remain closed until coil is de-energized.
9. Process control contacts (ON/OFF control); for example, level, pressure, or analytical process variable.
10. Push-to-test motor starter relay circuit leg (if MS coil and OLs OK, pump will start); used for troubleshooting purposes.
11. Main control contacts of relay R1, energizes MS coil to start pump.
12. "Lock-Out Stop" switch to prevent motor operation while being serviced, key lockable in open position.
13. Motor Starter ("Mag starter") relay coil.
14. Manually (push button) resettable overload relay contacts, opened by "heaters" (see 20).
15. Auxiliary contacts in MS relay.
16. Pump running light (red) and elapsed time (total hours run) meter.
17. Enclosure heater in panel (keeps components dry) with 1 amp circuit breaker and thermostat switch.
18. 100 amp 480 V AC cartridge-type fuses.
19. Main 480 V AC manual disconnect switch ("Main switch") local at the motor.
20. Motor circuit thermal overload "heaters" (set for maximum load motor current).
21. Motor starter relay contactors (large contacts).
22. Motor power, 480 volts AC, 60 cycle, 3 phase, 100 amp service.
23. Control power transformer (for control circuit above), 120 volts AC.

DESCRIPTION OF TYPICAL MOTOR CONTROL CIRCUIT OPERATION
(Refer to Figure 12.4)

Control circuit is shown Off; the H-O-A function switch is at "O." To start motor manually, place H-O-A in "H" and depress "Start" push button (PB) to energize main control relay R1. With power through coil, "holding contacts" R1-a close around the start PB to keep R1 energized (and motor running) when the operator's finger is removed from the button (such spring-loaded PBs are termed "momentary" contacts).

When R1 is energized, it also "pulls-in" (closes) contacts R1-b to energize motor starter relay MS in turn. With control power to the MS coil, its contactors MS-1, 2, and 3 "make" (close) to supply 480 V AC to the motor—assuming resettable thermal overloads H1, H2, and H3 are all closed (and the main disconnect is closed and 100A fuses are good, of course).

To stop the motor manually, with the H-O-A still in the "H" position, one pushes the "Stop" PB to break the circuit to the R1 coil, allowing contacts R1-a to "fall out" (open) so the control relay is de-energized (even after this momentary open PB is released).

For automatic operation, H-O-A switch is placed in "A" so motor will start and stop with the open and close, respectively, of the process controller contacts. Note that the PBs cannot start or stop the motor with H-O-A switch in the Auto position.

The "Test" PB, in Leg 1 of the schematic, will start stopped pump as long as it is depressed (being a momentary close PB), by bypassing the control contacts R1-b. This feature permits isolation of a no-start problem to Leg 1 circuitry, in that the motor will start if Leg 2 is OK.

In Leg 3, the "motor on" light and "elapsed time" meter are turned on when the MS relay is energized, through its auxiliary contacts (integral to the relay). Leg 4 consists of a small fused and thermostated space heater to keep all electrical panel components dry.

120 V AC control power for the control circuit (L1 and L2) is transformed from the main 480 V AC 3Ø motor power service. A main disconnect switch kills all power to the motor, for servicing, and all three phases are protected with 100-amp cartridge fuses.

This typical motor control circuitry has two types of motor overheating protection:

1. In-line heater elements H1, H2, and H3, which open any or all of their respective contacts above when the motor draws higher currents for a longer time than it is designed for. These interrupt the power to the MS relay (Leg 2).

2. Small, heat-sensitive contacts embedded in all three stator motor windings, which open at a predetermined maximum temperature to interrupt the control circuit in Leg 1, stopping the motor before it gets too hot.

Only the former (overload heaters) are manually resettable using the PB shown; the latter (stator-imbedded) must be allowed to cool before they remake contact.

must be *directly controlled by the value or values of some variable*, not merely by a device such as an ON/OFF timer. In other words, it must be turned on and off as a result of a *measurement* of a level, pressure, flow, chemical concentration, or other variable as it reaches a predetermined setting or settings. In the automatic mode (A on the H-O-A switch), its operation thus is, in fact, automatic in the sense that the *variable is automatically controlled*, even though the limits of its value may be quite wide compared to those attainable with a modulating controller as previously described. Whereas a basin level modulating controller may allow only an inch or so of water level change, an ON/OFF system might operate within a few feet of level difference. In many applications, however, such wide control is of no particular disadvantage, and sometimes is even desirable (such as a tank level where regular exchange of contents is desirable).

Terms used in control practice can now be defined operationally: FEEDBACK and CONTROL LOOP have been mentioned previously, however the term CONTROL LOOP needs some qualification. An OPEN-LOOP control system controls one variable on the basis of another. A good example of this is a chlorinator paced by process flow signals (rather than by a chlorine residual analyzer). CLOSED-LOOP control remains as discussed previously, the true control system with measurement and control of the same variable (and feedback). PROPORTIONAL-BAND, RESET, and DERIVATIVE actions are adjustments of a controller that relate to the effectiveness and speed of its control action. OFFSET is the difference between the desired value of the variable (the SET POINT) and the controlled (actual) value. DETECTION LAG, common to chlorinator control systems, refers to the time period between the moment when a change in control is made and the moment when such change is finally sensed by the associated measuring element. Long detection lags can result in unstable processes and poor control.

QUESTIONS

Write your answers in a notebook and then compare your answers with those on page 662.

12.0H What kinds of wastewater treatment process variables do control systems control?

12.0I List three examples of modulating control systems.

12.0J What is the essential purpose of a motor control station or panel?

12.1 SAFETY HAZARDS OF INSTRUMENTATION AND CONTROL SYSTEMS

12.10 General Precautions

The general principles for safe performance on the job, summed up as *ALWAYS* avoiding unsafe acts and correcting unsafe conditions, apply as much to instrumentation work as to other plant operations. However, there are some special dangers associated with instrument systems, namely electric shock hazards, that merit special mention in this section. Repetition is well justified here for the sake of safe practice.

12.11 Electrical Hazards

A hidden aspect of energized electrical equipment is that it looks normal; that is, there are no moving parts or other obvious signs that tend to discourage one from touching its components. In fact, there seems to be a peculiar fascination to "see if it is really live" by quickly touching circuit components with a tool, often a screwdriver (usually having an insulated handle, fortunately) but sometimes even with one's finger. Only training (coupled with bad experience, at times) is effective in squelching this morbid curiosity. Even so, most practicing electricians' tools still have the "arc-mark trademark" or two, evidence of the need for continuing self-discipline in this area. Though such mention may conjure up humorous images of the maintenance person's surprise and shock upon such an incident, one only need consider that electric shock can and does regularly cause serious burns and even death (by asphyxiation due to paralysis of the muscles used in breathing) to bring the problem into sober perspective. Also, the expense and effort caused by a needless shorting-out of an electrical device could be significant. The point is, *RESIST THE URGE* to test any electrical device with a hand tool or part of your body.

If there is *ANY* doubt in your mind that *ALL* sources of voltage (not merely the local switch) to a device have been switched off or disconnected, then *DO NOT TOUCH*, except possibly with the insulated probes of a test meter. Remember, you cannot see even the highest voltage, and assuming that a circuit is dead can be very hazardous, maybe even deadly.

Do not simulate an electrical action, for example by pressing down a relay armature, within an electrical panel without a positive understanding of the circuitry. Your innocent action may cause an electrical "explosion" to shower you with molten metal, startle you into a bumped head or elbow, or cause a bad fall. Remember, *WHEN IN DOUBT—DO NOT* when it comes to electricity.

Usually, a plant operator does not have the test equipment nor the technical knowledge to correct an electrical malfunction, other than possibly resetting a circuit breaker, regardless of how critical the device's function is to plant operations. In addition to the shock hazards of motor control centers, there also may be a shock hazard within measurement instrument cases. Expensive instrument components can easily be damaged or destroyed when a foolhardy operator uses the tool-touch-test method.

Most panels have an *INTERLOCK* on the door that interrupts all (local) power to a panel or device when the door is opened or the circuit is exposed for service. Never disconnect or disable interlocks that interrupt control power. Warning labels, insulating covers (over "hot" terminals), safety switches, lockouts, and other safety provisions on electrical equipment must remain functional at all times. Your attention to this crucial aspect of your workplace may save a life, and, as the saying goes, it could be your own.

Operators often use hand power tools around electrical equipment. All such power tools (even double-insulated ones) can present a shock hazard, treatment plants being damp places, at times. Power tools can be a mechanical hazard as well. For your own safety, use power tools only when you can have an observer on hand in case of an accident. *NEVER* stand in water with a power tool, even when it is turned off. Brace yourself, if necessary, in such a way that electric current cannot flow from arm to arm in case of a faulty power tool. Shocks through the upper body could involve your heart or your head, whose importance to you is self-evident.

QUESTIONS

Write your answers in a notebook and then compare your answers with those on page 662.

12.1A What are the general principles for safe performance on the job?

12.1B How can electric shock cause death?

12.1C What could happen to you as a result of an electrical "explosion"?

12.1D Why should you brace yourself when operating power equipment so that electric current cannot flow from arm to arm in case of a faulty tool?

12.12 Mechanical and Pneumatic Hazards

There exists a special danger when working around powered mechanical equipment, such as electric motors, valve operators, and chemical feeders that are operated remotely or by an automatic control system. Directly stated, the machinery may suddenly start or move when you are not expecting it. Most devices are powered by motors with enough torque or RPM to severely injure anyone in contact with a moving part. Even when the exposed rotating or meshing elements are fitted with guards in compliance with safety regulations, a danger may exist. A motor started remotely may catch a shirt sleeve, finger, or tool hanging near a loose or poorly fitted shaft or gear train guard.

The sudden, automatic operation of equipment, even when half-expected, may startle a nearby operator into a fall or slip. Signs indicating that "This Equipment May Start At Any Time" tend to be ignored after a while. Accordingly, you must stay alert to the fact that any automatic device may begin to operate at any time. Thus, you must stay well clear of active automatic equipment, especially when it is *NOT* operating.

Lockout devices on electric switches must be respected at all times. The electrician who attaches a lockout device to physically prevent the operation of an electric circuit is, in effect, trusting his or her life and health to the device. Once the lockout device is attached to the switch (whether the switch is tagged off or actually locked with lock and key), the electrician will consider the circuit and its connected equipment de-energized and safe and will feel free to work on it. Consider the potential consequences, then, of a careless operator who removes a lockout to place needed equipment back into service, *presuming* the electrician is finished (as might occur after several hours). The point cannot be overstressed:

RESPECT ALL LOCKOUT DEVICES AND ALL TAGGED-OFF EQUIPMENT AS IF A LIFE IS ENTRUSTED TO YOU ... IT MAY WELL BE.

PNEUMATIC HAZARDS

Working with and around pneumatic instrumentation presents an additional hazard associated with high-pressure air. Pneumatic devices are powered by air at supply pressures from about 50 to 100 psi (350 to 700 kPa). Serious injury can be caused by the high air (and trapped particle) velocities that can be produced. A pressure regulator normally reduces the high air pressures to a safer 30 psi (210 kPa) or so for each device, but even these lower pressures can cause injury if directed toward the eye or other delicate tissues.

Before disconnecting any air line, put on your safety goggles (not just glasses, but wraparound goggles). Then valve the line off, from both directions ideally, and crack (open slightly) a fitting on the pipe or tubing to permit the pressurized air to bleed out slowly. If it is necessary to purge a line of moisture after inspection or repair, do so through the filter trap valve (pointing down, if standard) on each of a supply line's connected components. If a very dirty, oily, or moist line must be purged with supply air, set up a temporary purge line to a low area and tape a sock or closed rag to the open end to trap particles and minimize dirt pick-up by the exhausting air. *NEVER* direct a pressurized air stream toward any part of your body or anyone else's body. High-velocity air can easily penetrate the body's tissues, even the skin of the hands. The tiny particles that are always present, even in filtered air (picked up from the inside of piping), can enter the eye's delicate tissues and cause an irritation at the least, or corneal damage and infection at worst. Breathing oil-laden air (many compressors seal with oil) from extensive line and filter purging can cause "chemical pneumonia" or even bacterial lung infections in sensitive individuals. Wearing a simple painter's (filter) mask will keep the suspended oil droplets and other particulates out of your respiratory system.

Though it is recognized that plant operators do little, if any, corrective maintenance around instrumentation and control components, it seems there is always some exposure to electrical, mechanical, and pneumatic hazards. Operators often do preventive maintenance tasks, and even minor fixing of many devices. Therefore, always play it smart and safe around all instrumentation.

12.13 Vaults and Other Confined Spaces

Many measurement and control systems include remotely installed sensors and control valves. Quite often, these are found in vaults or other closed concrete structures. Take a few minutes to review the special precautions for confined space work described in Chapter 13, "Safety." Be sure the ventilation equipment is working properly before entering a closed space.

12.14 Falls and Associated Hazards

All the general safety measures to guard the operator against falls, a leading cause of industrial lost-time accidents, are covered in Chapter 13, "Safety," and need not be repeated in this chapter. However, keep in mind that an electric shock of even minor intensity can result in a serious fall. When working above ground on a ladder, use the proper, nonconductive type of ladder (fiberglass rails) and position it safely. Even an alert, safety-conscious operator could be knocked off balance by a slight electric shock. When required to do preventive maintenance from a ladder, turn off the power to the equipment being serviced, if at all possible. If this is not feasible, take special care to stay out of contact with any component inside the enclosure of an operating mechanism, and well away from terminal strips, unconduited wiring, and electrical black boxes. Though not commonly considered essential, wearing thin rubber or plastic gloves can greatly reduce your chances of electric shock (whether on a ladder or off).

Make provisions for carrying tools or other required objects on an electrician's belt rather than in your hands when climbing up or down ladders. Finally, never leave tools or any object on a step or platform of the ladder when you climb down, even temporarily. You might be the one upon whom they fall if the ladder is moved or even steadied from below. In this regard, it is always a good idea (even if not required) for preventive maintenance personnel to wear a hard hat whenever working on or near equipment, especially when a ladder must be used.

QUESTIONS

Write your answers in a notebook and then compare your answers with those on page 662.

12.1E Why should operators be especially careful when working around powered mechanical equipment?

12.1F What is the purpose of an electrical lockout device?

12.1G What kind of specific protective clothing could be worn to protect you from electric shock?

12.2 MEASURED VARIABLES AND TYPES OF SENSORS/TRANSDUCERS

12.20 General Principles of Sensors/Transducers

A measured variable is any factor that is sensed and quantified (reduced to a reading of some kind) by a primary element or sensor. In industrial waste treatment, the variables of pressure, level, and flow are the most common ones measured; at times, chemical feed rates and some physical or chemical process characteristics are also sensed.

The sensor is usually a transducer of some type, in that it converts energy of one kind into some other form to produce a readout or signal. For example, one common type of flowmeter converts the hydraulic action of the water into the mechanical motion (for example, propeller) necessary to drive a meter indicator, and also produces an electrical signal for a remote readout device. If such a signal is produced, be it electric or pneumatic, the sensor is also then considered a transmitter.

The signal produced is not necessarily a continuous one proportional to the variable (that is, an analog signal), but often merely a switch set to detect when the variable goes above or below preset limits. In this type of ON/OFF control, the predetermined setting is called a control point. This distinction between continuous and ON/OFF operation relates to the two types of controllers discussed previously. The remainder of this section discusses each of the common variables sensed and measured in industrial wastewater treatment plants.

12.21 Pressure Measurements

Since pressure is defined as a force per unit of area (pound per square inch, or kiloPascal), you might expect that sensing pressure would thus entail the small movement of some flexible element subjected to a force. In fact, that is how pressure is measured in practice. There are many classes and brands of sensors, but the most common types contain mechanically deformable components such as the Bourdon tube (Figures 12.5 and 12.6), bellows, or diaphragm arrangements. The slight motion each exhibits, directly proportional to the applied force, is then amplified mechanically by levers or gears to position a pointer on a scale or provide an input for an associated transmitter. A so-called "blind transmitter" for pressure, or any variable, has no local readout.

Some pressure sensors are fitted with surge or overrange protection (snubbers) to limit the effect pressure spikes or water hammer have on the instrument. In most cases, such protection devices function by restricting flow into the sensing element. Surge protection devices thus prevent sudden pressure surges, which can easily damage most pressure sensors.

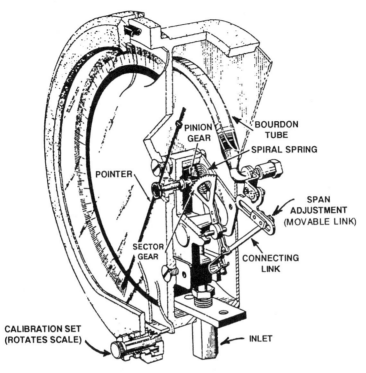

A precision industrial pressure gauge with a Bourdon-tube pressure element.

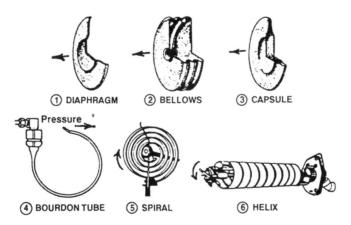

Elastic deformation elements. Pressure tends to expand or unroll elements to indicate pressure as shown by arrows.

Fig. 12.5 Bourdon tube and other pressure sensing elements
(Permission of Heise Guage, Dresser Industries)

NOTE: Diaphragm (round disk) below gauge protects sensing element from corrosion by chlorine gas. The Bourdon tube and all other fluid portions of the gauge above the diaphragm are filled with fluorolube.

Fig. 12.6 Photo of a chlorine pressure gauge

A snubber (Figure 12.7) consists of a restriction through which the pressure-producing fluid must flow. A more elaborate mechanical snubber responds to surges by moving a piston or plunger that effectively controls the size of an orifice. Some snubbers are subject to clogging or being adjusted so tight as to prevent any response at all to pressure changes. If a pressure sensor is not performing properly, look first for such clogging or adjustment that has become too restrictive.

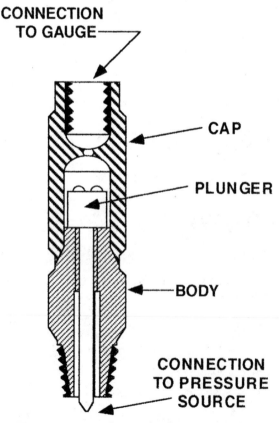

Fig. 12.7 Internal parts of a plunger-type snubber for surge protection

Another dampening device is an air cushion chamber (Figure 12.8), which is simply constructed yet very effective. The top part of the chamber contains air; water flows into the bottom part. A sudden change in water pressure compresses the air within the chamber, taking the shock. The rate of response can be further dampened by also placing a snubber into the air chamber line.

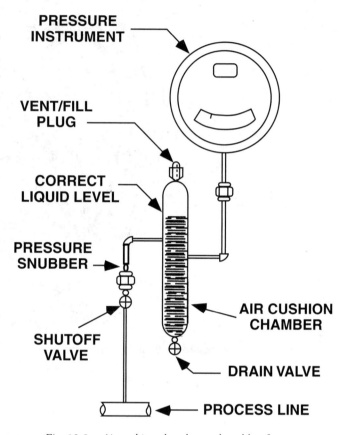

Fig. 12.8 Air cushion chamber and snubber for surge protection of pressure gauge

QUESTIONS

Write your answers in a notebook and then compare your answers with those on page 662.

12.2A What is a sensor?

12.2B How is pressure measured?

12.2C Why are some pressure sensors fitted with surge or overrange protection?

12.22 Level Measurements

Systems for sensing the level of water or any other liquid, either continuously or at a single point, are very common in municipal and industrial wastewater treatment plants. Pumps are controlled, filters operated, basins and tanks filled and emptied, chemicals fed (and ordered), sumps emptied, and many other variables controlled on the basis of liquid levels. Fortunately, level sensors usually are simple devices. A float, for example, can be a very reliable liquid level sensor both for single-point and continuous level sensing. Other types of liquid level sensing devices include displacers, electrical probes, direct hydrostatic pressure, pneumatic bubbler tubes, and ultrasonic (sonar) devices. Single-point detection of level is very common for levels controlled by pumps or valves within fairly wide limits.

Variations on mechanical ways to measure basin liquid levels use a fine cable riding over a grooved drum or a perforated steel tape riding on a toothed pulley. The cable or tape transmits the level sensed as an electrical signal proportional to pulley or drum rotation (Figures 12.9a and 12.9b). Most of these devices use a counterweight on the end of the tape or cable opposite the float to ensure tautness. All types of float-operated sensors work best in basins with the float traveling within a long tube called a stilling well, which dampens out unwanted liquid turbulence or waves.

Another system of level sensing is the displacer type (Figure 12.9c). By its nature only single-point determinations of level can be obtained, but this type of sensing for ON/OFF control is adequate for many purposes. The displacer is a weight, usually of a noncorroding, dense material such as porcelain, which hangs down on a cable into the liquid within a stilling well. The cable is supported by a spring that is sized so as to keep an electric switch (a mercury vial, usually) in one position with the displacer immersed, but allowing it to switch to another position when the displacer is partially out of the liquid. The basic principle is that the displacer is buoyed up by the liquid when immersed, thus weighing less. Accordingly, the motion of the displacer is very slight, typically less than one inch (25 mm), so this design is more reliable than a float device, which may be subject to sticking in its stilling well. A float switch, however, is a commonly used variation of the displacer-type switch. The float containing a mercury switch is fixed at the desired depth. When the water level is below the float, the float hangs down; when the level rises, the float rotates upward and the switch changes position.

An alternative to a float or displacer, both of which are mechanical systems, is the use of an electrical probe to sense liquid level (Figure 12.9d). Again, only single-point determinations can be made this way, though several probes (or one with several sensors along its length) can be set up to detect several different levels. Level probes are used where a mechanical system is impractical, such as within sealed or pressurized tanks or with chemically active liquids.

The probes can be small-diameter, stainless-steel rods inserted into a tank through a fitting, usually through the top but at times in the side of a vessel. Each rod is cut to length corresponding to a specific liquid level in the case of the top-entering probes; in the side-entering setup, a short rod merely enters the vessel at the appropriate height or depth. One problem encountered with probes is the accumulation of scum or caking (such as grease) on the surface of the rods.

A small voltage is applied to the probe(s) by the system's power supply, with current flowing only when the probe "sees" liquid, that is, just becomes immersed. When current flow is sensed, a switch activates a pump/valve control or alarm(s) at as many control points as necessary through an electrical relay.

Though at times only a single probe is used, with the metal tank completing the circuit as a ground, usually at least two probes are found. The ground probe extends to within a short distance of the bottom of the tank so as to be in constant contact with the electrically conducting liquid (a liquid ground, as it were).

Levels can be sensed continuously by measurement of liquid hydrostatic pressure near the bottom of a vessel or basin. The pressure elements used for level sensing must be quite sensitive to the low pressures created by liquid level (23 feet of water column equals only 10 psi, or 70 kPa). Therefore, simple off-the-shelf pressure gauges such as are found on pumps are not used to measure water levels. Instead, very sensitive water level sensors are used to measure levels of filter basins or in process tanks where control or monitoring must be close, continuous, and positive. Rather than being calibrated in units of pressure (psi or kPa), these gauges read directly in units of liquid level (feet or meters). One or more single-point control/alarm contacts can be made a part of this, or any, continuous type of level sensing system.

A very precise method of measuring liquid level is the bubbler tube with its associated pneumatic instrumentation (Figure 12.10). The pressure created by the liquid level is sensed, but not directly as with a liquid pressure element. A bubbler measures the level of a liquid by sensing the air pressure necessary to cause bubbles to just flow out the end of a submerged tube. Air pressure is created in a bubbler tube to just match the pressure applied by the liquid above the open end of the tube when it is immersed to a precisely determined depth in a tank or basin. The air pressure in the tube is then measured as proportional to the liquid level above the end of the tube. This indirect determination of level using air permits the placement of the instrumentation anywhere above or below the liquid's surface, whereas direct pressure-to-level gauges must be installed at the very point (or very close to) where liquid pressure must be sensed.

Bubbler systems are adjusted so air just begins to bubble slowly out of the submerged end of the sensing tube. They automatically compensate for changes in liquid level by providing a small, constant flow of air through the bubbler tube by means of the constant air flow (also called constant differential) regulator. There is no advantage to turning up the amount of air to create more intense bubbling because the back pressure will still mainly depend on the water level. In fact, increasing the air flow may create a sizeable measuring error in the system, so any air flow changes should be left to qualified instrument service personnel.

634 Treatment Plants

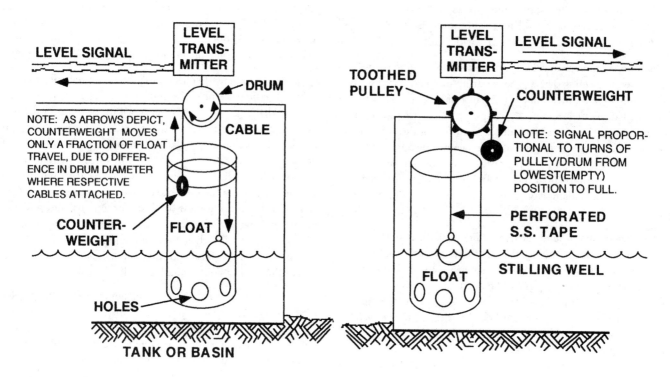

Fig. 12.9a Float and cable (continuous)

Fig. 12.9b Float and tape (continuous)

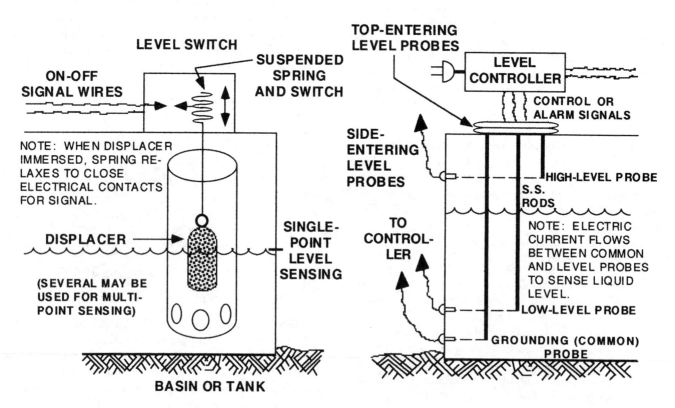

Fig. 12.9c Displacer (single point)

Fig. 12.9d Electrical probes (multipoint)

*Fig. 12.9 Liquid level sensing systems
(level measurement and single/multipoint sensing)*

Instrumentation 635

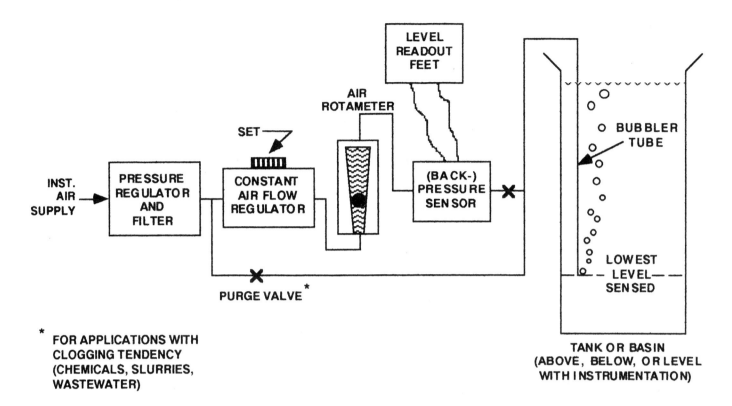

Constant-flow regulator and rotameter on left, back-pressure sensor (D.P. cell) to right.

Fig. 12.10 Diagram and photo of a bubbler tube system for measuring liquid level

Bubbler tube systems are common in basin-level controllers that must maintain water levels within a range of a few inches (centimeters) and for measurement of flow in open channels or over weirs. Usually, the level transmitter for basin-level controllers is blind since it only controls liquid level and need not provide a local readout of the level.

QUESTIONS

Write your answers in a notebook and then compare your answers with those on pages 662 and 663.

12.2D List the different types of liquid level sensing devices.

12.2E How can a continuous level signal be generated by a float element?

12.2F Under what circumstances can probes be used to measure liquid level?

12.2G How does a bubbler measure the level of a liquid?

12.23 Flow (Rate of Flow and Total Flow)

The term "flow" can be used to refer to either *RATE OF FLOW*, such as MGD, CFS, and GPM or m^3/sec or liters/sec (volume per unit of time), or to *TOTAL FLOW*, in simple units of volume (no time unit) such as the corresponding million gallons, cubic feet, or gallons (liters or cubic meters). Total flow volumes are usually obtained as a running total, with a comparatively long time period for the flow delivery, such as a day or month. Understanding this distinction is important because most flow instrumentation provides both values (for rate of flow and total flow).

While it is possible in principle to measure process flows directly, as described previously, with pressure and most level-sensing devices, it is quite impractical. Direct measurement would involve the constant filling and emptying of, say, a gallon container with water flowing from a pipe on a timed basis. This method is obviously not practical. Therefore, sensing of process flows in waterworks practice is done inferentially, that is by inferring what the flow is from the observation of some associated hydraulic action or effect of the water. The inferential flow-sensing techniques that are used in flow measurement are (1) velocity, (2) differential pressure, (3) magnetic, and (4) ultrasonic. First, let us look at one device, the rotameter, used in flow sensing for some specialized applications before studying the devices for process stream flows.

ROTAMETERS (Figures 12.11 and 12.12) are transparent (usually) tubes with a tapered bore containing a ball or float. The float rises up within the tube to a point corresponding to a particular rate of flow. The rotameter tube is set against, or has etched upon it, marks calibrated in whatever flow rate unit is appropriate. Rotameters are used to indicate approximate liquid or gas flow. For example, the readout for a gas chlorinator could be a rotameter. Sometimes a simple rotameter is installed merely to indicate a flow or no-flow condition in a pipe, for example, on a chlorinator injector supply line.

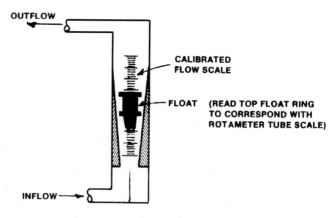

Fig. 12.11 Rotameter

For process flows, as mentioned, *VELOCITY-SENSING* meters measure water speed within a pipeline or channel. One way of doing this is by sensing the rate of rotation of a special impeller (Figure 12.13) placed within the flowing stream; the rate of flow is directly proportional to impeller RPM (within certain limits). Since normal water velocities in pipes and channels are under 10 feet per second (about 7 MPH or 3 m/sec), the impeller turns rather slowly. This rotary motion drives a train of gears, which indicates *RATE* of flow, in the same way a speedometer indicates the rate of travel for a car. *TOTAL* flow then appears as the cumulative number of revolutions, denoting total volume flowed, like the odometer on your car's speedometer indicates total mileage traveled.

Rotation of the velocity-sensing element is not always transferred by gears, but may be picked up as magnetic or electric pulses by the transducer. Nor is velocity always sensed mechanically; it may also be detected or measured purely electrically (the thermistor type), or hydraulically (the pitot tube). In each case, the principle of equating water velocity with rate of flow (within a constant flow area) is the same. Of course, all such flowmeters are calibrated to read out in an appropriate unit of flow rate, rather than velocity units.

Typically, a velocity-sensing element transmits its reading to a remote site as electric pulses, although other devices can be used in order to convert to any standard electrical or pneumatic signal.

Preventive maintenance of impeller-type flowmeters centers around regular lubrication of rotating parts, at least for the older types. Propeller meters, as they are called, have a long history of reliability and acceptable accuracy in both municipal and industrial applications. When propeller meters become old, they can become susceptible to under-registration (read low) due to bearing wear and gear train friction. Accordingly, annual teardown for inspection is indicated. Over-registration is a physical impossibility, but a partially full pipeline, wrong gears installed, or a malfunctioning transmitter or receiver can cause high readings.

DIFFERENTIAL PRESSURE SENSING TUBES (Figures 12.14 and 12.15), also called Venturi or differential meters (or flow tubes), depend for their operation upon a basic principle of

Rotameters (gas **rate**-of-flow)

Fig. 12.12   Flow-sensing devices for fluids

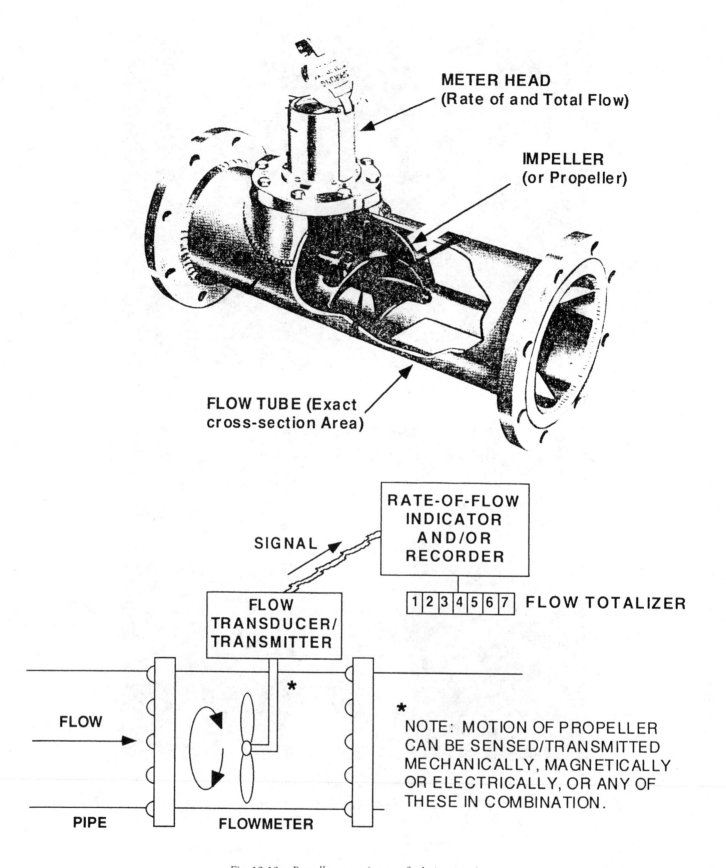

Fig. 12.13 Propeller meter (a type of velocity meter)

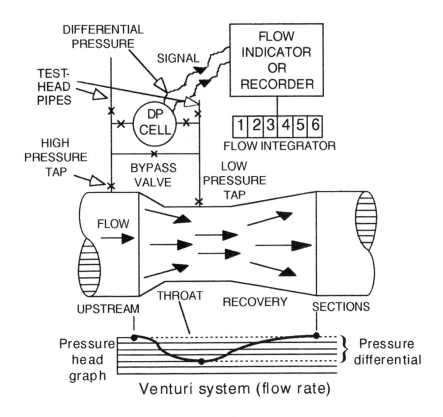
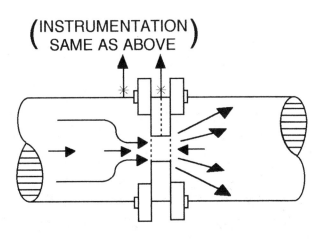

Fig. 12.14 Schematic diagrams of differential pressure flow measuring devices

54-inch Venturi tube

24-inch orifice plate

Fig. 12.15 Photos of differential pressure flow tubes

hydraulics, the Bernoulli Effect: When a liquid is forced to go faster in a pipe or channel, its internal pressure drops. If a carefully sized restriction is placed within the pipe or flow channel, the flowing water must speed up to get through it. In doing so, its pressure drops a little, and it drops an exact amount for a given flow rate. This small pressure drop, the "pressure differential," is the difference between the water pressure *before* the restriction and *within* the restriction. This difference is proportional in a certain way (but not directly proportional) to the rate of flow. The difference in pressure is measured very precisely by the instrumentation associated with the particular flow tube installed. Typically, a difference of only a few psi (kPa) is required. This small value of pressure difference is often described in inches (centimeters) of water (head).

Measuring flow by the differential pressure method removes a little hydraulic energy from the water. However, the modern flow tube, with its carefully tapered form, allows recovery of well over 95 percent of the original pressure throughout its range of flows. Other ways of constricting the flow, such as installation of orifice plates, do not allow such high recoveries of pressure, nor the accuracy possible with other modern flow tubes.

An orifice plate (Figures 12.14 and 12.15) is a steel plate with a precisely sized hole (orifice) in it. The plate is inserted between flanges in a pipe. The pressure drop is sensed right at the orifice, or immediately downstream to yield a less accurate flow indication than a Venturi meter. This drop in pressure is not recovered; that is, a permanent pressure loss occurs with orifice plate installations, unlike Venturi flow tubes.

Differential devices require little, if any, preventive maintenance by the operator since there are no moving parts. Occasional flushing of the hydraulic sensing lines is good practice. However, flushing should only be done by a qualified person. When dealing with an instrument sensitive to fractions of a psi or pascals, opening the wrong valve can instantly damage the internal parts severely. Also, if an older DP (differential pressure) cell containing mercury is used, this toxic (and expensive) metal can easily be blown out of the device and into the process pipeline. Thus, all valve manipulations must be understood and done deliberately after careful planning by a qualified person.

In nearly all cases, the signals from larger flow tubes are transmitted to a remote readout station. Local readout is also provided (sometimes inside the case only) for purposes of calibration. Differential pressure transmitters may be electrical or pneumatic types. The signal transmitted is proportional to the square root of the differential pressure.

Venturi meters have been in use for many decades and can produce very close accuracies year after year. Older flow tubes are quite long physically (to yield maximum accuracy and pressure recovery). Newer units are much shorter but have very good pressure recovery and even better accuracy. With no moving parts, the Venturi-type meter is not subject to mechanical failure as is the comparable propeller meter. Flow tubes, however, must be kept internally clean and without obstructions upstream (and even downstream) to provide the designed accuracies.

The piping design for flowmeters must provide for adequate lengths of straight pipe runs upstream and downstream of the meter. Flowmeters in pipes will produce accurate flowmeter readings when the meter is located at least five pipe diameters distance downstream from any pipe bends, elbows, or valves and also at least two pipe diameters distance upstream from any pipe bends, elbows, or valves. Flowmeters also should be calibrated in place to ensure accurate flow measurements.

In summary, all of the flowmeters described in this section provide rate of flow indication. The rate of flow can also be (and usually is) continuously totalized to give a reading of total flow past the measuring point. Total flow is usually indicated at the readout instrument in units of gallons or cubic feet (liters or cubic meters).

QUESTIONS

Write your answers in a notebook and then compare your answers with those on page 663.

12.2H What are two basic types of flow readings?

12.2I List the inferential techniques that are used in flow measurement.

12.2J How do velocity-sensing devices measure flows?

12.2K Flows measured with Venturi meters take advantage of what hydraulic principle?

12.24 Chemical Feed Rate

Chemical feed rate indicators are often an integral part of a particular chemical feed system and thus are usually not considered instrumentation as such. For example, a dry feeder for lime may be provided with an indicator for feed rate in units of weight per time, such as pounds/hr or grams/minute. In a fluid (liquid or gas) feeder, the indication of quantity per unit of time, such as gallons/hour or pounds/day (liters/hour or kilograms/day), may be provided by use of a rotameter (Figure 12.11, page 636) or built-in calibrated pump with indicated output settings.

12.25 Process Instrumentation

Process instrumentation, by definition, provides for continuous analysis of physical or chemical indicators of process variables in a municipal or industrial plant. This does not include laboratory "bench" instruments (unless set up to measure sample water continuously), although the operating principles are usually quite similar. The process variables of dissolved oxygen and pH are often monitored closely in a wastewater treatment plant (Figure 12.16). Very frequently, chlorine residuals are also continuously measured or controlled. These variables are usually measured at several locations. Additionally, other indicators of process status may be sensed on a continuous basis, such as electrical conductivity (for Total Dissolved Solids, TDS), Oxidation-Reduction Potential (ORP), water alkalinity, turbidity, and temperature. In every case, the instrumentation is specific as to operating principle, standardization procedures, preventive maintenance, and operational checks. The manufacturer's technical manual describes routine procedures to check and operate this sensitive type of equipment.

Operators must realize that most process instrumentation is quite delicate and thus requires careful handling and special training to service. No adjustments should be made without a true understanding of the specific device. Generally speaking, this category of instrumentation must be maintained by the plant's instrument specialist, or the factory representative, rather than by an operator (unless specially instructed).

12.26 Signal Transmitters/Transducers

Common system installations measure a variable at one location and provide a readout of the value at a remote location, such as a main panel board. Except in the case of a blind transmitter, local indication is provided at the field site as well as being presented at the remote site. Associated with the remote (panel) readout system are quite often the control and alarm set points. Usually, recorders are found only on a main panel, along with all other recorders remote from their respective sensors in the plant. These system components will be discussed further in Section 12.3, "Categories of Instrumentation."

In order to transmit a measured value to a remote location for readout, it is necessary to generate a signal directly proportional to the value measured. This signal is then transmitted to a remote receiver, which provides a reading based upon the signal. Also, a controller may use the signal to control the measured variable, and a totalizer to integrate it.

Presently, two general systems for transmission of signals are used in most industrial situations: electrical and pneumatic. Electricity, of course, requires wiring (although radio transmission or microwave are possible) and pneumatic systems require small-diameter tubing (usually 1/4 inch or 6 mm) between transmitter and receiver. When the transmitter is quite far removed from the receiving station, a special terminology is used for the electrical link between the two; this is called telemetry. Generally, when signals have to be transmitted from outside a plant site, for example from a very remote flowmeter, telemetry is used.

Electrical signals used within an industrial treatment plant are either voltage (1 to 5 volts DC), current (4 to 20 milliamps DC), or pulse types. Milliamp signals (4 to 20 mA) are the most common electrical signals for most instrumentation and control systems in recent years. In any of these, a very low voltage is applied so no severe shock hazard exists (although shorting signal wires can still destroy electrical components). Signal transmission is limited to several hundred feet, with signal voltages set up for the specific connecting lines and transmitters/receivers.

A power supply to generate the required electric energy may be located at the transmitter, at the receiver, or at another location in the signal loop. The transmitter may be an integral part of the measurement or readout/transducer, or separately housed. In any case, the transmitter adjusts the signal to a corresponding value of the measured variable, and the receiver, in turn, converts this signal to a visible indication, the readout.

Pneumatic signal systems are restricted to comparatively short distances. Components include a compressor to provide air under pressure, as well as the necessary air filters, and often an air dryer. The precision of signal transmission by pneumatics is comparable to electrical signals, so both systems are found about equally in industrial plants, although pneumatic systems are safer if explosive hazards exist.

Compressed air presents no shock hazard and most plants must have compressors available for other purposes, so pneumatic systems are very common. Also, pneumatic systems seem to be more understandable to operating personnel and thus easier to keep functioning as desired. As with electrical signal circuits, the transmitter and receivers perform as their names imply. Pneumatic controllers, and all other types of pneumatic equipment, are as available and effective as their electrical counterparts. In most installations, pneumatic signals are generated by creating air pressures from 3 to 15 psi (20 to 100 kPa) proportional to the variable, with 9 psi (62 kPa) then representing a 50 percent signal.

Preventive maintenance of pneumatic components centers around ensuring a clean dry air supply at all times, which requires alert operators.

QUESTIONS

Write your answers in a notebook and then compare your answers with those on page 663.

12.2L What is one way to measure liquid chemical feed rates?

12.2M What process variables are commonly monitored or controlled by process instrumentation?

12.2N By what means can values measured at one site be read out at a remote location?

12.2O What are the two general systems for transmission of measurement signals?

END OF LESSON 1 OF 2 LESSONS

on

INSTRUMENTATION

Please answer the discussion and review questions next.

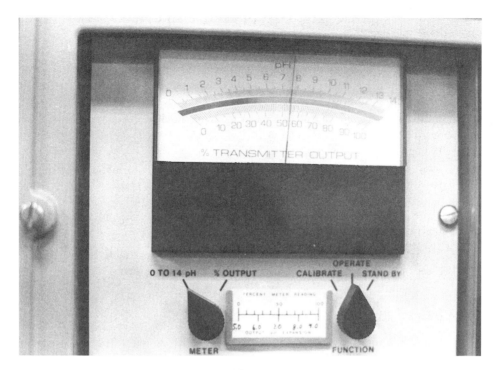

pH meter

Chlorine residual analyzer

Fig. 12.16   Industrial waste treatment plant process instrumentation

DISCUSSION AND REVIEW QUESTIONS
Chapter 12. INSTRUMENTATION
(Lesson 1 of 2 Lessons)

At the end of each lesson in this chapter you will find some discussion and review questions. The purpose of these questions is to indicate to you how well you understand the material in the lesson. Write the answers to these questions in your notebook.

1. Why should operators understand measurement and control systems?

2. How can measurement and control systems make an operator's job easier?

3. What is the difference between precision and accuracy?

4. Why should a screwdriver not be used to test an electric circuit?

5. What precautions should an operator take before entering a vault?

6. How can water levels be measured?

7. What problems can develop with propeller meters when they become old or worn?

CHAPTER 12. INSTRUMENTATION

(Lesson 2 of 2 Lessons)

12.3 CATEGORIES OF INSTRUMENTATION

12.30 Measuring Elements

Measuring (or primary) elements, or sensors, are those devices that make the actual measurement of the variable. Transducers are usually associated with them to convert the sensor's minute actions to a usable indication or a signal. If remote transmission of the value is required, a transmitter may be part of the transducer. An illustrative example of these three components is the typical Venturi meter (shown on Figure 12.14, page 639): (1) the flow tube is the primary element, (2) the differential pressure sensing device (DP cell) is the transducer, and (3) the signal-producing component is the transmitter. An understanding of the separate functions of each section of such a flowmeter is important to the proper understanding of equipment problems.

12.31 Panel Instruments (Figure 12.17)

12.310 Instrumentation

The components of measurement and control systems found on an industrial treatment plant's main panel board (often called the main control panel) are generally thought of as *the* plant instrumentation. These particular components are important to the operator and, hence, to plant operation itself, because they display or control the variable directly. The main control panel devices can also produce alarm signals to indicate if a variable is outside its range of expected values. The controllers are often installed on (or behind) the main panel along with the operating buttons, switches, and indicator lights for the plant's equipment.

In this age of cybernetics (a fancy term for instrumentation), an operator can easily be lulled into an overdependence on automation to operate the plant processes more or less blindly. However, there is no substitute for critical evaluation and informed judgment when operating important plant processes, no matter how sophisticated the instrumentation is. You must not rely solely upon the readings of any single instrument to ensure proper plant operation, but must consult other instruments and closely watch the other indicators of plant operation. Even the most sophisticated and expensive instrument systems do malfunction at times, even when well maintained.

12.311 Indicators

The major components found on a plant's main panel are indicators and recorders. Indicators give a visual presentation of

Fig. 12.17 Main control panel of an industrial treatment plant

a variable's present value, either as an analog or as a digital display (Figures 12.18 and 12.19). The analog display uses some type of pointer or graphical display (or other indicator) against a scale. A digital display is a direct numerical readout. Recorders, which by nature also serve as indicators, give a permanent record of how the variable changes with time by way of a moving chart or a printout of a graphical display on a computer. Whereas there are usually several indicators out in the plant or field, recorders are usually housed at a central location in the plant, as previously mentioned. Both indicators and recorders are discussed in the following paragraphs.

Since there are two main types of signal transmission means available, panel indicators may be of the electric or the pneumatic type. The digital readout has both advantages and disadvantages. Digitals may be read more quickly and precisely from a longer distance, and can respond virtually instantly to variable changes. But analog indicators are cheaper, more rugged, and may not even require electrical power (the pneumatic type), an advantage during a power failure or in explosive environments.

Another advantage of the electric or pneumatic analog display is that an incorrect readout may be more recognizable than with a digital system, and also is more easily corrected by the operator. For example, the pointer on a flowmeter gauge may merely

Fig. 12.18 Analog chlorine residual indicator

be stuck, as evidenced by a perfectly constant reading. With a digital reading, there is no practical way for the operator to see whether a problem actually does exist, nor is there any way for the operator to attempt a repair, such as freeing a pointer. Erratic or unreliable operation, while always a problem, seems to be worse when digitals are involved since there is no way for the operator to fix it. You often cannot tell if the problem is real, and you cannot do anything to get by until 8 am, as is often required on a night shift, until a technician comes on duty.

With all-electronic instrumentation, as advantageous as it may seem from a technical and economic standpoint, the operator has little recourse in case of malfunction of critical instrumentation. Temporary power failures, tripped panel circuit breakers, voltage surges (lightning) resulting in blown fuses, static electricity, and excessive heat, can all cause electronic instrument problems. However, electromechanical or pneumatic instruments may keep operating, or recover operation readily, after such power or heat problems. Electronic systems require the services of an instrument technician, or even the factory technician, to become fully operable again. Accordingly, the operator should insist upon some input into the design phase of instrument systems to ensure that the plant is still operable during power outages, hot weather, and other contingencies. Standby power generators or batteries can be used to keep plants operating during commercial power outages, but even they have shortcomings if plant operations depend entirely on electrical power.

QUESTIONS

Write your answers in a notebook and then compare your answers with those on page 663.

12.3A What is the purpose of instrumentation indicators?

12.3B Describe an analog display.

12.3C Where are recorders usually found at an industrial wastewater treatment plant?

12.3D What factors can cause electronic instrument problems?

12.312 Recorders

Recorders are indicators designed to show how the value of the variable has changed with time (Figure 12.19). Usually, this is done by attaching a pen (or stylus) to an indicator's arm, which then marks or scribes the value of the variable onto a continuously moving chart. The chart is marked on a horizontal, vertical, or circular scale in time units.

The chart is driven along at a precise speed under the pen to correspond with the time markings on the chart. Chart speeds range from several inches to a fraction of an inch (centimeter) per hour, with the pen and drying time of the ink specific to a given range of speeds.

There are two main types of recorder charts and recorders: the horizontal or vertical strip-chart type and the circular-chart type. The strip-chart type carries its chart on a roll or as folded stock, with typically several weeks' supply of chart available. Several hours of charted data are usually visible, or easily available, for the operator to read. On a circular recorder, the chart makes one revolution every day, week, or month, with the advantage that the record of the entire elapsed time period is visible at any time.

Changing of charts is usually the operator's duty. It is easier with circular recorders, though not that difficult with most strip-chart units with some practice.

Recorders may be electric or pneumatic. Pneumatic models frequently have electric chart drives. Purely mechanical units, useful at remote, unpowered stations, have hand-wound chart drives and are usually of the circular-chart type. Other models are battery-powered. Recorders are most commonly described by the nominal size of the strip-chart width or circular-chart diameter (for example, a 4-inch (100-mm) strip-chart, or a 10-inch (250-mm) circular-chart recorder). Figure 12.19 shows a combination indicator/recorder and Figure 12.20 presents two models of recorders.

12.313 Totalizers

Rate of flow, as a variable, is a time rate; that is, it involves time directly, such as in gallons per minute, or million gallons per day, or cubic feet per second (cubic meters per second or liters per minute). Flow rate units become units of volume with the passage of time. For example, flow in gallons per minute accumulates as total gallons during an hour or day. The process of calculating and presenting an ongoing running total of flow volumes passing through a meter is termed "integration" or totalizing.

A totalizer or integrator continually adds up gallons or cubic feet (liters or cubic meters) as a cumulative total up to that point in time. Virtually all flow indicators and recorders are equipped with totalizers, though sometimes as separate units (Figure 12.19).

Large quantities of water (or liquid chemical) are commonly read out in units of hundreds or thousands of gallons (liters). On the face of a totalizer you may find a multiplier such as

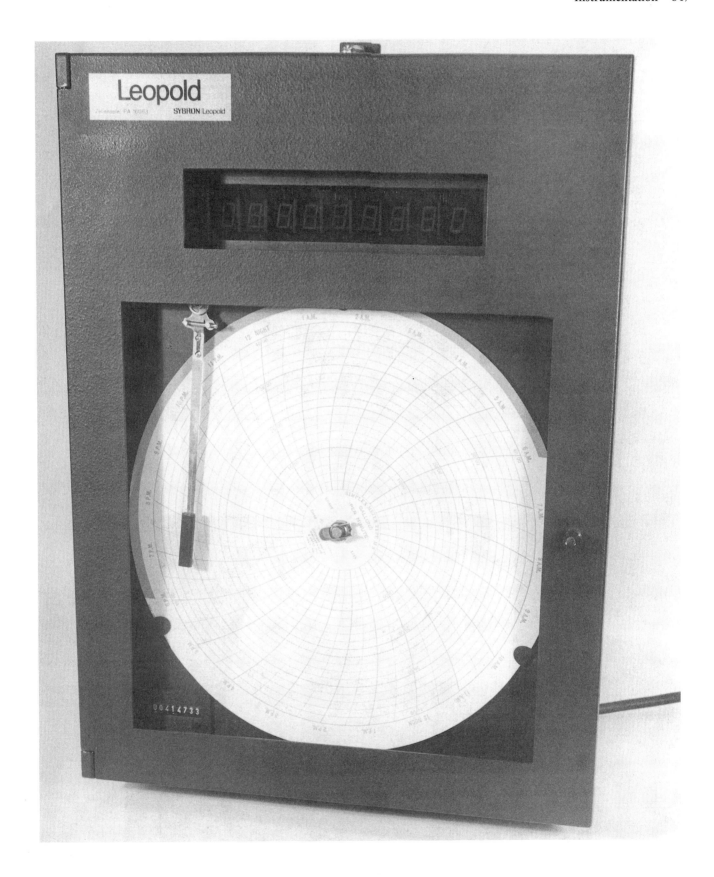

Fig. 12.19   Digital indicator/recorder combination (24-hour circular chart)
(Permission of Leopold Company, Division of Sybron Corporation)

648   Treatment Plants

Strip-chart recorders

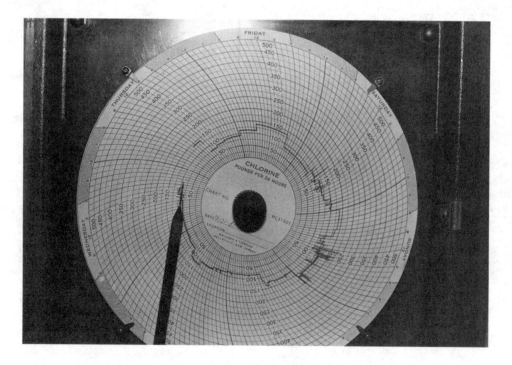

Seven-day circular-chart recorder

Fig. 12.20   Recorders, strip-chart and circular-chart

× 100 or × 1,000. This indicates that the reading is to be multiplied by this (or another) factor to yield the full amount of gallons or cubic feet. If the readout uses a large unit, such as mil gal, a decimal will appear between appropriate numbers on the display, or a fractional multiplier (× 0.001, for example) may appear on the face of the totalizer.

Every operator should personally be able to calculate total flow for a given time period in order to verify that the integrator is actually producing the correct value. Accuracy to one or two parts in a hundred (1 or 2 percent) is usually acceptable in a totalizer. There are methods to integrate (add up) the area under the flow-rate curve on a recorder chart, to check for long-term accuracy of total flow calculations, but it is cumbersome and rarely necessary.

12.314 Alarms

Alarms are visual or audible signals that a variable is out of bounds, or that a condition exists in the plant requiring the operator's attention. For noncritical conditions, the glow of a small lamp or LED (light-emitting diode) on (or outside of) an indicating instrument is sufficient notice. For more important variables or conditions, an attention-getting annunciator panel (Figure 12.21) with flashing lights and an unmistakable and penetrating alarm horn is commonly used.

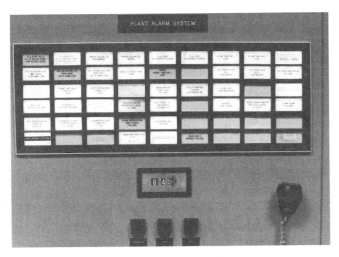

*Fig. 12.21 Annunciator (alarm) panel
(each rectangle represents a monitored location)*

Annunciator panels should all have "acknowledge" and "reset" features to allow the operator to squelch the alarm sound (leaving the visible indication alone), and then to reset the system after the alarm condition is corrected. Annunciator panels should also have a "test" button so that an operator can confirm that no alarm lamps have burned out. The alarm contacts (switches) activating the system commonly are within an indicator or recorder on the main control panel, or are wired in from remote alarm sensors in the plant or telemetered in from an off-site location. The operator is usually responsible for setting these alarm contacts and must use judgment as to the actual limits of the particular variables that will ensure meeting proper operational goals. Each system is different so no attempt will be made here to instruct operating personnel in alarm resetting procedures.

Sometimes operators fail to reset alarm limits as conditions and judgments change in the plant. It is not uncommon to see a plant's annunciator panel lit up, with the operator ignoring all the alarm conditions as the normal status quo. Such practice is not advised because a true alarm condition requiring immediate operator attention may be lost in the resulting general indifference to the alarm system. For some operators, acknowledging an alarm sound to get rid of the noise is second nature, without due attention to each and every activating condition. All alarm contact limits should be reset (or deactivated) as necessary to ensure that the operator is as attentive to the alarm system as necessary to handle real emergencies. The system design should include provisions for disconnecting an annunciator window when the associated instrument or device is out of service for long periods of time.

QUESTIONS

Write your answers in a notebook and then compare your answers with those on page 663.

12.3E What is the purpose of recorders?

12.3F What are the two general types of charts used on recorders?

12.3G How are charts driven in remote locations where electricity is not available?

12.3H List the two kinds of warnings that are produced by alarms.

12.32 Automatic Controllers

Section 12.0 explains the nature of control systems as they are used in industrial wastewater treatment plant operations. Indications of proper and improper control need to be recognized by operators, but actual adjustment of the controller is often left to a qualified instrument technician. By shifting to the manual mode, the operator can bypass the operation of any controller, whether electric or pneumatic. Learn how to shift all your controllers to manual operation. This will allow you to take over control of a critical system when necessary in an emergency, as well as at any other time it suits your purposes. For example, you may be able to quickly correct a cycling or sluggish variable by using manual control rather than waiting for the controller to correct the condition in time (if it ever does).

A controller, while seemingly superhuman in some of its abilities, is still limited. It can only do what it has been programmed to do. You, as the operator, can exercise judgment based on your experience and observations, so do not hesitate to intervene if a controller is not exercising control within sensible limits. Of course, you must be sure of your conclusions and competent to take over control if you decide to operate manually.

To repeat a few of the more important operational control considerations, remember that ON/OFF control is quite different in operation from proportional control. Both methods can exercise close control of a variable; however, proportional control is better suited for close control. Attempting to set up an ON/OFF control system to maintain a variable within too close a tolerance may result in rapid ON/OFF operation of equipment. Such operation can damage both the equipment and the switching devices. Therefore, do not attempt to set a level controller, for instance, to cycle the pump or valve more often than actually necessary for plant operations.

In the case of a modulating controller, it too may begin to cycle its final control element (pump or valve) through a wide range if any of the internal settings, namely proportional band or reset, are adjusted so as to attempt closer control of the variable than is reasonable. Accordingly, it may be better to accommodate to a small offset (difference between set point and control point) than risk an upset in control by attempting too close control.

12.33 Pump Controllers

Control of pumping systems can be achieved, as we have seen, by an ON/OFF type of controller starting and stopping pumps according to a level, pressure, or flow measurement.

Usually, an ON/OFF pump control system responds to level changes in a tank of some type. Water level can be sensed directly with a float or by a pressure change at the tank or pump site. The pump is thus turned off or on as the tank level rises above or falls below predetermined level or pressure limits. Control is rather simple in this case.

However, such systems may include several extra features to ensure fail-safe operation. To prevent the pump from running after a loss of level signal, electrical circuitry should be designed so the pump will turn OFF on an open signal circuit and ON only with a closed circuit. (Ideally, the controller would be able to distinguish between an open or closed remote level/pressure contact and an open or shorted signal line.) Larger pump systems also will often have a low-pressure cutoff switch on the suction side to prevent the pump from running when no water is available, such as with an empty tank or closed suction valve.

Controllers may also protect against overheating a pump (as happens when continuing to pump against a closed discharge valve) by a high-pressure (or low-flow) cutoff switch on the discharge piping. Both the high- and low-pressure switches should shut off a pump through a time delay circuit so that short-term pressure surges (dips and spikes) in the pump's piping can be tolerated. Ideally the low- or high-pressure switches also key alarms to notify the operator of the condition. For remote stations, a plant's main panel may include indicator lights to show the pumps' operating conditions. Figure 12.22 shows a simplified diagram of pump control circuitry.

Pump control panels (Figure 12.23) may also include automatic or manual alternators (two pumps) or sequencers (more than two pumps). This provision allows the total pump operating time required for the particular system to be distributed equally among all the pumps at a pump station. A manual switch for a two-pump station, for example, may read "1-2" in one position and "2-1" in the other position. In the first position, pump #1 is the lead pump (which runs most of the time) and #2 the lag pump (which runs less). When the operator changes the switch to "2-1," the lead-lag order of pump operation is reversed, as it should be periodically, to keep the running time (as read on the elapsed time meters, or as estimated) of both pumps to about the same number of total hours. In a station with multiple pumping units, an automatic alternator or sequencer regularly changes the order of the pumps' start-up to maintain similar operating times for all pumps. In order to protect a pump's electric motor from overheating, level controls should be set so that the pump starts no more often than six times per hour (average one start every 10 minutes).

QUESTIONS

Write your answers in a notebook and then compare your answers with those on page 663.

12.3I Under what conditions might an operator decide to bypass a modulating-type controller? How could this be done?

12.3J What basic principle should guide you in setting up the frequency of ON/OFF operation of a piece of equipment?

12.3K How can pumps be prevented from running upon loss of a remote control signal?

12.3L How can you ensure that both pumps in a duplex (two) pump station operate for roughly equal lengths of time?

12.34 Air Supply Systems

Pneumatic instrumentation depends upon a constant source of clean, dry, pressurized air for reliable operation. Given a quality air supply, pneumatic devices can operate for long periods without significant problems. Without a quality air supply, operational problems can be frequent. The operator of a plant is usually assigned the task of ensuring that the instrument air is always available and dry, so operators must learn how to accomplish this; it cannot be assumed that clean air is there automatically.

Instrumentation 651

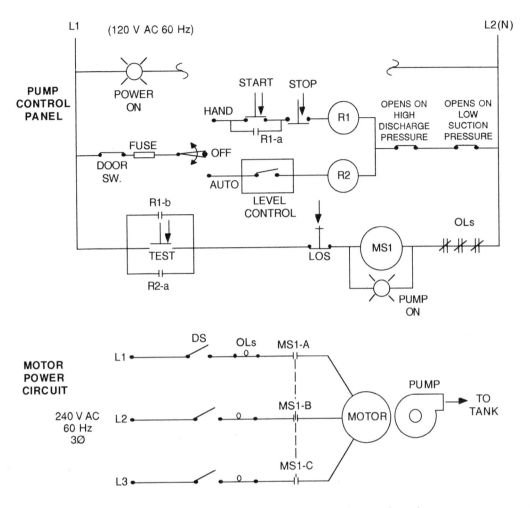

Refer to Figure 12.4 (page 626), and page 627 for parts legend.

Fig. 12.22   Pump control station ladder diagram (ON/OFF control) (simplified schematic)

Fig. 12.23   Photo of pump control station

The plant's instrument air supply system consists of a compressor with its own controls, master air pressure regulator, air filter, and air dryer, as well as the individual pressure regulator/filters in the line at each pneumatic plant instrument (Figure 12.24). Only the instrument air is filtered and dried; the plant air usually does not require such measures since it is being used only for other purposes.

As air passes through a compressor, it not only can pick up some oil but the air's moisture content is concentrated by the compression process. Special measures must be taken to remove both of the liquids. Oil is removed by filtering the air through special oil-absorbent elements. A process called *DESICCATION*[2] can be used to remove the water. This is simply a matter of either passing the moisture-laden air through desiccant columns, which regenerate their moisture-absorption capacity periodically through heating, or by refrigerating the instrument air. The refrigeration method is based upon the principle that cold air can hold comparatively little moisture within it. You must recognize that the capacity of any of these systems of oil or water removal is limited to amounts of liquid encountered under normal conditions.

If the compressor is worn so as to pass more oil than usual, the oil separation process may permit troublesome amounts of oil to pass into the air supply. If the air source contains excessive humidity (due to a rainy day, perhaps, or a wet compressor room), the air drying system may not handle the excess moisture. Learn enough about the instrument air system to be able to open the drain valves, cycle the desiccator, or even bypass the tank, in order to prevent instrumentation problems due to an oily, moisture-saturated air supply.

Operators should regularly crack the regulator/filter drain valves at each of the plant's process instruments. An unusual quantity of liquid drainage may indicate an overloading or failure in the instrument air filter/drying parts. Also, pneumatic indicators/recorders should be watched for erratic pointer/pen movements, which usually are indicative of air quality problems.

The seriousness of plant power or compressor failures can be lessened if you temporarily turn off all nonessential usages of compressed air in the plant. The air storage tank is usually sized so that there is enough air on hand to last for several hours, if conserved. Knowing this, you may be able to wait out a power failure without undue drastic action by conserving the remaining available pressure of the air supply.

QUESTIONS

Write your answers in a notebook and then compare your answers with those on page 663.

12.3M What are the essential qualities of the air supply needed for reliable operation of pneumatic pressure instrumentation systems?

12.3N How are moisture and oil removed from instrument air?

12.35 Laboratory Instruments

This category of instrumentation includes those analytical units typically found in industrial wastewater treatment plant laboratories (Figure 12.25). Some examples are turbidimeters, colorimeters and comparators, pH, conductivity (TDS), and dissolved oxygen (DO) meters. We have already seen that process models of each of the units monitor these same variables out in the plant. The models used in the laboratory are usually referred to as "bench" models rather than "process" instrumentation.

Operators often are required to make periodic readings from lab instruments, and periodic standardizing of particular instruments is often required before determinations are made. Preventive, and certainly corrective, maintenance is handled by the lab staff, factory rep, or instrument technician since each unit can be quite complex. Some of these countertop instruments or devices are very delicate and replacement parts, such as glass vessels or pH electrodes, are quite expensive. Moreover, the use of some of these instruments requires the regular handling of laboratory glassware and other breakable items. The operator who, through carelessness, lack of knowledge, or simple hurrying, consistently breaks glassware or "finds the darn meter broken again" does not become popular with the chemist, supervisor, or other operators. The byword in the lab is *WORK WITH CAUTION*. Protect valuable and essential instrumentation and supplies.

12.36 Test and Calibration Equipment

Plant measuring systems must be periodically calibrated to ensure accurate measurements. In most larger industrial wastewater treatment facility operations, the plant operating staff has little occasion to use testing and calibration devices on the plant instrumentation systems. A trained technician will usually be responsible for using such equipment. There are, however, some general considerations the operator should understand concerning the testing and calibration of plant measuring and control systems. With this basic knowledge, you may be able to discuss needed repairs or adjustments with an instrument technician and perhaps assist with that work. A better understanding of your plant's instrument systems may also enable you to analyze the effects of instrument problems on continued plant operation, and to handle emergency situations created by instrument failure. Your skills in instrument testing and calibration may even eventually result in a job promotion or pay raise.

The most useful piece of general electrical test equipment is the V-O-M, that is the Volt-Ohm-Milliammeter, commonly called the multimeter (Figure 12.26). To use this instrument, you will need a basic understanding of electricity, but once you learn to use it, the V-O-M has potential for universal usage in instrument and general electrical work. Local colleges and other educational institutions may offer courses in basic electricity, which undoubtedly include practice with a V-O-M. You, as a professional plant operator, are unlikely to find technical training of greater practical value than this type of course or program. Your future use of test and calibration equipment, in

---

[2] *Desiccation* (dess-uh-KAY-shun). A process used to thoroughly dry air; to remove virtually all moisture from air.

Instrumentation 653

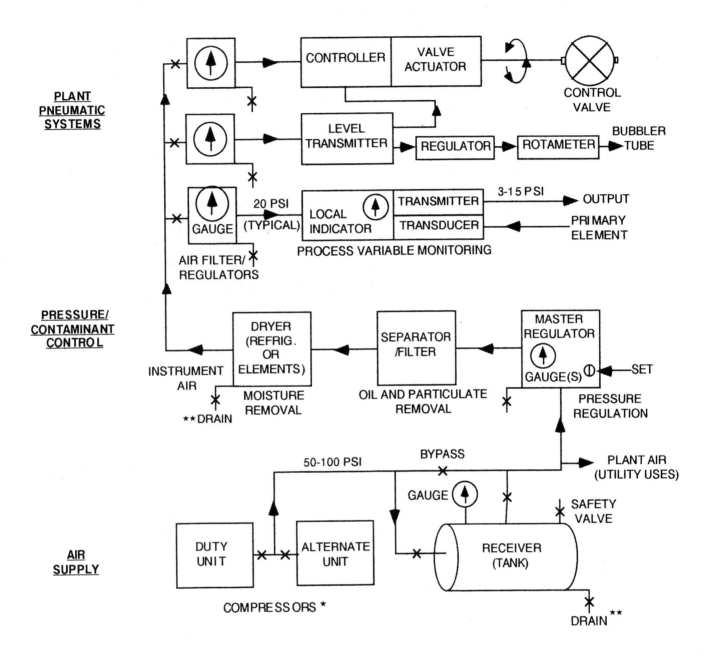

Fig. 12.24 *Typical plant instrument air system functional diagram (simplified, not all valves and piping shown)*

Fig. 12.25   Industrial waste treatment plant laboratory

SPECIFICATIONS

Voltage Ranges
0–199.9/750 V AC   15 kV AC
0–1.999/19.99/199.9/1,000 V DC   15 kV DC
0–1,999 mV DC

Resistance Ranges
0–199.9/1,999 ohms
0–19.99/199.9/1,999 K ohms

Current Ranges
0–1,999 µA DC
0–19.99/199.9/1,999 mA DC
0–10 Amps DC
0–19.99/199.9/1,999 mA AC
0–10 Amps AC
NOTE: AC accuracy may be affected by outside interference.

Accuracy
DC V: ± 0.5% of rdg ± 2LSD
AC V: ± 1.5% of rdg ± 2LSD
DC Amps: All ranges ± 1.0% of rdg ± 2LSD except 10 Amp range, which is ± 1.5% of rdg ± 3LSD.
AC Amps: All ranges ± 1.5% of rdg ± 2LSD except 10 Amp range, which is ± 2.0% of rdg ± 3LSD.
Ohms: All ranges ± 0.75% of rdg ± 2LSD except 2 megohm range, which is ± 1% of rdg ± 2LSD.
15 kV AC/DC high voltage probe: add up to ± 2% of rdg.

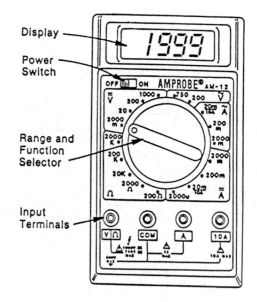

Fig. 12.26   Digital multimeter (V-O-M)

general, certainly should be preceded by instruction in the fundamentals of electricity.

QUESTIONS

Write your answers in a notebook and then compare your answers with those on page 663.

12.3O Why should an operator be especially careful when working in a laboratory?

12.3P Why should an operator become familiar with the testing and calibration of plant measuring and control systems?

12.3Q What is a V-O-M?

12.37 Computerization

12.370 Basic Principles of Computerization

There are about as many special definitions of computers as there are general applications; that is, business concerns, science, industry, and private users, so each would describe what a computer is—or more to the point, what it does—in its own terms. Actually, a broad technical definition of computers serves little purpose to most people using them. In other words, to make effective use of computers requires little, if any, detailed knowledge of just how they work. This fact is just as true for most of today's sophisticated technology, including most modern machines. For our purposes here, computers are devices we use to collect, store, display, and process data of all kinds. So, instead of recorder charts, logs of tabulated data, and other records on paper for files, a computer keeps everything in its electronic memory available for instant display on a screen when required.

The term "processing," as used above, means that computers can organize data, plot data, and perform calculations on the data they have collected. In fact, this is one of the computer's most useful features in industry; in this application they are often called micro (meaning very small) processors, and are extensively used in industrial plants. In effect, we can tell a computer exactly what we want it to do with certain process data, so the microprocessor not only collects the information but also processes it and then exercises some control over the plant process based on its calculations. In this sense, computers behave like the controllers (both motor control circuitry and true controllers) previously discussed, but they do the job with far more versatility and better reliability than electromechanical or pneumatic controllers. In addition, most computers are much smaller in size and require less power than conventional controllers.

The best way for an unfamiliar operator to get comfortable with plant computers is to work with them regularly, always keeping a receptive mind as to their advantages. Although computers are unlike the older mechanical instrumentation in that one cannot tweak this and that in case the device does not work, these black boxes are inherently less susceptible to malfunctions so do not require the preventive and corrective maintenance of mechanical equipment. It is safe to say that once operators get used to the computerization of their plant operations, none would return to the chart changing, manual data gathering, hand calculating, and files searching of traditional operations. Once even the most reluctant operator learns how to plot a graph and "hard copy" a critical process variable on the plant's desk computer to show how well his or her shift controlled the process, that operator will be hooked.

12.371 Video Display of Data

Process variables and other plant data can be organized and displayed on the operator-station computer(s) in different formats. Whatever arrangement of operating data is most useful to plant operations should be, and usually is, programmed into the unit. Often a spreadsheet depiction is set up, similar to a Plant Operations Log, whereby all relevant process data are shown on one or more screens. A screen is a particular display on a computer's monitor. (Screen also means the physical window on the front of the monitor where the picture is displayed.) As the data are gathered by the computer itself, through remote instrumentation, or hand-entered into its database, the display can be changed so as to continually represent the plant's operations. This ongoing updating is sometimes referred to as the computer operating in real time. In contrast, some systems can only indicate what happened well after process changes occur—much as is the case in older plants where the operator has to check several instruments to determine what the process was doing some minutes ago (often too late).

Screens can be scrolled up or down to continue the spreadsheet or to find the new one needed, just as a paper scroll could be rolled up or down to show the portion of interest. To enter (or delete) a particular item of data, the operator moves the cursor (the short blinking line) to the proper location on the video data display and enters in the correct numeral or letter. Cursors are moved by certain keyboard strokes or by moving a mouse to quickly pinpoint items or areas on a screen for special attention.

As mentioned, a very useful function of the plant operations computer is to plot data on a graph against time or another variable to show process trends as conditions change in the plant; for example, chemical feeds or equipment speeds. Such trend plots can be used in the same way as strip charts to facilitate troubleshooting of instrumentation and process problems. Also, the computer can very quickly do repetitive calculations for ongoing comparisons of many process and other variables, so mathematical relationships can instantly be determined; for example, which aeration blower combination is most cost-efficient. Instead of the hand-calculated results requiring hours or days as in the past, the computer can produce such results in a matter of minutes.

This discussion has covered only a few of the most common aspects of computerization. Each industrial wastewater treatment facility will have its own computer program and methods. With a little practice you will become proficient on your own plant's unit, as you must be in the current operations of an industrial wastewater treatment facility.

12.372 Process Control Using Microprocessors

The microprocessors used in industrial wastewater treatment facilities were previously described as controllers (functioning in the same way as pump control circuits) or modulating devices (functioning as true process controllers). Formerly, pump control circuits were conventional electromechanical relay logic panels with dozens of relays, timers, and other devices, each one having an inherent potential for developing electromechanical problems. Now, a tiny computer chip duplicates the necessary control logic exactly. In this application, the microprocessor is usually called a Programmable Logic Controller or PLC. (Sometimes the abbreviation PC is used but, unfortunately, PC is also the popular abbreviation for Personal Computer.) PLCs are very common devices produced under many brand names. They can be programmed in the field for the simplest or the most complex motor control operations, or "logic," as this is called, and can be retrofitted fairly easily to replace existing pump control panels. PLCs can be readily reprogrammed to modify their internal logic (unlike hard-wired relay panels) for such purposes as adding functions (for example, another motor overload protection feature) or to change time-delay settings. Also, a spare PLC (or just its replaceable logic component) can be plugged in to replace a malfunctioning unit to keep plant equipment online (this should only be done by someone trained for the task).

The microprocessor used as a true proportional controller also finds wide use, but usually as original plant equipment rather than a retrofit. For all practical purposes, it functions just as the older model electronic (versus pneumatic) controllers do, except that typically its setting ranges are wider.

Computerized controllers can be preprogrammed to precisely tailor their control modes to optimize a particular process. Some even have the ability to learn, that is, their memory can store the results of past control strategies and modify control in anticipation of process problems—just as an experienced operator does. Again, however, only fully qualified individuals should make any but the simplest controller adjustments, such as changing a set point and possibly the proportional-band setting. An unqualified person could easily damage the equipment and possibly upset the plant processes.

QUESTIONS

Write your answers in a notebook and then compare your answers with those on page 664.

12.3R What is meant by a computer "processing" data?

12.3S How is a mouse used with a computer?

12.3T What is a Programmable Logic Controller?

12.4 OPERATION AND PREVENTIVE MAINTENANCE

12.40 Proper Care of Instruments

Usually, instrumentation systems are remarkably reliable year after year, assuming proper application, set-up, operation, and maintenance. Reliable measurement systems, though outdated by today's standards, are still found in regular service at some plants up to 50 years after installation. To a certain extent, good design and application account for such long service life, but most important is the careful operation and regular maintenance of the instruments' components. The key to proper operation and maintenance (O & M) is the operator's practical understanding of the system. Operators must know how to (1) recognize malfunctioning instruments so as to prevent prolonged damaging operation, (2) shut down and prepare devices for seasonal or other long-term nonoperation, and (3) perform preventive (and minor corrective) maintenance tasks to ensure proper operation in the long term. A sensitive instrumentation system can be ruined in short order with neglect in any one of these three areas.

Operators should be familiar with the Technical Manual (also called the Instruction Book or Operating Manual) of each piece of equipment and instrument encountered in a plant. Each manual will have a section devoted to the operation of a certain component of a complete measuring or control system (although frequently not for the entire system). Detailed descriptions of maintenance tasks and operating checks will usually be found in the manual. Depending on the general type of instrument (electromechanical, pneumatic, or electronic), the suggested frequency of the operation and maintenance/checking tasks can range from none to monthly. Accordingly, this section of the course only describes those common and general tasks an operator might be expected to perform to operate and maintain instrumentation systems. These general tasks can be summed up from the operations standpoint as learning and constant attention to what constitutes normal function, and from a maintenance standpoint, ensuring proper and continuing protection and care of each component.

QUESTIONS

Write your answers in a notebook and then compare your answers with those on page 664.

12.4A List the three areas of operator responsibility that are the keys to proper instrument O & M.

12.4B What generally is expected of an operator of instrumentation systems from (1) an operations standpoint, and (2) a maintenance standpoint?

12.41 Indications of Proper Function

The usual pattern of day-to-day operation of every measuring and control system in a plant should become so familiar to operators that they almost unconsciously sense any significant change. This will be especially evident and true for systems with recorders where the pen trace is visible. An operator should thus watch indicators, recorders, and controllers for their characteristic actions. With analog instruments, each pen or pointer may display its own characteristics (although some may be virtually the same). Thus, the pen for "Flow Recorder A" may normally scribe out a one-eighth-inch (3-mm) wide track due to inherent sensitivity to flow variations, whereas "Level Recorder B" may normally produce a trace as steady as a rock. However, if the flow A pen is noticed as steady one day, or the level B indication widens, then the operator should suspect a problem (Figure 12.27). In this regard, signs of possible improper function (though not *necessarily* so) include:

1. Very flat or steady pen trace (is the system working at all, or is the variable really that constant?).

2. Excessive pen/pointer quiver (causes undue wear on parts, can usually be adjusted out).

3. Constant or periodic hunting, or spikes, in a pen/pointer (improper adjustment, control, or other problem).

Additionally, it is not that uncommon for a pointer or pen to become stuck at some position on its scale, often at the extreme limits of its movement. Pens are particularly prone to sticking or getting hung up on the chart edge or in a tear or hole. Therefore, operators should become observant not only of unusual pointer/pen movement, but unusual lack of movement by indicators and recorders. In the case of recorders, you may lightly tap an instrument to check on the pointer/pen motion. If a gentle tap does not cause a very slight jiggle, a problem may well exist. Anyone hitting or shaking a delicate instrument hard, however, in an attempt to check it out, only reveals a lack of good sense or training in this area.

At times, firmly pushing an instrument into its case, or closing the door completely, may close the interlock switch and switch the system on, as designed. However, jamming the device into its case, or slamming a door, is never considered proper action. If a device still does not begin to work, check the power connection and instrument fuse, if any.

For pneumatic systems, an unnoticed failure of the instrument air supply is the most common reason for an inoperable instrument. Such a failure of the air supply extends the inoperable situation to all pneumatic systems in the plant. Complete functional loss of a single pneumatic instrument is rather rare, but erratic operation is not uncommon (due to previously mentioned water or oil in the air supply).

One of the surest indications of a serious electrical problem in instrument or power circuits is, of course, smoke or a burning odor. Such signs of a problem should *never* be ignored. Smoke/odor means heat, and no device can operate long at unduly high temperatures. Any electrical equipment that begins to show signs of excessive heating must be shut down *immediately*, regardless of how critical it is to plant operation. Overheated equipment will very likely fail very soon anyway, with the damage being aggravated by continued usage. Fuses and circuit breakers do not always de-energize circuits before damage occurs, so cannot be relied upon to do so.

Finally, operators frequently forget to reset an individual alarm, either after an actual occurrence or after a system test. This is especially prevalent when an annunciator panel is allowed to operate day after day with lit-up alarm indicators (contrary to good practice) and one light (or more) is not easily noticeable. Also, when a plant operator must be away from the main duty station, the system may be set so the audible part of the alarm system is temporarily squelched. When the operator returns, the audible system may inadvertently not be reactivated. In both instances (individual or collective loss of audible alarm), the consequences of such inattention can be serious. Therefore, develop the habit of checking your annunciator system often.

QUESTIONS

Write your answers in a notebook and then compare your answers with those on page 664.

12.4C List three possible signs of an improperly functioning flow recorder.

12.4D Where or how are recording pens most likely to become stuck or hung up?

12.4E What is a common reason for nonoperation of pneumatic systems?

12.4F What are the immediate indications of a serious problem in an electrical instrument or power circuit?

12.42 Start-Up/Shutdown Considerations

The start-up and periodic or prolonged shutdown of instrumentation equipment usually requires very little extra work by the operator. Start-up is limited mainly to undoing or reversing the shutdown measures taken.

When shutting down any pressure, flow, or level measuring system, valve off the liquid to the measuring element. Exercise particular care, as explained previously, regarding the order in which the valves are manipulated for any flow tube installation. Also, the power source of some instruments may be shut off,

PROCESS MEASURING/MONITORING SYSTEMS

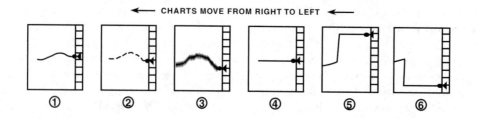

1. Normal function: Ink trace dark and steady; variable within expected range.
2. Pen skips: Pen dirty, dry, or not on chart; clean pen/tubing, re-ink, check contact with chart paper.
3. Wide trace: "Noisy" system, too sensitive; causes inking problems, can possibly be adjusted out.
4. Flat trace (upscale): OK if usual for system; otherwise check sensor or process.
5. Trace to max. scale: Instrumentation problem (sudden or constant 100% variable unlikely).
6. Trace to zero or min. scale: Process down, or sensor problem; also may be signal loss.

PROCESS MODULATING CONTROL SYSTEMS

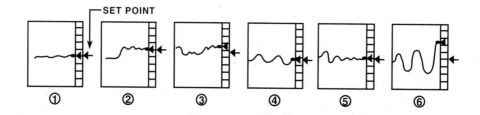

1. Normal control: Pen trace steady, process or set point changes controlled well.
2. Normal control: Small oscillations normal with process or set point change.
3. Abnormal control: Excessive departure from set point, "sloppy" control.
4. Cycling or hunting: Unstable control, controller settings need adjustment.
5. Damped oscillations: Process upset, control OK if acceptable for process.
6. Worsening oscillations: System out of control due to process or set point change, service required. Do not use "auto," switch to "manual" control.

Fig. 12.27 Indications of proper and abnormal function (systems with strip-chart recorders; circular-chart indications are similar)

unless the judgment is made that keeping an instrument case warm (and thus dry) is in order. Constantly moving parts, such as chart drives, should be turned off. With an electrical panel room containing instrumentation, it is good practice to leave some power components on (such as a power transformer) to provide space heat for moisture control. In a known moist environment, sealed instrument cases may be protected for a while with a container of *DESICCANT*[3] (indicating silica gel, which is blue if OK and pink when the moisture-absorbent capacity is exhausted).

Although preventing the access of insects and rodents into any area is difficult, general cleanliness seems to help considerably. Rodenticides are available to control mice; this is good preventive maintenance practice in any electrical space. Mice will chew off wire and transformer insulation, and may urinate on other insulator material, leading to serious problems for equipment (not to mention the rodent).

Nest-building activities of some birds can also be a problem. Screening some buildings and equipment against entry by birds has become a design practice of necessity. Insects and spiders are not generally known to cause specific functional problems, but start-up and operation of systems invaded by ants, bees, or spiders should await cleanup of each such component of the system. All of these pests can bite or sting, so take care.

With pneumatic instrumentation, it is desirable to purge each device with dry air before shutdown. This measure helps rid the individual parts of residual oil and moisture to minimize internal sticking and corrosion while standing idle. As before, periodic blow-off of air receivers and filters keeps these liquids out of the instruments to a large degree. Before shutdown, however, extra attention should be paid to instrument air quality for purging. Before start-up, each filter/receiver should again be purged.

Finally, pay attention to the pens and chart drives of recorders upon shutdown. Ink containers (capsules) may be removed if deemed necessary, and chart drives turned off. A dry pen bearing against one track (such as zero) of a chart for weeks on end is an invitation to start-up problems. Re-inking and chart replacement at start-up are easy if the proper shutdown procedures were followed.

QUESTIONS

Write your answers in a notebook and then compare your answers with those on page 664.

12.4G How can moisture be controlled in an instrument?

12.4H Why should pneumatic instrumentation be purged with dry air before shutdown?

12.43 Preventive Maintenance

Preventive maintenance (PM) means that attention is given periodically to equipment in order to prevent future malfunctions. Corrective maintenance involves actual, significant repairs, which are beyond the scope of this chapter and, in most cases, are not the responsibility of the operator. Routine operational checks are part of all PM programs in that a potential problem may be discovered and thereby corrected before it becomes serious.

PM duties for instrumentation should be included in the plant's general PM program. If your plant has no formal, routine PM program, it should have. Such a program must be set up on paper (or on a computer program). That is, the regular duties required are printed on forms or cards (or appear on a computer screen) that the operator (or technician) uses as a reminder, guide, and record of PM tasks performed. Without such explicit measures, experience shows that preventive maintenance will almost surely be put off indefinitely. Eventually, the press of critical corrective maintenance (often due to lack of preventive maintenance) and even equipment replacement projects may well eliminate forever any hope of a regular PM program. The fact that instrumentation is usually very reliable (being of quality design) may keep it running long after pumps and other equipment have failed. Nevertheless, instrumentation does require proper attention periodically to maximize its effective life. PM tasks and checks on modern instrument systems are quite minimal (even virtually nonexistent on some), so there are no valid reasons for ever failing to perform these tasks.

The technical manual for each item of instrumentation in your plant should be available so you can refer to it for O & M purposes. When a manual cannot be located, contact the manufacturer of the unit. Be sure to give all relevant serial/model numbers in your request for the manual. Request two manuals, one to use and one to put in reserve. All equipment manuals should be kept in one protected location, and signed out as needed. Become familiar with the sections of these manuals related to O & M, and follow their procedures and recommendations closely.

---

[3] *Desiccant* (DESS-uh-kant).   A drying agent that is capable of removing or absorbing moisture from the atmosphere in a small enclosure.

A good practice is to have on hand any supplies and spare parts that are or may be necessary for instrument operation (such as charts) or service (such as pens, pen cleaners, and ink). Some technical manuals contain a list of recommended spare parts that you could use as a guide. Try to obtain these supplies/parts for your equipment. A new pen on hand for a critical recorder can be a lifesaver, at times.

Since PM measures can be so diverse for different types, brands, and ages of instrumentation, only the few general considerations applicable to all will be covered in this section.

1. Protect all instrumentation from moisture (except as needed by design), vibration, mechanical shock, vandalism (a very real problem in the field), and unauthorized access.

2. Keep instrument components clean on the outside, and closed/sealed against inside contamination (for example, spider webs and rodent wastes).

3. *DO NOT* presume to lubricate, tweak, fix, calibrate, free up, or modify any component of a system arbitrarily. If you are not qualified to take any of these measures, then do not do it.

4. *DO* keep recorder pens and charts functioning as designed by frequent checking and service, bleed pneumatic systems regularly, as instructed, ensure continuity of power for electrical devices, and do not neglect routine analytical instrument cleanings and standardizing duties as required by your plant's established procedures.

As a final note, it is a good idea to get to know and cooperate fully with your plant's instrument service person. Good communication between this person and the operating staff can only result in a better all-around operation. If your agency is too small to staff such a specialist, it may be a good idea to enter into an instrumentation service contract with an established company or possibly even with the manufacturer of the majority of the components. With rare exceptions, general maintenance persons (even journeyman electricians) are *not* qualified to perform extensive maintenance on today's instrumentation. Be sure that someone takes good care of your instruments and they will take good care of you.

QUESTIONS

Write your answers in a notebook and then compare your answers with those on page 664.

12.4I Why should regular preventive maintenance duties be printed on forms or cards?

12.4J How can the technical manual for an instrument be obtained if the only copy in a plant is lost?

12.4K What instrument supplies and spare parts should always be available at your plant?

12.44 Operational Checks

Operational checks are most efficiently made by always observing each system for its continuing signs of normal operation. However, some measuring systems may be cycled within their range of action as a check on the responsiveness of components. For instance, if a pressure-sensing system indicates only one pressure for months on end, and some doubt arises as to whether it is working or not, the operator may bleed off a little pressure at the primary element to produce a small fluctuation. Or, if a flow has appeared constant for an overly long period, the bypass valve in the DP (Differential Pressure) cell piping[4] may be cracked open briefly to cause a drop in reading. Be sure you crack the *bypass* valve, not one of the others on the piping. If you open the wrong valve, the pressure may be excessive and be beyond the range of the DP cell, which could cause problems. A float suspected of being stuck (very constant level indication) may be freed by jiggling its cable, or other measures could be taken to cause a slight fluctuation in the reading.

Whenever an operator or a technician disturbs normal operation during checking or for any reason, plant process operating personnel must be informed—ideally prior to the disturbance. If a recorder trace is altered from its usual pattern in the process, the person causing the upset should initial the chart appropriately and note the time. Some plants, incidentally, require operators to mark or date each chart at midnight (or noon) of each day for easy reference and filing.

In the case where a pen/pointer is thought to be stuck mechanically, that is, it does not respond at all to simulated or actual change in the measured variable, it is normally permissible to open an instrument's case and try to move the pointer/pen, but only to the minimum extent necessary to free it. Further deflection may well bend or break the device's linkage. A dead pen often is due only to loss of power or air to the readout mechanism. Any hard or repeated striking of an instrument to make it work identifies the striker as ignorant of good operational practice and can ruin the equipment. Insertion of tools into an instrument case in a random fix-it attempt can easily damage an instrument. Generally speaking, any extensive operating check of instrumentation should be performed by the instrument technician during routine PM program activities.

QUESTIONS

Write your answers in a notebook and then compare your answers with those on page 664.

12.4L How are operational checks performed on instrumentation equipment?

12.4M What should be done if a recorder trace is altered from its usual pattern during the process of checking an instrument?

---

[4] There is no similar easy way to check a propeller meter's response.

12.5 ADDITIONAL READING

1. *INSTRUMENTATION—HANDBOOK FOR WATER AND WASTEWATER TREATMENT PLANTS*, Robert G. Skrentner. Obtain from CRC Press LLC, Attn: Order Entry, 6000 Broken Sound Parkway, NW, Suite 300, Boca Raton, FL 33487. ISBN 0873711262. Order No. L126. Price, $179.95, includes shipping and handling.

2. *INSTRUMENTATION IN WASTEWATER TREATMENT FACILITIES* (MOP 21). Obtain from Water Environment Federation (WEF), Publications Order Department, 601 Wythe Street, Alexandria, VA 22314-1994. Order No. MO2021. Price to members, $35.00; nonmembers, $55.00; plus shipping and handling.

3. *HANDS-ON WATER AND WASTEWATER EQUIPMENT MAINTENANCE*. Obtain from American Public Works Association (APWA) Bookstore, PO Box 802296, Kansas City, MO 64180-2296. Order No. PB.XHOW. Price to members, $110.00; nonmembers, $115.00; plus shipping and handling.

4. *OPERATION OF MUNICIPAL WASTEWATER TREATMENT PLANTS* (MOP 11). Obtain from Water Environment Federation (WEF), Publications Order Department, 601 Wythe Street, Alexandria, VA 22314-1994. Order No. M05110. Price to members, $120.00; nonmembers, $148.00; plus shipping and handling.

5. *INSTRUMENTATION AND CONTROL* (M2). Obtain from American Water Works Association (AWWA), Bookstore, 6666 West Quincy Avenue, Denver, CO 80235. Order No. 30002. Price to members, $84.00; nonmembers, $126.00; price includes cost of shipping and handling.

END OF LESSON 2 OF 2 LESSONS

on

INSTRUMENTATION

Please answer the discussion and review questions next.

DISCUSSION AND REVIEW QUESTIONS

Chapter 12. INSTRUMENTATION

(Lesson 2 of 2 Lessons)

Write the answers to these questions in your notebook. The question numbering continues from Lesson 1.

8. What are the advantages and limitations of analog versus digital indicators?

9. Why is it poor practice to ignore the lamps that are lit up (alarm conditions) on an annunciator panel?

10. How should the constantly lit up lamps (alarm conditions) on an annunciator panel be handled?

11. What electrical control features are available to protect pumps from damage?

12. What problems are created by oil and moisture in instrument air, and how can these contaminants be removed?

13. Why should plant measuring systems be periodically calibrated?

14. What could cause erratic operation of pneumatic instruments?

15. Why should insects and rodents be kept out of instruments?

16. How could you tell if a float for a level recorder might be stuck, and how would you determine if it was actually stuck?

SUGGESTED ANSWERS
Chapter 12. INSTRUMENTATION

ANSWERS TO QUESTIONS IN LESSON 1

Answers to questions on page 622.

12.0A An industrial wastewater treatment plant operator must have a good working familiarity with measurement instrumentation and associated control systems to properly monitor and control treatment processes.

12.0B This chapter contains information about instrument standardization and preventive maintenance measures.

12.0C Measurement instruments can serve as an extension of and improvement to an operator's senses of vision, touch, hearing, and even smell.

Answers to questions on page 623.

12.0D Our senses provide us with qualitative information but instrumentation measurements provide us with exact quantitative data.

12.0E The definition of a measurement is the comparison of a quantity with a standard unit of measure.

12.0F *ACCURACY* refers to how closely an instrument measures the true or actual value of a process variable, while *PRECISION* refers to how closely the instrument can reproduce a given reading (or measurement) time after time.

12.0G *ANALOG* instruments yield a reading as a pointer against a marked scale and *DIGITAL* instruments provide direct numerical readouts.

Answers to questions on page 628.

12.0H Control systems are the means by which such process variables as pressure, level, weight, or flow are controlled.

12.0I Examples of modulating control systems include: (1) chlorine residual analyzers/controllers; (2) flow-paced (open-loop) chemical feeders; (3) pressure- or flow-regulating valves; (4) continuous level control of process basins; and (5) variable-speed pumping systems for flow/level control.

12.0J A motor control station or panel and the related circuitry essentially provide ON/OFF operation of an electric motor or other device.

Answers to questions on page 629.

12.1A The general principles for safe performance on the job are to *ALWAYS* avoid unsafe acts and correct unsafe conditions immediately.

12.1B Electric shock can cause serious burns and even death (by asphyxiation due to paralysis of the muscles used in breathing).

12.1C An electrical "explosion" could shower you with molten metal, startle you into a bumped head or elbow, or cause a bad fall.

12.1D If electric current flows through your upper body, electric shock could harm your heart or your head.

Answers to questions on page 630.

12.1E Operators should be especially careful when working around powered mechanical equipment because the equipment could start unexpectedly and cause serious injury.

12.1F The purpose of an electrical lockout device is to positively prevent the operation of an electric circuit, or to de-energize the circuit temporarily.

12.1G Wearing thin rubber or plastic gloves can greatly reduce your chances of electric shock.

Answers to questions on page 632.

12.2A A sensor or primary element directly measures a variable. The sensor is often a transducer of some type that converts energy of one kind into some other form to produce a readout or signal.

12.2B Pressure is measured by the movement of a flexible element or a mechanically deformable device subjected to the force of the pressure being measured.

12.2C Some pressure sensors are fitted with surge and over-range protection to limit the effect of pressure spikes or water hammer on the device.

Answers to questions on page 636.

12.2D Different types of liquid level sensing devices include floats, displacers, electrical probes, direct hydrostatic pressure, pneumatic bubbler tubes, and ultrasonic (sonar) devices.

12.2E A continuous level signal can be generated by a float element by attaching the float to a steel tape or cable that is wrapped around a drum or pulley. The level sensed is transmitted as a signal (electrical) proportional to the rotation (position) of the pulley or drum.

12.2F Probes can be used instead of mechanical systems to measure liquid levels in sealed or pressurized tanks or with chemically active liquids.

12.2G A bubbler measures the level of a liquid by sensing the air pressure necessary to cause bubbles to just flow out the end of a submerged tube.

Answers to questions on page 641.

12.2H The two basic types of flow readings are (1) rate of flow, and (2) total flow (volume units).

12.2I The inferential techniques that are used in flow measurement are (1) velocity, (2) differential pressure, (3) magnetic, and (4) ultrasonic flow sensing.

12.2J Velocity-sensing devices measure flows by sensing the rate of rotation of a special impeller placed within the flowing stream of wastewater.

12.2K Flows measured with Venturi meters take advantage of a basic principle of hydraulics, the Bernoulli Effect: When a liquid is forced to go faster in a pipe or channel, its internal pressure drops.

Answers to questions on page 642.

12.2L In a fluid (liquid or gas) chemical feeder, the indication of quantity per unit of time, or liquid chemical feed rate, may be provided by use of a rotameter.

12.2M Process variables commonly monitored or controlled by process instrumentation include DO, pH, chlorine residuals, electrical conductivity, ORP, alkalinity, turbidity, and temperature.

12.2N In order to transmit a measured value to a remote location for readout, it is necessary to generate a signal directly proportional to the value measured. This signal is then transmitted to a remote receiver, which provides a reading based upon the signal.

12.2O Two general systems are used for transmission of measurement signals: electrical and pneumatic.

ANSWERS TO QUESTIONS IN LESSON 2

Answers to questions on page 646.

12.3A The purpose of instrumentation indicators is to give a visual presentation of a variable's present value, either as an analog or as a digital display.

12.3B An analog display uses some type of pointer (or other indicator) against a scale.

12.3C Recorders are usually found in a central location at an industrial wastewater treatment plant.

12.3D Factors that can cause electronic instrument problems include temporary power failures, tripped panel circuit breakers, and voltage surges, resulting in blown fuses, static electricity, and excessive heat.

Answers to questions on page 649.

12.3E Recorders are indicators designed to show (and produce a permanent record of) how the value of the variable has changed with time.

12.3F Recorder charts may be circular or strip types.

12.3G In remote locations where no electricity is available, charts are driven by hand-wound drives or batteries.

12.3H Alarms may produce visual or audible signals.

Answers to questions on page 650.

12.3I An operator might bypass a modulating-type controller in an emergency or when, in the judgment of the operator, the controller is not exercising control within sensible limits. To bypass a controller, switch to the manual mode of operation.

12.3J ON/OFF controls should be set to operate or cycle associated equipment on and off no more often than actually necessary for plant operation.

12.3K Pumps can be prevented from running upon loss of a remote control signal by electrical circuitry designed so the pump will turn *OFF* on an *OPEN* signal circuit and *ON* only with a *CLOSED* circuit.

12.3L Pumps in a duplex pump station can be operated for similar periods of time by the use of manual or automatic sequencers that periodically switch different pumps to the lead pump position and the other(s) to the lag position.

Answers to questions on page 652.

12.3M Pneumatic pressure instrumentation systems must have a constant source of clean, dry, pressurized air for reliable operation.

12.3N Oil is removed by filtration through special oil-absorbent elements, and a dryer/desiccator or refrigeration is used to remove moisture from instrument air.

Answers to questions on page 655.

12.3O Care must be exercised when working in the laboratory so as not to break sensitive instruments, delicate equipment, or fragile glassware.

12.3P Operators should become familiar with the testing and calibration of plant measuring and control systems in order to assist instrument technicians, and to better understand the plant's instrumentation systems. Also, development of skills in instrumentation testing and calibration may result in a job promotion or pay raise.

12.3Q V-O-M stands for Volt-Ohm-Milliammeter, commonly referred to as a multimeter.

Answers to questions on page 656.

12.3R "Processing" data means that computers can organize data, plot data, and perform calculations.

12.3S To enter (or delete) a particular item of data, the operator uses a mouse to move the computer's cursor to the proper location on the video data display and enters the correct numeral or letter.

12.3T A Programmable Logic Controller (PLC) is a microprocessor that can be programmed in the field for the simplest or the most complex motor control operations, or "logic," as this is called, and can be retrofitted fairly easily to replace existing pump control panels.

Answers to questions on page 656.

12.4A The three areas of operator responsibility that are the keys to proper instrument O & M are:

1. Recognizing malfunctioning instruments so as to prevent prolonged damaging operation.
2. Shutting down and preparing devices for seasonal or other long-term nonoperation.
3. Performing preventive (and minor corrective) maintenance tasks to ensure proper operation in the long term.

12.4B General tasks expected of operators of instrumentation systems can be summed up (1) from an operations standpoint, as learning and constant attention to what constitutes normal function, and (2) from a maintenance standpoint, as ensuring proper and continuing protection and care of each component.

Answers to questions on page 657.

12.4C Three signs that a flow recorder may not be functioning properly are:

1. Very flat or steady pen trace (is the system working at all, or is the variable really that constant?).
2. Excessive pen/pointer quiver (causes undue wear on parts, can usually be adjusted out).
3. Constant or periodic hunting, or spikes, in a pen/pointer (improper adjustment, control, or other problem).

12.4D Recording pens are most likely to become stuck or hung up on the chart edge or in a tear or hole.

12.4E A common reason for nonoperation of a pneumatic system is the failure of the instrument air supply caused by water or oil.

12.4F An indication of a serious problem in an electrical instrument or power circuit is the presence of smoke or a burning odor.

Answers to questions on page 659.

12.4G Moisture can be controlled in instruments by a space-heat source (such as a power transformer) and by inserting a container of desiccant into the instrument case.

12.4H Pneumatic instrumentation should be purged with dry air before shutdown to rid the individual parts of residual oil and moisture and to minimize internal sticking and corrosion.

Answers to questions on page 660.

12.4I Regular preventive maintenance duties should be printed on forms or cards (or appear on computer screens) for use by operators as a reminder, guide, and record of preventive maintenance tasks.

12.4J To obtain a technical manual for an instrument, write to the manufacturer. Be sure to provide all relevant serial/model numbers in your request to the manufacturer for a manual.

12.4K Instrument supplies and spare parts that should always be available include charts, pens, pen cleaners, and ink, and any other parts listed in the technical manual as necessary for instrument operation or service.

Answers to questions on page 660.

12.4L Operational checks on instrumentation equipment are performed by always observing each system for its continuing signs of normal operation, and cycling some indicators by certain test methods.

12.4M If a recorder trace is altered from its usual pattern during the process of checking an instrument, the operator causing the upset should initial the chart appropriately, with the time noted.

CHAPTER 13

SAFETY

by

Robert Reed

Revised by

Russ Armstrong

TABLE OF CONTENTS
Chapter 13. SAFETY

	Page
OBJECTIVES	670
WORDS	671

LESSON 1

13.0	WHY SAFETY?	675
13.1	TYPES OF HAZARDS	675
	13.10 Physical Injuries	676
	13.11 Infections and Infectious Diseases	676
	13.12 Confined Spaces	676
	13.13 Oxygen Deficiency or Enrichment	681
	13.14 Toxic or Suffocating Gases or Vapors	681
	13.15 Toxic and Harmful Chemicals	681
	13.16 Radiological Hazards	682
	13.17 Explosive Gas Mixtures	682
	13.18 Fires	684
	13.19 Other Hazards	684
	13.190 Electric Shock/Stored Energy	684
	13.191 Noise	685
	13.192 Dusts, Fumes, Mists, Gases, and Vapors	688
13.2	SPECIFIC HAZARDS	689
	13.20 Collection Systems	689
	13.200 Traffic Hazards	689
	13.201 Manholes	689
	13.202 Excavations	696
	13.203 Sewer Cleaning	700
	13.204 Acknowledgment	700
DISCUSSION AND REVIEW QUESTIONS		700

LESSON 2

	13.21 Pumping Stations	701
	13.22 Treatment Plants	701
	13.220 Headworks	701
	13.221 Grit Channels	703

		13.222	Clarifiers or Sedimentation Basins	704
		13.223	Digesters and Digestion Equipment	704
		13.224	Trickling Filters	706
		13.225	Aerators	707
		13.226	Ponds	707
		13.227	Chemical Treatment	708
			13.2270 Chlorine and Sulfur Dioxide	708
			13.2271 Polymers	709
		13.228	Applying Protective Coatings	709
		13.229	Housekeeping	710
	13.23	Industrial Waste Treatment		710
		13.230	Fuels	710
		13.231	Toxic Gases	711
		13.232	Amines	711
		13.233	Surface-Active Agents	711
		13.234	Biocides	711
		13.235	High or Low pH	712
		13.236	Summary	712

DISCUSSION AND REVIEW QUESTIONS ... 712

LESSON 3

13.3	SAFETY IN THE LABORATORY		713
	13.30	Sampling Techniques	713
	13.31	Equipment Use and Testing Procedures	713
13.4	FIRE PREVENTION		715
	13.40	Ingredients Necessary for a Fire	715
	13.41	Fire Control Methods	715
	13.42	Fire Prevention Practices	716
13.5	WATER SUPPLIES		716
13.6	SAFETY EQUIPMENT AND INFORMATION		717
13.7	"TAILGATE" SAFETY MEETINGS		718
13.8	HOW TO DEVELOP SAFETY TRAINING PROGRAMS		718
	13.80	Conditions for an Effective Safety Program	718
	13.81	Start at the Top	718
	13.82	The Supervisor's Role	719
	13.83	Plan for Emergencies	719
	13.84	Promote Safety	719
	13.85	Hold Safety Drills and Train for Safety	719
	13.86	Purchase the Obvious Safety Equipment First	720

	13.87 Safety Is Important for Everyone	720
	13.88 Necessary Paperwork	720
	13.89 Summary	720
13.9	HAZARD COMMUNICATION (WORKER RIGHT-TO-KNOW LAWS)	721
13.10	SAFETY SUMMARY	729
13.11	ADDITIONAL READING	730
DISCUSSION AND REVIEW QUESTIONS		731
SUGGESTED ANSWERS		731
APPENDIX: REGULATORY INFORMATION		735

OBJECTIVES
Chapter 13. SAFETY

Following completion of Chapter 13, you should be able to:

1. Identify the types of hazards you may encounter operating an industrial wastewater treatment facility.
2. Recognize unsafe conditions and correct them whenever they develop.
3. Organize regular "tailgate" safety meetings.
4. Develop the habit of always "thinking safety and working safely."

NOTE: Special safety information is given in other chapters because of the importance of safety considerations at all times.

S SAFETY FIRST
A ACCIDENTS COST LIVES
F FASTER IS NOT ALWAYS BETTER
E EXPECT THE UNEXPECTED
T THINK BEFORE YOU ACT
Y YOU CAN MAKE THE DIFFERENCE

ACCIDENTS DO NOT JUST HAPPEN... THEY ARE CAUSED!

WORDS
Chapter 13. SAFETY

ACUTE HEALTH EFFECT

An adverse effect on a human or animal body, with symptoms developing rapidly.

AIR GAP

An open, vertical drop, or vertical empty space, between a drinking (potable) water supply and potentially contaminated water. This gap prevents the contamination of drinking water by backsiphonage because there is no way potentially contaminated water can reach the drinking water supply.

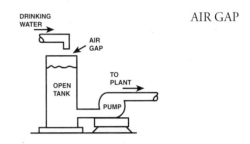

CFR

Code of Federal Regulations. A publication of the US government that contains all of the proposed and finalized federal regulations, including safety and environmental regulations.

CARCINOGEN (kar-SIN-o-jen)

Any substance that tends to produce cancer in an organism.

CHRONIC HEALTH EFFECT

An adverse effect on a human or animal body with symptoms that develop slowly over a long period of time or that recur frequently.

COMPETENT PERSON

A competent person is defined by OSHA as a person capable of identifying existing and predictable hazards in the surroundings, or working conditions that are unsanitary, hazardous, or dangerous to employees, and who has authorization to take prompt corrective measures to eliminate the hazards.

CONFINED SPACE

Confined space means a space that:

(1) Is large enough and so configured that an employee can bodily enter and perform assigned work; and

(2) Has limited or restricted means for entry or exit (for example, manholes, tanks, vessels, silos, storage bins, hoppers, vaults, and pits are spaces that may have limited means of entry); and

(3) Is not designed for continuous employee occupancy.

Also see DANGEROUS AIR CONTAMINATION and OXYGEN DEFICIENCY.

CONFINED SPACE, NON-PERMIT

A non-permit confined space is a confined space that does not contain or, with respect to atmospheric hazards, have the potential to contain any hazard capable of causing death or serious physical harm.

CONFINED SPACE, PERMIT-REQUIRED (PERMIT SPACE)

A confined space that has one or more of the following characteristics:

(1) Contains or has a potential to contain a hazardous atmosphere

(2) Contains a material that has the potential for engulfing an entrant

(3) Has an internal configuration such that an entrant could be trapped or asphyxiated by inwardly converging walls or by a floor that slopes downward and tapers to a smaller cross section

(4) Contains any other recognized serious safety or health hazard

DANGEROUS AIR CONTAMINATION

An atmosphere presenting a threat of causing death, injury, acute illness, or disablement due to the presence of flammable and/or explosive, toxic, or otherwise injurious or incapacitating substances.

(1) Dangerous air contamination due to the flammability of a gas, vapor, or mist is defined as an atmosphere containing the gas, vapor, or mist at a concentration greater than 10 percent of its lower explosive (lower flammable) limit (LEL).

(2) Dangerous air contamination due to a combustible particulate is defined as a concentration that meets or exceeds the particulate's lower explosive limit (LEL).

(3) Dangerous air contamination due to the toxicity of a substance is defined as the atmospheric concentration that could result in employee exposure in excess of the substance's permissible exposure limit (PEL).

NOTE: A dangerous situation also occurs when the oxygen level is less than 19.5 percent by volume (OXYGEN DEFICIENCY) or more than 23.5 percent by volume (OXYGEN ENRICHMENT).

DECIBEL (DES-uh-bull)

A unit for expressing the relative intensity of sounds on a scale from zero for the average least perceptible sound to about 130 for the average level at which sound causes pain to humans. Abbreviated dB.

ENGULFMENT

Engulfment means the surrounding and effective capture of a person by a liquid or finely divided (flowable) solid substance that can be aspirated to cause death by filling or plugging the respiratory system or that can exert enough force on the body to cause death by strangulation, constriction, or crushing.

ENTRAIN

To trap bubbles in water either mechanically through turbulence or chemically through a reaction.

FLAME POLISHED

Melted by a flame to smooth out irregularities. Sharp or broken edges of glass (such as the end of a glass tube) are rotated in a flame until the edge melts slightly and becomes smooth.

HYDRAULIC JUMP

The sudden and usually turbulent abrupt rise in water surface in an open channel when water flowing at high velocity is suddenly retarded to a slow velocity.

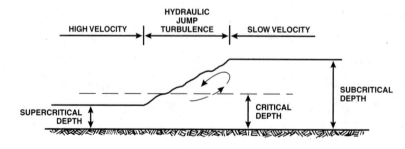

IDLH

Immediately Dangerous to Life or Health. The atmospheric concentration of any toxic, corrosive, or asphyxiant substance that poses an immediate threat to life or would cause irreversible or delayed adverse health effects or would interfere with an individual's ability to escape from a dangerous atmosphere.

LOWER EXPLOSIVE LIMIT (LEL)

The lowest concentration of a gas or vapor (percent by volume in air) that explodes if an ignition source is present at ambient temperature. At temperatures above 250°F (121°C) the LEL decreases because explosibility increases with higher temperature.

MATERIAL SAFETY DATA SHEET (MSDS)

A document that provides pertinent information and a profile of a particular hazardous substance or mixture. An MSDS is normally developed by the manufacturer or formulator of the hazardous substance or mixture. The MSDS is required to be made available to employees and operators or inspectors whenever there is the likelihood of the hazardous substance or mixture being introduced into the workplace. Some manufacturers are preparing MSDSs for products that are not considered to be hazardous to show that the product or substance is not hazardous.

NONSPARKING TOOLS

These tools will not produce a spark during use. They are made of a nonferrous material, usually a copper-beryllium alloy.

OSHA (O-shuh)

The Williams-Steiger Occupational Safety and Health Act of 1970 (OSHA) is a federal law designed to protect the health and safety of workers, including the operators of water supply and treatment systems and wastewater collection and treatment systems. The Act regulates the design, construction, operation, and maintenance of water and wastewater systems. OSHA regulations require employers to obtain and make available to workers the Material Safety Data Sheets (MSDSs) for chemicals used at industrial facilities and treatment plants. OSHA also refers to the federal and state agencies that administer the OSHA regulations.

OLFACTORY (all-FAK-tore-ee) FATIGUE

A condition in which a person's nose, after exposure to certain odors, is no longer able to detect the odor.

OXYGEN DEFICIENCY

An atmosphere containing oxygen at a concentration of less than 19.5 percent by volume.

OXYGEN ENRICHMENT

An atmosphere containing oxygen at a concentration of more than 23.5 percent by volume.

SET POINT

The position at which the control or controller is set. This is the same as the desired value of the process variable. For example, a thermostat is set to maintain a desired temperature.

SEWER GAS

(1) Gas in collection lines (sewers) that results from the decomposition of organic matter in the wastewater. When testing for gases found in sewers, test for oxygen deficiency, oxygen enrichment, and also for explosive and toxic gases.

(2) Any gas present in the wastewater collection system, even though it is from such sources as gas mains, gasoline, and cleaning fluid.

SPECIFIC GRAVITY

(1) Weight of a particle, substance, or chemical solution in relation to the weight of an equal volume of water. Water has a specific gravity of 1.000 at 4°C (39°F). Particulates with specific gravity less than 1.0 float to the surface and particulates with specific gravity greater than 1.0 sink.

(2) Weight of a particular gas in relation to the weight of an equal volume of air at the same temperature and pressure (air has a specific gravity of 1.0). Chlorine gas has a specific gravity of 2.5.

SPOIL

Excavated material, such as soil, from the trench of a water main or sewer.

SURFACE-ACTIVE AGENT

The active agent in detergents that possesses a high cleaning ability. Also called a SURFACTANT.

TAILGATE SAFETY MEETING

Brief (10 to 20 minutes) safety meetings held every 7 to 10 working days. The term comes from the safety meetings regularly held by the construction industry around the tailgate of a truck.

TIME-WEIGHTED AVERAGE (TWA)

The average concentration of a pollutant based on the times and levels of concentrations of the pollutant. The time-weighted average is equal to the sum of the portion of each time period (as a decimal, such as 0.25 hour) multiplied by the pollutant concentration during the time period divided by the hours in the workday (usually 8 hours). 8TWA PEL is the time-weighted average permissible exposure limit, in parts per million, for a normal 8-hour workday and a 40-hour workweek to which nearly all workers may be repeatedly exposed, day after day, without adverse effect.

VISCOSITY (vis-KOSS-uh-tee)

A property of water, or any other fluid, that resists efforts to change its shape or flow. Syrup is more viscous (has a higher viscosity) than water. The viscosity of water increases significantly as temperatures decrease. Motor oil is rated by how thick (viscous) it is; 20 weight oil is considered relatively thin while 50 weight oil is relatively thick or viscous.

CHAPTER 13. SAFETY

(Lesson 1 of 3 Lessons)

13.0 WHY SAFETY?

A cat may have nine lives, but you have only one. Protect it. Others may try, but only your efforts in thinking about safety and acting safely can ensure you the opportunity of continuing to live your one life.

You are working at an occupation that has an accident frequency rate second only to that of the mining industry. This is not a very desirable record. This chapter is intended to make you aware of the many hazards that may be encountered at your plant. Guidelines for working safely are provided. Precise requirements for your plant and work may vary depending on the specific design of your plant and the equipment and processes being used. Site-specific safety procedures should be confirmed with your local safety regulatory agency.

Your employer has the responsibility of providing you with a safe place to work. But you, the operator, who has overall responsibility for your treatment facility, must accept the task of seeing to it that your facility is maintained in such a manner as to continually provide a safe place to work. This can only be done by constantly "thinking safety and working safely."

You have the responsibility of protecting yourself and other facility personnel or visitors by establishing safety procedures for your facility and then by seeing that they are followed. Train yourself to analyze jobs, work areas, and procedures from a safety standpoint. Learn to recognize potentially hazardous actions or conditions. When you do recognize a hazard, take immediate steps to eliminate it by corrective action. Corrective actions can range from temporary isolation of the hazard through the placement of barricades or warning signs to engineering redesign and physical modifications to permanently eliminate the hazard(s). As an individual, you can be held liable for injuries or property damage that result from an unsafe condition, act, or situation that you knew of, or through reasonable diligence, could have known of.

REMEMBER: Accidents do not just happen—they are caused! Behind every accident there was a chain of events that led to an unsafe act, unsafe condition, or a combination of both. THINK SAFETY!

Accidents may be prevented by using good common sense, applying a few basic rules, and, particularly, by acquiring a good knowledge of the hazards unique to your job as a plant operator.

The Bell System had one of the best safety records of any industry. A variation of their successful policy statement is:

"There is no job so important nor emergency so great that we cannot take time to do our work safely."

Although this chapter is intended primarily for the industrial wastewater treatment facility operator, the operators of many industrial plants also have the responsibility of industrial wastewater collection system maintenance. Also, some industrial plant operators have the responsibility of collecting and treating the domestic wastewater from their facility. Therefore, the safety aspects of both collection system maintenance and facility operation will be discussed.

13.1 TYPES OF HAZARDS

You are equally exposed to accidents whether working on the collection system or working in a treatment facility. As an operator, you may be exposed to:

1. Physical injuries
2. Infections and infectious diseases
3. Confined spaces
4. Oxygen deficiency or enrichment
5. Toxic or suffocating gases or vapors
6. Toxic and harmful chemicals
7. Radiological hazards
8. Explosive gas mixtures
9. Fires
10. Electric shock/stored energy
11. Noise
12. Dusts, fumes, mists, gases, and vapors

13.10 Physical Injuries

The most common physical injuries are cuts, bruises, strains, and sprains. Injuries can be caused by many things including moving machinery, improper lifting techniques, or slippery surfaces. Falls from or into tanks, wet wells, catwalks, or conveyors can also be disabling or fatal.

If you work in an area six feet (1.8 meters) or more above a lower level and you are not protected by guardrails or other fall protection, you must use a personal fall arrest system. The fall arrest system may consist of a combination of anchorage points, connectors, a body harness, and a lanyard, deceleration device, or a lifeline. Connectors used in the fall restraint system must be the locking type with a self-closing, self-locking keeper that stays closed and locked until physically unlocked and pushed open for connecting and disconnecting. Lanyards, lifelines, and anchorage points must be rated at 5,000 pounds (2,273 kilograms) breaking and tensile strength. The fall arrest system must be rigged to prevent you from contacting the lower level or free falling more than six feet (1.8 meters). Fall arrest systems must be inspected prior to each use to detect damage, deterioration, or wear that could compromise the user's safety. Never use any suspect equipment. If the system is subjected to impact loading (someone falls), it must be removed from service until a COMPETENT PERSON[1] inspects it and approves it for reuse. Additional information and training on fall arrest systems can be obtained from your local safety regulatory agency and safety equipment vendor.

Working with ladders can also be very hazardous. Do not overreach when on a ladder and make sure that the ladder is positioned properly, is secured, and is appropriate for the job. Portable ladders are classified according to the weight they can sustain and should be labeled to identify the manufacturers' rated capacity. Make sure the ladder you use is rated for your weight and the weight of any tools or equipment it may have to support. Do not use a conductive (metal) ladder when working around electrical equipment.

Most injuries can be avoided by the proper use of ladders, hand tools, and safety equipment, and by following established safety procedures.

13.11 Infections and Infectious Diseases

Although treatment facilities and facility personnel are not expected to be pristine, personal hygiene is the best protection against the risk of infections and infectious diseases such as typhoid fever, dysentery, hepatitis, and tetanus. Immunization shots for protection against tetanus, polio, and hepatitis B are recommended and are often available free or for a minor charge from your local health department. *REMEMBER*, many pathogenic organisms can be found in wastewater. Some diseases that may be transmitted by wastewater are anthrax, tuberculosis, paratyphoid fever, cholera, and polio. Tapeworms and the organisms associated with food poisoning may also be present.

The possibility that Acquired Immune Deficiency Syndrome (AIDS), which is caused by a virus, can be contracted from exposure to raw wastewater has been discounted by researchers who have found that although the AIDS virus is present in the wastes from AIDS victims, the raw wastewater environment is hostile to the virus itself and has not been identified as a mode of transmission to date. Needle sticks from potentially contaminated syringes should remain a concern to operators and maintenance personnel. Fluids in or on syringes may provide a less severe environment than raw wastewater where dilution and chlorination significantly reduce infection potential.

Make it a habit to thoroughly wash your hands before eating or smoking, as well as before and after using the restroom. *ALWAYS* wear proper protective gloves when you may contact wastewater or sludge in any form. Bandages covering wounds should be changed frequently.

Do not wear your work clothes home because diseases may be transmitted to your family. Provisions should be made in your plant for a locker room where each employee has a locker. Work clothes should be placed in lockers and not thrown on the floor. Your work clothes should be cleaned as often as necessary. If you are required to wear protective clothing because of the possibility of contamination with toxic materials, you should store your street clothes and your protective clothing in separate lockers. If your employer does not supply you with uniforms and laundry service, investigate the availability of disposable clothing for "dirty" jobs. If you must take your work clothes home, launder them separately from your regular family wash. All of these precautions will reduce the possibility of you or your family becoming ill because of your contact with wastewater.

What is wrong with the above sketch? *NEVER* stick objects in your mouth that you do not intend to eat.

13.12 Confined Spaces

This section outlines procedures for preventing personal exposure to dangerous air contamination and/or oxygen deficiency/enrichment when working within such spaces as tanks, channels, boilers, sewers, or manholes. If you enter confined spaces, you must develop and implement written, understandable proce-

---

[1] *Competent Person.* A competent person is defined by OSHA as a person capable of identifying existing and predictable hazards in the surroundings, or working conditions that are unsanitary, hazardous, or dangerous to employees, and who has authorization to take prompt corrective measures to eliminate the hazards.

dures in compliance with OSHA standards and you must provide training in the use of these procedures for all persons whose duties may involve confined space entry. *The procedures presented here are intended as guidelines. Exact procedures for work in confined spaces may vary with different agencies and geographical locations and must be confirmed with the appropriate regulatory safety agency.*

A confined space may be defined as any space that: (1) is large enough and so configured that an employee can bodily enter and perform assigned work; and (2) has limited or restricted means for entry or exit (for example, manholes, tanks, vessels, silos, storage bins, hoppers, vaults, and pits are spaces that may have limited means of entry); and (3) is not designed for continuous employee occupancy. One easy way to identify a confined space is by whether or not you can enter it by simply walking while standing fully upright. In general, if you must duck, crawl, climb, or squeeze into the space, it is considered a confined space.

A major concern in confined spaces is whether the existing ventilation is capable of removing DANGEROUS AIR CONTAMINATION[2] and/or oxygen deficiency/enrichment that may exist or develop. In industrial treatment, we are concerned primarily with oxygen deficiency (less than 19.5 percent oxygen by volume), oxygen enrichment (greater than 23.5 percent by volume), methane (explosive), hydrogen sulfide (toxic), and other gases as identified in Table 13.1.

The potential for buildup of toxic or explosive gas mixtures and/or oxygen deficiency/enrichment exists in all confined spaces. The atmosphere must be checked with reliable, calibrated instruments before every entry. When testing the atmosphere, first test for oxygen deficiency/enrichment, then combustible gases and vapors, and then toxic gases and vapors. The oxygen concentration in normal breathing air is 20.9 percent. The atmosphere in the confined space must not fall below 19.5 percent or exceed 23.5 percent oxygen. Engineering controls are required to prevent low or high oxygen levels. However, personal protective equipment is necessary if engineering controls are not possible. In atmospheres where the oxygen content is less than 19.5 percent, supplied air or self-contained breathing apparatus (SCBA) is required. SCBAs are sometimes referred to as scuba gear because they look and work much like the air tanks used by divers.

Entry into confined spaces is never permitted until the space has been properly ventilated using specially designed forced-air ventilators. These blowers force all the existing air out of the space, replacing it with fresh air from outside. This crucial step must *ALWAYS* be taken even if atmospheric monitoring instruments show the atmosphere to be safe. Because some of the gases likely to be encountered in a confined space are combustible or explosive, the blowers must be specially designed so that the blower itself will not create a source of ignition that could cause an explosion.

There are two general classifications of confined spaces: (1) non-permit confined spaces, and (2) permit-required confined spaces (permit spaces).

A *NON-PERMIT CONFINED SPACE* is a confined space that does not contain or, with respect to atmospheric hazards, have the potential to contain any hazard capable of causing death or serious physical harm. The following steps are recommended *PRIOR* to entry into *ANY* confined space:

1. Ensure that all employees involved in confined space work have been effectively trained.

2. Identify and close off or reroute any lines that may carry harmful substance(s) to, or through, the work area.

3. Empty, flush, or purge the space of any harmful substance(s) to the extent possible.

4. Monitor the atmosphere at the work site and within the space to determine if dangerous air contamination and/or oxygen deficiency/enrichment exists.

5. Record the atmospheric test results and keep them at the site throughout the work period.

6. If the space is interconnected with another space, each space must be tested and the most hazardous conditions found must govern subsequent steps for entry into the space.

7. If an atmospheric hazard is noted, use portable blowers to further ventilate the area; retest the atmosphere after a suitable period of time. Do not place the blowers inside the confined space.

8. If the *ONLY* hazard posed by the space is an actual or potential hazardous atmosphere and the preliminary ventilation has eliminated the atmospheric hazard or continuous forced ventilation *ALONE* can maintain the space safe for entry, entry into the area may proceed.

---

[2] *Dangerous Air Contamination.* An atmosphere presenting a threat of causing death, injury, acute illness, or disablement due to the presence of flammable and/or explosive, toxic, or otherwise injurious or incapacitating substances.
   (1) Dangerous air contamination due to the flammability of a gas, vapor, or mist is defined as an atmosphere containing the gas, vapor, or mist at a concentration greater than 10 percent of its lower explosive (lower flammable) limit (LEL).
   (2) Dangerous air contamination due to a combustible particulate is defined as a concentration that meets or exceeds the particulate's lower explosive limit (LEL).
   (3) Dangerous air contamination due to the toxicity of a substance is defined as the atmospheric concentration that could result in employee exposure in excess of the substance's permissible exposure limit (PEL).
   NOTE: A dangerous situation also occurs when the oxygen level is less than 19.5 percent by volume (OXYGEN DEFICIENCY) or more than 23.5 percent by volume (OXYGEN ENRICHMENT).

TABLE 13.1 COMMON DANGEROUS GASES ENCOUNTERED IN WASTEWATER COLLECTION SYSTEMS AND AT WASTEWATER TREATMENT PLANTS[a]

Name of Gas and Chemical Formula	8TWA PEL[b]	Specific Gravity or Vapor Density[c] (Air = 1)	Explosive Range (% by volume in air)		Common Properties (Percentages below are percent in air by volume)	Physiological Effects (Percentages below are percent in air by volume)	Most Common Sources in Sewers	Method of Testing[d]
			Lower Limit	Upper Limit				
Oxygen, O_2 (in Air)		1.11	Not flammable		Colorless, odorless, tasteless, nonpoisonous gas. Supports combustion.	Normal air contains 20.93% of O_2. If O_2 is less than 19.5%, do not enter space without respiratory protection.	Oxygen depletion from poor ventilation and absorption or chemical consumption of available O_2.	Oxygen monitor.
Gasoline Vapor, C_5H_{12} to C_9H_{20}	300	3.0 to 4.0	1.3	7.0	Colorless, odor noticeable in 0.03%. Flammable. Explosive.	Anesthetic effects when inhaled. 2.43% rapidly fatal. 1.1% to 2.2% dangerous for even short exposure.	Leaking storage tanks, discharges from garages, and commercial or home dry-cleaning operations.	Combustible gas monitor.
Carbon Monoxide, CO	50	0.97	12.5	74.2	Colorless, odorless, nonirritating. Tasteless, Flammable. Explosive.	Hemoglobin of blood has strong affinity for gas causing oxygen starvation. 0.2 to 0.25% causes unconsciousness in 30 minutes.	Manufactured fuel gas.	1. CO monitor. 2. CO tubes.
Hydrogen, H_2		0.07	4.0	74.2	Colorless, odorless, tasteless, nonpoisonous, flammable. Explosive. Propagates flame rapidly; very dangerous.	Acts mechanically to deprive tissues of oxygen. Does not support life. A simple asphyxiant.	Manufactured fuel gas.	Combustible gas monitor.
Methane, CH_4		0.55	5.0	15.0	Colorless, tasteless, odorless, nonpoisonous. Flammable. Explosive.	See hydrogen.	Natural gas, marsh gas, manufactured fuel gas, gas found in sewers.	Combustible gas monitor.
Hydrogen Sulfide, H_2S	10	1.19	4.3	46.0	Rotten egg odor in small concentrations, but sense of smell rapidly impaired. Odor not evident at high concentrations. Colorless. Flammable. Explosive. Poisonous.	Death in a few minutes at 0.2%. Paralyzes respiratory center.	Petroleum fumes, from blasting, gas found in sewers.	1. H_2S monitor. 2. H_2S tubes.
Carbon Dioxide, CO_2	5,000	1.53	Not flammable		Colorless, odorless, nonflammable. Not generally present in dangerous amounts unless there is already a deficiency of oxygen.	10% cannot be endured for more than a few minutes. Acts on nerves of respiration.	Issues from carbonaceous strata. Gas found in sewers.	Carbon dioxide monitor.
Nitrogen, N_2		0.97	Not flammable		Colorless, tasteless, odorless. Nonflammable. Nonpoisonous. Principal constituent of air (about 79%).	See hydrogen.	Issues from some rock strata. Gas found in sewers.	Oxygen monitor.
Ethane, C_2H_4		1.05	3.1	15.0	Colorless, tasteless, odorless, nonpoisonous. Flammable. Explosive.	See hydrogen.	Natural gas.	Combustible gas monitor.
Chlorine, Cl_2	0.5	2.5	Not flammable Not explosive		Greenish yellow gas, or amble color liquid under pressure. Highly irritating and penetrating odor. Highly corrosive in presence of moisture.	Respiratory irritant, irritating to eyes and mucous membranes. 30 ppm causes coughing. 40–60 ppm dangerous in 30 minutes. 1,000 ppm apt to be fatal in a few breaths.	Leaking pipe connections. Overdosage.	1. Chlorine monitor. 2. Strong ammonia on swab gives off white fumes.
Sulfur Dioxide, SO_2	2	2.3	Not flammable Not explosive		Colorless compressed liquified gas with a pungent odor. Highly corrosive in presence of moisture.	Respiratory irritant, irritating to eyes, skin, and mucous membranes.	Leaking pipes and connections.	1. Sulfur dioxide monitor. 2. Strong ammonia on swab gives off white fumes.

[a] Originally printed in Water and Sewage Works, August 1953. Adapted from "Manual of Instruction for Sewage Treatment Plant Operators," State of New York.
[b] 8TWA PEL is the Time Weighted Average permissible exposure limit, in parts per million, for a normal 8-hour workday and a 40-hour workweek to which nearly all workers may be repeatedly exposed, day after day, without adverse effect.
[c] Gases with a specific gravity less than 1.0 are lighter than air; those more than 1.0, heavier than air.
[d] The first method given is the preferable testing procedure.

A *PERMIT-REQUIRED CONFINED SPACE* (permit space) is a confined space that has one or more of the following characteristics:

1. Contains or has the potential to contain a hazardous atmosphere
2. Contains a material that has the potential for engulfing an entrant
3. Has an internal configuration such that an entrant could be trapped or asphyxiated by inwardly converging walls or by a floor that slopes downward and tapers to a smaller cross section
4. Contains any other recognized serious safety or health hazard

OSHA regulations require that a confined space entry permit be completed for each permit-required confined space entry (Figure 13.1). The permit must be renewed each time the space is left and re-entered, even if only for a break or lunch, or to go get a tool. The confined space entry permit is "an authorization and approval in writing that specifies the location and type of work to be done, certifies that all existing hazards have been evaluated by a competent person, and that necessary protective measures have been taken to ensure the safety of each worker." A competent person, in this case, is a person designated in writing as capable, either through education or specialized training, of anticipating, recognizing, and evaluating employee exposure to hazardous substances or other unsafe conditions in a confined space. This person is authorized to specify control procedures and protective actions necessary to ensure worker safety.

The following procedures must be observed before entry into a permit-required confined space:

1. Ensure that personnel are effectively trained.
2. If the confined space has both side and top openings, enter through the side opening if it is within 3½ feet (1.1 meters) of the bottom.
3. Wear appropriate, approved, respiratory protective equipment.
4. Ensure that written operating and rescue procedures are at the entry site.
5. Wear an approved harness with an attached line. The free end of the line must be secured outside the entry point.
6. Test for atmospheric hazards as often as necessary to determine that acceptable entry conditions are being maintained.
7. Station at least one person to stand by on the outside of the confined space and at least one additional person within sight or call of the standby person.
8. Maintain effective communication between the standby person and the entry person.
9. The standby person, equipped with appropriate respiratory protection, should only enter the confined space in case of emergency.
10. If the entry is made through a top opening, use a hoisting device with a harness that suspends a person in an upright position. A mechanical device must be available to retrieve personnel from vertical spaces more than five feet (1.5 meters) deep.
11. If the space contains, or is likely to develop, flammable or explosive atmospheric conditions, do not use any tools or equipment (including electrical) that may provide a source of ignition.
12. Wear appropriate protective clothing when entering a confined space that contains corrosive substances or other substances harmful to the skin.
13. At least one person trained in first aid and cardiopulmonary resuscitation (CPR) should be immediately available during any confined space job.

Individuals designated to provide first aid or CPR should be included in a Bloodborne Pathogens (BBP) program. These employees may be exposed to contact with blood or other potentially infectious materials from the performance of their duties. The BBP program includes training in exposure potential determination, engineering and work practice controls, personal protective equipment (PPE), and the availability of the hepatitis B vaccination series (29 CFR 1910.1030). If the operator(s) must enter confined spaces to perform rescue services, they must be trained specifically to perform the assigned rescue duty and to use required personal protective equipment (PPE) and rescue equipment. Rescue practice sessions must be held at least once every 12 months.

If you arrange to have a contractor perform work in confined spaces at your facility or within your collection system, you must inform the contractor:

- That the contractor must comply with confined space regulations.
- Of hazards that you have identified and your experience with the space(s).
- Of precautions or procedures you have implemented for the protection of employees in or near the space where the contractor's personnel will be working.
- That a debriefing must occur at the conclusion of the entry operations regarding the confined space program followed and any hazards encountered or created during the entry operations.

To enhance safety, communications, and coordination of confined space activities, the contractor is also required to obtain available information from you and to inform you of the confined space program the contractor will follow. This exchange of information must occur before the confined space is entered by any operator.

Confined space work can present serious hazards if you are uninformed or untrained. The procedures presented are only guidelines and exact requirements for confined space work for

Confined Space Pre-Entry Checklist/Confined Space Entry Permit

Date and Time Issued: _____ Date and Time Expires: _____ Job Site/Space I.D.: _____

Job Supervisor: _____ Equipment to be worked on: _____ Work to be performed: _____

Standby personnel: _____ _____ _____

1. Atmospheric Checks: Time _____ Oxygen _____ %   Toxic _____ ppm
   Explosive _____ % LEL   Carbon Monoxide _____ ppm

2. Tester's signature: _____

3. Source isolation: (No Entry)   N/A   Yes   No
   Pumps or lines blinded,
     disconnected, or blocked   ( )   ( )   ( )

4. Ventilation Modification: N/A   Yes   No
   Mechanical   ( )   ( )   ( )
   Natural ventilation only   ( )   ( )   ( )

5. Atmospheric check after isolation and ventilation: Time _____
   Oxygen _____ % > 19.5%   < 23.5%   Toxic _____ ppm < 10 ppm H$_2$S
   Explosive _____ % LEL   < 10%   Carbon Monoxide _____ ppm < 35 ppm CO

Tester's signature: _____

6. Communication procedures: _____

7. Rescue procedures: _____
   _____

8. Entry, standby, and backup persons   Yes   No
   Successfully completed required training?   ( )   ( )
   Is training current?   ( )   ( )

9. Equipment: N/A   Yes   No
   Direct reading gas monitor tested   ( )   ( )   ( )
   Safety harnesses and lifelines for entry and standby persons   ( )   ( )   ( )
   Hoisting equipment   ( )   ( )   ( )
   Powered communications   ( )   ( )   ( )
   SCBAs for entry and standby persons   ( )   ( )   ( )
   Protective clothing   ( )   ( )   ( )
   All electric equipment listed for Class I, Division I,
     Groups A, B, C, and D, and nonsparking tools   ( )   ( )   ( )

10. Periodic atmospheric tests:
    Oxygen: ____% Time ____; ____% Time ____; ____% Time ____; ____% Time ____;
    Explosive: ____% Time ____; ____% Time ____; ____% Time ____; ____% Time ____;
    Toxic: ____ppm Time ____; ____ppm Time ____; ____ppm Time ____; ____ppm Time ____;
    Carbon Monoxide: ____ppm Time ____; ____ppm Time ____; ____ppm Time ____; ____ppm Time ____;

We have reviewed the work authorized by this permit and the information contained herein. Written instructions and safety procedures have been received and are understood. Entry cannot be approved if any brackets ( ) are marked in the "No" column. This permit is not valid unless all appropriate items are completed.

Permit Prepared By: (Supervisor) _____ Approved By: (Unit Supervisor) _____

Reviewed By: (CS Operations Personnel) _____
                              (Entrant)   (Attendant)   (Entry Supervisor)

This permit to be kept at job site. Return job site copy to Safety Office following job completion.

Fig. 13.1 Confined space pre-entry checklist/confined space entry permit

your locale may vary. Contact your local regulatory safety agency for specific requirements.

13.13 Oxygen Deficiency or Enrichment

Low oxygen levels may exist in any poorly ventilated, low-lying structure where gases such as hydrogen sulfide, gasoline vapor, carbon dioxide, or chlorine may be produced or may accumulate (see Table 13.1, "Common Dangerous Gases Encountered in Wastewater Collection Systems and at Wastewater Treatment Plants"). Oxygen in a concentration above 23.5 percent (oxygen enrichment) also can be dangerous because it speeds up combustion.

Oxygen deficiency is most likely to occur when structures or channels are installed below grade (ground level). Several gases (including hydrogen sulfide and chlorine) have a tendency to collect in low places because they are heavier than air. The specific gravity of a gas indicates its weight as compared to an equal volume of air. Since air has a specific gravity of exactly 1.0, any gas with a specific gravity greater than 1.0 may sink to low-lying areas and displace the air from that area or structure. (On the other hand, methane may rise out of a manhole because it has a specific gravity of less than 1.0, which means that it is lighter than air.) You should never rely solely on the specific gravity of a gas to tell you where it is. Air movement or temperature differences within a confined space may affect the location of atmospheric hazards. The only effective way of ensuring safe atmospheric conditions prior to entering a confined space is to test the atmosphere with an appropriate monitor(s) at various levels and locations throughout the space.

When oxygen deficiency or enrichment is discovered, the area should be ventilated with fans or blowers and checked again for oxygen deficiency/enrichment before anyone enters the area to work. Ventilation may be provided by fans or blowers. Follow confined space procedures before entering and during occupancy of any suspect area. *ALWAYS* get air into the confined space *BEFORE* you enter to work and maintain the ventilation until you have left the space. Equipment is available to measure oxygen concentration as well as toxic and combustible atmospheric conditions. You must use this equipment whenever you encounter a potential confined space situation. Ask your local safety regulatory agency or wastewater association about sources of this type of equipment in your area.

> NEVER ENTER AN ENCLOSED, POORLY VENTILATED AREA, WHETHER A MANHOLE, SUMP, OR OTHER STRUCTURE WITHOUT FIRST FOLLOWING CONFINED SPACE ENTRY PROCEDURES.

13.14 Toxic or Suffocating Gases or Vapors

The most common toxic gas in industrial waste discharges is hydrogen sulfide, which is produced during the anaerobic decomposition of certain materials containing sulfur compounds. Hydrogen sulfide will tend to accumulate in the lower voids of sewers, tanks, channels, and manholes because it is heavier than air, with a specific gravity of 1.19. The only reliable method of detecting hydrogen sulfide is by atmospheric monitoring since it has the unique ability to affect your sense of smell. You will lose the ability to smell hydrogen sulfide's "rotten egg" odor after only a short exposure. Your loss of ability to smell hydrogen sulfide is known as *OLFACTORY FATIGUE*.[3]

Other toxics, such as carbon monoxide, chlorinated solvents, and industrial toxins, may enter your plant as a result of industrial discharges, accidental spills, or illegal disposal of hazardous materials. You must become familiar with the waste discharges into your system. Table 13.1, "Common Dangerous Gases Encountered in Wastewater Collection Systems and at Wastewater Treatment Plants," contains information on methods of testing for several gases.

13.15 Toxic and Harmful Chemicals

Strong acids, bases, and liquid mercury are examples of toxic and harmful chemicals that operators may encounter working in and around industrial treatment facilities and laboratories. Be very careful when handling and using these chemicals. Chemical hazards may be present in many forms—vapors, dusts, mists, liquids, gases, and particles. The seriousness of a chemical hazard depends on exposure time and the concentration of the chemical, as well as which chemical you are exposed to. To avoid injury, be sure all hazardous chemicals are clearly labeled; obtain and read the health and safety data about the chemical *BEFORE* using the chemical; and learn and practice safe handling procedures and precautions.

A *MATERIAL SAFETY DATA SHEET*[4] (MSDS) (see Section 13.9, Figure 13.16) is your best source of information about hazardous chemicals and can be provided by your chemical supplier. The MSDS will provide at least the following information:

1. Identification of composition, formula, and synonyms—chemicals that are classified as health hazards must be identified if they are present in concentrations at 1.0 percent or greater. Carcinogens (cancer-causing chemicals) must also be identified if they are present at concentrations of 0.1 percent or greater.

2. Physical and chemical properties such as specific gravity.

3. Incompatible substances and decomposition products.

---

[3] *Olfactory* (all-FAK-tore-ee) *Fatigue.* A condition in which a person's nose, after exposure to certain odors, is no longer able to detect the odor.

[4] *Material Safety Data Sheet (MSDS).* A document that provides pertinent information and a profile of a particular hazardous substance or mixture. An MSDS is normally developed by the manufacturer or formulator of the hazardous substance or mixture. The MSDS is required to be made available to employees and operators or inspectors whenever there is the likelihood of the hazardous substance or mixture being introduced into the workplace. Some manufacturers are preparing MSDSs for products that are not considered to be hazardous to show that the product or substance is not hazardous.

4. *ACUTE*[5] and *CHRONIC*[6] health hazards.

5. Environmental impacts.

6. Personal protective measures and engineering/administrative controls.

7. Safe handling, storage, disposal, and cleanup procedures.

No chemical should be received, stored, or handled without essential safety information being provided to those who may come into contact with the substance.

Containers of hazardous chemicals at your plant must be labeled, tagged, or marked to identify the hazardous contents and appropriate hazard warnings (Figure 13.2). Exceptions are that you are not required to label portable containers that you transfer hazardous chemicals into for your personal use during your shift and individual stationary process containers can be identified by using signs or other written materials instead of attaching labels.

Operator exposure to chemical contaminants must be kept to a minimum. OSHA requires that engineering controls be initiated in preference to administrative controls or the use of personal protective equipment to minimize employee exposure. Engineering controls are considered the most effective means of protecting operators. Frequently, a combination of controls, training, safe work practices, and the use of personal protective equipment may be required to adequately reduce the potential for operator exposure.

Employers must develop, implement, and maintain a written hazard communication program describing how requirements for labeling and other forms of warning, material safety data sheets, and employee information and training will be met. The program must also include a list of the hazardous chemicals at your plant.

Training must include (1) methods and observations that can be used to detect the presence or release of a hazardous chemical, (2) the physical and health hazards of the chemicals, (3) measures that operators can take to protect themselves from those hazards, and (4) the details of the plant's specific hazard communication program, including labeling, material safety data sheets, and how employees can obtain and use the hazard information. Additional information concerning the Hazard Communication Standard can be found in Section 13.9.

Remember, do not work with a chemical unless you understand the hazards involved and are using the protective equipment necessary to protect yourself. Contact your local safety regulatory agency about specific chemicals you may deal with if there is any doubt in your mind about safe procedures.

13.16 Radiological Hazards

Creation of another hazard to industrial plant operators is a result of the use of radioactive isotopes in hospitals, research labs, and various industries. Check your sewer service area for the possible use of these materials. The routine handling and disposal of radioactive materials is stringently controlled by the US Nuclear Regulatory Commission (NRC). If you are receiving a discharge that may contain a radioactive substance, the NRC should be contacted. The only legal sources of radioactive wastes to a treatment facility are those specifically licensed by the NRC. Any personnel dealing with radioactive wastes must be monitored for exposure levels by pocket dosimeters or film badges. Special protective clothing must be worn and all work must be done under the direction of qualified staff.

Some plants use radioactive substances in control elements (for example, density elements or level indicators) or in various laboratory equipment. If your plant uses radioactive isotopes in these or other applications, you may be required to comply with NRC regulations and Title 10 of the Code of Federal Regulations, Parts 19 and 20. The requirements include, but are not limited to, licensing, training, recordkeeping, testing source integrity, and performing radiation surveys to determine potential exposure levels for personnel in the area of each source. You may also be required to have a "Radiation Safety Officer" on site. Consult the Title 10 regulations or contact the NRC to confirm the requirements for your radiation sources and to obtain information on how to safely deal with maintenance activities and accidents involving radioactive substances.

13.17 Explosive Gas Mixtures

Explosive gas mixtures may develop in many areas of a treatment facility from mixtures of air and methane, natural gas, manufactured fuel gas, hydrogen, or gasoline vapors. Table 13.1 lists the common dangerous gases that may be encountered in a treatment facility and identifies their explosive range where appropriate. The upper explosive limit (UEL) and lower explosive limit (LEL) indicate the range of concentrations at which combustible or explosive gases will ignite when an ignition source is present at ambient temperature. No explosion or ignition occurs when the concentration is outside these ranges. Gas concentrations below the LEL are too lean to ignite; there is not enough flammable gas or vapor to support combustion. Gas concentrations higher than the UEL are too rich to ignite; there is too much flammable gas or vapor and not enough oxygen to support combustion (Figure 13.3).

Explosive ranges can be measured by using a combustible gas detector calibrated for the gas of concern. Do not rely on your nose to detect gases. The sense of smell is absolutely unreliable for evaluating the presence of dangerous gases. Some gases have no smell and hydrogen sulfide can paralyze the sense of smell.

Avoid explosions by eliminating all sources of ignition in areas potentially capable of developing explosive mixtures. Only explosion-proof electrical equipment and fixtures should be used in these areas (influent/bar screen rooms, gas compressor areas, digesters, battery charging stations). Provide adequate ventilation in all areas that have the potential to develop an explosive atmosphere.

---

[5] *Acute Health Effect.* An adverse effect on a human or animal body, with symptoms developing rapidly.
[6] *Chronic Health Effect.* An adverse effect on a human or animal body with symptoms that develop slowly over a long period of time or that recur frequently.

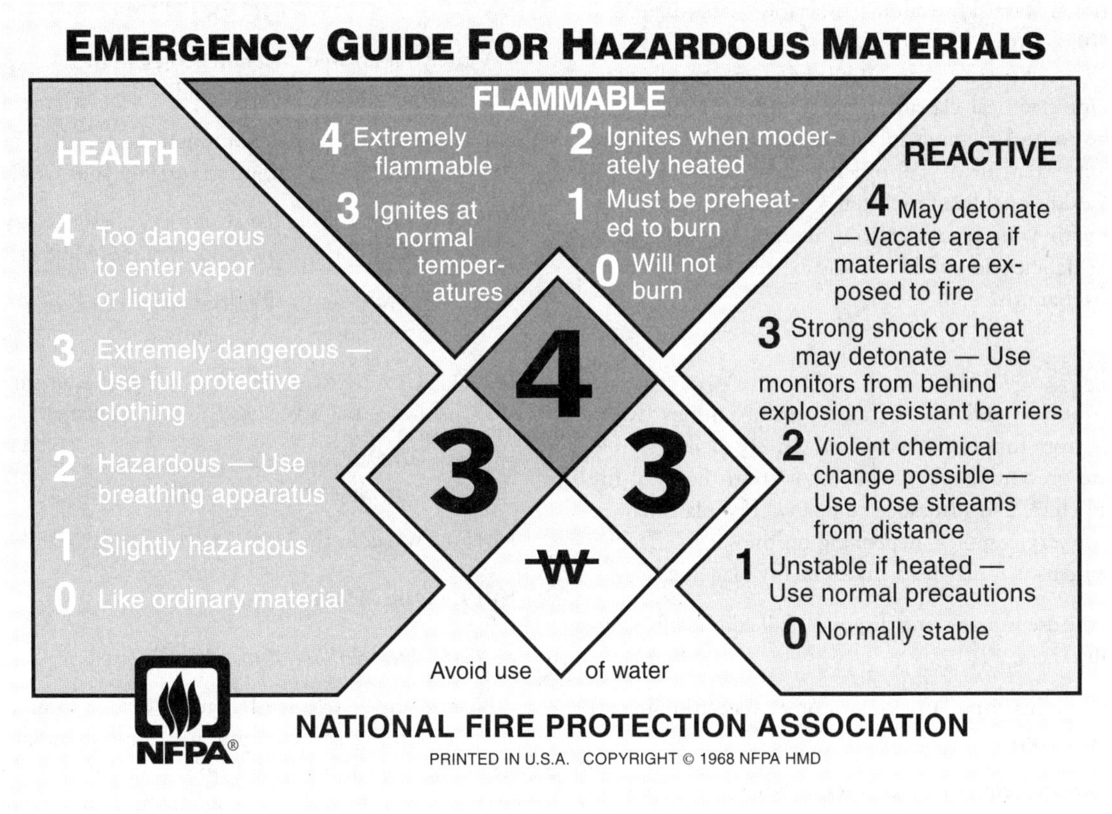

EMERGENCY GUIDE FOR HAZARDOUS MATERIALS
FLAMMABLE
4 Extremely flammable
3 Ignites at normal temperatures
2 Ignites when moderately heated
1 Must be preheated to burn
0 Will not burn
HEALTH
4 Too dangerous to enter vapor or liquid
3 Extremely dangerous — Use full protective clothing
2 Hazardous — Use breathing apparatus
1 Slightly hazardous
0 Like ordinary material
REACTIVE
4 May detonate — Vacate area if materials are exposed to fire
3 Strong shock or heat may detonate — Use monitors from behind explosion resistant barriers
2 Violent chemical change possible — Use hose streams from distance
1 Unstable if heated — Use normal precautions
0 Normally stable
4
3
3
W
Avoid use of water
NFPA®
NATIONAL FIRE PROTECTION ASSOCIATION
PRINTED IN U.S.A. COPYRIGHT © 1968 NFPA HMD

Fig. 13.2 NFPA hazard warning label
(Permission of National Fire Protection Association)

CONDITION 1
TOO LEAN
CONDITION 2
EXPLOSIVE ATMOSPHERE
CONDITION 3
TOO RICH
% OF LOWER EXPLOSIVE LIMIT
LOWER EXPLOSIVE LIMIT (LEL)
UPPER EXPLOSIVE LIMIT (UEL)
0%
50%
100%
ATMOSPHERIC TESTING EQUIPMENT - ALARM SET @ 10% LEL
THREE ATMOSPHERIC CONDITIONS CAN EXIST
1. TOO LEAN TO SUPPORT COMBUSTION
2. MIXTURE JUST RIGHT, EXPLOSION OCCURS
3. MIXTURE TOO RICH TO SUPPORT COMBUSTION

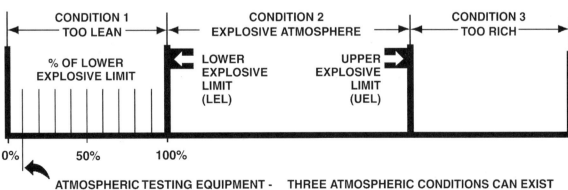

Fig. 13.3 Relationship between the lower explosive limit (LEL) and the upper explosive limit (UEL) of a mixture of air and gas

The National Fire Protection Association Standard 820 (NFPA 820), *FIRE PROTECTION IN WASTEWATER TREATMENT AND COLLECTION FACILITIES*, lists requirements for electrical classifications, ventilation, gas detection, and fire control methods in various wastewater treatment and collection system areas. (For ordering information, see Section 13.11, "Additional Reading," item 7.) Comparing these requirements with your plant's existing design and equipment may indicate deficiencies, which should be remedied to minimize potential hazards.

13.18 Fires

Burns can be very serious and cause painful injuries. Structural damage from fires can be very costly. Every facility should develop a fire prevention plan, with input from the local fire marshall, fire chief, and insurance company. The plan may be very simple or very complex, depending on the specific facility needs. Some items that may be included in any plan are:

1. Regulate the use, storage, and disposal of all combustible materials/substances.

2. Provide periodic cleanup of weeds or other vegetation in and around the facility.

3. Develop written response procedures for reacting to a fire situation, to include evacuation.

4. Provide required service on all fire detection and response equipment (inspection, service, hydrostatic testing).

5. Routinely inspect fire doors to ensure proper operation and unobstructed access.

6. Immediately repair, remove, or replace any defective wiring.

7. Restrict the use of any equipment that may provide a source of ignition in areas where combustible gases may exist.

8. Maintain clear access to fire prevention equipment at all times.

9. Develop a written hot work procedure and permit to provide written authorization to perform operations (for example, welding, cutting, burning, and heating) that involve a source of ignition.

Regardless of the size of the facility, each operator should know the location of fire protection equipment in their work area and must be trained in the proper use of fire extinguishing equipment and methods of extinguishing fires. Training must be provided upon initial employment and at least annually thereafter.

A portable extinguisher must also be visually inspected monthly and must receive an annual maintenance check. Maintenance checks must be documented and the records must be retained for one year. Hydrostatic testing of extinguishers is also required every 5 to 12 years, depending on the type of extinguisher. Remember, always have extinguishers serviced promptly after use so they will be ready if you need them again.

QUESTIONS

Write your answers in a notebook and then compare your answers with those on pages 731 and 732.

13.1A How can you help prevent the spread of infectious diseases from your job to you and your family?

13.1B When testing the atmosphere before entry in any confined space, what procedure should be used?

13.1C According to OSHA regulations, when is a confined space entry permit required?

13.1D How do toxic and suffocating gases or vapors enter the industrial wastewater treatment plant?

13.1E Why is your sense of smell not a reliable method of detecting hydrogen sulfide?

13.1F What do the initials UEL and LEL stand for?

13.19 Other Hazards

13.190 Electric Shock/Stored Energy

Electric shock frequently causes serious injury. Do not attempt to repair electrical equipment unless you know what you are doing. You must be qualified and authorized to work on electrical equipment before you attempt any troubleshooting or repairs. Ordinary 120 volt electricity may be fatal; 12 volts may, on good contact, cause injury. Any electrical system, regardless of voltage, should be considered dangerous unless you know positively that it is de-energized. Remember these basic safety rules when working around electrical equipment:

1. Keep your mind on the potential hazard at all times.

2. Always lock out and tag out any electrical equipment being serviced. *NEVER* remove anyone else's lock or tag.

3. Do not use portable ladders with conductive side rails.

4. Never override any electrical safety device.

5. Inspect extension cords for abrasion, insulation failure, and evidence of possible internal damage.

6. Use only grounded or insulated (Underwriter's Laboratory (UL) approved) electrical equipment.

7. Take care not to accidentally ground yourself when in contact with electrical equipment or wiring.

8. Do not alter or connect attachment plugs and receptacles in a manner that could prevent proper grounding.

9. Do not use flexible electrical cords connected to equipment to raise or lower the equipment.

10. Wear nonconductive head protection if there is a danger of head injury from contact with exposed energized parts.

11. Use a ground-fault circuit interrupter in damp locations.

12. Do not wear conductive articles of jewelry or clothing if they might contact exposed energized parts (unless they are covered, wrapped, or otherwise insulated).

EMERGENCY PROCEDURES

In the event of electric shock, the following steps should be taken:

1. Survey the scene and see if it is safe to enter.

2. If necessary, free the victim from a live power source by shutting power off at a nearby disconnect, or by using a dry stick or some other nonconducting object to move the victim.

3. Send for help, calling 911 or whatever the emergency number is in your community. Check for breathing and pulse. Begin CPR (cardiopulmonary resuscitation) immediately, if needed.

Remember, only trained and qualified individuals working in pairs should be allowed to service, repair, or troubleshoot electrical equipment and systems.

Whenever replacement, repair, renovation, or modification of equipment is performed, OSHA laws require that all equipment that could unexpectedly start up or release stored energy must be locked out or tagged out to protect against accidental injury to personnel. Some of the most common forms of stored energy are electrical and hydraulic energy. The energy isolating devices (switches, valves) for the equipment should be designed to accept a lockout device. A lockout device (Figure 13.4) uses a positive means such as a lock to hold a switch or valve in a safe position and prevent the equipment from becoming energized or moving. In addition, prominent warnings, such as the tags illustrated in Figure 13.5, must be securely fastened to the energy isolating device and the equipment (in accordance with an established written procedure) to indicate that both it and the equipment being controlled may not be operated until the tag and lockout device are removed by the person who installed them.

For the safety of all personnel, each plant should develop a standard operating procedure that must be followed whenever equipment must be shut down or turned off for repairs. If every operator follows the same procedures, the chances of an accidental start-up injuring someone will be greatly reduced. The following procedures are intended as guidelines and can be used as a model for developing your own written standard operating procedure for lockout/tagout.

Training must be provided to ensure that the purpose and function of the lockout/tagout program is understood and that the knowledge and skill required for the safe use of energy controls are gained. Each employee using lockout/tagout must be aware of applicable energy sources and the methods necessary for their effective isolation and control.

If you hire a contractor, you must inform each other of your respective lockout/tagout procedures. You must ensure that you and your employees or co-workers comply with the restrictions and prohibitions of the contractor's program.

Periodic inspections of the lockout/tagout program are also required (at least annually) to ensure that the requirements of the program are being followed. The inspection(s) must be done by someone other than the one(s) using the procedure.

BASIC LOCKOUT/TAGOUT PROCEDURES

1. Notify all affected employees that a lockout or tagged system is going to be used and the reason why. The authorized employee shall know the type and magnitude of energy that the equipment uses and shall understand the hazard thereof.

2. If the equipment is operating, shut it down by the normal stopping procedure.

3. Operate the switch, valve, or other energy isolating device(s) so that the equipment is isolated from its energy source(s). Stored energy such as that in springs; elevated machine members; rotating flywheels; hydraulic systems; and systems using air, gas, steam, or water pressure must be dissipated or restrained by methods such as repositioning, blocking, or bleeding down.

4. Lock out or tag out the energy isolating device with assigned individual lock or tag. If a tagout device is used, it must be substantial enough to prevent accidental removal. The attachment means for a tagout device must be a non-reusable type, attachable by hand, self-locking, and non-releasable with a minimum unlocking strength of no less than 50 pounds (22.7 kg).

5. After ensuring that no personnel are exposed, and as a check that the energy source is disconnected, operate the push button or other normal operating controls to make certain the equipment will not operate. *CAUTION: Return operating controls to the neutral or off position after the test.*

6. The equipment is now locked out or tagged out and work on the equipment may begin.

7. After the work on the equipment is complete, all tools have been removed, guards have been reinstalled, and employees are in the clear, remove all lockout or tagout devices. Operate the energy isolating devices to restore energy to the equipment.

13.191 Noise

Industrial wastewater treatment facilities contain some equipment that produces high noise levels, intermittently or continuously. Operators must be aware of this and use safeguards such as hearing protectors that eliminate or reduce noise to acceptable levels. In general, if you have to shout or cannot hear someone talking to you in a normal tone of voice, the noise level is excessive. Prolonged or regular daily exposure to high noise levels can produce at least two harmful, measurable effects: hearing damage and masking of desired sounds such as speech or warning signals.

Noise source monitoring should be conducted to measure noise levels in all suspect areas of the treatment plant and to identify excessive tool or equipment noise generation. Follow-up testing should be performed when any equipment or process is placed into service or modified.

The ideal method of dealing with any high-noise environment is the elimination or reduction of all sources through

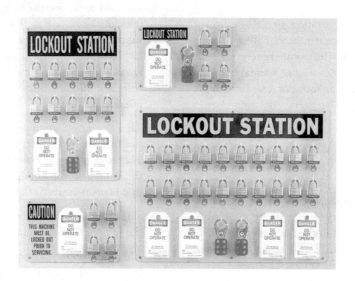

Fig. 13.4   Typical lockout devices
(Courtesy of Brady Worldwide)

Fig. 13.5   Typical lockout warning tags

feasible engineering or administrative controls. This approach is frequently not possible; therefore, employers are required to identify and monitor operators whose normal noise exposure might equal or exceed an 8-hour *TIME-WEIGHTED AVERAGE*[7] (TWA) of 85 *DECIBELS*[8] (A-scale).

To ensure that the welfare of operators is not compromised and to comply with federal regulations (29 CFR 1910.95), a comprehensive hearing conservation program should be implemented. All individuals whose normal noise exposure equals or exceeds the 8-hour TWA of 85 dBA must be included in this program. The primary elements of the program are monitoring, audiometric testing, hearing protection, training in the use of protective equipment and procedures, access to noise level information, and recordkeeping. The purpose of the conservation program is to prevent hearing loss that might affect an operator's ability to hear and understand normal speech.

A selection of hearing protection devices must be available to operators; however, a certain degree of confusion can arise concerning the adequacy of a particular protector. To estimate the adequacy of a hearing protector, use the noise reduction rating (NRR) shown on the hearing protector package. Subtract 7 dB from the NRR and subtract the remainder from the individual's A-weighted TWA noise environment to obtain the estimated A-weighted TWA under the ear protector. To provide adequate protection, the value under the ear protector should be 85 dB or less, the lower the better.

It is essential that individuals use properly rated protective devices in high-noise areas or during high-noise activities. Employee training must include information on (1) the effects of noise on hearing; (2) the purpose of hearing protectors and the advantages, limitations, and effectiveness of various types, and instruction on selection, fitting, use, and care; and (3) the purpose of the audiometric testing and an explanation of the test procedures. The training must be repeated annually. Audiometric test records must be retained for as long as the affected employee works at the plant. Contact your local health department or safety regulatory agency for assistance in the development of a program for your specific treatment facility.

13.192 Dusts, Fumes, Mists, Gases, and Vapors

The ideal way to control occupational diseases caused by breathing air contaminated with harmful dusts, fumes, mists, gases, and vapors is to prevent atmospheric contamination from occurring. This can sometimes be accomplished through engineering control measures. Remember, OSHA requires that engineering controls be implemented whenever feasible to eliminate or reduce operator exposure to a hazard. When effective engineering controls are not feasible, however, appropriate respirators must be used.

Respirators must be provided by the employer when they are necessary to protect the health of the operator. The respirators must be appropriate and suitable for the purpose intended. The four most common respirators are (1) Self-Contained Breathing Apparatus (SCBA), (2) Supplied Air Respirators (SAR), (3) Powered Air Purifying Respirators (PAPR), and (4) Air Purifying Respirators (APR).

SARs and SCBAs, known as air supplying respirators, supply clean air to the wearer from an independent source and provide the highest levels of protection. The air source may either be remotely located tanks or a tank carried by the user.

APRs and PAPRs take air from the immediate surroundings and purify it by passing it through filters, cartridges, or canisters. APRs and PAPRs provide lower levels of protection than air supplying types of respirators. These respirators (APRs and PAPRs) *ARE NOT* suitable for potentially oxygen-deficient atmospheres.

Positive pressure respirators, in which the pressure inside the face piece during inhalation (breathing in) remains higher than the pressure outside, help prevent contaminants from entering and offer the greatest protection. Conversely, negative pressure respirators, which allow the interior face piece pressure to drop below the outside pressure during the inhalation cycle, may not prevent contaminants from leaking into an improperly sealed face piece.

Positive pressure SCBAs and full face piece positive pressure SARs with an escape SCBA can be used in oxygen-deficient atmospheres (containing less than 19.5 percent oxygen) and atmospheres that are considered immediately dangerous to life or health (IDLH) as defined in OSHA regulations.

Negative pressure respirators (PAPRs and APRs) are approved only for atmospheres that are not immediately dangerous to life or health.

Selection of a respirator is based on the type of hazard and the contaminant concentration. Each respirator type is given an assigned protection factor (APF) by OSHA indicating the maximum contaminant level for which the respirator can be used. OSHA rates contaminants according to their permissible exposure limit (PEL). A contaminant's PEL is the legally established maximum time-weighted average level of contaminant to which an operator can be exposed during a work shift. The proper respirator is, therefore, chosen according to its APF and the PEL of the contaminant. For example, a respirator with an APF of 100, approved for a given contaminant, can be used in atmospheres containing 100 times the PEL of the contaminant.

---

[7] *Time-Weighted Average (TWA).* The average concentration of a pollutant based on the times and levels of concentrations of the pollutant. The time-weighted average is equal to the sum of the portion of each time period (as a decimal, such as 0.25 hour) multiplied by the pollutant concentration during the time period divided by the hours in the workday (usually 8 hours). 8TWA PEL is the time-weighted average permissible exposure limit, in parts per million, for a normal 8-hour workday and a 40-hour workweek to which nearly all workers may be repeatedly exposed, day after day, without adverse effect.

[8] *Decibel* (DES-uh-bull). A unit for expressing the relative intensity of sounds on a scale from zero for the average least perceptible sound to about 130 for the average level at which sound causes pain to humans. Abbreviated dB.

Remember, you, the operator, must use the provided respiratory protection in accordance with instructions and training provided to you.

Employers are also responsible for establishing and maintaining a respiratory protection program. Some of the basic elements of a respiratory protection program are:

1. Written standard operating procedures (SOPs) governing the selection and use of respirators,

2. Instruction and training in the proper use of respirators and their limitations (to include annual fit testing).

3. Physical assessment of individuals assigned tasks requiring the use of respirators.

These are only a few requirements for the safe use of respiratory protection. Specific requirements for a respiratory protection program for your application MUST BE CONFIRMED with your local regulatory safety agency.

QUESTIONS

Write your answers in a notebook and then compare your answers with those on page 732.

13.1G What is a lockout device and when is it used?

13.1H What types of equipment or systems are potentially hazardous for operators due to their stored energy?

13.1I Identify the primary elements of the hearing conservation program.

13.1J What types of respirators must be used in oxygen-deficient atmospheres?

13.2 SPECIFIC HAZARDS

The remainder of this chapter will acquaint you with the specific hazards, by location or type of work, that you may expect to encounter in the field of industrial wastewater collection and treatment.

13.20 Collection Systems

Good design and the use of safety equipment will not prevent physical injuries in sewer work unless safety practices are understood by the *entire crew* and are enforced.

Never attempt to do a job unless you have sufficient training, assistance, the proper tools, and the necessary safety equipment. *There are no shortcuts to safety.*

13.200 Traffic Hazards

Before starting any job in a street or other traffic area, even if you are just going to open a manhole, study the work area, and plan your work. Your task must be regulated to provide maximum safety.

Working in a roadway represents a significant hazard to an operator as well as pedestrians and drivers. Drivers can be seen applying makeup, shaving, talking on cellular phones, and changing tapes, CDs, or radio stations rather than concentrating on driving. The control of traffic is necessary to reduce the risk of injury or death while working in this hazardous area. The purpose of traffic control is to provide safe and effective work areas and to warn, control, protect, and expedite vehicular and pedestrian traffic. This can be accomplished by appropriate and prudent use of traffic control devices (Figures 13.6 and 13.7).

Upon arrival at the job site, look for a safe place to park your vehicle. If it must be parked in the street to do the job, route traffic around the job site *before* parking your vehicle in the street. If practical, park your vehicle in the work area between oncoming traffic and the job site to serve as a warning barricade and to discourage reckless drivers from plowing into you. The use of flashing warning lights is an excellent method of alerting traffic to your presence. Remember, you need protection from the drivers as well.

Traffic may be warned by high-level signs and flags far enough ahead (500 feet or 150 meters) of the job to adequately alert the driver, by traffic cones (fluorescent cones do an excellent job) arranged to guide traffic around your work area, by signs or barricades to direct traffic, by a flagger to direct and control traffic, or by any combination of these. Traffic warning devices must be placed to avoid causing confusion and congestion. Figure 13.8 provides two examples of appropriate placement of traffic control devices and a legend (key) for interpreting the diagrams. Your State Department of Transportation can provide you with the required patterns that you must use when setting up warning or traffic control devices for your specific applications. They can also advise you about the number of personnel required to control traffic in your proposed work area.

For additional information about traffic control and working safely in streets, see *OPERATION AND MAINTENANCE OF WASTEWATER COLLECTION SYSTEMS*, Volume I, Chapter 4, "Safe Procedures," in this series of operator training manuals.

13.201 Manholes[9]

Manholes are confined spaces and the requirements for entry into a confined space and the atmospheric hazards that may be encountered in a confined space have been discussed earlier in

---

[9] Also see OSHA standards regarding safe procedures for entry into confined spaces.

690 Treatment Plants

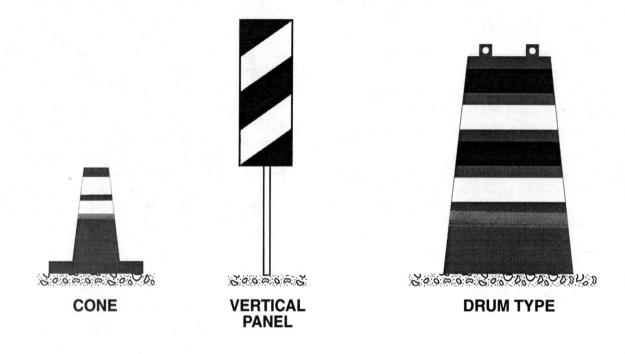

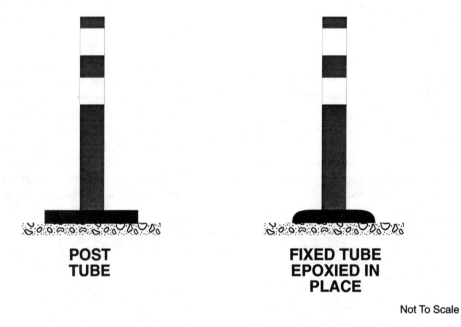

Fig. 13.6   Delineating and channelizing devices

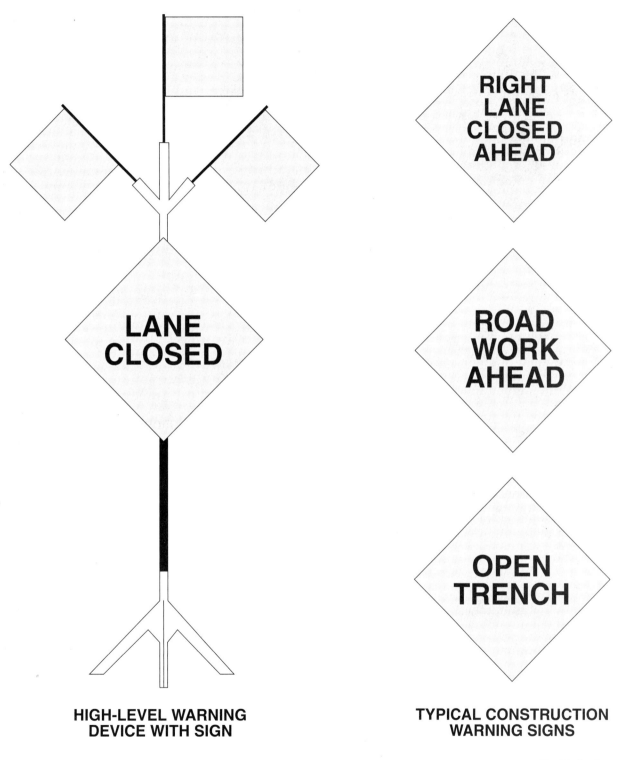

Fig. 13.7 *Warning devices and signs*

692 Treatment Plants

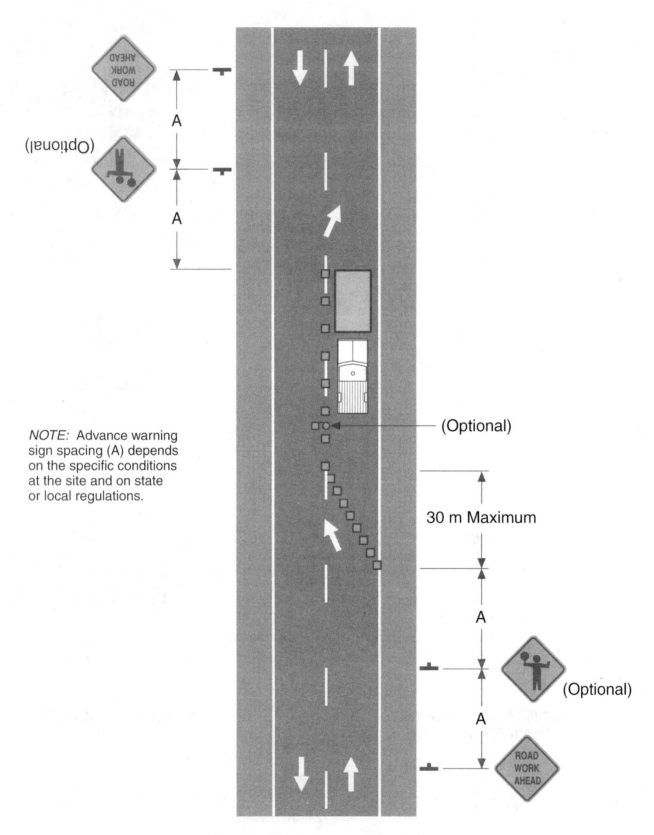

Fig. 13.8 Typical placement of traffic control devices
(Source: California Department of Transportation)

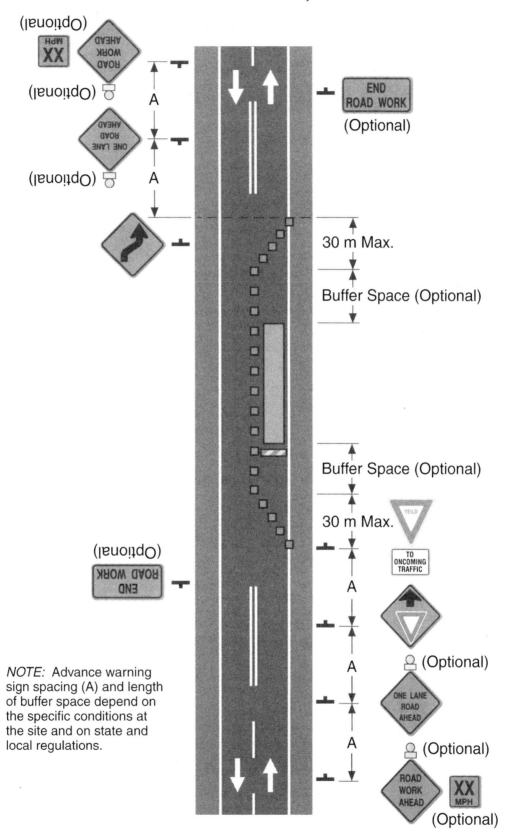

Fig. 13.8 *Typical placement of traffic control devices (continued)*
(Source: California Department of Transportation)

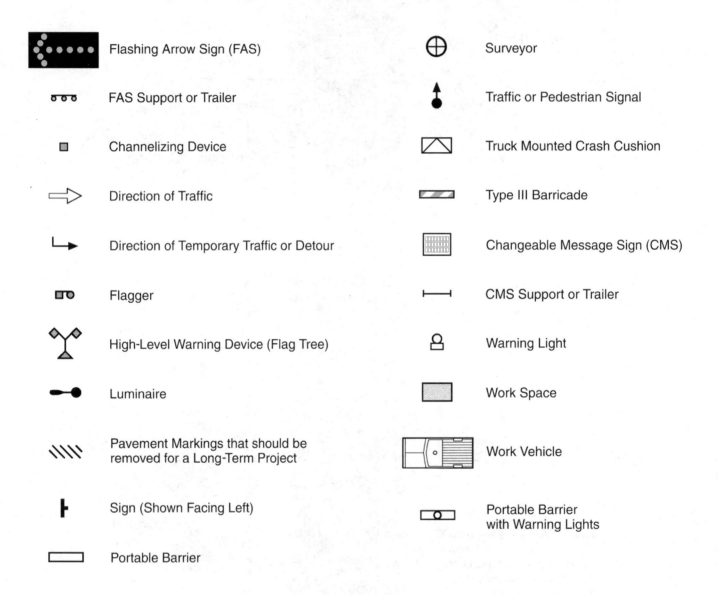

Fig. 13.8 Typical placement of traffic control devices (continued)
(Source: California Department of Transportation)

this chapter in Section 13.12. The following items are examples of other hazards that may exist in manholes, depending on the design and use of the manhole:

- Mechanical hazards, such as hot or cold surfaces, steam leaks, or rotating equipment.
- Electrical hazards involving conduit, energized circuits, lights, portable power tools, or moisture/water accumulation.
- Engulfment hazards, such as flooding from wastewater, sludge, or chemicals.
- Physical injury, such as sprains, strains, scrapes, and cuts due to uneven footing, inadequate working room, poor balance, or awkward positioning.
- Infections and diseases from bacteria, parasites, and viruses found in the wastewater stream.
- Bites from insects and rodents.
- Toxic exposure from substances illegally or accidentally discharged into the collection system or plant drainage system.

Manhole work will usually require job site protection, either a manhole safety enclosure by itself or in conjunction with traffic controls. Job site protection when working in manholes should include barricades and traffic warning devices for the safety of vehicles, bikes, pedestrians, and workers. In addition to following confined space procedures for entry and occupancy of the manhole, the atmosphere within the manhole should be tested prior to removing the manhole cover. A spark created by removing the cover could cause an explosion if a combustible atmosphere exists within the manhole. Also, it may be helpful to evaluate the conditions within the manhole before any ventilation occurs.

Remove the manhole cover with a tool specifically designed for the purpose, not your fingers. You have only 10 fingers. Protect them. When lifting a lid, the use of the rule "Lift with your legs, not with your back" will help eliminate back strains (Figure 13.9). Once the lid is removed, leave it flat on the ground and far enough away from the manhole to provide adequate room for a working area. This is usually at least two to three feet (0.6 to 1 m).

REMEMBER: Never enter a manhole without fully complying with the requirements for work in confined spaces.

BACK INJURIES ARE A FREQUENT INJURY

> BE CAREFUL...
> THE BACK YOU SAVE MAY BE YOUR OWN!
>
> Think about the following steps when lifting:
> 1. Size up the load.
> 2. Bend your knees.
> 3. Get a firm grip.
> 4. Lift with your legs... gradually.
> 5. Keep the load close to you.
>
> IT IS HOW YOU LIFT as well as WHAT YOU LIFT.

Fig. 13.9 Lifting guidelines

Be very cautious when using ladder rungs or steps installed in the side of the manhole. Be alert for loose or corroded steps. Always test each step individually before placing your weight upon it. If possible, it is much safer to use a portable ladder as a means of entering a manhole. Be certain, however, that the bottom feet are properly placed so that the ladder will not slip or twist when your weight is placed upon it. The top of the ladder should extend three feet (1 m) above the ground level to facilitate getting on and off the ladder. A mechanical lifting device is the safest possible way to be lowered into a manhole, wet well, or other below-grade work area or to be lifted to an elevated work area.

If you are working in wastewater, be sure to wear properly fitted rubber gloves and boots, or approved substitutes that will provide protection from infection. Be aware of possible needle sticks and other puncture wounds.

Tools and equipment should be lowered into a manhole by means of a bucket or basket. Do not drop them into the manhole for a person to catch. Attempting to carry tools in one hand while climbing up or down a ladder is an unsafe practice.

13.202 Excavations[10]

If it becomes necessary for you to excavate a sewer line, remember to contact utility companies to locate underground telephone, gas, fuel, electric, cable, and water lines *before* opening the excavation. If you cannot establish the exact location of the underground utilities, you must proceed with caution and you should use detection equipment, if possible, to locate the service lines. These lines can present a very significant safety hazard to the operator(s) and to the public during excavation activities.

Become familiar with the fundamentals of excavating and the proper, safe approach for shoring, shielding, and sloping and benching before excavating. Without a proper protective system the bank (wall) of a trench or excavation can cave in and kill you. It is strongly recommended that some type of adequate cave-in protection be provided when the trench or excavation is four feet (1.2 meters) deep or deeper. OSHA requirements state that adequate protection is absolutely required if the trench is five feet (1.5 meters) or more in depth. Types of adequate protection include shoring, shielding, and sloping and benching.

SHORING is a complete framework of wood or metal that is designed to support the walls of a trench (see Figure 13.10). Sheeting is the solid material placed directly against the side of the trench. Either wooden sheets or metal plates might be used. Uprights are used to support the sheeting. They are usually placed vertically along the face of the trench wall. Spacing between the uprights varies depending on the stability of the soil. Stringers (or walers) are placed horizontally along the uprights. Trench braces are attached to the stringers and run across the excavation. The trench braces must be adequate to support the weight of the wall to prevent a cave-in. Examples of different types of trench braces include solid wood or steel, screw jacks, or hydraulic jacks.

The space between the shoring and the sides of the excavation should be filled in and compacted in order to prevent a cave-in from getting started. If properly done, shoring may be the operator's best choice for cave-in protection because it actually prevents a cave-in from starting and does not require additional space.

SHIELDING is accomplished by using a two-sided, braced steel box that is open on the top, bottom, and ends (Figure 13.11). This "drag shield," as it is sometimes called, is pulled through the excavation as the trench is dug out in front and filled in behind. Operators using a drag shield must always work only within the walls of the shield and are not allowed in the shield when it is being installed, removed, or moved. If the trench is left open behind or in front of the shield, the temptation could be high to wander outside of the shield's protection. Shielding does not actually prevent a cave-in as the space between the trench wall and the drag shield is left open, allowing a cave-in to start. There have been cases where a drag shield was literally crushed by the weight of a collapsing trench wall.

SLOPING and BENCHING are practices that simply remove the trench wall itself (Figure 13.12). The amount of soil needed to be removed will vary, depending on the stability of the soil. A good rule of thumb is to always slope or bench at least one foot back for every one foot (0.3 m) of depth on *both* sides of the excavation. Exact sloping angles and benching dimensions largely depend on the type of soil that is being excavated. OSHA has established three types of soil classifications, Type A, Type B, and Type C, in decreasing order of stability. The type of soil dictates sloping/benching requirements. A competent person must examine the work site and determine the soil classification in order to determine requirements for a specific site.

Certain soil conditions can contribute to the chances of a cave-in. These conditions include low cohesion, high moisture content, freezing conditions, or a recent excavation at the same site. Other factors to be considered are the depth of the trench, the soil weight, the weight of nearby equipment, and vibration from equipment or traffic. It is worth repeating that regardless of the presence or absence of any or all of the above factors, the trench must still have proper cave-in protection if it is five feet (1.5 meters) or more deep. Excavations less than five feet (1.5 meters) deep may also require protection if a competent person determines that there are indications of a potential cave-in. The spoil (dirt removed from the trench) must be placed at least two feet (0.6 meter) back from the trench and should be placed on one side of the trench only. A stairway, ramp, or ladder is required in the trench if it is four feet (1.2 meters) or more deep. The means of leaving the trench must be placed so that no more than 25 feet (7.5 meters) of travel is required to exit the trench.

Atmospheric monitoring prior to entering excavations more than four feet (1.2 meters) deep is required if a hazardous atmosphere can reasonably be expected to exist. Excavations in landfill areas, in areas where hazardous substances are stored nearby, and in which a connection(s) is being made to an in-service manhole are examples of situations where a hazardous atmosphere could exist or develop. Emergency rescue equipment such as breathing apparatus, safety harness, and lifeline must also be available *AND* attended.

Excavations and adjacent areas must be inspected on a daily basis by a competent person for evidence of potential cave-ins, protective system failures, hazardous atmospheres, or other hazardous conditions. The inspections are only required if an employee exposure is anticipated. Walkways must also be provided where personnel must cross the excavation. If the excavation is six feet (1.8 meters) or more in depth at the crossing point, guardrails must also be provided on the walkways.

Accidents at the site of trenching and shoring activities are all too common. In addition to protecting workers from the danger of a cave-in, safety precautions must also be taken to protect them from traffic hazards if the work is performed in a street. Check with your local safety regulatory agency. They can

---

[10] Also see *OPERATION AND MAINTENANCE OF WASTEWATER COLLECTION SYSTEMS*, Volume I, Chapter 4, in this series of manuals.

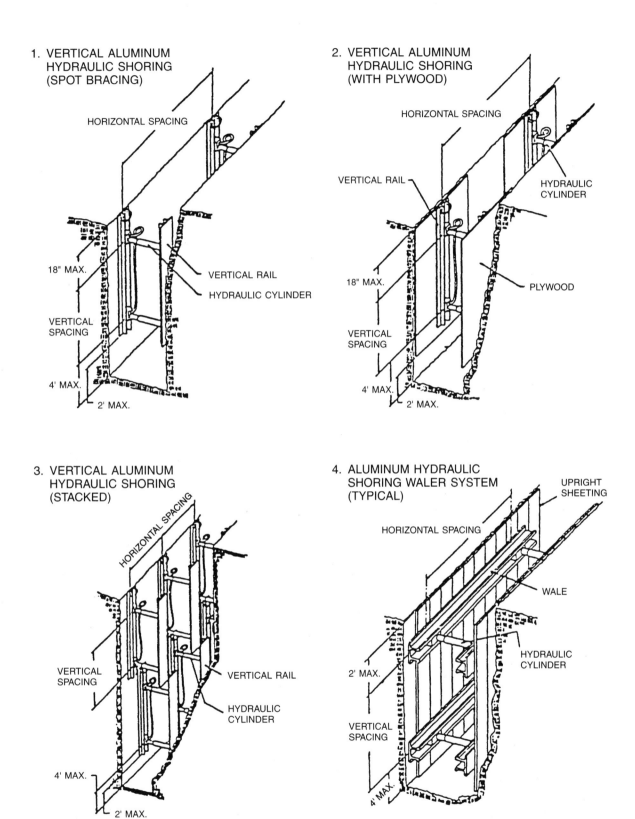

NOTE: Multiply inches × 2.5 to obtain centimeters; multiply feet × 0.3 to obtain meters.

Fig. 13.10 Typical installations of aluminum hydraulic shoring
(Source: 29 Code of Federal Regulations)

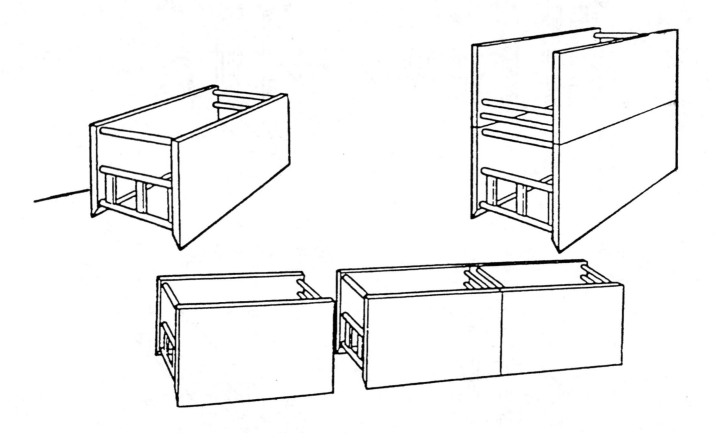

Fig. 13.11 Trench shields
(Source: 29 Code of Federal Regulations)

EXCAVATIONS MADE IN TYPE B SOIL

1. All simple slope excavations 20 feet or less in depth shall have a maximum allowable slope of 1:1.

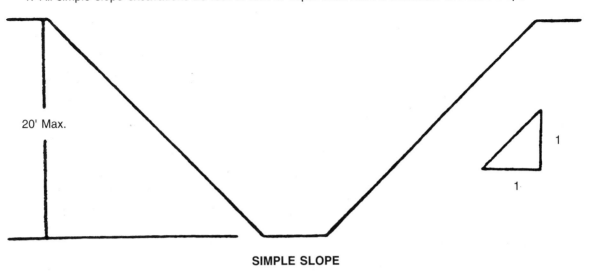

SIMPLE SLOPE

2. All benched excavations 20 feet or less in depth shall have a maximum allowable slope of 1:1 and maximum bench dimensions as follows:

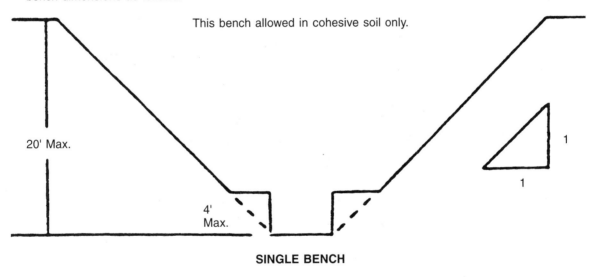

SINGLE BENCH

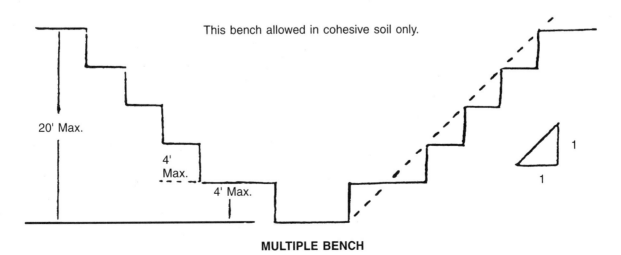

MULTIPLE BENCH

NOTE: Multiply inches × 2.5 to obtain centimeters; multiply feet × 0.3 to obtain meters.

Fig. 13.12 Typical sloping/benching schemes
(Source: 29 Code of Federal Regulations)

provide you with the appropriate regulations. Do not wait until an emergency arises to obtain the information.

13.203 Sewer Cleaning

Never use a tool or piece of equipment unless you have received training in its proper use or operation. Require the equipment vendor to provide you with this training. Know your limitations and the limitations and capabilities of your tools and equipment. Do not use tools or equipment improperly; you could be seriously injured.

If you use chemicals of any kind for root or grease control in your system, become thoroughly familiar with their use and, specifically, with any hazards involved. See Section 13.15, "Toxic and Harmful Chemicals."

13.204 Acknowledgment

Major portions of Section 13.12, "Confined Spaces," Section 13.190, "Electric Shock/Stored Energy," and Section 13.202, "Excavations," were adapted from "Wastewater System Operations Certification Study Guide," developed by the State of Oklahoma. Permission to use this material is greatly appreciated.

QUESTIONS

Write your answers in a notebook and then compare your answers with those on page 732.

13.2A   List three ways to warn traffic that you are working in a street or traffic area.

13.2B   List at least four hazards that may be encountered when entering a manhole, excluding atmospheric hazards.

13.2C   List three protective systems for excavations.

13.2D   At what excavation depth does OSHA require a protective system?

13.2E   From whom should you receive training regarding the proper use or operation of sewer cleaning tools and equipment?

END OF LESSON 1 OF 3 LESSONS

on

SAFETY

Please answer the discussion and review questions next.

DISCUSSION AND REVIEW QUESTIONS

Chapter 13. SAFETY

(Lesson 1 of 3 Lessons)

At the end of each lesson in this chapter you will find some discussion and review questions. The purpose of these questions is to indicate to you how well you understand the material in the lesson. Write the answers to these questions in your notebook.

1. What is the operator's responsibility with regard to safety?
2. Accidents do not just happen—they are _____!
3. How can an operator avoid physical injuries?
4. Immunization shots protect against which infectious diseases?
5. What precautions should you take to avoid transmitting disease to your family?
6. What should you do when you discover an area with an oxygen deficiency or enrichment?
7. What kind of job site protection is usually required when you are working in a manhole?
8. Lift with your legs, not with your _____.
9. How should tools and equipment be transported to the bottom of a manhole?

CHAPTER 13. SAFETY

(Lesson 2 of 3 Lessons)

13.21 Pumping Stations

Pumping stations may vary from small, telemetered lift stations that are visited monthly to large pumping stations that have operators on duty 24 hours a day. Regardless of the size, type, or complexity of the pumping station, safe procedures must be followed at all times. Safety precautions discussed in Section 13.12, "Confined Spaces," are very important for work in any potential confined space, including both wet wells and dry wells.

Always be aware of the possibility of oxygen deficiency/enrichment, explosive or flammable gases, and toxic gases that could be generated in, or be discharged into, the sewers. A properly operating ventilation system is essential, particularly in below-grade pumping stations where heavier-than-air gases may collect. Before entering a potential confined space in an unattended lift station, test the atmosphere for oxygen deficiency/enrichment, explosive or flammable conditions (lower explosive limit), and toxic gases (hydrogen sulfide).

Do not work on electrical systems or controls unless you are qualified and authorized to do so. Even if you are qualified and authorized, use caution when operating and maintaining electrical controls, circuits, and equipment. Operate only those switches and electrical controls installed for the purpose of your job. Do not open or work inside electrical cabinets or switch boxes unless you are an authorized and qualified electrician.

Be aware of moving equipment, especially reciprocating equipment and rotating shafts. Moving parts that create a contact hazard to employees must be guarded. Do not wear loose clothing, rings, or other jewelry around machinery. Long hair must be secured. Wear gloves and use appropriate tools when cleaning pump casings to protect your hands from dangerous sharp objects.

When starting equipment, everyone should stand away from rotating parts. Dust and oil or loose metal may be thrown from shafts and couplings, or sections of a long, vertical shaft could come loose and whip around, especially during start-up of equipment.

If stairs are installed in a pumping station, they should have handrails and nonslip treads. Where space limitations prevent the installation of stairs, a spiral stairway, ship's ladder, or vertical ladder should be provided. Vertical ladders must be provided with a cage enclosure if they exceed 20 feet (6.1 meters) in unbroken length. If the ladder is not provided with a safety device or cage, an intermediate platform or landing should be provided for each 20 feet (6.1 meters) of height or fraction thereof.

Lighting should be sufficient to avoid glare and shadows.

Fire extinguishers must be provided in the station, properly located and maintained. They should be of a type that may be used on electrical equipment as well as on solid material or power overload-type fires. The use of liquid-type fire extinguishers should be avoided. All-purpose, ABC chemical-type fire extinguishers are recommended.

Good housekeeping is a necessity in a pumping station. Common housekeeping problems include water and oil on the floor, dirty or oily rags from cleanup operations, dirty lighting fixtures leading to poor visibility, and stairways with dirt or grease carried in by maintenance or repair crews.

Always be sure you have properly secured the pumping station when you leave. This procedure is necessary to prevent injury to neighborhood children and possible vandalism to the station. Both of these problems can be very costly to your employer if proper precautions are not taken.

QUESTIONS

Write your answers in a notebook and then compare your answers with those on page 732.

13.2F What types of moving equipment should you be aware of in a pumping station and what safety precautions should be taken to protect yourself from injury?

13.2G What type of fire extinguisher should be available in a pumping station?

13.22 Treatment Plants

Because hazards found in pumping stations are similar to those found in treatment plants, the items discussed below may be applied to both situations. Safety precautions outlined in the previous section apply both to pumping stations and treatment plants.

13.220 *Headworks*

Structures and equipment in this category may consist of bar screens, racks, comminuting or grinding equipment, pump rooms, wet pits, and chlorination facilities.

1. *BAR SCREENS OR RACKS.* These may be either manually or mechanically cleaned. When manually cleaning screens or racks, be certain that you have a clean, firm surface to stand on. Remove all slimes, rags, greases, or other material that may create slip or trip hazards. *Good housekeeping in these areas is absolutely necessary.*

When raking screens, leave plenty of room for the length of your rake handle so you will not be thrown off balance if the handle strikes a wall, railing, or light fixture. Wear gloves to avoid slivers from the rake handle or scraping your knuckles on concrete. Injury may allow an infection to enter your body.

Place all material in a container that may be easily removed from the structure. Be careful lifting containers full of heavy material such as grit. Use of containers sized to prevent excessive weight is recommended. It is better to make more than one trip than to risk injury. Do not allow material to build up on the working surface.

If your rack area is provided with railings, check to see that they are properly anchored before you lean against them. If removable safety chains are provided, never use these to lean against or as a means of providing extra leverage for removing large amounts of material.

A hanging or mounting bracket should be used to hold the rake when not in use. Do not leave it lying on the deck.

If mechanically raked screens or racks are installed, never work on the electrical or mechanical part of this equipment without locking out the unit. Always open, lock out, and tag the main circuit breaker before you begin repairs.

The tag (Figure 13.5) should be securely fastened to the breaker handle or to the lockout device to notify others that you are working on the equipment and that it must remain de-energized.

The time and date the unit was locked out should be noted on the tag, as well as the reason it was locked out. The tag should be signed by the person who locked out the unit. No one should then close the main breaker and start the unit until the tag and lock have been removed by the person who placed them there, or until specific instructions are received from the person who tagged the breaker. Your local safety equipment supplier can obtain tags and lockout devices for you.

2. *COMMINUTING OR GRINDING EQUIPMENT.* This equipment may consist of barminutors, comminutors, grinders, or disintegrators.

NEVER work on the mechanical or electrical parts of the unit without first locking out the unit at its main circuit breaker. Be certain the breaker is properly tagged as explained in the previous section.

Good housekeeping is essential in the area of comminuting equipment. Keep all walking areas clean and free of slimes, oils, greases, or other materials. Clean up all spills immediately. Provide a storage place for equipment and tools used in this area.

See that proper guards are installed and kept in place around moving parts that create a contact hazard to employees. De-energize, lock out, and tag equipment before removing guards to make adjustments. Replace all of the guards prior to returning the equipment to service. Use extension tools to make adjustments if the equipment must be operated unguarded. *DO NOT* expose yourself to the hazard of moving parts.

3. *PUMP ROOMS.* The same basic precautions apply here as to any type of enclosed room or pit where wastewater or gases may enter and accumulate.

Always provide adequate ventilation to remove atmospheric hazards. If the room is below ground level and provided with forced-air ventilation, be certain the fan is operating properly before entering the area. Use confined space procedures (refer to Section 13.12) when entering pits, wet wells, or tanks.

The tops of all stairwells or ladders should be protected by a removable safety chain. Keep this chain in place when the stairwell or ladder is not being used.

Guards should be installed around all moving parts that create a contact hazard to employees. Never remove equipment guards without first locking out and tagging the equipment at the main breaker. Always replace all guards before starting units.

Maintain good housekeeping in pump rooms. Remove all oil and grease, and clean up spills immediately.

If you have a multilevel pump building, never remove and leave off equipment access hatches unless you are actually removing or replacing equipment. Be sure to provide barricades or posts with safety chains around the opening to prevent falls. Be extremely cautious when working around openings that have raised edges. These are hazardous because you can easily stumble over them.

Never start a positive displacement pump against a closed valve. Positive displacement pumps can create extreme pressures that can cause equipment or piping failures and endanger operators in the vicinity. Positive displacement pumps should be equipped with a pressure switch to shut the unit off before dangerous pressures are encountered.

Lighting fixtures, ventilation units, electric motors, and other electrical devices in pump rooms may be required to be rated for use in hazardous atmospheres (explosion proof), depending on the design and contents of the specific area. The integrity of electrical devices in potentially hazardous atmospheres must be maintained. Any defective or inoperative equipment must be repaired or replaced immediately by a qualified person. Only qualified electricians should work on an electrical control panel because unqualified persons may be seriously injured by electric shock and they could damage the equipment. Errors in reassembly of explosion-proof equipment can create a significant hazard potential. Sources of ignition must be controlled if the pump room has the potential to contain an explosive/flammable atmosphere. Hot work such as welding or soldering and work involving standard portable electrical tools should only be allowed in the area if atmospheric conditions can be maintained in a safe state while the ignition source(s) exist.

> **CAUTION**
> UNLESS YOU ARE A QUALIFIED ELECTRICIAN, STAY OUT OF THE INSIDE OF ALL ELECTRICAL PANELS. IF YOU ARE NOT TRAINED, AUTHORIZED, OR QUALIFIED TO WORK WITH THE EQUIPMENT, LEAVE IT ALONE!

4. *WET PITS—SUMPS.* Covered wet pits or sumps are potential death traps. *THEY ARE CONFINED SPACES.* Always strictly follow confined space procedures (refer to Section 13.12) when you work in these areas. Requirements for work in confined spaces include, but are not limited to, written procedures and permits, monitoring, rescue and respiratory protective equipment, training, atmospheric monitoring, and ventilation. In addition to the atmospheric, engulfment, and entrapment hazards common to many confined spaces, many other hazards may also exist.

Extreme care must be used when climbing up and down access ladders to pit areas. The application of a nonslip coating on ladder rungs is helpful. If available, a hoisting mechanism is safer than a ladder for entering pit and sump areas. Watch your footing on the floor of pits and sumps. They are very slippery.

Never attempt to carry tools or equipment up or down ladders into pits or sumps. Always use a bucket and handline or a sling for this purpose.

Only explosion-proof lights and equipment should be used in these areas if the potential for a flammable/explosive atmosphere exists.

A good safety practice is to effectively block the entry of potentially hazardous contaminants into the wet pit or sump. Isolate sources such as wastewater or sludge flows, chlorination, or plant drainage, whether they discharge upstream, downstream, or directly into the wet pit or sump. This exercise, coupled with forced ventilation, will minimize the potential for an atmospheric hazard to develop while the space is occupied.

For additional information about chlorination safety, see *OPERATION OF WASTEWATER TREATMENT PLANTS,* Volume I, Chapter 10, "Disinfection and Chlorination."

QUESTIONS

Write your answers in a notebook and then compare your answers with those on page 732.

13.2H Why should slimes, rags, or greases be removed from around bar screens or racks?

13.2I What precautions would you take when working on electrical or mechanical parts of equipment?

13.2J What parts of equipment should have guards installed around them?

13.2K How would you transport tools or equipment into or out of pits or sumps?

13.221 Grit Channels

Grit channels may be of various designs, sizes, and shapes, but they all have two things in common: they are confined spaces and they get dirty. Good housekeeping is needed. Keep walking surfaces free of grit, grease, oil, slimes, or other material that may create a slip or trip hazard.

Before working on mechanical or electrical equipment, be certain that it is locked out and properly tagged (Figure 13.5). Install and maintain guards over moving parts that could be contacted accidentally.

If it becomes necessary to enter the channel for cleaning or other work, do so with extreme caution. Follow all confined space procedures (refer to Section 13.12). Always provide sufficient ventilation to maintain safe atmospheric conditions within the channel. Before any entry occurs, use a high-pressure water hose to agitate any material that may have accumulated on the bottom of the channel. Agitate the material sufficiently to release any *ENTRAINED*[11] gases. Test the atmosphere first for oxygen deficiency/enrichment, then combustible gases, and then toxic gases.

Be sure of your footing when working in these structures. Rubber boots with steel safety toes and a nonskid, cleat-type sole should be worn. Use safety sole inserts in your boots to prevent puncture wounds. Step slowly and cautiously as there is usually an accumulation of slippery material or slimes on the bottom. Use hand holds and railings; if none are available, install them now.

Use ladders, whether vertical or ships' ladders, cautiously. If possible, apply nonslip material or coatings to ladder rungs. Keep handrails free of grease and other slippery substances.

When using a portable ladder, the horizontal offset should not be less than one-quarter of the ladder distance between supports. The ladder should extend three feet (one meter) beyond the access point to facilitate easy access to and from the ladder at the top.

---

[11] *Entrain.* To trap bubbles in water either mechanically through turbulence or chemically through a reaction.

13.222 Clarifiers or Sedimentation Basins

The greatest hazard involved in working on or in a clarifier is the danger of slipping. If possible, maintain a good, nonskid surface on all stairs, ladders, and catwalks. This may be done by using nonskid strips or coating. Be extremely cautious during freezing weather. A small amount of ice can be very dangerous. Be careful and do not fall in.

Your housekeeping program should include the brushing or cleaning of effluent weirs and launders (effluent troughs). Effluent weirs and launders on clarifiers should be brushed or cleaned to avoid the accumulation of slippery slimes and solids. Uneven flow over an effluent weir can encourage short-circuiting. When it is necessary to actually climb down into the launder, always use a fall arrest system as discussed in Section 13.10, page 676. A fall may result in a very serious injury. If you work in areas where the danger of drowning exists, you must wear a US Coast Guard-approved life jacket or buoyant work vest.

Remember that a clarifier is a confined space and confined space procedures must be followed for entry and occupancy (see Section 13.12). Be cautious when working on the bottom of a clarifier. When hosing down, always hose a clean path to walk upon. Avoid walking on the remaining sludge whenever possible.

Always turn off, lock out, and tag the clarifier breaker before working on the drive unit. Adjustments should not be made on flights or scrapers while the unit is in operation. Keep in mind that, although these are moving quite slowly, there is tremendous power behind their movement. Stay clear of any situation where your body or the tools you are using may get caught under one of the flights or scrapers.

Guards must be installed over moving parts that could be contacted accidentally. Keep these in place whenever the unit is in operation.

Railings must be installed along the tank side of all normal walkways. If the unit is elevated above ground, railings should be installed along the *outside* of all walkways, also. Standard guardrails usually consist of a top rail, an intermediate rail, and posts or uprights. The top rail should be 42 inches (107 cm) high, measured from the top of the rail to the surface of the walkway. The guardrail system is typically designed to support a load of at least 200 pounds (91 kilograms) applied in any direction at any point on the rail. Check with your state safety office for requirements on railing installation.

13.223 Digesters and Digestion Equipment

Digesters and their related equipment include many hazardous areas and potential dangers. The National Fire Protection Association Standard 820 (NFPA 820), *FIRE PROTECTION IN WASTEWATER TREATMENT AND COLLECTION FACILITIES*, lists requirements for electrical classifications, ventilation, gas detection, and fire control methods in sludge and gas handling areas. For ordering information, see Section 13.11, "Additional Reading," item 7.

NO SMOKING AND NO OPEN FLAMES should be allowed in the vicinity of digesters, in digestion control buildings, or in any other areas or structures used in the sludge digestion system. This includes pipe galleries, compressor or heat exchanger rooms, and others. All these areas should be posted with noticeable signs that forbid smoking and open flames. *Methane gas produced by anaerobic conditions is explosive when mixed with the proper proportion of air. For this reason, NFPA 820 prohibits the use of standard electrical systems on anaerobic digesters.*

Electrical systems on or within five feet (1.5 meters) of an anaerobic digester must be rated for use in Class I, Division I locations. Class I, Division I locations are those areas where a hazardous concentration of digester gas may exist (1) under normal operating conditions, or (2) frequently because of leakage repair/maintenance activities, or (3) because of a breakdown or faulty operation that might release hazardous concentrations of digester gas.

All rooms, galleries, and tunnels in the sludge and gas-handling areas should be provided with continuous forced ventilation. Required air changes per hour are specified in NFPA 820. Certain areas such as gas compressor rooms are also required to be fitted with classified electrical devices and gas detection systems.

Digesters are confined spaces. Follow all confined space procedures if it becomes necessary to work in a digester. In addition to confined space procedures, several additional measures can enhance the safety of digester entry. Stopping sludge addition several days before removing the digester from service will drastically reduce gas production. Piping that could introduce hazards to the digester such as gas and sludge lines should be mechanically separated from the digester (or pancaked by inserting a solid disk between flanges) to eliminate potential hazards from valve leakage. Purging of gas lines and equipment is also recommended. Purging is the displacement of hazardous gas with an inert substance such as nitrogen, carbon dioxide, or water. Purged gases are typically discharged at a controlled rate to a safe outside area where no ignition source is present and where no employee exposure will occur.

Explosion-proof ventilation, lighting, and "non-sparking" tools must also be used when working around digesters and gas-handling equipment unless safe atmospheric conditions can be maintained. Continuous atmospheric monitoring at the immediate work site should be performed throughout the work period.

> NEVER ALLOW SMOKING, OPEN FLAMES, OR OTHER SOURCES OF IGNITION WITHIN AN EMPTY DIGESTER UNLESS SAFE ATMOSPHERIC CONDITIONS CAN BE ENSURED AND MAINTAINED! (FIGURE 13.13)

Guardrails should be installed along the edges of the digester roof or cover in areas where it is necessary to work close to the edge. As previously discussed in Section 13.10, page 676, a fall

An explosion blew off the top of a digester …

… and it landed on top of a pickup truck.

Fig. 13.13   Blown-up digester

arrest system must be used if you work in an unguarded elevated area six feet (1.8 meters) or more above a lower level. A fall from the top of a digester could be fatal. Toe boards should also be installed to prevent objects/tools from being kicked over the edge.

When working on equipment such as draft tube mixers, compressors, and diffusers, be certain that the unit that operates or supplies gas to the equipment is properly valved or locked out and appropriately tagged (Figure 13.5).

If you have a heated digester, read and heed the manufacturer's instructions before working on the boiler or heat exchanger. Know that the main gas valve is turned off before attempting to light the pilot. Be certain that the fire box has been ventilated according to the manufacturer's instructions before lighting the pilot. Do not work on the boiler unless you are authorized and qualified to do so.

When it becomes necessary to clean tubes or coils in a heat exchanger, turn off the hot water supply far enough in advance to allow the heat exchanger to cool. Never open the unit without ensuring that sludge and water flows have been effectively isolated and that water and sludge temperatures have cooled down to body temperature or lower.

> **CAUTION**
> WASTE GAS BURNERS ARE NOTED FOR BLOWING OUT IN A MODERATE WIND. BEFORE YOU ATTEMPT TO RELIGHT THE UNIT, BE CERTAIN THAT THE MAIN VALVE HAS BEEN TURNED OFF AND THE STACK ALLOWED TO VENT ITSELF FOR A FEW MINUTES. THESE UNITS HAVE BEEN KNOWN TO "BACKFLASH" WHEN RE-IGNITED!

Before working on any sludge pump, whether it is centrifugal or positive displacement, be certain that the unit is effectively isolated from process piping and that it is locked out and properly tagged (Figure 13.5).

Positive displacement pumps should be equipped with a pressure switch to shut the unit off at a preset pressure. *NEVER* start a positive displacement pump against a closed discharge valve because pressure could build up and burst a line or damage the pump. If you have closed this valve to inspect or clean the pump, double check to be sure that it is open before starting the unit.

Sludge pump rooms should be well ventilated to remove any gases that might accumulate from leakage, spillage, or from a normal pump cleaning. If you spill digesting sludge, clean it up immediately to prevent the possible accumulation of gases or a slippery walkway.

Provide thorough, regularly scheduled inspections and maintenance of your gas collection system. Inspect drip traps regularly. The so-called "automatic" drip trap is known to jam open frequently, allowing gas to escape, and, therefore, should not be installed in any area where leakage could create a hazard.

Routine effective maintenance of safety equipment in gas handling systems is essential to maintain a safe working environment. Flame arresters, vacuum and pressure relief valves, and thermal valves in gas piping cannot prevent injury or property damage if they do not function properly. Equipment interlocks, automatic isolation valves, and gas detection and alarm systems all require regular testing and calibration to ensure that they will operate properly if and when they are needed.

Additional safety information for working in and around digesters is also contained in *OPERATION OF WASTEWATER TREATMENT PLANTS,* Volume II, Chapter 12, "Sludge Digestion and Solids Handling."

> **QUESTIONS**
> Write your answers in a notebook and then compare your answers with those on page 732.
>
> 13.2L How can the danger of slipping be reduced on slippery surfaces?
>
> 13.2M Why should no smoking or open flames be allowed in the vicinity of digesters?
>
> 13.2N What safety precautions would you take before entering a recently emptied digester?
>
> 13.2O What would you do before relighting a waste gas burner?
>
> 13.2P Why should you never start a positive displacement pump against a closed discharge valve?

13.224 Trickling Filters

When it becomes necessary to inspect or service a rotating distributor, stop the flow of wastewater to the unit and allow it to come to rest. Lock out and tag the influent valve or pump supplying flow and secure the distributor arm so no one can start the distributor while you are working.

> NEVER STAND OR WALK ON THE FILTER MEDIA WHILE THE ROTATING DISTRIBUTOR IS IN MOTION.

Provide an approved ladder or stairway for access to the media surface. Be positive this is free from obstructions such as hose bibs and valve stems.

Use extreme caution when walking on the filter media. The biological slimes make the media very slippery. Move cautiously and be certain of your footing.

> NEVER ALLOW ANYONE TO RIDE A ROTATING DISTRIBUTOR.

Although a rotating distributor moves fairly slowly, the force behind it is powerful. If you fall off and are dragged by a distributor, you will be fortunate if you can walk away under your own power.

When inspecting underdrains, remember that these are dangerous confined spaces and, if entry is necessary, confined space procedures must be used. Check to determine that the channels or conduits are adequately ventilated. Gases are not normally a problem here, but may be if there is a buildup of solids that have become septic.

If it becomes necessary to jack up a distributor mechanism for inspection or repair, always try to provide a firm base off the media or drainage system for the jack plate. A firm base may be provided by wooden planks, which will spread the weight over a large area. However, sometimes the only way to obtain firm support is to remove the media and use the drainage system as a firm base. Again, wooden planks should be placed over the drainage system to distribute the weight better. Remember, you are lifting a heavy weight. Do not attempt inspection or repair work until the distributor has been adequately and properly blocked in its raised position.

13.225 Aerators

Guardrails must be installed on the tank side of usual work areas or walkways. If the tank is elevated above ground, guardrails should also be installed on the ground side of the tank as discussed in Section 13.222.

When working on Y-walls, or other unguarded areas where work is done infrequently, at least two people should do the work. US Coast Guard-approved life preservers with permanently attached handlines should be accessible at strategic locations around the aerator. The life preservers must be provided with at least 90 feet (27 meters) of handline and be located not more than 200 feet (61 meters) apart. If you work in areas where the danger of drowning exists, you must wear a US Coast Guard-approved life jacket or buoyant work vest.

An experiment in England found that if an operator fell into a diffused aeration tank, the operator should be able to survive because air will collect in the clothing and tend to help keep the operator afloat.[12] Drownings apparently occur when a person is overcome by the initial shock or there is nothing to grab hold of to keep afloat or to pull oneself out of the aerator. In aerators where diffused air is supplied only along the walls, strong currents develop, which could pull anyone but a very strong swimmer under water. *ALWAYS wear a Coast Guard-approved flotation device when working unguarded over water.*

When removing or installing diffusers, be aware of the limitations of your working area. Inspect and properly position hoists and other equipment used in servicing swing diffusers.

Remember, an aerator is a confined space. When it is necessary to work in an empty aerator, follow confined space procedures and comply with all relevant safety requirements. Portable ladders can be awkward but they are safe if positioned and used properly. A good practice is to use a fall arrest system when climbing up or down a ladder and to secure the top of the ladder so that it cannot slip. Be extremely careful when walking in an aerator; the floor of the aerator can be very slippery.

If your plant is in an area subject to freezing weather, be aware of possible ice conditions around these units and use caution accordingly.

Additional safety information for working in and around aerators is contained in *OPERATION OF WASTEWATER TREATMENT PLANTS,* Volume II, Chapter 11, "Activated Sludge."

13.226 Ponds

Ponds of any kind present the same hazards. Therefore, the following safety measures will apply to ponds in general.

If it is necessary to drive a vehicle on top of the pond levees, maintain the roadway in good driving condition by surfacing it with gravel or asphalt. Do not allow chuckholes or the formation of ruts. Be extremely cautious in wet weather. The material used in the construction of most levees becomes very slippery when wet. Slippery conditions should be corrected using crushed rock or other suitable material.

Never go out on the pond for sampling or other purposes when you are by yourself. Someone should be standing by on the bank in case you get into trouble. Always wear a US Coast Guard-approved life jacket when working from a boat or raft. And, as in any boating activity, do not stand up in the boat while performing work.

QUESTIONS

Write your answers in a notebook and then compare your answers with those on page 733.

13.2Q How would you stop the rotating distributor on a trickling filter?

13.2R Why should you never work alone on the center Y-wall of an aerator?

13.2S What precautions should you take when using a boat to collect samples from a pond?

---

[12] Kershaw, M. A., "Buoyancy of Aeration Tank Liquid," Journal Water Pollution Control Federation, Vol. 33, No. 11, p. 1151 (Nov. 1961) and Stevens, Patrick L., "I Fell into an Aeration Tank," Operations Forum, Vol. 3, No. 5, p. 21 (May 1986).

13.227 Chemical Treatment

13.2270 CHLORINE AND SULFUR DIOXIDE

Chlorine and sulfur dioxide are very similar when considering safety and health aspects of the two chemicals. Because of this similarity, the following information is applicable to both substances. Additional safety considerations and information are contained in OPERATION OF WASTEWATER TREATMENT PLANTS, Volume I, Chapter 10, "Disinfection and Chlorination."

Chlorine and sulfur dioxide are usually purchased on a bid contract. Contracts should specify the conditions under which containers delivered to your plant will be accepted. The container should be stamped with the last date when it was pressure tested. Do not accept containers that have not been pressure tested within five years of the delivery date. Cover cap should be securely in place over the valve mechanism. All threaded connections should be clean and not worn or cross threaded. New, approved gaskets should be provided with each container. Do not accept containers not meeting these standards. A few minutes spent inspecting containers when they are delivered can prevent serious problems in the future.

> CHLORINE GAS AND SULFUR DIOXIDE GAS ARE HIGHLY IRRITATING AND CORROSIVE GASES.
> DANGER... HANDLE WITH CAUTION!

The most common causes of accidents involving chlorine and sulfur dioxide are leaking pipe connections and excessive dosage rates.

Containers should be stored in a cool, dry place, away from direct sunlight or from heating units. Some heat is needed to cause desired evaporation and to control moisture condensation on containers. Containers should never be dropped or allowed to strike each other with any force. Cylinders should be stored in an upright position and secured with a chain, wire rope, or clamp. They should be moved only by hand truck and should be well secured during moving. One-ton containers should be blocked so that they cannot move. They should be lifted only by an approved lifting bar with hooks over the ends of the containers. NEVER lift a container with an improvised sling.

Always wear a face shield when changing chlorine or sulfur dioxide containers. Escape supplied air respirators are also recommended. Connections to containers should be made only with approved fittings. Always inspect all surfaces and threads of the connector before connecting. If you are in doubt as to their condition, do not use the connector. Always use a *new*, approved-type gasket when making a connection. The reuse of gaskets very often will result in a leak. Check for leaks as soon as the connection is completed. Use aqua ammonia vapor to check for chlorine or sulfur dioxide leaks. The ammonia vapor will combine with any chlorine or sulfur dioxide present to form a white cloud. Do not apply ammonia solution directly to piping or connections; corrosion will result. Never wait until you smell chlorine or sulfur dioxide. If you discover even the slightest leak, correct it immediately, as leaks will get worse rather than better. Like accidents, leaks generally are caused by faulty procedure or carelessness.

Obtain from your supplier and post in a conspicuous place (*outside* the chlorination and sulfonation room and by the telephone) the name and telephone number of the nearest emergency service in case of a severe leak.

Storage and equipment rooms must be provided with forced ventilation. As chlorine and sulfur dioxide are approximately two and a half times heavier than air, vents should be provided at floor level. Some installations have a blower mounted on the roof to blow air into the room and are vented at the floor level to allow escaped gas to be blown out of the building. Other installations may have the ventilation system and duct work interlocked to a leak detection system to "bottle up" the room when a leak occurs. This prevents uncontrolled dispersion of chlorine or sulfur dioxide to the atmosphere.

The Uniform Fire Code, Article 80, Hazardous Materials, requires that exhaust ventilation for chlorine and sulfur dioxide areas must be taken from a point within 12 inches (30.5 centimeters) of the floor. Mechanical ventilation should be at a rate of not less than one cubic foot of air per minute per square foot (0.00508 cubic meter of air per second per square meter) of floor area of storage handling area. Normally, ventilation from these areas is discharged to the atmosphere, but when a leak occurs, the ventilated air containing the chlorine or sulfur dioxide should be routed to a treatment system to remove the chemical. A caustic scrubbing system can be used to treat the air from a leak. Treatment systems must be designed to reduce the maximum allowable discharge concentration of chlorine or sulfur dioxide to one-half the IDLH (Immediately Dangerous to Life or Health[13]) at the point of discharge to the atmosphere. The IDLH for chlorine is 10 ppm. For sulfur dioxide the IDLH is 100 ppm. A secondary, standby source of power is required for the detection, alarm, ventilation, and treatment systems.

Always enter storage or equipment rooms with caution. If you smell chlorine or sulfur dioxide when opening the door to the area, immediately close the door, leave ventilation on, and seek assistance.

Never attempt to enter an atmosphere containing chlorine or sulfur dioxide when you are by yourself. Operators engaged in emergency response to a hazardous materials release must be effectively trained in hazard recognition, repair techniques, use of personal protective equipment such as clothing and respiratory protective equipment, decontamination procedures, if required,

---

[13] *IDLH.* Immediately Dangerous to Life or Health. The atmospheric concentration of any toxic, corrosive, or asphyxiant substance that poses an immediate threat to life or would cause irreversible or delayed adverse health effects or would interfere with an individual's ability to escape from a dangerous atmosphere.

and use of the buddy system. In a buddy system, each operator in the work group is designated to observe another operator in the group, with the purpose of providing assistance in an emergency. Responses to chlorine or sulfur dioxide releases should be coordinated with your local fire department and hazardous materials (HAZMAT) response agency.

Excellent booklets may also be obtained from PPG Industries, Inc. (*CHLORINE SAFE HANDLING*[14]) and from The Chlorine Institute, Inc. (*CHLORINE MANUAL*[15]). Safety information on chlorine handling is also contained in OPERATION OF WASTEWATER TREATMENT PLANTS, Volume I, Chapter 10, "Disinfection and Chlorination." Your chlorine and sulfur dioxide suppliers will probably provide you with all the information you need to handle and use chlorine and sulfur dioxide safely. It is your responsibility to obtain, read, and understand safety information and to practice safety.

If you are the manager of a wastewater treatment facility that uses large quantities of chlorine or sulfur dioxide, you may be required to develop and implement written safety plans for dealing with the effects of both on-site and off-site chemical releases. See OPERATION OF WASTEWATER TREATMENT PLANTS, Volume II, Chapter 20, Section 20.10, "Emergency Response," for information about Process Safety Management (PSM) programs and Risk Management Programs (RMPs).

13.2271 POLYMERS

The primary use for polymers in wastewater treatment is the conditioning of sludge to facilitate removal of water in subsequent treatment processes such as belt filter presses, centrifuges, gravity belt filters, and dissolved air flotation thickeners. Polymers are available in either liquid or dry forms. The type of polymer you use depends on the application, product performance, volume of use, space availability, and equipment capabilities.

Dry polymers come in various types of powder, crystals, and beads. Most dry polymers have a carrier in them and are about 94 percent to 96 percent active. The powder form is the most economical, but it may release a fine dust when handled, which can present a safety hazard.

Liquid polymers are available in either solution or emulsion/dispersion types. Both types consist of a polymer material in a medium, either water or oil, respectively. Emulsion/dispersion polymers range from 25 percent to 50 percent concentration and must be mixed with a large volume of water to achieve full activation before use. When water is added to emulsion/dispersion polymers, there is a dramatic increase in *VISCOSITY*.[16] Solution polymers range from 4 percent to 50 percent actual polymer and are extremely viscous and difficult to pump if not mixed with water. A widely used solution polymer is the **mannich reaction** polymer, which typically contains 4 percent to 6 percent active polymer. Mannich reaction polymers are produced by using formaldehyde as a catalyst. Since vapors from these polymers pose a safety hazard (formaldehyde is a *CARCINOGEN*[17]), they should be stored and used only by trained personnel in well-ventilated areas.

The use of polymers exposes operators to a number of safety hazards that require appropriate precautions to prevent injury or illness. The hazards generally fall into two areas, slipping hazards and personal exposure hazards from contact or inhalation:

- Slipping hazards—polymers have a moisture attracting property. Even a thin film of dry polymer can combine with moisture and become extremely slippery. Liquid polymers are inherently slippery. Polymer spills must be cleaned up immediately. Gather up as much of the spilled material as possible by using an appropriate method, such as gentle sweeping, vacuuming, soaking it up with rags, or using kitty litter as an absorbent. Water flushing and the use of household bleach can remove remaining polymer material. The proper disposal of cleaned-up material should be confirmed with your local environmental regulatory agency.

- Personal exposure hazards—irritation of skin, eyes, or lungs can result from contact with polymer, polymer dust, or polymer fumes. Personal protective equipment (PPE), such as chemical-resistant gloves, splash-proof goggles, and an apron, may be required. If exposure to polymer dust or fumes from mannich reaction polymer may occur, appropriate respiratory protection should be used. Environmental monitoring should be performed to determine the extent of exposure as well as the required level of protection.

Areas where polymers are handled or stored should be equipped with sufficient continuous ventilation and emergency eye wash stations and deluge showers. Remember, read and heed the safety requirements identified on your polymer's Material Safety Data Sheet (MSDS).

13.228 Applying Protective Coatings

CAUTION: When applying protective coatings in a clarifier or any other tank, channel, or pit, whether enclosed or open topped, you can create an atmospheric hazard. These areas are confined spaces and the coating system used can create additional hazards by generating flammable or toxic vapors during application and curing periods. Whenever possible, use nonhazardous coating materials. The control of application rates and methods such as

---

[14] PPG Industries, Inc., Monroeville Chemical Center, 440 College Park Drive, Monroeville, PA 15146. No charge.

[15] The Chlorine Institute, Inc., Bookstore, PO Box 1020, Sewickley, PA 15143-1020. Pamphlet 1. Price to members, $28.00; nonmembers, $70.00; plus $6.95 shipping and handling.

[16] *Viscosity* (vis-KOSS-uh-tee). A property of water, or any other fluid, that resists efforts to change its shape or flow. Syrup is more viscous (has a higher viscosity) than water. The viscosity of water increases significantly as temperatures decrease. Motor oil is rated by how thick (viscous) it is; 20 weight oil is considered relatively thin while 50 weight oil is relatively thick or viscous.

[17] *Carcinogen* (kar-SIN-o-jen). Any substance that tends to produce cancer in an organism.

brushing or rolling rather than spraying can also minimize potential hazards. Always attempt to find an appropriate atmospheric monitor to ensure that a hazard is not being created; however, with the sophisticated coating products available now, this may not be possible. Industrial hygiene calculations can be made to control hazards if atmospheric monitors are not available for the specific hazardous component(s) in the coating material. These calculations consider characteristics of the coating material, application rates, ventilation rates, and a safety factor that can be increased as desired. Consult your local safety regulatory agency for assistance.

In addition to the atmospheric hazards encountered during coatings operations, some coatings (asphaltic or bitumastic) can also cause skin burns if you are exposed to higher vapor concentrations. Engineering controls such as ventilation and a combination of personal protective equipment (protective clothing and creams) may be required to prevent exposure and injury or illness.

Check with your paint suppliers for any hazards involved in using their products. Study the appropriate Material Safety Data Sheet (MSDS) and follow the safety guidelines provided.

13.229 *Housekeeping*

Good housekeeping can and has prevented many accidents. Housekeeping tasks that will keep a treatment facility a safer, cleaner place to work include:

1. Have a place for your tools and equipment. When they are not being used, see that they are kept in their proper place.

2. Clean up all spills of oil, grease, chemicals, polymers, wastewater, and sludge. Keep walkways and work areas clean. A clean facility will reduce the possibility of physical injuries and infections.

3. Provide proper containers for wastes, oily rags, and papers. Empty these frequently.

4. Store flammable substances in an approved storage cabinet.

5. Remove snow and ice in areas where a person may slip and fall.

QUESTIONS

Write your answers in a notebook and then compare your answers with those on page 733.

13.2T   How should one-ton chlorine containers be lifted?

13.2U   Why are chlorine vents placed at floor level?

13.2V   What should you do if you open a door and smell chlorine or sulfur dioxide?

13.2W   List four housekeeping tasks that will keep a treatment facility a safer, cleaner place to work.

13.23 Industrial Waste Treatment

If your wastewater treatment plant treats only industrial wastes or a mixture of both industrial and municipal wastes, you must be aware of the industries and the types of wastes that are discharged. Also, in spite of an effective sewer-use ordinance, you must be alert for accidental spills and other unanticipated discharges. These discharges can be toxic to you and the organisms in your treatment processes and corrosive to your equipment. From a safety viewpoint, if you know how to identify the types of wastes that may reach your facility, you can be prepared to take the proper action and safety precautions. This section will discuss some of the hazardous or toxic substances that could reach your treatment facility.

13.230 *Fuels*

Fuels may be dumped or drained into storm sewers connected to sanitary sewers or they may enter the collection system as a result of a leaking underground fuel line or tank. Fuel oil and gasoline usually float on the surface of wastewater and are not diluted by mixing. Therefore, most floating fuels tend to collect in wet wells and can create both explosive and toxic conditions or atmospheres.

To reduce the chances of an explosion in a wet well and downstream enclosed structures, a combustible-gas detector should be installed in wet wells to sound an alarm and transmit a signal to the main control panel *BEFORE* explosive conditions are reached. Explosion-proof wiring, equipment, and fixtures will help to prevent fires and explosions in hazardous areas such as wet wells. Adequate ventilation also is essential. Oil skimmers should be installed in wet wells to remove floating fuel oil and gasoline. This equipment may be rarely used, but can remove explosive fuels without exposing operators to hazardous conditions.

Fuels may be detected by permanent or portable devices that measure either hydrocarbons or the lower explosive limit (LEL). These devices must be installed in treatment plants using pure oxygen treatment processes (activated sludge). They are located in the collection system upstream from the plant, in the headworks, and at the oxidation reactors (aerators). If hydrocarbons or explosive conditions reach a given *SET POINT*,[18] an alarm is usually generated. If the concentrations increase beyond this point, the oxygen gas flow to the processes is automatically shut off. The system is purged with air to prevent a possible explosion. These detection devices must be properly located and maintained at frequent intervals to provide reliable service.

If gasoline reaches the wet well in the headworks of your treatment plant:

1. Try to remove the gasoline from the surface of the wet well with skimmers (if available) or with a portable pump.

2. Apply as much ventilation as possible to prevent an explosive atmosphere from developing.

---

[18] *Set Point.* The position at which the control or controller is set. This is the same as the desired value of the process variable. For example, a thermostat is set to maintain a desired temperature.

3. Monitor the atmosphere for toxic and explosive conditions.

4. Keep personnel away from the area.

5. Do not allow any flames, sparks, or other sources of ignition in the area (use nonsparking tools and explosion-proof equipment and wiring).

13.231 Toxic Gases

Hydrogen sulfide (H_2S) is the most common toxic gas encountered by operators because it is produced in collection systems by decomposing organic matter under anaerobic conditions. Hydrogen sulfide and other toxic gases can be discharged into sewers or produced by chemical and biological reactions in sewers, in pretreatment facilities, or at the wastewater treatment plant. A serious hydrogen sulfide problem can develop when a discharge of sodium sulfide (Na_2S) from one industry mixes with the discharge of an acid waste from another industry. This mixture can produce extremely hazardous concentrations of hydrogen sulfide that are not only toxic, but explosive, flammable, and very odorous. Sodium sulfide also reacts violently with oxidants to form sulfur dioxide, which is toxic and may accumulate in below-grade areas.

Other toxic gases include phosgene ("war or mustard gas"), chlorine, and tear-producing substances (lacrimators). See Table 13.1, page 678, "Common Dangerous Gases Encountered in Wastewater Collection Systems and at Wastewater Treatment Plants."

Phosgene is produced in sewers when discharges of alcohol saturated with phosgene-wasted chloroformates is back-hydrolyzed (reverse chemical reaction) to phosgene. If an industry or laundry quickly dumps a few hundred to one thousand gallons of bleach, a slug of chlorine can occur in the plant's influent. Naphtha is used as both a fuel and a solvent and can create hazardous conditions.

Tear-gas type substances (lacrimators) occasionally may reach treatment plants. Certain organic insecticide wastes, when only partially chlorinated at an industrial wastewater pretreatment plant, can form tear-gas type substances.

Toxic gases may be detected by probes that measure the concentration of a particular gas, such as hydrogen sulfide, chlorine, or sulfur dioxide. Most instruments are capable of detecting the lower explosive limit (LEL), an oxygen deficiency/enrichment, and hydrogen sulfide. Portable sensors and monitors for toxic gases usually require daily calibration. Permanent systems usually require weekly calibration. Both systems require regular maintenance.

13.232 Amines

Amines are compounds formed from ammonia. Some of these compounds may react with other substances in wastewater to form nitrosamines, some of which are considered carcinogens (capable of causing cancer in humans). A remote possibility exists that nitrosamines from industrial dumps or chemical-biological reactions in sewers could contaminate the air space around treatment plants. If this problem is discovered at a treatment plant, all treatment process structures (wet wells, clarifiers, aeration tanks) could have to be covered. Exhaust air from these sources would have to be treated to remove harmful contaminants and offensive odors. Operators must test the atmosphere for harmful contaminants before entering the area. The type of testing equipment needed will depend on the contaminants that are expected to be present.

13.233 Surface-Active Agents

Concentrated industrial SURFACE-ACTIVE AGENTS,[19] either accidentally spilled or dumped into a collection system, can upset wastewater treatment processes. Certain industrial wastes can serve as super floc agents that produce a much denser sludge than usually pumped. This sludge can be too thick to pump, which then requires special handling to reslurry the sludge so it can be pumped. The opposite can occur when antifloc agents lower the capture of suspended solids in primary clarifiers. Excessive solids can reach aeration tanks and even flow out in the plant effluent. Rejected batches of detergents can contain antifloc agents.

Foaming agents often cause treatment plant operators very serious problems. Foam in wet wells can prevent inspection and operation of screens and grinders. Foam on aeration tanks that blows into neighbors' yards and on the surface of receiving water is objectionable to the public. If foam is formed in a chlorination chamber by a HYDRAULIC JUMP,[20] the foam bubbles could contain chlorine gas. Detection devices are available that are capable of detecting excessive levels of foaming agents. These foam detection devices also can be programmed to automatically feed an antifoaming agent to keep foaming under control.

13.234 Biocides

Poisons or biocides from industries can be harmful to operators as well as toxic to organisms purifying the wastewater in treatment processes. Unfortunately, biocides frequently cannot be detected until after the organisms in the treatment processes have been killed. Not only can the activated sludge process and the digesters be put out of action, but the sludge can be so contaminated that it cannot be disposed of on land or in landfills.

---

[19] *Surface-Active Agent.* The active agent in detergents that possesses a high cleaning ability. Also called a SURFACTANT.
[20] *Hydraulic Jump.* The sudden and usually turbulent abrupt rise in water surface in an open channel when water flowing at high velocity is suddenly retarded to a slow velocity.

13.235 High or Low pH

Highly acidic or alkaline wastes can be very hazardous. They are dangerous to personnel, treatment processes, and equipment. pH probes installed in the headworks can detect abnormal pH levels. A low pH caused by an acid can be increased by the addition of sodium hydroxide (NaOH). Sulfuric acid (H_2SO_4) can be added at the headworks to lower the pH of an alkaline waste. Study the appropriate Material Safety Data Sheet(s) (MSDS) and follow the safety guidelines provided.

13.236 Summary

Your main concerns as an industrial wastewater treatment facility operator are your own personal survival and that of your co-workers, and the preservation of your plant and the organisms in the treatment processes. Industrial dumps can produce especially serious hazards. Effective sewer-use ordinances and industrial pretreatment facilities can help greatly to reduce the frequency and severity of industrial dumps. This section only covers a few of the many potential hazards that can be created by industrial wastes.

QUESTIONS

Write your answers in a notebook and then compare your answers with those on page 733.

13.2X List the major types of discharges from industrial plants that could create safety hazards.

13.2Y What would you do if a gasoline truck was in an accident and the fire department washed the spilled gasoline into a storm drain? The storm drain conveys wastewater to the wet well in the headworks of your treatment plant.

13.2Z What types of problems can be caused by surface-active agents?

END OF LESSON 2 OF 3 LESSONS

on

SAFETY

Please answer the discussion and review questions next.

DISCUSSION AND REVIEW QUESTIONS
Chapter 13. SAFETY
(Lesson 2 of 3 Lessons)

Write the answers to these questions in your notebook. The question numbering continues from Lesson 1.

10. When cleaning racks or screens, on what kind of surface should the operator stand?

11. Never lean against a removable safety chain. True or False?

12. Why should only qualified electricians work on an electrical control panel?

13. Why should effluent weirs and launders on clarifiers be brushed or cleaned?

14. Why should no smoking or open flames be allowed in the vicinity of the digester or sludge digestion system?

15. Why should you never go out on a pond for sampling or other purposes by yourself?

16. Where should the name and telephone number of the nearest emergency chlorine leak repair service be posted?

17. What safety precautions should be taken when applying protective coatings?

18. Why are discharges of amines into collection systems considered dangerous?

CHAPTER 13. SAFETY

(Lesson 3 of 3 Lessons)

13.3 SAFETY IN THE LABORATORY [21]

In addition to all safety practices and procedures mentioned in the previous sections of this chapter, the collecting of samples and the performance of laboratory tests require that you be aware of the specific hazards involved in this type of work.

Laboratories use many hazardous chemicals. These chemicals should be kept in limited amounts and used with respect. Your chemical supplier may be able to supply you with a safety manual. Everyone who may handle, use, or dispose of hazardous chemicals must be trained in appropriate methods to ensure their safety. See Section 13.15, "Toxic and Harmful Chemicals."

13.30 Sampling Techniques

Wear disposable, impervious gloves if your hands may come in contact with wastewater or sludge. When you have finished sampling, dispose of the gloves and wash your hands thoroughly, using a disinfectant-type soap, to prevent the spread of disease.

NEVER COLLECT ANY SAMPLES WITH YOUR BARE HANDS. IF YOU HAVE ANY BROKEN SKIN AREAS, SUCH AS CUTS OR SCRATCHES, YOU MAY EASILY BECOME INFECTED.

Do not climb over or go beyond guardrails or chains when collecting samples. Do not lean on safety chains. Use sample poles, ropes, and other devices as necessary to collect samples.

13.31 Equipment Use and Testing Procedures

The following are some basic procedures to keep in mind when working in the laboratory:

NEVER LOOK INTO THE OPEN END OF A CONTAINER DURING A REACTION OR WHEN HEATING A CONTAINER.

---

1. Use proper safety goggles or a face shield in all tests where there is danger to the eyes.

2. Use care in making rubber-to-glass connections. Lengths of glass tubing should be supported while they are being inserted into rubber. The ends of the glass should be *FLAME POLISHED* [22] to smooth them out, and a lubricant such as water should be used. Never use grease or oil. Wear gloves or some other form of protection for the hands when making such connections. Hold the tubing as close to the end being inserted as possible to prevent bending or breaking. Never try to force rubber tubing or stoppers from glassware. Cut the rubber as necessary to remove it.

3. Always check labels on bottles to make sure that the proper chemical is selected. Never permit unlabeled or undated containers to accumulate around or in the laboratory. Unlabeled containers are only allowed if the person using the chemical dispenses and uses the chemical during the shift the chemical was placed in the container. Keep storage areas well organized to prevent mistakes when selecting chemicals for use. Clean out old or excess chemicals. Separate flammable, explosive, or special hazard items for storage in an approved manner. See Section 13.11, "Additional Reading," Reference 11.

ALL CHEMICAL CONTAINERS SHOULD BE CLEARLY LABELED, INDICATING CONTENTS AND DATE OPENED OR SOLUTION PREPARED. ALL CHEMICALS MUST BE LABELED WITH APPROVED WARNING LABELS.

---

4. Never handle chemicals with your bare hands. Use a spoon or spatula for this purpose.

5. Be sure that your laboratory is adequately ventilated.

ALWAYS WORK IN A FUME HOOD IF WORKING WITH CHEMICALS OR SAMPLES HAVING TOXIC FUMES.

Even mild concentrations of fumes or gases can be dangerous.

---

[21] Also see *FISHER SAFETY CATALOG*. Obtain from Fisher Scientific Company, Safety Division, 4500 Turnberry Drive, Suite A (Customer Service), Hanover Park, IL 60103, or phone (800) 772-6733.

[22] *Flame Polished.* Melted by a flame to smooth out irregularities. Sharp or broken edges of glass (such as the end of a glass tube) are rotated in a flame until the edge melts slightly and becomes smooth.

There are minimum requirements for air velocity through laboratory hoods. A "normal hood" must have an average "face" velocity of 100 fpm (feet per minute) (30 mpm (meters per minute)) with a minimum of 70 fpm (21 mpm) at any point. If carcinogens are handled in the hood, a face velocity of 150 fpm (39 mpm) with a minimum of 125 fpm (38 mpm) is required. You must also provide a method of indication that the air flow is active at each hood. Contact your local safety regulatory agency to verify specific requirements.

6. Never use laboratory glassware for a cup or food dish. This is particularly dangerous when dealing with wastewaters.

7. When handling hot equipment of any kind, always use tongs, insulated gloves, or other suitable tools. Burns can be painful and can cause more problems (encourage spills, fire, and shock).

8. When working in the lab, do not smoke or eat except in prescribed coffee break areas or in the lunch areas.

> **ALWAYS THOROUGHLY WASH YOUR HANDS BEFORE SMOKING OR EATING TO PREVENT THE SPREAD OF DISEASE.**

9. Do not pipet chemicals or wastewater samples by mouth. Always use a suction bulb or an automatic buret.

10. Handle all chemicals and reagents with care to protect your body from serious injuries and possible poisoning. Read and become familiar with all precautions or warnings on labels. Know and have available the antidote for all chemicals in your lab.

11. A short section of rubber tube on each water outlet is an excellent water flusher to wash away harmful chemicals from the eyes and skin. It is easy to reach and can quickly be directed on the exposed area. Eyes and skin can be saved if dangerous materials are washed away quickly. Provide an emergency eye wash and deluge shower in any area where corrosive, caustic, or otherwise hazardous substances are handled and stored. This equipment requires monthly operation and flushing to ensure that it will operate when needed.

12. Unsafe glassware is the largest single cause of accidents in the laboratory. Chipped glassware may still be used if it is possible to fire polish the chip in order to eliminate the sharp edges. This may be done by slowly heating the chipped area until it reaches a temperature at which the glass will begin to melt. At this point remove from flame and allow to cool.

> **NEVER HOLD ANY PIECE OF GLASSWARE OR EQUIPMENT IN YOUR BARE HANDS WHILE HEATING. ALWAYS USE A SUITABLE GLOVE OR TOOL!**

13. Dispose of all broken or cracked glassware immediately. A special receptacle for broken glass should be available and well identified.

14. Wear a protective smock or apron when working in the lab. This may save you the cost of replacing your work clothes or uniform and prevent serious injury to yourself. Protective gloves and goggles should be worn when working with corrosive/dangerous chemicals.

> **REMEMBER TO ADD ACID TO WATER, BUT NEVER THE REVERSE.**

15. Electrical equipment must be properly grounded and safeguards provided to prevent insertion of improper plugs into the equipment.

16. Do not keep your lunch in a refrigerator that is used for samples or chemical storage.

17. Where cylinders of compressed gas are used or stored, the area must be well ventilated and heat sources eliminated. Ensure that the cylinders are properly secured.

18. Carbon dioxide (CO_2) or all-purpose, ABC chemical-type extinguishers should be mounted in readily accessible locations throughout the laboratory. A D-type extinguisher should be available where burning metals might be encountered.

19. Maintain spill control stations in sufficient quantity and locations so that chemical spills can be cleaned up in a safe and timely manner.

20. Provide appropriate first-aid kits at accessible and strategic locations within the laboratory.

21. Properly store and dispose of hazardous chemicals and wastes. Requirements for hazardous waste storage include, but are not limited to:

    - Do not place incompatible wastes in the same container.
    - Store hazardous waste containers in a secondary container to prevent uncontrolled leaks in the event the primary container fails.
    - Use a container made of a material that will not react with the hazardous material being stored. Containers should be DOT (Department of Transportation) approved for ultimate transport off site.
    - Keep the hazardous waste container closed unless it is necessary to add or remove waste.
    - Do not open, handle, or store a hazardous waste container in a manner that may cause it to rupture or leak.
    - At least weekly, inspect areas where hazardous waste containers are stored; look for evidence of leaks and container deterioration. Take corrective action as required.
    - Label containers with a hazardous waste label describing the waste contained and the date that accumulation started.

There are many more requirements for dealing with hazardous waste generation and disposal. Most areas limit the amount of time that a hazardous waste can be stored on site to 90 days, however, there are exceptions for certain operations. Your plant may also be required to obtain a federal EPA identification number that must be used on hazardous waste manifests that are required when you transport, or arrange transport of, hazardous wastes from your plant. Contact your local or state Department of Environmental Management for guidance and requirements for your particular plant and wastestreams. Remember, items such as lead-acid batteries, used oil and oil filters, spent or discarded laboratory chemicals, contaminated containers, and asbestos gaskets can all be classified as hazardous wastes.

Treatment plants that have a laboratory must develop and implement a written Chemical Hygiene Plan. Some of the items that must be included in the plan are:

- Standard operating procedures to be followed when using hazardous chemicals.

- Criteria that will be used to determine and implement control measures to reduce employee exposure, including engineering controls, personal protective equipment, and hygiene practices.

- A requirement that the operation of fume hoods will comply with regulatory requirements and that specific measures will be taken to ensure that all protective equipment will function properly.

- Provisions for employee training that includes, but is not limited to, hazard communication (see Section 13.9).

- The circumstances under which a particular laboratory operation or activity will require prior approval.

- Provisions for medical consultation and medical examinations, if required.

- Assignment of a chemical hygiene officer.

- Provisions for additional employee protection for work with particularly hazardous substances such as select carcinogens and reproductive toxins.

This listing does not include all of the detailed requirements of a chemical hygiene plan. You should contact your local safety regulatory agency if you need assistance in the development of your program.

QUESTIONS

Write your answers in a notebook and then compare your answers with those on page 733.

13.3A What safety precautions would you take when collecting laboratory samples from a plant influent?

13.3B Why should you always wash your hands before eating?

13.3C Why should chemicals and reagents be handled with care?

13.4 FIRE PREVENTION

Fires are a serious threat to the health and safety of the operator and to the buildings and equipment in a treatment plant. Fires may injure or cause the death of an operator. Equipment damaged by fire may no longer function properly, and your treatment facility may have difficulty adequately treating the influent wastewater.

Good safety practices with respect to fire prevention require a knowledge of:

1. Ingredients necessary for a fire
2. Fire control methods
3. Fire prevention practices

13.40 Ingredients Necessary for a Fire

The three essential ingredients of all ordinary fires are:

1. FUEL—paper, wood, oil, solvents, and gas.
2. HEAT—the degree necessary to vaporize fuel according to its nature.
3. OXYGEN—normally at least 15 percent of oxygen in the air is necessary to sustain a fire. The greater the concentration, the brighter the blaze and more rapid the combustion.

13.41 Fire Control Methods

To extinguish a fire, it is necessary to remove only one of the essentials by:

1. Cooling (temperature and heat control)
2. Smothering (oxygen control)
3. Isolation (fuel control)

Fire classifications are important for determining the type of fire extinguisher needed to control the fire. Classifications also aid in recordkeeping. Fires are classified as A, B, C, or D fires based on the type of material being consumed: A, ordinary combustibles; B, flammable liquids and vapors; C, energized electrical equipment; and D, combustible metals. Fire extinguishers are also classified as A, B, C, or D to correspond with the class of fire each will extinguish.

Class A fires: ordinary combustibles, such as wood, paper, cloth, rubber, many plastics, dried grass, hay, and stubble. Use foam, water, soda-acid, carbon dioxide gas, or almost any type of extinguisher.

Class B fires: flammable and combustible liquids, such as gasoline, oil, grease, tar, oil-based paint, lacquer, and solvents, and also flammable gases. Use foam, carbon dioxide, or dry chemical extinguishers.

Class C fires: energized electrical equipment, such as starters, breakers, and motors. Use carbon dioxide or dry chemical extinguishers to smother the fire; both types are nonconductors of electricity.

Class D fires: combustible metals, such as magnesium, sodium, zinc, and potassium. Operators rarely encounter this type of fire. Use a Class D extinguisher or use fine dry soda ash, sand, or graphite to smother the fire. Consult with your local fire department about the best methods to use for specific hazards that exist at your facility.

Multipurpose extinguishers are also available, such as a Class BC carbon dioxide extinguisher that can be used to smother Class B and Class C fires. A multipurpose, ABC carbon dioxide extinguisher will handle most laboratory fire situations. (When using carbon dioxide extinguishers, remember that the carbon dioxide can displace oxygen—take appropriate precautions.)

There is no single type of fire extinguisher that is effective for all fires so it is important that you understand the class of fire you are trying to control. You must be trained in the use of the different types of extinguishers, and the proper type should be located near the area where that class of fire may occur.

13.42 Fire Prevention Practices

You can prevent fires by:

1. Maintaining a neat and clean work area, preventing accumulation of debris and combustible materials.
2. Putting oil- and paint-soaked rags in covered, metal containers and regularly disposing of them in a safe manner.
3. Observing all "no smoking" signs.
4. Keeping fire doors, exits, stairs, fire lanes, and firefighting equipment clear of obstructions.
5. Keeping all combustible materials away from furnaces and other sources of ignition.
6. Reporting any fire hazards you see, especially electrical hazards, which are the source of many fires.

Finally, here again are the things to remember:

1. Prevent fires by good housekeeping and proper handling of flammables.
2. Make sure that everyone obeys "no smoking" signs in all classified areas of your facility.
3. In case of fire, turn in the alarm immediately and make sure that the fire department is properly directed to the place of the fire.
4. Action during the first few seconds of ignition generally means the difference between destruction and control. Use the available portable firefighting equipment to control the fire until help arrives only if you have been trained and are qualified to do so.
5. Use the proper extinguisher for that fire.
6. Learn how to operate the extinguishers BEFORE an emergency arises.

If it is necessary, evacuate the building; do not stop to get anything—just leave!

Can you prevent fires? You can if you try, so let us see what we can do to preserve our well-being and the water pollution control system.

If you guard against fires, you will be protecting your lives and your community.

Additional recommendations can be found in Section 13.18.

QUESTIONS

Write your answers in a notebook and then compare your answers with those on page 733.

13.4A What are the necessary ingredients of a fire?

13.4B How should oil- and paint-soaked rags be handled?

13.5 WATER SUPPLIES

Inspect your plant to see if there are any cross connections between your potable (drinking) water and items such as water seals on pumps, feed water to boilers, hose bibs below grade where they may be subject to flooding with wastewater or sludges, or any other location where wastewater could contaminate a domestic water supply.

If any of these or other existing or potential cross connections are found, be certain that your drinking water supply source is properly protected by the installation of an approved backflow prevention device. Many treatment plants use an *AIR GAP DEVICE*[23] (Figure 13.14) to protect their drinking water supply.

It is a good practice to have your drinking water tested at least monthly for coliform group organisms. Sometimes the best of backflow prevention devices do fail.

Never drink from outside water connections such as faucets and hoses. The hose you drink from may have been used to carry effluent or sludge.

You may find in your plant that it will be more reliable and more economical to use bottled drinking water. If so, be sure to

---

[23] *Air Gap.* An open, vertical drop, or vertical empty space, between a drinking (potable) water supply and potentially contaminated water. This gap prevents the contamination of drinking water by backsiphonage because there is no way potentially contaminated water can reach the drinking water supply.

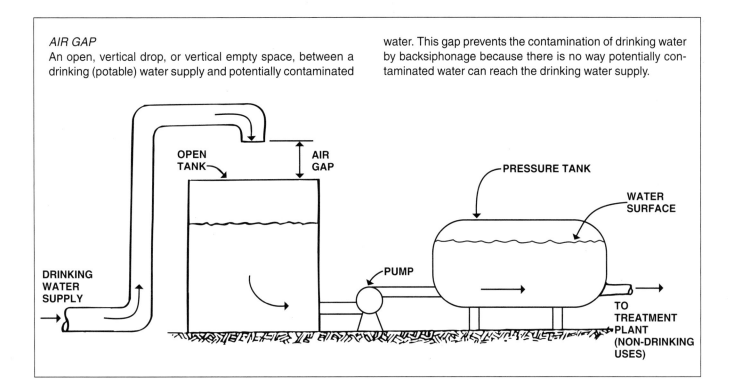

Fig. 13.14 Air gap device

post conspicuous signs that your plant water is not potable at all outlets. This also applies to all hose bibs in the plant from which you may obtain water other than a potable source. This is a must in order to inform visitors or absent-minded or thirsty employees that the water from each marked location is not for drinking purposes. This practice reduces the possibility of the spread of disease from unknown cross connections or defective devices installed to prevent contamination by backflows.

13.6 SAFETY EQUIPMENT AND INFORMATION

Conspicuously post on your bulletin board the location and types of safety equipment available at your plant (such as first-aid kits, respiratory protective devices, and atmospheric monitors). You, as the plant operator, should be thoroughly familiar with the operation and maintenance of each piece of equipment. You should review these at fixed intervals to be certain that you can safely use the piece of equipment as well as to be sure that it is in good operating condition.

Contacts should be made with your local fire and police departments to acquaint them with hazards at your plant as well as to inform them of the safety equipment that is necessary to cope with problems that may arise. Arrange a joint training session with these people in the use of safety equipment and the handling of emergencies. They also should know access routes to and around the treatment facility.

If you have any specific problems of a safety nature, do not hesitate to contact officials in your state safety agency. They can be of great assistance to you. And do not forget your equipment manufacturers; their familiarity with your equipment will be of great value to you.

Also, posted in conspicuous places in your facility should be such information as the phone numbers of your fire and police departments, ambulance service, chemical supplier or repair service, and the nearest doctor who has agreed to be available on call. Having these immediately available at telephone sites may save your or a fellow worker's life. Check and make sure these numbers are listed at your plant. If they are not listed, *add them now*. Also see OPERATION OF WASTEWATER TREATMENT PLANTS, Volume II, Chapter 15, Section 15.042, "Emergencies," for a list of emergency response agencies.

Prepare an emergency medical information sheet for each operator. Keep all of these sheets together in one binder. Send a copy of the appropriate sheet with the ambulance that takes an injured operator to the hospital.

QUESTIONS

Write your answers in a notebook and then compare your answers with those on page 733.

13.5A Why do some wastewater treatment plants use bottled water for drinking purposes?

13.6A What emergency phone numbers should be listed in a conspicuous place in your facility?

13.7 "TAILGATE" SAFETY MEETINGS [24]

Safety is crucial. Accidents cost money. No one can afford to lose time from their job due to injury. The purpose of tailgate safety meetings is to remind operators of the need for safety, to explain and discuss safe procedures and safe conditions, and to review potential hazards and how to correct or avoid dangerous situations.

In some states, you are required by law to conduct safety meetings at fixed intervals with employees. Whether this is required or not, it certainly is a good practice. Invite police and fire personnel to participate from time to time so you get to know them and they become acquainted with you and your facilities. Once every 7 to 10 working days is a good frequency. These meetings should usually be confined to one topic, and should be from 10 to 20 minutes long. It will be worthwhile to review any accidents that have occurred since the last meeting. Do not use this meeting to fix blame. Try to determine the cause and what can be or has been done to prevent a similar accident in the future.

To help you conduct tailgate safety meetings, this chapter was arranged to discuss the safety aspects of different plant operations. The material in some sections was deliberately repeated to cover the topic and to remind you of dangers. Some plants select topics for their tailgate safety meetings from a "safety goof box." The box is placed in a convenient location. Whenever anyone sees an unsafe situation or sees someone perform a hazardous act without proper safety precautions, this person places a note in the box identifying the situation or act. The box is opened at each safety meeting, and the cause of the "goof" and the steps that can be taken to correct and prevent it from happening again are discussed.

Your state safety agency, your insurance company, equipment and material suppliers, and the Water Environment Federation are all excellent sources of literature and aids that may help you in conducting tailgate safety meetings. Some of these agencies may be able to supply you with posters, signs, and slogans that are very effective safety reminders. You may wish to create some reminders of your own.

QUESTIONS

Write your answers in a notebook and then compare your answers with those on page 734.

13.7A What is the purpose of tailgate safety meetings?

13.7B How frequently should safety meetings be held for treatment facility operators?

13.8 HOW TO DEVELOP SAFETY TRAINING PROGRAMS

13.80 Conditions for an Effective Safety Program

Effective safety programs rely on many techniques to help workers recognize hazards and learn safe procedures. These safety programs can range from highly organized meetings to tailgate safety sessions to informal get-togethers. Safety programs of all types have proven very effective and usually stress the following points:

1. Basic safety concepts and practices are thoroughly understood by all.

2. Everyone participates and accepts personal responsibility for their own safety and that of their fellow workers (participation at all levels is absolutely essential to the continued success of any safety program).

3. Adequate safety equipment is available and its capabilities are thoroughly understood. Responsible individuals must regularly review and drill in the actual use of the equipment in order to safely respond to an emergency and hazardous conditions.

4. Everyone realizes that safety is a continuing learning and relearning process—a way of life that must become habit.

5. Accidents are studied step by step and thoroughly reviewed with the attitude that "they are caused and do not just happen." Every reasonable step will be taken to reduce the chance of an accident happening again to as near zero as is practical.

6. Every detail of work is a subject for discussion to the extent it will improve safety.

7. Operators realize they should stop anyone performing an unsafe act and remind the person that they are not following safe procedures, why, and how the job can be done safely.

8. Before starting a job, operators ensure that they can do the job without injury. If assigned work they are not qualified to perform, they do not just blindly do it, but bring it to the attention of their supervisor.

9. Before starting a job, operators thoroughly understand the work to be done, the job, and the safety rules that apply. Tailgate safety sessions or pre-job discussions will help promote safe operations.

10. Management actively supports the safety training program and demands that safe equipment and procedures be used at all times.

13.81 Start at the Top

An effective safety training program must start at the top. The person who controls the purse strings and makes final decisions must not only support safety but must promote it from the start

---

[24] *Tailgate Safety Meeting.* Brief (10 to 20 minutes) safety meetings held every 7 to 10 working days. The term comes from the safety meetings regularly held by the construction industry around the tailgate of a truck.

and continuously promote safety on a day-to-day basis. The safety director must have direct access to this person. Without this type of organization, a safety program may be put off, watered down, or even eliminated in the name of urgency, time, and cost.

Top management's attitude and approach toward safety will probably be reflected in the attitude of the supervisors, and the operator's attitude will most likely be the same as that of the supervisors. Thus, if management is not committed to safety, no one else is likely to be. Management must issue a safety policy statement that makes it clear that safety will take precedence over expediency or shortcuts and that the safety of employees, the public, and plant operations is a top priority.

13.82 The Supervisor's Role

Supervisors are the key people in a safety program because they are in constant contact with operators. Supervisors can actively support the plant's safety policy. This sets an example for their operators and gives safety the emphasis it must have. First-line supervisors should be responsible to see that operators:

- Understand the hazards of the various processes at the plant.

- Observe necessary precautions when performing operations, maintenance, or laboratory tasks, including the use of safe work techniques and appropriate personal protective equipment.

- Understand and properly follow established work procedures for their safety and the safety of others.

Supervisors are the plant's safety representatives in the field and their actions and attitudes will directly affect the entire safety program.

13.83 Plan for Emergencies

Start where you are. Nothing is going to stop while you get your safety program organized. Emergencies, accidents, and injuries can happen at any time and usually at the wrong time. Try to minimize the impact of accidents while trying to develop or improve your plan for prevention. The first step is to prepare emergency procedures for your treatment facilities, collection system, and vehicles. These plans should include:

1. What to do and what not to do for the injured.

2. How to contact the nearest fire department, rescue squad, or ambulance service.

3. Identification of the injured and notification of relatives.

4. Directions for emergency vehicles to reach the scene and to locate the victim.

5. Prevention of further damage to people and property.

6. Names of persons and authorities to be notified after the emergency.

All employees must be interested and trained in these procedures and copies should be posted in prominent places in all plant areas, pumping stations, and vehicles. The fact that employees are preparing for emergencies will have a positive effect in reducing accidents.

13.84 Promote Safety

Start early with your promotion of safety. Be proactive, not reactive. Make safety a part of discussions and work procedures. All types of on-the-job training should emphasize the importance of learning and practicing safe work procedures. Develop a written Standard Operating Practice (SOP) for routine duties or equipment operation and have regular training sessions over each SOP. This will not only point out safety aspects of the job, but will also be a way to train people in the most efficient way to work. Follow safe practices yourself—this applies especially to supervisory personnel. Example is a powerful incentive.

13.85 Hold Safety Drills and Train for Safety

No matter how well safety is engineered into your plant or a job, much of the safety of operators depends on their own conduct. Some people work safely in dangerous surroundings while others have accidents on jobs that seem very safe.

The training of operators begins the day they start work. Whether or not your plant has a formal safety orientation program, the operators start to learn about their jobs and to form attitudes about many things, including safety, the first day.

Use every opportunity to give safety instruction from 10-minute, on-the-spot chats to supervisory safety meetings. Vary the techniques and timing with chats, meetings, drills, exercises, workshops, and seminars. Cover all the subjects. Match discussions to incidents, such as slips and falls during the slippery season, defensive driving if a bad accident has occurred in the area, and chlorine safety if there have been problems with leaks. Make your point about safety while details of specific situations are fresh in everyone's mind.

Remember those fire drills in school? Try drilling on how employees should respond during emergencies, including evacuation of facilities. All facilities should have the necessary safety equipment (fire extinguishers, self-contained breathing apparatus, and atmospheric monitors), written procedures, and appropriate personnel trained for emergency response. Do not wait for an emergency before trying to learn how to use this equipment. Get proper instructions and conduct practice sessions.

First aid and chemical safety are important steps in organizing others to assist in your safety program. You can use the Red Cross multimedia program to train first-aid teams and instructors. Chemical manufacturers or distributors provide excellent instruction in chemical hazards and safety precautions.

When developing your training courses, try to emphasize the most hazardous tasks that are likely to cause accidents. Studies have shown that injuries most often occur when doing activities that are not routine. In your course discussions try to identify how hazards can cause injuries, how bad the injuries can be, and ways to avoid injury.

13.86 Purchase the Obvious Safety Equipment First

Hard hats, safety footwear, eye protection, and hearing protection apply to all personnel in designated areas and specific jobs. Purchase this equipment and post the areas where it must be used. The purchase of more specific and expensive equipment such as atmospheric monitors, light meters, and noise monitoring devices, is also very important. As your safety program develops, the benefits of more specific and more sophisticated safety equipment will become clear; the need and time to purchase them will be obvious.

13.87 Safety Is Important for Everyone

As your safety program develops, you will realize that safety is the responsibility of everyone, from managers to workers. Everyone must be involved. Organize safety committees and meetings from top to bottom, as well as from bottom to top. If you do it well, then safety practices will progress from top to bottom. Ideas and suggestions will come if they are recognized and implemented.

13.88 Necessary Paperwork

When you start to develop your safety program, concentrate your efforts on programs that apply generally to all employees. Paperwork can be helpful to identify the causes of accidents and to develop corrective procedures.

1. Accident report forms (Figure 13.15). Use these forms to record and analyze the causes of accidents and to prevent future accidents.
2. Safety policy. The plant manager must establish a safety policy and repeatedly demonstrate support of the policy.
3. Safety rules. Safety rules are as important as work rules and they should be implemented and enforced in the same manner. Most people perform better and with more confidence if they know the rules of the game. These rules must apply to everyone. Supervisors should serve as examples to the operators.
4. Supervisors' guidelines. Supervisors must have guidelines on how to promote and implement a safety program and enforce the rules.
5. Facility plans and specifications; plant inspection reports. State and OSHA regulations must be used when reviewing plans and specifications. Checklists are a tremendous aid during plant inspections.

```
CITY OF _____ WASTEWATER TREATMENT
              PLANT ACCIDENT REPORT
Date of this report _____ Name of person injured _____
                                              a.m.
Date of injury _____ Time _____ p.m. Occupation _____
Home address _____ Age _____ Sex _____
Check ___ First aid case, or ___ disabling (lost time) injury
___ Employee _____ on duty, or _____ off duty
___ Visitor injury
Date last worked _____ Date returned to work _____
Person reporting _____

              DESCRIPTION OF ACCIDENT
1. Description of Accident _____
   (Describe in detail what happened) (Name machine,
   tool, appliance, _____
   gas or liquid involved—if machine or vehicle—name
   part, gears, pulley, etc.)
2. Accident occurred where? _____
   If vehicle accident,
   make simple sketch of
   scene of accident.
3. Describe nature of injury and part of body affected ___
   _____
   (Amputation of finger, laceration of leg, back strain, etc.)
4. Were other persons involved? _____
        (If yes, give names and addresses.)
   _____
5. Names and addresses of witnesses _____
   _____
6. If property damage involved, give brief description ___
   _____
7. If hospitalized, name of hospital _____
8. Name and address of physician _____
9. Treatment given for injuries _____
```

Fig. 13.15 Typical accident report form

13.89 Summary

All types of safety programs are helpful. If variety is the spice of life, let variety add spice to your safety program. Informal chats on safety do not replace formal safety meetings or vice versa. Every type of safety meeting can help you develop a more effective safety program.

Your safety program should include the following items:

1. Get your top official to support and promote safety (have the official issue a written safety policy statement).
2. Give your safety officer direct access to the plant manager.
3. Direct your program from general topics to the more specific.
4. Organize from top to bottom.

5. Establish rules and implement and enforce them.

6. Train at all levels from employment to retirement.

7. *MAKE SAFETY A HABIT.*

QUESTIONS

Write your answers in a notebook and then compare your answers with those on page 734.

13.8A List three types of safety meetings.

13.8B What is the role of management in an effective safety training program?

13.8C Why should safety drills be held regularly?

13.8D What types of paperwork are necessary for an effective safety training program?

13.9 HAZARD COMMUNICATION (WORKER RIGHT-TO-KNOW LAWS)

In the past several years, there has been an increased emphasis nationally on hazardous materials and wastes. Much of this attention has focused on hazardous and toxic waste dumps, and the efforts to clean them up after the long-term effects on human health were recognized. Each year, thousands of new chemical compounds are produced for industrial, commercial, and household use. Frequently, the long-term effects of these chemicals are unknown. Exposure to the wastewater treatment plant operator can occur from one or more of the following:

1. Use of the collection system as an intentional or accidental disposal method for hazardous materials, for example:

    a. Industrial solvents

    b. Acids

    c. Flammable/explosive compounds

    d. Caustics

    e. Toxics

2. Chemicals and chemical compounds that we use every day for operation and maintenance in the treatment facility, for example:

    a. Solvents

    b. Degreasers

    c. Acids

    d. Chlorine

    e. Industrial cleaners

Consequently, federal and state laws have been enacted to control all aspects of hazardous materials handling and use. These laws are more commonly known as Worker Right-To-Know (R-T-K) laws. Every state is covered by one or more laws regarding Worker Right-To-Know. The Federal Occupational Safety and Health Administration (OSHA) Standard 29 CFR 1910.1200—Hazard Communication forms the basis of most laws. Although the federal standards were originally directed at the manufacturing sector, they now include all industries including the public sector and, therefore, collection systems and treatment plants.

In many cases, the individual states have the authority under the OSHA standard to develop their own state Worker Right-To-Know laws and most states have adopted their own laws. Unfortunately, state laws vary significantly from state to state. The state laws that have been passed are at least as stringent as the federal standard and, in most cases, are even more stringent and already apply to treatment facility operators.

State laws are also under continuous revision and, because a strong emphasis is being placed on hazardous materials and worker exposure, state laws can be expected to be amended in the near future to apply to virtually everybody in the workplace.

The purpose of this section is to familiarize you with general requirements of Worker RTK elements so that you are better prepared as a treatment plant operator to minimize risk to yourself and your co-workers from hazardous materials, and to comply with state and federal laws as well. Because of the wide diversity of existing laws, these guidelines may or may not meet your state's requirements. This section will give you an overview of the basic elements of a hazard communication program (see Table 13.2), which can be particularly useful if your agency currently does not have one. An extremely effective program can be developed in house through your safety program and committee. By cutting through the bureaucratic/legal language and applying some common sense to what the law is trying to accomplish (protection of the worker in the workplace from hazardous materials), an effective program can be developed.

Basically, the different elements of the program are as follows:

1. Identify Hazardous Materials

While there are thousands and thousands of chemical compounds that would fall under this definition in a technical sense, treatment plant operators should be concerned, first of all, with the materials they use in their everyday operation and maintenance activities; and also with materials that can be introduced to the collection system through intentional or accidental spills. Information on materials that could be introduced into the collection system can be obtained from industrial pretreatment inspectors. (Also see the manual, *PRETREATMENT FACILITY INSPECTION*, in this series of manuals.)

In particular, you should be familiar with the industries in your area and be aware of the materials they routinely discharge into the collection system as well as the types of hazardous materials they might accidentally discharge. In some cases, special precautions may have to be taken. For example, when your collection system operators are performing routine maintenance with a high-velocity cleaner (jet machine) in a section of line, you may wish to draw up an agreement with an industrial discharger upstream to halt discharges on the day the sewer main-

TABLE 13.2 ELEMENTS OF A HAZARD COMMUNICATION PROGRAM

I. *WRITTEN HAZARD COMMUNICATION PROGRAM*

   A. Hazard Determination (Chemical manufacturers and importers only)

      1. Person(s) responsible for evaluating the chemical(s)
      2. List of sources to be consulted
      3. Criteria to be used to evaluate studies
      4. A plan for reviewing information to update MSDSs if new and significant information is found

   B. Labels and Other Forms of Warning

      1. Person(s) responsible for labeling in-plant containers, if used
      2. Person(s) responsible for labeling shipped containers
      3. Description of labeling system
      4. Description of written alternatives to labeling of in-plant containers, if used
      5. Procedure to review and update label information

   C. Material Safety Data Sheets (MSDSs)

      1. Person(s) responsible for obtaining or maintaining the MSDSs
      2. Description of how the MSDSs will be made available to employees
      3. Procedure to follow when the MSDS is not received at time of first shipment
      4. Procedure to review and update MSDS information
      5. Description of alternatives to actual data sheets in the workplace, if used

   D. Training

      1. Person(s) responsible for conducting training
      2. Format of the program to be used
      3. Elements of the training program

          a. Requirements of the OSHA standard
          b. Operations, where hazardous chemicals are present in routine tasks, non-routine tasks, and foreseeable emergencies
          c. Location and availability of

               i. Written hazard communication program
              ii. List of hazardous chemicals
             iii. MSDSs

          d. Methods and observations that may be used to detect the presence or release of hazardous chemicals in the work area
          e. Physical and health hazards of chemicals in the work area
          f. Measures employees can take to protect themselves from hazards
          g. Details for the Hazard Communication Program developed

      4. Procedure to train new employees, as well as current employees, when a new chemical hazard is introduced into the workplace

   E. List of hazardous chemicals

   F. Procedure to inform employees of the hazards of chemicals in unlabeled pipes

   G. Procedure to inform on-site contractors

II. *LABELS AND OTHER FORMS OF WARNING*

   A. Labels or other markings on each container of hazardous chemicals

      1. Process vessels
      2. Storage tanks
      3. Compressed gas cylinders
      4. Product containers
      5. Tank truck and tank car labels

III. *MATERIAL SAFETY DATA SHEETS (MSDSs)*

   A. MSDSs developed or obtained for all hazardous chemicals

   B. Employees have access on each shift

   C. MSDS completed appropriately

   D. When no MSDS is available, the documentation requesting an MSDS from supplier is maintained

TABLE 13.2 ELEMENTS OF A HAZARD COMMUNICATION PROGRAM (continued)

IV. *TRAINING*

    A. Employee training files

    B. Employee questioning:

        1. Are they aware of the Hazard Communication Program (HCP) and its requirements?
        2. Have they received training?
        3. Are they able to locate the MSDSs?
        4. Do they have a general familiarity with the hazardous properties of the chemicals in their workplace?

---

tenance is being performed. Hazardous materials can be broken down into general categories as follows:

1. Corrosives
2. Toxics
3. Flammables and explosives
4. Asphyxiants
5. Harmful physical agents
6. Infectious agents

A complete inventory of materials in use will produce a list similar to the following:

1. Corrosives
   a. Sodium hydroxide
   b. Calcium oxide (lime)
   c. Hydrochloric acid
   d. Ferric chloride
2. Toxics
   a. Hydrogen sulfide
   b. Chlorine
   c. Carbon monoxide
3. Flammables and Explosives
   a. Methane
   b. Acetylene
   c. Gasoline
   d. Solvents
4. Asphyxiants
   a. Carbon Dioxide
   b. Nitrogen
5. Harmful Physical Agents
   a. Noise
   b. Temperature
   c. Radiation
6. Infectious Agents

2. Obtain Chemical Information and Define Hazardous Conditions

Once the inventory is complete, the next step is to obtain specific information on each of the chemicals and hazardous conditions. This information is generally incorporated into a standard format form called the "Material Safety Data Sheet" (MSDS). Figure 13.16 is the Material Safety Data Sheet (OSHA 174, September, 1985) produced by the US Department of Labor, Occupational Safety and Health Administration. This information is commonly available from manufacturers. Many agencies request an MSDS when the purchase order is generated and will refuse to accept delivery of the shipment if the MSDS is not included.

The purpose of the MSDS is to have a readily available reference document that includes complete information on common names, safe exposure level, effects of exposure, symptoms of exposure, flammability rating, type of first-aid procedures, and other information about each hazardous substance.

Operators must be trained to read and understand the MSDS forms. The forms themselves must be stored in a convenient location where they are readily available for reference.

3. Properly Label Hazards

Once the physical, chemical, and health hazards have been identified and listed, a labeling and training program must be implemented. To meet labeling requirements on hazardous materials, specialized labeling is available from a number of sources, including commercial label manufacturers. Exemptions to labeling requirements do exist, so consult your local safety regulatory agency for specific details.

Each hazardous substance container in the workplace must be labeled, tagged, or marked with the name of the hazardous substance and appropriate warning labels. There are a number of acceptable labeling systems available, including private labeling systems such as the one illustrated in Figure 13.17, produced by the J. T. Baker Chemical Company. The second illustration shows a label produced by a commercial label maker (Figure 13.18), and the third, in Figure 13.19, is one designed by the National Fire Protection Association (NFPA).

These are standardized formats that use a combination of pictographs, a numbering system, and colors to indicate various levels of conditions. In some cases, the MSDS sheet can be incorporated into the labeling requirements by locating the appropriate MSDS in close proximity to drums or storage areas. The labeling requirements offer the treatment facility operator virtually

Material Safety Data Sheet

May be used to comply with
OSHA's Hazard Communication Standard,
29 CFR 1910.1200 Standard must be
consulted for specific requirements.

U.S. Department of Labor
Occupational Safety and Health Administration
(Non-Mandatory Form)
Form Approved
OMB No. 1218-0072

IDENTITY (As Used on Label and List)

Note: Blank spaces are not permitted. If any item is not applicable, or no information is available, the space must be marked to indicate that.

Section I

Manufacturer's Name	Emergency Telephone Number
Address (Number, Street, City, State, and ZIP Code)	Telephone Number for Information
	Date Prepared
	Signature of Preparer (optional)

Section II—Hazardous Ingredients/Identity Information

Hazardous Components (Specific Chemical Identity: Common Name(s))	OSHA PEL	ACGIH TLV	Other Limits Recommended	%(optional)

Section III—Physical/Chemical Characteristics

Boiling Point		Specific Gravity (H_2O = 1)	
Vapor Pressure (mm Hg.)		Melting Point	
Vapor Density (AIR = 1)		Evaporation Rate (Butyl Acetate = 1)	
Solubility in Water			
Appearance and Odor			

Section IV—Fire and Explosion Hazard Data

Flash Point (Method Used)	Flammable Limits	LEL	UEL
Extinguishing Media			
Special Fire Fighting Procedures			
Unusual Fire and Explosion Hazards			

(Reproduce locally)

OSHA 174, Sept. 1985

Fig. 13.16 Material Safety Data Sheet

Section V—Reactivity Data

Stability	Unstable		Conditions to Avoid
	Stable		

Incompatibility *(Materials to Avoid)*

Hazardous Decomposition or Byproducts

Hazardous Polymerization	May Occur		Condition to Avoid
	Will Not Occur		

Section VI—Health Hazard Data

Route(s) of Entry: Inhalation?   Skin?   Ingestion?

Health Hazards *(Acute and Chronic)*

Carcinogenicity   NTP?   IARC Monographs?   OSHA Regulated?

Signs and Symptoms of Exposure

Medical Conditions
Generally Aggravated by Exposure

Emergency and First Aid Procedure

Section VII—Precautions for Safe Handling and Use

Steps to Be Taken in Case Material is Released or Spilled

Waste Disposal Method

Precautions to be Taken in Handling and Storing

Other Precautions

Section VIII—Control Measures

Respiratory Protection (Specify Type)

Ventilation	Local Exhaust		Special
	Mechanical *(General)*		Other

Protective Gloves	Eye Protection

Other Protective Clothing or Equipment

Work/Hygienic Practices

Fig. 13.16   Material Safety Data Sheet (continued)

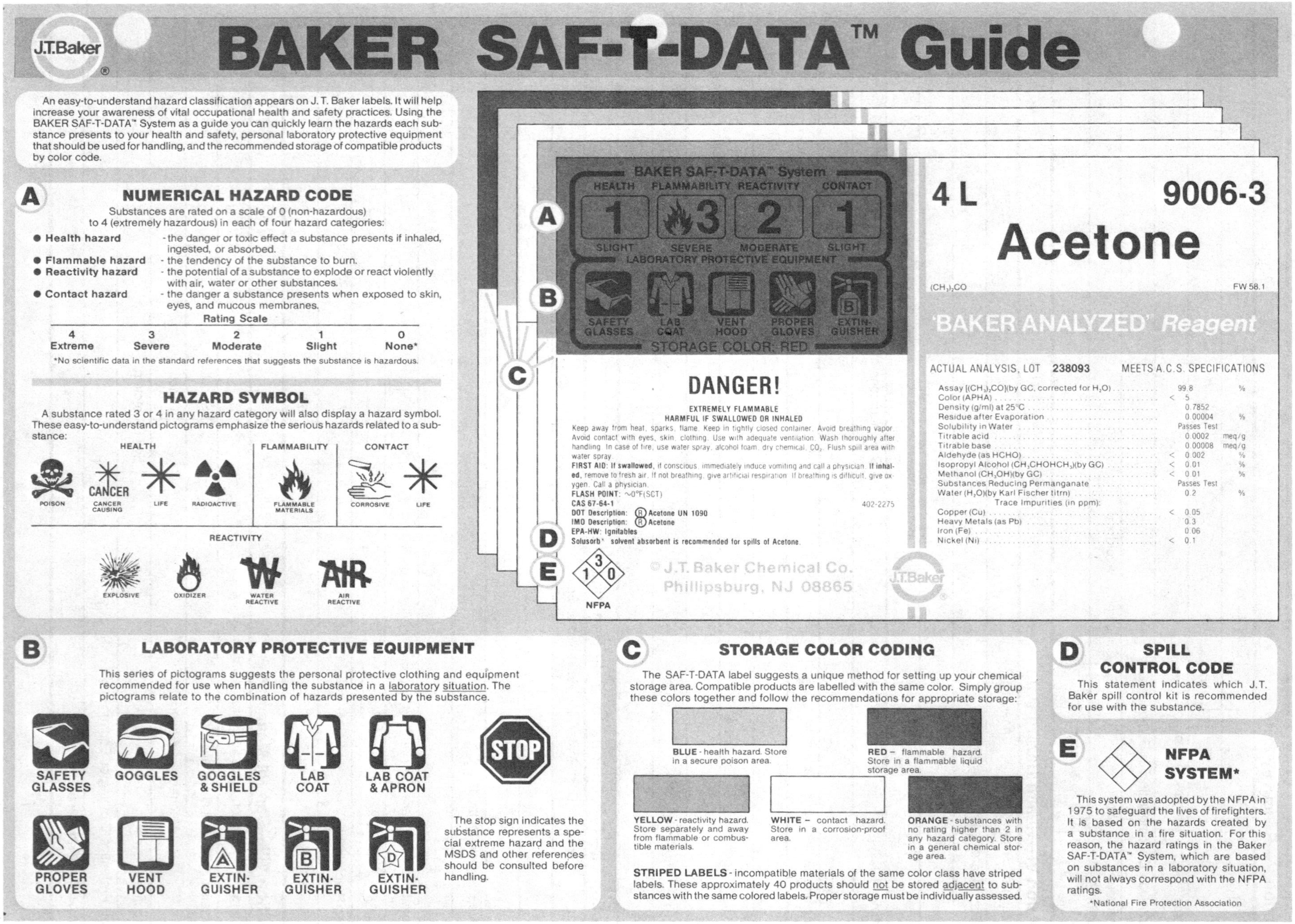

Fig. 13.17 J. T. Baker hazardous substance labeling system (Permission of J. T. Baker Chemical Company)

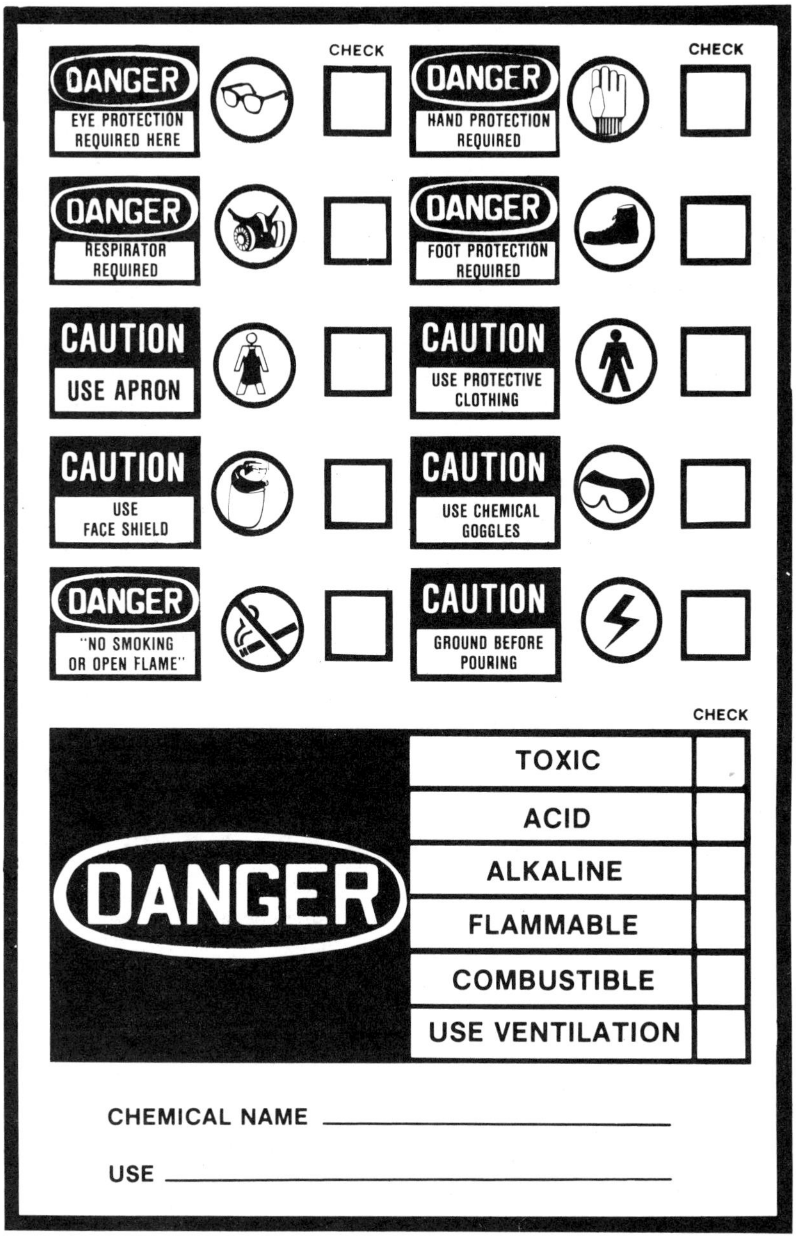

Fig. 13.18 Commercial warning labels
(Permission of the SIGNMARK Division, W. H. Brady Co.)

728 Treatment Plants

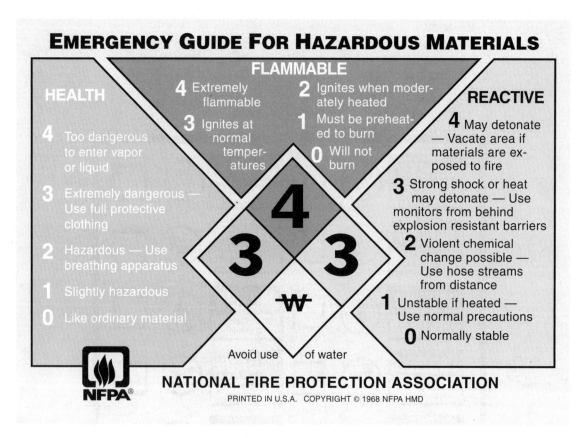

Fig. 13.19 NFPA hazard warning label
(Permission of National Fire Protection Association)

instant recognition of the hazards in dealing with specific substances, protective equipment required, and other information.[25]

4. Train Operators

The last element in the Hazard Communication Worker Right-To-Know Program is training and making information available to the collection system and treatment facility operators. A common-sense approach eliminates the confusing issue of which of the thousands of substances operators should be trained for, and concentrates on those that they will be exposed to or use in everyday maintenance routines. Obviously, the protection from hazardous materials is tied in with the confined space policy, since average domestic wastewater found in collection systems and treatment plants does not contain sufficient concentrations of hazardous materials to address each of the compounds that may be found. The use of ventilating equipment, atmospheric testing instrumentation, and the other precautions defined in the confined space procedure assist in protecting the collection system and treatment plant operators from hazardous materials.

Industrial dischargers can and do discharge significant concentrations of hazardous materials, either intentionally or unintentionally. An industrial treatment facility operator should be familiar with each industry that discharges to the collection system, since some special precautions may need to be taken to minimize exposure to specific hazardous materials.

The last element is a formal training program provided by qualified personnel designed to accomplish the following:

1. Make employees aware of the hazard communication standard and its requirements.

2. Familiarize employees with potentially dangerous operations and hazardous substances that are present in their routine and non-routine operation and maintenance tasks and how to deal with emergency situations involving those materials and conditions.

3. Inform employees of the location and availability of the written hazard communication program.

4. Train employees in the methods and observations that they may use to detect the presence or release of a hazardous substance in the work area such as visual appearance, odor, and monitoring.

5. Inform employees of the measures they can take to protect themselves from the physical and health hazards of the chemicals in the work area.

---

[25] The United States and Europe are working on an internationally acceptable labeling system, which ultimately could lead to standardized labeling for 575,000 chemical mixtures.

6. Explain how to read and interpret MSDS forms and have an MSDS file readily available for reference.

7. Train employees to use the labeling format.

8. Document what training has been performed.

The hazard communication standard and the individual state requirements are a very complex set of regulations. Remember, however, the ultimate goal of these regulations and other treatment plant procedures is to provide additional employee protection. These standards and regulations, once the intent is understood, are relatively easy to implement and can certainly be accomplished in house by a safety conscious organization.

When all is said and done, common sense, knowledge, awareness, and commitment are the keys to complying with the hazardous material and the Worker Right-To-Know regulations.

Depending on what state you are in, your agency could be in violation of federal and state regulations if it does not currently have a Worker Right-To-Know policy. As more states that are covered by the federal standard revise their laws, it is anticipated that in the near future all employees will be covered by Worker Right-To-Know laws.

QUESTIONS

Write your answers in a notebook and then compare your answers with those on page 734.

13.9A List the four basic elements of a hazard communication program.

13.9B List the general categories of hazardous materials.

13.9C What is the purpose of the Material Safety Data Sheet (MSDS)?

13.9D What information must be on the label of a hazardous material container?

13.10 SAFETY SUMMARY

Following is a summary of the safety precautions that have been discussed in the previous sections.

1. Good design without proper safety precautions will not prevent accidents. *ALL PERSONNEL MUST BE INVOLVED IN A SAFETY PROGRAM AND PROVIDED WITH FREQUENT SAFETY REMINDERS.*

2. Never attempt to do a job unless you have sufficient help, adequate training and skills, the proper tools, and necessary safety equipment.

3. Remove a manhole cover or heavy grate using the proper tool, not your fingers.

4. "Lift with your legs, not with your back" to prevent back strains.

5. Use ladders of any kind with caution. Be certain that portable ladders are positioned so they will not slip or twist and that they are placed at the appropriate angle. Whenever possible, secure the top of a ladder used to enter below-grade structures. Do not use metal ladders near electrical boards or appliances.

6. Never enter a manhole, pit, sump, or other enclosed area without complying with all confined space requirements.

7. Always test manholes, pits, sumps, and other enclosed areas for atmospheric hazards. When testing the atmosphere, first test for oxygen deficiency/enrichment, then combustible gases and vapors, and then toxic gases and vapors. Before entering, thoroughly ventilate with a forced air blower.

8. Wear or use personal protective equipment (PPE) such as safety harnesses, gas detectors, and rubber gloves to prevent infections and injuries.

9. Never use a tool or piece of equipment unless you have been trained and are thoroughly familiar with its use or operation and know its limitations.

10. When working in traffic areas, always provide:

    a. Adequate advance warning to traffic by signs and flags.

    b. Traffic cones, barricades, or other approved items for channeling the flow of traffic around your work area.

    c. Protection to workers by placing your vehicle between traffic and the job area, if practical, or by use of flashing or revolving lights, or other devices.

    d. Flaggers, when necessary, to direct and control flow of traffic.

11. Before starting a job, be certain that the work area is of adequate size. If not, make allowances for this. Keep all working surfaces free of material that may cause slip or trip hazards.

12. See to it that all guardrails and chains are properly installed and maintained.

13. Provide and maintain guards on all chains, sprockets, gears, shafts, and other similar moving pieces of equipment that can be accidentally contacted.

14. Before working on mechanical or electrical equipment, properly lock out and tag breakers and other sources of energy to prevent the accidental starting or movement of the equipment while you are working on it. Wear approved, insulated gloves and boots wherever you may contact live electric circuits.

15. Never enter a launder, channel, conduit, or other slippery area when by yourself. Use a fall arrest system if working in an unguarded, elevated area six feet (1.8 meters) or more above a lower level.

16. Do not allow smoking, open flames, or other sources of ignition in the area of, on top of, or in any structure in your digestion and gas-handling system. Post all these areas with warning signs in conspicuous places.

17. Never enter a chlorine or sulfur dioxide atmosphere by yourself or without proper protective equipment. Seek the cooperation of your local fire department in supplying support when responding to a release.

18. Obtain and post in a conspicuous location the name and telephone number of the nearest chlorine and sulfur dioxide emergency service. Acquaint your police and fire department with this service.
19. Inspect all chlorine and sulfur dioxide connectors and lines before using. Replace any of these that appear defective.
20. Keep all chlorine and sulfur dioxide containers secure to prevent falling or rolling. Use only approved methods of moving and lifting containers.
21. Maintain a good housekeeping program. This is a proven method of preventing many accidents.
22. Conduct an effective safety awareness and training program.

These are the highlights of what has been previously discussed. Whenever in doubt about the safety of any piece of equipment, structure, operation, or procedure, contact the equipment manufacturer, your city or county safety officer, or your state safety office. One of these should be able to supply you with an answer to your questions.

ACCIDENTS DO NOT JUST HAPPEN... THEY ARE CAUSED!

You can be held personally liable for injuries or damages caused by an accident as a result of your negligence.

Can you afford the price of one?

Can you afford the loss of one or more operators?

Can your family afford to lose YOU?

REMEMBER: SAFETY IS NO ACCIDENT!

13.11 ADDITIONAL READING

1. *MOP 11*, Chapter 5,* "Occupational Safety and Health," New Edition of MOP 11, Volume I, *MANAGEMENT AND SUPPORT SYSTEMS.*
2. *NEW YORK MANUAL*, Chapter 13,* "Safety."
3. *TEXAS MANUAL*, pages 689–706.*
4. *CHLORINE SAFE HANDLING BOOKLET.* Obtain from PPG Industries, Inc., Monroeville Chemical Center, 440 College Park Drive, Monroeville, PA 15146. No charge.
5. *SAFETY AND HEALTH IN WASTEWATER SYSTEMS* (MOP 1). Obtain from Water Environment Federation (WEF), Publications Order Department, 601 Wythe Street, Alexandria, VA 22314-1994. Order No. MO2001. Price to members, $20.00; nonmembers, $30.00; plus shipping and handling.
6. *CHLORINE MANUAL*, Sixth Edition. Obtain from the Chlorine Institute, Inc., Bookstore, PO Box 1020, Sewickley, PA 15143-1020. Pamphlet 1. Price to members, $28.00; nonmembers, $70.00; plus $6.95 shipping and handling.
7. National Fire Protection Association Standard 820 (NFPA 820), *FIRE PROTECTION IN WASTEWATER TREATMENT AND COLLECTION FACILITIES.* Obtain from National Fire Protection Association (NFPA), 11 Tracy Drive, Avon, MA 02322. Item No. 82003. Price to members, $27.00; nonmembers, $30.00; plus $7.95 shipping and handling.
8. *SAFETY IN THE CHEMICAL LABORATORY*, edited by Norman V. Steere. Out of print.
9. *GENERAL INDUSTRY, OSHA SAFETY AND HEALTH STANDARDS* (CFR, Title 29, Labor Pt. 1900–1910.999 (most recent edition)). Obtain from the US Government Printing Office, Superintendent of Documents, PO Box 371954, Pittsburgh, PA 15250-7954. Order No. 869-056-00108-8. Price, $61.00.
10. *FISHER SAFETY CATALOG.* Obtain from Fisher Scientific Company, Safety Division, 4500 Turnberry Drive, Suite A (Customer Service), Hanover Park, IL 60103, or phone (800) 772-6733.
11. *SAFETY RULES*, published by National Environmental Training Association (NETA), 5320 N. 16th Street, Suite 114, Phoenix, AZ 85016-3241. Out of print.
12. Safety-related training products are available from Communication Arts Multimedia, Inc., 226 Scenic View Lane, Ligonier, PA 15658. The titles are:
    a. Confined Space Safety/Manhole Entry. Price for video or DVD, $150.00.
    b. Shoring and Trenching. Price for video or DVD, $150.00.

The cost of shipping and handling is $2.00 per item.

* Depends on edition.

END OF LESSON 3 OF 3 LESSONS

on

SAFETY

Please answer the discussion and review questions next.

DISCUSSION AND REVIEW QUESTIONS
Chapter 13. SAFETY
(Lesson 3 of 3 Lessons)

Write the answers to these questions in your notebook. The question numbering continues from Lesson 2.

19. How can samples for lab tests be collected without going beyond guardrails or chains?

20. What should be done with the jagged ends of glass tubes?

21. How should hot lab equipment be handled?

22. How can a fire be extinguished?

23. Fires can be prevented by good housekeeping and proper handling of flammables. True or False?

24. Why should plant water supplies be checked monthly for coliform group bacteria?

25. Why should safety equipment be checked periodically?

26. Where would you look for safety posters, signs, and slogans to aid in tailgate safety meetings?

27. Carefully study the illustration below. List the safety hazards and indicate how each one can be corrected.

SUGGESTED ANSWERS
Chapter 13. SAFETY

ANSWERS TO QUESTIONS IN LESSON 1

Answers to questions on page 684.

13.1A   You can help prevent the spread of infectious diseases from your job to you and your family by immunization shots, thoroughly washing your hands before eating or smoking, wearing proper protective gloves when you may contact wastewater or sludge in any form, not wearing your work clothes home, keeping your work clothes clean, storing your street clothes and your work clothes in separate lockers, and, if your employer does not supply you with uniforms and laundry service, washing your work clothes separately from your regular family wash.

13.1B   When testing the atmosphere before entry in any confined space, first test for oxygen deficiency/enrichment, then combustible gases and vapors, and then toxic gases and vapors.

13.1C   OSHA regulations require that a confined space entry permit be completed for each permit-required confined space entry. The permit must be renewed each time the space is left and re-entered, even if only for a break or lunch, or to go get a tool.

13.1D   The most common toxic gas in industrial waste discharges is hydrogen sulfide, which is produced during the anaerobic decomposition of certain materials containing sulfur compounds. Other toxics, such as carbon monoxide, chlorinated solvents, and industrial toxins, may enter your plant as a result of industrial discharges, accidental spills, or illegal disposal of hazardous materials.

13.1E Hydrogen sulfide has the unique ability to affect your sense of smell. You will lose the ability to smell hydrogen sulfide's "rotten egg" odor after only a short exposure. Your loss of ability to smell hydrogen sulfide, and certain other odors, is known as olfactory fatigue. *NEVER* depend on your nose for detection, regardless of the situation.

13.1F UEL stands for upper explosive limit and LEL stands for lower explosive limit. UEL and LEL indicate the range of concentrations at which combustible or explosive gases will ignite when an ignition source is present at ambient temperature.

Answers to questions on page 689.

13.1G A lockout device uses a positive means such as a lock to hold a switch or valve in a safe position and prevent a piece of equipment from becoming energized or moving. Lockout devices are used whenever equipment must be repaired or replaced to ensure that the equipment will not start up or move unexpectedly and possibly injure workers.

13.1H Examples of equipment or systems that are potentially dangerous for operators due to stored energy include: springs; elevated machine members; rotating flywheels; hydraulic systems; and systems using air, gas, steam, or water pressure.

13.1I The primary elements of the hearing conservation program are monitoring, audiometric testing, hearing protection, training, access to information, and recordkeeping.

13.1J Positive pressure self-contained breathing apparatus (SCBAs) and full face piece positive pressure supplied air respirators (SARs) with an escape SCBA can be used in oxygen-deficient atmospheres.

Answers to questions on page 700.

13.2A Traffic may be warned that you are working in a street or traffic area by high-level signs and flags, by fluorescent traffic cones, by signs or barricades to direct traffic, by a flagger to direct and control traffic, or by any combination of these.

13.2B Mechanical, electrical, engulfment, and toxic exposure hazards may be encountered when entering a manhole. Physical injury, infections and diseases, and insect and rodent bites are other potential hazards.

13.2C Three protective systems for excavations are shoring, shielding, and benching and sloping.

13.2D OSHA requires a protective system when excavations are five feet (1.5 m) or more in depth.

13.2E You should receive training regarding the proper use or operation of sewer cleaning tools and equipment from the equipment vendor.

ANSWERS TO QUESTIONS IN LESSON 2

Answers to questions on page 701.

13.2F Be aware of moving equipment in pumping stations, especially reciprocating equipment and rotating shafts. Moving parts that create a contact hazard to employees must be guarded. Do not wear loose clothing, rings, or other jewelry around machinery. Long hair must be secured.

13.2G Fire extinguishers in pumping stations should be of a type that may be used on electrical equipment as well as on solid material or power overload-type fires. The use of liquid-type fire extinguishers should be avoided. All-purpose, ABC chemical-type fire extinguishers are recommended.

Answers to questions on page 703.

13.2H Slimes, rags, or greases should be removed from any area because they create slip and trip hazards.

13.2I When working on mechanical or electrical parts of equipment, you should lock out the main circuit breaker and fasten a tag to the handle or to the lockout device to notify others that you are working on the equipment and that it must remain de-energized.

13.2J Guards should be placed around all moving parts that create a contact hazard to employees.

13.2K Tools and equipment should not be carried, but should be transported in and out of pits and sumps by the use of buckets and a handline or sling.

Answers to questions on page 706.

13.2L Slippery surfaces, such as walkways, stairs, ladders, and catwalks, can be made less dangerous by keeping them free of grease, oil, slimes, or other materials and by applying nonslip strips or coatings. To reduce slipping risks, wear safety shoes with nonslip soles and install hand holds and railings.

13.2M Smoking and open flames should not be allowed in the vicinity of digesters because methane may be present and when methane gas is mixed with the proper proportion of air it forms an explosive mixture.

13.2N Follow all confined space procedures before entering and during occupancy of a recently emptied digester. *DO NOT* enter unless you are certain that sources of contaminants, such as sludge and gas lines, have been effectively isolated and that safe atmospheric conditions exist and can be maintained before entry and during occupancy.

13.2O Before relighting a waste gas burner, the main gas valve should be turned off and the stack allowed to vent itself for a few minutes.

13.2P If a positive displacement pump is started against a closed discharge valve, pressures could build up and break a pipe or damage the pump.

Answers to questions on page 707.

13.2Q The rotating distributor should be stopped by turning off the flow of water. Extreme care must be taken because of the powerful force behind the distributor.

13.2R You should never work alone on the center Y-wall of an aerator because you could fall into the aerator and need help getting out.

13.2S When using a boat to collect samples from a pond, have someone standing by in case you get into trouble, wear a Coast Guard-approved life jacket, and do not stand up in the boat while performing work.

Answers to questions on page 710.

13.2T One-ton chlorine containers should only be lifted by an approved lifting bar with hooks over the ends of the container.

13.2U Chlorine gas is two and a half times heavier than air and is best removed when leaks occur by blowing the gas out of the room at floor level.

13.2V If you open a door and smell chlorine or sulfur dioxide, immediately close the door, leave ventilation on, and seek help.

13.2W Housekeeping tasks that will keep a treatment facility a safer, cleaner place to work include:

1. Designate a place for tools and equipment and, when they are not being used, keep them in their proper places.
2. Clean up spills of oils, grease, chemicals, polymers, wastewater, and sludge. Keep walkways and work areas clean.
3. Provide proper containers for wastes, oily rags, and papers and empty the containers frequently.
4. Store flammable substances in an approved storage cabinet.
5. Remove snow and ice in areas where a person might slip and fall.

Answers to questions on page 712.

13.2X Major types of discharges from industrial plants that could create safety hazards include:

1. Fuels
2. Toxic gases
3. Amines
4. Surface-active agents
5. Biocides
6. Highly acidic or alkaline wastes

13.2Y If gasoline reaches the wet well in the headworks of your treatment plant:

1. Try to remove the gasoline from the surface of the wet well with skimmers (if available) or with a portable pump.
2. Apply as much ventilation as possible to prevent an explosive atmosphere from developing.
3. Monitor the atmosphere for toxic and explosive conditions.
4. Keep personnel away from the area.
5. Do not allow any flames, sparks, or other sources of ignition in the area (use nonsparking tools and explosion-proof equipment and wiring).

13.2Z Surface-active agents can cause three types of problems:

1. Super floc agents can produce a much denser sludge than usually pumped. This sludge can be too thick to pump, which then requires special handling to reslurry the sludge so it can be pumped.
2. Antifloc agents can lower the capture of suspended solids in primary clarifiers and cause a solids overload on downstream treatment processes.
3. Foaming agents can cause foam to cover treatment processes and prevent inspection and maintenance. Foam that blows into neighbors' yards or forms on the surface of receiving waters is objectionable to the public.

ANSWERS TO QUESTIONS IN LESSON 3

Answers to questions on page 715.

13.3A When collecting influent samples, disposable, impervious gloves should be worn to protect the operator's hands if there is any chance of direct contact with the wastewater. If possible, sample poles or other similar types of samplers should be used.

13.3B Hands should always be washed before eating to prevent the spread of disease.

13.3C Chemicals and reagents should be handled with care to protect your body from serious injuries and possible poisoning.

Answers to questions on page 716.

13.4A The necessary ingredients of a fire are fuel, heat, and oxygen.

13.4B Oil- and paint-soaked rags should be placed in covered metal containers and regularly disposed of in a safe manner.

Answers to questions on page 717.

13.5A Some treatment plants use bottled drinking water because it is an economical and reliable source of potable water. This practice reduces the possibility of the spread of disease from unknown cross connections or defective devices installed to prevent contamination by backflows.

13.6A The following phone numbers should be conspicuously posted in your facility: fire department, police department, ambulance, chemical supplier or repair service, and physician. Check your list to be sure they are all listed and the numbers are correct.

Answers to questions on page 718.

13.7A The purpose of tailgate safety meetings is to remind operators of the need for safety, to explain and discuss safe procedures and safe conditions, and to review potential hazards and how to correct or avoid dangerous situations.

13.7B Safety meetings should be held every 7 to 10 working days.

Answers to questions on page 721.

13.8A Safety meetings could be:

1. Formal, organized meetings
2. Tailgate safety sessions
3. Informal get-togethers

13.8B Management must be committed to safety and actively support a safety training program. An effective safety training program must start at the top.

13.8C Safety drills should be held regularly so everyone knows how to respond during emergencies. Do not wait for an emergency before trying to learn how to use safety equipment. Get proper instructions and conduct practice sessions.

13.8D Paperwork necessary for an effective safety training program includes:

1. Accident report forms
2. Safety policy
3. Safety rules
4. Supervisors' guidelines
5. Checklists for review of facility plans and specifications and also for plant inspections

Answers to questions on page 729.

13.9A The four basic elements of a hazard communication program include:

1. Identify hazardous materials
2. Obtain chemical information and define hazardous conditions
3. Properly label hazards
4. Train operators

13.9B The general categories of hazardous materials include:

1. Corrosives
2. Toxics
3. Flammables and explosives
4. Asphyxiants
5. Harmful physical agents
6. Infectious agents

13.9C The purpose of the Material Safety Data Sheet (MSDS) is to have a readily available document to be used for training and as an immediate reference, since it includes information on common names, safe exposure level, the effects of exposure, the symptoms of exposure, whether it is flammable, what type of first aid should be administered, and other information.

13.9D The label on a hazardous material container must indicate the name of the hazardous substance and appropriate warnings.

APPENDIX

REGULATORY INFORMATION
APPLICABLE OSHA REGULATIONS

REGULATORY INFORMATION

Safety regulations are generally developed and enforced under the overall jurisdiction of the Federal Government's Department of Labor Occupational Safety and Health Administration (OSHA). In general, coverage of the OSHA Act extends to all employers and their employees in the 50 states, the District of Columbia, Puerto Rico, and all other territories under Federal Government jurisdiction. Coverage is provided either directly by federal OSHA or through an OSHA-approved state program.

OSHA has developed a comprehensive set of safety regulations, many of them with specific applications to wastewater collection and treatment systems. Where OSHA has not developed specific standards, employers are responsible for following the Act's general duty clause, which states that each employer "shall furnish … a place of employment which is free from recognized hazards that are causing or are likely to cause death or serious physical harm to his employees."

States with OSHA-approved occupational safety and health programs must set standards that are at least as effective as the federal standards. Many states adopt standards identical to the federal standards. Where states adopt and enforce their own standards under state law, copies of state standards may be obtained from the individual states.

The Federal Register is one of the best sources of information on standards, since all OSHA standards are published there when adopted, as are all amendments, corrections, insertions, or deletions. The Federal Register is available in many public libraries. Annual subscriptions are available from the US Government Printing Office, Superintendent of Documents, PO Box 371954, Pittsburgh, PA 15250-7954, phone (866) 512-1800.

Each year the Office of the Federal Register publishes all current regulations and standards in the Code of Federal Regulations (CFR), available at many libraries and from the Government Printing Office. OSHA's regulations are collected in Title 29 of the Code of Federal Regulations (CFR), Part 1900–1999. Copies of the CFR may also be purchased from the Superintendent of Documents at the address above.

APPLICABLE REGULATIONS

OSHA regulations fill several hundred pages of text, which cannot be printed in this manual. Of the regulations, Part 1910, OCCUPATIONAL SAFETY AND HEALTH STANDARDS, contains the regulations most applicable to the work we do in wastewater collection and treatment systems. If your state is one of the 25 that has a state program, there will be a comparable set of regulations similar to Part 1910. Listed next are some of the OSHA regulations that apply.

PART 1910—OCCUPATIONAL SAFETY AND HEALTH STANDARDS

Subpart D — Walking/Working Surfaces

1910.21	Definitions.
1910.22	General requirements.
1910.23	Guarding floor and wall openings and holes.
1910.24	Fixed industrial stairs.
1910.25	Portable wood ladders.
1910.26	Portable metal ladders.
1910.27	Fixed ladders.
1910.28	Safety requirements for scaffolding.
1910.29	Manually propelled mobile ladder stands and scaffolds (towers).
1910.30	Other working surfaces.
1910.31	Sources of standards.
1910.32	Standards organizations.

Subpart E — Means of Egress

1910.35	Definitions.
1910.36	General requirements.

1910.37	Means of egress, general.	1910.110	Storage and handling of liquefied petroleum gases.
1910.38	Employee emergency plans and fire prevention plans.	1910.115	Sources of standards.
1910.39	Sources of standards.	1910.116	Standards organizations.
1910.40	Standards organizations.	1910.119	Process safety management of highly hazardous chemicals.
1910.40A	Appendix to Subpart E—Means of Egress.	1910.119A	Appendix A to §1910.119—List of Highly Hazardous Chemicals, Toxics and Reactives (Mandatory).

Subpart F — Powered Platforms, Manlifts, and Vehicle-Mounted Work Platforms

1910.66	Power platforms for exterior building maintenance.	1910.119B	Appendix B to §1910.119—Block Flow Diagram and Simplified Process Flow Diagram (Nonmandatory).
1910.66A	Appendix A to §1910.66—Guideline (Advisory)	1910.119C	Appendix C to §1910.119—Compliance Guidelines and Recommendations for Process Safety Management (Nonmandatory).
1910.66B	Appendix B to §1910.66—Exhibits (Advisory)		
1910.66C	Appendix C to §1910.66—Personal Fall Arrest System (Section I—Mandatory; Sections II and III—Nonmandatory)	1910.119D	Appendix D to §1910.119—Sources of Further Information (Nonmandatory).
1910.66D	Appendix D to §1910.66—Existing Installations (Mandatory)	1910.120	Hazardous waste operations and emergency response.
1910.67	Vehicle-mounted elevating and rotating work platforms.	1910.120A	Appendix A to §1910.120—Personal Protective Equipment Test Methods.
1910.68	Manlifts.	1910.120B	Appendix B to §1910.120—General Description and Discussion of the Levels of Protection and Protective Gear.
1910.69	Sources of standards.		
1910.70	Standards organizations.	1910.120C	Appendix C to §1910.120—Compliance Guidelines.

Subpart G — Occupational Health and Environmental Control

1910.94	Ventilation.	1910.120D	References.

Subpart I — Personal Protection Equipment

1910.95	Occupational noise exposure.	1910.132	General requirements.
1910.96	Ionizing radiation.	1910.133	Eye and face protection.
1910.97	Nonionizing radiation.	1910.134	Respiratory protection.
1910.98	Effective dates.	1910.135	Occupational head protection.
1910.99	Sources of standards.	1910.136	Occupational foot protection.
1910.100	Standards organizations.	1910.137	Electrical protective devices.

Subpart H — Hazardous Materials

Subpart J — General Environmental Controls

1910.101	Compressed gases (general requirements).	1910.141	Sanitation.
1910.102	Acetylene.	1910.144	Safety color code for marking physical hazards.
1910.103	Hydrogen.	1910.145	Specifications for accident prevention signs and tags.
1910.104	Oxygen.		
1910.106	Flammable and combustible liquids.	1910.145A	Appendix A to §1910.145(f)—Recommended Color Coding.
1910.107	Spray finishing using flammable and combustible materials.	1910.145B	Appendix B to §1910.145(f)—References for further information.
1910.108	Dip tanks containing flammable or combustible liquids.	1910.146	Permit-required confined spaces.

1910.146A	Appendix A to §1910.146—Permit-required Confined Space Decision Flow Chart.	1910.178	Powered industrial trucks.
		1910.179	Overhead and gantry cranes.
1910.146B	Appendix B to §1910.146—Procedures for Atmospheric Testing.	1910.180	Crawler locomotive and truck cranes.
		1910.181	Derricks.
1910.146C	Appendix C to §1910.146—Examples of Permit-required Confined Space Programs.	1910.182	Effective dates.
		1910.183	Helicopters.
1910.146D	Appendix D to §1910.146—Sample Permits.	1910.184	Slings.
1910.146E	Appendix E to §1910.146—Sewer System Entry.	1910.189	Sources of standards.
1910.147	The control of hazardous energy (lockout/tagout).	1910.190	Standards organizations.

Subpart O — Machinery and Machine Guarding

1910.147A	Appendix A to §1910.147—Typical Minimal Lockout Procedure.	1910.211	Definitions.
		1910.212	General requirements for all machines.
1910.148	Standards organizations.	1910.213	Woodworking machinery requirements.
1910.149	Effective dates.	1910.214	Cooperage machinery.

Subpart K — Medical and First Aid

1910.150	Sources of Standards.	1910.215	Abrasive wheel machinery.
1910.151	Medical services and first aid.	1910.216	Mills and calenders in the rubber and plastics industries.
1910.153	Sources of standards.	1910.217	Mechanical power presses.

Subpart L — Fire Protection

PORTABLE FIRE SUPPRESSION EQUIPMENT

1910.157	Portable fire extinguishers.	1910.218	Forging machines.
1910.158	Standpipe and hose systems.	1910.219	Mechanical power-transmission apparatus.
		1910.220	Effective dates.

FIXED FIRE SUPPRESSION EQUIPMENT

1910.159	Automatic sprinkler systems.	1910.221	Sources of standards.
1910.160	Fixed extinguishing systems, general.	1910.222	Standards organizations.

Subpart P — Hand and Portable Powered Tools and Other Hand-Held Equipment

1910.161	Fixed extinguishing systems, dry chemical.	1910.241	Definitions.
1910.162	Fixed extinguishing systems, gaseous agent.	1910.242	Hand and portable powered tools and equipment, general.
1910.163	Fixed extinguishing systems, water spray and foam.	1910.243	Guarding of portable powered tools.

OTHER FIRE PROTECTION SYSTEMS

1910.164	Fire detection systems.	1910.244	Other portable tools and equipment.
1910.165	Employee alarm systems.	1910.245	Effective dates.

Subpart M — Compressed Gas and Compressed Air Equipment

		1910.246	Sources of standards.
1910.169	Air receivers.	1910.247	Standards organizations.

Subpart Q — Welding, Cutting, and Brazing

1910.170	Sources of standards.	1910.251	Definitions.
1910.171	Standards organizations.	1910.252	General requirements.

Subpart N — Materials Handling and Storage

		1910.253	Oxygen-fuel gas welding and cutting.
1910.176	Handling materials — general.	1910.254	Arc welding and cutting.
1910.177	Servicing multi-piece and single piece rim wheels.	1910.255	Resistance welding.

1910.256	Sources of standards.		
1910.257	Standards organizations.		
1910.269	Electric power generation, transmission, and distribution.		
1910.269A	Appendix A to §1910.269—Flow Charts.		
1910.269A-1	Appendix A-1 to §1910.269—Application of Section 1910.269 and Subpart S of this Part to Electrical Installations.		
1910.269A-2	Appendix A-2 to §1910.269—Application of Section 1910.269 and Subpart S of this Part to Electrical Safety-Related Work Practices.		
1910.269A-3	Appendix A-3 to §1910.269—Application of Section 1910.269 and Subpart S of this Part to Tree-Trimming Operations.		
1910.269A-4	Appendix A-4 to §1910.269—Application of Section 1910.147, Section 1910.269, and Section 1910.333 to Hazardous Energy Control Procedures (Lockout/Tagout).		
1910.269A-5	Appendix A-5 to §1910.269—Application of Section 1910.146 and Section 1910.269 to Permit-Required Confined Spaces.		
1910.269B	Appendix B to §1910.269—Working on Exposed Energized Parts.		
1910.269C	Appendix C to §1910.269—Protection from Step and Touch Potentials.		
1910.269D	Appendix D to §1910.269—Methods of Inspecting and Testing Wood Poles.		
1910.269E	Appendix E to §1910.269—Reference Documents.		

Subpart S — Electrical

GENERAL

1910.301	Introduction.

DESIGN SAFETY STANDARDS FOR ELECTRICAL SYSTEMS

1910.302	Electric utilization systems.
1910.303	General requirements.
1910.304	Wiring design and protection.
1910.305	Wiring methods, components, and equipment for general use.
1910.306	Specific purpose equipment and installations.
1910.307	Hazardous (classified) locations.
1910.308	Special systems.

SAFETY-RELATED WORK PRACTICES

1910.331	Scope.
1910.332	Training.
1910.333	Selection and use of work practices.
1910.334	Use of equipment.

In addition to the regulations listed above, 29 CFR Part 1926, SAFETY AND HEALTH FOR CONSTRUCTION, defines the regulations for Excavations and Trenches under Subpart P. There are other federal regulations that may also apply to wastewater collection and treatment systems operation and maintenance, such as Department of Transportation (DOT), as well as state and local regulations for traffic control and protection of utilities during excavations using "one-call" notification systems.

CHAPTER 14

MAINTENANCE

GENERAL PROGRAM
by
Norman Farnum

MECHANICAL MAINTENANCE
by
Stan Walton

UNPLUGGING PIPES, PUMPS, AND VALVES
by
John Brady

FLOW MEASUREMENT
by
Roger Peterson

Revised by
Malcolm Carpenter and
Rick Arbour

TABLE OF CONTENTS
Chapter 14. MAINTENANCE

Page

OBJECTIVES 745

WORDS 746

LESSON 1

14.0 TREATMENT PLANT MAINTENANCE—GENERAL PROGRAM 749

    14.00   Preventive Maintenance Records 749

    14.01   Building Maintenance 750

    14.02   Plant Tanks and Channels 751

    14.03   Plant Grounds 751

    14.04   Library 752

DISCUSSION AND REVIEW QUESTIONS 752

LESSON 2

14.1 MECHANICAL EQUIPMENT 753

    14.10   Repair Shop 753

    14.11   Pumps 753

            14.110   Centrifugal Pumps 753

            14.111   Propeller Pumps 759

            14.112   Vertical Wet Well Pumps 759

            14.113   Reciprocating or Piston Pumps 759

            14.114   Incline Screw Pumps 768

            14.115   Progressive Cavity (Screw-Flow) Pumps 768

            14.116   Pneumatic Ejectors 768

    14.12   Pump Lubrication, Seals, and Bearings 768

            14.120   Lubrication 768

            14.121   Mechanical Seals 768

            14.122   Bearings 774

    14.13   Starting a New Pump 774

    14.14   Pump Shutdown 775

    14.15   Pump-Driving Equipment 775

    14.16   Electrical Controls 775

				Page
	14.17		Variable-Speed, Alternating Current (AC) Motors	776
			by Ken Peschel	
		14.170	Description	776
		14.171	Troubleshooting	777
		14.172	Preventive Maintenance	777
	14.18		Operating Troubles	781
	14.19		Starting and Stopping Pumps	782
		14.190	Centrifugal Pumps	783
		14.191	Positive Displacement Pumps	786
DISCUSSION AND REVIEW QUESTIONS				787

LESSON 3

				Page
14.2			BEWARE OF ELECTRICITY	788
14.3			ELECTRICAL EQUIPMENT MAINTENANCE	788
	14.30		Introduction	788
	14.31		Volts, Amps, Watts, and Power Requirements	790
	14.32		Tools, Meters, and Testers	791
	14.33		Electrical System Equipment	795
		14.330	Need for Maintenance	795
		14.331	Equipment Protective Devices	796
		14.332	Fuses	796
		14.333	Circuit Breakers	798
		14.334	Ground	801
	14.34		Motor Control/Supervisory Control and Electrical System	802
14.4			MOTORS	803
	14.40		Types	803
	14.41		Nameplate Data	803
	14.42		Causes of Failure	806
	14.43		Insulation	806
		14.430	Types and Specifications	806
		14.431	Causes of Failure	806
		14.432	Increasing Resistance Value	810
	14.44		Starters	810
	14.45		Safety	813
	14.46		Other Motor Considerations	814
		14.460	Alignment	814
		14.461	Changing Rotation Direction	817
		14.462	Allowable Voltage and Frequency Deviations	817
		14.463	Maximum Vibration Levels Allowed by NEMA	817
		14.464	Lubrication	817

	14.47	Troubleshooting	820
		14.470 Step-By-Step Procedures	820
		14.471 Troubleshooting Guide for Electric Motors	821
		14.472 Troubleshooting Guide for Magnetic Starters	823
		14.473 Trouble/Remedy Procedures for Induction Motors	825
14.5	RECORDS		826
14.6	ADDITIONAL READING		826
DISCUSSION AND REVIEW QUESTIONS			826

LESSON 4

14.7	MECHANICAL MAINTENANCE	829

The format of this section differs slightly from the others. The arrangement of procedures was designed specifically to assist you in planning an effective preventive maintenance program. The contents are at the beginning of the section, and the paragraphs are numbered for easy reference on equipment service record cards.

Paragraph

1. Pumps, General (Including Packing)	830
2. Reciprocating Pumps, General	835
3. Propeller Pumps, General	839
4. Progressive Cavity Pumps, General	839
5. Pneumatic Ejectors, General	839
6. Float and Electrode Switches	840

DISCUSSION AND REVIEW QUESTIONS ... 840

LESSON 5

7. Electric Motors	841
8. Belt Drives	843
9. Chain Drives	844
10. Variable-Speed Belt Drives	845
11. Couplings	846
12. Shear Pins	849

DISCUSSION AND REVIEW QUESTIONS ... 849

LESSON 6

13. Gate Valves	850
14. Check Valves	853
15. Plug Valves	853
16. Sluice Gates	853
17. Dehumidifiers	862
18. Air Gap Separation Systems	862
19. Plant Safety Equipment	862
Acknowledgment	864

DISCUSSION AND REVIEW QUESTIONS ... 864

LESSON 7

14.8	UNPLUGGING PIPES, PUMPS, AND VALVES		865
	14.80	Plugged Pipelines	865
	14.81	Scum Lines	865
	14.82	Sludge Lines	865
	14.83	Digested Sludge Lines	865
	14.84	Unplugging Pipelines	865
		14.840 Pressure Methods	865
		14.841 Cutting Tools	866
		14.842 High-Velocity Pressure Units	866
		14.843 Last Resort	866
	14.85	Plugged Pumps and Valves	866
14.9	REVIEW OF PLANS AND SPECIFICATIONS		867
	14.90	Examining Prints	867
	14.91	Reading Specifications	867
14.10	SUMMARY		867
14.11	ADDITIONAL READING		868
DISCUSSION AND REVIEW QUESTIONS			868
SUGGESTED ANSWERS			869

OBJECTIVES

Chapter 14. MAINTENANCE

Following completion of Chapter 14, you should be able to:

1. Develop a maintenance program for your plant, including equipment, buildings, grounds, channels, and tanks.

2. Start a maintenance recordkeeping system that will provide you with information to protect equipment warranties, to prepare budgets, and to satisfy regulatory agencies.

3. Schedule maintenance of equipment at proper time intervals.

4. Perform maintenance as directed by manufacturers.

5. Recognize symptoms that indicate equipment is not performing properly, identify the source of the problem, and take corrective action.

6. Start and stop pumps.

7. Unplug pipes, pumps, and valves.

8. Explain the operation and maintenance of sensors, transmitters, receivers, and controllers.

9. Determine when you need assistance to correct a problem.

NOTE: Special maintenance information is given in the previous chapters on treatment processes where appropriate.

WORDS
Chapter 14. MAINTENANCE

AIR GAP

An open, vertical drop, or vertical empty space, between a drinking (potable) water supply and potentially contaminated water. This gap prevents the contamination of drinking water by backsiphonage because there is no way potentially contaminated water can reach the drinking water supply.

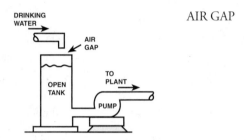

ANALOG READOUT

The readout of an instrument by a pointer (or other indicating means) against a dial or scale. Also, the continuously variable signal type sent to an analog instrument (for example, 4–20 mA). Also see DIGITAL READOUT.

AXIAL TO IMPELLER

The direction in which material being pumped flows around the impeller or flows parallel to the impeller shaft.

AXIS OF IMPELLER

An imaginary line running along the center of a shaft (such as an impeller shaft).

BRINELLING (bruh-NEL-ing)

Tiny indentations (dents) high on the shoulder of the bearing race or bearing. A type of bearing failure.

CAVITATION (kav-uh-TAY-shun)

The formation and collapse of a gas pocket or bubble on the blade of an impeller or the gate of a valve. The collapse of this gas pocket or bubble drives water into the impeller or gate with a terrific force that can cause pitting on the impeller or gate surface. Cavitation is accompanied by loud noises that sound like someone is pounding on the impeller or gate with a hammer.

CROSS CONNECTION

(1) A connection between drinking (potable) water and an unapproved water supply.

(2) A connection between a storm drain system and a sanitary collection system.

(3) Less frequently used to mean a connection between two sections of a collection system to handle anticipated overloads of one system.

DATEOMETER (day-TOM-uh-ter)

A small calendar disk attached to motors and equipment to indicate the year in which the last maintenance service was performed.

DIGITAL READOUT

The readout of an instrument by a direct, numerical reading of the measured value or variable. The signal sent to such instruments is usually an analog signal.

JOGGING

The frequent starting and stopping of an electric motor.

MEGOHM (MEG-ome)

Millions of ohms. Mega- is a prefix meaning one million, so 5 megohms means 5 million ohms.

MULTISTAGE PUMP
A pump that has more than one impeller. A single-stage pump has one impeller.

NAMEPLATE
A durable, metal plate found on equipment that lists critical operating conditions for the equipment.

POLE SHADER
A copper bar circling the laminated iron core inside the coil of a magnetic starter.

PROGRAMMABLE LOGIC CONTROLLER (PLC)
A small computer that controls process equipment (variables) and can control the sequence of valve operations.

PRUSSIAN BLUE
A blue paste or liquid (often on a paper like carbon paper) used to show a contact area. Used to determine if gate valve seats fit properly.

RADIAL TO IMPELLER
Perpendicular to the impeller shaft. Material being pumped flows at a right angle to the impeller.

SINGLE-STAGE PUMP
A pump that has only one impeller. A multistage pump has more than one impeller.

STATOR
That portion of a machine that contains the stationary (non-moving) parts that surround the moving parts (rotor).

STETHOSCOPE
An instrument used to magnify sounds and carry them to the ear.

CHAPTER 14. MAINTENANCE

(Lesson 1 of 7 Lessons)

14.0 TREATMENT PLANT MAINTENANCE— GENERAL PROGRAM

An industrial treatment plant operator has many duties. Most of them have to do with the efficient operation of the plant. An operator has the responsibility to discharge an effluent that will meet all the requirements established for the plant. By doing this, the operator develops a good working relationship with the regulatory agencies, water recreationists, water users, plant neighbors, and the downstream wastewater treatment plant.

Another duty an operator has is that of plant maintenance. A good maintenance program is essential for a wastewater treatment plant to operate continuously at peak design efficiency. A successful maintenance program will cover everything from mechanical equipment, such as pumps, valves, scrapers, and other moving equipment, to the care of the plant grounds, buildings, and structures.

Mechanical maintenance is of prime importance as the equipment must be kept in good operating condition for the plant to maintain peak performance. Manufacturers provide information on the mechanical maintenance of their equipment. You should thoroughly read their literature on your plant equipment and understand the procedures. Contact the manufacturer or the local representative if you have any questions. Follow the instructions very carefully when performing maintenance on equipment. You also must recognize tasks that may be beyond your capabilities or repair facilities, and you should request assistance when needed.

For a successful maintenance program, your supervisors must understand the need for and benefits from equipment that operates continuously as intended. Disabled or improperly working equipment is a threat to the quality of the plant effluent, and repair costs for poorly maintained equipment usually exceed the cost of maintenance.

14.00 Preventive Maintenance Records

Preventive maintenance programs help operating personnel keep equipment in satisfactory operating condition and aid in detecting and correcting malfunctions before they develop into major problems.

A frequent occurrence in a preventive maintenance program is the failure of the operator to record the work after it is completed. When this happens, the operator must rely on memory to know when to perform each preventive maintenance function. As days pass into weeks and months, the preventive maintenance program is lost in the turmoil of everyday operation.

The only way an operator can keep track of a preventive maintenance program is by good recordkeeping. A good recordkeeping system tells when maintenance is due and also provides a record of equipment performance. Poor performance is a good justification for replacement or new equipment. Good records also help keep your warranty in force. Whatever recordkeeping system is used, it should be kept up to date on a daily basis and not left to memory for some other time. Equipment service cards and service record cards (Figure 14.1) are easy to set up and require little time to keep up to date.

An *EQUIPMENT SERVICE CARD* (master card) should be filled out for each piece of equipment in the plant. Each card should have the equipment name on it, such as Sludge Pump No. 1, Primary Clarifier.

1. List each required maintenance service with an item number.

2. List maintenance services in order of frequency of performance. For instance, show daily service as items 1, 2, and 3 on the card; weekly items as 4 and 5; monthly items as 6, 7, 8, and 9; and so on.

3. Describe each type of service under work to be done.

Make sure all necessary inspections and services are shown. For reference data, list paragraph or section numbers as shown in the mechanical maintenance section of this chapter (Section 14.7, page 829). Also list frequency of service as shown in the time schedule columns of the same section. Under time, enter day or month service is due. Service card information may be changed to fit the needs of your plant or particular equipment as recommended by the equipment manufacturer. Be sure the information on the cards is complete and correct.

The *SERVICE RECORD CARD* should have the date and work done, listed by item number and signed by the operator who performed the service. Some operators prefer to keep both cards clipped together, while others place the service record card near the equipment.

When the service record card is filled, it should be filed for future reference and a new card attached to the master card. The equipment service card tells what should be done and when, while the service record card is a record of what you did and when you did it. Many plants keep this information on a computer.

EQUIPMENT SERVICE CARD

EQUIPMENT: #1 Raw Wastewater Lift Pump

Item No.	Work To Be Done	Reference	Frequency	Time
1	Check water seal and packing gland	Par. 1	Daily	
2	Operate pump alternately	Par. 1	Weekly	Monday
3	Inspect pump assembly	Par. 1	Weekly	Wed.
4	Inspect and lube bearings	Par. 1	Quarterly	1-4-7-10*
5	Check operating temperature of bearings	Par. 1	Quarterly	1-4-7-10
6	Check alignment of pump and motor	Par. 1	Semiannually	4 & 10
7	Inspect and service pump	Par. 1	Semiannually	4 & 10
8	Drain pump before shutdown	Par. 1		

* 1-4-7-10 represent the months of the year when the equipment should be serviced—1. January, 4. April, 7. July, and 10. October.

SERVICE RECORD CARD

EQUIPMENT: #1 Raw Wastewater Lift Pump

Date	Work Done (Item No.)	Signed	Date	Work Done (Item No.)	Signed
1-5-04	1 & 2	J.B.			
1-6-04	1	J.B.			
1-7-04	1-3-4-5-	R.W.			

Fig. 14.1 Equipment service card and service record card

QUESTIONS

Write your answers in a notebook and then compare your answers with those on page 869.

14.0A Why should you plan a good maintenance program for your treatment plant?

14.0B What general items would you include in your maintenance program?

14.0C Why should your maintenance program be accompanied by a good recordkeeping system?

14.0D What is the difference between an equipment service card and a service record card?

14.01 Building Maintenance

Building maintenance is another program that should be maintained on a regular schedule. Buildings in a treatment plant are usually built of sturdy materials to last for many years. Buildings must be kept in good repair. In selecting paint for a treatment plant, it is always a good idea to have a painting expert help the operator select the types of paint needed to protect the buildings from deterioration. The expert also will have some good ideas as to color schemes to help blend the plant in with the surrounding area. Consideration should also be given to the quality of paint. A good quality, more expensive material will usually give better service over a longer period of time than the economy-type products.

Building maintenance programs depend on the age, type, and use of a building. New buildings require a thorough check to be certain essential items are available and working properly. Older buildings require careful watching and prompt attention to keep ahead of leaks, breakdowns, replacements when needed, and changing uses of the building. Attention must be given to the maintenance requirements of many items in all plant buildings, such as electrical systems, plumbing, heating, cooling, ventilating, floors, windows, roofs, and drainage around the buildings. Regularly scheduled examinations and necessary maintenance of these items can prevent many costly and time-consuming problems in the future.

In each plant building, periodically check all stairways, ladders, catwalks, and platforms for adequate lighting, head clearance, and sturdy and convenient guardrails. Protective devices should be around all moving equipment. Whenever any repairs, alterations, or additions are built, avoid building accident traps such as pipes laid on top of floors or hung from the ceiling at head height, which could create serious safety hazards.

Organized storage areas should be provided and maintained in an accessible and neat manner.

Keep all buildings clean and orderly. Janitorial work should be done on a regular schedule. All tools and plant equipment should be kept clean and in their proper place. Floors, walls, and windows should be cleaned at regular intervals to maintain a neat appearance. A treatment plant kept in a clean, orderly condition makes a safer place to work and aids in building good public and employee relations.

14.02 Plant Tanks and Channels

Wastewater channels, wet wells, and tanks such as clarifiers and grit tanks should be drained and inspected at least once a year. Be sure that the groundwater level is down far enough so the tanks will not float on the groundwater when empty or develop cracks from groundwater pressure. Most of the tanks in recently constructed facilities contain relief valves in the floor of the tank to prevent tank flotation if it is dewatered under high groundwater conditions.

Schedule inspections and maintenance of tanks and channels during periods of low inflow to minimize the load on other plant treatment units. Route flows through alternate units, if available; otherwise provide the best possible treatment with remaining units not being inspected or repaired.

All metal and concrete surfaces that come in contact with wastewater and covered surfaces exposed to fumes should have a good protective coating. The coating should be reapplied where necessary at each inspection.

Digesters should also be drained and cleaned on a regular basis. Once every five years (actual times range from three to eight years) has been accepted as an approximate interval for this operation. Most digesters have a sludge inlet box on one side and a supernatant box on the opposite side. A sludge sampler can be lowered through the pipes into both of these boxes to check for sand and grit buildup. To determine the amount of grit buildup, you must know the sidewall depth of the digester.

If the sludge sampler will only drop to within four feet (1.2 m) of the bottom, you can assume that you have a four-foot (1.2-m) buildup of sand and grit. By measuring the depth of sand and grit at periodic intervals, you can determine how fast the buildup is accumulating. In digesters, all metal and concrete surfaces should be inspected for deterioration.

On surfaces where the protective coatings are dead and flake off, it is necessary to sandblast the entire surface before new coatings are applied. Usually, two or more coats are needed for proper protection. Be aware that sandblasting old coatings may result in the production of a hazardous waste. Many coating materials used in wastewater facilities contain lead and the resulting spent blast material may contain sufficient lead levels to be classified as a hazardous waste. Test kits are available to determine if the coating is lead based. Very stringent regulations exist that deal with abrasive blasting and disposal of lead-based coating materials. Check with your local environmental health and safety agencies.

The protective coatings used on these types of tanks and channels can also be black asphaltic-type paint. These coatings should be used wherever practical. In areas where fumes and moisture are not severe, aluminum coating or a color scheme may be desirable. In these areas, a rubber-base paint or some similar material may be used. Follow the recommendations of a paint expert.

> **CAUTION**
>
> Periodic drainage, inspection, and repair of tanks and channels is essential. Failure to do so may result in complete disruption of operations during the critical low-flow/high-temperature season. Select a time for maintenance (removing obstructions and repairing gates, concrete pipes, and pumps) when you can minimize the discharge of harmful wastes to receiving waters. Schedule as many concurrent events as possible during a shutdown to minimize the times the plant or parts of the plant must be taken out of service.

14.03 Plant Grounds

Plant grounds that are well groomed and kept in a neat condition will greatly add to the overall appearance of the plant area. Well-groomed and neat grounds are important because many people judge the ability of the operator and the plant performance on the basis of the appearance of the plant. Management, also, tends to view well-kept grounds as evidence of an operator's ability and competence.

If the plant grounds have not been landscaped, it is sometimes the responsibility of the operator to do so. This may consist of planting shrubs and lawns or just keeping the grounds neat and weed free. Some plant grounds may be entirely paved. In any case, they should be kept clean and orderly at all times.

Control rodents and insects so they will not spread diseases or cause nuisances.

For the convenience of visitors and new operators, signs directing people to the plant, indicating the way to different plant facilities, identifying plant buildings, and the direction of flow and contents flowing in a pipe can all be very helpful. Well-lighted and well-maintained walks and roadways are very important. Plant grounds should be fenced to prevent unauthorized persons and animals from entering the area. Keep seldom-used items and old, discarded equipment neatly stored to avoid the appearance of a cluttered junkyard. Take pride in your plant grounds and you will be amazed at the favorable impression your facility will convey to the public and administrators.

QUESTIONS

Write your answers in a notebook and then compare your answers with those on page 869.

14.0E What items should be included in a building maintenance program?

14.0F When plant tanks and channels are drained, what items would you inspect?

14.0G Why are neat and well-groomed grounds important?

14.04 Library

A plant library can contain helpful information to assist in plant operation. Material in the library should be cataloged and filed for easy use. Items in the library should include:

1. Plant operation and maintenance manual.
2. Plant plans and specifications.
3. Manufacturers' equipment instructions.
4. Reference books on wastewater treatment, including manuals in this series.
5. Professional journals and publications.
6. Manuals of Practice and safety literature published by the Water Environment Federation (WEF), 601 Wythe Street, Alexandria, VA 22314-1994.
7. First-aid book.
8. Reports from other plants.
9. A dictionary.
10. Local, state, and federal safety regulations.

Please answer the discussion and review questions next.

DISCUSSION AND REVIEW QUESTIONS
Chapter 14. MAINTENANCE
(Lesson 1 of 7 Lessons)

At the end of each lesson in this chapter you will find some discussion and review questions. The purpose of these questions is to indicate to you how well you understand the material in the lesson. Write the answers to these questions in your notebook.

1. Why should the operator thoroughly read and understand manufacturers' literature before attempting to maintain plant equipment?

2. Why must administrators or supervisors be made aware of the need for an adequate maintenance program?

3. What is the purpose of a maintenance recordkeeping program?

4. What kinds of maintenance checks should be made periodically of stairways, ladders, catwalks, and platforms?

5. When should inspection and maintenance of the underwater portions of plant structures such as clarifiers and digesters be scheduled?

6. Why should rodents and insects be controlled?

7. What items should be included in a plant library?

CHAPTER 14. MAINTENANCE

(Lesson 2 of 7 Lessons)

14.1 MECHANICAL EQUIPMENT

Mechanical equipment commonly used in treatment plants is described and discussed in this section. Equipment used with specific treatment processes, such as clarifiers or aeration basins, are not discussed. You must be familiar with equipment and understand what it is intended to do before developing a preventive maintenance program and maintaining equipment.

14.10 Repair Shop

Many large plants have fully equipped machine shops staffed with competent mechanics. But for smaller plants, adequate machine shop facilities often can be found in the community. In addition, most pump manufacturers maintain pump repair departments where pumps can be fully reconditioned.

The pump repair shop in a large plant commonly includes such items as welding equipment, lathes, drill press and drills, power hacksaw, flame-cutting equipment, micrometers, calipers, gauges, portable electric tools, grinders, a hydraulic press, metal-spray equipment, and sandblasting equipment. You must determine what repair work you can and should do and when you need to request assistance from an expert. Help may be obtained from the manufacturer, the local representative, a consulting engineer, or another operator.

14.11 Pumps

Pumps serve many purposes in wastewater collection systems and treatment plants. They may be classified by the character of the material handled: raw wastewater, grit, effluent, activated sludge, raw sludge, or digested sludge. Or, they may relate to the conditions of pumping: high lift, low lift, recirculation, or high capacity. They may be further classified by principle of operation, such as centrifugal, propeller, reciprocating, and turbine.

The type of material to be handled and the function or required performance of the pump vary so widely that the design engineer must use great care in preparing specifications for the pump and its controls. Similarly, the operator must conduct a maintenance and management program adapted to the specific characteristics of the equipment.

14.110 Centrifugal Pumps

A centrifugal pump is basically a very simple device: an impeller rotating in a casing. The impeller is supported on a shaft which is, in turn, supported by bearings. Liquid coming in at the center (eye) of the impeller (Figure 14.2) is picked up by the vanes and by the rotation of the impeller and then is thrown out by centrifugal force into the discharge.

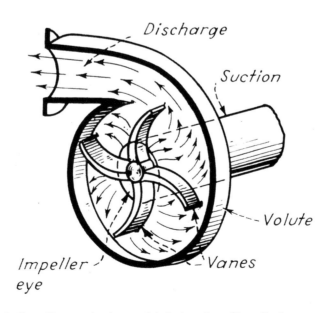

Refer to Figure 14.3 (page 761) for location of impeller in pump.

Fig. 14.2 Diagram showing details of centrifugal pump impeller

(Source: *CENTRIFUGAL PUMPS* by Karassik and Carter of Worthington Corporation)

754 Treatment Plants

To help you understand how pumps work and the purpose of the various parts, a section titled "Let's Build a Pump" has been included on the following pages. This material, reprinted with the permission of Allis-Chalmers Corporation, has been edited for style. Originally the material was printed in Allis-Chalmers Bulletin No. OBX62568.

Let's Build a Pump!

A student of medicine spends long years learning exactly how the human body is built before attempting to prescribe for its care. Knowledge of pump anatomy is equally basic in caring for centrifugal pumps.

But, whereas the medical student must take a body apart to learn its secrets, it will be far more instructive to us if we put a pump together (on paper, of course). Then we can start at the beginning—adding each new part as we need it in logical sequence.

As we see what each part does, how it does it … we will see how it must be cared for.

Another analogy between medicine and maintenance: there are various types of human bodies, but if you know basic anatomy, you understand them all. The same is true of centrifugal pumps. In building one basic type, we will learn about all types.

Part of this will be elementary to some maintenance people … but they will find it a valuable "refresher" course, and, after all, maintenance just cannot be too good.

So with a side glance at the centrifugal principle on page 755, let us get on with building our pump …

FIRST WE REQUIRE A DEVICE TO SPIN LIQUID AT HIGH SPEED …

That paddle-wheel device is called the "impeller" (Figure 14.2) … and it is the heart of our pump.

Note that the blades curve out from its hub. As the impeller spins, liquid between the blades is impelled outward by centrifugal force.

Note, too, that our impeller is open at the center—the "eye." As liquid in the impeller moves outward, it will suck more liquid in behind it through this eye … provided it is not clogged.

Maintenance Rule No. 1: If there is any danger that foreign matter (sticks, refuse, etc.) may be sucked into the pump—clogging or wearing the impeller unduly—provide the intake end of the suction piping with a suitable screen.

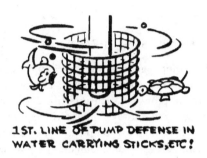

NOW WE NEED A SHAFT TO SUPPORT AND TURN THE IMPELLER …

Our shaft looks heavy—and it is. It must maintain the impeller in precisely the right place.

But that ruggedness does not protect the shaft from the corrosive or abrasive effects of the liquid pumped … so we must protect it with sleeves slid on from either end.

What these sleeves—and the impeller, too—are made of depends on the nature of the liquid we are to pump. Generally, they are bronze, but various other alloys, ceramics, glass, or even rubber-coating are sometimes required.

CENTRIFUGAL FORCE IN ACTION—

ALL MOVING BODIES TEND TO TRAVEL IN A STRAIGHT LINE. WHEN FORCED TO TRAVEL IN A CURVE, THEY CONSTANTLY *TRY* TO TRAVEL ON A TANGENT...

...IN AN "AIRPLANE RIDE"

Centrifugal force pushes dummy planes swung in a circle *away* from center of rotation.

...IN A WHIRLPOOL

Centrifugal force tends to push swirling water outward... forming vortex in center.

Maintenance Rule No. 2: Never pump a liquid for which the pump was not designed.

Whenever a change in pump application is contemplated and there is any doubt as to the pump's ability to resist the different liquid, check with your pump manufacturer.

WE MOUNT THE SHAFT ON SLEEVE, BALL, OR ROLLER BEARINGS ...

As we will see later, clearances between moving parts of our pump are quite small.

If bearings supporting the turning shaft and impeller are allowed to wear excessively and lower the turning units within a pump's closely fitted mechanism, the life and efficiency of that pump will be seriously threatened.

Maintenance Rule No. 3: Keep the right amount of the right lubricant in bearings at all times. Follow your pump manufacturer's lubrication instructions to the letter.

Main points to keep in mind are ...

1. Although too much oil will not harm sleeve bearings, too much grease in antifriction type bearings (ball or roller) will promote friction and heat. The main job of grease in antifriction bearings is to protect steel elements against corrosion, not friction.

2. Operating conditions vary so widely that no one rule as to frequency of changing lubricant will fit all pumps. So play it safe: if anything, change lubricant *before* it is too worn or too dirty.

TO CONNECT WITH THE MOTOR, WE ADD A COUPLING FLANGE ...

Some pumps are built with pump and motor on one shaft and, of course, offer no alignment problem.

But our pump is to be driven by a separate motor ... and we attach a flange to one end of the shaft through which bolts will connect with the motor flange.

Use a straightedge or a dial indicator to ensure shaft alignment (see pages 847 and 848, Figures 14.53 and 14.54).

Maintenance Rule No. 4: See that pump and motor flanges are parallel vertically and axially ... and that they are kept that way.

If shafts are eccentric or meet at an angle, every revolution throws tremendous extra load on bearings of both pump and motor. Flexible couplings will not correct this condition, if excessive.

Checking alignment should be a regular procedure in pump maintenance. Foundations can settle unevenly, piping can change pump position, bolts can loosen. Misalignment is a major cause of pump and coupling wear.

NOW WE NEED A "STRAW" THROUGH WHICH LIQUID CAN BE SUCKED ...

Notice two things about the suction piping: (1) the horizontal piping slopes upward toward the pump; (2) any reducer that connects between the pipe and pump intake nozzle should be horizontal at the top—(eccentric, not concentric).

This up-sloping prevents air pocketing in the top of the pipe ... where trapped air might be drawn into the pump and cause loss of suction.

Maintenance Rule No. 5: Any down-sloping toward the pump in suction piping (as exaggerated in the previous diagrams) should be corrected.

This rule is VERY important. Loss of suction greatly endangers a pump ... as we will see shortly.

WE CONTAIN AND DIRECT THE SPINNING LIQUID WITH A CASING ...

We got a little ahead of our story in the previous paragraphs ... because we did not yet have the casing to which the suction piping bolts. And the manner in which it is attached is of great importance.

Maintenance Rule No. 6: See that piping puts absolutely no strain on the pump casing.

When the original installation is made, all piping should be in place and self-supporting before connection. Openings should meet with no force. Otherwise, the casing is apt to be cracked ... or sprung enough to allow closely fitted pump parts to rub.

It is good practice to check the piping supports regularly to see that loosening, or settling of the building, has not put strains on the casing.

NOW OUR PUMP IS ALMOST COMPLETE, BUT IT WOULD LEAK LIKE A SIEVE ...

We are far enough along now to trace the flow of water through our pump. It is not easy to show suction piping in the cross-section view above, so imagine it stretching from your eye to the lower center of the pump.

Our pump happens to be a "double suction" pump, which means that water flow is divided inside the pump casing ... reaching the eye of the impeller from either side.

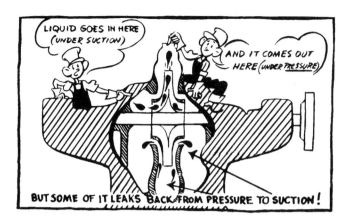

As water is drawn into the spinning impeller, centrifugal force causes it to flow outward ... building up high pressure at the outside of the pump (which will force water out) and creating low pressure at the center of the pump (which will draw water in). This situation is diagrammed in the upper half of the pump, above.

So far so good ... except that water tends to be drawn back from pressure to suction through the space between impeller and casing—as diagrammed in the lower half of the pump, above—and our next step must be to plug this leak, if our pump is to be very efficient.

SO WE ADD WEARING RINGS TO PLUG INTERNAL LIQUID LEAKAGE ...

You might ask why we did not build our parts closer fitting in the first place—instead of narrowing the gap between them by inserting wearing rings (Figure 14.3, page 761).

The answer is that those rings are removable and replaceable ... when wear enlarges the tiny gap between them and the impeller. (Sometimes rings are attached to impeller rather than casing—or rings are attached to both so they face each other.)

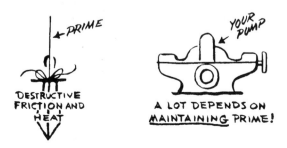

Maintenance Rule No. 7: Never allow a pump to run dry (either through lack of proper priming when starting or through loss of suction when operating). Water is a lubricant between rings and impeller.

Maintenance Rule No. 8: Examine wearing rings at regular intervals. When seriously worn, their replacement will greatly improve pump efficiency.

TO KEEP AIR FROM BEING DRAWN IN, WE USE STUFFING BOXES . . .

We have two good reasons for wanting to keep air out of our pump: (1) we want to pump water, not air; (2) air leakage is apt to cause our pump to lose suction.

Each stuffing box we use consists of a casing, rings of packing, and a gland at the outside end. A mechanical seal may be used instead.

Maintenance Rule No. 9: Packing should be replaced periodically—depending on conditions—using the packing recommended by your pump manufacturer. Forcing in a ring or two of new packing instead of replacing worn packing is bad practice. It is apt to displace the seal cage.

Put each ring of packing in separately, seating it firmly before adding the next. Stagger adjacent rings so the points where their ends meet do not coincide. See Figure 14.48, "How to pack a pump," on pages 836 and 837.

Maintenance Rule No. 10: Never tighten a gland more than necessary . . . as excessive pressure will wear shaft sleeves unduly.

Maintenance Rule No. 11: If shaft sleeves are badly scored, replace them immediately . . . or packing life will be entirely too short.

TO MAKE PACKING MORE AIRTIGHT, WE ADD WATER SEAL PIPING . . .

In the center of each stuffing box is a "seal cage." By connecting it with piping to a point near the impeller rim, we bring liquid under pressure to the stuffing box.

This liquid acts both to block out air intake and to lubricate the packing. It makes both packing and shaft sleeves wear longer . . . providing it is clean liquid.

Maintenance Rule No. 12: If the liquid being pumped contains grit, a separate source of sealing liquid should be obtained (for example, it may be possible to direct some of the pumped liquid into a container and allow the grit to settle out).

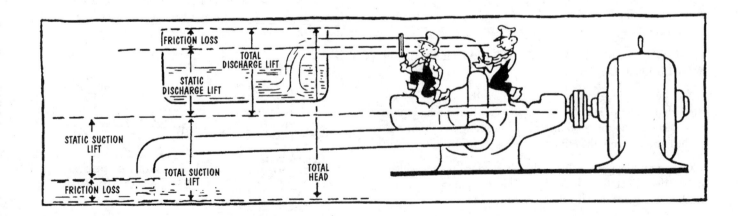

To control liquid flow, draw up the gland just tight enough so a thin stream (approximately one drop per second) flows from the stuffing box during pump operation.

DISCHARGE PIPING COMPLETES THE PUMP INSTALLATION—AND NOW WE CAN ANALYZE THE VARIOUS FORCES WE ARE DEALING WITH …

SUCTION. At least 75 percent of centrifugal pump troubles trace to the suction side. To minimize them …

1. Total suction lift (distance between centerline of pump and liquid level when pumping, plus friction losses) generally should not exceed 15 feet (4.5 meters).

2. Piping should be at least a size (diameter) larger than pump suction nozzle.

3. Friction in piping should be minimized . . . use as few and as easy bends as possible . . . avoid scaled or corroded pipe.

DISCHARGE lift, plus suction lift, plus friction in the piping from the point where liquid enters the suction piping to the end of the discharge piping equals total head.

Pumps should be operated near their rated heads. Otherwise, the pump is apt to operate under unsatisfactory and unstable conditions, which reduce efficiency and operating life of the unit, or "cavitation" could occur. Note the description of cavitation on page 760. Cavitation can seriously damage your pump.

PUMP CAPACITY generally is measured in gallons per minute (liters per second or m^3 per second). A new pump is guaranteed to deliver its rating in capacity and head. But whether a pump retains its actual capacity depends to a great extent on its maintenance.

Wearing rings must be replaced when necessary—to keep internal leakage losses down.

Friction must be minimized in bearings and stuffing boxes by proper lubrication … and misalignment must not be allowed to force scraping between closely fitted pump parts.

POWER of the driving motor, like capacity of the pump, will not remain at a constant level without proper motor maintenance.

Starting load on motors can be reduced by throttling or closing the pump discharge valve (*NEVER* the suction valve!) … but the pump must not be operated for long with the discharge valve closed. Power then is converted into friction, overheating the water with serious consequences.

Centrifugal pumps designed for pumping wastewater (Figures 14.3 and 14.4) usually have smooth channels and impellers with large openings to prevent clogging.

Impellers may be of the open or closed type (Figure 14.5). Submersible pumps (Figure 14.6) usually have open impellers and are frequently used to pump wastewater from wet wells in lift stations.

14.111 Propeller Pumps

There are two basic types of propeller pumps (Figure 14.7), axial-flow and mixed-flow impellers. The axial-flow propeller pump is one having a flow parallel to the *AXIS*[1] of the impeller (Figure 14.8). The mixed-flow propeller pump is one having a flow that is both *AXIAL*[2] and *RADIAL*[3] to the impeller (Figure 14.8).

14.112 Vertical Wet Well Pumps

A vertical wet well pump is a vertical shaft, diffuser-type centrifugal pump with the pumping element suspended from the discharge piping (Figure 14.9). The needs of a given installation determine the length of discharge column. The pumping bowl assembly may connect directly to the discharge head for shallow sumps, or may be suspended several hundred feet for raising water from wells. Vertical turbine pumps are used to pump water from deep wells, and may be of the *SINGLE-STAGE* or *MULTISTAGE TYPE*.[4]

14.113 Reciprocating or Piston Pumps

The word "reciprocating" means moving back and forth, so a reciprocating pump is one that moves water or sludge by a piston that moves back and forth. A simple reciprocating pump is shown in Figure 14.10 (below). If the piston is pulled to the left, Check Valve A will be open and sludge will enter the pump and fill the casing.

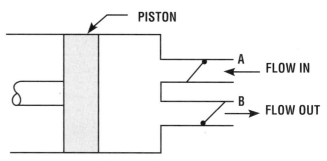

Fig. 14.10   Simple reciprocating pump
(See page 838 for pump details)

When the piston reaches the end of its travel to the left and is pushed back to the right, Check Valve A will close, Check Valve B will open, and wastewater will be forced out the exit line.

A piston pump is a positive-displacement pump. Never operate it against a closed discharge valve or the pump, valve, or

---

[1] *Axis of Impeller.*   An imaginary line running along the center of a shaft (such as an impeller shaft).
[2] *Axial to Impeller.*   The direction in which material being pumped flows around the impeller or flows parallel to the impeller shaft.
[3] *Radial to Impeller.*   Perpendicular to the impeller shaft. Material being pumped flows at a right angle to the impeller.
[4] Single-stage type pumps have only one impeller while multistage type pumps have more than one impeller.

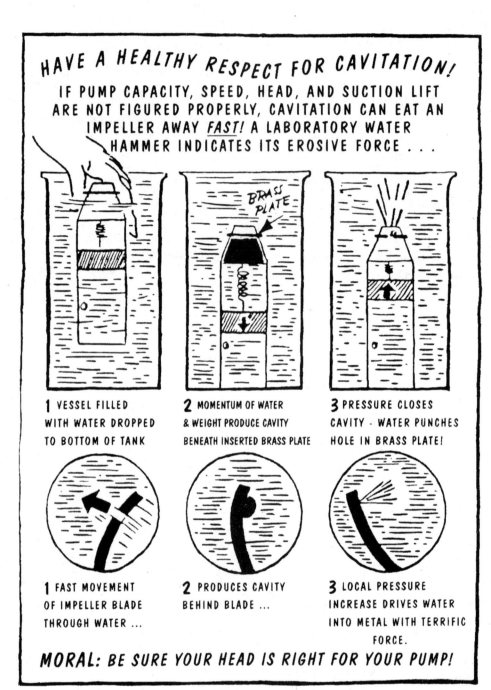

Cavitation is a condition that can cause a drop in pump efficiency, vibration, noise, and rapid damage to the impeller of a pump. Cavitation occurs due to unusually low pressures within a pump. These low pressures can develop when pump inlet pressures drop below the design inlet pressures or when the pump is operated at flow rates considerably higher than design flows. When the pressure within the flowing water drops very low, the water starts to boil and vapor bubbles form. These bubbles then collapse with great force, which knocks metal particles off the pump impeller. This same action can and does occur on pressure reducing valves and partially closed gate and butterfly valves.

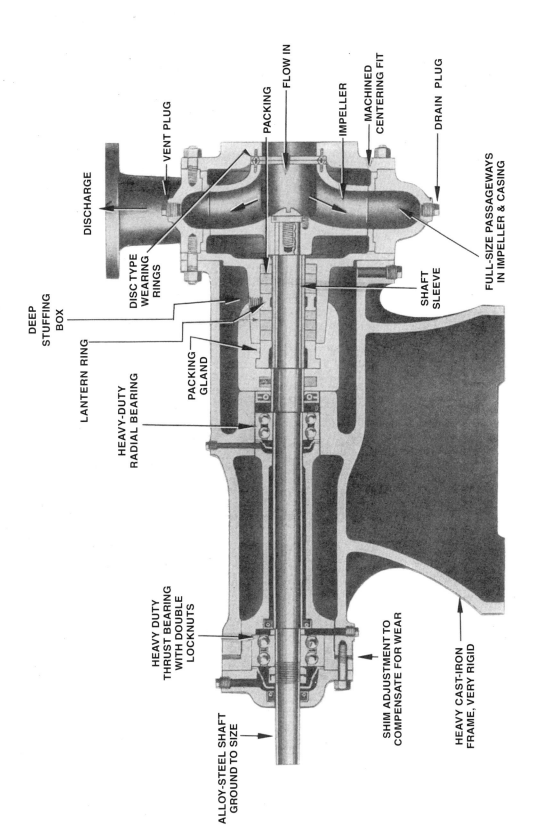

Fig. 14.3 Horizontal nonclog wastewater pump with open impeller
(Source: War Department Technical Manual TM5-666)

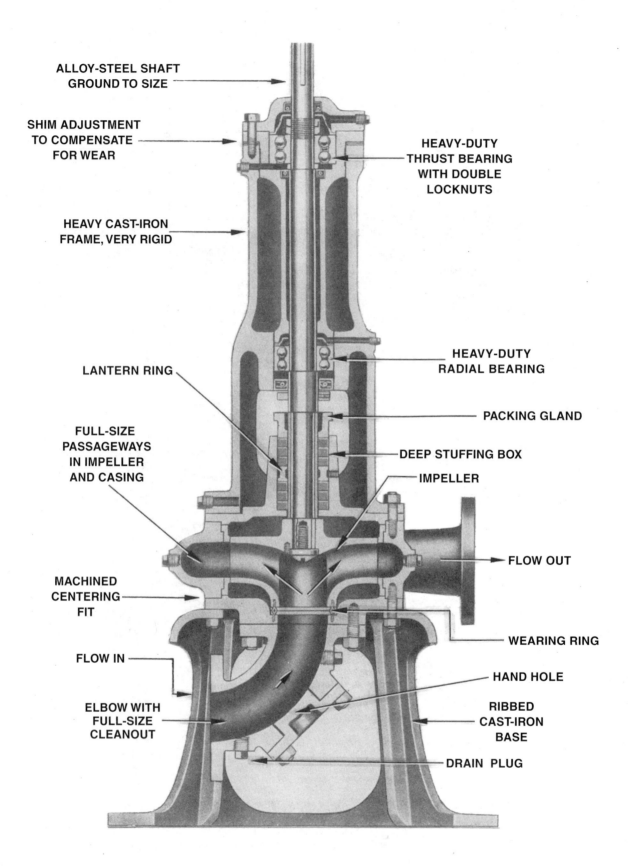

Fig. 14.4   Vertical ball bearing-type wastewater pump
(Source: War Department Technical Manual TM5-666)

Maintenance 763

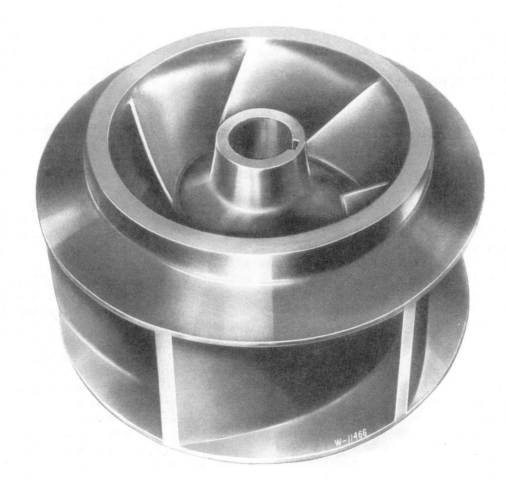

Closed Radial
(Closed radial impellers are used in wastewater treatment plants.)

Open Radial

Fig. 14.5   Impellers
(Source: *CENTRIFUGAL PUMPS* by Karassik and Carter of Worthington Corporation)

764   Treatment Plants

1. LIFTING HANDLE
2. JUNCTION CHAMBER WITH WATERTIGHT CABLE ENTRIES
3. ANTIFRICTION BEARINGS
4. SHAFT
5. STATOR WITH TEMPERATURE SENSING THERMISTORS
6. ROTOR
7. STATOR HOUSING LEAKAGE SENSOR
8. BEARING TEMPERATURE THERMISTOR
9. SHAFT SEAL
10. OIL CHAMBER
11. VOLUTE
12. NONCLOG IMPELLER
13. COOLING JACKET
14. SLIDING BRACKET
15. AUTOMATIC DISCHARGE CONNECTION

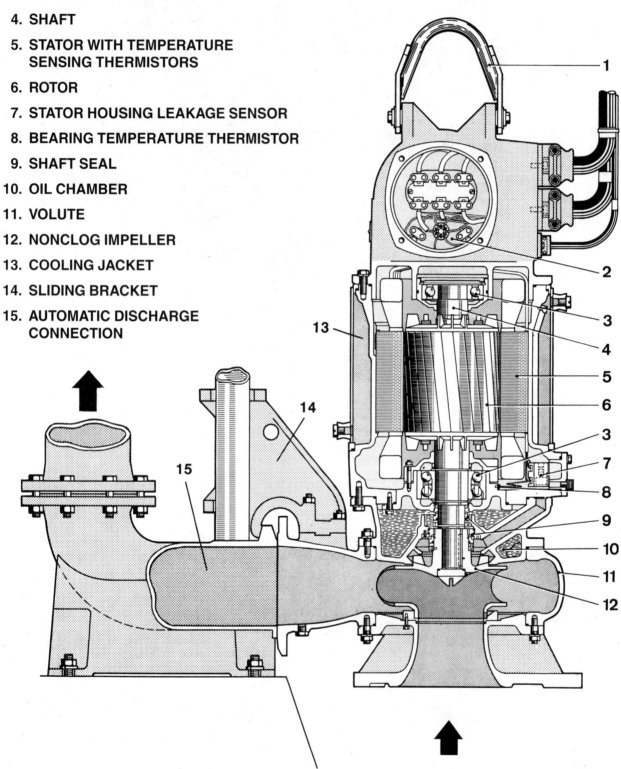

Fig. 14.6 Submersible wastewater pump
(Courtesy of Flygt Corporation)

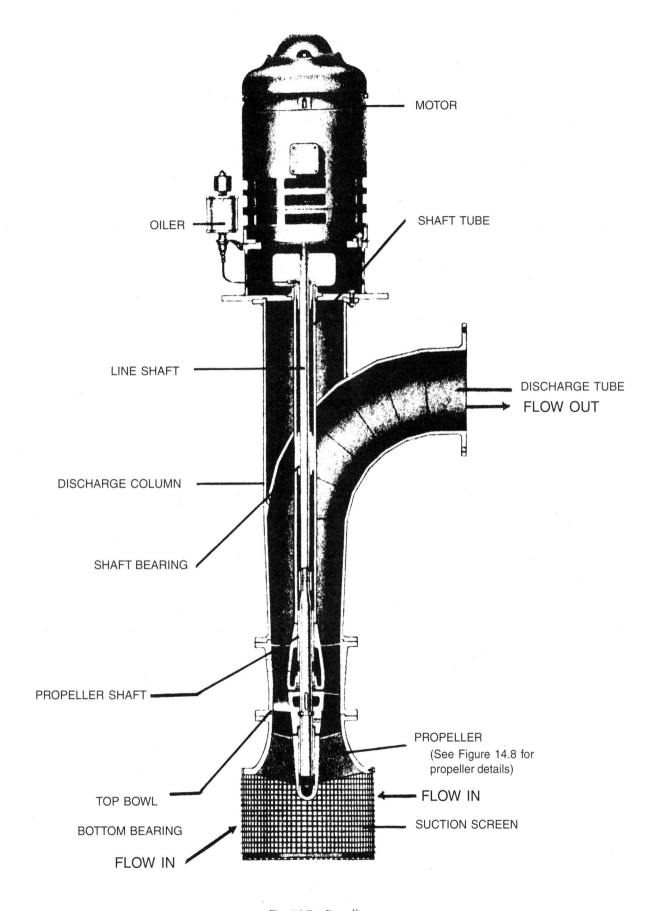

Fig. 14.7 Propeller pump

766 Treatment Plants

Axial-Flow

Mixed-Flow

Fig. 14.8 Propeller-type impellers
(Source: *CENTRIFUGAL PUMPS* by Karassik and Carter of Worthington Corporation)

Maintenance 767

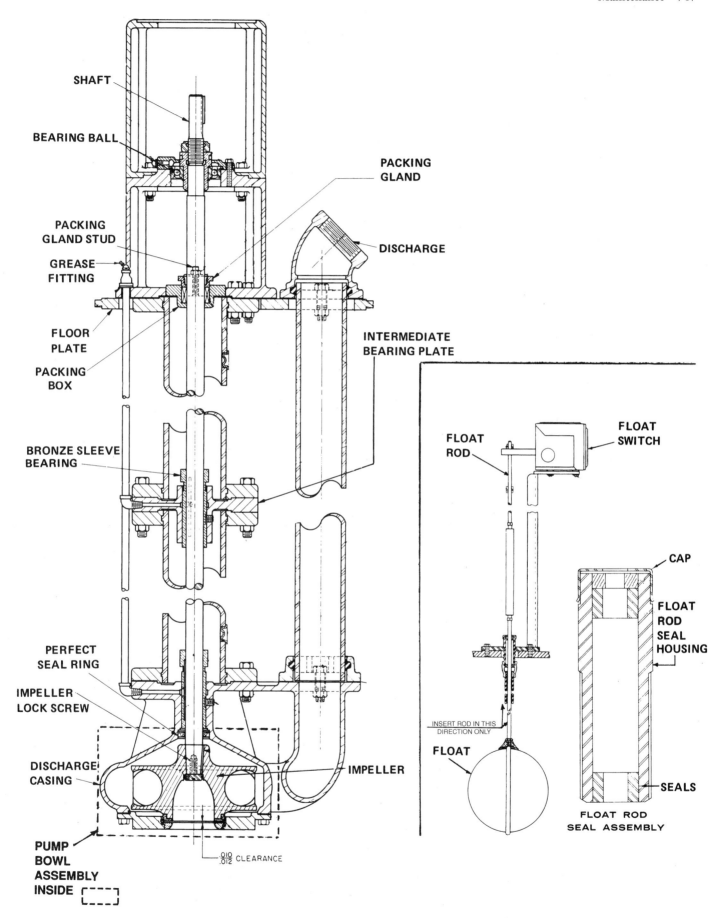

Fig. 14.9 Vertical wet well pump (motor not shown)
(Courtesy Chicago Pump)
Note: Figure 14.10 is on page 759

pipe could be damaged by excessive pressures. Also, the suction valve should be open when the pump is started. Otherwise, an excessive suction or vacuum could develop and cause problems.

14.114 Incline Screw Pumps

Incline screw pumps (Figure 14.11) consist of a screw operating at a constant speed within a housing or trough. When the screw rotates, it moves the wastewater up the trough to a discharge point. The screw is supported by two bearings, one at the top and one at the bottom.

14.115 Progressive Cavity (Screw-Flow) Pumps (Figure 14.12)

Operation of a progressive cavity pump is similar to that of a precision incline screw pump (Figure 14.11). The progressive cavity pump consists of a screw-shaped rotor snugly enclosed in a non-moving *STATOR*[5] or housing (Figure 14.13). The threads of the screw-like rotor (commonly manufactured of chromed steel) make contact along the walls of the stator (usually made of synthetic rubber). The gaps between the rotor threads are called "cavities." When wastewater is pumped through an inlet valve, it enters the cavity. As the rotor turns, the waste material is moved along until it leaves the conveyor (rotor) at the discharge end of the pump. The size of the cavities along the rotor determines the capacity of the pump.

All progressive cavity pumps operate on the basic principle described above. To further increase capacity, some models have a shaped inside surface of the stator (housing) with a similarly shaped rotor. In addition, some models use a rotor that moves up and down inside the stator as well as turning on its axis (Figure 14.12). This allows a further increase in the capacity of the pump.

These pumps are recommended for materials that contain higher concentrations of suspended solids. They are commonly used to pump sludges. Progressive cavity pumps should never be operated dry (without liquid in the cavities), nor should they be run against a closed discharge valve.

14.116 Pneumatic Ejectors (Figure 14.14)

Pneumatic ejectors are used when it is necessary to handle limited flows. Centrifugal pumps are highly efficient for pumping large flows; however, when scaled down for lower flows, they tend to plug easily. Unstrained solids will tend to block the small impeller opening (if less than two inches (5.1 cm)) of a centrifugal pump and will quickly reduce the flow. Pneumatic ejectors, on the other hand, are capable of passing solids up to the size of the inlet and discharge valves, and there is nothing on the inside of the ejector-receiver to restrict the flow. The ejector may be useless, however, if a stick or another object gets stuck under the inlet or discharge check valve preventing it from closing.

14.12 Pump Lubrication, Seals, and Bearings

14.120 Lubrication

Pumps, motors, and drives should be oiled and greased in strict accordance with the recommendations of the manufacturer. Cheap lubricants may often be the most expensive in the end. Oil should not be put in the housing while the pump shaft is rotating because the rotary action of the ball bearings will pick up and retain a considerable amount of oil. When the unit comes to rest, an overflow of oil around the shaft or out of the oil cup will result.

14.121 Mechanical Seals (Figure 14.15)

Many pumps use mechanical seals in place of packing. Mechanical seals serve the same purpose as packing; that is, they prevent leakage between the pump casing and shaft. Like packing, they are located in the stuffing box where the shaft goes through the volute; however, they should not leak. Mechanical seals are gaining popularity in the wastewater field.

Mechanical seals have two faces that mate tightly and prevent water from passing through them. One half of the seal is mounted in the pump or gland with an O-ring or gasket, thus providing sealing between the housing and seal face. This prevents water from going around the seal face and housing. The other half of the mechanical seal is installed on the pump shaft. This part also has an O-ring or gasket between the shaft and seal to prevent water from leaking between the seal part and shaft. There is a spring located behind one of the seal parts that applies pressure to hold the two faces of the seal together and keeps any water from leaking out. One half of the seal is stationary and the other half is revolving with the shaft.

Materials used in the manufacture of mechanical seal parts are carbon, stainless steel, ceramic, tungsten carbide, brass, and many others. The different type materials are selected for their best application. Some of the variables are:

1. Liquid and solids being pumped
2. Shaft speed
3. Temperature
4. Corrosion resistance
5. Abrasives

Initially, mechanical seals are more expensive than packing to install in a pump. This cost is gained back in maintenance savings during a period of time.

Some of the advantages of mechanical seals are as follows:

1. They last from three to four years without having to touch them, resulting in labor savings.
2. Usually there is not any damage to the shaft sleeve when they need replacing (expensive machine work is not needed).

---

[5] *Stator.* That portion of a machine that contains the stationary (non-moving) parts that surround the moving parts (rotor).

Maintenance 769

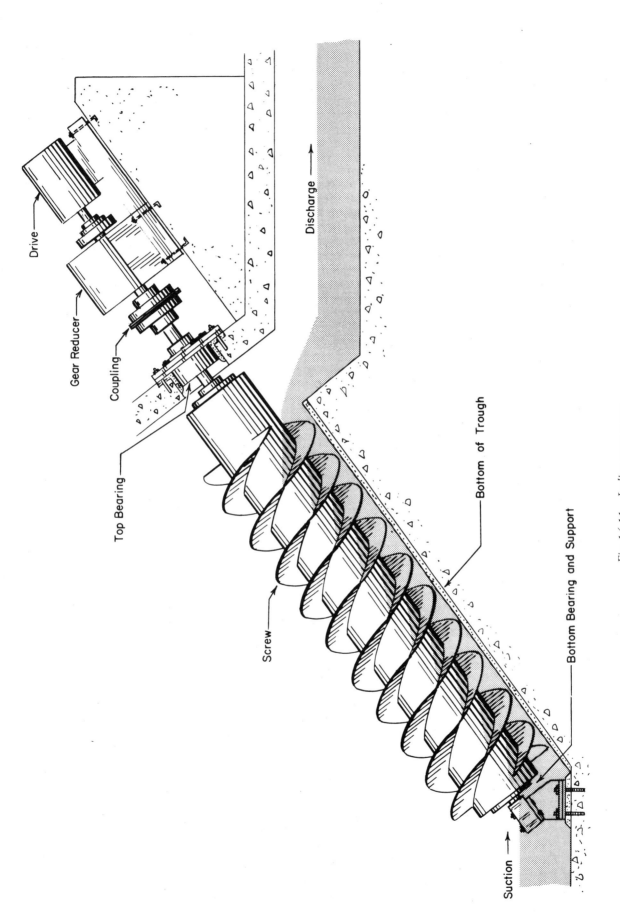

Fig. 14.11 Incline screw pump
(Courtesy of FMC Corporation, Environmental Equipment Division)

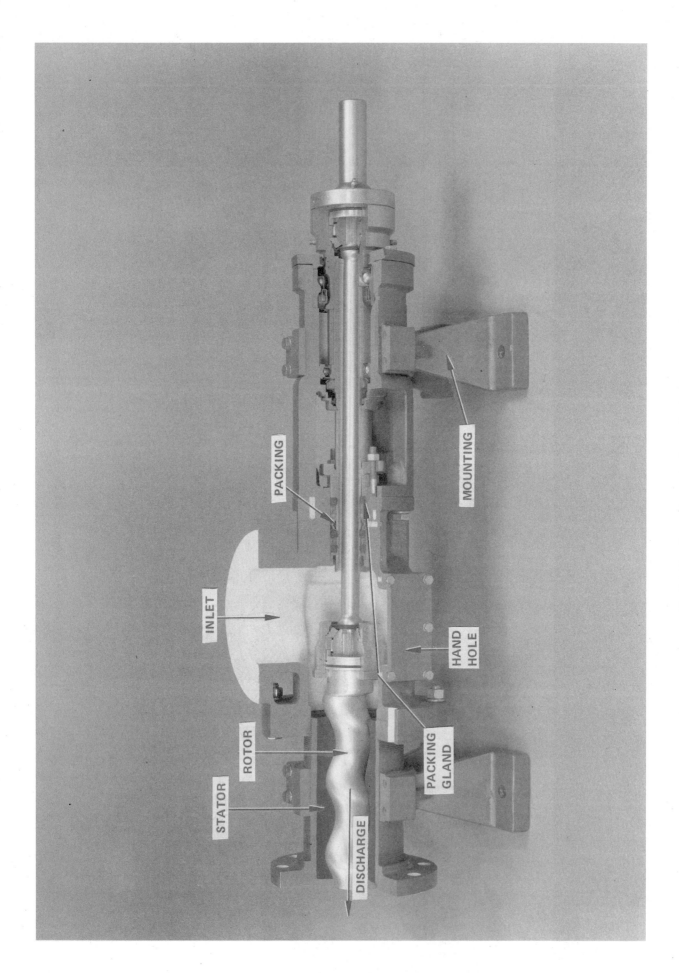

Fig. 14.12 Progressive cavity (screw-flow) pump
(Permission of Moyno Pump Division, Robbins & Meyer, Inc.)

Maintenance 771

Pumping principle

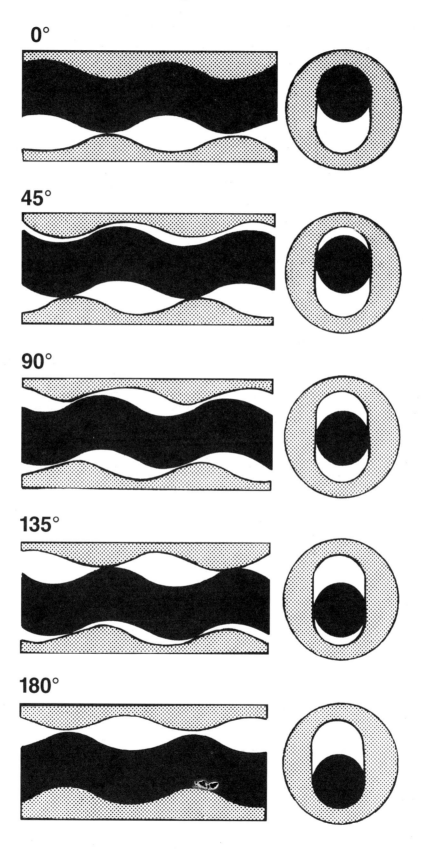

Fig. 14.13 Pumping principle of a progressive cavity pump
(Permission of Allweiler Pumps, Inc.)

772 Treatment Plants

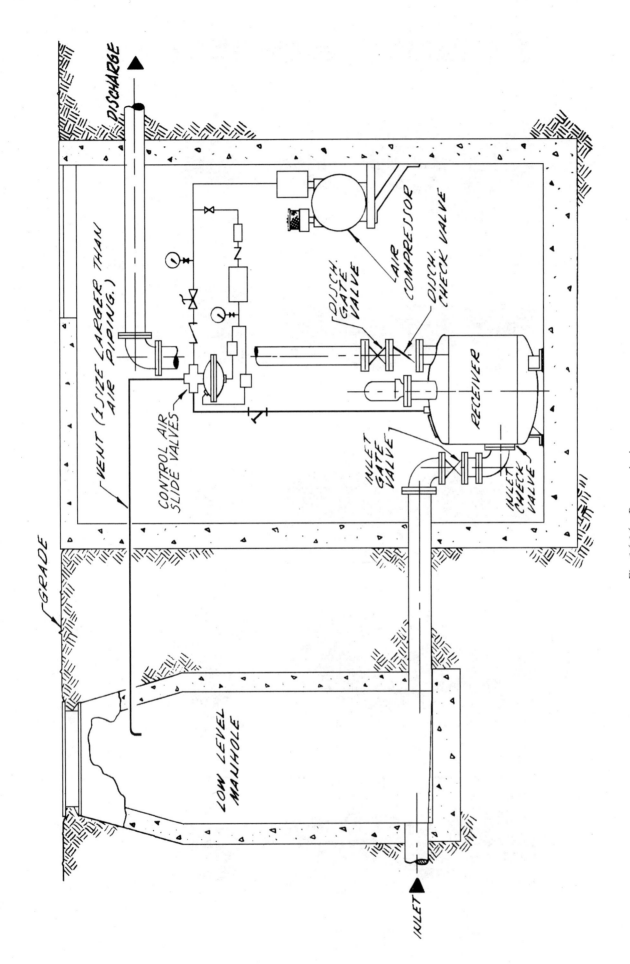

Fig. 14.14 Pneumatic ejector system
(Courtesy of James Equipment and Manufacturing Company)

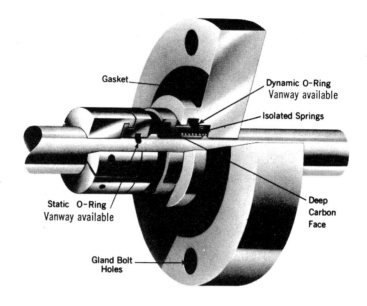

Fig. 14.15   Mechanical seals
(Courtesy A. W. Chesterton Co.)

3. Continual adjusting, cleaning, or repacking is not required.

4. The possibility of flooding a lift station because a pump has thrown its packing is eliminated; however mechanical seals can fail and flood a lift station too.

Some of the limitations of mechanical seals are as follows:

1. High initial cost.

2. Competent mechanic required for installation.

3. When they fail, the pump must be shut down.

4. Pump must be dismantled to repair.

Mechanical seals are always flushed in some manner to lubricate the seal faces and minimize wear. This may be the liquid being pumped and is referred to as "source lubrication." If fresh water is used, it is connected from the high-pressure side and back to the stuffing box low-pressure side. In some lift stations where fresh water is not available, the wastewater is used in this manner but it must be filtered. Another method is to use fresh water from an air gap system. Still another way of lubricating the seal is to use a spring-loaded grease cup.

Whatever method is used, the mechanical seal must be inspected frequently. The seal water is adjusted to five psi (0.35 kg/sq cm) above maximum discharge pressure to keep the wastewater and grit from entering the seal housing and contaminating the seal faces. Grease cups must be kept full at all times and inspected to make sure they are operating properly.

When a pump is fitted with a mechanical seal, it must never run dry or the seal faces will be burned and ruined. Mechanical seals are not supposed to have any leakage from the gland. If a leak develops, the seal may require resurfacing or it may have to be replaced.

Repair or replacement of mechanical seals requires the pump to be removed and dismantled. Seals are quite delicate and special care must be taken when installing them. Mechanical seals differ widely in their construction and installation, and individual manufacturer's instructions must be followed. Due to the complexity of installing mechanical seals, special tools and equipment are needed as well as a qualified pump mechanic.

The following is a checklist for cost-effective installation and use of mechanical seals:

1. Use cartridge-mounted seals. They are easy to install and do not damage either the seal or the pump.

2. Purchase standard, off-the-shelf seals to ensure seals will be available from suppliers, even in an emergency.

3. Return seals to the original manufacturer for repair.

4. Obtain bearing isolators (labyrinths) or magnetic seals to protect the bearing from intrusion of moisture or debris.

5. Use synthetic lubricants in all pumps, motors, bearings, and compressors. Synthetic lubricants are very cost-effective because they lower the operating temperature, tolerate water intrusion better, lower high-frequency vibration, and last two or three times longer than the low-priced mineral oils.

6. Every time the motor or pump is worked on, be sure to perform precision coupling alignment by using a dial indicator or laser to ensure long-term service life of bearings and mechanical seals.

14.122 Bearings

Pump bearings usually should last for years if serviced properly and used in their proper application. There are several types of bearings used in pumps, such as ball bearings, roller bearings, and sleeve bearings. Each bearing has a special purpose, such as thrust load, radial load, and speed. The type of bearing used in each pump depends on the manufacturer's design and application. Whenever a bearing failure occurs, the bearing should be examined to determine the cause and, if possible, to eliminate the problem. Many bearings are ruined during installation or start-up. Bearing failures may be caused by:

1. Fatigue failure

2. Contamination

3. *BRINELLING*[6]

4. False brinelling

5. Thrust failures

6. Misalignment

7. Electric arcing

8. Lubrication failure

9. Cam failure

14.13 Starting a New Pump

The initial start-up work described in this section should be done by a qualified and trained person, such as a manufacturer's representative, consulting engineer, or an experienced operator. The operator can learn a lot about pumps and motors by accompanying and helping an equipment representative put new equipment into operation.

Before starting a pump, lubricate it according to the lubrication instructions. Turn the shaft by hand to see that it rotates freely. Then check to see that the shafts of the pump and motor are aligned and the flexible coupling adjusted. (Refer to Section 14.7, Paragraph 11, "Couplings," page 846.) If the unit is belt driven, sheave (pulley) alignment and belt adjustment should be checked. (Refer to Paragraph 8, "Belt Drives.")

Check the electric current characteristics with the motor characteristics and inspect the wiring. See that thermal units in the starter are set properly. Turn on the motor just long enough to see that it turns the pump in the direction indicated by the rotational arrows marked on the pump. If separate water seal

---

[6] *Brinelling* (bruh-NEL-ing). Tiny indentations (dents) high on the shoulder of the bearing race or bearing. A type of bearing failure.

units or vacuum primer systems are used, these should be started. Finally, make sure lines are open. Sometimes there is an exception (see following paragraph) in the case of the discharge valve.

A pump should not be run without first having been primed. To prime a pump, the pump must be completely filled with water or wastewater. In some cases, automatic primers are provided. If they are not, it is necessary to vent the casing. Most pumps are provided with a valve to accomplish this. Allow the trapped air to escape until water or wastewater flows from the vent; then replace the vent cap. In the case of suction lift applications, the pump must be filled with water unless a self-primer is provided. In nearly every case, you may start a pump with the discharge valve open. Exceptions to this, however, are where water hammer or velocity disturbances might result, or where the motor does not have sufficient margin of safety or power. Sometimes there are no check valves in the discharge line. In this case (with the exception of positive displacement pumps), it is necessary to start the pump and then open the discharge lines. Where there are common discharge headers, it is essential to start the pump and then open the discharge valve. A positive displacement pump (reciprocating or progressive cavity types) should never be operated against a closed discharge line.

After starting the pump, again check to see that the direction of rotation is correct. Packing gland boxes (stuffing boxes) should be observed for slight leakage (approximately 60 drops per minute), as described in Paragraph 1, "Pumps, General (Including Packing)." Check to see that the bearings do not overheat from over- or underlubrication. The flexible coupling should not be noisy; if it is, the noise may be caused by misalignment or improper clearance or adjustment. Check to be sure pump anchorage is tight. To find out if a new pump is delivering design flows and pressures, measure the flows and pressures and compare them with the pump performance curves supplied by the manufacturer. (See OPERATION OF WASTEWATER TREATMENT PLANTS, Volume I, Appendix, "How To Solve Wastewater Treatment Plant Arithmetic Problems," Section A.8, PUMPS.) If pump delivery falls below performance curves, look for obstructions in the pipelines. Inspect piping for leaks.

14.14 Pump Shutdown

When shutting down a pump for a long period, the motor disconnect switch should be opened, locked out, and tagged with the reason for the shutdown noted on the tag. All valves on the suction, discharge, and water seal lines should be tightly shut. Completely drain the pump by removing the vent and drain plugs. Do not permit sludge to remain in pumps or piping for any length of time; there are cases on record in which the gas produced has created sufficient pressure to rupture pipes and sludge pumps.

Inspect the pump and bearings thoroughly so that all necessary servicing may be done during the inactive period. Drain the bearing housing and then add fresh lubricant. Follow any additional manufacturer's recommendations.

14.15 Pump-Driving Equipment

Driving equipment used to operate pumps includes electric motors and internal combustion engines. In rare instances, pumps are driven with steam turbines, steam engines, air, and hydraulic motors.

In all except the large installations, electric motors are used almost exclusively, with synchronous and induction types being the most commonly used. Synchronous motors operate at constant speeds and are used chiefly in large sizes. Three-phase, squirrel-cage induction motors are most often used in treatment plants. These motors require little attention and, under average operating conditions, the factory lubrication of the bearing will last approximately one year. (Check with the manufacturer for average number of operating hours for bearings.) When lubricating motors, remember that too much grease may cause bearing trouble or damage the winding.

Clean and dry all electrical contacts. Inspect for loose electrical contacts. Only qualified and authorized electricians should be permitted to work on electrical equipment and circuitry. Make sure that hold-down bolts on motors are secure. Check voltage while the motor is starting and running. Examine bearings and couplings.

14.16 Electrical Controls

A variety of electrical equipment is used to control the operation of wastewater pumps or to protect electric motors. The simplest type of control unit consists of a counterweighted float that triggers a switch. When the float is raised by the wastewater to a predetermined level, a switch is tripped to start the pump. When the wastewater level falls to the cutoff level, the float switch stops the pump. The time required for each cycle and the length of time between cycles depend on the pumping rate and the quantity of wastewater flow.

If starters, disconnect switches, and cutouts are used, they should be installed in accordance with the local regulations (city or county codes) regarding this equipment. In the case of larger motors, the power company often requires starters that do not overload the power lines.

The electrode-type, bubbler-type, and diaphragm-type water level control systems are all similar in effect to the float-switch system. Scum is a problem with most water level controls that operate pumps and it must be removed on a regular basis.

QUESTIONS

Write your answers in a notebook and then compare your answers with those on page 869.

14.1A   Where would you find out how to lubricate a pump?

14.1B   Why should a pump with a mechanical seal never be allowed to run dry?

14.1C   What should be done when a mechanical seal develops a leak?

14.1D   What problems can develop if too much grease is used in lubricating a motor?

14.17 Variable-Speed, Alternating Current (AC) Motors
by Ken Peschel

14.170 Description

Many of today's collection system pumping stations incorporate a wet well design that uses variable-speed pumping equipment. Lower building costs and the smaller (or nonexistent) wet well both contribute to the growing popularity of this type of installation. A variable-speed pump facility can maintain a predetermined level of flow in the incoming line during minimal flows in dry weather as well as during maximum flows in the peak wet weather season. Still, all components of this type of facility require a set preventive maintenance program to avoid costly breakdowns. Experience has shown that the motors themselves are most often overlooked during preventive maintenance.

A standard, three-phase, single-speed, synchronous AC motor can be satisfactorily maintained with a standard preventive maintenance program consisting primarily of lubrication, cleaning, and testing of electric circuits. In contrast, a standard, variable-speed, three-phase, AC motor incorporates three copper or copper alloy slip rings or "collectors" attached to the shaft of a wound rotor. These three slip rings receive a secondary electrical current through sets of carbon brushes from the rotor windings. The slip rings and brushes on such a motor require additional preventive maintenance procedures.

To properly maintain a standard, three-phase, variable-speed, synchronous AC motor, you must have some idea of what to look for when examining the slip rings and brushes. Examine these components before initial start-up, if you possibly can. You must also keep a close watch during the "filming" or seating period immediately after start-up; this may take anywhere from a few hours to several hundred hours, but proper formation of the film is crucial to the operation of the motor.

The film that forms on the slip rings during operation is the conducting medium by which the electric current is carried from the carbon brush to the metal slip ring. Without it, satisfactory operation of the sliding contact is impossible. This film is a form of corrosion caused by the chemical reaction between the metals and the atmosphere. The first layers protect the metal and the rate of film growth will decrease with time. This film is actually an oxide that starts forming when the ring surface is machined. Initial oxidation takes place in a matter of seconds and will have a thickness of about 20 molecules of oxide. This film must be deposited uniformly on the ring by some means throughout the life of the ring and brush. Film formation is accelerated by the heat of machining the ring surface, seating of brushes, friction while running, electrical load, and by mechanical burnishing during operation.

A uniform film of oxidation is essential to the flow of electricity, but it serves another purpose as well by protecting the metal surfaces from wear. The powerful forces of attraction cause rapid wear of the surfaces. In some cases, these forces may cause seizure and result in motor failure. Fortunately, these forces are effective only over short distances, and a boundary film separating and lubricating the components is sufficient to prevent seizure and wear.

The brush is the other main component of the variable-speed motor that needs periodic care and attention. The four basic types of brush composition are: carbongraphite, electrographite, graphite, and metal graphite. Complete details of these various types, together with characteristics, shunts, connections, and shunt insulation, may be obtained from brush manufacturers' publications.

It is extremely important to maintain a regular examination and cleaning schedule for all types of graphite brushes. The normal wear of brushes may cause a buildup of conductive particles along the creepage paths of the winding, the slip ring assembly, and the brushholder assembly. Some particles will be carried to various parts of the machine by ventilating air. They may clog the vent ducts, or settle between the brush and the pocket causing the machine to overheat or the brushes to bind. The buildup of particles could contribute to ring threading or grooving, as well as sparking, burning, and flashovers if they are allowed to accumulate. The importance of keeping these areas clean to avoid flashovers and breakdown cannot be overemphasized.

Satisfactory operation of brushes on AC slip rings requires that mechanical conditions be as perfect as possible. Intimate contact between the two surfaces must be maintained; the brushholder design and assembly are vital in this respect. The brushholder assembly normally provides for two or more brushes, which are mounted radially with respect to the ring surface. When mounting the brushholder assembly, position the brushes on the center of the slip ring. The holder should clear the ring by approximately $1/16$ to $1/8$ inch (1.6 to 3.2 mm) when the rings are new, and not over $1/4$ inch (6.4 mm) as the rings wear. The brushholder must have solid, rigid mountings to minimize vibration.

The brushholder assembly also provides a means of maintaining the proper pressure between the ring and brush surface—a most important factor in brush and ring performance. The rate of brush wear varies with the brush pressure. At light pressures, electrical wear is dominant because the brush can jump off the rings, sparking occurs, and the filming action on the ring becomes erratic. At higher pressures, mechanical wear is dominant because of high friction losses, needless heating, and needless abrasion. Check the machine manufacturer's recommended pressure for the type and grade of brush you are using.

Pressures of all brushes on a given ring must be as equal as possible. Check the brush pressures at regular intervals as the brushes wear to be sure that pressures stay equal at the proper value. If constant-pressure springs are used, adjustment may not be necessary. However, springs sometimes lose their temper (resilience) and require replacing—another reason to schedule regular maintenance.

It is also important to remember that the brushholders respond to machine vibration caused by any unbalanced condition. Brushholder mountings cannot eliminate the vibration that disrupts the contact surface between brush and ring. The causes of machine vibration are beyond the scope of this section, but everything possible should be done to eliminate vibration that disrupts brush contact.

It is highly recommended that you check slip ring runout with a dial indicator. Excessive runout of 0.003 of an inch (0.076 mm) or more on any slip ring will result in poor motor performance, rapid brush wear, and eventual motor failure. As Murphy's Law dictates, motors fail when they are needed most. A motor that has slip rings with excessive runout must be taken out of service as soon as possible. Send it to a reliable motor service shop to have the rings resurfaced to tolerances as close to zero runout as possible.

14.171 Troubleshooting

Table 14.1 lists the indications and sources of unsatisfactory performance of AC slip rings and brushes. Indications of unsatisfactory performance are divided into those that appear at the brushes, at the ring surface, and as heating. The immediate causes of a particular indication of unsatisfactory performance are listed in the middle column. The primary fault, responsible for a specific immediate cause, is listed by a number in the third column.

Table 14.2 lists the primary sources of unsatisfactory AC brush and ring performance. Sources discussed in this section are listed together with others that are beyond the scope of this section. They are numbered for use with Table 14.1 and, in addition, are broken down into basic groups of association as follows:

A. Preparation and care of machine
B. Machine adjustment
C. Mechanical fault in machine
D. Electrical fault in machine
E. Machine design
F. Load or service condition
G. Disturbing external condition
H. Wrong brush grade
I. Wrong brush shunting

By using the two summary tables, an indication of unsatisfactory AC slip ring and brush performance can be traced to the primary source of the trouble. For example, from Table 14.1, one immediate cause of flashover at the brushes is "lack of attention." The primary fault would be items 3 and 7 of Table 14.2. Looking now at Table 14.2, these items indicate a need for periodic cleaning and for cleaning after seating brushes (item 3) or incorrect spring tension (item 7). The basic fault is improper care of the machine (Part A of Table 14.2) or incorrect machine adjustment (Part B).

The two summary tables list only the general troubles most frequently experienced with the wound-rotor machines.

14.172 Preventive Maintenance

Accumulation of dry or oily dirt containing carbon from the brushes should be kept to a minimum by a regular preventive maintenance schedule as often as operating conditions may require. Flashover at the rings is apt to occur at start-up because the voltage between rings is highest at the moment when power is first applied. This secondary voltage is reduced to nearly zero as the motor reaches full speed. A dirty, low-resistance leakage path may withstand the lower secondary voltage during normal running operation, but may flash over at the next start-up. Therefore, it is recommended that the critical areas shown in Figure 14.16 be cleaned after the seating of new brushes, or after prolonged periods of operation or shutdown.

Critical leakage paths (refer to Figure 14.16) between phases and from phase to ground along the insulated surfaces are:

A. *Support-ring* surfaces and edges, between coils, and between coils and metal surfaces.

B. *Slip ring sleeve* surfaces and edges, between rings, and from rings to shafts.

C. *Brushholder-stud* surfaces and ends, between holders, and from holder to bracket.

D. *Through the air* separating the coils, the rings, and the brushholders at A, B, and C, but especially in the area of the brush shunts D where misplaced shunt reduces creepage through the air.

Recommended methods for cleaning the critical leakage paths are:

A. *For areas designated A.* Blow out the dirt with clean, dry, compressed air at 30 to 50 psi (2.1 to 3.5 kg/sq cm). Excessive oil or oil accumulations may require wiping with a clean, dry cloth or washing with an approved insulation cleaner (such as VM & P Naphtha, Stoddard's Solvent, or chloroethylene) or dry steam. Washing, blowing, or wiping may have to be repeated to ensure a clean leakage path of high resistance. *CAUTION: Use eye protection when cleaning with compressed air. Study appropriate Material Safety Data Sheets (MSDSs) and follow all safety requirements.*

B. *For areas designated B.* Wipe thoroughly with a clean, dry cloth and repeat as required to ensure a clean leakage path. Avoid accumulations of dirt containing carbon at the end of slip ring sleeve, along leads, and at junctions of a lead and its ring.

C. *For areas designated C.* Wipe surface as for B above.

D. *For areas designated D.* Here it is most important to maintain maximum creepage through the air. Do not allow the shunts to droop over. Keep them in line with the holder.

The critical leakage paths A, B, C, and D should be inspected for cleanliness before each start-up. If cracks appear in the insulated surfaces, check the insulation value with a megger. The megohm reading should exceed kV+1 megohms, where kV is the secondary voltage (given on the nameplate) divided by 1,000. If the megohm reading is low, clean the cracks, fill and seal with an insulating varnish such as glyptol; retest the insulation resistance value. In some instances, replacement with new insulation may be advantageous.

TABLE 14.1 INDICATIONS AND SOURCES OF UNSATISFACTORY PERFORMANCE OF AC SLIP RINGS AND BRUSHES

Indications	Immediate Causes	Primary Faults (See Table 14.2)
APPEARING AT BRUSHES:		
Sparking	Ring surface condition	1, 2, 26, 27, 28, 29, 32, 40, 41
	Faulty machine adjustment	5, 7
	Mechanical fault in machine	8, 9, 10, 11, 12, 14
	Electrical fault in machine	18, 19
	Bad load conditions	23, 24, 25
	Vibration	34, 35
	Chattering of brushes	See "Chattering and noisy brushes"
	Wrong brush grade	38, 40
Pitting of brush face	Glowing	See "Glowing at brush face"
	Embedded copper	See "Copper in brush face"
Rapid brush wear	Ring surface condition	See specific indication under "APPEARING AT RING SURFACE"
	Severe sparking	See "Sparking"
	Imperfect contact with ring	7, 8, 9, 10, 34, 35
	Wrong brush grade	39
Glowing at brush face	Embedded copper	See "Copper in brush face"
	Severe load condition	23, 24, 25
	Bad service condition	29, 30
	Wrong brush grade	38, 40, 42
Copper in brush face	Ring surface condition	2
	Bad service condition	26, 29, 30, 31, 32
	Wrong brush grade	40, 42
Flashover at brushes	Machine condition	3, 8
	Bad load condition	23, 24, 36, 41
	Lack of attention	3, 7
Chattering and noisy brushes	Ring surface condition	See specific indication under "APPEARING AT RING SURFACE"
	Looseness in machine	9, 10, 11
	Faulty machine adjustment	6, 7
	High friction	26, 28, 32, 35, 39, 40
	Wrong brush grade	39, 40
Brush chipping or breakage	Ring surface condition	See specific indication under "APPEARING AT RING SURFACE"
	Looseness in machine	9, 10, 11
	Vibration	35
	Sluggish brush movement	8
Burned shunts		See "Heating at shunts"
APPEARING AT RING SURFACE:		
Rough or uneven surface		1, 2, 11
Dull or dirty surface		3, 27, 33, 41
Eccentric surface		1, 12, 14, 35
Grooving or threading of surface	Sparking	26, 27, 28, 29, 32, 40
	Copper or foreign material in brush face	2, 29, 30, 31, 42
	Glowing	See "Glowing at brush face"
Ring burning	Sparking	2, 3, 7, 8, 23, 24, 25, 36, 41
	Flashover	3, 7, 8, 23, 24, 25, 36, 41
Brush outline appearing on ring at brush spacing	Sparking	33

TABLE 14.1 INDICATIONS AND SOURCES OF UNSATISFACTORY PERFORMANCE OF AC SLIP RINGS AND BRUSHES (continued)

Indications	Immediate Causes	Primary Faults (See Table 14.2)
APPEARING AT RING SURFACE: (continued)		
Flat spot	Sparking	1, 2, 36
	Flashover	3, 7, 8, 23, 24, 25, 36
	Lack of attention	1, 3, 7
Discoloration of surface	High temperature	See "Heating at ring"
	Atmospheric condition	27, 29
	Wrong brush grade	41
Raw-material surface	Embedded copper	See "Copper in brush face"
	Bad surface condition	26, 28, 30, 32
	Wrong brush grade	40, 42
Rapid ring wear with blackened surface	Burning	2, 7, 8
	Severe sparking	See "Sparking"
Rapid ring wear with bright surface	Foreign material in brush face	26, 28, 30, 32
	Wrong brush grade	42
APPEARING AS HEATING:		
Heating in windings	Severe load condition	23, 25, 36
	Unbalanced magnetic field	12, 18, 19, 20
	Unbalanced currents in the windings	12, 14, 16, 18, 19, 20, 22
	Lack of ventilation	4
Heating at ring	Severe load condition	23, 25
	Severe sparking	5, 13, 28, 38
	High friction	6, 7, 21, 26, 28, 32, 39, 40
	Poor ring surface	See specific indication under "APPEARING AT RING SURFACE"
	Degradation of ring material	15
	High contact resistance	37
Heating at brushes	Severe load condition	23, 25
	Faulty machine adjustment	6, 7, 17
	Severe sparking	See "Sparking"
	Raw streaks on ring surface	See "Grooving or threading of surface"
	Embedded copper	See "Copper in brush face"
	Wrong brush grade	38, 39, 40, 42, 43
Heating at shunts	Severe load condition	36, 38
	Selective action	8, 17, 43, 45, 46, 49
	Inadequate heat transfer	4, 17, 44, 46, 47, 48
	Corroded shunts	27, 28

TABLE 14.2 PRIMARY SOURCES OF UNSATISFACTORY AC BRUSH AND RING PERFORMANCE

A. PREPARATION AND CARE OF MACHINE
1. Poor preparation of ring surface
2. Porous ring material (inclusions, blowholes)
3. Need for periodic cleaning and for cleaning after seating brushes
4. Clogged ventilating ducts, poor ventilation in general

B. MACHINE ADJUSTMENT
5. Poor alignment of brushholders
6. Incorrect brush angle
7. Incorrect spring tension

C. MECHANICAL FAULT IN MACHINE
8. Brushes tight in holders
9. Brushes too loose in holders
10. Brushholders loose at mountings
11. Ring(s) loose
12. Loose or worn bearings
13. Unequal air gap around stator
14. Dynamic unbalance
15. Ring diameter too small

D. ELECTRICAL FAULT IN MACHINE
16. Open or high-resistance connection at ring
17. Poor connection of brush shunt terminal
18. Short circuit in primary or secondary winding
19. Ground in primary or secondary winding
20. Reversed coil(s) in primary or secondary winding

E. MACHINE DESIGN
21. High ratio of brush contact to ring surface area
22. Insufficient cross section of parallel rings in winding

F. LOAD OR SERVICE CONDITION
23. Overload
24. Rapid change in load
25. Plugging
26. Low average current density in brushes
27. Contaminated atmosphere
28. "Contact poisons"
29. Oil on ring or oil mist in air
30. Abrasive dust in air
31. Humidity too high
32. Humidity too low
33. Down for extended time periods

G. DISTURBING EXTERNAL CONDITION
34. Loose or unstable foundation
35. External source of vibration
36. External short circuit or very heavy load surge

H. WRONG BRUSH GRADE
37. Contact drop of brushes too high
38. Contact drop of brushes too low
39. Coefficient of friction too high
40. Lack of film-forming properties in brush
41. Lack of polishing action in brush
42. Brushes too abrasive
43. Lack of carrying capacity
44. Wrong thermal conductivity

I. WRONG BRUSH SHUNTING
45. Cross section too small
46. Poor shunt connection at brush
47. Insufficient dissipating surface area
48. Wrong shunt insulation
49. Improper temperature-resistance relationship

The other areas of preventive maintenance are:

E. *Brushes.* Brushes may be lifted off the rings if the machine is to be idle for a period extending beyond one week, especially if the absolute humidity is above one grain per cubic foot. An alternative way to protect brushes during a period of shutdown involves carefully placing a sheet of polyethylene film between the brushes and the ring. This will minimize chances of chipping brushes, changing brush pressure setting, and disturbing the film on the ring. Be sure to remove the polyethylene before the next start-up and check for freedom of brush movement in the brush pocket, for proper seating of brushes, and for correct, uniform brush pressure.

The brushes are centered on the rings at the factory for operation with the rotor on magnetic center. If the rotor is pulled off magnetic center in normal operation, determine if the brushes are positioned on the rings so that they will not override the ring at any positions of axial "float" during start-up or shutdown. It may be necessary to relocate the brushholders on the studs.

F. *Slip ring surfaces.* Unless the rings are refinished, the film that develops in the brush track during normal operation should not be removed or disturbed. This is the film that is vital and necessary for satisfactory operation. Remove any loose foreign material by blowing or wiping lightly with a clean, dry cloth. If the machine is to be shut down for several

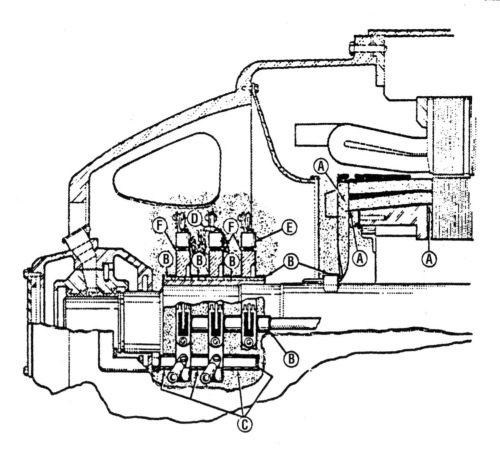

Fig. 14.16 Critical leakage paths

months or placed in storage, it is desirable to cover the rings with polyethylene film to prevent contamination of the ring surfaces. Do not paint, grease, or oil the ring surface.

QUESTIONS

Write your answers in a notebook and then compare your answers with those on page 869.

14.1E Why do wet wells use variable-speed pumping equipment?

14.1F What preventive maintenance is required by a standard, three-phase, single-speed, synchronous AC motor and what additional maintenance is required for a variable-speed, three-phase AC motor?

14.1G The rate of brush wear is influenced by what factors?

14.1H Indications of unsatisfactory performance of AC slip rings and brushes appear at or as what items?

14.18 Operating Troubles

The following list of operating troubles includes most of the causes of failure or reduced operating efficiency. The remedy or cure is either obvious or may be identified from the description of the causes. The lists of causes are not arranged in priority.

SYMPTOM A—Pump Will Not Start

Causes:

1. Blown fuses or tripped circuit breakers due to:

    a. Rating of fuses or circuit breakers not correct

    b. Switch (breakers) contacts corroded or shorted

    c. Terminal connections loose or broken somewhere in the circuit

    d. Automatic control mechanism not functioning properly

    e. Motor shorted or burned out

    f. Wiring hookup or service not correct

    g. Switches not set for operation

    h. Contacts of the control relays dirty and arcing

    i. Fuses or thermal units too warm

    j. Wiring short-circuited

    k. Shaft binding or sticking due to rubbing impeller, tight packing glands, or clogging of pump

2. Loose connection, fuse, or thermal unit

SYMPTOM B—Reduced Rate of Discharge

Causes:

1. Pump not primed
2. Mixture of air in the wastewater
3. Speed of motor too low
4. Improper wiring
5. Defective motor
6. Discharge head too high
7. Suction lift higher than anticipated
8. Impeller clogged
9. Discharge line clogged
10. Pump rotating in wrong direction
11. Air leaks in suction line or packing box
12. Inlet to suction line too high, permitting air to enter
13. Valves partially or entirely closed
14. Check valves stuck or clogged
15. Incorrect impeller adjustment
16. Impeller damaged or worn
17. Packing worn or defective
18. Impeller turning on shaft because of broken key
19. Flexible coupling broken
20. Loss of suction during pumping may be caused by leaky suction line, ineffective water or grease seal
21. Belts slipping
22. Worn wearing ring

SYMPTOM C—High Power Requirements

Causes:

1. Speed of rotation too high
2. Operating heads lower than rating for which pump was designed, resulting in excess pumping rates
3. Check valves open, draining long force main back into well
4. Specific gravity or viscosity of liquid pumped too high
5. Clogged pump
6. Sheaves on belt drive misaligned or maladjusted
7. Pump shaft bent
8. Rotating elements binding
9. Packing too tight
10. Wearing rings worn or binding
11. Impeller rubbing

SYMPTOM D—Noisy Pump

Causes:

1. Pump not completely primed
2. Clogged inlet
3. Inlet not submerged
4. Pump not lubricated properly
5. Worn impellers
6. Strain on pumps caused by unsupported piping fastened to the pump
7. Insecure foundation
8. Mechanical defects in pump
9. Misalignment of motor and pump where connected by flexible shaft
10. Rags or sticks bound (wrapped) around impeller

QUESTIONS

Write your answers in a notebook and then compare your answers with those on page 870.

14.1I What items would you check if a pump will not start?

14.1J How would you attempt to increase the discharge from a pump if the flow rate is lower than expected?

14.19 Starting and Stopping Pumps

The operator must determine what treatment processes will be affected by either starting or stopping a pump. The pump discharge point must be known and valves either opened or closed to direct flows as desired by the operator when a pump is started or stopped.

14.190 Centrifugal Pumps

Figure 14.17 illustrates a typical wet well and pumping system. The purpose of each part is explained in Table 14.3. Basic rules for the operation of centrifugal pumps include the following items.

1. Do not operate the pump when safety guards are not installed over or around moving parts.

2. Do not start a pump that is locked or tagged out for maintenance or repairs; serious injury to the operator or damage to the equipment could result.

3. Never run a centrifugal pump when the impeller is dry. Always be sure the pump is primed.

4. Never attempt to start a centrifugal pump whose impeller or shaft is spinning backward.

5. Do not operate a centrifugal pump that is vibrating excessively after start-up. Shut unit down and isolate pump from system by closing the pump suction and discharge valves. Look for a blockage in the suction line and the pump impeller.

There are several situations in which it may be necessary to start a centrifugal pump against a closed discharge valve until the pump has picked up its prime and developed a satisfactory discharge head for that operating system. Once the pump is primed, slowly open the pump discharge valve until the pump is fully on line. This procedure is used with treatment processes or piping systems with vacuums or pressures that cannot be dropped or allowed to fluctuate greatly while an alternate pump is put on the line.

Most centrifugal pumps used in wastewater treatment plants are designed so that they can be easily started even if they have not been primed. This is accomplished with a positive static suction head or a low suction lift. On most of these arrangements, the pump will not require priming as long as the pump and the piping system do not leak. Leaks would allow the water to drain out of the pump volute. When pumps in wastewater systems lose their prime, the cause is often a faulty check valve on the pump discharge line. When the pump stops, the discharge check valve will not seal (close) properly. Wastewater previously pumped then flows back through the check valve, down

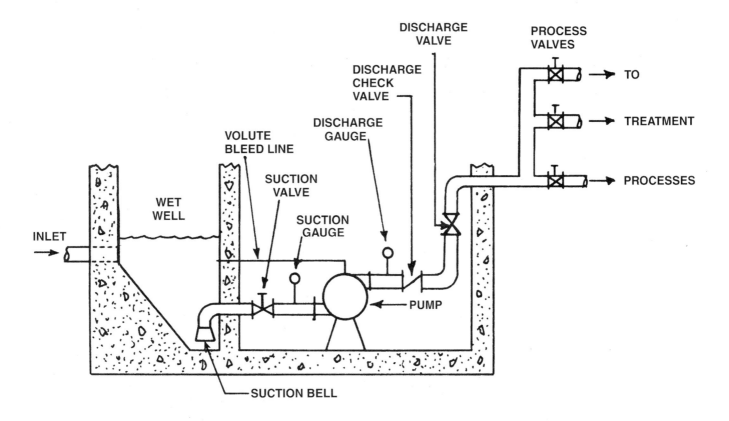

Fig. 14.17 Wet well and pump system

TABLE 14.3 PURPOSE OF WET WELL AND PUMP SYSTEM PARTS

Part	Purpose
1. Inlet	Carries wastewater to wet well.
2. Wet Well	Stores wastewater for removal by pump.
3. Suction Bell	Guides wastewater into pump suction pipe and reduces pipe entrance energy losses.
4. Suction Valve	Isolates pump and piping from wet well.
5. Suction Gauge	Indicates suction head or lift on suction side of pump.
6. Volute (not shown in Figure 14.17)	Collects wastewater discharged by pump impeller and directs flow to pump discharge.
7. Volute Bleed Line	Keeps pump primed for automatic operation by allowing entrapped gases (air) to escape from the pump volute.
8. Discharge Gauge	Indicates discharge head (energy imparted to wastewater by pump).
9. Discharge Valve	Isolates pump from discharge system.
10. Discharge Check Valve	Prevents discharge pipe and treatment process tanks from draining back through pump and into wet well.
11. Process Valves	Direct flow.

through the pump rotating the pump impeller and shaft backward, and back into the wet well. The pump is drained and has lost its prime.

The other danger associated with a faulty check valve is that if the pump wet well has a high inflow, plus inflow from the water running back through the pump, the wet well water level will rise to the elevation set to turn on the pump. If the pump attempts to start while the pump, motor, and shafting are rotating in the opposite direction, very serious damage can occur to the pumping equipment. Many pumps are equipped with anti-rotational devices that minimize this potential problem.

About 95 percent of the time, the centrifugal pumps in wastewater treatment plants are ready to operate with suction and discharge valves open and seal water turned on. When the automatic start or stop command is received by the pump from an air or electronic controller, the pump is ready to respond properly.

When the pumping equipment must be serviced, take it off the line by locking and tagging out the pump main breaker until all service work is completed.

QUESTIONS

Write your answers in a notebook and then compare your answers with those on page 870.

14.1K Why should a pump that has been locked or tagged out for maintenance or repairs not be started?

14.1L Under what conditions might a centrifugal pump be started against a closed discharge valve?

STOPPING PROCEDURES

This section contains a typical sequence of procedures to follow to stop a centrifugal pump.

1. Inspect process system affected by pump, start alternate pump if required, and notify supervisor or log action.

2. Before stopping the operating pump, check its operation. This will give an indication of any developing problems, required adjustments, or problem conditions of the unit. This procedure only requires a few minutes. Items to be inspected include:

   a. Pump packing gland

      (1) Seal water pressure

      (2) Seal leakage (too much, sufficient, or too little leakage)

      (3) Seal leakage drain flowing clear

      (4) Mechanical seal leakage (if equipped)

   b. Pump operating pressures

      (1) Pump suction (Pressure-Vacuum)

         A higher vacuum than normal may indicate a partially plugged or restricted suction line. A lower vacuum may indicate a higher wet well level or a worn pump impeller or wearing rings.

      (2) Pump discharge pressure

         System pressure is indicated by the pump discharge pressure. Lower than normal discharge pressures can be caused by:

         (a) Worn impeller or wearing rings in the pump

         (b) A different point of discharge can change discharge pressure conditions

         (c) A broken discharge pipe can change the discharge head

NOTE: To determine the maximum head a centrifugal pump can develop, slowly close the discharge valve at the pump. Read the pressure gauge between the pump and the discharge valve when the valve is fully closed. This is the maximum pressure the pump is capable of developing. Do not operate the pump longer than a few minutes with the discharge valve closed completely because the energy from the pump is converted to heat and water in the pump can become hot enough to damage the pump.

   c. Motor temperature and pump bearing temperature

   If motor or bearings are too hot to touch, further checking is necessary to determine if a problem has developed or if the temperature is normal. High temperatures may be measured with a thermometer.

   d. Unusual noises, vibrations, or conditions about the equipment

   If any of the above items indicate a change from the pump's previous operating condition, additional service or maintenance may be required during shutdown.

3. Actuate stop switch for pump motor and lock out switch. If possible, use switch next to equipment so that you may observe the equipment stop. Observe the following items:

   a. Check valve closes and seats.

   Valve should not slam shut, or discharge piping will jump or move in their supports. There should not be any leakage around the check valve shaft. If check valve is operated automatically, it should close smoothly and firmly to the fully closed position.

   NOTE: If the pump is not equipped with a check valve, close discharge valve before stopping pump.

   b. Motor and pump should wind down slowly and not make sudden stops or noises during shutdown.

   c. After equipment has completely stopped, pump shaft and motor should not start backspinning. If backspinning is observed, close the pump discharge valve slowly. Repair faulty check or foot valve.

4. Go to power control panel containing the pump motor starters just shut down and open motor breaker switch, lock out, and tag.

5. Return to pump and close:

   a. Discharge valve

   b. Suction valve

   c. Seal water supply valve

   d. Pump volute bleed line (if so equipped)

6. If pump is to be left out of service more than two days, drain pump volute and leave volute empty. Indicate "volute empty" on lockout tag.

7. If required, close and open appropriate valves along piping system through which pump was discharging.

STARTING PROCEDURES

This section contains a typical sequence of procedures to follow to start a centrifugal pump.

1. Check motor control panel for lock and tags. Examine tags to be sure that no item is preventing start-up of equipment.

2. Inspect equipment.

   a. Be sure stop switch is locked out at equipment location

   b. Guards over moving parts must be in place

   c. Cleanout on pump volute and drain plugs should be installed and secure

   d. Valves should be in closed position

   e. Pump shaft must rotate freely

   f. Pump motor should be clean and air vents clear

   g. Pump, motor, and auxiliary equipment lubricant level must be at proper elevations

   h. Determine if any special considerations or precautions are to be taken during start-up

3. Follow pump discharge piping route. Be sure all valves are in the proper position and that the pump flow will discharge where intended.

4. Return to motor control panel.

   a. Remove tag

   b. Remove padlock

   c. Close motor main breaker

   d. Place selector switch to manual (if you have automatic equipment)

5. Return to pump equipment.

   a. Open seal water supply line to packing gland. Be sure seal water supply pressure is adequate.

   b. Open pump suction valve slowly.

   c. Bleed air out of top of pump volute to prime pump. Some pumps are equipped with air relief valves or bleed lines back to the wet well for this purpose.

d. When pump is primed, slowly open pump discharge valve and recheck prime of pump. Be sure no air is escaping from volute.

e. Unlock stop switch and actuate start switch. Pump should start.

6. Inspect equipment.

   a. Motor should come up to speed promptly. If ammeter is installed on instrument panel, then read or watch for excessive draw of power (amps) during start-up and normal operation. If ammeter is not installed on panel and power leads are exposed from conduit or motor connection box, then a portable ammeter may be used to observe draw of power.

   b. No unusual noise or vibrations should be observed during start-up.

   c. Check valve should be open and no chatter or pulsation should be observed.

   d. Pump suction and discharge pressure readings should be within normal operating range for this pump.

   e. Packing gland leakage should be normal.

   f. If a flowmeter is on the pump discharge, record pump output.

7. If the unit is operating properly, return to the motor control panel and place the motor mode of operation selector in the proper operation position (manual-auto-off).

8. The pump and auxiliary equipment should be inspected routinely after it has been placed back into service.

QUESTIONS

Write your answers in a notebook and then compare your answers with those on page 870.

14.1M   What should be done before stopping an operating pump?

14.1N   What could cause a pump shaft or motor to spin backward?

14.1O   Why should the position (open or closed) of all valves be checked before starting the pump?

14.191 Positive Displacement Pumps

Steps for starting and stopping positive displacement pumps are outlined in this section. There are two basic differences in the operation of positive displacement pumps as compared with centrifugal pumps. Centrifugal pumps (due to their design) will permit an operator error, but a positive displacement pump will not and someone will have to pay for correcting the damages.

Important rules for operating positive displacement pumps include:

1. **Never operate a positive displacement pump against a closed valve, especially a discharge valve.** The pipe, valve, or pump could rupture from excessive pressure. The rupture will damage equipment and possibly seriously injure or kill someone standing nearby.

2. Positive displacement pumps are used to pump solids (sludge) and certain precautions must be taken to prevent injury or damage. If the valves on both ends of a sludge line are closed tightly, the line becomes a closed vessel. Gas from decomposition can build up to a pressure that will rupture pipes or valves.

3. Positive displacement pumps also are used to meter and pump chemicals. Care must be exercised to avoid venting chemicals to the atmosphere.

4. Never operate a positive displacement pump when it is dry or empty, especially the progressive cavity types that use rubber stators. A small amount of liquid is needed for lubrication in the pump cavity between the rotor and the stator.

In addition to never closing a discharge valve on an operating positive displacement pump, the only other difference (when compared with a centrifugal pump) may be that the positive displacement pump system may or may not have a check valve in the discharge piping after the pump. Installation of a check valve depends on the designer and the material being pumped.

Other than the specific differences mentioned in this section, the starting and stopping procedures for positive displacement pumps are similar to the procedures for centrifugal pumps.

QUESTIONS

Write your answers in a notebook and then compare your answers with those on page 870.

14.1P   What is the most important rule regarding the operation of positive displacement pumps?

14.1Q   What could happen if a positive displacement pump is started against a closed discharge valve?

14.1R   Why should both ends of a sludge line never be closed tight?

END OF LESSON 2 OF 7 LESSONS

on

MAINTENANCE

Please answer the discussion and review questions next.

DISCUSSION AND REVIEW QUESTIONS
Chapter 14. MAINTENANCE
(Lesson 2 of 7 Lessons)

Write the answers to these questions in your notebook. The question numbering continues from Lesson 1.

8. What should you do if you cannot understand the manufacturer's instructions?

Select the correct word:

9. Cheap lubricants may be the (1) *MOST*, or (2) *LEAST* expensive in the end.

10. Start-up of a new pump should be done by (1) *A NEW OPERATOR*, or (2) *A TRAINED PERSON*.

11. How can you determine if a new pump will turn in the direction intended?

12. How can you tell if a new pump is delivering design flows and pressures?

13. When shutting down a pump for a long period, what precautions should be taken with the motor disconnect switch?

14. What is a maintenance problem with water level float controls?

CHAPTER 14. MAINTENANCE

(Lesson 3 of 7 Lessons)

14.2 BEWARE OF ELECTRICITY

RECOGNIZE YOUR LIMITATIONS

In the wastewater collection and treatment maintenance departments of all cities, there is a need for maintenance operators to know something about electricity. Duties could range from repairing a taillight on a trailer or vehicle to repairing complex pump controls and motors. Very few maintenance operators do the actual electrical repairs or troubleshooting because this is a highly specialized field and unqualified people can seriously injure themselves and damage costly equipment. For these reasons you must be familiar with electricity, know the hazards, and recognize your own limitations when you must work with electrical equipment.

Most municipalities employ electricians or contract with a commercial electrical company that they call when major problems occur. However, the maintenance operator should be able to explain how the equipment is supposed to work and what it is doing or is not doing when it fails.

After this lesson, you should be able to tell an electrician what appears to be the problem with electrical panels, controls, circuits, and equipment. Even though operators may only be assigned maintenance tasks on electrical systems, the more operators know about electricity, the better equipped they are to diagnose electrical problems, especially during emergency situations.

When an outside electrical contractor performs maintenance on your system, ask the contractor to assign the same person each time. This allows one person to become familiar with the pump station controls and electrical systems and, therefore, troubleshoot and repair much more rapidly during failures. This person should be accompanied by an assistant so a backup person will be available with some knowledge of your system when the regular person is not available.

The need for safety should be apparent. If proper safe procedures are not followed in operating and maintaining the various electrical equipment used in wastewater collection and treatment facilities, accidents can happen that cause injuries, permanent disability, or loss of life. Some of the serious accidents that have happened, and could have been avoided, occurred when machinery was not shut off, locked out, and tagged properly (Figure 14.18) as required by OSHA. Possible accidents include:

1. Maintenance operator could be cleaning pump and have it start, thus losing an arm, hand, or finger.

2. Electrical motors or controls not properly grounded could lead to possible severe shock, paralysis, or death.

3. Improper circuits created by mistakes, such as wrong connections, bypassed safety devices, wrong fuses, or improper wire, can cause fires or injuries due to incorrect operation of machinery.

Another consideration for having a basic working knowledge of electricity is to prevent financial losses resulting from motors burning out and from damage to equipment, machinery, and control circuits. Additional costs result when damages have to be repaired, including payments for outside labor.

WARNING

Never work in electrical panels or on electrical controls, circuits, wiring, or equipment unless you are qualified and authorized. By the time you find out what you do not know about electricity, you could find yourself too dead to use the knowledge.

QUESTIONS

Write your answers in a notebook and then compare your answers with those on page 870.

14.2A Why must unqualified or inexperienced people be extremely careful when attempting to troubleshoot or repair electrical equipment?

14.2B What could happen when machinery is not shut off, locked out, and tagged properly?

14.3 ELECTRICAL EQUIPMENT MAINTENANCE

14.30 Introduction

This section contains a basic introduction to electrical terms and information plus directions on how to troubleshoot problems with electrical equipment.

DANGER

OPERATOR WORKING ON LINE

DO NOT CLOSE THIS SWITCH WHILE THIS TAG IS DISPLAYED

TIME OFF: _____

DATE: _____

SIGNATURE: _____

This is the ONLY person authorized to remove this tag.

INDUSTRIAL INDEMNITY/INDUSTRIAL UNDERWRITERS/
INSURANCE COMPANIES

4E210—R66

Fig. 14.18 Typical warning tag
(Source: Industrial Indemnity/Industrial Underwriters/Insurance Companies)

Most electrical equipment used in wastewater collection systems and treatment plants is labeled with *NAMEPLATE*[7] information indicating the proper voltage and allowable current in amps.

14.31 Volts, Amps, Watts, and Power Requirements

VOLTS

Voltage (E) is also known as Electromotive Force (EMF), and is the electrical pressure available to cause a flow of current (amperage) when an electric circuit is closed.[8] This force can be compared with the pressure or force that causes water to flow in a pipe. Some pressure in a water pipe is required to make the water move. The same is true of electricity. A force is necessary to push electricity or electric current through a wire. This force is called voltage.

There are two types of voltage: Direct Current (DC) and Alternating Current (AC).

DIRECT CURRENT

Direct current (DC) is flowing in one direction only and is essentially free from pulsation. Direct current is seldom used in wastewater treatment plants except in motor-generator sets, some control components of pump drives, and standby lighting. Direct current is used exclusively in automotive equipment, certain types of welding equipment, and a variety of portable equipment. Direct current is found in various voltages, such as 6 volts, 12 volts, 24 volts, 48 volts, and 110 volts. All batteries are direct current. DC is tested by holding the multimeter leads on the positive and negative poles on a battery. These poles are usually marked Positive (+) and Negative (–). Direct current usually is not found in higher voltages (over 24 volts) around plants and lift stations unless in motor-generator sets. Care must be taken when installing battery cables and wiring that Positive (+) and Negative (–) poles are connected properly to wires marked (+) and (–). If not properly connected, you could get an arc across the unit that could cause an explosion.

ALTERNATING CURRENT

Alternating current (AC) is periodic current that has alternating positive and negative values. In other words, it goes from zero to maximum strength, back to zero and to the same strength in the opposite direction, which comprises a cycle. Our AC voltage is 60-cycle frequency, or "Hertz," which means that this happens 60 times per second. Alternating current is classified as:

a. Single Phase

b. Two Phase

c. Three Phase or Polyphase

The most common of these are single phase and three phase. The various voltages you probably will find on your job are 110 volts, 120 volts, 208 volts, 220 volts, 240 volts, 277 volts, 440 volts, 460 volts, 480 volts, and 550 volts.

Single-phase power is found in lighting systems, small pump motors, various portable tools, and throughout our homes. It is usually 120 volts or 240 volts. Single phase means that only one phase of power is supplied to the main electrical panel at 240 volts and the power supply has three wires or leads. Two of these leads have 120 volts each, the other lead is neutral and usually is coded white. The neutral lead is grounded. Many appliances and power tools have an extra ground (commonly a green wire) on the case for additional protection.

Three-phase power is generally used with motors and transformers found in lift stations and wastewater treatment plants. This power generally is 208, 220, 240 volts, or 440, 460, 480, and 550 volts. Higher voltages are used in some lift stations. Three phase is used when higher power requirements or larger motors are used because efficiency is usually higher and motors require less maintenance. Generally, all motors above two horsepower are three phase unless there is a problem with the power company getting three phase to the installations. Quite a few residential lift stations are on single-phase power due to their remote locations. Three-phase power usually is brought in to the point of use with three leads and there is power in all three leads.

When taking a voltage check between any two of the three leads, you measure 208, 220, 240 volts, or 440, 460, 480 volts depending on the supply voltage. There are some instances where you might measure a difference between the leads due to the use of different transformers. If there is power in three leads and if a fourth lead is brought in, it is a neutral lead.

Two-phase and polyphase systems will not be discussed because they generally are not found in wastewater collection and treatment facilities.

AMPS

An ampere (I) is the practical unit of electric current. This is the current produced by a pressure of one volt in a circuit having a resistance of one ohm. Amperage is the measurement of current or electron flow and is an indication of work being done or "how hard the electricity is working."

In order to understand amperage, one more term must be explained. The ohm is the practical unit of electrical resistance (R). "Ohm's Law" states that in a given electrical circuit the amount of current in amperes (I) is equal to the pressure in volts (E) divided by the resistance (R) in ohms. The following three formulas are given to provide you with an indication of the rela-

---

[7] *Nameplate.* A durable, metal plate found on equipment that lists critical operating conditions for the equipment.
[8] Electricians often talk about closing an electric circuit. This means they are closing a switch that actually connects circuits together so electricity can flow through the circuit. Closing an electric circuit is like opening a valve on a water pipe.

tionships among current, resistance, and EMF (electromotive force).

Current, amps $= \dfrac{\text{EMF, volts}}{\text{Resistance, ohms}}$ $I = \dfrac{E}{R}$

EMF, volts $= $ (Current, amps)(Resistance, ohms) $E = IR$

Resistance, ohms $= \dfrac{\text{EMF, volts}}{\text{Current, amps}}$ $R = \dfrac{E}{I}$

These equations are used by electrical engineers for calculating circuit characteristics. If you memorize the following relationship, you can always figure out the correct formula.

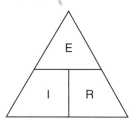

To use the above triangle you cover up with your finger the term you do not know or are trying to find out. The relationship between the other two known terms will indicate how to calculate the unknown. For example, if you are trying to calculate the current, cover up I. The two knowns (E and R) are shown in the triangle as E/R. Therefore, I = E/R. The same procedure can be used to find E when I and R are known or to find R when E and I are known.

WATTS

The watt and kilowatt (one thousand watts) are measures of power units used to rate electrical machines or motors or to indicate power in an electric circuit. The watts or power (P) required by a machine is obtained by multiplying the amps (I) required (or pulled) times the potential drop in volts (E) or the electromotive force (EMF). The amount of watts in a circuit is equal to the voltage (E) times the amperage (I). The total power in a circuit, measured in watts, at any given moment is similar to the power produced by a motor, measured in horsepower, at any given moment.

Power, watts = (Electromotive Force, volts)(Current, amps)

or P, watts = (E, volts)(I, amps)

For the sake of comparisons,

1 horsepower = 746 watts = 0.746 kilowatt = 550 ft-lbs/sec

The actual horsepower output of a motor also depends on the power factor (a number less than one) and the efficiency of the motor.

Output, HP $= \dfrac{(P, \text{watts})(\text{Power Factor})(\text{Efficiency, \%})}{(746 \text{ watts/horsepower})(100\%)}$

POWER REQUIREMENTS

Power requirements (PR) are expressed in kilowatt hours: 500 watts for two hours or one watt for 1,000 hours equals one kilowatt hour (kW-hr or kWh). The power company charges so many cents per kilowatt hour.

Power Req, kW-hr = (Power, kilowatts)(Time, hours)

PR, kW-hr = (P, kW)(T, hr)

CONDUCTORS AND INSULATORS

A material, like copper, that permits the flow of electric current is called a conductor. Material that will not permit the flow of electricity, like rubber, is called an insulator. When such a material is wrapped or cast around a wire, it is called insulation. Insulation is commonly used to prevent the loss of electrical flow by two conductors coming into contact with each other.

QUESTIONS

Write your answers in a notebook and then compare your answers with those on page 870.

14.3A   How can you determine the proper voltage and allowable current in amps for a piece of equipment?

14.3B   What are two types of voltage?

14.3C   Amperage is a measurement of what?

14.32   Tools, Meters, and Testers

WARNING

Never enter any electrical panel or attempt to troubleshoot or repair any piece of electrical equipment or any electric circuit unless you are qualified and authorized.

A wide variety of instruments are used to maintain lift station electrical systems. These instruments measure current, voltage, and resistance. They are used not only for troubleshooting, but for preventive maintenance as well. These instruments may have either an *ANALOG READOUT*,[9] which uses a pointer and scale, or a *DIGITAL READOUT*,[10] which gives a numerical reading of the measured value.

To check for voltage, a *MULTIMETER* is needed. There are several types on the market and all of them work. They are designed to be used on energized circuits and care must be exercised when testing. By holding one lead on ground and the other on a power lead, you can determine if power is available. You also can tell if it is AC or DC and the intensity or voltage (110, 220, 480, or whatever) by testing the different leads.

---

[9] *Analog Readout.* The readout of an instrument by a pointer (or other indicating means) against a dial or scale. Also, the continuously variable signal type sent to an analog instrument (for example, 4–20 mA). Also see DIGITAL READOUT.

[10] *Digital Readout.* The readout of an instrument by a direct, numerical reading of the measured value or variable. The signal sent to such instruments is usually an analog signal.

A multimeter can also be used to measure voltage, current, and resistance. A digital multimeter is shown in Figure 14.19 and an analog clamp-on multimeter is shown in Figure 14.20.

Fig. 14.19   Digital multimeter
(Reproduced with permission of Fluke Corporation)

Fig. 14.20   Analog clamp-on multimeter
(Permission of Simpson Electric)

Do not work on any electric circuits unless you are qualified and authorized. Use a multimeter and other circuit testers to determine if the circuit is energized, or if all voltage is off. This should be done after the main switch is turned off to make sure it is safe to work inside the electrical panel. Always be aware of the possibility that even if the disconnect to the unit you are working on is off, the control circuit may still be energized if the circuit originates at a different distribution panel. Check with a multimeter before and during the time the main switch is turned off to have a double-check. This procedure ensures that the multimeter is working and that you have good continuity to your tester. Use circuit testers to measure voltage or current characteristics to a given piece of equipment for making sure that you have or do not have a "live" circuit. *WARNING: Switches can fail and the only way to ensure that a circuit is dead is to test the circuit.*

In addition to checking for power, a multimeter can be used to test for open circuits, blown fuses, single phasing of motors, grounds, and many other uses. Some examples are illustrated in the following paragraphs.

In the circuit shown in Figure 14.21, test for power by holding one lead of the multimeter on point "A" and the other at point "B." If no power is indicated, the switch is open or faulty. The sketch shows the switch in the "open" position.

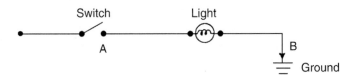

Fig. 14.21   Single-phase circuit

To test for power at point "A" and point "B" in Figure 14.22, open the switch as shown. Using a multimeter with clamp-on leads, clamp a lead on L1 and a lead on L2 between the fuses and the load. Bring the multimeter and leads out of the panel and close the panel door as far as possible without cutting or damaging the meter leads. Some switches cannot be closed if the panel door is open. The panel door is closed when testing because hot copper sparks could seriously injure you when the circuit is energized and the voltage is high. Close the switch.

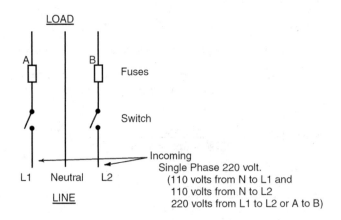

Fig. 14.22   Single-phase, three-lead circuit

1. Multimeter should register at 220 volts. If there is no reading at points "A" and "B," the fuse or fuses could be "blown."

2. Move multimeter down below fuses to line 1 and line 2 (L1, L2). If there is still no reading on the multimeter, check for an open switch in another location, or call the power company to find out if power is out.

3. If a 220-volt reading is registered at line 1 and line 2, move the test leads to point "A" and the neutral lead. If a reading of 110-volts is observed, the fuse on line "A" is OK. If there is no voltage reading, the fuse on line "A" is blown. Move the lead from line "A" to line "B." Observe the reading. If 110-volt power is recorded, the fuse on line "B" is OK. If there is no voltage reading, the fuse on line "B" is blown. Another possibility to consider is that the neutral line could be broken.

> **WARNING**
>
> Turn off power and be sure that there is no voltage in either power line before changing fuses. Use a fuse puller. Test circuit again in the same manner to make sure fuses or circuit breakers are OK. 220 volts power or voltage should be present between points "A" and "B." If fuse or circuit breaker trips again, shut off and determine the source of the problem.

Test for power at points "A," "B," and "C" in a three-phase circuit (Figure 14.23). Place multimeter leads on lines "A" and "B." Close all switches. 220 volts should register on multimeter. Check between lines "A" and "C," and between lines "B" and "C." 220 volts should be recorded between all of these points. If voltage is not present, one or all of the fuses are blown or the circuit breaker has been tripped. First, check for voltage above the fuses at all of these points, "A" to "B," "A" to "C," and "B" to "C," to make sure power is available (see 220 readings in Figure 14.23). If voltage is recorded, move leads back down to bottom of fuses. If voltage is present from "A" to "B," but not at "A" to "C" and "B" to "C," the fuse on line "C" is blown. If there were no voltage readings at any of the test points, all the fuses could be blown.

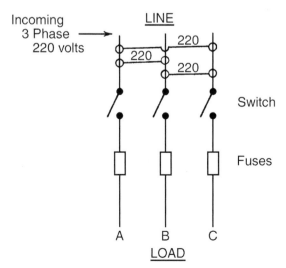

Fig. 14.23 Three-phase circuit, 220 volts

Another way of checking the fuses on this three-phase circuit would be to take your multimeter and place one lead on the bottom, and one lead on the top of each fuse. You should not get a voltage reading on the multimeter. This is because electricity takes the path of least resistance. If you get a reading across any of the fuses (top to bottom), that fuse is bad.

Always make sure that when you use a multimeter it is set for the proper voltage. If voltage is unknown and the meter has different scales that are manually set, always start with the highest voltage range and work down. Otherwise, the multimeter could be damaged. Look at the equipment instruction manual or nameplate for the expected voltage. Actual voltage should not be much higher than given unless someone goofed when the equipment was wired and inspected.

Voltage readings are important because they determine how you connect motor leads, relays, and transformers. Low or high voltages can drastically affect motors. Operators of small wastewater systems should also be aware of the common but little understood problem of unbalanced current. Operating a pumping unit with unbalanced current can seriously damage three-phase motors and cause early motor failure. Unbalanced current reduces the starting torque of the motor and can cause overload tripping, excessive heat, vibrations, and overall poor performance. (Section 14.431, Problem 7, describes how to test circuits for unbalanced current.)

Another meter used in electrical maintenance and testing is the *AMMETER*. The ammeter records the current or amps in the circuit. There are several types of ammeters, but only two will be discussed in this chapter. The ammeter generally used for testing is the clamp-on type (Figures 14.24 and 14.25). The

Fig. 14.24 Analog clamp-on type ammeter
(Permission of Amprobe)

Fig. 14.25 Digital clamp-on type ammeter
(Permission of Amprobe)

term "clamp-on" means that it can be clamped around a lead or each lead supplying a motor lead, and no direct electrical connection needs to be made. These are used by clamping the meter over only one of the power leads to the motor or other apparatus and taking a direct reading. Each "leg" or lead on a three-phase motor must be checked by itself.

The first step should be to read the motor nameplate data and find what the amperage reading should be for the particular motor or device you are testing. After you have this information, set the ammeter to the proper scale. Set it on a higher scale than necessary if the expected reading is close to the top of the meter scale. Place the clamp around one lead at a time. Record each reading and compare with the nameplate rating. If the readings do not compare with the nameplate rating, find the cause, such as low voltage, bad bearings, poor connections, plugging, or excessive load. If the ammeter readings are higher than expected, the high current could produce overheating and damage the equipment. Try to find the problem and correct it.

When using a clamp-on ammeter, be sure to set the meter on a high enough range or scale for the starting current if you are testing during start-up. Starting currents range from 110 to 150 percent higher than running currents and using too low a range can ruin an expensive and delicate instrument. Newer, clamp-on ammeters automatically adjust to the proper range and can measure both starting or peak current and normal running current.

Another type ammeter is one that is connected in line with the power lead or leads. Generally, they are not portable and are usually installed in a panel or piece of equipment. They require physical connections to put them in series with the motor or apparatus being tested. These ammeters are usually more accurate than the clamp-on type and are used in motor control centers and pump panels.

A *MEGGER* or *MEGOHMMETER* is used for checking the insulation resistance on motors, generators, feeders, bus bar systems, grounds, and branch circuit wiring. This device actually applies a DC test voltage, which can be as high as 5,000 volts DC, depending on the megohmmeter selected. The one shown in Figure 14.26 is a hand-held, hand-cranked system that applies 500 volts DC and is particularly useful for testing motor insulation. Battery-operated and instrument-style meggers are also available in both analog and digital models.

> **WARNING**
>
> Turn off circuit breaker when using a megger.

To use a megger there are two leads to connect. One lead is clamped to a ground lead and the other to the lead you are testing. The readings on the megger will range from "0" (ground) to infinity (perfect), depending on the condition of your circuit.

Fig. 14.26 Hand-cranked megohmmeter
(Permission of AVO International)

The megger is usually connected on the motor terminals, one at a time, at the starter, and the other lead to the ground lead. Results of this test indicate if the insulation is deteriorating or cut.

If a low reading is obtained, disconnect motor leads from power or line leads. Meg motor and if low reading is observed, the motor winding insulation is breaking down. If a good reading is obtained, meg the circuit or branch wiring. If this reading is low, the wiring to the motor is bad. A rule of thumb is not to run a motor if it is less than one *MEGOHM*[11] per horsepower. This means a 5 horsepower motor should have a 5 megohm reading; 10 horsepower, a 10 megohm reading; 20 horsepower, 20 megohm reading; and so forth.

Motors and wirings should be megged at least once a year, and twice a year, if possible. The readings taken should be recorded and plotted to determine when insulation is breaking down. Meg motors and wirings after a pump station has been flooded. If insulation is wet, excessive current could be drawn and cause pump motors to "kick out" (shut off).

OHMMETERS, sometimes called circuit testers, are valuable tools used for checking electric circuits. An ohmmeter is used only when the electric circuit is OFF, or de-energized. The ohmmeter supplies its own power by using batteries. An ohmmeter is used to measure the resistance (ohms) in a circuit. These are most often used in testing the control circuit components, such as coils, fuses, relays, resistors, and switches. They are used also to check for continuity. An ohmmeter has several scales that can be used. Typical scales are: $R \times 1$, $R \times 10$, $R \times 1,000$, and $R \times 10,000$. Each scale has a level of sensitivity for measuring different resistances. To use an ohmmeter, set the scale, start at the low point ($R \times 1$), and put the two leads across the part of the circuit to be tested, such as a coil or resistor, and read the resistance in ohms. A reading of infinity would indicate an open circuit, and a "0" would in-

---

[11] *Megohm* (MEG-ome). Millions of ohms. Mega- is a prefix meaning one million, so 5 megohms means 5 million ohms. For additional information, see *A STITCH IN TIME: THE COMPLETE GUIDE TO ELECTRICAL INSULATION TESTING*. Available from AVO Training Institute, 4271 Bronze Way, Dallas, TX 75237. Order No. AVOB001. Price, $7.00.

dicate no resistance. Ohmmeters usually would be used only by skilled technicians because they are very delicate instruments.

The motor rotation indicator illustrated in Figure 14.27 is another specialized instrument used in electrical maintenance. This device is useful for determining the phase rotation of utility power when connecting 3-phase motors to ensure that the motor is connected properly for correct rotation.

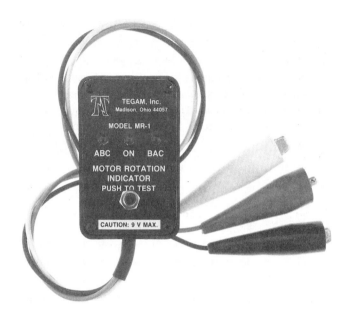

Fig. 14.27   Motor rotation indicator
(Permission of Tegam Inc.)

All meters should be kept in good working order and calibrated periodically. They are very delicate, susceptible to damage, and should be well protected during transportation. When readings are taken, they should always be recorded on a machinery history card for future reference. Meters are a good way to determine pump and equipment performance. *CAUTION: Never use a meter unless you are qualified and authorized.* The risk of electric shock or electrocution is very high if meters are used by unqualified personnel.

QUESTIONS

Write your answers in a notebook and then compare your answers with those on page 870.

14.3D   How can you determine if there is voltage in a circuit?

14.3E   What are some of the uses of a multimeter?

14.3F   What precautions should be taken before attempting to change fuses?

14.3G   How do you test for voltage with a multimeter when the voltage is unknown?

14.3H   What could be the cause of amp readings different from the nameplate rating?

14.3I   How often should motors and wirings be megged?

14.3J   An ohmmeter is used to check the ohms of resistance in what control circuit components?

14.33   Electrical System Equipment

14.330   Need for Maintenance

Electrical system equipment is frequently the least understood and, therefore, most neglected equipment in a pump station. Usually, we do not think of it as a system that requires frequent inspection or maintenance; however, the reverse of this is actually true. Electrical equipment can be damaged more readily by operating conditions than almost any other kind of equipment. Water, dust, heat, cold, humidity, corrosive atmospheres, and vibration are all common pump station conditions that can affect the performance and the life of electrical equipment.

This section will discuss various elements in the electrical system of a pump station including motor control center components, such as circuit breakers, contactors, protective devices, transformers, and control relays. The section will also explain in detail important maintenance and operating aspects of the AC induction motor, the most common pump driver used in pump stations.

Electrical equipment should be inspected and maintained on at least an annual basis or more frequently, depending on the equipment and the application. Inspection should include a thorough examination, replacement of worn and expendable parts, and operational checks and tests.

Listed below are examples of electrical equipment maintenance tasks. Check and inspect each of these items annually.

1. All switch gear and distribution equipment. Look for worn parts and note general condition.

2. Wiring integrity.

3. Terminal connections. Tighten, if necessary (often needed with aluminum wire conductors).

4. Interlocking devices that prevent unauthorized entry.

5. Control circuits. Check operation and verify sequencing, including actual sequencing of pump motors as a function of wet well level.

6. All panel instruments. Clean and check for accuracy. Permanently installed, panel-mounted current and multimeters can be added to existing electrical systems and should be specified on new projects. This eliminates the need for routine access to the inside of the electrical system to obtain voltage current readings and exposure to hazardous voltages. Select a multimeter that measures both phase-to-phase and phase-to-neutral voltage. Current meters should also allow switching to measure current in each of the three phases.

7. Mechanical disconnect switches. Service and lubricate all switches, fuses, disconnects, and transfer switches.

8. Fuses. Verify proper application, size, and general condition. Check inventory of spare fuses.

9. Circuit breakers. Cycle each breaker and check for proper response and performance.

10. Contacts. Check for response (especially those in motor starters that carry high switching current); replace, if necessary.

11. Enclosures. Clean and vacuum.

If these basic maintenance procedures are done at least annually, it will help minimize pump station electrical system failures including:

1. Current imbalances or unbalances, which ultimately result in motor failure.

2. Loose contacts or terminals in control and power circuits causing high-resistance contacts.

3. Overheating resulting in arcing, fire, and electrical system damage.

4. Dirty enclosures and components, which allow a conductive path to build up between incoming phases causing phase-to-phase shorts.

5. Corrosion that causes high-resistance contacts and heating.

14.331 Equipment Protective Devices

Electricity, by its very nature, is extremely hazardous and safety devices are needed to protect operators and equipment. Water systems have pressure valves, pop-offs, and different safety equipment to protect the pipes and equipment. So must electricity have safety devices to contain the voltage and amperage that come in contact with the wiring and equipment.

Pump station electrical equipment protective devices may consist of fuses, circuit breakers, motor and circuit overload devices, and grounds. These devices usually are found in combination so that all three are present. For example, a fused disconnect switch (see Figure 14.28) may protect the entire electrical distribution system within the pump station; circuit breakers may be used to protect branch motor circuits from short circuits; and overload elements would be installed to protect the motor from overloads.

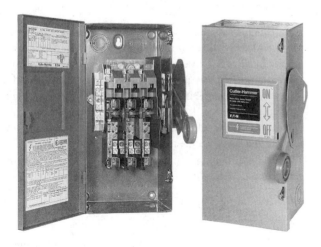

Fig. 14.28 Fused disconnect switch
(Permission of Eaton)

Figure 14.29 is a block diagram of motor circuit elements showing a typical arrangement of control and protective devices used to apply 3-phase line voltage to the motor. The first component in the circuit is some means of disconnecting the motor. This can be either a safety switch, a fused disconnect switch, or a circuit breaker. When using a fused disconnect switch or circuit breaker (the second element in the block diagram), the motor branch circuit overcurrent protection is incorporated. The next element in the block diagram is the motor controller, also referred to as a contactor. This is a 3-phase switch that connects 3-phase line voltage from the supply to the motor terminals. The final element in the diagram is the motor running overcurrent protection device. This device can be one that uses thermal elements that sense motor running current and trip under overload conditions or one of the newer solid-state motor control devices that also sense motor running current and provide overload protection. When the motor controller, or contactor, and the motor overload device are combined in a single unit, they are referred to as a motor starter.

The motor branch circuit breaker may provide only short-circuit fault current protection if it is a magnetic only breaker or, in the case of a thermal magnetic circuit breaker, it may provide overload protection as well. Typically, the motor branch circuit breaker is a molded-case circuit breaker (described later in Section 14.333).

Regardless of which type of protective device you are working with, if it blows or trips, the source of the problem should be investigated, identified, and corrected. All too frequently, a fuse is simply replaced or a circuit breaker simply reset without any investigation into the problem. If the protective device has operated reliably for long periods of time (for example, no nuisance tripping), then tripping is almost a sure indication that a problem exists somewhere in the circuit and the device is trying to protect the circuit. Simply resetting the circuit breaker or replacing a fuse and then restarting the system may result in damage or more severe damage to the equipment that is being protected.

Once again, if you are not qualified to perform electrical troubleshooting, you should not attempt to proceed any further than providing diagnostic analysis. Maintenance must always be performed by trained personnel. It does not take high voltage or extreme currents to seriously injure an operator or cause a fatality.

14.332 Fuses (Figure 14.30)

The power company installs fuses on their power poles to protect their equipment from damage. We also must install something to protect the main control panel and wiring from damage due to excessive voltage or amperage.

A fuse is a protective device having a strip or wire of fusible metal which, when placed in a circuit, will melt and break the electric circuit when subjected to an excessive temperature. This temperature will develop in the fuse when a current flows through the fuse in excess of what the circuit will carry safely. This means that the fuse must be capable of de-energizing the circuit before any damage is done to the wiring it is safely pro-

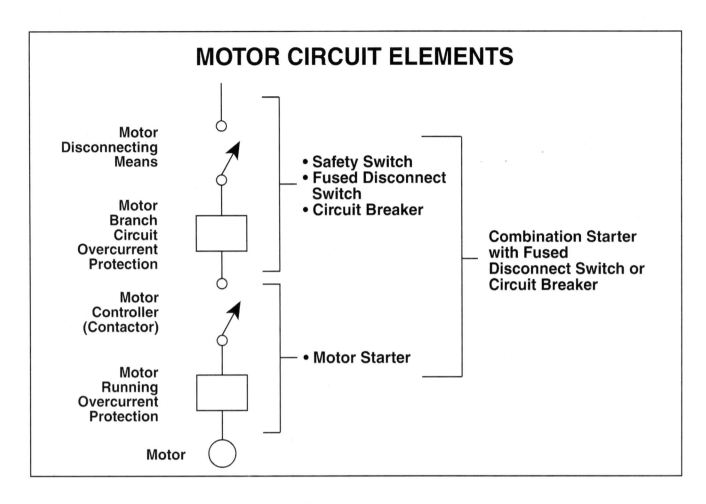

Fig. 14.29 Block diagram of motor circuit elements

tecting. Fuses are used to protect operators, wiring, main circuits, branch circuits, heaters, motors, and various other electrical equipment. A fuse must never be bypassed or jumped because the fuse is the only protection the circuit has. Without it, serious damage to equipment and possible injury to operators can occur.

There are several types of fuses, each being used for a certain type of protection. Some of these are:

1. Current-Limiting Fuses: Used to protect against current in circuits.
2. Dual-Element Fuses: Used for motor protection.
3. Time-Delay Fuses: Used in electronic and motor starting circuits.
4. Sand-Filled Fuses: Used on high voltage.
5. Phase Fuses: Used to protect phase sequence.
6. Voltage-Sensitive Fuses: Used where close voltage control is needed.

Since fuses are one-time-use devices, little can be done to check them during maintenance procedures; however, the following tasks should be performed at least annually:

1. Inspect bolted connections at the fuse clip or fuse holder for signs of looseness.
2. Check connections for any evidence of corrosion from moisture or atmosphere (air pollution).
3. Tighten connections.
4. Check fuse for obvious overheating.
5. Inspect insulation on the conductors coming into the fuses on the line side and out of the fuses on the load side for evidence of discoloration or bubbling, which would indicate overheating of the conductors.

Definition Line Side/Load Side

Line side/load side are terms frequently used to describe the incoming and outgoing conductors of circuit breakers, motor

798 Treatment Plants

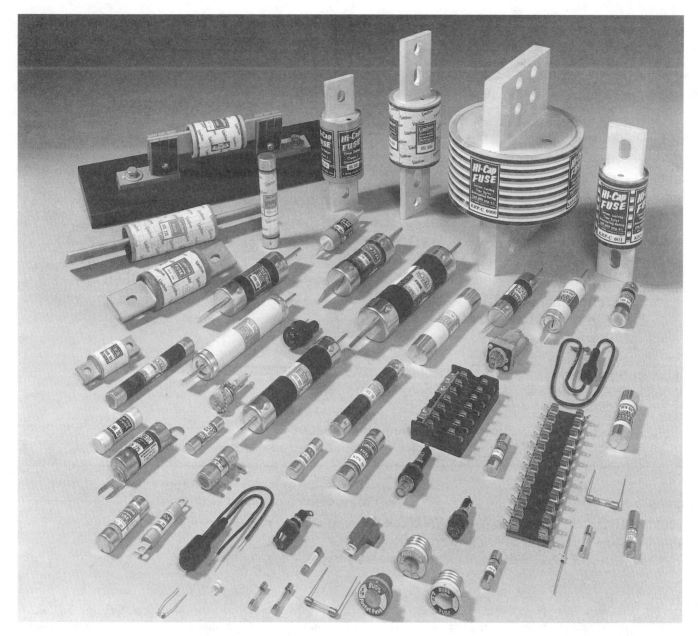

Fig. 14.30 Fuses
(Courtesy of Bussman Manufacturing, McGraw-Edison Company Division)

starters, and other devices. The line side of the device is where incoming power is fed into the device. The load side is the terminal where power is fed to the load, for example, a motor.

Usually, when a fuse blows it is not visibly apparent and it is necessary to check the fuse with a meter. The following procedure should be followed to determine whether or not a fuse has blown:

1. Ensure the main disconnect or circuit breaker is in the OFF position and locked out.

2. Check for live voltage in the panel (power may feed into a control system from other sources).

3. Remove fuse with a fuse-pulling device.

4. Test resistance of fuse using an ohmmeter. An open circuit (infinite resistance) indicates a blown fuse, whereas, a short circuit indicates a good fuse).

5. Only in the event it is not possible to disconnect incoming power should the fuse be checked with power applied. In this case the procedure as outlined in Section 14.32, "Tools, Meters, and Testers," may be used.

14.333 Circuit Breakers (Figure 14.31)

A circuit breaker is a switch that is opened automatically when the current or the voltage exceeds or falls below a certain limit. Unlike a fuse that has to be replaced each time it blows, a circuit breaker can be reset after a short delay to allow time for cooling.

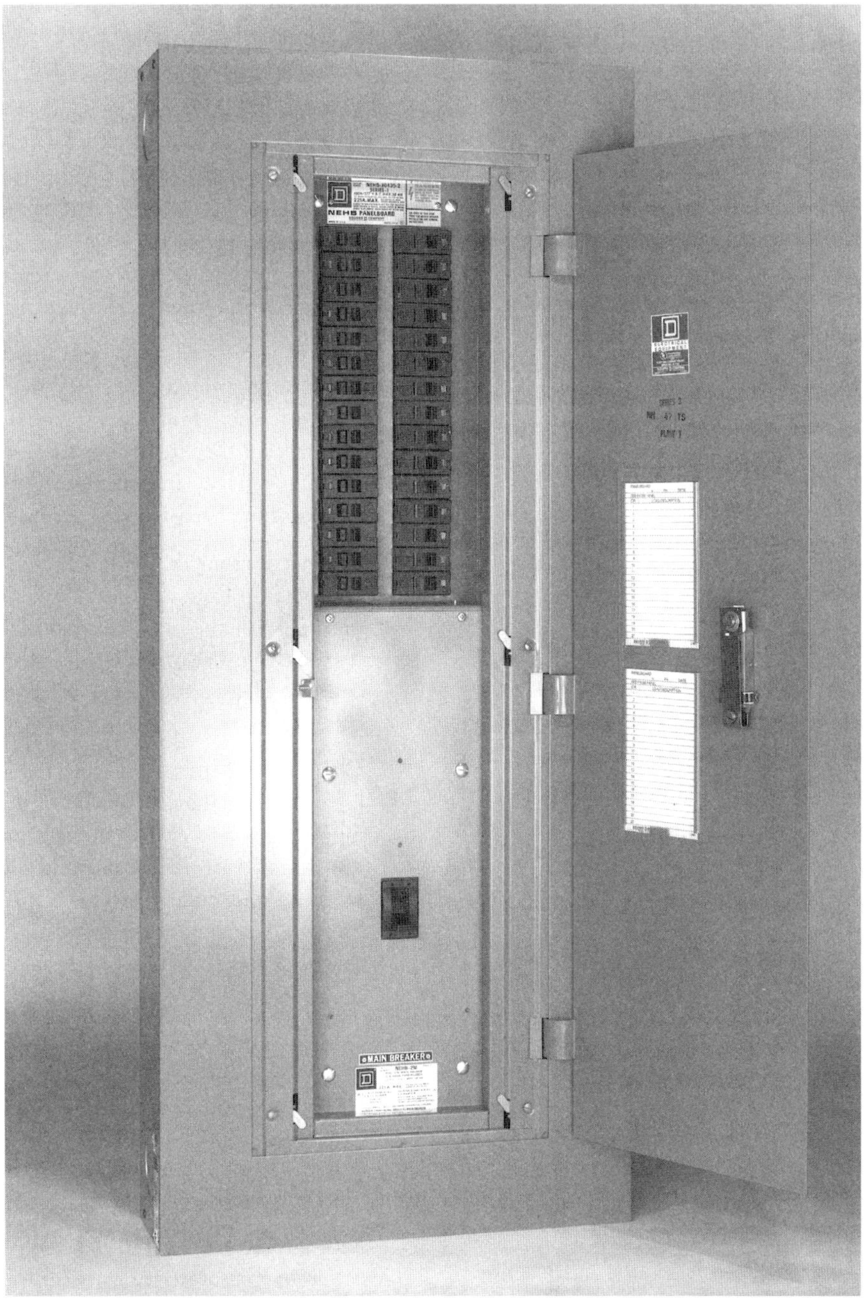

Fig. 14.31   Circuit breaker panelboard
(Permission of Square D Company)

800 Treatment Plants

Circuit breakers are used as disconnecting devices and to protect electrical systems primarily from short-circuit conditions that may occur on the load side of the circuit breaker. In some cases, circuit breakers can also be used in conjunction with motor starters to provide motor overload protection. In most cases, however, they are sized to protect the entire system as opposed to specific components in the circuit from short-circuit fault currents.

Molded-Case Circuit Breakers (Figure 14.32)

Pump station motor control center circuit breakers are typically housed in a molded case unless the motor or other electrical equipment is high horsepower. Molded-case circuit breakers can be thermal-magnetic trip or magnetic only.

Thermal-magnetic trip circuit breakers provide both overload protection and short-circuit protection. If the circuit is overloaded, a thermal sensing element will detect this and cause the circuit breaker to open or trip. In addition, the magnetic sensing element of the circuit breaker rapidly senses extremely high currents, which flow under a short-circuit condition, and will also trip the circuit breaker.

The amount of current that can flow during a short-circuit condition is a function of many things. For this reason, circuit breaker replacement should be done very carefully and by personnel who are qualified to evaluate circuit characteristics.

Fig. 14.32 Molded-case circuit breaker
(Permission of Eaton Corporation)

Molded-case circuit breakers require little maintenance other than:

1. Manually trip and operate the mechanism.
2. Check connections for tightness.
3. Inspect for evidence of overheating on the line and load side conductors.
4. Inspect the circuit breaker case for evidence of overheating.

The following additional tests can be performed on molded-case circuit breakers, but require specialized equipment.

1. Insulation resistance test. This test uses a high-voltage DC megger device (described previously in this chapter) to check the internal resistance of the circuit breaker.

    a. Phase-to-phase check. Apply high voltage between the three phases of the circuit breaker, which ultimately could lead to a phase-to-phase short. This check is performed by applying the high voltage to each combination of two phases of the three phases (1 and 2, 1 and 3, and 2 and 3) one at a time. At least one megohm of resistance in the insulation is adequate. Inadequate insulation resistance could lead to a phase-to-phase short.

    b. Phase-to-ground. To check the insulation resistance between each individual phase to ground, connect one phase on the circuit breaker at a time to ground, apply a high voltage, and measure the resistance. At least one megohm of resistance in the insulation is desirable.

2. Contact resistances. With a DC power supply, measure the voltage drop from the line side to the load side of the circuit breaker with the circuit breaker closed, but power disconnected. An excessive voltage drop indicates a high-resistance condition between the contacts, and the circuit breaker must be replaced.

3. Overload trip test. This test requires a specialized circuit breaker tester that is capable of generating currents in the range of the trip devices of the circuit breaker. Connect the test device between the line and the load and run a current equal to the circuit breaker capacity through the circuit breaker. A current in the range of the trip devices that causes them to trip will indicate that the circuit breaker is functioning properly.

Because the interior components of molded-case circuit breakers are not accessible, repair or replacement of internal parts is not possible. The entire circuit breaker must be replaced when it fails.

Motor Protection Devices

Since circuit breakers are normally used as a disconnecting device and as a protective device for the entire circuit, including the conductors that are connected to the breaker, the circuit breaker must be sized for the current carrying capacity of all components or equipment connected to it. For example, a thermal-magnetic circuit breaker must have a continuous current rating of at least 115 percent of motor full-load amps and the rating may be as great as 250 percent. This means that the thermal overload sensing device will not trip the breaker unless there is 115 percent overload present. In most cases, by the time this occurs, the motor will have been seriously damaged.

To overcome this problem and to provide further protection for motors, a device called a motor starter is usually installed in pump station motor control centers. The device operates through the supervisory control system responding automatically to the level in the wet well. A typical configuration is referred to as a magnetic motor starter and can be operated manually as well as automatically. Section 14.44 of this chapter contains a complete discussion of motor starters.

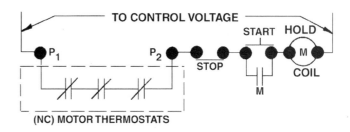

Fig. 14.33 Closed thermostats connected in series with the motor starter coil

CAUTION: Never increase the rating of the overload heaters because of tripping. You should find the problem and repair it.

Internal thermal protection devices are another form of motor and insulation protection. Such devices are especially desirable on motors that are likely to experience occasional overload conditions or voltage imbalances. Thermal protection devices may include the following features:

1. Built-in thermal switches that automatically open when the motor overheats. This type of device will not reset until the motor cools, at which time the switch resets automatically. (In some cases, there may be a reset button on the motor enclosure.) This type of protection is normally limited to small, single-phase motors.

2. Three-phase motors with internal thermal protection have thermostats bedded in each phase of the three-phase windings. The thermostats are connected in series and brought out into the motor conduit box and labeled P1 and P2. The switches are connected into the motor control circuit, which will open the magnetic starter contacts when an overtemperature condition occurs in the windings.

   Two circuit configurations are available:

   a. One in which the control circuit automatically restarts when the thermal switches and the motor reset.

   b. A circuit that is locked out and must be manually reset in order to restart.

Figure 14.33 illustrates a typical control circuit in which normally closed thermostats are connected in series with the motor starter coil in the control circuit. *NOTE:* This is the type of control circuit that will automatically reset when the motor thermostat's switches are reset.

There are many other protective devices for electricity, such as motor winding thermostats, phase protectors, low voltage protectors, and ground-fault circuit interrupters. Each has its own special applications and should never be tampered with or jumped (bypassed).

14.334 Ground

"Ground" is an expression representing an electrical connection to earth or a large conductor that is at the earth's potential or neutral voltage. Motor frames and all electrical tools and equipment enclosures should be connected to ground. This is generally referred to simply as grounding, or equipment ground.

Connecting motor frames, tools, and electrical equipment to ground is a safety precaution that protects you and the motors, tools, and equipment. If one of the conductors opens up and is not connected to ground, then *YOU* become the ground, the current could flow through you, and you could receive a severe or fatal electric shock. If the current flows through motors, tools, or equipment, severe damage could occur.

The third prong on cords from electric hand tools is the equipment ground and must never be removed. When an adapter is used with a two-prong receptacle, the green wire on the adapter should be connected under the center screw on the receptacle cover plate. Many times equipment grounding, especially at home, is achieved by connecting onto a metal water pipe or drain rather than a rod driven into the ground. This practice is not recommended when plastic pipes and other non-conducting pipe materials are used. Also, corrosion can be accelerated if pipes of different metals are used. A rod driven into dry ground is not very effective as a ground.

QUESTIONS

Write your answers in a notebook and then compare your answers with those on page 871.

14.3K What is the most common pump driver used in pump stations?

14.3L Basic pump station maintenance procedures performed at least annually will help minimize what types of pump station failures?

14.3M What should be done when a fuse or circuit breaker blows or trips?

14.3N What are two types of safety devices found in main electrical panels or control units?

14.3O What are fuses used to protect?

14.3P Why must a fuse never be bypassed or jumped?

14.3Q What types of annual maintenance should be performed with regard to fuses?

14.3R How does a circuit breaker work?

14.34 Motor Control/Supervisory Control and Electrical System

The motor and supervisory control systems are composed of the auxiliary electrical equipment, such as relays, transformers, lighting panels, pump control logic, alarms, and other electrical equipment typically found in a pump station electrical system, over and above the protective devices and the motor starters.

In general, annual maintenance should be performed on all these systems as follows:

1. Control Transformers
   a. Check primary, secondary, and ground connections
   b. Check for loose windings/coils
   c. Inspect insulation for signs of overheating as a result of overloading
   d. Check mounting for tightness
   e. Check primary/secondary fusing and fuse clips for tightness

2. Motor Control Centers
   a. Check panel lights for operation
   b. Check control knobs/switches for freedom of movement and contact condition
   c. Check horizontal and vertical bus and supports for evidence of heating or arcing and tighten. (Bus refers to the copper or aluminum bars that run horizontally and vertically in the motor control center; they feed the three-phase power to the branch circuits.)

3. Control Relays
   a. Check mounting for looseness
   b. Tighten all screw terminal connections
   c. Check for evidence of overheating or arcing indicated by carbon buildup or discoloration of plastic housing

4. Clean and vacuum enclosure

The use of aluminum wire as a conductor has become very common because of its economic advantage over copper. Copper traditionally has been used for virtually all wiring applications in wastewater lift stations, including the conductors feeding the station from the utility transformers, bus bar, motor control centers, control wiring, as well as power wiring to motors.

The greatest advantage of copper is that it oxidizes very slowly. Even though this reaction with air or moisture results in a surface layer of impurities, the surface is soft and is easily penetrated by the connecting device. With the exception of the annual tightening of the terminals, copper requires very little maintenance.

Aluminum is much softer and reacts much more rapidly with the air so aluminum begins oxidizing almost immediately when exposed to air. This oxide, as opposed to that found on copper, tends to form an insulating layer over the aluminum wire. Aluminum also expands and contracts 36 percent more than copper, which will result in loose connections, high-resistance contacts, heat formation, and failure of the connection. When using aluminum conductors, the following rules must be observed:

1. The connecting terminal must be specifically designed to accommodate aluminum conductors (connectors that are rated for this use are stamped with the letters CU/AL indicating that they are suitable for use with copper (CU) or aluminum (AL) conductors).

2. The termination point of the aluminum conductor must be coated with a compound to prevent the formation of the insulating oxides and should be wire-pressure scraped before the compound is applied.

Because of these limitations, aluminum conductors and bus bars should be inspected and maintained more frequently than the traditional copper conductors. Conduct a visual inspection of the terminal, conductor, and insulation for any evidence of discoloration and overheating.

Three basic factors contribute to the reliable operation of electrical systems found in pump stations. They are:

1. An adequate preventive maintenance program must be implemented.

2. A knowledge of the system by the operator, even though the operator does not perform the actual maintenance.

3. Adherence to three principles: KEEP IT CLEAN! KEEP IT DRY! KEEP IT TIGHT!

QUESTIONS

Write your answers in a notebook and then compare your answers with those on page 871.

14.3S Motor and supervisory control systems are composed of what types of auxiliary electrical equipment?

14.3T What are the three basic factors that contribute to the reliable operation of electrical systems found in pump stations?

14.4 MOTORS

14.40 Types

Electric motors are the machines most commonly used to convert electrical energy into mechanical energy. Although a multitude of different types of motors are produced today, the most common pump motor is the AC (Alternating Current) induction motor.

Two types of induction motor construction are typically encountered when dealing with AC induction motors:

1. Squirrel Cage Induction Motor (SCIM, by far the most numerous)
2. Wound Rotor Induction Motor (WRIM, used for variable-speed applications)

Figure 14.34 illustrates a horizontal squirrel cage induction motor with a drip-proof enclosure. Major components of the motor are:

1. Stator winding with connection to the three-phase power supply
2. Shaft end and opposite end ball bearings
3. Rotor assembly, including rotor, shaft, and fans on the rotor cage
4. Enclosure

Stator construction is generally the same in both types of motors. The primary difference is that the squirrel cage induction motor has no electrical connections to the rotor circuit. In contrast, a wound rotor motor has a slip ring assembly and brushes that connect the rotor circuit to an external electric circuit. This arrangement varies the resistance, thus causing the speed characteristics of the wound rotor motor to change. Therefore, a wound rotor is usually found in a variable-speed pump station application. This section will deal specifically with the squirrel cage induction motor since this is the type you will encounter most often.

14.41 Nameplate Data (Figure 14.35)

As part of your preventive maintenance program, record the motor nameplate information in the file for each motor or piece of equipment as recommended in NEMA (National Electrical Manufacturers Association) publication MG1, Section 10.3A. Motor nameplate data must be recorded and filed so the information is available when needed to repair the motor or to obtain replacement parts.

Following is a brief description of the information you could find on a nameplate and how it relates to motor performance.

Serial Number. This is a unique number assigned by the manufacturer based on the manufacturer's numbering system to identify that specific motor. (This number should be available and used whenever it is necessary to communicate with the manufacturer.)

Type. This may be a combination of letters and numbers, established by the manufacturer to identify the type of enclosure and any modifications to it.

Model Number. The manufacturer's model number.

Horsepower. Rated horsepower is the horsepower that the motor is designed to produce at the shaft when power is applied at the rated frequency and voltage and the motor is operating at a service factor of 1.0.

Frame. This identifies the frame size in accordance with established NEMA Standards and does not vary from manufacturer to manufacturer. Therefore, the motor from one manufacturer with a NEMA frame is dimensionally identical to that same frame size from another manufacturer.

Service Factor. Service factors of 1.0 and 1.15 are commonly found on pump station motors. A service factor of 1.0 indicates that the motor may be run continuously at its rated horsepower without causing damage to the insulation system. A 1.15 service factor indicates that the motor may occasionally be run at a horsepower equal to the rated horsepower times the service factor without serious injury to the insulation system. This allows for intermittent variations in voltage, which will cause some internal heating on the motor windings but not enough to cause damage. A service factor above 1.0 should never be relied upon to accommodate continuous loads because such use will quickly degrade the insulation system.

Amps. The current drawn by the motor at rated voltage and frequency at rated horsepower.

Volts. This is the voltage that would be measured at the terminals of the motor as opposed to the voltage of the supply.

Class of Insulation. Various classes of insulation material (described in Section 14.43) are available and the listed class will determine the operating temperatures at which the motor can safely operate.

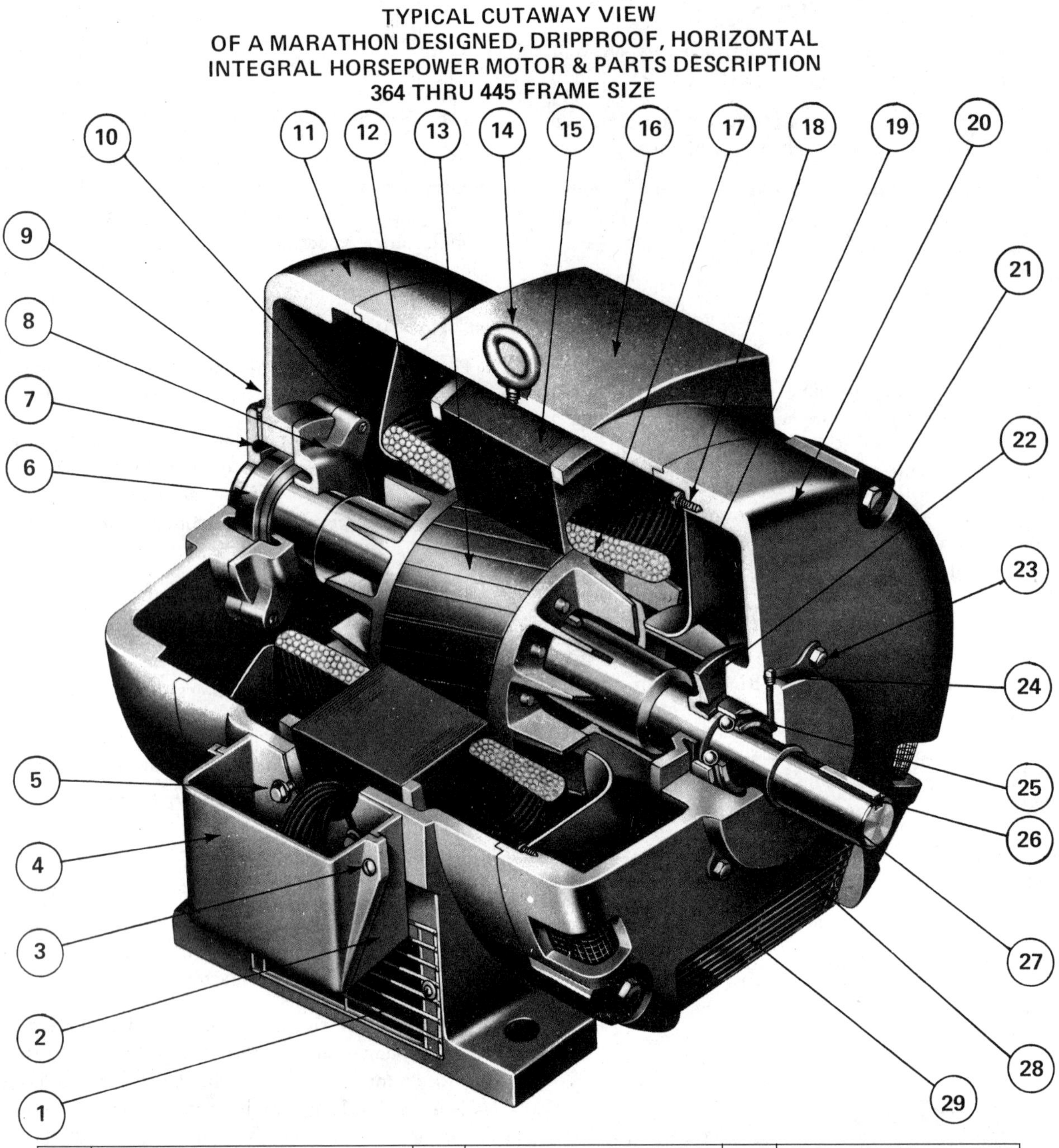

804 Treatment Plants

**TYPICAL CUTAWAY VIEW
OF A MARATHON DESIGNED, DRIPPROOF, HORIZONTAL
INTEGRAL HORSEPOWER MOTOR & PARTS DESCRIPTION
364 THRU 445 FRAME SIZE**

ITEM	DESCRIPTION	ITEM	DESCRIPTION	ITEM	DESCRIPTION
1.	*Frame Vent Screen	11.	Bracket O.P.E.	21.	Bracket Holding Bolt
2.	Conduit Box Bottom	12.	Baffle Plate O.P.E.	22.	Inner Bearing Cap P.E.
3.	Conduit Box Top-Holding Screw	13.	Rotor Core	23.	Inner Bearing Cap Bolt
4.	Conduit Box Top	14.	Lifting Eye Bolt	24.	Grease Plug
5.	Conduit Box Bottom Holding Bolt	15.	Stator Core	25.	**Ball Bearing P.E.
6.	**Ball Bearing O.P.E.	16.	Frame	26.	Shaft Extension Key
7.	Pre-Loading Spring	17.	Stator Winding	27.	Shaft
8.	Inner Bearing Cap O.P.E.	18.	Baffle Plate Holding Screw	28.	Drain Plug (grease)
9.	Grease Plug	19.	Baffle Plate P.E.	29.	*Bracket Screen
10.	Inner Bearing Cap Bolt	20.	Bracket P.E.		

P.E. = Pulley End
O.P.E. = Opposite Pulley End
* = Bracket and frame screens are optional.
** = Bearing Numbers are shown on motor nameplate. When requesting information or parts, always give complete motor description, model and serial numbers.

Fig. 14.34 Horizontal squirrel cage induction motor
(Courtesy of Marathon Electric)

Maintenance 805

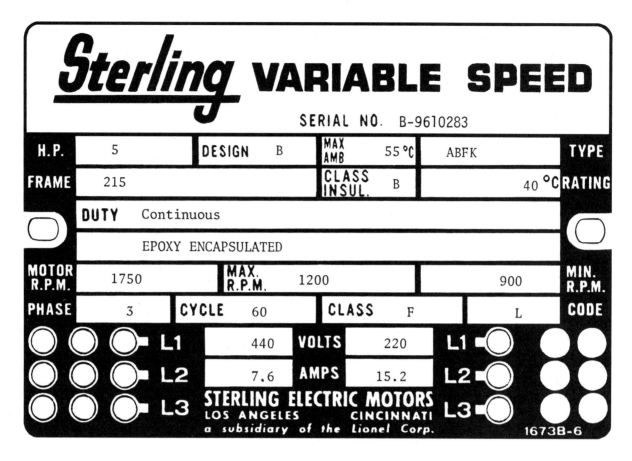

NOTE:
1. The motor for this unit is rated at 1,750 RPM and the maximum speed for the variable drive unit is 1,200 RPM.

2. The 40°C rating is the allowable operating temperature above ambient temperature.

Fig. 14.35 Typical nameplate
(Courtesy of Sterling Power Systems, Inc.)

RPM. Speed of the motor shaft, in revolutions per minute, at rated horsepower with rated voltage and frequency being supplied.

Hertz. Frequency, in cycles per second, of the utility company power.

Duty. Normally stamped "continuous," which means the motor can operate 24 hours a day, 365 days a year. In some cases, the motor nameplate will indicate "intermittent duty" for a specified time interval. This means the motor can operate at full load for the time interval specified, and then must be shut down and allowed to cool before restarting.

Ambient Temperature. Specifies the maximum ambient (surrounding) temperature at which the motor can safely operate. If the ambient temperature is exceeded, a corresponding increase of operating temperatures in the winding will degrade the insulation system and cause premature failure of the motor.

Phase. This indicates the number of phases; for example, 3 or three-phase, which indicates the phase for which the motor is designed.

KVA Code. Starting inrush current, which relates KiloVolt Amps to HorsePower (KVA/HP).

Design. This letter indicates the electrical design characteristics and, therefore, torque, speeds, inrush current, and slip values (these values and codes are also specified by NEMA).

Bearings. This number and letter sequence is designated by the Anti-Friction Bearing Manufacturers' Association (AFBMA) standards. It specifies looseness of the bearing fit, type of retainer, degree of protection, and dimension of the bearings. In most cases, unless the bearing is an extremely specialized type, the standards allow the use of bearings of different manufacturers.

Efficiency. This indicates the nominal operating efficiency of the motor at full load (Power Out/Power In)(100%).

14.42 Causes of Failure

Over the years, a considerable amount of research has been (and continues to be) conducted by various associations and manufacturers of electrical equipment to identify types of motor failures and the reasons for failures in AC induction motors. This section will deal with some of the common causes of motor failure and will suggest what you can do to minimize motor failures in pump stations.

In a survey of 9,000 motor failures by the Electrical Research Association, Letterhead, England, the following causes and percentages of motor failures were identified:

Causes of Motor Malfunction	Frequency, %
1. Overload (Thermal)	30
2. Contaminants	19
3. Single phasing	14
4. Bearing failures	13
5. Old age	10
6. Miscellaneous	9
7. Rotor failures	5

A study conducted by the Dymac Division of Scientific Atlanta identifies the frequency with which specific components failed.

Failed Component	Frequency, %
1. Stator windings	70
2. Rotor windings	10
3. Bearings	10
4. Other	10

This information suggests that a large percentage of motor failures could be prevented. Sixty-three percent of the failures examined were primarily due to overload, single phasing, and contaminants, and all of these factors can be controlled by the operator. In addition, since 70 percent of the failures occur in the stator, we can assume that many of the failures are related to the insulation system used to protect the motor stator windings. For this reason, the next section will focus on the motor insulation system and show how a variety of conditions can lead to excessive heat buildup that will eventually result in motor failure.

14.43 Insulation

14.430 Types and Specifications

Four classes or levels of insulation systems are available:

1. Class A
2. Class B
3. Class F
4. Class H

Each class of insulation is defined by the temperature limitation of the insulation itself. If the temperature limitation is exceeded, the insulation will deteriorate and, ultimately, cause a premature motor failure. The temperature rating consists of three components:

1. Ambient operating temperature—the air temperature where the motor is operating, for example, air temperature inside the pump station

2. Temperature rise—the maximum inside temperature of the windings during normal operation

3. Hot-spot allowance—because of inconsistencies in the manufacturing process and the winding materials, a 10°C hot-spot allowance is included

Older motors, those over 20 to 25 years old, are likely to have Class A insulation systems. A standard motor today has a Class B system or, in some cases, Class F. In extreme environments, a Class H insulation system may be used.

Table 14.4, AC Motor Temperature Limits, lists the temperature limitations for the four classes of insulation as they relate to the type of motor construction. For example, for all motors with a 1.15 or higher service factor, the total temperature that the insulation can withstand is 140°C for Class F insulation, or 165°C with Class H insulation. Table 14.5 lists typical materials used for motor insulation.

14.431 Causes of Failure

Obviously, induction motors are a critical and integral part of the operating ability, efficiency, and reliability of the lift station. When failures occur, they can be extremely serious because the loss of pumping capacity creates critical operating problems. In addition, repair or replacement of a motor is an expensive, labor-intensive job.

If a motor is operated in a clean, dry environment and within its specified nameplate load and operating characteristics, there is no reason why the motor should not operate for years and years without major maintenance. Unfortunately, a number of different factors can cause a motor to operate outside its specified ranges. When this occurs, the life of the motor is shortened significantly.

TABLE 14.4 AC MOTOR TEMPERATURE LIMITS[a]

	Temperature (Degrees C)			
	Class A	Class B	Class F	Class H
1.0 Service Factor				
Drip-Proof				
Ambient Temperature	40 40	40	40	40
Rise by Thermometer	40	..	..	..
Rise by Resistance	.. 50	80	105	125
Service-Factor Margin	10 10	..	..	..
Hot-Spot Allowance	15 5	10	10	15
Total Temperature	105 105	130	155	180
TEFC				
Ambient Temperature	40 40	40	40	40
Rise by Thermometer	55	..	..	..
Rise by Resistance	.. 60	80	105	125
Hot-Spot Allowance	10 5	10	10	15
Total Temperature	105 105	130	155	180
TENV[b]				
Ambient Temperature	40 40	40	40	40
Rise by Thermometer	55	85	..	..
Rise by Resistance	.. 65	..	110	135
Hot-Spot Allowance	10 0	5	5	5
Total Temperature	105 105	130	155	180
Encapsulated[c]				
Ambient Temperature			40	40
Rise by Thermometer			85	110
Hot-Spot Allowance			5	5
Total Temperature			130	155
1.15 or Higher Service Factor				
All Motors				
Ambient Temperature			40	40
Rise by Thermometer[d]			90	115
Hot-Spot Allowance			10	10
Total Temperature			140	165

[a] Adapted from National Electrical Manufacturers Association (NEMA) publication MG 1-12.39 and 12.40.
[b] Including all fractional-horsepower totally enclosed motors and fractional-horsepower motors smaller than frame 42.
[c] Enclosed.
[d] At service-factor load.

TABLE 14.5 MOTOR INSULATION MATERIALS

	Class A Systems-105°C	Class B Systems-130°C	Class F Systems-155°C	Class H Systems-180°C
Varnish	Modified phenolic Modified asphalt Alkyd polyester	Unmodified polyester Epoxy Modified phenolic	Modified polyester Epoxy	Silicone Polyimide
Wire Insulation	Vinyl acetal enameled	Modified polyester enameled Epoxy enameled Enamel plug glass yarn	Modified polyester enameled Epoxy enameled	Glass yarn, silicone varnish covered Polyimide
Other	Rag paper Kraft paper Polyester film Acetate film Varnished cambric Wood Fiber Cotton cord	Polyester film Polyester mat Varnished glass Mica flake or mica paper Polyester glass Laminated glass Asbestos Melamine Glass cord	Varnished glass Laminated glass Mica flake or mica paper Glass cord Polyester glass Tetrafluoroethylene resin	Glass cord Mica flake or mica paper Silicone varnished glass Laminated glass Polyimide film Polyimide varnished glass Polyimide filament

In the paragraphs that follow, these conditions will be discussed in terms of problems/solutions. You will see that, in one way or another, failure to adhere to motor nameplate guidelines frequently results in excessive heat buildup, damaged motor insulation, and eventual motor failure.

1. *PROBLEM:* Contaminants. Insulation failure can occur when deposits of dust, grease, or other foreign material accumulate on the windings and prevent the dissipation of the heat generated in the motor winding during normal operation. This causes local hot spots in the winding and, when the insulation breaks down completely, it will cause a phase-to-phase short characterized by arcing and melting between the phases' windings.

Vertical motors are susceptible to improper greasing methods and materials. Frequently, grease escapes from the bearing and the bearing housing and then contaminates the upper end turns on the stator winding, resulting in winding failure.

1. *SOLUTION:* Keep motors clean and free of dirt or grease accumulations. Follow the manufacturer's recommended methods and materials for greasing the equipment.

2. *PROBLEM:* Short cycling or excessive starts. Short cycling occurs when the automatic control system triggers frequent starts and stops of the pump and motor in response to fluctuating wet well elevations or because of a failure in the control system.

When induction motors start, the current (called locked rotor amps) required to magnetize the windings and start the rotor can be five to eight times the normal running current. For example, with a motor rated at 100 amps for full-load current, the starting sequence will require anywhere from 500 to 800 amps to start the motor rotating.

When frequent starts occur, the heat generated by the locked rotor currents never has a chance to dissipate and the internal winding temperature increases with each successive start of the motor. Typically, motors up to 100 horsepower (75 kilowatts) should not exceed more than four or five starts per hour and as the horsepower increases, motor starting limitations may reduce the number to as few as one start per hour.

2. *SOLUTION:* To overcome excessive heat buildup from frequent starts and stops of a motor, use a reduced-voltage method of starting the motor. Table 14.6 compares different reduced-voltage starting methods.

808 Treatment Plants

TABLE 14.6 REDUCED-VOLTAGE STARTING[a]

Type of Starting	Relative Starting Current	Relative Starting Torque
1. Across-the-Line	100%	100%
2. Resistors/reactor (at 65% voltage)	65%	65%
3. Auto Transformer (at 65% voltage)	42%	42%
4. Wye-Delta Winding	33%	33%
5. Two-Part Winding	50%	50%

[a] Compared with full-voltage, across-the-line starting, which typically draws 6.5 times the full-load current.

3. *PROBLEM:* High ambient operating temperature. If the ambient operating temperature exceeds that specified on the nameplate, it will contribute to a higher internal operating temperature. For all classes of insulation, the maximum ambient temperature is specified as 40°C (approximately 104°F).

3. *SOLUTION:* Provide adequate ventilation in lift stations and outdoor motor installations, particularly in southern climates where temperatures can be expected to exceed 104°F.

4. *PROBLEM:* Obstructed enclosure vents. All motor enclosures are designed for maximum dissipation of internally generated heat.

4. *SOLUTION:* In open, drip-proof motors, do not allow obstructions of the ventilation openings. Similarly, in totally enclosed, fan-cooled motors, a buildup of dirt, grease, or dust on the ribs of the enclosure will decrease the ability of the enclosure to dissipate heat. Keep the enclosure clean.

In outside installations, construct the enclosure to prevent entry by rodents. Frequently check rodent screens, if present, to ensure they are not clogged with foreign material. In areas of high humidity, treat the enclosure with a fungicide to prevent formation of fungus on the insulation surface.

5. *PROBLEM:* Single phasing. Single phasing refers to the condition that occurs when one phase of the power source to the motor is lost, either from the utility company or from a fuse blowing in one phase in the motor control center. Under a single-phase condition, an induction motor that is already rotating will continue to rotate; however, it will be characterized by increased noise and vibration. If the motor is not rotating, a single-phase condition will not start the motor, will cause excessive noise and vibration, and will continue to do so until the protective device senses the condition and trips the overloads. Single-phase condition causes unbalanced currents to circulate in the rotor causing increases in internal motor heating.

5. *SOLUTION:* To correct a single-phasing problem, determine why one phase is missing. Did the utility company lose a phase (a common problem) or is a fuse blown or a circuit breaker tripped in the motor control center? Once the source of the problem is identified, then it can be corrected by notifying the utility company, replacing the blown fuse, or resetting the tripped circuit breaker.

6. *PROBLEM:* Motor overloading. This is operation of the motor in a way that causes it to draw current in excess of the motor nameplate current value. Overloading can happen inadvertently through improper operation on the pump curve by changing impeller diameters or through a change in the dynamic operating conditions of the pump, which changes the total dynamic head (TDH). Other conditions that can cause motor overloading include bearing problems and jamming of material between the rotating impeller and the stationary pump housing. More energy is required to operate the pump when rags, rocks, or timbers interfere with free rotation of the impeller.

The heat generated by continuous operation of a motor above its design rating is extremely damaging to the insulation. For example, a relatively minor overload of 6 percent will cause a 10°C increase in temperature; continuous operation at this elevated temperature will reduce the insulation life by 50 percent. A continuous 12 percent overload cuts insulation life to one-quarter of its design life.

6. *SOLUTION:* A thorough understanding of hydraulics and the existing operating conditions is required before changes in the pump can be made since improper changes can have disastrous effects on the motor. Know what you are doing before making any changes. Also, avoid continuous operation of a motor above its design rating. For additional protection, consider installing an overload protective device (described in Section 14.331, "Equipment Protective Devices").

7. *PROBLEM:* Voltage imbalance. A common problem found in pump stations with a high rate of motor failure is voltage imbalance or unbalance. Unlike a single-phase condition, all three phases are present but the phase-to-phase voltage is not equal in each phase.

Voltage imbalance can occur in either the utility side or the pump station electrical system. For example, the utility company may have large, single-phase loads (such as residential services) that reduce the voltage on a single phase. This same condition can occur in the pump station if a large number of 120/220 volt loads are present. Slight differences in voltage can cause disproportional current imbalance; this may be six to ten times as large as the voltage imbalance. For example, a two percent voltage imbalance can result in a 20 percent current imbalance.

A 4.5 percent voltage imbalance will reduce the insulation life to 50 percent of the normal life. This is the reason a dependable voltage supply at the motor terminals is critical. Even relatively slight variations can greatly increase the motor operating temperatures and burn out the insulation.

It is common practice for electrical utility companies to furnish power to three-phase customers in open delta or wye configurations. An open delta or wye system is a two-transformer bank that is a suitable configuration where lighting loads are *large* and three-phase loads are *light*. This is the exact opposite of the configuration needed by most pumping facilities where three-phase loads are *large*. (Examples of three-transformer banks include Y-delta, delta-Y, and Y-Y.) In most cases, three-phase motors should be fed from three-transformer banks for proper balance. The capacity of a two-transformer bank is only 57 percent of the capacity of a three-transformer bank. The two-transformer configuration can cause one leg of the three-phase current to furnish higher amperage to one leg of the motor, which will greatly shorten its life.

Operators should acquaint themselves with the configuration of their electric power supply. When an open delta or wye configuration is used, operators should calculate the degree of current imbalance existing between legs of their polyphase motors. If you are unsure about how to determine the configuration of your system or how to calculate the percentage of current imbalance, always consult a qualified electrician. *Current imbalance between legs should never exceed 5 percent under normal operating conditions* (NEMA Standards MGI-14.35).

Loose connections will also cause voltage imbalance as will high-resistance contacts, circuit breakers, or motor starters.

Another serious consideration for operators is voltage fluctuation caused by neighborhood demands. A pump motor in near perfect balance (for example, 3 percent unbalance) at 9:00 am could be as much as 17 percent unbalanced by 4:00 pm on a hot day due to the use of air conditioners by customers on the same grid. Also, the hookup of a small market or a new home to the power grid can cause a significant change in the degree of current unbalance in other parts of the power grid. Because energy demands are constantly changing, wastewater system operators should have a qualified electrician check the current balances between legs of their three-phase motors at least once a year.

7. *SOLUTION:* Motor connections at the circuit box should be checked frequently (semiannually or annually) to ensure that the connections are tight and that vibration has not caused the insulation to rub through on the conductors. Measure the voltage at the motor terminals and calculate the percentage imbalance (if any) using the procedures below.

Do not rely entirely on the power company to detect unbalanced current. Complaints of suspected power problems are frequently met with the explanation that all voltages are within the percentages allowed by law and no mention is made of the percentage of current unbalance, which can be a major source of problems with three-phase motors. A little research of your own can pay large benefits. For example, a small water company in Central California configured with an open delta system (and running three-phase unbalances as high as 17 percent as a result) was routinely spending $14,000 a year for energy and burning out a 10-HP motor on the average of every 1.5 years (six 10-HP motors in 9 years). After consultation, the local power utility agreed to add a third transformer to each power board to bring the system into better balance. Pump drop leads were then rotated, bringing overall current unbalances down to an average of 3 percent; heavy-duty, three-phase capacitors were added to absorb the prevalent voltage surges in the area; and computerized controls were added to the pumps to shut them off when pumping volumes got too low. These modifications resulted in a saving in energy costs the first year alone of $5,500.00.

FORMULAS

Percentage of current unbalance can be calculated by using the following formulas and procedures:

$$\text{Average Current} = \frac{\text{Total of Current Value Measured on Each Leg}}{3}$$

$$\% \text{ Current Unbalance} = \frac{\text{Greatest Amp Difference From the Average}}{\text{Average Current}} \times 100\%$$

PROCEDURES

1. Measure and record current readings in amps for each leg (Hookup 1). Disconnect power.

2. Shift or roll the motor leads from left to right so the drop cable lead that was on terminal 1 is now on 2, lead on 2 is now on 3, and lead on 3 is now on 1 (Hookup 2). Rolling the motor leads in this manner will not reverse the motor rotation. Start the motor, measure and record current reading on each leg. Disconnect power.

3. Again shift drop cable leads from left to right so the lead on terminal 1 goes to 2, 2 goes to 3, and 3 to 1 (Hookup 3). Start pump, measure and record current reading on each leg. Disconnect power.

4. Add the values for each hookup.

5. Divide the total by 3 to obtain the average.

6. Compare each single leg reading to the average current amount to obtain the greatest amp difference from the average.

	Step 1 (Hookup 1)	Step 2 (Hookup 2)	Step 3 (Hookup 3)
(T_1)	DL_1 = 25.0 amps	DL_3 = 25 amps	DL_2 = 25.0 amps
(T_2)	DL_2 = 23.0 amps	DL_1 = 24 amps	DL_3 = 24.5 amps
(T_3)	DL_3 = 26.5 amps	DL_2 = 26 amps	DL_1 = 25.5 amps
Step 4	Total = 75 amps	Total = 75 amps	Total = 75 amps
Step 5	Average Current =	$\frac{\text{Total Current}}{3 \text{ readings}}$ =	$\frac{75}{3}$ = 25 amps
Step 6	Greatest amp difference from the average:	(Hookup 1) = 25 − 23 = 2 (Hookup 2) = 26 − 25 = 1 (Hookup 3) = 25.5 − 25 = .5	
Step 7	% Unbalance	(Hookup 1) = 2/25 × 100 = 8 (Hookup 2) = 1/25 × 100 = 4 (Hookup 3) = 0.5/25 × 100 = 2	

7. Divide this difference by the average to obtain the percentage of unbalance.
8. Use the wiring hookup that provides the lowest percentage of unbalance.

CORRECTING THE THREE-PHASE POWER UNBALANCE

Example: Check for current unbalance for a 230-volt, 3-phase, 60-Hz submersible pump motor, 18.6 full load amps.

Solution: Steps 1 to 3 measure and record amps on each motor drop lead for Hookups 1, 2, and 3 (Figure 14.36).

As can be seen, Hookup 3 should be used since it shows the least amount of current unbalance. Therefore, the motor will operate at maximum efficiency and reliability on Hookup 3.

By comparing the current values recorded on each leg, you will note the highest value was always on the same leg, L_3. This indicates the unbalance is in the power source. If the high current values were on a different leg each time the leads were changed, the unbalance would be caused by the motor or a poor connection.

If the current unbalance is greater than 5 percent, contact your power company for help.

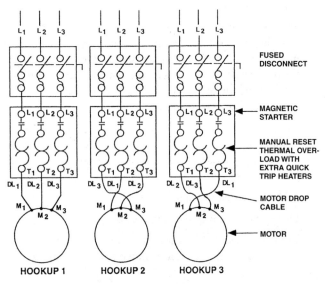

Fig. 14.36 Three hookups used to check for current unbalance

Acknowledgment

Material on unbalanced current was provided by James W. Cannell, President, Canyon Meadows Mutual Water Company, Inc., Bodfish, California. His contribution is greatly appreciated.

14.432 Increasing Resistance Value

The minimum insulation value in megohms is calculated by the following procedures. Divide the rated voltage by 1,000 and add 1 to the result. For example, for a motor operating at 460 volts, the minimum resistance value is 1.46 megohms.

$$\frac{460}{1,000} + 1 = 1.46$$

Measure the actual insulation resistance using a megger and compare the measured megger value with the calculated value; then any one of the three following procedures can be used to increase the actual (measured) insulation resistance value. The time required to increase the insulation resistance value depends on the wetness of the insulation and the size of the motor.

1. Remove the motor and bake in an oven at a temperature of not more than 194°F (90°C) until the resistance reaches an acceptable level.

2. Cover motor with a canvas or tarp and insert heating units or lamps. Heat until the resistance reaches an acceptable level. The heating time depends on the number and size of the heating units.

3. Provide a low-voltage current at the motor terminals that will generate heat within the windings. Heat until the resistance reaches an acceptable level.

14.44 Starters

A motor starter is a device or group of devices that are used to connect the electrical power to a motor. These starters can be either manually or automatically controlled.

Manual and magnetic starters range in complexity from a single ON/OFF switch to a sophisticated automatic device using timers and coils. The simplest motor starter is used on single-phase motors where a circuit breaker is turned on and the motor starts. This type of starter also is used on three-phase motors of smaller horsepower as well as on fan motors, machinery motors, and other motors, usually where it is not necessary to have automatic control.

Magnetic starters (Figure 14.37) are usually used to start pumps, compressors, blowers, and anything where automatic or remote control is desired. They permit low-voltage circuits to energize the starter of equipment at a remote location or to start larger starters (Figure 14.38). A magnetic starter is operated by electromagnetic action. It has contactors and these operate by energizing a coil that closes the contact, thus starting the motor. The circuit that energizes the starter is called the control circuit and it is usually operated on a lower voltage (115 volts) than the motor. Whenever a starter is used as a part of an integrated

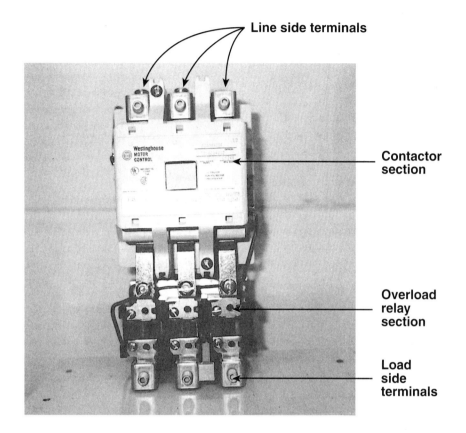

Fig. 14.37   Magnetic starter

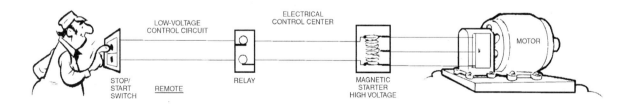

Fig. 14.38   Application of magnetic starter

circuit (such as for flow, pressure, or temperature control), a magnetic starter or controller is necessary.

Magnetic starters are sized for their voltage and horsepower ratings. Additional information can be found in electrical catalogs, manuals, and manufacturers' brochures.

A magnetic starter actually consists of two distinct sections:

1. The contacts, which connect and disconnect the power to the motor
2. Overload protection

Figure 14.37 illustrates a three-phase magnetic starter. The replaceable contacts (in the upper portion of the illustration) close when the motor is required to start, thus closing the electric circuit to the motor. Similarly, when the wet well level gets to a point where the pump is no longer required, a signal is sent to the motor starter and the contacts open, thus breaking the electric circuit to the motor. Each time this occurs, an arc takes place between the movable contact and the stationary contact and pitting occurs. This is why contacts must be replaced as a regular part of your preventive maintenance program.

The control coil usually uses a lower voltage than the line voltage to the motors. The magnetic coil is the device that actually causes the contacts to energize and de-energize.

In addition, each phase has an overload protection device that operates as a function of the length of the overload and the amount of the overload. Two types of devices are used:

1. A bimetallic strip that is precisely calibrated to open under higher temperature conditions to de-energize the coil
2. A small solder pot within the coil that melts because of the heat and will de-energize the system

A more common term for the overload protection devices is "heater elements." They are replaceable and can be selected and changed to correspond to the desired trip setting. Typically, the overload heater is selected for a trip setting that will de-energize the contactor when a 10 percent overload occurs.

Figure 14.39 shows a typical schematic diagram (referred to as a ladder diagram) for the control of one pump. This diagram is intended only for illustrative purposes. It does not include required grounding. Electrical systems must conform with the National Electrical Code in all cases.

Three-phase, 480-volt power is fed into terminals L1, L2, and L3 on the line side of the main circuit breaker. The main circuit breaker is normally located in the motor control center and provides circuit protection for all electrical equipment on the load side of the circuit breaker. On the load side of the circuit breaker, connections are made for a branch circuit breaker for the motor circuit. The load side of the branch circuit breaker feeds into motor starter 1M, which is a combination contactor and thermal overload. The load side of the motor starter 1M is then connected to the motor terminals.

Figure 14.39 shows the circuit components needed to get the 3-phase, 480-volt power to the motor terminals. However, some means must be provided for turning the motor on and off in response to wet well levels. The 24-volt AC, low-voltage control circuit illustrates how this is accomplished. Two phases of the 480-volt power are fed to a low-voltage control circuit breaker. The load side of the circuit breaker is connected to transformer T1, which reduces the line voltage to 24 volts AC. Power is then supplied to the low-voltage control circuit through fuse F1. Under a rising wet well condition, float switch FS1 is energized (as shown in Figure 14.39). This means that the float has tipped and the contacts are made in the float switch.

FS1 is the low-level or pump shutoff switch. As the level rises in the wet well, the pump start float switch FS2 tips. The contacts close and the circuit is completed through the hand-off-auto (HOA) switch; this energizes control relay CR1 and the motor starter contactor coil 1M. In a non-overload condition, thermal overload relay contacts OL1 are in the closed position allowing the circuit to be completed. When CR1 energizes, the contacts in parallel with FS2 close. This is necessary because as the wet well is pumping down, float switch FS2 will de-energize; however, we want the pump to continue operating until float switch FS1 de-energizes. When 1M energizes, this closes the contacts in the motor starter connecting the 3-phase power to the motor terminals. In the event of an overload condition, which would be sensed by the thermal overloads in motor starter 1M, the overload relay contacts OL1 in the low-voltage control circuit will open, thereby shutting off the motor. As the wet well continues to drop, float switch FS1 will return to the normal position when the water level drops below the float. This opens the control circuit and de-energizes relay coil 1M, which, in turn, opens the contacts in motor starter 1M.

If more than one motor is installed in the pump station, which is usually the case, additional branch circuit breakers, motor starters, float switches, and control devices are required. Additional control functions are easily added, such as indicating lights, alarms, and other control elements. As previously discussed, a *PROGRAMMABLE LOGIC CONTROLLER (PLC)*[12] could be installed to control pump functions. The PLC would essentially replace the control circuit elements, thus reducing the number of relays and the wiring required in pump motor control circuits.

Figure 14.40 illustrates how a motor contactor operates. The coil, when energized from the low-voltage control circuit, pulls the armature up in a vertical movement. The armature is attached to a bell crank lever that translates the vertical motion into horizontal motion, moving the contacts to the right where they make contact with the stationary contacts. Three-phase voltage is brought in on the line side connection and, in the closed position, the contacts allow that voltage to be applied through the load side connection to the motor. Both the stationary contacts and the moving contacts are replaceable. *NOTE:* This diagram does not illustrate the thermal overload protection.

---

[12] *Programmable Logic Controller (PLC).* A small computer that controls process equipment (variables) and can control the sequence of valve operations.

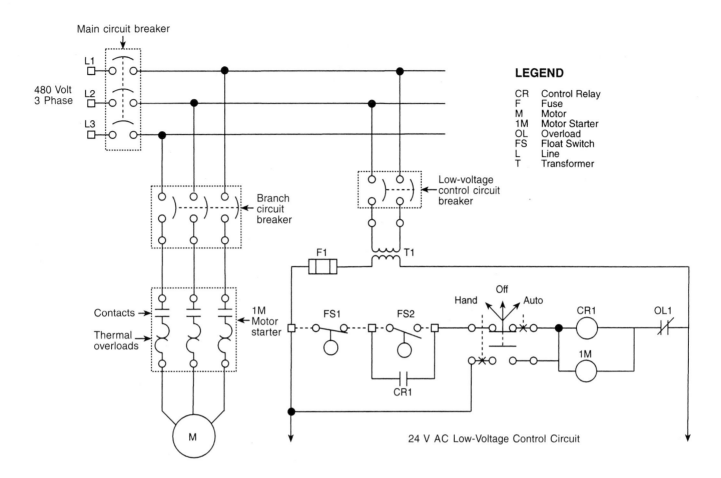

Fig. 14.39 Schematic (ladder) diagram for control of one pump

Starters are also available without the overload function. They provide only the motor disconnecting means and are referred to as contactors rather than motor starters. Manual contactors are available as well as magnetic contactors, which are capable of automatic control.

Magnetic starter maintenance consists of at least an annual inspection of the equipment, including the following tasks:

1. Inspect line and load conductors for evidence of high temperatures as indicated by bubbling or discoloring of the wire insulation.

2. Tighten all line and load connection terminals, including all low-voltage terminals (control, auxiliary switches).

3. Inspect and replace, if necessary, the stationary and movable contacts. See Figure 14.41, which illustrates the appearance of new contacts, contacts that are used but still suitable for use, and contacts that are used and should be replaced. A troubleshooting guide for magnetic starters is presented in Section 14.472. *CAUTION: No attempt should be made to file down contacts to restore the surface to a new condition, since this will result in an uneven surface and uneven distribution of electric energy across the face of the contacts.*

14.45 Safety

1. The eye bolt used for lifting and moving motors is designed for the weight of the motor alone without other equipment attached.

2. Whenever physically working on rotating equipment, including the pump and the motor, open, lock out, and tag the electrical disconnect switch. This will prevent accidental energization of the motor, either by a careless operator or as a result of motor thermal switches automatically resetting and restarting the motor.

3. Ground all motors in accordance with requirements of National Electrical Code, Article 340.

4. Always keep hands and clothing away from moving parts.

5. Discharge all capacitors including power factor direction capacitors before servicing motor or motor controller.

6. Be sure that required safety guards are always in place.

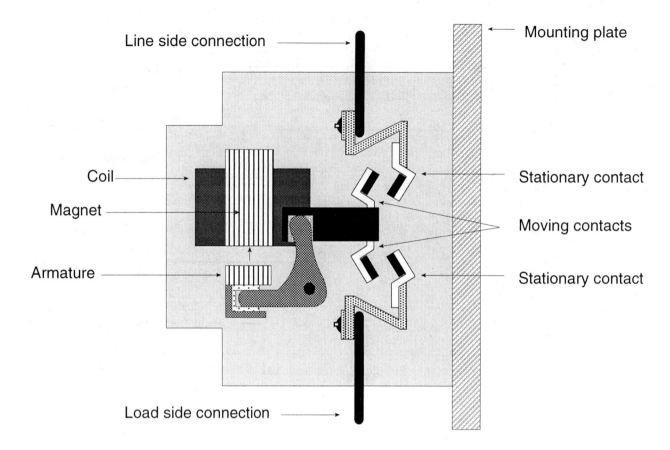

Fig. 14.40 How a motor contactor operates

14.46 Other Motor Considerations

14.460 Alignment

Horizontal motors should be mounted so that all four mounting feet are aligned to within 0.010 inch (0.25 mm) of each other for NEMA Frame 56 to NEMA Frame 210 and within 0.015 inch (0.38 mm) for NEMA Frame 250 to NEMA Frame 680. This alignment ensures a good, rigid foundation and also will make alignment of the motor and pump easier.

Whenever two pieces of rotating equipment, such as a pump and a motor, are used, there must be some means of transmitting the torque from the motor to the pump. Couplings are designed to do this. To function as intended, the equipment must be properly aligned at the couplings. Misalignment of the pump and the motor, or any two pieces of rotating equipment, can seriously damage the equipment and shorten the life of both the pump and the motor. Misalignment can cause excessive bearing loading as well as shaft bending, which will cause premature bearing failure, excessive vibration, or permanent damage to the shaft. Remember that the purpose of the coupling is to transmit power and, unless the coupling is of special design, it is not to be used to compensate for misalignment between the motor and the pump.

When connecting a pump and a motor, there are two important types of misalignment: (1) parallel, and (2) angular. Parallel misalignment occurs when the centerlines of the pump shaft and the motor shaft are offset. Figure 14.42 (page 816) illustrates parallel misalignment. The pump and the motor shafts remain parallel to each other but are offset by some amount. Parallel misalignment can be detected very easily by holding a straightedge on one hub of the coupling and measuring the gap between the straightedge and the other hub of the coupling. Feeler gauges can be used to measure the amount of offset misalignment.

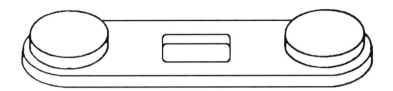

NEW

Smooth surface. May be bright or dull and somewhat discolored due to oxidation or tarnishing.

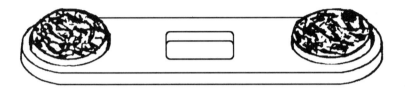

USED

Surface may be pitted and have discolored areas of black, brown, or may have blue (heat) tint. If half of the thickness (mass) of the silver points is still intact they are usable. This is the time to order a backup set.

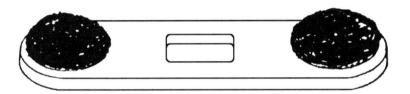

SEVERE OR LONG-TIME USE

Surface badly pitted and eroded with badly feathered and lifting edges. Replace entire contact set.

Fig. 14.41   Visual inspection of contact points

The second type of alignment problem is angular misalignment, also shown in Figure 14.42. In this case, the shaft centerlines are not parallel but instead form an angle, which represents the amount of angular misalignment. This type of misalignment can also be detected with a feeler gauge, calipers, or more sophisticated laser alignment equipment by measuring the distance between the coupling hubs at the point of maximum and minimum openings in the hubs, which would be 180° from each other. When using a dial indicator to measure angular misalignment, a general rule of thumb is that angular misalignment of the shafts must not exceed a total indicator reading of 0.002 inch (0.05 mm) for each inch (mm) of diameter of the coupling hub. To check for angular misalignment, mount a dial indicator on one coupling hub as shown in Diagram 1, Figure 14.43, with the finger or the button of the indicator against the finished face of the other hub and the dial set at zero. While rotating the shaft, note the reading on the indicator dial at each revolution.

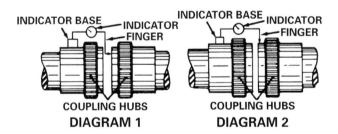

Fig. 14.43   Use of dial indicator to check for shaft angular alignment and trueness

In reality, misalignment usually includes both parallel and angular misalignment. The goal when aligning machines is to reduce both types of misalignment to a minimum. The purpose of this is two-fold: (1) couplings are not designed to accommodate large differences in parallel or angular misalignment, and

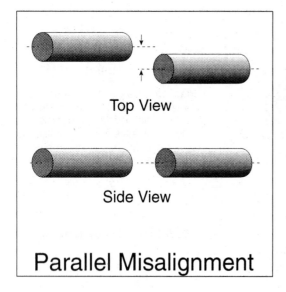
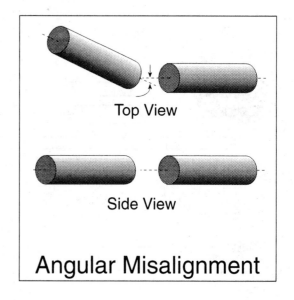
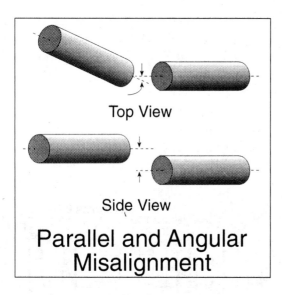
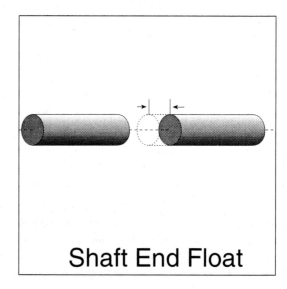

Fig. 14.42 *Types of shaft misalignment and end float*

(2) most alignment takes place when the machines being aligned are cold. However, all metal has a coefficient of expansion, which means that the metal expands as it heats up during operation. Since various types of metals expand at different rates and to different degrees, some additional misalignment may occur during operation. Couplings are designed to accommodate this type of misalignment.

In addition to misalignment, we are also concerned with end float in the shafts on the pump and the motor and with runout. End float is an in-and-out movement of the shaft along the axis of the shaft (see Figure 14.42).

Runout should also be checked. This checks the trueness or straightness of the shaft. Diagram 2, Figure 14.43, illustrates how the trueness is checked with the dial indicator mounted, again, on one coupling hub and the finger or button mounted on the outside surface of the second hub. As the shaft is rotated, note the indicator reading. The reading should not exceed 0.002 inch (0.05 mm). A reading in excess of 0.002 inch (0.05 mm) indicates a bent shaft or a shaft that is eccentrically machined.

Misalignment is one of the most frequent causes of vibration problems in lift station motors and pumps. It frequently causes premature failure of bearings, mechanical seals, and packing. While these failures in and of themselves are expensive and affect equipment availability and reliability, the failures can also cause more catastrophic related failures, such as broken shafts, destruction of the rotor element in the motor, or damage to motor windings.

14.461 Changing Rotation Direction

Changing the rotation direction of a three-phase motor is accomplished simply by changing any two of the power leads. Generally, this is done on the load side of a magnetic starter. The direction of motor rotation must be changed if a pump is rotating in the wrong direction.

14.462 Allowable Voltage and Frequency Deviations

1. Voltage 10 percent above or below the value stamped in the nameplate.

2. Frequency 5 percent above or below the value stamped in the nameplate.

3. Voltage and frequency together within 10 percent providing frequency is less than 5 percent above or below the value stamped in the nameplate.

As mentioned previously, both voltage and frequency have a direct effect on the performance and life of the motor, as illustrated in Tables 14.7 and 14.8. A 10 percent increase in the rated voltage, for example, results in a 0 to 17 percent increase in temperature, which will have a significant effect on the insulation life of the motor (see Table 14.8, Voltage DP, 110% of Rated Voltage, 1-200 HP, Temperature Rise, Full-Load, 0 to 17%).

14.463 Maximum Vibration Levels Allowed by NEMA

Speed, RPM	Maximum Vibration Amplitude, inches
3,000–4,000	0.001
1,500–2,999	0.002
1,000–1,499	0.0025
0–999	0.003

14.464 Lubrication

The correct procedure for lubricating motors is as follows:

1. Stop motor, lock out, and tag.

2. Wipe all grease fittings.

3. Remove filler and drain plugs. *CAUTION: Zerk fittings should not be installed in both the filler and the drain holes.*

4. Free drain hole of any hard grease using a piece of wire, if necessary.

5. Add appropriate amount and type of grease using low-pressure grease gun.

6. Start motor and let run for approximately 30 minutes. (This allows excess grease to drain out. If this is not done, pressure buildup in the bearing can blow out the bottom seal, allowing grease to run down the shaft onto the top of the motor windings.)

7. Stop motor, wipe off any drained grease, replace filler and drain plugs.

TABLE 14.7 TYPICAL MOTOR PERFORMANCE VARIATIONS DUE TO POWER SUPPLY VARIATIONS

POLYPHASE · INTEGRAL HORSEPOWER

				TORQUE			Speed Full-Load	Power Factor Full-Load
				Starting (Locked Rotor)	DIP	Break-Down	Full-Load		
			1% Unbalance	Slight −	Slight −	Slight −	Slight +	Slight −	−5.5%
			2% Unbalance	Slight −	Slight −	Slight −	Slight +	Slight −	−7.1%
V O L T A G E	D P	110% of Rated Volt.	1-10 HP[a]	+21 to 23%	+21 to 23%	+21 to 23%	−1.0%	+1.0%	−13 to 10%
			15-30 HP				−0.6%	+0.6%	−9 to 8%
			40-75 HP				−0.5%	+0.5%	−8 to 6%
			100-200 HP				−0.3%	+0.3%	−6 to 4%
		90% of Rated Volt.	1-10 HP	−17 to 19%	−17 to 19%	−17 to 19%	+1.5%	−1.5%	+11 to 7%
			15-30 HP				+1.0%	−1.0%	+6 to 3%
			40-75 HP				+0.6%	−0.6%	+3 to 2%
			100-200 HP				+0.3%	−0.3%	+2 to 3%
	T E F C	110% of Rated Volt.	1-10 HP	+21 to 23%	+21 to 23%	+21 to 23%	−1.0%	+1.0%	−13 to 6%
			15-30 HP				−0.6%	+0.6%	−5 to 3%
			40-75 HP				−0.5%	+0.5%	−3 to 2%
			100-200 HP				−0.3%	+0.3%	+2 to 0%
		90% of Rated Volt.	1-10 HP	−17 to 19%	−17 to 19%	−17 to 19%	+1.5%	−1.5%	+11 to 4%
			15-30 HP				+1.0%	−1.0%	+2 to 0%
			40-75 HP				+0.6%	−0.6%	+1 to 0%
			100-200 HP				+0.3%	−0.3%	+1 to 0%
F R E Q	DP & TEFC	105% of Rated Freq.	1 to 200 HP	−10%	−10%	−10%	−5%	+5%	Slight +
		95% of Rated Freq.	1 to 200 HP	+11%	+11%	+11%	+5%	−5%	Slight −

[a] Multiply HP × 0.746 to obtain kilowatts.

\+ Increase
− Decrease

Reprinted with permission of Marathon Electric, Wausau, WI

TABLE 14.8 TYPICAL MOTOR PERFORMANCE VARIATIONS DUE TO POWER SUPPLY VARIATIONS

				AMPS			Effic. Full-Load	Temp. Rise Full-Load
				Starting (Locked Rotor)	Full-Load	No-Load		
		1% Unbalance		+1.5%	+8%	+13%	−2%	+2%
		2% Unbalance		+3%	+17%	+27%	−8%	+8%
V O L T A G E	D P	110% of Rated Volt.	1-10 HP[a]	+10 to 12%	+8 to 4%	+25 to 37%	−6 to 1%	+17 to 9%
			15-30 HP		+4 to 1%	+32 to 37%	−1 to 0%	+7 to 3%
			40-75 HP		−0 to 2%	+37%	No Change	+2 to 0%
			100-200 HP		−3 to 5%	+36 to 30%	+0 to 0.3%	−0 to 4%
		90% of Rated Volt.	1-10 HP	−10 to 12%	−3 to +5%	−20 to 19%	+1 to 0%	−6 to +12%
			15-30 HP		+6 to 9%	−19 to 18%	−0 to 0.4%	+15 to 19%
			40-75 HP		+9 to 10%	−18 to 17%	−0.4 to 0.1%	+19 to 20%
			100-200 HP		+9 to 10%	−17 to 16%	−0.1%	+19 to 20%
	T E F C	110% of Rated Volt.	1-10 HP	+10 to 12%	+8 to −4%	+37 to 27%	−6 to 1%	+17 to −5%
			15-30 HP		−5 to 6%	+26 to 25%	−1 to 0%	−7 to 9%
			40-75 HP		−6 to 7%	+24 to 21%	No Change	−9 to 10%
			100-200 HP		−7 to 9%	+20 to 13%	+0 to 0.3%	−10 to 11%
		90% of Rated Volt.	1-10 HP	−10 to 12%	−3 to +5%	−20 to 16%	+1 to 0%	−6 to +15%
			15-30 HP		+6 to 11%	−14 to 12%	−0 to 0.4%	+19 to 23%
			40-75 HP		+12%	−12 to 10%	−0.4 to 0.1%	+24 to 23%
			100-200 HP		+11 to 9%	−10%	−0.1%	+24 to 23%
F R E Q	D P & T E F C	105% of Rated Freq.	1 to 200 HP	−5 to 6%	Slight −	−5 to 6%	Slight +	Slight −
		95% of Rated Freq.	1 to 200 HP	+5 to 6%	Slight +	+5 to 6%	Slight −	Slight +

[a] Multiply HP x 0.746 to obtain kilowatts.

+ Increase
− Decrease

Reprinted with permission of Marathon Electric, Wausau, WI

Figure 14.44 shows a cross section of the bearing cap and filler/drain plugs.

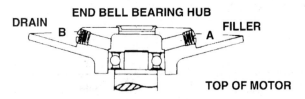

Fig. 14.44 Cross section of bearing cap and filler/drain plugs

Excessive greasing can cause as much damage to motor bearings as undergreasing. Therefore, manufacturer's recommendations should be followed without exception. The amount and frequency of greasing depends on a number of factors, including RPM, operating temperature, duty cycle, and environmental conditions.

QUESTIONS

Write your answers in a notebook and then compare your answers with those on page 871.

14.4A What are the two types of induction motor construction typically encountered when dealing with AC induction motors?

14.4B List the five most common causes of electric motor failure.

14.4C List the three components of an insulation temperature rating.

14.4D How are motor starters controlled?

14.4E When are magnetic starters used?

14.4F What two types of overload protection devices are used on magnetic starters?

14.4G How can the direction of rotation of a three-phase motor be changed?

14.47 Troubleshooting

14.470 Step-By-Step Procedures

The key to effective troubleshooting is the use of practical, step-by-step procedures combined with a common-sense approach.

"NEVER TAKE ANYTHING FOR GRANTED"

1. Gather preliminary information. The first step in troubleshooting any motor control that has developed trouble is to understand the circuit operation and other related functions. In other words, what is supposed to happen, operate, and so forth when it is working properly? Also, what is it doing now? The qualified maintenance operator should be able to do the following:

   a. *KNOW WHAT SHOULD HAPPEN WHEN A SWITCH IS PUSHED:* When switches are pushed or tripped, what coils go in, contacts close, relays operate, and motors run?

   b. *EXAMINE ALL OTHER FACTORS:* What other unusual things are happening in the pump station (facility) now that this circuit does not work properly? Lights dimmed, other pumps ran faster, lights went out when it broke, everything was flooded, operators were hosing down area, and many other possible factors.

   c. *ANALYZE WHAT YOU KNOW:* What part of it is working correctly? Is switch arm tripped? Everything but this is all right, except pump gets plugged with rags frequently. Is it a mechanical failure or an electrical problem caused by a mechanical failure?

   d. *SELECT SIMPLE PROCEDURES:* To localize the problem, select logical ways that can be simply and quickly accomplished.

   e. *MAKE A VISUAL INSPECTION:* Look for burned wires, loose wires, area full of water, coil burned, contacts loose, or strange smells.

   f. *CONVERGE ON SOURCE OF TROUBLE:* Mechanical or electrical. Motor or control, whatever it might be. Electrical problems result from some type of mechanical failure.

   g. *PINPOINT THE PROBLEM:* Exactly where is the problem and what do you need for repair?

   h. *FIND THE CAUSE:* What caused the problem? Moisture, wear, poor design, voltage, or overloading.

   i. *REPAIR THE PROBLEM AND ELIMINATE THE CAUSE, IF POSSIBLE:* If the problem is inside switch gear or motors, call an electrician. Give the electrician the information you have regarding the equipment. Do not attempt electrical repairs unless you are qualified and authorized, otherwise you could cause excessive damage to yourself and to the equipment.

2. Some of the things to look for when troubleshooting are given in the remainder of this section.

14.471 Troubleshooting Guide for Electric Motors

Symptom	Cause	Result[a]	Remedy
1. Motor does not start (switch is on and not defective)	a. Incorrectly connected	a. Burnout	a. Connect correctly per diagram on motor.
	b. Incorrect power supply	b. Burnout	b. Use only with correctly rated power supply.
	c. Fuse out, loose or open connection	c. Burnout	c. Correct open circuit condition.
	d. Rotating parts of motor may be jammed mechanically	d. Burnout	d. Check and correct: 1. Bent shaft. 2. Broken housing. 3. Damaged bearing. 4. Foreign material in motor.
	e. Driven machine may be jammed	e. Burnout	e. Correct jammed condition.
	f. No power supply	f. None	f. Check for voltage at motor and work back to power supply.
	g. Internal circuitry open	g. Burnout	g. Correct open circuit condition.
2. Motor starts but does not come up to speed	a. Same as 1-a, b, c above	a. Burnout	a. Same as 1-a, b, c above.
	b. Overload	b. Burnout	b. Reduce load to bring current to rated limit. Use proper fuses and overload protection.
	c. One or more phases out on a 3-phase motor	c. Burnout	c. Look for open circuits.
3. Motor noisy (electrically)	a. Same as 1-a, b, c above	a. Burnout	a. Same as 1-a, b, c above.
4. Motor runs hot (exceeds rating)	a. Same as 1-a, b, c above	a. Burnout	a. Same as 1-a, b, c above.
	b. Overload	b. Burnout	b. Reduce load.
	c. Impaired ventilation	c. Burnout	c. Remove obstruction.
	d. Frequent starts or stops	d. Burnout	d. 1. Reduce number of starts or reversals. 2. Secure proper motor for this duty.
	e. Misalignment between rotor and stator laminations	e. Burnout	e. Realign.
5. Motor noisy (mechanically)	a. Misalignment of coupling or sprocket	a. Bearing failure, broken shaft, stator burnout due to motor drag	a. Correct misalignment.
	b. Mechanical unbalance of rotating parts	b. Same as 5-a	b. Find unbalanced part, then balance.
	c. Lack of or improper lubricant	c. Bearing failure	c. Use correct lubricant, replace parts as necessary.

14.471 Troubleshooting Guide for Electric Motors (continued)

Symptom	Cause	Result[a]	Remedy
5. Motor noisy (mechanically) (continued)	d. Foreign material in lubricant	d. Bearing failure	d. Clean out and replace bearings.
	e. Overload	e. Bearing failure	e. Remove overload condition. Replace damaged parts.
	f. Shock loading	f. Bearing failure	f. Correct causes and replace damaged parts.
	g. Mounting acts as amplifier of normal noise	g. Annoying	g. Isolate motor from base.
	h. Rotor dragging due to worn bearings, shaft, or bracket	h. Burnout	h. Replace bearings, shaft, or bracket as needed.
6. Bearing failure	a. Same as 5-a, b, c, d, e	a. Burnout, damaged shaft, damaged housing	a. Replace bearings and follow 5-a, b, c, d, e.
	b. Entry of water or foreign material into bearing housing	b. Burnout, damaged shaft, damaged housing	b. Replace bearings and seals and shield against entry of foreign material (water, dust, etc.). Use proper motor.

Symptom	Caused By	Appearance
1. Shorted motor winding	a. Moisture, chemicals, foreign material in motor, damaged winding	a. Black or burned coil with remainder of winding good.
2. All windings completely burned	a. Overload	a. Burned equally all around winding.
	b. Stalling	b. Burned equally all around winding.
	c. Impaired ventilation	c. Burned equally all around winding.
	d. Frequent reversal or starting	d. Burned equally all around winding.
	e. Incorrect power	e. Burned equally all around winding.
3. Single-phase condition	a. Open circuit in one line. The most common causes are loose connection, one fuse out, loose contact in switch.	a. If 1,800 RPM motor—four equally burned groups at 90° intervals.
		b. If 1,200 RPM motor—six equally burned groups at 60° intervals.
		c. If 3,600 RPM motor—two equally burned groups at 180° intervals.
		NOTE: If Y connected, each burned group will consist of two adjacent phase groups. If delta connected, each burned group will consist of one-phase group.
4. Other	a. Improper connection	a. Irregularly burned groups or spot burns.
	b. Ground	

[a] Many of these conditions should trip protective devices rather than burn out motors. Also, many burnouts occur within a short period of time after motor is started up. This does not necessarily indicate that the motor was defective, but usually is due to one or more of the above-mentioned causes. The most common causes of failure shortly after start-up are improper connections, open circuits in one line, incorrect power supply, or overload.

14.472 Troubleshooting Guide for Magnetic Starters

Trouble	Possible Cause	Remedy
CONTACTS		
Contact chatter	1. Broken *POLE SHADER*[13]	1. Replace.
	2. Poor contact in control circuit	2. Improve contact or use holding circuit interlock.
	3. Low voltage	3. Correct voltage condition. Check momentary voltage drop.
Welding or freezing	1. Abnormal surge of current	1. Use larger contactor and check for grounds, shorts, or excessive motor load current.
	2. Frequent *JOGGING*[14]	2. Install larger device rated for jogging service or caution operators.
	3. Insufficient contact pressure	3. Replace contact spring; check contact carrier for damage.
	4. Contacts not positioning properly	4. Check for voltage drop during start-up.
	5. Foreign matter preventing magnet from seating	5. Clean contacts.
	6. Short circuit	6. Remove short fault and check that fuse and breaker are right.
Short contact life or overheating of tips	1. Contacts poorly aligned, poorly spaced, or damaged	1. Do not file silver-faced contacts. Rough spots or discoloration will not harm contacts. Replace.
	2. Excessively high currents	2. Install larger device. Check for grounds, shorts, or excessive motor currents.
	3. Excessive starting and stopping of motor	3. Caution operators. Check operating controls.
	4. Weak contact pressure	4. Adjust or replace contact springs.
	5. Dirty contacts	5. Clean with approved solvent.
	6. Loose connections	6. Check terminals and tighten.
Coil overheated	1. Starting coil may not kick out	1. Repair coil.
	2. Overload will not let motor reach minimum speed	2. Remove overload.
	3. Overvoltage or high ambient temperature	3. Check application and circuit.
	4. Incorrect coil	4. Check rating; if incorrect, replace with proper coil.
	5. Shorted turns caused by mechanical damage or corrosion	5. Replace coil.

---

[13] *Pole Shader.* A copper bar circling the laminated iron core inside the coil of a magnetic starter.
[14] *Jogging.* The frequent starting and stopping of an electric motor.

14.472 Troubleshooting Guide for Magnetic Starters (continued)

Trouble	Possible Cause	Remedy
CONTACTS (continued)		
Coil overheated (continued)	6. Undervoltage, failure of magnet to seal it	6. Correct system voltage.
	7. Dirt or rust on pole faces increasing air gap	7. Clean pole faces.
Overload relays tripping	1. Sustained overload	1. Check for grounds, shorts, or excessive motor currents. Mechanical overload.
	2. Loose connection on all or any load wires	2. Check, clean, and tighten.
	3. Incorrect heater	3. Replace with correct size heater unit.
	4. Fatigued heater blocks	4. Inspect and replace.
Failure to trip	1. Mechanical binding, dirt, or corrosion	1. Clean or replace.
	2. Wrong heater, or heaters omitted and jumper wires used	2. Check ratings. Apply heaters of proper rating.
	3. Motor and relay in different temperatures	3. Adjust relay rating accordingly, or install temperature compensating relays.
MAGNETIC AND MECHANICAL PARTS		
Noisy magnet (humming)	1. Broken shading coil	1. Replace shading coil.
	2. Magnet faces not mating	2. Replace magnet assembly or realign.
	3. Dirt or rust on magnet faces	3. Clean and realign.
	4. Low voltage	4. Inspct system voltage and voltage dips or drops during start-up.
Failure to pick up and seal	1. Low voltage	1. Inspect system voltage and correct.
	2. Coil open or shorted	2. Replace.
	3. Wrong coil	3. Check coil number and voltage rating.
	4. Mechanical obstruction	4. With power off, check for free movement of contact and armature assembly. Repair.
Failure to drop out	1. Gummy substance on pole	1. Clean with solvent.
	2. Voltage not removed from coil	2. Check coil circuit.
	3. Worn or rusted parts causing binding	3. Replace or clean parts as necessary.
	4. Residual magnetism due to lack of air gap in magnet path	4. Replace worn magnet parts or align, if possible.
	5. Welded contacts	5. Shorted circuit, grounded, overloaded.

14.473 Trouble/Remedy Procedures for Induction Motors

1. Motor will not start.

    Overload control tripped. Wait for overload to cool, then try to start again. If motor still does not start, check for the causes outlined below.

    a. Open fuses: test fuses.

    b. Low voltage: check nameplate values against power supply characteristics. Also, check voltage at motor terminals when starting motor under load to check for allowable voltage drop.

    c. Wrong control connections: check connections with control wiring diagram.

    d. Loose terminal-lead connection: turn power off and tighten connections.

    e. Drive machine locked: disconnect motor from load. If motor starts satisfactorily, check driven machine.

    f. Open circuit in stator or rotor winding: check for open circuits.

    g. Short circuit in stator winding: check for short.

    h. Winding grounded: test for grounded wiring.

    i. Bearing stiff: free bearing or replace.

    j. Overload: reduce load.

2. Motor noisy.

    a. Three-phase motor running on single phase: stop motor, then try to start. It will not start on single phase. Check for open circuit in one of the lines.

    b. Electrical load unbalanced: check current balance.

    c. Shaft bumping (sleeve-bearing motor): check alignment and conditions of belt. On pedestal-mounted bearing, check for play and axial centering of rotor.

    d. Vibration: driven machine may be unbalanced. Remove motor from load. If motor is still noisy, rebalance.

    e. Air gap not uniform: center the rotor and, if necessary, replace bearings.

    f. Noisy ball bearing: check lubrication. Replace bearings if noise is excessive and persistent.

    g. Rotor rubbing on stator: center the rotor and replace bearings, if necessary.

    h. Motor loose on foundation: tighten hold-down bolts. Motor may possibly have to be realigned.

    i. Coupling loose: insert feelers at four places in coupling joint before pulling up bolts to check alignment. Tighten coupling bolts securely.

3. Motor at higher than normal temperature or smoking. (Measure temperature with thermometer or thermister and compare with nameplate value.)

    a. Overload: measure motor loading with ammeter. Reduce load.

    b. Electrical load imbalance: check for voltage imbalance or single-phasing.

    c. Restricted ventilation: clean air passage and windings.

    d. Incorrect voltage and frequency: check nameplate values with power supply. Also, check voltage at motor terminals with motor under full load.

    e. Motor stalled by driven tight bearings: remove power from motor. Check machine for cause of stalling.

    f. Stator winding shorted or grounded: test windings by standard method.

    g. Rotor winding with loose connection: tighten, if possible, or replace with another rotor.

    h. Belt too tight: remove excessive pressure on bearings.

    i. Motor used for rapid reversing service: replace with motor designed for this service.

4. Bearings hot.

    a. End shields loose or not replaced properly: make sure end shields fit squarely and are properly tightened.

    b. Excessive belt tension or excessive gear side thrust: reduce belt tension or gear pressure and realign shafts. See that thrust is not being transferred to motor bearing.

    c. Bent shaft: straighten shaft or send to motor repair shop.

5. Sleeve bearings.

    a. Insufficient oil: add oil—if supply is very low, drain, flush, and refill.

    b. Foreign material in oil or poor grade of oil: drain oil, flush, and relubricate using industrial lubricant recommended by a reliable oil manufacturer.

    c. Oil rings rotating slowly or not rotating at all: oil too heavy; drain and replace. If oil ring has worn spot, replace with new ring.

    d. Motor tilted too far: level motor or reduce tilt and realign, if necessary.

    e. Rings bent or otherwise damaged in reassembling: replace rings.

    f. Rings out of slot (oil-ring retaining clip out of place): adjust or replace retaining clip.

    g. Defective bearings or rough shaft: replace bearings. Resurface shaft.

6. Ball bearings.
   a. Too much grease: remove relief plug and let motor run. If excess grease does not come out, flush and relubricate.
   b. Wrong grade of grease: flush bearing and relubricate with correct amount of proper grease.
   c. Insufficient grease: remove relief plug and grease bearing.
   d. Foreign material in grease: flush bearing, relubricate; make sure grease supply is clean (keep can covered when not in use).

14.5 RECORDS

Records are a very important part of electrical maintenance. They must be accurate and complete. Pages 827 and 828 are examples of typical record sheets. Most of the information you will need to complete these forms can be found on the manufacturer's data sheet or in the instruction manual.

Whenever a piece of equipment is changed, repaired, or tested, the work performed should be recorded on an equipment history card of some type. Complete, up-to-date equipment records will enable you to evaluate the reliability of your equipment and will provide the basis for a realistic preventive maintenance program.

14.6 ADDITIONAL READING

1. *BASIC ELECTRICITY* by Van Valkenburgh, Nooger & Neville, Inc. Obtain from Sams Technical Publishing Company, 9850 East 30th Street, Indianapolis, IN 46229, or call 800-428-7267. ISBN 0-7906-1041-8. Price, $38.95.

2. "Instrumentation" by Leonard Ainsworth, Chapter 9 in *ADVANCED WASTE TREATMENT*. Obtain from the Office of Water Programs, California State University, Sacramento, 6000 J Street, Sacramento, CA 95819-6025. Price, $45.00.

3. "Maintenance" by Parker Robinson, Chapter 18 in *WATER TREATMENT PLANT OPERATION*, Volume II. Obtain from the Office of Water Programs, California State University, Sacramento, 6000 J Street, Sacramento, CA 95819-6025. Price, $45.00.

4. *MAINTENANCE ENGINEERING HANDBOOK* by L. Higgins and K. Mobley. Obtain from the McGraw-Hill Companies, Order Services, PO Box 182604, Columbus, OH 43272-3031. ISBN 0-07-028819-4. Price, $150.00, plus nine percent of order total for shipping and handling.

QUESTIONS

Write your answers in a notebook and then compare your answers with those on pages 871 and 872.

14.4H What is the key to effective troubleshooting?

14.4I What are some of the steps that should be taken when troubleshooting magnetic starters?

14.5A What kind of information should be recorded regarding electrical equipment?

END OF LESSON 3 OF 7 LESSONS

on

MAINTENANCE

Please answer the discussion and review questions next.

DISCUSSION AND REVIEW QUESTIONS

Chapter 14. MAINTENANCE

(Lesson 3 of 7 Lessons)

Write the answers to these questions in your notebook. The question numbering continues from Lesson 2.

15. Why should inexperienced, unqualified, or unauthorized persons and even qualified and authorized persons be extremely careful around electrical panels, circuits, wiring, and equipment?

16. What is the difference between direct current (DC) and alternating current (AC)?

17. What meters and testers are used to maintain, repair, and troubleshoot electric circuits and equipment? Discuss the use of each meter and tester.

18. What protective or safety devices are used to protect operators and equipment from being harmed by electricity?

19. Why must motor nameplate data be recorded and filed?

20. How would you attempt to find the cause when a pump motor will not start?

PUMP RECORD CARD

NAME_____ MAKE_____ MODEL_____

TYPE_____ SIZE_____ SERIAL #_____

ORDER NUMBER_____ SUPPLIER_____ DATE PURCHASED_____

DATE INSTALLED_____ APPLICATION_____ PLANT #_____

Nameplate Data and Pump Info **Stuffing Box Data** **Motor Data**

GPM _____ Diameter____ Depth____ Name_____ Serial #_____

TDH _____ Pack. Size____ Type____ H.P._____ Speed_____

RPM _____ Length____ No. Rings____ Ambient°_____

Gauge Press Disc ____ Lantern Ring____ Flushed____ RPM____ Frame____

Gauge Press Suc ____ Mech. Seal Name____ Size____ Volts____ Amps____

Shut off Press ____ Type_____ Phase____ Cycle____

Suction Head ____ Shaft Size____ Key____

Pump Materials

Rotation ____ Casing_____ Bearing Front____

Impeller Type_____ Shaft_____ Rear____

Impeller Dia._____ Wearing Rings Casing_____ Code____ Type____

Impeller Clear_____ Wearing Rings Impeller_____ Amps @ Max Speed____

Coupl Type & Size_____ Shaft Sleeve_____ Amps @ Shut Off____

Front Brg #_____ Slinger_____ Control Data Info

Rear Brg #_____ Shims_____ Starter____

Lub Interval_____ Gaskets_____ NEMA Size____

Lubricant_____ "O" Rings_____ Cat. #____

Wearing Rings_____ Brg. Seals Front_____ Heater Size____

Shaft Sleeve Size_____ Rear_____ Rated @____

Pump Shaft Size_____ Casing Wear Ring Size ID____ Control Voltage____

Pump Keyway_____ OD____ Variable Speed

Type____

Width____ Speed Max____

Other Related Information:

Impeller Wear Ring ID____ Speed Min____

OD____

Width____

MOTOR STARTERS   Number_____

Title:_____

Mfg.:_____Address_____

Style:_____Class_____Size_____

Type:_____ _____ _____

O.L. HEATERS   O.L. TRIP UNITS

Style_____Code_____   Mfg:_____Style:_____

Amps_____   Type:_____ _____

_____   Amps Range:_____ _____

CIRCUIT BREAKER

Mfg:_____Address_____

Style:_____Frame:_____Volts_____Amps Setting_____

Cat. No._____ _____ _____ _____ _____

MOTOR   Number_____

TITLE_____

Mfg:_____Address_____

HP:_____Volts:_____Ser. No._____Duty:_____

Phase:_____Amps:_____Frame:_____Temp:_____

Cycles:_____RPM:_____Type_____Class:_____

Code:_____S.F.:_____Model_____Spec.:_____

SO#_____S#_____Style:_____CSA App:_____

Form_____Spec._____Shft. Brg._____Rear Brg._____

50 Cycle Data_____

Suitable for 208V Network:_____   Connection Diagram

Additional data_____   (6) (5) (4)   (6) (5) (4)

_____   (7) (8) (9)   (7) (8) (9)

_____   (1) (2) (3)   (1) (2) (3)

_____

The format of this section differs from the other chapters. The table of contents is outlined below and the paragraphs are numbered for easy reference when you use the Equipment Service Cards and Service Record Cards mentioned in Section 14.00, pages 749 and 750.

14.7 MECHANICAL MAINTENANCE

Paragraph	Page
(Lesson 4 of 7 Lessons)	
1. Pumps, General (Including Packing)	830
2. Reciprocating Pumps, General	835
3. Propeller Pumps, General	839
4. Progressive Cavity Pumps, General	839
5. Pneumatic Ejectors, General	839
6. Float and Electrode Switches	840
(Lesson 5 of 7 Lessons)	
7. Electric Motors	841
8. Belt Drives	843
9. Chain Drives	844
10. Variable Speed Belt Drives	845
11. Couplings	846
12. Shear Pins	849
(Lesson 6 of 7 Lessons)	
13. Gate Valves	850
14. Check Valves	853
15. Plug Valves	853
16. Sluice Gates	853
17. Dehumidifiers	862
18. Air Gap Separation Systems	862
19. Plant Safety Equipment	862
Acknowledgment	864

CHAPTER 14. MAINTENANCE

(Lesson 4 of 7 Lessons)

The format of this section differs from the other chapters. This format was designed specifically to assist you in planning an effective preventive maintenance program. The table of contents is outlined on the preceding page and the paragraphs are numbered for easy reference when you use the Equipment Service Cards and Service Record Cards mentioned in Section 14.00, pages 749 and 750. You can also use this paragraph numbering system when your maintenance program is on a computer.

An entire book could be written on the topics covered in this section. Step-by-step details for maintaining equipment are not provided because manufacturers are continually improving their products and these details could soon be out of date. You are assumed to have some familiarity with the equipment being discussed. For details concerning a particular piece of equipment, you should contact the manufacturer. This section indicates to you the kinds of maintenance you should include in your program and how you could schedule your work. Carefully read the manufacturer's instructions and be sure you clearly understand the material before attempting to maintain and repair equipment. If you have any questions or need any help, do not hesitate to contact the manufacturer or your local representative.

A glossary is not provided in this section because of the large number of technical words that require familiarization with the equipment being discussed. The best way to learn the meaning of these new words is from manufacturers' literature or from their representatives. Some new words are described in the lessons where necessary.

Preventive Maintenance

The following paragraphs list some general preventive maintenance services and indicate frequency of performance.

There are many makes and types of equipment and the wide variation of functions cannot be included; therefore, you will have to use some judgment as to whether the services and frequencies will apply to your equipment. If something goes wrong or breaks in your plant, you may have to disregard your maintenance schedule and fix the problem now.

NOTE: If you need to shut a unit down, make sure it is also locked out and tagged properly.

Paragraph 1: Pumps, General (Including Packing)

This paragraph lists some general preventive maintenance services and indicates frequency of performance. Typical centrifugal pump sections are shown in Figures 14.3 and 14.4, pages 761 and 762.

Frequency
of
Service

D   1. CHECK WATER-SEAL PACKING GLANDS FOR LEAKAGE. See that the packing box is protected with a clear-water supply from an outside source; make sure that water seal pressure is at least 5 psi (0.35 kg/sq cm) greater than maximum pump discharge pressure. See that there are no *CROSS CONNECTIONS*.[15] Check packing glands for leakage during operation. Allow a slight seal leakage when pumps are running to keep packing cool and in good condition. The proper amount of leakage depends on equipment and operating conditions. Sixty drops of water per minute is a good rule of thumb. If excessive leakage is found, hand tighten gland nuts evenly, but not too tight. After adjusting packing glands, be sure shaft turns freely by hand. If serious leakage continues, renew packing, shaft, or shaft sleeve.

---

[15] *Cross Connection.* (1) A connection between drinking (potable) water and an unapproved water supply. (2) A connection between a storm drain system and a sanitary collection system. (3) Less frequently used to mean a connection between two sections of a collection system to handle anticipated overloads of one system.

Frequency of Service	
D	2. CHECK GREASE-SEALED PACKING GLANDS. When grease is used as a packing gland seal, maintain constant grease pressure on packing during operation. When a spring-loaded grease cup is used, keep it loaded with grease. Force grease through packing at a rate of about one ounce (30 gm) per day. Never allow the seal to run dry.
W	3. OPERATE PUMPS ALTERNATELY. If two or more pumps of the same size are installed, alternate their use to equalize wear, keep motor windings dry, and distribute lubricant in bearings.
W	4. INSPECT PUMP ASSEMBLY. Check float controls noting how they respond to rising water level. See that unit starts when float switch makes contact and that pump empties basin at a normal rate. Apply light oil to moving parts.
D	5. CHECK MOTOR CONDITION. See Paragraph 7.
W	6. CLEAN PUMP. First lock out power and tag switch (Figure 14.45). Cleanout handholes are provided on the pump volute. To clean pump, close all valves, drain pump, remove handhole cover, and remove all solids. Wear gloves to protect your hands from sharp objects.
W	7. CHECK PACKING GLAND ASSEMBLY. Check packing gland, the unit's most abused and troublesome part. If stuffing box leaks excessively when gland is pulled up with mild pressure, remove packing and examine shaft sleeve carefully. Replace grooved or scored shaft sleeve because packing cannot be held in stuffing box with roughened shaft or shaft sleeve. Replace the packing a strip at a time, tamping each strip thoroughly and staggering joints. (See Figure 14.46.) Position lantern ring (water-seal ring) properly. If grease sealing is used, completely fill lantern ring with grease before putting remaining rings of packing in place. The type of packing used (Figure 14.47) is less important than the manner in which packing is placed. Both types of packing wrap around and score the shaft sleeve or are thrown out against outer wall of stuffing box, allowing wastewater to leak through and score the shaft. The proper size of packing should be available in your plant's equipment files. See pages 836 and 837, Figure 14.48, for illustrated steps on how to pack a pump.
W	8. CHECK MECHANICAL SEALS. Mechanical seals usually consist of two subassemblies: (1) a rotating ring assembly, and (2) a stationary assembly (Figure 14.15, page 773). Inspect seal for leakage and excessive heat. If any part of the seal needs replacing, replace the entire seal (both subassemblies) with a new seal that has been provided by the manufacturer. Before installing a new seal, be sure that there are no chips or cracks on the sealing surface. Keep a new mechanical seal clean at all times. Always be sure that a mechanical seal is surrounded with water before starting and running the pump.
Q	9. INSPECT AND LUBRICATE BEARINGS. Unless otherwise specifically directed for a particular pump model, drain lubricant and wash out oil wells and bearing with solvent. Check sleeve bearings to see that oil rings turn freely with the shaft. Repair or replace, if defective. Refill with proper lubricant. Check bearings by feeling for rough spots and looking for signs of binding and excessive movement up, down, back, and forth when rotated. Replace those bearings that are worn excessively. Generally, allow clearance of 0.002 inch plus 0.001 inch for each inch or fraction of inch of shaft-journal diameter (0.05 mm plus 0.025 mm for each 25 mm or fraction of 25 mm of shaft-journal diameter).
Q	10. CHECK OPERATING TEMPERATURE OF BEARINGS. Check bearing temperature with thermometer, not by hand. If antifriction bearings are running hot, check for overlubrication and relieve, if necessary. If sleeve bearings run too hot, check for lack of lubricant. If proper lubrication does not correct condition, disassemble and inspect bearing. Check alignment of pump and motor if high temperatures continue.
S	11. CHECK ALIGNMENT OF PUMP AND MOTOR. For method of aligning pump and motor, see Paragraph 11. If misalignment recurs frequently, inspect entire piping system. Unbolt piping at suction and discharge nozzles to see if it springs away, indicating strain on casing. Check all piping supports for soundness and effective support of load. Vertical pumps usually have flexible shafting, which permits slight angular misalignment; however, if solid shafting is used, align exactly. If beams carrying intermediate bearings are too light or are subject to contraction or expansion, replace beams and realign intermediate bearings carefully.

DANGER

OPERATOR WORKING ON LINE

DO NOT CLOSE THIS SWITCH WHILE THIS TAG IS DISPLAYED

TIME OFF: _____

DATE: _____

SIGNATURE: _____

This is the ONLY person authorized to remove this tag.

INDUSTRIAL INDEMNITY/INDUSTRIAL UNDERWRITERS/
INSURANCE COMPANIES

4E210—R66

Fig. 14.45 *Typical warning tag*
(Source: Industrial Indemnity/Industrial Underwriters/Insurance Companies)

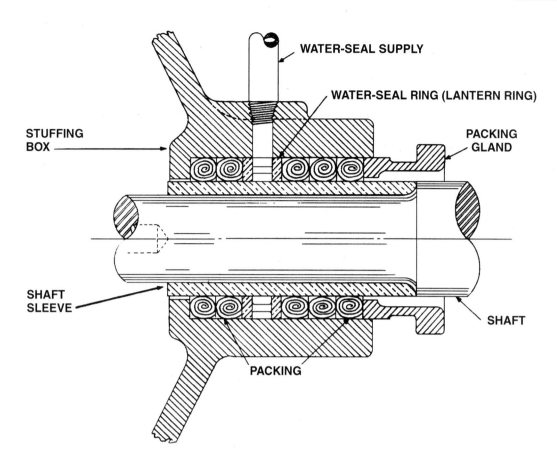

Fig. 14.46 Method of packing shaft
(Source: War Department Technical Manual TM5-666)

Frequency
of
Service

S   12. **INSPECT AND SERVICE PUMPS.**

    a. Remove rotating element of pump and inspect thoroughly for wear. Order replacement parts, where necessary. Check impeller clearance between volute.

    b. Remove any deposit or scaling. Clean out water-seal piping.

    c. Determine pump capacity by pumping into empty tank of known size or by timing the draining of pit or sump.

$$\text{Pump Capacity, GPM} = \frac{\text{Volume, gallons}}{\text{Time, minutes}}$$

or

$$\text{Pump Capacity,} \frac{\text{liters}}{\text{sec}} = \frac{\text{Volume, liters}}{\text{Time, seconds}}$$

    d. Test pump efficiency. Refer to pump manufacturer's instructions on how to collect data and perform calculations.

    e. Measure total dynamic suction lift and discharge head to test pump and pipe condition. Record figures for comparison with later tests.

    f. Inspect foot and check valves, paying particular attention to check valves, which can cause water hammer when pump stops. (See Paragraph 14 also.) Foot valves are used when pumping fresh water or plant effluent. Wet wells must be dewatered before foot valves can be inspected.

    g. Examine wearing rings. Replace seriously worn wearing rings to improve efficiency. Check wearing ring clearances, which generally should be no more than 0.003 inch per inch of wearing diameter (0.003 mm per mm of wearing diameter). *CAUTION: To protect rings and casings, never allow pump to run dry through lack of proper priming when starting or loss of suction when operating.*

Teflon Packing

Graphite Packing

Fig. 14.47 Packing
(Courtesy A. W. Chesterton Co.)

Frequency of Service	
A	13. DRAIN PUMP FOR LONG-TERM SHUTDOWN. When shutting down pump for a long period, open motor disconnect switch; shut all valves on suction, discharge, water-seal, and priming lines; drain pump completely by removing vent and drain plugs. This procedure protects pump against corrosion, sedimentation, and freezing. Inspect pump and bearings thoroughly and perform all necessary servicing. Drain bearing housings and replenish with fresh oil, purge old grease and replace. When a pump is out of service, run it monthly to warm it up and to distribute lubrication so the packing will not "freeze" to the shaft. Resume periodic checks after pump is put back in service.

QUESTIONS

Write your answers in a notebook and then compare your answers with those on page 872.

14.7A What is a cross connection?

14.7B Is a slight water-seal leakage desirable when a pump is running? If so, why?

14.7C How would you measure the capacity of a pump?

14.7D Estimate the capacity of a pump (in GPM) if it lowers the water in a 10-foot wide × 15-foot long wet well 1.7 feet in five minutes.

14.7E What should be done to a pump before it is shut down for a long time, and why?

Paragraph 2: Reciprocating Pumps, General (See Figure 14.49, page 838)

The general procedures in this paragraph apply to all reciprocating sludge pumps described in this section.

Frequency
of
Service

W 1. CHECK SHEAR PIN ADJUSTMENT. Set eccentric by placing shear pin through proper hole in eccentric flanges to give required stroke. Tighten the two $5/8$- or $7/8$-inch (1.5- or 2.2-cm) hexagonal nuts on connecting rods just enough to take spring out of lock washers. (See Paragraph 12.) When a shear pin fails, eccentric moves toward neutral position, preventing damage to the pump. Remove cause of obstruction and insert new shear pin. Shear pins fail because of one of three common causes:

  a. Solid object lodged under piston

  b. Clogged discharge line

  c. Stuck or wedged valve

D 2. CHECK PACKING ADJUSTMENT. Give special attention to packing adjustment. If packing is too tight, it reduces efficiency and scores piston walls. Keep packing just tight enough to keep sludge from leaking through gland. Before pump is installed or after it has been idle for a time, loosen all nuts on packing gland. Run pump with sludge suction line closed and valve covers open for a few minutes to break in the packing. Turn down gland nuts no more than necessary to prevent sludge from getting past packing. Tighten all packing nuts uniformly.

When packing gland bolts cannot be taken up farther, remove packing. Remove old packing and thoroughly clean cylinder and piston walls. Place new packing into cylinder, staggering packing-ring joints, and tamp each ring into place. Break in and adjust packing as explained above. When chevron type packing is used, tighten gland nuts only finger tight because excessive pressure ruins packing and scores plunger.

Q 3. CHECK BALL VALVES. When valve balls are so worn that diameter is $5/8$-inch (1.5-cm) smaller than original size, they may jam into guides in valve chamber. Check size of valve balls and replace if badly worn.

Q 4. CHECK VALVE CHAMBER GASKETS. Valve chamber gaskets on most pumps serve as a safety device and blow out under excessive pressure. Check gaskets and replace, if necessary. Keep additional gaskets on hand for replacement.

A 5. CHECK ECCENTRIC ADJUSTMENT. To take up babbitt bearing, remove brass shims provided on connecting rod. After removing shims, operate pump for at least one hour and check to see that eccentric does not run hot.

D 6. NOTE UNUSUAL NOISES. Check for noticeable water hammer when pump is operating. This noise is most pronounced when pumping water or very thin sludge; it decreases or disappears when pumping heavy sludge. Eliminate noise by opening the $1/4$-inch (0.6-cm) petcock on the discharge air chamber slightly; this draws in a small amount of air, keeping the discharge air chamber full of air at all times. This must be done with the pump locked out, discharge line isolated, and line pressure relieved to allow air to enter the chamber.

D 7. CHECK CONTROL VALVE POSITIONS. Because any plunger pump may be damaged if operated against closed valves in the pipeline, especially the discharge line, make all valve setting changes with pump shut down; otherwise pumps that are installed to pump from two sources or to deliver to separate tanks at different times may be broken if all discharge line valves are closed simultaneously for a few seconds or discharge valve directly above pump is closed.

W 8. GEAR REDUCER. Check oil level by removing plug on the side of the gear case. Unit should not be in operation.

Q 9. CHANGE OIL AND CLEAN MAGNETIC DRAIN PLUG.

836   Treatment Plants

 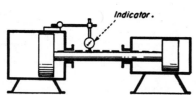 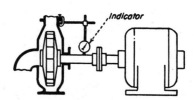

1 Remove *all* old packing. Aim packing hook at bore of the box to keep from scratching the shaft. Clean box thoroughly so the new packing won't hang up

2 Check for bent shaft, grooves or shoulders. If the neck bushing clearance in bottom of box is great, use stiffer bottom ring or replace the neck bushing

3 Revolve rotary shaft. If the indicator runs out over 0.003-in., straighten shaft, or check bearings, or balance rotor. Gyrating shaft beats out packing

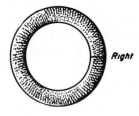 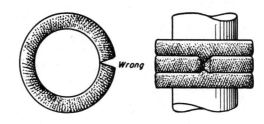

6 Cutting off rings while packing is wrapped around shaft will give you rings with parallel ends. This is very important if packing is to do job

7 If you cut packing while stretched out straight, the ends will be at an angle. With gap at angle, packing on either side squeezes into top of gap and ring cannot close. This brings up the question about gap for expansion. Most packings need none. Channel-type packing with lead core may need slight gap for expansion

HOW
TO PACK
A PUMP

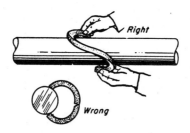 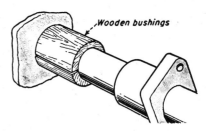

11 Open ring joint sidewise, especially lead-filled and metallic types. This prevents distorting molded circumference—breaking the ring opposite gap

12 Use split wooden bushing. Install first turn of packing, then force into bottom of box by tightening gland against bushing. Seat each turn this way

(*Editor's Note:* This step-by-step illustration of a basic maintenance duty was brought to our attention by Anthony J. Zigment, Director, Municipal Training Division, Department of Community Affairs.)

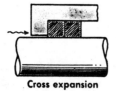 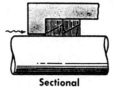 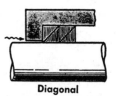

Cross expansion   Sectional   Diagonal

15 Always install cross-expansion packing so plies slope toward the fluid pressure from housing. Place sectional rings so slope between inside and outside ring is toward the pressure. Diagonal rings must also have slope toward the fluid pressure. Watch these details for best results when installing new packing in a box

Fig. 14.48   How to pack a pump

(Source: Water Pollution Control Association of Pennsylvania Magazine. January-February, 1976)

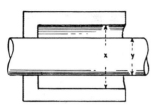 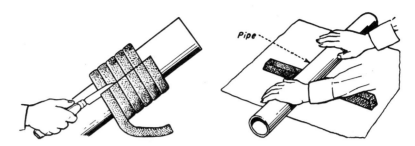

4 To find the right size of packing to install, measure stuffing-box bore and subtract shaft diameter, divide by 2. Packing is too critical for guesswork.

5 Wind packing, needed for filling stuffing box, snugly around shaft (or same size shaft held in vise) and cut through each turn while coiled, as shown. If the packing is slightly too large, never flatten with a hammer. Place each turn on a clean newspaper and then roll out with pipe as you would with a rolling pin

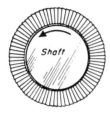 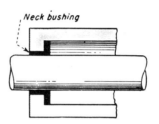 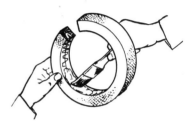

8 Install foil-wrapped packing so edges on inside will face direction of shaft rotation. This is a must; otherwise, thin edges flake off, reduce packing life

9 Neck bushing slides into stuffing box. Quick way to make it is to pour soft bearing metal into tin can, turn and bore for sliding fit into place

10 Swabbing new metallic packings with lubricant supplied by packing maker is OK. These include foil types, leadcore, etc. If the shaft is oily, don't swab it

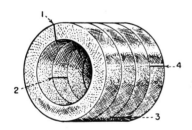 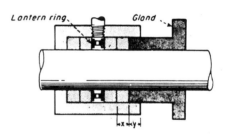

13 Stagger joints 180 degrees if only two rings are in stuffing box. Space at 120 degrees for three rings, or 90 degrees if four rings or more are in set

14 Install packing so lantern ring lines up with cooling-liquid opening. Also, remember that this ring moves back into box as packing is compressed. Leave space for gland to enter as shown. Tighten gland with wrench—back off finger-tight. Allow the packing to leak until it seats itself, then allow a slight operating leakage.

Hydraulic-packing pointers

First, clean stuffing box, examine ram or shaft. Next, measure stuffing box depth and packing set—find difference. Place 1/8-in. washers over gland studs as shown. Lubricate ram and packing set (if for water). If you can use them, endless rings give about 17% more wear than cut rings. Place male adapter in bottom, then carefully slide each packing turn home—don't harm lips. Stagger joints for cut rings. Measure from top of packing to top of washers, then compare with gland. Never tighten down new packing set until all air has chance to work out. As packing wears, remove one set of washers after more wear, remove other washer.

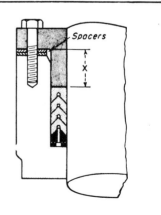

Fig. 14.48 How to pack a pump (continued)

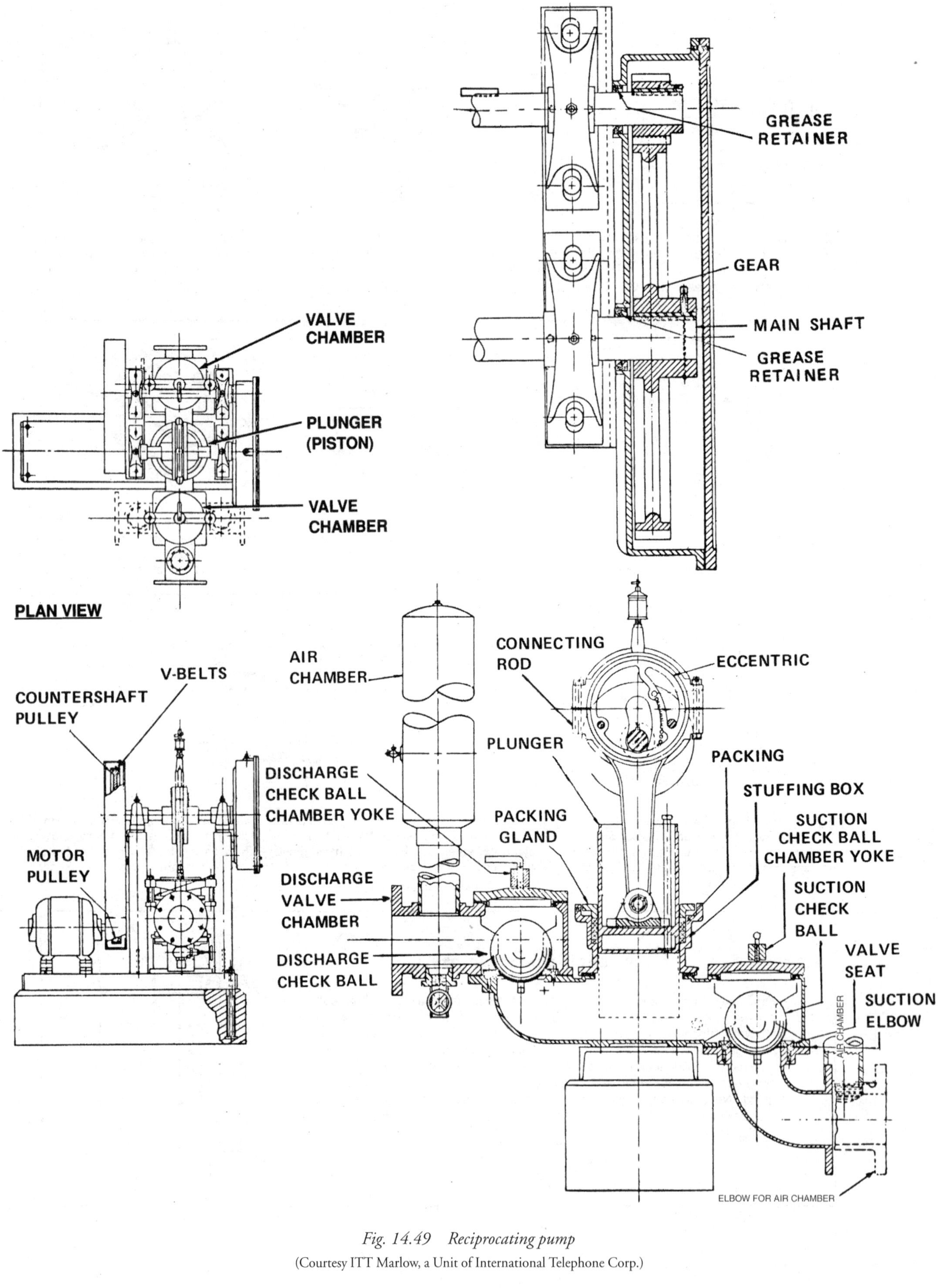

838　Treatment Plants

Fig. 14.49　Reciprocating pump
(Courtesy ITT Marlow, a Unit of International Telephone Corp.)

Frequency of Service

W 10. CONNECTING RODS. Set oilers to dispense two drops per minute.

W 11. PLUNGER CROSSHEAD. Fill plunger as required to half cover the wrist pin with oil.

D 12. PLUNGER TROUGH. Keep small quantity of oil in trough to lubricate the plunger.

M 13. MAIN SHAFT BEARING. Grease bearings monthly. Pump should be in operation when lubricating to avoid excessive pressure on seals.

14. CHECK ELECTRIC MOTOR. See Paragraph 7.

Paragraph 3: Propeller Pumps, General (Figure 14.7)

D 1. CHECK MOTOR CONDITION. See Paragraphs 7.1 and 7.2.

D 2. CHECK PACKING GLAND ASSEMBLY. See Paragraph 1.7.

W 3. INSPECT PUMP ASSEMBLY. See Paragraph 1.4.

W 4. LUBE LINE SHAFT AND DISCHARGE BOWL BEARING. Maintain oil in oiler at all times. Adjust feed rate to approximately four drops per minute.

W 5. LUBE SUCTION BOWL BEARING. Lube through pressure fitting. Usually three or four strokes of gun are enough.

W 6. OPERATE PUMPS ALTERNATELY. See Paragraph 1.3.

A 7. LUBE MOTOR BEARINGS. See Paragraph 7.3.

Paragraph 4: Progressive Cavity Pumps, General (Figure 14.12)

D 1. CHECK MOTOR CONDITION. See Paragraphs 7.1 and 7.2.

D 2. CHECK PACKING GLAND ASSEMBLY. See Paragraph 1.7.

D 3. CHECK DISCHARGE PRESSURE. A higher than normal discharge pressure may indicate a line blockage or a closed valve downstream. An abnormally low discharge pressure can mean reduced rate of discharge.

S 4. INSPECT AND LUBRICATE BEARINGS—GREASE. If possible, remove bearing cover and visually inspect grease. When greasing, remove relief plug and cautiously add 5 or 6 strokes of the grease gun. Afterward, check bearing temperature with thermometer. If over 220°F (104°C), remove some grease.

S 5. LUBEFLUSH MOTOR BEARINGS. See Paragraph 7.3.

S 6. CHECK PUMP OUTPUT. Check how long it takes to fill a vessel of known volume or quantity; or check performance against a meter, if available.

A 7. SCOPE MOTOR BEARINGS. See Paragraph 7.4.

A 8. SCOPE PUMP BEARINGS. See Paragraph 7.4.

Paragraph 5: Pneumatic Ejectors, General (Figure 14.14)

D 1. INSPECT UNIT. Check unit through a complete cycle. Look for air or water leaks. Keep units clean.

W 2. BLOW DOWN RECEIVERS.

M 3. CHECK VALVES. Keep check valve packing tight enough to prevent air or water leaks.

S 4. CLEAN AIR STRAINERS. Isolate air inlet line to ejector pot. Remove strainer and clean with water and wire brush ensuring free strainer openings. In case of different types of filters, consult particular manufacturer's literature.

A 5. CLEAN RECEIVER. After completely isolating ejector and power to it, remove and clean electrodes. Open ejector pot inspection plate and scrape inside walls of the pot.

A 6. INSPECT CHECK VALVES. Check operation by closing discharge valve, and actuating ejector on "hand" until pressure relief valve operates continuously for several minutes. Non-operation, or intermittent operation of relief valve in this mode indicates inlet check valve not holding. Discharge check valve can be checked by automatic ejection of pot, followed by locking out controls, closing inlet and discharge valves, removing top inspection plate, and checking for fluid leaking by the check valve. Some check valves can be checked by manually operating and feeling if they seat. At times, gauges, located in strategic locations, can indicate proper or improper operation. Inspect all check valves for leakage around packing.

A 7. INSPECT MAIN AIR VALVE. Check packing nut for lubrication. Check for smooth operation.

A 8. INSPECT PILOT VALVE. Time the ejection and check pressure gauge for maximum steady pressure.

840 Treatment Plants

Paragraph 6: Float and Electrode Switches

To ensure the best operation of the pump, a systematic inspection of the water level controls should be made at least once a week. Check to see that:

Frequency of Service		
W	1.	CHECK CONTROLS. Controls respond to a rising water level in the wet well.
W	2.	START-UP. The unit starts when the float switch or electrode system makes contact, and the pump stops at the prescribed level in the wet well.
W	3.	MOTOR SPEED. The motor speed comes up quickly and is maintained.
W	4.	SPARKING. A brush-type motor does not spark profusely in starting or running.
W	5.	INTERFERENCE WITH CONTROLS. Grease and trash are not interfering with controls. Be sure to remove scum from water-level float controls.
W	6.	ADJUSTMENTS. Any necessary adjustments are properly completed.

QUESTIONS

Write your answers in a notebook and then compare your answers with those on page 872.

14.7F What are some of the common causes of shear pin failure in reciprocating pumps?

14.7G What may happen when water or a thin sludge is being pumped by a reciprocating pump?

Please answer the discussion and review questions next.

DISCUSSION AND REVIEW QUESTIONS
Chapter 15. MAINTENANCE
(Lesson 4 of 7 Lessons)

Write the answers to these questions in your notebook. The question numbering continues from Lesson 3.

21. What would you do if considerable water was leaking from the water-seal of a pump?

22. When two or more pumps of the same size are installed, why should they be operated alternately?

23. What should be checked if pump bearings are running hot?

24. What happens when the packing is too tight on a reciprocating pump?

25. Why should changes in control valves for reciprocating pumps be adjusted when the pump is shut down?

CHAPTER 14. MAINTENANCE

(Lesson 5 of 7 Lessons)

Paragraph 7: Electric Motors (Figure 14.50)

To ensure the proper and continuous function of electric motors, the items listed in this paragraph must be performed at the designated intervals. If operational checks indicate a motor is not functioning properly, these items will have to be checked to locate the problem.

Frequency
of
Service
———

D 1. CHECK MOTOR CONDITIONS.

    a. Keep motors free from dirt, dust, and moisture.

    b. Keep operating space free from articles that may obstruct air circulation.

    c. Check for excessive grease leakage from bearings.

W 2. NOTE ALL UNUSUAL CONDITIONS.

    a. Unusual noises in operation.

    b. Motor failing to start or come to speed normally, sluggish operation.

    c. Motor or bearings that feel or smell hot.

    d. Continuous or excessive sparking commutator or brushes. Blackened commutator.

    e. Intermittent sparking at brushes.

    f. Fine dust under coupling having rubber bushings or pins.

    g. Smoke, charred insulation, or solder whiskers extending from armature.

    h. Excessive humming.

    i. Regular clicking.

    j. Rapid knocking.

    k. Brush chatter.

    l. Vibration.

    m. Hot commutator.

A 3. LUBRICATE BEARINGS (Figure 14.51).

    a. Check grease in ball bearings and replenish when necessary.

    Follow instructions below when preparing bearings for grease.

    b. Wipe pressure gun fitting, bearing housing, and relief plug to make sure that no dirt gets into bearing with grease.

    c. Before using grease gun, always remove relief plug from bottom of bearing to prevent excessive pressure in housing, which might rupture bearing seals.

    d. Use clean screwdriver or similar tool to remove hardened grease from relief hole and permit excess grease to run freely from bearing.

    e. While motor is running, add grease with hand-operated pressure gun until it flows from relief hole, purging housing of old grease. If there is no bottom or relief plug on bearing housing, insert grease cautiously through upper plug. Usually, four or five strokes of gun are enough. If bearing is overlubricated, seal may be ruptured. If lubricating a running motor is dangerous, follow above procedure with motor at a standstill. Lubricating a running motor is dangerous if you must remove protective gear such as guards over moving parts in order to lubricate the motor. Extension tubes and appropriate fittings should be installed wherever practical to facilitate lubrication with the guards in place.

    f. Allow motor to run for five minutes or until all excess grease has drained from bearing house.

    g. Stop motor and replace relief plug tightly with wrench.

A 4. USING A *STETHOSCOPE*,[16] CHECK BOTH BEARINGS. Listen for whines,

---

[16] *Stethoscope.* An instrument used to magnify sounds and carry them to the ear.

DRIP PROOF

ITEM NO.	PART NAME
1	Wound Stator w/ Frame
2	Rotor Assembly
3	Rotor Core
4	Shaft
5	Bracket
6	Bearing Cap
7	Bearings
8	Seal, Labyrinth
9	Thru Bolts/Caps
10	Seal, Lead Wire
11	Terminal Box
12	Terminal Box Cover
13	Fan
14	Deflector
15	Lifting Lug

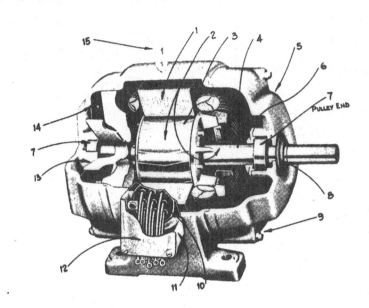

TOTALLY ENCLOSED FAN COOLED

ITEM NO.	PART NAME
1	Wound Stator w/ Frame
2	Rotor Assembly
3	Rotor Core
4	Shaft
5	Brackets
6	Bearings
7	Seal, Labyrinth
8	Thru Bolts/Caps
9	Seal, Lead Wire
10	Terminal Box
11	Terminal Box Cover
12	Fan, Inside
13	Fan, Outside
14	Fan Grill
15	Fan Cover
16	Fan Cover Bolts
17	Lifting Lug

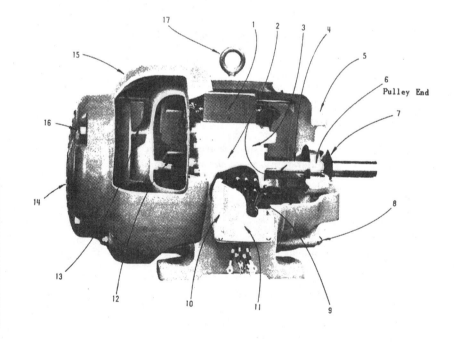

Fig. 14.50 Typical motors
(Courtesy of Sterling Power Systems, Inc.)

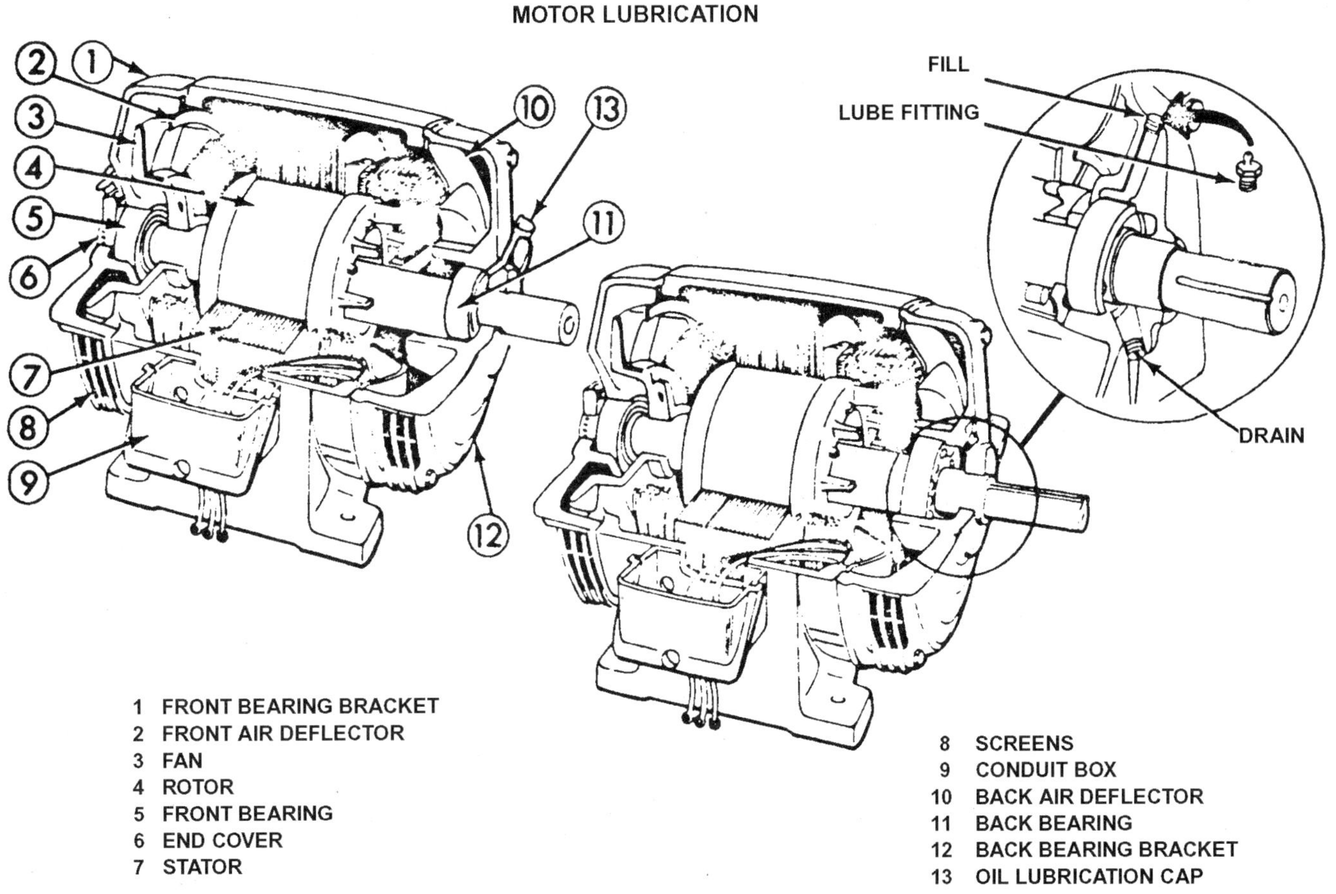

MOTOR LUBRICATION

1 FRONT BEARING BRACKET
2 FRONT AIR DEFLECTOR
3 FAN
4 ROTOR
5 FRONT BEARING
6 END COVER
7 STATOR
8 SCREENS
9 CONDUIT BOX
10 BACK AIR DEFLECTOR
11 BACK BEARING
12 BACK BEARING BRACKET
13 OIL LUBRICATION CAP

Fig. 14.51 Electric motor lubrication

Frequency of Service

gratings, or uneven noises. Listen all around the bearing and as near as possible to the bearing. Listen while the motor is being started and shut off. If unusual noises are heard, pinpoint the location.

5. IF YOU THINK THE MOTOR is running unusually hot, check with a thermometer. Place the thermometer in the casing near the bearing, holding it there with putty or clay. Magnetic thermometers are also available for this purpose.

A 6. *DATEOMETER.*[17] If there is a dateometer on the motor, after changing the oil in the motor, loosen the dateometer screw and set to the corresponding year.

QUESTION

Write your answer in a notebook and then compare your answer with the one on page 872.

14.7H What are the major items you would include when checking an electric motor?

Paragraph 8: Belt Drives

1. GENERAL. Maintaining a proper tension and alignment of belt drives ensures long life of belts and sheaves. Incorrect alignment causes poor operation and excessive belt wear. Inadequate tension reduces the belt grip, causes high belt loads, snapping, and unusual wear.

   a. Cleaning belts. Keep belts and sheaves (pulleys) clean and free of oil, which causes belts to deteriorate. To remove oil, take belts off sheaves and wipe belts and sheaves with a rag moistened in a non-oil-base sol-

---

[17] *Dateometer* (day-TOM-uh-ter). A small calendar disk attached to motors and equipment to indicate the year in which the last maintenance service was performed.

Frequency of Service		
		vent. Follow the requirements on the solvent's MSDS concerning ventilation and personal protective equipment (PPE).
	b.	Installing belts. Before installing belts, replace worn or damaged sheaves, then slack off on adjustments. Do not try to force belts into position. Never use a screwdriver or similar lever to get belts onto sheaves. After belts are installed, adjust tension; recheck tension after eight hours of operation. (See Table 14.9.)
	c.	Replacing belts. Replace belts as soon as they become frayed, worn, or cracked. *NEVER REPLACE ONLY ONE V-BELT ON A MULTIPLE DRIVE.* Replace the complete set with a set of matched belts, which can be obtained from any supplier. All belts in a matched set are machine-checked to ensure equal size and tension.
	d.	Storing spare belts. Store spare belts in a cool, dark place. Tag all belts in storage to identify them with the equipment on which they can be used.
	2.	V-BELTS. A properly adjusted V-belt has a slight bow in the slack side when running; when idle it has an alive springiness when thumped with the hand. An improperly tightened belt feels dead when thumped.

If the slack side of the drive is less than 45° from the horizontal, the vertical sag at the center of the span may be adjusted in accordance with Table 14.9 below:

TABLE 14.9 HORIZONTAL BELT TENSION

Span (inches)		10	20	50	100	150	200
Vertical Sag (inches)	From	.01	.03	.20	.80	1.80	3.30
	To	.03	.09	.58	2.30	4.90	8.60
Span (millimeters)		250	500	1,250	2,500	3,750	5,000
Vertical Sag (millimeters)	From	0.25	0.75	5.00	20.0	45.0	82.5
	To	0.75	2.25	14.50	57.5	122.5	215.0

M	a.	Check tension. If tightening belt to proper tension does not correct slipping, check for overload, oil on belts, or other possible causes. Never use belt dressing to stop belt slippage. Rubber wearings near the drive are a sign of improper tension, incorrect alignment, or damaged sheaves.
M	b.	Check sheave (pulley) alignment. Lay a long straightedge or string across outside faces of pulley, and allow for differences in dimensions from centerlines of grooves to outside faces of the pulleys being aligned. Be especially careful in aligning drives with more than one V-belt on a sheave, as misalignment can cause unequal tension.

Paragraph 9: Chain Drives

	1.	GENERAL. Chain drives may be designated for slow, medium, or high speeds.
	a.	Slow-speed drives. Because slow-speed drives are usually enclosed, adequate lubrication is difficult. Heavy oil applied to the outside of the chain seldom reaches the working parts; in addition, the oil catches dirt and grit and becomes abrasive. For lubricating and cleaning methods, see 5 and 6 below.
	b.	Medium- and high-speed drives. Medium-speed drives should be continuously lubricated with a device similar to a sight-feed oiler. High-speed drives should be completely enclosed in an oil-tight case and the oil maintained at proper level.
D	2.	CHECK OPERATION. Check general operating condition during regular tours of duty.
Q	3.	CHECK CHAIN SLACK. The correct amount of slack is essential to proper operation of chain drives. Unlike other belts, chain belts should not be tight around the sprocket; when chains are tight, working parts carry a much heavier load than necessary. Too much slack is also harmful; on long centers particularly, too much slack causes vibrations and chain whip, reducing life of both chain and sprocket. A properly installed chain has a slight sag or looseness on the return run.
S	4.	CHECK ALIGNMENT. If sprockets are not in line or if shafts are not parallel, excessive sprocket and chain wear and early chain failure result. Wear on inside of chain, side walls, and sides of sprocket teeth are signs of misalignment. To check alignment, remove chain and place a straightedge against sides of sprocket teeth.
S	5.	CLEAN. On enclosed types, flush chain and enclosure with a petroleum solvent (kerosene). On exposed types, remove chain and soak and wash it in solvent. Clean sprockets, install chain, and adjust tension.

Frequency of Service	

S   6. CHECK LUBRICATION. Soak exposed-type chains in oil to restore lubricating film. Remove excess lubricant by hanging chains up to drain.

Do not lubricate underwater chains that operate in contact with considerable grit. If water is clean, lubricate by applying waterproof grease with brush while chain is running.

Do not lubricate chains on elevators or on conveyors of feeders that handle dirty or gritty materials. Dust and grit combine with lubricants to form a cutting compound that reduces chain life.

S   7. CHANGE OIL. On enclosed types only, drain oil and refill case to proper level.

S   8. INSPECT. Note and correct abnormal conditions before serious damage results. Do not put a new chain on worn sprockets. Always replace worn sprockets when replacing a chain because out-of-pitch sprockets cause as much chain wear in a few hours as years of normal operation.

9. TROUBLESHOOTING. Some common symptoms of improper chain drive operation and their remedies follow:

   a. Excessive noise. Correct alignment, if misaligned. Adjust centers for proper chain slack. Lubricate in accordance with aforementioned methods. Be sure all bolts are tight. If chain or sprockets are worn, reverse or renew, if necessary.

   b. Wear on chain, side walls, and sides of teeth. Remove chain and correct alignment.

   c. Chain climbs sprockets. Check for poorly fitting sprockets and replace, if necessary. Make sure tightener is installed on drive chain.

   d. Broken pins and rollers. Check for chain speed that may be too high for the pitch, and substitute chain and sprockets with shorter pitch, if necessary. Breakage also may be caused by shock loads.

   e. Chain clings to sprockets. Check for incorrect or worn sprockets or heavy, tacky lubricants. Replace sprockets or lubricants, if necessary.

   f. Chain whip. Check for too-long centers or high, pulsating loads and correct cause.

   g. Chains get stiff. Check for misalignment, improper lubrication, or excessive overloads. Make necessary corrections or adjustments.

Paragraph 10: Variable-Speed Belt Drives (See Figure 14.52)

W   1. CLEAN DISKS. Remove grease, acid, and water from disk faces.

D   2. CHECK SPEED-CHANGE MECHANISM. Shift drive through entire speed range to make sure shafts and bearings are lubricated and disks move freely in lateral direction on shafts.

W   3. CHECK V-BELT. Make sure it runs level and true. If one side rides high, a disk is sticking on shaft because of insufficient lubrication or wrong lubricant. In this case, stop the drive at once, remove V-belt, and clean disk hub and shaft thoroughly with petroleum solvent until disk moves freely. Relubricate with soft, ball-bearing grease and replace V-belt in opposite direction from that in which it formerly ran.

M   If drive is not operated for 30 days or more, shift unit to minimum speed position, placing spring on variable-speed shaft at minimum tension and relieving belt of excessive pressure.

4. LUBRICATE DRIVE. Make sure to apply lubricant at all the six force-feed lubrication fittings (Figure 14.52: A, B, D, E, G, and H) and the one cup-type fitting (C). *NOTE: If the drive is used with a reducer, fitting E is not provided.*

W   a. Once every ten days to two weeks, use two or three strokes of a grease gun through fittings A and B at ends of shifting screw and variable-speed shaft, respectively, to lubricate bearings of movable disks. Then, with unit running, shift drive from one extreme speed position to the other to ensure thorough distribution of lubricant over disk-hub bearings.

Q   b. Add two or three shots of grease through fittings D and E to lubricate frame bearing on variable-speed shaft.

Q   c. Every 90 days, add two or three cupfuls of grease to Cup C, which lubricates thrust bearing on constant-speed shaft.

Q   d. Every 90 days, use two or three strokes of grease gun through fittings G and H to lubricate motor frame bearings.

CAUTION: Be sure to follow manufacturer's recommendation on type of grease. After lubricating, wipe excess grease from sheaves and belt.

846   Treatment Plants

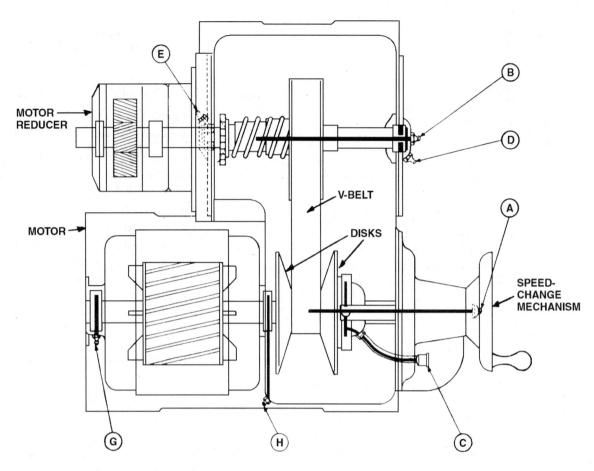

NOTE: A, B, D, E, G, and H are force-feed lubrication fittings. C is a cup-type lubrication fitting.

Fig. 14.52   Reeves varidrive (variable-speed belt drive)
(Source: War Department Technical Manual TM5-666)

QUESTIONS

Write your answers in a notebook and then compare your answers with those on page 872.

14.7I   How can you tell if a belt on belt-drive equipment has proper tension and alignment?

14.7J   Why should sprockets be replaced when replacing a chain in a chain-drive unit?

Paragraph 11: Couplings

Frequency
   of
Service

1. GENERAL. Unless couplings between the driving and driven elements of a pump or any other piece of equipment are kept in proper alignment, breaking and excessive wear results in either or both the driven machinery and the driver. Burned-out bearings, sprung or broken shaft, and excessively worn or ruined gears are some of the damages caused by misalignment. To prevent outages and the expense of installing replacement parts, check the alignment of all equipment before damage occurs.

a. Improper original installation of the equipment may not necessarily be the cause of the trouble. Settling of foundations, heavy floor loadings, warping of bases, excessive bearing wear, and many other factors cause misalignment. A rigid base is not always security against misalignment. The base may have been mounted off level, which could cause it to warp.

b. Flexible couplings permit easy assembly of equipment, but they must be aligned as ex-

Maintenance 847

Frequency of Service	

actly as flanged couplings if maintenance and repair are to be kept to a minimum. Rubber-bushed types cannot function properly if the bolts cannot move in their bushings.

S 2. CHECK COUPLING ALIGNMENT (straightedge method). Excessive bearing and motor temperatures caused by overload, noticeable vibration, or unusual noises may all be warnings of misalignment. Realign when necessary (Figure 14.53) using a straightedge and thickness gauge or wedge. To ensure satisfactory operation, level up to within 0.005 inch (0.13 mm) as follows:

a. Remove coupling pins.

b. Rigidly tighten driven equipment; slightly tighten bolts holding drive.

c. To correct horizontal and vertical misalignment, shift or shim drive to bring coupling halves into position so no light can be seen under a straightedge laid across them. Place straightedge in four positions, holding a light in back of straightedge to help ensure accuracy.

d. Check for angular misalignment with a thickness or feeler gauge inserted at four places to make certain space between coupling halves is equal.

e. If proper alignment has been secured, coupling pins can be put in place easily using only finger pressure. Never hammer pins into place.

f. If equipment is still out of alignment, repeat the procedure.

S 3. CHECK COUPLING ALIGNMENT (dial indicator method). Dial indicators also are used to measure coupling alignment. This method produces better results than the straightedge method. The dial indicates very small movements or distances, which are measured in mils (one mil equals $1/1000$ of an inch). The indicator consists of a dial with a graduated face (with "plus" and "minus" readings), a pedestal, and a rigid indicator bar (or "fixture") as shown in Figure 14.54.

The dial indicator is attached to one coupling at the fixture and adjusted to the zero position or reading. When the shaft of the machine is rotated, misalignment will cause the pedestal to compress (a "plus" reading), or extend (a "minus" reading). Literature provided by the manufacturer of machinery usually will indicate maximum allowable tolerances or movement.

Carefully study the manufacturer's literature provided with your dial indicator before attempting to use the device.

A 4. CHANGE OIL IN FAST COUPLINGS. Drain out old oil and add gear oil to proper level. Correct quantity is given on instruction card supplied with each coupling.

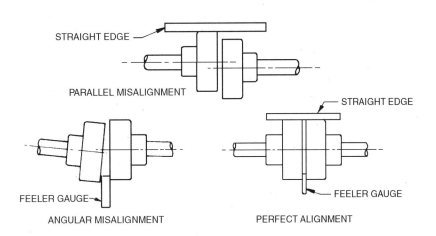

Fig. 14.53 *Testing alignment, straightedge*

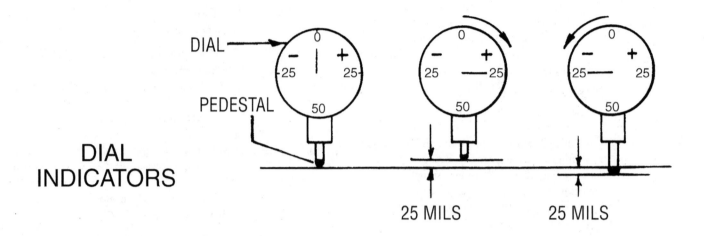

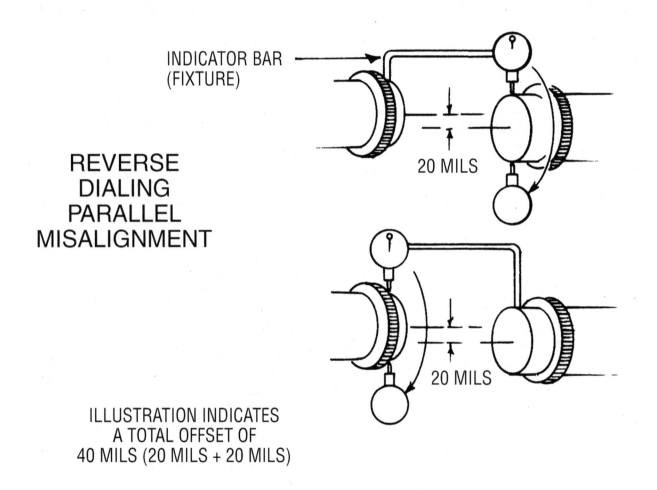

Fig. 14.54 Use of a dial indicator
(Permission of DYMAC, a Division of Spectral Dynamics Corporation)

Paragraph 12: Shear Pins

Many wastewater treatment units use shear pins as protective devices to prevent damage in case of sudden overloads. To serve this purpose, these devices must be in operational condition at all times. Under some operating conditions, shearing surfaces of a shear pin device may freeze together so solidly that an overload fails to break them.

Manufacturers' drawings for particular installations usually specify shear pin material and size. If this information is not available, obtain the information from the manufacturer, giving the model, serial number, and load conditions of unit. When necessary to determine shear pin size, select the lowest strength that does not break under the unit's usual loads. When proper size is determined, never use a pin of greater strength, such as a bolt or a nail.

If necked pins are used, be sure the necked-down portion is properly positioned with respect to shearing surfaces. When a shear pin breaks, determine and remedy the cause of failure before inserting new pin and starting drive in operation.

Frequency of Service		
M	1.	GREASE SHEARING SURFACES.
Q	2.	REMOVE SHEAR PIN. Operate motor for a short time to smooth out any corroded spots.
A	3.	CHECK SPARE INVENTORY. Make sure an adequate supply is on hand, properly identified and with record of proper pin size, necked diameter, and longitudinal dimensions.

QUESTIONS

Write your answers in a notebook and then compare your answers with those on page 872.

14.7K What factors could cause couplings to become out of alignment?

14.7L What is the purpose of shear pins?

Please answer the discussion and review questions next.

DISCUSSION AND REVIEW QUESTIONS

Chapter 14. MAINTENANCE

(Lesson 5 of 7 Lessons)

Write the answers to these questions in your notebook. The question numbering continues from Lesson 4.

26. Why would you use a stethoscope to check an electric motor?

27. How would you determine if a motor is running unusually hot?

28. How would you clean belts and sheaves (pulleys) on a belt drive?

29. Why should you never replace only one belt on a multiple-drive unit?

30. What do rubber wearings near a belt drive indicate?

31. How can you determine if a chain in a chain-drive unit has the proper slack?

32. What happens when couplings are not in proper alignment?

CHAPTER 14. MAINTENANCE

(Lesson 6 of 7 Lessons)

Paragraph 13: Gate Valves (Figures 14.55 and 14.56)

Gate valves require the following common maintenance: oiling, tightening, and replacing the stem stuffing box packing.

Frequency
of
Service

A 1. REPLACE PACKING. Modern gate valves can be repacked without removing them from service. Before repacking, open valve wide. This prevents excessive leakage when the packing or the entire stuffing box is removed. It draws the stem collar tightly against the bonnet on a non-rising stem valve, and tightly against the bonnet bushing on a rising stem valve.

   a. Stuffing box. Remove all old packing from stuffing box with a packing hook. Clean valve stem of all adhering particles and polish it with fine emery cloth. After polishing, remove the fine grit with a clean cloth to which a few drops of oil have been added.

   b. Insert packing. Insert new split-ring packing in stuffing box and tamp it into place with packing gland. Stagger ring splits. After stuffing box is filled, place a few drops of oil on stem, assemble gland, and tighten it down on packing.

S 2. OPERATE VALVE. Operate inactive gate valves to prevent sticking.

A 3. LUBRICATE GEARING. Lubricate gate valves as recommended by manufacturer. Lubricate thoroughly any gearing in large gate valves. Wash open gears with solvent and lubricate with grease.

S 4. LUBRICATE RISING STEM THREADS. Clean threads on rising stem gate valves and lubricate with grease.

A 5. LUBRICATE BURIED VALVES. If a buried valve works hard, lubricate it by pouring oil down through a pipe that is bent at the end to permit oiling the packing follower below the valve nut.

A 6. REFACE LEAKY GATE VALVE SEATS. If gate valve seats leak, reface them immediately, using the method discussed below. A solid wedge disk valve is used for illustration, but the general method also applies to other types of repairable gate valves. Proceed as follows:

   a. Remove bonnet and clean and examine disk and body thoroughly. Carefully determine extent of damage to body rings and disk. If corrosion has caused excessive pitting or eating away of metal, as in guide ribs in body, repairs may be impractical.

   b. Check and service all parts of valve completely. Remove stem from bonnet and examine it for scoring and pitting where packing makes contact. Polish lightly with fine emery cloth to put stem in good condition. Use soft jaws if stem is put in vise.

   c. Remove all old packing and clean out stuffing box. Clean all dirt, scale, and corrosion from inside of valve bonnet and other parts.

   d. Do not salvage an old gasket. Remove it completely and replace with one of proper quality and size.

   e. After cleaning and examining all parts, determine whether valve can be repaired by removing cuts from disk and body seat faces or by replacement of body seats. If repair can be made, set disk in vise with face leveled, wrap fine emery cloth around a flat tool, and rub or lap off entire bearing surface on both sides to a smooth, even finish. Remove as little metal as possible.

   f. Repair cuts and scratches on body rings, lapping with an emery block small enough to permit convenient rubbing all around

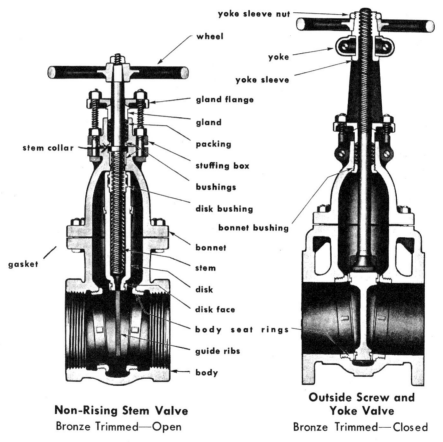

Fig. 14.55 *Wedge gate valve*
(Source: Crane Co.)

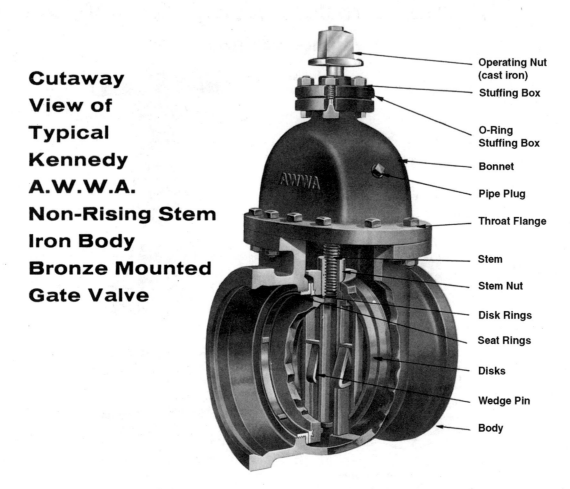

Fig. 14.56 Non-rising stem gate valve
(Permission of Kennedy Valve, Division of ITT Grinnell Valve Co., Inc.)

Frequency of Service	

rings. Work carefully to avoid removing so much metal that disk will seat too low. When seating surfaces of disk and seat rings are properly lapped in, coat faces of disk with *PRUSSIAN BLUE*[18] and drop disk in body to check contact. When good, continuous contact is obtained, the valve is tight and ready for assembly. Insert stem in bonnet, install new packing, assemble other parts, attach disk to stem, and place assembly in body. Raise disk to prevent contact with seats so bonnet can be properly seated on body before tightening the joint.

    g. Test repaired valve before putting it back in line to ensure that repairs have been properly made.

    h. If leaky gate valve seats cannot be refaced, remove and replace seat rings with a power lathe. Chuck up body with rings vertical to lathe and use a strong steel bar across ring lugs to unscrew them. They can be removed by hand with a diamond point chisel if care is taken to avoid damaging threads. Drive new rings home tightly. Use a wrench on a steel bar across lugs when putting in rings by hand. Always coat threads with a good lubricant before putting threads into the valve body. This helps to make the threads easier to remove the next time the seats have to be replaced. Lap in rings to fit disk perfectly.

Paragraph 14: Check Valves (Figure 14.57)

A   1. INSPECT DISK FACING. Open valves to observe condition of facing on swing check valves equipped with leather or rubber seats on disk. If metal seat ring is scarred, dress it with a fine file and lap with fine emery paper wrapped around a flat tool.

A   2. CHECK PIN WEAR. Check pin wear on balanced disk check valve, since disk must be accurately positioned in seat to prevent leakage.

Paragraph 15: Plug Valves (Figures 14.58, 14.59, 14.60, 14.61, and 14.62)

M   1. ADJUST GLAND. The adjustable gland holds the plug against its seats in body and acts through compressible packing, which functions as a thrust cushion. Keep gland tight enough at all times to hold plug in contact with its seat. If this is not done, the lubricant system cannot function properly and solid particles may enter between the body and plug and cause damage.

M   2. LUBRICATE ALL VALVES. Apply lubricant by removing lubricant screw and inserting stick of plug valve lubricant for stated temperature conditions. The check valve fitting within the shank prevents line pressure from blowing out when lubricant screw is removed. Inject lubricant into valve by turning screw down to keep valve in proper operating condition. If lubrication has been neglected, several sticks of lubricant may be needed before lubricant system is refilled to operating condition. Be sure to lubricate valves that are not used often to ensure that they are always in operating condition. Leave lubricant chamber nearly full so extra supply is available by turning screw down. Use lubricant regularly to increase valve efficiency and service, promote easy operation, reduce wear and corrosion, and seal valve against internal leakage.

Paragraph 16: Sluice Gates (Figure 14.63)

There are two general types of light-duty sluice gates: those that seat with the pressure, and those that seat against the pressure. Both are maintained similarly. Heavy-duty sluice gates (Figure 14.63) can seat under seating and unseating pressure.

M   1. TEST FOR PROPER OPERATION. Operate inactive sluice gates. Oil or grease stem threads.

A   2. CLEAN AND PAINT. Clean sluice gate with wire brush and paint with proper corrosion-resistant paint.

A   3. ADJUST FOR PROPER CLEARANCE. For gates seating against pressure, check and adjust top, bottom, and side wedges until in closed position each wedge applies nearly uniform pressure against gate (Figure 14.64).

QUESTION

Write your answer in a notebook and then compare your answer with the one on page 872.

14.7M   What maintenance is required by:

1. Gate valves?
2. Sluice gates?

---

[18] *Prussian Blue.* A blue paste or liquid (often on a paper like carbon paper) used to show a contact area. Used to determine if gate valve seats fit properly.

 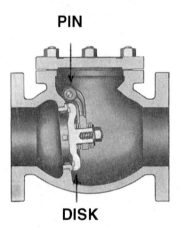

Fig. 14.57 Check valves
(Source: Crane Company)

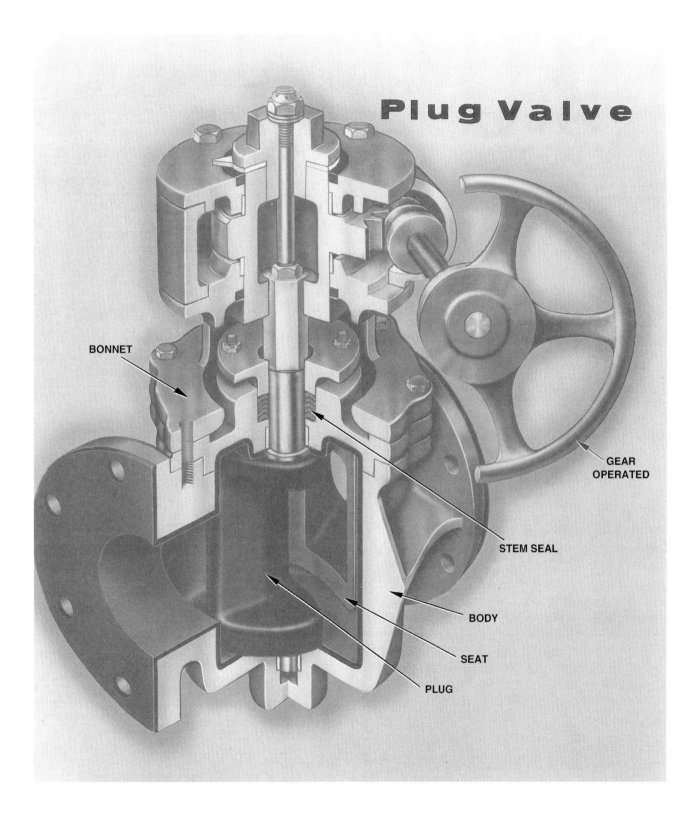

Fig. 14.58 Eccentric plug valve
(Permission of DeZurik, a Unit of General Signal)

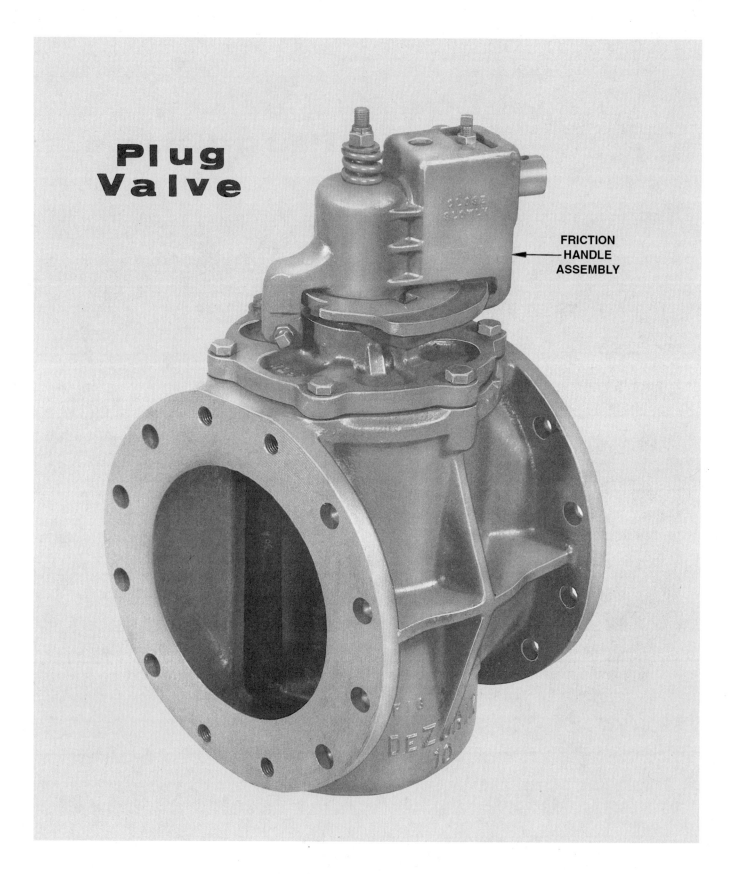

Fig. 14.59 Plug valve, lever operated
(Permission of DeZurik, a Unit of General Signal)

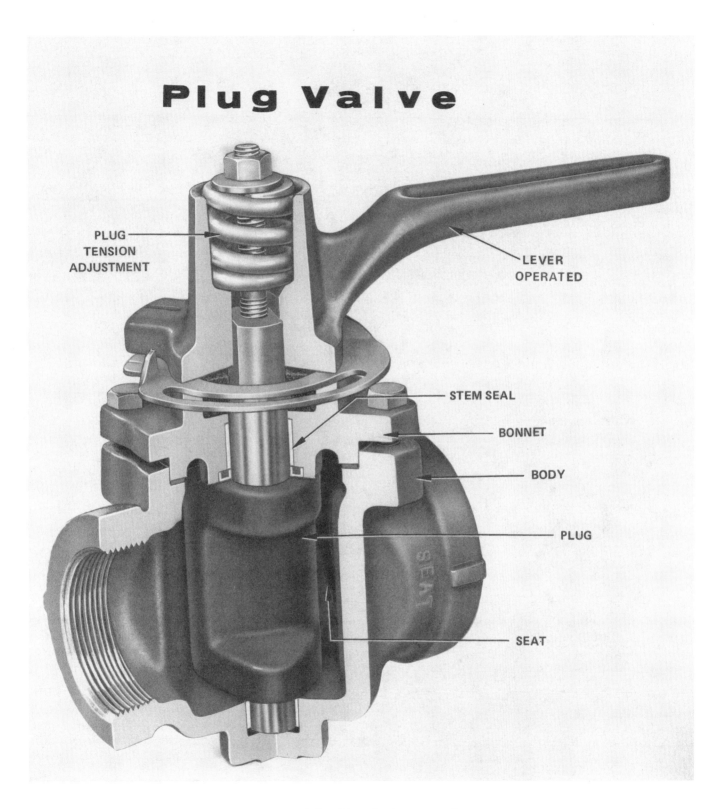

Fig. 14.60 Plug valve, lever operated
(Permission of DeZurik, a Unit of General Signal)

858 Treatment Plants

Fig. 14.61   Plug valve, gear operated
(Permission of DeZurik, a Unit of General Signal)

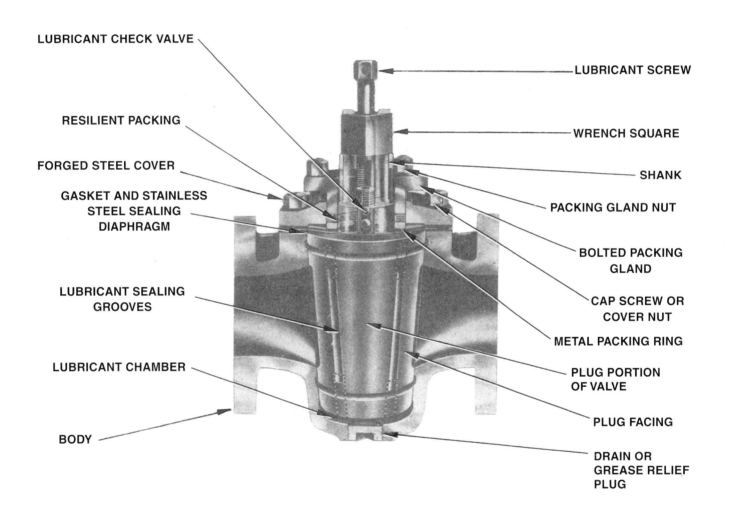

Fig. 14.62 Plug valve
(Source: War Department Technical Manual TM5-666)

Fig. 14.63 Heavy-duty sluice gate
(Permission of ARMCO)

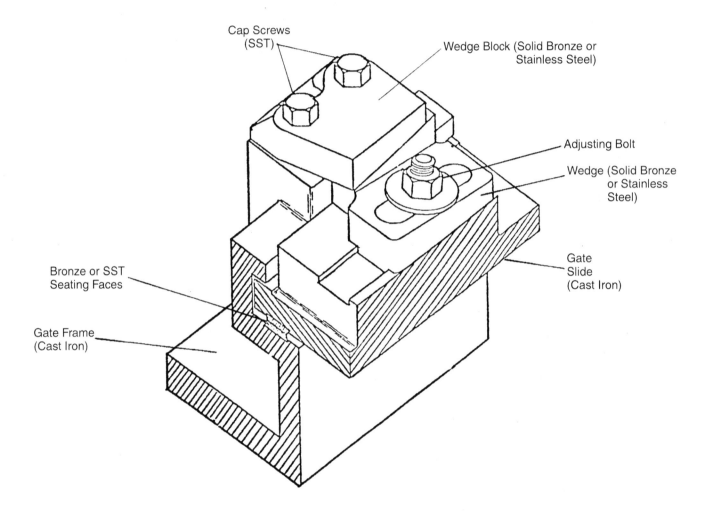

Fig. 14.64 Adjustment of sluice gate side wedges
(Permission of ARMCO)

Paragraph 17: Dehumidifiers

The job of the dehumidifier is to remove moisture from the surrounding atmosphere. Moisture can accumulate in locations that are below grade since the surrounding air temperature will remain relatively constant. This is especially true of pump stations. In the summertime, warm, humid air will enter the pump stations and, unless the air is dehumidified, moisture will form on pump controls, circuits, and other essential equipment. This moisture will cause malfunctions if allowed to exist, due to corrosion and oxidation.

Frequency of Service		
D	1.	Check dehumidifier performance by noting condensation on the walls of the station. If condensation continues, set the dehumidifier to a drier setting. Inspect dehumidifier coils. If frost is on coils, turn unit off until it defrosts.
S	2.	To lube dehumidifier, disassemble the unit and clean it thoroughly, including the drip pan and drain hose. Lube the fan with a few drops of oil. After the unit is reassembled and installed, check condensation buildup daily.
Q	3.	Be sure that the drain remains clear, since buildup of water will make the unit operate less efficiently.

Paragraph 18: Air Gap Separation Systems

An air gap separation system provides a physical break between the wastewater treatment plant's municipal or freshwater supply (well) and the plant's treatment process systems. The purpose of this system is to protect the potable (drinking) water supply in case wastewater backs up from a treatment process. For example, wastewater could travel back through a pump seal and cause contamination of a drinking water system.

A	1.	Installation of air gap systems is controlled by health regulations and periodically should be inspected by the local health department or the public water supply agency.
W	2.	Precautions must be taken to ensure that the air gap separation from the discharge of the water supply pipe is at least two pipe diameters above the rim of the air gap tank (see Figure 14.65). This separation prevents water from being drawn back down into the main water supply line under any circumstances because the water can never reach the elevation of the discharge pipe above the rim of the tank. Wind guards may be necessary around the rim to prevent water loss or damage by wind spray. The device must be easily accessible for weekly inspections.

An examination of Figure 14.65 shows how the plant receives its water supply in the receiving tank. If a plant power outage or other occurrence causes negative pressures in the plant water lines, it would be physically impossible for any contaminated water to enter the drinking water system.

3. Preventive maintenance should include the following regularly scheduled items:

D	a.	Pump and motor maintenance
W	b.	Servicing of float and control valve
A	c.	Periodic draining and cleaning of air gap tank
W	d.	Routine operational inspections should be conducted at least once a week

Paragraph 19: Plant Safety Equipment

Plant safety equipment must be maintained on a regular basis so it will always be ready for use when needed.

After use or W	1.	AIR BLOWERS (for ventilation of confined space). Examine carefully each time used. Inspect hose carefully. If used infrequently, check weekly.
After use or W	2.	SAFETY HARNESS. Examine carefully each time used. Inspect stitching and rings.
W	3.	PROTECTIVE CLOTHING. Inspect for rips or holes.
W	4.	GAS DETECTORS (PORTABLE). Examine carefully before each use. Ensure battery is charged. Calibrate using known standard gas concentrations obtained from detector manufacturer.
M	5.	FIRST-AID KITS. Inventory contents. Replace contents whenever used or contents discovered missing during inventory.
M	6.	FIRE EXTINGUISHERS. Inspect and inventory. Check extinguishers for adequate pressure, pin retainers, and service dates.
A		An authorized service representative must check/recharge and recertify portable extinguishers.
M	7.	Inspect all emergency respiratory protective equipment.
M	8.	Inspect and flush all emergency eye wash stations and deluge showers.
M	9.	Check face velocities of laboratory fume hoods.

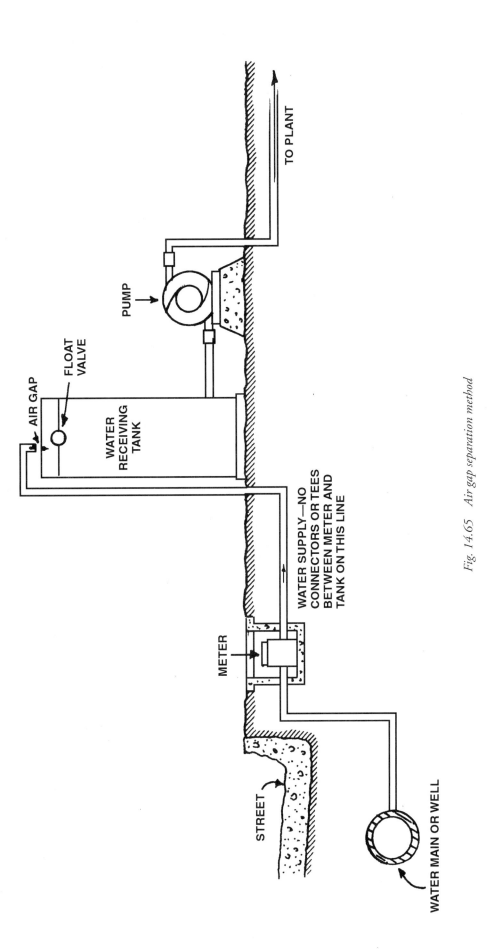

Fig. 14.65 Air gap separation method

Acknowledgment

Major portions and basic concepts in this section on mechanical maintenance are from the War Department Technical Manual, TM5-666, "Inspections and Preventive Maintenance Services, Sewage Treatment Plants and Sewer Systems at Fixed Installations," War Department.

QUESTIONS

Write your answers in a notebook and then compare your answers with those on page 873.

14.7N What is the job of the dehumidifier?

14.7O What happens if the dehumidifier does not do its job?

14.7P What is the purpose of an air gap separation system?

14.7Q How would you maintain a portable gas detector?

END OF LESSON 6 OF 7 LESSONS

on

MAINTENANCE

Please answer the discussion and review questions next.

DISCUSSION AND REVIEW QUESTIONS

Chapter 14. MAINTENANCE

(Lesson 6 of 7 Lessons)

Write the answers to these questions in your notebook. The question numbering continues from Lesson 5.

33. Why should inactive gate valves be operated periodically?

34. Why should plug valves that are not used very often be lubricated regularly?

35. Why is regular maintenance of plant safety equipment necessary?

CHAPTER 14. MAINTENANCE

(Lesson 7 of 7 Lessons)

14.8 UNPLUGGING PIPES, PUMPS, AND VALVES

14.80 Plugged Pipelines

Plugged pipelines are encountered in lines transporting scum, raw sludge, digested sludge, or grit. The frequency of a particular line plugging depends on the type of material passing through the line, the construction material of the line, the type of pumps or system used to move the material, and the routine maintenance performed on the line. This section outlines the preventive maintenance measures to reduce plugging problems in the different lines in a wastewater treatment plant and the methods of unplugging pipes, pumps, and valves.

14.81 Scum Lines

Scum will cause more problems in pipelines than any other substance pumped in a wastewater treatment plant. Problems are more frequent and more severe in colder weather when grease tends to harden quickly.

Preventive maintenance includes:

1. Hose down scum troughs, hoppers, and flush lines to scum box at least every two hours when an operator is on duty and problems are occurring.

2. Clean lines monthly using:

    a. Rods equipped with cutters

    b. High-pressure hydraulic pipe cleaning units

    c. Steam cleaning units

    d. Chemicals such as "Sanfax" or "HotRod" (strong hydroxides). This method is least desirable because of costs and the possibility that the chemicals could be harmful to biological treatment processes. *CAUTION: Study appropriate Material Safety Data Sheets (MSDSs) and follow all safety requirements.*

14.82 Sludge Lines

Sludge lines will plug more often when scum and raw sludge are pumped through the same line, or when stormwaters carry in grit and silt that are not effectively removed by the grit removal facilities.

Preventive maintenance includes:

1. Flush lines monthly with plant effluent or wastewater.

2. If possible, recirculate warm, digested sludge through the line for an hour each week if grease tends to build up on pipe walls.

3. Rod or high-pressure clean lines monthly or quarterly, depending on severity of problem.

4. If possible, force cleaning tool (pig) through line using pressures produced by pump. Line must be equipped with valves and wyes to insert and remove pig. Pumps must be located to allow pig to be forced through the line. A plastic bag full of ice cubes makes an excellent cleaning tool or pig. Force the bag down the line with hot water. If the line plugs, the ice will melt to the point where the bag will continue down the line.

14.83 Digested Sludge Lines

Problems develop in digested sludge lines of small plants from infrequent use, ineffective grit removal, and failure to remove sludge from the line after withdrawing sludge to a drying bed.

Preventive maintenance includes checking:

1. Condition of pipeline for wear or obstructions, such as sticks and rags.

2. Pump impellers for wear. A worn impeller will not maintain desired velocity and pressure in the line.

14.84 Unplugging Pipelines

Selection of a method to unplug a pipe depends on the location of the blockage and access to the plugged line. Pressure methods and cutting tools are the most common techniques used to clear stopped lines.

14.840 Pressure Methods

REQUIREMENTS:

1. Must be able to valve off or plug one end of pipeline in order to move obstruction or blockage down the line and out other end to a free discharge.

2. Pressure may be developed using water or air pressure. Maximum available pressures are usually less than 80 psi (5.6 kg/sq cm).

3. Pipeline must have tap and control valves to control applied water or air pressure.

PRECAUTIONS:

1. Never use water connected to a domestic water supply because you may contaminate the water supply.
2. Do not exceed pipeline design pressures, usually 125 psi (8.8 kg/sq cm).
3. Never attempt to use a positive displacement pump by overriding the safety cut-out pressure switches. This practice may damage the pump.

PROCEDURE:

1. Plug or valve off one end of pipe, but leave other end open. For example, (1) close valve to digester but open line to the drying beds, or a raw sludge line, or (2) close suction valve on raw sludge pump, and open pipe back to primary clarifier hopper.
2. Connect hose from pressure supply to tap and valve on pipeline as close as possible to the plugged or valved-off end.
3. Apply pressure to supply hose and then slowly open control tap valve and allow pressure to build up until obstruction is moved.

DO NOT EXCEED PIPELINE WORKING DESIGN PRESSURE.

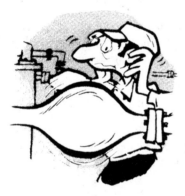

14.841 Cutting Tools

Cutting tools are usually available from sewer maintenance crews and may consist of hand rods, power rods, or snakes, which are capable of cutting or breaking up material causing a stoppage.

REQUIREMENTS:

1. One end of the line must be open and reasonably accessible.
2. Cutting tools should be able to remove material causing stoppage when line is cleared.

LIMITATIONS:

1. Most of these units cannot clean lines with sharp bends or pass through some of the common types of plug valves used in sludge lines.
2. A 4-inch (10-cm) cutter tool may have to be used on a 6-inch (15-cm) line due to 90-degree bends.
3. A part of the line may have to be dismantled to use a cutting tool.
4. Rods are difficult to hand push over 300 feet (90 m). The operator must have firm footing and room to work.

HAND RODS:

1. Use sufficient sections to clean full length of line.
2. Insert cutter in the open end of the pipeline and twist rods as they are pushed up the line.
3. If rods start to twist up due to torque, pull back and let rod unwind.

POWER RODS:

1. Power drive unit must be located over plugged line. Do not attempt to run 40 feet (12 m) across a clarifier and then into sludge line.
2. Do not run rods into line too fast. You may hit an obstruction or valve and break the cutter off of the rods. They will be very difficult to recover.

14.842 High-Velocity Pressure Units

This unit is very good for removing grease, sludge, or grit from pipelines.

PROCEDURE:

1. Insert nozzle and hose 3 feet (1 m) into line.
2. Increase pressure in cleaning system to 600 to 1,000 psi (42 to 70 kg/sq cm) and slowly unreel the hose into the pipeline.
3. Keep track of how much hose is in the line to prevent the nozzle from attempting to go through an open valve. The nozzle and hose may catch on the valve and require taking apart the valve to free the nozzle.
4. Run water through nozzle while reeling in hose.

14.843 Last Resort

If the methods described in this section fail, the only solution is to attempt to locate the position of the stoppage, drain the line, take apart the plugged section of pipe, and remove the obstruction.

14.85 Plugged Pumps and Valves

Isolate the plugged pump or valve from the remainder of the treatment plant by valving off the plugged section and tagging and locking out the power supply to the pump. Remove the pump inspection plate or dismantle the valve and remove the material causing the blockage. When removing the pump inspection plate, loosen the bolts and allow to drain BEFORE removing the bolts in case any of the pipes have not been properly valved off. Exercise caution when removing materials to avoid damaging the pump or valve.

QUESTIONS

Write your answers in a notebook and then compare your answers with those on page 873.

14.8A What methods are available for clearing plugged pipelines?

14.8B How would you clear a plugged pump?

14.9 REVIEW OF PLANS AND SPECIFICATIONS

This section covers pumps and lift stations. Many large municipal agencies do their own design work for pumps and lift stations. Smaller agencies usually rely on a consulting engineering firm for design. In either case, operators should be given the opportunity to review the prints and specifications of a new lift station or pumping facility before the award of a contract for construction and installation. This review is very important to be sure adequate provisions have been made for the station pumps, equipment, and instrumentation to be easily and properly operated and maintained.

14.90 Examining Prints

When examining the prints, operators should look for accessibility not only for equipment, but for operators to get to the facilities. Is there sufficient space for vehicles to park and not restrict vehicles passing on streets or pedestrians on sidewalks? Is there room to use hydrolifts, cranes, and high-velocity cleaners, as needed, at the lift station? Are overhead clearances of power lines, trees, and roofs adequate for a crane to remove large equipment? Is there sufficient room to set up portable pumping units or other necessary equipment in cases of major station failures or disasters? Are floors sloped to provide drainage, where needed, and are drains located in low spots? Are station doors and access hatches large enough to remove the largest piece of equipment? Are lifting eyes or overhead rails available, where needed, in the structure? Has sufficient overhead and work room been provided around equipment and control panels to work safely? Is lighting inside and outside the facility provided and is it adequate? Does the alarm system signal high water levels in the wet well and water on the floor of the dry well? Is there sufficient water at a high enough pressure to adequately wash down the wet well? Are there any low-hanging projections or pipes, or other hazards, such as unprotected holes, or unsafe stairs or platforms? Is there access to the wet well? If you have to clean out incoming lines, it may be necessary to put a temporary pump in the wet well. All of these questions must be answered satisfactorily if the lift station or pumping facility is to be easily operated and maintained.

Equipment should be laid out orderly with sufficient work room and access to valves and other station equipment, controls, wiring, pipes, and valves. If there is any possibility of future growth and the station may be enlarged, be sure provisions are made to allow for pumping units to be changed or for the installation of additional pumping units. If additional pumping units will be necessary, be sure spools and valves are built now for ease of expansion. Be sure there is sufficient room to add electrical switchgear for future units. If stationary standby power units are not provided, make certain there are external connections and transfer switches for a portable generator.

14.91 Reading Specifications

Plans and specifications should be carefully reviewed before construction and installation because changes are easily made on paper but are much more difficult in the field.

Review the specifications for the acceptability of the equipment, piping, electrical system, instrumentation, and auxiliary equipment. Determine if the equipment is familiar to your agency and if its reliability has been proven. Find out what warranties, guarantees, and operation and maintenance aids will be provided with the equipment and the lift station. Require a list of names, addresses, and phone numbers of persons to contact in case help is needed regarding supplies or equipment during start-up and shakedown runs. Be sure that the equipment brochures and other information apply to the equipment supplied. Sometimes new models are installed and you are provided with old brochures.

Be sure the painting and the color coding on pipes and electric circuits meet with your agency's practice. Try to standardize electrical equipment and components as much as possible so one manufacturer cannot blame the other when problems develop. Standardization also can help to reduce the inventory of spare parts necessary for replacement. A few hours spent reviewing plans and specifications will save many days of hard and discouraging labor in the future when it is a major job to make a change.

QUESTIONS

Write your answers in a notebook and then compare your answers with those on page 873.

14.9A Why should operators be given an opportunity to review the prints and specifications of a new lift station before the award of a construction contract?

14.9B Why is accessibility to pumping facilities important?

14.9C Why should plans and specifications be carefully reviewed?

14.9D What information should be determined for equipment to be installed in a pumping facility?

14.10 SUMMARY

1. Establish and follow a regular maintenance program.

2. Thoroughly read and understand manufacturers' maintenance instructions. Ask for assistance if you do not understand them. Follow the manufacturers' instructions in your maintenance program.

3. Critically evaluate the maintenance and repair capabilities of yourself and your facilities. Request the help of an expert, when necessary.

14.11 ADDITIONAL READING

1. *MOP 11*, Chapter 12,* "Maintenance." New edition of MOP 11, Volume I, *MANAGEMENT AND SUPPORT SYSTEMS.*

2. *NEW YORK MANUAL*, Chapter 12,* "Maintenance of Plant and Equipment."

3. *TEXAS MANUAL*, Chapter 7,* "Lift Stations and Sewage Pumps," Chapter 8,* "Measurement of Wastewater Flow," and Chapter 24,* "Plant Maintenance."

4. *OPERATION AND MAINTENANCE OF WASTEWATER COLLECTION SYSTEMS*, Kenneth D. Kerri and John Brady. Obtain from the Office of Water Programs, California State University, Sacramento, 6000 J Street, Sacramento, CA 95819-6025. See Chapter 6, "Pipeline Cleaning and Maintenance Methods," Chapter 8, "Lift Stations," and Chapter 9, "Equipment Maintenance," Volumes I and II, $45.00 each.

5. *PLANNED MANAGEMENT SYSTEMS FOR MUNICIPAL WASTEWATER TREATMENT PLANTS*, US Environmental Protection Agency. Obtain from National Information Technical Information Service (NTIS), 5285 Port Royal Road, Springfield, VA 22161. Order No. PB-233111. EPA No. 600-2-73-004. Price, $42.50, plus $5.00 shipping and handling per order.

6. *MAINTENANCE MANAGEMENT SYSTEMS FOR MUNICIPAL WASTEWATER FACILITIES*, US Environmental Protection Agency. Obtain from National Information Technical Information Service (NTIS), 5285 Port Royal Road, Springfield, VA 22161. Order No. PB-256611. EPA No. 430-9-74-004. Price, $42.50, plus $5.00 shipping and handling per order.

7. "Blue Plains Manages Maintenance" (implementation of a computerized system), *OPERATIONS FORUM*, November 1987.

8. *PUMP HANDBOOK*, Third Edition, edited by Igor Karassik, Joseph Messina, Paul Cooper, and Charles Heald. Obtain from the McGraw-Hill Companies, Order Services, PO Box 182604, Columbus, OH 43272-3031. ISBN 0-07-034032-3. Price, $135.00, plus nine percent of order total for shipping and handling.

* Depends on edition.

Please answer the discussion and review questions next.

DISCUSSION AND REVIEW QUESTIONS

Chapter 14.   MAINTENANCE

(Lesson 7 of 7 Lessons)

Write the answers to these questions in your notebook. The question numbering continues from Lesson 6.

36. How can scum lines be kept from plugging?

37. Which methods can be used by operators to unplug pipelines?

38. Why should operators review plans and specifications for a new facility?

SUGGESTED ANSWERS
Chapter 14. MAINTENANCE

ANSWERS TO QUESTIONS IN LESSON 1

Answers to questions on page 750.

14.0A A good maintenance program is essential for a wastewater treatment plant to operate continuously at peak design efficiency.

14.0B The most important item is maintenance of the mechanical equipment—pumps, valves, scrapers, and other moving equipment. Other items include plant grounds, buildings, and structures.

14.0C A good recordkeeping system tells when maintenance is due and also provides a record of equipment performance. Poor performance is a good justification for replacement or new equipment. Good records help keep your warranty in force.

14.0D An equipment service card tells what service or inspection work should be done on a piece of equipment and when, while the service record card is a record of what was done and when.

Answers to questions on page 752.

14.0E A building maintenance program will keep the building in good shape and includes painting, when necessary. Attention also must be given to electrical systems, plumbing, heating, cooling, ventilating, floors, windows, roofs, and drainage around the buildings. The building should be kept clean, tools should be stored in their proper place, and essential storage should be available. In each plant building, periodically check all stairways, ladders, catwalks, and platforms for adequate lighting, head clearance, and sturdy and convenient guardrails.

14.0F When plant tanks and channels are drained, the operator should check surfaces for wear and deterioration from wastewater or fumes. Protective coatings should be applied where necessary to prevent further damage.

14.0G Well-groomed and neat grounds are important because many people judge the ability of the operator and the plant performance on the basis of the appearance of the plant.

ANSWERS TO QUESTIONS IN LESSON 2

Answers to questions on page 775.

14.1A Pumps must be lubricated in accordance with the manufacturers' recommendations. Quality lubricants should be used.

14.1B A pump with a mechanical seal (or any other type of seal) should never be run dry because the seal faces will be burned and ruined.

14.1C Mechanical seals are not supposed to have any leakage from the gland; if a leak develops, the seal may require resurfacing or it may have to be replaced.

14.1D In lubricating motors, too much grease may cause bearing trouble or damage the winding.

Answers to questions on page 781.

14.1E Wet wells use variable-speed pumping equipment because of lower building costs and the smaller (or non-existent) wet well. A variable-speed pump facility can maintain a predetermined level of flow in the incoming line during minimal flows in dry weather as well as during maximum flows in the peak wet weather season.

14.1F A standard, three-phase, single-speed, synchronous AC motor requires a preventive maintenance program consisting primarily of lubrication, cleaning, and testing of electric circuits. The slip rings and brushes of a variable-speed motor require additional preventive maintenance procedures.

14.1G The rate of brush wear varies with the brush pressure. At light pressures, electrical wear is dominant because the brush can jump off the rings, sparking occurs, and the filming action on the ring becomes erratic. At higher pressures, mechanical wear is dominant because of high friction losses, needless heating, and needless abrasion.

14.1H Indications of unsatisfactory performance of AC slip rings and brushes are those indicators that appear at the brushes, at the ring surface, and as heating.

Answers to questions on page 782.

14.1I If a pump will not start, check for blown fuses or tripped circuit breakers and the cause. Also, check for a loose connection, fuse, or thermal unit.

14.1J To increase the rate of discharge from a pump, you should look for something causing the reduced rate of discharge, such as pumping air, motor malfunction, plugged lines or valves, impeller problems, or other factors.

Answers to questions on page 784.

14.1K If a pump that has been locked or tagged out for maintenance or repairs is started, an operator working on the pump could be seriously injured and also equipment could be damaged.

14.1L Normally, a centrifugal pump should be started after the discharge valve is opened. Exceptions are treatment processes or piping systems with vacuums or pressures that cannot be dropped or allowed to fluctuate greatly while an alternate pump is put on the line.

Answers to questions on page 786.

14.1M Before stopping an operating pump:

1. Start another pump (if appropriate).
2. Inspect the operating pump by looking for developing problems, required adjustments, and problem conditions of the unit.

14.1N A pump shaft or motor will spin backward if wastewater being pumped flows back through the pump when the pump is shut off. This will occur if there is a faulty check valve or foot valve in the system.

14.1O The position of all valves should be checked before starting a pump to ensure that the wastewater being pumped will go where intended.

Answers to questions on page 786.

14.1P The most important rule regarding the operation of positive displacement pumps is to *NEVER* operate the pump against a closed valve, especially a discharge valve.

14.1Q If a positive displacement pump is started against a closed discharge valve, the pipe, valve, or pump could rupture from excessive pressure. The rupture will damage equipment and possibly seriously injure or kill someone standing nearby.

14.1R Both ends of a sludge line should never be closed tight because gas from decomposition can build up and rupture pipes or valves.

ANSWERS TO QUESTIONS IN LESSON 3

Answers to questions on page 788.

14.2A Unqualified or inexperienced people must be extremely careful when attempting to troubleshoot or repair electrical equipment because they can be seriously injured and damage costly equipment if a mistake is made.

14.2B When machinery is not shut off, locked out, and tagged properly, the following accidents could occur:

1. Maintenance operator could be cleaning pump and have it start, thus losing an arm, hand, or finger.
2. Electrical motors or controls not properly grounded could lead to possible severe shock, paralysis, or death.
3. Improper circuits created by mistakes, such as wrong connections, bypassed safety devices, wrong fuses, or improper wire, can cause fires or injuries due to incorrect operation of machinery.

Answers to questions on page 791.

14.3A The proper voltage and allowable current in amps for a piece of equipment can be determined by reading the nameplate information or the instruction manual for the equipment.

14.3B The two types of voltage are Direct Current (DC) and Alternating Current (AC).

14.3C Amperage is a measurement of current or electron flow and is an indication of work being done or "how hard the electricity is working."

Answers to questions on page 795.

14.3D You test for voltage by using a multimeter.

14.3E A multimeter can be used to test for voltage, open circuits, blown fuses, single phasing of motors, and grounds.

14.3F Before attempting to change fuses, turn off power and check both power lines for voltage. Use a fuse puller.

14.3G If the voltage is unknown and the multimeter has different scales that are manually set, always start with the highest voltage range and work down. Otherwise, the multimeter could be damaged.

14.3H Amp readings different from the nameplate rating could be caused by low voltage, bad bearings, poor connections, plugging, or excessive load.

14.3I Motors and wirings should be megged at least once a year, and twice a year, if possible.

14.3J An ohmmeter is used to test the control circuit components, such as coils, fuses, relays, resistors, and switches.

Answers to questions on page 802.

14.3K The most common pump driver used in pump stations is the AC induction motor.

14.3L Types of pump station failures that can occur due to lack of proper maintenance include:

1. Current unbalances, which ultimately result in motor failure.
2. Loose contacts or terminals in control and power circuits causing high-resistance contacts.
3. Overheating resulting in arcing, fire, and electrical system damage.
4. Dirty enclosures and components, which allow a conductive path to build up between incoming phases causing phase-to-phase shorts.
5. Corrosion that causes high-resistance contacts and heating.

14.3M If a fuse or circuit breaker blows or trips, the source of the problem should be investigated, identified, and corrected.

14.3N The two types of safety devices in main electrical panels or control units are fuses or circuit breakers.

14.3O Fuses are used to protect operators, wiring, circuits, heaters, motors, and various other electrical equipment.

14.3P A fuse must never be bypassed or jumped because the fuse is the only protection the circuit has. Without it, serious damage to equipment and possible injury to operators can occur.

14.3Q Annual maintenance that should be performed with regard to fuses includes:

1. Inspect bolted connections at the fuse clip or fuse holder for signs of looseness.
2. Check connections for any evidence of corrosion from moisture or atmosphere (air pollution).
3. Tighten connections.
4. Check fuse for obvious overheating.
5. Inspect insulation on the conductors coming into the fuses on the line side and out of the fuses on the load side for evidence of discoloration or bubbling, which would indicate overheating of the conductors.

14.3R A circuit breaker is a switch that is opened automatically when the current or the voltage exceeds or falls below a certain limit. Unlike a fuse that has to be replaced each time it blows, a circuit breaker can be reset after a short delay to allow time for cooling.

Answers to questions on page 803.

14.3S The motor and supervisory control systems are composed of the auxiliary electrical equipment, such as relays, transformers, lighting panels, pump control logic, alarms, and other electrical equipment typically found in a pump station electrical system.

14.3T The three basic factors that contribute to the reliable operation of electrical systems found in pump stations include:

1. An adequate preventive maintenance program must be implemented.
2. A knowledge of the system by the operator, even though the operator does not perform the actual maintenance.
3. Adherence to three principles: KEEP IT CLEAN! KEEP IT DRY! KEEP IT TIGHT!

Answers to questions on page 820.

14.4A The two types of induction motor construction that are typically encountered when dealing with AC induction motors are:

1. Squirrel cage induction motor
2. Wound rotor induction motor

14.4B The five most common causes of electric motor failure are (1) overload (thermal), (2) contaminants, (3) single phasing, (4) bearing failures, and (5) old age.

14.4C The three components of an insulation temperature rating consist of:

1. Ambient operating temperature
2. Temperature rise
3. Hot-spot allowance

14.4D Motor starters can be either manually or automatically controlled.

14.4E Magnetic starters are usually used to start pumps, compressors, blowers, and anything where automatic or remote control is desired.

14.4F The two types of overload protection devices used with magnetic starters are:

1. A bimetallic strip that is precisely calibrated to open under higher temperature conditions to de-energize the coil.
2. A small solder pot within the coil that melts because of the heat and will de-energize the system.

14.4G The direction of rotation of a three-phase motor can be changed by changing any two of the power leads.

Answers to questions on page 826.

14.4H The key to effective troubleshooting is practical, step-by-step procedures combined with a common-sense approach.

14.4I When troubleshooting magnetic starters:

1. Gather preliminary information
2. Inspect:
   a. Contacts
   b. Mechanical parts
   c. Magnetic parts

14.5A Types of information that should be recorded regarding electrical equipment include every:

1. Change
2. Repair
3. Test

ANSWERS TO QUESTIONS IN LESSON 4

Answers to questions on page 835.

14.7A A cross connection is a connection between a drinking (potable) water system and an unapproved water supply. For example, if you have a pump moving nonpotable water and hook into the drinking water system to supply water for the pump seal, a cross connection or mixing between the two water systems can occur. This mixing may lead to contamination of the drinking water.

14.7B Yes. A slight water-seal leakage is desirable when the pumps are running to keep the packing cool and in good condition.

14.7C To measure the capacity of a pump, measure the volume pumped during a specific time period.

$$\text{Capacity, GPM} = \frac{\text{Volume, gallons}}{\text{Time, minutes}}$$

or

$$\text{Capacity, } \frac{\text{liters}}{\text{sec}} = \frac{\text{Volume, liters}}{\text{Time, sec}}$$

14.7D Estimate the capacity of a pump (in GPM) if it lowers the water in a 10-foot wide × 15-foot long wet well 1.7 feet in five minutes.

$$\text{Capacity, GPM} = \frac{\text{Volume, gallons}}{\text{Time, minutes}}$$

$$= \frac{10 \text{ ft} \times 15 \text{ ft} \times 1.7 \text{ ft} \times 7.5 \text{ gal/cu ft}}{5 \text{ minutes}}$$

$$= 382.5 \text{ GPM}$$

or

$$\text{Capacity, } \frac{\text{liters}}{\text{sec}} = \frac{\text{Volume, liters}}{\text{Time, sec}}$$

$$= \frac{3 \text{ m} \times 5 \text{ m} \times 0.5 \text{ m} \times 1{,}000 \text{ } L/m^3}{5 \text{ minutes} \times 60 \text{ sec/min}}$$

$$= 25 \text{ liters/sec}$$

14.7E When shutting down a pump for a long period, open the motor disconnect switch; shut all valves on suction, discharge, water-seal, and priming lines; drain the pump completely by removing the vent and drain plugs. This procedure protects the pump against corrosion, sedimentation, and freezing.

Answers to questions on page 840.

14.7F Shear pins commonly fail in reciprocating pumps because of (1) a solid object lodged under piston, (2) a clogged discharge line, or (3) a stuck or wedged valve.

14.7G A noise may develop when pumping thin sludge due to water hammer, but will disappear when heavy sludge is pumped.

ANSWERS TO QUESTIONS IN LESSON 5

Answer to question on page 843.

14.7H When checking an electric motor, the following items should be checked periodically, as well as when trouble develops:

1. Motor condition
2. Note all unusual conditions
3. Lubrication of bearings
4. Listen to motor
5. Check temperature
6. Set dateometer

Answers to questions on page 846.

14.7I A properly adjusted horizontal belt has a slight bow in the slack side when running. When idle, it has an alive springiness when thumped with the hand. To check for proper alignment, place a straightedge or string against the pulley face or faces and allow for differences in dimensions from centerlines of grooves to outside faces of the pulleys being aligned.

14.7J Always replace sprockets when replacing a chain because old, out-of-pitch sprockets cause as much chain wear in a few hours as years of normal operation.

Answers to questions on page 849.

14.7K Improper original installation of equipment, settling of foundations, heavy floor loadings, warping of bases, and excessive bearing wear could cause couplings to become out of alignment.

14.7L Shear pins are designed to fail if a sudden overload occurs that could damage expensive equipment.

ANSWERS TO QUESTIONS IN LESSON 6

Answer to question on page 853.

14.7M Maintenance required by (1) gate valves is oiling, tightening, and replacing the stem stuffing box packing. Maintenance required by (2) sluice gates is testing for proper operation, cleaning and painting, and adjusting for proper clearance.

Answers to questions on page 864.

14.7N The job of the dehumidifier is to remove moisture from the atmosphere.

14.7O If the dehumidifier does not do its job, moisture will form on pump controls, circuits, and other essential equipment and cause malfunctions due to corrosion and oxidation.

14.7P The purpose of an air gap separation system is to protect the potable (drinking) water supply in case wastewater backs up from a treatment process.

14.7Q To maintain a portable gas detector:

1. Be sure battery is charged
2. Calibrate using known standard gas concentrations obtained from detector manufacturer

ANSWERS TO QUESTIONS IN LESSON 7

Answers to questions on page 867.

14.8A Plugged pipelines may be cleared by the use of pressure methods, cutting tools, high-velocity pressure units and, as a last resort, dismantling the plugged section and removing the obstruction.

14.8B To clear a plugged pump, isolate the pump from the remainder of the plant by valving off the plugged section and tagging and locking out the power supply to the pump. Loosen the pump inspection plate, allow the lines to drain, and remove the material causing the blockage.

Answers to questions on page 867.

14.9A Operators should review prints and specifications of a new lift station before construction to be sure adequate provisions have been made for the facility to be easily and properly operated and maintained.

14.9B Accessibility to a pumping facility is important not only for equipment but also because operators need to be able to easily reach the facility and have room to park vehicles and work.

14.9C Plans and specifications should be carefully reviewed before construction and installation because changes are easily made on paper but are much more difficult in the field.

14.9D Before equipment is installed in a pumping facility, determine if the equipment is familiar to your agency and if its reliability has been proven. Find out what warranties, guarantees, and operation and maintenance aids will be provided with the equipment and the lift station.

APPENDIX

INDUSTRIAL WASTE TREATMENT
(VOLUME I)

Comprehensive Review Questions and
Suggested Answers

Industrial Waste Words

Subject Index

COMPREHENSIVE REVIEW QUESTIONS
VOLUME I

This section was prepared to help you review the material in Volume I. The questions are divided into five types:

1. True-False
2. Best Answer
3. Multiple Choice
4. Short Answer
5. Problems

To work this section:

1. Write the answer to each question in your notebook.

2. After you have worked a group of questions (you decide how many), check your answers with the suggested answers at the end of this section.

3. If you missed a question and do not understand why, reread the material in the manual.

You may wish to use this section for review purposes when preparing for civil service and certification examinations.

Since you have already completed this course, please *DO NOT SEND* your answers to California State University, Sacramento.

True-False

1. Historically, Americans have shown a great lack of interest in the protection of their water resources.

    1. True
    2. False

2. Historically, Americans have been content to think that the solution to pollution is dilution.

    1. True
    2. False

3. Unfortunately, too many operators take safety for granted.

    1. True
    2. False

4. An important basic concept is that water in the environment belongs to all of us.

    1. True
    2. False

5. Noncompatible pollutants are defined as those pollutants that are normally removed by the POTW.

    1. True
    2. False

6. Hydrogen sulfide presents a toxic gas problem to sewer maintenance personnel and to the IWTS operator.

    1. True
    2. False

7. The discharge of flammables is potentially the most damaging industrial discharge to the collection system.

    1. True
    2. False

8. Each point source discharger is required to obtain an NPDES permit.

    1. True
    2. False

9. In place of monitoring for TTO (total toxic organics), the control authority may permit the industrial user to submit a solvent management plan (also known as a toxic management plan (TOMP)) for the POTW's review and approval.

    1. True
    2. False

10. A Significant Industrial User (SIU) must have a program in place to reduce the volume and toxicity of hazardous wastes generated to the degree it has determined to be economically practical.

    1. True
    2. False

11. Waste minimization is a term primarily used in connection with hazardous wastes.

    1. True
    2. False

12. Equipment cleaning using water-based cleaners should be scheduled frequently on a regular basis.

    1. True
    2. False

13. Accurate use of the appropriate flow measurement devices is an essential skill for an industrial wastewater treatment plant operator.
    1. True
    2. False

14. Most accurate weir flow measurements occur when the nappe clings to the crest of the weir.
    1. True
    2. False

15. Any deposits that raise the surface of the water at the measuring point for a Palmer-Bowlus-type flume would cause the head measurement to indicate lower than actual.
    1. True
    2. False

16. The variation of industrial waste flows and concentrations of pollutants affects the operation of the industrial wastewater treatment system (IWTS) and could adversely affect the effluent quality from the plant.
    1. True
    2. False

17. The more variable the flow and concentration coming into the industrial wastewater treatment system (IWTS), the less equalization will help.
    1. True
    2. False

18. Rotating screens represent an advance in fine screening in that they are self-cleaning or largely self-cleaning.
    1. True
    2. False

19. In practice, industrial wastewaters are rarely truly neutralized to a pH of 7.
    1. True
    2. False

20. Filters are often installed after chemical treatment to produce a highly polished effluent.
    1. True
    2. False

21. Both dry dust and liquid forms of alum are irritating to the skin and mucous membranes and can cause serious eye injury.
    1. True
    2. False

22. Liquid alum becomes very slick upon evaporation and therefore spills and leaks should be avoided.
    1. True
    2. False

23. If the detention time is greater than the settling rate, then there will be a carryover of particles into the effluent.
    1. True
    2. False

24. Most gravity filters operate on a continuous basis.
    1. True
    2. False

25. Manual backwashings of a filter are inconsistent.
    1. True
    2. False

26. In a turbidity meter, the amount of light absorbed is proportional to the turbidity of the water.
    1. True
    2. False

27. A batch process is the least efficient mode in terms of membrane area requirements and system design.
    1. True
    2. False

28. Each chemical has a similar tendency to transfer from the liquid to the gaseous phase.
    1. True
    2. False

29. In an air stripping tower, flow instrumentation should be provided for both the water flow and the air flow.
    1. True
    2. False

30. The better the quality of water that enters the carbon column reactors, the more efficient the adsorption process will be.
    1. True
    2. False

31. The hydroxide form of most metals is relatively soluble.
    1. True
    2. False

32. Very high or very low pH industrial wastestreams may originate at cleaning, manufacturing, or waste treatment processes.
    1. True
    2. False

33. Both pH and ORP probes are cleaned using the same procedures.
    1. True
    2. False

34. Voltage can be viewed as the "pressure," while current can be viewed as the "flow."
    1. True
    2. False

35. Samples should be mixed thoroughly before compositing and at the time of the test.
    1. True
    2. False

36. Measurement instruments can be considered extensions of and improvements to an operator's senses of vision, touch, hearing, and even smell.
    1. True
    2. False

37. Simple "off-the-shelf" pressure gauges such as are found on pumps are used to measure water levels.
    1. True
    2. False

38. The piping design must provide for adequate lengths of straight pipe runs upstream and downstream of the flowmeter.
    1. True
    2. False

39. Any electrical equipment that begins to show any signs of excessive heating must be shut down immediately.
    1. True
    2. False

40. Operators must constantly be thinking of safety and working safely.
    1. True
    2. False

41. An explosion will occur when a gas concentration is higher than the upper explosive limit (UEL) or below the lower explosive limit (LEL).
    1. True
    2. False

42. Never attempt to do a job unless you have sufficient training, assistance, the proper tools, and the necessary safety equipment.
    1. True
    2. False

43. A treatment plant kept in a clean, orderly condition makes a safer place to work and aids in building good public and employee relations.
    1. True
    2. False

44. Vertical turbine pumps are used to pump water from deep wells, and may be of the single-stage or multi-stage type.
    1. True
    2. False

45. Unqualified people doing electrical work can seriously injure themselves and damage costly equipment.
    1. True
    2. False

46. Regardless of which type of electrical equipment protective device you are working with, if it blows or trips, the potential source of the problem should be investigated, identified, and corrected.
    1. True
    2. False

47. Industrial wastes often cause shock loads.
    1. True
    2. False

48. The work of laboratory technicians must be better than the samples that have been collected for analysis.
    1. True
    2. False

49. Cleanliness is a very important part of getting accurate test results.
    1. True
    2. False

Best Answer (Select only the closest or best answer.)

1. What can operators learn from this course about their industrial wastewater treatment plant?
    1. How to avoid getting caught for noncompliance
    2. How to increase company profits
    3. How to manipulate the plant so it operates at maximum efficiency
    4. How to read manufacturers' equipment warranties

2. Why must special care and safety be practiced when visitors are taken through your treatment plant?
    1. A visitor might collect a sample and obtain a noncompliance test result
    2. A visitor might spot a safety violation
    3. An accident could spoil all of your public relations efforts
    4. Someone might decide they should replace you

3. Before industrial wastes are discharged, they should have a pH similar to that of which item?
    1. Initial industrial waste
    2. Langelier Index
    3. Neutral
    4. Receiving sewer or receiving water

4. What is drag out?
   1. A process used to reduce the amount of dissolved salts in the recirculated cooling water
   2. Regulated substances that must be removed from wastestreams or neutralized prior to discharge to POTW sewers
   3. The liquid film (plating solution) that adheres to the workpieces and their fixtures as they are removed from any given process solution or their rinses
   4. The removal of accumulated solids in boilers to prevent plugging of boiler tubes and steam lines

5. What is the purpose of chelation?
   1. To form a thin, protective coating on metal
   2. To keep precipitates in suspension
   3. To prevent the corrosion of concrete
   4. To prevent the precipitation of metals (copper)

6. What is bright dipping?
   1. An electrochemical process that converts the metal surface to a coating of an insoluble oxide
   2. The process of coating a metallic workpiece with another metal by immersion in a molten bath to provide a protective film
   3. The process of forming a protective film on metals by immersion in an acid solution, usually nitric acid or nitric acid with sodium dichromate
   4. The process used to remove oxide and tarnish from ferrous and nonferrous materials

7. What is a statute?
   1. A comprehensive source of regulations
   2. A printed record of all activity that takes place in the US Congress
   3. An act of the legislature declaring, commanding, or prohibiting something
   4. Rules issued by various governmental departments to carry out the intent of the statute

8. What are indirect dischargers?
   1. Facilities that are funded directly by user charges
   2. Facilities that are funded indirectly by public funds
   3. Facilities that discharge wastewater directly into US streams or lakes
   4. Facilities that discharge wastewater to a POTW and are subject to pretreatment standards

9. What is the basis for the concentration-based standards?
   1. The actual mass of pollutants in a categorical wastewater stream per unit of production
   2. The maximum actual mass of pollutants in a categorical wastestream per unit of production during a typical shift
   3. The maximum concentration measured during a typical shift, usually expressed as mg/L
   4. The relative strength of a pollutant in a wastestream, usually expressed as mg/L

10. What is the focus of pollution prevention?
    1. On minimizing the amount of waste generated
    2. On neutralizing toxic wastes
    3. On recycling wastes after generation
    4. On treating wastes after generation

11. What is recycling?
    1. The elimination of a manufacturing process
    2. The reduction or elimination of the generation of the waste at the source
    3. The reuse of materials in the original process or in another process
    4. The use of less hazardous materials

12. What is process modification?
    1. A change in process operation or equipment
    2. Changing a product in such a way that the process for the new product generates less wastewater or less-toxic wastewater
    3. Procedures and policies that reduce the quantity or toxicity of wastewater
    4. The replacement of one raw material with another raw material

13. What do concentration-based limits require? The amount of pollutant in a discharge
    1. Does not cause a blockage in the pipes
    2. Does not cause spills or overflows
    3. Does not exceed a certain mass per unit volume, such as mg/L
    4. Is of a concentration low enough to flow in pipes

14. What is a flume?
    1. A specially shaped flowmeter that uses a pressure differential to measure flow rate in the channel
    2. A specially shaped flowmeter that uses a propeller to measure flow rate
    3. A specially shaped open channel section that can be installed in an open channel or an unpressurized pipe line
    4. A specially shaped weir that can be placed in an open channel

15. What is a good practice to prevent flow measurement problems caused by deposits upstream from a flow measurement device?
    1. Sewer closed-circuit television (CCTV) inspection program
    2. Sewer flowmeter calibration program
    3. Sewer flushing schedule
    4. Sewer preventive maintenance program

16. For what flow measurement task is the dye dilution method especially useful?
    1. Continuous flow monitoring
    2. Measuring fluctuating flow rates
    3. Measuring large flow rates
    4. Measuring low flow rates

17. How can equalization be accomplished within the wastewater collection system?

    1. By using a waste discharge fee structure that discourages industries from dumping batch loads
    2. By using flowmeters to regulate flows from large dischargers
    3. By using minimum-sized pipes to prevent maximum or peak flows
    4. By using sumps and pumping stations where flows could be discharged at controlled rates

18. Why are static screens popular in industrial wastewater treatment?

    1. Because they can be moved around the plant
    2. Because they can remove most particles
    3. Because they have no moving parts
    4. Because they never plug

19. What is a doctor blade?

    1. A blade used to apply solids to the outside of a fixed screen
    2. A blade used to apply solids to the outside of a rotating screen
    3. A blade used to remove any excess solids that may cling to the outside of a fixed screen
    4. A blade used to remove any excess solids that may cling to the outside of a rotating screen

20. What happens as the velocity past the face of a pH electrode increases?

    1. Poor mixing past the electrode results
    2. The boundary layer thickness increases
    3. The electrode will tend to become coated
    4. The response time to a pH change becomes almost instantaneous

21. What is coagulation?

    1. The addition of chemicals to adjust the pH to be within acceptable limits
    2. The clumping together of very fine particles into larger particles caused by the use of chemicals
    3. The gathering together of fine particles to form larger particles by a process of gentle mixing
    4. The separation of suspended particles into a clarified liquid and a more concentrated suspension

22. Why is liquid alum preferred by operators? Because of its

    1. Ease of handling
    2. Lack of corrosivity
    3. Low cost
    4. Relatively small volume

23. What is the objective of the preliminary screening test?

    1. To confirm the performance and fine-tune the dosage rates for the selected chemical(s)
    2. To determine the chemical dosages required to achieve the desired treatment performance
    3. To keep treatment process performance within the desired limits while minimizing the chemical costs involved
    4. To select a limited number of potentially suitable chemical treatment products from a large number of choices

24. Why are chemical feeders (metering equipment) required?

    1. To accurately control the desired dosage
    2. To properly adjust the concentration of solution in the storage tank
    3. To provide the desired degree of mixing
    4. To record the total amount of chemical applied

25. What is the surface loading rate? An expression of the

    1. Length of time it would take a plug of water to enter a clarifier and to exit the effluent
    2. Quantity of solids being treated in relation to available clarifier surface
    3. Quantity of water being treated in relation to available clarifier surface
    4. Quantity of water that passes out of the clarifier in relation to the lineal feet of weir available

26. When should a filter be backwashed?

    1. After a known volume of water has passed through the filter
    2. After an assumed amount of solids have been removed by the filter
    3. After the solids capacity of the media has been reached, but before solids breakthrough occurs
    4. When other filters are available to meet the demand for water during the backwash cycle

27. What is the purpose of a direct, spring-loaded pressure relief valve installed on the top of pressure vessels?

    1. To allow use of treated wastewater for side stream purposes
    2. To ensure treatment of all industrial wastewater flows
    3. To prevent escape from vessels containing contaminated waters
    4. To prevent vessel rupture in case effluent flow is restricted or stopped while influent flow continues

28. What is one of the most important determinants for controlling the operation of a continuous backwash, upflow, deep-bed silica sand media filter?

    1. Coagulation chemical dose
    2. Depth of the sand filter
    3. Flocculant aid chemical dose
    4. Reject water flow rate leaving the filter

29. Under what conditions is incineration of the vapors selected for the control of volatile organic compounds (VOCs) from air strippers?

    1. When the air pollution control regulations are insignificant
    2. When the concentration of VOCs in the off gas is high enough that the fuel value of the VOCs becomes significant
    3. When the recovery of heat is considered inappropriate
    4. When the source of natural gas is limited

30. What should be done when the head loss is too high through a carbon adsorber?

    1. Backflushing of the screen
    2. Reduction of the flow through the carbon adsorber
    3. Regeneration of the carbon
    4. Replacement of the screen

31. Why can carbon dust be particularly hazardous if allowed to accumulate in a confined space?

    1. The dust can discolor recordkeeping forms and make the forms difficult to read
    2. The dust can ignite with explosive force if a spark or even a hot surface is present
    3. The dust can plug ventilation vents
    4. The dust can render clothing dirty and difficult to wash

32. Why are chelating agents added to plating solutions?

    1. To cause metals to remain in solution
    2. To create a shiny finished surface
    3. To improve the quality of the plated surfaces
    4. To prevent surfaces from reflecting bright lights

33. What is the first step in hydroxide precipitation of common metals?

    1. To determine the optimum pH for precipitating each type of metal you wish to remove
    2. To evaluate sludge (precipitates) collection and disposal methods
    3. To investigate the sources of metals in the wastestream
    4. To select the proper acid for pH adjustment

34. Where may toxic organics be found in industrial wastestreams?

    1. With collected hazardous wastes from laboratories
    2. With effluents from bioassay toxicity tests
    3. With equalization tanks to smooth out flows
    4. With oily wastes and solvents used for cleaning

35. Why should the intake lines of automatic sampling devices be cleaned out regularly?

    1. To prevent bacterial degradation of the wastestream being sampled
    2. To prevent buildup with precipitates, which may slough off and contaminate the sample
    3. To prevent high head losses in the intake lines
    4. To prevent loss of wastestream extreme events

36. What is first aid?

    1. Aid for persons who are unable to help themselves
    2. Emergency treatment for injury or sudden illness before regular medical treatment is available
    3. The first corrective action applied to an injured person
    4. The response of the first person to arrive at the scene of an accident

37. What is a measurement?

    1. A determination made by an operator
    2. A unit consisting of the appropriate symbol
    3. The comparison of a quantity with a standard unit of measurement
    4. The recording of the result of an experiment or test

38. What does precision refer to?

    1. A physical or chemical quantity that is measured or controlled in an individual waste treatment process
    2. How closely an instrument can reproduce a given reading time after time
    3. How closely an instrument measures the true or actual value of a process variable
    4. The smallest change in a process variable able to produce an appropriate response from an instrument

39. What is the purpose of surge protection devices on pressure sensors?

    1. To desensitize the sensor from low pressures
    2. To prevent sudden pressure surges from damaging the pressure sensor
    3. To prevent the pressure sensor from recording surge pressures
    4. To protect the pressure sensor from cavitation

40. What are the most common physical injuries in wastewater treatment facilities?

    1. Cuts, bruises, strains, and sprains
    2. Electric shock
    3. Infections
    4. Needle sticks from potentially contaminated syringes

41. Why should trenches or excavations have proper shoring?

    1. A trench or excavation can cave in and kill you
    2. Other operators or pedestrians can fall into the trench or excavation
    3. Vehicles can drive into the trench or excavation
    4. Water can flood the trench or excavation

42. What can be done if excess foaming begins to cause serious problems with wastewater treatment processes?

    1. Cause the foaming to occur upstream
    2. Encourage rapid mixing
    3. Feed an antifoaming agent
    4. Install a hydraulic jump to reduce foaming

43. What is a cross connection?

    1. A connection between a potable clear well and a hydropneumatic tank
    2. A connection between a potable (drinking) water and any other location where wastewater could contaminate a domestic water supply
    3. A connection between a potable well and a water supply pump
    4. A connection where a potable well supply crosses over a wastewater discharge from a pump

44. Why is mechanical maintenance of prime importance in an industrial wastewater treatment plant? Equipment must be kept

    1. In good operating condition for the plant to maintain peak performance
    2. In proper condition for company sales literature
    3. Looking like new to impress visitors
    4. New to convince regulators that the plant is in compliance

45. What is the purpose of wearing rings on a pump? To keep the pump

    1. Cool
    2. From losing suction
    3. Impeller rotating
    4. Primed

46. What is a fuse? A protective device that

    1. Can be lit and cause a circuit to break if overloaded too long
    2. Has a strip or wire of fusible metal that, when placed in a circuit, will melt and break the electric circuit when subjected to an excessive temperature
    3. Measures the flow of electricity
    4. Prevents inefficient operation

47. What are the types of pump and motor misalignment?

    1. Impeller and rotor
    2. Noise and vibration
    3. Parallel and angular
    4. Volute and stator

48. What is the purpose of a dehumidifier?

    1. To increase the ease of working in a confined space
    2. To obtain water to prime a pump
    3. To provide moisture for the workplace
    4. To remove moisture from the surrounding atmosphere

49. Why should fines and extra charges for accidental spills, dumps, or additional loadings be greater than the extra costs associated with resulting problems and a poor plant effluent?

    1. To discourage future problems resulting from negligence
    2. To ensure a profitable operation of the POTW
    3. To keep industrial dischargers happy
    4. To provide adequate and competitive salaries for operators

50. What is a good sampling point?

    1. One that has ample parking space for vehicles transporting samples to the lab
    2. One that is easily accessible and located where a composite sample may be obtained
    3. One that is easily accessible and located where a representative sample may be obtained
    4. One that is located near the laboratory to facilitate short transportation times to the lab

51. Where should the search begin to trace an industrial discharge causing problems at a wastewater treatment plant?

    1. At a point in the treatment plant or collection system where the discharge has been discovered
    2. At the most remote points in the collection system above the industrial discharger
    3. At the point of discharge into the collection system of the largest industrial discharger
    4. At the point of discharge into the collection system of the newest discharger

Multiple Choice (Select all correct answers.)

1. The industrial wastewater treatment plant operator physically performs which tasks?

    1. Collects samples
    2. Interviews reporters
    3. Lubricates equipment
    4. Pushes switches
    5. Turns valves

2. Why must industrial wastewater receive adequate treatment? To protect

    1. Game fishing
    2. Health of downstream neighbors
    3. Property values
    4. Recreational uses
    5. Water for water-using industries

3. Which of the following items are included in definitions of pollution?

    1. Any discharge of waste to a body of water
    2. Any failure to meet water quality requirements
    3. Any interference with the beneficial reuse of water
    4. Any time operators fail to properly treat wastes
    5. Any treatment process that fails due to budget limitations

4. What are possible alternatives for treatment of industrial wastewaters?

   1. Direct discharge to the receiving water
   2. Discharge to a POTW for treatment and return to the environment
   3. No treatment, after obtaining a variance
   4. Pretreatment by industry, followed by discharge to a POTW
   5. Treatment by industry to the extent required before discharge to a receiving water

5. Why does an operator need an understanding of the concentration and the mass of a pollutant in an industrial waste? To determine the effects on the

   1. Industry's pretreatment system
   2. POTW's collection system
   3. POTW's disposal systems
   4. POTW's treatment system
   5. Sampling of the industry's discharge

6. Acids will corrode which types of facilities?

   1. Concrete and cast-iron sewers
   2. Concrete wet wells and tanks
   3. Internal steel equipment in clarifiers
   4. Internal steel equipment in pumps
   5. Stainless-steel pipes

7. Which measurements of wastewater are used to detect changes in the influent to the industrial waste treatment system (IWTS) or the publicly owned treatment works (POTW)?

   1. BOD
   2. Conductivity
   3. Flow
   4. pH
   5. Temperature

8. Where do industrial facilities discharge treated wastewater?

   1. Directly to the environment
   2. To evapotranspiration systems
   3. To regulatory compliance facilities
   4. To the atmosphere
   5. To wastewater collection systems and treatment plants (POTWs)

9. The pretreatment standards for the metal finishing category list the contaminant and which concentrations?

   1. Maximum for any hour, mg/L
   2. Maximum for any one day, mg/L
   3. Monthly average limit, mg/L
   4. Quarterly average limit, mg/L
   5. Weekly average limit, mg/L

10. EPA has promulgated only production-based (mass-based) pretreatment standards for which industrial categories?

    1. Battery manufacturing
    2. Coil coating
    3. Electroplating
    4. Metal finishing
    5. Pesticides

11. What are the benefits of pollution prevention?

    1. A process change could eliminate a whole wastestream
    2. Good operating practices could minimize the volume of waste entering the treatment system
    3. Material substitution could eliminate a pollutant altogether
    4. Operators could be required to handle and treat a more hazardous waste
    5. Particular pollutants could be minimized or even eliminated from the wastestream

12. What is the role of industrial treatment personnel in pollution prevention? To be aware of pollution prevention opportunities in the areas of

    1. Budget presentation and justification
    2. Good housekeeping
    3. Implementation of pollution prevention strategies
    4. Improved operation and maintenance
    5. Process modification

13. Economic benefits and cost savings that could be achieved by a pollution prevention program include which items?

    1. Lower health and safety costs
    2. Lower long-term environmental liability and insurance costs
    3. Lower manifesting and reporting costs
    4. Lower risks of emergency spills and accidents
    5. Reduced transportation and disposal costs for sludges

14. What are the keys to pollution prevention in the chemicals formulating industry?

    1. Avoiding the use of chlorinated solvents
    2. Better scheduling to minimize changeover
    3. Increasing water usage to dilute wastestreams
    4. Using alternative cleaning methods to minimize cleaning wastes
    5. Using better cleaning methods to minimize cleaning wastes

15. Which items are examples of open channel flow?

    1. Canals
    2. Chlorine contact chambers
    3. Force mains
    4. Gravity sewer lines
    5. Pump discharge lines

16. In closed pipe flow, the primary element generally consists of which types of devices?

    1. Flow tube
    2. Flume
    3. Orifice plate
    4. Venturi
    5. Weir

17. Which factors may cause measurement errors when using bubblers?

    1. Clogs in the end of the bubbler tube from solids in the wastestream
    2. Density of the liquid
    3. Kinks in the bubbler tubing
    4. Lag effects
    5. Splits in the bubbler tubing

18. How can a manual measurement be made when a staff gauge is unavailable? By the use of

    1. A Doppler technique
    2. A measuring tape
    3. A pitot meter
    4. A thin steel rule
    5. An air bubbler

19. Which types of batch discharges may cause the influent concentration to vary?

    1. Discharge from the equalization basin
    2. Flow through screens
    3. pH adjustment
    4. Release of process wastewater
    5. Spent plating baths

20. Which types of industrial wastewater treatment system (IWTS) recycle flows need to be included in the calculation of the volume of the equalization tank?

    1. Chlorine contact basin flows
    2. Decant water from the digester or sludge holding tank
    3. Filter backwash
    4. Primary effluent
    5. Screening flows

21. What kinds of problems may be discovered when troubleshooting automatically cleaned bar screens?

    1. Excessive noise or vibration
    2. Faulty bar rack engagement by cleaning rake
    3. Proper cleaning rake wiping
    4. Screenings dropping off cleaning rake shelf
    5. Too high temperature

22. Which safety precautions must be practiced when working around screens?

    1. Be careful to avoid tripping or slipping and falling into the moving machinery
    2. Be careful to avoid tripping or slipping and falling into the wastewater
    3. Keep walkways and guardrails clean, in good repair, and free of obstructions
    4. Operate screening equipment without guardrails
    5. Stay behind guardrails whenever possible

23. Which items are important process control guidelines for the chemical treatment processes?

    1. Controlling the chemical dose
    2. Controlling the intensity of mixing during flocculation
    3. Handling the chemicals in a safe manner
    4. Providing enough energy to completely mix the chemicals with the wastewater
    5. Providing storage space for the chemicals

24. What are the uses of hydrated lime (calcium hydroxide) in treating industrial wastewaters?

    1. To adjust the pH to improve the coagulation process
    2. To coagulate solids
    3. To dissolve solids and precipitates
    4. To enhance filtration
    5. To increase flocculation

25. What information does an operator obtain from a jar test?

    1. Best chemical storage procedures
    2. Best chemical to use
    3. Best dosage to feed
    4. Best time to collect samples
    5. Safe chemical handling procedures

26. When reviewing chemical feed system designs and specifications, the operator should check which items?

    1. Be sure provisions are made for standby equipment in order to maintain uninterrupted dosages during equipment maintenance
    2. Determine if sampling points are provided to measure chemical feeder output
    3. Examine plans for valving to allow flushing the system with water before removing from service
    4. Look for adequate valving to allow bypassing or removing equipment for maintenance without interrupting the chemical dosage
    5. Review the results of the pre-design tests to determine the chemical feed rate for both the present and future

27. What are the purposes of a filter underdrain system?

    1. To collect the filtered water uniformly throughout the bed
    2. To drain stormwater runoff away from the filter
    3. To prevent high groundwater levels from seeping into the filter
    4. To remove water that infiltrates past the filters
    5. To uniformly apply the backwash water during backwashing

28. Which items should be included in a log of the filtering operation?

    1. Backwash water flow rates and duration
    2. Chemicals added as filter aids
    3. Head loss at start and end of filter run
    4. Influent and effluent turbidity
    5. Remarks of special observations and maintenance

29. If there are problems with the performance of a continuous backwash, upflow, deep-bed silica sand media filter, which questions should an operator investigate?

    1. Are the chemical feed systems functioning properly?
    2. Do the influent suspended solids exceed the design solids loading rate?
    3. Does the influent flow rate exceed the design flow and loading rates?
    4. Is sand being pumped to the top of the airlift and at the proper rate?
    5. Is there enough reject water flow?

30. In the air stripping process, the amount of liquid chemical (pollutant) transferred from the water to the air depends on which factors?

    1. Air pressure
    2. Properties of the liquid chemical
    3. Temperature
    4. Toxicity of the liquid chemical
    5. Volume of container

31. How can flow rates through the carbon system be controlled?

    1. By controlling production processes
    2. By controlling the pumping rate
    3. By controlling waste production processes
    4. By throttling the effluent valve on an individual adsorber
    5. By throttling the flow of carbon

32. Why should operators sample both the influent and the effluent of the activated carbon process?

    1. To determine BOD removal efficiencies
    2. To determine corrosiveness of the effluent
    3. To determine organic contaminant removal efficiencies
    4. To determine the quantities of carbon fines remaining in the effluent from the process
    5. To determine the toxicity of the effluent

33. An important aspect of physical–chemical treatment is precise and accurate control of which operating factors?

    1. Chemical doses
    2. Chlorine residual
    3. Dissolved oxygen
    4. ORP
    5. pH

34. Why should operators be very careful whenever strongly acidic and basic solutions are mixed? Because of the potential for

    1. Generation of heat
    2. Neutralization
    3. Precipitation
    4. Production of toxic gases
    5. Splattering

35. What are the primary problems operators encounter with ion exchange systems?

    1. Excessive energy costs
    2. Excessive pressure drop across the ion exchange system
    3. Failure to produce specified water quality
    4. High level of maintenance required
    5. Throughput capacity loss

36. Which common methods are used to dewater metallic sludges?

    1. Anaerobic digesters
    2. Bag filters
    3. Centrifuges
    4. Filter presses
    5. Vacuum filters

37. How can industrial treatment plant operators obtain information regarding the treatment processes they are required to operate? Sources of information include

    1. Instrument service company representatives
    2. Manufacturers' technical representatives
    3. Operators who treat wastewater for other companies or shops
    4. System construction drawings
    5. System O & M manuals

38. After starting a pump, which items should be checked?

    1. Check to be sure the pump anchorage is tight
    2. Check to see that the bearings do not overheat from over- or underlubrication
    3. Check to see that the direction of rotation is correct
    4. Inspect the piping for leaks
    5. Observe the packing gland boxes (stuffing boxes) for slight leakage

39. Operators should know how to perform first aid for which situations?

    1. Contact with acids
    2. Contact with bases
    3. Eye burns
    4. Hypochlorite solution contact
    5. Skin burns

40. In industrial waste treatment, which variables are the most common ones measured?

    1. Density
    2. Flow
    3. Level
    4. Pressure
    5. Viscosity

41. Which panel instruments could be located in the main control panel of a typical industrial treatment plant?

    1. Alarms
    2. Controllers
    3. Indicators
    4. Recorders
    5. Totalizers

42. Which tasks must be performed to ensure instrumentation systems are reliable year after year?

    1. Proper application
    2. Proper operation and maintenance
    3. Proper recording of readings
    4. Proper set-up
    5. Proper storage of data

43. Operators may be exposed to which types of hazards?

    1. Confined spaces
    2. Infections and infectious diseases
    3. Physical injuries
    4. Toxic and harmful chemicals
    5. Toxic or suffocating gases or vapors

44. What information is included on a confined space entry permit?

    1. Certification that all existing hazards have been evaluated by a competent person
    2. Certification that necessary protective measures have been taken to ensure the safety of each worker
    3. Dimensions of the confined space
    4. Location of work to be done
    5. Type of work to be done

45. Which types of hazardous or toxic substances could reach an industrial wastewater treatment facility?

    1. Amines
    2. Biocides
    3. Fuels
    4. Surface-active agents
    5. Toxic gases

46. What are the essential ingredients necessary for a fire?

    1. Fuel
    2. Heat
    3. Observers
    4. Oxygen
    5. Water

47. Which points are stressed in an effective safety program?

    1. Accidents are studied step by step and thoroughly reviewed with the attitude that "they are caused and do not just happen"
    2. Adequate safety equipment is available and its capabilities are thoroughly understood
    3. Basic safety concepts and practices are thoroughly understood by all
    4. Everyone participates and accepts personal responsibility for their own safety and that of their fellow workers
    5. Everyone realizes that safety is a continuing learning and re-learning process

48. What are the benefits of a good maintenance recordkeeping program?

    1. Program helps to keep equipment warranty in force
    2. Program indicates when maintenance is due
    3. Program keeps operators busy during off hours
    4. Program monitors plant effluent and compliance
    5. Program provides a record of equipment performance

49. Which items are included in the total head in a pump system from where the water enters the suction piping to the end of the discharge piping?

    1. Cavitation head
    2. Discharge lift
    3. Friction in the piping
    4. Suction lift
    5. Water hammer head

50. Which factors could cause lower than normal discharge pressures in a centrifugal pump?

    1. A broken discharge pipe
    2. A change of the source of electrical power
    3. A different point of discharge
    4. Worn impeller in the pump
    5. Worn wearing rings in the pump

51. Which items are pump station operating conditions that can affect the performance and the life of electrical equipment?

    1. Corrosive atmospheres
    2. Dust
    3. Heat
    4. Humidity
    5. Vibration

52. What is the purpose of a magnetic starter? Magnetic starters are used to start
    1. Anything where automatic or remote control is desired
    2. Blowers
    3. Compressors
    4. Equipment when the commercial power is unavailable
    5. Pumps

53. What treatment plant problems could occur as a result of a discharge of improperly pretreated industrial wastes (shock load)?
    1. Hydraulic overloads could "wash out" the microorganisms in the treatment process
    2. Odors could result from upset biological treatment processes
    3. Solids carryover to the effluent could result from a process upset
    4. Toxic materials could destroy or affect the metabolism of microorganisms that are part of the treatment process
    5. Toxic materials could pass through the treatment process and be present in the plant effluent

54. What equipment is used to obtain composite samples?
    1. Conveyors with small "buckets" attached
    2. Scoops
    3. Small pumps
    4. Test tubes
    5. Vacuum chambers

55. Containers for samples collected from the discharge of industrial wastewaters are made of what kinds of materials?
    1. Cardboard
    2. Glass
    3. Polyethylene
    4. Stainless steel
    5. Wood

56. Which activities are required before entering a confined space?
    1. Collect a sample of the flowing wastewater
    2. Notify upstream dischargers to eliminate hazardous discharges
    3. Provide for monitoring of the atmosphere whenever anyone is in the space
    4. Provide for two people to be standing by topside to assist in an emergency
    5. Wear a safety harness

Short Answer

1. Describe in general terms the types of activities that make up an industrial treatment plant operator's job.
2. Why is recordkeeping an important part of an operator's job?
3. What is the main job of an industrial wastewater treatment plant operator?
4. Identify one advantage and one limitation of the use of chlorine as a disinfectant.
5. Explain how compatible pollutants can sometimes exhibit the characteristics of noncompatible pollutants, and vice versa.
6. What problems can be caused in sewers when an industry discharges out-of-pH-range wastewater?
7. What processes are used to clean and prepare the surface of a workpiece for a surface coating?
8. What are the advantages of EPA's delegation of implementation authority to local agencies?
9. What are categorical pretreatment standards?
10. How can an industrial discharger avoid monitoring for TTO?
11. What is the meaning of "waste minimization"?
12. How should employees be trained with regard to off-specification process batches?
13. How can a representative sample be obtained from an industry with a fluctuating or unusual waste discharge flow rate?
14. How does a differential pressure meter measure flow in a pipe?
15. Why is in-line equalization more effective than side-line equalization in leveling influent concentrations?
16. What types of downstream processes and equipment are protected by screening?
17. Why do some pH adjustment systems use three pH electrodes?
18. What are the three most important process control guidelines in any chemical treatment process?
19. Why is paddle speed important during flocculation?
20. Why should extensive laboratory testing be conducted before treating the entire plant effluent with a particular polyelectrolyte?
21. Explain how to run the jar test.
22. What economic factors should be considered when selecting a chemical feeder?
23. Why are multi-media filters used?
24. How are floatable and settleable solids removed from a holding tank?
25. What are the advantages of continuous backwash, upflow, deep-bed silica sand media filters over other types of granular media filters?
26. How can fouled membranes be cleaned?
27. What usually happens to water flux with time? Explain.

28. What are the most common symptoms of operational problems in an air stripper?

29. What industries use activated carbon to treat industrial wastestreams?

30. What tests should be run on the effluent of the activated carbon process and why?

31. How can the costs and complexity of metal waste treatment facilities be reduced?

32. Why should a spent process solution (a pickle, for example) be treated as a batch process and not dumped directly into the rinse water for treatment?

33. How can the ion exchange process be used to treat metal wastes?

34. Why should an operator have a fundamental knowledge of electrical power and how it is controlled?

35. When handling acids, what type of protective clothing would you wear?

36. Why should operators understand measurement and control systems?

37. Why should a screwdriver not be used to test an electric circuit?

38. Why is it poor practice to ignore the lamps that are lit up (alarm conditions) on an annunciator panel?

39. What problems are created by oil and moisture in instrument air, and how can these contaminants be removed?

40. What is the operator's responsibility with regard to safety?

41. What precautions should you take to avoid transmitting disease to your family?

42. What kind of job site protection is usually required when you are working in a manhole?

43. Why should only qualified electricians work on an electrical control panel?

44. What safety precautions should be taken when applying protective coatings?

45. How can samples for lab tests be collected without going beyond guardrails or chains?

46. Why should safety equipment be checked periodically?

47. Why should the operator thoroughly read and understand manufacturers' literature before attempting to maintain plant equipment?

48. What is the purpose of a maintenance recordkeeping program?

49. How can you determine if a new pump will turn in the direction intended?

50. What protective or safety devices are used to protect operators and equipment from being harmed by electricity?

51. What should be checked if pump bearings are running hot?

52. Why would you use a stethoscope to check an electric motor?

53. Why should inactive gate valves be operated periodically?

54. How can scum lines be kept from plugging?

55. Why should operators review plans and specifications for a new facility?

56. What problems could a wastewater treatment plant experience if improperly pretreated industrial wastes are discharged?

57. Why must representative samples be collected?

58. How would you attempt to determine the peak concentration and time of occurrence of a toxic waste discharge?

59. How would you measure the flow from an industry?

Problems

1. Determine the mass emission rate in pounds of copper per day from an electroplater if the effluent flow is 5,000 gallons per day containing 6 milligrams per liter of copper.

2. An equalization basin stores 150,000 gallons of wastewater from an eight-hour work shift. The pretreatment system is designed to treat a flow of 120 GPM. How long will it require to process the wastewater from the eight-hour production period?

3. An industrial pretreatment facility receives a flow of 200 GPM containing a BOD of 450 mg/L for a 24-hour period. Estimate the BOD loading on the facility in pounds of BOD per day. What would be the BOD loading if the flow was for only one eight-hour shift during 24 hours?

4. An 80-foot diameter clarifier drops 4.0 inches in 20 minutes during a calibration test on the RAS pump. Calculate the RAS pumping rate in gallons per minute (GPM).

5. Estimate the average flow in gallons per minute from an equalization basin if the total flow volume is 40,000 gallons and the release time is from 8 am to 5 pm.

6. Determine the chemical application rate in pounds per day and pounds per hour to treat a flow of 200 GPM with a chemical dose of 6 mg/L.

7. Jar test results indicate an optimum chemical dose of 6 mg/L to treat a flow of 0.3 MGD. If the chemical feed solution strength is 0.5 percent, what is the chemical pumping rate in gallons per hour?

8. Estimate the polymer dose in milligrams per liter if 80 pounds of polymer per day are delivered to a flow of 5,000 GPM.

9. Polymer is supplied to your plant at a concentration of 0.6 pound polymer per gallon. The polymer feed pump delivers a flow of 0.12 GPM and the flow being treated is 2,500 GPM. What is the polymer dose in mg/L in the water being treated?

10. Liquid alum is supplied at a concentration of 5.4 pounds alum per gallon. If the alum feed pump delivers 80 mL per minute to a flow of 2,000 GPM, calculate the dose of alum in mg/L to the water being treated.

11. A membrane filtration system is operating with 500 square feet of membrane area and the permeate flow is 10 gallons per minute. What is the flux (J) in GFD (gallons/sq ft-day)?

SUGGESTED ANSWERS TO COMPREHENSIVE REVIEW QUESTIONS
VOLUME I

True-False

1. True — Historically, Americans have shown a great lack of interest in the protection of their water resources.

2. True — Historically, Americans have been content to think that the solution to pollution is dilution.

3. True — Unfortunately, too many operators take safety for granted.

4. True — An important basic concept is that water in the environment belongs to all of us.

5. False — Noncompatible pollutants are defined as those pollutants that are *NOT* normally removed by the POTW.

6. True — Hydrogen sulfide presents a toxic gas problem to sewer maintenance personnel and to the IWTS operator.

7. True — The discharge of flammables is potentially the most damaging industrial discharge to the collection system.

8. True — Each point source discharger is required to obtain an NPDES permit.

9. True — In place of monitoring for TTO, the control authority may permit the industrial user to submit a solvent management plan for the POTW's review and approval.

10. True — A SIU must have a program in place to reduce the volume and toxicity of hazardous wastes generated to the degree determined to be economically practical.

11. True — Waste minimization is a term primarily used in connection with hazardous wastes.

12. False — Equipment cleaning using water-based cleaners should be scheduled as needed, *NOT* frequently on a regular basis.

13. True — Accurate use of the appropriate flow measurement devices is an essential skill for an industrial wastewater treatment plant operator.

14. False — Most accurate weir flow measurements occur when the nappe springs free of the crest (*NOT* when the nappe clings to the crest) of the weir.

15. False — Any deposits that raise the surface of the water at the measuring point for a Palmer-Bowlus-type flume would cause the head measurement to indicate higher (*NOT* lower) than actual.

16. True — The variation of industrial waste flows and concentrations of pollutants affects the operation of the IWTS and could adversely affect the effluent quality from the plant.

17. False — The more variable the flow and concentration coming into the industrial wastewater treatment system (IWTS), the more (*NOT* less) equalization will help.

18. True — Rotating screens represent an advance in fine screening in that they are self-cleaning or largely self-cleaning.

19. True — In practice, industrial wastewaters are rarely truly neutralized to a pH of 7.

20. True — Filters are often installed after chemical treatment to produce a highly polished effluent.

21. True — Both dry dust and liquid forms of alum are irritating to the skin and mucous membranes and can cause serious eye injury.

22. True — Liquid alum becomes very slick upon evaporation and therefore spills and leaks should be avoided.

23. False — If the detention time is less (*NOT* greater) than the settling rate, then there will be a carryover of particles into the effluent.

24. False — Most gravity filters operate on a batch (*NOT* continuous) basis.

25. True — Manual backwashings of a filter are inconsistent.

26. False — In a turbidity meter, the amount of light scattered (*NOT* absorbed) is proportional to the turbidity of the water.

27. False — A batch process is the most (*NOT* least) efficient mode in terms of membrane area requirements and system design.

28. False — Each chemical has a different (*NOT* similar) tendency to transfer from the liquid to the gaseous phase.

29. True — In an air stripping tower, flow instrumentation should be provided for both the water flow and the air flow.

30. True — The better the quality of water that enters the carbon column reactors, the more efficient the adsorption process will be.

31. False — The hydroxide form of most metals is relatively insoluble (NOT soluble).

32. True — Very high or very low pH industrial wastestreams may originate at cleaning, manufacturing, or waste treatment processes.

33. True — Both pH and ORP probes are cleaned using the same procedures.

34. True — Voltage can be viewed as the "pressure," while current can be viewed as the "flow."

35. True — Samples should be mixed thoroughly before compositing and at the time of the test.

36. True — Measurement instruments can be considered extensions of and improvements to an operator's senses of vision, touch, hearing, and even smell.

37. False — Simple "off-the-shelf" pressure gauges such as are found on pumps are NOT used to measure water levels.

38. True — The piping design must provide for adequate lengths of straight pipe runs upstream and downstream of the flowmeter.

39. True — Any electrical equipment that begins to show any signs of excessive heating must be shut down immediately.

40. True — Operators must constantly be thinking of safety and working safely.

41. False — An explosion will NOT occur when a gas concentration is higher than the upper explosive limit (UEL) or below the lower explosive limit (LEL).

42. True — Never attempt to do a job unless you have sufficient training, assistance, the proper tools, and the necessary safety equipment.

43. True — A treatment plant kept in a clean, orderly condition makes a safer place to work and aids in building good public and employee relations.

44. True — Vertical turbine pumps are used to pump water from deep wells, and may be of the single-stage or multi-stage type.

45. True — Unqualified people doing electrical work can seriously injure themselves and damage costly equipment.

46. True — Regardless of which type of electrical equipment protective device you are working with, if it blows or trips, the potential source of the problem should be investigated.

47. True — Industrial wastes often cause shock loads.

48. False — The work of laboratory technicians CANNOT be better than the samples that have been collected for analysis.

49. True — Cleanliness is a very important part of getting accurate test results.

Best Answer

1. 3 Operators can learn from this course how to manipulate their industrial wastewater treatment plant so it operates at maximum efficiency.

2. 3 Special care and safety must be practiced when visitors are taken through your treatment plant because an accident could spoil all of your public relations efforts.

3. 4 Before industrial wastes are discharged, they should have a pH similar to that of the receiving sewer or receiving water.

4. 3 Drag out is the liquid film (plating solution) that adheres to the workpieces and their fixtures as they are removed from any given process solution or their rinses.

5. 4 The purpose of chelation is to prevent the precipitation of metals (copper).

6. 4 Bright dipping is the process used to remove oxide and tarnish from ferrous and nonferrous materials.

7. 3 A statute is an act of the legislature declaring, commanding, or prohibiting something.

8. 4 Indirect dischargers are facilities that discharge wastewater to a POTW and are subject to pretreatment standards.

9. 4 The basis for the concentration-based standards is the relative strength of a pollutant in a wastestream, usually expressed as mg/L.

10. 1 The focus of pollution prevention is on minimizing the amount of waste generated.

11. 3 Recycling is the reuse of materials in the original process or in another process.

12. 1 Process modification is a change in process operation or equipment.

13. 3 Concentration-based limits require that the amount of pollutant in a discharge does not exceed a certain mass per unit volume, such as mg/L.

14. 3 A flume is a specially shaped open channel section that can be installed in an open channel or an unpressurized pipe line.

15. 3 A sewer flushing schedule is a good practice to prevent flow measurement problems caused by deposits upstream from a flow measurement device.

16. 3 The dye dilution method is especially useful for measuring large flow rates.

17. 4 Equalization can be accomplished within the wastewater collection system by using sumps and pumping stations where flows could be discharged at controlled rates.

18. 3 Static screens are popular in industrial wastewater treatment because they have no moving parts.

19. 4 A doctor blade is a blade used to remove any excess solids that may cling to the outside of a rotating screen.

20. 4 As the velocity past the face of a pH electrode increases, the response time to a pH change becomes almost instantaneous.

21. 2 Coagulation is the clumping together of very fine particles into larger particles caused by the use of chemicals.

22. 1 Liquid alum is preferred by operators because of its ease of handling.

23. 4 The objective of the preliminary screening test is to select a limited number of potentially suitable chemical treatment products from a large number of choices.

24. 1 Chemical feeders (metering equipment) are required to accurately control the desired dosage.

25. 3 The surface loading rate is an expression of the quantity of water being treated in relation to available clarifier surface.

26. 3 A filter should be backwashed after the solids capacity of the media has been reached, but before solids breakthrough occurs.

27. 4 The purpose of a direct, spring-loaded pressure relief valve is to prevent vessel rupture in case effluent flow is restricted or stopped while influent flow continues.

28. 4 One of the most important determinants for controlling the operation of a continuous backwash, upflow, deep-bed silica sand media filter is the reject water flow rate leaving the filter.

29. 2 Incineration of vapors is selected for the control of volatile organic compounds (VOCs) from air strippers when the concentration of VOCs in the off gas is high enough that the fuel value of the VOCs becomes significant.

30. 1 When the head loss is too high through a carbon adsorber, backflush the screen.

31. 2 Carbon dust can be particularly hazardous if allowed to accumulate in a confined space because the dust can ignite with explosive force if a spark or even a hot surface is present.

32. 3 Chelating agents are added to plating solutions to improve the quality of the plated surfaces.

33. 1 The first step in hydroxide precipitation of common metals is to determine the optimum pH for precipitating each type of metal you wish to remove.

34. 4 Toxic organics may be found in industrial wastestreams with oily wastes and solvents used for cleaning.

35. 2 The intake lines of automatic sampling devices should be cleaned out regularly to prevent buildup with precipitates, which may slough off and contaminate the sample.

36. 2 First aid is emergency treatment for injury or sudden illness before regular medical treatment is available.

37. 3 A measurement is the comparison of a quantity with a standard unit of measurement.

38. 2 Precision refers to how closely an instrument can reproduce a given reading time after time.

39. 2 The purpose of surge protection devices on pressure sensors is to prevent sudden pressure surges from damaging the pressure sensor.

40. 1 Cuts, bruises, strains, and sprains are the most common physical injuries in wastewater treatment facilities.

41. 1 Trenches or excavations should have proper shoring because without proper shoring a trench or excavation can cave in and kill you.

42. 3 Feed an antifoaming agent if excess foaming begins to cause serious problems with wastewater treatment processes.

43. 2 A cross connection is a connection between a potable (drinking) water and any other location where wastewater could contaminate a domestic water supply.

44. 1 Mechanical maintenance is of prime importance in an industrial wastewater treatment plant because equipment must be kept in good operating condition for the plant to maintain peak performance.

45. 2 The purpose of wearing rings on a pump is to keep the pump from losing suction.

46. 2 A fuse is a protective device that has a strip or wire of fusible metal that, when placed in a circuit, will melt and break the electric circuit when subjected to an excessive temperature.

47. 3 Parallel and angular are the types of pump and motor misalignment.

48. 4 The purpose of a dehumidifier is to remove moisture from the surrounding atmosphere.

49. 1 Fines and extra charges for accidental spills, dumps, or additional loadings should be greater than the extra costs associated with resulting problems and a poor plant effluent to discourage future problems resulting from negligence.

50. 3 A good sampling point is one that is easily accessible and located where a representative sample may be obtained.

51. 1 Begin the search to trace an industrial discharge causing problems at a wastewater treatment plant at a point in the treatment plant or collection system where the discharge has been discovered.

Multiple Choice

1. 1, 3, 4, 5 — The industrial wastewater treatment plant operator physically collects samples, lubricates equipment, pushes switches, and turns valves.

2. 1, 2, 3, 4, 5 — Industrial wastewater must receive adequate treatment to protect game fishing, health of downstream neighbors, property values, recreational uses, and water for water-using industries.

3. 2, 3 — Definitions of pollution include any failure to meet water quality requirements or any interference with the beneficial reuse of water.

4. 2, 4, 5 — Possible alternatives for treatment of industrial wastewaters include discharge to a POTW for treatment and return to the environment; pretreatment by industry, followed by discharge to a POTW; and treatment by industry to the extent required before discharge to a receiving water.

5. 1, 2, 3, 4, 5 — An operator needs an understanding of the concentration and the mass of a pollutant in an industrial waste to determine the effects on the industry's pretreatment system; the POTW's collection system, disposal systems, and treatment system; and the sampling of the industry's discharge.

6. 1, 2, 3, 4 — Acids will corrode concrete and cast-iron sewers, concrete wet wells and tanks, internal steel equipment in clarifiers, and internal steel equipment in pumps.

7. 2, 3, 4, 5 — Measurements of wastewater used to detect changes in the influent to the industrial waste treatment system (IWTS) or the publicly owned treatment works (POTW) include conductivity, flow, pH, and temperature.

8. 1, 5 — Industrial facilities discharge treated wastewater directly to the environment or to wastewater collection systems and treatment plants (POTWs).

9. 2, 3 — The pretreatment standards for the metal finishing category list the contaminant and the contaminant concentrations including the maximum for any one day, mg/L, and the monthly average limit, mg/L.

10. 1, 2, 5 — EPA has promulgated only production-based pretreatment standards for the battery manufacturing industry, the coil coating industry, and the pesticides industry.

11. 1, 2, 3, 5 — The benefits of pollution prevention include a process change that could eliminate a whole wastestream, good operating practices that could minimize the volume of waste entering the treatment system, material substitution that could eliminate a pollutant altogether, and the potential of minimizing or even eliminating particular pollutants from the wastestream.

12. 2, 3, 4, 5 — The role of industrial treatment personnel in pollution prevention is to be aware of pollution prevention opportunities in the areas of good housekeeping, implementation of pollution prevention strategies, improved operation and maintenance, and process modification.

13. 1, 2, 3, 4, 5 — The economic benefits and cost savings that could be achieved by a pollution prevention program include lower health and safety costs, lower long-term environmental liability and insurance costs, lower manifesting and reporting costs, lower risks of emergency spills and accidents, and reduced transportation and disposal costs for sludges.

14. 1, 2, 4, 5 — The keys to pollution prevention in the chemicals formulating industry include avoiding the use of chlorinated solvents, better scheduling to minimize changeover, using alternative cleaning methods to minimize cleaning wastes, and using better cleaning methods to minimize cleaning wastes.

15. 1, 2, 4 — Canals, chlorine contact chambers, and gravity sewer lines are examples of open channel flow.

16. 1, 3, 4 — In closed pipe flow, the primary element generally consists of either a flow tube (such as a Venturi) or an orifice plate.

17. 1, 2, 3, 4, 5 — Factors that may cause measurement errors when using bubblers include clogs in the end of the bubbler tube from solids in the wastestream, density of the liquid, kinks in the bubbler tubing, lag effects, and splits in the bubbler tubing.

18. 2, 4 — When a staff gauge is unavailable, a measuring tape or a thin steel rule can be used to make a manual measurement.

19. 4, 5 — Releases of process wastewater and spent plating baths are types of batch discharges that may cause the influent concentration to vary.

20. 2, 3 — Industrial wastewater treatment system (IWTS) recycle flows that need to be included in the calculation of the volume of the equalization tank include the decant water from the digester or sludge holding tank and the filter backwash.

21. 1, 2, 4, 5 — When troubleshooting automatically cleaned bar screens, problems that may be discovered include excessive noise or vibration, faulty bar rack engagement by cleaning rake, screenings dropping off cleaning rake shelf, and too high temperature.

22. 1, 2, 3, 5 — Safety precautions that must be practiced when working around screens include being careful to avoid tripping or slipping and falling into the moving machinery or the wastewater; keeping walkways and guardrails clean, in good repair, and free of obstructions; and staying behind guardrails whenever possible.

23. 1, 2, 4 — Important process control guidelines for the chemical treatment processes include controlling the chemical dose, controlling the intensity of mixing during flocculation, and providing enough energy to completely mix the chemicals with the wastewater.

24. 1, 2 — Hydrated lime (calcium hydroxide) is used in treating industrial wastewaters to adjust the pH to improve the coagulation process and to coagulate solids.

25. 2, 3 — Operators can obtain information on the best chemical to use and the best dosage to feed from a jar test.

26. 1, 2, 3, 4, 5 — When reviewing chemical feed system designs and specifications, be sure provisions are made for standby equipment in order to maintain uninterrupted dosages during equipment maintenance; determine if sampling points are provided to measure chemical feeder output; examine plans for valving to allow flushing the system with water before removing from service; look for adequate valving to allow bypassing or removing equipment for maintenance without interrupting the chemical dosage; and review the results of the pre-design tests to determine the chemical feed rate for both the present and future.

27. 1, 5 — The purposes of a filter underdrain system are to collect the filtered water uniformly throughout the bed and to uniformly apply the backwash water during backwashing.

28. 1, 2, 3, 4, 5 — A log of the filtering operation should include backwash water flow rates and duration, chemicals added as filter aids, head loss at start and end of filter run, influent and effluent turbidity, and remarks of special observations and maintenance.

29. 1, 2, 3, 4, 5 — If there are problems with the performance of a continuous backwash, upflow, deep-bed silica sand media filter, an operator should investigate whether (1) the chemical feed systems are functioning properly; (2) the influent suspended solids exceed the design solids loading rate; (3) the influent flow rate exceeds the design flow and loading rates; (4) sand is being pumped to the top of the airlift and at the proper rate; and (5) there is enough reject water flow.

30. 1, 2, 3 — In the air stripping process, the amount of liquid chemical (pollutant) transferred from the water to the air depends on the air pressure, the properties of the liquid chemical, and the temperature.

31. 2, 4 — Flow rates through the carbon system can be controlled by controlling the pumping rate and by throttling the effluent valve on an individual adsorber.

32. 3, 4 — Operators sample both the influent and the effluent of the activated carbon process to determine organic contaminant removal efficiencies and the quantities of carbon fines remaining in the effluent from the process.

33. 1, 4, 5 — An important aspect of physical–chemical treatment is precise and accurate control of chemical doses, ORP, and pH.

34. 1, 4, 5 — Operators should be very careful whenever strongly acidic and basic solutions are mixed because of the potential for the generation of heat, the production of toxic gases, and splattering.

35. 2, 3, 5 — The primary problems encountered with ion exchange systems include excessive pressure drop across the ion exchange system, the failure to produce specified water quality, and throughput capacity loss.

36. 2, 3, 4, 5 — Common methods used to dewater metallic sludges include bag filters, centrifuges, filter presses, and vacuum filters.

37. 1, 2, 3, 4, 5   Industrial treatment plant operators can obtain information regarding the treatment processes they are required to operate from (1) instrument service company representatives; (2) manufacturers' technical representatives; (3) operators who treat wastewater for other companies or shops; (4) system construction drawings; and (5) system O & M manuals.

38. 1, 2, 3, 4, 5   After starting a pump, check to be sure the pump anchorage is tight; check to see that the bearings do not overheat from over- or underlubrication; check to see that the direction of rotation is correct; inspect the piping for leaks; and observe the packing gland boxes (stuffing boxes) for slight leakage.

39. 1, 2, 3, 4, 5   Operators should know how to perform first aid for the following situations: (1) contact with acids; (2) contact with bases; (3) eye burns; (4) hypochlorite solution contact; and (5) skin burns.

40. 2, 3, 4   In industrial waste treatment, the variables most commonly measured are flow, level, and pressure.

41. 1, 2, 3, 4, 5   Panel instruments that could be located in the main control panel of a typical industrial treatment plant include alarms, controllers, indicators, recorders, and totalizers.

42. 1, 2, 4   Proper application, proper operation and maintenance, and proper set-up will help to ensure instrumentation systems are reliable year after year.

43. 1, 2, 3, 4, 5   Operators may be exposed to the following types of hazards: (1) confined spaces; (2) infections and infectious diseases; (3) physical injuries; (4) toxic and harmful chemicals; and (5) toxic or suffocating gases or vapors.

44. 1, 2, 4, 5   A confined space entry permit includes certification that all existing hazards have been evaluated by a competent person; certification that necessary protective measures have been taken to ensure the safety of each worker; the location of work to be done; and the type of work to be done.

45. 1, 2, 3, 4, 5   The types of hazardous or toxic substances that could reach an industrial wastewater treatment facility include amines, biocides, fuels, surface-active agents, and toxic gases.

46. 1, 2, 4   The three essential ingredients necessary for a fire are fuel, heat, and oxygen.

47. 1, 2, 3, 4, 5   An effective safety program stresses the following points: (1) accidents are studied step by step and thoroughly reviewed with the attitude that "they are caused and do not just happen"; (2) adequate safety equipment is available and its capabilities are thoroughly understood; (3) basic safety concepts and practices are thoroughly understood by all; (4) everyone participates and accepts personal responsibility for their own safety and that of their fellow workers; and (5) everyone realizes that safety is a continuing learning and re-learning process.

48. 1, 2, 5   The benefits of a good maintenance record-keeping program include helping to keep the equipment warranty in force, indicating when maintenance is due, and providing a record of equipment performance.

49. 2, 3, 4   Items that are included in the total head in a pump system from where the water enters the suction piping to the end of the discharge piping include discharge lift, friction in the piping, and suction lift.

50. 1, 3, 4, 5   Factors that could cause lower than normal discharge pressures in a centrifugal pump include a broken discharge pipe, a different point of discharge, a worn impeller in the pump, and worn wearing rings in the pump.

51. 1, 2, 3, 4, 5   Pump station operating conditions that can affect the performance and the life of electrical equipment include corrosive atmospheres, dust, heat, humidity, and vibration.

52. 1, 2, 3, 5   Magnetic starters are used to start blowers, compressors, pumps, and anything where automatic or remote control is desired.

53. 1, 2, 3, 4, 5   Treatment plant problems that could occur as a result of a discharge of improperly pretreated industrial wastes (shock load) include (1) hydraulic overloads could "wash out" the microorganisms in the treatment process; (2) odors could result from upset biological treatment processes; (3) solids carryover to the effluent could result from a process upset; (4) toxic materials could destroy or affect the metabolism of microorganisms that are part of the treatment process; and (5) toxic materials could pass through the treatment process and be present in the plant effluent.

54. 1, 2, 3, 5   Composite samples can be obtained by using conveyors with small "buckets" attached, scoops, small pumps, or vacuum chambers.

55. 2, 3   Glass and polyethylene are used to contain samples collected from the discharge of industrial wastewaters.

56. 3, 4, 5   Before entering a confined space, provisions must be made to monitor the atmosphere whenever anyone is in the space; provisions must be made for two people to be standing by topside to assist in an emergency; and you must be wearing a safety harness.

Short Answer

1. An operator keeps an industrial wastewater treatment plant working by operating and maintaining the equipment, collecting samples, and recording data. The operator also communicates with supervisors and management, and explains the operations and importance of the facility to the public.

2. Adequate, reliable records of every phase of plant operation are needed to document the effectiveness of the operation and the plant's compliance with regulatory requirements.

3. The main job of an industrial wastewater treatment plant operator is to protect the many users of receiving waters. Also, if the industrial wastewater treatment plant discharges to the municipal wastewater collection systems, toxic industrial wastes must be treated to prevent toxic wastes from upsetting municipal treatment plants, flowing through the municipal plants, and killing aquatic life in receiving waters.

4. One advantage of the use of chlorine as a disinfectant is that chlorine is economical and effective. A limitation is that when chlorine combines with certain other chemicals it may cause long-term disease.

5. Compatible pollutants sometimes exhibit the characteristics of noncompatible pollutants, and vice versa. Soluble BOD from a food industry may have some harmful effects on a POTW's secondary treatment system. The accidental discharge of ammonia by a fertilizer manufacturer may disrupt the nitrification/denitrification or stripping tower processes used by the POTW to treat ammonia. On the other hand, some of the heavy metals (usually classified as noncompatible pollutants) are used as micronutrients to aid in the production of biological mass and the reduction of BOD. Certain organic chemical wastes, such as acetone and isopropanol, are biodegradable and, in dilute solutions, are removed by biological action in secondary treatment.

6. The discharge of out-of-pH range wastewater by an industry will result in damage to the sewer, which, if left undetected, can eventually corrode the pipe completely, causing exfiltration and contamination of the groundwater. Infiltration of groundwater could occur in areas where the groundwater level is above the depth of the sewer. Industrial discharge violations of pH will also increase the maintenance requirements on pumps in the pumping stations. The damage to pumps could eventually cause their failure, resulting in sewer backups and raw wastewater overflows.

7. Processes used to clean and prepare the surface of a workpiece for a surface coating include (1) alkaline cleaning, (2) acid cleaning, (3) paint stripping, and (4) solvent degreasing.

8. By delegating its implementation and enforcement authority to local agencies, EPA utilizes local industrial source control programs already in effect. The local POTWs are familiar with their industrial dischargers, may already have developed an extensive database, and may have ongoing wastewater permit and administration mechanisms. Local authorities are also better able to respond promptly and effectively to wastewater collection system or treatment plant emergencies; they are better able to resolve problems with industrial dischargers; and greater resources are usually available to local agencies to conduct a pretreatment program.

9. Categorical pretreatment standards were established by EPA in conjunction with its program to regulate the direct discharge to the environment of wastewaters from certain categories of industrial facilities. The standards apply to both direct discharges to the environment and to indirect discharges passing through a POTW. Categorical pretreatment standards are technology-based standards, meaning they are based upon the available treatment technologies that could be used to remove pollutants.

10. An industrial discharger can avoid monitoring for TTO by submitting a Solvent Management Plan for the POTW's review and approval.

11. Waste minimization is the reduction of hazardous waste that is generated or subsequently treated, stored, or disposed of. It includes any activity that results in the reduction of the total volume or quantity of hazardous waste, or the reduction of toxicity of the hazardous waste.

12. Employees should be trained regarding proper process operations and procedures so that off-specification process batches will not be generated or will be reworked rather than disposed of through the wastewater treatment system.

13. For industries with fluctuating or unusual waste discharge flow rates, composite samples taken over time are most representative if they are acquired using flow proportioning techniques.

14. Differential pressure devices measure flow in a pipe by measuring the change in pressure that occurs when the diameter of the pipe is reduced, comparing it to the original line pressure, and relating the pressure drop to the known change in pipe area. The velocity can then be determined and converted to a flow measurement.

15. In-line equalization is more effective in leveling the influent concentration variations because the entire flow is blended with the entire contents of the holding tank.

16. Screening protects downstream processes and equipment such as pumps, valves, aeration headers, and heat exchanges, which often become clogged with nuisance solids.

17. Some pH adjustment systems use three pH electrodes and use the middle electrode value because of the variability of pH electrodes.

18. The three most important process control guidelines for any chemical treatment process are:

    1. Providing enough energy to completely mix the chemicals with the wastewater.
    2. Controlling the intensity of mixing during flocculation.
    3. Controlling the chemical(s) dose.

19. Paddle speed is important during flocculation because it must be sufficient to keep the floc from settling while at the same time it must not be so great as to shear and break up the floc formed.

20. Extensive laboratory testing should be conducted before treating the entire plant effluent because of the wide selection of available polyelectrolyte products and the different and changing chemical characteristics of water being treated.

21. To run the jar test, various types of chemicals or different doses of a single chemical are added to sample portions of wastewater and all portions of the sample are rapidly mixed. After rapid mixing, the samples are slowly mixed to approximate the conditions in the plant. Mixing is then stopped and the floc formed is allowed to settle. The appearance of the floc, the time required to form a floc, and the settling conditions are recorded. The supernatant (liquid above the settled sludge) is analyzed for turbidity, suspended solids, and pH. With this information, the operator selects the best chemical or best dosage to feed on the basis of clarity of effluent and minimum cost of chemicals.

22. Economic factors that should be considered when selecting a chemical feeder include costs of purchase, installation, operation, maintenance, and replacement and energy requirements.

23. Multi-media filters are used to take advantage of solids removal by both surface straining and depth filtration. Depth filtration has a slower buildup of head loss, but solids will break through more readily than with surface straining. To reduce breakthrough yet retain depth filtering, the mixed-media design is used.

24. Floatable and settleable solids are removed from holding tanks similar to any other clarifier. Flights or scrapers move settled solids to a sludge hopper and floatable solids are removed by a pan-type skimmer.

25. Continuous backwash, upflow, deep-bed silica sand media filters have two main advantages over other types of granular media filters:

    1. They do not need to be taken out of service for backwashing.
    2. They operate continuously so fewer filters or smaller-sized filters may be enough to provide full design flow processing capacity.

26. Fouled membranes can be cleaned with caustic, acids, surfactants, chlorine, or hydrogen peroxide and cleaning will return the membrane flux very nearly to original values.

27. Even under ideal conditions (pure feedwater and no fouling of the membrane surface), there is a decline in water flux with time. This decrease in flux is due to membrane compaction. Flux decline also results from foulants and bacterial growth.

28. The most common symptoms of operating problems in an air stripper are high differential pressure readings or low removal efficiencies.

29. Industries using activated carbon to treat industrial wastestreams include the petroleum, chemical, and textile industries.

30. COD and turbidity tests should be run on the effluent of the activated carbon process to determine process efficiency. Determination of the quantities of carbon fines remaining in the effluent gives an indication of the status of the carbon in the carbon column contactors.

31. The costs and complexity of metal waste treatment facilities can be reduced by an effective source control program. Plating processes can be changed or altered to reduce the volumes of process wastewaters and to control amounts of wastes that must be treated. Volumes of wastewater to be treated can be reduced by countercurrent rinsing processes. Good housekeeping practices, prevention of accidental spills, and control of leaks and drag out are other procedures used to reduce the amount of waste that must be treated.

32. A spent process solution (a pickle, for example) should be treated as a batch process and not dumped directly into rinse waters because the dump can cause significant increases in the contaminant level. The control system for rinse water treatment cannot respond to this surge without losing the close control required for normal discharge of effluent.

33. Ion exchange is the process in which ions, held by electrostatic forces to charged functional groups on the surface of an ion exchange resin, are exchanged for ions of similar charge from the solution in which the resin is immersed. The ion exchange process may be used to recover precious metals from metal wastes, including copper, molybdenum, cobalt, nickel, gold, and silver. Metal plating chemicals recovered by ion exchange include chromium, nickel, phosphate solution, sulfuric acid from anodizing, and chromic acid. Ion exchange units also have been used to treat metal wastes containing aluminum, arsenic, cadmium, chromium (hexavalent and trivalent), cyanide, iron, lead, manganese, selenium, tin, and zinc.

34. An operator should have a fundamental knowledge of electrical power and how it is controlled for the operator's safety, to ensure the continuous operation of the plant, to analyze electrical problems, and to communicate effectively with electricians. An operator should also pay constant attention to electrical safety.

35. When handling acids, wear protective clothing and equipment to prevent body contact with the acid. Wear rubber gloves, and safety goggles or a face shield for eye protection against splashing. Also wear a rubber apron, rubber boots, and a long-sleeved polyester shirt.

36. Operators must have a good working familiarity with measurement and control systems to properly monitor and control the treatment processes.

37. A screwdriver should not be used to test an electric circuit because electric shock can and does cause serious burns and even death (by asphyxiation due to paralysis of the muscles used in breathing).

38. It is poor practice to ignore the lamps that are lit up (alarm conditions) on an annunciator panel because a true alarm condition requiring immediate operator attention may be lost in the resulting general indifference to the alarm system.

39. Oil and moisture in instrument air can cause frequent operational problems. Oil and moisture in instrument air can be removed by filtering and drying. Oil is removed by filtering the air through special oil-absorbent elements. A process called desiccation can be used to remove the water.

40. The operator's responsibility with regard to safety includes making certain that the facility is maintained in such a manner as to continually provide a safe place to work.

41. You can avoid transmitting disease to your family by wearing proper protective gloves when you may contact wastewater or sludge in any form, not wearing your work clothes home, keeping your work clothes clean, storing your street clothes and your work clothes in separate lockers, and, if your employer does not supply you with uniforms and laundry service, washing your work clothes separately from your regular family wash.

42. Job site protection when working in manholes should include barricades and traffic warning devices for the safety of vehicles, bikes, pedestrians, and workers.

43. Only qualified electricians should work on an electrical control panel because unqualified persons may be seriously injured by electric shock and they could damage the equipment.

44. When applying protective coatings, evaluate potential atmospheric hazards, provide adequate ventilation, and use protective equipment (protective clothing and creams). Other precautions include using nonhazardous coating materials (if possible) and brushing or rolling rather than spraying coatings.

45. Samples for lab tests should be collected by using sample poles or ropes attached to sample collectors and by staying behind guardrails and chains.

46. Safety equipment should be checked periodically to be sure the operators know how to operate the equipment and that the equipment works properly.

47. The operator should thoroughly read and understand manufacturers' literature before attempting to maintain plant equipment in order to do the job required to keep the equipment operating.

48. A maintenance recordkeeping program indicates the frequency of the various maintenance jobs for each piece of equipment and also indicates what was done, when, and who did the work. The information is very valuable in revealing when major repairs or replacements are necessary.

49. To determine if a pump will turn in the direction intended, momentarily start the motor by a quick electrical contact and check to be sure the motor will turn the pump in the direction indicated by the rotational arrows marked on the pump.

50. Protective or safety devices used to protect operators and equipment from being harmed by electricity include fuses, circuit breakers, motor and circuit overload devices, and grounds.

51. If pump bearings are running hot, check for under- or over-lubrication. If proper lubrication is observed, check alignment of pump and motor.

52. A stethoscope is used to listen to an electric motor for whines, gratings, or uneven noises. (A large screwdriver is used by some operators to magnify noises in electric motors.)

53. Inactive gate valves should be operated periodically to prevent sticking.

54. Scum lines can be kept from plugging by an effective preventive maintenance program. Remove scum from troughs, and clean lines at regular intervals (monthly).

55. Plans and specifications should be carefully reviewed before construction and installation because changes are easily made on paper but are much more difficult in the field. A few hours spent reviewing plans and specifications will save many days of hard and discouraging labor in the future when it is a major job to make a change.

56. If improperly pretreated industrial wastes are discharged, a wastewater treatment plant could experience the following problems:

    1. Failure to meet the discharge requirements set by the NPDES permit, the regulating agency, or other authority.
    2. A process upset occurring due to toxic materials that destroy or affect the metabolism of microorganisms that are part of the treatment process.
    3. Toxic materials that pass through the treatment process and are present in the plant effluent.
    4. Hydraulic overloads upsetting treatment processes by "washing out" the microorganisms in the treatment process.
    5. Process upsets resulting in solids carryover to the effluent and odors resulting from upset biological treatment processes.

57. Representative samples must be collected in order to know the actual characteristics of an industrial waste discharge.

58. To determine the peak concentration and time of occurrence of a toxic waste discharge, collect grab samples hourly or use a composite sampler that collects samples at regular time intervals and does not mix the samples. Analyze the individual samples—do not mix them.

59. The method used to measure flow from an industry will depend on each particular situation. For larger industries, open-channel flowmeters, such as Parshall flumes, generally are used. For smaller enterprises, portable metering equipment often is installed in existing collection systems. The standard approach is to use a compact flume, such as a Palmer-Bowlus flume, as the primary flow-measuring device.

Problems

1. Determine the mass emission rate in pounds of copper per day from an electroplater if the effluent flow is 5,000 gallons per day containing 6 milligrams per liter of copper.

Known	Unknown
Effluent Flow, GPD = 5,000 GPD	Mass Emission Rate, lbs/day
or MGD = 0.005 MGD	
Copper Conc, mg/L = 6 mg/L	

    Calculate the mass emission rate in pounds per day.

    $$\text{Mass Emission Rate, lbs/day} = (\text{Flow, MGD})(\text{Conc, mg}/L)(8.34 \text{ lbs/gal})$$
    $$= (0.005 \text{ MGD})(6 \text{ mg}/L)(8.34 \text{ lbs/gal})$$
    $$= 0.25 \text{ lb/day}$$

2. An equalization basin stores 150,000 gallons of wastewater from an eight-hour work shift. The pretreatment system is designed to treat a flow of 120 GPM. How long will it require to process the wastewater from the eight-hour production period?

Known	Unknown
Wastewater Volume, gal = 150,000 gallons	Process Time, hours
Pretreatment Flow, GPM = 120 GPM	

    Calculate the process time in hours.

    $$\text{Process Time, hours} = \frac{\text{Wastewater Vol, gal}}{(\text{Pretreatment Flow, GPM})(60 \text{ min/hr})}$$
    $$= \frac{150,000 \text{ gallons}}{(120 \text{ gal/min})(60 \text{ min/hr})}$$
    $$= 20.8 \text{ hours}$$

3. An industrial pretreatment facility receives a flow of 200 GPM containing a BOD of 450 mg/L for a 24-hour period. Estimate the BOD loading on the facility in pounds of BOD per day. What would be the BOD loading if the flow was for only one eight-hour shift during 24 hours?

Known	Unknown
Flow, GPM = 200 GPM	a. BOD Loading, lbs/day for 24 hours
BOD, mg/L = 450 mg/L	b. BOD Loading, lbs/day for 8 hours

    1. Calculate the BOD loading in pounds per day for 24 hours.

    $$\text{BOD, lbs/day} = \frac{(\text{Flow, GPM})(\text{BOD, mg}/L)(8.34 \text{ lbs/gal})}{694 \text{ GPM/MGD}}$$
    $$= \frac{(200 \text{ GPM})(450 \text{ mg}/L)(8.34 \text{ lbs/gal})}{694 \text{ GPM/MGD}}$$
    $$= 1,082 \text{ lbs/day}$$

    2. Calculate the BOD loading in pounds per day for one 8-hour shift during 24 hours.

    $$\text{BOD, lbs/day} = \frac{(\text{Flow, GPM})(\text{BOD, mg}/L)(8.34 \text{ lbs/gal})(\text{Time, hr/day})}{(694 \text{ GPM/MGD})(24 \text{ hr/day})}$$
    $$= \frac{(200 \text{ GPM})(450 \text{ mg}/L)(8.34 \text{ lbs/gal})(8 \text{ hr/day})}{(694 \text{ GPM/MGD})(24 \text{ hr/day})}$$
    $$= 361 \text{ lbs/day}$$

4. An 80-foot diameter clarifier drops 4.0 inches in 20 minutes during a calibration test on the RAS pump. Calculate the RAS pumping rate in gallons per minute (GPM).

Known	Unknown
Diameter, ft = 80 ft	Pumping Rate, GPM
Drop, in = 4.0 in	
Time, min = 20 min	

    Calculate the RAS pumping rate in GPM.

    $$\text{Pumping Rate, GPM} = \frac{\text{Volume Pumped, gal}}{\text{Time Pumped, min}}$$
    $$= \frac{(0.785)(\text{Diameter, ft})^2(\text{Drop, in})(7.48 \text{ gal/cu ft})}{(\text{Time, min})(12 \text{ in/ft})}$$
    $$= \frac{(0.785)(80 \text{ ft})^2(4.0 \text{ in})(7.48 \text{ gal/cu ft})}{(20 \text{ min})(12 \text{ in/ft})}$$
    $$= 626 \text{ GPM}$$

5. Estimate the average flow in gallons per minute from an equalization basin if the total flow volume is 40,000 gallons and the release time is from 8 am to 5 pm.

Known	Unknown
Total Volume, gal = 40,000 gal	Flow, GPM
Release Time, hr = 8 am to 5 pm	
= 9 hours	

Estimate the average flow in gallons per minute.

$$\text{Flow, GPM} = \frac{\text{Total Volume, gal}}{(\text{Release Time, hr})(60 \text{ min/hr})}$$

$$= \frac{40,000 \text{ gal}}{(9 \text{ hr})(60 \text{ min/hr})}$$

$$= 74 \text{ GPM}$$

6. Determine the chemical application rate in pounds per day and pounds per hour to treat a flow of 200 GPM with a chemical dose of 6 mg/L.

Known		Unknown
Flow Rate, GPM	= 200 GPM	a. Chemical Application Rate, lbs/day
Chemical Dose, mg/L	= 6 mg/L	b. Chemical Application Rate, lbs/hr

1. Calculate the chemical application rate in pounds per day.

$$\text{Application Rate, lbs/day} = \frac{(\text{Flow, MGD})(\text{Chem Dose, mg}/L)(8.34 \text{ lbs/gal})}{}$$

$$= \frac{(200 \text{ GPM})(6 \text{ mg}/L)(8.34 \text{ lbs/gal})}{694 \text{ GPM/MGD}}$$

$$= 14.42 \text{ lbs/day}$$

2. Convert the application rate from lbs/day to lbs/hr.

$$\text{Application Rate, lbs/hr} = \frac{\text{Application Rate, lbs/day}}{24 \text{ hr/day}}$$

$$= \frac{14.42 \text{ lbs/day}}{24 \text{ hr/day}}$$

$$= 0.60 \text{ lb/hr}$$

7. Jar test results indicate an optimum chemical dose of 6 mg/L to treat a flow of 0.3 MGD. If the chemical feed solution strength is 0.5 percent, what is the chemical pumping rate in gallons per hour?

Known		Unknown
Chemical Dose, mg/L	= 6 mg/L	Chemical Pumping Rate, gal/hr
Flow Rate, MGD	= 0.3 MGD	
Solution Strength, %	= 0.5%	

Calculate the chemical pumping rate for a 0.5% solution.

$$\text{Chem Pumping Rate, gal/hr} = \frac{(\text{Flow, MGD})(\text{Dose, mg}/L)(8.34 \text{ lbs/gal})(100\%)}{(24 \text{ hr/day})(8.34 \text{ lbs/gal})(\text{Sol Strength, \%})}$$

$$= \frac{(0.3 \text{ MGD})(6 \text{ mg}/L)(8.34 \text{ lbs/gal})(100\%)}{(24 \text{ hr/day})(8.34 \text{ lbs/gal})(0.5\%)}$$

$$= 15 \text{ gal/hr}$$

8. Estimate the polymer dose in milligrams per liter if 80 pounds of polymer per day are delivered to a flow of 5,000 GPM.

Known		Unknown
Polymer Delivered, lbs/day	= 80 lbs/day	Polymer Dose, mg/L
Flow, GPM	= 5,000 GPM	

Estimate the polymer dose in milligrams per liter.

$$\text{Polymer Dose, mg}/L = \frac{\text{Polymer Delivered, lbs polymer/day}}{\text{Flow, million lbs water/day}}$$

$$\text{or} \quad \frac{\text{Polymer Delivered, lbs polymer/day}}{(\text{Flow, GPM})(8.34 \text{ lbs/gal})(60 \text{ min/hr})(24 \text{ hr/day})}$$

$$= \frac{80 \text{ lbs polymer/day}}{(5,000 \text{ gal/min})(8.34 \text{ lbs/gal})(60 \text{ min/hr})(24 \text{ hr/day})}$$

$$= \frac{80 \text{ lbs polymer/day}}{60,048,000 \text{ lbs water/day}} \times 1,000,000/\text{M}$$

$$= \frac{80 \text{ lbs polymer/day}}{60 \text{ M lbs water/day}}$$

$$= 1.3 \text{ mg polymer/liter water}$$

$$= 1.3 \text{ mg}/L$$

9. Polymer is supplied to your plant at a concentration of 0.6 pound polymer per gallon. The polymer feed pump delivers a flow of 0.12 GPM and the flow being treated is 2,500 GPM. What is the polymer dose in mg/L in the water being treated?

Known		Unknown
Polymer Conc, lbs/gal	= 0.6 lb/gal	Polymer Dose, mg/L
Polymer Pump, GPM	= 0.12 GPM	
Flow, GPM	= 2,500 GPM	

Calculate the polymer dose, mg/L.

$$\text{Dose, mg}/L = \frac{(\text{Pump, GPM})(\text{Conc, lbs/gal})}{(\text{Flow, GPM})(8.34 \text{ lbs/gal})}$$

$$= \frac{(0.12 \text{ GPM})(0.6 \text{ lb/gal})}{(2,500 \text{ GPM})(8.34 \text{ lbs/gal})}$$

$$= \frac{0.072 \text{ lb polymer}}{20,850 \text{ lbs/water}} \times 1,000,000/\text{M}$$

$$= \frac{3.45 \text{ lbs polymer}}{1 \text{ M lbs water}}$$

$$= \frac{3.45 \text{ mg polymer}}{1 \text{ M mg water}}$$

$$= 3.45 \text{ mg}/L$$

10. Liquid alum is supplied at a concentration of 5.4 pounds alum per gallon. If the alum feed pump delivers 80 mL per minute to a flow of 2,000 GPM, calculate the dose of alum in mg/L to the water being treated.

Known	Unknown
Alum Conc, lbs/gal = 5.4 lbs/gal	Alum Dose, mg/L
Alum Pump, mL/min = 80 mL/min	
Flow, GPM = 2,000 GPM	

Calculate the alum dose in mg/L.

$$\text{Dose, mg}/L = \frac{(\text{Pump, m}L/\text{min})(\text{Conc, lbs/gal})(0.00026 \text{ gal/m}L)}{(\text{Flow, gal/min})(8.34 \text{ lbs/gal})}$$

$$= \frac{(80 \text{ m}L/\text{min})(5.4 \text{ lbs/gal})(0.00026 \text{ gal/m}L)}{(2,000 \text{ gal/min})(8.34 \text{ lbs/gal})}$$

$$= \frac{0.112 \text{ lbs alum}}{16,680 \text{ lbs water}} \times 1,000,000/\text{M}$$

$$= \frac{6.7 \text{ lbs alum}}{1\text{M lbs water}}$$

$$= 6.7 \text{ mg}/L$$

11. A membrane filtration system is operating with 500 square feet of membrane area and the permeate flow is 10 gallons per minute. What is the flux (J) in GFD (gallons/sq ft-day)?

Known	Unknown
Membrane Area, sq ft = 500 sq ft	J (Flux), GFD
Flow, gal/min = 10 gal/min	

Calculate the flux in GFD.

$$J, \text{GFD} = \frac{(\text{Flow, gal/min})(60 \text{ min/hr})(24 \text{ hr/day})}{\text{Membrane Area, sq ft}}$$

$$= \frac{(10 \text{ gal/min})(60 \text{ min/hr})(24 \text{ hr/day})}{500 \text{ sq ft}}$$

$$= 29 \text{ GFD}$$

INDUSTRIAL WASTE WORDS

A Summary of the Words Defined

in

INDUSTRIAL WASTE TREATMENT

PROJECT PRONUNCIATION KEY

by Warren L. Prentice

The Project Pronunciation Key is designed to aid you in the pronunciation of new words. While this key is based primarily on familiar sounds, it does not attempt to follow any particular pronunciation guide. This key is designed solely to aid operators in this program.

You may find it helpful to refer to other available sources for pronunciation help. Each current standard dictionary contains a guide to its own pronunciation key. Each key will be different from each other and from this key. Examples of the difference between the key used in this program and the *WEBSTER'S NEW WORLD COLLEGE DICTIONARY*[1] "Key" are shown below.

In using this key, you should accent (say louder) the syllable that appears in capital letters. The following chart is presented to give examples of how to pronounce words using the Project Key.

	SYLLABLE				
WORD	1st	2nd	3rd	4th	5th
acid	AS	id			
coliform	KOAL	i	form		
biological	BUY	o	LODGE	ik	cull

The first word, *ACID*, has its first syllable accented. The second word, *COLIFORM*, has its first syllable accented. The third word, *BIOLOGICAL*, has its first and third syllables accented.

We hope you will find the key useful in unlocking the pronunciation of any new word.

---

[1] The *WEBSTER'S NEW WORLD COLLEGE DICTIONARY*, Fourth Edition, 1999, was chosen rather than an unabridged dictionary because of its availability to the operator. Other editions may be slightly different.

INDUSTRIAL WASTE WORDS

>GREATER THAN

DO >5 mg/*L* would be read as DO GREATER THAN 5 mg/*L*.

<LESS THAN

DO <5 mg/*L* would be read as DO LESS THAN 5 mg/*L*.

40 CFR 403

EPA's General Pretreatment Regulations appear in the Code of Federal Regulations under 40 CFR 403. 40 refers to the numerical heading for the environmental regulations portion of the Code of Federal Regulations. 403 refers to the section that contains the General Pretreatment Regulations. Significant amendments to the General Pretreatment Regulations include the PIRT Amendments (*FEDERAL REGISTER,* October 18, 1988) and the DSS Amendments (*FEDERAL REGISTER,* July 24, 1990).

A

ACEOPS

See ALLIANCE OF CERTIFIED OPERATORS, LAB ANALYSTS, INSPECTORS, AND SPECIALISTS.

atm

The abbreviation for atmosphere. One atmosphere is equal to 14.7 psi or 100 kPa.

ABSORPTION (ab-SORP-shun)

The taking in or soaking up of one substance into the body of another by molecular or chemical action (as tree roots absorb dissolved nutrients in the soil).

ACCURACY

How closely an instrument measures the true or actual value of the process variable being measured or sensed.

ACTIVATED SLUDGE

Sludge particles produced in raw or settled wastewater (primary effluent) by the growth of organisms (including zoogleal bacteria) in aeration tanks in the presence of dissolved oxygen. The term "activated" comes from the fact that the particles are teeming with bacteria, fungi, and protozoa. Activated sludge is different from primary sludge in that the sludge particles contain many living organisms that can feed on the incoming wastewater.

ACTIVATED SLUDGE PROCESS

A biological wastewater treatment process that speeds up the decomposition of wastes in the wastewater being treated. Activated sludge is added to wastewater and the mixture (mixed liquor) is aerated and agitated. After some time in the aeration tank, the activated sludge is allowed to settle out by sedimentation and is disposed of (wasted) or reused (returned to the aeration tank) as needed. The remaining wastewater then undergoes more treatment.

ACUTE HEALTH EFFECT

An adverse effect on a human or animal body, with symptoms developing rapidly.

ADSORPTION (add-SORP-shun)

The gathering of a gas, liquid, or dissolved substance on the surface or interface zone of another material.

AERATION (air-A-shun) LIQUOR

Mixed liquor. The contents of the aeration tank, including living organisms and material carried into the tank by either untreated wastewater or primary effluent.

AERATION (air-A-shun) TANK

The tank where raw or settled wastewater is mixed with return sludge and aerated. The same as aeration bay, aerator, or reactor.

AEROBES

Bacteria that must have dissolved oxygen (DO) to survive. Aerobes are aerobic bacteria.

AEROBIC (air-O-bick)

A condition in which atmospheric or dissolved oxygen is present in the aquatic (water) environment.

AEROBIC BACTERIA (air-O-bick back-TEER-e-uh)

Bacteria that will live and reproduce only in an environment containing oxygen that is available for their respiration (breathing), namely atmospheric oxygen or oxygen dissolved in water. Oxygen combined chemically, such as in water molecules (H_2O), cannot be used for respiration by aerobic bacteria.

AEROBIC (air-O-bick) DIGESTION

The breakdown of wastes by microorganisms in the presence of dissolved oxygen. This digestion process may be used to treat only waste activated sludge, or trickling filter sludge and primary (raw) sludge, or waste sludge from activated sludge treatment plants designed without primary settling. The sludge to be treated is placed in a large aerated tank where aerobic microorganisms decompose the organic matter in the sludge. This is an extension of the activated sludge process.

AGGLOMERATION (uh-glom-er-A-shun)

The growing or coming together of small scattered particles into larger flocs or particles, which settle rapidly. Also see FLOC.

AGRONOMIC RATES

Sludge application rates that provide the amount of nitrogen needed by the crop or vegetation grown on the land while minimizing the amount that passes below the root zone.

AIR BINDING

The clogging of a filter, pipe, or pump due to the presence of air released from water. Air entering the filter media is harmful to both the filtration and backwash processes. Air can prevent the passage of water during the filtration process and can cause the loss of filter media during the backwash process.

AIR GAP

An open, vertical drop, or vertical empty space, between a drinking (potable) water supply and potentially contaminated water. This gap prevents the contamination of drinking water by backsiphonage because there is no way potentially contaminated water can reach the drinking water supply.

AIR LIFT PUMP

A special type of pump consisting of a vertical riser pipe submerged in the wastewater or sludge to be pumped. Compressed air is injected into a tail piece at the bottom of the pipe. Fine air bubbles mix with the wastewater or sludge to form a mixture lighter than the surrounding water, which causes the mixture to rise in the discharge pipe to the outlet.

ALARM CONTACT

A switch that operates when some preset low, high, or abnormal condition exists.

ALGAE (AL-jee)

Microscopic plants containing chlorophyll that live floating or suspended in water. They also may be attached to structures, rocks, or other submerged surfaces. Excess algal growths can impart tastes and odors to potable water. Algae produce oxygen during sunlight hours and use oxygen during the night hours. Their biological activities appreciably affect the pH, alkalinity, and dissolved oxygen of the water.

ALKALI (AL-kuh-lie)

Any of certain soluble salts, principally of sodium, potassium, magnesium, and calcium, that have the property of combining with acids to form neutral salts and may be used in chemical processes such as water or wastewater treatment.

ALKALINITY (AL-kuh-LIN-it-tee)

The capacity of water or wastewater to neutralize acids. This capacity is caused by the water's content of carbonate, bicarbonate, hydroxide, and occasionally borate, silicate, and phosphate. Alkalinity is expressed in milligrams per liter of equivalent calcium carbonate. Alkalinity is not the same as pH because water does not have to be strongly basic (high pH) to have a high alkalinity. Alkalinity is a measure of how much acid must be added to a liquid to lower the pH to 4.5.

ALLIANCE OF CERTIFIED OPERATORS, LAB ANALYSTS, INSPECTORS, AND SPECIALISTS (ACEOPS)

A professional organization for operators, lab analysts, inspectors, and specialists dedicated to improving professionalism; expanding training, certification, and job opportunities; increasing information exchange; and advocating the importance of certified operators, lab analysts, inspectors, and specialists. For information on membership, contact ACEOPS, 808 1st Avenue SE, Le Mars, IA 51031, phone (712) 548-4281, or e-mail: Info@aceops.org.

AMBIENT (AM-bee-ent) TEMPERATURE

Temperature of the surroundings.

AMPHOTERIC (AM-fuh-TUR-ick)

Capable of reacting chemically as either an acid or a base.

ANAEROBES

Bacteria that do not need dissolved oxygen (DO) to survive.

ANAEROBIC (AN-air-O-bick)

A condition in which atmospheric or dissolved oxygen (DO) is *NOT* present in the aquatic (water) environment.

ANAEROBIC BACTERIA (AN-air-O-bick back-TEER-e-uh)

Bacteria that live and reproduce in an environment containing no free or dissolved oxygen. Anaerobic bacteria obtain their oxygen supply by breaking down chemical compounds that contain oxygen, such as sulfate (SO_4^{2-}).

ANAEROBIC (AN-air-O-bick) SELECTOR

Anaerobic refers to the practical absence of dissolved and chemically bound oxygen. Selector refers to a reactor or basin and environmental conditions (food, lack of DO) intended to favor the growth of certain organisms over others. Also see SELECTOR.

ANALOG READOUT

The readout of an instrument by a pointer (or other indicating means) against a dial or scale. Also, the continuously variable signal type sent to an analog instrument (for example, 4–20 mA). Also see DIGITAL READOUT.

ANALYZER

A device that conducts periodic or continuous measurement of some factor such as chlorine, fluoride, or turbidity. Analyzers operate by any of several methods including photocells, conductivity, or complex instrumentation.

ANHYDROUS (an-HI-drous)

Very dry. No water or dampness is present.

ANION (AN-EYE-en)

A negatively charged ion in an electrolyte solution, attracted to the anode under the influence of a difference in electrical potential. Chloride ion (Cl⁻) is an anion.

ANNULAR (AN-yoo-ler) SPACE

A ring-shaped space located between two circular objects. For example, the space between the outside of a pipe liner and the inside of a pipe.

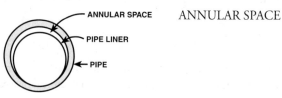

ANODIZING

An electrochemical process that deposits a coating of an insoluble oxide on a metal surface. Aluminum is the most frequently anodized material.

ANOXIC (an-OX-ick)

A condition in which the aquatic (water) environment does not contain dissolved oxygen (DO), which is called an oxygen deficient condition. Generally refers to an environment in which chemically bound oxygen, such as in nitrate, is present. The term is similar to ANAEROBIC.

ANOXIC DENITRIFICATION (dee-nye-truh-fuh-KAY-shun)

A biological nitrogen removal process in which nitrate nitrogen is converted by microorganisms to nitrogen gas in the absence of dissolved oxygen.

APPURTENANCE (uh-PURR-ten-nans)

Machinery, appliances, structures, and other parts of the main structure necessary to allow it to operate as intended, but not considered part of the main structure.

ARCH

(1) The curved top of a sewer pipe or conduit.

(2) A bridge or arch of hardened or caked chemical that will prevent the flow of the chemical.

ASPIRATE (AS-per-rate)

Use of a hydraulic device (aspirator or eductor) to create a negative pressure (suction) by forcing a liquid through a restriction, such as a Venturi tube. An aspirator may be used in the laboratory in place of a vacuum pump; sometimes used instead of a sump pump.

AUTOTROPHIC (auto-TROF-ick)

Describes organisms (plants and some bacteria) that use inorganic materials for energy and growth.

AVAILABLE EXPANSION

The vertical distance from the sand surface to the underside of a trough in a sand filter. This distance is also called FREEBOARD.

AXIAL TO IMPELLER

The direction in which material being pumped flows around the impeller or flows parallel to the impeller shaft.

AXIS OF IMPELLER

An imaginary line running along the center of a shaft (such as an impeller shaft).

B

BOD (pronounce as separate letters)

Biochemical Oxygen Demand. The rate at which organisms use the oxygen in water or wastewater while stabilizing decomposable organic matter under aerobic conditions. In decomposition, organic matter serves as food for the bacteria and energy results from its oxidation. BOD measurements are used as a surrogate measure of the organic strength of wastes in water.

BOD$_5$

BOD$_5$ refers to the five-day biochemical oxygen demand. The total amount of oxygen used by microorganisms decomposing organic matter increases each day until the ultimate BOD is reached, usually in 50 to 70 days. BOD usually refers to the five-day BOD or BOD$_5$.

BACTERIAL (back-TEER-e-ul) CULTURE

In the case of activated sludge, the bacterial culture refers to the group of bacteria classified as AEROBES and FACULTATIVE BACTERIA, which covers a wide range of organisms. Most treatment processes in the United States grow facultative bacteria that use the carbonaceous (carbon compounds) BOD. Facultative bacteria can live when oxygen resources are low. When nitrification is required, the nitrifying organisms are obligate aerobes (require oxygen) and must have at least 0.5 mg/L of dissolved oxygen throughout the whole system to function properly.

BAFFLE

A flat board or plate, deflector, guide, or similar device constructed or placed in flowing water, wastewater, or slurry systems to cause more uniform flow velocities, to absorb energy, and to divert, guide, or agitate liquids (water, chemical solutions, slurry).

BASE

(1) A substance that takes up or accepts protons.

(2) A substance that dissociates (separates) in aqueous solution to yield hydroxyl ions (OH$^-$).

(3) A substance containing hydroxyl ions that reacts with an acid to form a salt or that may react with metals to form precipitates.

BASELINE MONITORING REPORT (BMR)

All industrial users subject to categorical pretreatment standards must submit a baseline monitoring report (BMR) to the control authority (POTW, state, or EPA). The purpose of the BMR is to provide information to the control authority to document the industrial user's current compliance status with a categorical pretreatment standard.

BATCH PROCESS

A treatment process in which a tank or reactor is filled, the water (or wastewater or other solution) is treated or a chemical solution is prepared, and the tank is emptied. The tank may then be filled and the process repeated. Batch processes are also used to cleanse, stabilize, or condition chemical solutions for use in industrial manufacturing and treatment processes.

BENCH-SCALE ANALYSIS (TEST)

A method of studying different ways or chemical doses for treating water or wastewater and solids on a small scale in a laboratory. Also see JAR TEST.

BEST AVAILABLE TECHNOLOGY (BAT)

A level of technology represented by a higher level of wastewater treatment technology than required by Best Practicable Technology (BPT). BAT is based on the very best (state-of-the-art) control and treatment measures that have been developed, or are capable of being developed, and that are economically achievable within the appropriate industrial category.

BEST PRACTICABLE TECHNOLOGY (BPT)

A level of technology represented by the average of the best existing wastewater treatment performance levels within the industrial category.

BIOASSAY (BUY-o-AS-say)

(1) A way of showing or measuring the effect of biological treatment on a particular substance or waste.

(2) A method of determining the relative toxicity of a test sample of industrial wastes or other wastes by using live test organisms, such as fish.

BIOCHEMICAL OXYGEN DEMAND (BOD)

See BOD.

BIOCHEMICAL OXYGEN DEMAND (BOD) TEST

A procedure that measures the rate of oxygen use under controlled conditions of time and temperature. Standard test conditions include dark incubation at 20°C for a specified time (usually five days).

BIODEGRADABLE (BUY-o-dee-GRADE-able)

Organic matter that can be broken down by bacteria to more stable forms that will not create a nuisance or give off foul odors is considered biodegradable.

BIOMASS (BUY-o-mass)

A mass or clump of organic material consisting of living organisms feeding on wastes, dead organisms, and other debris. Also see ZOOGLEAL MASS and ZOOGLEAL MAT (FILM).

BIOSOLIDS

A primarily organic solid product produced by wastewater treatment processes that can be beneficially recycled. The word biosolids is replacing the word sludge when referring to treated waste.

BLINDING

The clogging of the filtering medium of a microscreen or a vacuum filter when the holes or spaces in the media become clogged or sealed off due to a buildup of grease or the material being filtered.

BLOWDOWN

The removal of accumulated solids in boilers to prevent plugging of boiler tubes and steam lines. In cooling towers, blowdown is used to reduce the amount of dissolved salts in the recirculated cooling water.

BOUND WATER

Water contained within the cell mass of sludges or strongly held on the surface of colloidal particles. One of the causes of bulking sludge in the activated sludge process.

BREAKPOINT CHLORINATION

Addition of chlorine to water or wastewater until the chlorine demand has been satisfied. At this point, further additions of chlorine will result in a free chlorine residual that is directly proportional to the amount of chlorine added beyond the breakpoint.

BRINELLING (bruh-NEL-ing)

Tiny indentations (dents) high on the shoulder of the bearing race or bearing. A type of bearing failure.

BUFFER

A solution or liquid whose chemical makeup neutralizes acids or bases without a great change in pH.

BUFFER CAPACITY

A measure of the capacity of a solution or liquid to neutralize acids or bases. This is a measure of the capacity of water or wastewater for offering a resistance to changes in pH.

BULKING

Clouds of billowing sludge that occur throughout secondary clarifiers and sludge thickeners when the sludge does not settle properly. In the activated sludge process, bulking is usually caused by filamentous bacteria or bound water.

C

CFR

Code of Federal Regulations. A publication of the US government that contains all of the proposed and finalized federal regulations, including safety and environmental regulations.

COD (pronounce as separate letters)

Chemical Oxygen Demand. A measure of the oxygen-consuming capacity of organic matter present in wastewater. COD is expressed as the amount of oxygen consumed from a chemical oxidant in mg/L during a specific test. Results are not necessarily related to the biochemical oxygen demand (BOD) because the chemical oxidant may react with substances that bacteria do not stabilize.

CALIBRATION

A procedure that checks or adjusts an instrument's accuracy by comparison with a standard or reference.

CARCINOGEN (kar-SIN-o-jen)

Any substance that tends to produce cancer in an organism.

CATEGORICAL STANDARDS (LIMITS)

Industrial waste discharge standards (limits) developed by EPA that are applied to the effluent from any industry in any category anywhere in the United States that discharges to a Publicly Owned Treatment Works (POTW). These are standards based on the technology available to treat the wastestreams from the processes of the specific industrial category and normally are measured at the point of discharge from the regulated process. The standards are listed in the Code of Federal Regulations.

CATHODIC (ka-THOD-ick) PROTECTION

An electrical system for prevention of rust, corrosion, and pitting of metal surfaces that are in contact with water, wastewater, or soil. A low-voltage current is made to flow through a liquid (water) or a soil in contact with the metal in such a manner that the external electromotive force renders the metal structure cathodic. This concentrates corrosion on auxiliary anodic parts, which are deliberately allowed to corrode instead of letting the structure corrode.

CATION (KAT-EYE-en)

A positively charged ion in an electrolyte solution, attracted to the cathode under the influence of a difference in electrical potential. Sodium ion (Na^+) is a cation.

CAUTION

This word warns against potential hazards or cautions against unsafe practices. Also see DANGER, NOTICE, and WARNING.

CAVITATION (kav-uh-TAY-shun)

The formation and collapse of a gas pocket or bubble on the blade of an impeller or the gate of a valve. The collapse of this gas pocket or bubble drives water into the impeller or gate with a terrific force that can cause pitting on the impeller or gate surface. Cavitation is accompanied by loud noises that sound like someone is pounding on the impeller or gate with a hammer.

CENTRATE

The water leaving a centrifuge after most of the solids have been removed.

CENTRIFUGE

A mechanical device that uses centrifugal or rotational forces to separate solids from liquids.

CERTIFICATION EXAMINATION

An examination administered by a state agency or professional association that operators take to indicate a level of professional competence. In the United States, certification of operators of water treatment plants, wastewater treatment plants, water distribution systems, and small water supply systems is mandatory. In many states, certification of wastewater collection system operators, industrial wastewater treatment plant operators, pretreatment facility inspectors, and small wastewater system operators is voluntary; however, current trends indicate that more states, provinces, and employers will require these operators to be certified in the future. Operator certification is mandatory in the United States for the Chief Operators of water treatment plants, water distribution systems, and wastewater treatment plants.

CERTIFIED OPERATOR

A person who has the education and experience required to operate a specific class of treatment facility as indicated by possessing a certificate of professional competence given by a state agency or professional association.

CHAIN OF CUSTODY

A record of each person involved in the handling and possession of a sample from the person who collected the sample to the person who analyzed the sample in the laboratory and to the person who witnessed disposal of the sample.

CHELATING (KEY-LAY-ting) AGENT

A chemical used to prevent the precipitation of metals (such as copper).

CHELATION (key-LAY-shun)

A chemical complexing (forming or joining together) of metallic cations (such as copper) with certain organic compounds, such as EDTA (ethylene diamine tetracetic acid). Chelation is used to prevent the precipitation of metals (copper). Also see SEQUESTRATION.

CHEMICAL OXYGEN DEMAND (COD)

A measure of the oxygen-consuming capacity of organic matter present in wastewater. COD is expressed as the amount of oxygen consumed from a chemical oxidant in mg/L during a specific test. Results are not necessarily related to the biochemical oxygen demand (BOD) because the chemical oxidant may react with substances that bacteria do not stabilize.

CHRONIC HEALTH EFFECT

An adverse effect on a human or animal body with symptoms that develop slowly over a long period of time or that recur frequently.

CILIATES (SILLY-ates)

A class of protozoans distinguished by short hairs on all or part of their bodies.

CLARIFICATION (klair-uh-fuh-KAY-shun)

Any process or combination of processes the main purpose of which is to reduce the concentration of suspended matter in a liquid.

CLARIFIER (KLAIR-uh-fire)

A tank or basin in which water or wastewater is held for a period of time during which the heavier solids settle to the bottom and the lighter materials float to the surface. Also called settling tank or SEDIMENTATION BASIN.

COAGULANT (ko-AGG-yoo-lent)

A chemical that causes very fine particles to clump (floc) together into larger particles. This makes it easier to separate the solids from the liquids by settling, skimming, draining, or filtering.

COAGULANT (ko-AGG-yoo-lent) AID

Any chemical or substance used to assist or modify coagulation.

COAGULATION (ko-agg-yoo-LAY-shun)

The clumping together of very fine particles into larger particles (floc) caused by the use of chemicals (coagulants). The chemicals neutralize the electrical charges of the fine particles, allowing them to come closer and form larger clumps.

CODE OF FEDERAL REGULATIONS (CFR)

A publication of the US government that contains all of the proposed and finalized federal regulations, including safety and environmental regulations.

COLIFORM (KOAL-i-form)

A group of bacteria found in the intestines of warm-blooded animals (including humans) and also in plants, soil, air, and water. The presence of coliform bacteria is an indication that the water is polluted and may contain pathogenic (disease-causing) organisms. Fecal coliforms are those coliforms found in the feces of various warm-blooded animals, whereas the term "coliform" also includes other environmental sources.

COLLOIDS (KALL-loids)

Very small, finely divided solids (particles that do not dissolve) that remain dispersed in a liquid for a long time due to their small size and electrical charge. When most of the particles in water have a negative electrical charge, they tend to repel each other. This repulsion prevents the particles from clumping together, becoming heavier, and settling out.

COLORIMETRIC MEASUREMENT

A means of measuring unknown chemical concentrations in water by measuring a sample's color intensity. The specific color of the sample, developed by addition of chemical reagents, is measured with a photoelectric colorimeter or is compared with color standards using, or corresponding with, known concentrations of the chemical.

COMMON METAL

Aluminum, cadmium, chromium, copper, iron, lead, nickel, tin, zinc, or any combination of these elements are considered common metals.

COMPATIBLE POLLUTANTS

Those pollutants that are normally removed by the POTW treatment system. Biochemical oxygen demand (BOD), suspended solids (SS), and ammonia are considered compatible pollutants.

COMPETENT PERSON

A competent person is defined by OSHA as a person capable of identifying existing and predictable hazards in the surroundings, or working conditions that are unsanitary, hazardous, or dangerous to employees, and who has authorization to take prompt corrective measures to eliminate the hazards.

COMPLIANCE

The act of meeting specified conditions or requirements.

COMPOSITE (PROPORTIONAL) SAMPLE

A composite sample is a collection of individual samples obtained at regular intervals, usually every one or two hours during a 24-hour time span. Each individual sample is combined with the others in proportion to the rate of flow when the sample was collected. Equal volume individual samples also may be collected at intervals after a specific volume of flow passes the sampling point or after equal time intervals and still be referred to as a composite sample. The resulting mixture (composite sample) forms a representative sample and is analyzed to determine the average conditions during the sampling period.

CONCENTRATION POLARIZATION

(1) A buildup of retained particles on the membrane surface due to dewatering of the feed closest to the membrane. The thickness of the concentration polarization layer is controlled by the flow velocity across the membrane.

(2) Used in corrosion studies to indicate a depletion of ions near an electrode.

(3) The basis for chemical analysis by a polarograph.

CONDUCTIVITY

A measure of the ability of a solution (water) to carry an electric current.

CONFINED SPACE

Confined space means a space that:

(1) Is large enough and so configured that an employee can bodily enter and perform assigned work; and

(2) Has limited or restricted means for entry or exit (for example, manholes, tanks, vessels, silos, storage bins, hoppers, vaults, and pits are spaces that may have limited means of entry); and

(3) Is not designed for continuous employee occupancy.

Also see DANGEROUS AIR CONTAMINATION and OXYGEN DEFICIENCY.

CONFINED SPACE, NON-PERMIT

A non-permit confined space is a confined space that does not contain or, with respect to atmospheric hazards, have the potential to contain any hazard capable of causing death or serious physical harm.

CONFINED SPACE, PERMIT-REQUIRED (PERMIT SPACE)

A confined space that has one or more of the following characteristics:

(1) Contains or has a potential to contain a hazardous atmosphere

(2) Contains a material that has the potential for engulfing an entrant

(3) Has an internal configuration such that an entrant could be trapped or asphyxiated by inwardly converging walls or by a floor that slopes downward and tapers to a smaller cross section

(4) Contains any other recognized serious safety or health hazard

CONING

Development of a cone-shaped flow of liquid, like a whirlpool, through sludge. This can occur in a sludge hopper during sludge withdrawal when the sludge becomes too thick. Part of the sludge remains in place while liquid rather than sludge flows out of the hopper. Also called coring.

CONTACT STABILIZATION

Contact stabilization is a modification of the conventional activated sludge process. In contact stabilization, two aeration tanks are used. One tank is for separate reaeration of the return sludge for at least four hours before it is permitted to flow into the other aeration tank to be mixed with the primary effluent requiring treatment. The process may also occur in one long tank.

CONTACTOR

An electric switch, usually magnetically operated.

CONTINUOUS PROCESS

A treatment process in which water is treated continuously in a tank or reactor. The water being treated continuously flows into the tank at one end, is treated as it flows through the tank, and flows out the opposite end as treated water.

CONTROL LOOP

The path through the control system between the sensor, which measures a process variable, and the controller, which controls or adjusts the process variable.

CONTROL SYSTEM

An instrumentation system that senses and controls its own operation on a close, continuous basis in what is called proportional (or modulating) control.

CONTROLLER

A device that controls the starting, stopping, or operation of a device or piece of equipment.

CONVENTIONAL POLLUTANTS

Those pollutants that are usually found in domestic, commercial, or industrial wastes, including suspended solids, biochemical oxygen demand, pathogenic (disease-causing) organisms, and oil and grease.

CO-PRECIPITATION

A treatment process that occurs when ferrous iron is added to metallic wastestreams and subsequently oxidized in an aerator. The oxidized iron, which is insoluble, precipitates along with other metallic contaminants present in the wastestream, thereby enhancing metals removal.

CRADLE TO GRAVE

A term used to describe a hazardous waste manifest system used by regulatory agencies to track a hazardous waste from the point of generation to the hauler and then to the ultimate disposal site.

CROSS CONNECTION

(1) A connection between drinking (potable) water and an unapproved water supply.

(2) A connection between a storm drain system and a sanitary collection system.

(3) Less frequently used to mean a connection between two sections of a collection system to handle anticipated overloads of one system.

CRYOGENIC (KRY-o-JEN-nick)

Very low temperature. Associated with liquified gases (liquid oxygen).

D

DPD METHOD

A method of measuring the chlorine residual in water. The residual may be determined by either titrating or comparing a developed color with color standards. DPD stands for N,N-diethyl-p-phenylenediamine.

DALTON

A unit of mass designated as one-sixteenth the mass of oxygen-16, the lightest and most abundant isotope of oxygen. The dalton is equivalent to one mass unit.

DANGER

The word *DANGER* is used where an immediate hazard presents a threat of death or serious injury to employees. Also see CAUTION, NOTICE, and WARNING.

DANGEROUS AIR CONTAMINATION

An atmosphere presenting a threat of causing death, injury, acute illness, or disablement due to the presence of flammable and/or explosive, toxic, or otherwise injurious or incapacitating substances.

(1) Dangerous air contamination due to the flammability of a gas, vapor, or mist is defined as an atmosphere containing the gas, vapor, or mist at a concentration greater than 10 percent of its lower explosive (lower flammable) limit (LEL).

(2) Dangerous air contamination due to a combustible particulate is defined as a concentration that meets or exceeds the particulate's lower explosive limit (LEL).

(3) Dangerous air contamination due to the toxicity of a substance is defined as the atmospheric concentration that could result in employee exposure in excess of the substance's permissible exposure limit (PEL).

NOTE: A dangerous situation also occurs when the oxygen level is less than 19.5 percent by volume (OXYGEN DEFICIENCY) or more than 23.5 percent by volume (OXYGEN ENRICHMENT).

DATEOMETER (day-TOM-uh-ter)

A small calendar disk attached to motors and equipment to indicate the year in which the last maintenance service was performed.

DECANT (de-KANT)

To draw off the upper layer of liquid (water) after the heavier material (a solid or another liquid) has settled.

DECANT (de-KANT) WATER

Water that has separated from sludge and is removed from the layer of water above the sludge.

DECIBEL (DES-uh-bull)

A unit for expressing the relative intensity of sounds on a scale from zero for the average least perceptible sound to about 130 for the average level at which sound causes pain to humans. Abbreviated dB.

DEFINING

A process that arranges the activated carbon particles according to size. This process is also used to remove small particles from granular contactors to prevent excessive head loss.

DELAMINATION (DEE-lam-uh-NAY-shun)

Separation of a membrane or other material from the backing material on which it is cast.

DELETERIOUS (DELL-eh-TEER-ee-us)

Refers to something that can be or is hurtful, harmful, or injurious to health or the environment.

DENITRIFICATION (DEE-nye-truh-fuh-KAY-shun)

(1) The anoxic biological reduction of nitrate nitrogen to nitrogen gas.

(2) The removal of some nitrogen from a system.

(3) An anoxic process that occurs when nitrite or nitrate ions are reduced to nitrogen gas and nitrogen bubbles are formed as a result of this process. The bubbles attach to the biological floc and float the floc to the surface of the secondary clarifiers. This condition is often the cause of rising sludge observed in secondary clarifiers or gravity thickeners. Also see NITRIFICATION.

DENSITY

A measure of how heavy a substance (solid, liquid, or gas) is for its size. Density is expressed in terms of weight per unit volume, that is, grams per cubic centimeter or pounds per cubic foot. The density of water (at 4°C or 39°F) is 1.0 gram per cubic centimeter or about 62.4 pounds per cubic foot.

DESICCANT (DESS-uh-kant)

A drying agent that is capable of removing or absorbing moisture from the atmosphere in a small enclosure.

DESICCATION (dess-uh-KAY-shun)

A process used to thoroughly dry air; to remove virtually all moisture from air.

DETECTION LAG

The time period between the moment a process change is made and the moment when such a change is finally sensed by the associated measuring instrument.

DETENTION TIME

(1) The time required to fill a tank at a given flow.

(2) The theoretical (calculated) time required for water to pass through a tank at a given rate of flow.

(3) The actual time in hours, minutes, or seconds that a small amount of water is in a settling basin, flocculating basin, or rapid-mix chamber. In septic tanks, detention time will decrease as the volumes of sludge and scum increase. In storage reservoirs, detention time is the length of time entering water will be held before being drafted for use (several weeks to years, several months being typical).

$$\text{Detention Time, hr} = \frac{(\text{Basin Volume, gal})(24 \text{ hr/day})}{\text{Flow, gal/day}}$$

or

$$\text{Detention Time, hr} = \frac{(\text{Basin Volume, m}^3)(24 \text{ hr/day})}{\text{Flow, m}^3/\text{day}}$$

DEWATERABLE

This is a property of sludge related to the ability to separate the liquid portion from the solid, with or without chemical conditioning. A material is considered dewaterable if water will readily drain from it.

DIAPHRAGM PUMP

A pump in which a flexible diaphragm, generally of rubber or equally flexible material, is the operating part. It is fastened at the edges in a vertical cylinder. When the diaphragm is raised, suction is exerted, and when it is depressed, the liquid is forced through a discharge valve.

DIATOMACEOUS (DYE-uh-toe-MAY-shus) EARTH

A fine, siliceous (made of silica) earth composed mainly of the skeletal remains of diatoms.

DIATOMS (DYE-uh-toms)

Unicellular (single cell), microscopic algae with a rigid, box-like internal structure consisting mainly of silica.

DIFFUSED-AIR AERATION

A diffused-air activated sludge plant takes air, compresses it, and then discharges the air below the water surface of the aerator through some type of air diffusion device.

DIFFUSER

A device (porous plate, tube, bag) used to break the air stream from the blower system into fine bubbles in an aeration tank or reactor.

DIGITAL READOUT

The readout of an instrument by a direct, numerical reading of the measured value or variable. The signal sent to such instruments is usually an analog signal.

DIRECT DISCHARGER

A point source that discharges a pollutant(s) to waters of the United States, such as streams, lakes, or oceans. These sources are subject to the National Pollutant Discharge Elimination System (NPDES) program regulations.

DISCHARGE HEAD

The pressure (in pounds per square inch (psi) or kilopascals (kPa)) measured at the centerline of a pump discharge and very close to the discharge flange, converted into feet or meters. The pressure is measured from the centerline of the pump to the hydraulic grade line of the water in the discharge pipe.

    Discharge Head, ft = (Discharge Pressure, psi)(2.31 ft/psi)

    or

    Discharge Head, m = (Discharge Pressure, kPa)(1 m/9.8 kPa)

DISINFECTION (dis-in-FECT-shun)

The process designed to kill or inactivate most microorganisms in water or wastewater, including essentially all pathogenic (disease-causing) bacteria. There are several ways to disinfect, with chlorination being the most frequently used in water and wastewater treatment plants. Compare with STERILIZATION.

DISSOLVED OXYGEN

Molecular oxygen dissolved in water or wastewater, usually abbreviated DO.

DOCTOR BLADE

A blade used to remove any excess solids that may cling to the outside of a rotating screen.

DRAG OUT

The liquid film (plating solution) that adheres to the workpieces and their fixtures as they are removed from any given process solution or their rinses. Drag-out volume from a tank depends on the viscosity of the solution, the surface tension, the withdrawal time, the draining time, and the shape and texture of the workpieces. The drag-out liquid may drip onto the floor and cause wastestream treatment problems. Regulated substances contained in this liquid must be removed from wastestreams or neutralized prior to discharge to POTW sewers.

DRIFT

The difference between the actual value and the desired value (or set point); characteristic of proportional controllers that do not incorporate reset action. Also called OFFSET.

DYNAMIC HEAD

When a pump is operating, the vertical distance (in feet or meters) from a point to the energy grade line. Also see ENERGY GRADE LINE, STATIC HEAD, and TOTAL DYNAMIC HEAD.

DYNAMIC PRESSURE

When a pump is operating, pressure resulting from the dynamic head.

Dynamic Pressure, psi = (Dynamic Head, ft)(0.433 psi/ft)

or

Dynamic Pressure, kPa = (Dynamic Head, m)(9.8 kPa/m)

E

EGL

See ENERGY GRADE LINE.

EDUCTOR (e-DUCK-ter)

A hydraulic device used to create a negative pressure (suction) by forcing a liquid through a restriction, such as a Venturi. An eductor or aspirator (the hydraulic device) may be used in the laboratory in place of a vacuum pump. As an injector, it is used to produce vacuum for chlorinators. Sometimes used instead of a suction pump.

EFFECTIVE RANGE

That portion of the design range (usually from 10 to 90+ percent) in which an instrument has acceptable accuracy. Also see RANGE and SPAN.

EFFLUENT (EF-loo-ent)

Water or other liquid—raw (untreated), partially treated, or completely treated—flowing *FROM* a reservoir, basin, treatment process, or treatment plant.

EFFLUENT LIMITS

Pollutant limitations developed by a POTW for industrial plants discharging to the POTW system. At a minimum, all industrial facilities are required to comply with federal prohibited discharge standards. The industries covered by federal categorical standards must also comply with the appropriate discharge limitations. The POTW may also establish local limits more stringent than or in addition to the federal standards for some or all of its industrial users.

ELECTROCOAGULATION

A treatment process that uses sacrificial anodes of iron, aluminum, or other ion-producing materials in place of flocculants such as aluminum sulfate (alum), ferric chloride, or polymers to provide the chemistry. Instead of energy from chemical bonds or ions, electric current is used. In the process, an electric current is incorporated to neutralize ionic, valence, and particulate charges thereby allowing contaminants, such as colloidal particulates, oils, and dissolved metals, to be precipitated and removed from stable suspensions and emulsions.

ELECTROLYSIS (ee-leck-TRAWL-uh-sis)

The decomposition of material by an outside electric current.

ELECTROLYTE (ee-LECK-tro-lite)

A substance that dissociates (separates) into two or more ions when it is dissolved in water.

ELUTRIATION (e-LOO-tree-A-shun)

The washing of digested sludge with either fresh water, plant effluent, or other wastewater. The objective is to remove (wash out) fine particulates and/or the alkalinity in sludge. This process reduces the demand for conditioning chemicals and improves settling or filtering characteristics of the solids.

EMULSION (e-MULL-shun)

A liquid mixture of two or more liquid substances not normally dissolved in one another; one liquid is held in suspension in the other.

ENCLOSED SPACE

See CONFINED SPACE.

ENDOGENOUS (en-DODGE-en-us) RESPIRATION

A situation in which living organisms oxidize some of their own cellular mass instead of new organic matter they adsorb or absorb from their environment.

ENERGY GRADE LINE (EGL)

A line that represents the elevation of energy head (in feet or meters) of water flowing in a pipe, conduit, or channel. The line is drawn above the hydraulic grade line (gradient) a distance equal to the velocity head ($V^2/2g$) of the water flowing at each section or point along the pipe or channel. Also see HYDRAULIC GRADE LINE.

[SEE DRAWING ON PAGE 920]

ENGULFMENT

Engulfment means the surrounding and effective capture of a person by a liquid or finely divided (flowable) solid substance that can be aspirated to cause death by filling or plugging the respiratory system or that can exert enough force on the body to cause death by strangulation, constriction, or crushing.

ENTRAIN

To trap bubbles in water either mechanically through turbulence or chemically through a reaction.

EQUIVALENCE POINT

The point on a titration curve where the acid ion concentration is equal to the base ion concentration.

EVAPOTRANSPIRATION (ee-VAP-o-TRANS-purr-A-shun)

(1) The process by which water vapor is released to the atmosphere from living plants. Also called TRANSPIRATION.

(2) The total water removed from an area by transpiration (plants) and by evaporation from soil, snow, and water surfaces.

EXFILTRATION (EX-fill-TRAY-shun)

Liquid wastes and liquid-carried wastes that unintentionally leak out of a sewer pipe system and into the environment.

EXISTING SOURCE

An industrial discharger that was already in operation when the proposed pretreatment standard for the industrial category was promulgated.

EXOTHERMIC (EX-o-THUR-mick)

A chemical change that produces heat.

F

F/M RATIO

See FOOD/MICROORGANISM RATIO.

FACULTATIVE (FACK-ul-tay-tive) BACTERIA

Facultative bacteria can use either dissolved oxygen or oxygen obtained from food materials such as sulfate or nitrate ions. In other words, facultative bacteria can live under aerobic, anoxic, or anaerobic conditions.

FACULTATIVE (FACK-ul-tay-tive) POND

The most common type of pond in current use. The upper portion (supernatant) is aerobic, while the bottom layer is anaerobic. Algae supply most of the oxygen to the supernatant.

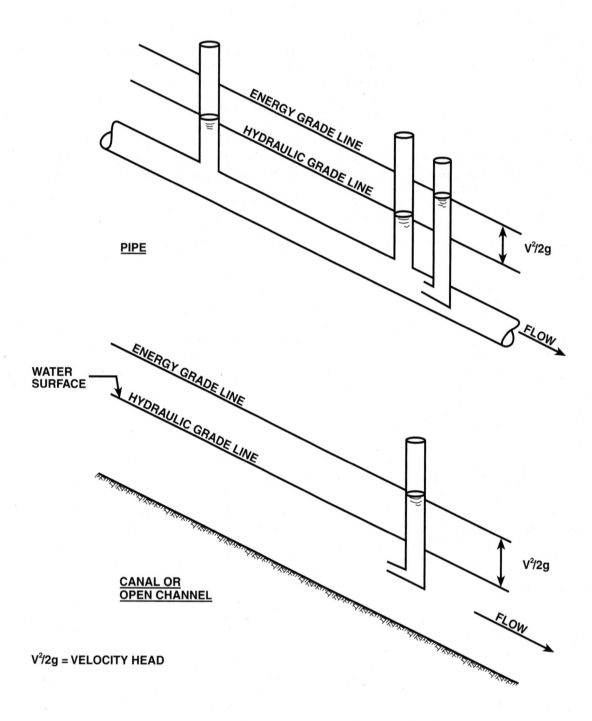

ENERGY GRADE LINE and HYDRAULIC GRADE LINE

FAIL-SAFE

Design and operation of a process control system whereby failure of the power system or any component does not result in process failure or equipment damage.

FEEDBACK

The circulating action between a sensor measuring a process variable and the controller that controls or adjusts the process variable.

FILAMENTOUS (fill-uh-MEN-tuss) ORGANISMS

Organisms that grow in a thread or filamentous form. Common types are *Thiothrix* and *Actinomycetes*. A common cause of sludge bulking in the activated sludge process.

FILTER AID

A chemical (usually a polymer) added to water to help remove fine colloidal suspended solids.

FLAME POLISHED

Melted by a flame to smooth out irregularities. Sharp or broken edges of glass (such as the end of a glass tube) are rotated in a flame until the edge melts slightly and becomes smooth.

FLOC

Clumps of bacteria and particles, or coagulants and impurities, that have come together and formed a cluster. Found in flocculation tanks, sedimentation basins, aeration tanks, secondary clarifiers, and chemical precipitation processes.

FLOCCULATION (flock-you-LAY-shun)

The gathering together of fine particles after coagulation to form larger particles by a process of gentle mixing. This clumping together makes it easier to separate the solids from the water by settling, skimming, draining, or filtering.

FLUIDIZED (FLOO-id-i-zd)

A mass of solid particles that is made to flow like a liquid by injection of water or gas is said to have been fluidized. In water and wastewater treatment, a bed of filter media is fluidized by backwashing water through the filter.

FLUX

A flowing or flow.

FOOD/MICROORGANISM (F/M) RATIO

Food to microorganism ratio. A measure of food provided to bacteria in an aeration tank.

$$\frac{\text{Food}}{\text{Microorganisms}} = \frac{\text{BOD, lbs/day}}{\text{MLVSS, lbs}}$$

$$= \frac{\text{Flow, MGD} \times \text{BOD, mg}/L \times 8.34 \text{ lbs/gal}}{\text{Volume, MG} \times \text{MLVSS, mg}/L \times 8.34 \text{ lbs/gal}}$$

or by calculator math system

$$= \text{Flow, MGD} \times \text{BOD, mg}/L \div \text{Volume, MG} \div \text{MLVSS, mg}/L$$

or metric

$$= \frac{\text{BOD, kg/day}}{\text{MLVSS, kg}}$$

$$= \frac{\text{Flow, M}L/\text{day} \times \text{BOD, mg}/L \times 1 \text{ kg/M mg}}{\text{Volume, M}L \times \text{MLVSS, mg}/L \times 1 \text{ kg/M mg}}$$

FREEBOARD

(1) The vertical distance from the normal water surface to the top of the confining wall.

(2) The vertical distance from the sand surface to the underside of a trough in a sand filter. This distance is also called AVAILABLE EXPANSION.

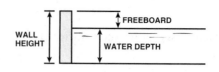

FRICTION LOSS

The head, pressure, or energy (they are the same) lost by water flowing in a pipe or channel as a result of turbulence caused by the velocity of the flowing water and the roughness of the pipe, channel walls, or restrictions caused by fittings. Water flowing in a pipe loses head, pressure, or energy as a result of friction losses. Also called HEAD LOSS.

G

GIS

See GEOGRAPHIC INFORMATION SYSTEM.

GAS, SEWER

See SEWER GAS.

GASIFICATION (gas-uh-fuh-KAY-shun)

The conversion of soluble and suspended organic materials into gas during aerobic or anaerobic decomposition. In clarifiers, the resulting gas bubbles can become attached to the settled sludge and cause large clumps of sludge to rise and float on the water surface. In anaerobic sludge digesters, this gas is collected for fuel or disposed of using a waste gas burner.

GEOGRAPHIC INFORMATION SYSTEM (GIS)

A computer program that combines mapping with detailed information about the physical locations of structures, such as pipes, valves, and manholes, within geographic areas. The system is used to help operators and maintenance personnel locate utility system features or structures and to assist with the scheduling and performance of maintenance activities.

GRAB SAMPLE

A single sample of water collected at a particular time and place that represents the composition of the water only at that time and place.

GROWTH RATE, Y

An experimentally determined constant to estimate the unit growth rate of bacteria while degrading organic wastes.

H

HGL

See HYDRAULIC GRADE LINE.

HARMFUL PHYSICAL AGENT or TOXIC SUBSTANCE

Any chemical substance, biological agent (bacteria, virus, or fungus), or physical stress (noise, heat, cold, vibration, repetitive motion, ionizing and non-ionizing radiation, hypo- or hyperbaric pressure) that:

(1) Is regulated by any state or federal law or rule due to a hazard to health

(2) Is listed in the latest printed edition of the National Institute of Occupational Safety and Health (NIOSH) Registry of Toxic Effects of Chemical Substances (RTECS)

(3) Has yielded positive evidence of an acute or chronic health hazard in human, animal, or other biological testing conducted by, or known to, the employer

(4) Is described by a Material Safety Data Sheet (MSDS) available to the employer that indicates that the material may pose a hazard to human health

Also see ACUTE HEALTH EFFECT and CHRONIC HEALTH EFFECT.

HAZARDOUS WASTE

A waste that possesses any one of the following four characteristics:

(1) Ignitability, which identifies wastes that pose a fire hazard during routine management. Fires not only present immediate dangers of heat and smoke, but also can spread harmful particles over wide areas. A liquid that has a flash point of less than 140°F (60°C).

(2) Corrosivity, which identifies wastes requiring special containers because of their ability to corrode standard materials, or requiring segregation from other wastes because of their ability to dissolve toxic contaminants. An aqueous solution with a pH less than or equal to 2 or a pH greater than or equal to 12.5.

(3) Reactivity (or explosiveness), which identifies wastes that, during routine management, tend to react spontaneously, to react vigorously with air or water, to be unstable to shock or heat, to generate toxic gases, or to explode.

(4) Toxicity, which identifies wastes that, when improperly managed, may release toxicants in sufficient quantities to pose a substantial present or potential hazard to human health or the environment.

HEAD

The vertical distance, height, or energy of water above a reference point. A head of water may be measured in either height (feet or meters) or pressure (pounds per square inch or kilograms per square centimeter). Also see DISCHARGE HEAD, DYNAMIC HEAD, STATIC HEAD, SUCTION HEAD, SUCTION LIFT, and VELOCITY HEAD.

HEAD LOSS

The head, pressure, or energy (they are the same) lost by water flowing in a pipe or channel as a result of turbulence caused by the velocity of the flowing water and the roughness of the pipe, channel walls, or restrictions caused by fittings. Water flowing in a pipe loses head, pressure, or energy as a result of friction losses. The head loss through a filter is due to friction losses caused by material building up on the surface or in the top part of a filter. Also called FRICTION LOSS.

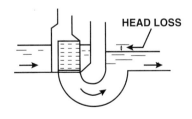

HEADER

A large pipe to which the ends of a series of smaller pipes are connected. Also called a MANIFOLD.

HERTZ (Hz)

The number of complete electromagnetic cycles or waves in one second of an electric or electronic circuit. Also called the frequency of the current.

HETEROTROPHIC (HET-er-o-TROF-ick)

Describes organisms that use organic matter for energy and growth. Animals, fungi, and most bacteria are heterotrophs.

HINDERED SOLIDS SEPARATION

Solids settling in a thickening rather than in a clarifying mode. Suspended solids settling (clarifying) velocities are strongly influenced by the applied solids concentration. The greater the applied solids concentration, the greater the opportunity for thickening (hindered) solids settling.

HYDRAULIC GRADE LINE (HGL)

The surface or profile of water flowing in an open channel or a pipe flowing partially full. If a pipe is under pressure, the hydraulic grade line is that level water would rise to in a small, vertical tube connected to the pipe. Also see ENERGY GRADE LINE.

[SEE DRAWING ON PAGE 920]

HYDRAULIC JUMP

The sudden and usually turbulent abrupt rise in water surface in an open channel when water flowing at high velocity is suddenly retarded to a slow velocity.

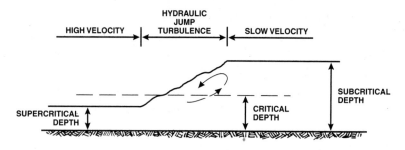

HYDROLYSIS (hi-DROLL-uh-sis)

(1) A chemical reaction in which a compound is converted into another compound by taking up water.

(2) Usually a chemical degradation of organic matter.

HYGROSCOPIC (hi-grow-SKOP-ick)

Absorbing or attracting moisture from the air.

I

IDLH

Immediately Dangerous to Life or Health. The atmospheric concentration of any toxic, corrosive, or asphyxiant substance that poses an immediate threat to life or would cause irreversible or delayed adverse health effects or would interfere with an individual's ability to escape from a dangerous atmosphere.

IMHOFF CONE

A clear, cone-shaped container marked with graduations. The cone is used to measure the volume of settleable solids in a specific volume (usually one liter) of water or wastewater.

IMMISCIBLE (im-MISS-uh-bull)

Not capable of being mixed.

INCINERATION

The conversion of dewatered wastewater solids by combustion (burning) to ash, carbon dioxide, and water vapor.

INDICATOR

(1) (Chemical indicator) A substance that gives a visible change, usually of color, at a desired point in a chemical reaction, generally at a specified end point.

(2) (Instrument indicator) A device that indicates the result of a measurement, usually using either a fixed scale and movable indicator (pointer), such as a pressure gauge, or a moving chart with a movable pen like those used on a circular flow-recording chart. Also called a RECEIVER.

INDIRECT DISCHARGER

A nondomestic discharger introducing pollutants to a POTW. These facilities are subject to the EPA pretreatment regulations.

INFILTRATION (in-fill-TRAY-shun)

The seepage of groundwater into a sewer system, including service connections. Seepage frequently occurs through defective or cracked pipes, pipe joints and connections, interceptor access risers and covers, or manhole walls.

INHIBITORY SUBSTANCES

Materials that kill or restrict the ability of organisms to treat wastes.

INORGANIC WASTE

Waste material such as sand, salt, iron, calcium, and other mineral materials that are only slightly affected by the action of organisms. Inorganic wastes are chemical substances of mineral origin; whereas organic wastes are chemical substances usually of animal or plant origin. Also see NONVOLATILE MATTER, ORGANIC WASTE, and VOLATILE SOLIDS.

INTEGRATOR

A device or meter that continuously measures and calculates (adds) a process rate variable in cumulative fashion; for example, total flows displayed in gallons, million gallons, cubic feet, or some other unit of volume measurement. Also called a TOTALIZER.

INTERFACE

The common boundary layer between two substances, such as water and a solid (metal); or between two fluids, such as water and a gas (air); or between a liquid (water) and another liquid (oil).

INTERFERENCE

Interference refers to the harmful effects industrial compounds can have on POTW operations, such as killing or inhibiting beneficial microorganisms or causing treatment process upsets or sludge contamination.

INTERLOCK

An electric switch, usually magnetically operated. Used to interrupt all (local) power to a panel or device when the door is opened or the circuit is exposed to service.

INTERSTICE (in-TUR-stuhz)

A very small open space in a rock or granular material. Also called a PORE, VOID, or void space. Also see VOID.

ION

An electrically charged atom, radical (such as SO_4^{2-}), or molecule formed by the loss or gain of one or more electrons.

ION EXCHANGE

A water or wastewater treatment process involving the reversible interchange (switching) of ions between the water being treated and the solid resin contained within an ion exchange unit. Undesirable ions are exchanged with acceptable ions on the resin or recoverable ions in the water being treated are exchanged with other acceptable ions on the resin.

ION EXCHANGE RESINS

Insoluble polymers, used in water or wastewater treatment, that are capable of exchanging (switching or giving) acceptable cations or anions to the water being treated for less desirable ions or for ions to be recovered.

J

JAR TEST

A laboratory procedure that simulates coagulation/flocculation with differing chemical doses. The purpose of the procedure is to estimate the minimum coagulant dose required to achieve certain water quality goals. Samples of water to be treated are placed in six jars. Various amounts of chemicals are added to each jar, stirred, and the settling of solids is observed. The lowest dose of chemicals that provides satisfactory settling is the dose used to treat the water.

JOGGING

The frequent starting and stopping of an electric motor.

K

(NO LISTINGS)

L

LAUNDERS

Sedimentation basin and filter discharge channels consisting of overflow weir plates (in sedimentation basins) and conveying troughs.

LINEAL (LIN-e-ul)

The length in one direction of a line. For example, a board 12 feet (meters) long has 12 lineal feet (meters) in its length.

LINEARITY (lin-ee-AIR-it-ee)

How closely an instrument measures actual values of a variable through its effective range; a measure used to determine the accuracy of an instrument.

LIPOPHILIC (lie-puh-FILL-ick)

Having a strong affinity for fats. Compounds that dissolve in fats, oils, and greases.

LOWER EXPLOSIVE LIMIT (LEL)

The lowest concentration of a gas or vapor (percent by volume in air) that explodes if an ignition source is present at ambient temperature. At temperatures above 250°F (121°C) the LEL decreases because explosibility increases with higher temperature.

LOWER FLAMMABLE LIMIT (LFL)

The lowest concentration of a gas or vapor (percent by volume in air) that burns if an ignition source is present.

LYSIMETER (lie-SIM-uh-ter)

A device containing a mass of soil and designed to permit the measurement of water draining through the soil.

M

MCRT

Mean Cell Residence Time. An expression of the average time (days) that a microorganism will spend in the activated sludge process.

$$\text{MCRT, days} = \frac{\text{Total Suspended Solids in Activated Sludge Process, lbs}}{\text{Total Suspended Solids Removed From Process, lbs/day}}$$

or

$$\text{MCRT, days} = \frac{\text{Total Suspended Solids in Activated Sludge Process, kg}}{\text{Total Suspended Solids Removed From Process, kg/day}}$$

NOTE: Operators at different plants calculate the Total Suspended Solids (TSS) in the Activated Sludge Process, lbs (kg), by three different methods:

1. TSS in the Aeration Basin or Reactor Zone, lbs (kg)
2. TSS in the Aeration Basin and Secondary Clarifier, lbs (kg)
3. TSS in the Aeration Basin and Secondary Clarifier Sludge Blanket, lbs (kg)

These three different methods make it difficult to compare MCRTs in days among different plants unless everyone uses the same method.

mg/*L*

See MILLIGRAMS PER LITER, mg/*L*.

MLSS

Mixed Liquor Suspended Solids. The amount (mg/*L*) of suspended solids in the mixed liquor of an aeration tank.

MLVSS

Mixed Liquor Volatile Suspended Solids. The amount (mg/*L*) of organic or volatile suspended solids in the mixed liquor of an aeration tank. This volatile portion is used as a measure or indication of the microorganisms present.

MSDS

See MATERIAL SAFETY DATA SHEET.

MACRONUTRIENT

A chemical element of which relatively large quantities are essential for the growth of an organism.

MAIN SEWER

A sewer line that receives wastewater from many tributary branches and sewer lines and serves as an outlet for a large territory or is used to feed an intercepting sewer. Also called TRUNK SEWER.

MANIFOLD

A large pipe to which the ends of a series of smaller pipes are connected. Also called a HEADER.

MANOMETER (man-NAH-mut-ter)

An instrument for measuring pressure. Usually, a manometer is a glass tube filled with a liquid that is used to measure the difference in pressure across a flow measuring device, such as an orifice or a Venturi meter. The instrument used to measure blood pressure is a type of manometer.

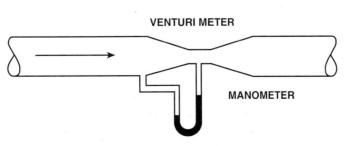

MASKING AGENTS

Substances used to cover up or disguise unpleasant odors. Liquid masking agents are dripped into the wastewater, sprayed into the air, or evaporated (using heat) with the unpleasant fumes or odors and then discharged into the air by blowers to make an undesirable odor less noticeable.

MASS EMISSION RATE

The rate of discharge of a pollutant expressed as a weight per unit time, usually as pounds or kilograms per day.

MATERIAL SAFETY DATA SHEET (MSDS)

A document that provides pertinent information and a profile of a particular hazardous substance or mixture. An MSDS is normally developed by the manufacturer or formulator of the hazardous substance or mixture. The MSDS is required to be made available to employees and operators or inspectors whenever there is the likelihood of the hazardous substance or mixture being introduced into the workplace. Some manufacturers are preparing MSDSs for products that are not considered to be hazardous to show that the product or substance is not hazardous.

MEAN CELL RESIDENCE TIME (MCRT)

See MCRT.

MEASURED VARIABLE

A factor (flow, temperature) that is sensed and quantified (reduced to a reading of some kind) by a primary element or sensor.

MECHANICAL AERATION

The use of machinery to mix air and water so that oxygen can be absorbed into the water. Some examples are: paddle wheels, mixers, or rotating brushes to agitate the surface of an aeration tank; pumps to create fountains; and pumps to discharge water down a series of steps forming falls or cascades.

MEG

(1) Abbreviation of MEGOHM.
(2) A procedure used for checking the insulation resistance on motors, feeders, bus bar systems, grounds, and branch circuit wiring. Also see MEGGER.

MEGGER (from megohm)

An instrument used for checking the insulation resistance on motors, feeders, bus bar systems, grounds, and branch circuit wiring. A megger reads in millions of ohms. Also see MEG.

MEGOHM (MEG-ome)

Millions of ohms. Mega- is a prefix meaning one million, so 5 megohms means 5 million ohms.

MERCAPTANS (mer-CAP-tans)

Compounds containing sulfur that have an extremely offensive skunk-like odor; also sometimes described as smelling like garlic or onions.

MESH

One of the openings or spaces in a screen or woven fabric. The value of the mesh is usually given as the number of openings per inch. This value does not consider the diameter of the wire or fabric; therefore, the mesh number does not always have a definite relationship to the size of the hole.

MESOPHILIC (MESS-o-FILL-ick) BACTERIA

Medium temperature bacteria. A group of bacteria that grow and thrive in a moderate temperature range between 68°F (20°C) and 113°F (45°C). The optimum temperature range for these bacteria in anaerobic digestion is 85°F (30°C) to 100°F (38°C).

METABOLISM

All of the processes or chemical changes in an organism or a single cell by which food is built up (anabolism) into living protoplasm and by which protoplasm is broken down (catabolism) into simpler compounds with the exchange of energy.

MICRON (MY-kron)

µm, Micrometer or Micron. A unit of length. One millionth of a meter or one thousandth of a millimeter. One micron equals 0.00004 of an inch.

MICRONUTRIENT

A trace element or an organic compound that is essential in tiny amounts for the growth of an organism.

MICROORGANISMS (MY-crow-OR-gan-is-ums)

Living organisms that can be seen individually only with the aid of a microscope.

MIL

A unit of length equal to 0.001 of an inch. The diameter of wires and tubing is measured in mils, as is the thickness of plastic sheeting.

MILLIGRAMS PER LITER, mg/L

A measure of the concentration by weight of a substance per unit volume in water or wastewater. In reporting the results of water and wastewater analysis, mg/L is preferred to the unit parts per million (ppm), to which it is approximately equivalent.

MISCIBLE (MISS-uh-bull)

Capable of being mixed. A liquid, solid, or gas that can be completely dissolved in water.

MIXED LIQUOR

When the activated sludge in an aeration tank is mixed with primary effluent or the raw wastewater and return sludge, this mixture is then referred to as mixed liquor as long as it is in the aeration tank. Mixed liquor also may refer to the contents of mixed aerobic or anaerobic digesters.

MIXED LIQUOR SUSPENDED SOLIDS (MLSS)

The amount (mg/L) of suspended solids in the mixed liquor of an aeration tank.

MIXED LIQUOR VOLATILE SUSPENDED SOLIDS (MLVSS)

The amount (mg/L) of organic or volatile suspended solids in the mixed liquor of an aeration tank. This volatile portion is used as a measure or indication of the microorganisms present.

MOTHER CIRCUIT BOARD

The base circuit board or the main circuit board.

MOVING AVERAGE

To calculate the moving average for the last 7 days, add up the values for the last 7 days and divide by 7. Each day add the most recent day's value to the sum of values and subtract the oldest value. By using the 7-day moving average, each day of the week is always represented in the calculations.

MUDBALLS

Material, approximately round in shape, that forms in filters and gradually increases in size when not removed by the backwashing process. Mudballs vary from pea-sized up to golf-ball-sized or larger.

MULTISTAGE PUMP

A pump that has more than one impeller. A single-stage pump has one impeller.

N

N or NORMAL

A normal solution contains one gram equivalent weight of reactant (compound) per liter of solution. The equivalent weight of an acid is that weight which contains one gram atom of ionizable hydrogen or its chemical equivalent. For example, the equivalent weight of sulfuric acid (H_2SO_4) is 49 (98 divided by 2 because there are two replaceable hydrogen ions). A one N solution of sulfuric acid would consist of 49 grams of H_2SO_4 dissolved in enough water to make one liter of solution.

NAICS

North American Industry Classification System. A code number system used to identify various types of industries. This code system replaces the SIC (Standard Industrial Classification) code system used prior to 1997. Use of these code numbers is often mandatory. Some companies have several processes, which will cause them to fit into two or more classifications. The code numbers are published by the US Government Printing Office, Superintendent of Documents, PO Box 371954, Pittsburgh, PA 15250-7954. Stock No. 041-001-00509-9; price, $33.00. There is no charge for shipping and handling.

NPDES PERMIT

National Pollutant Discharge Elimination System permit is the regulatory agency document issued by either a federal or state agency that is designed to control all discharges of potential pollutants from point sources and stormwater runoff into US waterways. NPDES permits regulate discharges into US waterways from all point sources of pollution, including industries, municipal wastewater treatment plants, sanitary landfills, large animal feedlots, and return irrigation flows.

NTU

Nephelometric Turbidity Units. See TURBIDITY UNITS.

NAMEPLATE

A durable, metal plate found on equipment that lists critical operating conditions for the equipment.

NAPPE (NAP)

The sheet or curtain of water flowing over a weir or dam. When the water freely flows over any structure, it has a well-defined upper and lower water surface.

NEPHELOMETRIC (neff-el-o-MET-rick)

A means of measuring turbidity in a sample by using an instrument called a nephelometer. A nephelometer passes light through a sample and the amount of light deflected (usually at a 90-degree angle) is then measured.

NEUTRALIZATION (noo-trull-uh-ZAY-shun)

Addition of an acid or alkali (base) to a liquid to cause the pH of the liquid to move toward a neutral pH of 7.0.

NEW SOURCE

Any building, structure, facility, or installation from which there is or may be a discharge of pollutants. Construction of the facility must have begun after promulgation of the applicable Pretreatment Standards. The building, structure, facility, or installation must also be constructed at a site at which no other source is located; or, must totally replace the existing process or production equipment producing the discharge at the site; or, must be substantially independent of an existing source of discharge at the same site.

NITRIFICATION (NYE-truh-fuh-KAY-shun)

An aerobic process in which bacteria change the ammonia and organic nitrogen in water or wastewater into oxidized nitrogen (usually nitrate).

NITRIFYING BACTERIA

Bacteria that change ammonia and organic nitrogen into oxidized nitrogen (usually nitrate).

NONBIODEGRADABLE (NON-buy-o-dee-GRADE-uh-bull)

Substances that cannot readily be broken down by bacteria to simpler forms.

NONCOMPATIBLE POLLUTANTS

Those pollutants that are normally *NOT* removed by the POTW treatment system. These pollutants may be a toxic waste and may pass through the POTW untreated or interfere with the treatment system. Examples of noncompatible pollutants include heavy metals, such as copper, nickel, lead, and zinc; organics, such as methylene chloride, 1,1,1-trichloroethylene, methyl ethyl ketone, acetone, and gasoline; or sludges containing toxic organics or metals.

NON-PERMIT CONFINED SPACE

See CONFINED SPACE, NON-PERMIT.

NONPOINT SOURCE

A runoff or discharge from a field or similar source, in contrast to a point source, which refers to a discharge that comes out the end of a pipe or other clearly identifiable conveyance. Also see POINT SOURCE.

NONSPARKING TOOLS

These tools will not produce a spark during use. They are made of a nonferrous material, usually a copper-beryllium alloy.

NONVOLATILE MATTER

Material such as sand, salt, iron, calcium, and other mineral materials that are only slightly affected by the actions of organisms and are not lost on ignition of the dry solids at 550°C (1,022°F). Volatile materials are chemical substances usually of animal or plant origin. Also see INORGANIC WASTE and VOLATILE SOLIDS.

NORMAL

See *N* or NORMAL.

NOTICE

This word calls attention to information that is especially significant in understanding and operating equipment or processes safely. Also see CAUTION, DANGER, and WARNING.

NUTRIENT

Any substance that is assimilated (taken in) by organisms and promotes growth. Nitrogen and phosphorus are nutrients that promote the growth of algae. There are other essential and trace elements that are also considered nutrients. Also see NUTRIENT CYCLE.

NUTRIENT CYCLE

The transformation or change of a nutrient from one form to another until the nutrient has returned to the original form, thus completing the cycle. The cycle may take place under either aerobic or anaerobic conditions.

O

ORP (pronounce as separate letters)

Oxidation-Reduction Potential. The electrical potential required to transfer electrons from one compound or element (the oxidant) to another compound or element (the reductant); used as a qualitative measure of the state of oxidation in water and wastewater treatment systems. ORP is measured in millivolts, with negative values indicating a tendency to reduce compounds or elements and positive values indicating a tendency to oxidize compounds or elements.

OSHA (O-shuh)

The Williams-Steiger Occupational Safety and Health Act of 1970 (OSHA) is a federal law designed to protect the health and safety of workers, including the operators of water supply and treatment systems and wastewater collection and treatment systems. The Act regulates the design, construction, operation, and maintenance of water and wastewater systems. OSHA regulations require employers to obtain and make available to workers the Material Safety Data Sheets (MSDSs) for chemicals used at industrial facilities and treatment plants. OSHA also refers to the federal and state agencies that administer the OSHA regulations.

OBLIGATE AEROBES

Bacteria that must have atmospheric or dissolved molecular oxygen to live and reproduce.

OCCUPATIONAL SAFETY AND HEALTH ACT OF 1970 (OSHA)

See OSHA.

OFFSET

(1) The difference between the actual value and the desired value (or set point); characteristic of proportional controllers that do not incorporate reset action. Also called DRIFT.

(2) A pipe fitting in the approximate form of a reverse curve or other combination of elbows or bends that brings one section of a line of pipe out of line with, but into a line parallel with, another section.

(3) A pipe joint that has lost its bedding support, causing one of the pipe sections to drop or slip, thus creating a condition where the pipes no longer line up properly.

OLFACTORY (all-FAK-tore-ee) FATIGUE

A condition in which a person's nose, after exposure to certain odors, is no longer able to detect the odor.

ORGANIC WASTE

Waste material that may come from animal or plant sources. Natural organic wastes generally can be consumed by bacteria and other small organisms. Manufactured or synthetic organic wastes from metal finishing, chemical manufacturing, and petroleum industries may not normally be consumed by bacteria and other organisms. Also see INORGANIC WASTE and VOLATILE SOLIDS.

ORIFICE (OR-uh-fiss)

An opening (hole) in a plate, wall, or partition. An orifice flange or plate placed in a pipe consists of a slot or a calibrated circular hole smaller than the pipe diameter. The difference in pressure in the pipe above and at the orifice may be used to determine the flow in the pipe. In a trickling filter distributor, the wastewater passes through an orifice to the surface of the filter media.

OVERFLOW RATE

One factor of the design flow of settling tanks and clarifiers in treatment plants used by operators to determine if tanks and clarifiers are hydraulically (flow) over- or underloaded. Also called SURFACE LOADING.

$$\text{Overflow Rate, GPD/sq ft} = \frac{\text{Flow, gallons/day}}{\text{Surface Area, sq ft}}$$

or

$$\text{Overflow Rate, } \frac{m^3/day}{m^2} = \frac{\text{Flow, } m^3/day}{\text{Surface Area, } m^2}$$

OXIDATION

Oxidation is the addition of oxygen, removal of hydrogen, or the removal of electrons from an element or compound; in the environment and in wastewater treatment processes, organic matter is oxidized to more stable substances. The opposite of REDUCTION.

OXIDATION STATE/OXIDATION NUMBER

In a chemical formula, a number accompanied by a polarity indication (+ or −) that together indicate the charge of an ion as well as the extent to which the ion has been oxidized or reduced in a REDOX REACTION.

Due to the loss of electrons, the charge of an ion that has been oxidized would go from negative toward or to neutral, from neutral to positive, or from positive to more positive. As an example, an oxidation number of 2+ would indicate that an ion has lost two electrons and that its charge has become positive (that it now has an excess of two protons).

Due to the gain of electrons, the charge of the ion that has been reduced would go from positive toward or to neutral, from neutral to negative, or from negative to more negative. As an example, an oxidation number of 2− would indicate that an ion has gained two electrons and that its charge has become negative (that it now has an excess of two electrons). As an ion gains electrons, its oxidation state (or the extent to which it is oxidized) lowers; that is, its oxidation state is reduced. Also see REDOX REACTION.

OXIDATION-REDUCTION POTENTIAL (ORP)

The electrical potential required to transfer electrons from one compound or element (the oxidant) to another compound or element (the reductant); used as a qualitative measure of the state of oxidation in water and wastewater treatment systems. ORP is measured in millivolts, with negative values indicating a tendency to reduce compounds or elements and positive values indicating a tendency to oxidize compounds or elements.

OXIDATION-REDUCTION (REDOX) REACTION

See REDOX REACTION.

OXYGEN DEFICIENCY

An atmosphere containing oxygen at a concentration of less than 19.5 percent by volume.

OXYGEN ENRICHMENT

An atmosphere containing oxygen at a concentration of more than 23.5 percent by volume.

P

POTW

Publicly Owned Treatment Works. A treatment works that is owned by a state, municipality, city, town, special sewer district, or other publicly owned and financed entity as opposed to a privately (industrial) owned treatment facility. This definition includes any devices and systems used in the storage, treatment, recycling, and reclamation of municipal sewage (wastewater) or industrial wastes of a liquid nature. It also includes sewers, pipes, and other conveyances only if they carry wastewater to a POTW treatment plant. The term also means the municipality (public entity) that has jurisdiction over the indirect discharges to and the discharges from such a treatment works.

PASSIVATING

A metal plating process that forms a protective film on metals by immersion in an acid solution, usually nitric acid or nitric acid with sodium dichromate.

PATHOGENIC (path-o-JEN-ick) ORGANISMS

Organisms, including bacteria, viruses, or cysts, capable of causing diseases (such as giardiasis, cryptosporidiosis, typhoid, cholera, dysentery) in a host (such as a person). Also called PATHOGENS.

PATHOGENS (PATH-o-jens)

See PATHOGENIC ORGANISMS.

PERISTALTIC (PAIR-uh-STALL-tick) PUMP

A type of positive displacement pump.

PERMEATE (PURR-me-ate)

(1) To penetrate and pass through, as water penetrates and passes through soil and other porous materials.

(2) The liquid (demineralized water) produced from the reverse osmosis process that contains a low concentration of dissolved solids.

PERMIT-REQUIRED CONFINED SPACE (PERMIT SPACE)

See CONFINED SPACE, PERMIT-REQUIRED (PERMIT SPACE).

pH (pronounce as separate letters)

pH is an expression of the intensity of the basic or acidic condition of a liquid. Mathematically, pH is the logarithm (base 10) of the reciprocal of the hydrogen ion activity. If $\{H^+\} = 10^{-6.5}$, then pH = 6.5. The pH may range from 0 to 14, where 0 is most acidic, 14 most basic, and 7 neutral.

PICKLE

An acid or chemical solution in which metal objects or workpieces are dipped to remove oxide scale or other adhering substances.

PLUG FLOW

A type of flow that occurs in tanks, basins, or reactors when a slug of water or wastewater moves through a tank without ever dispersing or mixing with the rest of the water or wastewater flowing through the tank.

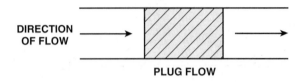

POINT SOURCE

A discharge that comes out the end of a pipe or other clearly identifiable conveyance. Examples of point source conveyances from which pollutants may be discharged include: ditches, channels, tunnels, conduits, wells, containers, rolling stock, concentrated animal feeding operations, landfill leachate collection systems, vessels, or other floating craft. A NONPOINT SOURCE refers to runoff or a discharge from a field or similar source.

POLE SHADER

A copper bar circling the laminated iron core inside the coil of a magnetic starter.

POLLUTION

The impairment (reduction) of water quality by agricultural, domestic, or industrial wastes (including thermal and radioactive wastes) to a degree that the natural water quality is changed to hinder any beneficial use of the water or render it offensive to the senses of sight, taste, or smell or when sufficient amounts of wastes create or pose a potential threat to human health or the environment.

POLYELECTROLYTE (POLY-ee-LECK-tro-lite)

A high-molecular-weight (relatively heavy) substance, having points of positive or negative electrical charges, that is formed by either natural or synthetic (manmade) processes. Natural polyelectrolytes may be of biological origin or obtained from starch products or cellulose derivatives. Synthetic polyelectrolytes consist of simple substances that have been made into complex, high-molecular-weight substances. Used with other chemical coagulants to aid in binding small suspended particles to larger chemical flocs for their removal from water. Often called a POLYMER.

POLYMER (POLY-mer)

A long-chain molecule formed by the union of many monomers (molecules of lower molecular weight). Polymers are used with other chemical coagulants to aid in binding small suspended particles to larger chemical flocs for their removal from water. Also see POLYELECTROLYTE.

POLYSACCHARIDE (poly-SAC-uh-ride)

A carbohydrate, such as starch, insulin, or cellulose, composed of chains of simple sugars.

PORE

A very small open space in a rock or granular material. Also called an INTERSTICE, VOID, or void space. Also see VOID.

POTTING COMPOUNDS

Sealing and holding compounds (such as epoxy) used in electrode probes.

PRECIOUS METAL

Metal that is very valuable, such as gold or silver.

PRECIPITATE (pre-SIP-uh-TATE)

(1) An insoluble, finely divided substance that is a product of a chemical reaction within a liquid.

(2) The separation from solution of an insoluble substance.

PRECISION

The ability of an instrument to measure a process variable and repeatedly obtain the same result. The ability of an instrument to reproduce the same results.

PRECOAT

Application of a free-draining, noncohesive material, such as diatomaceous earth, to a filtering medium. Precoating reduces the frequency of media washing and facilitates cake discharge.

PRIMARY ELEMENT

(1) A device that measures (senses) a physical condition or variable of interest. Floats and thermocouples are examples of primary elements. Also called a SENSOR.

(2) The hydraulic structure used to measure flows. In open channels, weirs and flumes are primary elements or devices. Venturi meters and orifice plates are the primary elements in pipes or pressure conduits.

PRIMARY TREATMENT

A wastewater treatment process that takes place in a rectangular or circular tank and allows those substances in wastewater that readily settle or float to be separated from the wastewater being treated. A septic tank is also considered primary treatment.

PROCESS VARIABLE

A physical or chemical quantity that is usually measured and controlled in the operation of a water, wastewater, or industrial treatment plant.

PROGRAMMABLE LOGIC CONTROLLER (PLC)

A small computer that controls process equipment (variables) and can control the sequence of valve operations.

PROPORTIONAL WEIR (WEER)

A specially shaped weir in which the flow through the weir is directly proportional to the head.

PROTEINACEOUS (PRO-ten-NAY-shus)

Materials containing proteins, which are organic compounds containing nitrogen.

PROTOPLASM

A complex substance (typically colorless and semifluid) regarded as the physical basis of life, having the power of spontaneous motion and reproduction; the living matter of all plant and animal cells and tissues.

PROTOZOA (pro-toe-ZOE-ah)

A group of motile, microscopic organisms (usually single-celled and aerobic) that sometimes cluster into colonies and generally consume bacteria as an energy source.

PRUSSIAN BLUE

A blue paste or liquid (often on a paper like carbon paper) used to show a contact area. Used to determine if gate valve seats fit properly.

PUG MILL

A mechanical device with rotating paddles or blades that is used to mix and blend different materials together.

PURGE

To remove a gas or vapor from a vessel, reactor, or confined space, usually by displacement or dilution.

PUTREFACTION (PYOO-truh-FACK-shun)

Biological decomposition of organic matter, with the production of foul-smelling and -tasting products, associated with anaerobic (no oxygen present) conditions.

PUTRESCIBLE (pyoo-TRES-uh-bull)

Material that will decompose under anaerobic conditions and produce nuisance odors.

Q

(NO LISTINGS)

R

RAS (pronounce as separate letters, or RAZZ)

Return Activated Sludge. Settled activated sludge that is collected in the secondary clarifier and returned to the aeration basin to mix with incoming raw or primary settled wastewater.

RCRA (RICK-ruh)

The Federal Resource Conservation and Recovery Act (10/21/76), Public Law (PL) 94-580, provides technical and financial assistance for the development of plans and facilities for recovery of energy and resources from discarded materials and for the safe disposal of discarded materials and hazardous wastes. This act introduces the philosophy of the "cradle-to-grave" control of hazardous wastes. RCRA regulations can be found in Title 40 of the Code of Federal Regulations (40 CFR) Parts 260-268, 270, and 271.

RABBLING

The process of moving or plowing the material inside a furnace by using the center shaft and rabble arms.

RADIAL TO IMPELLER

Perpendicular to the impeller shaft. Material being pumped flows at a right angle to the impeller.

RANGE

The spread from minimum to maximum values that an instrument is designed to measure. Also see EFFECTIVE RANGE and SPAN.

READOUT

The reading of the value of a process variable from an indicator or recorder (or on a computer screen).

REAGENT (re-A-gent)

A pure, chemical substance that is used to make new products or is used in chemical tests to measure, detect, or examine other substances.

RECALCINATION (re-kal-sin-NAY-shun)

A lime recovery process in which the calcium carbonate in sludge is converted to lime by heating at 1,800°F (980°C).

RECARBONATION (re-kar-bun-NAY-shun)

A process in which carbon dioxide is bubbled into the water being treated to lower the pH.

RECEIVER

A device that indicates the result of a measurement, usually using either a fixed scale and movable indicator (pointer), such as a pressure gauge, or a moving chart with a movable pen like those used on a circular flow-recording chart. Also called an INDICATOR.

RECEIVING WATER

A stream, river, lake, ocean, or other surface or groundwaters into which treated or untreated wastewater is discharged.

RECORDER
A device that creates a permanent record, on a paper chart or magnetic tape, of the changes in a measured variable.

RECYCLE
The use of water or wastewater within (internally) a facility before it is discharged to a treatment system. Also see REUSE.

REDOX (REE-docks) REACTION
A two-part reaction between two ions involving a transfer of electrons from one ion to the other. Oxidation is the loss of electrons by one ion, and reduction is the acceptance of electrons by the other ion. Reduction refers to the lowering of the OXIDATION STATE/OXIDATION NUMBER of the ion accepting the electrons.

In a redox reaction, the ion that gives up the electrons (that is oxidized) is called the reductant because it causes a reduction in the oxidation state or number of the ion that accepts the transferred electrons. The ion that receives the electrons (that is reduced) is called the oxidant because it causes oxidation of the other ion. Oxidation and reduction always occur simultaneously.

REDUCTION (re-DUCK-shun)
Reduction is the addition of hydrogen, removal of oxygen, or the addition of electrons to an element or compound. Under anaerobic conditions (no dissolved oxygen present), sulfur compounds are reduced to odor-producing hydrogen sulfide (H_2S) and other compounds. In the treatment of metal finishing wastewaters, hexavalent chromium (Cr^{6+}) is reduced to the trivalent form (Cr^{3+}). The opposite of OXIDATION.

REFERENCE
A physical or chemical quantity whose value is known exactly, and thus is used to calibrate instruments or standardize measurements. Also called a STANDARD.

REFRACTORY (re-FRACK-toe-ree) MATERIALS
Materials difficult to remove entirely from wastewater, such as nutrients, color, taste- and odor-producing substances, and some toxic materials.

REPRESENTATIVE SAMPLE
A sample portion of material, water, or wastestream that is as nearly identical in content and consistency as possible to that in the larger body being sampled.

RESINS
See ION EXCHANGE RESINS.

RESPIRATION
The process in which an organism takes in oxygen for its life processes and gives off carbon dioxide.

RETURN ACTIVATED SLUDGE (RAS)
Settled activated sludge that is collected in the secondary clarifier and returned to the aeration basin to mix with incoming raw or primary settled wastewater.

REUSE
The use of water or wastewater after it has been discharged and then withdrawn by another user. Also see RECYCLE.

RISING SLUDGE
Rising sludge occurs in the secondary clarifiers of activated sludge plants when the sludge settles to the bottom of the clarifier, is compacted, and then starts to rise to the surface, usually as a result of denitrification, or anaerobic biological activity that produces carbon dioxide and/or methane.

ROTAMETER (ROTE-uh-ME-ter)

A device used to measure the flow rate of gases and liquids. The gas or liquid being measured flows vertically up a tapered, calibrated tube. Inside the tube is a small ball or bullet-shaped float (it may rotate) that rises or falls depending on the flow rate. The flow rate may be read on a scale behind or on the tube by looking at the middle of the ball or at the widest part or top of the float.

ROTARY PUMP

A type of displacement pump consisting essentially of elements rotating in a close-fitting pump case. The rotation of these elements alternately draws in and discharges the water being pumped. Such pumps act with neither suction nor discharge valves, operate at almost any speed, and do not depend on centrifugal forces to lift the water.

ROTIFERS (ROTE-uh-fers)

Microscopic animals characterized by short hairs on their front ends.

S

SCFM

Standard Cubic Feet per Minute. Cubic feet of air per minute at standard conditions of temperature, pressure, and humidity (0°C, 14.7 psia, and 50 percent relative humidity).

SIC CODE

Standard Industrial Classification code. A code number system used to identify various types of industries. In 1997, the United States and Canada replaced the SIC code system with the North American Industry Classification System (NAICS); Mexico adopted the NAICS in 1998. Also see NAICS.

SPC CHART

Statistical Process Control chart. A plot of daily performance such as a trend chart.

SVI

Sludge Volume Index. A calculation that indicates the tendency of activated sludge solids (aerated solids) to thicken or to become concentrated during the sedimentation/thickening process. SVI is calculated in the following manner: (1) allow a mixed liquor sample from the aeration basin to settle for 30 minutes; (2) determine the suspended solids concentration for a sample of the same mixed liquor; (3) calculate SVI by dividing the measured (or observed) wet volume (mL/L) of the settled sludge by the dry weight concentration of MLSS in grams/L.

$$\text{SVI, m}L/\text{gm} = \frac{\text{Settled Sludge Volume/Sample Volume, m}L/L}{\text{Suspended Solids Concentration, mg}/L} \times \frac{1{,}000 \text{ mg}}{\text{gram}}$$

SACRIFICIAL ANODE

An easily corroded material deliberately installed in a pipe or tank. The intent of such an installation is to give up (sacrifice) this anode to corrosion while the water supply facilities remain relatively corrosion free.

SCALE

(1) A combination of mineral salts and bacterial accumulation that sticks to the inside of a collection pipe under certain conditions. Scale, in extreme growth circumstances, creates additional friction loss to the flow of water. Scale may also accumulate on surfaces other than pipes.

(2) The marked plate against which an indicator or recorder reads, usually the same as the range of the measuring system. See RANGE.

SECCHI (SECK-key) DISK

A flat, white disk lowered into the water by a rope until it is just barely visible. At this point, the depth of the disk from the water surface is the recorded Secchi disk transparency.

SECONDARY ELEMENT

The secondary measuring device or flowmeter used with a primary measuring device (element) to measure the rate of liquid flow. In open channels, bubblers and floats are secondary elements. Differential pressure measuring devices are the secondary elements in pipes or pressure conduits. The purpose of the secondary measuring device is to (1) measure the liquid level in open channels or the differential pressure in pipes, and (2) convert this measurement into an appropriate flow rate according to the known liquid level or differential pressure and flow rate relationship of the primary measuring device. This flow rate may be integrated (added up) to obtain a totalized volume, transmitted to a recording device, or used to pace an automatic sampler.

SECONDARY TREATMENT

A wastewater treatment process used to convert dissolved or suspended materials into a form more readily separated from the water being treated. Usually, the process follows primary treatment by sedimentation. The process commonly is a type of biological treatment followed by secondary clarifiers that allow the solids to settle out from the water being treated.

SEDIMENTATION (SED-uh-men-TAY-shun) BASIN

A tank or basin in which water or wastewater is held for a period of time during which the heavier solids settle to the bottom and the lighter materials float to the surface. Also called settling tank or CLARIFIER.

SEIZING or SEIZE UP

Seizing occurs when an engine overheats and a part expands to the point where the engine will not run. Also called freezing.

SELECTOR

A reactor or basin in which baffles or other devices create a series of compartments. The environment and the resulting microbial population within each compartment can be controlled to some extent by the operator. The environmental conditions (food, lack of dissolved oxygen) that develop are intended to favor the growth of certain organisms over others. The conditions thereby select certain organisms.

SELECTOR RECYCLE

The recycling of return sludge or oxidized nitrogen to provide desired environmental conditions for microorganisms to perform a desired function.

SENSITIVITY

The smallest change in a process variable that an instrument can sense.

SENSOR

A device that measures (senses) a physical condition or variable of interest. Floats and thermocouples are examples of sensors. Also called a PRIMARY ELEMENT.

SEPTIC (SEP-tick) or SEPTICITY

A condition produced by bacteria when all oxygen supplies are depleted. If severe, the bottom deposits produce hydrogen sulfide, the deposits and water turn black, give off foul odors, and the water has a greatly increased oxygen and chlorine demand.

SEQUESTRATION (SEE-kwes-TRAY-shun)

A chemical complexing (forming or joining together) of metallic cations (such as iron) with certain inorganic compounds, such as phosphate. Sequestration prevents the precipitation of the metals (iron). Also see CHELATION.

SET POINT

The position at which the control or controller is set. This is the same as the desired value of the process variable. For example, a thermostat is set to maintain a desired temperature.

SEWER GAS

(1) Gas in collection lines (sewers) that results from the decomposition of organic matter in the wastewater. When testing for gases found in sewers, test for oxygen deficiency, oxygen enrichment, and also for explosive and toxic gases.

(2) Any gas present in the wastewater collection system, even though it is from such sources as gas mains, gasoline, and cleaning fluid.

SHOCK LOAD

The arrival at a treatment process of water or wastewater containing unusually high concentrations of contaminants in sufficient quantity or strength to cause operating problems. Organic or hydraulic overloads also can cause a shock load.

(1) For activated sludge, possible problems include odors and bulking sludge, which will result in a high loss of solids from the secondary clarifiers into the plant effluent and a biological process upset that may require several days to a week to recover.

(2) For trickling filters, possible problems include odors and sloughing off of the growth or slime on the trickling filter media.

(3) For drinking water treatment, possible problems include filter blinding and product water with taste and odor, color, or turbidity problems.

SHORT-CIRCUITING

A condition that occurs in tanks or basins when some of the flowing water entering a tank or basin flows along a nearly direct pathway from the inlet to the outlet. This is usually undesirable since it may result in shorter contact, reaction, or settling times in comparison with the theoretical (calculated) or presumed detention times.

SIGNIFICANT INDUSTRIAL USER (SIU)

A significant industrial user includes all categorical industrial users, and any noncategorical industrial user that:

(1) Discharge 25,000 gallons per day or more of process wastewater ("process wastewater" excludes sanitary, noncontact cooling and boiler blowdown wastewaters), or

(2) Contribute a process wastestream that makes up five percent or more of the average dry weather hydraulic or organic (BOD, TSS) capacity of a treatment plant, or

(3) Have a reasonable potential, in the opinion of the control or approval authority, to adversely affect the POTW treatment plant (inhibition, pass-through of pollutants, sludge contamination, or endangerment of POTW workers).

SINGLE-STAGE PUMP

A pump that has only one impeller. A multistage pump has more than one impeller.

SLAKE

To mix with water so that a true chemical combination (hydration) takes place, such as in the slaking of lime.

SLIME GROWTH

See ZOOGLEAL MAT (FILM).

SLOUGHED or SLOUGHING (SLUFF-ing)

The breaking off of biological or biomass growths from the fixed film or rotating biological contactor (RBC) media. The sloughed growth becomes suspended in the effluent and is later removed in the secondary clarifier as sludge.

SLUDGE (SLUJ)

(1) The settleable solids separated from liquids during processing.

(2) The deposits of foreign materials on the bottoms of streams or other bodies of water or on the bottoms and edges of wastewater collection lines and appurtenances.

SLUDGE AGE

A measure of the length of time a particle of suspended solids has been retained in the activated sludge process.

$$\text{Sludge Age, days} = \frac{\text{Suspended Solids Under Aeration, lbs or kg}}{\text{Suspended Solids Added, lbs/day or kg/day}}$$

SLUDGE DENSITY INDEX (SDI)

This calculation is used in a way similar to the Sludge Volume Index (SVI) to indicate the settleability of a sludge in a secondary clarifier or effluent. The weight in grams of one milliliter of sludge after settling for 30 minutes. SDI = 100/SVI. Also see SLUDGE VOLUME INDEX.

SLUDGE VOLUME INDEX (SVI)

A calculation that indicates the tendency of activated sludge solids (aerated solids) to thicken or to become concentrated during the sedimentation/thickening process. SVI is calculated in the following manner: (1) allow a mixed liquor sample from the aeration basin to settle for 30 minutes; (2) determine the suspended solids concentration for a sample of the same mixed liquor; (3) calculate SVI by dividing the measured (or observed) wet volume (mL/L) of the settled sludge by the dry weight concentration of MLSS in grams/L.

$$\text{SVI, m}L/\text{gm} = \frac{\text{Settled Sludge Volume/Sample Volume, m}L/L}{\text{Suspended Solids Concentration, mg}/L} \times \frac{1{,}000 \text{ mg}}{\text{gram}}$$

SLUDGE/VOLUME (S/V) RATIO

The volume of sludge blanket divided by the daily volume of sludge pumped from the thickener.

SLUG

Intermittent release or discharge of wastewater or industrial wastes.

SLURRY

A watery mixture or suspension of insoluble (not dissolved) matter; a thin, watery mud or any substance resembling it (such as a grit slurry or a lime slurry).

SOFTWARE PROGRAM

Computer program; the list of instructions that tell a computer how to perform a given task or tasks. Some software programs are designed and written to monitor and control treatment processes.

SOLENOID (SO-luh-noid)

A magnetically operated mechanical device (electric coil). Solenoids can operate small valves or electric switches.

SOLVENT MANAGEMENT PLAN

A strategy for keeping track of all solvents delivered to a site, their storage, use, and disposal. This includes keeping spent solvents segregated from other process wastewaters to maximize the value of the recoverable solvents, to avoid contamination of other segregated wastes, and to prevent the discharge of toxic organics to any wastewater collection system or the environment. The plan should describe measures to control spills and leaks and to ensure that there is no deliberate dumping of solvents. Also known as a TOXIC ORGANIC MANAGEMENT PLAN.

SPAN

The scale or range of values an instrument is designed to measure. Also see RANGE.

SPECIFIC GRAVITY

(1) Weight of a particle, substance, or chemical solution in relation to the weight of an equal volume of water. Water has a specific gravity of 1.000 at 4°C (39°F). Particulates with specific gravity less than 1.0 float to the surface and particulates with specific gravity greater than 1.0 sink.

(2) Weight of a particular gas in relation to the weight of an equal volume of air at the same temperature and pressure (air has a specific gravity of 1.0). Chlorine gas has a specific gravity of 2.5.

SPOIL

Excavated material, such as soil, from the trench of a water main or sewer.

STABILIZATION

Conversion to a form that resists change. Organic material is stabilized by bacteria that convert the material to gases and other relatively inert substances. Stabilized organic material generally will not give off obnoxious odors.

STABILIZED WASTE

A waste that has been treated or decomposed to the extent that, if discharged or released, its rate and state of decomposition would be such that the waste would not cause a nuisance or odors in the receiving water.

STANDARD

A physical or chemical quantity whose value is known exactly, and thus is used to calibrate instruments or standardize measurements. Also called a REFERENCE.

STANDARDIZE

To compare with a standard.

(1) In wet chemistry, to find out the exact strength of a solution by comparing it with a standard of known strength. This information is used to adjust the strength by adding more water or more of the substance dissolved.

(2) To set up an instrument or device to read a standard. This allows you to adjust the instrument so that it reads accurately, or enables you to apply a correction factor to the readings.

STARTERS (MOTOR)

Devices used to start up large motors gradually to avoid severe mechanical shock to a driven machine and to prevent disturbance to the electrical lines (causing dimming and flickering of lights).

STATIC HEAD

When water is not moving, the vertical distance (in feet or meters) from a reference point to the water surface is the static head. Also see DYNAMIC HEAD, DYNAMIC PRESSURE, and STATIC PRESSURE.

STATIC PRESSURE

When water is not moving, the vertical distance (in feet or meters) from a specific point to the water surface is the static head. The static pressure in psi (or kPa) is the static head in feet times 0.433 psi/ft (or meters $\times$ 9.81 kPa/m). Also see DYNAMIC HEAD, DYNAMIC PRESSURE, and STATIC HEAD.

STATOR

That portion of a machine that contains the stationary (non-moving) parts that surround the moving parts (rotor).

STEP-FEED AERATION

Step-feed aeration is a modification of the conventional activated sludge process. In step-feed aeration, primary effluent enters the aeration tank at several points along the length of the tank, rather than at the beginning or head of the tank and flowing through the entire tank in a plug flow mode.

STERILIZATION (STAIR-uh-luh-ZAY-shun)

The removal or destruction of all microorganisms, including pathogens and other bacteria, vegetative forms, and spores. Compare with DISINFECTION.

STETHOSCOPE

An instrument used to magnify sounds and carry them to the ear.

STILLING WELL

A well or chamber that is connected to the main flow channel by a small inlet. Waves and surges in the main flow stream will not appear in the well due to the small-diameter inlet. The liquid surface in the well will be quiet, but will follow all of the steady fluctuations of the open channel. The liquid level in the well is measured to determine the flow in the main channel.

STRIPPED GASES

Gases that are released from a liquid by bubbling air through the liquid or by allowing the liquid to be sprayed or tumbled over media.

STRUVITE (STREW-vite)

A deposit or precipitate of magnesium ammonium phosphate hexahydrate found on the rotating components of centrifuges and centrate discharge lines. Struvite can be formed when anaerobic sludge comes in contact with spinning centrifuge components rich in oxygen in the presence of microbial activity. Struvite can also be formed in digested sludge lines and valves in the presence of oxygen and microbial activity. Struvite can form when the pH level is between 5 and 9.

SUBSTRATE

(1) The base on which an organism lives. The soil is the substrate of most seed plants; rocks, soil, water, or other plants or animals are substrates for other organisms.

(2) Chemical used by an organism to support growth. The organic matter in wastewater is a substrate for the organisms in activated sludge.

SUCTION HEAD

The positive pressure [in feet (meters) of water or pounds per square inch (kilograms per square centimeter) of mercury vacuum] on the suction side of a pump. The pressure can be measured from the centerline of the pump up to the elevation of the hydraulic grade line on the suction side of the pump.

SUCTION LIFT

The negative pressure [in feet (meters) of water or inches (centimeters) of mercury vacuum] on the suction side of a pump. The pressure can be measured from the centerline of the pump down to (lift) the elevation of the hydraulic grade line on the suction side of the pump.

SUPERNATANT (soo-per-NAY-tent)

Liquid removed from settled sludge. Supernatant commonly refers to the liquid between the sludge on the bottom and the scum on the surface.

SURCHARGE

Sewers are surcharged when the supply of water to be carried is greater than the capacity of the pipes to carry the flow. The surface of the wastewater in manholes rises above the top of the sewer pipe, and the sewer is under pressure or a head, rather than at atmospheric pressure.

SURFACE-ACTIVE AGENT

The active agent in detergents that possesses a high cleaning ability. Also called a SURFACTANT.

SURFACE LOADING

One factor of the design flow of settling tanks and clarifiers in treatment plants used by operators to determine if tanks and clarifiers are hydraulically (flow) over- or underloaded. Also called OVERFLOW RATE.

$$\text{Surface Loading, GPD/sq ft} = \frac{\text{Flow, gallons/day}}{\text{Surface Area, sq ft}}$$

or

$$\text{Surface Loading, } \frac{m^3/day}{m^2} = \frac{\text{Flow, } m^3/day}{\text{Surface Area, } m^2}$$

SURFACTANT (sir-FAC-tent)

Abbreviation for surface-active agent. The active agent in detergents that possesses a high cleaning ability.

SUSPENDED GROWTH PROCESSES

Wastewater treatment processes in which the microorganisms and bacteria treating the wastes are suspended in the wastewater being treated. The wastes flow around and through the suspended growths. The various modes of the activated sludge process make use of suspended growth reactors. These reactors can be used for BOD (biochemical oxygen demand) removal, nitrification, and denitrification.

SUSPENDED SOLIDS

(1) Solids that either float on the surface or are suspended in water, wastewater, or other liquids, and that are largely removable by laboratory filtering.

(2) The quantity of material removed from water or wastewater in a laboratory test, as prescribed in *STANDARD METHODS FOR THE EXAMINATION OF WATER AND WASTEWATER*, and referred to as Total Suspended Solids Dried at 103–105°C.

T

TOC (pronounce as separate letters)

Total Organic Carbon. TOC measures the amount of organic carbon in water.

TWA

See TIME-WEIGHTED AVERAGE.

TAILGATE SAFETY MEETING

Brief (10 to 20 minutes) safety meetings held every 7 to 10 working days. The term comes from the safety meetings regularly held by the construction industry around the tailgate of a truck.

TELEMETRY (tel-LEM-uh-tree)

The electrical link between a field transmitter and the receiver. Telephone lines are commonly used to serve as the electrical line.

THERMOCOUPLE

A heat-sensing device made of two conductors of different metals joined at their ends. An electric current is produced when there is a difference in temperature between the ends.

THERMOPHILIC (thur-moe-FILL-ick) **BACTERIA**

A group of bacteria that grow and thrive in temperatures above 113°F (45°C). The optimum temperature range for these bacteria in anaerobic decomposition is 120°F (49°C) to 135°F (57°C). Aerobic thermophilic bacteria thrive between 120°F (49°C) and 158°F (70°C).

THRUST BLOCK

A mass of concrete or similar material appropriately placed around a pipe to prevent movement when the pipe is carrying water. Usually placed at bends and valve structures.

TIME-WEIGHTED AVERAGE (TWA)

The average concentration of a pollutant based on the times and levels of concentrations of the pollutant. The time-weighted average is equal to the sum of the portion of each time period (as a decimal, such as 0.25 hour) multiplied by the pollutant concentration during the time period divided by the hours in the workday (usually 8 hours). 8TWA PEL is the time-weighted average permissible exposure limit, in parts per million, for a normal 8-hour workday and a 40-hour workweek to which nearly all workers may be repeatedly exposed, day after day, without adverse effect.

TIMER

A device for automatically starting or stopping a machine or other device at a given time.

TITRATE (TIE-trate)

To titrate a sample, a chemical solution of known strength is added drop by drop until a certain color change, precipitate, or pH change in the sample is observed (end point). Titration is the process of adding the chemical reagent in small increments (0.1–1.0 milliliter) until completion of the reaction, as signaled by the end point.

TOTAL DYNAMIC HEAD (TDH)

When a pump is lifting or pumping water, the vertical distance (in feet or meters) from the elevation of the energy grade line on the suction side of the pump to the elevation of the energy grade line on the discharge side of the pump. The total dynamic head is the static head plus pipe friction losses.

TOTALIZER

A device or meter that continuously measures and calculates (adds) a process rate variable in cumulative fashion; for example, total flows displayed in gallons, million gallons, cubic feet, liters, cubic meters, or some other unit of volume measurement. Also called an INTEGRATOR.

TOXIC

A substance that is poisonous to a living organism. Toxic substances may be classified in terms of their physiological action, such as irritants, asphyxiants, systemic poisons, and anesthetics and narcotics. Irritants are corrosive substances that attack the mucous membrane surfaces of the body. Asphyxiants interfere with breathing. Systemic poisons are hazardous substances that injure or destroy internal organs of the body. Anesthetics and narcotics are hazardous substances that depress the central nervous system and lead to unconsciousness.

TOXIC ORGANIC MANAGEMENT PLAN (TOMP)

A strategy for keeping track of all solvents delivered to a site, their storage, use, and disposal. This includes keeping spent solvents segregated from other process wastewaters to maximize the value of the recoverable solvents, to avoid contamination of other segregated wastes, and to prevent the discharge of toxic organics to any wastewater collection system or the environment. The plan should describe measures to control spills and leaks and to ensure that there is no deliberate dumping of solvents. Also known as a SOLVENT MANAGEMENT PLAN.

TOXIC POLLUTANT

Those pollutants or combinations of pollutants, including disease-causing agents, that cause death, disease, behavioral abnormalities, cancer, genetic mutations, physiological malfunctions (including malfunctions in reproduction), or physical deformations.

TOXIC SUBSTANCE

See HARMFUL PHYSICAL AGENT and TOXIC.

TRAMP OIL

Oil that comes to the surface of a tank due to natural flotation. Also called free oil.

TRANSDUCER (trans-DUE-sir)

A device that senses some varying condition measured by a primary sensor and converts it to an electrical or other signal for transmission to some other device (a receiver) for processing or decision making.

TRANSPIRATION (TRAN-spur-RAY-shun)

The process by which water vapor is released to the atmosphere by living plants. This process is similar to people sweating. Also see EVAPOTRANSPIRATION.

TRUE COLOR

Color of the water from which turbidity has been removed. The turbidity may be removed by double filtering the sample through a Whatman No. 40 filter when using the visual comparison method.

TRUNK SEWER

A sewer line that receives wastewater from many tributary branches and sewer lines and serves as an outlet for a large territory or is used to feed an intercepting sewer. Also called MAIN SEWER.

TURBID

Having a cloudy or muddy appearance.

TURBIDIMETER

See TURBIDITY METER.

TURBIDITY (ter-BID-it-tee)

The cloudy appearance of water caused by the presence of suspended and colloidal matter. In the waterworks field, a turbidity measurement is used to indicate the clarity of water. Technically, turbidity is an optical property of the water based on the amount of light reflected by suspended particles. Turbidity cannot be directly equated to suspended solids because white particles reflect more light than dark-colored particles and many small particles will reflect more light than an equivalent large particle.

TURBIDITY (ter-BID-it-tee) METER

An instrument for measuring and comparing the turbidity of liquids by passing light through them and determining how much light is reflected by the particles in the liquid. The normal measuring range is 0 to 100 and is expressed as nephelometric turbidity units (NTUs).

TURBIDITY (ter-BID-it-tee) UNITS (TU)

Turbidity units are a measure of the cloudiness of water. If measured by a nephelometric (deflected light) instrumental procedure, turbidity units are expressed in nephelometric turbidity units (NTU) or simply TU. Those turbidity units obtained by visual methods are expressed in Jackson turbidity units (JTU), which are a measure of the cloudiness of water; they are used to indicate the clarity of water. There is no real connection between NTUs and JTUs. The Jackson turbidimeter is a visual method and the nephelometer is an instrumental method based on deflected light.

TURBULENT MIXERS

Devices that mix air bubbles and water and cause turbulence to dissolve oxygen in the water.

U

(NO LISTINGS)

V

VARIABLE, MEASURED

A factor (flow, temperature) that is sensed and quantified (reduced to a reading of some kind) by a primary element or sensor.

VARIABLE, PROCESS

A physical or chemical quantity that is usually measured and controlled in the operation of a water, wastewater, or industrial treatment plant.

VECTOR

An insect or other organism capable of transmitting germs or other agents of disease.

VELOCITY HEAD

The energy in flowing water as determined by a vertical height (in feet or meters) equal to the square of the velocity of flowing water divided by twice the acceleration due to gravity ($V^2/2g$).

VENTURI (ven-TOOR-ee) METER

A flow measuring device placed in a pipe. The device consists of a tube whose diameter gradually decreases to a throat and then gradually expands to the diameter of the pipe. The flow is determined on the basis of the difference in pressure (caused by different velocity heads) between the entrance and throat of the Venturi meter.

NOTE: Most Venturi meters have pressure sensing taps rather than a manometer to measure the pressure difference. The upstream tap is the high pressure tap or side of the manometer.

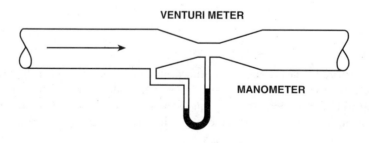

VISCOSITY (vis-KOSS-uh-tee)

A property of water, or any other fluid, that resists efforts to change its shape or flow. Syrup is more viscous (has a higher viscosity) than water. The viscosity of water increases significantly as temperatures decrease. Motor oil is rated by how thick (viscous) it is; 20 weight oil is considered relatively thin while 50 weight oil is relatively thick or viscous.

VOID

A pore or open space in rock, soil, or other granular material, not occupied by solid matter. The pore or open space may be occupied by air, water, or other gaseous or liquid material. Also called an INTERSTICE, PORE, or void space.

VOLATILE (VOL-uh-tull)

(1) A volatile substance is one that is capable of being evaporated or changed to a vapor at relatively low temperatures. Volatile substances can be partially removed from water or wastewater by the air stripping process.

(2) In terms of solids analysis, volatile refers to materials lost (including most organic matter) upon ignition in a muffle furnace for 60 minutes at 550°C (1,022°F). Natural volatile materials are chemical substances usually of animal or plant origin. Manufactured or synthetic volatile materials, such as plastics, ether, acetone, and carbon tetrachloride, are highly volatile and not of plant or animal origin. Also see NONVOLATILE MATTER.

VOLATILE ACIDS

Fatty acids produced during digestion that are soluble in water and can be steam-distilled at atmospheric pressure. Also called organic acids. Volatile acids are commonly reported as equivalent to acetic acid.

VOLATILE SOLIDS

Those solids in water, wastewater, or other liquids that are lost on ignition of the dry solids at 550°C (1,022°F). Also called organic solids and volatile matter.

VOLUTE (vol-LOOT)

The spiral-shaped casing that surrounds a pump, blower, or turbine impeller and collects the liquid or gas discharged by the impeller.

W

WAS

See Waste Activated Sludge.

WARNING

The word *WARNING* is used to indicate a hazard level between *CAUTION* and *DANGER*. Also see CAUTION, DANGER, and NOTICE.

WASTE ACTIVATED SLUDGE (WAS)

The excess quantity (mg/L) of microorganisms that must be removed from the process to keep the biological system in balance.

WASTEWATER

A community's used water and water-carried solids (including used water from industrial processes) that flow to a treatment plant. Stormwater, surface water, and groundwater infiltration also may be included in the wastewater that enters a wastewater treatment plant. The term sewage usually refers to household wastes, but this word is being replaced by the term wastewater.

WASTEWATER FACILITIES

The pipes, conduits, structures, equipment, and processes required to collect, convey, and treat domestic and industrial wastes, and dispose of the effluent and sludge.

WASTEWATER ORDINANCE

The basic document granting authority to administer a pretreatment inspection program. This ordinance must contain certain basic elements to provide a legal framework for effective enforcement.

WATER HAMMER

The sound like someone hammering on a pipe that occurs when a valve is opened or closed very rapidly. When a valve position is changed quickly, the water pressure in a pipe will increase and decrease back and forth very quickly. This rise and fall in pressures can cause serious damage to the system.

WEIR (WEER)

(1) A wall or plate placed in an open channel and used to measure the flow of water. The depth of the flow over the weir can be used to calculate the flow rate, or a chart or conversion table may be used to convert depth to flow. Also see PROPORTIONAL WEIR.

(2) A wall or obstruction used to control flow (from settling tanks and clarifiers) to ensure a uniform flow rate and avoid short-circuiting.

X

(NO LISTINGS)

Y

Y, GROWTH RATE

An experimentally determined constant to estimate the unit growth rate of bacteria while degrading organic wastes.

Z

ZOOGLEAL (ZOE-uh-glee-ul) FILM

See ZOOGLEAL MAT (FILM).

ZOOGLEAL (ZOE-uh-glee-ul) MASS

Jelly-like masses of bacteria found in both the trickling filter and activated sludge processes. These masses may be formed for or function as the protection against predators and for storage of food supplies. Also see BIOMASS.

ZOOGLEAL (ZOE-uh-glee-al) MAT (FILM)

A complex population of organisms that form a slime growth on the sand filter media and break down the organic matter in wastewater. These slimes consist of living organisms feeding on wastes, dead organisms, silt, and other debris. On a properly loaded and operating sand filter, these mats are so thin as to be invisible to the naked eye. Slime growth is a more common term.

SUBJECT INDEX

A

AIDS (Acquired Immune Deficiency Syndrome), 29, 676
Abbreviations
    instrumentation, 616
    regulatory requirements, 79–80
Abnormal conditions, carbon adsorption, 506
Abnormal operations
    gravity filters, 408
    pressure filters, 426, 427
    upflow filters, 440
Abrasion number, 515
Absorption, 495
Absorption, alkaline buffers, 293
Accident
    form, 720
    frequency, 675
    investigation, 718
    operators, 675, 720
    record, 686
    reports, 720
Accidental discharge, 201
Accuracies, flow measurement, 240–244
Acidity error, pH adjustment, 294
Acids, 182, 681
Acoustic flow measurement, 238
Activated carbon adsorption
    See Carbon adsorption
Activated sludge
    flow measurement, 244–246
    pumps, 753
Acts, pollution control, 81
Acute health effect, 35, 532, 682
Administration of a monitoring program
    database, 160
    dealing with industry, 161
    enforcement, 160
    organization, 160
Administrative fine penalties, 100
Adsorber loading, 505
Adsorption
    See Carbon adsorption
Aeration tanks, activated sludge, 707
Aerobic bacteria, 28
Agglomeration, 322
Agricultural use of water, 36
Air binding, 409, 425

Air chamber, pump, 835, 838
Air drying, metal sludges, 574
Air gap device, 716, 717, 862, 863
Air pockets, adsorption, 507
Air pressure, air stripping, 490
Air strainer, ejectors, 839
Air stripping
    air pressure, 490
    chemical characteristics, 490
    controlling discharge, 491
    discharge, 491
    equipment, 491
    incineration, 492
    maintenance, 493
    operation, 492
    organics, volatile, 489
    packed tower, 489
    principles, 490
    product recovery, 492
    purpose, 489
    safety, 494
    start-up, 492
    temperature, 490
    troubleshooting, 493
    types of systems, 489
    vapor phase, 491
    volatile organics, 489
Air supply instrumentation, 650
Air-vacuum relief valve, 502
Algae, 30, 32
Alarm instrumentation, 649
Alarms, carbon adsorption, 511
Alarms, gravity filters, 406
Algae, 410
Alignment
    motors, 814, 815, 816
    pumps, 756, 782, 831, 847, 848
Alkali conversion table, 288
Alkali neutralization graph, 289
Alkalinity error, pH adjustment, 294
Alternating current, 790
Alum, 327, 346, 418
Aluminum sulfate, 327, 346, 418
Amines, 711
Ammeter, 793
Amphoteric, 597
Amps, 790

Anaerobic bacteria, 29
Anaerobic sludge digestion
   drip traps, 706
   flame arresters, 706
Analog readout, 791
Analysis, carbon adsorption, 508
Anhydrous, 328
Anion, 294
Annular space, 436
Anodizing, 56
Apparent density tests, 515
Applying protective coating, 709
Aquatic vegetation, 36
Aspirator, 367
Audits, pollution prevention, 118
Authority, pollution control, 81
Automatic samplers, 163
Axial flow pumps, 759, 766

B

BMR, 87, 95
BOD, 268
Backflow prevention, 716
Backwashing
   gravity filter, 396, 400, 402, 404, 406
   pressure filters, 423–425
Bag filters, metal sludges, 569
Ball valves, 835, 838
Bar screens, 268–272
Bar screens, safety, 701
Baseline monitoring reports
   initial, 87
   requirements, 95
Bases, 681
Basins, flow equalization, 244, 260, 261–267
Batch processes, metals, 527, 533–536
Battery charging, 179, 206
Bearings, electric motor, 841
Bearings, pump, 753, 756, 761, 762, 764, 765, 767, 768, 769, 774, 775, 785, 831, 834, 839
Belt drives, pump maintenance, 843, 845, 846
Benching, excavations, 696, 699
Benefits
   flow equalization, 259
   pollution prevention, 117
Bernoulli effect, 641
Best Available Technology (BAT), 91
Best Practicable Technology (BPT), 91
Biochemical Oxygen Demand (BOD), 268
Biocides, 711
Biological contamination, 33, 35
Biological process, effects of industrial wastes
   See Toxic wastes
Biological treatment, pH adjustment, 301
Biomass, 493

Biosolids, 116, 260
Bloodborne Pathogens (BBP) program, 679
Blow down receiver, ejectors, 839
Blowdown, 40
Blowers
   safety, 681, 862
   ventilation, 677, 862
Bonnet valve, 855, 857
Bourdon tube, 630
   See Chapter 9, *ADVANCED WASTE TREATMENT*
Breathing apparatus, self-contained, 719
Bright dipping, 58
Brine, reverse osmosis, 474
Brinelling, 774
Bubble pipe, level measurement
   See Volume II, Chapter 15
Bubbler, level measurement, 633, 635
Bubblers, flow measurement, 232
Bubbler-type controls, 775
Budget administrator, 8
Buffer capacity, 286, 537
Building codes, 101
Buildings, maintenance, 750
Bulking, sludge, 322, 408

C

COD removal efficiencies, adsorption, 506
Calculations, reverse osmosis, 466–474
Calibration
   equipment, 652
   flowmeters, 240, 241, 243
   gas detectors, 862
   probes, 564
Cannell, James W., 810
Capacitance strips, 233
Capacity, pump, 759, 833
Carbon adsorption
   abnormal adsorption conditions, 506
   abrasion number, 515
   adsorber loading, 505
   air pockets, 507
   alarms, 511
   analysis, 508
   apparent density tests, 515
   COD removal efficiencies, 506
   carbon regeneration, 508
   carbon transfer, 509
   chemical loading, 505
   chemical warfare service (CWS) test, 515
   coatings, deterioration, 507
   collapsed screens, 507
   decolorizing index, 515
   disposal of carbon, 510
   dust control, 510
   effective size, 515

Carbon adsorption (continued)
  emergency conditions, 506
  equipment, 496
  fixed beds, 497
  fouling, 507
  hardness number, 515
  head losses, 507
  hydraulic loading, 505
  iodine number, 496, 515
  laboratory procedures, 515
  manufacture of carbon, 495
  methylene blue number, 516
  moisture, 516
  molasses number, 496, 516
  moving bed, 497
  operation, 505
  physical–chemical treatment, 481–516
  plans, 510
  plugged screens, 507
  principles, 496
  process, 496
  purpose, 495
  reactivation of carbon, 508
  regeneration, carbon, 508
  safety, 509
  sampling and analysis, 508
  shutdown, 506
  sieve analysis, 516
  specifications, 510
  spent activated carbon, 508, 510
  start-up, 498–505
  total ash of regenerated carbon, 516
  transfer of carbon, 509
  uniformity coefficient, 515
  unloading station, 510
  upflow column, 503, 504
  upstream processes, 507, 511
  ventilation, 511
Carbon regeneration, 508
Carbon transfer, 509
Carbon usage, management, 481–516
Carcinogens, 709, 711
Casing, pump, 753, 757, 761, 762
Categorical pretreatment standards, 88–97
Categories, pretreatment
  application, 93–97
  determination request, 92
  exempt, 92
  industrial, 89, 93
  local limits, 92
  modification, 96
  regulated, 88–93
Cation, 294
Caustic soda, 345
Caustic wastes, 182
Cave-ins, 696

Cavitation, 369, 595, 760
Cellulose acetate membranes, 465
Centrifugal force, 755
Centrifugal pumps
  description, 753, 759, 761–767
  maintenance, 783, 830
Centrifugation
  See Chapter 3, *ADVANCED WASTE TREATMENT*
Centrifuges, metal sludges, 568
Chain drives, 844
Chain of custody, 172
Change oil, pump, 835
Changed discharge, 96
Channels (flow), maintenance, 751
Characteristics of industrial wastes, 159
Charging batteries, 179, 206
Check valves, 783, 833, 839, 853, 854
Checklists, pollution prevention, 120, 123–150
Chelating agent, 56, 301, 493, 531, 532
Chelation, 48, 56
Chemical characteristics, wastewater, 28
Chemical conditioning
  See Chapter 3, *ADVANCED WASTE TREATMENT*
Chemical conditioning of sludge, dissolved air flotation (DAF) thickeners, 373
Chemical contamination, 35
Chemical feed
  filters, 418
  flow measurement, 246
  instrumentation, 641
  physical–chemical treatment, 309–382
Chemical loading, adsorption, 505
Chemical treatment
  agglomeration, 322
  alum, 327, 346
  aluminum sulfate, 327
  bulking, 322
  caustic soda, 345
  chemical feed, 342–358
  chemical mixing, 342
  chemical storage, 342
  chemicals, 327–331
  clarifiers, 358–366, 375, 376
  coagulation, 321, 358
  colloids, 322
  co-precipitation, 326
  day tank, 342, 343
  destabilization, 324, 325
  diaphragm pump, 349, 350
  equipment, 342–358
  feed equipment, 342–358
  ferric chloride, 328
  flocculation, 321, 358, 359
  gravimetric belt chemical feeder, 353
  iron salts, 324, 325, 328
  jar test, 331–341

952 Treatment Plants

Chemical treatment (continued)
   lime, 328, 344
   log, 356, 357
   MSDS, 340
   maintenance, 374
   material safety, 328, 340, 377
   material safety data sheet, 340
   metal salts, 324, 325, 328
   operation, 355, 358, 374–377
   phosphate monitoring, 340
   polishing process, 322, 323
   polymer, 324, 325, 347
   polymer map, 329, 330
   polymeric flocculants, 328
   pumps, chemical feed, 348, 349, 350
   rotary feeder, 352
   safety, 340, 377
   sedimentation, 321, 358–377
   shutdown, 355
   solids contact clarifier, 362, 363
   start-up, 355, 374, 375
   troubleshooting, 377, 378
   tube settlers, 364, 365
   volumetric screw feeder, 350
Chemicals
   feed, 342–358
   labeling, 722, 723, 727, 728
   mixing, 342
   safety, 681, 708, 713
   settling, 327–331
   storage, 342, 713
   toxic, 681
Chemicals, pH adjustment, 287
Chemical warfare service (CWS) test, 515
CHEMTREC [(800) 424-9300], 600
Chlorine monitor, 678
Chlorine safety, 678, 703, 708, 729
Cholera, 29
"Christmas Tree" arrangement, reverse osmosis, 474, 475
Chromium, 532
Chronic health effect, 35, 532, 682
Cipolletti weir, 223, 583
Circuit breakers, 795, 796, 797, 798–801
Clarifiers
   chemical treatment, 358–366, 375, 376
   safety, 377, 704
Clean Air Act regulations, 103
Clean Water Act regulations, 103
Cleaning pipelines, 865, 866
Cleaning probes, 560
Cleaning procedures, membranes, 460
Cleaning pumps, 831
Clearing plugged pipes, pumps, and valves
   costs, 865
   cutting tools, 866
   digested sludge lines, 865

   equipment, 865, 866
   high-velocity pressure units, 866
   methods of clearing, 865, 866
   pipes, 865
   pressure clearing methods, 865, 866
   pumps, 866
   scum lines, 865
   sludge lines, 865
   valves, 865, 866
Closed channel, flow measurement, 218, 233–238
Closed impeller, 763
Coagulation, 299, 321, 358, 436
Coarse screens, 268–272, 283
Coatings, deterioration, adsorption, 507
Code of Federal Regulations (CFR), 81, 83, 84
Codes
   See Ordinances, local
Collapsed screens, adsorption, 507
Collection system
   effects of discharges, 47–50
   flow equalization, 260
   pH adjustment, 285–304
Collection system, 689, 710
Colloids, 322
Color, true, 338
Combustible gas monitor, 678, 682, 710
Comminution, safety, 702
Common metals removal, 538–541
Comparative depth measurements, 240–242
Compatible pollutants, 38, 42
Competent person, 676
Complexed metals removal, 546
Compliance
   reports, 95
   schedule, 95
Composite samples, 163, 217, 246, 367, 580
Computerization, 655
Concentrated solutions, 43
Concentration-based standards, 93
Concentration factor, sludge, 372
Concentration polarization, 456, 474
Conditioning of sludges, chemical conditioning, 373
Conductivity, 51, 233
Conductors, 791
Cones, traffic, 178
Configurations
   filters, 397, 398
   membranes, 449
Confined space
   entry, 630
   upflow filters, 441, 445
Confined space entry permit, 680
Confined spaces, 179, 675, 676–681, 689, 703, 704, 707, 862
Connecting rods, 835, 838, 839
Construction, equalization tanks, 265
Contacts, starters, 810–813

Contamination
    biological, 33, 35
    chemical, 35
    radioactive, 35
Continuous discharges, 45
Continuous monitoring, 161
Continuous processes, metals, 533–536
Control systems
    flow equalization, 266
    instrumentation, 621–630
    metal wastestreams, 560–568
    pH adjustment, 298–299
Controllers
    pH adjustment, 298
    programmable, 567
Controllers, 649, 650
Controls, level, 775, 840
Conventional pollutants, 116
Cooking water, 35
Co-precipitation, 326
Cost savings, pollution prevention, 117
Costs
    preventive maintenance, 749
    repair, 749
    safety, 718
    unplugging pipelines, 865
Couplings, 756, 769, 782, 846, 847, 848
Criminal actions, 100
Cross connections, 716, 830
Current, electrical, 790, 791, 793, 794, 795, 796, 800, 801
Current measurements, 567
Cutting tools, plugged pipes, 866
Cycle
    natural purification, 32
    nutrient, 32
Cylinder, calibration, chemical pumps, 243

D

DPD method, 552
Dairy processing, 130–131
Dall flow tube, 235
Dalton, 447
Database, 160
Dateometer, 843
Day tank, 420
Day tank, chemicals, 342, 343
Decant, 416
Decant tank, filters, 425
Decant water, 264
Decibel, 688
Decolorizing index, 515
Dehumidifiers, 862
Delamination, 461
Delegation of federal authority, 82
Demineralization, reverse osmosis, 463

Department of Transportation (DOT) regulations, 104
Desiccation, 652, 659
Destabilization, chemical, 324, 325
Dial indicators, alignment, 847, 848
Diaphragm bulb, level measurement, 775
Diaphragm pump, 349, 350, 593
Diatomaceous earth, 569
Differential pressure devices, 233, 234, 583, 584
Digested sludge handling, unplugging pipelines, 865
Digested sludge, pumps, 753
Digester heating, heat exchanger, 704
Digestion, equipment safety, 704
Digital readout, 791
Dilute solutions, 42
Direct current, 790
Discharge
    changed, 96
    pH adjustment, 300
    standards, prohibited, 87
Discharge, pump, 753, 759, 782, 783, 784, 786, 839
Discharge table, weir, 221
Dischargers
    direct, 82
    indirect, 82
    regulated, 82
Discharges, industrial, 44, 45
Disconnect switch, 796
Displaced volume meter, 235, 237, 583, 584
Disposal of carbon, 510
Dissolved air flotation (DAF) thickeners
    age of sludge, 369
    air to solids (A/S) ratio, 370, 371, 373, 374
    biological flotation, 367
    biological sludges, 369
    blanket thickness, 371, 373, 374
    chemical conditioning, 373
    concentration factor, 372
    dispersed air flotation, 367
    efficiency, 373
    effluent, 373, 374
    factors affecting performance, 369, 372
    float characteristics, 373
    guidelines, operation, 369, 372
    hydraulic loading, 369, 373
    observations, 373
    operating guidelines, 369, 372
    operation, 369, 372
    performance, 369, 372
    pressure flotation, 367
    primary sludge thickening, 369
    recycle rate, 371
    rising sludge, 369
    shutdown, 372
    sludge blanket, 371, 374
    solids loading, 369, 373
    solids recovery, 372

Dissolved air flotation (DAF) thickeners (continued)
   start-up, 371
   thickened sludge characteristics, 371, 373
   troubleshooting, 373, 374
   vacuum flotation, 367
   variables, 369
   visual inspection, 373
   withdrawal of sludge, 369
Dissolved solids, in wastewater, 30
Doctor blade, 275, 569
Domestic sewage exemption, 102
Domestic wastes, 28
Domestic wastewater, characteristics, 28
Doppler method, flow measurement, 238
Dosimeter, 682
Drag out, 39, 119, 528
Drag shield, 696
Drain, backwash, 400
Draining pumps, 775, 785, 834
Drinking water, 33, 862
Drip traps, 706
Drowning, 707
Drums, traffic control, 178
Dry cleaning, 132
Drying sludge, 574
Dust control, carbon, 511
Dusts, 675, 688
Dye dilution method, 242
Dysentery, 29, 676

E

EPA
   authority, 81
   organizational structure, 82
   pollution control, 81
   regulated categories, 92
   regulation development, 91
   reporting requirements, 94
Ear protecting devices, 688
Eccentric
   pipe, 756
   pump, 835, 838
Economics, pollution prevention, 117
Eductor, 498
Effective size, 515
Effects, industrial wastewaters, 46–53
Efficiency, dissolved air flotation (DAF) thickeners, 373
Efficiency, pump, 833
Effluent
   flow measurement, 243
   pumps, 753
Effluent effects, industrial wastewater, 53
Effluent rate control valve, 405
Ejectors, 768, 772, 839
Electric motors, 775, 803–826, 841–843

Electric shock, 684
Electrical
   capacitance strips, 233
   instrumentation hazards, 628
   symbols, 616–620
Electrical controls, 775
Electrical equipment maintenance, 788–803
Electrical safety, 684
Electrical system
   control, 795
   starters, 810–813
   switch gear, 795
Electricity
   circuit breakers, 795, 796, 797, 798–801
   circuits, 788, 790, 792, 793, 794, 796, 800, 801
   fuses, 796, 798
   hazards, 788
   meters, 791–795
   starters, 810–813
   terms, 790
   testers, 791–795
   tools, 791–795
Electrode switches, 775, 840
Electrode troubleshooting, 302
Electrodes, pH adjustment, 291
Electroless plating solutions, 531
Electrolyte, 287, 328
Electromagnetic flowmeter, 235
Elements
   control systems, 298
   measuring, 645
   primary and secondary, 221–235
Emergency conditions, adsorption, 506
Emergency procedures
   electric shock, 685
   planning, 719, 867
Employers, operators, 5, 9
Emulsion, 260
Energy grade line, 219
Energy requirements, pumps, 759, 782, 786
Enforcement, 160
Entrain, 703
Equalization of flows
   basins, 244, 260, 261–267
   benefits, 259
   collection system, 260
   construction, tanks, 265
   controls, flow, 266
   flows, 44
   location, 260
   manufacturing process, 261
   mixing, 265
   need, 257
   operation strategy, 266
   pumps, 266
   sizing tank, 261–265

Equalization of flows (continued)
    strategy, operation, 266
    tanks, 244, 260, 261–267
    volume, 261–265
    when, 257
Equalization, pH adjustment, 294
Equipment
    air stripping, 491
    calibration, 652
    carbon adsorption, 496
    chemical treatment, 342–358
    records, 749, 750, 826, 827, 828
    service card, 749, 750
    test, 652
Equipment storage, 179
Equivalence point, 288
Errors, pH adjustment, 294
Etching and chemical milling, 58
Excavations, 696
Exemptions
    domestic sewage, 102
    on-site treatment, 102
Exfiltration, 49
Exothermic, 531
Explosive gases, 678, 682, 710, 711
Eye protection, 713, 714

F

Face shield, 708, 713
Face velocity, 293
Failure, motors, 806, 807, 820–826
Fall arrest system, 676, 706
Falling hazards, 630
Federal pollution control regulations, 81
Federal Register, 81
Federal statutes, 101–104
    Also see Regulations
Feedback control, 295, 296
Feed equipment, chemical, 342–358
Feed forward control, 295
Feeler gauge, alignment, 814
Felony criminal actions, 100
Ferric chloride, 328
Film badges, 682
Filter aid, 408
Filter backwash, flows, 246
Filter feed pumps, 418
Filter press, 569
Filters, gravity
    abnormal operation, 408
    alarms, 406
    backwashing, 396, 400, 402, 404, 406
    description, 395
    drain, backwash, 400
    effluent rate control valve, 405

head loss, 396, 405
inlet, 400
instrumentation, 405–406
log, operation, 413
media, 400
methods, 396
mudballs, 400
operation, 406–413
parts, 400–405
plans, 414
rapid sand filter, 396, 402
rate control valve, 405
safety, 414
scouring, media, 400
shutdown, 412
specifications, 414
strategy, 410
systems, 395
totalizer, 405
troubleshooting, 412
troughs, 400
turbidity, 405
types, 396
underdrains, 400, 403
wash water troughs, 400
Filters, membrane
    See Membrane filtration
Filters, pressure
    abnormal operation, 426, 427
    backwash, 423–425
    chemical feed, 418
    decant tank, 425
    facilities, 416–425
    filter feed pumps, 418
    filters, 422
    holding tank, 416
    maintenance, 426, 428
    operation, 426
    plans, 429
    pumps, 418
    recovery, backwash, 425
    safety, 427
    specifications, 429
    strategy, 426
    use, 416, 417
    wet well, 416
Filters, upflow
    abnormal operation, 440
    coagulation, 436
    equipment, 433–436
    flocculation, 436
    maintenance, 440, 443, 444
    metering, flow, 436
    metering, turbidity, 433–436
    operation, 436–440
    plans, 445

Filters, upflow (continued)
  safety, 445
  specifications, 445
  strategy, 439
  turbidity metering, 433–436
  use, 430
Filtration
  abnormal operation, 408, 426, 427, 440
  alarms, 406
  backwashing, 396, 400, 402, 404, 406, 423–425
  chemical feed, 418
  decant tank, 425
  description, 395
  differential pressure, 399
  drain, backwash, 400
  effluent rate control valve, 405
  equipment, upflow, 433–436
  feed pumps, 418
  gravity filters, 395–415
  head loss, 396, 405
  holding tank, 416
  inlet, 400
  instrumentation, 405–406
  maintenance, 426, 428, 440, 443, 444
  media, 400
  membrane filtration
    See Membrane filtration
  methods, 396
  mudballs, 400, 426, 427
  operation, 406–413, 426, 436–440, 456–463
  parts, 400–405, 430, 431
  plans, 414, 429, 445
  pressure, differential, 399
  pumps, pressure filters, 418
  rapid sand filter, 396, 402
  rate control valve, 405
  recovery, backwash, 425
  safety, 414, 427, 445, 463
  scouring, media, 400
  shutdown, 412
  specifications, 414, 429, 445
  strategy, 410, 426, 439
  systems, 395
  totalizer, 405
  troubleshooting, 412
  troughs, 400
  turbidity, 405, 410, 433–436
  types, 395
  underdrains, 400, 403
  wash water troughs, 400
  wet well, 416
  Also see Filters, gravity; Filters, pressure; Filters, upflow; and Membrane filtration
Final element, pH adjustment, 299

Fine screens, 273–282, 283
Fines, administrative, 100
Fire, 714, 715
Fire control, safety, 715
Fire drills, 719
Fire extinguishers, 701, 714, 715, 719, 862
First aid, 598
First-aid kit, 862
Fish, 36
Fixed beds, carbon adsorption, 497
Flame polished, 713
Flame trap, anaerobic digester, 706
Flammables, sewers, 49
Float control, 775, 831, 840
Float mechanism, level measurement, 775, 831, 840
Float switches, 775, 831, 840
Floats
  flow measurement, 230, 231
  level, 634
Flocculation, 299, 321, 358, 359, 436
Flow equalization
  industrial wastes, 180
  See Equalization of flows
Flow measurement
  accuracies, 240–244
  acoustic flow measurement, 238
  activated sludge, 244–246
  basics, 217–221
  bubblers, 232
  calibration, 240, 241, 243
  capacitance strips, 233
  chemical feed, 246
  Cipolletti weir, 223
  closed channel flow, 218, 233–238
  comparative depth measurements, 240–242
  comparisons, 243, 244
  cylinder, calibration, 243
  Dall flow tube, 235
  devices, 218
  differential pressure devices, 233, 234
  discharge, weir, 221
  displaced volume meter, 235, 237
  doppler methods, 238
  dye dilution method, 242
  effluent, 243
  electrical capacitance strips, 233
  electromagnetic flowmeter, 235
  elements, primary and secondary, 221–235
  equalization basin, 244
  filter backwash, 246
  floats, 230, 231
  flow nozzles, 226, 228, 233
  flumes, 218, 220, 224–226
  gauges, 230, 231

Flow measurement (continued)
   head, 218
   hydraulic calibration, 240
   influent, 243
   instrument calibration, 240
   Kennison nozzle, 226, 228
   low flow measurements, 238
   magnetic flowmeter, 235
   mechanical flow devices, 235
   need, 217
   nozzles, flow, 226, 228, 233
   open channel flow, 218, 221–233
   operation and maintenance, 226–230
   orifice plate, 218, 234, 235
   Palmer-Bowlus flume, 226, 227, 241
   Parshall flume, 224, 225, 241
   pipe flow, 218, 233–238
   pressure pipe flow, 218, 233, 234
   pressure transducers, 233
   primary elements, 221–230
   primary sludge, 244
   propeller meter, 238, 239
   return activated sludge, 245
   rotameter, 235, 236
   secondary elements/devices, 230–235
   staff gauge, 230, 231
   stilling well, 230, 231
   stormwater, 238
   submerged pressure transducers, 233
   transducers, pressure, 233
   transit time flow measurement, 238
   ultrasonic devices, 232
   velocity meter, 235
   Venturi nozzle, 218, 233, 234
   very low flow measurements, 238
   waste activated sludge, 245
   weirs, 218, 220, 221–224
   well, stilling, 230
Flow measurement (rate and total), 636
Flow measurement/totalization, 566, 582–584
Flow nozzles, 226, 228, 233, 583
Flow regulation, 175, 180
Flowmeters, level measurement, 775
Fluid milk processing, 133–134
Fluidized, filter, 424
Flumes, 218, 220, 224–226, 583
Flux decline, 466
Flux, membranes
   concentration dependent, 456
   membrane, 453
   water measurements, 461
Flux, reverse osmosis, 465, 466
Foam control, surface-active agents, 711
Foaming, safety hazard, 711

Foot valves, 833
Fouling
   carbon adsorption, 507
   membranes, 456
   pH sensors, 293
Free oil, 458
Fuels, safety, 710
Fume hood, 713
Fumes, 678, 681, 682, 709, 711, 713
Fundamentally different factors (FDF), 97
Fuses, 781, 796, 798

G

Gas, detection, 862
Gas extraction, 139
Gases
   characteristics, 678
   explosive, 678, 682, 710, 711
   explosive range, 678
   safe exposure, 678, 681
   testing methods, 678
   toxic, 678, 681, 682, 709, 711, 713
Gasket, pump, 835
Gasoline, 182
Gasoline vapors, 678, 681, 682
Gate valves, 850–853
Gauges, flow measurement, 230, 231
Gear pump, 348
Gear reducer, 835
Generation of wastewater, 39, 44, 54–62
Generators, 867
Gland
   pump, 758, 781, 784, 786
   valve, 853
Gold wastestreams, 532
Good housekeeping
   management responsibilities, 675, 718
   safety, 701, 702, 704, 710
Grab samples, 162, 580
Gravimetric belt chemical feeder, 353
Gravity filters, 395–415
   Also see Filters, gravity
Greasing, pump, 839
Grit channels, 703
Grit, pumps, 753
Ground, 801
Grounds, maintenance, 751

H

Handling of chemicals, 597
Hardness number, 515
Harness, safety, 676, 679, 862

Hazard communication, 103
Hazard communication program, 721–729
Hazardous chemicals and wastes, safety, 595
Hazardous waste
    disposal reports, 96
    monitoring, 160
    regulations, 104
Hazards
    See Safety hazards
Head, flow measurement, 218
Head loss
    carbon adsorption, 507
    gravity filter, 396, 405
Headworks, 701
Health effects
    acute, 35, 532
    chronic, 35, 532
    organic wastes, 532
Hearing protection devices, 688
Heat exchanger, anaerobic digester, 704
Heated wastewaters, sewers, 50
Heating (drying sludge), 574
Heavy metal removal, pH adjustment, 301
Hepatitis, 29, 676
Hexavalent chromium, 532, 547
High-temperature waste, 182
High-velocity pressure units, 866
History, pollution control regulations, 81
Holding tank, filters, 416
Hollow fiber membranes, 449
Hood, fume, 713
Housekeeping
    management responsibilities, 675, 718
    safety, 701, 702, 704, 710
Hydraulic calibration, flowmeters, 240
Hydraulic capacity, 47, 50
Hydraulic grade line, 219
Hydraulic jump, 226, 711
Hydraulic loading
    carbon adsorption, 505
    dissolved air flotation (DAF) thickeners, 369, 373
Hydraulic shock load, 183
Hydrocarbons, 182
Hydrogen peroxide, odor control
    See Chapter 1, *ADVANCED WASTE TREATMENT*
Hydrogen sulfide
    hazards, 29
    industrial wastes, 159
    monitor, 678
    problems, 29
    safety hazard, 678, 681, 701, 711
Hydrolysis, 469, 473
Hydroxide precipitation, metals, 534–536, 538–541, 551
Hygiene, safety considerations, 676
Hygroscopic, 598

I

IDLH, 584
IDLH (Immediately Dangerous to Life or Health), 708
Imbalance, voltage, 808, 810
Imhoff cone, 30
Immediately Dangerous to Life or Health (IDLH), 708
Immiscible, solvents, 49
Immunization shots, 676
Impacts of waste discharges
    algae, 32
    human health, 29
    nutrients, 30
    odors, 28
    oxygen depletion, 28
    pathogenic bacteria, 29
    sludge and scum, 28
    toxic substances, 30
Impellers, 753, 759, 761, 762, 763, 764, 766, 782, 784, 833
Incineration, 492
Incline screw pumps, 768, 769
Indicators, instruments, 645
Industrial pretreatment categories, 89, 93
Industrial use of water, 36
Industrial waste monitoring
    accidental discharge, 201
    administration, 160
    battery charging, 179, 206
    care of monitoring equipment, 166
    chain of custody, 172
    characteristics of industrial wastes, 159
    charging batteries, 179, 206
    composite samples, 163
    confined spaces, 179
    continuous monitoring, 161
    database, 160
    dealing with industry, 161
    enforcement, 160
    equipment storage, 179
    flow metering, 175
    flows, regulation, 180
    grab samples, 162
    hazardous wastes, 160
    hydrogen sulfide, 159
    identifying waste materials, 171
    importance, 159
    labeling samples, 172
    locating sources of discharges, 171
    maintenance, 170
    monitoring, 161, 179
    need, 159
    objectives, 159
    odors, 159
    ordinance, sewer-use, 197
    Palmer-Bowlus flume, 163, 175

Industrial waste monitoring (continued)
  Parshall flume, 175
  permit, sewer-use, 187
  portable sampling equipment, 163
  preservation of samples, 171, 173
  pretreatment inspection, 161
  records, 201
  refractory materials, 160
  regulation of high flows, 180
  representative samples, 162
  safety, 175, 206
  sample preservation and security, 171
  sampling points, 162
  security of samples, 171
  self-monitoring, 161
  sewer service charges, 160
  sewer-use ordinance, 160, 197
  sewer-use permit, 187
  shock loads, 159
  slug discharge, 171, 192
  standard industrial classification, 160
  storage of equipment, 179
  storage time and temperature of samples, 171
  strategy for monitoring, 179
  thermal wastes, 159
  toxic wastes, 159
  traffic safety, 175
  warning systems, 181
  water meters, 175
Industrial waste treatment, safety, 681, 682, 710, 711
Industrial waste treatment system, 40
Industrial wastes
  need to treat, 32, 33
  reasons for treatment, 33
  types, 28
Industrial wastewaters
  acid, 182
  caustic, 182
  collection system effects, 47–50
  color, 182
  compatible pollutants, 38, 42
  concentrated solutions, 42
  dilute solutions, 42
  discharges, 44, 45
  effects, 46–53
  effluent effects, 53
  frequency, 44
  gasoline, 182
  generation of wastewater, 39, 44, 54–62
  grease, 183
  heavy metals, 183
  high temperature, 182
  hydraulic shock load, 183
  hydrocarbons, 182
  importance, 38–41
  industrial waste treatment system, 40
  interference, 42, 51
  intermittent discharges, 45
  maintenance activities, 40
  manufacturing processes, 39, 54–62
  metal finishing, 54–60
  metals, 183
  nitrogen, 183
  noncompatible pollutants, 38, 42
  nutrients, 183
  odors, 183
  oil, 183
  operator's responsibility, 46
  organic solids, 182
  POTW, effects on, 53
  pathogens, 183
  pesticides, 183
  phosphorus, 183
  pollutants, 42
  printed circuit board manufacturing, 60–62, 63
  radioactive wastes, 183
  references, 62
  responsibility, operator's, 46
  sludge disposal, effects on, 53
  solids, 182
  solvent, 182
  tastes, 183
  thermal waste, 182
  toxic substances, 183
  treatment system, effects on, 50–54
  turbidity, 182
  variables, 42–46
  wastestreams, 42–46
  wastewater generation, 39, 44, 54–62
Infections, 676
Infectious diseases, 676
Infectious hepatitis, 29
Infiltration, 49
Influent, flow measurement, 243
Inlet, gravity filter, 400
Inorganic compounds, in wastewater, 31
Inorganic wastes, 28
Inspection of facilities, 751
Instrument calibration, flowmeters, 240
Instrumentation
  air supply, 650
  alarms, 649
  Bourdon tube, 630
  bubbler, 633, 635
  calibration, equipment, 652
  categories, 645
  chemical feed rate, 641
  computerization, 655
  control systems, 621–630
  controllers, 649, 650

Instrumentation (continued)
   electrical hazards, 628
   electrical symbols, 616–620
   elements, measuring, 645
   floats, 634
   flow measurements (rate and total), 636
   gravity filter, 405–406
   indicators, 645
   laboratory, 652
   level measurement, 632–635
   maintenance, 656–660
   measurement systems, 621, 622, 630–642
   measuring elements, 645
   mechanical hazards, 629
   microprocessors, 656
   motor control panel, 625
   operation and maintenance, 656–660
   orifice plate, 640, 641
   panel instruments, 645
   pneumatic hazards, 629
   pressure measurements, 630
   probes, 633, 634
   process analytical instrumentation, 642
   recorders, 646
   rotameter, 636
   safety, 628–630
   sensors/transducers, 630–642
   shutdown, 657–659
   start-up, 657–659
   symbols, 616–620
   test equipment, 652
   totalizers, 646
   transducers, 630–642
   Venturi meter, 640, 641
   VOM (Volt-Ohm-Milliammeter), 652, 654
Instrumentation and controls, 560–568, 592
Insulation
   materials, 806, 807
   motors, 806, 807
Insulators, 791
Interference, 42, 51
Intermittent discharges, 45
Iodine number, 496, 515
Ion, 286
Ion exchange, 553–557
Ion exchange troubleshooting, 555–557
Iron salts, 324, 325, 328

J

Jar test, 331–341, 408
Jobs for operators
   duties, 5
   locations, 5
   opportunities, 9
Jogging, 823

K

Kennison nozzle, 226, 228

L

Labeling samples, 172
Labels, chemicals, 722, 723, 727, 728
Laboratory equipment
   breathing apparatus, self-contained, 719
   face shield, 708, 713
Laboratory hazards
   hydrogen sulfide, 678, 681, 701, 711
   tetanus, 676
Laboratory, instrumentation, 652
Laboratory procedures, carbon adsorption, 515
Laboratory safety
   face shield, 708, 713
   tetanus, 676
Lagoons, safety, 707
Lamella settler, 541–543
Land use ordinances, 101
Landscaping, 751
Lantern ring, 831, 833
Launders, 436
Let's Build a Pump, 754
Level controls, 567
Level measurement instruments, 632–635
Library, plant, 752
Lifeline, 676
Lift stations, 701, 867
Lifting, safe practices, 695
Lime, chemical treatment, 328, 344
Lipophilic, 53
Livestock watering, 36
Loading of material and product, 118
Local
   limits, 92
   ordinances, 98
   Also see Ordinances, local
Location, flow equalization, 260
Lockout tag, 685, 702, 788, 789, 831, 832
Lockout/tagout procedures, 685
Log
   chemical feeder, 356, 357
   filter operation, 413
Low flow measurements, 238
Lower explosive limit, 682, 683, 701, 710
Lubrication, 756, 768, 774, 775, 782, 817, 820, 831, 839, 841, 843, 845, 847, 853

M

MSDS (Material Safety Data Sheet), 340, 414, 681, 710, 712, 723–725, 777, 865
Magnetic flowmeter, 235, 583, 584

Magnetic starters, 810
Maintenance
   activities, 40
   air stripping, 493
   chemical treatment, 374
   electric motors, 775, 803–826, 841–843
   electrical equipment, 788–803
   flowmeters, 226–230
   mechanical seals, 768, 773, 831
   metal wastestreams, 590
   motors, 803–826, 841–843
   pH adjustment, 302
   pressure filters, 426, 428
   pumps, 829
   screens, 282–284
   upflow filters, 440, 443, 444
   valves, 833, 835, 838, 839, 850, 853, 855–859
Maintenance activities, 40
Maintenance, instruments, 656–660
Maintenance program, pump, 830
Management support, 118
Manholes, 689, 695
Manometer, flow measurements, 235
Manufacture of carbon, 495
Manufacturing process
   flow equalization, 261
   troubleshooting pH changes, 304
Manufacturing processes, 39, 54–62
Map, polymer, 329, 330
Mass-based standards, 93
Mass emission rate, 43
Material distribution systems, 118
Material Safety Data Sheet (MSDS), 340, 414, 681, 710, 712, 723–725, 777, 865
Material storage and loading, 118
Materials, membrane, 449
Mean, 259
Measurement electrode, 291
Measurement of gases, 678, 679, 680, 681, 682, 683, 703
Measurement systems, 621, 622, 630–642
Measuring elements, 645
Meat packing, beef, 135
Mechanical bar screen, 269
Mechanical equipment
   axial flow pumps, 759, 766
   centrifugal pumps, 753, 759, 761–767, 783, 830
   electric motors, 775, 803–826, 841–843
   electrical controls, 775
   incline screw pumps, 768, 769
   lubrication, 756, 847
   multi-stage pumps, 759
   operation, 774, 782
   piston pumps, 759, 786, 835, 838
   pneumatic ejectors, 768, 772, 839
   positive displacement pump, 759, 786, 835, 838
   progressive cavity pumps, 768, 770, 771, 786, 839
   propeller pumps, 759, 765, 766, 839
   pump-driving equipment, 775
   radial flow pumps, 759, 766
   reciprocating pumps, 759, 786, 835, 838
   repair shop, 753
   screw-flow pumps, 768, 770, 771, 786
   shutdown, 775, 782, 784, 786
   single-stage pumps, 759, 767
   sludge pump, 759, 786, 835, 838
   start-up, 759, 774, 781, 782, 785
   submersible pump, 759, 764
   troubleshooting, 777–782, 845
   turbine pumps, 759, 767
   vertical wet well pump, 759, 767
   wet well pump, 759, 764, 767, 783, 784
   Also see Mechanical maintenance, Meter maintenance, and chapters 12 (607–664) and 14 (739–875)
Mechanical flow devices, 235
Mechanical hazards, 629
Mechanical maintenance
   air chamber, 835, 838
   air gap separation systems, 862, 863
   alignment, 756, 782, 814, 815, 816, 831, 847, 848
   ball valves, 835, 838
   bearings, 831, 839, 841, 843
   belt drives, 843–844, 845, 846
   capacity, pump, 759, 833
   centrifugal pumps, 783, 830
   chain drives, 844
   change oil, 835, 845
   check valves, 783, 833, 839, 853, 854
   cleaning pump, 831
   connecting rods, 835, 838, 839
   controls, 831
   couplings, 756, 769, 782, 846, 847, 848
   dehumidifiers, 862
   draining pump, 775, 785, 834
   eccentric, 835, 838
   efficiency, pump, 833
   ejectors, 768, 772, 839
   electric motors, 775, 803–826, 841–843
   electrode switches, 840
   equipment service card, 749, 750
   float switches, 831, 840
   frequency of service, 749
   gaskets, 835
   gate valves, 850–853
   gear, reducer, 835
   greasing, 839
   impeller, 833
   lantern ring, 831, 833
   lubrication, 756, 768, 774, 775, 782, 817, 820, 831, 839, 841, 843, 845, 847, 853
   mechanical seals, 831
   noises, 782, 835, 841, 845, 847
   oil change, 835, 845

Mechanical maintenance (continued)
  packing, 770, 781, 782, 786, 830, 831, 833, 834, 836–837, 838, 839, 850, 851
  piston pumps, 835, 838
  plug valves, 853, 855–859
  plunger pumps, 835, 838
  pneumatic ejectors, 768, 772, 839
  positive displacement pumps, 835, 838
  preventive maintenance, 830, 865
  progressive cavity pumps, 839
  propeller pumps, 839
  pumps, general, 830
  reciprocating pumps, 835, 838
  reducer, gear, 835
  rings, pump, 757, 761, 782, 784, 831, 833
  rods, 835, 838, 839
  safety equipment, 862
  seals, 768, 773, 782, 783, 784, 785, 830, 831, 833
  service record card, 749, 750
  shear pins, 835, 849
  sludge pump, 835, 838
  sluice gates, 853, 860, 861
  stuffing box, 758, 761, 831, 850, 851, 852
  switches, 775, 831, 840
  TDH, 833
  total dynamic head, 833
  troubleshooting, 777–782, 845
  valve chamber gasket, 835
  valves, 833, 835, 838, 839, 850, 853, 855–859
  variable-speed belt drives, 845
  vibrations, 783, 785, 817, 841, 847
  water seal, 831, 833
  wearing rings, 757, 761, 782, 784, 833
Mechanical operation, troubleshooting pH adjustment, 302, 303
Mechanical seals, 768, 773, 831
Media, gravity filter, 400
Megger, 794
Megohm, 794
Membrane filtration (cross flow)
  cleaning procedures, 460
  concentrating components, 453
  configurations, membranes, 449
  elements of membrane process, 453
  feed pretreatment, 458
  flux, concentration dependent, 456
  flux, membrane, 453
  flux, water measurements, 461
  fouling, 456
  hollow fiber membranes, 449
  materials, membrane, 449
  membrane life, 449
  microfiltration, 447
  nanofiltration, 449
  operation, 456–463
  permeate, 453
  plate and frame membranes, 452
  pressure, transmembrane, 455
  pretreatment, feed, 458
  recirculation flow, 455
  recordkeeping, 463
  reverse osmosis, 449
  safety, 463
  sampling, 462
  spiral membranes, 451
  temperature, 456
  transmembrane pressure, 455
  tubular membranes, 449
  types, 447
  ultrafiltraton, 447
Membrane life, 449
Membranes
  performance, 465
  properties, 465
  reverse osmosis, 463, 465, 466
Mercaptans, 47
Metal fabrication, 136–137
Metal finishing, 54–60, 120, 137–138
Metal finishing standards, 92
Metal removal, 301
Metal salts, 324, 325, 328
Metal wastestreams
  air drying, 574
  bag filters, 569
  batch processes, 527, 533–536
  calibration of probes, 564
  centrifuges, 568
  chelating agents, 531, 532
  chromium, 532
  cleaning probes, 560
  common metals removal, 538–541
  complexed metals removal, 546
  continuous processes, 533–536
  controllers, programmable, 567
  controls, 560–568
  current measurement, 567
  cyanide, 532, 549–552
  cyanide destruction, 549–552
  dewatering, sludge, 568–574
  dilute rinse waters, 532
  drag out, 528
  drying sludge, 574
  electroless plating solutions, 531
  filter press, 569
  first aid, 598
  flow measurement/totalization, 566, 582–584
  gold wastestreams, 532
  handling of chemicals, 597
  hazardous chemicals and wastes, safety, 595
  heating (drying sludge), 574
  hexavalent chromium, 532, 547
  hydroxide precipitation, 534–536, 538–541, 551

Index 963

Metal wastestreams (continued)
    instrumentation and controls, 560–568, 592
    ion exchange, 553–557
    Lamella settler, 541–543
    level controls, 567
    maintenance, 590
    metallic sludge dewatering, 568–574
    methods of treatment, 527
    neutralization, 537–538
    normal operation, 577–590
    ORP probes, 560–566, 592
    oily waste removal, 557
    operation, 576–590
    organically contaminated wastewaters, 532
    organics, toxic, 558
    pH adjustment, 537–538
    pH probes, 560–566, 591
    pickle, 533
    precious metals recovery, 553–557
    preservation of samples, 580
    probes, 560–566, 591, 592
    processes, treatment, 533–558
    programmable controllers, 567
    pumps, 585–590, 593
    reduction of hexavalent chromium, 547
    representative sampling, 579
    resistance measurements, 567
    reverse osmosis, 555
    rinse waters, 530, 532
    safety, 549, 590, 595–600
    sampling, 579–581
    sampling devices, 580
    sand filter, 545
    silver wastestreams, 532
    sludge dewatering, 568–574
    sludge drying, 574
    solvent control, 558
    source control, 528
    sources, 528
    spare parts inventory, 591
    spent baths, 531
    storage and handling of chemicals, 597
    sulfide precipitation, 546
    toxic organics, 558
    treatment processes, 533–558
    troubleshooting probes, 565
    troubleshooting processes, 591–595
    vacuum drying of sludge, 574
    vacuum filters, 569
    ventilation and exhaust systems, 584
    voltage measurements, 567
    waste minimization, 528
    wastestream types, 531–533
Metallic sludge dewatering, 568–574
Metals, 183
Meters, electrical, 791–795
Methane gas, 678, 704
Methylene blue number, 516
Microfiltration, 447
Micron, 273, 322, 496
Microprocessors, 656
Milk processing, 133–134
Mineral rejection, 466, 467, 468
Miscible, solvents, 40
Misdemeanor criminal actions, 100
Mixing
    flow equalization, 265
    pH adjustment, 296
Moisture, carbon adsorption, 516
Molasses number, 496, 516
Monitoring, industrial wastes
    accidental discharge, 201
    acids, 159
    alkalies, 159
    collection system, 159
    corrosion, 159
    cyanide, 159
    flammable materials, 159
    hazardous wastes, 160
    hydrogen sulfide, 159
    labeling samples, 171, 172
    Palmer-Bowlus flume, 163, 175
    Parshall flume, 175
    records, 201
    refractory materials, 160
    sewer service charges, 160
    sewer-use ordinance, 160, 197
    slug discharge, 171, 192
    standard industrial classification, 160
    thermal wastes, 159
Monitoring, TTO, 94
Mother circuit board, 593
Motor control center, 802
Motor control panel, 625
Motor protection devices, 796, 800, 801
Motor, pump, 785
Motors
    alignment, 814, 815, 816
    contacts, 810–813
    electric, 775, 803–826, 841–843
    failure, 806, 807, 820–826
    humming, 841
    induction, 825
    insulation, 806, 807
    lubrication, 817, 820
    performance, 818–819
    pump, 785
    records, 826, 828
    rotation, 817
    safety, 813
    starters, 810–813
    temperature limits, 785, 807

Motors (continued)
   troubleshooting, 820–826
   types, 803
   vibrations, 817, 841, 847
Moving bed, carbon adsorption, 497
Mudballs, 400, 426, 427
Multimeter, 791, 792
Multi-stage pumps, 759

N

NFPA 820 (National Fire Protection Association Standard 820), 684, 704
NPDES permit program, 83
NPDES permits, 85, 416
Nameplate, 355, 790, 803
Nanofiltration, 449
Nanofiltration, reverse osmosis, 469
Naphtha, 711
Nappe, 221, 222, 224
National Fire Protection Association Standard 820 (NFPA 820), 684, 704
National Pollutant Discharge Elimination System permits, 85
Natural purification, 5, 32
Nephelometric, 436
Net gross calculations, 97
Neutralization, 285, 537–538
Neutralizing reagents, 286
Nitrogen, cycle, 32
Nitrogen, industrial wastes, 183
Noise, safety, 685
Noisy chain drive, 845
Noisy electric motor, 841
Noisy pump, 782, 835
Noncategorical industries, local limits, 92
Noncompatible pollutants, 38, 42
Nonsparking tools, 704
Normal operation, metals treatment, 577–590
Notification of changed discharge, 96
Nozzles, flow, 226, 228, 233, 583
Nuclear power plants, 28
Nutrient cycle, 32
Nutrients, 30, 183

O

O & M (Operation and Maintenance) manual, 752
ORP, 51, 301
ORP probes, 560–566, 592
OSHA (Occupational Safety and Health Act), 179, 677, 688
OSHA regulations, 103
Objectives
   maintenance, 745
   safety, 670
Observations, dissolved air flotation (DAF) thickeners, 373
Occupational Safety and Health Act (OSHA), 179, 677, 688
Occupational Safety and Health Act (OSHA) regulations, 103
Odors
   industrial wastes, 159, 183
   sewers, 47
Ohmmeter, 794
Ohm's Law, 790
Oil and gas extracting, 139
Oil change, pump, 835
Oily waste removal, 557
Olfactory fatigue, 681
On-site treatment exemption, 102
Open channel flow, 218, 221–233
Open impeller, 763
Operation
   air stripping, 492
   carbon adsorption, 505
   chemical treatment, 355, 358, 374–377
   dissolved air flotation (DAF) thickeners, 369, 372
   gravity filters, 406–413
   instrumentation, 656–660
   membranes, 456–463
   metals treatment, 576–590
   pressure filters, 426
   pumps, 774, 782
   sludge thickeners, 371
   upflow filters, 436–440
Operating practices, pollution prevention, 117
Operation and maintenance (O & M) manuals, 752
Operation and maintenance, flowmeters, 226–230
Operation guidelines, dissolved air flotation (DAF) thickeners, 369, 372
Operation strategy, flow equalization, 266
Operational strategy
   industrial waste monitoring, 180
   monitoring programs, 179
Operator
   administrator, 8
   duties, 5
   employers, 5, 9
   employment, 5, 9
   job duties, 5
   job locations, 5
   job opportunities, 9
   pay, 5
   public relations, 8
   qualifications, 6, 7
   safety, 9
   training, 6, 10, 721, 728
Operator's responsibility, 46
Operator's role, pollution prevention, 116
Opportunities, pollution prevention, 117–119
Ordinance, sewer-use, 197
Ordinances, local
   building codes, 101
   land use, 101
   underground tank, 101
   wastewater, 98–100

Ordinances, sanitary sewer, 100
Organic compounds in wastewater, 31
Organic loading or overloads
    See specific process of interest
Organic, wastes, 182
Organically contaminated wastewaters, 532
Organics
    total toxic, 94
    toxic, 558
    volatile, 489
Orifice plate, 218, 234, 235, 424, 583, 584, 640, 641
Osmosis, 463, 464
Overload protection, 796
Oxidation, chemical, pH adjustment, 301
Oxidation-reduction potential (ORP), 51, 301
Oxygen consuming wastes, 36
Oxygen deficiency/enrichment, 675, 676, 678, 681
Oxygen depletion, 28

P

PLC (programmable logic controller), 458, 812
POTW, 7, 33
    effects, wastestreams, 53
    facilities, 98
Packed tower, air stripping, 489
Packing, 770, 781, 782, 786, 830, 831, 833, 834, 836–837, 838, 839, 850, 851
Paint, building and equipment, 750
Paints, 129
Palmer-Bowlus flume, 163, 175, 226, 227, 241, 583
Panel instruments, 645
Parshall flume, 175, 224, 225, 241, 583
Pathogens, 183
Pay for operators, 5
Penalties, administrative fines, 100
Permeability constants, 466
Permeate, 453, 466
Permeate projections, computerized, 470, 471
Permit, confined space entry, 680
Permit, sewer-use, 187
Permits, effluent discharge, 85
Permits NPDES, 83, 416
Personal hygiene, 676
Personal training, 6
Pesticides, 129, 183
Petroleum refining, 140–145
pH
    adjustment, 285–304, 537–538
    definition, 30
    probes, 290–294, 560–566, 591
    problems, sewers, 48
    safety, 712
pH adjustment
    absorption, alkaline buffers, 293
    acidity error, 294
    alkalinity error, 294
    biological treatment, 300
    chemicals used, 287
    control system, 298–299
    controller, 298
    dead time, 296
    definition, 286
    discharge, 300
    electrode troubleshooting, 302
    electrodes, 291
    elements, 298
    equalization, 294
    errors, 294
    face velocity, 293
    feedback control, 295, 296
    feed forward control, 295
    final element, 299
    fouling, 293
    heavy metal removal, 301
    maintenance, 302
    manufacturing process changes, troubleshooting, 304
    measurement electrode, 291
    mechanical operation, troubleshooting, 302, 303
    metal removal, 301
    mixing, 296
    need, 285
    oxidation, 301
    primary element, 298
    reduction, 301
    reference electrode, 291
    residence time, 296
    response time, 293
    sensors, 290–294
    strategy, 294
    temperature, 293
    titration curves, 287, 288, 290
    transmitter, 298
    troubleshooting, 302–304
    velocity, face, 293
pH control
    See pH adjustment
pH, effects on, reverse osmosis, 469, 473
pH probes, 290–294, 560–566, 591
Phosgene, 711
Phosphate monitoring, 340
Phosphorus, industrial wastewaters, 183
Photo processing, 146
Physical injuries, 676
Physical treatment, clarification
    equipment, 358–366
    operation, 374–377
    safety, 377
    troubleshooting, 377, 378
Physical treatment processes
    See Air stripping, Carbon adsorption, Filtration
Pickle, 533
Pipe flow measurements, 218, 233–238
Piping, maintenance, 865

Piston pumps, 759, 786, 835, 838
Pitot tube, 584
Plans
    carbon adsorption, 510
    gravity filters, 414
    pressure filters, 429
    upflow filters, 445
Plans and specifications
    electrical systems, 867
    equipment, 867
    lift stations, 867
    pump stations, 867
    safety, 720
Plant, appearance, 751
Plate and frame membranes, 452
Plug valves, 853, 855–859
Plugged screens, carbon adsorption, 507
Plugging sewers, 47
Plunger pumps, 348, 835, 838
Pneumatic ejectors, 768, 772, 839
Pneumatic hazards, 629
Pocket dosimeter, 682
Polarization, concentration, 456
Pole shader, 823
Policy, safety, 720
Polio, 29
Polishing process, chemical treatment, 322, 323
Pollutants
    conventional, 116
    priority toxic, 90
Pollutants from wastestreams, 42
Pollution
    solids, 36
    visible, 36
    water, 7
Pollution control regulations, 81
Pollution prevention
    audits, 118
    benefits, 117
    checklists, 120, 123–150
    chemical formulating, 121, 129
    chemical manufacturing, 124–128
    cost savings, 117
    dairy processing, 130–131
    dry cleaning, 132
    economics, 117
    fluid milk processing, 133–134
    gas extraction, 139
    industry specific, 120
    loading of material and product, 118
    management support, 118
    material distribution systems, 118
    material storage and loading, 118
    material substitution, 118
    meat packing, beef, 135
    metal fabrication, 136–137
    metal finishing, 120, 137–138
    milk processing, 133–134
    oil and gas extraction, 139
    operating practices, 117
    operator's role, 27, 116
    opportunities, 117–119
    paints, 129
    pesticides, 129
    petroleum refining, 140–145
    photo processing, 146
    printed circuit board manufacturing, 146–147
    printing, 148
    process modification, 119
    process operations, 118
    product reformulation, 119
    product storage and loading, 118
    pulp and paper manufacturing, Kraft segment, 149
    radiator repair, 150
    raw material purchasing, 118
    reformulation of product, 119
    regulatory requirements, 116
    segregation of wastes, 118
    storage of material and product, 118
    strategy, 115
    walk-through, 120
    waste segregation, 118
Polyelectrolyte, 328, 408, 419
Polymer, 324, 325, 347, 416, 419, 436, 709
Polymer map, 329, 330
Polymeric flocculants, 328
Ponds, safety, 707
Portable generators, 867
Portable samplers, 163
Positive displacement pumps
    sludge, 759, 786, 835, 838
    wastewater, 759, 786, 835, 838
Potable water, 716, 862
Potting compounds, 449, 564
Power
    requirements, 791
    supply, 818–819
Power outage, 867
Power, pump, 759, 782, 786
Precious metals recovery, 553–557
Precipitate, 287, 324
Preliminary treatment, 257
Preserving samples, 171, 173, 580
Pressure cleaning pipelines, 865, 866
Pressure measurements, 630
Pressure pipe flow measurement, 218, 233, 234
Pressure, transmembrane, 455
Pretreatment, 249–308
    application, 93–97
    categorical standards, 88–97
    feed, 458
    general regulations, 84
    industrial, 33
    industrial categories, 89, 93
    industrial wastes, 198
    inspection, 161

Pretreatment, 249–308 (continued)
   metal finishing standards, 92
   prohibited discharge standards, 87
   regulation development, 91
   regulations, 84, 92, 93–97
   standards, 92, 93
   Also see Chapter 4, *OPERATION OF WASTEWATER TREATMENT PLANTS*, Volume I
Preventing wastes
   See Pollution prevention
Preventive maintenance
   program, 749, 830, 865
   records, 749
   variable-speed AC motors, 777
Primary element, pH adjustment, 298
Primary elements, 221–230
Primary sludge flow measurement, 244
Prime, centrifugal pump, 757, 758, 775, 782, 783, 785
Printed circuit board manufacturing, 60–62, 63, 146–147
Printing, 148
Priority toxic pollutants, 90
Probes, 290–294, 560–566, 591, 592
Probes, electrical, 633, 634
Process analytical instrumentation, 642
Process modification, waste reduction, 119
Process operations, waste reduction, 118
Processes, treatment, metals, 533–558
Product recovery, air stripping, 492
Product reformulation, 119
Product storage and loading, 118
Programmable controllers, 567
Programmable logic controller (PLC), 458, 812
Programs
   NPDES permit, 83
   National pretreatment, 83, 84
Progressive cavity pumps, 768, 770, 771, 786, 839
Prohibited discharge standards, 87
Propeller meter, 238, 239, 583, 584
Propeller pumps, 759, 765, 766, 839
Protection devices, motor, 796, 800, 801
Protective clothing, 676, 679, 682, 688, 707, 710, 713, 714, 720, 862
Protective coating, 709
Prussian blue, 853
Public relations, 8, 750, 751, 752
Publicly Owned Treatment Works (POTW), 7, 33
Pulp and paper manufacturing, Kraft segment, 149
Pump capacity, 759, 833
Pumping stations, 701, 867
Pumps and pump parts
   air chamber, 835, 838
   alignment, 756, 782, 831, 847, 848
   axial flow, 759, 766
   ball valves, 835, 838
   bearings, 753, 756, 761, 762, 764, 765, 767, 768, 769, 774, 775, 785, 831, 834, 839
   belts, 782, 843–844, 845, 846
   capacity, 759, 833
   casing, 753, 757, 761, 762
   cavitation, 760
   centrifugal, 753, 759, 761–767, 783, 830
   chemical feed, 348, 349, 350
   circuit breakers, 795, 796, 797, 798–801
   cleaning pump, 831
   connecting rods, 835, 838, 839
   coupling, 756, 769, 782, 846, 847, 848
   cross connections, 716, 830
   description, 753
   discharge, 753, 759, 782, 783, 784, 786, 839
   drain, 775, 785, 834
   draining pumps, 775, 785, 834
   driving equipment, 775
   eccentric, 835, 838
   electric motors, 776, 803–826
   energy, 759, 782, 786
   flow equalization, 266
   friction, 759
   fuses, 781
   gasket, 835
   gear reducer, 835
   general, 753, 830
   gland, 758, 781, 784, 786
   greasing, 839
   head, 758
   impeller, 753, 759, 761, 762, 763, 764, 766, 782, 784, 833
   incline screw, 768, 769
   lantern ring, 831, 833
   Let's Build a Pump, 754
   lubrication, 756, 768, 774, 775, 782, 831, 839, 853
   maintenance, 829
   metal treatment, 585–590, 593
   motor, 785
   multi-stage, 759
   noisy, 782, 835
   oil change, 835
   operation, 774, 782
   packing, 770, 781, 782, 786, 830, 831, 833, 834, 836–837, 838, 839
   plunger, 835, 838
   pneumatic ejectors, 768, 772, 839
   positive displacement, 759, 786, 835, 838
   power, 759, 782, 786
   pressure filters, 418
   prime, 757, 758, 775, 782, 783, 785
   progressive cavity, 768, 770, 771, 786, 839
   propeller, 759, 765, 766, 839
   pump-driving equipment, 775
   radial flow, 759, 766
   reciprocating, 759, 786, 835, 838
   records, 826, 827
   reducer, gear, 835
   repair shop, 753
   rings, 757, 761, 782, 784, 831, 833
   rods, 835, 838, 839
   rotation, 775, 782
   rotor, 768, 770, 776

Pumps and pump parts (continued)
  screw-flow, 768, 770, 771, 786
  seal cage, 758
  seals, 768, 773, 782, 783, 784, 785, 830, 831, 833
  shaft, 753, 754, 761, 762, 781, 782
  shear pin, 835, 849
  shutdown, 775, 782, 784, 786
  single-stage, 759, 767
  sleeves, 754
  sludge pump, 759, 786, 835, 838
  start-up, 759, 774, 781, 782, 785
  stator, 768, 770, 786
  stuffing boxes, 758, 761, 831
  submersible pump, 759, 764
  suction, 753, 756, 759, 782, 784, 786
  troubleshooting, 781
  turbine pumps, 759, 767
  valve chamber gasket, 835
  vanes, 753
  vertical wet well, 759, 767
  vibrations, 783, 785
  volute, 753, 783, 784, 785
  wearing rings, 757, 761, 782, 784, 833
  wet well pumps, 759, 764, 767, 783, 784

Q

Qualifications for jobs, 6, 7

R

RCRA, 86, 101
Radial flow pumps, 759, 766
Radiator repair, 150
Radioactive contamination, 35
Radioactive wastes, 28, 183
Radiological hazards, 682
Rapid sand filter, 396, 402
Rate control valve, gravity filters, 405
Rate-of-flow controller, 423
Raw material purchasing, 118
Raw sludge, pumps, 753
Raw wastewater, pumps, 753
Reactivation of carbon, 508
Reagent, 56
Reagents, neutralizing, 286
Receiving waters
  algae, 32
  definition, 7
  human health, 29
  nutrients, 30
  odors, 28
  oxygen depletion, 28
  pathogenic bacteria, 29
  sludge and scum, 28
  toxic substances, 30

Reciprocating pumps, 759, 786, 835, 838
Recirculation flow, membranes, 455
Recorders, instrumentation, 646
Recordkeeping
  electrical equipment, 826, 827, 828
  membranes, 463
  motors, 826, 828
  preventive maintenance, 749
  pumps, 826, 827
Recovery, backwash, 425
Recovery, reverse osmosis, 469
Recreation, 35
Recycle, 115
Reduction, chemical, pH adjustment, 301
Reducer, gear, 835
Reduction of hexavalent chromium, 547
Reference electrode, pH, 291
References, industrial wastewaters, 62
Reformulation of product, 119
Refractory, 527
Refractory materials, 160
Regeneration of carbon, 508
Regulated dischargers, 82
Regulation of high flows, 180
Regulations
  abbreviations, 79–80
  Clean Air Act, 103
  Clean Water Act, 103
  Department of Transportation, 104
  development, 91
  EPA, 81, 94, 98
  federal, 101–104
  hazard communication, 103
  hazardous material regulations, 104
  history, 81
  NPDES permit program, 83
  Occupational Safety and Health Act (OSHA), 103
  pollution control, 81
  pretreatment, 84, 92, 93–97
  Resource Conservation and Recovery Act (RCRA), 86, 101
  statutes, 101–104
  stormwater, 103
  Superfund Amendment Reauthorization Act (SARA), 102
Regulatory requirements, 116
Rejection, mineral, 466, 467, 468
Removal credits, 97
Report forms, accident, 720
Reports
  baseline monitoring (BMR), 87, 95
  compliance, 95
  EPA, 94
  hazardous waste disposal, 96
  slug load, 95
Representative samples, 162
Representative sampling, metals, 579
Resampling, violation, 96

Residence time, pH adjustment, 296
Resistance measurements, 567
Resource Conservation and Recovery Act (RCRA), 86, 101
Response time, pH adjustment, 293
Responsibility, operator's, 46
Return activated sludge flow, 245
Reuse, 115
Reverse osmosis (RO)
   brine, 474
   calculations, 466–474
   "Christmas Tree" arrangement, 474, 475
   concentration polarization, 474
   definition, 463
   flow diagram, 464
   flux, 465, 466
   flux decline, 466
   hydrolysis, 469, 473
   membrane, 449, 463, 465, 466
   mineral rejection, 466, 467, 468
   osmosis, 463, 464
   permeate, 466
   pH effects, 469, 473
   recovery, 469, 555
   rejection, mineral, 466, 467, 468
   temperature effects, 469, 472, 473
   Also see Demineralization and Electrodialysis
Review of plans and specifications
   electrical systems, 867
   equipment, 867
   lift stations, 867
   pump stations, 867
   safety, 720
Right-To-Know (RTK) Laws, 721–729
Rings, pump, 757, 761, 782, 784, 831, 833
Rinse waters, 530, 532
Rising sludge, 369, 408
Rods, pump, 835, 838, 839
Rotameter, 235, 236, 419, 582, 583, 636, 637
Rotary chemical feeder, 352
Rotating screens, 275–282
Rotation, motors, 817
Rotation, pump, 775, 782
Rotor, 768, 770, 776, 843

S

SARA, 102
SPC, 462
SCFM, 370
Safety
   air stripping, 494
   carbon adsorption, 509
   chemical treatment, 340, 377
   clarifiers, 377
   cones, traffic, 178
   confined spaces, 441, 445, 704
   drums, traffic control, 178
   electricity, 788
   gravity filters, 414
   instrumentation, 628–630
   membrane filters, 463
   metal treatment, 549, 590, 595–600
   motors, 813
   OSHA, 677, 688
   operator, 9
   policy statement, 720
   pressure filters, 427
   screens, 284
   traffic cones, 178
   tubular markers, traffic control, 178
   upflow filters, 445
   vertical panels, traffic control, 178
   warning tag, 685, 702, 788, 789, 831, 832
Safety equipment
   air gap device, 716, 717, 862, 863
   air supply, 708
   blowers, 681, 862
   breathing apparatus, self-contained, 688, 719
   chlorine monitor, 678
   combustible gas monitor, 678, 682, 710
   dosimeter, 682
   ear protecting devices, 688
   eye protection, 713, 714
   face shield, 708, 713
   fall arrest system, 676, 704, 706, 729
   fans, 681, 862
   film badges, 682
   fire extinguishers, 701, 714, 715, 719, 862
   fire-fighting equipment, 684, 701, 715, 720, 862
   first-aid kit, 717, 862
   fume hood, 713
   gas testing, 678, 679, 680, 681, 682, 683
   gloves, 676, 713
   hard hat, 720
   harness, 676, 679, 862
   hearing protection devices, 688
   hydrogen sulfide monitor, 678
   lifeline, 676
   lockout tag, 685, 702, 788, 789, 831, 832
   measurement of gases, 678, 679, 680, 681, 682, 683, 703
   nonsparking tools, 704
   oxygen deficiency/enrichment indicator, 678
   pocket dosimeter, 682
   protective clothing, 676, 679, 682, 688, 707, 710, 713, 714, 720, 862
   purchase, 720
   respirators, 688
   safety glasses and goggles, 713, 714
   safety harness, 676, 679, 862
   self-contained breathing apparatus, 688, 719
   storage containers, 708, 710
   tag, 685, 702, 788, 789, 831, 832

Safety equipment (continued)
  traffic control devices, 689, 690–694
  ventilation, 681, 682, 702, 704, 708, 710
Safety harness, 676, 679, 862
Safety hazards
  acids, 681
  aeration tanks, 707
  amines, 711
  applying protective coatings, 709
  back strains, 695
  bar screens, 701
  biocides, 711
  blowers, 681, 862
  carcinogens, 709, 711
  chemicals, 681, 708, 713
  chlorine, 678, 703, 708, 729
  clarifiers, 704
  collection systems, 689, 710
  comminutors, 702
  confined spaces, 675, 676–681, 689, 703, 707, 862
  cross connections, 716, 830
  digesters, 704
  digestion equipment, 704
  drowning, 707
  dysentery, 676
  electrical, 684, 788
  emergencies, 719
  empty digesters, 751
  excavations, 696
  explosive gas mixtures, 678, 682, 710
  falls, 676
  fire, 714, 715
  flammable atmospheric condition, 679
  foam, 711
  fuels, 710
  gasoline vapors, 678, 681, 682
  grit channels, 703
  hepatitis, 676
  hydrogen sulfide, 678, 681, 701, 711
  ice, 704, 708, 710
  industrial waste sampling, 175
  industrial waste treatment, 681, 710, 711
  infections, 676
  infectious diseases, 676
  laboratory, 713
  lift station, 701
  manholes, 689, 695
  methane, 678, 704
  monitoring, 175
  natural gas, 682
  noise, 685
  oil and grease soaked rags, 710
  oil and grease spills, 710
  oxygen deficiency/enrichment, 675, 676, 678, 681
  pH, 712
  physical injuries, 676
  polymers, 709
  ponds, 707
  protective coatings, 709
  pumping stations, 701
  radiological, 682
  sampling, 175, 713
  sedimentation basins, 704
  sewer cleaning, 700
  slippery surfaces, 701, 703, 706, 707
  spills, 706, 710
  start-up, 701
  stored energy, 685
  suffocating gases or vapors, 675, 678, 681, 682, 709, 711, 713
  sulfur dioxide, 678, 708, 729
  surface-active agents, 711
  tetanus, 676
  toxic chemicals, 681
  toxic gases or vapors, 675, 678, 681, 682, 709, 711, 713
  traffic, 689
  treatment plants, 701
  trickling filters, 706
  typhoid fever, 676
  vapors, 675, 678, 681, 682, 709, 711, 713
  wet wells, 701, 702
Safety, OSHA standards, 735–738
Safety program
  accident report form, 720
  accidents, 718, 720
  backflow prevention, 716
  conditions for program, 718
  confined spaces, 676–681
  cross connections, 716
  development, 718
  drills, 719
  effectiveness, 718
  emergencies, 719
  entry of confined spaces, 676–681
  equipment use, 713, 717, 719
  fire control methods, 715
  fire drills, 719
  fire prevention, 714, 715, 716
  forms, accident, 720
  hazard communication program, 721–729
  hearing protection program, 688
  housekeeping, 701, 702, 704, 710
  immunization shots, 676
  importance of program, 718, 720
  information, 717–721
  inoculations (immunizations), 676
  laboratory, 713
  laundry service, 676
  library, 752
  locker room, 676
  management responsibility, 675, 718

Safety program (continued)
   manhole entry, 689, 695
   material safety data sheets, 681, 710, 712, 723–725, 777, 865
   meetings, 718
   need, 675
   operator responsibility, 675
   paper work, 720
   personal hygiene, 676
   planning, 719
   policy, 720
   procedures, 713, 714
   promotion, 719
   protective clothing, 676, 679, 682, 688, 707, 710, 713, 714, 720, 862
   purchase of equipment, 720
   purpose, 675
   report form, 720
   respiratory protection, 679
   review of plans and specifications, 720
   safety policy, 720
   safety rules, 720
   sampling techniques, 713
   sewer-use ordinance, 710
   starting program, 718, 720
   storage of chemicals, 713
   supervisor's role, 719
   tailgate safety meetings, 718
   testing procedures, 713
   training programs, 717, 718, 720
   uniforms, 676
   water supply protection, 716
   work clothes, 676
Safety rules, 720
Samplers, 163
Sampling
   activated carbon, 508
   chain of custody, 172
   chemical preservation of samples, 171
   composite samples, 163, 580
   continuous sampling, 161
   devices, 580
   equipment, 163, 166
   grab samples, 162, 580
   industrial wastes, 159, 161
   jar test, 335
   labeling samples, 172
   location of sampling points, 162
   maintenance of equipment, 170
   membrane filtration, 462
   portable equipment, 163
   preservation of samples, 171, 173, 580
   procedures, 579–581
   representative samples, 162
   representative sampling, 579
   safety, 713
   security of samples, 171
   storage of samples, 171
   techniques, 713
   temperature of stored samples, 171
   time of sample storage, 171
Sampling location, 162
Sand filter, metals, 545
Sanitary sewer codes, 100
Schedule, compliance, 95
Scouring, media, 400
Screening
   bar screens, 268–272
   coarse screens, 268–272, 283
   fine screens, 273–282, 283
   maintenance, screens, 282–284
   mechanical bar screen, 269
   purpose, 268
   rotating screens, 275–282
   safety, screens, 284
   shutdown procedures, 269, 281
   start-up procedures, 269, 281
   static screens, 273–274
   troubleshooting, 271–272
   types, 268–282
   wedgewire screen, 273
   why, 268
Screw-flow pumps, 768, 770, 771, 786
Screw-lift pumps, 768, 770, 771, 786
Scum
   problem, 775
   receiving water, 28
   unplugging pipes, 865
Seal cage, pump, 758
Seals, pump, 768, 773, 782, 783, 785, 830, 831, 833
Secondary elements/devices, flows, 230–235
Security of samples, 171
Sedimentation
   physical–chemical treatment, 321, 358–377
   safety, 704
Sedimentation basins, safety, 377
Segregation of wastes, 118
Self-contained breathing apparatus (SCBA), 719
Self-monitoring, 161
Sensors, pH adjustment, 290–294
Sensors/transducers, 630–642
Septic, 265, 366, 32
Service record card, 749, 750
Sewer cleaning, 700
Sewer gases, characteristics, 682
Sewer service charges, 160
Sewer-use ordinance, 160, 197, 710
Sewer-use permit, 187
Shaft, pump, 753, 754, 761, 762, 781, 782
Shear pins, 835, 849
Sheaves, 844

Sheeting, 696
Shielding, 696
Shock loads, 51, 159
Shockcor, Joe, 534–536, 551
Shoring, excavations, 696, 697
Short-circuiting
    clarifier, 366
    equalization tank, 266
    filter, 408
Shutdown
    carbon adsorption, 506
    chemical treatment, 355
    dissolved air flotation (DAF) thickeners, 372
    draining pump, 834
    gravity filters, 412
    instrumentation, 657–659
    pumps, 775, 782, 784, 786
    screens, 269, 281
Sieve analysis, 516
Signal transmitters/transducers, 642
Significant Industrial User (SIU), 86
Silicon controlled rectifier (SCR), 419
Silver wastestreams, 532
Single-stage pumps, 759, 767
Slake, 328
Sleeves, 754
Sloping, excavations, 696, 699
Sludge
    receiving waters, 28
    unplugging pipelines, 866
Sludge dewatering, 568–574
Sludge disposal effects, 53
Sludge drying, 574
Sludge load report, 95
Sludge pumps, 759, 786, 835, 838
Slug discharge, 171, 192
Slug loadings, 51
Sluice gates, 853, 860, 861
Slurry, 496
Small quantity generator, 102
Sodium sulfide, 711
Softening membrane, RO, 469
Solids
    dissolved, 30, 31
    floatable, 31
    inorganic, 31
    nonsettleable, 30
    organic, 31
    settleable, 30
    suspended, 30
    test in wastewater, 31
    total, 30
    wastewater characteristics, 30
Solids contact clarifier, 362, 363
Solids, industrial wastes, 182
Solids loadings, dissolved air flotation (DAF) thickeners, 369, 373
Solids pollution, 36
Solvent controls, 558
Solvent Management Plan, 94
Solvents, 182
Source control, 528
Sources, metal wastestreams, 528
Spare parts inventory, 591
Specific gravity, 407
Specifications
    carbon adsorption, 510
    gravity filters, 414
    pressure filters, 429
    upflow filters, 445
Spent activated carbon, 510
Spent baths, 531
Spiral membranes, 451
Spoil, 696
Squirrel-cage induction motor (SCIM), 775, 803, 804
Staff gauge, flow, 230, 231
Staffing needs, 9
Standard deviation, 259
Standard industrial classification, 160, 198
Starters, 810–813
Start-up
    air stripping, 492
    carbon adsorption, 498–505
    chemical treatment, 355, 374, 375
    dissolved air flotation (DAF) thickeners, 371
    instrumentation, 657–659
    pumps, 759, 774, 781, 782, 785
    safety hazards, 701
    screens, 269, 281
    water level controls, 840
Static screens, 273–274
Stator, 768, 770, 786, 843
Statutes
    See Regulations
Stethoscope, 841
Stilling well, 230, 231
Storage
    bins, 677
    chemicals, 713
    containers, 708, 710
    samples, 171
Storage and handling of chemicals, 597, 713
Storage of material and product, 118
Stored energy, 685
Stormwater, flow measurement, 238
Straightedge, alignment, 847
Strategy, pollution prevention, 115
Strategy
    gravity filter operation, 410
    operation, flow equalization, 266
    pH adjustment, 294
    pressure filters, 426
    upflow filters, 439
Strategy for monitoring, 179

Stuffing box, valve, 850, 851, 852
Stuffing boxes, pump, 758, 761, 831
Submerged pressure transducers, 233
Submersible pump, 759, 764
Suction, pump, 753, 756, 759, 782, 784, 786
Suffocating gases or vapors, 678, 681, 682, 709, 711, 713
Sulfide precipitation, 546
Sulfur dioxide, 678, 708, 729
Superfund Amendment Reauthorization Act (SARA), 102
Supervisors, safety, 719
Supervisory controls, 802
Surcharge, 238
Surface-active agents, safety, 711
Surfactants, 119, 453, 711
Suspended solids, 30
Switch gear, electrical, 795
Switches, level control, 775, 831, 840
Symbols, instrumentation, 616–620

T

TDH (total dynamic head), 833
TTO, 94
TWA (time weighted average), 688
Tailgate safety meetings, 718
Tanks, flow equalization, 244, 260, 261–267
Tanks, maintenance, 751
Tastes in water, 183
Temperature
   air stripping, 490
   pH adjustment, 293
Temperature effect, plugging pipelines, 865
Temperature effects, reverse osmosis, 469, 472, 473
Temperature limits, motors, 807
Test equipment, 652
Testers, electrical, 791–795
Tetanus, 676
Thermal wastes, 28, 159, 182
Thermophilic process, anaerobic digester
   See Chapter 12, *OPERATION OF WASTEWATER TREATMENT PLANTS*, Volume II
Thickening sludges, dissolved air flotation (DAF) thickeners, 367
Thickness gauge, alignment, 847
Thin film composite membrane, 465
Time weighted average (TWA), 688
Titration curves, pH adjustment, 287, 288, 290
Total ash of regenerated carbon, 516
Total dynamic head (TDH), 833
Total toxic organics (TTO), 94
Totalizer, gravity filters, 405
Totalizers, instrumentation, 646
Toxic
   chemicals, 35
   organics, 558
Toxic chemicals, 681
Toxic compounds, 183

Toxic gases or vapors, 678, 681, 682, 709, 711, 713
Toxic Organic Management Plan (TOMP), 94
Toxic organics, 558
Toxic pollutants, priority, 90
Toxic wastes
   gases, 711, 714
   general, 28
   industrial, 159
   materials, 711, 714
Traffic
   cones, 178
   drums, 178
   tubular markers, 178
   vertical panels, 178
Traffic control devices, 689, 690–694
Traffic hazards, 689
Training
   courses, 719
   hazard communication, 721–729
   management planning, 718
   safety, 717, 718, 719
Training for operators, 6, 10
Tramp oil, 458
Transducers, 233, 630–642
Transfer of carbon, 509
Transit time flow measurement, 238
Transmembrane pressure, 455
Transmitter, pH adjustment, 298
Treatment of industrial wastes, need, 32, 33
Treatment plant maintenance
   buildings, 750
   channels, 751
   costs, 749
   general program, 749
   grounds, 751
   inspection, 751
   landscaping, 751
   library, 752
   paint, 750
   preventive maintenance program, 749, 830, 865
   preventive maintenance records, 749
   public relations, 751, 752
   records, 749
   safety, 701, 751
   tanks, 751
   Also see Mechanical maintenance, Meter maintenance, and chapters 12 (607–604) and 14 (739–875)
Treatment plants, safety, 701
Treatment, preliminary, 257
Treatment processes, metals, 533–558
Treatment system effects, 50–54
Trenching, 696, 697, 698, 699
Trickling filters, safety, 707
Troubleshooting
   air stripping, 492
   bar screen, 271–272
   centrifugal pump, 593

Troubleshooting (continued)
  chain drive, 845
  chemical treatment, 377, 378
  diaphragm pump, 593
  dissolved air flotation (DAF) thickeners, 373, 374
  gravity filters, 412
  instrumentation, 592
  ion exchange, 555–557
  metals treatment processes, 591–595
  motors, 820–826
  ORP meter, 592
  pH adjustment, 302–304
  pH meter, 591
  probes, pH and ORP, 565
  pumps, 781
  rotating screen, 281
  static screen, 274
  upflow filter, 441–442
  variable-speed AC motors, 777–782
  water level controls, 840
Troughs, gravity filters, 400
True color, 338
Trunk sewer, 163
Tube settlers, 364, 365
Tubular markers, traffic control, 178
Tubular membranes, 449
Turbidity, 182
Turbidity meter, 433–436
Turbine pumps, 759, 767
Typhoid fever, 29, 676

U

Ultrafiltration, 447
Ultrasonic flow devices, 232
Unbalance, voltage, 808, 810
Underdrains, gravity filters, 400, 403
Underground tank laws, 101
Uniformity coefficient, 515
Uniforms, 676
Unloading station, carbon, 510
Unplugging pipes, pumps, and valves
  costs, 865
  cutting tools, 866
  digested sludge lines, 865
  equipment, 865, 866
  high-velocity pressure units, 866
  methods of unplugging, 865, 866
  pipes, 865
  pressure clearing methods, 865, 866
  pumps, 866
  scum lines, 865
  sludge lines, 865
  valves, 865, 866
Upflow carbon column, 503, 504
Upper explosive limit (UEL), 682, 683
Upstream processes, carbon adsorption, 507, 511

V

VOM (Volt-Ohm-Milliammeter), 652, 654
Vacuum drying of sludge, 574
Vacuum filters, 569
Valve chamber gaskets, 835
Valve parts
  bonnet, 855, 857
  gear, 850, 855
  gland, 853
  packing, 850, 851
  plug, 855, 857
  rising stem, 850
  seal, 855
  seats, 850, 853, 855, 857
  stuffing box, 850, 851, 852
  threads, 850
Valves
  ball valves, 835, 838
  check valves, 783, 833, 839, 853, 854
  description, 850, 853
  foot valves, 833
  gate valves, 850–853
  location, 783
  maintenance, 833, 835, 838, 839, 850, 853, 855–859
  plug valves, 853, 855–859
  unplugging, 865, 866
Vanes, pump, 753
Vapor phase carbon, 491
Vapors, toxic or suffocating, 678, 681, 682, 709, 711, 713
Variables, industrial wastewaters, 42–46
Variable-speed AC motors
  description, 776
  preventive maintenance, 777
  troubleshooting, 777–782
Variable-speed belt drives, 845
Variance from categorical standards, 97
Velocity, face, sensors, 293
Velocity meters, 235, 583, 584
Ventilation and exhaust, systems, 584
Ventilation, carbon adsorption, 511
Ventilation, safety, 681, 682, 702, 704, 708, 710
Venturi meter, 640, 641
Venturi nozzle, 218, 233, 234, 583
Vertical panels, traffic control, 178
Vertical wet well pumps, 759, 767
Very low flow measurements, 238
Vibration, motors, 817, 841, 847
Vibrations, pump, 783, 785
Viscosity, 336, 709
Visible pollution, 36
Visual inspection, chain drive, 845
Visual inspection, dissolved air flotation (DAF) thickeners, 373
Volatile liquids, 39
Volatile organics, 489
Voltage, imbalance, 808, 810
Voltage measurements, 567

Volts, 790
Volume, flow equalization, 261–265
Volumetric screw chemical feeder, 350
Volute, 753, 783, 784, 785

W

Walk-through, 120
Warning tag, 685, 702, 788, 789, 831, 832
Wash water troughs, gravity filters, 400
Waste activated sludge flow, meter, 245
Waste discharges, 28
Waste minimization, 528
Waste segregation, 118
Wastes
   biological, 33, 35
   chemical, 35
   dissolved, 33
   industrial, 32
   inorganic, 36
   need to treat, 32, 33
   oxygen consuming, 36
   suspended, 33
Wastestreams
   industrial, 42–46
   metal, 531–533
   types, 94
Wastewater
   definition, 5
   effluent, 9
   facilities, 98
   generation, 39, 54–62
   ordinances, 98–100
   pumps, 753, 759
Wastewater characteristics, 28
Wastewater treatment objectives, 27
Wastewaters, industrial
   See Industrial wastewaters
Water
   aquatic vegetation, 36
   contact recreation, 35
   fish, 36
   level controls, 775, 840
   non-body contact recreation, 36
   pollution, 7, 27
   receiving, 7
   supply protection, 716
   uses by industry, 32, 34, 36
   wildlife, 36
Water hammer, 407, 585, 835
Water meters, 175
Water quality protector, 6
Watts, 791
Wearing rings, pump, 757, 761, 782, 784, 833
Wedgewire screen, 273
Weirs, 218, 220, 221–224, 582, 583
Well, stilling, 230
Wet well, pressure filter, 416
Wet well pumps, 759, 764, 767, 783, 784
Wet wells, 701, 702, 783, 784
Wildlife, 36
Work clothes, 676
Worker Right-to-Know (RTK) Laws, 721–729

X, Y, Z

(NO LISTINGS)

NOTES

NOTES

NOTES

NOTES

NOTES

NOTES

NOTES